COLOSS
BEEBOOK

Volume I:
Standard Methods
for *Apis mellifera*
Research

Edited by
Vincent Dietemann
James D Ellis
Peter Neumann

INTERNATIONAL BEE
RESEARCH ASSOCIATION

Jointly published by: The International Bee Research Association, a
Company Limited by Guarantee, 91 Brinsea Road, Congresbury,
Bristol BS49 5JJ (UK) & Northern Bee Books, Scout Bottom Farm,
Mytholmroyd, Hebden Bridge HX7 5JS (UK).

Obtainable from:
www.ibrabee.org.uk & www.northernbeebooks.co.uk

© 2013 The International Bee Research Association.
Reprinted 2017

ISBN 978-0-86098-283-8

IBRA

INTERNATIONAL BEE
RESEARCH ASSOCIATION

From its origins as a mere idea borne out of frustrations caused by a lack of standardisation of laboratory techniques expressed at an early meeting of the COLOSS (Prevention of honey bee COlony LOSSes) Network in Bern, Switzerland in 2009, this project has grown to what you see today. This is the result of work by more than 234 authors from 34 countries, together with a large number of others who have reviewed the papers. Without the voluntary efforts of these, a large proportion of the world's bee scientists, the BEEBOOK would not have been possible. In particular our thanks are due to Vincent Dietemann, Jamie Ellis and Peter Neumann, who selected, approached and then cajoled the authors into writing the chapters. They then solicited reviews and edited the papers. The unique way that BEEBOOK has evolved also allowed constructive criticisms from many more people. Thanks to Pilar de la Rua for translating the summaries into Spanish, and to Shi Wei and Huo Qing Zheng for the Mandarin Chinese summary translations. The mammoth task of production of the papers for publication was carried out by Tony Gruba, Sarah Jones, Julian Rees and especially Diane Griffiths at the IBRA office in Cardiff. I must thank the scientific members of IBRA Council, most of whom have acted as authors or reviewers, and Ivor Davis, IBRA Vice-Chairman for realising the potential importance of the BEEBOOK from the outset, and for providing unwavering support for the project. Ralf Bünemann, COLOSS webmaster has produced the html version.

The COLOSS network was funded through the COST Action FA0803. COST (European Cooperation in Science and Technology) is a unique means for European researchers to jointly develop their own ideas and new initiatives across all scientific disciplines through trans-European networking of nationally funded research activities. Based on a pan-European intergovernmental framework for cooperation in science and technology, COST has contributed since its creation more than 40 years ago to closing the gap between science, policy makers and society throughout Europe and beyond. COST is supported by the EU Seventh Framework Programme for research, technological development and demonstration activities (Official Journal L 412, 30 December 2006). The European Science Foundation as implementing agent of COST provides the COST Office through an EC Grant Agreement. The Council of the European Union provides the COST Secretariat. The COLOSS network is now supported by the Ricola Foundation - Nature & Culture.

Norman L Carreck.

IBRA Science Director and Senior Editor, *Journal of Apicultural Research*.

July 2013.

COST is supported by the
EU Framework Programme

ESF provides the COST Office through
a European Commission contract

The COLOSS *BEEBOOK*, Volume I: Standard methods for *Apis mellifera* research.	
Foreword	Robert E Page Jr
Standard methods for *Apis mellifera* research	Vincent Dietemann, James D Ellis, Peter Neumann
Standard methods for *Apis mellifera* anatomy and dissection	Norman L Carreck, Michael Andree, Colin S Brent, Diana Cox-Foster, Harry A Dade, James D Ellis, Fani Hatjina, Dennis vanEnglesdorp
Standard methods for behavioural studies of *Apis mellifera*	Ricarda Scheiner, Charles I Abramson, Robert Brodschneider, Karl Crailsheim, Walter M Farina, Stefan Fuchs, Bernd Grünewald, Sybille Hahshold, Marlene Karrer, Gudrun Koeniger, Niko Koeniger, Randolf Menzel, Samir Mujagic, Gerald Radspieler, Thomas Schmickl, Christof Schneider, Adam J Siegel, Martina Szopek, Ronald Thenius
Standard methods for cell cultures in *Apis mellifera* research	Elke Genersch, Sebastian Gisder, Kati Hedtke, Wayne B Hunter, Nadine Möckel, Uli Müller
Standard methods for characterising subspecies and ecotypes of *Apis mellifera*	Marina D Meixner, Maria Alice Pinto, Maria Bouga, Per Kryger, Evgeniya Ivanova, Stefan Fuchs
Standard methods for chemical ecology research in *Apis mellifera*	Baldwyn Torto, Mark J Carroll, Adrian Duehl, Ayuka T Fombong, Tamar Katzav Gozansky, Francesco Nazzi, Victoria Soroker, Peter E A Teal
Standard methods for estimating strength parameters of *Apis mellifera* colonies	Keith S Delaplane, Jozef J M van der Steen, Ernesto Guzman-Novoa
Standard methods for research on *Apis mellifera* gut symbionts	Philipp Engel, Rosalind R James, Ryuichi Koga, Waldan K Kwong, Quinn S McFrederick, Nancy A Moran
Standard use of Geographic Information Systems (GIS) techniques in honey bee research	Stephanie R Rogers, Benno Staub
Standard methods for maintaining adult *Apis mellifera* in cages under *in vitro* laboratory conditions	Geoffrey R Williams, Cédric Alaux, Cecilia Costa, Támas Csáki, Vincent Doublet, Dorothea Eisenhardt, Ingemar Fries, Rolf Kuhn, Dino P McMahon, Piotr Medrzycki, Tomás E Murray, Myrsini E Natsopoulou, Peter Neumann, Randy Oliver, Robert J Paxton, Stephen F Pernal, Dave Shutler, Gina Tanner, Jozef J M van der Steen, Robert Brodschneider
Standard methods for artificial rearing of *Apis mellifera* larvae	Karl Crailsheim, Robert Brodschneider, Pierrick Aupinel, Dieter Behrens, Elke Genersch, Jutta Vollmann, Ulrike Riessberger-Gallé

The COLOSS *BEEBOOK*, Volume I: Standard methods for *Apis mellifera* research.	
Standard methods for instrumental insemination of *Apis mellifera* queens	Susan W Cobey, David R Tarpy, Jerzy Woyke
Miscellaneous standard methods for *Apis mellifera* research	Hannelie Human, Robert Brodschneider, Vincent Dietemann, Galen Dively, James D Ellis, Eva Forsgren, Ingemar Fries, Fani Hatjina, Fu-Liang Hu, Rodolfo Jaffé, Annette Bruun Jensen, Angela Köhler, Josef P Magyar, Asli Özkýrým, Christian W W Pirk, Robyn Rose, Ursula Strauss, Gina Tanner, David R Tarpy, Jozef J M van der Steen, Anthony Vaudo, Fleming Vejsnæs, Jerzy Wilde, Geoffrey R Williams, Huo-Qing Zheng
Standard methods for molecular research in *Apis mellifera*	Jay D Evans, Ryan S Schwarz, Yan Ping Chen, Giles Budge, Robert S Cornman, Pilar De la Rua, Joachim R de Miranda, Sylvain Foret, Leonard Foster, Laurent Gauthier, Elke Genersch, Sebastian Gisder, Antje Jarosch, Robert Kucharski, Dawn Lopez, Cheng Man Lun, Robin F A Moritz, Ryszard Maleszka, Irene Muñoz, M Alice Pinto
Standard methods for physiology and biochemistry research in *Apis mellifera*	Klaus Hartfelder, Márcia M G Bitondi, Colin S Brent, Karina R Guidugli-Lazzarini, Zilá L P Simões, Anton Stabentheiner, Érica D Tanaka, Ying Wang
Standard methods for pollination research in *Apis mellifera*	Keith S Delaplane, Arnon Dag, Robert G Danka, Brenno M Frietas, Lucas A Garibaldi, R Mark Goodwin, Jose I Hormoza
Standard methods for rearing and selection of *Apis mellifera* queens	Ralph Büchler, Sreten Andonov, Kaspar Bienefeld, Cecilia Costa, Fani Hatjina, Nikola Kezic, Per Kryger, Marla Spivak, Aleksandar Uzunov, Jerzy Wilde
Statistical guidelines for *Apis mellifera* research	Christian W W Pirk, Joachim R de Miranda, Matthew Kramer, Tomàs Murray, Francesco Nazzi, Dave Shutler, Jozef J M van der Steen, Coby van Dooremalen
Standard methods for toxicology research in *Apis mellifera*	Piotr Medrzycki, Hervé Giffard, Pierrick Aupinel, Luc P Belzunces, Marie-Pierre Chauzat, Christan Claßen, Marc E Colin, Thierry Dupont, Vincenzo Girolami, Reed Johnson, Yves Le Conte, Johannes Lückmann, Matteo Marzaro, Jens Pistorius, Claudio Porrini, Andrea Schur, Fabio Sgolastra, Noa Simon Delso, Jozef J M van der Steen, Klaus Wallner, Cédric Alaux, David G Biron, Nicolas Blot, Gherado Bogo, Jean-Luc Brunet, Frédéric Delbac, Marie Diogon, Hicham El Alaoui, Bertille Provost, Simone Tosi, Cyril Vidau

Journal of Apicultural Research 52(4): (2013)
DOI 10.3896/IBRA.1.52.4.02

Notes and Comments

The COLOSS *BEEBOOK* Volume I - Foreword

I B R A

INTERNATIONAL BEE
RESEARCH ASSOCIATION

Robert E Page Jr[1]*

[1]School of Life Sciences, PO Box 874701, Arizona State University, Tempe, AZ 85287-4701, USA.

*Corresponding author: Email: Robert.Page@asu.edu

Considering the COLOSS *BEEBOOK*, two things immediately come to mind. First is colossal, the size of the undertaking by the community of honey bee researchers that took on this project. The second is cooperation, the fabric of the honey bee societies we study and the fabric of this community of authors contributing to this work. Ten years ago I would have said that this book is the outcome of a unique collaborative effort extending across the honey bee research community. But that was before the honey bee genome sequencing project brought together an international community, showed how we could work together, and opened the door for many more to join. This is the second such effort. By sharing best practices and standard methods for conducting research, the way will be paved for new investigators to enter the field and for current investigators to broaden their interests and impact.

The organizers of this effort should be commended, as should the European Science Foundation, European Cooperation in Science and Technology, the European Union, and the Ricola Foundation for making this happen. Honey bee researchers for generations to come will benefit from and contribute to this undertaking.

Journal of Apicultural Research 52(4): (2013)
DOI 10.3896/IBRA.1.52.4.23

GUEST EDITORIAL

The COLOSS *BEEBOOK* Volume I, Standard methods for *Apis mellifera* research: Introduction

Vincent Dietemann[1,2*], James D Ellis[3] and Peter Neumann[4,2]

[1]Swiss Bee Research Centre, Agroscope Liebefeld-Posieux Research Station ALP-Haras, Bern, Switzerland.
[2]Social Insect Research Group, Department of Zoology & Entomology, University of Pretoria, Pretoria, South Africa.
[3]Honey bee Research and Extension Laboratory, Department of Entomology and Nematology, University of Florida, Gainesville, Florida, USA.
[4]Institute of Bee Health, Vetsuisse Faculty, University of Bern, Bern, Switzerland.

Received 29 July 2013, accepted for publication 31 July 2013.

*Corresponding author: Email: vincent.dietemann@agroscope.admin.ch

Keywords: COLOSS, *BEEBOOK,* honey bee, *Apis mellifera,* research, standard methods, laboratory, field

The COLOSS *BEEBOOK* is a practical manual compiling standard methods in all fields of research on the western honey bee, *Apis mellifera.* The COLOSS network was founded in 2008 as a consequence of the heavy and frequent losses of managed honey bee colonies experienced in many regions of the world (Neumann and Carreck, 2010). As many of the world's honey bee research teams began to address the problem, it soon became obvious that a lack of standardized research methods was seriously hindering scientists' ability to harmonize and compare the data on colony losses obtained internationally. In its second year of activity, during a COLOSS meeting held in Bern, Switzerland, the idea of a manual of standardized honey bee research methods emerged. The manual, to be called the COLOSS *BEEBOOK,* was inspired by publications with similar purposes for fruit fly research (Lindsley and Grell, 1968; Ashburner 1989; Roberts, 1998; Greenspan, 2004).

Production of the *BEEBOOK* began after recruiting international experts to lead the compilation of each research domain. These senior authors (first in the author list) were tasked with recruiting a suitable team of contributors to select the methods to be used as standards and then to report them in a user-friendly manner (Williams *et al.,* 2012).

The initial *BEEBOOK* project is divided into three volumes: The COLOSS *BEEBOOK,* Volume I: Standard methods for *Apis mellifera* research; The COLOSS *BEEBOOK,* Volume II: Standard methods for *Apis mellifera* pest and pathogen research; and The COLOSS *BEEBOOK,* Volume III: Standard methods for *Apis mellifera* product research.

Papers in the *BEEBOOK* are organized according to research topics. The authors have compiled those methods selected as the 'best' in each domain of research. These methods are for both laboratory and field research. We recognize that it is often necessary to use methods from several domains of research to complete a given experiment with honey bees. Whenever there is a need for multi-disciplinary approach, the manual describes the specific instructions

necessary for a given method, and cross references all general methods from other papers as necessary. For example, identifying a subspecies of honey bee can be done using genetic tools. The general instructions to use microsatellites are given in the molecular methods paper (Evans *et al.*, 2013), whereas the specific method appropriate for subspecies identification is described in the paper on ecotypes and subspecies identification (Meixner *et al.,* 2013). Consequently, one would visit the ecotypes paper to determine how to identify a given subspecies. That paper will then refer to the molecular methods paper when discussing microsatellites specifically.

The reader may wonder about the difference between the *BEEBOOK* and existing standards provided by the Office International des Epizooties (OIE), and the European Organisation for Economic Co-operation and Development (OECD). In the *BEEBOOK,* we often refer to OIE, OECD, and other standards, since they describe methods to diagnose pests and diseases (OIE) or to perform, for example, routine analyses for toxicity tests (OECD). The *BEEBOOK,* however, goes well beyond diagnosis and routine analyses by describing the methods to perform research on the honey bee and associated organisms. Where necessary, the *BEEBOOK* recognizes existing standards such as those provided by the OIE and OECD, and presents a harmonized compendium of research methods, written and reviewed by an international team of scientists.

In addition to producing a bench-friendly manual, and in an effort to make the methods broadly available, every paper forming the *BEEBOOK* is also available as open access articles in several special issues of the *Journal of Apicultural Research.*

To further build on the availability of digital media, a novel concept was developed around the manual. An online version of the manual was created, where each method can be discussed and improvements suggested. Development work on the online *BEEBOOK* platform started in 2009, and the current iteration can be found at www.coloss.org/beebook. On the platform, each webpage describing a method has a comment field, which can be used to suggest changes or additions. Users can thus assist with the improvement and further development of the *BEEBOOK*. Once sufficient updates have accumulated online, a new print version of the manual can then be edited and published. Such a Wiki-like tool is especially useful for following fast evolving fields, such as for molecular protocols.

The *BEEBOOK* is a tool for all who want to do research on honey bees. It was written in such a way that those new to honey bee research can use it to start research in a field with which they may not be familiar. Of course, such an endeavour is often limited by the availability of complex and expensive machinery and other equipment. However, provided access to and training on the necessary equipment are secured, the instructions provided in the *BEEBOOK* can be followed by everyone, from undergraduate student to experienced researcher. All details on how to implement instructions are given.

The editors and author team hope that the *BEEBOOK* will serve as a reference tool for honey bee and other researchers globally. As with the original *Drosophila* book that evolved into a journal where updates and new methods are published, we hope that the honey bee research community will embrace this tool and work to improve it. The online platform is open for everyone to use and further contribute to the development of our research field.

The study of honey bees is globally relevant and remarkably varied. In the applied sense, honey bees have been studied due to their use as producers of honey and suppliers of pollination services in agricultural situations. Furthermore, honey bees have been used as a model organism to address basic questions in multiple scientific disciplines. Consequently, the editorial and author team felt it necessary to develop a Volume focused solely on protocols related to studying the organism and the colony in which it resides. This effort resulted in the production of Volume I of the COLOSS *BEEBOOK*, Standard Methods for *Apis mellifera* Research.

In Volume I of the COLOSS *BEEBOOK*, 167 international scientists from 29 countries have collaborated to produce 18 chapters, including over a thousand protocols related to studying honey bees and their colonies. These chapters include research protocols associated with honey bees in the following scientific domains: anatomy (Carreck *et al.*, 2013), artificial rearing of *A. mellifera* larvae (Crailsheim *et al.*, 2013), behaviour (Scheiner *et al.*, 2013), cage studies (Williams *et al.*, 2013), cell cultures (Genersch *et al.*, 2013), characterization of subspecies and ecotypes (Meixner *et al.*, 2013), chemical ecology (Torto *et al.*, 2013), estimation of colony strength parameters (Delaplane *et al.*, 2013), endosymbionts (Engel *et al.*, 2013), geographic information system (Rogers *et al.*, 2013), instrumental insemination (Cobey *et al.*, 2013),

miscellaneous methods (Human *et al.*, 2013), molecular biology (Evans *et al.*, 2013), physiology and biochemistry (Hartfelder *et al.*, 2013), pollination (Delaplane *et al.*, 2013), rearing and selecting queens (Büchler *et al.*, 2013), statistics (Pirk *et al.*, 2013), and toxicology (Medrzycki *et al.*, 2013). It was our intention to be exhaustive when working with senior authors to develop the chapters included in Volume I. We hope that we have included all of the relevant research domains but recognize that, as with any undertaking of such size, we may have overlooked important topics, and new topics and research areas may well emerge in the future. If so, this can be addressed via the online *BEEBOOK* platform (www.coloss.org/beebook), leading to an improved version in the future. We hope that the information provided herein will assist everyone interested in investigating the honey bee.

The western honey bee is a fascinating research model, and one of timeless significance considering the bee's importance to food production and ecosystem sustainability. We hope that we and our team of international colleagues have produced a resource that will be useful into perpetuity. We also hope that you will find research on honey bees to be professionally rewarding and intellectually stimulating.

Acknowledgements

The COLOSS (Prevention of honey bee COlony LOSSes) network aims to explain and prevent massive honey bee colony losses. It was funded through the COST Action FA0803. COST (European Cooperation in Science and Technology) is a unique means for European researchers to jointly develop their own ideas and new initiatives across all scientific disciplines through trans-European networking of nationally funded research activities. Based on a pan-European intergovernmental framework for cooperation in science and technology, COST has contributed since its creation more than 40 years ago to closing the gap between science, policy makers and society throughout Europe and beyond. COST is supported by the EU Seventh Framework Programme for research, technological development and demonstration activities (Official Journal L 412, 30 December 2006). The European Science Foundation as implementing agent of COST provides the COST Office through an EC Grant Agreement. The Council of the European Union provides the COST Secretariat. The COLOSS network is now supported by the Ricola Foundation - Nature & Culture.

References

ASHBURNER, M (1989) *Drosophila: a laboratory manual*. Cold Spring Harbor Laboratory Press; Cold Spring Harbor, USA. 434 pp.

BÜCHLER, R; ANDONOV, S; BIENEFELD, K; COSTA, C; HATJINA, F; KEZIC, N; KRYGER, P; SPIVAK, M; UZUNOV, A; WILDE, J (2013) Standard methods for rearing and selection of *Apis mellifera* queens. In *V Dietemann; J D Ellis; P Neumann (Eds) The COLOSS BEEBOOK, Volume I: standard methods for* Apis mellifera *research. Journal of Apicultural Research* 52(1): http://dx.doi.org/10.3896/IBRA.1.52.1.07

CARRECK, N L; ANDREE, M; BRENT, C S; COX-FOSTER, D; DADE, H A; ELLIS, J D; HATJINA, F; VANENGELSDORP, D (2013) Standard methods for *Apis mellifera* anatomy and dissection. In *V Dietemann; J D Ellis; P Neumann (Eds) The COLOSS BEEBOOK, Volume I: standard methods for* Apis mellifera *research. Journal of Apicultural Research* 52(4): http://dx.doi.org/10.3896/IBRA.1.52.4.03

COBEY, S W; TARPY, D R ; WOYKE, J (2013) Standard methods for instrumental insemination of *Apis mellifera* queens. In *V Dietemann; J D Ellis; P Neumann (Eds) The COLOSS BEEBOOK, Volume I: standard methods for* Apis mellifera *research. Journal of Apicultural Research* 52(4): http://dx.doi.org/10.3896/IBRA.1.52.4.09

CRAILSHEIM, K; BRODSCHNEIDER, R; AUPINEL, P; BEHRENS, D; GENERSCH, E; VOLLMANN, J; RIESSBERGER-GALLÉ, U (2013) Standard methods for artificial rearing of *Apis mellifera* larvae. In *V Dietemann; J D Ellis; P Neumann (Eds) The COLOSS BEEBOOK, Volume I: standard methods for* Apis mellifera *research. Journal of Apicultural Research* 52(1): http://dx.doi.org/10.3896/IBRA.1.52.1.05

DELAPLANE, K S; DAG, A; DANKA, R G; FREITAS, B M; GARIBALDI, L A; GOODWIN, R M; HORMAZA, J I (2013) Standard methods for pollination research with *Apis mellifera*. In *V Dietemann; J D Ellis; P Neumann (Eds) The COLOSS BEEBOOK, Volume I: standard methods for* Apis mellifera *research. Journal of Apicultural Research* 52(4): http://dx.doi.org/10.3896/IBRA.1.52.4.12

DELAPLANE, K S; VAN DER STEEN, J; GUZMAN, E (2013) Standard methods for estimating strength parameters of *Apis mellifera* colonies. In *V Dietemann; J D Ellis; P Neumann (Eds) The COLOSS BEEBOOK, Volume I: standard methods for* Apis mellifera *research. Journal of Apicultural Research* 52(1): http://dx.doi.org/10.3896/IBRA.1.52.1.03

ENGEL, P; JAMES, R; KOGA, R; KWONG, W K; MCFREDERICK, Q; MORAN, N A (2013) Standard methods for research on *Apis mellifera* gut symbionts. In *V Dietemann; J D Ellis; P Neumann (Eds) The COLOSS* BEEBOOK, *Volume I: standard methods for* Apis mellifera *research. Journal of Apicultural Research* 52(1): http://dx.doi.org/10.3896/IBRA.1.52.1.07

EVANS, J D; SCHWARZ, R S; CHEN, Y P; BUDGE, G; CORNMAN, R S; DE LA RUA, P; DE MIRANDA, J R; FORET, S; FOSTER, L; GAUTHIER, L; GENERSCH, E; GISDER, S; JAROSCH, A; KUCHARSKI, R; LOPEZ, D; LUN, C M; MORITZ, R F A; MALESZKA, R; MUÑOZ, I; PINTO, M A (2013) Standard methodologies for molecular research in *Apis mellifera*. In *V Dietemann; J D Ellis; P Neumann (Eds) The COLOSS* BEEBOOK, *Volume I: standard methods for* Apis mellifera *research. Journal of Apicultural Research* 52(4): http://dx.doi.org/10.3896/IBRA.1.52.4.11

GENERSCH, E; GISDER, S; HEDTKE, K; HUNTER, W B; MÖCKEL, N; MÜLLER, U (2013) Standard methods for cell cultures in *Apis mellifera* research. In *V Dietemann; J D Ellis; P Neumann (Eds) The COLOSS* BEEBOOK, *Volume I: standard methods for* Apis mellifera *research. Journal of Apicultural Research* 52(1): http://dx.doi.org/10.3896/IBRA.1.52.1.02

HARTFELDER, K; GENTILE BITONDI, M M; BRENT, C; GUIDUGLI-LAZZARINI, K R; SIMÕES, Z L P; STABENTHEINER, A; DONATO TANAKA, É; WANG, Y (2013) Standard methods for physiology and biochemistry research in *Apis mellifera*. In *V Dietemann; J D Ellis; P Neumann (Eds) The COLOSS* BEEBOOK, *Volume I: standard methods for* Apis mellifera *research. Journal of Apicultural Research* 52(1): http://dx.doi.org/10.3896/IBRA.1.52.1.06

GREENSPAN, R J (2004) *Fly pushing: the theory and practice of* Drosophila *genetics (Second Ed.)*. Cole Spring Harbor Laboratory Press; Cold Spring Harbor, USA. 191 pp.

HUMAN, H; BRODSCHNEIDER, R; DIETEMANN, V; DIVELY, G; ELLIS, J; FORSGREN, E; FRIES, I; HATJINA, F; HU, F-L; JAFFÉ, R; JENSEN, A B; KÖHLER, A; MAGYAR, J; ÖZIKRIM, A; PIRK, C W W; ROSE, R; STRAUSS, U; TANNER, G; TARPY, D R; VAN DER STEEN, J J M; VAUDO, A; VEJSNÆS, F; WILDE, J; WILLIAMS, G R; ZHENG, H-Q (2013) Miscellaneous standard methods for *Apis mellifera* research. In *V Dietemann; J D Ellis; P Neumann (Eds) The COLOSS* BEEBOOK, *Volume I: standard methods for* Apis mellifera *research. Journal of Apicultural Research* 52(4): http://dx.doi.org/10.3896/IBRA.1.52.4.10

LINDSLEY, D L; GRELL, E H (1968) *Genetic variations of* Drosophila melanogaster. Carnegie Institute of Washington; Washington, USA. 472 pp.

MEDRZYCKI, P; GIFFARD, H; AUPINEL, P; BELZUNCES, L P; CHAUZAT, M-P; CLAßEN, C; COLIN, M E; DUPONT, T; GIROLAMI, V; JOHNSON, R; LECONTE, Y; LÜCKMANN, J; MARZARO, M; PISTORIUS, J; PORRINI, C; SCHUR, A; SGOLASTRA, F; SIMON DELSO, N; VAN DER STEEN, J J F; WALLNER, K; ALAUX, C; BIRON, D G; BLOT, N; BOGO, G; BRUNET, J-L; DELBAC, F; DIOGON, M; EL ALAOUI, H; PROVOST, B; TOSI, S; VIDAU, C (2013) Standard methods for toxicology research in *Apis mellifera*. In *V Dietemann; J D Ellis; P Neumann (Eds) The COLOSS BEEBOOK, Volume I: standard methods for* Apis mellifera research. *Journal of Apicultural Research* 52(4): http://dx.doi.org/10.3896/IBRA.1.52.4.14

MEIXNER, M D; PINTO, M A; BOUGA, M; KRYGER, P; IVANOVA, E; FUCHS, S (2013) Standard methods for characterising subspecies and ecotypes of *Apis mellifera*. In *V Dietemann; J D Ellis; P Neumann (Eds) The COLOSS BEEBOOK, Volume I: standard methods for* Apis mellifera research. *Journal of Apicultural Research* 52(4): http://dx.doi.org/10.3896/IBRA.1.52.4.05

NEUMANN, P; CARRECK, N L (2010) Honey bee colony losses. *Journal of Apicultural Research* 49(1): 1-6. http://dx.doi.org/ 10.3896/IBRA.1.49.1.01

PIRK, C W W; DE MIRANDA, J R; FRIES, I; KRAMER, M; MURRAY, T; PAXTON, R; NAZZI, F; SHUTLER, D; VAN DER STEEN, J J M; VAN DOOREMALEN, C (2013) Statistical guidelines for *Apis mellifera* research. In *V Dietemann; J D Ellis; P Neumann (Eds) The COLOSS BEEBOOK, Volume I: standard methods for* Apis mellifera research. *Journal of Apicultural Research* 52(4): http://dx.doi.org/10.3896/IBRA.1.52.4.13

ROBERTS, D B (Ed.) (1998) Drosophila: *a practical approach*. Oxford University Press; Oxford, UK. 389 pp.

ROGERS, S R; STAUB, B (2013) Standard use of Geographic Information System (GIS) techniques in honey bee research. In *V Dietemann; J D Ellis; P Neumann (Eds) The COLOSS BEEBOOK, Volume I: standard methods for* Apis mellifera research. *Journal of Apicultural Research* 52(4): http://dx.doi.org/10.3896/IBRA.1.52.4.08

SCHEINER, R; ABRAMSON, C I; BRODSCHNEIDER, R; CRAILSHEIM, K; FARINA, W; FUCHS, S; GRÜNEWALD, B; HAHSHOLD, S; KARRER, M; KOENIGER, G; KOENIGER, N; MENZEL, R; MUJAGIC, S; RADSPIELER, G; SCHMICKLI, T; SCHNEIDER, C; SIEGEL, A J; SZOPEK, M; THENIUS, R (2013) Standard methods for behavioural studies of *Apis mellifera*. In *V Dietemann; J D Ellis; P Neumann (Eds) The COLOSS BEEBOOK, Volume I: standard methods for* Apis mellifera research. *Journal of Apicultural Research* 52(4): http://dx.doi.org/10.3896/IBRA.1.52.4.04

TORTO, B; CARROLL, M J; DUEHL, A; FOMBONG, A T; GOZANSKY, K T; NAZZI, F; SOROKER, V; TEAL, P E A (2013) Standard methods for chemical ecology research in *Apis mellifera*. In *V Dietemann; J D Ellis; P Neumann (Eds) The COLOSS BEEBOOK, Volume I: standard methods for* Apis mellifera research. *Journal of Apicultural Research* 52(4): http://dx.doi.org/10.3896/IBRA.1.52.4.06

WILLIAMS, G R; ALAUX, C; COSTA, C; CSÁKI, T; DOUBLET, V; EISENHARDT, D; FRIES, I; KUHN, R; MCMAHON, D P; MEDRZYCKI, P; MURRAY, T E; NATSOPOULOU, M E; NEUMANN, P; OLIVER, R; PAXTON, R J; PERNAL, S F; SHUTLER, D; TANNER, G; VAN DER STEEN, J J F; BRODSCHNEIDER, R (2013) Standard methods for maintaining adult *Apis mellifera* in cages under *in vitro* laboratory conditions. In *V Dietemann; J D Ellis; P Neumann (Eds) The COLOSS BEEBOOK, Volume I: standard methods for* Apis mellifera research. *Journal of Apicultural Research* 52(1): http://dx.doi.org/10.3896/IBRA.1.52.1.04

WILLIAMS, G R; DIETEMANN, V; ELLIS, J D; NEUMANN P (2012) An update on the COLOSS network and the "*BEEBOOK*: standard methodologies for *Apis mellifera* research". *Journal of Apicultural Research* 51(2): 151-153. http://dx.doi.org/10.3896/IBRA.1.51.2.01

Journal of Apicultural Research 52(4): (2013)
DOI 10.3896/IBRA.1.52.4.03

REVIEW ARTICLE

Standard methods for *Apis mellifera* anatomy and dissection

Norman L Carreck[1,2*], **Michael Andree**[3], **Colin S Brent**[4], **Diana Cox-Foster**[5], **Harry A Dade**[†], **James D Ellis**[6], **Fani Hatjina**[7] and **Dennis vanEnglesdorp**[8]

[1]International Bee Research Association, Unit 6, Centre Court, Treforest, CF37 5YR, UK.
[2]Laboratory of Apiculture and Social Insects, School of Life Sciences, University of Sussex, Falmer, Brighton, East Sussex, BN1 9QG, UK.
[3]University of California Cooperative Extension, Oroville, CA 95965, USA.
[4]USDA, Arid Land Agricultural Research Center, Maricopa, AZ, USA.
[5]Department of Entomology, Penn State University, University Park, Pennsylvania PA 16802, USA.
[6]Honey Bee Research and Extension Laboratory, Department of Entomology and Nematology, University of Florida, Steinmetz Hall, Natural Area Dr., P.O. Box 110620, Gainesville, FL, 32611, USA.
[7]Hellenic Institute of Apiculture, Hellenic Agric. Organization 'DEMETER', Nea Moudania, Greece.
[8]Department of Entomology, University of Maryland, College Park, MD 20742 USA.
[†]Deceased.

Received 3 July 2013, accepted subject to revision 15 July 2013, accepted for publication 1 August 2013

*Corresponding author: Email: norman.carreck@btinternet.com

Summary

An understanding of the anatomy and functions of internal and external structures are fundamental to many studies on the honey bee *Apis mellifera*. Similarly, proficiency in dissection techniques is vital for many more complex procedures. In this paper, which is a prelude to the other papers of the COLOSS *BEEBOOK*, we outline basic honey bee anatomy and basic dissection techniques.

Métodos estandar para la disección y anatomía de *Apis mellifera*

Resumen

El conocimiento de la anatomía y las funciones de las estructuras internas y externas es fundamental para muchos estudios sobre la abeja de la miel *Apis mellifera*. Del mismo modo, el dominio de técnicas de disección es vital para muchos procedimientos más complejos. En este trabajo, que es un preludio de los demás documentos del BEEBOOK COLOSS, describimos la anatomía básica de abejas y las técnicas básicas de disección.

西方蜜蜂解剖学和解剖的标准方法

摘要

在西方蜜蜂的很多研究中都很有必要了解蜜蜂解剖学和内外结构的功能。同样地，熟练的解剖技术对于许多复杂的研究也很重要。作为COLOSS *BEEBOOK*的开篇，本文概述了基本的蜜蜂解剖学和基本的解剖技术。

Keywords: COLOSS, *BEEBOOK*, honey bee, *Apis mellifera*, anatomy, dissection, autopsy

Footnote: Please cite this paper as: CARRECK, N L; ANDREE, M; BRENT, C S; COX-FOSTER, D; DADE, H A; ELLIS, J D; HATJINA, F; VANENGELSDORP, D (2013) Standard methods for *Apis mellifera* anatomy and dissection. In *V Dietemann; J D Ellis; P Neumann (Eds) The COLOSS* BEEBOOK, *Volume I: standard methods for* Apis mellifera *research. Journal of Apicultural Research* 52(4): http://dx.doi.org/10.3896/IBRA.1.52.4.03

Table of Contents

1. Introduction

This paper is placed first in the COLOSS *BEEBOOK*, because an understanding of honey bee anatomy is essential for much of the work described in the other papers. Similarly, basic dissection techniques are also fundamental to many facets of the study of honey bees. Man has kept honey bees for many thousands of years, and they have long held a fascination for those keen on understanding natural history. The development of our understanding of the anatomy of honey bees has been outlined by Crane (1999). In the modern era, two textbooks have become standard, those by Snodgrass (1956; 2004), and Dade (1962; 2009), and these are still readily available. For the purposes of this paper, we have therefore tried to give essential information only, and we suggest that the reader seeking further information consults these. Much of the section on dissection here is taken from Dade's (1962) work, and for reference we have reproduced his fine plates, and retained the same numbering for these and the figures. The plates themselves are also available separately from the International Bee Research Association in enlarged and laminated form for use at the laboratory bench: http://ibrastore.org.uk/index.php?main_page=product_info&cPath=4&products_id=176

Those seeking further information about the functions of the structures shown here are suggested to consult two other books, those by Goodman (2003), and Stell (2012). Specific techniques of dissection for the diagnosis of nosema infection and tracheal mite infestation are given in the relevant *BEEBOOK* papers (Fries *et al.*, 2013; Sammataro *et al.*, 2013).

2. Anatomy

Table 1 summarises the external and internal anatomy of the honey bee. The list of structures is not exhaustive, but is presented based on the structures discussed in the text and shown in the figures and plates of Dade (1962), which are reproduced here for reference using his original numbering. Only notable muscles, tergites, sternites, ganglia, and other "minor" or "repeated" structures labelled in the figures or plates are included in the table. Most information regarding structure definition or function is from Dade (1962) but supplemented with information from Goodman (2003), the social insect anatomy glossaries at antbase.org (2008) and the Hymenoptera Anatomy Ontology Portal (2013). If column 2 shows "not labelled", the structure is not labelled in any plate or figure, though we felt it necessary to define. Though unlabelled, the plate or figure where the structure is shown is included in column 2. If "not shown", then the structure is not shown in any plate or figure, but is worth mentioning nonetheless. We list in column 3 the caste(s) in Dade (1962) figures or plates for which the structures are shown. "Generic" means that the image is presented stylized for two or more castes. For example, the abdomen is shown for the drone, worker and queen in different figures or plates. However it is shown generically (no specific caste) in other figures or plates. Columns 4 and 5 list the life stage (egg, embryo, larva, prepupa, pupa, adult) and structure location (head, thorax and/or abdomen and internal/external) shown in the figures or plates. That does not mean the structures are unique to that life stage or figure or plate, only that they were presented as such in Dade (1962).

3. Dissection

3.1 Apparatus and materials

3.1.1. The dissecting microscope

The type of instrument required for dissection is a prismatic dissecting microscope and suitable lighting source. Instruments of this kind embody prisms which erect the image, so that we can see our tools moving in the correct directions. A compound microscope is not suitable as it inverts the image. Binocular instruments give stereoscopic vision, enabling us to perceive depth in the object as well as to use both eyes. The most useful magnification is about x 20, and higher powers are neither useful nor desirable.

3.1.2. Dissecting instruments

Only a few instruments are needed (Fig. 36), but it is important that three of them should be of exactly the right kind.

- The scissors should be 'cuticle' scissors, not less than 90 mm in length, and not much longer, with very fine points which cut cleanly right up to their tips. Looked at sideways, they should be very slim.
- Forceps need to have very fine points and grip very firmly at their extreme tips.
- A very sharp and finely-pointed knife is the third important tool. The Swann-Morton scalpel No. 3 with replaceable Swann-Morton No. 11 blades of the correct shape is widely used.
- A pair of needles, mounted in metal handles.
- Pasteur pipettes.
- Coarse forceps.
- A stout wire, bent into an L-shape, its long limb being about 150 mm long and the short one 20 mm long. The best material is brass rod, 5 mm thick. This brass wire has to be heated (Section 3.1.5.; Plate 1A).
- Two or three dissecting dishes need to be made from flat round metal tins about 75 mm in diameter. These are to be filled with melted beeswax to within 6 mm of the top of the rim; the wax must then be allowed to solidify. The surface of the wax has to be re-melted frequently, and this is done most conveniently by turning a Bunsen burner flame downwards over the dish.

Table 1. External and internal anatomy of the honey bee. The figure and plate numbers apply both to this paper and to Dade (1962).

Structure: singular(plural)	Figures (F) or Plates (P) where labelled	Caste presented in Figures (F) or Plates (P)	Life stage shown in Figures (F) or Plates (P)	Structure location	Definition/function
abdomen(s)	F20, 22, P4, 5, 8-10, 14-17	drone (P14, 15), worker (P4, 8, 9, 10), queen (P16, 17) generic (F20, 22, P5)	adult	3rd main body section	Third tagma of the body. It is located after the thorax.
abdominal muscle(s)	P10	worker	adult	abdomen, internal	Muscles responsible for general abdominal movement and movement of internal abdominal organs.
acid gland(s) of sting	P10, 11	worker (P10), generic (P11)	adult	abdomen, internal	A long, forked tubule with slightly expanded trips. The single, proximal tube widens for form the venom sac, in which the gland secretion is stored. Also known as the "venom gland".
acinus(-ni)	not labelled				One of the small, sac-like dilations composing a compound gland. For example, they are present in the hypopharyngeal glands and look like a string of onions.
adult(s)	not labelled				The mature stage of honey bee.
alkaline gland(s) of sting	P10, 11	worker (P10), generic (P11)	adult	abdomen, internal	White, strap-shaped organ that empties its secretion into the sting chamber.
antenna(-ae)	F5, 28, P2, 18B, C	drone (F28), all (P2), generic (F5, P18B, C)	embryo (F5), larva (P18B, C), adult (F28, P2)	head, external	Important, paired sensory organs on bees. They aid in tactile, olfactory, and gustatory sensory perception.
antenna(-ae) cleaner	P6	generic	adult	foreleg, thorax, external	A notch with comb structure on the basitarsus of the foreleg of the honey bee. When the leg is bent, the fibula and notch form a structure through which the bee pulls its antennae to clean it.
antennal lobe(s)	F24, P9, 10, 13	worker (P9, 10, 13), generic (F24)	adult	head, internal	Areas in the deutocerebrum composed of bundles of nerve fibres that connect with the sense organs of the antennae.
antennal vesicle(s)	F20	generic	adult	head, internal	Small pulsating area under the base of the antennae. Small vessels run from the vesicles into the antennae to supply them with haemolymph.
anus(-ni)	P5, 16C, 19	generic (P5, 19), queen (P16C)	adult (5P, 16C), larva (P19)	abdomen, external	An orifice in the proctiger through which wastes are expelled.
aorta(-ae)	F20, P12	generic	adult	head, thorax, internal	Continuous with the heart, it is a blood vessel that runs through the thorax to the head, where its end opens below the brain, thus supplying haemolymph to the brain/head.
apodeme(s)	P12	worker	adult	abdomen, internal	Peg-like, ingrowth extensions of thickened plate edges of the exoskeleton. They serve as points of attachment for muscles and may support internal organs.
arcus(-ci)	P6	generic	adult	legs, thorax, external	An arc on the ventral surface of the "foot" or pretarsus to which the dorsally-located manubrium is attached.
arolium(-ia)	P6	generic	adult	legs, thorax, external	A pad on the pretarsus that is normally folded and raised between the claws. Bees unfold the arolium for adhesion when walking on a smooth surface.
auricle(s)	P7	worker	adult	hind legs, thorax, external	A sloping shelf on the end of the basitarsus (i.e. in the pollen press). It has a textured surface and a fringe of hairs to facilitate pollen processing.
basalare(s) and basalare muscle(s)	F9	generic	adult	thorax, external (basalare) and internal (basalare muscle)	A plate on the thorax at the base of the wing that is hinged to a pleurite and attached to a muscle. Contraction of the muscle (basalar muscle) causes the basalare to swing inward on its hinge, thus pulling the leading edge of the wing down during flight.
basement membrane(s)	not shown			all, internal	Supportive layer for epidermal cells composed of basal lamina and collagen fibres. It separates the haemocoel from the exoskeleton.
basitarsus(-rsi)	P6, 7	worker (P7), generic (P6)	adult	legs, thorax, external	The first, and largest, tarsomere (or subsection of the tarsus).
bulb(s) of endophallus(-lli)	P14, 15	drone	adult	abdomen, internal	An ovoid body on the distal end of the endophallus, being crescent-shaped and having roughly triangular sclerotized plates in its walls and trough-like internal projections. Dorsal and lateral plates of bulb are seen in P15.
bulb(s) of sting	F12, P11	generic	adult	abdomen, internal	Inflated organ between the ramus and shaft of sting. It is continuous with the stylet and full of venom.

Table 1. cont'd

Structure: singular(plural)	Figures (F) or Plates (P) where labelled	Caste presented in Figures (F) or Plates (P)	Life stage shown in Figures (F) or Plates (P)	Structure location	Definition/function
bursa(-ae)	F32, 33	queen	adult	abdomen, internal	A wide, membranous pouch at the anterior end of the sting chamber.
brain(s)	F20, 23, 24, P13	worker (P13), generic (F20, 23, 24)	larva (F23), adult (P13, F20, 24)	head, internal	The centre of sensory perception in the bee. It principally receives stimuli from the eye and antennae and transmits the nervous impulses to the motor centres (ganglia) of the ventral nerve cord.
cardo(-dines)	F3, 4A, P2D, 3	worker (P2D), generic (F3, 4A, P3)	adult	head, external	The basal hinge of the maxilla, located on the proximal end of the proboscis. The cardines are instrumental in extending the proboscis.
cervix(-ices)	not labelled but shown in P15	drone	adult	abdomen, internal	The part of the shaft of the endophallus in drone honey bees between the horns and bulb.
chitin(s)	not shown			all, external	Tough, protective polysaccharide that forms much of the insect exoskeleton.
cibarium(-ia)	F15, P13	worker (P13), generic (F15)	adult	head, internal	Food chamber in the bee mouth. Muscles attached to it cause it to act as a sucking pump, thus raising fluid through the proboscis.
clasper(s)	P15	drone	adult	abdomen, external	External part of the reproductive organs of drones. They are reduced to small sclerites attached to the sternite of A9.
claw(s)	P6	generic	adult	legs, thorax, external	Pair of strong, recurved structures present on the pretarsus (foot). They provide secure footing on rough surfaces. Also known as unguis.
clypeus(-pei)	P2C	worker	adult	head, external	A plate of the exoskeleton on the "face" of the bee.
compound eye(s)	P2C, 9, 13	worker	adult	head, external	The two, large eyes on the head of the bee. The external surface of both eyes is an elongated oval, strongly convex and consists of the lenses of thousands of ommatidia.
corbicula(-ae)	P7	worker	adult	hind leg, thorax, external	The "pollen basket" or pollen carrying apparatus on the hind leg of the worker bee.
corna(-nua)	See "horn"				Synonymous with "horn".
corpus (-pora) allatum (-ta)	P13	worker	adult	head, internal	Paired endocrine glands behind the brain. They produce juvenile hormone.
corpus (-pora) cardiacum (-iaca)	not shown			head, internal	Two small knots of tissue just in front of the corpora allata, on either side of the aorta. They pour the hormones they produce into the haemolymph, stimulating larval prothoracic endocrine glands to produce ecdysone.
corpus (-pora) pedunculatum (-ta)	not shown			head, internal	Also called "mushroom bodies". Buried in protocerebrum, under the ocelli, they contain small groups of nerve cells. They coordinate the actions of the insect according to information received from the sense organs.
coxa(-ae)	P5, 7	worker (P7), generic (P5)	adult	legs, thorax, external	Segment of the leg, closest to the thorax.
crop(s)	P8B, 9	worker	adult	abdomen, internal	Also known as the "honey stomach", the crop is a transparent bag in which foraging bees store nectar after it has been collected from flowers and while it is being transported to the hive. The crop is the last section of the foregut.
crystalline cone(s)	F26	generic	adult	head, internal	A transparent area behind the ommatidium lens that is surrounded by pigment cells.
cuticle(s)	not shown			all, external	The main layer of the exoskeleton. It is secreted by epidermal cells which lie beneath it. It has two layers: exocuticle and endocuticle. It is covered by the epicuticle.
deutocerebrum(s)	not shown			head, internal	Part of the brain. It is composed of bundles of nerve fibres connected with the sense organs of the antennae.
dorsal diaphragm(s)	F20, P8C	worker (P8C), generic (F20)	adult	abdomen, internal	A thin, transparent membrane spread over the roof of the abdomen and attached to the apodemes of the tergites and sternites. Responsible (with the ventral diaphragm) for setting up circulation inside the abdomen and for drawing blood from the thorax into the abdomen.
duct(s) of hypopharyngeal gland(s)	F15	generic	adult	head, internal	Ducts leading from the hypopharyngeal gland to the back of the hypopharynx. The duct carries the secretions of the hypopharyngeal gland to the mouth of the bee.

Table 1. cont'd

Structure: singular(plural)	Figures (F) or Plates (P) where labelled	Caste presented in Figures (F) or Plates (P)	Life stage shown in Figures (F) or Plates (P)	Structure location	Definition/function
duct(s) of salivary gland(s)	F15	generic	adult	head, internal	Ducts leading from the salivary glands of the thorax to the salivarium in the bee mouth, at the base of the glossa.
duct(s) of spermatheca(-ae)	F32, 33	queen	adult	abdomen, internal	Duct connecting the spermatheca to the vagina. At mating, the drones deposit spermatozoa into the oviducts. The spermatozoa then migrate up the spermathecal duct into the spermatheca.
duct(s) of venom gland(s)	F12	generic	adult	abdomen, internal	A tube that runs from the venom gland to the bulb of the sting.
egg(s)	P20	generic	egg		Also known as "ovum", the egg contains the embryo, the youngest juvenile form of the honey bee. They are about 1.5 mm long and 1/3 mm in diameter at the apex.
egg cell(s)	F31	queen	adult	abdomen, internal	The cells present in the ovarioles that later become the mature eggs.
ejaculatory duct(s)	P14	drone	adult	abdomen, internal	The duct in the endophallus through which spermatozoa pass at the moment of copulation.
embryo(s)	P20	generic	egg	egg, internal	The youngest juvenile form of the honey bee. It is present in the egg.
endocuticle(s)	not shown			all, external	The innermost layer of cuticle. It is a soft layer, composed mostly of chitin.
endophallus(-lli)	P15	drone	adult	abdomen, internal	The part of the drone reproductive organ that is inserted into the queen at copulation. It is stored, inverted, in the abdomen.
epicuticle(s)	not shown			all, external	The thin, waxy/waterproof layer covering the exoskeleton. It has the following layers (from outside to inside): cement, wax, polyphenol, and cuticulin. It reduces water loss and blocks invasion of foreign matter.
epidermis(es)	not shown			all, internal	Single layer of epithelial cells that secrete the rest of the layers of the exoskeleton. They are located below the endocuticle and are attached to the basement membrane.
epipharynx(-nges)	F4A, 15, P13	worker (P13), generic (F4A, 15)	adult	head, internal	A soft pad on the inner surface of the labrum that is shaped to fit closely (i.e. form an airtight seal) against the proboscis when the proboscis is in use.
exocuticle(s)	not shown			all, external	The outermost layer of cuticle. It is hard and composed mainly of sclerotin and chitin. It hardens in a process called sclerotization, thus forming the sclerites.
exoskeleton(s)	not shown			all, external	The protective outer shell of the bee. From outside to inside, it is composed of the following layers: epicuticle, cuticle (exocuticle and endocuticle,), epidermis, and basement membrane. It protects the internal organs and serves as locations for muscle attachment.
fat body(ies)	P10	worker	adult	abdomen, internal	A layer of conspicuous, creamy cells concentrated principally on the floor and roof of the abdomen. The fat body contains fat cells, which are mainly fat but also may contain protein and glycogen.
femur(-mora)	P6, 7	worker (P7), generic (P6)	adult	legs, thorax, external	From the thorax, the 3rd segment of each bee leg.
fibula(-ae)	P6	generic	adult	foreleg, thorax, external	A jointed spur, present on the foreleg tibia, that is part of the antenna cleaner complex.
flabellum(-lla)	P3	generic	adult	head, external	The small, rounded scoop at the end of the glossa. Also known as the labellum.
flagellum(-lla)	F28A	drone	adult	head, external	The largest part of the bee antenna. It contains 11 segments in queens and workers and 12 segments in drones.
fold(s), forewing(s)	P6	generic	adult	wing, external	Folds on the trailing edge of the forewing to which the hamuli on the leading edge of the hind wings connect.
follicle cell(s)	F31	queen	adult	abdomen, internal	Cells that cluster around the egg cells in the ovariole to form a sheath around each egg cell. They disappear once the egg is mature, thus leaving the network of markings on the exterior of the egg.

Table 1. cont'd

Structure: singular(plural)	Figures (F) or Plates (P) where labelled	Caste presented in Figures (F) or Plates (P)	Life stage shown in Figures (F) or Plates (P)	Structure location	Definition/function
food canal(s)	F4B	generic	adult	head, internal	A tube formed when the galeae and labial palps are brought close together. It surrounds the glossa and is used to suck nectar, honey, water or other liquids.
foramen(-mina)	P2D	worker	adult	head, external	A hole below the occiput on the posterior portion of the head through which the organs inside the head are connected to the thorax.
foregut(s)	not labelled			head, thorax, abdomen, internal	The mouth cavity, oesophagus, and crop of the honey bee. It is lined with cuticle. Also called stomodaeum in embryonic and larval bees.
foreleg(s)	F5, P6	generic	embryo (F5), adult (P6)	thorax, external	The front pair of the three pairs of legs of the honey bee. These are located on the prothorax and contain the antenna cleaner. Also called proleg.
forewing(s)	not labelled but seen in P6			thorax, external	The larger of the paired wings on both sides of the body. The forewing is located on the mesothorax.
fossa(-ae)	P2D, 3	worker (P2D), generic (P3)	adult	head, external	A U-shaped hollow on the posterior portion of the head.
frons(-ntes)	P2C	worker	adult	head, external	An area on the anterior portion of the head (face). It is the segment best seen below the ocelli but above and beside the antennae. Also called the "brow".
furca(-ae)	P9, 10, 12	worker (P9, 10), generic (P12)	adult	thorax, internal	Hardened processes of the thoracic sternites that reach inside the thorax. They may protect internal organs or have other functions.
furcula(-ae) of the sting(s)	P11	generic	adult	abdomen, internal	Small, recurved process rising from the base of the bulb as two branches which unite after curving over it dorsally. It provides sites to which muscles attach in the sting.
galea(-ae)	F4B, P2D, E, 3	worker (P2D, E), generic (F4B, P3)	adult	head, external	Paired parts of the proboscis that, with the labial palps, form the food canal.
ganglion(-ia)	F23, P10, 11, 12	worker (P10, 11), generic (F23, P12)	larva (F23), adult (P10-12)	thorax, abdomen, internal	Nerve centres located in the thorax and abdomen. They are responsible, primarily, for movement and organ control.
gena(-ae)	P2C, 13	worker	adult	head, external	The "cheek" region or lateral plates on the bee head.
glossa(-ae)	F4B, 15, P2D, E, 3	worker (P2D, E), generic (F4B, 15, P3)	adult	head, external	The bee "tongue", the distal tip of which contains the flabellum. It is a hollow tube of thin, tough membrane and is flattened and curled at its sides. It is covered with small hairs.
glossal rod(s)	F4B	generic	adult	head, internal	A slender rod that stiffens the glossa and which can be drawn backwards by muscles in the prementum.
hamulus(-li)	P6	generic	adult	wing, external	Hooks on the leading edge of the hind wing that latch onto a fold on the trailing edge of the fore-wing, thus joining the fore- and hind wings.
head(s)	P2	drone (P2B), queen (P2A), worker (PC-F)	adult	1st main body section	First tagma of the bee body. It is located before the thorax.
heart(s)	F20, P8C	worker (P8C), generic (F20)	adult	abdomen, internal	Elongated organ lying just under the roof of the abdomen and attached to the dorsal diaphragm. It has muscular walls and small holes (ostia) which allow haemolymph into the heart. The heart pumps the haemolymph forward through the abdomen and into the thoracic aorta.
haemocoel(s)	not shown			all, internal	The internal body cavity of the bee.
haemolymph	not shown			all, internal	Bee blood.
hind gut(s)	not labelled			abdomen, internal	The Malpighian tubules, small intestines, rectum and anus of the bee. Also called proctodeum in embryonic and larval bees.
hind leg(s)	F5, P7	worker (P7), generic (F5)	embryo (F5), adult (P7)	thorax, external	The last pair of the three pairs of legs of the honey bee. These are located on the metathorax and contain the pollen press, pollen brush, and corbicula (pollen basket).
hind wing(s)	not labelled but seen in P6			thorax, external	The smaller of the paired wings on both sides of the body. The hind wing is located on the metathorax.

Table 1. cont'd

Structure: singular(plural)	Figures (F) or Plates (P) where labelled	Caste presented in Figures (F) or Plates (P)	Life stage shown in Figures (F) or Plates (P)	Structure location	Definition/function
horn(s) of endophallus, also corna(-nua) or pneumophysis(-ses)	P15	drone	adult	abdomen, internal	Conspicuous projections on either side of the drone vestibule.
hypopharynx(-nges)	F15	generic	adult	head, internal	A plate on the floor of the cibarium. It is hardened and the front lobe bends downward.
hypopharyngeal gland(s)	P9, 13	worker	adult	head, internal	Paired glands in the head with ducts opening at the base of the hypopharynx. They produce components of brood food.
ileum(-lei)	not labelled but seen in F17, P8B, 9	worker (P8B, 9), generic (F17)	adult	abdomen, internal	Part of the hindgut. Also called the "small intestine". It is a narrow tube, surrounded by circumferential muscle fibres. It is pleated into 6 longitudinal folds.
imago(-gines)	not labelled				The adult or sexually developed insect.
instar(s)	P20	generic	larva and pupa		The developmental stage of the bee between each moult, until sexual maturity (adulthood) is reached.
Johnston organ(s)	not shown			head, internal	A sense organ located in the antennae pedicel. The cells are arranged around the nerve trunk and are believed to be speed-of-flight indicators and also sensitive to gravity and electromagnetic fields.
labellum(-lla)	not labelled as labellum but shown in P3				The small, rounded scoop at the end of the glossa. Also known as the flabellum.
labial palp(s)	F4B, P2D, E, 3	worker (P2D, E), generic (F4B, P3)	adult	head, external	Paired parts of the proboscis that, with the galea, form the food canal.
labium(-ia)	F5, P18B, C	generic	embryo (F5), larva (P18B, C)	head, external	The "lower lip" of the bee from which the inner members of the proboscis are all derived. These include the postmentum, prementum, labial palps, glossa, two paraglossae and the labellum.
labrum(-ra)	F5, P2C, E, 13, 18B, C	worker (P2C, E, 13), generic (F5, P18B, C)	embryo (F5), larva (P18B, C), adult (P2C, E, 13)	head, external	A sclerotized flap hinged to the clypeus on the anterior side of the head.
lacinia(-ae)	F4A, P3	generic	adult	head, external	Part of the maxillae. They press against the epipharynx when the proboscis is in use. This forms an airtight joint to facilitate sucking though the proboscis.
lancet(s)	F12, P11, 18D	generic	adult (F12, P11), prepupa (18D)	abdomen, external	Paired, hardened shafts that are part of the sting apparatus. They are pointed and barbed on the distal end. When the sting is used, muscles cause the lancets to dig into the victim at the sting site.
lancet track(s)	P5, 11	generic	adult	abdomen, internal	A semicircular path on which the ramus of the lancet runs in the process of deploying the sting.
larva(-ae)	P20	generic	larva		The immature stage of the honey bee that emerges from the egg. Larvae spend all their time feeding and growing. This stage immediately precedes the prepupal stage of bee development.
lateral pouch(es) of bursa(-ae)	F32	queen	adult	abdomen, internal	Bulbous sacs located on both sides of the bursa (1 sac per side).
lens(es)	F26	generic	adult	head, external	The outer layer of the eye, both on the ocelli and the ommatidium.
ligula(-ae)	not shown				The region where the base of the glossa and the small pair of paraglossal lobes join the prementum.
lip(s), proventriculus(-li)	F16A-D	generic	adult	abdomen, internal	4 triangular flaps on the apex of the proventriculus can be closed/opened. They are fringed with hairs. The hairs filter pollen and other particles from nectar.
longitudinal commissure(s)	P10	worker	adult	abdomen, internal	Twin nerve trunks that connect ganglia.
longitudinal muscle(s), thorax(-aces)	P8A, 12	worker (P8A), generic (P12)	adult	thorax, internal	Two bundles of muscles that run side by side from the mesothorax tergite and 1st phragma to the 2nd phragma. Contraction of this muscle squeezes the anterior and posterior ends of the thorax together, raising the roof of the thorax and forcing the wings down.
lorum(-ra)	P3	generic	adult	head, external	V-shaped submentum that join the stipites together. It is located between the cardines.

Table 1. cont'd

Structure: singular(plural)	Figures (F) or Plates (P) where labelled	Caste presented in Figures (F) or Plates (P)	Life stage shown in Figures (F) or Plates (P)	Structure location	Definition/function
lumen(-mina) of the proventriculus(-li)	F16B	generic	adult	abdomen, internal	The inside space of the proventriculus.
Malpighian tubule(s)	F17, P8A, B, 9, 19	worker (P8A, B, 9), generic (F17, P19)	adult (F17, P8A, B, 9), larva (P19)	abdomen, internal	Function as the "kidneys" in the bee. They filter nitrogenous wastes from the haemolymph and pass it through the small intestines to the rectum.
mandible(s)	F5, P2C, E, 3, 13, 18B, C	worker (P2C, E, P13), all (P3), generic (F5, P18B, C)	embryo (F5), larva (P18B, C), adult (P2C, E, 3, 13)	head, external	The "jaws" which are hinged to the genae. They are strong, spoon-shaped organs in the worker, concave and ridged on the inner side. The queen's mandibles are toothed, and, along with the drone's, are unspecialized. Mandibles have many functions in the worker bee.
mandible groove(s)	P3	generic	adult	head, external	A straight depression on the mandible down which secretions from the mandibular gland flow.
mandible orifice(s) of gland	P3	generic	adult	head, external	An opening on the mandible that leads to the mandibular gland.
mandibular gland(s)	P13	worker	adult	head, internal	A pair of glands above the mandibles. They are single, somewhat lobate sacs lying under the genae. They are developed in the worker, rudimentary in the drone, and very large in the queen. The secretion of the mandibular glands in queens is known as "queen substance", which has multiple, important functions in the colony.
manubrium(-ria)	P6	generic	adult	legs, thorax, external	A dorsal plate on the pretarsus. It has 5 or 6 long bristles and is attached to the arcus of the arolium.
maxilla(-ae)	F5, P2D, E, 18B, C	worker (P2D, E), generic (F5, P18B, C)	embryo (F5), larva (P18B, C), adult (P2D, E)	head, external	Part of the proboscis composed of the stipites, galeae, laciniae and the maxillary palps.
maxillary palp(s)	P3	generic	adult	head, external	Small appendages that are part of the maxilla of the bee. They are vestiges, with no clear function.
median oviduct(s)	F32, 33, P16C	queen	adult	abdomen, internal	The duct formed where the two lateral oviducts join. Eggs pass from the ovarioles into the lateral oviducts, and then into the median oviduct. The median oviduct opens into the vagina.
mesenteron(s)	labelled as ventriculus in P19	generic	larva	abdomen, internal	The midgut of the immature bee. It becomes the ventriculus in the adult bee.
mesothorax(-aces), T2	P4	generic	adult	thorax, external	The second thoracic segment. The middle legs and forewing are attached to this segment.
metathorax(-aces), T3	P4	generic	adult	thorax, external	The third thoracic segment. The hind legs and hind wing are attached to this segment.
micropyle(s)	not shown			egg, external	A location at the apex of the egg that is not covered by follicle cells but rather by a thin membrane. It is this area through which spermatozoa penetrate when the egg is fertilized.
middle leg(s)	F5	generic	embryo	thorax, external	The middle pair of legs on the thorax. It has a characteristic spine on the distal end of the tibia. Also called mesoleg.
midgut(s)	labelled as ventriculus and/or mesenteron P9			abdomen, internal	Also called ventriculus or mesenteron, the latter in embryonic and larval bees.
mucus gland(s)	P14, 15	drone	adult	abdomen, internal	Club-shaped sacs in the drone, associated with the endophallus. It produces mucus and increases in size as the drone matures. The mucus may be deposited in the mated queen to prevent the escape of spermatozoa.
Nasanov gland(s)	not labelled				Synonymous with scent gland.
nerve(s)	P10	worker	adult	head, thorax, abdomen, legs, antennae, internal	Cells responsible for sensory transmission and processing to/in the ganglia and brain and for muscle movement.
nurse cell(s)	F31	queen	adult	abdomen, internal	Clusters of cells that follow each egg cell in the ovarioles. The nurse cells provide nutrients to the growing eggs and are absorbed as the egg approaches full size.
oblong plate(s)	F12, P5, 10, 18D	generic	adult (F12, P5, 10), prepupa (P18D)	abdomen, external	One of three pairs of plates in the sting which are moved by muscles and cause the sting to deploy.
occiput(s)	P2D	worker	adult	head, external	The upper posterior region of the head.

Table 1. cont'd

Structure: singular(plural)	Figures (F) or Plates (P) where labelled	Caste presented in Figures (F) or Plates (P)	Life stage shown in Figures (F) or Plates (P)	Structure location	Definition/function
ocellus(-lli)	F24, P2C, 9, 13	worker (P2C, 9, 13), generic (F24)	adult	head, external	The three simple eyes on the dorsal surface of the head, between the two compound eyes. They consist of a lens above a layer of simple, elongated retinal cells. Used to measure light intensity.
oenocyte(s)	not shown				Type of cell in the fat body. Function possibly linked to wax production.
oesophagus(-gi)	P9, 12, 13	worker (P9, 13), generic (P12)	adult	head, thorax, internal	The tube connecting the mouth to the crop. It is part of the bee foregut.
ommatidium(-ia)	F26, P13	worker (P13), generic (F26)	adult	head, internal	Elongated bodies of the compound eyes, tapering towards their inner ends. It consists of a lens, crystalline cone (surrounded by pigment cells), a bundle of eight retinula cells (also surrounded by pigment cells).
optic lobe(s)	F24, P9, 13, 18B	worker (P9, 13), generic (F24, P18B)	adult (F24, P9, 13), larva (P18B)	head, internal	Bundles of crossing nerve fibres (chiasmata) connected with the thousands of units in the compound eyes. The lobes are part of the protocerebrum, which is one of the brain's three component parts.
ostium(-ia)	P8C	worker	adult	abdomen, internal	One-way valves in the heart that allow haemolymph to enter the heart when it is dilated, but confined it and force it forward when the heart contracts.
ovariole(s)	F31	queen	adult	abdomen, internal	A long tubule located in the ovary. It contains the egg cells, nurse cells, and follicle cells. Groups of ovarioles (150 or more) coalesce to form the ovary.
ovary(-ies)	P10, 16A-C	worker (P10), queen (P16A-C)	adult	abdomen, internal	Paired organs in the queen, each consisting of a bundle of about 150 ovarioles. They are greatly reduced in the worker.
oviduct(s), lateral	F32, 33, P16B, C	queen	adult	abdomen, internal	A broad tube that is attached to both ovaries. The two lateral oviducts connect directly to ovarioles on their posterior side and to one another to form the median oviduct on their anterior side.
ovum(-va)	not labelled				Another term for "egg".
paraglossa(-ae)	P3	generic	adult	head, external	Part of the proboscis, they are a pair of small appendages of the ligula.
pecten(-tines)	not labelled				Another term for "rastellum".
pedicel(s)	F28A	drone	adult	head, external	The second segment of the bee antenna.
peduncle(s)	not labelled				Another term for "petiole".
petiole(s)	not labelled but shown in P4, 5			thorax, abdomen external	The restricted (narrowed) area linking the thorax to the abdomen. Often called the "waist" or "peduncle".
phallotreme(s)	not shown			abdomen, internal	Genital opening of endophallus.
pharynx(-nges)	F3, 15	generic	adult	head, internal	The gullet-cavity behind the cibarium.
phragma(ae)	P12	generic	adult	thorax, internal	Fence-like ridges that are ingrowths of the exoskeleton. They serve as strong places of attachment for muscles, and struts to stiffen the shell in places where added strength is needed.
pigment cell(s)	F26	generic	adult	head, internal	Cells in the ommatidium that are located around the crystalline cone and the retinula cells. They appear to exclude the light which enters neighbouring ommatidia, thus ensuring stimulation is applied only by the light entering a single ommatidium.
pit(s), anterior head	P2C	worker	adult	head, external	Two depressions at the edges of the clypeus that mark the position of the upper ends of the tentoria.
pit(s), posterior head	P2D	worker	adult	head, external	Two depressions adjoining the foramen that indicate the position of the ends of the tentoria.
planta(-ae)	P6	generic	adult	legs, thorax, external	A sclerite of the pretarsus between the unguitractor and arolium.
pleurite(s)	P4, 5	worker (P4), generic (P5)	adult	thorax, external	In honey bees, this mainly refers to lateral plates (sclerites) of the thoracic exoskeleton. The legs on the pro-, meso-, and metathorax are articulated to the pleurites located on the respective thoracic segments. Pleurites are absent in abdomen.
pneumophysis(-ses)	labelled as horn in P15				Synonymous with "horn".
pollen brush(es)	P7	worker	adult	hind legs thorax, external	On the hind leg, it is the inner side of the flat, broad basitarsi. It is covered with rows of closely set, stiff hairs. It is used for brushing pollen from the abdomen after visits to flowers.

Table 1. cont'd

Structure: singular(plural)	Figures (F) or Plates (P) where labelled	Caste presented in Figures (F) or Plates (P)	Life stage shown in Figures (F) or Plates (P)	Structure location	Definition/function
pollen press(es)	P7	worker	adult	hind leg, thorax, external	The tibio-tarsal joint on the hind leg of the worker bee, modified for use as a pollen manipulator.
postcerebral gland(s)	P9, 13	worker	adult	head, internal	Glands that lie behind the brain. Produces enzymes that are a component of bee saliva.
postmentum(-ta)	P3	generic	adult	head, external	Part of the proboscis. It is articulated to the middle of the lorum and between the cardines.
pouch(es), proventriculus(-li)	F16B-D	generic	adult	abdomen, internal	Pockets behind the proventricular lips in which pollen is collected after being filtered from nectar. Pollen masses pass from here into the ventriculus when full.
prementum(-ta)	F15, P2D, E, 3	worker (P2D, E), generic (F15, P3)	adult	head, external	A sclerite of the proboscis to which is joined the labial palps, glossa, and two paraglossae.
prepupa(ae)	P20	generic	prepupa		The immature stage of bee development between the larval and pupal stages. The prepupal stage occurs in capped cells.
pretarsus(-si)	P6	generic	adult	legs, thorax, external	The bee "foot".
proboscis(es)	P2C, 3	worker (P2), generic (P3)	adult	head, external	An anatomical cluster composed of the labium and maxillae. It has many component parts. It use used by the bee to suck in liquids (nectar, water, honey, etc.), for exchanging food with other bees, and for removing water from nectar.
proctiger(-ra)	P5, 11	generic	adult	abdomen, external	The remains of the 10th abdominal segment. It carries the anus and is fixed to the sting.
proctodeum(-ea)	P19	generic	larva	abdomen, internal	The hindgut of the embryo and larva. It does not connect with the midgut (mesenteron) until the full grown larva has taken its last meal and is ready to pupate. It becomes the Malpighian tubules, small intestine, rectum, and anus in the adult bee.
propodeum(-ea), A1	P4	worker	adult	thorax, external	The first abdominal segment located immediately behind the metathorax. Its tergite (A1, tg) nearly encircles the rear part of the thorax and narrows to fit around the petiole. Its sternite (A1, st) is a small strap underneath the petiole. The precursor for this is shown for larvae in P18A and for prepupae in P18 E (in both, labelled A1).
prothorax(-aces), T1	P4	worker	adult	thorax, external	The first thoracic segment. The two forelegs are attached to the pleurites of this segment. Its tergite encircles the bee neck like a collar. A lobe on either side projects backwards to cover the first spiracle.
protocerebrum(-ra)	F24	generic	adult	head, internal	Part of the brain. It is primarily composed of the optic lobes.
proventriculus(-li)	F16, P9	worker (P9), generic (F16)	adult	abdomen, internal	Valve between the crop and ventriculus that prevents the collected nectar from running into the stomach. It also comprises a filtering apparatus for extracting pollen.
pupa(-ae)	P20	generic	pupa		The immature stage of bee development that occurs after the prepupal stage and before the adult stage. This stage occurs in capped cells and is when the grub-like body of the prepupa begins to develop into that of an adult.
pyloric valve(s)	F17	generic	adult	abdomen, internal	Valve that regulates the passage of material from the ventriculus into the intestine.
quadrate plate(s)	F12, P11	generic	adult	abdomen, external	One of three pairs of plates in the sting which are moved by muscles and cause the sting to deploy.
ramus(-mi)	F12	generic	adult	worker	Proximal end of the lancet. It is flexible and runs on a semicircular track. Pressure by the triangular plate pushes the ramus round its track, thus forcing the lancet in the same direction.
rastellum(-lla)	P7	worker	adult	hind legs, thorax, external	A row of wide and pointed spines on the distal end of the tibia. It is used to rake pollen out of pollen brushes and prevent the pollen mass from escaping the pollen press. Also known as "pecten".
rectal pad(s)	P8A, 9	worker	adult	abdomen, internal	Six partly chitinized pads arranged around the rectum. They reabsorb ions and water (used to collect wastes by the Malpighian tubules) from the rectum.

Table 1. cont'd

Structure: singular(plural)	Figures (F) or Plates (P) where labelled	Caste presented in Figures (F) or Plates (P)	Life stage shown in Figures (F) or Plates (P)	Structure location	Definition/function
rectum(-ta)	P5, 8A, B, 9, 16A	generic (P5) worker (P8A, B, 9), queen (P16A)	adult	abdomen, internal	Part of the hind gut into which the contents of the small intestines empty and out of which wastes pass through the anus. It can expand greatly to hold wastes when bees are unable to leave the colony to defecate.
retaining hair(s) or auricle(s)	P7	worker	adult	hind legs, thorax, external	Hairs on the fringe of the auricle that keep pollen from falling out of the pollen press.
retina(-ae) of ocellus(-lli)	P9, 13	worker	adult	head, internal	Part of the ocellus that detects relative intensity of flight which falls on the lens.
retinula(-ae) cell(s)	F26	generic	adult	head, internal	A bundle of 8 cells in the ommatidium. They are surrounded by pigment cells. The edges of the retinula cells, which meet in the axis of the ommatidium, combine to form a long, narrow rhabdom.
rhabdom(s)	F26	generic	adult	head, internal	A transparent rod that is formed by the hollow left when the edges of 8 retinula cells meet in the axis of the ommatidium. It is striated and likely diffuses light laterally into the retinula cells.
salivarium(-ia)	F15	generic	adult	head, internal	A pouch under the hypopharynx into which opens the common duct of the postcerebral and thoracic salivary glands.
salivary gland(s), thorax	P9	worker (P9)	adult	thorax, internal	Glands in the thorax which are partially responsible for producing enzymes in the saliva.
scape(s)	F28A	drone	adult	head, external	The first antennal segment.
scent canal(s)	P12	worker	adult	abdomen, external	Structure on the 7th abdominal tergite onto which the pheromone of the scent gland is released.
scent gland(s)	P4	worker	adult	abdomen, internal	Lies under the front part of the 7th abdominal tergite. Secretes a pheromone into the scent canal, which is dissipated by the bees when they fan their wings. Also called Nasanov gland.
sclerite(s)	not labelled			all, external	Hardened plates of cuticle.
sclerotin(s)	not labelled			all, external	A tanned protein present in the exocuticle.
scopa(-ae)	not labelled			abdomen, external	Pollen-carrying apparatus on bee body. Also known as "corbicula" or "pollen basket" for honey bees.
scutal fissure(s)	P5	generic	adult	thorax, external	A fissure (line that divides a sclerite) between the scutum and scutellum.
scutellum(-lla), T3, tg	P4, 5	worker (P4), generic (P5)	adult	thorax, external	The tergite of the 3rd thoracic segment. It is visible as a prominent "roll" behind the scutum.
scutum(-ta), T2, tg	P4, 5	worker (P4), generic (P5)	adult	thorax, external	The tergite of the 2nd thoracic segment. It is strongly domed and covers most of the thorax.
semen	P15	drone	adult	abdomen, internal	A mixture of spermatozoa and glandular secretions produced by the drone reproductive organs.
seminal vesicle(s)	P14, 15	drone	adult	abdomen, internal	Two curved, sausage-shaped organs which increase in length and girth as they receive spermatozoa from the testes. Their walls are muscular and lined with glandular tissue. When the drone is sexually mature the vesicles are packed with spermatozoa and glandular secretions.
sense hair(s)	F28C, 29B	drone (F28C), generic (F29B)	adult	head, external	Sensilla with fine bristles projecting from the cuticle on a smooth surface or from in a pit. They are tactile organs, with many located on the antenna.
sense peg(s)	not shown				Similar to sense hairs, they are sensilla with short, stout pegs rather than a bristle, projecting from the cuticle on a smooth surface or from in a pit. They are tactile organs (mechanoreceptors).
sense plate(s)	F28C, 29A	drone (F28C), generic (F29A)	adult	head, antennae, external	Sensilla that consist of a hollow in the cuticle, capped by a thin plate, level with the surrounding surface. They are chemoreceptors.
sensillum(-lla)	F29	generic	adult	head, external	A sense cell or cells with a nerve fibre connected with the central nervous system and its distal end in close connection to the cuticle. There are several types and they are involved in various types of sensory perception (principally mechano- and chemoreceptors).
seta(-ae)	not labelled			all, external	A sensillum that is multicellular, consisting of trichogen (hair shaft), tormogen (the "socket" holding the shaft), and sense cells.

Table 1. cont'd

Structure: singular(plural)	Figures (F) or Plates (P) where labelled	Caste presented in Figures (F) or Plates (P)	Life stage shown in Figures (F) or Plates (P)	Structure location	Definition/function
silk gland(s)	F5, P19	generic	embryo (F5), larva (P19)	abdomen, internal	Long, kinked glands that extend through most of the larva's body. They unite in a common duct which opens in the spinneret on the labium. They become the thoracic salivary glands in the adult bee. The larva uses the glands to produce a silk cocoon while entering the prepupal phase.
sinus(es), dorsal and ventral	not labelled			abdomen, internal	The space between the dorsal/ventral diaphragms and the body wall. It is important for haemolymph circulation in the bee's body.
small intestine(s)	F17; P8A, 16A	worker (P8), queen (P16A), generic (F17)	adult	abdomen, internal	Derived from the proctodeum, it is a narrow tube surrounded by circumferential muscle fibres. The tube is pleated into six longitudinal folds and is coiled. It follows the ventriculus and Malpighian tubules and precedes the rectum in the alimentary tract. Also called ileum.
spermatheca(-ae)	F33, P16B, C	queen	adult	abdomen, internal	A spherical sac which holds the spermatozoa in a mated queen. The spermatozoa can be stored and kept alive here for the life of the queen.
spermathecal gland(s)	F33	queen	adult	abdomen, internal	A gland with two branches that loop over the dorsal surface of the spermatheca. The common duct of these two branches joins the spermathecal duct and the gland is believed to produce a nutrient secretion for the spermatozoa.
spermathecal valve and pump	F33	queen	adult	abdomen, internal	A fixture on the duct of the spermatheca that has muscles which draw spermatozoa out of the spermatheca and forces it into the vagina where it is available to fertilize an egg.
spermatozoon(-zoa)	F30	drone	adult	abdomen, internal	The male gametes. They are also called sperm. They are slender threads, about ¼ mm long. They have head and tail regions, using their tails to swim.
spiracle(s), sp	F5, P5	generic	embryo (F5), adult (P5)	thorax/abdomen, external	The bee's breathing holes. Spiracles are arranged on the lateral sides of the various body segments. They open to the tracheal system and facilitate air exchange into and out of the body.
spiracle plate(s) of sting, A8, tg	P5	generic	adult	abdomen, external	The remains of the 8th abdominal tergite. It composes part of the sting.
sternite(s), st	P4, 5	worker (P4), generic (P5)	adult	thorax/abdomen, external	Ventral plates of the exoskeleton. There is one for each visible body segment of the thorax and abdomen. They are labelled 1-3 on the thorax and 1-7 on abdomen.
sting apparatus(es)	P10, 11, 16B, C	worker (P10), queen (P16B, C), generic (P11)	adult	abdomen, internal	A number of abdominal parts that collectively compose the honey bee sting. These include modified sclerites, muscles, glands, etc. The honey bee sting is used in defence.
sting chamber(s)	P5, 16C	generic (P5), queen (P16C)	adult	abdomen, external	The area covered by the 7th abdominal segment that contains the sting apparatus.
sting shaft(s)	P4, 5, 10	worker (P4, 10), generic (P5)	adult	abdomen, external	The part of the sting that is deployed into the stung victim. It is composed of three parts, the stylet and two lancets.
sting sheath(s)	P5, 11, 18D	generic (P5, 11, 18D)	adult (P5, 10), prepupa (P18D)	abdomen, external	Two soft extensions of the oblong plates about the same length of the sting shaft. It may produce a glandular secretion that is involved in the alarm response.
stipes(-pites)	P3	generic	adult	head, external	Part of the maxillae, which, in turn, is part of the proboscis. The two stipites are joined together by the transverse lorum.
stomodaeum(-ea)	F5, P19	generic	embryo (F5), larva (P19)	head, thorax, abdomen, internal	The foregut of the immature honey bee embryo and larva. It becomes the mouth cavity, oesophagus, and crop in the adult bee.
stylet(s)	F12, P18D	generic	prepupa	abdomen, external	A slender, rigid and sharply pointed rod that, with the two lancets, compose the sting shaft. The stylet is barbed at its point. The two lancets and stylet form a canal, down which the venom flows when the sting is deployed.
subalare(s) and subalare muscle (s)	F9	generic	adult	abdomen, external (subalare) and internal (subalare muscle)	A plate on the thorax that is hinged to the notched pleurite of the mesothorax. Inside the body, this plate is attached to the subalare muscle, the other end of which is anchored to the coxa of the second leg. Contraction of the subalare muscle pulls down the subalare and with it, the trailing edge of the wing.

Table 1. cont'd

Structure: singular(plural)	Figures (F) or Plates (P) where labelled	Caste presented in Figures (F) or Plates (P)	Life stage shown in Figures (F) or Plates (P)	Structure location	Definition/function
suboesophageal ganglion(-ia)	F23, 24, P13	worker (P13), generic (F23, 24)	larva (F23), adult (F24, P13)	head, internal	The ganglion in the head that sends nerves to the mandibles and proboscis.
taenidium(-ia)	not shown			all, internal	Spiral thickenings of the cuticle in the walls of trachea. They keep the trachea open or dilated.
tagma(-mata)				head, thorax, abdomen	The three major body regions of the bee: the head, thorax and abdomen.
tarsomere(s)	P6	generic	adult	legs, thorax, external	The five subdivisions of the tarsus, of which the first and largest is the basitarsus.
tarsus(-si)	P7	worker	adult	legs, thorax, external	A part of the bee leg that contains 5 subdivisions. It follows the tibia.
tegula(-ae)	P4, 5	worker (P4), generic (P5)	adult	thorax, external	A large scale on the mesothorax that overlaps and protects the root of the forewing
tentorium(-ia)	F3, 4A, P3, 13	generic (F3, 4A, P3), worker (P13)	adult	head, internal	Struts which strengthen the head framework.
tergite(s), tg	P4, 5, 8C	worker (P4, 8C), generic (P5)	adult	thorax/abdomen, external	Dorsal plates of the exoskeleton. There is one for each visible body segment of the thorax and abdomen. They are labelled 1-3 on the thorax and 1-7 on abdomen.
testis(-tes)	P14, 15, 19	drone (P14, 15), generic (P19)	adult (P14, 15), larva (P19)	abdomen, internal	Two bundles of tubules in which spermatozoa are produced and matured. They occupy much of the abdomen in newly-emerged drones but they shrink to a smaller size in sexually-mature drones. This occurs when the testes empty their contents into the seminal vesicles.
thoracic collar(s), T1, tg	P4, 5	worker (P4), generic (P5)	adult	thorax, external	A notable thoracic tergite located on the prothorax. It encircles the neck like a collar. A lobe ("lobe" in P4 and "lobe covering spiracle" in P5) on each side projects backwards to cover the spiracle through which tracheal mites (*Acarapis woodi*) enter.
thorax(-aces), T1-3	P4, 5	worker (P4), generic (P5)	adult	2nd main body section	The 2nd tagma of the body. It occurs after the head and before the abdomen. It is the centre of bee locomotion, being the tagma where the wings and legs are attached.
tibia(-ae)	P6, 7	worker (P7), generic (P6)	adult	legs, thorax, external	The 4th segment of the bee leg. It is proximal to the tarsus and distal to the femur.
tormogen(s)	not labelled			all, external	Cell that secretes the cuticle of the "socket" holding the trichogen of the seta.
trachea(-ae) or tracheal trunk (s)	F21, 22, P9	worker (P9), generic (F21, 22)	larva (F21), adult (F22, P9)	all, internal	Tubes in the body connected to spiracles through which respiratory air is delivered to organs. They are maintained in a dilated state by spiral thickenings of cuticle (taenidia) in their walls. They resemble the tracheal of terrestrial vertebrates. They expand into tracheal sacs.
tracheal sac(s)	not labelled but seen in F22			all, internal	Large air bags formed by longitudinal tracheal trunks. They serve as bellows to move air into and out of the body.
tracheoles(s)	not labelled			all, internal	The smallest of the branches that radiate from the tracheal trunks. They have no spiral thickening of the cuticle (taenidia). They are closely applied to the tissues. Here, oxygen brought to the tracheoles is dissolves in the haemolymph and carbon dioxide is removed.
triangular plate(s)	F12, P5, 11, 18D	generic	adult (F12, P5, 11), prepupa (P18D)	abdomen, external	One of three pairs of plates in the sting which are moved by muscles and cause the sting to deploy.
trichogen(s)	not labelled			all, external	The "hair shaft" of the seta.
tritocerebrum(-ra)	not labelled			head, internal	The smallest of the three component parts of the brain. It is obscured by the other parts and sends nerves to the labrum and frons (for the cibarium).
trochanter(s)	P7	worker	adult	thorax, external	The second leg segment. It is proximal to the femur and distal to the coxa.
unguis(-ues)				thorax, external	Synonymous with "claws".
unguitractor plate(s)	P6	generic	adult	thorax, external	The sclerite that corresponds to the site of insertion of the tendon which flexes the pretarsus.
valvefold(s)	F33, P16C	queen	adult	abdomen, internal	A muscular fold in the floor of the vagina that projects upwards as a flap. It may be used to press eggs against the opening of the duct as sperm are released.

Table 1. cont'd

Structure: singular(plural)	Figures (F) or Plates (P) where labelled	Caste presented in Figures (F) or Plates (P)	Life stage shown in Figures (F) or Plates (P)	Structure location	Definition/function
vagina(-ae)	F32	queen	adult	abdomen, internal	The sac-like pouch that follows the median oviduct. Eggs pass through the vagina and the queen can release spermatozoa from the spermatheca, through a duct, to fertilize the egg.
vaginal orifice(s)	F32	queen	adult	abdomen, internal	Opening leading to the vagina.
vas(-sa) deferens(-ntia)	P14, 15	drone	adult	abdomen, internal	Coiled tubes leading from the testes to the seminal vesicles. Spermatozoa pass through the vas deferentia.
venom canal(s)	F12	generic	adult	abdomen, internal	The canal formed by the pair of lancets and the stylet. This is the canal through which venom passes on its way to the victim.
venom gland(s)	labelled as "acid gland"				Synonymous with "acid gland".
venom sac(s)	P10, 11	worker (P10), generic (P11)	adult	abdomen, internal	A single tube proximal to the venom glands (acid glands) that widens to form a large, club-shaped sac in which the secretion of the venom gland is stored. The sac tapers to a narrow duct which connects with and opens into the bulb of the sting.
ventral diaphragm(s)	F20	generic	adult		A thin, transparent membrane spread over the floor of the abdomen and attached to the apodemes of the sternites. Responsible (with the dorsal diaphragm) for setting up circulation inside the abdomen and for drawing blood from the thorax into the abdomen.
ventriculus(-li)	F17, P8A, B, 9, 16A, 19	worker (P8A, B, 9), queen (P16A), generic (F17, P19)	adult (F17, P8A, B, 9, 16A), larva (P19)	abdomen, internal	Also known as mid-gut or stomach. It is derived from the embryonic mesenteron. It is a long, wide tube, lying in a loop in the abdomen. It is constricted at intervals by contracted circumferential muscles. It is lined with epithelium cells which aid in digestion and nutrient uptake.
vertex(-tices)	P2C	worker	adult	head, external	The crown or top of the head.
vertical muscle(s), thorax(-aces)	P8A, 12	worker (P8A), generic (P12)	adult	thorax, internal	Two bundles of muscles that run side by side from the domed scutum of the mesothorax to the sternite of the same. When they contract, the roof of the thorax is pulled down, thus raising the wings.
vestibule(s)	not labelled but shown in P15	drone	adult	abdomen, internal	The part of the shaft of the endophallus in drone honey bees between the horns and the genital opening at A9.
wax gland(s), w	P4	worker	adult	abdomen, internal	Glands located on inner sides of abdominal sternites 4-7 that secrete wax.
wax mirror(s)	P12	worker	adult	abdomen, external	Areas on abdominal sternites 4-7 onto which wax is secreted by the wax gland.

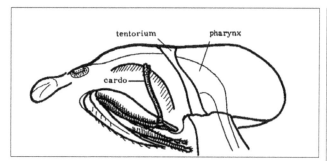

Fig. 3. The proboscis folded, cardines swung back.

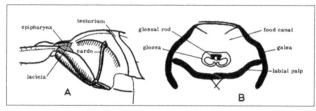

Fig. 4. A. The proboscis extended, cardines swung forward, laciniae pressed against the epipharynx. **B.** cross-section of proboscis.

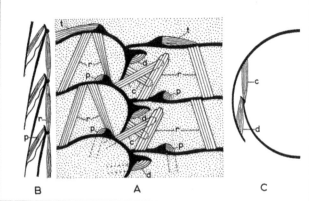

Fig. 11. The principal abdominal muscles. **A.** the diagram shows half of some abdominal segments, laid out; **B.** section through the sternites, showing how the protractor and retractor muscles act on the plates; **C.** half of cross-section of the abdomen, showing how the compressor and dilator muscles work. p. protractors; r, retractors; c. compressors; d. dilators; t. muscles in anterior segments which produce torsion.

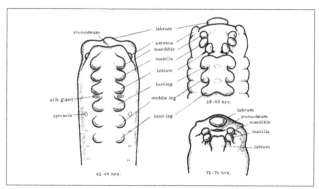

Fig. 5. Development of embryonic mouthparts, showing their progressive arrangement round the mouth, and the final fusion of the two parts of the labium.

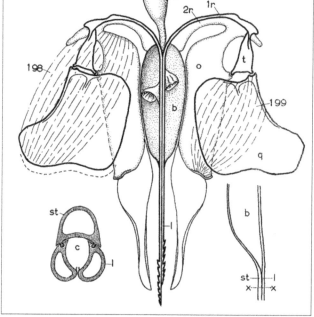

Fig. 12. The sting mechanism: movements of the plates and lancets. The diagram shows the sting flattened out, the moving parts in thick lines. On the left side, the protractor muscle (198) contracts, pulling the quadrate plate forward and thus causing the triangular plate to swing on its articulation with the oblong plate; the movement is transmitted to the ramus of the lancet (1r), which slides on the ramus of the oblong plate (2r) and causes the lancet to be protracted. On the right hand side the lancet is simultaneously retracted by the reverse movements produced by the contraction of the retractor muscle (199). The small drawings show: on the right, a longitudinal section through part of the bulb and shaft; on the left, a cross-section through the shaft. v, duct of venom gland; b, bulb; l, lancet; o, oblong plate; q, quadrate plate; t, triangular plate; 1r, first ramus; 2r, second ramus; st, stylet; c, venom canal.

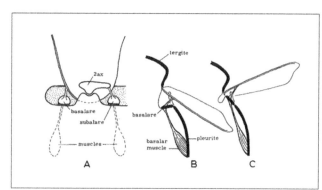

Fig. 9. Diagram, greatly simplified, showing the feathering action of the direct wing muscles. **A.** left forewing, seen from outside the body: the 2nd axillary sclerite (2ax) of the wing is articulated in a notch on the pleurite; the basalare and subalare are hinged to the pleurite and float on the membrane (dotted); their direct muscles are inside the thorax. **B.** cross-section of part of thorax: basalare pulled down by its muscle, drawing down with it the leading edge of the wing.

C. basalare muscle relaxed, while the subalare muscle (hidden) pulls down the subalare and with it the trailing edge of the wing.

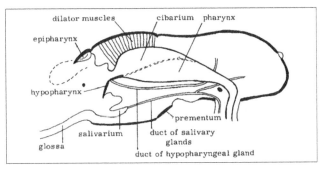

Fig. 15. The cavity of the mouth and the associated structures. A longitudinal section through the head, diagrammatic.

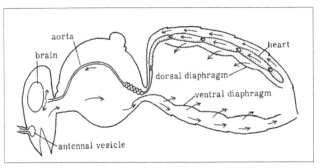

Fig. 20. Diagram illustrating the action of the heart and diaphragms.

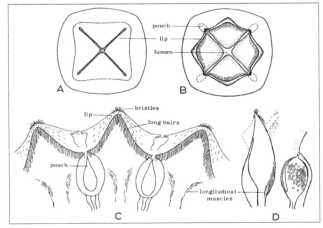

Fig. 16. The proventriculus. **A.** anterior aspect, lips closed; **B.** ditto, lips open to show the short spines and long hairs of the lips, and the lumen partly closed by the muscles below the lips, also the pouches which open into the lumen. **C.** part of the proventriculus laid out after slitting up on one side; three of the four lips are shown, with the pollen pouches between them. **D.** sketch of a longitudinal section, on the left through a lip, on the right through a pouch.

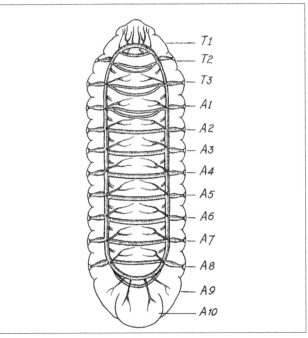

Fig. 21. The larval tracheal system.

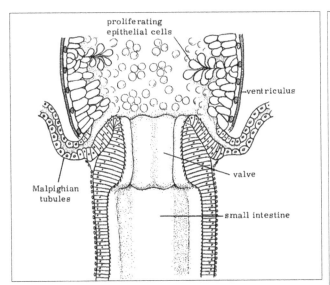

Fig. 17. The pyloric region. Half of the canal is cut away, showing the interior of the other half, with the junction of the ventriculus and the small intestine, and the valve, also the insertion of the Malpighian tubules. Note the small recurved setae lining the valve, also the proliferating digestive cells of the ventricular epithelium.

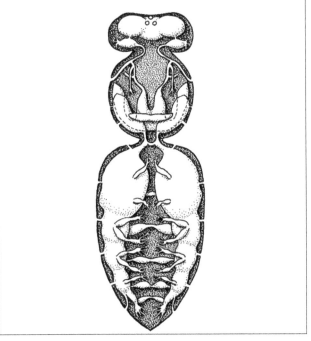

Fig. 22. The principal tracheal sacs and trunks of the adult bee.

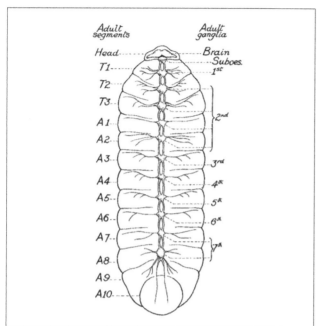

Fig. 23. The larval nervous system. The six ganglia of the head are already combined to form the brain and suboesophageal ganglion. Those of T2 and T3, and A1 and A2, are still discrete; that of A7 is still discrete, but those of A8 to A10 are already fused.

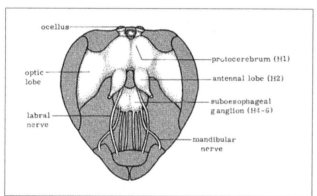

Fig. 24. The brain and principal nerves of the head, anterior aspect. The labral nerves come from the very small tritocerebrum (H3), concealed behind the antennal lobes. The mandibular nerve comes from H4. Two other pairs of nerves (shown but not flagged) come from H5 and H6, and go to the maxillae and labium respectively.

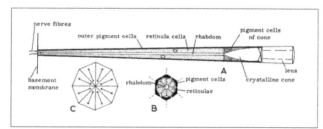

Fig. 26. An ommatidium. **A.** longitudinal section; **B.** transverse section through the rhabdom and retinula cells; **C.** von Frisch's polaroid model of B, consisting of 8 triangles with axes of polarization shown by double arrows.

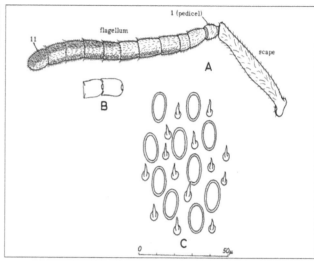

Fig. 28. A. antenna of worker; **B.** jointing of segments; **C.** sense plates and sense hairs on the antenna.

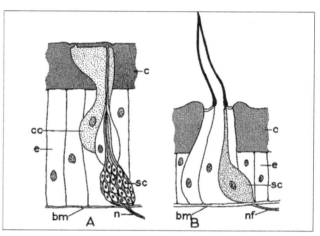

Fig. 29. Sensilla. **A.** a sense plate in section; **B.** a sense hair. c, cuticle; cc, cap cell; sc, sense cells; e, epidermis; bm, basement membrane; n, nerve; nf, nerve fibre.

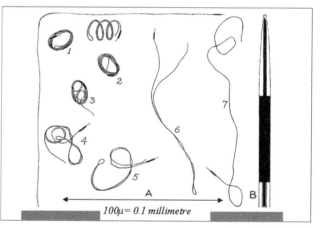

Fig. 30. A. spermatozoa, as they appear in a stained smear: 1, 2, coiled, inactive; 3 to 7, stages in uncoiling. The total length of a spermatozoon is about 0.25 mm; the head is about 10 μm long and 0.5 μm in diameter. **B.** structure of the head and part of the tail. (A, drawn from smear; B, simplified after Rothschild.). Bar represents 100 μm.

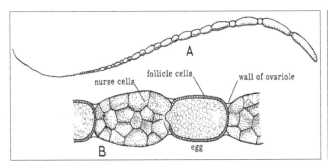

Fig. 31. A. an ovariole with eggs and nurse cells, in all stages of development. **B.** an egg with its nurse cells, drawn from a microtome section; a plug of the egg's cell plasm is in direct contact with the nurse cells through an opening in the layer of follicle cells.

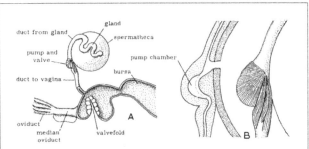

Fig. 33. A. the spermatheca and vagina, with adjoining organs, of a queen. **B.** the spermathecal valve and pump, in section, and external view, showing muscles.

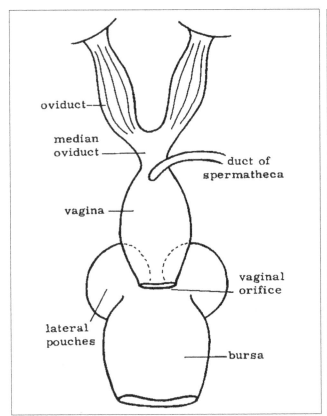

Fig. 32. The reproductive tract of a queen, dorsal aspect. Diagrammatic.

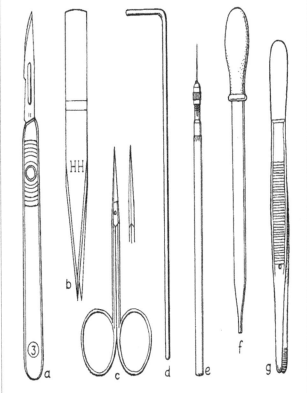

Fig. 36. Dissecting instruments. **a.** scalpel with No. 11 blade; **b.** watchmaker's forceps, HH pattern; **c.** cuticle scissors; **d.** bent wire; **e.** needle in chuck-top holder; **f.** pipette with teat; **g.** coarse forceps.

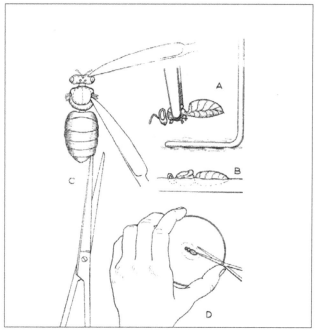

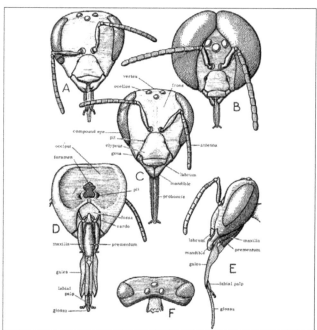

Plate 1. Preparation for dissection and use of instruments.
A, anchoring bee in pool of melted wax. **B**, bee fixed in wax.
C, methods of using scissors and scalpel. **D**, holding dissecting dish and steadying scissors against thumb.

Plate 2. External anatomy of the head. Faces of queen (A), drone (B), worker (C). Other aspects of worker's head: posterior (D), lateral (E), dorsal (F).

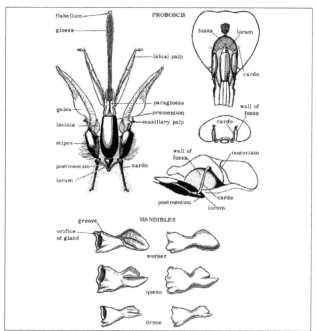

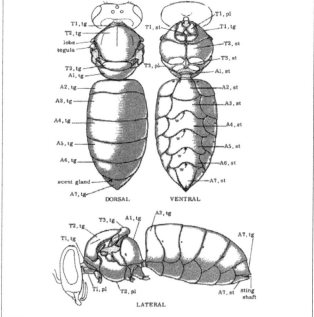

Plate 3. Mouthparts: proboscis, parts and suspension; mandibles, all castes.

Plate 4. Thorax and abdomen of worker, external features from all aspects. (For details of lateral aspect of thorax, see Plate 5.)

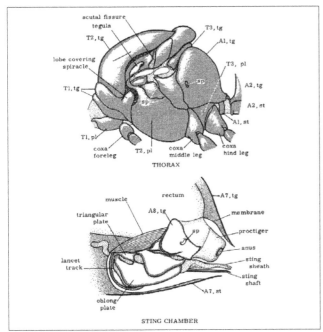

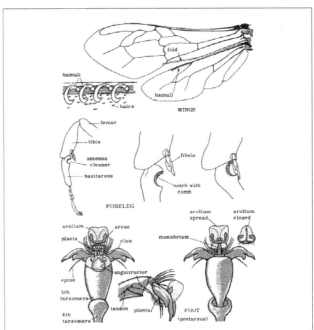

Plate 5. Above: lateral aspect of thorax. Below: lateral dissection of sting chamber, from the left side.

Plate 6. Above: the wings, with details of the hamuli. Centre: the foreleg with the antenna cleaner; the antenna cleaner open and closed. Below: the foot (pretarsus); left to right, ventral, lateral, dorsal aspects.

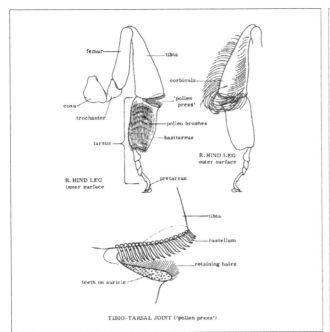

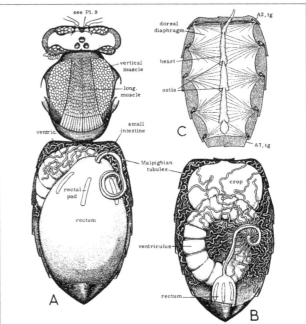

Plate 7. The hind leg, inner and outer surfaces, and details of the tibio-tarsal joint ('pollen press').

Plate 8. Dissection of the worker from the dorsal aspect, Stage 1.
A, roofs of the head, thorax, and abdomen removed, and underlying organs undisturbed; abdomen shows condition of bee confined to hive with full rectum; for details of head, see Plate 9. **B**, abdomen shows condition of bee returning to the hive, rectum empty, crop full of nectar. **C**, roof of abdomen inverted, showing heart and dorsal diaphragm attached to tergites.

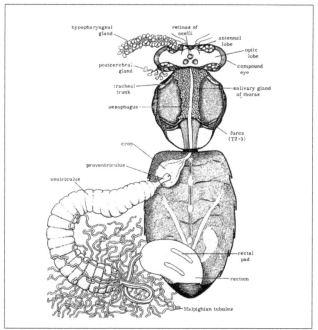

Plate 9. Dissection of worker, Stage 2. Glands of head lifted out, indirect flight muscles removed from thorax to expose underlying organs, and alimentary canal displayed.

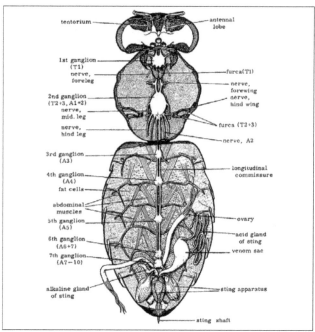

Plate 10. Dissection of worker, Stage 3. Head canted back to show some parts more clearly, glands removed; alimentary canal removed to expose nervous system, sting apparatus, and floor of abdomen.

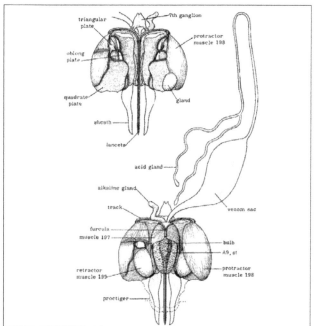

Plate 11. The sting apparatus. Below: dorsal aspect, as in Plate 10. Above: ventral aspect, revealed when apparatus is lifted out and turned over.

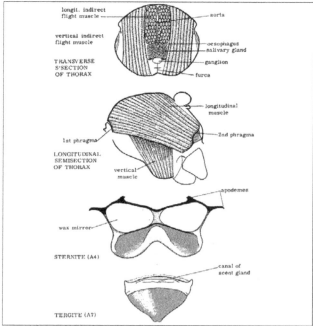

Plate 12. Above: transverse and longitudinal 'semisections' of the thorax, to show the indirect flight muscles, etc. Below: the fifth abdominal sternite with 'wax mirrors', and the seventh abdominal tergite, with the canal of the scent gland.

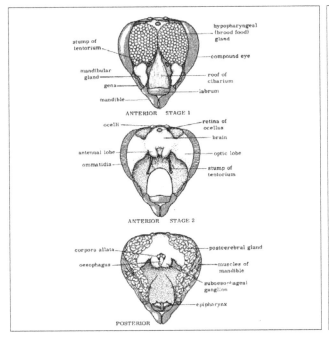

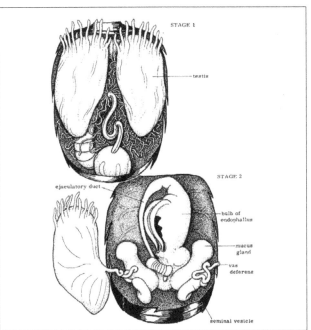

Plate 13. Dissections of head of worker. Above: from the anterior aspect, Stages 1 and 2. Below: from the posterior aspect.

Plate 14. Dissection of immature (newly emerged) drone. Stage 1, viscera undisturbed. Stage 2, testes laid out and alimentary canal removed to expose the complete reproductive apparatus.

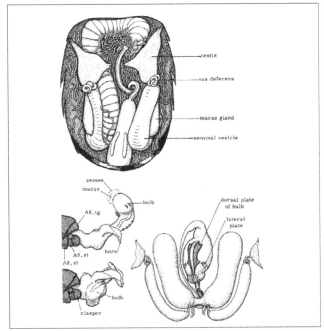

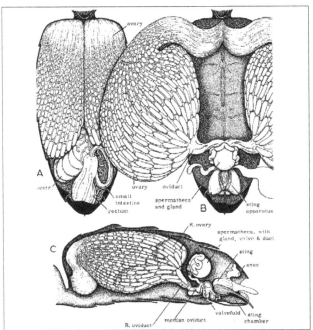

Plate 15. Dissection of maturing drone, viscera undisturbed. Below, right: the reproductive apparatus removed and laid out. Below, left: two stages in induced eversion of the endophallus.

Plate 16. Dissection of the fertile queen. **A**, Stage 1, viscera undisturbed. **B**, ovaries laid out and alimentary canal removed. **C**, longitudinal 'semisection' right side, viewed from the left.

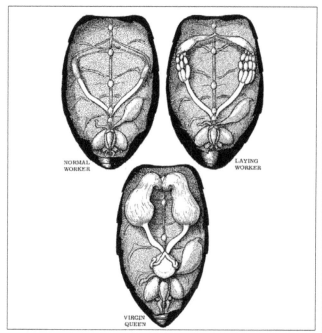

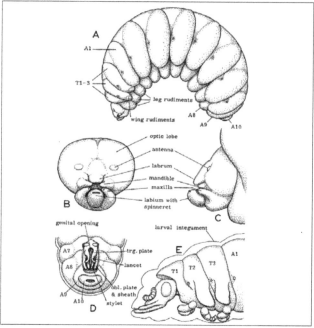

Plate 17. Reproductive organs of normal worker, laying worker, and virgin queen. Compare with Plate 16.

Plate 18. External anatomy of larva and prepupa. **A**, larva, lateral aspect. **B**, face of larva. **C**, head, from left side. **D**, posterior segments of prepupa, ventral aspect, showing sting initials. **E**, head and thorax of prepupa, from left side, showing the rapid reorganization of thoracic segments and head.

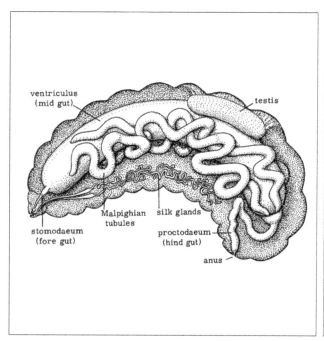

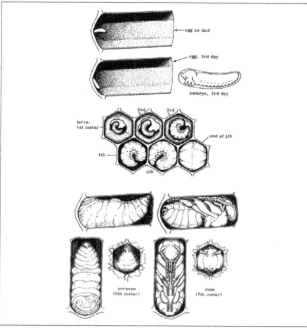

Plate 19. Dissection of larva from lateral aspect.

Plate 20. Stages in development from egg to pupa.

If a Bunsen is not available, a hand gas tool may be used or a butane blow-lamp; otherwise the whole of the wax will have to be melted. When dissections are done to prepare glands (e.g. mandibular glands) for chemical analyses, wax contamination can be a problem. In this instance a clean glass Petri dish can be used. There the bee cannot be fixed, but with some training, this is not a problem.

3.1.3. Reagents

- Alcohol has a variety of uses. The stock should be 95% Industrial Denatured Alcohol (IDA). The best dissecting fluid is 30 or 50%, made by diluting the stock with water (30 parts spirit to 65 parts water by volume, or 50 to 45, respectively). Water cannot be used by itself because it will not wet the hairy bodies of bees.
- Formalin (38-40% solution of formaldehyde) will be needed as preserving fluid. Caution: toxic compound.
- Bees can be killed with many volatile substances such as ethyl acetate or chloroform, but freezing is also a good method of killing bees (see Human *et al.*, 2013).
- Glacial acetic acid will be required if Carl's solution is preferred for preservation; it has many advantages (see section 3.1.4.).
- Glycerol, diluted with water, is an excellent preserving fluid for finished dissections.
- Insect saline solution is a good preservative for dissections: 7.5 g NaCl, 2.38 g Na_2HPO_4, 2.72 g KH_2PO_4 in 1 l of distilled water. Insect saline is preferable to alcohol as it preserves the tissues for the time of dissection, but it does not harden the tissues as alcohol does.
- Either saline solution or Grace's insect tissue culture medium should be used when observing organs *in vivo*. These solutions are isotonic relative to the organisms body fluid. Sterile bee saline is appropriate for honey bees and can be easily made. Sterile bee saline is composed of 130 mM NaCl, 6 mM KCl, 4 mM MgCl2, 5 mM CaCl2, 160 mM sucrose, 25 mM glucose, 10 mM 4-(2-hydroxyethyl)-1-piperazineethane-sulfonic acid in distilled water (pH 6.7, 500 mOsmol) (Richard *et al.*, 2008). Alternatively, Grace's Insect Tissue culture medium can be purchased from various biological supply companies or made according to Grace (1962).

3.1.4. Material and its preservation

For guidance on how to collect bees, see the *BEEBOOK* paper on miscellaneous methods (Human *et al.*, 2013). Freshly killed bees will be available for dissection at all times, and even in the depths of winter can be collected from the cluster. With the exception of queens and drones, preserved bees are more satisfactory for dissection. Bees outside of preservation fluid deteriorate quickly, so bees that are intended for autopsies need to be collected alive and appropriately

preserved. Black bees, or dark hybrids, are easier to dissect than pale races, the internal sclerotized parts being more deeply coloured and therefore more conspicuous. Queens and drones must be dissected immediately after killing; their reproductive organs become brittle when they are preserved by any method, and thus break up when they are disturbed.

Alcohol is commonly supposed to be a good preservative. This is a fallacy. It does not penetrate sufficiently rapidly to prevent the onset of decomposition, which ensues very rapidly after death, and it renders tissues very brittle. It is therefore necessary to add to the alcohol some substance which has great penetrating power and which tends to toughen tissues. Two such fluids are commonly used, formalin and acetic acid.

Formol alcohol is prepared by mixing the following:
- 95% alcohol 70 parts by volume
- formalin 5 parts
- water 25 parts.

The formalin penetrates quickly, and hardens soft tissues, making them leathery and easy to handle. This fluid can be used not only for preserving material, but also for fixing and hardening finished dissections of freshly killed bees, prior to mounting them in glycerol.

For preserving material, however, many prefer Carl's solution. This is made up of:
- 95% alcohol 17 parts
- formalin 6 parts
- water 28 parts
- glacial acetic acid 2 parts.

The acetic acid should be added just before using, i.e. 2 parts of acetic acid to 51 parts of the remaining mixed constituents.

Larvae and pupae can be extracted from their cells without damage by floating them. This is quite easy if a jet of water from a fine nozzle is directed into the cells. They should be carefully handled, using a small spoon or section lifter, and at once put into formol alcohol or Carl's solution. They should be dissected two days later (see section 3.6.1.).

3.1.5. Mounting bees for dissection

Insects are usually dissected with their backs uppermost, though in some cases another posture may be more appropriate for some special reason.

1. Take a freshly killed or preserved worker, and cut off its wings, legs and proboscis with scissors.
2. If the bee has been preserved, dry it as well as possible by rolling it gently on blotting paper.
3. Consult Plate 1.

4. Seize the bee by the thorax, back uppermost with coarse forceps.

5. Take the bent wire with the other hand.

6. Heat its short limb in a flame.

7. Apply the hot wire to the wax in the middle of the dissecting dish, thus forming a small pool of melted wax somewhat bigger than the bee.

8. Place the bee quickly in the pool, hold it there with the cool end of the wire, and withdraw the forceps.

9. Reheat the wire and melt a little of the wax near the sides and ends of the bee

10. Push this melted wax against the body, so that it piles up slightly and makes good contact.

This will ensure that the specimen is firmly anchored and will not come adrift during dissection. The bee should be sunk nearly halfway in the wax. Plate 1A shows this operation, and B the embedded bee. All this must be done quickly and without overheating the insect. A better posture is obtained if, when lowering the bee into the pool of wax, the tip of the abdomen touches the wax first; it will adhere, and then the body can be drawn forward slightly, thus stretching the abdomen a little. Whole insects, or parts like the head, can be prepared in any posture that may be desired.

11. Pour insect saline or dissecting fluid (diluted alcohol) on at once, enough to cover the bee.

The fluid floats and supports the internal organs and of course prevents them from drying.

12. Place the dish under the microscope, it then is ready for work.

13. Focus the microscope and adjust the spot lamp.

14. Instruments should be laid out ready for use.

3.2. Dissection of the worker bee

3.2.1. The general dissection

Plates 8 to 10 illustrate the dissection of the whole body, from the dorsal aspect, in three stages. In practice it will be found convenient to begin with the abdomen, and to complete its examination before starting on the thorax and head, and the directions which follow are arranged in that order.

For the study of the alimentary canal, the heart, the tracheal sacs, and the ovaries in both normal and laying workers, bees should be dissected immediately after killing. These organs become brittle in preserved specimens. For all other purposes, preserved bees are much more satisfactory. Bees with fully distended abdomina will show the heart and nervous system to the best advantage. The specimen is prepared, fixed in the dissecting dish, back uppermost, and covered with dissecting fluid, as described in Section 3.1. and illustrated in Plate 1.

3.2.2. The abdomen

Turn the dish so that the head of the insect points to 10 o'clock. During operations, remember to move the dish to suit the convenience of the right hand and scissors (see section 3.4.).

3.2.2.1. Exposing the viscera (Plates 1, and 8 A and B)

To open the abdomen:

1. Steady the dish with the left hand, and the scissors by resting them against the left thumb on the edge of the dish (Plate 1).

2. Insert one point of the scissors under the overlapping edge of the tergite of A5, on the right side of the body.

3. Cut through the body wall.

4. Continue snipping through the right side, working forwards towards the thorax.

Keep the inner blade of the scissors as far as possible parallel with the side of the insect, thus avoiding thrusting it in deeply and damaging the viscera.

5. Turn the dish clockwise to suit the scissors hand when the corner at the front end of the abdomen is reached.

6. Cut across the broad front of the abdomen to the opposite corner.

7. Then again turn the dish and work down the left side.

8. Turn the dish again when the tergite of A6 is reached.

9. Cut across the tergite, taking great care not to damage the soft organs underneath, and so complete the circuit at the beginning of the first incision.

10. Lift off the roof of the abdomen gently with the point of a needle; it should be free and easily removed.

If it resists, one or more of the infolded parts of the tergites have not been severed. These uncut parts must be found and cut through, using the inner scissors point as a probe while gently lifting the roof with the needle. When it is clear that the roof is free, do not pull it off roughly, but lift it gently with two needles. There may be slight resistance from tracheae, but these will break without doing any damage. If, however, it seems that the internal organs are being pulled out or disturbed, take the roof by its edge with the fine forceps, and with the needle in the other hand break the tracheae, which will show as fine threads stretching between the roof and the organs below. Finally lift off the roof and turn it over. If the work has been done neatly, it will come off in one piece.

11. Examine the underside of the roof (Plate 8C) as it lies in the dissecting fluid.

Observe:

11.1. the heart, with its closed posterior chamber and its ostioles (there are five pairs, but the anterior pair may have been lost), in the mid-line of the roof;

11.2. the dorsal diaphragm, transparent, but clearly visible, and its attachments to the apodemes of the tergites;

11.3. the pericardial fat cells, large numbers of small, creamy bodies clustered against the heart;

11.4. the dorsal sheet of the fat body forming a pad between the heart and the body wall;

11.5. some of the abdominal muscles may be seen as flat, nearly transparent bands stretched across the tergites.

In preserved bees the heart and dorsal diaphragm occasionally adhere to the viscera and thus tear away from the roof. Having examined all these organs, lay aside the roof, and

12. look at the contents of the abdomen.

12.1. The appearance of the undisturbed viscera is very variable, depending on the state of the alimentary canal (Plate 8, A and B). In a bee which has been confined to the hive for some time, or a young bee which has not yet flown, the rectum is greatly distended by accumulated faeces, the bulk of which are yellow pollen husks (A). If the rectum has been damaged by instruments during the opening operations, some of the faeces will have escaped, and will litter the dissection. If the bee has just returned to the hive after a flight, the rectum will be empty and shrunken to very small proportions; if she has brought home a load of nectar or water, the crop (honey stomach) will be expanded into a large, transparent globe (B); if it is empty it will appear as a small, semi-opaque, pear-shaped body;

12.2. part at least of the ventriculus will be visible as a broad, corrugated tube;

12.3. a loop of the small intestine will be found connected to the forward end of the rectum; its other end, which joins the ventriculus, may not be visible;

12.4. the slender, tangled threads which spread all over the abdomen are the Malpighian tubules;

12.5. in a freshly killed bee, the tracheal sacs will be seen as large bags, silvery with included air (which escapes when a needle point is inserted), obscuring parts of the other organs. In preserved bees the sacs are almost invisible, filmy membranes, the air having been dissolved by the preserving fluid. When air-filled sacs obscure the view, they should be pulled out with forceps.

12.6. Tracheae in large numbers appear as silvery tubules in all parts of the body.

13. Clear away debris (faeces, fragments of tissues, etc.) which collects in the abdominal cavity from time to time during dissection.

This is done by irrigation with clear dissecting fluid, a jet of which is directed into the cavity with the pipette.

3.2.2.2. Displaying the alimentary canal (Plate 9)

1. Take a needle in each hand.
2. Pass them under the rectum and ventriculus
3. Lift up the alimentary canal, gently tearing away the network of investing tracheae in which it hangs.
4. Lay it over to the left side, as shown in Plate 9.
5. Carefully tease out the tracheae and Malpighian tubules to permit the canal to lie loosely, showing all its parts.
6. Identify and examine the parts. Notice:

6.1. the six rectal pads, which appear as whitish bars on the wall of the rectum;

6.2. the small intestine, as a narrow coiled tube with six longitudinal pleats;

6.3. at its junction with the ventriculus, about one hundred Malpighian tubules are inserted. This is the pyloric region of the canal;

6.4. the ventriculus, in which food masses in the course of digestion can usually be seen, showing as dark areas where the corrugations of the ventriculus are smoothed out. If the ventriculus is torn with a needle, this food mass will exude as a brownish gelatinous substance.

6.5. the proventriculus, which will be visible through the walls of the crop if it is full of nectar. In any case, tear open the wall of the crop, using needles, and turn up the proventriculus so that its four triangular lips may be clearly seen. If they are closed, the lips meet to form a cross. If they are partly open, an aperture like a four pointed star is seen;

6.6. the forward end of the crop, which narrows into

6.7. the oesophagus, which enters the thorax through the petiole.

3.2.2.3. The underlying organs (Plate 10)

1. Grasp the alimentary canal with forceps.
2. Stretch it.
3. Cut through the oesophagus with scissors.
4. Treat the rectum in the same way, cutting through it as far back as possible.
5. Carefully cut away the remaining small triangle of the roof which was left at the tip of the abdomen.
6. Lower the side walls with scissors, giving a better view of the floor.
7. Flush out the cavity with the pipette.
8. Compare with Plate 10.
9. Identify and examine the parts. Notice:

9.1. the ventral diaphragm, which may not be noticed at first, but closer inspection will reveal it as a transparent film which very slightly obscures the view of the chain of ganglia and other underlying features. In a later dissection, the diaphragm may be studied more carefully; it is attached to the apodemes of the sternites; its anterior end extends into the thorax and is attached to the furca of T2 and T3, while its posterior end is anchored to the spiracle plate of A8.

10. Tear out the diaphragm with the fine forceps, taking care not to damage other structures in the process.

11. Observe the now more clearly visible organs lying on the floor of the abdomen:

  11.1. the chain of five ganglia is the most conspicuous, it is connected by twin longitudinal commissures. The last, the 7th ganglion, is attached to the sting apparatus, and comes away with the latter when it is torn out of the worker's body after stinging;

  11.2. the main lateral nerves which spring from the ganglia can be seen running out to right and left; those of the 7th may be seen passing to the muscles of the sting;

  11.3. the fat body spreads widely over the floor of the abdomen, being particularly well developed over the wax glands of the sternites of A4 to A7. Smaller clusters of fat cells occur along the sides of the abdomen. The fat body is highly developed in young bees and winter bees, where the cells are large and plump, but in old foragers they are shrunken;

  11.4. the abdominal muscles show clearly, some of the larger sets being very conspicuous as broad V-shaped pairs of bands stretching between the thickened forward margins of adjacent sternites;

  11.5. the ovaries are difficult to see, and since they encircle the alimentary canal they are torn away when it is lifted out. To prevent this, after removing the roof of the abdomen,

    11.5.1. lift the alimentary canal slightly from the right-hand side,

    11.5.2. look sideways under it, the right ovary with its oviduct will be seen as an almost transparent, narrow, flat tube running to the root of the sting,

    11.5.3. gently disengage the ovary from the tracheae which tie it down and attach it to the other viscera,

    11.5.4. repeat this operation from the other side, thus freeing the left ovary,

    11.5.5. go on with the dissection, removing the canal. The ovaries will then be seen lying or floating in the abdominal cavity, their oviducts disappearing behind the sting, their distal ends separated. In the undisturbed abdomen, the tips of the ovaries are joined and attached to the heart (Plate 17).

  11.6. the sting, if not wholly visible, can be examined *in situ* by removing more of the wall of the abdomen at the tip (Plate 10).

Identify the parts flagged in Plate 11, dorsal aspect. The whole apparatus can be removed intact very easily by passing needles below it and lifting it out, the small muscles which hold it giving way without offering noticeable resistance. The extracted apparatus can now be turned over, as it lies in clear fluid, and its ventral aspect (Plate 11) can be examined. Very rarely, the sting apparatus is laterally reversed, the only evidence of this being that the positions of the venom gland and the alkaline gland are reversed. The powerful muscles of the sting apparatus conceal the plates which constitute the system of levers actuating the lancets. The plates can be exposed by removing the muscles by maceration. Note that the sting apparatus is arched; it can be flattened by tearing away the proctiger, which is firmly attached to the oblong plates, and it is then easier to examine and also to mount as a microscopical preparation.

3.2.2.4. Rapid removal of alimentary canal and associated glands in the abdomen

There may be occasions when rapid removal of abdominal organs (alimentary canal, and sting glands) is needed. For instance, one may want to quantify the percentage of bees exhibiting a certain pathology such as scaring of the pyloric valve (Bailey, 1981). In these cases it may be more economical and efficient to expose the abdominal organs without mounting the bee in wax. Instead, after blotting individual bees on a paper towel, the intact bee can be held by the thorax, ventral side uppermost. Using scissors, two shallow incisions are made along the lateral sides of the sclerites (Fig. 42). The incisions should start at the posterior end of the abdomen (between tergite and sternite A6, Plate 4) and end near the petiole. It is helpful to lay the scissors flat against the sclerites while cutting between the tergites and sternites. When the incisions are complete, remove the abdomen from the thorax, and placing the abdomen in a petri dish

Fig. 42. An individual bee being held for rapid removal of abdominal organs.

Photo: Michael Andree.

containing the appropriate preservation solution to submerge the
abdomen. With the aid of a compound microscope, remove the
sclerites by grabbing the edge of the most anterior sternite (A2 st,
Plate 4) with forceps and peel the sternites (st A2 - A6, Plate 4) back,
exposing the gastro-intestinal tract and other abdominal organs
(Fig. 43). The digestive tract and sting gland organs can be removed
from the tergites by gently teasing them away from the integument
beginning at the anterior end and working back toward the sting
gland (Fig. 44). Examination of the desired organs can then
commence.

3.2.3. The thorax
In a freshly killed bee it is not possible to see the structure of the
indirect wing muscles, or to remove them without great difficulty, for
they are extremely soft and elastic. In bees which have been
preserved, however, these great muscles are hardened and tough,
very easy to examine and to handle. Therefore preserved bees should
always be used for work on the thorax.

3.2.3.1. Exposing the flight muscles (Plates 8A and 12)
The roof of the thorax is best taken off with the knife:
1. Insert the extreme point only, as shown in Plate 1.
2. Make a short slit in the body wall by an outward and forward
   stroke.
3. Continue this along the dotted line in the Plate, all round the
   domed roof of T2.
4. Make a longitudinal slit along the mid-line.

Usually the roof is very firmly attached to the flight muscles, and must
be detached, again with the point of the knife.
5. Keep the blade in a horizontal position, pass its point under
   the body wall, through the longitudinal slit, and separate the
   roof from the muscles by small forward movements, gradually
   working the point further under the body wall. When the first
   half of the roof is nearly free, steady it with forceps while
   completing the separation. If this is done carefully, the
   muscles will be undisturbed and undamaged, and will have
   the appearance shown in Plate 8.
6. Remove the other half of the roof.
7. Take off the remainder of the roof, along the second dotted
   line in Plate 1.

This is not attached to the muscles, and will come off easily.
8. If necessary, remove more of the side walls of the thorax,
   down to the level of the wings.

The indirect flight muscles are now exposed.

3.2.3.2. Oesophagus and glands (Plate 9)
1. Remove the flight muscles: simply grip bunches of them with
   the forceps and pull them out (this is virtually impossible in a
   freshly-killed bee).

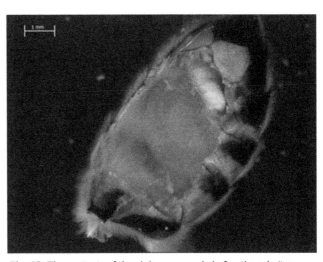

Fig. 43. The contents of the abdomen revealed after the sclerites
have been removed.   Photo: Michael Andree.

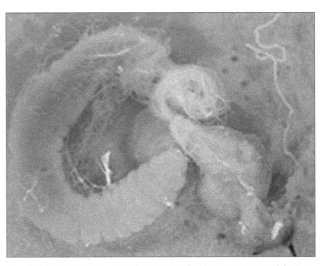

Fig. 44. The alimentary canal and sting gland organs removed from a
worker bee using the rapid autopsy procedure.

Photo: Michael Andree.

2. Observe:
   2.1. the attachment of the longitudinal muscles to the 2nd
        phragma, which is an extension inwards of the tergite of
        the second segment;
   2.2. the oesophagus below the longitudinal muscles, it passes
        from abdomen to head;
   2.3. the salivary glands of the thorax (derived from the silk
        glands of the larva);
   2.4. the aorta is a delicate tube which is destroyed by the
        removal of the indirect flight muscles; it can be found by
        careful lateral dissection.

3.2.3.3. The nervous system in the thorax (Plate 10)
1. Remove the oesophagus and salivary glands with forceps.
2. Observe:
   2.1. the combined furcae of T2 and T3, which are now conspicuous.

They form a strong strut and bridge across the thorax and protect:

    2.2. the great second ganglion, which can be seen below the furcae;

    2.3. the first ganglion lies in front of and partly concealed by the furca of T1, at the anterior end of the thorax. To see this, it will probably be necessary to remove more of the body wall in this region. Both ganglia can be exposed by removing the overlying parts of the furcae, using the point of the knife and forceps. Sheets of semi-transparent muscles attached to the furca will also have to be removed; this must be done cautiously, for the ganglia are very easily damaged. Notice the thick twin commissures joining the ganglia;

    2.4. commissures also run forward from the first ganglion to the suboesophageal in the head, and from the second ganglion to the abdomen, where they join the abdominal chain;

    2.5. nerves from the second ganglion can also be seen passing to the propodeum and into the abdomen, where they serve A2.

3.2.4. The head (Plates, 8, 9, 10)
3.2.4.1. Exposing the brain and other structures.

    1. Cut along the dotted line of Plate 1C, using the point of the knife. The isolated portion of the wall will come away easily. Conspicuous on the summit of the brain are the purple retinae of the three ocelli.

    2. Extend the opening to each side, cutting away part of the compound eyes (Plate 8A).

Both knife and scissors can be used for this part of the work.

    3. In a freshly killed bee, clear away the large tracheal sac filled with air, which obscures the brain.

    4. Observe:

        4.1. the protocerebrum, bearing the ocelli;

        4.2. the optic lobes connected with the compound eyes.

The pigmented parts of the compound eyes indicate the radiating ommatidia;

        4.3. the hypopharyngeal glands in front of the brain.

They are conspicuous in young nurse bees and winter bees, in which the acini are large and white, but in foragers they are much shrunken and may be difficult to find;

        4.4. another branch of the hypopharyngeals;

        4.5. a small part of the postcerebral glands behind the brain.

3.2.4.2. Displaying the glands and antennal lobes (Plates 9, 10)

    1. Lift out the glands with the point of a needle.

    2. Note:

        2.1. the form of the hypopharyngeals, like strings of onions;

        2.2. the quite different branched structure of the postcerebrals;

        2.3. the mandibular glands.

Beware of rupturing them and thus losing their contents.

    3. Cut away the frons down to the level of the antennae.

    4. Observe:

        4.1. the antennal lobes,

        4.2. nerves running into the antennae.

They can be seen if the head is lying in a favourable position, as in Plate 10.

3.2.5. Lateral dissection of the worker

Mount a preserved worker on its right side.

3.2.5.1. The head

The dissection is chiefly useful to demonstrate the suspension of the proboscis, but also shows different aspects of the organs which are the principal subjects of other dissections.

    1. Take off the left gena.

    2. Work carefully downwards, lowering the walls.

    3. Note:

        3.1. the tentoria, two powerful struts running from the foramen to the clypeus, and joined at their feet by a small transverse bridging piece;

        3.2. the fossa, of which the upturned wall lies just below

        3.3. the gena;

        3.4. the cardo, its end will be found articulated to a knob on the edge of this wall;

        3.5. the oesophagus, it enters the head through the foramen and then expands into the pharynx;

        3.6. the hypopharyngeal plate, of which the long extensions embrace

        3.7. the pharynx;

        3.8. the cibarium, its dilator muscles connects it to

        3.9. the clypeus.

The organs revealed by this dissection will be recognized from the diagrams in Figs. 4 and 15.

3.2.5.2. The thorax

    1. Remove the left side of the thorax.

    2. Excavate, removing the flight muscles and lowering the walls as the work proceeds.

    3. Observe:

        3.1. the attachment of the vertical muscles to the floor and roof of the thorax;

        3.2. the attachment of the longitudinal muscles to the roof and first phragma in front, and to the second phragma in the rear;

        3.3. the aorta, which can be found when approaching the mid-line of the thorax, between the right and left sets of longitudinal muscles, and below it

        3.4. the oesophagus;

3.5. the ganglia;

3.6. the commissures;

3.7. the anterior end of the ventral diaphragm, which enters the thorax through the petiole.

4. Remove everything else.

5. Observe:

5.1. the direct wing muscles on the inner surface of the right hand wall of the thorax.

A special preparation to show the direct muscles can easily be made by slicing off one side of a thorax with a razor blade and trimming away debris.

3.2.5.3. The abdomen

1. Examine the sting chamber.

2. Cut a window in the side of the abdomen, starting at the rear edge of the tergite of A6.

3. Enlarge the opening towards the posterior end of the abdomen, severing the muscles which hold the isolated parts of the body wall by passing the point of the knife under them.

4. Observe:

4.1. the spiracle plate of A8, it is found overlying the sting apparatus, and has something of the appearance of a dog's head (Plate 5). The spiracle plate is bound to

4.2. the quadrate plate, near the dog's nose;

4.3. the proctiger, attached to the sting apparatus and also to

4.4. the rectum.

5. Remove the contents of the abdomen, lower the walls.

6. Survey the abdominal muscles of the right side (some of them are shown in Fig. 11).

3.2.6. Dissection of the laying worker (Plate 17)

See note on obtaining material in section 3.1.4. Laying workers should be dissected when freshly killed, or within two or three days after preservation, or the ovaries will become too brittle to handle.

1. Remove the roof of the abdomen.

If the crop is full and the ovaries are well developed, the anterior extremities of the ovaries may be seen crossing the crop, but this is not usual.

2. Proceed as though dissecting out the ovaries of a normal worker (see section 3.9.2.3.). The laying worker's ovaries are much more easily seen, as the ovarioles contain strings of eggs.

3. Break the tracheae binding the ovaries to the alimentary canal.

4. With care, remove the alimentary canal without tearing the ovaries apart, and they are then displayed as shown in Plate 17.

3.3. Dissection of the head, all castes

3.3.1. Preparation

Heads are removed and mounted for dissection from both anterior and posterior aspects. Beeswax alone is too soft to hold the head firmly, and a stiffer cement must be used. A wax-resin mixture is

suitable, and may be used to mount the heads on small pieces of acetate or Perspex® sheet, these in turn being temporarily anchored in the dissecting dish by melting the wax in the dish at their corners with the toe of the bent wire. The heads should be embedded deeply, with their upper surfaces almost level with the surface of the cement.

3.3.2. Worker's head, from the anterior aspect (Plate 13)
3.3.2.1. The glands

1. Cut through the wall of the mask with the point of the knife, across the vertex, round the margins of the compound eyes, and round the edges of the mask, excluding the mandibles, labrum, and clypeus.

2. Snip off the antennae near to their insertion.

3. Lift off the mask which remains connected by tentoria.

4. Cut around the small pits in the suture surrounding the clypeus. The mask will then lift off.

5. Hold the clypeus down firmly by the cibarial muscles (dilators of the cibarium).

6. Disengage the cibarial muscles by using the point of the knife in the same way as when taking the roof off the thorax.

7. Notice:

7.1. the hypopharyngeal glands. In a young bee, five or six days old, or in a winter bee which has not yet nursed brood, they will have the appearance shown in the Plate, the acini being plump and creamy white; they will almost fill the space in front of the brain, as well as sending branches to the back of the brain. They can be lifted out to show their string-of-onions structure. In foraging bees which have completed their nursing duties, these glands are greatly shrunken, almost to the point of disappearance, leaving only thin thread-like remains;

7.2. the mandibular glands, beware of rupturing them and thus losing their contents.

3.3.2.2. The brain

1. Remove the hypopharyngeal glands to expose the brain.

2. Identify the structures flagged in the plate 13.

3. Cut down the compound eyes to the level of the optic lobes, and note the indications of the radiating ommatidia.

4. Remove the roof of the cibarium.

5. Examine the floor of the mouth cavity, finding:

5.1. the hypopharynx, with the two pores marking the ends of

5.2. the ducts of the brood-food glands.

3.3.3. Worker's head, from the posterior aspect (Plate 13)
3.3.3.1. The glands

1. Remove the proboscis.

2. Remove the wall of the occiput and postgenae, by the same means as in the preceding dissection.

3. Lift off the back wall of the head cut down the tentoria, to give a clear view.

4. Clear the tentoria in the normal way.

5. Identify the structures flagged in the plate 13. The postcerebral glands are more translucent than the hypopharyngeals; note their branching form.

3.3.3.1. The brain and suboesophageal ganglion

1. Remove the glands.

The cut end of the oesophagus will be seen projecting from the space between the brain and the suboesophageal ganglion.

2. If the corpora allata are not visible, pull the oesophagus out a little way with forceps, when they should come into view.

3. Trace:

3.1. the principal nerves (Fig. 24);

3.2. the ducts of the glands (Fig. 15).

3.3.4. Isolation of the retrocerebral complex

The retrocerebral complex (RCC) is a set of neurosecretory organs that sit at the base of the brain and play a crucial role in controlling insect behaviour and physiology. The complex consists of the corpora allata (CA) and corpora cardiaca (CC), both of which are paired organs. In honey bees, the complex is located close to the subesophageal ganglion (Fig. 45.3), and forms an incomplete ring around the aorta and oesophagus (AO-ES) (Hannan, 1955).

The CA are responsible for the production and release of juvenile hormone (Goodman and Cusson, 2012). The CC are a conduit for neurosecretory cells of the brain, responsible for storing and releasing factors such as prothoracicotropic hormone, adipokinetic hormone and other regulators of metabolism and muscular activity (Woodring *et al.*, 1994; Lorenz *et al.*, 1999; Takeuchi *et al.*, 2003; Audsley and Weaver, 2006; Boerjan *et al.*, 2010).

Isolation of the RCC permits quantification of these various hormonal factors, including the *in vitro* measurement of the rate of synthesis and release of juvenile hormone (Rachinsky and Hartfelder, 1990; Huang *et al.*, 1991; Hartfelder *et al.*, 2013).

3.3.4.1. Dissection of the RCC from the head capsule

1. Sedate a live adult bee by chilling.

2. Decapitate the bee by gently pulling on the head to elongate the neck.

3. Sever the neck as close to the thorax as possible.

4. Secure the head, face side down, to a wax surface using a pair of pins through the eyes.

Angle the pins away from the head to maximize accessibility (Fig. 45.1).

5. Using a micro-scalpel or –scissors, cut the area (postgena) that encircles the occipital foramen.

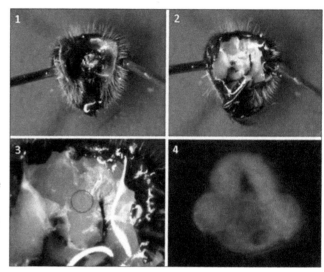

Fig. 45. Dissection of the retrocerebral complex of an adult honey bee. **1.** Head capsule viewed from behind; **2.** cuticle removed, giving a posterior view of the brain; **3.** enlarged image of Fig. 45.2, with the position of the retrocerebral complex marked by a red circle; **4.** the isolated retrocerebral complex showing the spherical corpora allata connected to the corpora cardiaca, which sits like a saddle over the foregut.   Photos: Colin Brent.

6. Widen the circle out to the inner margins of the eyes.

7. Gently tilt up the cuticle from the anterior end to gain access to any connective tissue bound to the cuticle.

8. Sever these connections.

Be sure to minimally disturb the posterior end and leave the oesophagus intact.

9. Remove the cuticle plate.

10. Add a droplet of incubation medium if dissecting to measure the rate of juvenile hormone biosynthesis by the paired corpora allata (see section 3.5 of the *BEEBOOK* paper on physiology and biochemistry (Hartfelder *et al.*, 2013)) or another appropriate physiological liquid (e.g. Ringer solution, see Table 1 of the *BEEBOOK* paper on cell cultures (Genersch *et al.*, 2013)) to ensure the interior of the head stays moist (Fig. 45.2.).

11. Examine the posterior end of the brain where the suboesophogeal ganglion completes a circle around the aorta and oesophagus; the retrocerebral complex will be found in this area, also tightly associated with AO-ES.

12. If the glands are not immediately visible, gently pull on the oesophagus and the complex should come into view.

The RCC has a slightly blue colour that is distinct from the whiter nervous tissue around it (Fig. 45.3.).

3.3.4.2. Isolation of the RCC and CA

1. Sever the section of AO-ES just anterior to where the complex is attached.

2. Remove the whole mass from the head.

3. Grasp the AO-ES rather than the RCC to avoid damaging the CA and CC.

4. Use forceps to gently push the complex off of the AO-ES, leaving an isolated RCC.

5. Transfer the AO-ES/RCC to an incubation medium if dissecting to measure the rate of juvenile hormone biosynthesis by the paired corpora allata (see section 3.5. of the *BEEBOOK* paper on physiology and biochemistry (Hartfelder *et al.*, 2013)) or another appropriate physiological liquid (e.g. Ringer solution, see Table 1 of the *BEEBOOK* chapter on cell cultures (Genersch *et al.*, 2013)) to prevent desiccation.

For some assays, it is desirable to leave attached tracheal elements intact to enhance the buoyancy of the complex (Fig. 45.4.). Avoid grasping the glands with forceps as the endocrine tissue is readily damaged, and often the glands will adhere to the forceps and become difficult to dislodge. The CA or entire RCC can be moved between different media by suspending the glands in a droplet of liquid held by a small wire hoop.

3.3.4.3. Dissection of the brain from the head capsule

1. Flash freeze the bees in liquid nitrogen.

2. Disconnect the head from the body.

3. Place the head on dry ice to ensure it stays frozen.

4. Cut away the face plate using a micro-scalpel or scissors.

5. Make the incision from just above the epistomal sulcus, along the inner margin of the eyes and over the ocelli.

6. Gently lift the face away, severing any underlying connections to reveal the brain below.

7. Cut the connections between the brain and optic lobes.

8. Lift the brain away from the head capsule.

If the brain is to be further subdivided prior to analysis, ensure that it stays chilled during the dissection.

3.3.5. The head of the drone

Proceed as with the worker. The hypopharyngeal glands are absent, and the postcerebrals are reduced to rudiments adhering to the posterior wall of the head. The mandibular glands are also vestigial. Clusters of fat cells in some parts of the head may be mistaken for glands. The brain proper (part of the protocerebrum under the ocelli) is smaller than the worker's, though the very large optic lobes give a misleading appearance of size.

3.3.6. The head of the queen

Proceed as with the worker. The hypopharyngeal glands are lacking (vestiges of their ducts may sometimes be found). The postcerebrals are like the worker's. The mandibular glands (source of queen substance)

are much larger than the worker's. The brain is somewhat smaller than that of the worker.

3.4. Dissection of the drone

3.4.1. Preparation for dissection

Immature drones which have just emerged from their cells, or which can be caught actually emerging, should be killed and dissected at once. Mature drones cannot be killed without partial eversion of the endophallus ensuing. Partly matured drones, if their development has not proceeded too far, can usually be killed without inducing eversion, though ejaculation often begins. Preserved material is useless for dissection of the reproductive organs, but preserved drones may be used for the dissection of the head, of the thorax to show the powerful flight muscles, and of the abdomen to show the immensely powerful abdominal muscles.

3.4.2. Immature drone, dissection from the dorsal aspect (Plate 14)

3.4.2.1 Exposing the viscera

1. Proceed, as in the case of the worker (see section 3.2.2.1.), to take off the roof of the abdomen.

In crossing over at the anterior end it will be difficult to avoid damage to the testes, which will then be slightly frayed out in that region.

2. Notice the resistance to the scissors offered by the greatly developed abdominal muscles.

3. Take care in all dissections of the drone to avoid damage to the mucus glands.

If these are pierced, mucus will exude, coagulate, and obscure the work.

4. Compare the undisturbed viscera with the plate 14 and identify the organs.

Note the enormous testes, composed of bundles of tubules which can be shown by teasing out one of the testes.

3.4.2.2. The reproductive apparatus

1. Lay out the testes, as shown in Plate 14 (where only one testis is drawn).

If they are not very gently handled, they will break off at the vasa deferentia.

2. Grasp the ventriculus with forceps and draw it out backwards until the crop appears.

3. Cut through the oesophagus with scissors, also through the rectum.

4. Remove the alimentary canal.

5. Flush out the abdominal cavity with clean dissecting fluid from the pipette.

This exposes the rest of the reproductive apparatus.

6. Identify the parts flagged in Plate 14.

The whole of the apparatus can be removed for more detailed examination by cutting through the body wall, round the genital aperture, with the point of the knife. It can then be lifted out, very gently to avoid damage, fixed in formol-alcohol, and preserved.

3.4.2.3. Maturing drone (Plate 15)

Treat as in the preceding dissection. The testes are reduced in size, finally becoming thin, triangular, translucent scales. The seminal vesicles, having received spermatozoa, are increased in size, so are the mucus glands. If eversion has not occurred, there may yet be partial ejaculation, the evidence of which may be seen in the swelling of the ejaculatory duct near the bulb.

3.4.2.4. Mature drone

1. Catch a flying drone on the alighting board of a hive or at a DCA (see section 13.4. of the *BEEBOOK* paper on behavioural methods (Scheiner *et al.*, 2013)).
2. Kill it (see section 2.1.2. on immobilising, killing and storing adult *Apis mellifera* in the *BEEBOOK* paper on miscellaneous methods (Human *et al.*, 2013)).
3. Cut off the everted endophallus.
4. Mount the insect for dissection.
5. Remove the roof of the abdomen.
6. Look for the testes, now shrunken to mere yellowish or greenish scales.
7. In the second stage of the dissection, compare the condition of the rest of the apparatus (except the endophallus, which is now outside the body) with that of the immature and maturing drone. The ejaculatory duct will be seen passing out through the genital aperture.

3.4.2.5. The everted endophallus (Plate 15)

1. Kill a mature drone.
2. Examine the partially everted endophallus, and compare it with the drawing in Plate 15.
3. Take the abdomen between the thumb and forefinger and compress it laterally; eversion will continue until the bulb reaches the tip of the organ, when sperm and mucus will be liberated.

In living drones, notice how easily eversion is induced by the slightest pressure, and the 'snap' when this occurs.

3.5. Dissection of the queen

3.5.1. Preparation

Queens must be dissected immediately after killing. If they are preserved in alcohol the ovaries become extremely brittle and cannot be handled for proper examination. They can, however, be placed at -20°C alive, without alcohol, and they can be dissected even months afterwards. The remainder of the body may be preserved, however, and can then be dissected at leisure (see section 3.10.5.).

3.5.2. Dissection of the fertile queen (Plate 16)

3.5.2.1. Exposing the viscera (Plate I6A)

1. Mount the queen for dissection, dorsal surface uppermost.

2. Open the abdomen carefully, avoiding damage to the underlying organs when taking off the roof.
3. When the roof appears to be quite free, raise the front edge of it slightly with the forceps.
4. Gently detach, with a needle held in the right hand, the soft organs which cling to the roof.

If this is not done very carefully the ovaries will be pulled out and spoiled. The undisturbed viscera will then have the appearance shown in Plate 16A. At their anterior ends, the ovarioles of the two ovaries are joined, and at this point were attached to the heart, from which they were detached before lifting the roof.

5. Identify the parts flagged in the plate, and note:
   5.1. the enormous ovaries that fill about two-thirds of the abdomen;
   5.2. the rectum that is always empty.

3.5.2.2. Displaying the reproductive organs (Plate 16B)

1. Grip the ventriculus with forceps and stretch it, drawing it towards the rear end of the abdomen.
2. Cut the canal in front of the crop and through the rectum.
3. Remove the alimentary canal.

This must be done without damaging the ovaries.

4. Very carefully slip a needle under one of the ovaries and lift it, turning it slightly outwards.

The tracheae which bind and suspend the ovary will now be seen stretched.

5. Break the tracheae with a needle held in the other hand, until the ovary is freed sufficiently to lie over to the side, as shown in the plate.
6. Repeat this operation with the other ovary.
7. Lower the sides of the body wall near the tip of the abdomen.
8. Flush out the body cavity with the pipette.
9. Note:
   9.1. the paired oviducts, which converge to the root of the sting, where they join
   9.2. the median oviduct;
   9.3. the spermatheca;
   9.4. the spermathecal gland and investment of silvery tracheae;
   9.5. the acid gland of the sting, which is much longer than the worker's;
   9.6. the sting apparatus is firmly anchored and cannot be lifted out as it can from a worker dissection. The shaft of the sting is curved, not straight like the worker's.

Queens used for dissection are often discarded ones, a year or two old, sometimes older. In these old queens the fat body and Malpighian tubules are discoloured by the accumulation of waste products (this can be seen in the pericardial fat cells; it is not necessary to spoil a good dissection by removing the viscera to expose the fat cells on the floor of the abdomen).

3.5.2.3. Weight of ovaries

The ovaries can be weighed immediately after dissection (wet weight), or after dehydration in an incubator (in separate Eppendorf tubes at 60°C for 24 hours (dry weight). Dehydration requires time and it does not give an immediate result.

3.5.2.4. Number of ovarioles

The number of ovarioles (in one ovary) can vary among honey bee races. For example it has been reported that there are an average of 130-155 ovarioles/ovary in *A. m. macedonica* (Hatjina *et al.*, 2013). In *A. m. ligustica*, there are between 135-175 (Woyke, 1971; Casagrande-Ialoretto *et al.*, 1984) and in *A.m. carnica* they range between 145-160 (Hatjina *et al.*, 2013). The number of ovarioles can be evaluated at any time during the life of a fertilized queen but preferably a few months after mating. Left and right ovaries contain almost the same number of ovarioles, so there is no need to perform the count for both ovaries. It is recommended to always count ovaries on the same side of all queens. The number of ovarioles can be estimated by two main methods: histological preparations or real-time counting. The steps for each method are described below (Table 2; Fig. 46). These two methods have advantages and disadvantages as described in Table 3.

3.5.2.5. Diameter of the spermatheca

The diameter of the spermatheca is evaluated with or without the tracheal net which surrounds it. The diameter is a direct estimation of the volume and an indirect estimation of the number of spermatozoa. Digital photographs and measurements with the use of an Image Analysis system give very accurate results. Without this system, measurements of the diameter can be performed using an eye-piece ocular micrometer.

1. Measure, two cross diameters because spermathecae do not always have a perfect spherical shape.
2. Calculate average value in mm.

Before the tracheal net is removed, the full spermatheca should have a diameter > 1.2 mm. One can use this measure for spermatheca volume to calculate the theoretical maximum storage capacity and percentage filled for spermatheca (see Tarpy *et al.*, 2011).

3. Remove the tracheal net.

The spermatheca of a mated queen is white milky colour, while the one of an unfertilized queen is transparent (Fig. 47).

3.5.3. Dissection of the fertile queen from the ventral aspect

This more difficult operation will demonstrate that part of the reproductive tract which was concealed by the sting apparatus in the preceding dissection. This part of the tract, including the median oviduct, vagina, and bursa, together with the sting apparatus which lies above them, are firmly attached to the floor of the abdomen, from which they can be detached only by careful, patient use of the point of the knife.

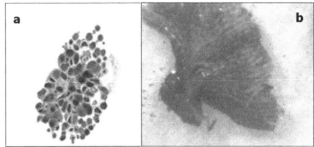

Fig. 46. Two different methods of seeing the ovarioles: **a.** the histological preparation (transversal section from Jackson *et al.*, 2011), each ellipse represents and ovariole; and **b.** real-time stereomicroscopy counting (from Hatjina, 2012).

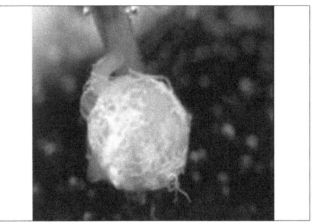

Fig. 47. Spermatheca with the tracheal net.   Photo: Fani Hatjina.

1. Mount the insect with the ventral surface upwards.
2. Cut away the floor of the abdomen using scissors, starting at the anterior end, and working backwards, removing the sternites until the lateral oviducts are uncovered.
3. Continue cutting through the body wall at the extreme sides of the abdomen, towards the tip.
4. Grip the floor with forceps, at the forward edge.
5. Insert the point of the knife and very carefully separate the sternites from the soft parts beneath them.
6. Remove the floor piecemeal, as it becomes possible to cut away parts of it to facilitate the work.

When the dissection is finished, it will be possible easily to identify the bursa and lateral pouches; the latter, in this position of the queen, are uppermost. The remaining structures are more difficult to identify, but will become clearer after soaking in glycerol.

3.5.4. Dissection of the virgin queen (Plate 17)

Dissect from the dorsal aspect, and compare with the fertile queen. The ovaries are small and undeveloped, the ovarioles containing no eggs. The spermatheca contains fluid, and collapses easily. Compare also with the laying worker.

Table 2. Estimating the number of ovarioles using histological preparations or real-time counting.

Steps of procedure	Histological preparations	Real-time counting
1	The ovaries are dissected in saline solution and then are immediately transferred to alcoholic Bouin's fixative prepared by combining 150 ml 80% ethanol, 60 ml commercially prepared formaldehyde, 15 ml glacial acetic acid, and 1 g of saturated picric acid solution for fixation. Jackson *et al.* (2011).	The ovaries are dissected in saline solution.
2	After overnight fixation at room temperature, dehydration is completed by transferring the ovaries to 100% ethanol (two changes, 15 minutes each). Dehydrated ovaries are then immersed into a 1:1 solution of 100% ethanol: xylene for one hour, then transferred to xylene for storage until infiltration with paraffin wax in a 60°C oven (two changes of paraffin, approximately 60 minutes each in duration).	Ovaries are removed, dyed by electrophoresis gel stain, and then rinsed with alcohol and transferred to a microscope slide. (Rodes and Somerville, 2003). A transverse section cuts the ovary into two parts. Most ovarioles contain eggs at the same stage of maturation.
3	Finally they are embedded in paraffin wax and the wax blocks containing the embedded ovaries are sectioned using a rotary microtome. Sections of 10 µm-thickness should be mounted in a pool of water at 48°C and collected on slides. The slides can stay on a slidewarmer overnight and then stored at room temperature in a covered box until staining. Immediately prior to staining, paraffin is removed from sections by immersion in xylene (three changes, 5 minutes each) and rehydrated in a graded series of ethanol of descending concentrations, 5 minutes per change (100%, 100%, 95%, 70% with lithium carbonate, 50%, and 30%). After a brief dip in distilled water, sections can then stained for 2 minutes in trichrome stain prepared by adding phosphotungstic acid (1 g), orange G (2 g), aniline blue WS (1 g), and acid fuchsin (3 g) to 200 ml distilled water and differentiated in 95% ethanol prior to dehydrating in 100% ethanol, clearing in CitriSolv and coverslipping with Permount (Fisher Scientific). Slides are stored flat for a week at room temperature prior to microscopy to ensure adequate drying of the Permount mounting medium.	The upper part of the ovary is then used for separation and counting of the ovarioles under a stereomicroscope at 200x or 250x magnification (Hatjina, 2012).
4	Ovarioles in several tissue sections from each ovary are counted and the average number is noted.	Ovariole numbers are noted only once per ovary (left and right).

Table 3. Advantages and disadvantages of estimating the number of ovarioles using histological preparations or real-time counting.

	Histological preparations	Real-time counting
Advantages	• The preparations are permanent. • Photographs can be taken and the ovarioles can be counted at any time. • Greater accuracy in the estimation can be achieved.	• The procedure is very quick and simple. • It does not require special chemicals or equipment such as a microtome. • You count each ovary only once.
Disadvantages	• The method requires special chemicals and equipment such as a microtome. • The preparation of the samples is vey long. • Handling the ovary so many times can be destructive.	• The method is destructive for the ovary. • The ovariole count needs to be done at the time of the dissection. • You cannot have more than two different estimations of the same ovary as the ovarioles are destroyed.

3.5.5. Spermatozoa in the spermatheca

To obtain spermatozoa or check their presence:

1. Anaesthetize the queen, and kill her (see section 2.1.2. on immobilising, killing and storing adult *Apis mellifera* in the *BEEBOOK* paper on miscellaneous methods (Human *et al.*, 2013)).
2. Dissect quickly in 0.15% salt solution.
3. Remove the spermatheca.
4. Put the spermatheca into a drop of the salt solution in a watch glass or on a microscope slide.
5. Tear open the spermatheca with needles.

If spermatozoa are present they will emerge as a creamy mass with something of the appearance of a tuft of cotton wool.

6. Spread this out with needles and examine under a microscope with 16 mm objective or higher power. See Fig. 30.

Alternatively:

1. Pinch off the top of the abdomen.
2. Pick out the spermatheca (easily seen with the naked eye).
3. Burst it under a coverslip.
4. Examine under a microscope with 16 mm objective or higher power. See Fig. 30.

3.6. Dissection of the juvenile forms

3.6.1. Dissection of the larva (Plate 19)

1. Choose large coiled larvae.

In the fresh state they are too fragile to handle, while if they are fully hardened the fat cells form a solid mass which cannot be broken up without destroying the viscera. They must therefore be hardened:

2. Immerse in formol-alcohol for only two or three days before dissection.

In any case, the removal of the fat is a rather tedious process, requiring care and patience to take it away piecemeal.

3. Lay the coiled larva on the side since it cannot be straightened out for dissection.
4. Remove the body wall using scissors.
5. Disengage the fat in small fragments.

If this is done successfully, the structures shown in Plate 19 will be revealed. The plate shows a drone larva, more convenient for this work on account of its larger size. The ovaries are very small in worker larvae, much larger in queen larvae; they occupy the same position as the testes, which are still larger.

3.6.2. Dissection of the prepupa

Prepupae show the same structures as larvae, but after the faeces are discharged the ventriculus is reduced to a narrow, flat strap, hardly recognizable as the same organ as the enormously distended ventriculus of the feeding larva.

3.6.3. Pupae

In the early stages pupae may be dissected like larvae, but from the dorsal aspect. Older pupae can be treated like adult insects.

4. Acknowledgements

Mention of trade names or commercial products in this article is solely for the purpose of providing specific information and does not imply recommendation or endorsement by the US Department of Agriculture. USDA is an equal opportunity provider and employer. The COLOSS (Prevention of honey bee COlony LOSSes) network aims to explain and prevent massive honey bee colony losses. It was funded through the COST Action FA0803. COST (European Cooperation in Science and Technology) is a unique means for European researchers to jointly develop their own ideas and new initiatives across all scientific disciplines through trans-European networking of nationally funded research activities. Based on a pan-European intergovernmental framework for cooperation in science and technology, COST has contributed since its creation more than 40 years ago to closing the gap between science, policy makers and society throughout Europe and beyond. COST is supported by the EU Seventh Framework Programme for research, technological development and demonstration activities (Official Journal L 412, 30 December 2006). The European Science Foundation as implementing agent of COST provides the COST Office through an EC Grant Agreement. The Council of the European Union provides the COST Secretariat. The COLOSS network is now supported by the Ricola Foundation - Nature & Culture.

5. References

ANTBASE.ORG (2008) American Museum of Natural History: http://antbase.org/index.htm.

AUDSLEY, N; WEAVER, R J (2006) Analysis of peptides in the brain and corpora cardiaca–corpora allata of the honey bee, *Apis mellifera* using MALDI-TOF mass spectrometry. *Peptides* 27: 512-520. http://dx.doi.org/10.1016/j.peptides.2005.08.022.

BAILEY, L (1981) *Honey bee pathology*. Academic Press; London, UK. 193 pp.

BOERJAN, B; CARDOEN, D; BOGAERTS, A; LANDUYT, B; SCHOOFS, L; VERLEYEN, P (2010) Mass spectrometric profiling of (neuro)-peptides in the worker honey bee, *Apis mellifera*. *Neuropharmacology* 58: 248-258. http://dx.doi.org/10.1016/j.neuropharm.2009.06.026.

CASAGRANDE-JALORETTO, D C; BUENO, O C; STORT, A C (1984) Numero de ovariolos em rainhas de *Apis mellifera*. *Naturalia* 9: 73-79.

CRANE, E E (1999) *The world history of beekeeping and honey hunting*. Duckworth ; London, UK. pp 559-561. ISBN 0 7156 2827 5

DADE, H A (1962) *Anatomy and dissection of the honey bee*. International Bee Research Association; Cardiff, UK. 178 pp. ISBN 0-86098-214-9

DADE, H A (2009) *Anatomy and dissection of the honey bee (revised Edition)*. International Bee Research Association; Cardiff, UK. 196 pp. ISBN 0-86098-214-9

FRIES, I; CHAUZAT, M-P; CHEN, Y-P; DOUBLET, V; GENERSCH, E; GISDER, S; HIGES, M; MCMAHON, D P; MARTÍN-HERNÁNDEZ, R; NATSOPOULOU, M; PAXTON, R J; TANNER, G; WEBSTER, T C; WILLIAMS, G R (2013) Standard methods for *nosema* research. In *V Dietemann; J D Ellis; P Neumann (Eds) The COLOSS BEEBOOK, Volume II: Standard methods for* Apis mellifera *pest and pathogen research. Journal of Apicultural Research* 52(1): http://dx.doi.org/10.3896/IBRA.1.52.1.14

GENERSCH, E; GISDER, S; HEDTKE, K; HUNTER, W B; MÖCKEL, N; MÜLLER, U (2013) Standard methods for cell cultures in *Apis mellifera* research. In *V Dietemann; J D Ellis; P Neumann (Eds) The COLOSS* BEEBOOK, *Volume I: standard methods for* Apis mellifera *research. Journal of Apicultural Research* 52(1): http://dx.doi.org/10.3896/IBRA.1.52.1.02

GOODMAN, L J (2003) *Form and function in the honey bee.* International Bee Research Association; Cardiff, UK. 220 pp. ISBN 0-86098-243-2

GOODMAN, W G; CUSSON M (2012) The juvenile hormones. In *L I Gilbert (Ed.). Insect Endocrinology.* Academic Press; Oxford, UK. pp. 310-365. http://dx.doi.org/10.1016/B978-0-12-384749-2-10008-1.

GRACE, T D C (1962) Establishment of four strains of cells from insect tissues grown *in vitro. Nature* 195(4843): 788.

HANAN, B B (1955) Studies of the retrocerebral complex in the honey bee: Part I: Anatomy and histology. *Annals of the Entomological Society of America* 48: 315-320.

HARTFELDER, K; GENTILE BITONDI, M M; BRENT, C; GUIDUGLI-LAZZARINI, K R; SIMÕES, Z L P; STABENTHEINER, A; DONATO TANAKA, É; WANG, Y (2013) Standard methods for physiology and biochemistry research in *Apis mellifera.* In *V Dietemann; J D Ellis; P Neumann (Eds) The COLOSS* BEEBOOK, *Volume I: standard methods for* Apis mellifera *research. Journal of Apicultural Research* 52(1): http://dx.doi.org/10.3896/IBRA.1.52.1.06

HATJINA, F (2012) Greek honey bee queen quality certification. *Bee World* 89: 18-20.

HATJINA, F; BIEŃKOWSKA, M; CHARISTOS, L; CHLEBO, R; COSTA, C; DRAŽIĆ, M; FILIPI, J; GREGORC, A; IVANOVA, E N; KEZIC, N; KOPERNICKY, J; KRYGER, P; LODESANI, M; LOKAR, V; MLADENOVIC, M; PANASIUK, B; PETROV, P P; RAŠIĆ, S; SMODIS-SKERL, M I; VEJSNÆS, F; WILDE, J (2013) Examples of different methodology used to access the quality characteristics of honey bee queens. *Journal of Apicultural Research* (in press).

HUANG, Z Y; ROBINSON, G E; TOBE, S S; YAGI, K J; STRAMBI, C; STRAMBI, A; STAY B (1991) Hormonal regulation of behavioural development in the honey bee is based on changes in the rate of juvenile hormone biosynthesis. *Journal of Insect Physiology* 37: 733-741 http://dx.doi.org/10.1016/0022-1910(91)90107-B.

HUMAN, H; BRODSCHNEIDER, R; DIETEMANN, V; DIVELY, G; ELLIS, J; FORSGREN, E; FRIES, I; HATJINA, F; HU, F-L; JAFFÉ, R; KÖHLER, A; PIRK, C W W; ROSE, R; STRAUSS, U; TANNER, G; TARPY, D R; VAN DER STEEN, J J M; VEJSNÆS, F; WILLIAMS, G R; ZHENG, H-Q (2013) Miscellaneous standard methods for *Apis mellifera* research. In *V Dietemann; J D Ellis; P Neumann (Eds) The COLOSS* BEEBOOK, *Volume I: standard methods for* Apis mellifera *research. Journal of Apicultural Research* 52(4): http://dx.doi.org/10.3896/IBRA.1.52.4.10

HYMENOPTERA ANATOMY ONTOLOGY PORTAL (2013) Hymenoptera Anatomy Ontology: http://portal.hymao.org/projects/32/public/ontology/.

JACKSON, J T; TARPY, D R; FAHRBACH, S E (2011) Histological estimates of ovariole number in honey bee queens, *Apis mellifera,* reveal lack of correlation with other queen quality measures. *Journal of Insect Science* 11: 82. http://dx.doi.org/10.1016/S0093-691X(99)00094-1

LORENZ, M W; KELLNER, R; WOODRING, J; HOFFMANN, K H; GADE, G (1999) Hypertrehalosaemic peptides in the honeybee (*Apis mellifera*): purification, identification and function. *Journal of Insect Physiology* 45: 647–53. http://dx.doi.org/10.1016/S0022-1910(98)00158-9.

RACHINSKY, A; HARTFELDER K (1990) Corpora allata activity, a prime regulating element for caste-specific juvenile hormone titre in honey bee larvae (*Apis mellifera carnica*). *Journal of Insect Physiology* 36: 189-194. http://dx.doi.org/10.1016/0022-1910(90)90121-U.

RICHARD, F J; AUBERT, A; GROZINGER, C M (2008) Modulation of social interactions by immune stimulation in honey bee, *Apis mellifera,* workers. *BMC Biology* 6: 50.

SAMMATARO, D; DE GUZMAN, L; GEORGE, S; OCHOA, R (2013) Standard methods for tracheal mites research. In *V Dietemann; J D Ellis; P Neumann (Eds) The COLOSS* BEEBOOK, *Volume II: Standard methods for* Apis mellifera *pest and pathogen research. Journal of Apicultural Research* 52(4): http://dx.doi.org/10.3896/IBRA.1.52.4.20

SCHEINER, R; ABRAMSON, C I; BRODSCHNEIDER, R; CRAILSHEIM, K; FARINA, W; FUCHS, S; GRÜNEWALD, B; HAHSHOLD, S; KARRER, M; KOENIGER, G; KOENIGER, N; MENZEL, R; MUJAGIC, S; RADSPIELER, G; SCHMICKLI, T; SCHNEIDER, C; SIEGEL, A J; SZOPEK, M; THENIUS, R (2013) Standard methods for behavioural studies of *Apis mellifera.* In *V Dietemann; J D Ellis; P Neumann (Eds) The COLOSS* BEEBOOK, *Volume I: standard methods for* Apis mellifera *research. Journal of Apicultural Research* 52(4):http://dx.doi.org/10.3896/IBRA.1.52.4.04

SNODGRASS, R E (1956) *Anatomy of the honey bee.* Cornell University Press; Ithica, USA. 334 pp. ISBN 1-904846-05-X

SNODGRASS, R E (2004) *Anatomy of the honey bee*. Northern Bee Books; Mytholmroyd, UK. 334 pp. ISBN 1-904846-05-X

STELL, I (2012) *Understanding bee anatomy: a full colour guide.* The Catford Press; Catford, UK. 200 pp. ISBN 978-0-9574228-0-3

TAKEUCHI, H; YASUDA, A; YASUDA-KAMATANI, Y; KUBO, T; NAKAJIMA, T (2003) Identification of a tachykinin-related neuropeptide from the honey bee brain using direct MALDI-TOF MS and its gene expression in worker, queen and drone heads. *Insect Molecular Biology* 12: 291–298. http://dx.doi.org/10.1046/j.1365-2583.2003.00414.x

TARPY, D R; KELLER, J J; CAREN, J R; DELANEY, D A (2011) Experimentally induced variation in the physical reproductive potential and mating success in honey bee queens. *Insectes Sociaux* 58(4): 569-574. http://dx.doi.org/10.1007/s00040-011-0180-z

WOODRING, J; DAS, S; GÄDE, G (1994) Hypertrehalosemic factors from the corpora cardiaca of the honey bee (*Apis mellifera*) and the paper wasp (*Polistes exclamans*) *Journal of Insect Physiology* 40: 685-692. http://dx.doi.org/10.1016/0022-1910(94)90095-7.

WOYKE, J (1971) Correlations between the age at which honeybee brood was grafted, characteristics of the resultant queens, and results of insemination. *Journal of Apicultural Research* 10(1): 45-55.

Journal of Apicultural Research 52(4): (2013)
DOI 10.3896/IBRA.1.52.4.04

REVIEW ARTICLE

Standard methods for behavioural studies of *Apis mellifera*

Ricarda Scheiner[1]†*, Charles I Abramson[2], Robert Brodschneider[3], Karl Crailsheim[3], Walter M Farina[4], Stefan Fuchs[5], Bernd Grünewald[5], Sybille Hahshold[3], Marlene Karrer[3], Gudrun Koeniger[5], Niko Koeniger[5], Randolf Menzel[6], Samir Mujagic[7], Gerald Radspieler[3], Thomas Schmickl[3], Christof Schneider[5], Adam J Siegel[8], Martina Szopek[3] and Ronald Thenius[3]

[1]Universität Potsdam, Institut für Biochemie und Biologie, Karl-Liebknecht-Str. 24-25, 14476 Potsdam, Germany.
[2]Oklahoma State University, Laboratory of Comparative Psychology and Behavioural Biology, 116 N. Murray, 74078, Stillwater, OK, USA.
[3]Department of Zoology, Karl-Franzens-University Graz, Universitätsplatz 2, 8010 Graz, Austria.
[4]Universidad de Buenos Aires, Departamento de Biodiversidad y Biología Experimental, IFIBYNE-CONICET, Facultad de Ciencias Exactas y Naturales, Pab. II, Ciudad Universitaria, C1428EHA Buenos Aires, Argentina.
[5]Institut für Bienenkunde, FB Biowissenschaften, Goethe-Universität, Frankfurt am Main, Karl-von-Frisch-Weg 2, 61440 Oberursel, Germany.
[6]Freie Universität Berlin, Institut Biologie - Neurobiologie - , Königin-Luise-Str. 28/30, 14195 Berlin, Germany.
[7]Samir Mujagic, Bielefeld University, Faculty of Biology, Department of Biological Cybernetics, Universitätsstr. 25, 33615 Bielefeld, Germany.
[8]The Department of Ecology, Evolution, and Behaviour, The Hebrew University of Jerusalem, Jerusalem 91904, Israel.

Received 29 April 2012, accepted subject to revision 11 September 2012, accepted for publication 7 March 2013.

†All authors except the first are listed alphabetically.

*Corresponding author: Email: rscheine@uni-potsdam.de

Summary

In this *BEEBOOK* paper we present a set of established methods for quantifying honey bee behaviour. We start with general methods for preparing bees for behavioural assays. Then we introduce assays for quantifying sensory responsiveness to gustatory, visual and olfactory stimuli. Presentation of more complex behaviours like appetitive and aversive learning under controlled laboratory conditions and learning paradigms under free-flying conditions will allow the reader to investigate a large range of cognitive skills in honey bees. Honey bees are very sensitive to changing temperatures. We therefore present experiments which aim at analysing honey bee locomotion in temperature gradients. The complex flight behaviour of honey bees can be investigated under controlled conditions in the laboratory or with sophisticated technologies like harmonic radar or RFID in the field. These methods will be explained in detail in different sections. Honey bees are model organisms in behavioural biology for their complex yet plastic division of labour. To observe the daily behaviour of individual bees in a colony, classical observation hives are very useful. The setting up and use of typical observation hives will be the focus of another section. The honey bee dance language has important characteristics of a real language and has been the focus of numerous studies. We here discuss the background of the honey bee dance language and describe how it can be studied. Finally, the mating of a honey bee queen with drones is essential to survival of the entire colony. We here give detailed and structured information how the mating behaviour of drones and queens can be observed and experimentally manipulated.

The ultimate goal of this chapter is to provide the reader with a comprehensive set of experimental protocols for detailed studies on all aspects of honey bee behaviour including investigation of pesticide and insecticide effects.

Footnote: Please cite this paper as: SCHEINER, R; ABRAMSON, C I; BRODSCHNEIDER, R; CRAILSHEIM, K; FARINA, W; FUCHS, S; GRÜNEWALD, B; HAHSHOLD, S; KARRER, M; KOENIGER, G; KOENIGER, N; MENZEL, R; MUJAGIC, S; RADSPIELER, G; SCHMICKLI, T; SCHNEIDER, C; SIEGEL, A J; SZOPEK, M; THENIUS, R (2013) Standard methods for behavioural studies of *Apis mellifera*. In *V Dietemann; J D Ellis; P Neumann (Eds) The COLOSS BEEBOOK, Volume I: standard methods for* Apis mellifera *research*. *Journal of Apicultural Research* 52(4):http://dx.doi.org/10.3896/IBRA.1.52.4.04

Métodos estándares para el estudio del comportamiento de *Apis mellifera*

Resumen

En este artículo BEEBOOK presentamos un conjunto de métodos establecidos para cuantificar el comportamiento de la abeja de la miel. Empezamos con métodos generales para la preparación de las abejas para ensayos de comportamiento. A continuación presentamos ensayos para cuantificar la respuesta sensorial a estímulos gustativos, visuales y olfativos. La presentación de comportamientos más complejos como el apetivo y el aprendizaje aversivo bajo condiciones controladas de laboratorio y el aprendizaje de paradigmas en condiciones de vuelo libre le permitirá al lector investigar una amplia gama de habilidades cognitivas de las abejas de la miel. Las abejas son muy sensibles a los cambios de temperatura. Por lo tanto, presentamos experimentos que tienen como objetivo el análisis de la locomoción de la abeja de la miel en gradientes de temperatura. El complejo comportamiento de vuelo de las abejas puede ser investigado bajo condiciones controladas en el laboratorio o con tecnologías sofisticadas como radar armónico o RFID (son sus siglas en ingles) en el campo. Estos métodos se explicarán en detalle en diferentes secciones. Las abejas son organismos modelo para la biología del comportamiento dada su compleja y a la vez plástica división del trabajo. Para observar el comportamiento diario de abejas individuales en una colonia, las colmenas de observación clásicos son muy útiles. La puesta en marcha y el uso de colmenas típicas de observación serán el foco de otra sección. El lenguaje de la danza de la abeja de la miel tiene características importantes como un idioma real y ha sido el foco de numerosos estudios. Nosotros aquí discutimos la base del lenguaje de la danza de la abeja de miel y describimos la forma en que se puede estudiar. Por último, el apareamiento de la reina de la abeja de la miel con los zánganos es esencial para la supervivencia de toda la colonia. Damos aquí, información detallada y estructurada sobre cómo se puede observar y manipular experimentalmente el comportamiento de apareamiento de los zánganos y las reinas.

El objetivo final de este capítulo es ofrecer al lector un conjunto completo de protocolos experimentales para estudios detallados sobre todos los aspectos del comportamiento de las abejas de la miel, incluyendo la investigación sobre los efectos de los plaguicidas y los insecticidas.

西方蜜蜂行为研究的标准方法

摘要

本章我们介绍一系列已经确立的量化蜜蜂行为的方法。我们从行为试验中准备蜜蜂的通用方法开始。然后介绍对味觉、视觉和嗅觉刺激的感官反应定量试验。对更加复杂的行为介绍帮助读者广泛地研究蜜蜂的认知能力，比如蜜蜂在实验室控制条件下的奖励性和惩罚性学习以及在自由飞行条件下的学习范式。蜜蜂对温度变化十分敏感。因此我们介绍用于分析蜜蜂在温度梯度下的运动情况的实验。蜜蜂复杂的飞行行为可以在实验室控制的条件下或在室外应用复杂的技术如谐波雷达和无线射频识别（RFID）进行研究。这些方法将分别在不同小节做详细解释。因为蜜蜂复杂且可塑的劳动分工，它们被用作行为生物学研究的模式生物。传统的观察箱对观察单只蜜蜂在蜂巢内的日常行为十分有用。传统观察箱的设置和使用将于另一小节作重点介绍。蜜蜂的舞蹈语言具备一个真正语言所需的重要特征，也成为众多研究的焦点。我们在这里讨论了蜜蜂舞蹈语言的背景并介绍如何对其开展研究。最后，蜂王与雄蜂的交配是整个蜂群存活的要素。我们在这里给出详细有组织的资料介绍如何观察和实验操作雄蜂与蜂王的交配行为。本章节的最终目的是提供给读者一系列综合的实验方案，以详细研究蜜蜂所有的行为，包括农药和杀虫剂对蜜蜂行为的影响。

Keywords: : COLOSS, *BEEBOOK*, honey bee, behaviour, gustatory responsiveness, olfactory responsiveness, phototaxis, non-associative learning, associative learning, appetitive learning, aversive learning, locomotion, temperature sensing, honey bee flight, observation hive, honey bee dance, honey bee navigation, harmonic radar, BeeScan, RFID, honey bee mating, free-flying honey bees

Table of Contents

Table of Contents cont'd

1. General introduction

The honey bee is a powerful model system for dissecting the molecular and genetic mechanisms underlying complex behaviours such as learning and memory, navigation or dance communication. In addition, the honey bee has long served as a model for the effective organisation of group living. The huge behavioural repertoire of the honey bee has been explored extensively over the last century. We here have selected a range of standard methods for quantifying the most frequently studied behaviours in honey bee research laboratories world wide. The emphasis lies on methods which can be adopted easily by researchers not specialized in this field. The wide range of behaviours we cover in this article can also be employed to investigate the effects of pesticides, insecticides or pathogens on all areas of honey bee behaviour. We start this chapter with some general methods for preparing bees for behavioural assays and then describe in detail experimental protocols for quantifying sensory responsiveness, learning, locomotion, navigational skills, communication and mating in honey bees.

2. Methods for preparing bees for behavioural tests

2.1. Capturing free-flying bees

For many behavioural studies, free-flying foragers need to be taken into the laboratory. Foragers can be caught when they leave the hive (Friedrich *et al.*, 2004; Matsumoto *et al.*, 2012) or when they return to the hive (Page *et al.*, 1998; Scheiner *et al.*, 1999, 2001b, 2002, 2003; Tsuruda and Page, 2009).

2.1.1. How to catch leaving bees

- Use a pyramid (height: 24.5 cm, apex: 3.5 cm x 3.5 cm, base: 18 cm x 18 cm).
- The pyramid should be made of UV-translucent Plexiglas® to enable the bees to perceive celestial cues and to lure them into the pyramid (Matsumoto *et al.*, 2012).
- Hold the base of the pyramid in front of the hive (at 20 cm distance) with the base open and the apex closed.
- Avoid blocking the hive entrance and arousing guard bees by standing laterally to the hive.
- After catching the required number of bees, quickly close the base and take the pyramid to the laboratory.
- There, darken the pyramid except for its apex.
- Forager bees are phototactically positive (Erber *et al.*, 2006) and will move towards he uncovered apex.
- Catch them individually while they try to leave the pyramid at the apex (Matsumoto *et al.*, 2012).

2.1.2. How to catch returning foragers

- Block the hive entrance temporarily with a wire mesh.
- Wait for the returning bees to land on the mesh and collect them individually in 10 ml glass vials.
- Close the vials with foam plugs.

The advantage of catching bees on their return to the hive is that the researcher can differentiate between pollen foragers (filled pollen baskets) and nectar or water foragers (empty pollen baskets and distended abdomen). In addition, the researcher does not need a glass pyramid when catching returning bees. However, catching returning bees might in some weather conditions take a much longer time than catching departing bees. In this situation, and when the foraging role of the bees is of minor importance, leaving bees should be caught.

Sometimes, bees are collected directly from their frames according to their age and independent of their social role (Pankiw and Page, 1999; Scheiner *et al.*, 2001a; Behrends *et al.*, 2007; Rueppell *et al.*, 2007; Scheiner and Amdam, 2009). For this, use soft forceps and collect individual bees in glass vials. If necessary, use smoke to calm bees down. More details on selecting colonies and workers for laboratory experiments can be found in the section 'Obtaining adult workers for laboratory experiments' of the *BEEBOOK* paper on standard methods for maintaining adult *Apis mellifera* in cages under *in vitro* laboratory conditions (Williams *et al.*, 2013)

2.2. Narcotizing bees

Sometimes, bees need to be narcotized, for example to mark them individually, to fix plates or transponders to them or to mount them in holders for measuring their gustatory responsiveness. To do this,

different options are available. Most frequently, individuals are briefly cooled to immobility on ice (Friedrich *et al.*, 2004; Iqbal and Müller, 2007; Rueppell *et al.*, 2007) or in a refrigerator (Scheiner *et al.*, 1999; 2001a, b; Erber *et al.*, 2006). But CO_2 narcosis is also possible. For more details on narcotizing bees refer to the section 'Standard methods for immobilising, killing and storing adult *Apis mellifera* in the laboratory' in the *BEEBOOK* paper on miscellaneous methods (Human *et al.*, 2013).

2.3. Marking individual bees

Several systems for individually marking honey bees have been developed. Karl von Frisch used a complex paint-dot system where paint-colour and location represented numeral value (1, 2, 3, etc.) as well as positional notation (ones place, tens place, hundreds place; von Frisch, 1950). Today, numbered, coloured plastic discs are available that can be glued to the thorax, and make individual tagging of bees much simpler to conduct and easier to read (Honig Müngersdorff; Fig. 1). Additionally, a paint-dot can be placed on the tag when marking a very large number of bees. Seeley and Kolmes found that an additional paint mark on the tip of the abdomen is essential for avoiding data bias due to the inability to see the thorax tag of a bee with its head and thorax concealed in a comb cell (Seeley and Kolmes, 1991).

Fig. 1. Workers tagged on the thorax with numbered coloured disks. Paint mark on disc increases number of unique tags available. Note that these bees do not have an abdomen paint mark.

Photo: A J Siegel.

To tag newly emerged bees (< 24-hour-old adults) using plastic tags:

1. Remove all adult bees from combs of mature pupae.
2. Place the combs in individual sealed cages in an incubator (34.5ºC, 60-70% relative humidity) overnight (see the *BEEBOOK* paper on Standard methods for maintaining adult *Apis mellifera* in cages under *in vitro* laboratory conditions (Williams *et al.*, 2013) for more details).
3. After 12-24 hours, shake all adult bees found on the combs into plastic bins with a thin layer of petroleum jelly applied around the upper edge of the bin (at this age, most workers are unable to fly or sting; petroleum jelly will keep them in the bin).
4. Paint-mark bees near the tip of the abdomen using enamel paint applied with a small brush, stick, or dropper (do not glue down the wings or glue shut the anus with the paint).
5. Gently pick up individuals using soft forceps or fingers, and glue a numbered tag to the centre of the thorax (Fig. 1) using fine forceps or a small-moistened stick.
6. Wood-glue (Elmer's Glue) and shellac (Honig Müngersdorff) are appropriate glues for bees.
7. Do not impede neck or wing movement with the tag.
8. Do not damage the bee while handling; the abdomen is particularly delicate at this age.
9. Orient tags in the same direction.

When tagging older workers, it is easier to work if bees are narcotized briefly prior to tagging to inhibit movement. Alternatively, place an individual bee in a tube with a grid at one end, through which the tag is glued on the thorax (see also section 10; Figure 24B). After receiving tags, bees can be temporarily stored in cups with a thin layer of petroleum jelly around the edges (one-day-old bees) or a ventilated cover (older bees). A sugar cube in the bottom prevents the bees from starvation. A section on marking honey bee queens can be found in the *BEEBOOK* paper on miscellaneous methods (Human *et al.*, 2013).

2.4. Harnessing individual bees

For a number of behavioural assays, honey bees need to be harnessed individually. Metal or plastic holders (Fig. 2A) have proved very useful for mounting bees, and they can be produced easily. Bees are placed in a holder while being narcotized. A strip of textile tape between head and thorax fixes the bee. A second strip of tape fixed over the abdomen prevents the experimenter from getting stung. If bees are supposed to walk freely after their confinement to the holder, a plaster should be used instead of textile tape to save the wings and legs from sticking to the tape. For more details on immobilizing bees, refer to the section 'Standard methods for immobilising, killing and storing adult *Apis mellifera* in the laboratory' in the *BEEBOOK* paper on miscellaneous methods (Human *et al.*, 2013).

3. Quantifying sensory responsiveness
3.1. Introduction

Honey bees rely on their olfactory, visual, gustatory and tactile senses to communicate with their environment. Odours, for example, play an important role in nest mate recognition (Hölldobler and Wilson, 2008; Ratnieks *et al.*, 2011) and in locating food sources (Menzel *et al.*, 1993; Dobson *et al.*, 1996). Not surprisingly, honey bees are capable of discriminating a large variety of different odours (Guerrieri *et al.*, 2005). For orientation and navigation, honey bees rely to a large part on their visual senses (Dyer, 1996; Horridge, 2009). The perception of gustatory stimuli is decisive for evaluating nectar sources by foragers (Seeley *et al.*, 1991) and for the preparation of food by nurse bees (Winston, 1987). Tactile cues are decisive for orientation in the dark hive, for comb building and for locating the way to a food source on a flower (Kevan and Lane, 1985; Winston, 1987).

A honey bee colony depends on the accurate sensory perception of its individuals. Measuring the sensory responsiveness of a bee can therefore not only be indicative of the individual's physiological state and health but of the state of the entire colony. Diseases or pesticides can obviously affect the vitality of a colony by changing the sensory responsiveness of some of its members. Analysing the sensory responsiveness for different stimulus modalities is therefore an important tool for studying the physiology and behaviour of honey bees.

A number of different assays are available for measuring responsiveness for gustatory, visual and olfactory stimuli in individual honey bees (see Table 1). For testing visual or olfactory responsiveness, bees do not need to be fixed, but for analysing gustatory responsiveness, individuals need to be mounted in holders (see section 2.4.). To reduce the mobility of the bees and to prevent them from stinging, they are usually narcotized (see section 2.2.). Once the bee has adapted to the new situation, different tests for measuring sensory responsiveness can be employed.

3.2. Gustatory responsiveness

Gustatory responsiveness, i.e. responsiveness to water and sucrose solutions of different concentrations, can be easily measured using the proboscis extension response (PER):

1. Present water and a series of sucrose concentrations (i.e. 0.1%; 0.3%; 1.0%; 3.0%; 10%; 30%) to the antennae of a fixed bee. When the concentration of the sucrose solution exceeds the individual response threshold of a bee, it extends its proboscis in expectation of food (Fig. 2A).
2. Report on each stimulation whether the bee extended its proboscis or not.
3. Use the PERs to water and the different sucrose concentrations (Fig. 2B) to calculate an individual gustatory response score (GRS, Fig. 2C). This score is the sum of

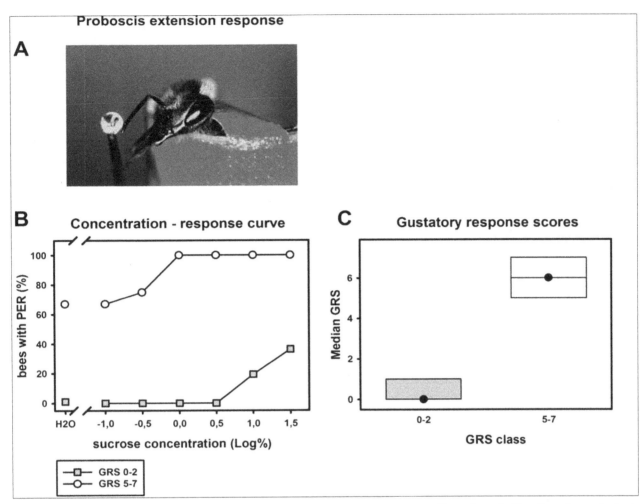

Fig. 2. Use of the proboscis extension response (PER) for measuring gustatory responsiveness of honey bees. **A.** When the antenna of a fixed bee is stimulated with a droplet of sucrose solution of sufficient concentration, the bee extends its proboscis. This reflex behaviour is termed "proboscis extension response". The number of proboscis extension responses following stimulation with a series of sucrose concentrations can be counted. This number constitutes the gustatory response score (GRS) of a bee. **B.** Sucrose-concentration-response curves of unresponsive bees with low gustatory response scores (GRS 0-2) and of highly responsive bees with high gustatory scores (GRS 5-7). The x-axis displays the sucrose concentration. The y-axis shows the percentage of bees extending their probosces at stimulation with that sucrose concentration. **C.** Gustatory response scores are a measure for the overall gustatory responsiveness and are well suited for correlation analyses. Because gustatory response scores are usually not distributed normally, median GRS (circles) and quartiles (upper and lower lines) are displayed.

proboscis extension responses following a series of sucrose stimulations at the antennae. Bees with a score of "1" usually only respond to 30% sucrose, while bees with a score of "7" respond to all sucrose stimulations with proboscis extension.

The data gained from measuring the occurrence of proboscis extension are bimodal. Either a bee extends its proboscis or it does not. Thus, frequencies of responding and non-responding animals can be used to compare different populations of bees. Alternatively, the researcher can compare the gustatory response scores of groups of bees (for review see Scheiner *et al.*, 2003). The GRSs of a population of bees usually correlate with the gustatory response threshold of the group (Page *et al.*, 1998), i.e. the lowest sucrose concentration that the bees can distinguish from water.

Gustatory responsiveness is an excellent indicator of the physiological state of a bee. It correlates with responsiveness to light and to odours, with division of labour and with non-associative and associative learning performance (for review see Scheiner *et al.*, 2004). The PER assay is an inexpensive and high-throughput procedure and allows far-reaching conclusions on related behaviours such as learning and memory (Scheiner *et al.*, 1999). It has been used frequently in neurobiological and behavioural physiological studies on honey bees (Page *et al.*, 1998; Scheiner *et al.*, 1999; Amdam *et al.*, 2006; Rueppell *et al.*, 2007). In addition, this assay can be used for examining the effects of diseases and pesticides on honey bee behaviour (Iqbal and Müller, 2007; Aliouane *et al.*, 2009).

Table 1. Experimental designs for measuring sensory responsiveness to gustatory, olfactory and visual stimuli in honey bees.

	gustatory responsiveness	olfactory responsiveness	visual responsiveness (phototaxis)
fix bee in holder	yes	no	no
devices needed	syringes with sucrose solutions of different concentrations	olfactometer, i.e. an arena which allows observation of the bee's behaviour towards odorant sources of different intensities; system for continuously removing air (i.e. exhaust)	dark arena which can be illuminated by LEDs of different intensities; infrared illumination, infrared-sensitive camera which can be mounted on the arena
parameters to be measured	occurrence of proboscis extension when the antennae are stimulated with different sucrose solutions	time a bee spends in an odour-containing arm	time a bee needs for reaching a light source

3.3. Responsiveness to light (phototaxis)

Phototaxis or the movement of bees towards light can be compared between individuals by measuring the time a bee needs to reach a light source in an otherwise dark arena (Fig. 3). The dark arena (a bucket standing upside down to form a light-proof arena) is illuminated by green light emitting diodes with a maximum light intensity of 5600 mcd. The diodes are driven by 4 V power supply. Different relative light intensities are achieved by using neutral density filters. Twelve LEDs are fixed in the arena with opposite LEDs having the same light intensity. The LEDs are covered by a ground glass. The following logarithmic order of relative light intensities is suggested: 100%, 50%, 25%, 12.5%, 6.25% and 3.125%. At the maximum (100%) intensity, the 520 nm stimulus has an intensity of 560 lx measured directly at the ground glass.

The phototactic behaviour of bees naturally depends on their locomotor behaviour. Therefore, it is necessary to study this factor independent of illumination. The length of the walking path the bee covers in a specific time in complete darkness can be measured and an average walking speed can be calculated (Erber *et al.*, 2006). These measures should be compared between individuals to test whether differences in phototactic behaviour are causally linked to differences in locomotor behaviour. An example of how to determine the phototaxis of honey bees is given here:

1. Release an individual bee into the dark arena.
2. Videotape its behaviour till you remove the bee from the arena.
3. Measure the walking path length of the bee during 30 seconds in the dark arena.

If you do not possess a computer program for doing this, simply follow the replayed walking path with a pen on the monitor after covering it with transparent film.

4. Switch on one of the light sources with the lowest intensity.
5. As soon as the bee has reached it, switch on the opposite light source of equal intensity.
6. Take the time the bee needs to get from the first LED to the second LED of the same intensity.
7. Repeat this procedure three times, so that you can average the walking times of a bee to one light intensity over four trials.
8. When the bee has reached the lowest light intensity for the final time, switch on the next higher light intensity and so forth.
9. Make sure that you switch off one light source before the next one is turned on.
10. To analyse phototaxis of different groups, compare walking times between groups or treatments.

While the walking speed of a bee towards different light sources strongly correlates with its gustatory responsiveness, locomotion was shown to be independent of gustatory responsiveness (Erber *et al.*, 2006; Tsuruda and Page, 2009). Using this assay, Rueppel *et al.* (2007) demonstrated that bees of different ages do not differ in their phototactic behaviour, if behavioural role was not accounted for. Tsuruda and Page (2009) compared the visual responsiveness of bees of the two genetic strains of Page and Fondrk (1995) which differ in their foraging preferences. This assay is ideally suited to study visual responsiveness separately from locomotor responses in honey bees and will reveal impairments in the processing of visual stimuli. However, one has to bear in mind that using this assay one cannot make statements about the bee's perception of the visual stimuli, because behavioural responses are recorded, not the activity of photoreceptors.

3.4. Responsiveness to odours

Gustatory responsiveness correlates with responsiveness to odours, which can be tested in an olfactometer (Fig. 4; Pham-Delègue *et al.*, 1991; Scheiner *et al.*, 2004). In this assay, bees are placed in the centre of a four-armed maze. One of the arms contains an odorant of a certain concentration. If the bee can smell the odorant and if it is attracted by it, the bee will soon walk towards the odorant and remain close to the odour source. Most bees prefer a high odorant intensity over a low intensity. Responsiveness to different odorant intensities can be tested by placing different concentrations of an odorant alternating in the four arms of the olfactometer. An example of an olfactometer experiment is given below:

1. Place an individual bee in the centre of the arena.
2. Add one odour to the end of one arm of the four-armed olfactometer.

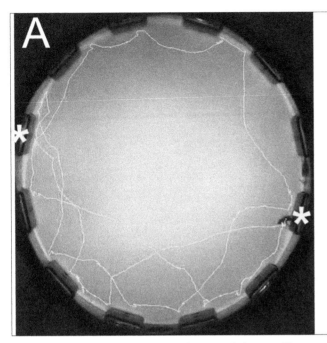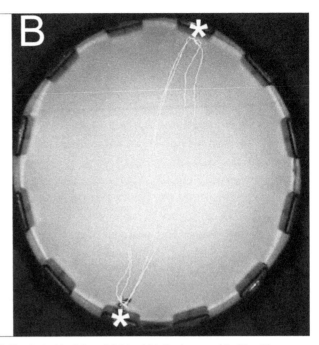

Fig. 3. Phototactic behaviour of honey bees in a dark arena. The arena is illuminated by infrared light, which the bees are blind for. The phototaxis arena can be illuminated by twelve light sources. Two opposing light sources (light emitting diodes, LED) are of the same relative light intensity. The walking paths of a bee (yellow) are displayed for four walks between LEDs of low light intensity (**A**) and for four walks between LEDs of high intensity (**B**). The location of the respective light sources is indicated by asterisks. While the bee takes many detours when walking towards an LED of low intensity, its walking paths towards a high light intensity are direct.

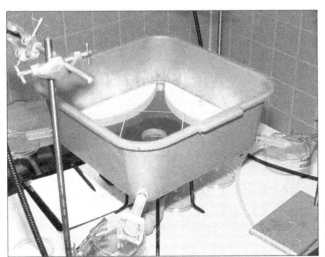

Fig. 4. Olfactometer for measuring responsiveness to odours. In this arena, the bee is placed in the centre. An odorant of a specific concentration is placed in front of one of the four arms of the olfactometer. While the air is drawn from the arena from its middle and therefore pulls in the odour trough the arm, the bee can smell it. If the bee is attracted by the odour, it will spend a longer time in the odour-containing arm of the arena than in the other arms.

3. Determine the time the bee spends in the centre of the arena and in each arm of the olfactometer.

4. Once the bee has reached the odorant source, remove the odorant quickly.

5. Allow a 30-second pause without olfactory stimulation.

6. Place the next odorant or the next higher concentration of the same odorant in a different arm of the olfactometer.

7. Measure the time the bee spends in each part of the olfactometer, which corresponds to the preference of the bee for that particular odour.

8. At the end, calculate the time the bee spends in an odorant-containing arm and compare it with the times the bee spent in the other parts of the arena using T Tests (for further details see the *BEEBOOK* paper on statistical methods (Pirk *et al.*, 2013)).

Individuals with low gustatory responsiveness only walk directly towards high odorant intensities, while bees with high gustatory responsiveness are also attracted by lower odorant concentrations (Scheiner *et al.*, 2003). If a bee is not attracted by an odour or if it is unable to perceive the odour, it spends the time randomly in the four arms of the olfactometer.

Using an olfactometer it was shown, for example, that bees display an age-dependent preference for queen mandibular pheromone with five-day-old bees exhibiting the strongest attraction to this pheromone (Pham-Delègue *et al.*, 1991). Also, it was tested using an olfactometer how well bees discriminate between combs based on their odours (Breed and Stiller, 1992). These examples suggest that the olfactometer assay is also well suited to detect impairments of olfactory perception. One problem with this assay is that bees might not be attracted by an odour, even though they may be able to detect differences in the odour concentration. Therefore, odours to be used in this assay need to be tested carefully for their behavioural effects on honey bees prior to use in the olfactometer.

4. Quantifying non-associative learning

Habituation and sensitization are two well-known forms of non-associative learning, which change the response probability of an animal towards a stimulus following repeated exposure to the same stimulus (habituation) or exposure to a strong yet different stimulus (sensitization). Habituation is a response decrement to a monotonously repeated stimulus (Braun and Bicker, 1992), while sensitization implies that a strong or particularly salient stimulus enhances the response of the bee to a test stimulus (Menzel *et al.*, 1991).

A honey bee extends its proboscis when its antennae are stimulated with a sucrose solution whose concentration exceeds its sucrose response threshold (see section 3.2.). When low-concentrated sucrose solutions are repeatedly applied to the antennae of a bee with a short interval between two stimulations ("inter-trial interval"), the proboscis extension response habituates (Braun and Bicker, 1992; Scheiner, 2004). The bee no longer shows proboscis extension when its antennae are stimulated with this sucrose stimulus.

To perform a habituation experiment, adapt the following standard protocol to your needs:

1. Harness the bee (see section 2.4).
2. Test for each bee whether it shows the PER to antennal stimulation with 1% sucrose.
3. Stimulate both antennae of each bee repeatedly with 1% sucrose (for example 30 times), leaving a break of 3 seconds in between the stimulations ("inter-trial interval").

Take care not to stimulate the extended proboscis with sucrose solution.

4. For each individual, count the number of trials that the bee needs to habituate the PER, i.e. count how often the bee shows the PER before it stops extending its proboscis.

Alternatively, calculate a habituation score (i.e. the sum of PERs during the 30 habituation trials). You may also calculate response frequencies over all trials across groups.

5. Compare scores between two groups using T Tests or non-parametric equivalents, if necessary. More than two groups can be compared using analysis of variance (ANOVA) or non-parametric equivalents with respective post-hoc tests. Response frequencies can be compared using Chi square tests or multiple-measurement ANOVA (for further details see the *BEEBOOK* paper on statistical methods (Pirk *et al.*, 2013)).

One disadvantage of this method is that it is difficult to interpret why a bee has stopped responding. Because the researcher does not know whether a bee has become exhausted of extending its proboscis, whether it really habituated or whether it may show peripheral adaptation, a dishabituation test is necessary at the end of the experiment. Here, a high-concentrated sucrose solution (30% or 50% for example) is applied to the antennae. If the bee shows dishabituation, it responds again both to the high-concentrated sucrose solution and to a subsequent stimulation with the low sucrose concentration used for habituation. Therefore, you need to add these steps to your protocol:

6. Apply a 30% sucrose solution to both antennae 3 seconds after your final habituation trial.
7. Three seconds later, test whether the bee shows the PER to the 1% sucrose solution used in the habituation session.
8. Only compare habituation data of bees which responded in the tests for dishabituation.

The course of habituation is determined by individual gustatory responsiveness and the sucrose concentration used as habituating stimulus. Bees with low gustatory responsiveness and satiated bees habituate faster than bees that are more responsive to gustatory stimuli (Braun and Bicker, 1992; Scheiner *et al.*, 2004). High-concentrated sucrose concentrations lead to slower habituation than low sucrose concentrations (Scheiner, 2004).

This paradigm is very simple and allows a high throughput of bees in a short time. For that reason, habituation tests have been used repeatedly to study effects of pesticides or viruses on honey bee behaviour. Guez *et al.* (2001) showed, for example, how the insecticide imidacloprid affected habituation of the proboscis extension response. Iqbal and Müller (2007) demonstrated that DWV infection has no effect on habituation in honey bees and Kralj *et al.* (2007) analysed whether infection with *Varroa* mites affects habituation.

Sensitization can be easily tested in honey bees using an odorant and a sucrose stimulus. Most bees do not show spontaneous proboscis extension when their antennae are stimulated with an odorant such as carnation, citral or geraniol. When their antennae are briefly stimulated with a high-concentrated sucrose solution (30% or 50% for example) and with an odorant immediately afterwards, the bees become sensitized to the odorant (Menzel *et al.*, 1993; Hammer *et al.*, 1994). Now they display proboscis extension in response to antennal stimulation with the odorant which previously did not evoke a visible response. Note that sensitization to an odour is different from classical olfactory conditioning (see section 5.), because: (1) It does not require repeated pairings of odour and sucrose solution; (2) It only occurs within up to two minutes after olfactory stimulation.

Satiated bees can also be sensitized to gustatory stimuli which do not elicit a PER such as low sucrose concentrations (1%) or water. Normally, satiated bees rarely extend their probosces when their antennae are stimulated with these solutions. But when their antennae are briefly stimulated with a high sucrose concentration and subsequently with a low sucrose concentration, bees can become sensitized to the low sucrose concentration and now respond to it. To analyse different groups, compare the response frequencies to the sensitized odour or low sucrose stimulus using Chi square tests (for further details see the *BEEBOOK* paper on statistical methods (Pirk *et al.*, 2013)).

Similar to habituation, the degree of sensitization depends on the gustatory responsiveness of the bee and the strength (i.e. the sucrose concentration) of the sensitizing stimulus (Menzel *et al.*, 1993; Scheiner 2004). Sensitization to odours or low sucrose concentrations is a very effective and high-throughput assay, which allows the experimenter to draw conclusions on the non-associative learning abilities of a bee or a population of bees and which includes different stimulus modalities (i.e. gustatory and olfactory).

5. Quantifying associative appetitive learning and memory in the laboratory

The honey bee is well-known for its excellent associative learning capacities in a wide range of assays (Menzel and Müller, 1996; Menzel, 2001; Giurfa, 2007; Menzel, 2012). Both in the field and under the controlled conditions of a laboratory, bees reliably learn and remember odours, shapes and surface structures. Learning of colours is much more complicated under laboratory conditions (Gerber and Smith, 1998; Mota *et al.*, 2011). Bees can be trained in classical and operant conditioning paradigms. In classical conditioning, the honey bee learns to associate an originally neutral stimulus (conditioned stimulus, CS) with a biologically relevant stimulus (unconditioned stimulus, US), while in operant conditioning, a bee evaluates its own behaviour and its consequences (Menzel, 2012).

5.1. Classical conditioning

Most studies on learning and memory employ classical olfactory conditioning of the proboscis extension response. Here, the bee learns to associate an odorant with proboscis extension (Fig. 5A, B; Takeda, 1961; Bitterman *et al.*, 1983; Hammer and Menzel, 1995; Sandoz *et al.*, 1995; Scheiner *et al.*, 2003; Matsumoto *et al.*, 2012). A simple olfactory conditioning experiment with one odorant as conditioned stimulus and a 30% sucrose solution as unconditioned stimulus and reward could be performed like this:

1. Place a harnessed bee (see section 2.4.) in a constant airstream.
2. Add the odorant (for example 2 μl citral on a filter paper) to the airstream for a few seconds (1-5 seconds).
3. Observe whether the bee shows spontaneous proboscis extension.
4. Consider discarding bees with spontaneous response.
5. Test whether the bee shows proboscis extension at stimulation of its antennae with the sucrose solution.
6. Discard bees without PER in response to antennal stimulation with sucrose.
7. Start each conditioning trial by exposing a bee to the odorant for 1-3 seconds.
8. Elicit the proboscis extension response by touching the antennae of the bee with the sucrose solution.
9. Allow the bee to lick from the sucrose solution for up to one second.

Make sure that the odorant is turned off while the bee consumes the reward.

10. Once the bee has completed drinking, proceed with conditioning the next bee.
11. Use an interval of 5 minutes before you train your first bee again.

Most bees learn quickly to associate odorant and reward and will soon extend their probosces when their antennae are only stimulated with the odorant, which demonstrates associative learning (Fig. 5B). Although bees learn better if they are allowed to drink from the sucrose solution, they also learn to associate the odorant with proboscis extension when only their antennae are stimulated with sucrose (Sandoz *et al.*, 2002). Honey bees need few trials for olfactory learning and establish long-lasting memories (Bitterman *et al.*, 1983; Menzel and Müller, 1996; Sandoz *et al.*, 1995; Matsumoto *et al.*, 2012). As three conditioning trials usually suffice to measure the learning ability of a population of bees, this assay allows a comparatively high throughput. Classical olfactory conditioning is also a good example of appetitive learning, where the behaviour of the bee is rewarded. Later on (see section 6.), aversive learning paradigms will be introduced to the reader. In those paradigms, the behaviour of a bee is punished.

Classical olfactory conditioning has not only been used to study the mechanisms underlying different forms of learning and memory processes (Menzel *et al.*, 1993; Hammer and Menzel, 1995; Giurfa, 2007; Haehnel and Menzel, 2012) but also to analyse the effects of diverse pesticides on honey bee behaviour. Taylor *et al.* (1987) showed that exposure to different pyrethroids prior to conditioning differentially impairs associative learning ability in bees. Short-term treatment with imidacloprid had differential effects on olfactory acquisition and memory (Decourtye *et al.*, 2004). Abramson *et al.* (1999) tested the effects of a whole set of products on classical olfactory learning and extinction and demonstrated differential effects, depending on the time of application. Ramirez-Romero *et al.* (2007) demonstrated an effect of genetically modified crops containing the soil bacterium *Bacillus thuringiensis* on feeding behaviour and proboscis extension learning.

5.2. Differential olfactory conditioning

In differential olfactory conditioning, a second odorant is introduced into the conditioning protocol, which, however, is unrewarded (Giurfa, 2003). While both odorants are given repeatedly, one of them is always followed by a sucrose reward, while the other odorant is not. This procedure usually produces a stronger conditioned response to the conditioned odorant than simple classical olfactory conditioning. Outside neurobiological studies this assay has, for example, been used successfully to analyse effects of dicofol on learning behaviour of bees (Stone *et al.*, 1997).

Fig. 5. Classical olfactory conditioning (**A, B**) and operant tactile conditioning (**C, D**) of the honey bee. Naïve bees do not show proboscis extension when their antennae are stimulated with a certain odour like citral or geraniol (**A**) or when their antennae can scan a tactile stimulus (**C**). After pairing the odour or the tactile stimulus with a sucrose droplet, which is first applied to the antennae and then to the proboscis (not shown), the bees show conditioned proboscis extension to the odour (**B**) and to the tactile stimulus (**D**). Note that for tactile conditioning, the eyes are painted over with black colour to improve perception of tactile stimuli.

5.3. Tactile conditioning

Similar to olfactory conditioning, bees can be trained to associate a tactile stimulus with proboscis extension (Fig. 5C, D). In this paradigm, the bee learns to associate the characteristics of a tactile stimulus with a sucrose reward. In contrast to olfactory conditioning, the eyes of the bee have to be occluded by black paint before training it to tactile cues. This improves tactile scanning behaviour. The tactile cue, which can be a small plate (3 x 4 mm, Scheiner *et al.*, 1999), is first presented to the bee for analysis of spontaneous behaviour like for olfactory conditioning (Fig. 5C). As soon as the plate is brought into the scanning range of the bee antennae, the animal starts scanning the tactile stimulus with its antennae. This behaviour is normally not associated with proboscis extension. During training, the plate is repeatedly brought into the scanning range of the bee's antennae and while the bee touches the plate, proboscis extension is elicited by also applying a small droplet of sucrose solution to its antennae (see olfactory conditioning, sections 5.1.) and the bee is

allowed to drink from the sucrose solution. While the bee is licking the sucrose, the plate is removed from the scanning range. Like in classical olfactory conditioning, bees learn to show proboscis extension to tactile stimuli after one or two trainings (Erber *et al.*, 1998; Scheiner *et al.*, 1999, 2001a, b) and retain long-lasting tactile memories (Scheiner *et al.*, 2001a).

5.4. Mechanosensory conditioning

Similar to tactile learning, honey bees can be trained classically to mechanosensory stimulation of their fixed antennae. Here, bees are rewarded with sucrose, while their antennae are stimulated with a tactile stimulus, even though the bees themselves are passive (Giurfa and Malun, 2004). Unlike tactile conditioning, this procedure employs classical conditioning, because the bee does not need to demonstrate active antennal scanning movements.

Appetitive olfactory learning and tactile learning strongly depend on individual gustatory responsiveness. Bees with high responsiveness learn fast, while bees that display a low gustatory responsiveness need more training to form an association (Scheiner *et al.*, 1999; 2001a, b, 2003, 2005; Amdam *et al.*, 2010). If bees differing in their gustatory responsiveness are rewarded with equal *subjective* rewards, depending on their individual gustatory responsiveness, learning performance does not differ between individuals (Scheiner *et al.*, 2005). This underlines the importance of measuring individual gustatory responsiveness prior to testing learning abilities.

5.5. Pitfalls

Despite the usefulness of these learning paradigms, one has to bear in mind that the interpretation of why a bee does not extend its proboscis can be difficult. The bee might not find the reward motivating enough or it might have forgotten the association of conditioned and unconditioned stimuli, to name just a few examples. This is certainly a drawback of this method, which cannot be overcome completely by experimental controls. Nevertheless, a large number of bees tested (around 30 – 40 bees per group) will yield robust results for interpretation.

6. Quantifying associative aversive learning in the laboratory

6.1. Introduction

Compared to studies using appetitive sucrose stimuli, the use of aversive stimuli has been rather neglected in honey bee research (Wells, 1973; Srinivasan, 2010), although there are several reasons for studying aversive learning in honey bees. First, aversive conditioning paradigms can help in teasing apart theories in the comparative analysis of learning (Abramson *et al.*, 2010b). Second, honey bees must be able to predict dangers (McNally and Westbrook, 2006). Under field conditions, they face numerous challenges related to the escape and avoidance of aversive stimuli including those related to predators, pesticides, and repellents. This behaviour can be tested in laboratory assays on aversive conditioning (Abramson *et al.*, 2006a). A third reason to study aversive conditioning is to provide new paradigms that can further advance the neurobiological and genomic understanding of learning mechanisms. Two of the most well-known examples using aversive stimuli are based on conditioning of sting extension (Giurfa *et al.*, 2009) and on place learning (Agarwal *et al.*, 2011).

6.2. Aversive conditioning in a shuttle box

6.2.1. Example protocols

The most popular and versatile apparatus for studying aversive conditioning is the shuttle box (also known as a choice chamber). A shuttle box (Fig. 6) is a two-compartment chamber between which an

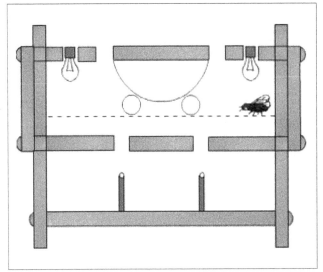

Fig. 6. Overview of a shuttle box. In this version, an aversive odour such as formic acid is used. The odour enters the device through the tubes shown in the bottom of the drawing. The bee walks on a screen floor and its movement is detected with infrared photo-detectors represented by the open circles. Lights located on the ceiling provide a cue signalling the onset of the aversive stimulus. The aversive stimulus can be terminated or avoided when the bee activates the photo-detector farthest away from it at any given time. To cross from one side of the chamber to another, the bee passes through a "tunnel" represented by the semi-circle.

unrestrained bee moves back and forth ("shuttles"). The shuttle box can accommodate all types of aversive conditioning designs in the areas of escape, punishment, and avoidance (see Table 2). Molecular studies of learning in the honey bee can also be performed in a shuttle box (Agarwal *et al.*, 2011). If a researcher is interested in the effect of reward on learning, escape conditioning is a good alternative to an appetitive reward. To perform escape conditioning:

1. Select the parameters for the training variables. Table 3 provides a list.
2. Place the honey bee in the shuttle box for a 5 minute adaptation period.
3. Present an electric shock to the bee while it stays in one compartment. As a response to this shock, the bee will escape from the shock and enter a different compartment. The shock-free time in this compartment constitutes the reward.
4. Measure the time the bee takes to terminate (escape) the shock.

Escape conditioning is an especially useful aversive conditioning technique for bee researchers, because it is simple to use and there is no concern for motivational issues associated with food rewards in situations where the bee is confined and unable to unload. In addition, the results of escape conditioning are similar to those found with food rewards (see Macintosh, 1974 and Campbell and Church, 1969).

Table 2. Different categories of aversive conditioning in honey bees using a shuttle box.

category of aversive conditioning	escape	punishment	avoidance	fear conditioning
application of electric shock	shock remains on until bee leaves the current compartment	shock turns on when bee enters the other compartment	shock is omitted when bee enters compartment	response-independent pairings of light with shock
relationship between requisite response and shock	yes	yes	yes	no
references	Abramson, 1986; Agarwal *et al.*, 2011	Agarwal *et al.*, 2011	Abramson, 1986; Agarwal *et al.*, 2011	

Table 3. Variables affecting aversive conditioning.

training variable	effects
- **type of aversive stimuli** (electric shock, olfactory, temperature, vibration, noise, agrochemicals) - **type of shock** (alternating or direct current) - **subspecies**	affects all responses
- **intensity of warnings and aversive stimuli** - **timing of aversive stimuli**	function is an inverted "U" shape
- **duration of aversive stimulus**	short durations are often more effective than longer durations
- **schedule of aversive stimuli**	continuous reinforcement schedules are more effective than intermittent schedules; a single presentation has different effects from a series of short bursts
- **prior exposure to aversive stimuli**	prior habituation or sensitization influence performance
- **interval between CS and US stimuli**	the shorter the interval, the more effective the association between a conditioned stimulus (CS) and an aversive unconditioned stimulus (US)
- **delay between response and consequence**	the shorter the delay the more effective the conditioning
- **control of aversive stimuli by the bee or independent of the bee**	the aversive stimulus can be terminated either by the bee or independently depending upon the paradigm
- **magnitude of a negative reinforcer**	frequency of escape responses is a function of reinforcement magnitude
- **shock-shock (S-S) and response-shock (R-S) intervals in unsignalled avoidance**	if the S-S and R-S intervals are short the response rate will be high, if these intervals are long, the response rate will be low
- **type of extinction procedure**	if the aversive event is omitted during extinction, performance will

The shuttle box can also be used to study punishment, another simple type of aversive conditioning. Here, the honey bee receives an aversive stimulus, such as shock, as a consequence of entering a compartment. As a result of this contingency, the bee decreases the probability of shuttling. This situation is ideal for bee researchers looking for a situation that produces the opposite effect of reward training. If a researcher is looking for a complex conditioning situation, one protocol to use with a shuttle box is signalled avoidance. Signalled avoidance has the interesting property that it contains elements of both Pavlovian and instrumental conditioning. The experiment begins with a conditioned stimulus (CS) predicting an impending aversive event such as electric shock. If the honey bee does not respond to the CS by entering the shuttle box compartment farthest away from it at the time of CS presentation, the shock is presented (the Pavlovian component). After a number of CS-shock pairings, the honey bee makes the instrumental response of entering the compartment during the presentation of the CS, but before the onset of shock, thereby avoiding or postponing the shock (instrumental component). To perform a signalled avoidance experiment:

1. Select the parameters for the training variables. Table 3 provides a list.

2. Place the honey bee in the shuttle box for a 5 minute adaptation period.

3. Present the CS, if the honey bee does not respond by entering the compartment farthest away from it at any given time, present the shock. Do not present the shock if the honey bee responds during the presentation of the CS.

4. Measure on each CS presentation whether or not the bee entered the compartment farthest away from it (i.e., an avoidance response).

6.2.2. The study of "emotions"

For researchers interested in studying "emotions" in bees, the signalled avoidance paradigm is ideal. One protocol is known as "fear conditioning." Here a conditioned stimulus is paired with an aversive event in a response-independent fashion. As applied to the shuttle box, the bee receives response-independent pairings of light with shock. Fear conditioning is measured indirectly by the effect of light only trials on some on-going behaviour (known as conditioned suppression) or directly by observing the conditioned response to light such as sting extension.

6.2.3. Standard two-way shuttle box construction

An example of a honey bee shuttle box is shown in Fig. 6. The apparatus is made from plastic. Infra-red photocell-detector circuits automatically monitor shuttle responses. The photocells must be in the infra-red range, so not to attract the bee. The aversive stimulus is typically electric shock but odours also are used. An olfactory shuttle box for insects is described by Abramson *et al.* (1982) and modified for honey bees (Abramson, 1986). The honey bee can be placed in the shuttle box in several ways. The easiest is to capture bees (see section 2.1.) and narcotize them (section 2.2.). When the bee is less active it can be safely placed in the apparatus. Alternatively, an active bee placed inside of a container or vial can be lured into the shuttle box by light. Light can also be used to induce the bees to leave the apparatus when the experimental session is completed. Luring the bee to the shuttle box eliminates the need to narcotize it.

When constructing and using a shuttle box there are some important issues you must consider. For example, great care must be given to the dimensions of the apparatus, how stimulus cues are presented if punishment and signalled avoidance paradigms are used, and the placement of photocell-detector pairs. A disadvantage of the shuttle box is that it represents an unnatural situation for the honey bee and the bee is often highly active when placed inside the shuttle box.

For technical details, the reader is advised to look at the Abramson (1986) and Agarwal *et al.*, (2011) papers before constructing their own shuttle box. If shock is used rather than odours, it is critical that the bee stays in contact with the shocking surface. If the ceiling of the apparatus is too high, the bee will quickly learn to walk on the ceiling and/or keep a few legs on the shocking surface and the rest on

the walls of the apparatus. In both situations the bee is not receiving the programmed intensity and duration of shock. This situation is easily eliminated by having a ceiling just a few millimetres above the bee.

Another issue related to apparatus size is that there must be enough room for the bee to turn around and re-enter the compartment it just came from. The bee will be highly active making many responses and often moving from end to end. If the width of the shuttle box is to narrow, the bee will find it difficult to shuttle between compartments. This is readily seen, because the bee will try and walk backwards. This problem can be eliminated by using a suitable width between the walls.

For the shuttle box to function effectively, the bee must have a cue that it entered a new compartment. The cue can be as simple as small strips of plastic that it steps over, a sloping roof that requires the bee to "duck under" or visual stimuli. One method to present visual stimuli is to set the shuttle box on a flat screen computer monitor. However, this will only work if you are using shock grids. An alternative is to place coloured plastic strips under the grid that are manually changed at the appropriate time (Agarwal *et al.*, 2011).

The final issue to consider when using a shuttle box is the photocells and their placement. A photocell-detector pair should be placed on either side of the mid-line of the apparatus approximately at bee's length. This length ensures that the experimenter has a clear definition that the bee fully entered a compartment. The photocell circuit should also be carefully designed. It is not a trivial problem to get a bee to trip a photo beam. Consideration must also be given to the background illumination in the area where the shuttle box is to be placed. If the level of infra-red in the background illumination is too high, it might activate the detector without a bee being present.

6.2.4. One-way shuttle box

One strategy to eliminate some of the problems associated with a two way shuttle box is to use a "one-way" shuttle box. A one-way shuttle box gets its name because the animal is always moving in the same direction. This device eliminates the problem of shuttle box dimensions, bees potentially walking on the ceiling of the apparatus, and accommodates a variety of bee species without much effort. In one version of this device a bee is tethered to a pole in the centre of a 50.8 mm square grid using a stiff wire. One end of the wire is attached to the thorax with wax and the other end to the pole. The wire is secured to the pole by inserting it into a small hole. The location of the hole is approximately the height of a honey bee enabling the bee to easily turn the pole while it circumnavigates the space. To provide a cue for the bee that it enters a new section of the space (i.e. a new compartment), small, thin 2 mm plastic strips can be secured to surface that the bee must step over. For example, if a 4 section shuttle box is required for a 50.8 mm space, plastic hurdles are placed every 12.7 mm. Infrared photocells can be placed near the plastic hurdles to automate response detection. To provide electric shock a grid can be used or antistatic foam (Abramson *et al.*, 2004).

6.3. Aversive conditioning in harnessed bees

In the shuttle box (Fig. 6) situation bees are able to move within the confines of the chamber. It is also possible to study bees using aversive stimuli when their movement is restricted by harnessing. Harnessing has a number of advantages over the shuttle box in that the training variables such as stimulus intensity can be more precisely controlled and it is easier to measure physiological responses. It is also easier to put a bee in a harness than to put it in a shuttle box. However, the harnessing technique is not as versatile as the shuttle box technique, as you cannot perform escape experiments.

6.3.1. Aversive conditioning of proboscis extension

In one option, electric shock is applied to the extended proboscis when it makes contact with the tip of a syringe. Smith, Abramson, and Tobin (1991) trained bees to withhold proboscis extension in a situation in which two conditioned stimuli were paired with sucrose, but response to one of them was followed by a shock to the extended proboscis. Bees readily learned to make the discrimination by responding to the odour not associated with shock. Although this procedure has not been used to investigate the molecular basis of learning in the honey bee, it can certainly be used so.

6.3.2. Aversive conditioning of sting extension

In the second situation, a conditioned stimulus is paired with a shock known to elicit sting extension. The bee is harnessed between two electrodes and after a number of odour-shock pairings the sting extends to the odour (Vergoz *et al.*, 2007; Carcaud *et al.*, 2009; Giurfa *et al.*, 2009; Mota *et al.*, 2011). In contrast to the shuttle box conditioning studies, which involve some type of consequence consistent with operant conditioning (escape, punishment, avoidance), the sting extension procedure is an application of classical conditioning and can be classified as an example of "fear" conditioning. It is important to note that at this time the learning curves associated with the classical conditioning of sting extension are quite low compared to classical conditioning with appetitive stimuli and when appetitive stimuli are combined with shock to an extended proboscis. In contrast to other aversive conditioning protocols, a start has been made investigating the effect of such training variables as number of trials, inter-stimulus interval, and inter-trial interval on sting extension conditioning (Giurfa *et al.*, 2009). Recently, the sting response assay was used to demonstrate that bees learn well to differentiate between different colours under laboratory conditions, which otherwise inhibit visual learning (Mota *et al.*, 2011).

6.4. Aversive conditioning in free-flying bees

In addition to protocols that use harnessed bees or bees confined to a shuttle box, there is a technique in which free-flying bees receive aversive stimulation (instructions for building the device is available in Abramson, 1986). In the situation shown in Fig. 7, a bee is trained to visit a target. The target is modified in such a way as to administer electric shock when the bee is standing on the plate and extending its proboscis into a well containing high concentrated sucrose solution. The target can also be modified to either vibrate and/or administer an air puff. The unconditioned response to shock in the vast majority of bees is to fly off the target. A bee can also avoid the shock by retracting its proboscis from the well. When vibration or air puff precedes shock, the bee will learn to use these warning stimuli to avoid the shock. Bees can also learn to discriminate between the air puff and vibration when one of them is followed by shock and the other is not (Abramson and Bitterman, 1986a, b). Shock is also known to improve discrimination learning in free-flying bees (Avargues-Weber *et al.*, 2010).

The free-flying technique is highly versatile. Studies on punishment and avoidance conditioning can easily be done. What cannot be done is escape conditioning. Bees will not readily land on a target which is electrified.

6.5. Training variables and aversive stimuli

Aversive conditioning is influenced by many training variables. These variables should be considered in the design and interpretation of aversive conditioning experiments (Abramson, 1994). Very few of these variables have been systematically manipulated in honey bee aversive conditioning studies (Giurfa *et al.*, 2009). An overview of important training variables and their effects is given in Table 3.

6.5.1. Electric shock as an aversive stimulus

Electric shock is the most frequently used aversive stimulus, because it is easy to control. It is advisable to begin with the lowest effective intensity. If the shock is too intense, the bee will be damaged.

Fig. 7. Free-flying aversive conditioning situation for honey bees. A honey bee lands on a conductive plate and when the proboscis comes in contact with sucrose solution a shock can be delivered. The plate is attached to a vibrator that can provide a warning stimulus. A second warning stimulus is air puff which can be seen above the bee.

Alternating current (AC) gives good results and direct current (DC) is also effective. There are no studies that test whether AC or DC current is more effective for honey bees. In addition, resistance measurements of several bees should be taken prior to conducting the formal experiments, for example by using a Volt meter. These measurements will provide an estimate of the amount of current that will flow through the bees. Shock can be applied in several ways depending upon the requirements of the experiment. For example, in a place learning assay, a grid based on the design of Kolmes *et al.* (1989) was effective (Agarwal *et al.* 2011). If the researcher wants to prepare grids from metal rods, the rods need to be cleaned, because oil or other material on the surface of the rods will interfere with the flow of current. If the use of a grid is not practical, antistatic foam is useful (Abramson *et al.*, 2004). It has a number of advantages over other shocking surfaces including ease of construction and standardisation. The major disadvantage is that the material is not transparent. If visual (or olfactory) stimuli are going to be used with antistatic foam, the stimuli cannot be presented beneath the bee.

Shock can also be applied to free-flying bees by using a modified landing surface constructed from conductive material. When the bee lands on the surface and extends its proboscis into an electrically isolated feeding well containing sucrose, the circuit is completed and the bee receives a shock (Fig. 7; Abramson, 1986). Another method to administer the shock that does not need the feeding well is to use the antistatic foam.

6.5.2. Nonstandard aversive stimuli

Electric shock is the aversive stimulus most often used in learning experiments because of its ease of use, controllability, and standardization However, there are other, less standardized aversive stimuli that can be used. Calcium chloride, citric acid and rapid changes in temperature have been shown to act as aversive stimuli for bees (Voskresenskaja and Lopatina, 1953; Lopatina, 1971). Various agrochemicals and repellents have also been used as aversive stimuli in conditioning experiments (Abramson *et al.*, 2006a, b; Abramson *et al.*, 2010a) as have formic acid (Abramson, 1986), GSM cellular phone radiation (Mixson *et al.*, 2009) and "virtual reality" (Abramson *et al.*, 1996).

6.6. Perspectives for aversive conditioning

When conducting aversive conditioning studies control groups are critical. In situations where the bee must discriminate between two stimuli, one of which is associated with an aversive stimulus, some way of randomizing the two stimuli must be used. One way is to present a pseudo-random sequence consisting of A, B, B, A, B, A, A, B with A being the stimulus followed by the aversive event and B the stimulus not followed by with the aversive event. If A, for example, is the conditioned stimulus associated with the shock (also known as the CS+) and B is the conditioned stimulus not followed by shock (also

known as the CS-) the sequence would be CS+, CS-, CS-, CS+, CS-, CS+, CS+, CS-. If more trials are needed, the sequence is repeated. This pseudo-random sequence will also work with discriminative stimuli. The intervals between the stimuli should be 10 minutes to minimize any carryover effects between the stimuli.

When exploring aversive conditioning in honey bees, researchers should consider several points:

- Punishment leads to a decrease in behaviour, while avoidance leads to an increase (Abramson, 1997; Abramson *et al.*, 2011)
- There is a large need for individual data, because there are few examples of individual learning curves of aversive conditioning with invertebrates, which are important for modelling aversive learning (Abramson and Stepanov, 2012)
- Researchers are encouraged to develop new aversive conditioning protocols and to provide parametric data on the effect of training variables for comparative studies

7. Investigating honey bee locomotion in complex and dynamic temperature gradient fields

7.1. Introduction

7.1.1. Temperature and thermotactic orientation in a honey bee hive

Honey bees live in a complex environment inside their hive and perform different tasks in different locations of the hive, depending on their age (Lindauer, 1952; Sakagami, 1953; Seeley, 1982). These locations differ from each other with respect to the stimuli they present to the bees (i.e. temperature: Fahrenholz *et al.*, 1989; Crailsheim *et al.*, 1999a; Tautz *et al.*, 2003; Stabentheiner *et al.*, 2003; Groh *et al.*, 2004; Kernbach *et al.*, 2009; Becher *et al.*, 2010; Schmickl and Hamann, 2011). Several important regions of a honey bee colony (Fig. 8) differ significantly in their temperature, ranging from the actively temperature-controlled brood nest (33-36°C; Kleinhenz *et al.*, 2003) to the cooler regions of the dance floor and the honey storage area (30-32°C). Newly emerged honey bees, which have a temperature preference for approximately 36°C (Heran, 1952), tend to locate themselves in the brood nest area. These bees are not yet fully developed physiologically. Stabentheiner *et al.* (2010) suggest that by searching for places with higher brood nest temperature they ensure the proper development of their flight muscles.

Studies have shown that bees are able to find and stay in areas of their preferred temperature (Heran, 1952; Ohtani, 1992; Grodzicki and Caputa, 2005). Therefore, the concept of actively-controlled motion by honey bees in heterogeneous temperature fields is widely accepted.

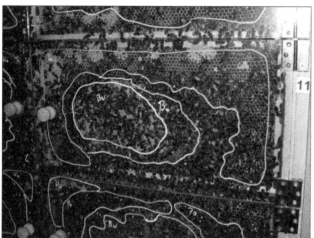

Fig. 8. Combs in an observation hive. Different regions of the combs are marked as honey (Ho) or pollen storage area (Po) or as brood area with open (Bo) and sealed brood (Bv). Some areas have a characteristic temperature (e.g. 33-36°C in the brood nest), which appears to contribute to the bees' navigation within the hive.

7.1.2. Investigating thermotactic behaviour

One approach to investigate the dynamics of the honey bees' winter cluster, for example, is to model honey bee thermoregulation. These modelling studies readily assume a direct uphill/downhill approach of honey bees towards a temperature optimum (Sumpter and Broomhead, 2000). Although studies on the location of sensory organs necessary for such a direct gradient ascent/descent were performed decades ago (Lacher, 1964), little is known about the proximate mechanisms of thermotaxis of honey bees in temperature fields. Early experiments in a so-called "temperature organ" (Fig. 9; Herter, 1924; Heran, 1952) have indicated that the outer two segments of the bee antenna are the most important locations for the temperature sense of the honey bee. However, such "temperature organs" are only a very rough approximation of the inner climate of a honey bee colony. They can only generate simple steep gradients in an almost one-dimensional environment. In contrast to that, the inner space of a honey bee colony is a complex three-dimensional structure and every comb the bees crawl on can be interpreted as a two-dimensional space.

Fig. 9. The classical temperature organ according to Herter. Bees move in a one-dimensional thermal gradient in the central tunnel (l = 53 cm) until they find the spot of their preferred temperature. To establish the gradient, the tunnel is heated by a hot water bath at the left side and cooled by ice contained in the box to the right. The thermometers at the back survey the gradient.

7.1.3. Emulation and application of two-dimensional temperature fields

Thus, investigating the trajectories of bees in two-dimensional, complex and dynamic gradient fields is an important field of research to understand the navigation principles of single honey bees as well as the process of collective aggregation. In addition to investigating the locomotion of single and multiple bees in such two-dimensional temperature fields for the sake of fundamental research, such methods could also be used as simple bio-assays to measure some important fitness variables (e.g. based on social capabilities) of bees reared under different conditions (incubation temperature of brood, diet, insecticide exposure, genetic traits etc.).

Single bees or groups of honey bees can be studied in an arena as they move through a well-defined two-dimensional thermal gradient (Fig. 10) or form aggregations at optimal temperature spots (Fig. 11). A number of behavioural parameters such as moving speed or resting durations of individuals or aggregation time and cluster stability can be evaluated to quantify the locomotion behaviour of the bees and to determine their efficiency in finding an area of preferred temperature. This efficiency can be reduced by environmental factors that affect the thermotactic capabilities of the bees (e.g. chemicals or parasites). By comparing the behaviour between healthy and irritated individuals or between groups with different ratios of irritated to healthy individuals, the extent can be assessed to which different types of irritations compromise the bees at the level of individuals or at colony level.

7.2 Arena construction

The heart of the setup is an adaptable circular arena with an exchangeable horizontal floor (Fig. 12A) and a plastic wall. The wall, which is available in various diameters, is coated with Teflon® to prevent the bees from climbing out of the arena. The arena is designed to allow for fast and easy adaptation of arena size and floor

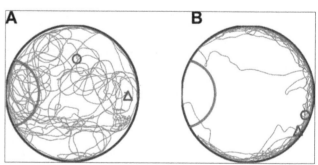

Fig. 10. Two bees exhibiting different forms of locomotion behaviour. The trajectories of two bees moving in the arena for 20 min are shown along with their starting (triangle) and stopping positions (circle). A thermal gradient spans the arena, whose area of optimum temperature is designated by a red border. The bees exhibit different locomotion behaviours. While **A** performs a slightly biased random walk, **B** follows the arena wall most of the time.

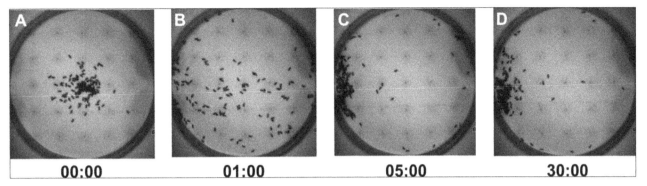

Fig. 11. Timeline of an aggregation experiment. 128 bees move in a thermal gradient with the optimum spot (36°C) at the left side and the coolest spot (32.5°C) to the right. After being released at the centre of the arena (**A**) the bees quickly disperse across the arena (**B**). After 5 min. most of the bees are aggregated in a cluster at the optimal area (**C**), where they remain till the end of the experiment (**D**).

material. A temperature field (usually with a thermal gradient) can be established in the arena with the help of heat bulbs (*p1*, consult Table 4 for specifications of referenced parts) above the arena. The temperature field in the arena is measured by an array of up to 61 temperature sensor modules mounted beneath the floor (Fig. 12B-D). The sensors are addressed individually with the help of two 8-channel multiplexers (*p4*) and the sensor voltage is relayed to an instrumentation amplifier (*p5*) to adjust the measuring range. The digitization of the sensor voltage is facilitated by a standard PC I/O board (*p6*), which is also used to control the heat bulbs. This is done via DC controlled dimmers (*p7*) which allow a nearly continuous regulation of heat bulb power (see Fig. 13 for wiring scheme) and thus serve as the actuators of a software-controlled feedback system for the stabilization of the thermal gradient. The setup of the arena, especially the design of the control elements for the two multiplexers, was strongly influenced by the work of Becher (2007) and Becher and Moritz (2009).

The behavioural observations are performed in an environment which is virtually dark for the bees due to infra-red lighting. The IR-light is produced by a set of modules consisting of low power (12 V) halogen lamps (*p8*), which emit a considerable part of their energy in the far red and IR-spectrum ($\lambda \geq 700$ nm), and an IR filter (*p9*) which blocks out light visible to the bees ($\lambda < 630$ nm). The bees in the arena can be observed using an IR sensitive CCTV camera (*p10*) with VGA resolution (720 x 576 px). The video stream is recorded on an HD video recorder (*p11*) in MPEG2 HQ format.

The arena floor is made of standard wax foundations which are easily available for purchase and approximate the natural floor in the bee hive. In preliminary tests bees showed a clear preference for wax ground over alternative materials (Fig. 14). In experiments with a vertically split arena floor consisting of wax in one and soft tissue in the other half, bees are found almost exclusively on the wax floor and seem to avoid the soft tissue in a thermal gradient that extends

equally into both halves. In thermal gradients with equal optimum spots on both halves, more than 70% of the individuals of groups released on the soft tissue side switched to the wax side during a 30 min experiment, while nearly none switched to the soft tissue side after being released on the wax side (Fig. 14). Groups released on the split line quickly gather at the wax side, and more than 80% of the individuals are assembled there after 30 min.

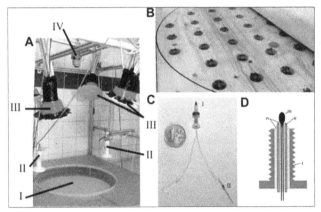

Fig. 12. Arena and technical details. **A.** Image of the arena setup: I: Wax floor with sensor array, II: Heat bulbs, III: IR-light emitters, IV: IR-light sensitive camera. **B.** Array of sensor modules. Holes in the acrylic glass have an M4-screw thread (fitting to the sensor modules) to allow for adjustment of sensor module protrusion. **C.** Picture of a sensor module, consisting of the actual temperature sensor (I, p2) and a diode (II, p3), which is essential for the functionality. **D.** Scheme of a single sensor module: I: Polyamide screw (size M4), drilled through along the length axis, II: carbon pipe for mechanical stabilization of the sensor unit (outer diam. / inner diam.: 2.5 mm / 1.7 mm). III: Temperature sensor (p2), wires of the sensor are fed through the hollow screw (diode not depicted) IV: Epoxy Glue.

Table 4. List of devices and parts used in the construction of the arena (identified by their reference number in the text).

#	Device type	Specification	Manufacturer
p1	heat bulb	ReptilHeat 60W	JBL GmbH & Co. KG, Dieselstraße 3, 67141 Neuhofen, Germany [www.jbl.de]
p2	temperature sensor (RTH)	NTC SEMI833-ET	Hygrosens Instruments GmbH, Maybachstr. 2, 79843 Löffingen, Germany [www.hygrosens.com]
p3	diode	1N4148	Diotec Semiconductor AG, Kreuzmattenstrasse 4, 79423 Heitersheim, Germany [diotec.com]
p4	multiplexer	MAX4051	Maxim Integrated Products Inc., 120 San Gabriel Drive, Sunnyvale, CA 94086, USA [www.maxim-ic.com]
p5	Instrumentation amplifier	LT1789-1	Linear Technology Corporate, 1630 McCarthy Blvd., Milpitas, CA 95035-7417, USA [www.linear.com]
p6	PC I/O-board	K8055	Velleman® nv, Legen Heirweg 33, B-9890 GAVERE, Belgium [www.velleman.eu]
p7	digital dimmer	K8064	Velleman® nv, Legen Heirweg 33, B-9890 GAVERE, Belgium [www.velleman.eu]
p8	halogen lamp	Decostar 35S	Osram GmbH, Hellabrunner Strasse 1, 81543 Munich, Germany [www.osram.com]
p9	infrared filter	Schott & Gen. IR Filter 22cm	Schott AG, Hattenbergstrasse 10, 55122 Mainz, Germany [www.schott.com]
p10	CCTV camera	WV-BP330/GE	Panasonic Corporation, 1006, Oaza Kadoma, Kadoma-shi, Osaka 571-8501, Japan [www.panasonic.com]
p11	HD video recorder	ME 1000sMM	Gerhard Witter GmbH, Kirchplatz 16, 94513 Schönberg, Germany (vendor) [www.witter-gmbh.de]

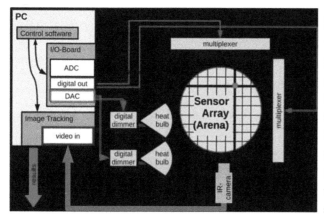

Fig. 13. Wiring scheme of the arena electronics. A USB I/O-board (p6) connects the arena electronics to a PC. A temperature sensor is selected for read-out via the multiplexers and its voltage is amplified (not depicted), digitized and relayed to the PC. Based on these sensor readings, a software-controlled feedback loop ensures the stability of the thermal gradient by continuously adjusting the heat lamp power. The I/O-board's DAC outputs are connected to an amplifier (p5, not depicted), which amplifies its output voltage (0-5 V) to the control voltage (0-12 V) required by the digital dimmers (p7), which in turn control the heat bulb power.

7.3. Generating a thermal gradient

The thermal gradient is usually characterized by the temperature and location of the temperature pessimum (usually at the coolest spot) and one or more global (36°C) or local temperature optima (32-34°C). While the coolest temperature in the arena depends on ambient temperature, the global and local optima are generated by heat lamps. A special computer program polls the temperature sensors in a given time interval and determines their temperature. Based on these measurements, a heat map is calculated for the entire arena (Fig. 15) which can be logged and used as a feedback for the maintenance of a stable thermal gradient. Depending on the desired temperature and the actual temperature of deliberately designated control sensors, the control software turns the heat lamps on or off individually to achieve a simple yet effective ($\Delta T = 0.1°C$) automatic control over the thermal gradient.

7.4. Data analysis

Experiments with the temperature arena produce a large amount of data in the form of temperature logs and video recordings. A suite of software programs is available to reduce the analysis effort to a minimum:

- Freely available video editing software is used to extract individual frames from the video recordings at a deliberate rate.

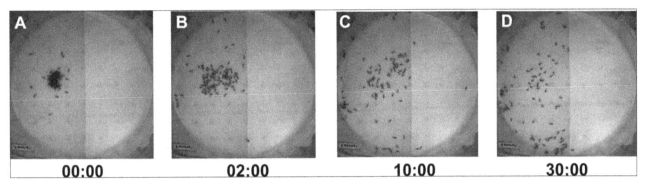

Fig. 14. Time line of a wax / tissue choice experiment. We released 100 bees on the wax side (left) of an arena with split floor. During the entire experiment of 30 min, only a few bees entered the soft tissue side (right). The thermal gradient with two optima (one to the left and one to the right) is too weak to induce the formation of notable aggregations.

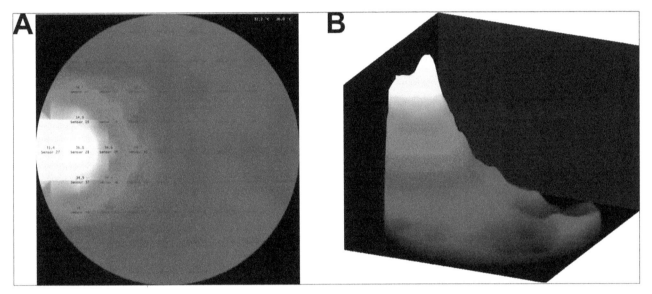

Fig. 15. Temperature distribution in the arena. The figure shows a heat map of the temperature distribution as a false colour representation (**A**) and the corresponding 3D-representation (**B**). The temperature optimum is located to the left, the pessimum to the right.

- A MatLab® script (available at http://zool33.uni-graz.at/artlife/ sites/default/files/track_single_agent.zip) iterates over these frames, identifies and extracts the positions of single bees or groups of bees and stores them in a Microsoft Excel® file.
- The number of individuals in a group can be determined with sufficient accuracy from the area covered by the group.
- A Microsoft Excel® script (available on request from http:// zool33.uni-graz.at/artlife/contact) gathers the sequence of positions from all experiments, reads the temperature logs and links the positions to the local temperature.
- The script establishes the trajectories of single bees (e.g. Fig. 10) and analyses a number of behavioural parameters (e.g. distribution of moving speeds, turning angles, moving and resting behaviour or the time spent in distinct arena zones like optimum, sub-optimum or wall zone) in dependence of local temperature and gradient steepness.

This analysis can serve as a bio-assay to assess the health status of a colony by comparing the clustering performance of a sample group from the compromised hive to the performance of a comparable group from a healthy colony (for data analysis see the *BEEBOOK* paper on statistical methods (Pirk *et al.*, 2013)). The clustering performance is determined by the time it takes until a given percentage of the individuals are clustered in the optimum, the ratio of individuals in the optimum after a given time and the distribution of individuals between the global and a local optimum. Groups from colonies with a compromised health status are expected to have a decreased clustering performance, which can be detected using established statistical methods.

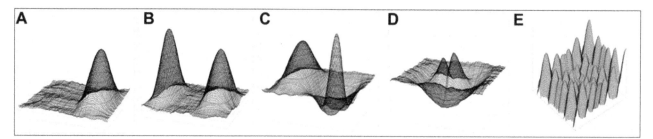

Fig. 16. Assorted optimization problems. The 3D-mesh depicts the temperature distribution in the arena. The highest peak represents the temperature optimum. The bees' task is to find the global optimum while avoiding getting stuck in local optima. The image sequence shows tasks which increase in difficulty. They range from a simple gradient with a single optimum area (**A**) via various gradients with one or more local optima (**B-D**) to the quite complicated Schwefel's problem (**E**).

7.5. Application of thermal gradients

The ability to discriminate a local optimum (e.g. a temperature spot which is less than optimal, but closer to the optimum than its immediate vicinity) from the global optimum (e.g. an area with 36°C) is a well-established measure for the quality of search and optimization algorithms. These algorithms can be benchmarked by applying them to specifically tailored test problems which are best visualized as 3D-landscapes in which the third dimension represents the quality to be optimized (e.g. temperature in the case of the thermal gradient).

The bee arena allows us to establish various forms of simple and complex thermal gradients to test the capability of the bees to find the temperature optimum. To study the success of the bees in local optimization tasks (Fig. 16):

1. Maintain a static thermal gradient while the bees roam the arena.
2. Determine the ratio of bees aggregated at the optimum (most likely the majority) to bees aggregated at one or more of the sub-optima.
3. Measure the success of the group by counting the number of bees ultimately aggregated in the optimum area or by taking the time it takes until a given number of bees find the optimum.

For temporal optimization tasks:

4. Create dynamically-changing gradients to test the dynamic components of the aggregation behaviour of the bees, e.g. by shifting the location of the global optimum or switching the location of the global and local optimum gradually (Fig. 17) or instantly.
5. Measure the latency between the optimum shift and the bees' re-aggregation at the new optimum site (this latency also depends on group size).

Dynamic gradients are convenient to test the ability of honey bees to revise a group decision and to react to changes in the environment. Note that larger clusters tend to be more stable and thus take longer to dissolve (Kernbach *et al.*, 2009).

7.6. Advantages and disadvantages of the method

With this setup the researcher can observe the behaviour of individual honey bees or of groups with respect to their temperature optimum. The setup allows automated and reproducible experiments with dynamically changing thermal gradients. Examples of the questions which can be investigated with the described hardware are:

- The natural temperature preferendum and thermotactic behaviour of honey bees.
- The responses of individuals and groups to temperature changes on a local level.
- The influence of environmental factors (incubation temperature, diet, parasites, pesticides etc.) on individual and social (thermotactic) behaviour.
- The learning abilities of honey bees regarding circadian temperature changes.

In light of the increasing environmental stress honey bee colonies have to deal with it is especially important to have a method available for the standardized investigation of the bees' natural (healthy) behaviour and its modulation by compromising environmental factors. The bee arena facilitates such investigations with a focus on thermotactic behaviour, both on the individual level (temperature preferendum, locomotion and search behaviour) and on the collective level (aggregation behaviour, social interactions).

The advantages of the setup are:

- The very low price of the setup: the total price for the setup is approximately 10% of the price for a commercial thermovision camera.
- Due to the "do it yourself" nature of the setup it is easy to adapt it to new tasks (e.g. new bioassay).
- The given setup allows the researcher to easily replace the given sensors (temperature sensors in the described setup) by other off-the-shelf sensors like sensors for light, gas, sound-pressure or magnetic fields.
- The present actuators (e.g. heat lamps) can be replaced by light- or gas-emitters, vibrators, electromagnets and others

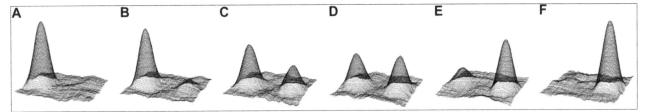

Fig. 17. Example of a dynamic gradient. The experiment starts with a simple gradient with the temperature optimum at the left (**A**). Once a deliberate ratio or number of bees has aggregated in this area, the temperature of the optimum is decreased while a new local optimum is generated at the opposite side of the arena (**B-D**). Subsequently, the former local optimum becomes the global (**E**) and ultimately the only optimum (**F**). After a certain delay, the aggregation at the left side will dissolve and reappear at the right side.

for use in different types of bioassays.

- It is possible to establish different gradients of different environmental parameters (e.g., temperature AND light) within one setup to investigate effects of combined environmental influences on the behaviour of animals; this is especially interesting for behavioural changes or behavioural disorders based on several unknown parameters (e.g. colony collapse disorder).
- The effect of the given environmental conditions can be investigated by observing easy-to-measure (behavioural) parameters like moving speed, orientation, average or relative position of the animals over time, the area covered by groups of animals or the probability of animal-to-animal interactions within specific zones and during different experimental phases; these measurements can be evaluated both manually or automatically.

The given setup allows experiments under well-defined and controlled laboratory experimental conditions. The extent to which the method is adaptable and usable for outdoor experiments is yet to be investigated.

A shortcoming of the current setup is the fact that the heat bulbs have to be positioned at the arena wall lest they obscure parts of the floor, as the camera is mounted on top of the arena. This implementation confines the heat spots to the wall area. If a heat spot needs to be established at a central position, alternative filming methods must be considered. For instance, the camera could be mounted at a flatter angle or two cameras could be used to create a confluent image of the arena floor. Both methods entail additional post-processing of video data (i.e. correction of image distortion or blending of partial images). This is, however, a minor effort to overcome this current limitation when more freedom in the positioning of the heat spots is required. Another shortcoming is the necessity to have basic knowledge in electronics (e.g. reading a datasheet) and soldering techniques (e.g. usage of a soldering iron), which can take some time (several days) to learn.

The setup can be easily scaled for use with different species (e.g. ants, wasps, bumble bees or even mammals) by simply re-adjusting the size of the arena and the density of the temperature sensors. If the sensors are integrated into pre-structured honey bee hives it is even possible to perform the experiments inside a colony. The use of the arena is not restricted to the investigation of thermal gradients. In fact, the quality of the gradient can easily be adapted by replacing the heat-bulbs by other types of actuators, e.g. Peltier-elements, wind machines, pheromone evaporators, magnetic field generators, sound emitters, food dispensers and so on. The sensor array can be adapted accordingly to measure the gradient and provide the feedback for its automatic maintenance.

In combination with online image tracking even a reactive experimental setup is possible. In this setup, the experimenter can pre-define a set of rules according to which the temperature gradient changes once a deliberate number of bees have aggregated at a given place. This way, experiments regarding reaction time and adaptation to environmental fluctuations can be performed.

8. Testing honey bee flight capability in the laboratory

8.1. Introduction

Flight is an indispensable ability of honey bees. Flight physiology of adults may be altered by larval and adult conditions like nutrition, developmental temperature, pests, pathogens or pesticides but also by external factors such as temperature or dust. The flight capability and flight performance can be measured in tethered flight using a flight mill or roundabout. These devices were originally constructed to investigate thermoregulation or flight metabolism (sugar utilization, oxygen consumption) of honey bees and other insects (Sotavalta, 1954; Jungmann *et al.*, 1989).

8.2. Roundabout construction

Several issues need to be considered, when a researcher wants to construct a roundabout for measuring honey bee flight behaviour:

- The arm of the roundabout should have a length of about 14-16 cm, so that one revolution equals roughly one meter.

- The arm should be connected to the stand as frictionless as possible (Fig. 18A).
- A circa 10 mm long spur pointing downwards is needed at the end of the arm to attach the bee via a connecting tube.
- The spur must be moveable in all directions to adjust the bee's position (angle) before flight.
- The roundabout should be placed in a containment with regular (2.5 cm) vertical black and white stripes to provide a homogeneous optical flow during flight and an opening that allows manipulation of the bee.
- Mount a 60 W bulb on top of the roundabout as a light source and for adjusting temperature (ambient temperature affects metabolism and flight speed (Hrassnigg and Crailsheim, 1999)).
- Maintain the temperature (e.g. 28°C) throughout the experiment and when testing different groups.
- Register every rotation of the roundabout electronically (e.g. by a light beam).
- Collect data of the duration of every round in seconds and possibly of temperature at regular intervals.
- Flight time can either be derived automatically (computer program that clocks the time when the arm of the roundabout is in rotation) or manually.
- Because the arm may continue to rotate for a few rounds after the bee has stopped, we recommend to clock manually.

8.3. Preparation and treatment of bees

Forager-aged bees are needed for flight tests because they will most likely perform long and persisting flights. Care should be taken to use same-aged bees both in the treatment and control groups (see the section on 'Obtaining brood and adults of known age' in the *BEEBOOK* paper on miscellaneous methods (Human *et al.*, 2013)). When bees are specially treated or manipulated at earlier age, they should be marked and introduced into a nucleus or observation hive until experiments, to allow the ontogeny from hive bees to foragers including orientation and defecation flights. Otherwise, bees from differentially treated colonies can be readily used at foraging age (see the *BEEBOOK* paper on standard methods for maintaining adult *Apis mellifera* in cages under *in vitro* laboratory conditions (Williams *et al.*, 2013)). Caged bees may not fly as well as those from colonies. You should prepare the bees for the further experiment like this:

1. Collect the bees.
2. Immobilize them with needles on balsa wood (or similar material).
3. Glue the connecting tube (about 4 mm long, ca. 1.5 mm in diameter) to the thorax with Pattex (Henkel); super glue is not suitable because bees will die quickly.
4. Allow a few minutes for glue drying before using the bees for flight experiments.

Alternatively, they may also be kept in hoarding cages until usage.

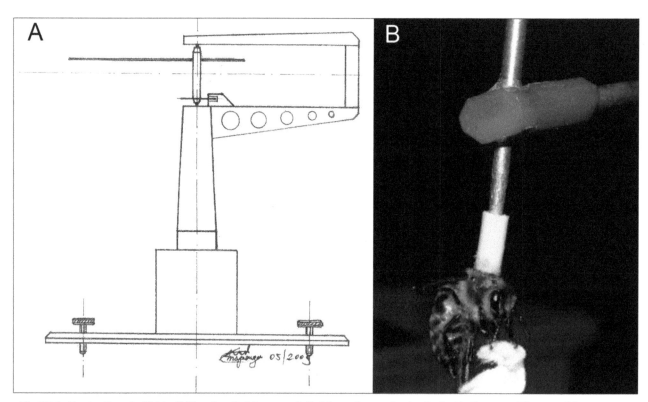

***Fig. 18.* A.** Construction of a flight mill (Drawing: Kurt Ansperger). **B.** A honey bee worker is attached to the spur of the arm of the roundabout, holding a ball of paper. Note that the spur can be moved in all directions to find the best position and angle for the flying bee.

The inexperienced experimenter should keep the bees in cages, because the longer the connecting tube is fixed to the bees, the more naturally the bees behave, and the more likely they will fly successfully in the roundabout.

5. Attach the bees to the spur of the arm of the roundabout without prior feeding.
6. Perform an "emptying" flight to deplete all of their energy reserves.

    6.1. Stimulate flight by removing the small ball of paper or styrofoam that the bee holds with its legs (Fig. 18B) and which prevents the bee from flying.

    6.2. "Restart" the bees several times, in case they cease flying after a few rounds. Stimulate them again to completely deplete their reserves until the movements of the wings are very weak, and the "state of 'apparent exhaustion' cannot be prolonged but leads quickly to the state of complete exhaustion" (Sotavalta, 1954). During experimental flights with defined feeding, the arm must be prevented manually from swinging when the bee stops flying to avoid false counts of revolutions; we recommend implementing an acoustic signal at every rotation of the arm for aural control of flight (a metronome-like frequency confirms constant flight). Once the bees have started flying, they will continue to fly until they are exhausted (depending on their sugar reserves in the honey crop; they will first become slower and then stop when they have no more sugar solution in their honey crops and hardly any sugar reserves in their haemolymph).

7. Weigh the bees.
8. Feed them 10 µl of 1 M or 2 M glucose solution with a pipette within few minutes after the emptying flight, bees are now ready for experiments.
9. Weigh the bees once more.

Weighing before and after feeding allows controlling for complete ingestion of food (Brodschneider *et al.*, 2009). Ingestion of 10 µl of glucose solution corresponds to an increase of about 10 mg of weight.

10. Attach the bees to the roundabout.
11. Place a small paper ball between their legs.
12. Allow exactly 5 minutes of rest after feeding before starting experimental flights; to prevent them from flying during this time, you should give them a ball again. An acoustic signal helps monitoring that bees do not drop the ball and fly. If they do so, they should be stopped immediately and given a ball again.

It takes some experience to find a good flight position which allows optimal power transmission of the bees to the roundabout and hence maximum speed.

According to the construction of the flight mill, bees fed with 10 µl of a 1 M glucose solution fly for about 15 minutes (Brodschneider *et al.*, 2009). The flight is characterized by an increase in flight speed in the first three minutes followed by flight with constant speed and a decrease in flight speed, leading to the total exhaustion of the bee. At this point, further flight experiments with feeding of e.g. 2 M glucose solution can be conducted, because bees quickly recover when provided with energy. Flight time, distance covered (the perimeter of one round is known), maximum speed, average speed and other parameters can be calculated and compared among groups (Gmeinbauer and Crailsheim, 1993). A good verification of the method is to demonstrate that feeding bees with 2 M glucose solutions increases their maximum flight speed compared to that of bees fed with 1 M glucose solution (Gmeinbauer and Crailsheim, 1993; Brodschneider *et al.*, 2009).

Testing the flight performance of honey bees has been applied to compare sexes, castes, the value of energetic substrates and to identify physiological deficiencies such as poor larval nutrition (Jungmann *et al.*, 1989; Gmeinbauer and Crailsheim, 1993; Hrassnig *et al.*, 2005; Brodschneider *et al.*, 2009). Studying the effect of physiological deficiencies on flight performance requires the comparison of treated versus untreated groups in defined conditions as described above. Other applications might include testing the capability of bees of a certain age or treatment to perform long and persisting flights within a given period (e.g. 20 minutes). However, the rate of successful flights should be reported for both types of experiments and can be compared among groups using chi^2 tests.

9. Working with honey bee observation hives

9.1. Introduction

Honey bee (genus *Apis*) nest architecture can be easily utilized for visual studies of eusocial insect behaviour. Nests are primarily comprised of flat combs of uniformly sized and shaped hexagonal cells laid out on a vertical plane (Fig. 19). Combs can be constructed to utilize cavities of a variety of sizes and shapes. The uniformity of honey bee combs and flexibility of bees to accept a variety of cavity shapes allowed agricultural researchers to develop hives with removable, uniformly sized "frames" of comb. The hive designed by Langstroth in the 19th century is still utilized today, worldwide (Graham *et al.*, 1992). The observation hive is a valuable tool for investigating the behaviour and life history of bees inside the nest. Rösch began using this type of hive to study honey bee lifespan and behaviour in the 1920s (Rösch, 1925, 1927, 1930) and the design has changed little over the last century (von Frisch, 1950; Lindauer, 1952; Seeley, 1982; Calderone and Page, 1991; Seeley and Kolmes, 1991; Wang *et al.*, 2010). The topics that observation hives can be used to investigate are extensive (e.g. dance language, food transfer between nest mates, disease response, thermoregulation, temporal polyethism).

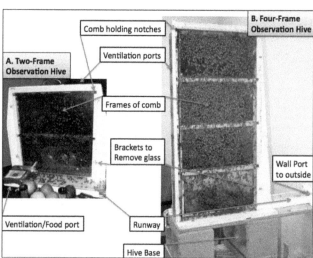

Fig. 19. Newly built honey bee comb cells with thorax-tagged bee inspecting a cell (no abdomen paint mark). Note the uniform size and shape of the cells.   Photo: A. J. Siegel

Fig. 20. A. A two-frame observation hive from the Starks Laboratory at Tufts University. **B.** A four-frame observation hive from the Page Laboratory at Arizona State University.   Photos: A J Siegel

This tool takes advantage of the natural flat quality of honey bee comb. The hive is constructed so that all comb surfaces can be observed. To facilitate this, combs are stacked with only the edges, and not the comb surfaces (as would be found in commercial or natural hives; Winston, 1987), facing each other. Additionally, the hive walls facing the combs are constructed of clear glass, Plexiglas®, or plastic (Fig. 20). Honey bees are behaviourally flexible and will readily accept this atypical colony shape as long as the queen has space to lay eggs, the workers can forage, and the hive is kept warm. Individual bees can be tagged using paint, or coloured-numbered plastic discs and placed in the hive (see section 2.3.; Fig. 1). The visibility of all surfaces allows researchers to categorize the behaviour of individually marked bees using a behavioural catalogue (Table 5; Seeley, 1982, 1995; Winston and Punnett, 1982; Kolmes, 1985; Robinson, 1987; Calderone and Page, 1991, 1996; Seeley and Kolmes, 1991). Observations can be conducted by following a single tagged bee over an extended period of time, or conducting daily behavioural surveys of many tagged workers to study group behaviour. Another advantage of observation hives is the ability to manipulate environmental conditions such as temperature and food storage (Riessberger and Crailsheim, 1997; Crailsheim *et al.*, 1999b). The remainder of this section will detail the setup and use of an observation hive for a study of temporal polyethism (change in task performance in the hive over time). This is one of the most commonly studied topics that employs an observation hive setup. However, the general setup described below could be used for many topics with slight modification.

9.2. Hive setup

Modern experimental observation hives usually consist of either two or four standard Langstroth "deep super" frames. To construct the hive:

1. Build a rectangular hive framing of parallel wooden bars with notches to hold the Langstroth comb-frame lips.
2. Space the comb frames ~0.95 cm apart; this leaves enough space to allow bees to pass between the edges of the frames, but not so much that the bees are encouraged to build additional comb between frames.
3. Leave the same space between the sides of the comb frames and the hive side bars.
4. Include a wooden hive top bar to connect the hive sidebars together.
5. Include wire-mesh covered ventilation holes in the top and/or side bars of the hive to allow free flow of fresh air in the nest.
6. When bees cover these holes with propolis from time to time, clean it away.
7. Attach removable glass or Plexiglas® front and back walls to the parallel wooden side-bars to allow for viewing of the frame surfaces; again, leave ~0.95 cm space between the comb surface and the viewing glass to allow bees to move freely on the comb surface, but minimize additional comb construction, which can obstruct comb viewing (Graham *et al.*, 1992).
8. Complete the hive with a tube, port, or runway connecting the lower part of the hive to the outside in order to allow bees to forage naturally.

Table 5. Honey bee worker behavioural catalogue.

Task	Description
cell cleaning	removing debris from used brood cells (cocoons, larvae excretion), cleaning cell walls. Takes place in a cell not currently being used
general nest sanitation	removing debris from nest (mouldy pollen, old cappings, dead brood, and dead adults)
brood care	feeding larvae (head in brood cell > 1.3 min), attending queen
construction	smoothing wooden hive parts with mandibles and manipulating wax and propolis in cracks and corners of the hive
fanning wings	flapping wings while standing in hive/at entrance
food care	insertion of head into a cell containing nectar, receiving nectar-on bridge
grooming a nestmate	running nest mate body parts through mandibles
grooming self	running own body parts through mandibles
inspecting a cell	momentary insertion of the anterior portion of the head into an empty cell
nest care	manipulating wax of cells (not cappings), building new empty cells
patrolling	walking around nest
standing and chaining	standing stationary or hanging while stationary on nestmates
brood cap manipulation	trimming or smoothing wax cappings on brood cells and capping brood with wax
honey cap manipulation	trimming or smoothing wax cappings on cells of honey and capping honey with wax
trophallaxis	nestmate exchange of food (not near entrance), receiver thrusts tongue at donators mouthpart, donator opens mouthparts pushes tongue forward, and regurgitates a drop which is lapped up
vibrating	fast rhythmic body vibrations (non-dance)
head in pollen	insertion of head into a cell containing pollen
inspecting brood	head in brood cell, < 1.3 min
dancing	dancing without/with pollen
washboarding/ plaining	standing and rocking back and forth with mouthparts open
attending dance	dance attendance without/with pollen

Hives can also be connected to enclosed flight rooms; however, flight rooms increase forager mortality.

9. Install a "background" colony in the observation hive approximately one week before beginning an experiment to give the bees time to adjust to their new environment.

The size of your observation hive depends on the available space; note however, that larger hives can be used for longer studies.

When using a two-frame observation hive design (Fig. 20A):

10. Include one comb frame containing brood (50%-75% capped) and pollen, and one empty frame (drawn or foundation) for honey storage.
11. Add a queen and 1,330-1,530 workers from a queen-right colony to the hive (Siegel *et al.*, 2005; Starks *et al.*, 2005).

When using a four-frame observation hive (Fig. 20B):

10. Include 3 frames consisting of ~75% brood in all stages of development and ~25% pollen, and one frame of honey and empty comb cells.
11. Add a queen and 4,500-5,000 workers from a queen-right colony (Calderone and Page, 1991).

Regardless of size, a hive should be installed in a highly regulated environment:

- Maintain the observation room at 21-30°C.
- Because natural hives are dark inside, the room should be kept dark or the hives should be kept covered when no observation is being performed.
- Red light can simulate a dark environment, as honey bees do not perceive red light (Backhaus, 1993).

Focus bees must be tagged individually (see section 2.3.). Tagged bees are introduced to a colony by carefully removing the wire mesh from a ventilation hole in the observation hive and gently pouring the bees in, or by placing tagged bees on a platform in front of the hive entrance and gently blowing smoke on the bees to "herd" them into the hive.

9.3. Behavioural observations

There are two main strategies for observing honey bee worker behaviour over time. First, a researcher can tag one individual and make frequent observations of the focal bee over the course of its life (Lindauer, 1952). Alternatively, a large number of bees can be tagged and a random subset of these bees can be observed at less frequent time intervals (Seeley, 1982; Calderone and Page, 1991; Seeley and Kolmes, 1991). Ultimately, the basic techniques are the same. The primary differences between the two techniques are the number of bees observed and the frequency of observations. After introducing tagged bees, the transparent walls of the observation hive are covered with sampling grids of equal-sized numbered boxes. 60 mm x 60 mm quadrants can be utilized for four-frame observation hive studies using a large quantity of tagged bees (Calderone and Page, 1991). A random number generator is frequently used to determine the order of sampled grid boxes (Seeley, 1982). For each data point collected, the observer needs to record:

- The identity of the observed bee (tag number).
- The location of the bee in the hive (grid square).
- The activity performed by the focus bee.

If few tagged bees are used, the majority of tagged bees can be sampled during an observation period. Additionally, a small number of tagged bees allows for multiple observations of each quadrant throughout the day (Seeley, 1982). If a larger sample size is necessary, a subset of quadrants (and sometimes a subset of bees within each quadrant) can be observed once a day over an extended period of time. Regardless of whether a researcher observed a single focal bee many times over the course of a day or a subset of hundreds of experimental bees once each day, behaviour can be categorized based on a list of stereotyped tasks and activities. Table 5 comprises a behavioural catalogue of recognizable tasks and activities, which was compiled from several sources (Seeley, 1982, 1995; Winston and Punnett, 1982; Kolmes, 1985; Robinson, 1987; Calderone and Page, 1991, 1996; Seeley and Kolmes, 1991). Note that if a tagged worker is observed with its head in a comb cell (Fig. 19), it is necessary to wait for the bee to exit, so that it is possible to read the tag and look inside the cell. A small flashlight can be helpful with this procedure.

9.4. Other observation hives

Observation hive studies are not limited to honey bees. Bumble bees can be kept in a wooden cube with a transparent top for easy viewing (Jandt and Dornhaus, 2009; Jandt *et al.*, 2009). Many species of ants can easily be kept in wooden, glass or plaster nest boxes with transparent walls or tops (Holbrook *et al.*, 2011; Sasaki and Pratt, 2011; Schneider, *et al.* 2011). *Polistes* wasp nests can be placed in wooden boxes with wire mesh fronts *(Liebert et al.,* 2010). With access to a basic wood shop and a bit of creativity, the behaviour of hundreds of eusocial insect species could be observed using an observation hive in a controlled laboratory setting.

10. The honey bee dance: background, knowledge and observation

10.1. Introduction

When a honey bee forager discovers a profitable floral patch in the field it returns to the nest to share the collected resource with its nest mates. Either during or after unloading the food, the honey bee can perform a series of repetitive waggle movements on the wax comb (Fig. 21). The first person to analyse and to decode this form of honey bee communication was Karl von Frisch. He was awarded the Nobel Prize in Medicine and Physiology in 1973 for his contributions to

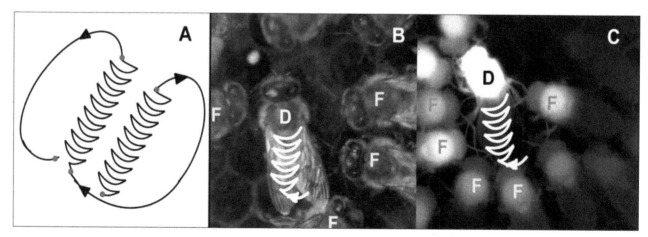

Fig. 21. **A**. The scheme represents the trajectory of a typical waggle dance, which encodes a far-distance goal. Two successive waggle runs are separated by round walks in which no waggle movements are made (the circle phase). The duration of the waggle runs correlates with the distance of the goal (food source or nest site) and the gravity-axis correlates with the orientation of the goal. The length of the circle phase has a negative correlation with the profitability of the food source. **B**. A photograph and **C** a thermogram of a successful forager (**D**) dancing on the wax comb while a group of nest mates (**F**) follow it. Whiter body parts in **C** indicate higher body temperatures. Note that the highest temperatures correspond to the dancer's thorax and head and to the different thoracic temperatures of the followers around the dancer.

behavioural sciences. More than six decades ago, Karl von Frisch began to train honey bees to forage at an artificial feeder by using sucrose syrup as reward (see von Frisch, 1967 for review; see also section 14). While the bees that arrived at the feeder fed from the liquid food he marked them with acrylic paint. This procedure allowed von Frisch to follow the behaviour of individual bees not only at the artificial feeder but also inside the nest. In-hive observations were possible because he had built a two-comb hive (see also section 10). In this way, he was able to observe and describe that the marked foraging bees returned to the hive, unloaded their crop and often performed a complex motor-coordinated pattern on the vertical wax combs. If von Frisch located the feeder 100 m or more away from the hive, the marked bees displayed a figure-of-eight shaped movement which he defined as the *waggle dance* (*Schwänzeltanz* in German). During the waggle dance, the bee releases intense vibrations and moves its body laterally as it runs straight forward on the comb (the waggle-run phase). The duration of each run during the waggle phase conveys information about the distance of the discovered resource whilst the orientation of the dancer's body relative to gravity during the waggle runs provides information on the direction of the goal (von Frisch, 1967). This complex motor pattern (see Fig. 21A) takes place not only when bees discover a profitable nectar source, but also when other resources such as pollen or water (von Frisch, 1967) or even potential nest sites (Lindauer, 1955) are found. If, on the other hand, the goal discovered is close to the nest, the display becomes a succession of running circuits in which the bee suddenly changes its direction after completing each round. This dance display, which indicates resources at a short distance, is known as the *round dance* (von Frisch, 1967).

10.2 Parameters to consider when studying honey bee dance

Some dance parameters can vary according to the profitability of the food source. For instance, if the sugar concentration or the solution flow rate (sugar mass per time) change, the dance probability, the dance time and the dance tempo (the number of waggles or reversals per 15 s running on the comb) increase according to the energetic value of the exploited resource (von Frisch, 1967; Farina, 1996; De Marco and Farina, 2001). In the specific case of the waggle dance, the interval between two successive waggle runs (defined as the circle phase) is shorter for higher food source profitability (Seeley *et al.*, 2000; Hrncir *et al.*, 2011). Thus, an enhanced recruitment for higher food source profitability might be explained by a larger number of waggle runs per unit time inside the hive (von Frisch, 1967; Seeley, 1986; Fernández *et al.*, 2003). A larger number of waggle runs per unit time attracts the attention of more bees located in the surrounding of the dance floor (Wells and Wenner, 1973) and motivates them to fly out. Once in the field, bees use the spatial

information acquired while following the dance to arrive at the goal (Riley *et al.*, 2005). Therefore, the higher the redundancy of the signal (higher number of waggle runs displayed) the higher the probability of transmitting the spatial information to nest mates and consequently the higher the recruitment level observed at the exploited food source.

However, the parameters recorded during the dance not only depend on the profitability of the resource discovered by the forager. These dance parameters also depend on the forager's own experience in the field (von Frisch, 1967; Gil and Farina, 2002), on the predictability of the food source in terms of profitability (Raveret, Richter and Waddington, 1993; De Marco *et al.*, 2005) and on the nutritional state of the hive (Lindauer, 1954; Seeley, 1989), amongst other factors (see Dyer, 2002 for a more detailed review).

The waggle dance of the honey bee *Apis mellifera* is, without any doubt, one of the best studied communication systems in the animal kingdom. Inexperienced recruits who follow the dances are capable of deciphering the vector information provided by the dancers' movements (the waggle runs) and can use this information when searching for the advertised patch (Esch *et al.*, 2001; Riley *et al.*, 2005). Furthermore, the information obtained from decoding the waggle dance is broad. This implies that other stimuli belonging to different sensory modalities are possibly being used. One possible additional source of information used by dance followers are the thoracic vibrations pulsed by the dancer while waggling (frequently referred to as "dance sounds"; Esch, 1961a, b; Wenner, 1962; Hrncir *et al.*, 2011). The volatile compounds released by the dancers, which act as a "recruiting pheromone", are another additional source of information (Thom *et al.*, 2007). The olfactory cues incidentally acquired while bees forage at flowers (Farina *et al.*, 2005) are yet another source of information available to dance followers.

In other words, the honey bee dance should be considered a multi-component signal, i.e. a signal comprised of more than one informational component. These components can be redundant or not and can lead to an enhanced response, because the receivers (the dance followers) might acquire more than a single type of information. Thus, these signals can even provide multiple messages (for details see Grüter and Farina, 2009). Within this framework, the function of the waggle dance is not only to indicate the location of food sources but to attract surrounding bees, so that they can receive other types of information, i.e. it informs bees of the presence of good food sources, it activates self-acquired navigational information (if present), and it facilitates the acquisition of information about food odours.

10.3 Measuring dance-related parameters

Audio and video recordings (Fig. 21B) have been the most common procedures for registering in-hive behaviours. Nowadays, however, alternative visual techniques such as thermographic recordings (Stabentheiner and Hagmüller, 1991; Farina and Wainselboim, 2001) and high-speed video analysis (Tautz *et al.*, 1996) are much more common and accessible. Thermal recordings provide the instantaneous temperature of the experimental bee, even if it is inside the nest. This data could give valuable information on the activity level and/or on the motivation of the dancer and its followers (Fig. 21C).

A suitable procedure to analyse focal bees (dancers/followers) is to mark them individually (see section 2.3). Bees can be marked directly at the feeder. Data can be processed and different behaviours can be quantified by using commercial or free software packages. Once quantified, some dance-related parameters can be used to infer the responsiveness of the dancers, the effect of the social surrounding and the feeding sites exploited by the colony. What follows is a list of suggested parameters useful to approach these questions from different angles.

10.3.1 Dancer's behaviour

Some common variables used to estimate the internal state or motivation of the active forager are:

- The occurrence and duration of dances performed after returning to the hive.
- The latency of the first reversal or waggle run, i.e. the time the bee spends from its arrival to the hive until it begins to dance.
- The number of runs per unit time (e.g. Waddington, 1982; Seeley, 1986; Farina, 1996; De Marco and Farina, 2001).

10.3.2 Food source location

To estimate the location of the food source, video tape the in-hive behaviour of the bees returning from the outside, some of them are successful foragers.

1. Focus on the waggle-run phase of each dancer, i.e. the part of the dance display during which the incoming foraging bee moves its body laterally while running straight forward on the comb.
2. Estimate the distance to the food source by calculating the average time taken in consecutive dancing bouts (waggle runs) in an uninterrupted dance sequence.
3. Once calculated, compare this average with the values corresponding to a calibration curve in which the mean time of the successive waggle-runs is represented against the distance to the food source. You can build your own calibration curve by measuring the length of the waggle-runs for bees trained at feeders located at different distances, while maintaining a constant direction of the food source.

Alternatively, use the calibration curve presented in von Frisch's book (1967), which allows for fairly good predictions on distances to unknown feeding sites to be made.

However, slight deviations are possible, mainly among different honey bee races (von Frisch 1967).

10.3.3 To estimate the direction of the food source:

1. Focus on the orientation of the waggle-runs in relation to the gravity-axis.
2. Video tape dance behaviour under diffuse light conditions to avoid changes in dance orientation due variation in light intensity (Edrich, 1975).

This is because when you compare dances performed at different times of the day, you need to account for the changes in the sun's azimuth. To estimate the direction of the advertised foraging site measure first the average angles of 4 waggle-runs by using a radial protractor over the screen. Such average angle was added to the sun azimuth angle at the solar time when the dance was captured. The sun azimuth angle for each dance can be calculated using the world-wide web (e.g. http://aa.usno.navy.mil/data/docs/AltAz.php#formb).

The foraging sites exploited at a social scale can be estimated by a systematic recording of the dance orientation in the whole colony. If video recordings are made systematically, this parameter is especially useful for analysing collective honey bee foraging under natural conditions.

10.3.4 Dance precision

The precision of the directional information encoded in the dances might change either with the hive-feeder distance (Towne and Gould, 1988) or with changes in the profitability of the food source (De Marco *et al.*, 2005). A simple measure to estimate the precision of the information is to consider the directional scatter of the total waggle runs measured for a single dance. Another possibility is to measure the divergence between the angles of consecutive waggle runs, a parameter that allows quantifying changes in the scatter along the dance display. Divergence is calculated as the direction of a particular waggle run minus the direction of the previous one (with a minimum of 0° to a maximum of 180°). Thus, the first divergence angle in a given dance is associated with the second waggle run displayed by the dancer.

10.3.5 Dance follower's behaviour

Followers might preferably contact certain body parts of the dancer during the waggle phases in order to obtain specific information. For instance, the best position to perceive the acoustic near field produced by the waggle is located right behind the dancer (Michelsen, 2003). The best position to obtain pollen load information is near the dancer's posterior legs (Díaz *et al.*, 2007). Additionally, the preferred contact to obtain information on floral scents diluted in the nectar is to touch the dancer's head or mouth parts (Díaz *et al.*, 2007). This analysis of contact preferences requires a frame-by-frame study of video recordings. It might not be possible to distinguish the movements of the dancer's antennae (its olfactory organs) due to the temporal resolution of the video. The number of head contacts, however, might be easier to quantify. These recordings guarantee that the hive bee's antennae indeed contact the dancer's body. Head contacts of hive bees with the dancers' body should not be confused with trophallaxis events (mouth-to-mouth food exchange). These social interactions should only be considered as trophallaxis when they are recorded for more than one frame in the analysis. The number of contacts recorded onto the dancer's body parts can be registered and pooled to analyse frequency of contacts (Michelsen *et al.*, 1987; Díaz *et al.*, 2007).

10.3.6 Dance attractiveness

A snapshot of the number of simultaneous followers can give an idea of the attractiveness of the dance recorded. One suggestion to measure the attractiveness of the dance is to quantify the number of followers 10 s after the beginning of dancing. The duration of the dance is relevant for this measurement, because it is strongly related to forager motivation (von Frisch, 1967).

11. Honey bee navigation: tracking bees with harmonic radar

Bees are too small to be observed over the range of their natural flight behaviour. Under optimal conditions one may be able to follow a bee by sight over a distance of up to 30 m, a small distance indeed given an action radius of kilometres. These limitations have forced researchers interested in navigation of bees to limit their observations to the initial flight path in catch-and-release experiments.

11.1. Recording initial flight path

To perform this kind of experiment:
1. Train foragers to an artificial feeder.
2. Catch them at one of the four foraging motivations (outbound at the hive, inbound out the hive, arrival at the feeder, departure from the feeder)
3. Transport the bees to a remote release site.
4. Release the bees.

When the bee is released, it usually performs a few circling flights and then vanishes along an often rather straight flight in a particular direction.
5. Note their flight directions.

For this a compass and a map can be used.

11.2. Recording whole flight path

Since a forager trained along a route to a feeder chooses the direction it would have taken if not being transported to an unexpected site, it was concluded that the bee applies an egocentric reference in navigation. The bee would be lost if it did not find back to the hive by steering towards a beacon at the hive or some other procedures of localizing the hive (Wehner and Menzel, 1990; Dyer, 1998; Collett

et al., 2002). Since bees are not lost even if such beacons are not available and if they were not trained along a route (Menzel *et al.*, 2000), it is necessary to follow their whole flight path after they have escaped from direct view and deduce from their initial flight path the spatial reference they may apply.

Harmonic radar provides the option to track bees over the range of several hundred meters (up to 3 km depending on the direction of flight relative to the radius of radar scan). The range of the radar scan depends strongly on the local conditions and can reach a radius of 1.5 km under ideal conditions. The working of the harmonic radar device has been described by several authors (Riley *et al.*, 1996; Osborne *et al.*, 1997; Riley and Smith, 2002).

11.2.1. Radar

The radar pulse (for example generated by a commercial 9.4 GHz radar transceiver, Raytheon Marine GmbH, Kiel, NSC 2525/7 XU) is emitted by a parabolic antenna providing approx. 44 dBi (Fig. 22A). The pulse is received by a transponder fixed to the thorax of the bee (see section 11.2.3.) and emits the second harmonic component of the signal (18.8 GHz). The receiving unit consists of a 18.8 GHz parabolic antenna, with a low-noise pre-amplifier, directly coupled to a mixer (18.8 GHz oscillator), a preamplifier with low noise and a downstream amplifier with a 90 MHz ZF-Filter. A 60 MHz ZF-Signal is used for signal recognition. The two antennae are mounted on a common vertical pole which rotates the two antennae in synchrony (Fig. 22A).

11.2.2. Transponder

The transponder consists of a dipole antenna with a Low Barrier Schottky Diode HSCH-5340 of centred inductivity. The transponder is 10.5 mg in weight and 12 mm in length. It consists of a silver or gold wire with a diameter of 0.3 mm and a loop inductance of 1.3 nH (Fig. 22B, 23A). The dimensions of the wires, the loop and the conductive glue (two component conductive epoxy, Chemtronics, Kennesaw, GA, 30152, USA) used to connect the diode across the loop to the wire are very important and define the strength of the harmonic signal. The loop is stabilized with nail polish. A simple device can be used (Fig. 23B) to stretch and stiffen the silver wire in order to produce a precise loop both with respect to loop diameter (0.35 mm) and distance of the loop at the location where the diode is glued across the loop (Fig. 23A). A laboratory-based calibration radar unit may be helpful for estimating the quality of the transponders, but such a unit is not essential. The calibration can also be performed in the field. The price for such a custom-made transponder is about 20 €.

11.2.3. Attaching the transponder

The transponders can be attached to the bee thorax in the following way, which is a further development of the method reported by Capaldi *et al.* (2000):

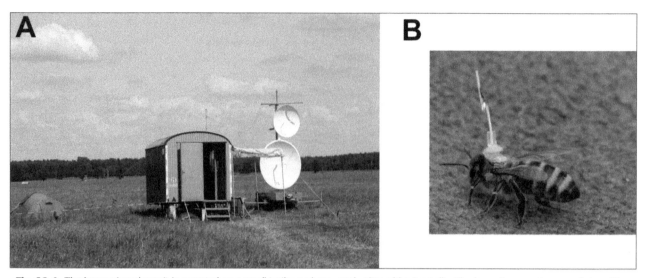

Fig. 22. A. The harmonic radar unit is mounted on a small trailer and accurately aligned horizontally. The lower larger antenna radiates off the main radar pulse, the higher smaller antenna received the harmonic signal. Both antennae are attached to the vertical rotating pole. Notice the displacement between the two antennae, which requires calibration of the transponders localization by a side-wise correction. **B.** A honey bee carrying a transponder. Notice the two number tags glued together with a double sticky tape.   Photo B: Dr Christoph Grüter.

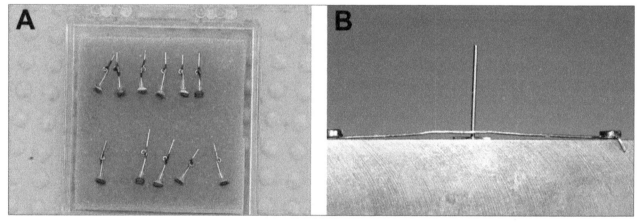

Fig. 23. A. A collection of transponders. Notice the loop bridged by the diode, the nail polish for stabilizing the loop, the PVC tubing into which the wire is inserted, the number tag and the double sticky tape.   Photos: Christoph Grüter.
B. A simple device for producing a loop with an inner diameter of 0.35 mm and a defined distance between the distance of the wire in the loop. One end of the wire is pulled below the horizontal wire, bent around the vertical filament above the horizontal wire and then strongly stretched.

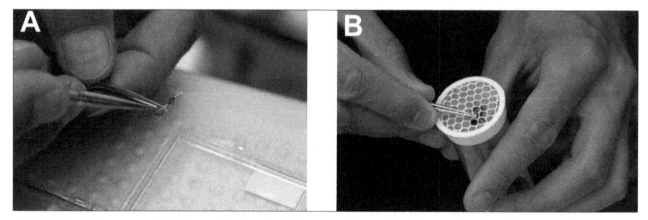

Fig. 24. Fixing the transponder to the thorax of a test bee. **A.** First the cover of the double sticky tape at the base of the transponder is removed. **B.** Then the bee is caught with a marking device and pushed with a stopper against a net. The transponder is caught with fine forceps at the tube and pushed onto the number tag of the bee.   Photos: Christoph Grüter.

1. Tag individual bees with a number plate (see 2.4; Fig. 1).
2. The base of the transponder is also a number tag which carries in its centre a tiny piece of a tubing (*Micro-katheter PVC Chemies Schlauch*, www.rct-online.de) into which the wire fits tightly (Fig. 23A).
3. Glue this tube to the number tag with acrylic superglue.
4. Fix the other side of the number tag to a fitting piece of double sided sticky tape (Pattex, Henkel).
5. Keep the protecting cover of the double sided sticky tape on the other side.
6. Remove the cover immediately before fixing the transponder to the number tag on the thorax of the bee (Fig. 24).
7. Now push the experimental bee with a stopper against the net of a queen marking device (Fig. 24B) and align it such that the number tag on the thorax appears in a window of the net.
8. Finally, push gently the fresh gluing surface of the double sticky tape against the number tag on the bee.

11.2.4. Field studies

The optimal area for tracking bees with the harmonic radar is an open, flat and horizontal landscape without any radar reflecting obstacles (trees, houses). The reason for these requirements is that the radar beam should not be reflected either from the ground or from an object because the reflected radar pulse with its energy multiple times stronger than the second harmonic signal radiated off from the transponder would interfere with this small signal. In such an ideal area, transponder-carrying bees can be detected within a radius of up to 3 km. Examples are given in Fig. 25. The transponder is detected every 3 s defined by the revolution of the radar unit. The circular map resulting from the radar scans is transformed into a Cartesian map with a custom-written program. Bees travelling with 5 – 20 km/h are therefore detected about every 5 – 15 m. The length and width of the radar paint (the technical term for a radar signal) depends strongly on the strength of the harmonic signal, but usually lies in the range of about 15 m at a distance of 600 m. Additional processing of the radar signals allows to substantially improve spatial/temporal resolution. Signal strength of the harmonic radar depends also on the animal's flight height over ground. The unit described above allows for detecting bees reliably between 70 cm and 9 m height for distances up to 1,200 m straight away from the radar. Bees learn quickly to fly with the transponder. Usually, they first land after being released but then start and fly equally fast and reliably as those without a transponder. However, they appear to be more sensitive to wind speeds above > 20 km/h. At such wind speeds they also fly low, making it more difficult to detect them.

The harmonic radar system allowed proving von Frisch's proposal (von Frisch, 1967) that bees communicate distance and direction in their waggle dances (Riley *et al.*, 2005; see section 10). In addition, their navigation strategies could be characterized (Menzel *et al.*, 2005,

2011b). The experiments performed led to the conclusion that navigating bees integrate multiple learning strategies which provide them with the capacity to localize themselves and several goals in such a way that they can perform novel short-cutting flights between them (Menzel *et al.*, 2011a). Such a capacity has been related to a memory structure best conceptualized as a cognitive map (Tolman, 1948).

12. Monitoring honey bee flight – BeeScan and RFID technologies

12.1. Introduction

Foraging honey bees are exposed to various environmental stressors. Their flight performance can therefore indicate noxious influences. Basic parameters for estimating stress are forager longevity and foraging success as indicated by foraging activity, time spent outside the hive and ability to return to the colony. Most of these parameters require detailed observation and bees often need to be labelled individually. These methods are time-consuming and expensive. However, a number of approaches register honey bee flight automatically or use simplified test paradigms. We here discuss how to monitor flight activity of unmarked or individually marked honey bees.

12.2. Tracking unmarked bees

Although foraging activity can be inferred from overall colony weight gains or losses measured with precision balances (Meikle *et al.*, 2008), hive entrance activity is a more reliable parameter. The advantages of this method are:

- Registration of most departing or arriving bees.
- Straightforward measuring of daily flight rhythms.

The major disadvantages, however, are

- Not all of the bees entering the hive have foraged (some bees do not even take off).
- Accumulation of large datasets during a relatively brief observation period.

A straightforward approach for automatic registration uses photoelectric relays which are connected to a single or to multiple tunnels. Bees are forced to pass through the tunnels, where they are detected while leaving or entering the hive. Various constructions have been developed during the last five decades. A commercial counter is produced by Lowland Electronics bvba (Belgium). This device is attached to the hive entrance and uses 32 parallel tunnels such that it can cope with bee trafficking in full colonies. The registration of bees is direction-specific, because two relays are positioned in sequence within each tunnel. Thus, the system can differentiate between bees leaving the colony and those entering the colony. It allows estimating the number of bees outside the colony at each given moment, so that bee losses can be calculated at the end of the day.

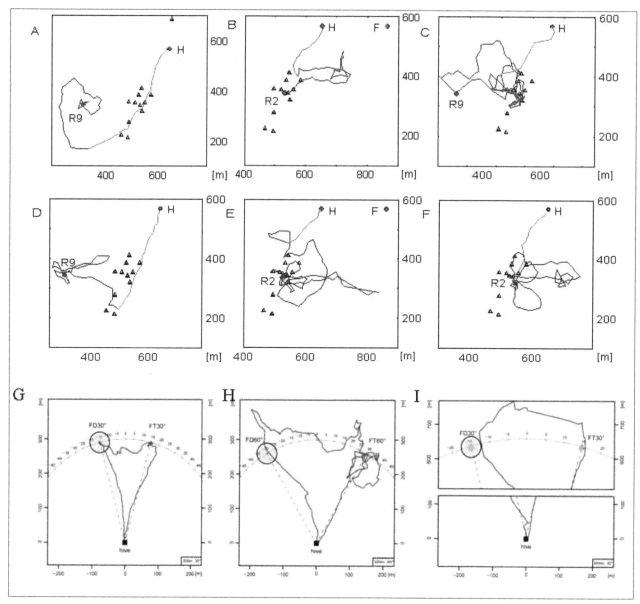

Fig. 25. Nine examples of flight tracks. In the subfigures **A – F** three phases of the homing flights are marked in colour. Vector flight component in red, search flight component in blue, straight homing component in green. H marks the location of the hive, R2 and R9 two different release sites. These bees followed a waggle dance that indicated a location 200 m to the east. The thin blue line in **C** marks a portion of the flight which could not be detected most likely because the bee flew very close to the ground. **G – I**: Each of these three bees was first trained to a location indicated as FT30° or FT60°. Then feeding at this site was terminated, motivating the bees to stay inside the hive and attend dances. A single dancing bee indicated a location FD30" or FD60°. Each bee flew first to the place indicated by the dance and then flew directly to the learned place along a novel shortcut. The green area around the place indicated by the dance marks the visual catchment area as defined by the person sitting there and the spatial resolution of the bee eye. Since the spatial resolution of the bee eye is 1.5° it can be calculated at which distance the bee will see the person sitting in front of the hive. There was no visual mark at the trained place during the test (after Menzel *et al.*, 2011b).

The device has been used in a number of studies (Struye *et al.*, 1994; Struye, 2001). It was generally found to be reliable if it was attended to carefully, particularly by frequent checking and cleaning of the tunnels. This method allows for correlating hive entrance activity with different external parameters such as precipitation, temperature, sunshine etc. In addition, it can detect substantial bee losses. However, the system is unable to track individuals or groups of individuals. Therefore, in experimental studies, the basic statistical unit is a single colony. This requires a sufficient number of colony replicates.

12.3. Tracking individual bees

Measuring the performance of individual bees greatly increases the precision with which various influences on forager activity can be studied. To recognize individuals, they need to be tagged (see section 2.3). Because number tags are not well-suited for automatic identification of bees during hive entrance passage, a combination of optical recognition software and high-resolution video technique is recommended for these applications. Recently, radio-frequency identification (RFID) chips became available which are light and small enough to be used on bees. In the following, a simple method to assess forager performance using "classical" labelling will be described. Afterwards, a detailed account of the RFID method will be provided, which has been employed successfully in several studies.

Flights from individual bees back to the colony are fairly predictable. They can be measured relatively easily, in particular at seasons when feeder trainings are limited to more attractive natural nectar flow. It can be determined reliably whether bees return to the colony and the time they need to return by releasing the bees at some distance (10 m to several 100 m) and registering their arrival time at the colony.

To do this:

1. Catch returning foragers at the hive entrance or from within the hive (see section 2.1) and carry them in glass vials to the release location (Fig. 26).
2. Restrict the observation times to e.g. 30 min after release.
3. Determine returning success to small nucleus colonies late in the evening or early morning.

To be successful, these tests require some precaution. The duration of a return flight is influenced by various variables such:

- Temperature
- Weather condition
- Time of the day
- Age of the bees, and others

In order not to drown the searched-for effects in a vast general variability, one needs to compare carefully matched pairs of individuals or small groups of bees having the same age, and flying at the same time. Accordingly, appropriate statistics for observations on paired samples need to be applied (e.g. Wilcoxon paired sample test, see also the *BEEBOOK* paper on statistics (Pirk *et al.*, 2013)).

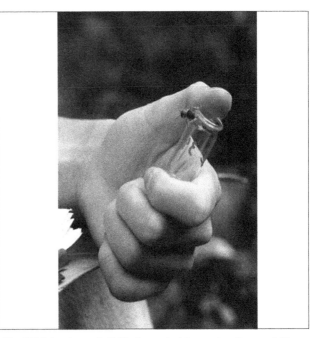

Fig. 26. Releasing an individually marked forager bee from a vial to measure its return time and ability to return to its home colony.

In addition, experimental treatment differences can easily be overrun by any form of general stress applied to the foragers, which requires the experimenter to handle bees with great care, and to release them from the vials without delay. This method may be extended by observing waggle dances of labelled bees in order to determine the danced vector and distance where the foragers found rewarding food sources (see section 10). They may then be released either along the indicated flight path or at certain angles from this vector.

Even low stress levels, due for example to parasitism by *Varroa destructor* or *Nosema spp.*, can be detected during return flight experiments (Kralj and Fuchs, 2006, 2010), because they affect the duration of return flights and the rate of successful returns. Although these experiments are laborious, often require two experimenters, and are restricted to small numbers of bees, they are reliable and very useful when technically sophisticated methods are unavailable.

12.4. Advanced methods: radio frequency identification

12.4.1. Equipment

Radio frequency identification (RFID) systems were initially developed for tagging and identifying stored goods. They are non-contact identification systems consisting of two components: (1) The transponder (tag), which is attached to the object that needs to be identified; (2) The reading device (scanner), which is required to retrieve the information stored on the transponder (Finkenzeller, 2002).

The tag is the data storage medium of the RFID system and usually has no on-board power supply, allowing maximum miniaturization. The energy needed to retrieve the information stored on the tag is provided by the scanner via inductive coupling. As soon as the tag enters the operating distance of the scanner it will be activated, enabling the transfer of the identity data to the scanner (Finkenzeller, 2002).

The majority of RFID tags consist of a spatially separated transponder coil (antenna) and transponder chip (hybrid technology; Finkenzeller, 2002). This design limits the level of miniaturization and their use in honey bee studies. A different tag design in which the antenna is planarly coiled on the surface of the chip (Coil-on-Chip; Jurisch, 2001; Finkenzeller, 2002) results in dimensions of 0.8 or 1.6 mm^2 and weights of 2 or 4 mg. These devices are much more appropriate for use on honey bees. The technique was first used by Streit *et al.* (2003). Recently, Decourtye *et al.* (2011), Schneider *et al.* (2012), and Henry *et al.* (2012) determined effects of insecticides on foraging activity with RFID technologies.

The RFID equipment (tags, scanners, software) in the aforementioned studies was provided by microsensys GmbH, Erfurt, Germany. The tag belongs to the mic 3-series (mic 3-64) with a 64 bit read-only (RO) memory containing the ID-number. It works at a frequency of 13.56 MHz and the reading distance is limited to 2 – 4 mm (depending on the size of the tag). Different scanner models are available. Among these are the 2k6 HEAD used in the studies of Decourtye *et al.* (2011) and Schneider *et al.* (2012). A new scanner model has recently been released. To our knowledge, no studies have been published with this model yet.

A special characteristic of the 2k6 HEAD readers is their direction sensitivity. Two separate, cascaded reader-antennae provide the information on the movement direction of an RFID-tagged honey bee. This yields information about the time a given tagged bee spent outside or inside the colony. To retrieve data collected by the readers, they are equipped with a serial interface to establish a connection to a computer. A USB to serial adapter (e.g. ATEN USB to SERIAL ADAPTER, chipset PL 2303) can be used for the data transfer.

12.4.2. Location of scanners

This RFID system provides a limited reading distance. Thus, the tag needs to be aligned in the correct position for the scanner to register it effectively. The setup requires custom-made bee tunnels in front of the hive entrance. Above these tunnels, the scanners are mounted. Despite this setup, registration errors may occur due to non-detections of bees during the scanner passage. They should be estimated for each experiment. It is essential that the tunnel design is adjusted to ensure that bees pass directly below the scanner with the dorsal side up. This helps to reduce non-detections and yields consistent data files. However, occasionally single readings may be missed, necessitating software filter routines for passage interpretations. Additional care is required if absolute precision is essential. A sufficient number of tunnels, depending on the size of the colony, should be used in order to avoid overly crowding in times of high traffic, each tunnel being monitored by a separate scanner.

12.4.3. Attaching the RFID tags to the thorax of a bee

1. Immobilize the bee (see section 2.2) or trap it in a queen-marking tube (see section 11; Fig. 24B).
2. Glue the tags onto the bees either using gum arabic (Decourtye *et al.*, 2011), or using a non-toxic shellac-based, viscous adhesive (Streit *et al.*, 2003; Schneider *et al.*, 2012).
3. Apply a thin layer of liquid shellac (e.g. Lumberjack Schellack-Streichlack-natur, Alfred Clouth Lackfabrik GmbH&Co., Offenbach, Germany) with a fine brush; this coats the surface of the thorax and removes hair and wax residues which might impede the adhesion.
4. Apply a drop of the viscous shellac-based adhesive onto the coated thorax and immediately press the tag gently on the glue.
5. Allow about 15-20 minutes for the adhesive to dry.
6. Remove the plunger and release the bee into the marking tube.
7. Make sure the bee is unable to remove the tag on its own before you release it from the marking tube.

12.4.4. Monitoring activity

There are two different paradigms using RFID. The first includes life-long monitoring of honey bee activity starting with the day of emergence and ending at the last day of registration. The second is an analysis of foraging behaviour using an artificial feeder training (see section 14).

12.4.4.1. Life-long monitoring

During life-long monitoring three main parameters can be analysed:

- Transition from in-hive tasks to outside tasks.
- Lifespan of every tagged bee as defined by the period from emergence until the day of the last registration.
- Times spent outside / inside the hive and their changes over time, e.g. in relation to the age of the bees or to circadian activity patterns, foraging conditions, or in response to various treatments.

Mostly, these procedures require individually tagged bees of known age (see section 2.3 and the section on obtaining adult and brood of known age in the *BEEBOOK* paper on miscellaneous methods (Human *et al.*, 2013)). All newly emerged bees need to be tagged with transponders and introduced into a registration colony with RFID-scanners installed at the hive entrance.

12.4.4.2. Monitoring foraging behaviour

To analyse foraging behaviour, you need tents to:

- Refrain bees from foraging in unknown terrain.
- Guarantee the attractiveness of the offered sugar solution.

Alternatively, the experiments can be performed during mid/late European summer, when natural forage is declining. In order to analyse foraging behaviour, bees have to be trained to forage from an artificial food source (see section 14).

For short distances up to 20 meters:

1. Catch departing worker bees of unknown age at the hive entrance, using e.g. sample glass bottles with snap caps (70 x 25 mm, 20 ml, neoLab, Heidelberg, Germany).
2. Release these bees at the feeding site by placing the opening of the sample bottle as close to the feeder as possible, so that the bees can collect sugar solution if it is attractive for them.

For mid/to long-distance experiment:

3. Train bees stepwise to the feeder as described in section 14.
4. Colour-mark them on the abdomen.
5. Attach the RFID tags to the thorax of each bee (see section 2.3; Fig. 27A)
6. Provide the feeder with a compartment whose entrance is monitored by RFID scanners (similar to the hive entrance) to automatically identify RFID-tagged bees (Fig. 27B).
7. Unlike the registration tunnels in front of the beehive, which the bees learn to pass on their own when leaving and entering the colonies, you have to guide the foragers stepwise by laying a trace of sucrose along the tunnel tubes which leads the bees to the feeder within compartment; these training procedures incur some cost in labour, thereby restricting the numbers of bees which can be investigated simultaneously.

12.4.5. Parameters of foraging flight

Multiple parameters of a foraging flight can be measured from individual bees when the scanners are positioned at the hive entrance and at the feeder compartment. In order to leave the hive for a foraging flight, a tagged bee has to pass the scanner at the hive entrance in an outward bound direction (Fig. 28, point "A"). After departure, the bees either forage from the artificial feeder or choose a random honey flow source (not shown in Fig. 28) around the perimeter of the hive. When the tagged bees chose to forage from the feeder, they pass the scanner at the feeder compartment entrance (Fig. 28, point "B"). This allows the researcher to determine the frequency of feeder visits from passage sequences of "A" followed by "B". Additionally, the time elapsed from leaving "A" to reaching "B" gives a measure of the flight time to the feeder. The time between "B" and "C" is a measure of the time a given bee spent at the feeder. Finally, the time interval between "C" and "D" yields a measure of the duration of the homing flight. The total time needed for a complete foraging flight is then given by the time between "A" and "D", while the time spent inside the hive between flights is given by the interval between "D" and "A". Finally, failures to return are determined if no "D" registration follows "C". Foraging flights to food source other than the artificial feeder show no "B" and "C" registrations between "A" and "D".

12.5. Data handling

12.5.1. Transponder identification

Before mounting the transponders on bees at the start of an experiment, all transponder-IDs to be used have to be identified using a USB-Reader-Pen (iID® PEN mini USB + software, microsensys, Erfurt, Germany) and stored in a text-file. This allows an overview of all tags present in the experiment and their comparison with the tag IDs detected by the scanner. The scanner data can be read out with software supplied by microsensys GmbH and are stored as a text-file. The next step is to import this file into a statistic program datasheet e.g. SPSS Statistics (SPSS Inc.; Chicago, Illinois, USA), differentiating the four variables: transponder-ID, date, time, scanner ID, and antenna. From here the processing of the data starts using self-programmed algorithms (Syntax-files).

12.5.2. Data filtering

To extract the relevant data, the files first need to be filtered. A custom-made filter algorithm erases rapid-succession-readings of the same transponder-ID at antennas of the same scanner, which occur when a tagged bee lingers beneath the scanners for too long e.g. by being blocked by a flow of opposing traffic. Another custom-made filter programmed to search only for times that take longer than 30 seconds helps to erase many short-duration stays outside the hive, which possibly occur during guard duties and quick movements into the hive or out of the hive. To calculate the time needed from hive to feeder and back again, scanner data at the hive entrance and at the feeder need to be synchronized. Three different algorithms are required:

- For calculation of foraging trip duration and flight times back and forth.
- For feeder stay duration.
- For in-hive intervals.

It is advisable to introduce additional variables which include entry numbering to avoid errors during data resorting in same-second entries. If this is not done, same-second entries might be inverted by the program, leading to errors in data evaluation.

12.5.3. Study design

The RFID technology allows detailed analysis and comparison of various foraging flight parameters (see section 11) between different groups of bees. It is a very sensitive method for registering even subtle effects of stressors on forager behaviour, vitality and viability. Thus, in designing experimental groups, great care is required to randomize any treatment influence in the preparatory steps. In addition, handling of the bees should be performed very carefully to avoid unnecessary stress during treatment administration, which might cover treatment effects.

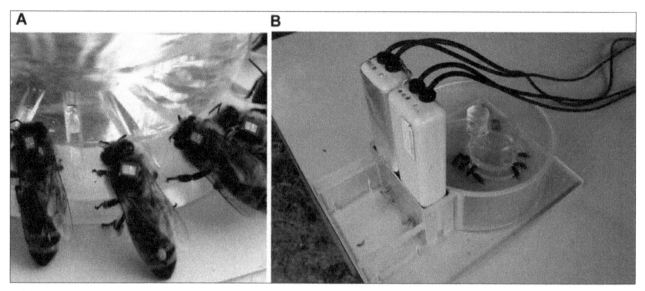

Fig. 27. A. RFID tagged bees foraging from a feeder filled with sugar solution. **B.** Feeder compartment to automatically detect bees at an artificial feeder site.

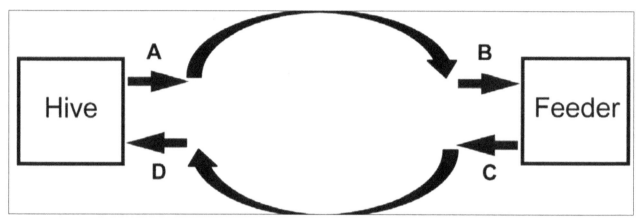

Fig. 28. Schematic drawing of the various parameters that can be obtained from an RFID feeder training experiment. See text for details.

12.6. Examples of application

A wide range of influences on bee performance can be investigated by this method. Treatments may include inoculation or infestation of bees with diseases or parasites and their influence on the behaviour of the bees. We describe in more detail two application examples below. The RFID method is of particular significance for investigations of short- and long-term effects of exposure to agricultural plant protection products or acaricides used in the hive (Decourtye *et al.*, 2011; Schneider *et al.*, 2012). RFID techniques may even have the potential to be included in standard protocols for pesticide registration.

12.6.1. Example 1: *Nosema* infection for life-long monitoring

To study the effects of *Nosema* infections,

1. Mark 300 -500 newly emerged bees at the thorax (see section 2.3.).

2. Introduce the young bees into a nucleus bee hive which serves as the test hive.

3. After four days, collect 140 of the colour-marked bees from the combs of the registration hive.

4. Divide the bees in the following way:

   4.1. Label 60 bees with RFID-tags.

   4.2. Divide these bees into two subgroups; each group of 30 bees is confined to a cage.

5. Bulk-feed both groups with honey solution (33% weight/volume). The control group (*Nosema*-) receives pure sucrose solution, while the *Nosema* group (*Nosema*+) receives honey solution contaminated with *Nosema* spores (5 x 10^6 spores/millilitre, for experimental infestation, see the *BEEBOOK* paper on Nosema; Fries *et al.*, 2013).

6. The remaining 80 bees are needed for subsequent determination of a successful *Nosema* infection. These bees are only required for infection control and therefore are not RFID-tagged. Instead, label these bees with an additional, different colour-marking: e.g. blue for *Nosema*- and red for *Nosema*+ group.

7. Distribute these bees for infection control evenly over the cages of the RFID-tagged bees, so that each cage now contains 70 bees.

8. When all of the bees in each cage have consumed the honey solution offered completely, re-introduce them into the registration colony with a delay of 15-30 minutes between the cages to avoid trophallaxis between inoculated and non-inoculated bees.

9. Twelve days later, collect the colour-marked positive and negative control bees from the hive and examine them for *Nosema* infection (see the *BEEBOOK* paper on *Nosema*; Fries *et al.*, 2013).

10. Leave the RFID-labelled bees in the colony and do not disturb them again. Their foraging behaviour can be studied throughout their entire live as described above.

12.6.2. Example 2: Acute effects of sublethal doses of insecticides or acaricides on foraging

To investigate whether the acute oral administration of insecticides or acaricides alters foraging parameters

1. Use queen marking tubes (Fig. 24B) for catching foragers.

2. Tag foragers with an RFID transponder (see section 11.2.3.).

3. Release bees to forage freely until the next day.

4. On the following day, catch previously tagged bees at the feeder entrance and identify them with the USB-Reader-Pen (iID® PEN mini USB, microsensys, Erfurt, Germany) directly after landing (for this, the feeder needs to be removed from the feeder compartment).

5. Allocate the tagged bees to different experimental groups.

6. Dissolve the substance (insecticide or acaricide) in sugar solution.

7. Offer the substance solution to the treated group in a cap removed from a 1.5 ml Rotilabo® micro centrifuge tube (Carl Roth GmbH, Karlsruhe, Germany), embedded in the foam plastic of the plunger of the queen marking tube.

8. Offer a control sugar solution containing an equivalent of the solvent (if any is used) to the control group in the same manner as described in step 5.

9. Keep bees isolated in the marking tubes for approx. 20 min to avoid trophallaxis with other bees and observe potential regurgitation.

10. After the treatment, the bees can be released at the feeding site or into the registration hive.

11. Determine their subsequent foraging activity for any desired experimental period.

13. Equipment and devices for experiments on mating behaviour

13.1. Introduction

As a member of the Hymenoptera, honey bees have haplo-diploid sex determination. This system of sex determination depends on a single sex locus. Thus inbreeding has a severe negative impact on the viability of the offspring. Thus, natural selection has favoured several behavioural mechanisms to prevent mating among related drones and queens. Altogether, the "general aim" of honey bee mating can best be understood as an optimized mode "of avoiding relatives".

Queens are able to regulate fertilization of an egg (Bresslau's sperm pump) and thus, together with worker bees, regulate the number of drones per season. At the same time, worker bees can regulate the number of daughter queens by feeding them with royal jelly. Investment in queens, which need large numbers of worker bees for swarming (colony fission), is costly while drone production does not involve more than rearing and maintaining individual drones. The estimated sex ratio is about 1 queen to up to 2,000 drones. The absolute control of the production of "full" females and males combined with the high susceptibility of inbreeding is a unique character for *Apis* and results in a mating behaviour which is exceptional in several aspects.

For mating, drones and queens leave the hive and meet quite far away at special congregation areas high in the air, where they copulate in free flight. So far, mating cannot be achieved in confinement. Consequently, natural mating behaviour cannot be observed directly or controlled for. Nevertheless, several aspects of mating behaviour have been uncovered by indirect conclusions from flight behaviour at the flight entrance, dissection of queens and drones, watching drone behaviour towards fixed queens at a drone congregation area (DCA) and analysis of offspring. Drones are attracted by the queen's sex pheromones whose main component is 9-oxodec-2-enoic acid (9 ODA; Gary, 1962). While flying at the DCA the queen's sex pheromones are dispersed not only because of the flight pattern of the queen, but also by the flock of drones which pursue the queen from a short distance (Koeniger *et al.*, 2011). Therefore, it is not only the sex pheromones that are important for mating behaviour, but the shape, colour and pattern of the queen dummy also play an important role (Gary and Marston, 1971; Koeniger, 1990; Vallet and Coles, 1993; Gries and Koeniger, 1996). Drones of all *Apis* species tested so far have reacted to a black wooden stick, 3.2 cm long, 0.5 cm in diameter, which was impregnated with 1 mg synthetic 9 ODA. Smaller black objects and a queen carcass also elicit drone pursuit when 9 ODA is present. Sex pheromones seem to indicate the presence of a queen but close-by navigation is greatly influenced by the shape of the dummy.

13.2. Time and weather restrictions for studying mating behaviour

Reproduction in *Apis* only takes place during a season with ample nectar flow and high availability of pollen. Then the number of worker bees increases rapidly and the colonies soon become strong enough for fission. In addition, conditions for the survival of swarms are optimal. Shortly before swarming season, colonies start producing some thousands of drones and later they also produce about ten queens. These numbers and the number of swarms vary according to the strength of the colonies and depend on species and races. This dependency on season restricts the time for studying mating behaviour to a short period per year. In addition, the reproductive season cannot be pinpointed to fixed dates but varies according to climatic conditions in different years and geographical and environmental conditions. Additional restrictions occur because of the short daily mating flight periods. In regions were several *Apis* species coexist like in South East Asia, a fixed daily mating flight period guarantees reproductive isolation (Koeniger and Wijayagunesekera, 1976; Koeniger and Koeniger, 2000; for review see Otis *et al.*, 2001). In regions with several species, the daily period is about 90 minutes. In regions where only one species occurs, the mating flight period is extended to about 3 hours. The time of flight is genetically fixed in queens and drones and not regulated by the colony (Koeniger *et al.*, 1994). Drone flight starts between 15 to 30 minutes before queens start and also end later (for review see Koeniger *et al.*, 2011).

Generally, queens and drones fly at temperatures above 20°C, blue or partly clouded sky, and low wind velocity. But this behaviour is greatly influenced by the weather of the days before. When there was a rainy period of some days, queens and drones fly under suboptimal conditions and drones react to queen dummies more frequently than after a period of good weather conditions.

It is therefore advisable to observe the hive entrance of some bee hives and start observations and experiments on mating behaviour only when the exiting and returning of drones indicates full mating flight activity. This is important in particular when one wants to detect or work at DCAs.

13.3. Data collection at the flight entrance: daily mating flight period, duration of individual flights, total life time duration and age-dependent flight behaviour

13.3.1. Determining mating flight periods

Daily mating flight periods of drones can be easily monitored by observing the flight activity at the hive entrance by a person with a watch and a tally counter. The number of leaving drones per minute gives data on time of beginning, peak and cessation of drone flight. Further counting is useful for records on the influence of weather conditions during mating flights. This simple method indicates the chances to do experiments on a DCA. It can be performed in any country and is even useful to prove differences in flight times of races and species, especially when they occur sympatrically.

13.3.2. Monitoring individual flight behaviour

For data collection of individual flight behaviour individual number or colour tags (see section 2.3.) or small transponders for RFID recognition (see section 11.) can be glued to the dorsal side of the thorax of queens and drones. Monitoring individual flight duration further allows for differentiating between orientation flights or cleansing flights (duration less than 10 min) and mating flights (duration longer than 10 min).

It is useful to use a flat transparent tunnel at the hive entrance to simplify reading the labels of the drones. The height should be 1 cm at maximum, so that each drone can be detected. The width depends on the type of the hive. The depth should range between 5 cm to 10 cm, depending on the strength of the colony. The RFID technique (see section 11.) is more useful than personal observation for long-term data such as total daily flight duration, total lifetime flight duration or influence of age on flight behaviour in drones.

13.3.3. Observing and monitoring queen flights

1. Use a transparent flight tunnel at the hive entrance (Fig. 29); it should prevent the queen from leaving on her own.
   1.1. The width of the tunnel depends on the type of the hive; its depth should be about 5 cm.
   1.2. Put a movable glass lid on top for observations and a movable piece of queen excluder in front.

Fig. 29. Mating nuc with veranda in front of the flight entrance. The top is covered with a movable glass lid (depth 5cm) to observe the queen. The proximal opening is blocked by a movable piece of queen excluder to control starts and returns of the queen.

2. If a queen wants to leave for mating, open either the glass lid or the queen excluder and observe the start of the queen.

3. After her departure, close the excluder again.

4. The returning queen will "wait" until the researcher removes the excluder and allows it to enter the colony again.

The RFID technique can be used for queens (see section 11). Returning from a mating flight, the queen often does not enter the hive right away but flies around searching for the right entrance, lands and takes off again after some seconds; sometimes the queen even takes a rest nearby on the wall of the hive.

In this case, personal observation yields more information than a RFID reader.

13.4. Assembling equipment for the DCA

For finding a drone congregation area (DCA), catching drones on a DCA for typifying, and estimating number of drones visiting a DCA you will need:

13.4.1. Material for balloons and drone traps

- Weather balloons (natural colour), of 200 g or 300 g weight (depending on the weight of the drone trap that is to be lifted (Fig. 30).
- A 50 m kite line and a kite winder (for about 50 kg) (Fig. 31A).
- A helium tank with a regulator filler valve including a gauge to inflate balloons (Fig. 31B).
- A hand truck for moving helium tanks securely (Fig. 31C).
- Several ball-bearing snap swivels (for about 50 kg, shops for kite or fishing supplies).
- Queen in cage and/or queen dummies with queen pheromones, bee boost.
- Fly catcher paper ribbon to trap drones for dead drone samples.
- White nylon net to trap live drones.
- Wedding veil with a wide mesh diameter to limit wind resistance (in craft or fabric stores). Stainless steel wire (diameter 0.014 cm).
- Lead containing white cotton bead (drapery store).
- Blackened cigarette filters or black wooden sticks of about the same size as drone decoys.
- Clear nylon monofilament fishing line.

13.4.2. Components for a stable platform in the height of flying drones

The components for constructing the basic mast to hold a rotating platform are given below. Many parts fixed to the rotating platform differ depending on the research, the maximal flight speed of drones, the choice tests (visual stimuli, pheromones etc.), the drone competition, etc.

Fig. 30. Weather balloon with drone trap.

You will need the following components:

- Collapsible mast 8 to 12 m high: 4 to 6 steel pipes each 200 cm by 10 cm (Fig. 32A).
- Base to fix the mast.
- Tilt mechanism.
- Electric motor.
- Several aluminium pipes, 200 by 1 cm as rotating bars.
- Rotating platform on top of the motor to fix aluminium pipes.
- Camera or video cameras at your choice (Fig. 32B).
- Battery.
- Electric cables.
- Tent guys, tent pegs, screws or other tools for safety device of the mast.

A locksmith (metal worker) is needed to construct the different parts of the mast.

Any live queen older than 10 days, including those that are several years old, will attract drones. The queens are maintained in nukes near the laboratory. For the duration of an experiment, they can be kept in a queen cage with candy and 3 worker bees. Various queen models (black wood, queen carcass etc.) can be affixed to the cage. (Gary and Marston, 1971).

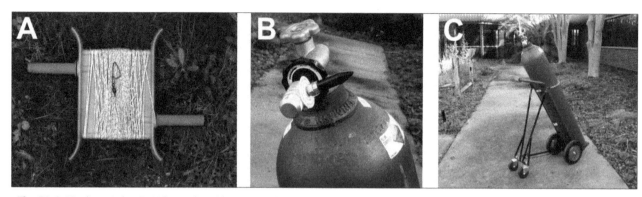

Fig. 31. **A.** Kite line winder. **B.** Helium valve with gauge. **C.** helium tank with hand truck.

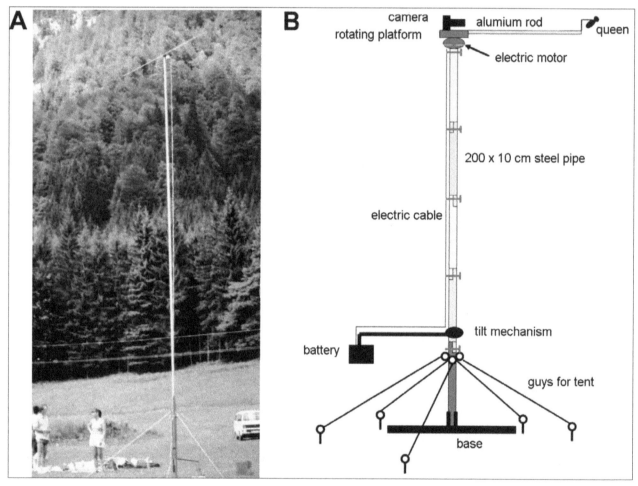

Fig. 32. **A.** mast to measure flight speed of drones. **B.** basic equipment for photographing queens and pursuing drones.

13.4.3. Experiments on copulation

For experiments on copulation with live queens special preparations are required:

1. Keep the sting chamber of the queen open to allow copulation with the drone.

   To achieve this:

   1.1. Fix the queen in an apparatus for instrumental insemination.

   1.2. Open the sting chamber with the hooks.

   1.3. Deposit a drop of elastic glue (Pattex) at the outside of the joint of tergite and sternite of the last segment of the abdomen and allow it to dry.

   Thus the queen cannot close its sting chamber (Fig. 33A).

2. Glue the queen to a wire at the thorax and the first and last tergites of the abdomen to prevent the queen from bending its abdomen when touched by a drone (Fig. 33B; Koeniger, 1981); the glue will be removed by the worker bees later without wounding the queen.

A queen carcass (last abdominal segment removed; Fig. 33C) or a

Fig. 33. A. Elastic glue (Pattex) must be dropped at the outside of the joint between tergite and sternite of the last segment of the queen to prevent it from closing its sting chamber **B.** Living queen glued to a thin wire. **C.** Queen carcass. The wire is pierced through the thorax to the last but one abdominal segment.

wooden queen model with an artificial sting chamber (diameter from 3.2 to 4.0 mm and a depth above 1.6 mm) will also result in copulation (Gary and Marston, 1971) when 9ODA is added or when near a caged living queen.

A dummy can be fixed on a wire bent by 90° (Fig. 34). On one end, a small loop can tie the thread holding the dummy. The other side is fixed at the kite line with tape. This way, the dummy is swinging back and forth. The thread must be shorter than the wire bar, so that the dummy does not entangle itself with the kite line.

Drones of all *Apis* species tested so far have been attracted to a black wooden stick, 3.2 by 0.8 cm, impregnated with 1 mg synthetic 9 ODA, the main component of the queen's sex pheromone. Blackened cigarette filters (2.4 by 0.7 cm) are also attractive, and Bee Boost can be used too. Bee Boost contains a precise blend of 5 pheromone biochemicals mimicking the queen mandibular pheromones. However, a black dummy has to be added since bee boost is sold in a transparent plastic release device.

13.4.4. Drone trap construction

For constructing a drone trap a modified design of a wind-directed pheromone trap of Williams (1987) has proved very effective (Fig. 35). It is constructed of white tulle for wedding veils which is supported by three wire rings of stainless steel.

- The diameter of the steel is 0.014 cm.
- The diameter of the rings are:
  50 cm for the lowest ring
  35 cm for the middle ring
  29 cm for the top ring.
- The net dimensions are:
  104 cm in height
  50 cm at the bottom (encompassing the bottom ring)
  22.5 cm at the top (encompassing the top ring).
- When sewn together, the shape is formed by the three steel rings and the top is closed by a disc of tulle.
- In the bottom ring and the middle ring a fishing line is fixed between two diametrically opposite points, with a small loop in the middle to fix the scented queen dummy at the DCA.
- It is advisable to use two dummies (each with 1 mg of 9 ODA):

one fixed at the lower diagonal fishing line – hanging on a 25 cm long thread.
one fixed beneath the middle fishing line.

- To imitate drone pursuit, drone decoys are tied at the opposite sides of the bottom ring with a thread and the others are distributed at different places.
- If there are sudden gusts of wind, the trap may be blown into a horizontal position, and many drones may escape. To prevent too much movement, white cotton bead containing tiny lead balls (such as used for draperies) can be fixed at the bottom ring. Alternatively, a heavier (thicker) stainless steel wire can be used for the bottom ring.

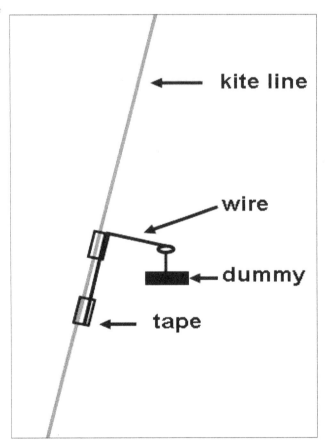

Fig. 34. Queen dummy bound to a wire, which is fixed with adhesive tape to the line.

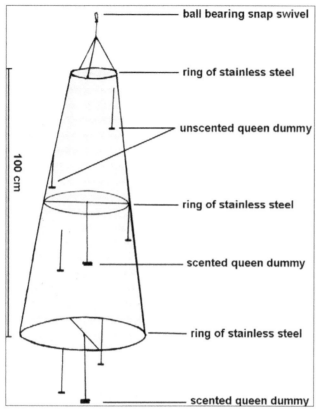

ball bearing snap swivel

ring of stainless steel

unscented queen dummy

100 cm

ring of stainless steel

scented queen dummy

ring of stainless steel

scented queen dummy

Fig. 35. Drone trap equipped with scented and unscented dummies.

13.4.5. Catching drones on fly catching paper

Drones that do not have to be kept alive can be caught by fly catching paper. One or even several pieces can be fixed lengthwise to the kite line, each underneath a dummy (Fujiwara *et al.*, 1994). If it moves freely, it may be fixed to a wire bar.

13.5 Starting the experiment

Before starting the experiment make sure that the temperatures are above 20°C, wind velocity is low and the sun shines most of the day. After a previous period with rainy days drones will fly for mating under less favourable conditions than after a period with fine weather. Also, drones will follow the queen dummies more intensely after some days of unfavourable weather. It is helpful to create a check list before starting for drone experiments in the field. The drone flight time is short and often the experimental drone site is far away from the lab.

13.5.1. Balloon preparation

1. Fix one end of the kite line to the winder.
2. Fix the ball bearing snap swivel to the other end.
   This will prevent any twist of the line.
3. Chose the balloon size adapted to your experiment.
   For carrying only a queen dummy (and optionally a fly paper), a smaller balloon with a diameter of about 150 cm (200 g) is adequate.

To carry a trap, the 300 g balloon inflated to a diameter of about 200 cm is more suitable.

4. Fill the balloon with helium only at the experimental site since transportation of inflated balloons is risky.
5. After inflating, close the balloon's mouth tight with some strong line leaving enough extra line to knot a loop.
6. Hook the snap swivel of the kite line into the loop.
7. Fix queens, dummies or traps 5 m below the balloon at minimum.
   You can do this by fixing a loop in the kite line at this distance.
8. Hook queens in cages and trap equipped with snap swivels in this loop.

To catch drones, the trap must be lifted from about 10 m to up to 40 m high. As mentioned above, the height of drone flight varies. On bright days, drones may fly preferably as high as 20 to 40 m, on a very warm day with an approaching thunderstorm they may fly between only 5 to 10 m high.

If the experiment requires that drones return to their colony, do not keep the drones inside the trap longer than 10 minutes (Fig. 36). Drones need enough fuel (honey reserve in their crop) for flying home. In this case, it is advisable to operate with two traps: one is lifted up while the drones from the other net can be marked and/or counted.

13.5.2. Locating DCAs

Drone congregation areas are normally in proximity to the apiaries. In hilly regions, drones start in the direction of depressions in the horizon and stop at a suitable area (Ruttner and Ruttner, 1965). Up to now, the physical characteristics of a DCA have not been determined, so there is no other way of finding a DCA than to lift a queen or a queen dummy into the flight height of drones and wander in direction of the depression in the horizon (lightest point). The flight height may differ

Fig. 36. Drones have to be released from the trap within 10 minutes to keep them strong enough for the flight to their home colony.

from 10 m to 40 m according to weather conditions and races of *A. mellifera* (Koeniger *et al.*, 1989). Single drones near the apiary are often attracted by a queen, but they stay only for a short while. At the DCA, a swarm of drones will assemble around a queen. In flat areas, the horizon is uniform. Up to now, no DCA has been described for this landscape. Within 10 minutes, drones are attracted everywhere at some distance to the apiary.

13.5.3 Experiments with the mast

For experiments on the mast, all of the items have to be composed according to the specific question, apart from the collapsible (portable) mast, built by a locksmith. To measure the potential flight speed of drones, for example, the electro motor must allow adjustable rotation speed. Only one aluminium pipe (rotating bar) is required. For choice tests including two stimuli (visual stimuli, pheromones etc.) a forked wire has to be attached to the rotating bar (Koeniger 1981). For analysing drone flight near the queen or drone competition, two video recorders have to be fixed on two rotating bars. The queen or dummy has to be fixed on a third bar in an identical angle to both recorders (Gries and Koeniger, 1996). Examples of possible research goals and their respective methods are given in Table 6.

14. Appetitive learning in the field

14.1. Introduction

Many assays employ the excellent learning capacities of honey bees under controlled laboratory conditions (see sections 4-6), where a single bee is restrained in a holder. Although these conditions are very helpful for specific analyses, for some questions like the behaviour of the honey bee in the social context or behavioural adaptations to environmental issues, a more natural learning situation is required. Honey bee foragers typically associate the characteristics of a food source with its location. Therefore, a learning situation is required which rewards the bee in the field. In the last century, several appetitive conditioning protocols were used to train free-flying bees in the field to different sensory cues like odours and colours (von Frisch, 1919; Menzel, 1967; Waller *et al.*, 1973; Greggers and Menzel, 1993), visual patterns and geometrical forms (Hertz, 1935; Giurfa and Menzel, 1997; Giurfa *et al.*, 1999; Horridge, 2000) or varying tactile cues (Martin, 1965; Erber *et al.*, 1998). The learning performance of a bee is strongly connected to the selection and combination of the sensory stimuli used. Bees, for example, learn odours faster and more reliably than colours (von Frisch, 1967; Kriston, 1973; Couvillon and Bitterman, 1988). Visual cues can be perceived at a much further distance than odours and tactile cues need a direct physical contact to be learned.

The majority of appetitive learning protocols uses sucrose as a reward. Several studies on PER-learning in the laboratory showed that the sucrose concentration of the reward and the individual

responsiveness to sucrose strongly correlate with the learning performance in honey bees (see Scheiner *et al.*, 1999, 2003, 2004, 2005). Bees that received a high-concentrated sucrose solution as reward learned faster than bees receiving a low sucrose concentration. Bees with high responsiveness to sucrose performed better than bees with low responsiveness. Similar rules between responsiveness to sucrose and appetitive learning performance in bees can be found under free-flying conditions in the field (Mujagic *et al.*, 2010). Bees accepting low sucrose concentrations from a feeder also show a significant better appetitive learning performance than bees accepting only high-concentrated sucrose solutions. The sucrose acceptance threshold of a bee is a reliable measure of its responsiveness to sucrose in the field and can be employed to select bees for training (von Frisch, 1927; Mujagic and Erber, 2009).

Table 6. Examples of some research goals concerning mating behaviour and some methods to achieve them.

Research goal	devices needed	parameter to measure
determine drone flight period	personal observation at flight entrance, tally counter video observation	time of start of first drones starts / min time of last drones returning
individual drone flight time	individually marked drones 1. Numbered tags, stop watches (personal observation) 2. RFID technique	age of drones at first flight duration of orientation and mating flight duration of life flight time life span
individual queen flight time	veranda covered with glass, entrance closed by removable queen excluder, stop watch	duration of orientation and mating flights
behaviour of drones pursuing a fixed queen	high mast with one or two cameras on top of the queen or dummy on a carrousel	drone competition: flight speed, overtakes, drone/drone distance copulation
finding a drone congregation area (DCA)	helium filled balloon, queen or queen dummy fixed to the line, testing height from 10 to 40 m	number of drones following queen or dummy
typifying drones on a DCA	helium filled balloon, drone traps or flypaper fixed at the line 5 m below the balloon, testing height from 10 to 40 m	flight distance (colour marks on drones, recover in apiaries) genetic diversity (molecular techniques)
estimation of drone number visiting a DCA	"capture-mark-recapture method": catch, mark and release about 500 drones and wait until drones are evenly mixed in the population (30 min!). then catch again and count marked and unmarked drones	estimate the total drone population size on the DCA by dividing the number of marked individuals by the proportion of marked individuals in the following catches.

14.2. Training bees to an artificial sucrose feeder in the field

Free-flying honey bees can be trained to forage on an artificial sucrose feeder in the field. In the following, one way is explained how bees can be prepared and trained to a sucrose feeder:

1. Fill a jar (e.g. a glass beaker of 250 ml volume) with a sucrose solution of a defined concentration (e.g. 30% weight/weight; sugar diluted in water).

    The exact sucrose concentration can be tested with a refractometer (e.g. N-50 E, Atago, Tokyo, Japan).

2. Turn the jar upside down on a 10 cm x 10 cm Plexiglas®-plate with radially engraved grooves (Fig. 37A).

    The sucrose solution accumulates inside each groove and bees landing on the platform can easily access the food with their probosces without drowning in the solution (Fig. 37B).

3. To attract bees to the artificial feeder, place it directly in front of the hive entrance; the Plexiglas®-plate builds a bridge between the entrance of the hive and the feeder.

    Thus bees can directly walk on the plate to the grooves filled with sucrose solution.

4. Mark 5 – 10 bees continuously foraging on the feeder with a colour spot on the abdomen or thorax (see section 2.3.) for better identification.

5. Move the feeder successively away from the hive in 2 m steps every 15 min to the final destination.

    Care should be taken that all marked bees are able to follow the moving food place; if a bee lost its track to the new location, the feeder must be returned to a former position until each single marked bee has found the food place again.

6. The marked bees will recruit new foragers to the feeder.

7. The number of newly recruited foragers can be controlled by increasing the sucrose concentration of the feeder.

    Increasing the concentration e.g. from 3% to 50% sucrose results in more bees being attracted to the feeder. Inversely, a decrease in concentration reduces the number of active foragers at this site.

14.3. Determining sucrose acceptance thresholds in the field

By reducing the concentration of sucrose in the feeder one can determine under free-flying conditions at which concentration an individual bee gives up foraging for sucrose. This concentration can be defined as the sucrose acceptance threshold of an individual.

1. Mark approximately 50 foragers visiting an artificial sucrose feeder on their abdomen or thorax (see section 2.3.).

2. Reduce the concentration in the feeder logarithmically in 20 min steps (30% → 10% → 3%; Fig. 37C).

3. Record for each individual marked bee the number of visits to each presented concentration.

4. After testing bees at each concentration for 20 min, assign each individual to one of two groups: (1) Bees accepting sucrose concentrations of 30% (w/w), but do not accept lower concentrations; (2) Bees accepting concentrations down to 3% (w/w).

When you train the two groups of bees in the field (see section 14.4.), you will notice that bees accepting only high sucrose concentrations (group 1) will perform poorly, while bees accepting low sucrose concentrations (group 2) will learn very well, even if the sucrose concentration used as reward is not very high.

14.4. Conditioning free-flying bees in the field

Honey bee foragers can be conditioned in the field to discriminate between different sensory cues such as odours, visual cues and tactile patterns. The basic procedures for conditioning bees are very similar. A crucial part is the right choice of the conditioning stimulus, because it has to be designed such that it can also be perceived by the bees, i.e. for visual cues for example one has to take care of the visual angle, the distance of perception, the wavelength spectrum, the green contrast, etc. In Fig. 38 we present one example of how to condition free-flying bees to a visual pattern in the field. In this example, a dual-choice test is used for quantifying the learning success. The test apparatus is a white box with three disk holders mounted horizontally to each other. The centre has an opening for the sucrose reward. The conditioned and tested visual cues used in this protocol are vertically and horizontally oriented black stripes on a white background placed onto rectangular discs (for more details see Fig. 38). The test apparatus and the stimuli used here are just one example to explain the conditioning procedure. The discs on the apparatus can be easily exchanged by other cues, such as colours, tactile cues or specific odours which are applied on filter papers, depending on the question of the study.

14.4.1. Preparing the bees

1. Mark a newly recruited forager from a sucrose feeder with a highly visible colour spot on the thorax or abdomen.

2. Train the marked bee to find a droplet of a high-concentrated sucrose solution (e.g. 50%) on the setup (for more details see Fig. 38).

3. Use landmarks (e.g. coloured plates) to guide the marked bee to the sucrose droplet or transfer the bee on a tooth pick from the feeder to the setup.

    In the following visits, the bee usually remembers the location of the high-sucrose reward.

4. After three visits to the setup, start with the conditioning procedure.

    A visit is defined as one foraging flight between the hive and the setup.

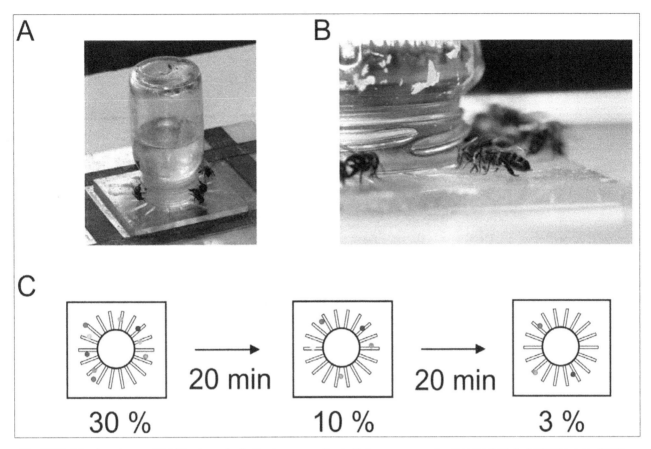

Fig. 37. Training bees to an artificial feeder and selecting bees according to their sucrose acceptance in the field. **A.** Artificial sucrose feeder. **B.** Honey bee foragers drinking sucrose solution from the feeder. **C.** Schematic drawing of the sucrose feeder with decreasing sucrose concentrations (30%, 10% and 3%) used for selecting bees with different sucrose acceptance thresholds in the field. Foragers visiting a 30% sucrose feeder are individually marked with a colour spot on the abdomen or thorax. After 20 min the concentration on the feeder is reduced to the next lower concentration (10%) and all individuals continuing to visit the feeder are registered. Again 20 min later, the sucrose concentration is reduced to 3%. 20 min later, the sucrose concentration is increased again to 30% sucrose. Bees visiting the feeder regardless of the sucrose concentration are assigned to the group with low acceptance thresholds. Individuals only foraging at the highest concentration are considered bees with high acceptance threshold.

14.4.2. Conditioning the bees

1. Place the conditioned stimulus (CS; disc with vertical pattern) into the central holder of the box.
2. Place a drop of the sucrose reward (e.g. 50% sucrose) in the opening of the disk (Fig. 38 C).
3. Let the marked bee visit the CS pattern with the reward for five times, before starting with the first dual-choice test.
4. After five conditioning trials, test the choice behaviour of the bee by:
   4.1. Cover the centre with a neutral white disc (Fig. 38C).
   4.2. Present the CS pattern and an alternative test stimulus (e.g. a horizontal pattern) to the left and to the right side of the box.
5. Register all approaches towards the left and the right stimulus for 1 min.
6. Then reverse the side of the tested stimuli and record the choice behaviour of the bee again for 1 min.
   This procedure helps to avoid artefacts due to a side preference of the bee.
7. Afterwards, condition the bee again five times to the CS pattern.
8. Repeat the dual choice test to both the vertical and horizontal stimuli.
9. Exchange each disc with the stimulus after one presentation to avoid scent marks of the bee; discs or substrates used for conditioning should not be used again for the testing.

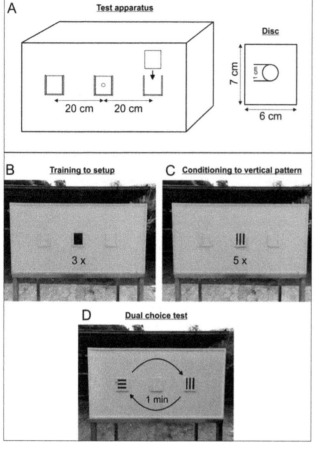

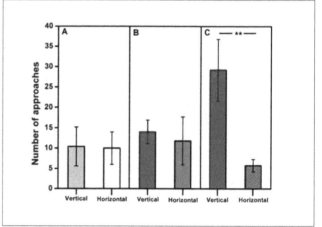

Figure 39. shows an example of the learning performance of five nectar foragers trained to a vertical pattern in the field using this protocol. The dual-choice test is just one method of quantifying the learning behaviour of bees. Another possibility is to determine the learning success after each conditioning trial, using individual learning curves as a measure. It is also important to hold the number of conditioning visits flexible, because the learning speed depends on the complexity of the learning task.

Fig. 39. Conditioning of free flying bees to a vertical visual pattern. The graph shows mean and SEM of the number of approaches towards a vertical and a horizontal visual pattern of five tested bees. The asterisks indicate significant differences between the number of approaches to the vertical and horizontal discs (** $P < 0.05$; paired sample *t*-test). **A.** The number of approaches of bees during the spontaneous choice test. **B.** First dual-choice test after 5 conditioning trials towards the rewarded vertical pattern and the alternative horizontal pattern. **C.** Second dual-choice test after 10 conditioning trials towards the rewarded vertical pattern and the alternative horizontal pattern. The tested bees showed a highly significant preference for this pattern ($P = 0.002$; *t*-statistic $= 7.101$).

Fig. 38. An example of a visual conditioning and testing protocol for free flying bees in the field. **A.** Schematic drawing of the test apparatus (80 cm x 40 cm x 40 cm) used for conditioning and dual-choice testing. The front side of the apparatus is used both for conditioning and testing. Three holders for discs are mounted 20 cm next to each other. The visual, olfactory or tactile stimuli can be mounted on the small rectangular discs (7 cm x 6 cm) that fit into the holders. The centre is used for conditioning and has a 1 cm opening. Bees can enter the small hole in the centre and receive the sugar reward. The left and right holders are for presentation of stimuli during the dual choice test. The discs for testing are closed and have no openings. **B – D.** Pictures of the conditioning procedure for visual discrimination in the field. **B.** Training to the setup. A droplet of a 50% sucrose solution is placed inside the opening of the centre. A black disc placed in the central holder is used as a landmark. The bee tested has to collect the reward during 3 visits before the conditioning procedure can begin. One visit is defined as one foraging flight between the hive and the setup. **C.** Conditioning procedure. The bee is conditioned to a vertical pattern (conditioning stimulus; CS) placed in the centre in 5 learning trials (= 5 visits). **D.** Dual-choice test. A vertical (CS) and a horizontal (alternative) stimulus pattern are presented on the left and right side. The centre is covered with a white disc. The approaches towards the vertical and the horizontal stimuli are registered for 1 min. Afterwards, the side of the tested stimuli is reversed and the choice behaviour is recorded again for 1 min.

14.5. Conclusions

The floral nectar plays a key role in the nutrition of bees and also serves as a reward in learning the characteristics of the food source (see Menzel and Müller, 1996). The concentration of the nectar varies during the season and foragers must adapt very fast to environmental changes. The selection of bees according to their individual sucrose acceptance threshold together with the free-flying appetitive conditioning protocol described here are suited well to analyse the temporal responsiveness to sucrose and the learning performance of a foraging group under more natural conditions. It further allows monitoring the behavioural adaptation of a group of foragers or a single bee to daily or seasonal changes in the environment. A recent challenge in honey bee science is to understand the influence of viruses and parasites on the complex behaviours of the honey bee, especially with respect to colony collapse disorder. Recent studies showed that the mite *V. destructor* influences the flight behaviour, orientation and returning success of forager honey bees (Kralj, 2004). The lower returning rate might be a result of reduced sensory and/or neural processing capabilities involved in navigation during foraging flights. In free-flying bumble bees, a strong correlation was found between the immune response caused by parasite infections and an impaired colour leaning behaviour (Alghamdi *et al.*, 2008). This learning protocol can be used to analyse differences in the learning and foraging behaviour of bees infected with parasites directly in the field.

15. General conclusions

Honey bees display a unique behavioural repertoire both under controlled laboratory conditions (see sections 3 – 8) and under free-flying conditions (see sections 9 – 14). Behaviours which can be quantified reliably under controlled laboratory conditions include sensory responsiveness to gustatory, visual and olfactory stimuli, appetitive and aversive learning performance, locomotion and flight performance of individual bees. The availability of diverse systems for videotaping behaviour enables us to perform long-term studies on honey bee behaviour inside the hive on issues such as division of labour and dance communication. These observation studies are particularly useful for analysing the influence of stressors such as pesticides or diseases on individual honey bee behaviour and on the complex system of division of labour. A range of modern technical devices and a long history of behavioural studies allow us today to investigate with high precision complex behaviours in the field. Honey bee navigation, flight performance and mating behaviour are just some examples. Observing the appetitive and aversive learning performance of free-flying honey bees can give us valuable information on how stress or other parameters affect cognitive abilities of foragers, thereby influencing the success of the entire colony.

Naturally, the number of available protocols for honey bee behaviour is much larger than what can be covered in this chapter. We have therefore selected a range of assays which 1) are well established in honey bee research laboratories world-wide, 2) have been used successfully in the past, 3) are comparatively simple to perform, 4) yield reliable results and 5) cover a wide range of exciting behaviours. The introduced methods and techniques are thus intended to serve as standard protocols for honey bee researcher and bee keepers.

16. Acknowledgements

The COLOSS (Prevention of honey bee COlony LOSSes) network aims to explain and prevent massive honey bee colony losses. It was funded through the COST Action FA0803. COST (European Cooperation in Science and Technology) is a unique means for European researchers to jointly develop their own ideas and new initiatives across all scientific disciplines through trans-European networking of nationally funded research activities. Based on a pan-European intergovernmental framework for cooperation in science and technology, COST has contributed since its creation more than 40 years ago to closing the gap between science, policy makers and society throughout Europe and beyond. COST is supported by the EU Seventh Framework Programme for research, technological development and demonstration activities (*Official Journal L 412, 30 December 2006*). The European Science Foundation as implementing agent of COST provides the COST Office through an EC Grant Agreement. The Council of the European Union provides the COST Secretariat. The COLOSS network is now supported by the Ricola Foundation - Nature & Culture.

15. References

ABRAMSON, C I (1986) Aversive conditioning in honey bees (*Apis mellifera*). *Journal of Comparative Psychology* 100: 108-116. http://dx.doi.org/10.1037/0735-7036.100.2.108

ABRAMSON, C I (1994) *A primer of invertebrate learning: the behavioural perspective.* American Psychological Association; Washington, DC, USA. http://dx.doi.org/10.1037/10150-000

ABRAMSON, C I (1997) Where have I heard it all before: Some neglected issues of invertebrate learning. In *Greenberg, G; Tobach, E (Eds). Comparative psychology of invertebrates the field and laboratory study of insect behaviour.* Garland Publishing; New York, USA. pp. 57-78. http://dx.doi.org/10.2466/pr0.97.3.721-731

ABRAMSON, C I; BITTERMAN, M E (1986a) Latent inhibition in honey bees. *Animal Learning and Behaviour* 14: 184-189. http://dx.doi.org/10.3758/BF03200054

ABRAMSON, C I; BITTERMAN, M E (1986b) The US-preexposure effect in honey bees. *Animal Learning and Behaviour* 14: 374-379. http://dx.doi.org/10.3758/BF03200081

ABRAMSON, C I; STEPANOV, I I (2012) The use of the first order system transfer function in the analysis of proboscis extension learning of honey bees, *Apis mellifera* L., exposed to pesticides. *Bulletin of Environmental Contamination and Toxicology.* http://dx.doi.org/10.10007/s00128-011-0512-8

ABRAMSON, C I; MILER, J; MANN, D W (1982) An olfactory shuttle box and runway for insects. *Journal of Mind and Behaviour* 3: 151-160. http://dx.doi.org/10.1037/0735-7036.100.2.108

ABRAMSON, C I; BUCKBEE, D A; EDWARDS, S; BOWE, K (1996) A demonstration of virtual reality in free-flying honey bees: *Apis mellifera. Physiology & Behaviour* 59: 39-43. http://dx.doi.org/10.1016/0031-9384(95)02023-3

ABRAMSON, C L; AQUINO, I S; RAMALHO, F S; PRICE, J M (1999) The effect of insecticides on learning in the Africanized honey bee (*Apis mellifera* L.). *Archives of Environmental Contamination and Toxicology* 37(4): 529-535.

ABRAMSON, C I; MORRIS, A W; MICHALUK, L M; SQUIRE, J (2004) Antistatic foam as a shocking surface for terrestrial invertebrates. *Journal of Entomological Science* 39: 562-566.

ABRAMSON, C I; SINGLETON, J B; WILSON, M K; WANDERLEY, P A; RAMALHO, F S; MICHALUK, L M (2006a) The effect of an organic pesticide on mortality and learning in Africanized honey bees (*Apis mellifera* L.) in Brazil. *American Journal of Environmental Science* 2: 37-44. http://dx.doi.org/10.3844/ajessp.2006.33.40

ABRAMSON, C I; WILSON, M K; SINGLETON, J B; WANDERLEY, P A; WANDERLEY, M J A; MICHALUK, L M (2006b) Citronella is not a repellent to Africanized honey bees *Apis mellifera* L. (Hymenoptera: Apidae). *BioAssay* 1(13): 1-7. http://dx.doi.org/www.bioassay.org.br/articles/1.13

ABRAMSON, C I; GIRAY, T; MIXSON, T A; NOLF, S L; WELLS, H; KENCE, A; KENCE, M (2010a) Proboscis conditioning experiments with honey bees (*Apis mellifera caucasica*) show butyric acid and DEET not to be repellents. *Journal of Insect Science* 10: 122. http://dx.doi.org/10.122/i1536-2442-10-122

ABRAMSON, C I; NOLF, S L; MIXSON, T A; WELLS, H (2010b) Can honey bees learn the removal of a stimulus as a conditioning cue? *Ethology* 116: 843-854. http://dx.doi.org/10.1111/j.1439-0310.2010.01796.x

ABRAMSON, C I; SOKOLOWSKI, M B C; WELLS, H (2011) Issues in the study of proboscis conditioning. In *Columbus, F (Ed.). Social insects: structure, function, and behaviour.* Nova Science Publishers; Hauppaug, NY, USA. pp. 25-49.

AGARWAL, M; GUZMAN, M G; MORALES-MATOS, C; DEL VALLE DIAZ, R; ABRAMSON, C I; GIRAY, T (2011) Dopamine and octopamine influence passive avoidance learning of honey bees in a place preference assay. *PLOS One* 6(9): e25371. http://dx.doi.org/10.1371/journal.pone.0025371

ALGHAMDI, A; DALTON, L; PHILLIS, A; ROSATO, E; MALLON, B (2008) Immune response impairs learning in free-flying bumble bees. *Biology Letters* 4(5): 479-481. http://dx.doi.org/10.1098/rsbl.2008.0331

ALIOUANE, Y; HASSANI, A K; GARY, V; ARMENGAUD, C; LAMBIN, M; GAUTHIER, M (2009) Subchronic exposure of honey bees to sublethal doses of pesticides: effects on behaviour. *Environmental Toxicology and Chemistry* 28: 113-122. http://dx.doi.org/10.1897/08-110.1

AMDAM, G V; CSONDES, A; FONDRK, M K; PAGE, R E (2006) Complex social behaviour derived from maternal reproductive traits. *Nature* 439: 76-78. http://dx.doi.org/10.1038/nature04340

AMDAM, G V; PAGE, R E; FONDRK, M K; BRENT, C S (2010) Hormone response to bidirectional selection on social behaviour. *Evolution and Development* 12(5): 428-436. http://dx.doi.org/10.1111/j.1525-142X.2010.00429.x

AVARGUES-WEBER, A; DE BRITO SANCHEZ, M G; GIURFA, M; DYER, A G (2010) Aversive reinforcement improves visual discrimination learning in free-flying honey bees. *PLOS One* 5(10): e15370. http://dx.doi.org/10.1371/journal.pone.0015370

BACKHAUS, W (1993) Color-vision and color choice behavior of the honey bee. *Apidologie* 24: 309-331.

BECHER, M A (2007) Measuring temperature distributions in the brood-comb of honey bees (*Apis mellifera*) in dependence of group size. In *20th meeting of the German speaking section of IUSSI, Bochum, Germany, 26-28 September 2007.*

BECHER, M A; MORITZ, R F A (2009) A new device for continuous temperature measurement in brood cells of honey bees (*Apis mellifera*). *Apidologie* 40(5): 577-584. http://dx.doi.org/10.1051/apido/2009031

BECHER, M A; HILDENBRANDT, H; HEMELRIJK, C K; MORITZ, R F A (2010) Brood temperature, task division and colony survival in honey bees: A model. *Ecological Modelling* 221(5): 769-776. http://dx.doi.org/10.1016/j.ecolmodel.2009.11.016

BEHRENDS, A; SCHEINER, R (2009) Evidence for associative learning in newly emerged honey bees (*Apis mellifera*). *Animal Cognition* 12: 249-255.

BEHRENDS, A; SCHEINER, R; BAKER, N; AMDAM, G V (2007) Cognitive aging is linked to social role in honey bees (*Apis mellifera*). *Experimental Gerontology* 42: 1146-1153. http://dx.doi.org/10.1016/j.exger.2007.09.003

BITTERMAN, M R; MENZEL, R; FIETZ, A; SCHÄFER, S (1983) Classical conditioning of proboscis extension in honey bees (*Apis mellifera*). *Journal of Comparative Psychology* 97(2): 107-119.

BRAUN, G; BICKER, G (1992) Habituation of an appetitive reflex in the honey bee. *Journal of Neurophysiology* 67(3): 588-598.

BREED, M D; STILLER, T M (1992) Honey bee, *Apis mellifera*, nestmate discrimination: hydrocarbon effects and the evolutionary implications of comb choice. *Animal Behaviour* 43: 875-883. http://dx.doi.org/10.1016/S0003-3472(06)80001-1

BRODSCHNEIDER, R; RIESSBERGER-GALLÉ, U; CRAILSHEIM, K (2009) Flight performance of artificially reared honey bees (*Apis mellifera*). *Apidologie* 40: 441-449. http://dx.doi.org/10.1051/apido/2009006

CALDERONE, N W; PAGE, R E (1991) Evolutionary genetics of division of labor in colonies of the honey-bee (*Apis mellifera*). *American Naturalist* 138: 69-92.

CALDERONE, N W; PAGE, R E (1996) Temporal polyethism and behavioural canalization in the honey bee, *Apis mellifera. Animal Behaviour* 51: 631-643. http://dx.doi.org/10.1006/anbe.1996.0068

CAMPBELL, B A; CHURCH, R M (Eds). (1969) *Punishment and aversive behaviour*. Appleton Century Crofts; New York, USA.

CAPALDI, E A; SMITH, A D; OSBORNE, J L; FAHRBACH, S E; FARRIS, S M; REYNOLDS, D R; EDWARDS, A S; MARTIN, A; ROBINSON, G E; POPPY, G M; RILEY, J R (2000) Ontogeny of orientation flight in the honey bee revealed by harmonic radar. *Nature* 403: 537-540. http://dx.doi.org/10.1038/35000564

CARCAUD, J; ROUSEL, E; GIURFA, M; SANDOZ, J C (2009) Odour aversion after olfactory conditioning of the sting extension reflex in honey bees. *Journal of Experimental Biology* 212: 620-626. http://dx.doi.org/10.1242/jeb.02664

COLLETT, M; HARLAND, D; COLLETT, T S (2002) The use of landmarks and panoramic context in the performance of local vectors by navigating honey bees. *Journal of Experimental Biology* 205: 807-814.

COUVILLON, P A; BITTERMAN, M E (1988). Compound-component and conditional discrimination of colours and odours by honey bees: further tests of a continuity model. *Learning and Behaviour* 16: 67-74. http://dx.doi.org/10.3758/BF03209045

CRAILSHEIM, C; EGGENREICH, U; RESSI, R; SZOLDERITS, M (1999a) Temperature preference of honey bee drones (Hymenoptera: Apidae). *Entomologia Generalis* 24: 37-47.

CRAILSHEIM, K; RIESSBERGER, U; BLASCHON, B; NOWOGRODZKI, R; HRASSNIGG, N (1999b) Short-term effects of simulated bad weather conditions upon the behaviour of food-storer honey bees during day and night (*Apis mellifera carnica* Pollmann). *Apidologie* 30(4): 299-310. http://dx.doi.org/10.1051/apido:19990406

DE MARCO, R J; FARINA, W M (2001) Changes in food source profitability affect the trophallactic and dance behaviour of forager honey bees (*Apis mellifera* L.). *Behavioural Ecology and Sociobiology* 50: 441-449. http://dx.doi.org/10.1007/s002650100382

DE MARCO, R J; GIL, M; FARINA, W M (2005) Does an increase in reward affect the precision of the encoding of directional information in the honey bee waggle dance? *Journal of Comparative Physiology* A191: 413-419. http://dx.doi.org/10.1007/s00359-005-0602-3

DECOURTYE, A; ARMENGAUD, C; RENOU, M; DEVILLERS, J; CLUZEAU, S; GAUTHIER, M; PHAM-DELEGUE, M H (2004) Imidacloprid impairs memory and brain metabolism in the honey bee (*Apis mellifera* L.). *Pesticide Biochemistry and Physiology* 78: 83-92. http://dx.doi.org/10.1016/j.pestbp.2003.10.001

DECOURTYE, A; DEVILLERS, J; AUPINEL, P; BRUN, F; BAGNIS, C; FOURRIER, J; GAUTHIER, M (2011) Honey bee tracking with microchips: a new methodology to measure the effects of pesticides. *Ecotoxicology London England* 20(2): 429-437. http://dx.doi.org/10.1007/s10646-011-0594-4

DIAZ, P C; GRÜTER, C; FARINA, W M (2007) Floral scents affect the distribution of hive bees around dancers. *Behavioural Ecology and Sociobiology* 61: 1589-1597. http://dx.doi.org/10.1007/s00265-007-0391-5

DOBSON, H E M; GROTH, I; BERGSTROM, G (1996) Pollen advertisement: chemical contrasts between whole-flower and pollen odours. *American Journal of Botany* 83(7): 877-885.

DYER, A G (1996). Reflection of near-ultraviolet radiation from flowers of Australian native plants. *Australian Journal of Botany* 44(4): 473-488.

DYER, F C (1998) Cognitive ecology of navigation. In *Dukas, R (Ed.)*. *Cognitive ecology*. Chicago: University of Chicago Press; USA. pp 201-260.

DYER, F C (2002). The biology of the dance language. *Annual Review of Entomology* 47: 917-949. http://dx.doi.org/10.1146/annurev.ento.47.091201.145306

EDRICH, W (1975) The waggle dance of the honey bee: a new formulation. *Fortschritte der Zoologie* 23: 20-30.

ERBER, J; KIERZEK, S; SANDER, E; GRANDY, K (1998) Tactile learning in the honey bee. *Journal of Comparative Physiology* A183: 737-744.

ERBER, J; HOORMAN, J; SCHEINER, R (2006) Phototactic behaviour correlates with gustatory responsiveness in honey bees (*Apis mellifera* L.). *Behavioural Brain Research* 174: 174-180. http://dx.doi.org/10.1016/j.bbr.2006.07.023

ESCH, H (1961a). Ein neuer Bewegungstyp im Schwänzeltanz der Bienen. *Naturwissenschaften* 48: 140-141.

ESCH, H (1961b) Über die Schallerzeugung beim Werbetanz der Honigbiene. *Zeitschrift für vergleichende Physiologie* 45: 1-11.

ESCH, H E; ZHANG, S; SRINIVASAN, M V; TAUTZ, J (2001) Honey bee dances communicate distances measured by optic flow. *Nature* 411: 581-583. http://dx.doi.org/10.1038/35079072

FAHRENHOLZ, L; LAMPRECHT, L; SCHRICKER, B (1989) Thermal investigations of a honey bee colony: thermoregulation of the hive during summer and winter and heat production of members of different bee castes. *Journal of Comparative Physiology* B159: 551 -560. http://dx.doi.org/10.1007/BF00694379

FARINA, W M (1996). Food-exchange by foragers in the hive – a means of communication among honey bees? *Behavioural Ecology and Sociobiology* 38: 59-64. http://dx.doi.org/10.1007/s002650100382

FARINA, W M; WAINSELBOIM, A J (2001) Changes in the thoracic temperature of honey bees while receiving nectar from foragers collecting at different reward rates. *Journal of Experimental Biology* 204: 1653-1658.

FARINA, W M; GRÜTER, C; DIAZ, P C (2005) Social learning of floral odours inside the honey bee hive. *Proceedings of the Royal Society of London* B272: 1923-1928. http://dx.doi.org/10.1098/rspb.2005.3172

FERNÁNDEZ, P C; GIL, M; FARINA, W M (2003) Reward rate and forager activation in honey bees: recruiting mechanisms and temporal distribution of arrivals. *Behavioural Ecology and Sociobiology* 54: 80-87.
http://dx.doi.org/10.1007/s00265-003-0607-2

FINKENZELLER, K (2002) *RFID-Handbuch: Grundlagen und praktische Anwendungen induktiver Funkanlagen, Transponder und kontaktloser Chipkarten.* Carl Hanser Verlag; München, Germany.

FRIEDRICH, A; THOMAS, U; MÜLLER, U (2004) Learning at different satiation levels reveals parallel functions for the cAMP–protein kinase a cascade in formation of long-term memory. *The Journal of Neuroscience* 24(18): 4460-4468.
http://dx.doi.org/10.1523/JNEUROSCI.0669-04.2004

FRIES, I; CHAUZAT, M-P; CHEN, Y-P; DOUBLET, V; GENERSCH, E; GISDER, S; HIGES, M; MCMAHON, D P; MARTÍN-HERNÁNDEZ, R; NATSOPOULOU, M; PAXTON, R J; TANNER, G; WEBSTER, T C; WILLIAMS, G R (2013) Standard methods for *nosema* research. In *V Dietemann; J D Ellis, P Neumann (Eds) The COLOSS BEEBOOK: Volume II: Standard methods for* Apis mellifera *pest and pathogen research. Journal of Apicultural Research* 52(1):
http://dx.doi.org/10.3896/IBRA.1.52.1.14

FUJIWARA, S; MIURA, H; KUMAGAI, T; SAWAGUCHI, T; NAYA, S; GOTO, K T; SUZUKI, K (1994) Drone congregation of *Apis cerana japonica* in an open area over larger trees *Zelkovia serrata*).
Apidologie 25: 331-337.

GARY, N E (1962) Chemical mating attractants in the queen honey bee. *Science* 136: 773-774.
http://dx.doi.org/10.1126/science.136.3518.773

GARY, N E; MARSTON, J (1971) Mating behaviour of drone honey bees with queen models (*Apis mellifera* L.). *Animal Behaviour* 19: 299-304.

GERBER, B; SMITH, B H (1998) Visual modulation of olfactory learning in honey bees. *The Journal of Experimental Biology* 201: 2213-2217.

GIL, M; FARINA, W M (2002) Foraging reactivation in the honey bee *Apis mellifera* L.: factors affecting the return to a known nectar source. *Naturwissenschaften* 89: 322-325.
http://dx.doi.org/10.1007/s00114-002-0323-1

GIURFA, M (2003) Cognitive neuroethology: dissecting non-elemental learning in a honey bee brain. *Current Opinion in Neurobiology* 13: 726-735. http://dx.doi.org/10.1016/j.conb.2003.10.015

GIURFA, M (2007) Behavioural and neural analysis of associative learning in the honey bee: a taste from the magic well. *Journal of Comparative Physiology* A193: 801-824.
http://dx.doi.org/10.1007/s00359-007-0235-9

GIURFA, M; MALUN, D (2004) Associative mechanosensory conditioning of the proboscis extension reflex in honey bees. *Learning and Memory* 11: 294-302.
http://dx.doi.org/10.1101/lm.63604

GIURFA, M; MENZEL, R (1997) Insect visual perception: complex abilities of simple nervous systems. *Current Opinion in Neurobiology* 7: 505–513. http://dx.doi.org/10.1016/S0959-4388 (97)80030-X

GIURFA, M; HAMMER, M; STACH, S; STOLLHOFF, N; MÜLLER-DEISIG, N; MIZYRYCKI, C (1999) Pattern learning by honey bees: conditioning procedure and recognition strategy. *Animal Behaviour* 57: 315-324.

GIURFA, M; FABRE, E; FLAVEN-POUCHON, J; GROLL, H; OBERWALLNER, B; VERGOZ, V; ROUSSEL, E; SANDOZ, J C (2009) Olfactory conditioning of the sting extension reflex in honey bees: memory dependence on trial number, interstimulus interval, intertrial interval, and protein synthesis. *Learning and Memory* 16: 761-765. http://dx.doi.org/10.1101/lm.1603009

GMEINBAUER, R; CRAILSHEIM, K (1993) Glucose utilisation during flight of honey bee workers, drones and queens. *Journal of Insect Physiology* 39: 959-967.
http://dx.doi.org/10.1016/0022-1910(93)90005-C

GRAHAM, J M; AMBROSE, J T; ATKINS, E L; AVITABILE, A; AYERS, G S; BLUM, M S (1992) *The Hive and the Honey Bee: A new book on beekeeping which continues the tradition of "Langstroth on the Hive and the Honey bee".* Dadant & Sons; Hamilton, Il, USA.

GREGGERS, U; MENZEL, R (1993) Memory dynamics and foraging strategies of honey bees. *Behavioural Ecology and Sociobiology* 32: 17–29.

GRIES, M; KOENIGER, N (1996) Straight forward to the queen: pursuing honey bee drones (*Apis mellifera* L.) adjust their body axis to the direction of the queen. *Journal of Comparative Physiology* A179: 539-545.

GRODZICKI, P; CAPUTA, M (2005) Social versus individual behaviour: a comparative approach to thermal behaviour of the honey bee (*Apis mellifera* L.) and the American cockroach (*Periplaneta americana* L.). *Journal of Insect Physiology* 51: 315 - 322.
http://dx.doi.org/10.1016/j.jinsphys.2005.01.001

GROH, C; TAUTZ, J; RÖSSLER, W (2004) Synaptic organization in the adult honey bee brain is influenced by brood-temperature control during pupal development. *Proceedings of the National Academy of Sciences of the United States of America* 101(12): 4268-4273.
http://dx.doi.org/10.1073/pnas.0400773101

GRÜTER, C; FARINA, W (2009). The honey bee waggle dance: can we follow the steps? *Trends in Ecology and Evolution* 24: 242-247.

GUERRIERI, F; LACHNIT, H; GERBER, B, GIURFA, M (2005) Olfactory blocking and odorant similarity in the honey bee. *Learning and Memory* 12: 86-95. http://dx.doi.org/10.1101/lm.79305

GUEZ, D; SUCHAIL, S; GAUTHIER, M; MALESZKA, R; BELZUNCES, L P (2001) Contrasting effects of imidacloprid on habituation in 7 – and 8 –day-old honey bees (*Apis mellifera* L.). *Neurobiology of Learning and Memory* 76(2): 183-191.
http://dx.doi.org/10.1006/nlme.2000.3995

HAEHNEL, M; MENZEL, R (2012) Long-term memory and response generalization in mushroom body extrinsic neurons in the honey bee *Apis mellifera*. *The Journal of Experimental Biology* 215: 559-565.

HAMMER, M; MENZEL, R (1995) Learning and memory in the honey bee. *The Journal of Neuroscience* 15(3): 1617-1630.

HAMMER, M; BRAUN, G; MAUELSHAGEN, J (1994) Food-induced arousal and nonassociative learning in honey bees – dependence of sensitization on the application site and duration of food stimulation. *Behavioural and Neural Biology* 62(3): 210-223. http://dx.doi.org/10.1016/S0163-1047(05)80019-6

HENRY, M; BEGUIN, M; REQUIER, F; ROLLIN, O; ODOUX, J F; AUPINEL, P; APTEL, J; TCHAMITCHIAN, S; DECOURTYE, A (2012) A common pesticide decreases foraging success and survival in honey bees. *Science* 336 (6079): 348-350. http://dx.doi.org/10.1126/science.1215039

HERAN, H (1952) Untersuchungen über den Temperatursinn der Honigbiene *Apis mellifica* unter besonderer Berücksichtigung der Wahrnehmung strahlender Wärme. *Zeitschrift für vergleichende Physiologie* 34: 179-206. http://dx.doi.org/10.1007/BF00339537

HERTER, K (1924) Untersuchungen über den Temperatursinn einiger Insekten. *Zeitschrift für vergleichende Physiologie* 1: 221-288. http://dx.doi.org/10.1007/BF00338213

HERTZ, M (1935) Die Untersuchungen über den Formensinn der Honigbiene. *Naturwissenschaften* 23: 618-624.

HOLBROOK,C T; BARDEN, P M; FEWELL, J H (2011) Division of labor increases with colony size in the harvester ant *Pogonomyrmex californicus*. *Behavioral Ecology* 22: 960-966. http://dx.doi.org/*10.1093/beheco/arr075*

HÖLLDOBLER, B; WILSON, E O (2008) *The superorganism: the beauty, elegance, and strangeness of insect societies*. W W Norton; New York, USA.

HORRIDGE, A (2000) Seven experiments on pattern vision of the honey bee, with a model. *Vision Research* 40(19): 2589-2603. http://dx.doi.org/10.1016/S0042-6989(00)00096-1

HORRIDGE, A (2009) *What does the honey bee see and how do we know? a critique of scientific reason*. ANU E Press; Canberra, Australia.

HRASSNIGG, N; CRAILSHEIM, K (1999) Metabolic rates and metabolic power of honey bees in tethered flight related to temperature and drag (*Hymenoptera: Apidae*). *Entomolgia Generalis* 24: 23-30.

HRASSNIGG, N; BRODSCHNEIDER, R; FLEISCHMANN, P H; CRAILSHEIM, K (2005) Unlike nectar foragers, honey bee drones (*Apis mellifera*) are not able to utilize starch as fuel for flight. *Apidologie* 36: 547-557. http://dx.doi.org/10.1051/apido:2005042

HRNCIR, M; MAIA-SILVA, C; MC CABE, S I; FARINA, W M (2011). The recruiter's excitement – thorax vibration features of the honey bee's waggle dance related to food source profitability. *Journal of Experimental Biology* 214: 4055-4064. http://dx.doi.org/10.1242/jeb.063149

HUMAN, H; BRODSCHNEIDER, R; DIETEMANN, V; DIVELY, G; ELLIS, J; FORSGREN, E; FRIES, I; HATJINA, F; HU, F-L; JAFFÉ, R; KÖHLER, A; PIRK, C W W; ROSE, R; STRAUSS, U; TANNER, G; TARPY, D R; VAN DER STEEN, J J M; VEJSNÆS, F; WILLIAMS, G R; ZHENG, H-Q (2013) Miscellaneous standard methods for *Apis mellifera* research. In *V Dietemann; J D Ellis; P Neumann (Eds) The COLOSS* BEEBOOK, *Volume I: standard methods for* Apis mellifera *research. Journal of Apicultural Research* 52(4): http://dx.doi.org/10.3896/IBRA.1.52.4.10

IQBAL, J; MUELLER, U (2007) Virus infection causes specific learning deficits in honey bee foragers. *Proceedings of the Royal Society B* 274: 1517-1521. http://dx.doi.org/10.1098/rspb.2007.0022

JANDT, J M; DORNHAUS, A (2009) Spatial organization and division of labour in the bumble bee *Bombus impatiens*. *Animal Behaviour* 77: 641-651. http://dx.doi.org/10.1016/j.anbehav.2008.11.019

JANDT, J M; HUANG, E; DORNHAUS, A (2009) Weak specialization of workers inside a bumble bee (*Bombus impatiens*) nest. *Behavioral Ecology and Sociobiology* 63: 1829-1836. http://dx.doi.org/10.1007/s00265-009-0810-x

JUNGMANN, R; ROTHE, U; NACHTIGALL, W (1989) Flight of the honey bee. I. Thorax surface temperature and thermoregulation during tethered flight. *Journal of Comparative Physiology B* 158(6): 711-718.

JURISCH, R (2001) Miniaturising transponders. *GID Magazine* 11: 50-51.

KERNBACH, S; THENIUS, R; KERNBACH, O; SCHMICKL, T (2009) Re-embodiment of honey bee aggregation behaviour in an artificial micro-robotic swarm. *Adaptive Behaviour* 17: 237-259. http://dx.doi.org/10.1177/1059712309104966

KEVAN, P G; LANE, M A (1985) Flower petal microtexture is a tactile cue for bees. *Ecology* 82: 4750-4752.

KLEINHENZ, M; BUJOK, B; FUCHS, S; TAUTZ, J (2003) Hot bees in empty brood nest cells: heating from within. *Journal of Experimental Biology* 206: 4217-4231. http://dx.doi.org/10.1242/jeb.00680

KOENIGER, G (1981) In which segment of the mating process of the queen bee does the induction oviposition occur? *Apidologie* 12: 329-343. [in German].

KOENIGER, G (1990) The role of mating sign in honey bees, *Apis mellifera* L.: does it hinder or promote multiple mating? *Animal Behaviour* 39: 444-449. http://dx.doi.org/10.1016/S0003-3472(05)80407-5

KOENIGER, N; KOENIGER, G (2000) Reproductive isolation among species of the genus *Apis*. *Apidologie* 31: 313–339.

KOENIGER, N; WIJAYAGUNESEKERA, H N P (1976) Time of drone flight in the three Asian honey bee species (*Apis cerana, Apis florea, Apis dorsata*). *Journal of Apicultural Research* 15: 67–71.

KOENIGER, G; KOENIGER, N; PECHHACKER, H; RUTTNER, F; BERG, S (1989) Assortative mating in a mixed population of European honey bees, *Apis mellifera ligustica* and *Apis mellifera carnica*. *Insectes Sociaux* 36: 129-138.

KOENIGER, G; KOENIGER, N; TINGEK, S (1994) Crossfostered drones of *Apis cerana* (Fabricius, 1793) and *Apis koschevnikovi* (v. Buttel-Reepen, 1906) fly at their species specific mating times. *Insectes Sociaux* 41: 73–78.

KOENIGER, G; KOENIGER, N; PHIANCHAROEN, M (2011) Comparative reproductive biology of honey bees. In *Hepburn, R; Radloff, S (Eds). Honey bees of Asia.* Springer; Germany. pp 159-206. http://dx.doi.org/10.1007/978-3-642-16422-4

KOLMES, S A (1985) A quantitative study of the division of labor among worker honey bees. *Zeitschrift Fur Tierpsychologie-Journal of Comparative Ethology* 68: 287-302. http://dx.doi.org/10.1111/j.1439-0310.1985.tb00130.x

KOLMES, S A; FERGUSSON-KOLMES, L A (1989) Stinging behaviour and residual value of worker honey bees (*Apis mellifera*). *Journal of the New York Entomological Society* 97: 218-231.

KRALJ, J (2004) Parasite - host interactions between *V. destructor* Anderson and Trueman and *Apis mellifera* L.: influence of parasitism on fight behaviour and on the loss of infested foragers. Dissertation, Fachbereich Biologie und Informatik der Johann Wolfgang Goethe Universität Frankfurt am Main, Germany.

KRALJ, J; FUCHS, S (2006) Parasitic *Varroa destructor* mites influence flight duration and homing ability of infested *Apis mellifera* foragers. *Apidologie* 37 5: 577-587. http://dx.doi.org/10.1051/apido:2006040

KRALJ, J; FUCHS, S (2010) *Nosema sp.* influences flight behaviour of infected honey bee (*Apis mellifera*) foragers. *Apidologie* 41: 21-28. http://dx.doi.org/10.1051/apido/2009046

KRALJ, J; BROCKMANN, A; FUCHS, S; TAUTZ, J (2007) The parasitic mite *Varroa destructor* affects non-associative learning in honey bee foragers, *Apis mellifera* L. *Journal of Comparative Physiology* A193: 363-370. http://dx.doi.org/10.1007/s00359-006-0192-8

KRISTON, I (1973) Die Bewertung von Duft- und Farbsignalen als Orientierungshilfen an der Futterquelle durch *Apis mellifera* L. *Jounal of Comparative Physiology* 84: 77–94.

LACHER, V (1964) Elektrophysiologische Untersuchungen an einzelnen Rezeptoren für Geruch, Kohlendioxyd, Luftfeuchtigkeit und Temperatur auf den Antennen der Arbeitsbiene und der Drohne (*Apis mellifica* L.). *Journal of Comparative Physiology* A48(6): 587-623. http://dx.doi.org/10.1007/BF00333743.

LIEBERT, A E; WILSON-RICH, N; JOHNSON, C E; STARKS, P T (2010) Sexual interactions and nestmate recognition in invasive populations of *Polistes dominulus* wasps. *Insectes Sociaux* 57:457-463. http://dx.doi.org/10.1007/s00040-010-0105-2

LINDAUER, M (1952) Ein Beitrag zur Frage der Arbeitsteilung im Bienenstaat. *Zeitschrift für vergleichende Physiologie* 34: 299–345. http://dx.doi.org/10.1007/BF00298048

LINDAUER, M (1954) Temperaturregulierung und Wasserhaushalt im Bienenstaat. *Zeitschrift für vergleichende Physiologie* 36: 391–432. http://dx.doi.org/10.1007/BF00345028

LINDAUER, M (1955) Schwarmbienen auf Wohnungssuche. *Zeitschrift für vergleichende Physiologie* 37: 263–324. http://dx.doi.org/10.1007/BF00303153

LOPATINA, N G (1971) *Signalnaya deyatyelnost v semye myedonosnoy pchely* (Apis mellifera *L.*) [Signal activities in honey bee colonies (*Apis mellifera* L.) (In Russian)]. Nauka; Leningrad, USSR.

MACKINTOSH, N J (1974) *The psychology of animal learning.* Academic Press; New York, USA.

MARTIN, H (1965) Leistungen des topochemischen Sinnes bei der Honigbiene. *Zeitschrift für vergleichende Physiologie* 50: 254-292. http://dx.doi.org/10.1007/BF00339481

MATSUMOTO, Y; MENZEL, R; SANDOZ, J-C; GIURFA, M (2012) Revisiting olfactory classical conditioning of the proboscis extension response in honey bees: A step toward standardized procedures. *Journal of Neuroscience Methods* 211: 159-167. http://dx.doi.org/10.1016/j.jneumeth.2012.08.018

MCNALLY, G P; WESTBROOK, R F (2006) Predicting danger: the nature, consequences, and neural mechanisms of predictive fear learning. *Learning and Memory* 13: 245-253. http://dx.doi.org/10.1101/lm.196606

MEIKLE, W G; RECTOR, B G; MERCADIER, G; HOLST, N (2008) Within-day variation in continuous hive weight data as a measure of honey bee colony activity. *Apidologie* 39: 694-707. http://dx.doi.org/10.1051/apido:2008055

MENZEL, R (1967) Untersuchungen zum Erlernen von Spektralfarben durch die Honigbiene. *Zeitschrift für vergleichende Physiologie* 56: 22–62. http://dx.doi.org/10.1007/BF00333562

MENZEL, R (2001) Searching for the memory trace in a mini-brain, the honey bee. *Learning and Memory* 8: 53-62. http://dx.doi.org/10.1101/lm.38801

MENZEL, R (2012) The honey bee as a model for understanding the basis of cognition. *Nature Reviews Neuroscience* 13: 758-768. http://dx.doi.org/ 10.1038/nrn3357

MENZEL, R; MÜLLER, U (1996) Learning and memory in honey bees: from behaviour to neural substrates. *Annual Review of Neuroscience* 19: 379-404. http://dx.doi.org/ 10.1146/annurev.ne.19.030196.002115

MENZEL, R HAMMER, M BRAUN, G MAUELSHAGEN, J SUGAWAM. (1991) Neurobiology of learning and memory in honey bees. In *Goodman, L J; Fisher, R C (Eds). The behaviour and physiology of bees.* CAB International; Wallingford, Oxon, USA. pp 323-353.

MENZEL, R; GREGGERS, U; HAMMER, M (1993) Functional organization of appetitive learning and memory in a generalist pollinator, the honey bee. In *Papaj, D R; Lewis, A C (Eds). Insect learning. Ecology and evolutionary perspectives.* Chapman & Hall; New York, London. pp 79-125.

MENZEL, R; BRANDT, R; GUMBERT, A; KOMISCHKE, B; KUNZE, J (2000) Two spatial memories for honey bee navigation. *Proceedings of the Royal Society of London* B267: 961-968. http://dx.doi.org/ 10.1098/rspb.2000.109

MENZEL, R; GREGGERS, U; SMITH, A; BERGER, S; BRANDT, R; BRUNKE, S; BUNDROCK, G; HUELSE, S; PLUEMPE, T; SCHAUPP, F; SCHUETTLER, E; STACH, S; STINDT, J; STOLLHOFF, N; WATZL, S (2005) Honey bees navigate according to a map-like spatial memory. *Proceedings of the National Academy of Sciences of the USA* 102(8): 3040-3045. http://dx.doi.org/ 10.1073/pnas.0408550102

MENZEL, R; FUCHS, J; KIRBACH, A; LEHMANN, K; GREGGERS, U (2011a) Navigation and communication in honey bees. In *Galizia, C G; Eisenhardt, D, Giurfa, M, (Eds). Honey bee neurobiology and behaviour. A tribute to Randolf Menzel.* Dordrecht; Heidelberg, Germany. pp 103-116. http://dx.doi.org/10.1007/978-94-007-2099-2

MENZEL, R; KIRBACH, A; HAASS, W D; FISCHER, B; FUCHS, J; KOBHLOFSKY, M; LEHMANN; K; REITER, L; MEYER, H; NGUYEN, H; JONES, S; NORTON, P; GREGGERS, U (2011b) A common frame of reference for learned and communicated vectors in honey bee navigation. *Current Trends in Biology* 21: 645-650. http://dx.doi.org/ 10.1016/j.cub.2011.02.039

MICHELSEN, A (2003). Signals and flexibility in the dance communication of honey bees. *Journal of Comparative Physiology* A189: 165-174. http://dx.doi.org/ 10.1007/s00359-003-0398-y

MICHELSEN, A; TOWNE, W F; KIRCHNER, W H; KRYGER, P (1987). The acoustic near field of a dancing honey bee. *Journal of Comparative Physiology* A161: 633-643. http://dx.doi.org/ 10.1007/BF00605005

MIXSON, T A; ABRAMSON, C I; NOLF, S L; JOHNSON, G A; SERRANO, E; WELLS, H (2009) Effects of GSM cellular phone radiation on the behaviour of honey bees (*Apis mellifera*). *The Science of Bee Culture* 1: 22-27.

MOTA, T; ROUSSEL, E; SANDOZ, J C; GIURFA, M (2011) Visual conditioning of the sting extension reflex in harnessed honey bees. *The Journal of Experimental Biology* 214: 3677-3687. http://dx.doi.org/ 10.1242/jeb.062026

MUJAGIC, S; ERBER, J (2009) Sucrose acceptance, discrimination and proboscis responses of honey bees (*Apis mellifera* L.) in the field and the laboratory. *Journal of Comparative Physiology* A195: 325–339. http://dx.doi.org/ 10.1007/s00359-008-0409-0

MUJAGIC, S; SARKANDER, J; ERBER, B; ERBER, J (2010) Sucrose acceptance and different forms of associative learning of the honey bee (*Apis mellifera* L.) in the field and laboratory. *Frontiers in Behavioural Neuroscience* 4: 46. http://dx.doi.org/10.3389/fnbeh.2010.00046

OHTANI, T (1992) Spatial distribution and age-specific thermal reaction of worker honey bees. *Humans and Nature* 1: 11-25.

OSBORNE, J L; WILLIAMS, I H; CARRECK, N L; POPPY, G M; RILEY, J R; SMITH, A D; REYNOLDS, D R; EDWARDS, A S (1997) Harmonic radar: a new technique for investigating bumble bee and honey bee foraging flight. VII. International Symposium for Pollination ISHS. *Acta Horticulturae* 437: 159-164.

OTIS, G W; KOENIGER, N; RINDERER, T E; HADISOESILO, S; YOSHIDA, T; TINGEK, S; WONGSIRI, S; MARDAN, M B (2001) Comparative mating flight times of Asian honey bees. In *Proceedings of 7th international conference on tropical bees, Chiang Mai, Thailand.* pp 137–141.

PAGE, R E; FONDRK, M K (1995) The effects of colony level selection on the special organization of honey bee (*Apis mellifera* L.) colonies: colony-level components of pollen hoarding. *Behavioural Ecology and Sociobiology* 36: 135-144.

PAGE, R E; ERBER, J; FONDRK, M K (1998) The effect of genotype on response thresholds to sucrose and foraging behaviour of honey bees (*Apis mellifera* L.). *Journal of Comparative Physiology* A182: 489-500.

PANKIW, T; PAGE, R E (1999) The effect of genotype, age, sex and caste on response thresholds to sucrose and foraging behaviour of honey bees (*Apis mellifera* L.). *Journal of Comparative Physiology* A185: 207-213.

PANKIW, T; PAGE, R E (2003) Effect of pheromones, hormones, and handling on sucrose response thresholds of honey bees (*Apis mellifera* L.). *Journal of Comparative Physiology* A189: 675-684. http://dx.doi.org/10.1007/s00359-003-0442-y

PHAM-DELÈGUE, M H; TROUILLER, J; BAKCHINE, E; ROGER, B; MASSON, C (1991) Age dependency of worker bee response to queen pheromone in a four-armed olfactometer. *Insectes Sociaux* 38: 283- 292. http://dx.doi.org/10.1007/BF01314914

PIRK, C W W; DE MIRANDA, J R; FRIES, I; KRAMER, M; PAXTON, R; MURRAY, T; NAZZI, F; SHUTLER, D; VAN DER STEEN, J J M; VAN DOOREMALEN, C (2013) Statistical guidelines for *Apis mellifera* research. In *V Dietemann; J D Ellis; P Neumann (Eds) The COLOSS BEEBOOK, Volume I: standard methods for* Apis mellifera *research. Journal of Apicultural Research* 52(4): http://dx.doi.org/10.3896/IBRA.1.52.4.13

RAMIREZ-ROMERO, R; DESNEUX, N; DECOURTYE, A; CHAFFIOL, A; PHAM-DELEGUE, M H (2007) Does cry1Ab protein affect learning performances of the honey bee *Apis mellifera* L. (Hymenoptera, Apidae)? *Ecotoxicology and Environmental Safety* 70: 327-333. http://dx.doi.org/10.1016/j.ecoenv.2007.12.002

RATNIEKS, F L W; KÄRCHER, M H; FIRTH, V; PARKS, D; RICHARDS, A; RICHARDS, P; HELANTERÄ, H (2011) Acceptance by honey bee guards of non-nestmates is not increased by treatment with nestmate odours. *Ethology* 117: 655-663. http://dx.doi.org/10.1111/j.1439-0310.2011.01918.x

RAVERET-RICHTER, M; WADDINGTON, K D (1993) Past foraging experience influences honey bee dance behaviour. *Animal Behaviour* 46: 123-128. http://dx.doi.org/10.1006/anbe.1993.1167

RIESSBERGER, U; CRAILSHEIM, K (1997) Short-term effect of different weather conditions upon the behaviour of forager and nurse honey bees (*Apis mellifera carnica* Pollmann). *Apidologie* 28 (6): 411–426. http://dx.doi.org/10.1051/apido:19970608

RILEY, J R; SMITH, A D (2002) Design considerations for a harmonic radar to investigate the flight of insects at low altitude. *Computers and Electronics in Agriculture* 35: 151-369. http://dx.doi.org/10.1016/S0168-1699(02)00016-9

RILEY, J R; SMITH, A D; REYNOLDS, D R; EDWARDS, A S; OSBORNE, J L; WILLIAMS, I H; CARRECK, N L; POPPY, G M (1996) Tracking bees with harmonic radar. *Nature* 379: 29-30. http://dx.doi.org/10.1038/379029b0

RILEY, J R, GREGGERS, U; SMITH, A D; REYNOLDS, D R; MENZEL, R (2005) The flight paths of honey bees recruited by the waggle dance. *Nature* 435(7039): 205-207. http://dx.doi.org/10.1038/nature03526

ROBINSON, G E (1987) Regulation of honey bee age polyethism by juvenile-hormone. *Behavioral Ecology and Sociobiology* 20: 329-338. http://dx.doi.org/10.1007/BF00300679

RÖSCH, G A (1925) Untersuchungen über die Arbeitsteilung im Bienenstaat, I. Teil: Die tätigkeiten im normalen Bienenstaate und ihre Beziehungen zum Alter der Arbeitsbienen. *Zeitschrift für vergleichende Physiologie* 2: 571-631.

RÖSCH, G A (1927) Über die Bautätigkeit im Bienenvolk und das alter der Baubienen. *Zeitschrift für vergleichende Physiologie* 6: 264-298.

RÖSCH, G A (1930) Untersuchungen über die Arbeitsteilung im Bienenstaat, II. Teil: Die Tätigkeiten der Arbeitsbienen unter experimentell veränderten Bedingungen. *Zeitschrift für vergleichende Physiologie* 12: 1-71.

RUEPPEL, O; BACHELIER, C; FONDRK, M K; PAGE, R E (2007) Regulation of life history determines lifespan of worker honey bees (*Apis mellifera* L.). *Experimental Gerontology* 42: 1020-1032. http://dx.doi.org/10.1016/j.exger.2007.06.002

RUTTNER, F; RUTTNER, H (1965) Untersuchungen ueber die Flugaktivitaet und das Paarungsverhalten der Drohnen. II. Beobachtungen an Drohnensammelplaetzen. *Zeitschrift für Bienenforschung* 8: 1–9. [in German].

SAKAGAMI S F (1953) Untersuchungen über die Arbeitsteilung in einem Zwergvolk der Honigbiene. *Japanese Journal of Zoology* 11: 117-185.

SASAKI, T; PRATT, S C (2011) Emergence of group rationality from irrational individuals in ants. *Integrative and Comparative Biology* 51: E122-E122. http://dx.doi.org/10.1093/beheco/arq198

SANDOZ, J C; ROGER, B; PHAM-DELEGUE, M H (1995) Olfactory learning and memory in the honey bee: comparison of different classical conditioning procedures of the proboscis extension response. *Comptes rendus de l'Académie de Sciences Paris, Sciences de la vie/Life sciences* 318: 749-755.

SANDOZ, J C; HAMMER, M; MENZEL, R (2002) Side-specificity of olfactory learning in the honey bee: US input side. *Learning and Memory* 9: 337-348. http://dx.doi.org/10.1101/lm.50502

SCHEINER, R (2004) Responsiveness to sucrose and habituation of the proboscis extension response in honey bees. *Journal of Comparative Physiology* A190: 727-733. http://dx.doi.org/10.1007/s00359-004-0531-6

SCHEINER, R; AMDAM, G V (2009) Impaired tactile learning is related to social role in honey bees. *The Journal of Experimental Biology* 212: 994-1002. http://dx.doi.org/10.1242/jeb.021188

SCHEINER, R; ERBER, J; PAGE, R E (1999) Tactile learning and the individual evaluation of the reward in honey bees (*Apismellifera* L.). *Journal of Comparative Physiology* A185: 1-10. http://dx.doi.org/10.1007/s003590050360

SCHEINER, R; PAGE, R E; ERBER, J (2001a) Responsiveness to sucrose affects tactile and olfactory learning in preforaging honey bees of two genetic strains. *Behavioural Brain Research* 120: 67-73. http://dx.doi.org/10.1016/S0166-4328(00)00359-4

SCHEINER, R; PAGE, R E; ERBER, J (2001b) The effects of genotype, foraging role, and sucrose responsiveness on the tactile learning performance of honey bees (*Apis mellifera* L.). *Neurobiology of Learning and Memory* 76: 138-150. http://dx.doi.org/10.1006/nlme.2000.3996

SCHEINER, R; PLÜCKHAHN, S; ÖNEY, B; BLENAU, W; ERBER, J (2002) Behavioural pharmacology of octopamine, tyramine and dopamine in honey bees. *Behavioural Brain Research* 136: 545-553. http://dx.doi.org/ 10.1016/S0166-4328(02)00205-X

SCHEINER, R; BARNERT, M; ERBER, J (2003) Variation in water and sucrose responsiveness during the foraging season affects proboscis extension learning in honey bees. *Apidologie* 34: 67-72. http://dx.doi.org/10.1051/apido:2002050

SCHEINER, R; PAGE, R E; ERBER, J (2004) Sucrose responsiveness and behavioural plasticity in honey bees (*Apis mellifera*). *Apidologie* 35: 133-142. http://dx.doi.org/10.1051/apido:2004001

SCHEINER, R; KURITZ-KAISER, A; MENZEL, R; ERBER, J (2005) Sensory responsiveness and the effects of equal subjective rewards on tactile learning and memory of honey bees, *Learning and Memory* 12: 626-635. http://dx.doi.org/10.1101/lm.98105

SCHMICKL, T; HAMANN, H (2011) BEECLUST: A swarm algorithm derived from honey bees. In *Xiao, Y; Hu F (Eds). Bio-inspired computing and communication networks.* Routledge; UK. pp 95-137.

SCHNEIDER, C W; TAUTZ, J; GRÜNEWALD, B; FUCHS, S (2012) RFID tracking of sublethal effects of two neonicotinoid insecticides on the foraging behaviour of *Apis mellifera*. *PLoS ONE* 7(1): e30023. http://dx.doi.org/10.1371/journal.pone.0030023

SCHNEIDER, S A; SCHRADER, C; WAGNER, A E; BOESCH-SAADATMANI, C; LIEBIG, J; RIMBACH, G; ROEDER, T (2011) Stress resistance and longevity are not directly linked to levels of enzymatic antioxidants in the Ponerine ant *Harpegnathos saltator*. *PLoS ONE* 6:e14601. http://dx.doi.org/10.1371/journal.pone.0014601

SEELEY, T D (1982) Adaptive significance of the age polyethism schedule in honey bee colonies. *Behavioural Ecology and Sociobiology* 11: 287-293. http://dx.doi.org/10.1007/BF00299306

SEELEY, T D (1986) Social foraging by honey bees: how colonies allocate foragers among patches of flowers. *Behavioural Ecology and Sociobiology* 19: 343-354. http://dx.doi.org/10.1007/BF00295707

SEELEY, T D (1989) Social foraging in honey bee: how nectar foragers assess their colony's nutritional status. *Behavioural Ecology and Sociobiology* 24: 181–199. http://dx.doi.org/ 10.1007/BF00292101

SEELEY, T D (1995) *The wisdom of the hive : the social physiology of honey bee colonies.* Harvard University Press; Cambridge, MA, USA. http://dx.doi.org/ 9780674043404

SEELEY, T D; KOLMES, S A (1991) Age poleythism for hive duties in honey bees - illusion or reality? *Ethology* 87: 284-297. http://dx.doi.org/ 10.1111/j.1439-0310.1991.tb00253.x

SEELEY, T D; CAMAZINE, S; SNEYD, J (1991) Collective decision-making in honey bees: how colonies choose among nectar sources. *Behavioural Ecology and Sociobiology* 28: 277-290. http://dx.doi.org/10.1007/BF00175101

SEELEY, T D; MIKHEYEV, A S; PAGANO, G J (2000) Dancing bees tune both duration and rate of waggle-run production in relation to nectar-source profitability. *Journal of Comparative Physiology* A186: 813-819. http://dx.doi.org/10.1007/s003590000134

SIEGEL, A J; HUI, J; JOHNSON, R N; STARKS, P T (2005) Honey bee workers as mobile insulating units. *Insectes Sociaux* 52: 242-246. http://dx.doi.org/10.1007/s00040-005-0805-1

SMITH B H; ABRAMSON C I; TOBIN, T R (1991) Conditioned withholding of proboscis extension in honey bees (*Apis mellifera*) during discriminative punishment. *Journal of Comparative Psychology* 105: 345-356. http://dx.doi.org/10.1037/0735-7036.105.4.345

SOTAVALTA, O (1954) On fuel consumption of the honey bee (*Apis mellifica* L.) in flight experiments. *Annales Zoologici Societatis Zoologicae Botanicae Fennicae Vanamo* 16: 1-22.

SRINIVASAN, M V (2010) Honey bees as a model for vision, perception, and cognition. *Annual Review of Entomology* 55: 267-284. http://dx.doi.org/ 10.1146/annurev.ento.010908.164537

STABENHEIMER, A; HAGMÜLLER, K (1991) Sweet food means hot dancing in honey bees. *Naturwissenschaften* 78: 471-473. http://dx.doi.org/ 10.1007/BF01134389

STABENTHEINER, A; PRESSL, H; PAPST, T; HRASSNIGG, N; CRAILSHEIM, K (2003) Endothermic heat production in honey bee winter clusters. *Journal of Experimental Biology* 206: 353-358. http://dx.doi.org/ 10.1242/jeb.00082

STABENTHEINER, A; KOVAC, H; BRODSCHNEIDER, R (2010) Honey bee colony thermoregulation - Regulatory mechanisms and contribution of individuals in dependence on age, location and thermal stress. *PLoS ONE* 5: e8967. http://dx.doi.org/ 10.1371/journal.pone.0008967

STARKS, P T; JOHNSON, R N; SIEGEL, A J; DECELLE, M M (2005) Heat-shielding: A task for youngsters. *Behavioral Ecology* 16: 128-132. http://dx.doi.org/*10.1093/beheco/arh124*

STONE, J C; ABRAMSON, C I; PRICE, J M (1997) Task-dependent effects of dicofol (kelthane) on learning in the honey bee (*Apis mellifera*). *Bulletin of Environmental Contamination and Toxicology* 58: 177-183. http://dx.doi.org/10.1007/s001289900317

STREIT, S; BOCK, F; PIRK, C W W; TAUTZ, J (2003) Automatic life-long monitoring of individual insect behaviour now possible. *Zoology* 106(3): 169-171. http://dx.doi.org/10.1078/0944-2006-00113

STRUYE, M H (2001) *Methodology. Possibilities and limitations of monitoring the flight activity of honey bees by means of BeeSCAN bee counters.* INRA; Paris, France.

STRUYE, M H; MORTIER, H J; ARNOLD, G; MINIGGIO, C; BORNECK, R (1994) Microprocessor-controlled monitoring of honey bee flight activity at the hive entrance. *Apidologie* 25(4): 384-395. http://dx.doi.org/10.1051/apido:19940405

SUMPTER D J T; BROOMHEAD D S (2000) Shape and dynamics of thermoregulating honey bee clusters. *Journal of Theoretical Biology* 204: 1-14. http://dx.doi.org/10.1006/jtbi.1999.1063

TAKEDA, K (1961) Classical conditioned response in the honey bee. *Journal of Insect Physiology* 6: 168-179. http://dx.doi.org/10.1016/0022-1910(61)90060-9

TAUTZ, J; ROHRSEITZ, K; DANDEMAN, D C (1996) One-strided waggle dance in bees. *Nature* 382: 32. http://dx.doi.org/ 10.1038/382032a0

TAUTZ, J; GROH, C; RÖSSLER, W; BROCKMANN, A (2003) Behavioural performance in adult honey bees is influenced by the temperature experienced during their pupal development. *Proceedings of the National Academy of Sciences of the United States of America* 100(12): 7343-7347. http://dx.doi.org/10.1073/pnas.1232346100

TAYLOR, K S; WALLER, G D; CROWDER, L A (1987) Impairment of classical conditioned response of the honey bee (*Apis mellifera* L.) by sublethal doses of synthetic pyrethroid insecticides. *Apidologie* 18: 243-252. http://dx.doi.org/10.1051/apido:19870304

THOM, C; GILLEY, D C; HOPPER, J; ESCH, H E (2007) The scent of the waggle dance. *PLoS Biology* 5: 1862-1867. http://dx.doi.org/10.1371/journal.pbio.0050228

TOLMAN, E C (1948) Cognitive maps in rats and men. *Psychological Revue* 55: 189-208. http://dx.doi.org 10.1037/h0061626

TOWNE, W F; GOULD, J L (1988) The spatial precision of the honey bee dance communication. *Journal of Insect Behaviour* 1(2): 129–156. http://dx.doi.org/10.1007/BF01052234

TSURUDA, J M; PAGE, R E (2009) The effects of foraging role and genotype on light and sucrose responsiveness in honey bees (*Apis mellifera* L.). *Behavioural Brain Research* 205: 132-137. http://dx.doi.org/10.1016/j.bbr.2009.07.022

VALLET, A M; COLE, J A (1993) The perception of small objects by the drone honey bee. *Journal of Comparative Physiology* A172: 183–188. http://dx.doi.org/10.1007/BF00189395

VERGOZ, V; ROUSEL, E; SANDOZ, J C; GIURFA, M (2007) Aversive learning in honey bees revealed by the olfactory conditioning of the sting extension reflex. *PLoS ONE* 2(3): e288. http://dx.doi.org/10.1371/journal.pone.0000288

VON FRISCH, K (1919) Ueber den Geruchsinn der Biene und seine blütenbiologische Bedeutung. *Zoologisches Jahrbuch* 37: 2–238.

VON FRISCH, K (1927) Versuche über den Geschmacksinn der Bienen, *Naturwissenschaften* 14: 1–20.

VON FRISCH, K (1950) *Bees: their vision, chemical senses and language.* Cornell University Press; Ithica, NY, USA.

VON FRISCH, K (1967) *The dance language and orientation of bees.* Harvard University Press; Cambridge, UK.

VOSKRESENSKAJA, A K; LOPATINA, N G (1953) Interrelationships between food and defensive conditioning in controlling the flight activity of honey bees [in Russian]. *Trudy Instituta Fiziologii imeni I. P. Pavlova Akademii Nauk SSR*, 2: 542-561.

WADDINGTON, K D (1982) Honey bee foraging profitability and round dance correlates. *Journal of Comparative Physiology* A148: 297-301. http://dx.doi.org/10.1007/BF00679014

WALLER, G D; LOPER, G M; BERDEL, R L (1973) A bioassay for determining bee responses to flower volatiles. *Environmental Entomology* 2(2): 255–259.

WANG, Y; KAFTANOGLU, O; SIEGEL, A J; PAGE, R E; AMDAM, G V (2010) Surgically increased ovarian mass in the honey bee confirms link between reproductive physiology and worker behavior. *Journal of Insect Physiology* 56: 1816-1824. http://dx.doi.org/10.1016/j.jinsphys.2010.07.013

WEHNER, R; MENZEL, R (1990) Do insects have cognitive maps? *Annual Revue of Neuroscience* 13: 403-414. http://dx.doi.org/10.1146/annurev.ne.13.030190.002155

WELLS, P H (1973) Honey bees. In *Corning, W C; Dyal, J A; Willows, A O D (Eds). Invertebrate learning Vol 2: arthropods and gastropod molluscs.* Plenum Press; New York, USA. pp 173-185. http://dx.doi.org/10.1016/0160-9327(75)90153-2

WELLS, P H; WENNER, A M (1973) Do bees have a language? *Nature* 241: 171–174. http://dx.doi.org/10.1038/241171a0

WENNER, A M (1962) Sound production during the waggle dance of the honey bee. *Animal Behaviour* 10: 79-95.

WILLIAMS, J L (1987) Wind-directed pheromone trap for drone honey bees (Hymenoptera: Apidae) *Journal of Economic Entomology* 80: 532-536.

WILLIAMS, G R; ALAUX, C; COSTA, C; CSÁKI, T; DOUBLET, V; EISENHARDT, D; FRIES, I; KUHN, R; MCMAHON, D P; MEDRZYCKI, P; MURRAY, T E; NATSOPOULOU, M E; NEUMANN, P; OLIVER, R; PAXTON, R J; PERNAL, S F; SHUTLER, D; TANNER, G; VAN DER STEEN, J J M; BRODSCHNEIDER, R (2013) Standard methods for maintaining adult *Apis mellifera* in cages under *in vitro* laboratory conditions. In *V Dietemann; J D Ellis; P Neumann (Eds) The COLOSS BEEBOOK, Volume I: standard methods for* Apis mellifera *research. Journal of Apicultural Research* 52(1): http://dx.doi.org/10.3896/IBRA.1.52.1.04

WINSTON, M L (1987) *The biology of the honey bee.* Harvard University Press; Cambridge, MA, USA.

WINSTON, M L; PUNNETT, E N (1982) Factors determining temporal division of Labor in honey bees. *Canadian Journal of Zoology* 60: 2947-2952. http://dx.doi.org/10.1139/z82-372

Journal of Apicultural Research 52(1): (2013)
DOI 10.3896/IBRA.1.52.1.02

© IBRA 2013

REVIEW ARTICLE

Standard methods for cell cultures in *Apis mellifera* research

Elke Genersch[1]*, Sebastian Gisder[1], Kati Hedtke[1], Wayne B Hunter[2], Nadine Möckel[1] and Uli Müller[3]

[1]Institute for Bee Research, Friedrich-Engels-Str. 32, 16540 Hohen Neuendorf, Germany.
[2]USDA, ARS, US Horticultural Research Lab, 2001 South Rock Road, Fort Pierce, FL 34945, USA.
[3]ZHMB (Center of Human and Molecular Biology), Dept. 8.3 Biosciences Zoology/Physiology-Neurobiology, Saarland University, D-66041 Saarbrücken, Germany.

Received 30 April 2012, accepted subject to revision 27 June 2012, accepted for publication 30 August 2012.

*Corresponding author: Email: elke.genersch@rz.hu-berlin.de

Summary

Cell culture techniques are indispensable in most, if not all life science disciplines to date. Wherever appropriate cell culture models are lacking, scientific development is hampered. Unfortunately this has been and still is the case in honey bee research, because permanent honey bee cell lines have not so far been established. To overcome this hurdle, protocols for the cultivation of primary honey bee cells and of non-permanent honey bee cell lines have been developed. In addition, heterologous cell culture models for honey bee pathogens based on non-*Apis* insect cell lines have recently been developed. To further advance this progress and to encourage bee scientists to enter the field of cell biology based research, here we present protocols for the cultivation of honey bee primary cells and non-permanent cell lines, as well as hints for the cultivation of permanent insect cell lines suitable for honey bee research.

Métodos estándar para cultivos celulares en *Apis mellifera*

Resumen

Las técnicas de cultivo celular son indispensables en la mayoría, si no en todas las disciplinas de ciencias de la vida hasta la fecha. Siempre que se carezca de modelos de cultivo celular apropiados, el desarrollo científico se ve obstaculizado. Desafortunadamente, esto ha sido y todavía es el caso de la investigación en la abeja de la miel, ya que hasta ahora, no se han establecido líneas celulares permanentes de abeja de la miel. Para superar este obstáculo, se han desarrollado protocolos para el cultivo de células primarias de la abeja de la miel y líneas celulares no permanentes de abejas. Además, también se han desarrollado recientemente modelos heterólogos de cultivo celular para patógenos de las abejas melíferas basados en líneas celulares de insectos que no pertenecen al género *Apis*. Para avanzar en este progreso y alentar a los científicos apícolas a entrar en el campo de la investigación basada en la biología celular, presentamos aquí los protocolos para el cultivo de células primarias de abejas y líneas celulares no permanentes, así como consejos para el cultivo de líneas celulares de insectos permanentes adecuadas para la investigación en la abeja de la miel.

西方蜜蜂细胞培养的标准方法

细胞培养技术在大多数生命科学研究中都是必不可少的。如果缺乏合适的细胞培养模型，相应学科的发展将受到阻碍。遗憾的是，这种情况在蜜蜂研究中一直存在，目前为止还未建立持久的蜜蜂细胞系。为了克服这种障碍，我们建立了初级蜜蜂细胞系和非持久蜜蜂细胞系的培养程序。此外，最近还建立了基于非蜂属昆虫细胞系的异种细胞培养模型。为了进一步推进研究发展、鼓励蜂学研究者进入细胞生物学方面的研究，在此列出了蜜蜂初级细胞系和非持久细胞系的培养程序，并给出了适于蜜蜂研究的持久昆虫细胞系培养方面的建议。

Keywords: honey bee cell, primary cell, non-permanent cell line, insect cell line, permanent cell line, cell culture, tissue culture, COLOSS, *BEEBOOK*

Footnote: Please cite this paper as: GENERSCH, E; GISDER, S; HEDTKE, K; HUNTER, W B; MÖCKEL, N; MÜLLER, U (2012) Standard methods for cell cultures in *Apis mellifera*. In *V Dietemann; J D Ellis; P Neumann (Eds) The COLOSS BEEBOOK, Volume I: standard methods for Apis mellifera research. Journal of Apicultural Research* 52(1) http://dx.doi.org/10.3896/IBRA.1.52.1.02

1. Introduction

In the beginning of the twentieth century, a new area of research began to develop. With the cultivation of tissues (explants) followed by the cultivation of primary cells and later by the development of immortalized, permanent cell lines Cell Culture / Cell Biology became a discipline of its own. These early works already included the use of invertebrates, and even Hymenoptera, when muscle explants from *Vespa* were cultivated to perform polarization optical experiments with reflected light (Pfeiffer, 1941). Since then reams of vertebrate and invertebrate permanent cell lines have been developed, and many of them are now commercially available from different cell culture collections. There also has been, and continues to be, a growing body of published work on the development and use of hymenopteran tissue and cell cultures (Giauffret, 1971; Kaatz *et al.*, 1985; Greany, 1986; Ferkovich *et al.*, 1994; Rocher *et al.*, 2004).

The lack of immortalized cell lines especially for honey bees, *Apis mellifera* L., continues, however, to be a major limiting factor of many studies trying to examine physiology and disease. Current studies which use bee cell cultures have thus relied on primary cultures or on non-permanent cell lines of low passage number (Lynn, 2001; Bergem *et al.*, 2006; Barbara *et al.*, 2008; Chan *et al.*, 2010; Hunter, 2010; Poppinga *et al.*, 2012). One of the main drawbacks of such primary cell cultures and non-permanent cell lines is that they are usually produced within the laboratory of origin and are thus not made available for widespread use by other researchers. They might also present problems with reproducibility. Even so, such bee cell cultures are and will be useful for e.g., examining bee cell physiology, host cell -pathogen interactions, or the effects of various chemicals on gene and protein expression in bee cells using modern technology and approaches like genomics or transcriptomics.

In contrast to these cell culture approaches, immortalized, permanent cell lines like those established from many non-hymenopteran insects (mainly Lepidoptera and Diptera) or vertebrates (mainly mammals) have several important advantages. They provide an excellent system to study cellular events, such as gene expression, DNA replication, pathogen interactions and more. They also provide a reproducible system which can be shared and replicated in many laboratories. Thus, as long as a permanent honey bee cell line is not available, insect cell cultures other than *Apis mellifera* should be screened for their suitability to examine aspects of honey bee biology and pathology. The usefulness of such an approach has been proven recently, when the first heterologous cell culture model for a honey bee pathogen has been reported (Gisder *et al.*, 2011). The cell line IPL-LD65Y, a permanent lepidopteran cell line established from the gypsy moth, *Lymantria dispar*, was shown to be susceptible to infection by honey bee pathogenic microsporidia (*N. apis* and *N. ceranae*) and to support the entire life cycle of *Nosema* spp. in cell culture providing a new model of microsporidiosis (Troemel, 2011).

To further cell culture based experiments in bee research, we here present protocols for both the isolation and cultivation of non-permanent honey bee cells and the cultivation of permanent insect cell lines proven to be suitable for honey bee research, most of them established from Lepidoptera. We hope that these protocols will foster progress in the development of techniques for the isolation and cultivation of permanent honey bee cell lines.

2. Working with non-permanent honey bee cells

2.1. Isolation and cultivation of primary neuronal cells

Protocols for the cultivation of honey bee neuronal cells were developed about twenty years ago (Gascuel *et al.*, 1991; Kreissl and Bicker, 1992; Devaud *et al.*, 1994; Gascuel *et al.*, 1994;). The original purpose at that time was to complement *in vivo* studies on the insect olfactory system with data from *in vitro* cell culture experiments. Unfortunately, these protocols did not find their way from bee neuroscience into bee pathology until recently, when they were adopted for cultivation of several cell types originating from pupal or adult brain and gut (Möckel *et al.*, 2009; Poppinga *et al.*, 2012). Although these primary cells proved to be useful, they have their limitations. Most primary cells stay viable only for a limited time period or, if it is possible to split the culture, they can be passaged several times only as a non-permanent cell line before they die. During these passages most cells change their characteristics to adapt to the artificial environment, which sometimes creates problems with reproducibility of results.

2.1.1 Protocol for pupal cells

For the isolation of neuronal cells from pupae:

1. Collect 13-14 day old pupae (red-eyed pupae, see *BEEBOOK* paper on miscellaneous research methods (Human *et al.*, 2013) for the method to obtain them).
2. Remove pupae carefully from brood cell with forceps. Make sure that the head is not even slightly turned and that the neck is not stretched.
3. Separate the head from thorax by means of a scalpel and pin it down on a wax, paraffin or silicon coated petri dish (35 mm in diameter) with micro pins near the antennae.
4. Make a cut axial around the head by means of a scissor. Start on one mandible, go over the backside with the developing ocelles and finish on the second mandible. Make sure to cut not only the cuticula of the pupae but also the head capsule underneath.
5. Cover the head with L15 medium (Table 1).
6. Remove the complete head capsule carefully without the brain.
7. Separate the brain from the head.
8. Remove the visible neurons (optical lobes) from the brain as well

Table 1. Recipes for media used for cultivation of primary neuronal and gut honey bee cells as well as non-permanent honey bee cell lines.

L 15 medium, pH 7.2	14.9 g L-15 powder, 4.0 g glucose, 2.5 g fructose, 3.3 g prolin, 30 g sucrose, dissolve in bi-distilled water and fill-up to 1000 ml with bi-distilled water; adjust pH 7.2 with NaOH
BM 3 medium, pH 6.7	1000 ml L 15 medium, 0.75 g Pipes, 30 ml FCS (heat inactivated), 12 g Yeastolate
AmWH5 medium (Hunter, 2010)	500 ml Grace's insect medium (supplemented), 500 ml Schneider's insect medium, 1000 ml 0.06 M L-histidine hydrochloride monohydrate (pH 6.5), 20 ml M199 medium (10X) with Hank's salts, 34 ml medium CMRL 1066, 66 ml Hank's balanced salts (1X), 52 ml 2 N glucose solution (filter sterilized, adjust osmolarity), 108 ml foetal bovine serum (FBS, heat inactivated) Add to final volume of medium (2280 ml): 3 ml L-glutamine (100X), 3 ml MEM (50X) amino acid solution, 3 ml gentamycin (10,000 U/ml), 5 ml PenStrep (100X) Note: 0.05 M HEPES buffer (pH 6.5) works as substitute for L-histidine monohydrate; final osmolarity is about 380 mOsm/l; can use Grace's insect medium as primary medium, with no Schneider's medium
ringer solution, Ca-free, pH 7.2 (147 mM NaCl, 5 mM KCl, 65 mM HEPES)	8.6 g NaCl, 0.36 g KCl, 15.6 g HEPES; dissolve in bi-distilled water and fill-up to 1000 ml with bi-distilled water; adjust pH 7.2 with NaOH
1 X PBS, pH 7.4 (137 mM NaCl, 2.7 mM KCl, 10 mM Na_2HPO_4, 2 mM KH_2PO_4)	8 g NaCl, 0.2 g KCl, 1.15 g Na_2HPO_4, 0.2 g KH_2PO_4; dissolve in bi-distilled water and fill-up to 1000 ml with bi-distilled water; pH will be between 7.2-7.6

as the ocelles and the eye-retina, otherwise growing of neurons will be suppressed.

9. Prepare a 24-well-plate (sterile tissue culture quality) with fresh cold (4°C – 10°C) L15 medium.

10. Choose the parts of the brain that will be used for cell culture and transfer them to a medium-filled well of the 24-well plate (see step 9).

11. Prepare as many brains as needed (a minimum of five is recommended to obtain enough cells) following the above outlined procedure.

12. After collecting enough brains, transfer them to a 1.5 ml reaction tube with calcium-free ringer solution (Table 1).

13. Incubate the brains for 10 min in the ringer solution.

14. Aspirate the ringer solution and add cold L15 medium (for 5 brains add 1000 µl L15).

15. Carefully resuspend the brains with a pipette (1 ml pipette tip) in the L15 medium to disintegrate the tissue.

16. Transfer the cell suspension to a poly-L-lysine coated cell culture plate ($10cm^2$, commercially available from several suppliers).

17. Let the cells attach for 20 min.

18. Carefully add 4 volumes of pre-warmed (27°C) BM3 medium (Table 1) supplemented with 10% antibiotic/antimycotic solution, pH 6.7 (Sigma-Aldrich, A5955).

19. Cultivate the cells at 27°C in an incubator suitable for insect cell culture [cooling incubator]; avoid desiccation of the cells by placing water filled bowls into the incubator.

20. If the medium becomes viscous, change the BM3 medium after a week.

The cells are vital for a minimum of 14 days.

2.1.2. Protocol for adult cells

For isolation of neurons from adult animals (age 1-3 days, see *BEEBOOK* paper on miscellaneous research methods (Human *et al.*, 2013) for the method to obtain them) the procedure follows a slightly modified protocol.

1. Collect the brain parts of interest, e.g., mushroom bodies, antennal or optical lobes (it is recommended to take at least 5 animals to obtain enough cells).

2. Incubate in accutase (PAA, #L11-007) for 30 min at RT. Using collagenase/dispase (Roche, 10269638001) (1mg/ml calcium-free ringer solution (Table 1)) for 30 min is also possible. However, this requires pre-tests to determine the temperature for optimal results (≈30-36°C).

3. Carefully resuspend the brain tissue 5 times with a pipette (1 ml pipette tip).

4. Incubate about 15-20 sec to allow for sedimentation of the neuropil parts.

5. Transfer the supernatant with the neurons into a 1.5 ml reaction tube.

6. Centrifuge at 1,100 rcf for 3 min.

7. Discard the accutase-supernatant and add calcium-free ringer solution.

8. Centrifuge at 1,100 rcf for 3 min.

9. Discard the supernatant.

10. Resuspend the cells in L15 medium (Table 1).

11. Transfer the suspension to the poly-L-lysine coated culture plates as described above (see 2.1.1, step 16).

2.2. Isolation and cultivation of primary gut cells

The gut epithelium provides a barrier or a first line defence against many honey bee pathogenic viruses, bacteria, and fungi (including microsporidia). It is therefore among the first tissues to be attacked and infected by several honey bee pathogens. Studying these interactions at the cellular level is best accomplished by using the appropriate target cells, which are gut epithelial cells. Hence, protocols for the cultivation of gut cells have been urgently needed. We here provide such a protocol recently developed for studying *Paenibacillus larvae* interactions with midgut cells (Poppinga *et al.*, 2012).

2.2.1. Protocol for primary gut cells

1. Briefly immerse 10 day old pupae (see *BEEBOOK* paper on miscellaneous research methods (Human *et al.*, 2013) for the method to obtain them) in 3% H_2O_2 for surface sterilization.
2. Wash pupae with 1 X PBS (Table 1).
3. Cut off heads.
4. Fix thorax on a petri dish (35 mm in diameter) with wax, paraffin or silicon coated dish.
5. Cut proximal abdomen lateral and dorsal.
6. Open abdomen carefully.
7. Carefully add cold L15 medium (Table 1) supplemented with 10% antibiotic/antimycotic solution (Sigma-Aldrich, A5955) to the opened abdomen.
8. Prepare a 24-well-plate with several wells filled with cold (4°C – 10°C) L15 medium.
9. Extract gut and place it in a medium-filled well of the 24-well plate(see step 8).
10. Prepare several guts following the above outlined procedure and place up to 10 guts into one well.
11. Remove medium carefully.
12. Add 1 ml of enzyme solution (L15 medium (Table 1), 0.05% trypsin (Invitrogen, 15400054) and 0.5% collagenase/dispase (Roche, 10269638001)) to disintegrate the tissue.
13. Incubate plate with gentle shaking at 4°C for 1 hour.
14. Incubate plate with gentle shaking at 30°C for 1 hour.
15. Incubate plate with gentle shaking at 4°C for 1 hour.
16. Transfer gut/cell-suspension to a 1.5 ml-reaction tube.
17. Centrifuge for 3 min with 300 rcf (Eppendorf 5415 R).
18. Remove supernatant and gently resuspend the pellet in L15 medium (40 µl per gut) to dissociate the cells.
19. Dispense 40 µl of cell suspension per well in a 96-well plate or per well of a chamber slide (8 well glass slide, VWR).
20. Incubate 20 min at 33°C to allow cell attachment.
21. Add 60 µl pre-warmed (37°C) BM3 medium (Table 1). supplemented with 10% antibiotic/antimycotic solution per well.
22. Incubate in an incubator suitable for insect cell culture [cooling incubator] for 24 h at 33°C.
23. Discard medium.
24. Add 100 µl fresh, pre warmed BM3 medium.

   Cells remain vital for several weeks or even months.

2.3. Isolation and cultivation of non-permanent cell lines

Recently, considerable progress has been made in the development of techniques for the isolation and cultivation of non-permanent honey bee cell lines. These cell lines have some advantages over primary cells because they can be passaged at least once and, therefore, can be cultivated for a longer time than primary cells. All life stages, from eggs to adult, and various tissues, appear to be capable of producing primary cell cultures, with isolations from bee brains (Goldberg *et al.*, 1999), antennae (Barbara *et al.*, 2008) embryos (Chan *et al.*, 2010), haemolymph (VanSteenkiste 1988; Sorescu *et al.*, 2003) fat bodies (Kaatz *et al.*, 1985; Hunter, 2010), with the most successful reports supporting use of 4-9 day old developing larvae (Sorescu *et al.*, 2003; Rocher *et al.*, 2004, Hunter, 2010). Even so, bee cell proliferation in culture is generally slow. Addition of foetal calf serum (FCS) at a concentration of about 5–20%, or various amounts of haemolymph or pollen do not appear to affect cell proliferation or rate of growth. Cells normally do not show any signs of differentiation over time, thus cell passages reported are few, up to 5 times over several months (3-8 months cultivation). Larvae with developing head capsule, white eyes, along with the light brown eye, early stage pupae appeared to produce more active cell cultures with diverse cell types, especially when a special culture medium AmWH5 (Table 1) developed for the establishment of non-permanent cell-lines from honey bee tissues was used (Hunter, 2010). An example of such a non-permanent honey bee cell line established from white-eyed pupae is shown in Fig. 1.

2.3.1. Protocol for preparing sterilized tissues

1. Submerge sample (eggs, larvae, pupae, adult) in 0.2% bleach for 3 min.
2. Rinse 3 times, with filter sterilized water, 1 minute each.
3. Rinse 3 times with 70% ethanol, 3-4 min each.
4. In sterile hood, remove ethanol.
5. Rinse twice with sterile water, 1 min each.
6. Remove water.
7. Place sample on sterile surface, i.e. top of tissue plate lid.

→ If eggs are used: place in depression well of sterilized slide or plate, i.e. black porcelain makes easier to see.

8. Add one drop of medium AmWH5 (Table 1).
9. Use sterile glass rod to gently crush eggs 2-3 taps per egg.
10. Add more medium AmWH5.
11. Using Pasteur glass pipette suck up medium AmWH5 with tissues.
12. Dispense into one well of a 24 multi-well tissue culture plate (3-6 eggs per well).

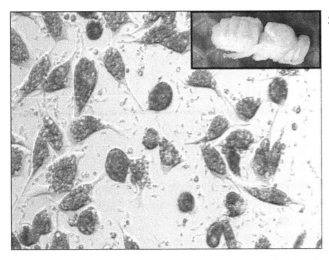

Fig. 1. Primary culture of honey bee, *A. mellifera*, pupae, white head, 17d post explanted. AmWH5 medium (Hunter, WB, USDA,ARS 2011.).

2.3.2. Care and observation of explanted material

1. On second day, transfer all floating material into a new plate (—➤ P1) with all previous information from P0 source plate (new date).
2. The P1 samples will all be transferred to a new plate, 4 days post being created (—➤P2).
3. The original plates P0 can be observed and cells should be visible attached to substrate. Floating material will look good, and viable.
4. Medium in wells should be above halfway full, and lids parafilmed around edges to reduce vapour loss.
5. Change half the medium AmWH5 once a week (but can push out to 10-12 days at first).
6. Once cells attached, or fatbody cells increasing, change medium AmWH5 once a week.
7. Cultures may be kept on counter top at 18-25°C. Increasing temperatures **did not** show any increase of cell growth (27-31°C).

2.4. Determining the viability of cultured cells

Cultured primary cells or non-permanent cell lines need to be tested for viability, because sometimes it is difficult to correctly differentiate between small cells and cell foci attached to the plate and cell debris or clumps of cell debris also adhering to the plate. This is especially important when no data on the healthy morphology of this cell type exist. Two commonly used methods are outlined below, the MTT test and the MitoTracker test, a fluorescence based viability test. In comparison to the MitoTracker test, the MTT test is less time consuming and less expensive. It can serve as a fast and reliable method to analyse cell proliferation of cell populations and it is suitable for identification of cytotoxic substances. With the help of the MitoTracker test the viability of cell populations can be analysed with special emphasis on single cell analysis and visualization.

2.4.1. MTT-viability test

To test the viability of cultured cells:

1. Collect a sample of adherent cells which were incubated at minimum for 24 h at 33°C in a microtitre plate.
2. Centrifuge the plate for 10 min at 210 rcf (Eppendorf 5415 R, rotor A-2-DWP).
3. Aspirate the medium using a vacuum pump.
4. Cover the cells with 100 µl of freshly prepared BM3 medium supplemented with 250 µg/ml penicillin/streptomycin-solution (Roth, HP10.1) and 2.5% antibiotic/antimycotic-solution (Sigma-Aldrich, A5955).
5. Incubate for 72 h at 33°C in a cooling incubator.
6. Centrifuge the microtitre plate for 10 min at 210xg to pellet the cells.
7. Aspirate the medium carefully without scratching the cells.

—➤ If larvae, pupae, or adults are used: then after surface sterilized, rinsed, and dry, sitting on sterile lid of plate in the hood:

8. Put medium AmWH5 into all the wells of four multi-well tissue culture plate, enough to cover bottom of each well.
9. Using sterilized fine tip metal forceps, grab dorsal surface of one bee abdomen, and tear small opening with second forceps. A **haemolymph** droplet will form.
10. Gently touch this droplet to the surface of medium to wick it from the bee's body, into the medium.
11. Readjust your forceps to gently squeeze the abdomen or thorax and a second and third **clear droplet** can be collected in similar fashion.
12. The first three droplets can all be placed into one well.
13. Squeeze the bee's body and **cloudy droplets** are now formed.
14. Put these one droplet per well, until you run out of haemolymph.
15. Tear the abdomen from the thorax and head, or if this is a larvae, tear in half.
16. Working with the abdomen first (as the head material often results in contamination): Dip the abdomen in a well, gently shake, move to next well, repeat until no more cells are observed to come off the material (this can fill 1-3 plates).
17. Next process the dorsal half, if an undifferentiated larvae, or work with the thorax if sample is pupae/adult: Tear sample in the medium in well, move to next well, tear and shake, repeat until material is used up (about ½ to 1 full plate).
18. Now the head is last, put into medium and tear apart, makes one to three wells.
19. Plates are labelled genus species abbreviated, Am, Date, body part [haemolymph, head, abdomen, thorax, head (He., ab, Tx, hd)], passage number (P0).

8. Add 100 µl of 0.5 mg/ml 3-(4,5-Dimethylthiazol-2-yl)-2,5-diphenyltetrazoliumbromid (MTT)-solution in BM3medium (Table 1) supplemented with 250 µg/ml penicillin/streptomycin-solution and 2.5% antibiotic/antimycotic-solution.

9. Incubate at 33°C for minimum 3 h.

10. Centrifuge the plate again at 210 rcf for 10 min.

11. Carefully aspirate the medium.

12. Add 100 µl of dimethylsulfoxid/acetic acid/sodium dodecyl sulphate (89.4%/0.6%/10%) and incubate the plate for 5 min at room temperature on a shaker (Heydolph, Polymax 1040) for cell lysis.

13. Analyse the viability-related colour in an ELISA reader (BioTek, Synergy HT) with 595 nm excitation wavelength.

Colorimetric intensity depends on individual cell type and doubling time. Percentage of viable cells should be between 80 and 100% depending on cell strain.

2.4.2. MitoTracker® Red FM-viability test for cultured cells

1. Let the cells adhere to the glass surface of a chamber slide for 24h.

2. Centrifuge the chamber slide for 10 min at 220 rcf (Eppendorf 5810 R, rotor A-2-DWP).

3. Aspirate the medium.

4. Add 100 µl of freshly prepared MitoTracker® Red FM (300nM) diluted in BM3 medium (Table 1) supplemented with 250 µg/ml penicillin/streptomycin-solution (Roth, HP10.1)) and 2.5% antibiotic/antimycotic-solution (Sigma-Aldrich, A5955).

5. Incubate for 1 hour at 27°C.

6. Centrifuge the chamber slide again for 10 min at 220 rcf.

7. Remove the medium carefully without scratching the surface.

8. Wash the cells with 1x phosphate buffered saline (1xPBS,Table 1)

9. Centrifuge again at 220 rcf for 10 min.

10. Aspirate the PBS-buffer.

11. Fix the cells in 4% formalin-solution (Roth, 4980.1) for 20 min at room temperature.

12. Centrifuge the chamber slide again for 10 min at 220 rcf.

13. Aspirate the formalin.

14. Wash the cells with 1xPBS.

15. Centrifuge again at 220 rcf for 10 min.

16. Aspirate the PBS-buffer.

17. Stain the nuclei with 250 µl DAPI (4',6-Diamidin-2-phenylindole, VWR, 1mg/ml in 99% methanol) for 5 min in the dark.

18. Aspirate the DAPI-solution and remove the chamber.

19. Wash the cells with 1xPBS-buffer.

20. Let the slide air dry.

21. Cover the cells with ProLong® Gold antifade reagent (Invitrogen, P36930) and a cover slip to preserve the fluorescent dyes.

22. Visualize viable cells under a fluorescence Microscope using a DAPI-filter or a TexasRed-filter.

MitoTracker probes label **active** mitochondria of **living** cells and, therefore, allow the identification of individual living cells amongst a cell population and to visually demonstrate the proportion of living cells within a cell population.

3. Working with permanent insect cell lines

3.1. Available and suitable cell lines

Working with permanent cell lines has several advantages over working with primary cells and non-permanent cell lines. One of the major advantages is that most of these cell lines can be propagated endlessly and without any restriction in the quantity of cells available for experiments. It may take several weeks to have hundreds of flasks with confluent cell layers, but it is possible to obtain them. In contrast, the establishment of primary cells and non-permanent cell lines depends on the availability of the organisms or organs used for cell isolation and the amount of cells depends on the size of the organ and the survival rate of the dissociated cells once they are in culture. In addition, permanent cell lines are advantageous when experiments need standardized conditions or when experiments need to be performed or reproduced at different locations. Reproducibility is much more difficult with primary cells and non-permanent cell lines. Even if the involved groups follow the very same protocol, they will have to use different animals for cell isolation, which might lead to deviation in results. A recent publication described the alleged immortalization of honey bee embryonic cells by gene transfer of the human c-myc proto-oncogene (Kitagishi *et al.*, 2011). Although this might be the first permanent honey bee cell line, this cell line can only be considered "of honey bee character" due to the expression of a central transcription factor of human origin known to change the entire cellular program by unregulating the expression of many genes (Nasi *et al.*, 2001; Pelengaris and Khan, 2003). Therefore, working with permanent cell lines in honey bee research is equivalent to working with heterologous (isolated from lepidopteran or dipteran insects or else) or aberrant (*in vitro* transformed) cell lines and special experimental precautions are necessary. Experiments need to be thoroughly conducted, and proper controls need to be included to avoid cell culture artifacts or artifacts due to the heterologous system. A list of cell lines which proved to be useful heterologous models in bee pathology (Gisder *et al.*, 2011) is given in Table 2. Which heterologous cell line is the best for the planned experimental approach needs to be tested by each researcher.

3.2. Cultivation of insect cell lines

Cultivation of insect cell lines is straightforward. Normally, they are maintained at room temperature (20-27°C) without CO_2 allowing

Table 2. List of commercially available, permanent cell lines established from lepidopteran or dipteran insects suitable for certain applications in honey bee research.

cell line	source organism	source tissue	cell morphology*
IPL-LD-65Y	*Lymantria dispar*	larval tissue	Large cells ; up to 30% grow adherent with processes; suspension cells are round to oval
MB-L2	*Mamestra brassicae*	larval tissue	Polymorphic round cells, partly adherent
MB-03	*Mamestra brassicae*	larval tissue	Polymorphic round cells, partly adherent
MB-L11	*Mamestra brassicae*	larval tissue	Polymorphic round cells, partly adherent
Schneider-2	*Drosophila melanogaster*	late embryo	Small adherent cells growing in monolayers, a small number of cells is also in suspension
Sf-9	*Spodoptera frugiperda*	pupal ovarian tissue	Polymorphic round cells, partly adherent
Sf 21	*Spodoptera frugiperda*	immature ovaries	90% round cells, 10% spindle shaped, adherent
Sf-158	*Spodoptera frugiperda*	pupal ovarian tissue	Polymorphic round cells, partly adherent
SPC-BM-36	*Bombyx mori*	larval tissue	Large, mostly adherent cells ; 90% round cells (singly or aggregates), 10% spindle-shaped cells with long processes
Tn-368	*Trichoplusia ni*	larval tissue	Spindle-shaped cells growing in suspension (90%); cells tend to cluster in aggregates

Note: *, all information according to cell line data sheets from DSMZ (Deutsche Sammlung von Mikroorganismen und Zellkultur)

cultivation of these cells "in the desk drawer". However, suitable cooling incubators (e.g. Heraeus BK6160 through Thermo Fisher Scientific) which keep a defined temperature are recommended. To avoid evaporation of the medium during cultivation, cooling should not be accomplished through air ventilation. Each cell line, when purchased from a cell culture collection, will be accompanied by a data sheet giving all necessary information concerning the medium for cultivation, how and when to subculture, doubling time, cell harvest, and storage conditions. It is advisable to first follow these instructions before adapting these protocols to experimental needs.

4. Conclusions and outlook

The knowledge base of modern infection biology has been built upon a foundation of cell culture systems, from which most cell culture techniques are now taken for granted in many scientific disciplines. Unfortunately, honey bee pathology is lacking an established, vigorously dividing, immortal cell line for use by the larger research community. The absence of permanent bee cell lines has motivated many researchers to develop alternative systems, or to use primary cultures within the short time frame of their viability. These approaches, however suitable, will be surpassed by development of continuous honey bee cell lines. The few reports of cultured honey bee cells (Barbara *et al.*, 2008; Bergem *et al.*, 2006; Hunter, 2010; Lynn, 2001) and those from other hymenopterans continues to increase. These advances in cell culture methodologies, as well as our increasing understanding of bee cell requirements and responses to current media components through genomic analyses continues to push the field towards the development of new cell lines for a wide range of hymenopteran species hopefully including *A. mellifera* in the near future.

5. References

BARBARA, G S; GRÜNEWALD, B; PAUTE, S; GAUTHIER, M; RAYMOND -DELPECH, V (2008) Study of nicotinic acetylcholine receptors on cultured antennal lobe neurons from adult honey bee brains. *Invertebrate Neuroscience* 8: 19-29.

BERGEM, M; NORBERG, N; AAMODT, R A (2006) Long-term maintenance of *in vitro* cultured honey bee (*Apis mellifera*) embryonic cells. *BMC Developmental Biology* 6: 17.

CHAN, M M Y; SHOI, S Y C; CHAN, Q W T; LI, P; GUARNA, M M; FOSTER, L J (2010) Proteome profile and lentiviral transduction of cultured honey bee (*Apis mellifera* L.) cells. *Insect Molecular Biology* 19: 653-658.

DEVAUD, J-M; QUENET, B; GASCUEL, J; MASSON, C (1994) A morphometric classification of pupal honey bee antennal lobe neurones in culture. *NeuroReport* 6: 214-218.

FERKOVICH, S M; OBERLANDER, H; DILLARD, C; LEACH, E (1994) Embryonic development of an endoparasitoid, *Microplitis croceipes* (Hymentopera: Braconidae) in cell line-conditioned media. In Vitro *Cellular and Developmental Biology* 30A: 279-282.

GASCUEL, J; MASSON, C; BEADLE, D J (1991) The morphology and ultrastructure of antennal lobe cells from pupal honey bees (*Apis mellifera*) growing in culture. *Tissue and Cell* 23: 547-559.

GASCUEL, J; MASSON, C; BERMUDEZ, I; BEADLE, D J (1994) Morphological analysis of honey bee antennal cells growing in primary cultures. *Tissue and Cell* 26: 551-558.

GIAUFFRET, A (1971) Cell culture of Hymenoptera. In *VAGO, C (Ed.), Invertebrate tissue culture*. Academic Press; New York, USA. pp. 295-305.

GISDER, S; MÖCKEL, N; LINDE, A; GENERSCH, E (2011) A cell culture model for *Nosema ceranae* and *Nosmea apis* allows new insights into the life cycle of these important honey bee-pathogenic microsporidia. *Environmental Microbiology* 13: 404-413.

GOLDBERG, F; GRÜNEWALD, B; ROSENBOOM, H; MENZEL, R (1999) Nicotinic acetylcholine currents of cultured Kenyon cells from the mushroom bodies of the honey bee *Apis mellifera. Journal of Physiology* 514: 759-768.

GREANY, P (1986) *In vitro* culture of hymenopterous larval endoparasitoids. *Journal of Insect Physiology* 32: 409.

HUMAN, H; BRODSCHNEIDER, R; DIETEMANN, V; DIVELY, G; ELLIS, J; FORSGREN, E; FRIES, I; HATJINA, F; HU, F-L; JAFFÉ, R; KÖHLER, A; PIRK, C W W; ROSE, R; STRAUSS, U; TANNER, G; VAN DER STEEN, J J M; VEJSNÆS, F; WILLIAMS, G R; ZHENG, H-Q (2013) Miscellaneous standard methods for *Apis mellifera* research. In *V Dietemann; J D Ellis; P Neumann (Eds) The COLOSS BEEBOOK, Volume I: standard methods for* Apis mellifera *research. Journal of Apicultural Research* 52(4): http://dx.doi.org/10.3896/IBRA.1.52.4.10

HUNTER, W B (2010) Medium for development of bee cell cultures (*Apis mellifera*: Hymenoptera: Apidae). In Vitro *Cellular and Developmental Biology - Animal* 46: 83-86.

KAATZ, H-H; HAGEDORN, H H; ENGELS, W (1985) Culture of honey bee organs: Development of a new medium and the importance of tracheation. In Vitro *Cellular and Developmental Biology* 21: 347-352.

KITAGISHI, Y; OKUMURA, N; YOSHIDA, H; NISHIMURA, Y; TAKAHASHI, J; MATSUDA, S (2011) Long-term cultivation of *in vitro Apis mellifera* cells by gene transfer of human c-myc proto-oncogene. In Vitro *Cellular and Developmental Biology - Animal* 47: 451-453.

KREISSL, S; BICKER, G (1992) Dissociated neurons of the pupal honey bee brain in cell culture. *Journal of Neurocytology* 21: 545-556.

LYNN, D E (2001) Novel techniques to establish new insect cell lines. In Vitro *Cellular and Developmental Biology* 37: 319-321.

MÖCKEL, N; GISDER, S; GENERSCH, E (2009) Establishment of an *in vitro* model for DWV-infections. *Apidologie* 40: 666.

NASI, S; CIARAPICA, R; JUCKER, R; ROSATI, J; SOUCEK, L (2001) Making decisions through Myc. *FEBS Letters* 490: 153-162.

PELENGARIS, S; KHAN, M (2003) The many faces of c-MYC. *Archives in Biochemistry and Biophysics* 416: 129-136.

PFEIFFER, H H (1941) Quantitative polarisationsoptische Versuche mit reflektiertem Licht. *Archiv für experimentelle Zellforschung besonders Gewebezüchtung* 24: 273-287.

POPPINGA, L; JANESCH, B; FÜNFHAUS, A; SEKOT, G; GARCIA-GONZALEZ, E; HERTLEIN, G; HEDTKE, K; SCHÄFFER, C; GENERSCH, E (2012) Identification and functional analysis of the S-layer protein SplA of *Paenibacillus larvae*, the causative agent of American foulbrood of honey bees. *PLoS Pathogens* 8: e1002716.

ROCHER, J; RAVALLEC, M; BARRY, P; VOLKHOFF, A-N; RAY, D; DEVAUCHELLE, G; DUONOR-CERUTTI, M (2004) Establishment of cell lines from the wasp *Hyposoter didymator* (Hym., Ichneumonidae) containing the symbiotic polydnavirus H. didymator ichnovirus. *Journal of General Virology* 85: 863-868.

SORESCU, I; TANSA, R; GHEORGHE, L; MARDARE, A; CHIOVEANU, G (2003) Attempts to *in vitro* cultivate honey bee (*Apis mellifera* L.) haemocytes. *Studies Research Veterinary Medicine (Bucharest)* 9: 123-131.

TROEMEL, E R (2011) New models of microsporidiosis: Infections in zebrafish, *C. elegans*, and honey bee. *PLoS Pathogens* 7: e1001243.

VAN STEENKISTE, D (1988) De hemocyten van de honingbij (*Apis mellifera* L.). Typologie, bloedbeeld en cellulaire verdedigingsreacties. Doctoraatsproefschrift, Rijksuniversiteit Gent, Belgium. 95-105.

Journal of Apicultural Research 52(4): (2013)
DOI 10.3896/IBRA.1.52.4.05

REVIEW ARTICLE

Standard methods for characterising subspecies and ecotypes of *Apis mellifera*

Marina D Meixner[1]*, Maria Alice Pinto[2], Maria Bouga[3], Per Kryger[4], Evgeniya Ivanova[5] and Stefan Fuchs[6]

[1]LLH, Bee Institute, Erlenstr. 9, 35274 Kirchhain, Germany.
[2]Mountain Research Centre (CIMO), Polytechnic Institute of Bragança, Campus de Sta. Apolónia, Apartado 1172, 5301-855 Bragança, Portugal.
[3]Laboratory of Agricultural Zoology and Entomology, Agricultural University of Athens, 75 Iera Odos St., Athens 11855, Greece.
[4]Aarhus University, Flakkebjerg, Forsøgsvej 1, 4200 Slagelse, Denmark.
[5]University of Plovdiv, Biological Faculty, Department of Developmental Biology, 24, Tzar Assen Str. Plovdiv 4000, Bulgaria.
[6]Institut für Bienenkunde, Goethe Universität Frankfurt am Main, FB Biowissenschaften, Karl-von-Frisch-Weg 2, 61440 Oberursel, Germany.

Received 19 March 2012, accepted subject to revision 4 June 2012, accepted for publication 23 January 2013.

*Corresponding author: Email: marina.meixner@llh.hessen.de

Summary

The natural diversity of honey bees in Europe is eroding fast. A multitude of reasons lead to a loss of both genetic diversity and specific adaptations to local conditions. To preserve locally adapted bees through breeding efforts and to maintain regional strains in conservation areas, these valuable populations need to be identified.

In this paper, we give an overview of methods that are currently available and used for recognition of honey bee subspecies and ecotypes, or that can be utilised to verify the genetic origin of colonies for breeding purposes. Beyond summarising details of morphometric, allozyme and DNA methods currently in use, we report recommendations with regard to strategies for sampling, and suggest methods for statistical data analysis. In particular, we emphasise the importance of reference data and consistency of methods between laboratories to yield comparable results.

Métodos estándar para la caracterización de las subespecies y ecotipos de *Apis mellifera*

Resumen

La diversidad natural de la abeja de la miel se está deteriorando rápidamente en Europa. Existen multitud de razones que conducen tanto a una pérdida de diversidad genética como de adaptaciones específicas a las condiciones locales. Se necesita identificar a estas valiosas poblaciones para preservar a las abejas adaptadas a nivel local, mediante esfuerzos para la mejora y el mantenimiento de variedades regionales en las áreas de conservación.

En este artículo, realizamos una revisión general de los actuales métodos disponibles que se utilizan para la determinación de subespecies y ecotipos de abejas melíferas, o que pueden ser utilizados para verificar el origen genético de las colmenas seleccionadas con fines de cría. Además, resumimos las características de los métodos morfométricos, de aloenzimas y de ADN, realizamos recomendaciones con respecto a las estrategias de muestreo, y sugerimos métodos para el análisis estadístico de los datos. En particular, destacamos la importancia de los datos de referencia y la coherencia de los métodos entre laboratorios para producir resultados comparables.

Footnote: Please cite this paper as: MEIXNER, M D; PINTO, M A; BOUGA, M; KRYGER, P; IVANOVA, E; FUCHS, S (2013) Standard methods for characterising subspecies and ecotypes of *Apis mellifera*. In *V Dietemann; J D Ellis; P Neumann (Eds) The COLOSS BEEBOOK, Volume I: standard methods for* Apis mellifera *research. Journal of Apicultural Research* 52(4): http://dx.doi.org/10.3896/IBRA.1.52.4.05

辨别西方蜜蜂亚种和生态型的标准方法

摘要

欧洲的蜜蜂自然多样性正在快速消失。导致遗传多样性和对局部条件的特定适应性丢失的原因有很多。不论是通过育种措施保存当地的适应蜂种还是维持保护区的局部种系，都需要辨别有价值的蜂种。

在本文中，我们概述蜜蜂亚种和生态型识别的一些已知并可用的方法，以及用于育种目的的蜂群遗传来源鉴定方法。除了总结目前使用的形态测定，异型酶和DNA方法的细节外，我们还提出了针对抽样策略的推荐规范，并推荐了统计数据分析使用的方法。尤其，我们强调参考数据的重要性和实验室之间方法的一致性，从而获得可比较的结果。

Keywords: COLOSS, *BEEBOOK*, *Apis mellifera*, honey bee, subspecies identification, ecotype, classical morphometry, geometric morphometry, allozymes, mitochondrial DNA, microsatellites, SNP analysis

Table of Contents

1. Introduction

The natural range of *Apis mellifera* includes Africa and Eurasia. In addition, this species has been introduced by humans to all other continents except Antarctica and is used intensively in pollination and honey production all over the world.

In this chapter, we place a stronger focus on the natural diversity within Europe than elsewhere for several reasons: (i) the diversity situation and the status of subspecies and ecotypes in Europe is known best, (ii) bee diversity research and breeding efforts in many parts of Africa and Asia are still in their infancy, (iii) most subspecies utilised for breeding (including the New World) originate from Europe. However, the approach and the methods discussed are of course valid for all populations of *A. mellifera* from its entire range.

Honey bees show considerable geographical variation, resulting in adaptation to regionally varying factors of climate and vegetation, but also to prevailing pests and pathogens. However, this natural heritage is increasingly subject to diffusion by human beekeeping efforts at a worrisome speed. The demand for high economic performance of bee colonies, combined with desirable behavioural characteristics, has led to considerable changes caused by systematic bee breeding. Thus, the original geographic distribution pattern is being dissolved EU-wide by mass importations and an increasing practice of queen trade and colony movements. These activities endanger regional races and ecotypes by promoting hybridisation (De la Rúa *et al.*, 2009; Meixner *et al.*, 2010), and by adding various breeder lines with distinct properties to the picture. Yet another dimension is added by the deliberate replacement of native subspecies in some regions by non-native bees with more desirable characters and greater commercial interest (for instance, the replacement of *A. m. mellifera* in northern and central Europe by *A. m. carnica or A. m. ligustica*) (Bouga *et al.*, 2011).

The downside of these economically driven processes is an increasing trend towards uniformity of honey bee populations across Europe, leading to a loss of both genetic diversity and specific adaptations to local conditions (reviewed in De la Rua *et al.*, 2009; Meixner *et al.*, 2010).

Honey bees are particularly sensitive to inbreeding (Seeley and Tarpy, 2007 and references therein). Therefore, the loss of genetic diversity is of grave concern. It has been shown that colonies with reduced genetic diversity are less capable of controlling hive temperature (Jones *et al.*, 2004) and more prone to develop diseases when challenged by parasites (e.g. Tarpy, 2003). This reduction in genetic diversity may also affect the capacity of honey bee populations to adapt to new threats, such as newly introduced parasites like varroa.

Thus, there is a widely recognised need to encourage regional breeding efforts to preserve local adaptation, and to maintain local strains in isolated conservation apiaries. To attain this goal, it is necessary to have a reference base to identify strains to be used for breeding. To provide a stable baseline, it is important that this reference reflects the natural variation of honey bees, since beekeepers and breeders are known to often work with non-native stock. However, although the rough outlines of the original patterns of honey bee diversity have more or less been clarified, many of the details are still "work in progress". Ten of the currently 27 described subspecies can be found in Europe, but considerable variation can be observed within many of them and several can be further subdivided into a diversity of "ecotypes". Several subspecies and ecotypes can be considered as endangered (De la Rúa *et al.*, 2009). Other than subspecies which can be assigned a formal trinomen according to the ICZN, there is no formal definition for ecotypes. However, the ecotype concept may include any regional natural bee type distinct enough to raise the need for a name. In addition to these refinements, the scientific description and recognition of honey bee diversity in Europe cannot be regarded as complete, since vast areas, predominantly in the eastern part of the continent, have not yet been studied systematically. Table 1 summarises the current state of knowledge regarding honey bee subspecies and ecotypic variation in Europe.

Historically, the current picture of honey bee geographic variation has emerged gradually over time from the results of numerous publications. Up to now, the most comprehensive compilation is still provided by the monograph of Ruttner (1988), based on the application of numerical taxonomy using characters of "classical" morphometry. Since then, many details to the contours of this basic picture have been added by others. However, as methods differ due to the varying preferences of different researchers, and the results achieved using different techniques are oftentimes not congruent, there is now a critical need to integrate data bases, make reference data sets generally accessible, and harmonise procedures to achieve sound and generally recognised "consensus reference data sets". To date, different laboratories keep independent reference samples, and there are still considerable gaps in relating molecular results to classical morphometric group definitions. In short, there is no agreement on a common procedure yet, raising a strong need to work towards an accessible data base of European subspecies and ecotypes, integrating records obtained with different methods and serving as reference for future research projects and identification needs.

Having pointed out these shortcomings and aims for improvement, nonetheless, methods for characterisation and identification of honey bees on the morphological and the molecular level have reached a fairly sophisticated stage. Our aim with this paper is to give recommendations and detailed descriptions for a variety of morphological and molecular methods in order to facilitate the acquisition of data compatible for integration towards a more coherent capture of the diversity of honey bees in Europe and the world.

2. General considerations

In exploring honey bee variation for practical or scientific reasons, researchers face two fairly different tasks. One is to investigate unknown variation to establish firm groupings. To complete this task,

one has to include a wide range of parameters of variation, in fact as many as available. This route has led from subspecies and ecotype definitions developed from morphological and behavioural characters to an increasing refinement and re-evaluation of this picture through the inclusion of molecular characterisations, and increased illumination of phylogenetic relations between populations.

The second task is to just identify unknown samples in the established picture. Here, few effective identifying characters are sufficient, and the main objective is to establish labour-effective diagnostic tools. These two aspects will consistently come up in the following sections introducing various morphometric and molecular methods, where they will be discussed in more detail. To facilitate the decision, which method(s) would be most appropriate in a given situation, we have compiled methods in a simple tree structure in Table 2.

2.1. Sampling and storing specimens for analysis

The scope of any sampling largely depends on the aim of the study. In case of identification, it may just comprise the samples of interest, e.g. from a specific apiary or bee line. However, because of inter-colony differences, a reasonable number of at least three colonies should be sampled. If one wants to investigate an entire population covering an area to detect regional differences, about 5 colonies should be sampled per location to keep sampling errors within tolerable limits (Radloff *et al.*, 2003). In particular, attention should be given to the sample coverage of a region, ideally with balanced spacing between locations and no apparent gaps in the sampling to avoid artificial groupings due to local sample clustering. Indeed, resolution of type changes strongly depends on sample density, where at low density large-scale differences tend to be more pronounced, but small-scale regional differences might get blurred (Radloff and Hepburn, 1998). Therefore, it may be necessary to differentiate between mesoscale and macroscale studies and adapt the sampling regime accordingly (Radloff *et al.*, 2010).

Ideally, samples should be collected in a way that ascertains their origin from a given hive under analysis. While the sampling of pupae would be ideal for this purpose, they are not suitable for morphometric analysis and their collection is time consuming. Therefore, we recommend sampling workers from inside the colonies, most desirably from the brood area. However, this is also not always possible, since hives may not easily be opened for a variety of reasons (traditional hives, natural comb, defensive bees, etc.). In these cases sampling from the flight entrance will have to suffice. However, one sample should always represent one colony, i.e. only contain workers sampled from one hive. The sampling of drones is more difficult, since they are only seasonally available and more prone to drifting than workers (Currie and Jay, 1988; Jay, 1969; Neumann *et al.*, 2000).

Table 1. Current status of honey bee subspecies and ecotypes in Europe (the references given here are meant as examples, and not as an exhaustive list of all studies performed).

Subspecies name	Status of knowledge	Morphometric lineage	Mitochondrial lineage	Morphometric references available?	mtDNA references available? Studies on the COI-COII intergenic region are printed in bold	Microsatellite references available?
A. m. mellifera	Despite the fact that the original description of *Apis mellifera* by Linnaeus was on this bee and this subspecies has the widest range of all *A. mellifera* sub-species, surprisingly little information is available. This race has been comparatively well studied in the western part of its range, especially in France and Britain, and also in southern Scandinavia. Almost nothing is known about the extent of its range and its variability towards the east, beyond Poland. Older literature from Russia often restricted to description of single characters (e.g. tongue length). Ecotypic variation has been described and confirmed in France (Louveaux *et al.*, 1966; Strange *et al.*, 2007), where also several conservation areas have been established	M	M	yes Ruttner, 1988 Meixner *et al.*, 2007	yes Smith and Brown, 1990 **Franck *et al.*, 1998** **Garnery *et al.*, 1998a** **Jensen *et al.*, 2005** **Rortais *et al.*, 2011**	yes Franck *et al.*, 1998 Garnery *et al.*, 1998b Jensen *et al.*, 2005
A. m. ligustica	Comparatively well studied Ecotypic variation not known, likely to be diffused by colony movements	C	C(M)	yes Ruttner, 1988 Nazzi, 1992	Yes **Franck *et al.*, 2000**	Yes Franck *et al.*, 2000 Dall'Olio *et al.*, 2007
A. m. carnica	Well studied in Austria, Slovenia and Croatia, few studies from eastern part of range Ecotypic variation postulated (e.g. Alpine and Pannonian ecotypes, Ruttner 1988) but few data	C	C	yes Ruttner, 1988 Meixner *et al.*, 1993 Dedej *et al.*, 1996 Mladenovic *et al.*, 2011 Marghitas *et al.*, 2008 Reka *et al.*, 2007	yes Smith and Brown, 1990 Meixner *et al.*, 1993 **Sušnik *et al.* 2004** **Muñoz *et al.*, 2008** **Nedić *et al.*, 2009**	Yes Muñoz *et al.*, 2008
A. m. macedonica	Few publications The subspecies extends from the southern Balkan to Ukraine, but the true range is not yet known Ecotypic variation hypothesised, but no internationally available description	C	C	yes Ruttner, 1988 Uzunov *et al.*, 2009	yes Bouga *et al.*, 2005 Ivanova *et al.*, 2010 Stevanovic *et al.*, 2010	yes Uzunov *et al.*, 2013
A. m. iberiensis	Well studied in most of Iberian Peninsula No distinct ecotypic variants found for the mainland, but ecotypes described for island populations	M	M/A	yes Ruttner, 1988 Arias *et al.*, 2006 Cornuet and Fresnaye, 1989 Padilla Alvarez *et al.*, 1998, 2001 Radloff *et al.*, 2001	yes Smith *et al.*, 1991 **De la Rúa *et al.*, 1998** **De la Rúa *et al.*, 2001a** **De la Rúa *et al.*, 2001b** **De la Rúa *et al.*, 2002** Arias *et al.*, 2006 **Canovas *et al.* 2008**	yes De la Rúa *et al.*, 2003 De la Rúa *et al.*, 2006 Miguel *et al.*, 2007 Miguel *et al.*, 2010 Canovas *et al.*, 2011
A. m. cecropia	Very few studies, current status questionable	C	C	yes Ruttner, 1988	yes Bouga et *al.*, 2005	no
A. m. siciliana	Little studied, near extinct	C/A	A	yes Ruttner, 1988	yes Sinacori *et al.*, 1998 **Franck *et al.*, 2000**	no
A. m. cypria	Few studies	C	C/Z	yes Ruttner, 1988 Kandemir *et al.*, 2000	yes **Kandemir *et al.*, 2006a**	yes Kandemir *et al.*, 2006a
A. m. anatoliaca	Some studies Possible ecotypic variation in western Anatolia (Aegean region), but few data available; more variation may be possible	O	C/Z	yes Ruttner, 1988 Kandemir *et al.*, 2000	yes Palmer *et al.*, 2000 Kandemir *et al.*, 2006b	yes Bodur *et al.*, 2007
A. m. caucasica	Despite its economic importance there are substantial gaps in our knowledge about this subspecies. Some data are available from the northern part (Ukraine, Russia), and very few from the southern Caucasus (Turkey). Ecotypic variation seems possible but not studied.	O	C	yes Ruttner, 1988	yes Kandemir *et al.*, 2006b	yes Bodur *et al.*, 2007
A. m. adami	Sparse information	C	C	yes Ruttner, 1980, 1988	yes Bouga *et al.*, 2005	no
A. m. ruttneri	Only one study. Current fate of population unknown	M/A	A	yes Sheppard *et al.*, 1997	yes Sheppard *et al.*, 1997	no

Table 2. Key to recommended methods, depending on the aim of the study.

1.a The population or sample under study originates from the native range of *Apis mellifera*	go to 3
1.b The population or sample under study does not originate from the native range of *A. mellifera*	go to 2
2.a Your main concern is to identify potential Africanization in the population or sample rapidly	Use geometric morphometrics or one of the fast identification systems (e.g. FABIS). See sections 3.1.1.2 and 3.1.3.2
2.b Your study aims at a characterisation of the population beyond the question of Africanization	Combine methods (morphometric, mtDNA or microsatellites) for a comprehensive characterisation. Include reference data from potential source populations in analysis. See sections 3.1, 3.2, and 3.3.2
3.a The population under study has been described and published reference data exist in the literature	go to 4
3.b The population under study has not been described or originates from a hybrid zone	Combine methods (morphometric, mtDNA or microsatellites) for a comprehensive characterisation. Include reference data from adjacent areas in analysis. See sections 3.1, 3.2, and 3.3.2
4.a A fast assignment is needed to confirm the origin of the samples in question	Use geometric morphometrics or classical wing characters; include reference data in the analysis. See sections 3.1.1.2 and 3.1.4
4.b The aim of your study is a comprehensive characterisation of the samples in question	Combine methods (morphometric, mtDNA or microsatellites) for a comprehensive characterisation. Include reference data in analysis. See sections 3.1, 3.2, and 3.3.2

2.2. Specific sampling recommendations

2.2.1. Sample size

It appears reasonable to collect enough bees to enable analysis of the sample with a combination of methods. Whereas for most molecular methods one bee per colony is sufficient for analysis, 15 bees are recommended for morphometric analysis and 10 should be analysed at the minimum (Table 3). In fact, the number of bees needed in morphometry depends on the degree of intracolonial variation, and is higher for variable traits such as colours, but can be reduced to 5 with more stable traits as wing venation (Alpatov, 1929, Radloff *et al.*, 2003). For all methods discussed here, the bees are destroyed during analysis and therefore sampling of about 30 to 40 workers per colony appears adequate.

2.2.2. Killing and storage

For molecular (DNA) methods, the method of killing does not matter very much, and storage in 95% ethanol is recommended. Diluted EtOH (70%) will work for a limited amount of time, when no 95% alcohol is available, but the bees should be transferred to higher concentrations as soon as possible. Freezing is also possible, but rarely an option when collecting in the field. Storage in any liquid containing formaldehyde or acetic acid destroys DNA and is not recommended (for more details, see also the section 'Standard methods for immobilising, terminating and storing adult *Apis mellifera*' in the *BEEBOOK* paper on miscellaneous methods (Human *et al.*, 2013)).

For analysis with "classical" morphometry, the bees should preferably be killed by immersion in hot (boiling) water or by ether

vapours, because no other method will surely lead to the extension of their proboscis which cannot be measured if not stretched out. In the field, water would keep hot over several hours in a common thermos bottle. The best storage method for morphometric analysis is 70% EtOH, where the chitin stays soft enough for dissection. However, dissection is also still possible after storage in 95% EtOH as recommended for DNA analysis.

For allozyme analysis, the bees should be transported to the laboratory alive, or frozen on site (dry ice, liquid nitrogen) (see also chapter the section 'Standard methods for immobilising, terminating and storing adult *Apis mellifera*' in the *BEEBOOK* paper on miscellaneous methods (Human *et al.*, 2013)).

3. Available methods and markers

3.1. Morphometry

There is no morphological "key" to honey bee subspecies, no simple logical tree based on a sequence of single discriminating characters. Instead, measurable morphometric characters show gradual changes and their ranges mostly overlap between subspecies. Thus, subspecies often differ only slightly in the mean values of several body characters, and therefore advanced statistical methods are required for discrimination of groups. The concept of numerical taxonomy was introduced into honey bee taxonomy by DuPraw (1964, 1965) and further elaborated by Ruttner *et al.* (1978).

Table 3. Morphometric characters in current use. For characters described in Ruttner (1988), character numbers are given, with abbreviations according to Ruttner *et al.* (1978) in brackets. Author numbers give sources of description, with original descriptions where possible. Characters used in the Dawino method are described in www.beedol.cz and Bouga *et al.* (2011), and abbreviations are given. Other authors: 1 Alpatov, 1928, 1929; 2 Goetze,1964; 3 Dupraw, 1964; 4 Ruttner *et al.*, 1978, 10 http://apiclass.mnhn.fr and Miguel *et al.*, 2010. Abbreviations suggested in the last column are merged from several sources to reflect common use, but in places have been modified to increase understandability and consistency. Characters recommended to measure are in printed in bold. Indices and wing angles maybe calculated from these for analyses.

Character	Ruttner	DAWINO	Other authors	Abbreviations	Character	Ruttner	DAWINO	Other authors	Abbreviations
HAIR					**Cubital vein, distance b**	20 (b)	Len B	2	*CUBB*
Length of cover hair on tergite 5	1 (h)		2	*HLT5*	**20 points at venation junctions**			10	
Width of tomentum on tergite 4	2 (a)		2	*TOM A*	Radial field		Radial 1		*RAD1*
Width of stripe posterior of tomentum	3 (b)		2	*TOM B*	Length C		Len C		*LENC*
Tomentum index	(2:3)		4	*TI*					
SIZE					Length D		Len D		*LEND*
Proboscis	4		1	*PROBL*	Inner wing length		Inn. len.		*INLEN*
Femur	5 (Fe)		1	*FEM*	Inner wing width		Inn. wid.		*INWID*
Tibia	6 (Ti)		1	*TIB*	Wing angle A1		A1	3	*A1*
Basitarsus length	7 (M_L)		1	*TAL*	Wing angle A4	21 (A4)	A1	3	*A4*
Basitarsus width	8 (M_T)		1	*TAW*	Wing angle B3		B3	3	*B3*
Tergite 3, longitudinal	9 (T_3)		1	*T3*	Wing angle B4	22 (B4)	B4	3	*B4*
Tergite 4, longitudinal	10 (T_4)		1	*T4*	Wing angle D7	23 (D7)	D7	3	*D7*
Sternite 3, longitudinal	11 (S_3)		1	*LS3*	Wing angle E9	24 (E9)	E9	3	*E9*
Wax mirror of sternite 3 longitudinal	12 (W_L)		1	*WML*	Wing angle G7		G7	3	*G7*
Wax mirror of stemite 3, transversal.	13 (W_T)		1	*WMT*	Wing angle G18	25 (G18)	J10	3	*J10*
Distance between wax mirrors st. 3	14 (W_D)		4	*WD*	Wing angle H12		H12	3	*H12*
Stemite 6, longitudinal	15 (L_6)		4	*S6L*	Wing angle J10	26 (J10)	J10	3	*J10*
Sternite 6, transversal	16 (T_6)		4	*S6T*	Wing angle J16	27 (J16)	J16	3	*J16*
Length of hind leg	(5+6+7)		4	*LEG*	Wing angle K19	28 (K19)	K19	3	*K19*
Body size	(9+6)		4	*BODY*	Wing angle L13	29 (L13)	L13	3	*L13*
Slenderness index	(15:16)		4	*SLENI*	Wing angle M17		M17	3	*M17*
Basitarsal index	(7:8)		1	*TAI*	Wing angle N23	30 (N23)	N23	3	*N23*
					Wing angle O26	31 (O26)	O26	3	*O26*
COLOUR					Wing angle Q21		Q21	3	Q21
Pigmentation of tergite 2	32		2	*PT2*	Discoidal shift		Disc. sh.	2	*DISCSH*
Pigmentation of tergite 3	33		2	*PT3*	Cubital index		Ci	2	*CI*
Pigmentation of tergite 4	34		2	*PT4*	Precubital index		Pci	2	*PCI*
Pigmentation of scutellum, Cupolla	35 (Sc)		4	*PSC1*	Dumb-bell index		Dbi	2	*DBI*
Pigmentation of scutellum, B and K	36 (B, K)		4	*PSC2*	Radial index		Ri	2	*RI*
					Area of 6 fields		Area6		*AREA6*
WING					*Hind wing*				
Fore wing					Hind wing length			1	HWL
Fore wing length	17 (F_L)		1	*FWL*	Hind wing width			1	HWL
Fore wing width	18 (F_W)		1	*FWW*	Number of Hamuli			1	HA
Cubital vein, distance a	19 (a)	Len A	2	*CUBA*					

ADDITIONAL CHARACTERS RARELY MEASURED:
Head capsule width and length, antenna length and number of segments, compound eyes length and width, interocellar and oculo-ocellar distance, Pigmentation of labrum, mandible length, proboscis postmentum, glossae, proximal and distal segment length, thorax width, pollen basket size and area, brush hair rows number and sting shaft length.

3.1.1. Character suites

3.1.1.1. Choice of character suite

Consequently, these subtle and gradual morphological differences lead to the questions of how many body characters should be measured and which ones should be chosen. Various body parts, falling into four main categories, have been used in morphometric analyses, starting with the first studies of Alpatov (1929): characters of body size, colouration patterns, wing venation characteristics, and characteristics of pilosity. Their descriptions are scattered over the literature, and the most commonly used characters are listed in Table 3. A core set of 36 characters selected for discriminative power is described in Ruttner (1988), containing the recognised measurements referred to as "classical morphometry".

Although most researchers utilise at least some characters from the above-mentioned set, the characters sets used in many studies vary considerably and deviate in the choice or number of these "standard" characters, and less than 10 are used in a number of studies. Occasionally, additional characters are introduced, but the selection of characters appears widely arbitrary. As comparisons between studies or to reference data rely on characters matching between data sets, these incongruities seriously compromise meaningful analysis. In consequence, to date it is difficult or impossible to piece together the patchwork of various comprehensive local studies that each must have involved significant amounts of labour, and much effort is wasted due to lack of consistency. As an orientation, in studies exploring unknown variation, we recommend to include the 25 characters proposed in Table 3, combined with wing shape analysis.

3.1.1.2. Wing venation pattern analysis

Particular discrepancies exist in the measurement and analysis of wing venation, whose pattern variation is of specific interest in insect taxonomy in general. Due to their key role and the current interest, they are discussed here in greater depth. At present, three different main approaches are in use: classical wing morphometry as defined by Ruttner (1988), the DAWINO (Discriminant Analysis With Numerical Output) method (www.beedol.cz), and geometric wing shape analysis (Bookstein, 1991).

3.1.1.2.1. Classical wing morphometry

Classical wing morphometry captures variation in wing shape by calculating 11 angles between 18 junctions in the wing venation (Fig. 1) which constitute a subset of a suite of 17 angles first introduced into bee morphometry by DuPraw (1965). The DAWINO method consists of the full set of DuPraw's angles, supplemented by 7 linear measurements, 5 indices, and one area (Table 3). All these angles and other parameters are considered as measurement characters in further analysis, where they can be combined with measurements of body characters. In the past decade, these somehow idiosyncratic

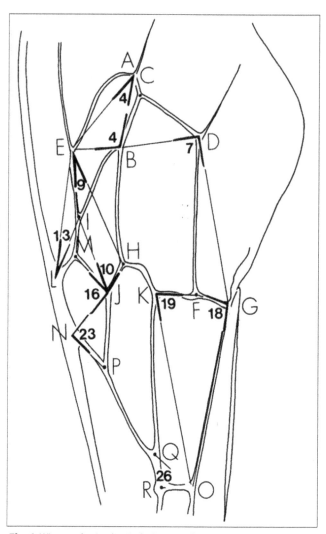

Fig. 1. Wing angles in classical wing morphometry (Ruttner, 1988).

morphometric methods for the bee-wing were increasingly replaced in a number of studies by "geometric morphometry", based on the theory of shape, which is explained here in more detail.

3.1.1.2.2. Geometric wing shape morphometry

The geometric morphometric method is based on the coordinates of landmarks located at vein intersections of the wing (Bookstein, 1991; Smith *et al.*, 1997). The fundamental advances of geometric morphometrics over traditional approaches include (i) the way the amount of difference between shapes can be measured (using Procrustes distance), (ii) the elucidation of the properties of the multidimensional shape space defined by this distance coefficient, (iii) the development of specialised statistical methods for the study of shape, and (iv) the development of new techniques for the graphical representations of the results (Bookstein, 1991; Rohlf, 2000; Mendes *et al.*, 2007).

In the measuring process, landmark coordinates are superimposed using a Generalized least Squares (GLS) Procrustes Superimposition (Rohlf and Slice, 1990): Specimens are centered, normalised to unit centred-size (Bookstein, 1991) and interactively rotated to minimise the sum-of squared distances between each location and its sample mean. Shape differences are then analysed by multivariate analysis of variance (MANOVA), Canonical Variate Analyses (CVA) and Multiple Discriminate Analyses, and shape patterns along the canonical axes are estimated by multivariate regression (Monteiro, 1999). For this kind of analysis, the software packages Morpheus (Slice, 2002), NTSYS (Rohlf, 1990), MORPHOJ (Klingenberg, 2011) and DrawWing (Tofilslki 2004) are commonly used. The new methods of automated measures and geometric morphometry have been used to distinguish Africanized honey bees from African and European subspecies, and to characterise the evolutionary lineages of *A. mellifera* (Francoy *et al.*, 2006, 2008; Baylac *et al.*, 2008; Tofilski, 2008; Miguel *et al.*, 2010; Kandemir *et al.*, 2011). This method has also been used to analyse differences between three honey bee subspecies in Poland: *A. m. mellifera*, *A. m. carnica,* and *A. m. caucasica* (Tofilski, 2004, 2008). The high rates of correct classification obtained with the geometric method indicate that forewings carry sufficient information to distinguish between different groups of bees. Discrimination results obtained with this method proved superior to classical wing angles, although the degree of improvement was moderate (Tofilski, 2008).

3.1.1.2.3. Classical wing angles, geometric wing analysis, and full body character suites

A main difficulty associated with this recent diversification of wing character suites is lack of downward compatibility. The morphometric definition of the currently recognised *Apis mellifera* subspecies has been based on traditional classical morphometry; consequently, studies based on a geometric character set cannot utilise reference data generated with traditional wing angle measurements accumulated in previous work by various authors, including the reference subspecies descriptions as given in Ruttner's (1988) monograph.

Fortunately, however, there is a high degree of consensus concerning the marking points between the different kinds of wing shape analysis. Fig. 2 shows the 20 landmarks predominantly used in geometric morphometry. Tofilski (2004, 2008) omitted landmark 15,

while Kandemir *et al.* (2011), following Zelditch et *al.* (2004), moved point 15 from the apex of the radial cell to the junction of Rs5 and the costa, and located one additional point at the end of the vannal fold. Classical morphometry and the DAWINO method use the same landmarks, omitting point 15. However, the methods disagree in the sequence of numbering these points, which is of no major concern.

To overcome the present sets of incompatible data and to avoid further parallel development of incompatible data sets in honey bee morphometry, our suggestion for a standardisation of wing measurements is to store all future data as point coordinates (instead of the format of derived characters such as angles) to facilitate data exchange between different studies and research teams. We suggest using the point format exemplified in the description of Apiclass (http://apiclass.mnhn.fr), shown in Fig. 2. From these coordinates, used in a majority of geometric studies, all 30 DAWINO characters can be calculated, which include the Ruttner (1988) wing angles as a subset. Storing the point coordinates instead of calculated characters will also keep all options open for any future progress in analysis techniques. Unfortunately, however, the coordinates cannot be recreated from classical wing angles, but first attempts have been made to re-measure reference samples with the geometric method (Kandemir *et al.*, 2011) to link geometric morphometry to subspecies characterisations obtained by the classical method.

As a further issue the question arises whether geometric morphometry should replace "classical" morphometry for good, meaning that the accurate, powerful and labour-effective shape analysis based on wing geometry alone should replace the full set of classical characters, including all traditional body characters. Phylogenetically, the wing venation is more informative compared to the more environment–sensitive character categories of size, colour or pilosity (Diniz -Filho *et al.*, 2000), and thus represents a character set somewhat comparable to molecular characters. A high degree of consistency between wing morphometry and molecular information has been demonstrated by Miguel *et al.* (2010). Therefore, wing geometry is particularly suitable to track phylogenetic relationships between subspecies, where the full "classical" character set can be misleading. However, aiming at an inventory of honey bee variation as a numerical account of ecotype morphology, it appears indispensable to maintain classical morphology with a broad character set to represent the actual features of subspecies or ecotypes, apart from and in addition to the question of their phylogeny. However, geometric wing venation morphometry might replace the classical wing angles even within the classical morphometry set, but no attempt has been made so far to combine these methods.

3.1.2. Preparation and measuring

1. Dissect each bee individually on a flat glass surface or in a petri dish (Fig. 3).
2. Mount the wings dry on microscope slides after rinsing in

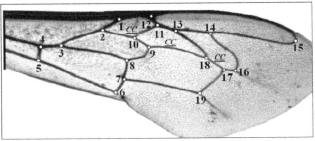

Fig. 2. Location of the nineteen landmarks on the fore-wing of honey-bee workers considered in the geometric morphometric analysis (CC = cubital cell) http://apiclass.mnhn.fr; Miguel *et al.*, 2010).

Fig. 3. Dissection of bees for the full character suite of classical morphometry. Photo: Institut für Bienenkunde Oberursel

ethanol with decreasing concentrations (70% to 20%) and distilled water. Care has to be taken to mount the wings flat on the slide, without any folds or distortions. Taking care that all wings face the same direction will facilitate measurements. Cover with second microscope slide and fix together with masking tape. Label with sample ID.

3. Carefully clean sternites and tergites from adhering tissue with forceps and/or a small spatula so as not to tear the chitin. If the chitin is very tender (e.g. because material is very old or has previously been subjected to freezing), tissue can be removed by immersion into concentrated KOH solution for a few hours (obey operator safety precautions: wear gloves and goggles) and subsequent rinsing in H_2O.

4. Embed flat body parts (legs, sternites, proboscis) in a solution of gum Arabic (not permanent), Euparal or Canada balsam (permanent) on microscope slides. Take care to avoid any air bubbles in the solution. Mounting body parts to face the same direction will facilitate measurements (Fig. 4).

5. Cover with a second microscope slide and label with sample ID.

6. When using Euparal or Canada balsam, keep the slides on a warming plate for several hours for the medium to harden.

7. Because of their natural rounded shape, tergites are best glued to 10 mm glass rods covered with gum Arabic, Euparal or Canada balsam. Cover rod and tergites with clear tape after mounting. Use a spacer to suspend rods from surface to avoid sticking.

After mounting, the body parts can be measured. There is a wide range of measuring methods, including microscopes with ocular scales, online video measurements, and photographing the slides and analysing them with commercial or freeware picture analysis programs. As long as devices are carefully calibrated, the specific methods do not matter too much. Precise calibration, however, is a critical prerequisite for a true representation of all measurements related to size and its importance cannot be emphasised enough!

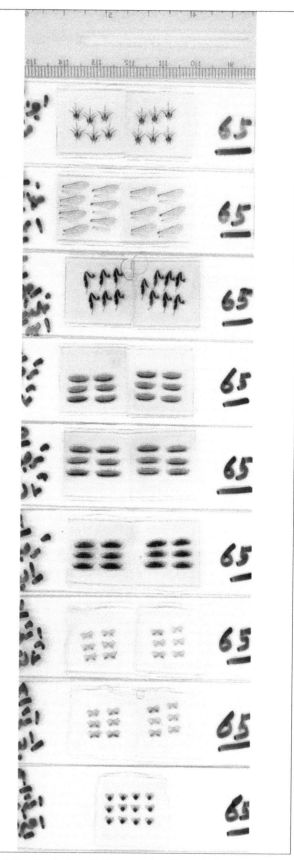

Fig. 4. Mounted body parts for measuring. Photo: Mohammed Al Sharhi.

- The calibration procedure greatly depends on the measuring method used (computer-aided techniques will mostly come with inbuilt calibration steps).
- Measure any distance using your own measuring device (microscope eyepiece, computer program etc.).
- Use a micrometre slide or similar exact measuring device to calculate the real distance from the units of the device used in step above.
- It is important to re-calibrate after each change of magnification.

Mostly, measurements are taken manually, but often with customised digital support as introduced by Daly (1982), which greatly improved precision. Any digitizing picture analysis program might be used, after verifying that it possesses sufficient precision. In geometric morphometry, a commonly used program is TPSdig (Rohlf, 2001), but efforts have been made to use automation for capturing wing morphometry. Quezada-Euán *et al.* (2003) and Steinhage *et al.* (1997) developed a semi-automated method to obtain wing measures that reduced the analysis time and improved precision for identifying bee species; this method was later improved to an automated system (Steinhage *et al.*, 2007). The coordinates of the eighteen wing landmarks can also be determined automatically using the DrawWing software (Tofilski, 2004).

In addition, to achieve the best possible compatibility of data obtained by different laboratories, great care has to be taken to always carry out measurements or assessments in exactly the same way. This is of essential importance, since particular personal habits or "measuring traditions" within different laboratories can introduce biases that can become troublesome and render data sets incompatible. It is thus essential for inexperienced researchers, before starting a new morphometric project, to seek advice from experienced laboratories and to use repeated measurements from the same samples for cross-checking. Even between laboratories involved in morphometry there is a need for crosschecking to standardize methods in detail.

3.1.3. Analysis and statistical methods
3.1.3.1. Documenting and naming honey bee variation
To detect, document and define previously unknown variation is a rather demanding task. First, it requires measuring a broad set of characters, in order not to miss the relevant morphological differences by not measuring a particular feature. In fact, even the conventional set already constitutes a compromise between desirable accuracy and work load, and is perhaps representing a dangerous restriction, because a (hitherto unknown) variant might be crucially different in some characters not considered. Therefore, although the logical recommendation would be to measure all characters possible, we recommend to measure at least a core set of the characters presented in Table 3 (bold type), together with the 20 point coordinates of wing

morphometry as shown in Fig. 2. From these measures, any indices or compound measures can be calculated. This minimum of common measurements would also ensure a sufficiently broad base for referencing the new variation against the known character distributions, and provide a sufficiently accurate account of features representing a numerical description of the morphological variation of the bees.

Procedures for analysis are supplied in common statistical software packages, such as SPSS, Systat, Statistica or others. Analysis should be based on the mean of 10 to 15 colony members rather than on single workers. Otherwise, because of the relatedness of workers to each other, some degree of pseudo-replication, difficult to account for, would be introduced. Only the measurements of the primary characters, not the derived compound measurements or indices are used for analysis. In the investigation of unknown variation, a first aim is to investigate the sample character sets, in reference to samples from already identified and confirmed groups.

There are two main methods (detailed below) to detect whether the samples under study represent one or more groups. The most common and recommended primary method is principal component analysis (PCA), which reduces the numbers of dimensions (each character is a dimension) to two or three main factor dimensions, into which the positioning of samples can be plotted and then inspected visually. In an easy situation, the new samples separate into one or few clusters distinct from all reference clusters. However, quite frequently, they only occupy distinct areas of a point cloud and are not clearly set apart, signifying some not unusual overlap with other groups. For interpretation, plotting several dimensions against each other, combined with local labelling of the samples can be helpful to clarify the positioning.

A second useful method to identify grouping patterns is k-means clustering, a procedure separating the samples under analysis into different, predefined, numbers of groups, and indicating by an F-statistics based goodness of fit test which number of groups fits best. In addition, k-means group memberships can be matched against the geographical distribution of the samples to investigate the consistency of grouping with local coherence or ecological zones. Hierarchical clustering procedures may also be used to supplement the analysis, as long as the number of samples is not too high. Distance matrices can be very helpful for clarifying relations to other groups.

If the above methods have led to a sensible group definition, this should finally be verified by discriminant analysis which determines the significance of group differences and the accuracy with which samples are re-allocated into their correct groups.

A major difficulty in finding and verifying groups is to differentiate gradual, clinal change in character distributions from truly distinct groupings. The main caveat is that gaps in the sampling pattern can easily give the impression that two or more distinct groups exist, which may even be verified statistically, although the distinctness would disappear if the whole range had been sampled evenly (Radloff and Hepburn, 1998). In contrast, true groups are characterised by sudden morphological changes in relation to their geographic origin, which can be verified by geographic plots and relations to physiogeography. However, even in the case of clinal changes over extended regions the necessity might arise to name the different ends of the cline differently, although they only represent parts of a continuous distribution. There is no obvious general solution to this problem.

3.1.3.2. Sample identification

In contrast to the above, sometimes sufficient information on the potential origin of the samples in question may already be available, and it is therefore just necessary to assign the samples to the correct subspecies from a few options. This is a very different issue, and a far less demanding task. It relates to fairly common situations: for example, somebody simply wants to know which subspecies, ecotype or possibly even breeder-line a given set of bee samples belongs to. Mostly, the question can even be asked more narrowly, as only a restricted set of possibilities needs to be considered, e.g., the allocation of a given sample to two, three or four subspecies.

Accordingly, few characters would already suffice to distinguish these possibilities, thus considerably reducing the labour involved. In specific cases, as in the replacement of *A. m. mellifera* by *A. m. carnica* in Germany during the past century, diagnosing a sample into one of these two subspecies could be based on hair length and cubital index alone, and for current characterisation of *A. m. carnica* the addition of two further characters, tergite colour and tomentum width, is sufficient. Similar simple methods have been applied in other countries, e.g. the three characters width of tergite IV, length of proboscis, and cubital index are used in Poland, and a similar set of length of proboscis, and cubital index together with wing length and width is applied in Slovakia (Bouga *et al.*, 2011). In particular, the derived characters, such as composed measures and indices which are useless in statistical evaluations, can be very useful to construct simple character sets for simplified distinctions which can be applied by bee breeders. In Switzerland, distinctions between *A.m. mellifera* and *A. m. carnica* are predominantly based on the cubital and dumb-bell indices.

Because the requirements for subspecies identification are predominantly designed by local demands (see Bouga *et al.*, 2011), together with the need for economizing the effort of character measuring and the lack of a complete catalogue of honey bee morphometric variation in Europe, there is no accredited identification system for European bees with general validity. Instead, several independent national systems exist, each considering a limited number of alternatives. In some countries (Italy, Greece), formal accreditations are given to laboratories for relevant identification (Bouga *et al.*, 2011).

However, there is no need to economise on the number of wing characters or the precision of shape analysis, once our suggestion has been followed and the standardised and generally accepted method has shifted to storing the point coordinates of the wing venation junctions. Presumably, automated geometric wing analysis may then prove to be the most efficient identification method in the future. However, it is not yet clear how reliably geometric wing morphometry alone will discriminate all European subspecies and known ecotypes from each other. Additional morphometric characters from other body parts may need to be included in such an envisioned all-European system still to be worked out.

The necessity to recognize Africanized bees in the New World may be regarded as a special case of sample identification, for instance when investigating if a swarm originates from an Africanized colony, or for certifying a potential breeder colony as free from African genes. In principle, the same criteria as above apply to this situation, but to facilitate achieving a fast result, specialised applications have been developed for this specific use, based on only few morphometric characters and the size differences between honey bees of African and European origin (e.g. FABIS, Rinderer *et al.*, 1986, 1987). However, as the purpose of the method is focused on discriminating "Africanized" from "European", the resolution is rather low and it is not suitable for further discrimination between subspecies. The USDA-ARS (http://www.ars.usda.gov/Research/docs.htm?docid=11053) uses FABIS for a preliminary test, followed by a full morphometric analysis, if possible Africanization is detected.

The most relevant and most powerful method to establish an allocation of "unknown" samples to a number of reference groups is discriminant analysis, based on the full set of characters. In a first step of this process, a large set of reference samples would be included, confirming their correct reallocation to their own groups. By entering the measured characters stepwise, the character subset relevant to separate the groups in question can be determined, and for further identification only these characters need to be considered. In the next step, the unknown samples would be entered into the analysis and allocated to the reference groups, including the probability of these allocations. It needs to be stressed again that this method is unsuitable to detect any new variation or ecotypes, as it forces a choice between a limited and predefined number of possible group allocations. Non-fitting types may not be allocated with high probability to any of these, but hybrids could not be told apart from genuinely new types.

3.1.4. Reference data

It is obvious that recognised sets of reference samples are of utmost importance not only in determining new ecotypes, but even more so for allocating unknown samples into the existing pattern. However, there is a regrettable lack of consistency between the reference sets

Table 4. PCR-RFLP assays that have been used to identify honey bee matrilineal origins. [1]Intergenic region also named tRNAleu-cox2. This intergenic region has generated over 100 distinct haplotypes. Therefore, a simple allocation into two categories, as given in this Table, is not possible. This assay is recommended for honey bee maternal identification.

Genes	Restriction Enzyme	Cleaved fragment	Uncleaved fragment	Authors
Cytochrome *b*	*Bgl*I	European	African	Crozier *et al.*, 1991
Ls rRNA	*Eco*RI	*A. m. ligustica*, *A. m. carnica*, *A. m. caucasica*	*A. m. mellifera*, *A. m. iberiensis* of lineage M	Hall and Smith, 1991
COI	*Hinc*II	*A. m. mellifera*, *A. m. iberiensis* of lineage M	*A. m. ligustica*, *A. m. carnica*, *A. m. caucasica*	Hall and Smith, 1991
	*Xba*I	*A. m. ligustica*, *A. m. carnica*, *A. m. caucasica*	*A. m. mellifera*, *A. m. iberiensis* of lineage M	Hall and Smith, 1991
	*Hinf*I	*A. m. lamarckii*	Non-*A. m. lamarckii*	Nielsen *et al.*, 2000
	*Nco*I	*A. m. macedonica*	*A. m. adami* *A. m. cecropia* *A. m. cypria*	Bouga *et al.*, 2005 Stevanovic *et al.*, 2010
	*Sty*I	*A. m. macedonica*	*A. m. adami* *A. m. cecropia* *A. m. cypria*	Bouga *et al.*, 2005 Stevanovic *et al.*, 2010
	*Ssp*I	Greek	Bulgarian	Ivanova *et al.*, 2010
ND5	*Alu*I	*A. m. macedonica*	*A. m. adami* *A. m. cecropia* *A. m. cypria*	Bouga *et al.*, 2005
	*Hinc*II	Greek	Bulgarian	Ivanova *et al.*, 2010
	*Fok*I	Greek	Bulgarian	Ivanova *et al.*, 2010
[1]COI-COII	*Dra*I			Garnery *et al.*, 1993

used, and frequently, except some general information on sampling location, no further specification of the origin of reference samples or their validation method is given. This is specifically worrisome as beekeeping might already have changed the properties of the population in question, or colonies might have been selected as source for breeder lines without previous confirmation of their congruency with the original subspecies. Thus, there is an urgent need to establish generally and widely accepted standards for subspecies reference samples, combined with the storage of the corresponding samples and data in accessible collections and data banks. To date, the only comprehensive and well documented compilation of data for morphometric characters for the majority of honey bee subspecies is given by Ruttner (1988), and complete sets of these measurements can be obtained from the Oberursel Institute, Germany.

3.2. Mitochondrial DNA

Mitochondrial DNA is a small (circa 16 000 bp), circular molecule that is transmitted intact by the queen to her offspring (workers and drones). Therefore, unlike the other markers discussed in this chapter, mtDNA is uniparentally (maternally) inherited, rather than biparentally. The non-recombining maternal inheritance renders interpretation of mtDNA data straightforward, which in combination with relatively simple and inexpensive assays has made mtDNA one of the most popular markers in honey bee genetic studies.

Variation in honey bee mtDNA has been detected using a variety of molecular methods, ranging from RFLPs (Restriction Fragment Length Polymorphisms), to PCR-RFLPs and direct sequencing. RFLP variation is revealed by digesting the entire mitochondrial genome with restriction endonucleases. These enzymes are isolated from bacteria and cut DNA at a constant position within a specific recognition site that is typically 4 or 6 bases long. A battery of 4 (*Hinf*I) and 6-base (*Acc*I, *Ava*I, *Bcl*I, *Bgl*I, *Eco*RI, *Hinc*II, *Hind*II, *Hind*III, *Nde*I, *Pst*I, *Pvu*II, *Xba*I) recognition site restriction enzymes were used in early studies of European and African honey bee subspecies in the native (Smith *et al.*, 1991; Garnery *et al.*, 1992; Arias *et al.*, 2006), and introduced ranges (Smith and Brown, 1988; Hall and Muralidharan, 1989; Smith *et al.*, 1989). While these RFLP surveys roughly discriminated and supported three of the major morphology-based lineages (M, A, and C, see also Annex) proposed by Ruttner (1988) and revealed the existence of some subspecies-specific haplotypes (Smith *et al.*, 1991), no single or composite set of restriction enzymes have proved to be diagnostic markers for subspecies identification.

The RFLP method has been replaced by PCR-RFLP mainly to overcome the disadvantage that it requires relatively large amounts of non-degraded DNA. In both methods, haplotypes are resolved by employing restriction enzymes. However, unlike RFLPs, which survey the whole mitochondrial genome, PCR-RFLP variation is revealed within a specific PCR-amplified region. Several PCR-RFLP assays (Table 4), spanning coding regions, have been developed and employed in identification of New World honey bee populations undergoing Africanization (Hall and Smith, 1991; Clarke *et al.*, 2002; Pinto *et al.*, 2003; Pinto *et al.*, 2004) and in Old World honey bees of eastern European ancestry (Smith *et al.*, 1997; Bouga *et al.*, 2005; Ivanova *et al.*, 2010; Stevanovic *et al.*, 2010). These assays have proved to be useful for discriminating variation in some eastern European subspecies, previously undetected using other mtDNA methods. Specifically, amplification of COI and ND5 genes followed by digestion with *NcoI*/*StyI* and *AluI*, respectively, produce diagnostic patterns characteristic of *A. m. macedonica* (Bouga *et al.*, 2005; Stevanovic *et al.*, 2010). Additionally, there are differences between Greek and Bulgarian honey bees in the digestion pattern of COI with *SspI*, and of ND5 with *HincII* and *FokI*, which do not recognize any site in Bulgarian honey bees (Ivanova *et al.*, 2010).

The PCR amplification of the non-coding region located between the tRNAleu and COII genes (originally named COI-COII intergenic region), followed by digestion with the *DraI* restriction enzyme, has been the most popular PCR-RFLP assay (Garnery *et al.*, 1993). This assay, which is commonly known as the **DraI test,** has been widely used in honey bee maternal identification in both Old World and New World populations. It has been helpful in (i) phylogeographical studies (De la Rúa *et al.*, 1998; Franck *et al.*, 2001; De la Rúa *et al.*, 2006), (ii) understanding the complexities underlying natural hybrid zones (Franck *et al.*, 1998; Franck *et al.*, 2000; Cánovas *et al.*, 2008), (iii) detecting introgression of foreign queens (Jensen *et al.*, 2005), (iv) tracking the temporal changes of maternal composition of honey bee populations undergoing Africanization (Clarke *et al.*, 2002), among others.

The *DraI* test has shown a high power of resolution which results from a combination of length variation with restriction site polymorphisms. The COI-COII intergenic region is composed of two distinct nucleotide sequences, named P and Q, where P can also appear in several variations (P_0, P_1, P_2). According to Garnery *et al.* (1993), each evolutionary lineage includes a variant of the P sequence, combined with a different copy number of the Q sequence, resulting in length polymorphisms of this mtDNA region. Honey bees of the eastern European lineage (C) have the shortest intergenic sequence, because they lack the P element and carry a single copy of the Q element (Cornuet *et al.*, 1991; Garnery *et al.*, 1993). Honey bees belonging to lineages M, A, Z and Y exhibit longer intergenic regions because they carry from one up to five Q elements (Rortais *et al.*, 2011) in addition to a variant of the P element (see the *BEEBOOK* paper on molecular methods (Evans *et al.*, 2013) for further details). Additional polymorphisms can be resolved following digestion of the length variants with the *DraI* restriction enzyme. Lineage Z was formerly known as (mitochondrial) O (see Annex).

The combination of length and restriction-site polymorphisms produced by the *DraI* test has resolved over 100 haplotypes (Franck *et al.*, 2001; De la Rúa *et al.*, 2005; Collet *et al.*, 2006; Shaibi *et al.*, 2009; Alburaki *et al.*, 2011; Rortais *et al.*, 2011; Pinto *et al.*, 2012),

which have been correctly assigned to evolutionary lineages. However, despite the higher resolving power of the *DraI* test compared to other assays, it is unable to identify honey bees at the subspecies level because it does not produce diagnostic haplotypes. For example, if there is no further information available, a given honey bee carrying a C1 haplotype could be identified as *A. m. carnica* or *A. m. ligustica,* because this haplotype is present in both populations, although at different frequencies (Muñoz *et al.*, 2009). The same applies to an A1 haplotype which could be carried by an *A. m. iberiensis*, *A. m. adansonii* or other African subspecies (Franck *et al.*, 2001; Cánovas *et al.*, 2008).

Despite its inherent limitations, the *DraI* test should be adopted as the standard for honey bee maternal identification, for several reasons. First, because among all PCR-based methods, it has proven to be the most powerful and informative. Second, because COI-COII/*DraI* variation has been widely documented across natural and introduced honey bee ranges. Consequently, a large catalogue of haplotype patterns is available in the literature (see restriction maps in Franck *et al.*, 2001; Collet *et al.*, 2006; Shaibi *et al.*, 2009; Pinto *et al.*, 2012), and sequence data for numerous described haplotypes have been deposited in GenBank. Additionally, as for the other PCR-based assays, colony identification using the *DraI* test can be performed in a small-sized laboratory equipped with basic equipment (Table 5), and it only takes two days for haplotype identification. A detailed protocol to implement the *DraI* test (from DNA extraction to digestion with *DraI*) is available in the *BEEBOOK* paper on molecular methods (Evans *et al.*, 2013).

Sequencing is the ultimate method for assessing mtDNA variation. Sequence data generate haplotypes that are identical by descent instead of identical by state, required for phylogenetic analysis (Arias and Sheppard, 1996). Sequencing has been employed mostly for characterising novel COI-COII/*DraI* haplotypes (Franck *et al.*, 2001; Collet *et al.*, 2006; Shaibi *et al.*, 2009; Pinto *et al.*, 2012). However, it should be employed more often in mtDNA surveys, because by providing further resolution it may reveal diversity patterns that would otherwise have gone undetected (Muñoz *et al.*, 2009; Martimianakis *et al.*, 2011).

While important insights into patterns and processes of honey bee maternal diversity across natural and introduced ranges have been gained studying mitochondrial DNA variation, the sole use of this molecule for inferring honey bee evolutionary and human-mediated contemporary history (e.g. introgression) and for subspecies identification is questionable. First, because there is no single mtDNA assay that is subspecies diagnostic. Second, because it only provides the maternal component of variation and is therefore unable to detect hybridisation or introgression events, which are becoming increasingly common with human-assisted gene flow through queen trading and colony transhumance. Therefore, a full and accurate identification requires employment of both mtDNA and nuclear markers (see also the *BEEBOOK* paper on molecular methods (Evans *et al.*, 2013)).

Table 5. Summary of characteristics of the different methods used for identification of honey bee subspecies.

Characteristic	Morphometrics	Allozymes	MtDNA	Microsatellites	SNPs
Number of individuals per colony	10-15	10	1	1 or more (depending on the goal)	1 or more (depending on the goal)
Characters/loci usually screened	Up to 41; wing venation	MDH, ME, EST, PGM, HK, ALP	COI-COII/*Dra*I, COI/ *Nco*I/ *Sty*I/ *Ssp*I, ND5/*Alu*I/ *Hin*dII/*Fok*I, 16s rDNA/ *Eco*RI	Hundreds available. However, for most studies less than 20 screened (e.g. A7, A24, A28, A88, A113, B124, Ap43, A14, A107, A35, Ap55, Ap66)	Hundreds (1536 for Golden Gate Assay of Illumina) to thousands (with the Infinium Assay of Illumina)
Inheritance	Biparental	Biparental	Maternal	Biparental	Biparental
Dominance		Co-dominant	N/A	Co-dominant	Co-dominant
Polymorphism		Low	Very high in COI-COII intergenic region, otherwise medium to low	Very high	Can be high
Number of alleles		Multi-allelic	Multi-allelic	Multi-allelic	Biallelic
Abundance in the genome		Low		Medium	Very high
Cross-lab/study comparisons	Cross-checking recommended	Easy	Easy	Requires special preparation and cross calibration	Easy
Time to complete lab protocol	Depends on character suite typically 1 sample per day for full suite	1 day	Depends on assay, up to 2 days	1 locus or one multiplex up to 2 days	3 days
Main software packages	SPSS, Systat, Statistica, Morpheus, NTSYS, MORPHOJ	GenAlex, Genepop, and others	GenAlex, Genepop, Network, Structure, and others	Genepop, Arlequin, Structure, GenAlex, GeneClass, Adegenet, and other R packages	Plink, Structure, Admixture
Main Equipment	Microscope, camera, measuring software, computer	Centrifuge, Electrophoresis Unit, incubator	Thermal Cycler, Electrophoresis Unit, Centrifuge, Water Bath	Thermal Cycler, Electrophoresis Unit, Centrifuge, Automated Sequencer	Thermal Cycler, Analyst Plate Reader, Hybridisation Oven, Bead Array Reader
Cost of equipment	Low	Low	Medium	High	Very high
Cost of genotyping	Low	Low	Medium	High	Very High

3.3. Nuclear markers

3.3.1. Allozymes

Allozyme analysis – the investigation of allelic variation at enzyme loci, accessible through gel electrophoresis - was the first type of molecular marker to be applied in the field of honey bee research (Mestriner, 1969; Mestriner and Contel, 1972). The majority of allozyme studies in honey bee diversity appeared in the 80's and 90's of the past century (Garstide, 1980; Nunamaker *et al.*, 1984; Sheppard and Berlocher, 1984, 1985; Badino *et al.*, 1988; Lobo *et al.*, 1989; Kandemir and Kence, 1995; Dedej *et al.*, 1996). Allozyme studies have contributed to our understanding of gene flow, population structure and hybridisation (Badino *et al.*, 1983, 1985, 1988; Sheppard and McPheron, 1986; Del Lama *et al.*, 1990; Meixner *et al.*, 1994) and revealed founder effects (Cornuet and Fresnaye, 1979; Sheppard, 1988).

One of the most polymorphic and most widely used enzyme loci in studies of population variation of honey bees has been the Mdh1 (malate dehydrogenase) locus. Its variation has been particularly useful, because its allele frequencies differ greatly between different the lineages of the honey bee. Therefore, Mdh variation has been

extensively used in studies about the spread of Africanized honey bees, or in studies on differentiation of European lineages (Cornuet, 1982; Cornuet *et al.*, 1986; Lobo *et al.*, 1989; Del Lama *et al.*, 1990). However, as there is evidence that variation of the Mdh1 locus can be interpreted as consequence of physiological adaptation to different climates and thus may not necessarily reflect gene flow (Coelho and Mitton, 1988; Nielsen *et al.*, 1994, Cornuet *et al.*, 1995), phylogeographic conclusions should be drawn with caution, since the Mdh1 Locus is obviously not selectively neutral.

Comparatively few loci are polymorphic in honey bees and there are no fixed allelic differences between subspecies; therefore, variation within *A. mellifera* consists exclusively of differences in allele frequencies between populations. This limitation renders allozymes less suitable for the purpose of identification of a small sampling or a single colony; however, the marker may still be employed usefully in population studies. As an advantage, the method is easy to use and scoring the data is comparatively unproblematic and straightforward. In addition, the method is economical in setup and running, since it does not require too much expensive laboratory equipment. The most

Table 6. Most polymorphic and most commonly used enzyme loci for allozyme studies. E.C. Enzyme commission number of the locus (www.chem.qmul.ac.uk/iubmb/enzyme).

Polymorphic locus	Known Alleles	References	Staining recommendation
MDH-1 EC 1.1.1.37	MDH65 MDH80 MDH87 MDH100 MDH116 MDH125 MDH133	Garstide, 1980; Nunamaker *et al.*, 1984; Badino *et al.*, 1983, 1985, 1988; Sheppard, 1988; Sheppard and Berlocher, 1984, 1985; Sheppard and McPheron, 1986; Lobo *et al.*, 1989; Meixner *et al.*, 1994; Kandemir and Kence, 1995; Smith and Glenn, 1995; Dedej *et al.*, 1996; Kandemir *et al.*, 2000; Bouga *et al.*, 2005; Ivanova, 2010; Ivanova *et al.*, 2007, 2011	NAD (0.035 g) NBT (0.03 g) PMS (0.0015 g) Na malate (10 ml) Phosphate buffer (pH 7.3) – 100 ml
ME EC 1.1.1.40	ME70 ME100 ME106 ME117	Sheppard and Berlocher, 1984, 1985; Sheppard and McPheron, 1986; Meixner *et al.*, 1994; Dedej *et al.*, 1996; Kandemir *et al.*, 2000, 2005; Bouga *et al.*, 2005; Ivanova, 2010, Ivanova *et al.*, 2011	NADP (0.035 g) NBT (0.03 g) PMS (0.0015 g) Na malate (10 ml) Phosphate buffer (pH 7.3) – 100 ml
EST 3 EC 3.1.1	EST70 EST80 EST88 EST94 EST100 EST105 EST118 EST130	Sheppard and McPheron, 1986; Kandemir and Kence, 1995; Bouga *et al.*, 2005; Ivanova, 2010; Ivanova *et al.*, 2010a, b, c, 2011	1-naphtil acetate Na salt (0.010 g) 2-naphtil acetate Na salt (0.020 g) Fast blue BB salt (0.050 g) Phosphate buffer (pH 7.0) – 100 ml
ALP EC 3.1.3.1	ALP80 ALP90 ALP100	Bouga *et al.*, 2005; Ivanova, 2010; Ivanova *et al.*, 2010a, b, c; Ivanova *et al.*, 2011	1-naphtil phosphate Na salt (0.020 g) Fast garnet (0.050 g) MgCl2 (0.120 g) 0.1 M TRIS-HCl buffer (pH 8.6) – 100 ml
PGM EC 5.4.2.2	PGM75 PGM80 PGM100 PGM114 PGM120	Mestriner and Contel, 1972; Brueckner, 1974; Nunamaker and Wilson, 1980; Badino *et al.*, 1983; Sheppard and Berlocher, 1985; Del Lama *et al.*, 1985; Meixner *et al.*, 1994; Smith and Glenn, 1995; Kandemir and Kence, 1995; Ivanova, 2010; Ivanova *et al.*, 2010a, b, c; Ivanova *et al.*, 2011	G-1-Phosphate Na salt (0.6 g) MgCl2.6H2O (0.2 g); NADP (0.01 g) NBT (0.020 g) PMS (0.001 g) G-6-P-D (80 U (0.035 mg)) 0.5 M TRIS-HCl buffer (pH 7.1) – 10 ml H2O – 90 ml
HK EC 2.7.1.1	HK77 HK87 HK100 HK110 HK120	Sheppard and McPheron, 1986; Badino *et al.*, 1988; Del Lama *et al.*, 1988, 1990; Smith and Glenn, 1995; Kandemir and Kence, 1995; Kandemir *et al.*, 2000; Ivanova *et al.*, 2010a, b, c; Ivanova *et al.*, 2011	Glucose (0.09 g) MgCl2.6H2O (0.02 g) ATP (0.025 g) NADP (0.025 g) NBT (0.020 g) PMS (0.003 g) G-6-P-D (80 un (0.035 mg)) 0.5 M TRIS-HCl buffer (pH 7.1) – 10 ml H2O – 90 ml

commonly studied enzyme systems and polymorphic loci, together with the most common alleles and recommendations for staining are summarized in Table 5.

3.3.1.1. Brief description of procedures

We here give a very brief overview over the steps involved.

1. Extraction:
   a. Dissect the bee and place thorax in a microtiter plate well.
   b. Grind the thorax with 20 µl of Tris-HCl buffer (for MDH, ME, EST, PGM and HK, see table 5). Use the whole body for the study of ALP (see Table 6).
2. Electrophoresis: good results can be obtained with the use of 7.5% polyacrylamide gel (Davis, 1964) with Tris-Glycine buffer pH 8.3.

Caution: acrylamide in unpolymerised form is a potent neurotoxin! Observe operator safety when handling (wear gloves, work under fume hood)! It is recommendable to cast the gels using acrylamide solution and we discourage the use of acrylamide powder. The liquid preparation is much easier to handle and presents less of a health

hazard. A good alternative to polyacrylamide alternative is the purchase and use of ready-cast cellulose acetate gels.

3. Staining:
   a. prepare locus-specific stain solution according to recommendations.
      Recommendations for substrate systems, stains and buffers have been published in: Boyer, 1961; Gahne, 1967; Shaw and Prasad, 1970; Harris and Hopkinson, 1976. We have included recommendations for stains in Table 6.
      Obey safety precautions, since many ingredients may pose a health hazard!
   b. immerse gel into solution,
   c. incubate in the dark according to recommendations (specific for different loci, no general recommendation can be given here).
4. Scoring: It is the convention to score the most common band as *locus*[100] and measure the mobility of all other bands in relation to this band (e.g., *Mdh*[100], *Mdh*[65]).

3.3.1.2. Data analysis

Allozyme data can be statistically analysed using similar approaches as those for DNA microsatellites outlined below (section 3.3.2).

3.3.2. DNA microsatellites

DNA microsatellites are a type of variable number of tandem repeats (VNTR), also known as short tandem repeats (STR) and simple sequence repeats (SSR). Microsatellites are segments of DNA with a sequence motif ranging from 1 to 6 bases, repeated from 4 to probably 100 or more times (Tautz, 1990). It seems that most eukaryotes carry these elements, whose role is not particularly clear. Honey bees and bumble bees were the first social insects for which DNA microsatellites were developed (Estoup *et al.*, 1993). The aim of the original project was to determine the zone of introgression between *A. m. mellifera* and *A. m. ligustica* in the region of the French-Italian Alps. For each microsatellite locus, a specific set of primers is used for PCR. A valuable source for loci in honey bees was published by Solignac *et al.* (2003), including information on subspecies variation and possible amplification in three other species of *Apis*.

DNA microsatellites offer variation useful for inferring population differentiation, originating from a high mutation rate (Levinson and Gutman, 1987) during replication, by adding or removing repeat units compared to the original microsatellite element. Estoup *et al.* (1995a) described the process of allelic evolution and used these markers to separate 9 populations of honey bees with seven microsatellite loci. The approaches used for inferring population differentiation at this early stage were mainly based on F-statistics and tree drawing methods from phylogeny. The first studies were done with a sample size of approximately 40 individuals per population, which since then has been found an effective number (Cornuet *et al.*, 1999). A detailed study of the west European honey bees used 11 loci to differentiate the M lineage bees from C lineage bees of Greek origin and to demonstrate a high amount of introgression into the French population (Garnery *et al.*, 1998b). The statistical analysis used were Fst values and phylogenetic trees, indicating a marked interest in grouping samples, based not on their sampling origin, but based on their similarity. This method does cluster the Iberian honey bees separate from the rest. The paper was important, as the data were used to suggest methods to define conservation strategies of local honey bee populations.

Franck *et al.* (1998) analysed the origin of west European honey bees with 8 microsatellite loci and mtDNA. The analysis demonstrated that the Iberian bees were rather pure, with limited influence from Northern Africa, drawing strength from the combination of two independent marker types. Similar methods were used to determine variation in honey bees from the Middle East that led to the hypothesis of a fourth lineage based on molecular data (Franck *et al.*, 2000) (see Annex on lineages). A time related study concerning the Africanization process of the Yucatan peninsula demonstrated the strength of DNA microsatellites in finding hybrids (Clarke *et al.*, 2002).

3.3.2.1. Data analysis

A method of analysis alternative to F-statistics and phylogenetic trees is assignment testing, which can be applied with several variations (Manel *et al.*, 2005). Two main types of assignment test can be distinguished:

Deterministic assignment compares the genotype of each individual, and groups are formed according to the sampling location or other likely categories. The assignment analysis then compares the probability for each sampled genotype being drawn at random from its own group of individuals, or from one or more alternative groups, based on the allele frequencies of each group. The population of origin is determined from the probability; however, it is also possible to reject the hypothesis that any of the reference populations are the source of origin, based on the calculated probabilities. The software package GENECLASS is the most advanced tool for this task (Piry *et al.*, 2004). For small sample sizes of less than 30 individuals, it is best to consider the individual genotypes as belonging to each population (as is), for large sample sizes it is better to remove the individual genotype from all subgroups ("leave-one-out" approach) to avoid self-assignment.

Alternative to the classical or deterministic assignment test, Bayesian assignment works without prior knowledge of the number of populations. Instead, it tries to determine the best assortment of the genotypes, while varying the number of clusters that the individuals are sorted into. The data from microsatellites are entered raw for analysis without designation of population origin, and the software varies the number of clusters in order to determine not only their numbers, but also for each individual from which cluster it most likely originates. The program STRUCTURE (Pritchard *et al.*, 2000) is the most commonly used software, but several other options exist. Ideally, the numbers of clusters found resembles the number of populations expected by the investigator. However, more objective methods exist to determine the optimal number of clusters for a given data set, based on the posterior probability calculated (Evanno *et al.*, 2005). The Bayesian method is sensitive and can assign populations at various levels, like closely related subspecies and more distantly related branches. However, it is important to avoid genotyping related individuals, as the software is clearly capable of picking up differences based on resemblance due to common ancestry. An example of this method in honey bees is a study of various levels of introgression of *A. m. ligustica* into populations of *A. m. mellifera* (Jensen *et al.*, 2005). Assignment tests have also been used to detect recent hybrids using the software NewHybrids, because individuals with intermediate probability are likely to have mixed origin (Soland-Reckeweg *et al.*, 2009).

Spatial methods have been developed for the use with DNA microsatellites. Currently there are studies underway with the methods in GENELAND (Guillot, 2005) and TESS (Durand, 2009), two software packages based on Bayesian assignment, and in ADEGENET (Jombart, 2008), a PCA based software package in the Statistical language R. We recommend analysis with ADEGENET, which produces interesting and rapid results, even without geographical data attached to the genotypes (Uzunov *et al.*, 2013), and it has fewer underlying assumptions than the other methods.

3.3.2.2. Problems and potential pitfalls with using microsatellites:

Homoplasy, the evolution of identity in state is one of the biggest problems when using DNA microsatellites. The rate of mutations in DNA microsatellites is high, due to replication slippage, and hence homoplasy is a factor of importance. Estoup *et al.* (1995b) demonstrated a method to quantify this process in honey bees, using interrupted microsatellite loci and bees of various subspecies. Their results show that alleles of identical length can in fact be differentiated by sequencing. Within the major (molecularly defined, see Annex) lineages A, C and M, alleles with identical length tend to also have identical sequences, but variation between subspecies is frequent. Detailed studies demonstrate that 7 different alleles all have the length of 224 in the microsatellite locus A113 (Viard *et al.*, 1998) in eight populations. Most studies on DNA microsatellites are limited to determining the length of each allele, which will lead to an under-estimation of allelic richness. This is an important caveat to consider when interpreting the results of studies between honey bee populations belonging to different lineages.

It is common in studies of bee populations to find a lower level of observed heterozygosity than expected. The term null alleles is used to describe alleles that will not yield a PCR product (Chapuis and Estoup, 2007). This is probably due to a mutation at the primer binding site. The bee will appear homozygous if a null allele occurs together with a normal allele. If the null allele occurs at a high frequency in a given population, the observed heterozygosity will decrease. However, in general there is little testing performed for the presence of null alleles. It is possible to detect the presence of null alleles directly, via the observation of parent offspring triads. This process is comparatively easy in honey bees due to the haploid state of the drone, however rarely pursued. Alternatively, a more indirect method was established in the software packages Genepop (Raymond and Rousset, 1995), and FreeNA (Chapuis and Estoup, 2007), using the presence of homozygous null alleles in individuals and the increase of homozygosity in various populations to estimate the frequency of null alleles. Not surprisingly, it is possible to show that the frequency of null alleles varies across a range of populations and loci (Kryger, unpublished data).

In order to determine the subspecies status of an individual honey bee, a honey bee colony, or a honey bee population, it is important to compare the results to reference material and published genotype information. Unfortunately, no standard reference material, such as a standard allelic ladder, is available for honey bees as there is for several other species (O'Reilly *et al.*, 1996; Schnabel *et al.*, 2000). No accepted source is available that would provide a standard set of

alleles of known length to compare to your results and calibrate your own fragment sizes against. Even the use of custom made oligos of known length is not an ideal remedy, since their run times may vary depending on the base composition of the fragment. The lack of organisation amongst the scientists studying honey bee populations has resulted in the use of a large variety of loci with differently labelled primers. As a result, the data from independent studies are at best difficult to compare. This is a serious limitation, and we strongly urge the community to develop commonly available standards, to increase reproducibility and comparability of data between labs and studies.

3.3.3. Single nucleotide polymorphisms (SNPs)

Single nucleotide polymorphism (SNP) markers are the most recent addition to the molecular toolkit for honey bee genetic analysis (see also the section on SNPs of the *BEEBOOK* paper on molecular methods (Evans *et al.*, 2013). A SNP is a change of a single base, usually by just one alternative nucleotide, in a given position of a DNA sequence. For example, in chromosome 5 of *A. m. mellifera* there is a DNA sequence in the gene that codes for subunit 1 of replication factor C that displays two alternative forms, either ...AACTT**A**TCAAA... or ...AACTT**G**TCAAA... (Pinto, unpublished data). In this case there are two alleles, A and G, created by a transition mutation. While there would be four possible nucleotides at each position of a sequence stretch, due to the low mutation rate, which is about 10^{-8} to 10^{-9} changes per nucleotide per generation (Brumfield *et al.*, 2003), SNPs are usually bi-allelic.

While SNPs have only been used in two evolutionary studies (Whitfield *et al.*, 2006; Zayed and Whitfield, 2008) and a QTL study (Spötter *et al.*, 2012), they have great potential for application in subspecies identification, for several reasons. At the analytical level, the genome-wide coverage (coding and non-coding regions), ubiquity, codominance, and conformation to infinite sites model of evolution (Vignal *et al.*, 2002) facilitate employment of more powerful and robust approaches, potentially leading to more reliable identification and more accurate estimates of introgression levels. At the technical level, the possibility of using new technologies enabling high throughput genotyping, data quality, and easy calibration among laboratories facilitate screening of large sample sizes (loci and individuals), data exchange among laboratories, and development of public databases.

Employment of SNPs for subspecies identification using high throughput technologies requires a SNP assay, which can be purchased, if commercially available. Otherwise, it must be developed (as in Whitfield *et al.*, 2006 and Spötter *et al.*, 2012), an expensive and time consuming endeavour requiring high tech equipment and expertise often only available in a core laboratory facility (see development details in Spötter *et al.*, 2012). Unlike for humans and other model organisms, there is only one commercial SNP assay for honey bees, available via AROS Applied Biotechnology AS (Denmark). This assay was designed by Spötter *et al.* (2012) for detection of *Varroa* tolerance in *A. m. carnica* and allows screening of 44,000 loci. Hence, its application for honey bee subspecies identification may not be appropriate.

Genotyping is also costly, but it will likely become increasingly affordable. As an example, AROS Applied Biotechnology AS company charges 261€ (2012 price) per individual honey bee (minimum number of analysed individuals is 95) for screening the 44,000 loci, which is inexpensive if we consider the per locus price. While contracting the services of a private company is expensive, purchasing the equipment and software for SNP genotyping is not affordable any longer for an academic laboratory performing medium-scale studies, unlike the standard equipment needed for mtDNA and microsatellite analysis (Table 6).

Other obstacles of working with thousands of SNP loci are related with the size of the datasets as they require more powerful computers, especially for analyses that are computationally intensive such as Bayesian Markov Chain Monte Carlo methods (used by the popular software Structure, for example). In addition, the software packages must be able to handle large input files. However, for many of the standard analyses, this is not a problem anymore as packages have been modified to deal with large datasets.

The sampling scheme (e.g. number of individuals per colony, number of colonies per apiary and population) in SNP surveys will depend on the research question (e.g., a population genetics-related question requires only one individual per colony), as for the other markers described in this chapter. While the number of individuals used for the other markers could be adopted for SNPs, genome saturation with thousands of SNP loci may lead to violation of the assumption of independent (unlinked) loci, assumed by many analytical approaches. Therefore, either linked loci are removed or other analytical methods (haplotype-based, for example) are employed. Most software packages used for microsatellites, such as Structure (Pritchard *et al.*, 2000), Arlequin (Excoffier *et al.*, 2005), NewHybrids (Anderson and Thompson, 2002), GenAlEx (Peakall and Smouse, 2006), Genepop (Raymond and Rousset, 1995), GeneClass (Cornuet *et al.*, 1999), FSTAT (Goudet, 1995), can also be applied to SNPs. However, for clustering analysis the recently developed software Admixture (Alexander *et al.*, 2009) is much faster than Structure (Pritchard *et al.*, 2000) and therefore more suited for large datasets.

In spite of the promising power of SNPs for subspecies identification, the cost of developing a SNP assay and genotyping will probably preclude widespread adoption of this cutting edge tool in the near future.

4. Annex: Evolutionary lineages of *Apis mellifera*

Besides serving the aims of identifying samples and discriminating subspecies or populations, the techniques discussed in this chapter have been used from the beginning (Ruttner *et al.*, 1978) to shed light on the evolutionary history of *Apis mellifera* and its many subspecies and regional populations. Innumerable publications have been produced on this subject, with the results produced with morphological and molecular techniques partially supporting each other, while being incongruent in several aspects (Garnery *et al.*, 1992, 1993; Franck *et al.*, 1998), leading, unfortunately, to substantial confusion about the number of "true" evolutionary subgroupings and the appropriate way of naming them.

Ruttner (1988) developed and introduced the concept of "branches", or evolutionary lineages within *Apis mellifera*, upon the notice that in statistical analyses of morphometric data samples from different subspecies grouped together according to their geographic origin. Based on morphometric analysis results, *A. mellifera* subspecies can therefore be grouped into four well differentiated lineages: a west Mediterranean and northwest European lineage M (*mellifera* and *iberiensis*, but originally also including *intermissa, sahariensis, siciliana*, as well as the related and later described *ruttneri* (Sheppard *et al.*, 1997)) that were considered as links between the tropical African and the west Mediterranean subspecies), lineage C from southeastern Europe and the eastern Mediterranean (*ligustica, carnica, macedonica, cecropia, cypria, adami*), lineage O in the Near East and western Asia (*caucasica, anatoliaca, syriaca, meda, armeniaca, jemenitica,* and the later described *pomonella* (Sheppard and Meixner, 2003)), and lineage A (*lamarckii, andansonii, scutellata, monticola, litorea, capensis, unicolor* and the newly described *simensis* (Meixner *et al.*, 2011)) from the African continent.

Following the development of molecular techniques and their application to the intraspecific variation of *A. mellifera*, it became evident that mitochondrial DNA variation reflected the lineages proposed by Ruttner in some cases, while in others it did not. While subspecies from Africa and north and west Europe grouped together consistently into lineages M and A based on both morphological and mtDNA results (with *intermissa, sahariensis* and *siciliana* belonging to a North African group within lineage A), this was not the case for the Mediterranean, east European and western Asian subspecies that formed the morphological lineages C and O. Regarding these groups, mitochondrial DNA patterns did not vary substantially and did not differentiate between them. This result was interpreted in a way that a mitochondrial lineage C was hypothesised inclusive of all subspecies belonging to the morphologically defined C and O lineages, with the existence of lineage O being questioned (Garnery *et al.*, 1992). Thus, in the subsequently published body of papers on mitochondrial variation within *A. mellifera*, a model of only three major evolutionary lineages (A, M, C) was generally assumed to reflect the evolutionary history of the species (Garnery *et al.*, 1993; Franck *et al.*, 1998), and the morphologically based model of four lineages was mostly rejected.

Later however, populations from the Near East were found to exhibit mitochondrial patterns different from the three lineages found so far, and the existence of a fourth mitochondrial lineage was postulated (Franck *et al.*, 2000; Palmer *et al.*, 2000). Unfortunately however, following the interpretation that the mitochondrial "associate" of the morphological lineage O had finally been detected, this newly found mitochondrial lineage was also named O (Franck *et al.*, 2000), now making confusion complete. Upon the inclusion of additional data and a new analysis, this lineage has recently been identified as a sublineage of the African lineage A and been renamed to Z (Alburaki *et al.*, 2011). Finally, yet another mitochondrial lineage, named Y, has been identified in northeastern Africa (Franck *et al.*, 2001) which also belongs into the context of the African lineages (Meixner *et al.*, unpublished).

Most interestingly, in recent comprehensive analyses based on nuclear markers (SNPs), and including representative samples of 14 subspecies (Whitfield *et al.*, 2006), the resulting groupings largely reflected the traditional four lineages postulated on the basis of morphology.

5. Acknowledgements

The COLOSS (Prevention of honey bee COlony LOSSes) network aims to explain and prevent massive honey bee colony losses. It was funded through the COST Action FA0803. COST (European Cooperation in Science and Technology) is a unique means for European researchers to jointly develop their own ideas and new initiatives across all scientific disciplines through trans-European networking of nationally funded research activities. Based on a pan-European intergovernmental framework for cooperation in science and technology, COST has contributed since its creation more than 40 years ago to closing the gap between science, policy makers and society throughout Europe and beyond. COST is supported by the EU Seventh Framework Programme for research, technological development and demonstration activities (*Official Journal L 412, 30 December 2006*). The European Science Foundation as implementing agent of COST provides the COST Office through an EC Grant Agreement. The Council of the European Union provides the COST Secretariat. The COLOSS network is now supported by the Ricola Foundation - Nature & Culture.

6. References

ALEXANDER, D H; NOVEMBRE, J; LANGE, K (2009) Fast model-based estimation of ancestry in unrelated individuals. *Genome Research* 19: 1655-1664. http://dx.doi.org/10.1101/gr.094052.109

ALPATOV, W W (1929) Biometrical studies on variation and races of the honey bee (*Apis mellifera* L.). *Quarterly Review of Biology* 4: 1–58.

ALBURAKI, M; MOULIN, S; LEGOUT, H; ALBURAKI, A; GARNERY, L (2011) Mitochondrial structure of Eastern honey bee populations from Syria, Lebanon and Iraq. *Apidologie* 42: 628-641. http://dx.doi.org/10.1007/s13592-011-0062-4

ANDERSON, E C; THOMPSON, E A (2002) A model-based method for identifying species hybrids using multilocus genetic data. *Genetics* 160: 1217-1229.

ARIAS, M C; SHEPPARD, W S (1996) Molecular phylogenetics of honey bee subspecies (*Apis mellifera*) inferred from mitochondrial DNA sequence. *Molecular Phylogenetics and Evolution* 5: 557–566. http://dx.doi.org/10.1006/mpev.1996.0050

ARIAS, M C; RINDERER, T E; SHEPPARD, W S (2006) Further characterization of honey bees from the Iberian Peninsula by allozyme, morphometric and mtDNA haplotype analyses. *Journal of Apicultural Research* 45(4): 188–196. http://dx.doi.org/10.3896/IBRA.1.45.4.04

BADINO, G; CELEBRANO, G; MANINO, A (1983) Population structure and Mdh-1 locus variation in *Apis mellifera ligustica*. *Journal of Heredity* 74: 443–446.

BADINO, G; CELEBRANO, G; MANINO, A; LONGO, S (1985) Enzyme polymorphism in the Sicilian honey bee. *Experientia* 41: 752–754. http://dx.doi.org/10.1007/BF02012580

BADINO, G; CELEBRANO, G; MANINO, A; IFANTIDIS, M D (1988) Allozyme variability in Greek honey bees (*Apis mellifera* L). *Apidologie* 19: 337-386. http://dx.doi.org/10.1051/apido:19880405

BAYLAC, M; GARNERY, L; THARAVY, D; PEDRAZA-ACOSTA, J; RORTAIS, A; ARNOLD, G (2008) ApiClass, an automaticwing morphometric expert system for honey bee identification, [online] http://apiclass.mnhn.fr

BODUR, C; KENCE, M; KENCE, A (2007) Genetic structure of honey bee, *Apis mellifera* L. (Hymenoptera: Apidae) populations of Turkey inferred from microsatellite analysis. *Journal of Apicultural Research* 46(1): 50–56. http://dx.doi.org/10.3896/IBRA.1.46.1.09

BOOKSTEIN, F L (1991) *Morphometric tools for landmark data. Geometry and biology.* Cambridge University Press; New York, USA. ISBN 0-521-38385.

BOUGA, M; KILIAS, G; HARIZANIS, P C; PAPASOTIROPOULOS, V; ALAHIOTIS, S (2005a) Allozyme variability and phylogenetic relationships in honey bee (Hymenoptera: Apidae: *A. mellifera*) populations from Greece and Cyprus. *Biochemical Genetics* 43: 471-484. http://dx.doi.org/10.1007/s10528-005-8163-2

BOUGA, M; HARIZANIS, P C; KILIAS, G; ALAHIOTIS, S (2005b)
Genetic divergence and phylogenetic relationships of honey bee
Apis mellifera (Hymenoptera: Apidae) populations from Greece
and Cyprus using PCR - RFLP analysis of three mtDNA Segments.
Apidologie 36: 335-344. http://dx.doi.org/10.1051/apido:2005021

BOUGA, M; ALAUX, C; BIENKOWSKA, M; BÜCHLER, R; CARRECK, N L;
CAUIA, E; CHLEBO, R; DAHLE, B; DALL'OLIO, R; DE LA RÚA, P;
GREGORC, A; IVANOVA, E; KENCE, A; KENCE, M; KEZIC, N;
KIPRIJANOVSKA, H; KOZMUS, P; KRYGER, P; LE CONTE, Y;
LODESANI, M; MURILHAS, A M; SICEANU, A; SOLAND, G;
UZUNOV, A; WILDE, J (2011): A review of methods for discrimination
of honey bee populations as applied to European beekeeping.
Journal of Apicultural Research 50: 51-84
http://dx.doi.org/10.3896/IBRA.1.50.1.06

BOYER, S H (1961) Alkaline phosphatase in human sera and placentae.
Science 134: 1002-1004.
http://dx.doi.org/10.1126/science.134.3484.1002

BRÜCKNER, D (1974) Reduction of biochemical polymorphism in
honey bees (*Apis mellifica*). *Experientia* 30: 618-619.
http://dx.doi.org/10.1007/BF01921504

BRUMFIELD, R T; BEERLI, P ; NICKERSON, D A; EDWARDS, S V (2003)
The utility of single nucleotide polymorphisms in inferences of
population history. *Trends in Ecology & Evolution* 18: 249-256.
http://dx.doi.org/10.1016/S0169-5347(03)00018-1

CÁNOVAS, F; DE LA RÚA, P; SERRANO, J; GALIÁN, J (2008)
Geographical patterns of mitochondrial DNA variation in *Apis
mellifera iberiensis* (Hymenoptera: Apidae). *Journal of Zoological
Systematics and Evolutionary Research* 46(1): 24–30.
http://dx.doi.org/10.1111/j.1439-0469.2007.00435.x

CÁNOVAS, F; DE LA RÚA, P; SERRANO, J; GALIÁN, J (2011)
Microsatellite variability reveals beekeeping influences on Iberian
honey bee populations. *Apidologie* 42: 235-251.
http://dx.doi.org/10.1007/s13592-011-0020-1

CHAPUIS, M-P; ESTOUP, A (2007) Microsatellite null alleles and
estimation of population differentiation. *Molecular Biology and
Evolution* 24(3): 621-631. http://dx.doi.org/10.1093/molbev/msl191

CLARKE, K E; RINDERER, T E; FRANCK, P; QUEZADA-EUÁN, J J G;
OLDROYD, B P (2002) The Africanization of honey bees (*Apis
mellifera* L.) of the Yucatan : a study of a massive hybridization
event across time. *Evolution* 56(7): 1462-1474.
http://dx.doi.org/10.1111/j.0014-3820.2002.tb01458.x

COELHO, J R; MITTON, J B (1988) Oxygen consumption during
hovering is associated with genetic variation of enzymes in honey
bees. *Functional Ecology* 2: 141-146.

COLLET, T; FERREIRA, K M; ARIAS, M C; SOARES, A E E; DEL LAMA,
M A (2006). Genetic structure of Africanized honey bee populations
(*Apis mellifera* L.) from Brazil and Uruguay viewed through
mitochondrial DNA COI–COII patterns. *Heredity* 97: 329-335.
http://dx.doi.org/10.1038/sj.hdy.6800875

CORNUET, J M (1982) The MDH polymorphism in some west
Mediterranean honey bee populations. In *Proceedings of the 9th
Congress of the International Union for the Study of Social Insects,
Boulder, Colorado, USA*. pp. 5-6.

CORNUET, J M; FRESNAYE, J (1979) Production de miel chez des
hybrides interraciaux d'abeilles (*Apis mellifica* L.) lors de
générations successives de rétrocroisement sur la race locale.
Apidologie 10: 3-15. http://dx.doi.org/10.1051/apido:19790101

CORNUET, J M; FRESNAYE, J (1989) Etude biométrique de colonies
d'abeilles d'Espagne et du Portugal. *Apidologie* 20(1): 93–101.
http://dx.doi.org/10.1051/apido:19890109

CORNUET, J M; DAOUDI, A; CHEVALET, C (1986) Genetic pollution
and number of matings in a black honey bee (*Apis mellifera
mellifera*) population. *Theoretical and Applied Genetics* 73: 223-227.
http://dx.doi.org/10.1007/BF00289278

CORNUET, J M; GARNERY, L; SOLIGNAC, M (1991) Putative origin
and function of the intergenic region between COI and COII of
Apis mellifera L. mitochondrial DNA. *Genetics* 128: 393-403.

CORNUET, J M; OLDROYD, B P; CROZIER, R H (1995) Unequal
thermostability of allelic forms of malate dehydrogenase in honey
bees. *Journal of Apicultural Research* 34: 45-47.

CORNUET, J M; PIRY, S; LUIKART, G; ESTOUP, A; SOLIGNAC, M (1999)
New methods employing multilocus genotypes to select or exclude
populations as origins of individuals. *Genetics* 153: 1989-2000.

CROZIER, Y C; KOULIANOS, S; CROZIER, R H (1991). An improved
test for Africanized honey bee mitochondrial DNA. *Experientia* 47:
968-969. http://dx.doi.org/10.1007/BF01929894

CURRIE, R W; JAY, S C (1988) The influence of a colony's queen
state, time of the year, and drifting behaviour, on the acceptance
and longevity of adult drone honey bees (*Apis mellifera* L.),
Journal of Apicultural Research 27: 219–226.

DALL'OLIO, R; MARINO, A; LODESANI, M; MORITZ, R F A (2007)
Genetic characterization of Italian honey bees, *Apis mellifera
ligustica*, based on microsatellite DNA polymorphisms. *Apidologie*
38: 207–217. http://dx.doi.org/10.1051/apido:2006073

DALY, H V; HOELMER, K; NORMAN, P; ALLEN, T (1982) Computer
assisted measurement and identification of honey bees. *Annals of
the Entomological Society of America* 75: 591–594.

DAVIS, B J (1964). Disc electrophoresis. II. Method and application to
human serum proteins. *Annals of the New York Academy of
Sciences* 121: 404 427.
http://dx.doi.org/10.1111/j.1749-6632.1964.tb14213.x

DEDEJ, S; BASIOLO, A; PIVA, R (1996) Morphometric and alloenzymatic
characterisation in the Albanian honey bee population *Apis mellifera* L.
Apidologie 27(3): 121-131. http://dx.doi.org/10.1051/apido:19960301

DE LA RÚA, P; SERRANO, J; GALIÁN, J (1998) Mitochondrial DNA
variability in the Canary Islands honey bees (*Apis mellifera* L.).
Molecular Ecology 7(11): 1543-1547.
http://dx.doi.org/10.1046/j.1365-294x.1998.00468.x

DE LA RÚA, P; GALIÁN, J; SERRANO, J; MORITZ, R F A (2001) Genetic structure and distinctness of *Apis mellifera* L: populations from the Canary Islands. *Molecular Ecology* 19: 1733-1742. http://dx.doi.org/10.1046/j.1365-294X.2001.01303.x

DE LA RÚA, P; GALIÁN, J; SERRANO, J; MORITZ, R F A (2001) Molecular characterization and population structure of the honey bees from the Balearic Islands (Spain). *Apidologie* 32: 417-427. http://dx.doi.org/10.1051/apido:2001141

DE LA RÚA, P; SERRANO, J; GALIÁN J (2002) Biodiversity of *Apis mellifera* populations from Tenerife (Canary Islands) and introgressive hybridization with East European races. *Biodiversity and Conservation* 11: 59-67. http://dx.doi.org/10.1023/A:1014066407307

DE LA RÚA, P; GALIÁN, J; SERRANO, J; MORITZ, R F A (2003) Genetic structure of Balearic honey bee populations based on microsatellite polymorphism. *Genetics, Selection, Evolution* 35: 339-350. http://dx.doi.org/10.1186/1297-9686-35-3-339

DE LA RÚA, P; HERNANDEZ-GARCIA, R; JIMÉNEZ, Y; GALIÁN, J; SERRANO, J (2005) Biodiversity of *Apis mellifera iberica* (Hymenoptera: Apidae) from northeastern Spain assessed by mitochondrial analysis. *Insect Systematics and Evolution* 36: 21-28.

DE LA RÚA, P; GALIÁN, J; PEDERSEN, B V; SERRANO, J (2006) Molecular characterization and population structure of *Apis mellifera* from Madeira and the Azores. *Apidologie* 37: 699-708. http://dx.doi.org/10.1051/apido:2006044

DE LA RÚA, P; JAFFÉ, R; DALL'OLIO, R; MUÑOZ, I; SERRANO, J (2009) Biodiversity, conservation and current threats to European honey bees. *Apidologie* 40(3): 263-284. http://dx.doi.org/10.1051/apido/2009027

DEL LAMA, M A; MESTRINER, M A; PAVIA, J C A (1985) Est-5 and Pgm-1: new polymorphism in *Apis mellifera*. *Brazilian Journal of Genetics* 8: 17-27.

DEL LAMA, M A; FIGUEIREDO, R A; SOARES, A E E; DEL LAMA, S N (1988) Hexokinase polymorphism in *Apis mellifera* and its use for Africanized honey bee identification. *Brazilian Journal of Genetics* 11: 287-292.

DEL LAMA, M A; LOBO, J A; SOARES, A E E; DEL LAMA, S N (1990) Genetic differentiation estimated by isozymic analysis of Africanized honey bee populations from Brazil and from Central America. *Apidologie* 21: 271-280. http://dx.doi.org/10.1051/apido:19900401

DINIZ-FILHO, J A F; HEPBURN, H R; RADLOFF. S; FUCHS, S (2000) Spatial analysis of morphological variation in African honey bees (*Apis mellifera* L.) on a continental scale. *Apidologie* 31: 191–204. http://dx.doi.org/10.1051/apido:2000116

DU PRAW, E J (1964) Non-Linnean taxonomy. *Nature* 202: 849-852.

DUPRAW, E J (1965) Non-Linnean taxonomy and the systematics of honey bees. *Systematic Zoology* 14: 1-24.

DURAND, E; JAY, F; GAGGIOTTI, O E; FRANÇOIS, O (2009) Spatial inference of admixture proportions and secondary contact zones. *Molecular Biology and Evolution* 26:1963-1973. http://dx.doi.org/10.1093/molbev/msp106

ESTOUP, A; SOLIGNAC, M; HARRY, M; CORNUET, J M (1993) Characterization of (GT)n and (CT)n microsatellites in two insect species *Apis mellifera* and *Bombus terrestris*. *Nucleic Acids Research* 21: 1427-1431. http://dx.doi.org/10.1093/nar/21.6.1427

ESTOUP, A; GARNERY, L; SOLIGNAC, M; CORNUET, J M (1995a) Microsatellite variation in honey bee (*Apis mellifera* L) populations: Hierarchical genetic structure and test of the infinite allele and stepwise mutation models. *Genetics* 140: 679-695.

ESTOUP, A; TAILLIEZ, C; CORNUET, J M; SOLIGNAC, M (1995b) Size homoplasy and mutational processes of interrupted microsatellites in two bee species, *Apis mellifera* and *Bombus terrestris* (Apidae). *Molecular Biology and Evolution* 12: 1074-1084.

EVANNO, G; REGNAUT, S; GOUDET, J (2005) Detecting the number of clusters of individuals using the software STRUCTURE: a simulation study. *Molecular Ecology* 14(8) 2611-2620. http://dx.doi.org/10.1111/j.1365-294X.2005.02553.x

EVANS, J D; CHEN, Y P; CORNMAN, R S; DE LA RUA, P; FORET, S; FOSTER, L; GENERSCH, E; GISDER, S; JAROSCH, A; KUCHARSKI, R; LOPEZ, D; LUN, C M; MORITZ, R F A; MALESZKA, R; MUÑOZ, I; PINTO, M A; SCHWARZ, R S (2013) Standard methodologies for molecular research in *Apis mellifera*. In *V Dietemann; J D Ellis; P Neumann (Eds) The COLOSS BEEBOOK, Volume I: standard methods for* Apis mellifera *research. Journal of Apicultural Research* 52(4): http://dx.doi.org/10.3896/IBRA.1.52.4.11

EXCOFFIER, L; LAVAL, G; SCHNEIDER, S (2005) Arlequin (version 3.0): An integrated software package for population genetics data analysis. *Evolutionary Bioinformatics* 1: 47-50.

FRANCK, P; GARNERY, L; SOLIGNAC, M; CORNUET, J M (1998) The origin of west European subspecies of honey bees (*Apis mellifera*): New insights from microsatellite and mitochondrial data. *Evolution* 52(4): 1119–1134. http://dx.doi.org/10.2307/2411242

FRANCK, P; GARNERY, L; CELEBRANO, G; SOLIGNAC, M; CORNUET, J M (2000) Hybrid origin of honey bees from Italy (*Apis mellifera ligustica*) and Sicily (*A. m. sicula*). *Molecular Ecology* 9: 907–921. http://dx.doi.org/10.1046/j.1365-294x.2000.00945.x

FRANCK, P; GARNERY, L; LOISEAU, A (2001) Genetic diversity of the honey bee in Africa: microsatellite and mitochondrial data. *Heredity* 86: 420–430. http://dx.doi.org/10.1046/j.1365-2540.2001.00842.x

FRANCK, P; GARNERY, L; SOLIGNAC, M; CORNUET, J M (2000) Molecular confirmation of a fourth lineage in honey bees from the Near East. *Apidologie* 31: 167-180. http://dx.doi.org/10.1051/apido:2000114

FRANCOY, T M; PRADO, P R R; GONÇALVES, L S; DA FONTOURA COSTA, L; DE JONG D (2006) Morphometric differences in a single wing cell can discriminate *Apis mellifera* racial types. *Apidologie* 37: 91-97. http://dx.doi.org/10.1051/apido:2005062

FRANCOY, T M; WITTMANN, D; DRAUSCHKE, S; MÜLLER. S; STEINHAGE, V (2008) Identification of Africanized honey bees through wing morphometrics: two fast and efficient procedures. *Apidologie* 39(5): 488-494. http://dx.doi.org/10.1051/apido:2008028

GAHNE, B (1967) Alkaline phosphatase isoenzyme in serum and seminal plasma. *Hereditas* 57: 83-99. http://dx.doi.org/10.1111/j.1601-5223.1967.tb02094.x

GARNERY, L; CORNUET, J M; SOLIGNAC, M (1992) Evolutionary history of the honey bee *Apis mellifera* inferred from mitochondrial DNA analysis. *Molecular Ecology* 1: 145-154. http://dx.doi.org/10.1111/j.1365-294X.1992.tb00170.x

GARNERY, L; SOLIGNAC, M; CELEBRANO, G; CORNUET, J M (1993) A simple test using restricted PCR-amplified mitochondrial-DNA to study the genetic-structure of *Apis mellifera* L. *Experientia* 49(11): 1016–1021. http://dx.doi.org/10.1007/BF02125651

GARNERY, L; FRANCK, P; BAUDRY, E; VAUTRIN, D; CORNUET, J M; SOLIGNAC, M (1998a) Genetic diversity of the west European honey bee (*Apis mellifera* and *A. m. iberica*). I. Mitochondrial DNA. *Genetics Selection Evolution* 30:S 31-47. http://dx.doi.org/10.1186/1297-9686-30-S1-S31

GARNERY, L; FRANCK, P; BAUDRY, E; VAUTRIN, D; CORNUET, J M; SOLIGNAC, M (1998b) Genetic biodiversity of the West European honey bee (*Apis mellifera mellifera* and *Apis mellifera iberica*). II. Microsatellite loci. *Genetics Selection Evolution* 30: 49– S74. http://dx.doi.org/10.1051/gse:19980703

GARSTIDE, D F (1980) Similar allozyme polymorphism in honey bees (*Apis mellifera*) from different continents. *Experientia* 36: 649-650.

GOUDET, J (1995) FSTAT (Version 12): A computer program to calculate F-statistics *Journal of Heredity* 86: 485-486.

GOETZE, G (1964) *Die Honigbiene in natürlicher und künstlicher Zuchtauslese Parey, Hamburg.*

GUILLOT, G; MORTIER, F; ESTOUP, A (2005) Geneland: A program for landscape genetics. *Molecular Ecology Notes* 5: 712-715. http://dx.doi.org/10.1111%2Fj.1471-8286.2005.01031.x

HALL, H G; MURALIDHARAN, K (1989) Evidence from mitochondrial DNA that African honey bees spread as continuous maternal lineages. *Nature* 339: 211-213. http://dx.doi.org/10.1038/339211a0

HALL, H G; SMITH, D R (1991) Distinguishing African and European honey bee matrilines using amplified mitochondrial DNA. *Proceedings of the National Academy of Sciences* 88: 4548-4552. http://dx.doi.org/10.1073/pnas.88.10.4548

HARRIS, H; HOPKINSON, D A (1976) *Handbook of enzyme electrophoresis in human genetics.* North-Holland Publishing Company; Amsterdam, Netherlands.

HEPBURN, H R; RADLOFF, S E (1996) Morphometric and pheromonal analysis of *Apis mellifera* L. along a transect from the Sahara to the Pyrenees. *Apidologie* 27: 35–45. http://dx.doi.org/10.1051/apido:19960105

HEPBURN, H R; RADLOFF, S E (1998) *Honey bees of Africa.* Springer; Berlin, Germany.

HUMAN, H; BRODSCHNEIDER, R; DIETEMANN, V; DIVELY, G; ELLIS, J; FORSGREN, E; FRIES, I; HATJINA, F; HU, F-L; JAFFÉ, R; KÖHLER, A; PIRK, C W W; ROSE, R; STRAUSS, U; TANNER, G; TARPY, D R; VAN DER STEEN, J J M; VEJSNÆS, F; WILLIAMS, G R; ZHENG, H-Q (2013) Miscellaneous standard methods for *Apis mellifera* research. In *V Dietemann; J D Ellis; P Neumann (Eds) The COLOSS BEEBOOK, Volume I: standard methods for* Apis mellifera *research. Journal of Apicultural Research* 52(4): http://dx.doi.org/10.3896/IBRA.1.52.4.10

IVANOVA, E (2010) Investigation on genetic variability in honey bee populations from Bulgaria, Greece and Serbia. *Biotechnology & Biotechnological Equipment* 24(2): 385-389.

IVANOVA, E; STAYKOVA, T; BOUGA, M (2007) Allozyme variability in honey bee populations from some mountainous regions in southwest of Bulgaria. *Journal of Apicultural Research* 46(1): 3-7. http://dx.doi.org/10.3896/IBRA.1.46.1.02

IVANOVA, E; PETROV, P; BOUGA, M; EMMANOUEL, N; IVGIN-TUNKA, R; KENCE, M (2010a) Genetic variation in honey bee (*Apis mellifera* L.) populations from Bulgaria. *Journal of Apicultural Science* 54(2): 49-60.

IVANOVA, E; STAYKOVA, T; PETROV, P (2010b). Allozyme variability in populations of local Bulgarian honey bee. *Biotechnology & Biotechnological Equipment* 24(2): 371–374.

IVANOVA, E; BOUGA, M; PETROV, P; MLADENOVIC, M; RASIC, S; CHARISTOS, L; HATJINA, F (2010c) Preliminary results from a study on Balkan honey bees' genetic variability using isoenzymic approach. In *Proceedings of 4th European Conference of Apidology, 7-9 September 2010, Ankara, Turkey.* p 45.

IVANOVA, E; BIENKOWSKA, M; PETROV, P (2011) Allozyme polymorphism and phylogenetic relationships *in Apis mellifera* subspecies selectively reared in Poland and Bulgaria. *Folia biologica* 59: 3-4. http://dx.doi.org/10.3409/fb59_3-4.121-126

JAFFÉ, R; DIETEMANN, V; ALLSOPP, M H; COSTA, C; CREWE, R M; DALL'OLIO R; DE LA RÚA, P; EL-NIWEIRI, M A A; FRIES, I; KEZIC, N; MEUSEL M S; PAXTON, R J; SHAIBI, T; STOLLE, E; MORITZ, R F A (2010) Estimating the density of honey bee colonies across their natural range to fill the gap in pollinator decline censuses. *Conservation Biology* 24: 583-593. http://dx.doi.org/10.1111/j.1523-1739.2009.01331.x

JAY, S C (1969) The problem of drifting in commercial apiaries. *American Bee Journal* 109: 178-179.

JENSEN, A B; PALMER, K A; BOOMSMA, J J; PEDERSEN, B V (2005) Varying degrees of *Apis mellifera ligustica* introgression in protected populations of the black honey bee, *Apis mellifera mellifera*, in northwest Europe. *Molecular Ecology* 14: 93–106. http://dx.doi.org/10.1111/j.1365-294X.2004.02399.x

JOMBART, T (2008) Adegenet: a R package for the multivariate analysis of genetic markers. *Bioinformatics* 24: 1403-1405. http://dx.doi.org/10.1093/bioinformatics/btn129

JONES, J C; MYERSCOUGH, M R; GRAHAM, S; OLDROYD, B P (2004) Honey bee nest thermoregulation: diversity promotes stability. *Science* 305: 402-404. http://dx.doi.org/10.1126/science.1096340

KANDEMIR, I; KENCE, A (1995) Allozyme variation in a central Anatolian honey bee (*Apis mellifera* L.) population. *Apidologie* 26: 503-510. http://dx.doi.org/10.1051/apido:19950607

KANDEMIR, I; KENCE, M; KENCE, A (2000) Genetic and morphometric variation in honey bee (*Apis mellifera* L.) populations of Turkey. *Apidologie* 31: 343-356. http://dx.doi.org/10.1051/apido:2000126

KANDEMIR, I; KENCE, M; KENCE, A (2005) Morphometric and electrophoretic variation in different honey bee (*Apis mellifera* L.) populations. *Turkish Journal of Veterinary and Animal Science* 29: 885-890.

KANDEMIR, I; MEIXNER, M D; ÖZKAN, A; SHEPPARD, W S (2006a) Genetic characterization of honey bee (*Apis mellifera cypria*, Pollmann 1879) populations in Cyprus. *Apidologie* 37: 547-555. http://dx.doi.org/10.1051/apido:2006029

KANDEMIR, I; KENCE, M; SHEPPARD, W S; KENCE, A (2006b) Mitochondrial DNA variation in honey bee (*Apis mellifera* L.) populations from Turkey. *Journal of Apicultural Research* 45(1): 33–38. http://dx.doi.org/10.3896/IBRA.1.45.1.08

KANDEMIR, I; ÖZKAN, A; FUCHS, S (2011) Reevaluation of honey bee (*Apis mellifera*) microtaxonomy: a geometric morphometric approach. *Apidologie* 42: 618–627. http://dx.doi.org/10.1007/s13592-011-0063-3

KLINGENBERG, C P (2011) Morpho J: an integrated software package for geometric morphometrics. *Molecular Ecology Resources* 11: 353-357.

LEVINSON, G; GUTMAN, G A (1987) Slipped-Strand Mispairing: A major mechanism for DNA sequence evolution. *Molecular Biology and Evolution* 4: 203-221.

LOBO, J A; DEL LAMA, M A; MESTRINER, M A (1989) Population differentiation and racial admixture in the Africanized honey bee (*Apis mellifera* L). *Evolution* 43: 794-802. http://dx.doi.org/10.2307/2409307

LOUVEAUX, J; ALBISETTI, M; DELANGUE, M; THEURKAUFF, J (1966) Les modalités de l'adaptation des abeilles (*Apis mellifica* L.) au milieu naturel. *Annales de l' Abeille* 9: 323-350. http://dx.doi.org/10.1051/apido:19660402

MANEL, S; GAGGIOTTI, O E; WAPLES, R E (2005) Assignment methods: matching biological questions with appropriate techniques. *Trends in Ecology and Evolution* 20(3): 136-142. http://dx.doi.org/10.1016/j.tree.2004.12.004

MĂRGHITAS, L A; PANITI-TELEKY, O; DEZMIREAN, D; MĂRGĂOAN, R; BOJAN, C; COROIAN, C; LASLO, L; MOISE, A (2008) Morphometric differences between honey bee (*Apis mellifera carpatica*) populations from transysIvanian area. *Lucrări stiinifice Zootehnie si Biotehnologii* 41(2): Timisoara.

MARTIMIANAKIS, S; KLOSSA-KILIA, E; BOUGA, M; KILIAS, G (2011) Phylogenetic relationships of Greek *Apis mellifera* subspecies based on sequencing of mtDNA segments (COI and ND5). *Journal of Apicultural Research* 50(1): 42-50. http://dx.doi.org/10.3896/IBRA.1.50.1.05

MEIXNER, M; SHEPPARD, W S; POKLUKAR, J (1993) Asymmetrical distribution of a mitochondrial DNA polymorphism between two introgressing honey bee races. *Apidologie* 24: 147–153. http://dx.doi.org/10.1051/apido:19930207

MEIXNER, M D; SHEPPARD, W S; DIETZ, A; KRELL, R (1994) Morphological and allozyme variability in honey bees from Kenya. *Apidologie* 25: 188-202. http://dx.doi.org/10.1051/apido:19940207

MEIXNER, M D; WOROBIK, M; WILDE, J; FUCHS, S; KOENIGER, N (2007) *Apis mellifera mellifera* in eastern Europe - morphometric variation and determination of range limits. *Apidologie* 38: 191-197. http://dx.doi.org/10.1051/apido:2006068

MEIXNER, M D; COSTA, C; KRYGER, P; HATJINA, F; BOUGA, M; IVANOVA, E; BÜCHLER, R (2010) Conserving diversity and vitality for honey bee breeding. *Journal of Apicultural Research* 49(1): 85 -92. http://dx.doi.org/10.3896/IBRA.1.49.1.12

MEIXNER, M D; LETA, M A; KOENIGER, N; FUCHS, S (2011) The honey bees of Ethiopia represent a new subspecies of *Apis mellifera* – *Apis mellifera simensis* n. ssp. *Apidologie* 42: 425-437. http://dx.doi.org/10.1007/s13592-011-0007-y

MENDES, M F M; FRANCOY, T M; NUNES-SILVA, P; MENEZES, C; IMPERATRIZ-FONSECA, V L (2007) Intra-populational variability of *Nannotrigona testaceicornis* Lepeletier 1836 (Hymenoptera, Meliponini) using relative warp analysis. *Bioscience Journal* 23: 147-152.

MESTRINER, M A (1969) Biochemical polymorphism in bees (*Apis m. ligustica*). *Nature* 223: 188-189.

MESTRINER, M A; CONTEL, E P B (1972) The P-3 and Est loci in the honey bee *Apis mellifera*. *Genetics* 72: 733-738.

MIGUEL, I; IRIONDO, M; GARNERY, L; SHEPPARD W S; ESTONBA, A (2007) Gene flow within the M evolutionary lineage of *Apis mellifera*: role of the Pyrenees, isolation by distance and post-glacial re-colonization routes in the western Europe. *Apidologie* 38: 141-155. http://dx.doi.org/10.1051/apido:2007007

MIGUEL, I; BAYLAC, M; IRIONDO, M; MANZANO, C; GARNERY, L; ESTONBA, A (2010) Both geometric morphometric and microsatellite data consistently support the differentiation of the *Apis mellifera* M evolutionary branch. *Apidologie* 42: 150–161. http://dx.doi.org/10.1051/apido/2010048

MONTEIRO, L R (1999) Multivariate regression models and geometric morphometrics: the search for causal factors in the analysis of shape. *Systematic Biology* 48: 192–199. http://dx.doi.org/10.1080/106351599260526

MLADENOVIĆ, M; RADOŠ, R; STANISAVLJEVIĆ, L Ž; RAŠIĆ, S (2011)
Morphometric traits of the yellow honey bee (*Apis mellifera carnica*)
from Vojvodina (Northern Serbia) *Archives of Biological Sciences,
Belgrade* 63(1): 251-257. http://dx.doi.org/10.2298/ABS1101251M

MUÑOZ, I; DALL´OLIO, R; LODESANI, M; DE LA RÚA, P (2008)
Sequence variation in the mitochondrial tRNAleu-cox2 intergenic
region of African and African-derived honey bee populations. In
*Proceedings of the 3ʳᵈ European Conference of Apidology, Belfast,
United Kingdom, 8-11 September 2008*. p 73.

MUÑOZ, I; DALL'OLIO, R; LODESANI, M; DE LA RÚA, P (2009)
Population genetic structure of coastal Croatian honey bees (*Apis
mellifera carnica*). *Apidologie* 40: 617-626.
http://dx.doi.org/10.1051/apido/2009041

NAZZI, F (1992) Morphometric analysis of honey bees from an area of
racial hybridization in northeastern Italy. *Apidologie* 23: 89–96.
http://dx.doi.org/10.1051/apido:19920201

NEUMANN, P; MORITZ, R F A; MAUTZ, D (2000) Colony evaluation is
not affected by drifting of drone and worker honey bees (*Apis
mellifera* L.) at a performance testing apiary. *Apidologie* 31: 67-79.
http://dx.doi.org/10.1051/apido:2000107

NIELSEN, D; PAGE, J R E; CROSLAND, M W J (1994) Clinal variation
and selection of MDH allozymes in honey bee populations.
Experientia 50: 867-871. http://dx.doi.org/10.1007/BF01956474

NIELSEN, D J; EBERT, P R; PAGE, R E; HUNT, G J; GUZMAN-NOVOA, E
(2000) Improved polymerase chain reaction-based mitochondrial
genotype assay for identification of the Africanized honey bee
(Hymenoptera: Apidae). *Annals of Entomological Society of
America* 93: 1-6.
http://dx.doi.org/10.1603/0013-8746(2000)093[0001:IPCRBM]2.0.CO;2

NIKOLOVA, S (2011) Genetic variability of local Bulgarian honey bees
Apis mellifera macedonica (*rodopica*) based on microsatellite DNA
analysis. *Journal of Apicultural Science* 55(2): 117-129.

NEDIĆ, N; STANISAVLJEVIĆ, L; MLADENOVIĆ, M; STANISAVLJEVIĆ, J
(2009) Molecular characterization of the honey bee *Apis mellifera
carnica* in Serbia. *Archives of Biological Sciences, Belgrade* 61(4):
587-598. http://dx.doi.org/10.2298/ABS0904587N

NUNAMAKER, R A; WILSON, W T; HALEY, B E (1984) Electrophoretic
detection of Africanized honey bees (*Apis mellifera scutellata*) in
Guatemala and Mexico based on malate dehydrogenase allozyme
patterns. *Journal of the Kansas Entomological Society* 57: 622-
631.

O'REILLY, P T; HAMILTON, L C; McCONNELL, S K; WRIGHT, J M (1996)
Rapid analysis of genetic variation in Atlantic salmon (*Salmo salar*)
by PCR9 multiplexing of dinucleotide and tetranucleotide
microsatellites. *Canadian Journal of Fisheries and Aquatic
Sciences* 53: 2292-2298.

PALMER, M R; SMITH, D R; KAFTANOĞLU, O (2000) Turkish honey
bees: genetic variation and evidence for a fourth lineage of *Apis
mellifera* mtDNA. *Journal of Heredity* 91: 42-66.
http://dx.doi.org/10.1093/jhered/91.1.42

PADILLA ÁLVAREZ, F; HERNÁNDEZ FERNÁNDEZ, R; REYES LÓPEZ, R;
PUERTA, F; FLORES SERRANO, J M; BUSTOS RUIZ, M (1998)
Estudio morfológico de las abejas melíferas del archipelago
Canario (Gran Canaria, Tenerife, la Palma, Gomera) [Morphological
study of honey bees on the Canary Islands (Gran Canaria,
Tenerife, la Palma, Gomera]. *Archivos de Zootecnia* 178: 451-459.

PADILLA ÁLVAREZ, F; VALERIO DA SILVA, M J; CAMPANO CABANES,
F; JIMÉNEZ VAQUERO, E; PUERTA PUERTA, F; FLORES SERRANO,
J M; BUSTOS RUIZ, M (2001) Biometric study of *Apis mellifera*
populations from central Portugal and Madeira. *Archivos de
Zootecnia* 50(189): 67-77.

PEAKALL, R; SMOUSE, P E (2006) GENALEX 6: genetic analysis in
Excel Population genetic software for teaching and research.
Molecular Ecology Notes 6: 288-295.
http://dx.doi.org/10.1111/j.1471-8286.2005.01155.x

PINTO, M A; JOHNSTON, J S; RUBINK, W L; COULSON, R N; PATTON,
J C; SHEPPARD, W S (2003) Identification of Africanized honey
bee (Hymenoptera: Apidae) mitochondrial DNA: validation of a
rapid PCR-based assay. *Annals of the Entomological Society of
America* 96: 679-684.
http://dx.doi.org/10.1603/0013-8746(2003)096[0679:IOAHBH]2.0.CO;2

PINTO, M A; RUBINK, W L; COULSON, R N; PATTON, J C; JOHNSTON,
J S (2004) Temporal pattern of Africanization in a feral honey bee
population from Texas inferred from mitochondrial DNA. *Evolution*
58: 1047–1055. http://dx.doi.org/10.1554/03-334

PINTO, M A; MUÑOZ, I; CHÁVEZ-GALARZA, J; DE LA RUA, P (2012)
The Atlantic side of the Iberian peninsula: a hot-spot of novel
African honey bee maternal diversity. *Apidologie* 43: 663-673.
http://dx.doi.org/10.1007/s13592-012-0141-1

PIRY, S; ALAPETITE, A; CORNUET, J. M; PAETKAU, D; BAUDOUIN, L;
ESTOUP, A (2004) GeneClass2: a software for genetic assignment
and first-generation migrant detection. *Journal of Heredity* 95:
536-539. http://dx.doi.org/10.1093/jhered/esh074

PRITCHARD, J K; STEPHENS, M; DONNELLY, P (2000) Inference of
population structure using multilocus genotype data. *Genetics* 155:
945–959.

QUEZADA-EUÁN, J G; PÉREZ-CASTRO, E E; MAY-ITZÁ, W J (2003)
Hybridization between European and African-derived honey bee
populations (*Apis mellifera*) at different altitudes in Peru.
Apidologie 34: 217-225. http://dx.doi.org/10.1051/apido:2003010

RADLOFF, S E; HEPBURN, H R (1997a) Multivariate analysis of honey
bees, *Apis mellifera* L. (Hymenoptera: Apidae) on the horn of
Africa. *African Entomology* 5: 57–64.

RADLOFF, S E; HEPBURN, H R (1997b) Multivariate analysis of honey
bees, *Apis mellifera* Linnaeus (Hymenoptera: Apidae) from
western central Africa. *African Entomology* 5: 195–204.

RADLOFF, S E; HEPBURN, H R (1998) The matter of sampling
distance and confidence levels in the subspecies classification of
honey bees, *Apis mellifera* L. *Apidologie* 29: 491-591.
http://dx.doi.org/10.1051/apido:19980602

RADLOFF, S E; HEPBURN, H R; HEPBURN, C; DE LA RÚA, P (2001) Morphometric affinities and population structure of honey bees of the Balearic Islands in the Mediterranean Sea. *Journal of Apicultural Research* 40: 97-103.

RADLOFF, S E; HEPBURN, H R; LINDSEY, J B (2003) Quantitative analysis of intracolonial and intercolonial morphometric variance in honey bees, *Apis mellifera* and *Apis cerana*. *Apidologie* 34: 339-351. http://dx.doi.org/10.1051/apido:2003034

RADLOFF, S E; HEPBURN, C; HEPBURN, H R; HADISOESILO, S; FUCHS, S; TAN, K; ENGEL, M S; KUZNETSOV, V (2010) Population structure and classification of *Apis cerana*. *Apidologie* 41: 589–601. http://dx.doi.org/10.1051/apido/2010008

RAYMOND, M; ROUSSET, F (1995) Genepop (version 12) - Population genetics software for exact tests and ecumenicism. *Journal of Heredity* 86: 248-249.

RÉKA, T O; BOJAN, C; MOISE, A; COROIAN, C; DEZMIREAN, D; MĂRGHITAS, L A (2007) Ecotypes differentiation within honey bees (*Apis mellifera carpatica*) from Transylvania. *Bulletin USAMV-CN*, 63-64.

RINDERER, T E; SYLVESTER, H A; BROWN, M A; VILLA, J D; PESANTE, D; COLLINS, A M (1986) Field and simple techniques for identifying Africanized and European honey bees. *Apidologie* 17: 33-48. http://dx.doi.org/10.1051/apido:19860104

RINDERER, T E; SYLVESTER, H A; BUCO, S M; LANCASTER, V A; HERBERT, E W; COLLINS, A M; HELLMICH, R L (1987) Improved simple techniques for Identifying Africanized and European honey bees. *Apidologie* 18: 179-196. http://dx.doi.org/10.1051/apido:19870208

ROHLF, F J (1990) *NTSYS-pc, Numerical taxomomy and multivariate analysis system.* Exeter Software; NY, USA.

ROHLF, F J (2000) Statistical power comparisons among alternative morphometric methods. *American Journal of Physical Anthropology* 111:463-478. http://dx.doi.org/10.1002/(SICI)1096-8644(200004)111:4<463::AID-AJPA3>3.0.CO;2-B

ROHLF, F J (2001) *TPSdig: digitize landmarks from image files, scanner, or video.* Department of Evolutionary Biology, University of New York; Stony Brook, NY, USA.

ROHLF, F J; SLICE, D. (1990) Extensions of the Procrustes method for the optimal superimposition of landmarks. *Systematic Zoology* 39: 40–59. http://dx.doi.org/10.2307/2992207

RORTAIS, A; ARNOLD, G; ALBURAKI, M; LEGOUT, H; GARNERY, L (2011) Review of the *DraI* COI-COII test for the conservation of the black honey bee (*Apis mellifera mellifera*). *Conservation Genetics Resources* 3: 383-391. http://dx.doi.org/10.1007/s12686-010-9351-x

RUTTNER, F (1980) *Apis mellifera adami* (n.ssp.), die Kretische Biene. *Apidologie* 11: 385-400. http://dx.doi.org/10.1051/apido:19800407

RUTTNER, F (1988) *Biogeography and taxonomy of honey bees.* Springer-Verlag; Berlin, Germany. ISBN 0387177817.

RUTTNER, F; TASSENCOURT, L; LOUVEAUX, J (1978) Biometrical-statistical analysis of the geographic variability of *Apis mellifera* L. *Apidologie* 9: 363-381. http://dx.doi.org/10.1051/apido:19780408

SEELEY, T D; TARPY, D R (2007) Queen promiscuity lowers disease within honey bee colonies. *Proceedings of the Royal Society London B* 274: 67–72. http://dx.doi.org/10.1098/rspb.2006.3702

SHAIBI, T; MUNOZ, I; DALL'OLIO, R; LODESANI, M; DE LA RUA, P; MORITZ, R F A (2009) *Apis mellifera* evolutionary lineages in Northern Africa: Libya, where orient meets occident. *Insectes Sociaux* 56: 293-300. http://dx.doi.org/10.1007/s00040-009-0023-3

SHAW, C R; PRASAD, R (1970) Starch-gel electrophoresis - a compilation of recipes. *Biochemical Genetics* 4: 297-320. http://dx.doi.org/10.1007/BF00485780

SHEPPARD, W S (1988) Comparative study of enzyme polymorphism in United States and European honey bee (Hymenoptera: Apidae) populations. *Entomological Society of America* 81: 886-889.

SHEPPARD, W S; BERLOCHER, S H (1984) Enzyme polymorphism in *Apis mellifera* from Norway. *Journal of Apicultural Research* 23: 64-69.

SHEPPARD, W S; BERLOCHER, S H (1985) New allozyme variability in Italian honey bees. *Journal of Heredity* 76: 45-48.

SHEPPARD, W S; MCPHERON, B A (1986) Genetic variation in honey bees from an area of racial hybridization in western Czechoslovakia. *Apidologie* 17: 21-32. http://dx.doi.org/10.1051/apido:19860103

SHEPPARD, W S; ARIAS, M C; GRECH, A; MEIXNER, M D (1997) *Apis mellifera ruttneri*, a new honey bee subspecies from Malta. *Apidologie* 28: 287-293. http://dx.doi.org/10.1051/apido:19970505

SHEPPARD, W S; MEIXNER, M D (2003) *Apis mellifera pomonella*, a new honey bee subspecies from the Tien Shan mountains of Central Asia. *Apidologie* 34: 367-375. http://dx.doi.org/10.1051/apido:2003037

SCHNABEL, R D; WARD, T J; DERR, J N (2000) Validation of 15 microsatellites for parentage testing in North American bison, *Bison bison* and domestic cattle. *Animal Genetics 31*: 360-366. http://dx.doi.org/10.1046/j.1365-2052.2000.00685.x

SINACORI, A; RINDERER, T E; LANCASTER, V; SHEPPARD, W S (1998) A morphological and mitochondrial assessment of *Apis mellifera* from Palermo, Italy. *Apidologie* 29: 481–490. http://dx.doi.org/10.1051/apido:19980601

SLICE, D E (2002) *Morpheus, for morphometric research software.* Department of Biomedical Engineering, Wake Forest University School of Medicine; Winston, Salem, USA.

SMITH, D R; BROWN, W M (1988) Polymorphisms in mitochondrial DNA of European and Africanized honey bees (*Apis mellifera*). *Experientia* 44: 257-260. http://dx.doi.org/10.1007/BF01941730

SMITH, D R; BROWN, W M (1990) Restriction endonuclease cleavage site and length polymorphisms in mitochondrial DNA of *Apis mellifera mellifera* and *A. m. carnica* (Hymenoptera: Apidae). *Annals of the Entomological Society of America* 83: 81-88.

SMITH, D R; GLENN, T (1995) Allozyme polymorphisms in Spanish honey bees. *Journal of Heredity* 86: 12-16.

SMITH, D R; TAYLOR, O R; BROWN, W M (1989) Neotropical Africanized honey bees have African mitochondrial DNA. *Nature* 339: 213-215. http://dx.doi.org/10.1038/339213a0

SMITH, D R; SLAYMAKER, A; PALMER, M; KAFTANOGLU, O (1997) Turkish honey bees belong to the east Mediterranean mitochondrial lineage. *Apidologie* 28: 269-274. http://dx.doi.org/10.1051/apido:19970503

SMITH, D R; PALOPOLI, M F; TAYLOR, B R; GARNERY, L; CORNUET, J M; SOLIGNAC, M; BROWN, W M (1991) Geographical overlap of two mitochondrial genomes in Spanish honey bees (*Apis mellifera iberica*). *Journal of Heredity* 82: 96-100.

SOLAND-RECKEWEG, G; HECKEL, G; NEUMANN, P; FLURI, P; EXCOFFIER, L (2009) Gene flow in admixed populations and implications for the conservation of the Western honey bee, *Apis mellifera*. *Journal of Insect Conservation* 13: 317-328. http://dx.doi.org/10.1007/s10841-008-9175-0

SOLIGNAC, M; VAUTRIN, D; LOISEAU, A; MOUGEL, F; BAUDRY, E; ESTOUP, A; GARNERY, L; HABERL, M; CORNUET, J M (2003) Five hundred and fifty microsatellite markers for the study of the honey bee (*Apis mellifera* L.) genome. *Molecular Ecology Notes* 3: 307-311. http://dx.doi.org/10.1046/j.1471-8286.2003.00436.x

SPÖTTER, A; GUPTA, P; NÜRNBERG, G; REINSCH, N; BIENEFELD, K (2012) Development of a 44K SNP assay focussing on the analysis of a varroa-specific defence behaviour in honey bees (*Apis mellifera carnica*). *Molecular Ecology Resources* 12: 323-332. http://dx.doi.org/10.1111/j.1755-0998.2011.03106.x

STEINHAGE, V; KASTENHOLZ, B; SCHRÖDER, S; DRESCHER, W (1997) A hierarchical approach to classify solitary bees based on image analysis. *Mustererkennung 19. DAGM-Symposium, Braunschweig, Sept. 15–17, 1997, Informatik Aktuell; Springer, Germany.* pp. 419-426.

STEINHAGE, V; SCHRÖDER, S; LAMPE, K H; CREMERS, A B (2007) Automated extraction and analysis of morphological features for species identification. In *N MacLeod (Ed.). Automated object identification in systematics: theory, approaches, and applications.* pp. 115–129

STEVANOVIC, J; STANIMIROVIC, Z; RADAKOVIC, M; KOVACEVIC, S R (2010) Biogeographic study of the honey bee (*Apis mellifera* L.) from Serbia, Bosnia and Herzegovina and Republic of Macedonia based on mitochondrial DNA analyses. *Russian Journal of Genetics* 46(5): 603–609. http://dx.doi.org/10.1134/S1022795410050145

STRANGE, J P; GARNERY, L; SHEPPARD, W S (2007) Persistence of the Landes ecotype of *Apis mellifera mellifera* in southwest France: confirmation of a locally adaptive annual brood cycle trait. *Apidologie* 38: 259-267. http://dx.doi.org/10.1051/apido:2007012

SUŠNIK, S; KOZMUS, P; POKLUKAR, J; MEGLI, V (2004) Molecular characterisation of indigenous *Apis mellifera carnica* in Slovenia. *Apidologie* 35: 623–636. http://dx.doi.org/10.1051/apido:2004061

TARPY, D R (2003) Genetic diversity within honey bee colonies prevents severe infections and promotes colony growth. *Proceedings of the Royal Society,* London B 270: 99-103. http://dx.doi.org/10.1098/rspb.2002.2199

TAUTZ, D (1990) Notes on the definition and nomenclature of tandemly repetitive DNA sequences. In *S D J PENA; R CHAKRABORTY; J T EPPLEN; J JEFFREYS (Eds). DNA fingerprinting state of the science.* Birkhauser; Basel, Switzerland. pp. 21-28.

TOFILSKI, A (2004) DrawWing, a program for numerical description of insect wings. *Journal of Insect Science* 4: 1-5.

TOFILSKI, A (2008) Using geometric morphometrics and standard morphometry to discriminate three honey bee subspecies. *Apidologie* 39: 558-563. http://dx.doi.org/10.1051/apido:2008037

UZUNOV, A; KIPRIJANOVSKA, H; ANDONOV, S; NAUMOVSKI, M; GREGORC, A. (2009) Morphological diversity and racial determination of the honey bee (*Apis mellifera* L.) population in the Republic of Macedonia. *Journal of Apicultural Research* 48(3): 196-203. http://dx.doi.org/10.3896/IBRA.1.48.3.08

UZUNOV, A; MEIXNER, M D; KIPRIJANOVSKA, H; ANDONOV, S; GREGORC, A; IVANOVA, E; BOUGA, M; DOBI, P; FRANCIS, R M; BUECHLER, R; KRYGER, P (2014) Genetic structure of *Apis mellifera macedonica* population based on microsatellite DNA polymorphism. *Journal of Apicultural Research* (in press).

VIARD, F; FRANCK, P; DUBOIS, M-P; ESTOUP, A; JARNE, P (1998) Variation of microsatellite size homoplasy across electromorphs, loci and population in three invertebrate species. *Journal of Molecular Evolution* 47: 42-51 http://dx.doi.org/10.1007/PL00006361

VIGNAL, A; MILAN, D; SANCRISTOBAL, M; EGGEN, A (2002) A review on SNP and other types of molecular markers and their use in animal genetics. *Genetics Selection Evolution* 34: 275-305. http://dx.doi.org/10.1051/gse:2002009

WHITFIELD, C W; BEHURA, S K; BERLOCHER, S H; CLARK, A G; JOHNSTON, J S; SHEPPARD, W S; SMITH, D R; SUAREZ, A V; WEAVER, D; TSUTSUI, N D (2006) Thrice out of Africa: ancient and recent expansions of the honey bee, *Apis mellifera. Science* 314: 642-645. http://dx.doi.org/10.1126/science.1132772

ZAYED, A; WHITFIELD, C W (2008) A genome-wide signature of positive selection in ancient and recent invasive expansions of the honey bee *Apis mellifera. Proceedings of the National Academy of Sciences of the United States of America* 105: 3421-3426. http://dx.doi.org/10.1073/pnas.0800107105

ZELDITCH, M L; SWIDERSKI, D L; SHEETS, H D; FINK, W L (2004) *Geometric morphometrics for biologists: a primer.* Elsevier Academic; New York, USA.

Journal of Apicultural Research 52(4): (2013)
DOI 10.3896/IBRA.1.52.4.06

REVIEW ARTICLE

Standard methods for chemical ecology research in *Apis mellifera*

Baldwyn Torto[1*], **Mark J Carroll**[2], **Adrian Duehl**[3], **Ayuka T Fombong**[1], **Tamar Katzav Gozansky**[4], **Francesco Nazzi**[5], **Victoria Soroker**[6] and **Peter E A Teal**[3]

[1]International Centre of Insect Physiology and Ecology, P.O. Box 30772-00100, Nairobi, Kenya.
[2]Carl Hayden Bee Research Center, USDA-ARS, 2000 E. Allen Rd., Tucson, AZ 85719, USA.
[3]Centre for Veterinary, Medical and Agricultural Entomology, USDA-ARS, 1600/1700 SW 23rd Dr. Gainesville, Fl. 32406, USA.
[4]Department of Natural and Life Sciences, The Open University of Israel, Raanana, Israel.
[5]Dipartimento di Scienze Agrarie e Ambientali, Università di Udine via delle Scienze 206, Udine, Italy.
[6]Department of Entomology, Agricultural Research Organization, The Volcani Center, Bet Dagan 50250, Israel.

Received 4 June 2012, accepted subject to revision 31 July 2012, accepted for publication 28 February 2013.

*Corresponding author: Email: btorto@icipe.org

Summary

This paper describes basic methods essential in elucidating chemically-mediated behavioural interactions among honey bees, and between honey bees and other arthropods. These range from bioassay methods used to demonstrate the role of specific behaviours, techniques and equipment used to collect and analyse semiochemicals (both volatiles and non-volatiles e.g. cuticular hydrocarbons) from individual honey bees, groups of bees or an entire colony in its native environments. This paper covers: collection and analysis of honey bee volatiles in the natural environment, collection and analysis of bee volatiles out of their natural environment and their antennal detection, collection and analysis of non-volatile cuticular hydrocarbons, bioassays with queen pheromone and finally a section focusing on *in vitro* bioassays as a tool for elucidation of mechanisms regulating pheromone gland activity.

Métodos estándar para la investigación en la ecología química en *Apis mellifera*

Resumen

Este artículo describe los métodos esenciales básicos para dilucidar las interacciones de comportamiento mediadas por la química entre las abejas y entre éstas y otros artrópodos. Estos van desde los métodos de bioensayo usados para demostrar el rol de comportamientos específicos, hasta las técnicas y equipamientos usados para colectar y analizar semioquímicos (tanto volátiles como no volátiles, por ejemplo hidrocarburos cuticulares) en abejas al nivel individual, en grupos de abejas o en una colonia entera en su ambiente natural. Este artículo engloba: colección y análisis de los volátiles de la abeja de la miel en su ambiente natural, colección y análisis de los volátiles de la abeja fuera de su ambiente natural y con su detección por las antenas, colección y análisis de los hidrocarburos cuticulares no volátiles, bioensayos con la feromona real y finalmente una sección enfocada a los bioensayos *in vitro* como herramienta para dilucidar los mecanismos que regulan la actividad de la glándula que produce feromonas.

Footnote: Please cite this paper as: TORTO, B; CARROLL, M J; DUEHL, A; FOMBONG, A T; GOZANSKY, K T; NAZZI, F; SOROKER, V; TEAL, P E A (2013) Standard methods for chemical ecology research in *Apis mellifera*. In *V Dietemann; J D Ellis; P Neumann (Eds) The COLOSS BEEBOOK, Volume I: standard methods for Apis mellifera research. Journal of Apicultural Research* 52(4): http://dx.doi.org/10.3896/IBRA.1.52.4.06

西方蜜蜂化学生态学研究的标准方法

摘要

本文介绍用于研究蜜蜂之间以及蜜蜂与其它节肢动物间化学介导的交互行为作用的基本方法。涵盖用于论证特定行为的作用的生物测定方法，以及从蜜蜂个体、群体或整个蜂群自然环境中收集和分析化学信息素的技术和装备，包括挥发性和非挥发性成分，如表皮烃类。本文包含：在自然条件下收集和分析蜜蜂挥发物；在非自然条件下收集和分析蜜蜂挥发物及这些挥发物的触角探测；收集和分析非挥发性表皮烃类；生物测定蜂王信息素；最后集中介绍用于阐明信息素腺体活性调控机理的体外生物测定方法。

Keywords: COLOSS *BEEBOOK*, semiochemicals, headspace, EAG, exocrine gland, *in vitro*, QMP, honey bee

西方蜜蜂化学生态学研究的标准方法

Table of Contents

Table of Contents cont'd

1. Introduction

Maintenance of integrity of a honey bee colony is highly dependent on a sophisticated communication system that is largely dependent on chemical cues. Due to the crowded environment within the hive, bees appear to communicate to large extent through chemicals, of low volatility, vibrations, and other near-contact or contact modalities (Blum and Fales, 1988; Breed *et al.*, 1988; Naumann *et al.*, 1991; Breed, 1998; Slessor *et al.*, 2005). However, volatiles form a small but important part of the signalling chemicals (semiochemicals), that mediate interactions between colony members (for reviews, see Free, 1987; Pankiw, 2004; Slessor *et al.*, 2005). Some volatiles are used as releaser pheromones to rapidly communicate information to the rest of the colony (Pankiw, 2004), including alarm pheromones for defence (Boch *et al.*, 1962; Hunt *et al.*, 2003) and Nasanov pheromones for colony cohesion (Boch and Shearer, 1964; Pickett *et al.*, 1980). Other volatile pheromones, such as (*Z*)-β-ocimene, have long-term primer effects on physiology, development and fertility (Maisonnasse *et al.*, 2009; Maisonnasse *et al.*, 2010). A few volatile pheromones, such as the alarm pheromone component isoamyl acetate (IAA), have both releaser effects on behaviour and primer effects on physiology (Alaux and Robinson, 2007). On a more localized scale, both volatiles and non-volatile contact cues (e.g. cuticular hydrocarbons) serve as signalling cues for intimate interactions of workers with their immediate colony environment such as hygienic behaviour (Masterman *et al.*, 2001; Gramacho and Spivak, 2003; Swanson *et al.*, 2009; Schöning *et al.*, 2012). Volatiles also mediate interactions of bees with non-nestmates.

Like many social hymenopterans, honey bees use non-volatile to volatile acquired colony chemicals to distinguish between nestmates and non-nestmates (Breed, 1998). Natural enemies of the honey bee such as the small hive beetle *Aethina tumida* and the parasitic mite *Varroa destructor* also use colony odours as kairomone cues (Nazzi *et al.*, 2004; Torto *et al.*, 2005; Torto *et al.*, 2007a; Nazzi *et al.*, 2009; see the *BEEBOOK* papers on small hive beetles (Neumann *et al.*, 2013) and varroa mites (Dietemann *et al.*, 2013) for more details on these organisms).

This paper describes basic methods essential in elucidating chemically-mediated behavioural interactions among honey bees, and between honey bees and other arthropods. These range from bioassay methods used to demonstrate the role of specific behaviours, techniques and equipment used to collect and analyse semiochemicals (both volatiles and non-volatiles) from individual honey bees, groups of bees or an entire colony in its native environments. This paper is subdivided into 5 main sections; collection and analysis of honey bee volatiles in the natural environment (Section 2), collection and analysis of bee volatiles out of their natural environment and their antennal detection (Section 3), collection and analysis of non-volatile semiochemicals (Section 4), bioassays with queen pheromone (Section 5) and the *in vitro* bioassays as a tool for elucidation of mechanisms regulating pheromone gland activity (Section 6). While this paper intends to provide simple easy to follow and replicate guidelines when working on the semiochemically-mediated interactions of honey bees, readers must bear in mind that chemical ecology requires basic understanding of behavioural biology and analytical chemistry, which are two very broad fields of study, that cannot be exhaustively dealt with in this *BEEBOOK* paper. Therefore, before embarking on any chemical ecology experiments, it is advised to consult a chemist (if a biologist) or a biologist (if a chemist) and the other *BEEBOOK* papers (e.g. for behaviour, Scheiner *et al.*, 2013).

2. *In situ* volatile collection of odours in the colony environment

2.1. Introduction

Semiochemicals play a vital role within and among colony interactions, and they also mediate interactions between honey bees and their parasites and predators. In this section, we describe methods used to collect and analyse volatiles from honey bees in their native environments. The discussion focuses on *in situ* collection of volatiles because natural emission rates are the most biologically relevant metric of volatile characterization. While volatile collection and analysis techniques are described broadly, these methods are specifically discussed in the context of working with a hive and its inhabitants. Given the complexity of honey bee chemical interactions, the authors strongly emphasize that detection of a compound in the gaseous phase does not confirm activity of the compound in that phase (see Keeling *et al.*, 2003, on contact activity of volatile queen retinue pheromone components). Researchers use bioassays to determine the mode of biological activity for volatiles collected from honey bees (Torto *et al.*, 2007b). *In situ* analytical methods are most informative when used in tandem with other methods to characterize volatiles. In particular, *ex situ* volatile collection (see section 3 of this paper) and chemical analysis of tissue extracts (see sections 4 and 6) can be used to specifically identify the odour source in the colony and within the bee itself.

2.2. Collection and analyses of honey bee volatiles

In situ volatile collection and analytical methods can be conceptually divided into four basic sections: headspace environment, volatile collection, volatile separation, and volatile detection and analysis (see Table 1).The reader should be aware that no one method provides all the information required to identify or characterize volatiles. Different methods are better suited for elucidation of compound identity, mass, quantification and sensitivity to trace volatiles (D'Alessandro and Turlings, 2006). In the following sections, the advantages and disadvantages of methods commonly used at each step of volatile collection and analysis are briefly discussed.

2.2.1. Volatiles in the headspace environment

Headspace environment refers to the techniques and devices used to manipulate the headspace, or air surrounding an odour source where volatiles are actively emitted. The volatile profile obtained from an odour source strongly depends on whether the headspace volatiles are contained (concentrated) or actively relayed to the collection device (via air flow) (for examples, see Heath and Manukian, 1994; Tholl *et al.*, 2006; Carroll and Duehl, 2012). Volatiles can be collected from an open or closed air space. Open sampling schemes are simple to carry out but result in variable losses of target emissions and contamination by background odours. Partially or completely enclosed air systems surround the odour source with a containment system (containers built of glass, metal and other odourless materials) to concentrate volatile emissions and control the sampling rate. For longer collections in an enclosed system, an air flow system is required to ventilate the bees.

Collection techniques can also be classified as dynamic or static. Dynamic flow systems use active air flow (push), vacuum flow (pull), or combined air and vacuum flow (push-pull) systems to move headspace volatiles through a filter containing an adsorbent material volatile trap. The benefit of dynamic collections is that quantification is easier because volatiles are trapped at a known rate. Static flow systems have little or no flow through the headspace. In a static airspace, there will be equilibrium between volatile compounds present in the solid or liquid source and the same volatiles in the gas phase. Changes in any volatile component will affect the equilibrium for all other components in the airspace blend. The main benefit of static technique is that it is very easy to use and does not require any expensive equipment or instrument modifications.

2.2.1.1. Volatile sampling in the headspace environment

The complete flow path for a closed push-pull flow system (described without connecting lines) is 1) air source, 2) bubbler humidifier, 3) air flowmeter array, 4) air line, 5) air port connector, 6) port connector extension tubes, 7) observation frame, 8) vacuum port connector, 9) absorbent filter volatile trap and 10) vacuum line (see Fig. 1). Each setup may not necessarily have all these features. A step-by-step guideline to the selection, assembly and sampling of headspace volatile is provided below.

- Collect volatiles in a headspace environment that mimics core colony conditions (Winston, 1987; Kraus and Velthuis, 1997).
- Use incubators and glass in-line bubblers to control temperature and humidity.
- To minimize contaminants, construct all enclosures, lines, and tubing before the volatile trap out of odourless glass, Teflon/ PFTE (polytetrafluoroethylene), copper, or stainless steel. All line materials under significant pressure or vacuum should be constructed of metal tubing. Materials after the volatile trap can be made of "dirty" (e.g. odiferous) materials such as Tygon plastics.
- Control the air and vacuum flow rates in closed and partially-closed collection systems with flowmeters. Flowmeters are necessary for quantitative comparisons of volatile emissions.
- Filter the main air source with gas filters (molecular sieve and activated charcoal) to eliminate water and oil contaminants from air pumps. Clean air can also be purchased in pressurized tanks.

Table 1. Headspace environments and methods commonly used for the collection and analysis of volatiles. Several of these methods have been well developed in other systems (e.g. plant-insect, microbial) but not applied to honey bee volatiles yet. Not all headspace environments and methods can be used interchangeably.

Headspace environment	Volatile collection	Volatile separation	Volatile detection
Enclosure type Open Partially-closed (excess air) Closed (balanced flow) **_Flow system_** *Static flow* *Dynamic flow* Push (air only) Pull (vacuum only) Push-pull (air and vacuum)	**_Adsorption method_** *Thermal desorption* SPME Tenax TA *Solvent desorption* Porapak HayeSepQ (SuperQ) Tenax TA Activated charcoal	**_Injector type_** Split Splitless On-column SPME (insert) Tenax TA (trap) **_Column type_** HP-1 (non-polar) DB-1 (non-polar) DB-35 (polar) Affinity Chiral	**_Detector type_** *Mass spectrum* Electronic ionization (EI – compound identity) Chemical ionization (CI – compound mass) *Ionization (detection/quantification)* Flame ionization (FID - organic compounds) Photoionization (PID - heteroatoms) Flame thermionic detector (FTD - heteroatoms) Electron capture (ECD - nitrogenous compounds) Flame photometric (FPD - sulfides) *Thermal conductivity (detection/quantification)* Thermal conductivity (TCD) *Electrophysiological response* GC-EAD (electrophysiology)

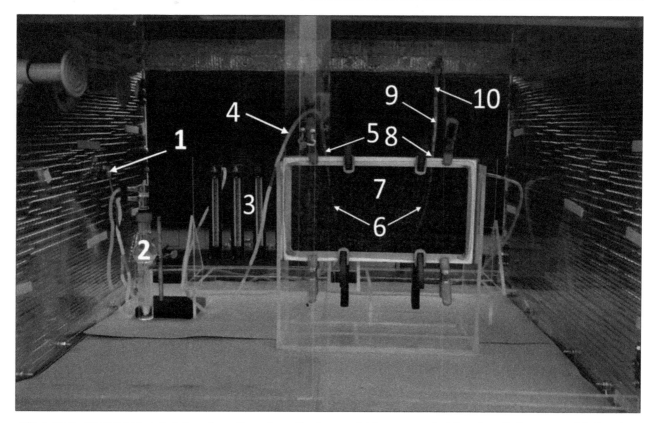

Fig. 1. *In situ* collection from a whole frame face with an observation frame containing pupae and adult workers. Air flows sequentially from the **(1)** air source through the **(2)** bubbler humidifier, **(3)** air flowmeter array, **(4)** air line, **(5)** air port connector, **(6)** port connector extension tubes into the **(7)** observation frame. Sample headspace is pulled through the **(6)** extension tubes through the **(8)** vacuum port connector, **(9)** adsorbent filter volatile trap, by the **(10)** vacuum line which is regulated by the vacuum flowmeters (hidden by the observation frame). The collection equipment is kept in a temperature controlled environmental chamber to keep the frame near colony temperature (32°C). Some flow system features such as the air source pressure reducer, air filter, air heater, vacuum pressure regulator and the vacuum source are not visible in this picture.

- Provide airflow to ventilate subjects without creating disturbance. A flow rate that exchanges the air in the container about once every 20 minutes is sufficient.
- Protect SPME fibres, adsorbent traps, and other sensitive objects from direct contact with bees by shielding the devices with a screen or Teflon shield.
- Collect appropriate odour "blanks" to account for background odours originating from the equipment, the air supply, and the surrounding environment. Use hive equipment of similar age (preferably less than two years old) whenever possible.
- Limit background odours by eliminating unnecessary odour sources. Avoid hive materials containing dead or diseased bees, rotten food stores, active small hive beetle infestations, or odiferous treatments such as essential oil patties. Limit the use of smoke. If use of smoke is unavoidable, pass clean air through the collection system for 30 minutes before volatile collection, or collect appropriate background samples.
- If needed, scale up the amount of volatiles sampled either by extending the sampling period or increasing the sample size by forming homogenous patches (e.g. single age brood cohorts, pollen patches, and simultaneous infection of brood).
- Avoid accidental contamination of collection equipment by household cleaners and detergents. Use GC grade solvents (ethanol, methanol, hexanes, and acetone), water, or unscented detergent to clean equipment. Equipment can be baked at 100°C for 1 hour to remove most volatile contaminants.

2.2.2. Collection and recovery (desorption) of volatiles

Volatiles are collected by trapping sample odours on exposed inert polymer matrix (adsorbent packing materials) which vary in their adsorbent properties and method of analysis. The trapped volatiles are then removed by using either heat (thermal desorption) or solvent (solvent desorption) to strip (desorb) the compounds from the packing material. Collected volatiles can be desorbed either directly into the analytical instrument or into a solvent that is later analysed. Each approach has distinct advantages and disadvantages in terms of information generated, ease-of-use, and cost to the researcher and are described below.

2.2.2.1. Solvent desorption of volatiles

The most common method for dynamic headspace collection is to use adsorbent materials that require a solvent wash to remove trapped volatiles (Heath and Manukian, 1994; D'Alessandro and Turlings, 2006; Tholl *et al.*, 2006; Carroll and Duehl, 2012). In general, adsorbent techniques that employ solvent extraction are best adapted for large scale sampling over longer periods of time. Solvent-based adsorbent techniques can also be used to quantify volatiles in closed airspace systems or when the relative capture rate can be calculated.

Sample headspace air is actively pulled through an adsorbent filter trap where the volatiles adhere onto the adsorbent matrix. Commonly used adsorbent materials include Super Q, Hayesep Q, Porapak Q, Tenax TA, and speciality adsorbents that target specific chemical groups (Núñez *et al.*, 1984; D'Alessandro and Turlings, 2006). Most of these adsorbents trap large quantities of volatiles without much bias toward specific chemical classes. The captured volatiles are desorbed off the matrix by a solvent (chosen on the basis of the polarity of the target compounds i.e. polar solvent for polar volatiles, non-polar solvents for non-polar volatile) rinse. An active vacuum source is required to draw the sample air through the tightly-packed adsorbent material. Refer to paragraphs 2.2.2.3. and 2.2.2.4. for application examples.

Pros: A distinct feature of solvent desorption is that only a small fraction of the sample solution is normally used during analysis. Thus, a single volatile sample can be analysed separately by different instruments and also tested for bioactivity.

Cons: On the negative side, analysing just a small fraction of the sample also decreases sample sensitivity. Another disadvantage of solvent desorption is that highly volatile compounds may co-elute with the solvent peak during separation (Núñez *et al.*, 1984).

2.2.2.2. Thermal desorption of volatiles

Thermal desorption is an approach that uses heat rather than solvents to remove trapped volatiles from the adsorbent packing material. During analysis, the packing material is heated in a temperature-controlled environment in the analytical instrument and the volatiles desorb off the collection surface into the column flow. Thermal desorption techniques are particularly advantageous for: 1) sampling volatiles emitted at very low concentrations, 2) identifying highly volatile chemicals that would co-elute with solvent during separation, 3) obtaining a rapid profile of the volatiles associated with an odour source, or 4) collection of volatiles in a static system with very limited or no air exchange. Thermal desorption also has certain disadvantages compared to solvent desorption. Because the entire sample is desorbed on column during GC analysis, each sample can be analysed only once. For various reasons, most thermal desorption techniques are not particularly well suited to quantification of volatile compounds. Methods of thermal desorption using SPME and Tenax cartridges are described in the following sections.

2.2.2.2.1. Thermal desorption of static headspace volatiles by SPME

SPME (solid phase microextraction), is the most common thermal desorption method (Augusto and Valente, 2002). The SPME technique includes exposing a fibre to odour source headspace in a static environment with little or no air flow. SPME is ideal for rapid analysis of volatiles emitted by small and strong odour sources, preferably in a closed container. Because of its ease of use, many researchers have begun to use SPME fibres in honey bee systems (Gilley *et al.* 2006; Schmitt *et al.*, 2007; Maisonnasse *et al.*, 2010). SPME has rather uniquely been used repeatedly to sample odours *in situ* from single bees in the open comb environment (Thom *et al.*, 2007).

Procedure:

1. Insert the SPME holder in the container with the sample. Volatiles should reach equilibrium (which should be in trial sampling and analysis) in the closed container before the SPME fibre is exposed to the headspace.
    2. Expose the SPME fibre to the headspace without the fibre touching the sample or container.

The fibre should ideally be allowed to reach equilibrium with headspace volatiles – typical adsorption times ranging from a few seconds to 30 minutes.

3. Retract the fibre into the protective sheath.
    4. Inject the fibre into a splitless GC injection port for analysis (see section 2.2.3.)

Trapped volatiles must be desorbed rapidly after collection since the trapped volatiles are exposed to heat and carrier gas flow and desorb off the fibre onto the column head as the fibre is heated.

The selection of SPME fibre type influences the sensitivity toward specific compound groups as well as the exposure time required to reach sorption equilibrium (Augusto and Valente, 2002). Polydimethylsiloxane (PDMS) fibres have been used to sample less polar volatiles and Carbowax/PDMS to capture more polar volatiles (Zabaras and Wyllie, 2002). Researchers should try a variety of related fibres to determine what works best for their system. For queen and worker volatiles, a number of fibre compositions were tested and PDMS/divinylbenzene was selected as the best (Gilley *et al.*, 2006). Consider the following guidelines for use of SPME:

- Use enclosed or partially enclosed static systems to limit dissipation of headspace volatiles (see Augusto and Valente, 2002 for a design). In general, the more static and concentrated the headspace volatiles, the more rapidly equilibrium is achieved.
- To limit background contaminants, only expose the fibre from its sheath when you are actively collecting volatiles. Protect the exposed fibre from the bees and hive materials with a Teflon jacket perforated with holes.
- Before sampling, bake the fibre in the GC port to remove residual volatiles left on the fibre. Follow the guidelines in the instructions that come with the SPME fibre.
- If unable to detect compounds of interest, try a longer equilibrium and exposure times first and different fibre materials next.

- Once specific chemicals of interest have been identified, optimize detection with chemical standards. Test equilibration time and sensitivity by exposing the fibre to each standard's headspace for different periods of time. Fibres have reached minimal equilibrium time when the volatile capture no longer increases with exposure time.

Pros: SPME is easy to use

Cons: Collected compounds cannot be stored on the fibre for longer period before analysis. Researchers should take caution in over-interpreting the volatile profiles obtained with SPME. Fibres are easily contaminated by background volatiles that may not be present in the target odour source. SPME is also poorly suited for volatile quantification because the fibres have different affinities for different chemical classes (Agelopoulos and Pickett, 1998). Unfortunately, the adsorbance rate of each volatile can be significantly influenced by the other compounds present in the headspace (Romeo, 2009). There are methods to quantify SPME samples, but given the variable chemical affinities, it is difficult to calculate amounts with confidence (Augusto and Valente, 2002). Volatile emission rates are often expressed as relative emission ratios rather than absolute amounts. For these reasons, researchers should use other methods to quantify volatile emission rates (see sections 2.2.2.3 and 2.2.2.4).

2.2.2.2.2. Thermal desorption of dynamic headspace volatiles by Tenax

A second set of thermal desorption techniques combines the sensitivity of thermal desorption with the controlled sampling of dynamic headspace collections. Collection of sample volatiles is similar to other dynamic headspace collection techniques. Unlike SPME, sample peaks collected and separated by thermal desorption can be readily quantified using techniques for solvent-desorbed collection systems (see section 2.2.2.3.).

Sample headspace is actively drawn by vacuum through a cartridge where volatiles are trapped in the adsorbent packing material. The packing material most commonly used is Tenax TA (replacing Tenax GC), which can be combined with activated carbon to increase capture of both non-polar and polar certain chemicals (Raguso and Pellmyr, 1998). Thermally-desorbed Tenax has a long history of use in honey bee systems and is still used today (Moritz and Crewe, 1991; Schöning *et al.*, 2012). The cartridge is later (immediately after volatile sampling) inserted into a modified GC injection port (which is larger than the normal syringe needle injection port to accommodate the Tenax filter) and rapidly heated to desorb all of the trapped volatiles from the packing material onto the column (see section 2.2.3.). Refer to sections 2.2.2.3. and 2.2.2.4. for application examples.

Pros: Like SPME, Tenax thermal desorption has a distinct advantage over solvent-desorbed samples in the absence of solvent peaks that may obscure highly-volatile compounds. Typically, samples are analysed immediately after volatile collection.

Cons: One major disadvantage of Tenax is that these thermal desorption methods require significant equipment and expertise compared with SPME and solvent desorption methods. Tenax collections make use of the volatile collection infrastructure used with solvent-extracted adsorbents as well as a modified injector port. Compared to solvent desorption, thermal desorption is a relatively slow sample injection technique that will lead to peak broadening for very volatile compounds during GC separations. These problems are partially corrected by cryofocusing techniques that use an automatic thermal inject system to rapidly heat and inject sample chemicals onto the column. Because cryotrap methods are advanced techniques, new researchers should use other collection techniques to initially sample volatiles (see sections 2.2.2.2.1 and 2.2.2.3).

2.2.2.3. Sampling odours at the whole colony scale

Whole colony volatiles can be collected from colonies using the hive equipment itself as a partial enclosure to concentrate colony volatiles. The colony must be well sealed to capture colony odours before volatiles escape to the outside atmosphere. Either replace leaky hive components or plug the gaps with wax. Collect a sample from air outside the colony as a control since this air replaces the colony headspace.

1. Add 200 µl of the elution solvent (dichloromethane or hexane) to the filter solvent reservoir (just above the adsorbent packing material).
2. Gently push the solvent through the packing material with a clean air or nitrogen flow.
3. Repeat steps 1 and 2 two times.

This procedure rinses residual contaminants from the adsorbent filter to prepare the adsorbent filter trap for volatile collection.

4. Place the filter into a sleeve jacket made of short interlocking sections of rigid Teflon tubing (0.635 cm OD, 0.794 cm OD, 0.952 cm OD) (Fig. 2) to protect the filter from the bees.

The top of the filter needs to attach tightly to the jacket tubing; otherwise, air will flow around, instead of through, the filter.

5. Construct a sampling tube out of 0.64 cm OD Teflon tubing that reaches from outside the colony to the centre of the colony (~ 30 cm for a Langstroth deep).

Cover the end with metal screen to prevent bees from entering the tube.

6. Carefully insert the sampling tube into the colony either through the entrance or a small hole in the equipment to the centre of the colony.

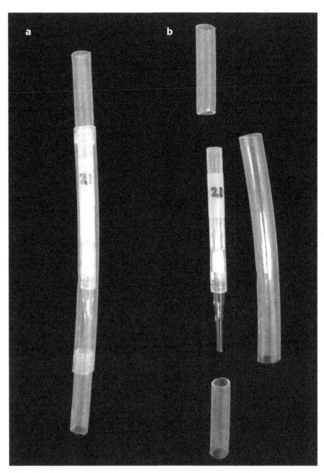

Fig. 2. **(a)** Assembled and **(b)** unassembled views of a SuperQ adsorbent filter enclosed in a protective Teflon tube jacket made of interlocking sections of tube.

7. Attach the adsorbent filter in-line between the sampling tube and a flowmeter-regulated vacuum line leading to the vacuum pump.
8. Collect colony volatiles by pulling colony headspace through the filter at 600 ml/min to 3l / min with flowmeter-regulated vacuum (exchange at least one volume of colony airspace every 20 minutes).

Most volatile collections require 3 to 12 hours to collect sufficient material for GC analysis.

9. End the volatile collection by removing the filter trap from the vacuum line.
10. Carefully remove the filter from its protective plastic jacket.

Secure the filter in a holder.

11. Add 5 µl of an internal standard solution (80 ng nonyl acetate/ µl) directly to the top of the adsorbent packing material with a syringe.

Avoid touching the packing material with the syringe.

12. Place a GC vial or vial with a glass insert directly underneath the tip of the filter.

13. Add 200 µl solvent (dichloromethane or hexane) to the solvent reservoir above the packing material to extract the trapped volatiles from the filter packing material.
14. Gently push the solvent through the packing material at a steady drip with clean air or nitrogen flow.
15. Cap and store the sample vials in a -80°C freezer until GC analysis (see section 2.2.3.).
16. Volatile emissions can be calculated by comparing compound peak areas to the known amount of internal standard added to the sample:

$$\text{Sample compound amount (ng)} = \frac{\text{Area of compound peak}}{\text{Area of internal standard peak (area counts)}} \times \text{internal standard amount (ng)}$$

Note that quantification may be difficult if large amounts of the whole colony volatiles escaped the colony headspace before collection due to a slow sampling rate (colony volumes/h) or poor seals between hive equipment.

2.2.2.4. Sampling odours at a whole frame scale

Volatiles can be sampled from bees enclosed on a single frame face with a partially-enclosed push-pull airflow system (Fig. 1; Carroll and Duehl, 2012). A metal and glass observation frame is pushed into the wax comb of a colony frame to enclose the bees and materials on the frame face inside. Controlled airflow from the enclosed headspace through a filter trap allows for ventilation of the bees and recovery of most volatile emissions. This approach provides a much more targeted method for *in situ* sampling of colony volatiles since background odours from most hive materials and outside air are excluded from the collection.

1. Weld together the observation frame from metal L-bar material. This rectangular frame should precisely fit the inside the inner perimeter of the wooden bars of the colony frame (23.2 cm x 51.1 cm for a standard Langstroth deep) out of 1.91 cm stainless steel or aluminium angle L-bar. The two edges of the L-bar extend perpendicularly down toward the wax comb and outward horizontally as a flat phalange.
2. To provide access for air and vacuum flow into the enclosure, drill two 0.65 cm diameter port holes through the top edge of the metal frame.

Drill the holes about 7.70 cm in from the ends of the frame. Position the holes as close to the angle of the metal frame as possible.

3. Provide a junction (port connector) through the port holes with 4 cm pieces of 0.635 cm outer diameter (OD) Teflon tubing. Air and vacuum lines attach to the outside of the port connector and port connector extension tubes attach to the inside of the port connector.
4. To enclose the bees in the metal frame, cut a piece of 0.47 cm

thick piece of rectangular plate glass that extends to the outer edges of the metal frame.

5. Form a partial gasket between the glass plate and metal frame of the observation frame by wrapping the perimeter of the glass plate 3 times with 0.635 cm wide Teflon tape. Secure the glass plate to the phalanges of the metal frame with four small (3.75 cm) spring clamps.
6. Connect air and vacuum lines consisting of 0.635 cm OD flexible Teflon tubing to the outside of the port connectors with slightly larger diameter pieces of Teflon tubing.
7. To direct flow into the centre of the frame airspace, attach an 11.6 cm long piece of 0.48 cm OD Teflon tubing to the inside of each port connector.

Orient these slightly curved port connector extension tubes toward the plate glass to avoid contact with the comb (Fig. 3).

8. The volatile collection system flow rates must be adjusted to final rates with all of the components in place except for the enclosed frame.

Removal of any part, especially the adsorbent filter traps, alters the resistance of the system to air flow. Likewise, changes in flow in one sampling line affects other sampling lines. Provide a slight excess of air flow (680 ml/min) to vacuum flow (515 ml/min) to ensure that no outside air enters the enclosed frame headspace. Check air and vacuum flow rates against a calibrated flowmeter placed in line after the air flowmeter and between the filter trap and the vacuum flowmeter.

9. Remove the adsorbent filter traps
10. Rinse adsorbent filter traps of residual contaminants 3 times with solvent as previously described in section 2.2.2.3.
11. Place the cleaned filters into protective tube jackets as previously described in section 2.2.2.3 (Fig. 2).
12. Select the frame face to be sampled.

The frame should have completely drawn comb without any major holes or gaps. Small cracks and gaps can be plugged with wax from the colony.

13. Carefully push the metal frame into the wax comb along the inner perimeter of the wooden colony frame.

If bees are present, move slowly to avoid crushing them.

14. Secure the observation frame to the colony frame with four C-clamps.
15. Remove any bees remaining on the outside of the frames.

Keep the enclosure out of direct sunlight to avoid overheating the bees.

16. Transfer the frame quickly to the volatile collection system site (Fig 3) and place the frame on a frame stand to hold the frame upright.
17. To ventilate the bees, connect one observation frame port connector to the air flow.
18. To collect volatiles, attach the filter trap in-line to the vacuum line and the other port connector. Check the air and vacuum

flow rates to see that they match the previous settings. Sampling times run from 30 minutes to 12 hours, with 3 hours being sufficient for most odour sources.

19. End the collection by detaching the filter trap from the vacuum line.

Maintain an air flow through the observation frame headspace to keep the bees ventilated until you return them to their colony.

20. Extract the trapped volatiles from the filter trap with solvent and analyse the samples as previously described in section 2.2.2.1.

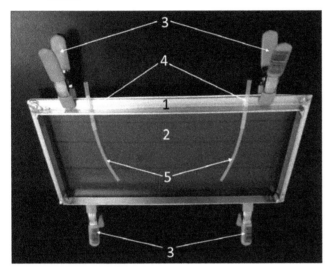

Fig. 3. Top inside view of assembled observation frame consisting of a **(1)** metal frame attached to a **(2)** glass plate by **(3)** four spring clamps and a Teflon tape gasket (hidden on other side). Air and vacuum enter across the frame through **(4)** two port connectors and **(5)** two port connector extension tubes.

2.2.3. Separation and analysis of volatiles by gas chromatography (GC)

Sample volatiles are separated by gas chromatography due to the greater sensitivity, easy reproducibility, and lower operating costs of this method (Handley and Adlard, 2005). Volatiles can be reliably separated under different chromatographic conditions by varying the injection conditions, the column, and the run parameters (temperature and flow). Here is a set of sample methods with comments on how certain parameters can be changed to evaluate chemicals with different volatilities.

2.2.3.1. Injector

GC injection ports are used to desorb trapped volatiles from the sample (solvent, surface, or adsorbent matrix) and channel sample volatiles into the column. Injectors differ in their temperature profiles and the proportion of sample volatiles directed to the column head. The following tips should serve as a guide when deciding on what kind of analysis and injector type to use.

1. For solvent desorption methods, the injection port temperature should be at least 10°C warmer than the maximum oven temperature used in the analysis. For thermal desorption methods such as SPME or Tenax, a slightly cooler injector temperature of 200°C or less is sufficient to desorb all of the absorbed volatiles.

2. The injection split refers to the proportion of the volatilized sample channelled onto the column. To maximize the amount of material injected on the column, use splitless injection.

3. Purge the injector at 1 minute to keep additional less volatile compounds from loading onto the column and degrading the column chromatography.

4. For thermal desorption methods such as Tenax TA, a modified injector improves the chromatography. By contrast, SPME thermal desorption and solvent desorption methods use an unmodified injection inlet with a special inner glass liner.

5. A serious drawback of most GC methods is that the extremely high temperatures commonly used in injector ports can thermally degrade unstable compounds. Destruction of these thermolabile compounds can be minimized through the use of Cool-on-column (COC) injector ports that allow sample volatiles to enter the column at low injector temperatures before the system is heated up.

2.2.3.2. Column

GC columns vary considerably in their polarity and affinity for different chemical functional groups. Compound separation is largely dependent on the affinity each sample compound has for the stationary phase (column coating material) relative to the mobile phase (nitrogen, hydrogen or helium carrier gas). A standard column is a DB-1 column (30 m long x 0.25 mm ID x 0.25 µm coating thickness), a column with a nonpolar stationary phase (a column whose inner surface is lined with a nonpolar matrix such as dimethylpolysiloxane) which is an excellent general purpose column that separates nonpolar and slightly polar compounds very well. Polar compounds such as organic acids chromatograph poorly on this column. Similar nonpolar dimethylpolysiloxane (PDMS) columns, such as HP-1 and SPB-1, also separate most volatiles associated with honey bees quite well. For polar compounds, a better alternative is the DB-35 ((35 % Phenyl)-methylpolysiloxane) column.

2.2.3.3. Run parameters:

2.2.3.3.1. Column flow parameters

Helium or hydrogen can be used as a carrier gas at a constant velocity of 20 cm/sec. The GC holds the carrier gas velocity constant as the temperature and pressure change. Check the system for leaks by placing small amounts of a column-compatible solvent (i.e. hexane, a mixture of isopropanol and water) around fittings, applying gas pressure to the system and observing for tiny bubbles. This should be done routinely especially after the installation of a new column to avoid chromatography degradation and accidental explosions (with hydrogen).

2.2.3.3.2. Oven temperature ramp
The oven temperature affects how compounds interact with the stationary phase (column coating). Generally, more volatile compounds with lower column coating affinities pass through the column before less volatile compounds with higher affinities for the column coating. Choosing the right temperature regime is important for GC analysis. The points below are basic guidelines which should the selection of the right temperature conditions for any given analysis.

- A standard method for solvent desorption starts at 35°C, holds for 2 minutes at this temperature, and then increases at 10°C per minute to 230°C, followed by a final temperature hold for 5 minutes.
- To separate highly volatile compounds, start the oven lower (30°C) and hold temperature for 5 minutes before starting the temperature ramp. A slower ramp (5°C per minute to 75°C, followed by 10°C per minute to 230°C) can also help to separate highly volatile compounds that would otherwise elute together early in the run.
- For thermal desorption, start oven temperatures at 30°C with similar hold and ramp parameters.
- To separate compounds with lower volatility, a higher maximum oven temperature with a longer final temperature hold is required. Increase the maximum oven temperature to 280°C and then hold the maximum temperature for 10 minutes.

2.2.3.3.3. MS detector parameters
The MS detector provides useful information about the chemical structure of compounds which is vital to their identification. When using a MS connected in tandem to a GC, ensure that:

1. The transfer line is generally held at the same temperature as the injector and should be a section of column that extends between the GC oven and the MS detector..
2. The ion source should also be held at a high temperature (220°C).
3. Scan parameters define the m/z (mass/charge) range of fragmented ions scanned by the detector. This range needs to contain the ion fragments of the compounds of interest while avoiding major low mass background ions such as nitrogen gas. Use a m/z range of 35 to 40 minimum to 400 to 600 maximum.

2.2.4. Detection and analysis of volatiles
The various types of GC detectors provide different information about volatile compound identity, emission rates, and activity (Table 1; Tholl *et al.*, 2006; D'Alessandro and Turlings, 2006; Skoog *et al.*, 2007).

- Mass spectrometry (MS) detection is used to provide information about peak identity and molecular mass.
- Electron ionization (EI) mass spectrometry is used to identify unknown compounds based on the comparison of their mass spectral fragment patterns with the fragment patterns of chemical standards found in mass spectra software libraries (i.e. NIST, WILEY, and other commercial libraries).
- Chemical ionization (CI) mass spectrometry is a much milder ionization technique than EI for mass spectrometry that provides information about the molecular mass of sample compounds. It is an excellent technique to quantify known compounds in complex mixtures, especially when combined with selected ion monitoring (SIM) to filter out all ions (m/z) except the selected ions from the chromatogram.
- SIM can also be used with EI ionization if characteristic mass fragments are known for the compounds of interest (Tholl *et al.*, 2006).

Unfortunately, EI-MS is less accurate for quantification due to differences in ionization rates between compounds. This is even more pronounced with SIM and CI where quantification is impossible unless standard curves are generated separately for each compound. For accurate compound quantification, detection methods are available that combine great sensitivity with linear responses over a wide range of concentrations (Table 1; Skoog *et al.*, 2007). Detectors differ considerably in their sensitivities and biases toward various compounds (Núñez *et al.*, 1984). Flame ionization detection (FID) is a technique commonly used for quantification because of its sensitive and relatively unbiased detection of organic compounds. However, FID often displays notably less sensitivity to some oxygen, nitrogen, and sulphur-containing compounds. Other less common, specialized ionization detectors have been developed to provide sensitive detection of these heteroatom-containing compounds. Researchers should generate standard curves of synthetic chemical standards to test the response of detectors to their compounds of interest over a range of concentrations.

2.2.4.1. Identification of volatile and non-volatile compounds
With advances in analytical techniques, the minimal criteria for conclusive identification of natural products have substantially increased. Compounds can no longer be identified solely on the basis of a retention time match with a synthetic chemical standard or a single mass spectrum library match. It is strongly suggested that the reader consult guidelines for natural product identification (Ducret *et al.*, 2008). These guidelines outline how confirmation of compound identity should involve multiple methods (mass spectral matches by GC-MS EI, molecular mass determination by GC-MS CI, structure elucidation by NMR) and comparison with synthetic chemical standards whenever possible. These guidelines may appear stringent to someone not familiar with chemical ecology but are important to assure the quality of chemical identifications.

2.2.4.1.1. Using GC-MS to identify sample peaks

1. Run the sample on EI GC-MS (section 2.2.4.). Examine a peak of interest and make sure that it is a single peak with a symmetrical shape and single apex.
2. Extract the mass spectrum associated with the peak.
3. Select a background mass spectrum from a section of baseline (area lacking peaks) near the peak of interest. Subtract the background mass spectrum from the peak spectrum – commands vary between software packages.
4. Search the mass spectral libraries (NIST, WILEY, and other self-defined libraries) for a match (minimally 80%, 90% for structurally similar compounds such as terpenes) between the peak spectrum and the mass spectra of known library standards.
5. Closely compare the mass spectrum of the library standard against the mass spectrum of the unknown peak. This peak's spectrum should have all of the ion fragments present in the library standard plus additional ion fragments from other minor compounds. If the spectrum is missing ions present in the library standard within the m/z sampling range, there is not a match.
6. Compare the ion fragment patterns of the compound of interest and the library standard to see if the fragment ratios are similar.
7. Test the tentative identity of the unknown peak by running a synthetic chemical standard under identical chromatographic conditions as the sample of interest. To be a match, the synthetic standard should have an identical retention time and similar ion patterns as the peak of interest.
8. Further test the match by running the sample and the synthetic standard separately under different chromatographic conditions (usually a different column type) to determine if the similarities hold.
9. Confirm the match in retention time by co-injecting the sample with a known amount of the synthetic standard. The two compound peaks should overlap completely.
10. Determine the mass of the molecular ion fragment (M+1, the molecular mass plus one additional mass unit) of the unknown peak and the synthetic standard peak by running the sample and standards on CI GC-MS.

2.2.4.1.2. Quantification of volatile and non-volatile compounds with internal standards

The goal of quantification is to obtain an accurate estimate of volatile emission rates from an odour source. This is readily applicable to volatiles sampled using Super Q and Haysep Q adsorbents, and extracts (see sections 4 and 6 of this paper). One approach to quantification is to add a known amount of a synthetic chemical to a sample before processing the sample through extraction and separation (Heath and Manukian, 1994). Researchers usually select a chemical that is not present in their sample but has similar separation properties as the chemicals of interest. One internal standard used with honey bee volatiles is nonyl acetate (Carroll and Duehl, 2012). The internal standard automatically scales the peak areas to known amounts of material. One useful aspect of this internal standard is that quantitative errors in sample processing (i.e. fraction of sample injected on the GC, compound concentration or losses, pipetting errors) are automatically factored out, as both the internal standard and the sample compounds experience similar changes.

To make nonyl acetate internal standard:

1. Add 92.6 µl (80.0 mg) of nonyl acetate to 9.907 ml of the same solvent used for filter extraction (dichloromethane or hexane).
2. Vortex for a few seconds.
3. Add 100 µl of this stock solution to 9.900 ml solvent.
4. Vortex for a few seconds.

The concentration of this internal standard solution is 80 ng/µl.

5. Aliquot the internal standard solution into working amounts of 500 to 1,000 µl.

For long term storage, aliquot the internal standard solution into glass ampules and seal.

It is very important that solvent evaporation of internal standard solutions be kept to an absolute minimum to maintain an accurate concentration. Store the internal standard solutions in a -80°C freezer between uses.

The sample chromatogram will have a nonyl acetate peak that represents the 400 ng of nonyl acetate (5 µl of 80 ng/µl internal standard solution) that was added to the original sample during rinsing with solvent containing internal standard. Run a sample containing only the internal standard to determine the retention time and peak characteristics of the internal standard. Because the amount of internal standard added to the sample is known, conversion rates can be made between peak areas and compound amounts. Sample compound amounts are calculated from peak areas against the standard as:

$$\textit{Sample compound amount (ng)} = \frac{\textit{Area of compound peak (area counts)}}{\textit{Area of internal standard peak (area counts)}} \times 400 \ \textit{ng}$$

2.3. Conclusion

To summarize, the most important features of a volatile collection and analysis design are to:

1. Isolate the sample of interest as much as possible without disturbing it.
2. Collect emitted volatiles with a technique that is as sensitive and efficient as possible while limiting the introduction of contaminants.

3. Collect background samples to help identify outside contaminants in the samples.
4. Use appropriate analytical techniques to detect and evaluate the biologically-important compounds of samples.
5. Confirm compound identities and quantities with appropriate chemical standards and techniques.
6. Test the biological activities of samples with bioassays on isolated native material. Then confirm the activities of suspected compounds with bioassays on synthetic chemical standards.

3. *Ex-situ* collection of honey bee odours and electrophysiology

3.1. Introduction

Honey bees produce airborne volatile organic compounds which serve as indicators of the 'status' of the colony. As such, the collection, analyses, bioactivity of identified honey bee odours is vital to understanding how social cohesion is maintained, regulated and influenced by various biotic factors such as foreign intruders (Torto *et al.*, 2005, 2007a, 2007b) and pathogens (Swanson *et al.*, 2009). Odours associated with honey bees have been used to improve colony vigour and to manage certain pest such as the small hive beetle *Aethina tumida*, (Teal *et al.*, 2006; Arbogast *et al.*, 2007; Torto *et al.*, 2007b).

A specialized method that provides information on honey bee detection and sensitivity to specific compounds is coupled Gas Chromatography-ElectroAntennographic Detection (GC-EAD) (Schneider, 1957; Baker *et al.*, 1985; Torto *et al.*, 2007a; Swanson *et al.*, 2009). This coupled system uses an insect antenna in tandem with a flame ionization detector (FID) to link the electrical activity stemming from neurons signalling receptor binding of chemicals to individual peaks. GC-EAD is an excellent analytical tool but works best for chemicals detected by many receptors thereby producing a strong electrical stimulation from the antenna. In this hybrid method, volatiles emerging from a GC column are split between a conventional GC detector (chemical sensor) and an antenna mounted between two electrodes (biological sensor). The interpretation of detected compounds requires an understanding of insect physiology, since antennal stimulation to chemicals represents activity which may indicate attraction or repellence. Some chemicals may be detected by a single receptor, transmitted by a single neuron and then amplified in the brain while others may stimulate several receptors and their associated receptor neurons. Honey bee chemosensory organs can be much more sensitive to bioactive compounds than analytical detectors, sometimes leading to strong electrophysiological responses to correspondingly weak chemically (FID) detected components. Although the honey bee antenna is most commonly used, other body

parts having chemosensory activity can be used as biosensors. Antennae and body parts with weak electrophysiological activity can have their activity amplified by mounting multiple parallel sensory organs in tandem across a single electrode.

This section focuses on methods to collect honey bee odours outside of the colony environment (*ex-situ* volatile collection) and to carry out electrophysiological recordings using antenna or other chemosensory body parts of the honey bee.

3.2. Volatile collection

3.2.1. Setup and volatile sampling

Laboratory collection of headspace odours can be carried out using a static or dynamic sampling method. A detailed description of the merits and disadvantages of these two sampling techniques, adsorbents and their associated desorption techniques have been dealt with in the previous section (see section 2.2.2. on *In-situ* collection of volatiles). Briefly:

- Dynamic headspace sampling requires an air supply, an air purification system, flowmeters, quickfit glass containers to hold odour sources with air entry and exit ports, and a mesh screen metallic canister with a tight fitting lid to hold honey bees, copper/Teflon tubing connectors, adsorbents (Super Q, Tenax, and SPME fibres), humidifier and a vacuum supply.
- Static headspace sampling on the other hand simply requires glass containers with lids fitted with air ports to contain odour sources, SPME adsorbent fibres (suited for collecting both polar and non-polar chemicals) or gas tight glass syringes for sampling head space. Sample the headspace directly with a gas tight syringe by pulling 50 or 100 ml of odour and analyse on a GC or GC-EAD or GC-MS.

An example of a laboratory volatile collection setup for honey bees has been described by Torto *et al.* (2005). A similar setup is shown in Fig. 4, illustrated using 6 components. In this setup, medical air from a pressurised air tank (not shown in figure) is passed through a copper tubing (component 1) and then through activated charcoal (component 2) to purify it and into to a humidifier containing double distilled water (component 4). The humidified air is pushed through a y- or t-split (for treatment and control) or multiple ports (manifold) (component 5) to which odour sources enclosed in glass jars are connected to in parallel. The vacuum supply pulls air from the glass jar at a specified flow rate set on the flow meter (component 6) through the adsorbent filters.

Odour collections can be made from adult worker honey bees only; an entire honey bee comb bearing adult workers, bee brood, pollen and honey; honey comb containing bee larvae; just to name a few depending on the hive odour source of interest and the research

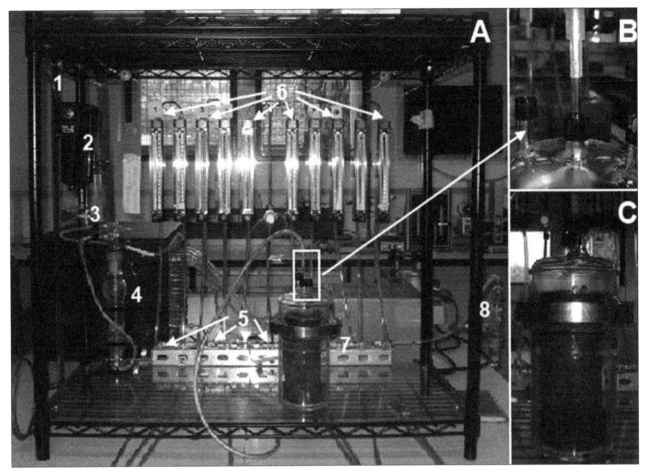

Fig. 4. **(A)** *Ex-situ* volatile collection setup, **(B)** Adsorbent filter fitted on lid and **(C)** glass jar with metallic canister containing worker honey bees. **(1)** copper tubes, **(2)** air filtering system (e.g. .activated charcoal), **(3)** air pressure regulator, **(4)** humidifier, **(5)** multiple air supply ports, **(6)** air flow meters, **(7)** glass jar containing odour source and 8-air exit monitor (water bubbler which give evidence of air flowing through the setup).

question being addressed. However, the experimenter should bear in mind that bee larvae usually become stressed in the absence of worker bees (see sections 2.2.2. and 2.2.3. on *in situ* volatile collection and analysis).

Once ready to collect odours in the laboratory and in possession of all the bits and pieces of equipment, following the steps below will ensure a successful process.

1. To collect odours from worker bees only, gently brush-off workers from a comb into a screen mesh metallic canister containing sugar source (a ball of cotton wool soaked in 50% sucrose solution or sucrose solution in a vial fitted with a dental cotton wick) and close it with its lid once the required population is obtained.

2. A setup consisting of a canister and sugar source without bees should also be prepared to serve as a control.

3. Return to the laboratory with the canisters and place them in an appropriate quickfit glass container (e.g. 2 or 5 l glass jar with a tight fitting lid).

4. To sample representative odours of the hive environment, gently push a honey bee comb or cut out a section of a comb with adult bees, brood, pollen and honey out of its frame.

5. Place in a clean quickfit glass jar with aeration ports and carry the confined comb and bees back to the laboratory for volatile collection.

6. Connect one aeration port attached to the glass jar to an air supply and the other to the a vacuum supply.

7. Pass clean air through the system for 15-20 min to purge out the alarm pheromone produced by the bees and any contaminants before collecting volatiles for a specified period.

3.2.2. Dynamic volatile collection

8. Briefly disconnect the vacuum line once the bees have become relatively calm.

9. Place a protective screen (Teflon sheath or screen mesh around the tip of an adsorbent (e.g. Porapak Q, Super Q or Tenax) filter to protect it from being blocked with wax by the bees.

10. Place the filter at the vacuum line's tip and reconnect it to the lid of the container.

Avoid placing the vacuum supply source such as a pump and odour source on the same bench to prevent alarming the bees from the vibrations/noise from the pump.

11. Pull air out of the system at a desired flow rate and for a specific duration (see section 2.2.2. and 2.2.3. on *in situ* volatile collection and analysis). The sucrose solution serves a dual role; as a source of food to keep bees alive and to calm them.

12. Remove the filter and seal the ends with Teflon tape after volatile collection is complete.

13. Switch-off the vacuum.

14. Disassemble the odour container.

15. Return the bees to their colony of origin.

16. Elute the volatiles adsorbed on the filter using an appropriate volume of solvent (e.g. 200 µl) such as dichloromethane or redistilled ether in to a sample vial using a gentle flow of N_2 gas.

17. Label the sample bottle and store at -80°C prior to analysis.

3.2.3. Static volatile collection

Static odour sampling can be carried out using several sampling techniques. The use of SPME fibres is the most convenient in a closed head space environment (see section 2.2.2.2.1). For this procedure:

8. Gently disconnect the air supply and vacuum pumps from the odour containment glass jars after purging for 15-20 min.

9. Immediately close all ports with screw caps lined with flexible Teflon linings.

10. Insert a SPME fibre holder or gas tight syringe by pushing its needle through one of the flexible Teflon linings of the screw caps until the needle is fully within the glass jar.

11a. Expose the fibre (adsorbent) through the needle tip by pushing down the SPME holder plunger. Maintain this experimental setup for 0.5 - 12 h (or for a desired duration which should not be lengthy to avoid stressing the bees).

11b. For the gas tight syringe, maintain a closed experimental setup for 0.5 - 12 h and afterwards sample the head space odour by pulling a desired volume of air from the head space and analyse directly on a GC, coupled GC-EAD or GC-MS (see section 2.2.4.).

12. Retract the SPME fibre into the needle before withdrawing the needle from the Teflon lining.

13. Analyse sample as described in the previous section (see section 2.2.3. on *in situ* volatile collection and analysis)

3.3. Electrophysiology
3.3.1. Setup
Electrophysiological studies require expensive specialized equipment including software, which can be purchased from commercially available sources. A basic electrophysiological setup consists of four elements (Syntech, 2004).

- A biological sensor which is usually an antenna or any other chemosensory organ mounted across a pair of electrodes.
- Amplifier and signal processing electronics specially built to minimize noise and to control the baseline signal.
- Signal display and recording system which makes use of computer software for display, record and analysis of signals.
- Stimulus application system which ensures a continuous or discontinuous release of test stimuli over the antenna while its electrical activity is being measured.

Equipment similar to that described by Torto *et al.* (2005) can be used to measure bee antennal responses to various chemicals. A stimulus source consisting of an inert metallic delivery tube (Fig. 5a) with a hole at its basal end originating from the side of a gas chromatograph serves as the channel via which volatile stimuli are applied over the mounted antenna on a micro-manipulator or gel probe (Syntech, 2004). The micro-manipulator has two terminals; an indifferent and a 10x amplification recording terminal. Both terminals containing silver wires (0.1 mm in diameter) are sheathed with capillary tubes tapered at one end by drawing them out as heated tubes in pipette pullers, filled with insect saline solution such as Ephrussi solution (consisting of an aqueous mixture of Na, K, Mg, Ca and Cl) (Christensen, 2004), thereby converting them into conducting electrodes (Christensen, 2004). The manipulator base is connected to an earth cable to minimize internal electrical interference while its recording electrode is connected to an AC/DC-EAG amplifier. Signals from the recording electrodes are amplified and digitized by data acquisition electronics. This electronic equipment acquires signals from the amplifier and transforms them into digital wave-like signal depicting antennal responses as peaks recorded in millivolts (mV). Signal visualisation, recording and analysis is carried out using a specific software (e.g. GC-EAD, Syntech) installed on a personal computer (PC). All the EAD equipment except for the gas chromatograph, data acquisition electronics and PC are enclosed within a Faraday's cage to reduce external electrical interference. For more details of the setup, see the system's manual (e.g. Syntech, 2004).

3.3.2. Types of electrodes
Two types of electrodes are often used to record EAGs (electroantennograms); glass capillary and probe electrodes.

3.3.2.1. Glass capillary electrodes

A glass capillary electrode usually consists of a borosilicate glass capillary pulled out to produce a tapering end ($\sim$ 1 µm in diameter) in a pipette puller and filled afterwards with sensillum haemolymph mimics e.g. Ringer (see the *BEEBOOK* paper on cell cultures by Genersch *et al.*, 2013 for a recipe) or Ephrussi solution (Christensen, 2004). The saline-filled glass capillary is then pushed over a silver wire fixed into a micro-manipulator (Fig. 5a). This electrode type can be used to record EAGs from intact honey bees, excised honey bee heads and excised antennae.

3.3.2.2. Probe electrodes

A probe electrode is similar to a two-pronged fork with blunt flat tips (Fig. 5b). The probe is metallic and often made of silver or gold. An excised antenna or chemosensory organ is mounted across the prongs and held in place using conducting gel (Spectra 360 Electrode Gel). Modified versions of this electrode include the:

- Tetraprobe with a single base electrode and four (4) different recording electrodes which can be used to mount several antennae of the same insect species in tandem (Syntech).
- Quadroprobe similar in design to the Tetraprobe but with a detachable probe end which is useful in mounting several antennae from different insects.

3.3.3. Sensory organ preparation and mounting

No universal standard exists for preparing honey bee sensory organs for electroantennographic studies (Syntech, 2004). Various mounting techniques exist and the choice of technique to use depends on the objectives and experience of the experimenter. For recording bee antennal responses over a long period of time (over one hour), whole insect preparations should be the ideal choice (Syntech, 2004). For shorter durations lasting up to an hour, excised antennal preparations can be used. Both whole insect and excised antenna preparations can be carried out using saline-filled glass or probe electrodes (Torto *et al.*, 2005, 2007b). To prepare bee antenna for an electrophysiological recording:

1. Collect insects i.e. forager, guard or nurse bees at the hive entrance or comb using an aspirator (manual or automated) as required.
2. Immobilise the collected bees by placing them on ice for about 1-2 min.
3. Pick individual immobilised bees and insert each at the base of a pipette tip (100-1000 µL).
4. Gently blow the bee towards the tip whose apical portion has been cut-off to allow only the head of the bee to go through it in order to restrain it.
5. Plug the base of the pipette tip with cotton wool or paper towel to prevent the insect from crawling back inside the restraining tube.

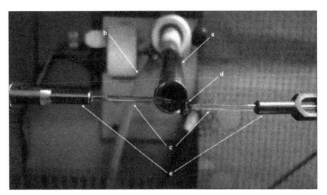

***Fig. 5a.* (a)** The stimulus delivery line for the column effluent, **(b)** humidified air supply used to flush column effluent in the background (blurred), **(c)** Saline-filled glass electrodes **(d)** a mounted bee antenna showing both head and antenna **(e)** mounted electrode holders.

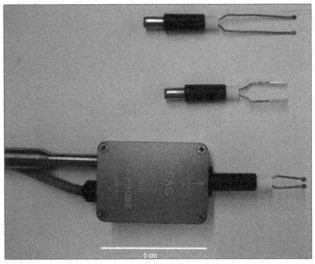

Fig. 5b. Gel electrodes and probe. Electrodes vary in their sizes fitting both large and small antennae or chemosensory organs.

6. Prepare the electrodes (saline-filled glass or gel electrodes) to be used for mounting. For saline electrodes, ensure that its drawn out tips have been cut to allow both antenna base and tip to fit through under a stereomicroscope. For gel electrodes, make sure that a sufficient amount of conducting gel is applied on the electrode tip.
7. Mount the excised antenna between glass micropipette electrodes on a micro-manipulator or probe electrodes covered with conducting gel (Syntech, 2004). For whole insect preparations, push the recording electrode into the distal end of the organ of interest (antennal tip) and the base or ground electrode into a body part close to the terminal end of the organ (which is often the eye for antennal preparations). This should always be done under a stereomicroscope.
8. Ensure that saline electrodes are free of air bubbles as they can interfere with a smooth recording of EAG.
9. Make sure that the mounted antenna has its basal end connected to the ground/base/indifferent electrode while its apical portion is connected to the recording/different electrode (usually the one connected to the amplifier).

10. Move either the antennal preparation closer to the source of the stimulus or the stimulus dispenser close to the mounting.

11. When the amplifier light signal changes from red to green, it indicates that a complete circuit has been established and the mounting process is complete.

3.3.4. EAG recording

After the insect antenna or sensory organ has been successfully mounted, start recording the signal on the computer monitor. A stimulus should only be applied when a relatively stable baseline signal is obtained.

Depending on the nature of the application of the stimulus, EAG recordings can either be described as continuous or discontinuous. A continuous recording involves the continuous application of the stimulus over the antennal preparation throughout the recording session while a discontinuous recording refers to the application of stimulus at intervals. Coupled gas chromatography electroantennographic detection (GC-EAD) is the most widely used technique for continuous recording.

Coupled GC-EAD recordings can be used to record antennal responses to both natural and synthetic mixtures of chemicals. The GC chemically separates the mixture into individual components which are then detected by the antenna, followed by identification of the antennally-active components by coupled gas chromatography-mass spectrometry (GC-MS). However, a coupled GC-EAD system is expensive to setup and run, bulky and requires lengthy recording times. Compared to continuous GC-EAD recording, discontinuous EAG recording is a simple, less bulky and fast way of testing individual chemicals at different doses for antennal activity. Discontinuous recordings are mostly used in dose response studies to compare antennal responses to identified chemicals. This usually involves quantification of responses to the various chemicals, averaging these responses and comparing them using various conventional statistical tests (see the *BEEBOOK* paper on statistics (Pirk *et al.*, 2013)). The drawback of this method is that it is not suitable for testing mixtures of chemicals, especially natural unidentified mixtures, and it cannot be used to identify unknown compounds.

Both GC-EAD and EAG recording systems complement each other and it is advantageous to have both systems in place in any insect electrophysiology laboratory. A coupled GC-EAD system can easily be converted into stimulus puff EAG recording system by switching-off the GC and using the opening at the side of the metallic stimulus delivery tube to puff odours over mounted antenna.

3.3.5. Coupled gas chromatography-electroantennographic detection (GC-EAD) recording

In this continuous recording system, the stimulus (a natural extract or synthetic mixture of chemicals is applied by injection into a gas chromatograph where individual components are separated based on

their physical and chemical properties while being carried through a column (see section 2.2.3.). One part of the column effluent is continuously flushed over the antenna while the other flows into the flame ionisation detector (FID).

1. Switch on the GC and programme it to run a specified separation method (see section 2.2.4. on *In situ* volatile collection) and wait for it to get ready.

2. Following successful antennal preparation, inject 1-5 µl of the extract depending upon its concentration into the GC through its injection port and run the method.

3. At the same time start the recording programme on the computer to synchronise the signal output of the GC's sensor (FID) with that of the EAD setup (antenna).

4. Allow the program method to run from start to finish. When the run is over, stop the recording and save the file on the PC. The file can then be retrieved later for analysis.

5. The recorded file contains two line tracings, the FID and EAD outputs (Fig. 6a).

Peaks on the FID trace represent the different chemical components separated on the column and detected by the FID, while those corresponding to EAD peaks indicate antennal detection to specific compounds.

3.3.6. Discontinuous EAG recording

Discontinuous EAG does not require a gas chromatograph, but rather a stimulus delivery system for delivery of odours over the antenna at intervals as puff (sometimes referred to as puff stimulation). Stimulus delivery systems are commercially available as stand-alone devices or as accessories which can be installed on to a GC (Fig. 5a). It also requires a preparation of the stimulus in a manner different from natural volatile extracts (Syntech, 2004) as follows:

1. Prepare a stock solution of 1 µg/µl (equivalent to 1 mg/ml) by dissolving 5 mg of a pure compound of interest to the experimenter in 5 ml of solvent (e.g. hexane, dichloromethane, acetone) and shake gently to dissolve the sample.

2. Serially dilute the stock solution to prepare a range of doses (e.g. 100 ng/µl, 10 ng/µl, 1 ng/µl, etc.) to be tested.

3. Place a cut piece of paper, preferably filter paper (3 cm long x 1 cm wide), in the wide end of a standard Pasteur pipette.

4. Apply a specified amount of the solution of the compound to be tested. For example apply 1 µl of a 1 µg/µl solution to test 1 µg of the compound, etc.

5. Allow the solvent to evaporate for 30 s to several min and gently push the filter paper completely into the Pasteur pipette. Seal both ends of the pipette using Teflon tape or parafilm.

6. Label the pipette by the side to indicate the stimulus type (code names for test compound and concentration).

7. Prepare three control stimuli consisting of:

    7a. a clean Pasteur pipette,

    7b. a pipette containing filter paper only,

    7c. a pipette with filter paper with only solvent applied. These three controls check for pipette, filter paper and solvent contamination respectively.

8. Place the smaller end of the pipette inside the hole on the side of the delivery tube (Fig. 5a) and connect its wider end to an air supply after removing the Teflon tape or parafilm.

9. Apply the stimulus over the antenna by puffing using the EAG puff pedal for about 0.5-3 s, in the order; control stimulus followed by the test stimuli and finally control stimuli again. You may randomize the application of the test and control stimuli on the antennae, and puff each test stimulus several times since only a fraction of the test compound is delivered in each puff.

Stimulus application should be done at intervals of 30-120 s to allow the antenna to recover from the previous stimulus.

10. Record antenna signals as described in section 3.3.5. and open files later for analysis (see Fig. 6b for an example of a stimulus-puff recording).

EAG recordings are limited in their sensitivity and specificity to certain chemicals because responses to these chemicals are a summation of the total depolarisations elicited by the chemicals across the mounted antenna at the level of the olfactory receptor neurons (ORNs) contained within each receptor cell (sensillum) (Christensen, 2004). As such, chemicals which elicit action potentials on very small number of ORNs are often not detected in EAG recordings. A newer technique, single-sensillum recording (SSR) partially overcomes EAG limitations by recording antennal electrical activity at the level of ORNs (Millar and Haynes, 1998; Christensen, 2004). Details on the equipment type, experimental protocols and analysis of results in SSR studies have been well described by Christensen (2004).

3.4. Chemical identification of electrophysiologically-active components

See section 2.2.4 on *In-situ* volatile collection for details of how volatile components can be identified. Briefly,

1. Collect volatiles on adsorbent and wash them-off the filters with an appropriate solvent (hexane, dichloromethane or redistilled ether).

2. Analyse aliquots of the volatile extracts using coupled GC-EAD and coupled GC-MS (linked gas chromatography mass spectrometry) using identical GC columns and oven conditions

in both equipment. Add a specific amount of Internal Standard (highly recommended to facilitate matching and identification of peaks when comparing traces obtained from the GC-EAD and GC-MS) (see section 2.2.4. on *In-situ* volatile collection) for qualitative and quantitative comparisons.

3. Compare retention times of separated components from both GC-EAD and GC-MS traces and identify peaks representing EAD-active components in the mixture on the GC-MS trace.

4. Tentatively identify chemical structures of EAD-active components (based on their representative peaks) using their fragmentation patterns (mass spectral data) while comparing it with those already identified and stored in a mass spectra database (e.g. NIST, ADAMS) (see section 2.2.4. on *In situ* volatile analysis).

5. Individually prepare 50 – 100 ng/µl solutions of the tentatively identified components in an appropriate solvent from authentic compounds obtained from commercial sources or synthesized by a chemist and analyse via linked GC-EAD and GC-MS to confirm identities of EAD-active components. Repeat the same procedure using a mixture of the tentatively identified components from premix authentic compounds obtained from commercial sources or synthesized and constituted by a chemist. Using a mixture of compounds is advantageous in that it saves time and resources in cases where many EAG-active components (> 20) are present in the natural extract. You may also analyse each component separately.

3.5. Summary

In a nutshell, the following steps should serve as a guide for planning experiments designed for *ex-situ* volatile collections and analysis:

1. Electrophysiology is much of practical science, it takes practice to consistently set up organ preparations and adapting methods to different organisms takes some trial and error. Take the time initially to get a good feel for consistently setting up an organ to detect known chemicals before evaluating novel detections.

2. Design and carry out appropriate bioassays to demonstrate the involvement of honey bee emitted volatiles in bee behaviour (see the *BEEBOOK* paper on methods for behavioural research by Scheiner *et al.* (2013)).

3. Use appropriate analytical techniques to detect (coupled GC-EAD), identify (GC, linked GC-MS) and quantify (GC, linked GC-MS) biologically relevant chemicals present in the emitted odours (see section 3.3. and 3.4. of this paper).

4. Confirm biological activities of identified chemicals by using their authentic (pure synthetic versions) equivalents in bioassays.

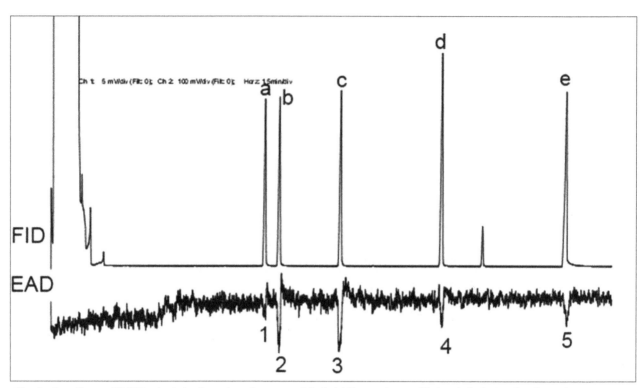

Fig. 6a. A continuous GC-EAD output showing antennal responses labelled 1-5 to a mixture of 6 chemicals. Notice that the peak representing a chemical (impurity) between peaks d and e is not detected by the antenna.

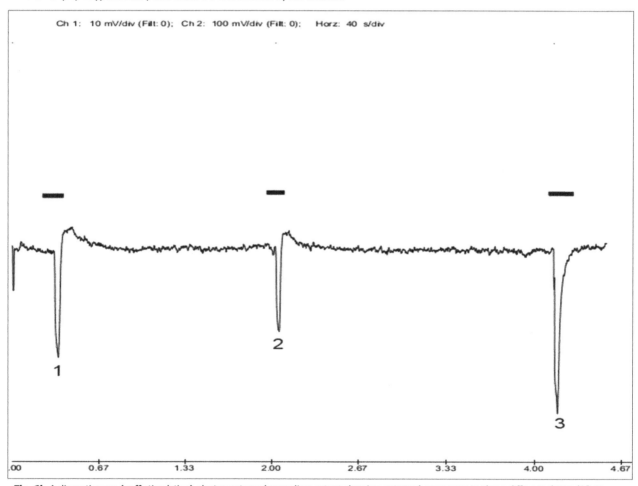

Fig. 6b. A discontinuous (puff stimulation) electro-antennal recording output showing antennal responses to three different doses (10 ng, 10 pg and 10 µg respectively) of a single chemical applied over the antenna at approximately 1 min intervals. Black horizontal bars above responses labeled 1 to 3 represent puff application.

4. Extraction and analysis of honey bee non-volatile cuticular hydrocarbons

4.1. Introduction

The honey bee cuticle is covered by a thin layer of non-volatile hydrocarbons that are used for several functions including prevention of dehydration and chemical communication. The study of honey bee cuticular hydrocarbons (CHC) has been carried out for chemotaxonomic characterization (Carlson and Bolton, 1984), nestmate recognition (Dani *et al.*, 2005) and elucidation of host-parasite relationships (Rickli *et al.*, 1994; Salvy *et al.*, 2001; Del Piccolo *et al.*, 2010).

Cuticular hydrocarbons of the honey bee include straight chain saturated and unsaturated hydrocarbons as well as branched saturated hydrocarbons. Chain length varies between 15 and 35, with odd numbered hydrocarbons being the most abundant; unsaturation is common at the 9 position in monoenes, but other positions are possible in longer chain hydrocarbons (Blomquist *et al.*, 1980; Francis *et al.*, 1985; Francis *et al.*, 1989; Carlson *et al.*, 1989). Some excellent reviews about insect CHC analysis have been published that can be used as a reference (e.g. Blomquist, 2010).

4.2. Techniques for analysing honey bee CHC

In general, the study of honey bee cuticular HC involves the following steps: extraction, sample preparation, identification and data-analysis, as detailed in the following sections.

4.2.1. Extraction

Modern analytical equipment easily allows for the analysis of CHC from single specimens.

The most used method for extracting CHC is, by far, solvent extraction. This can be done using the following protocol.

1. Transfer a single bee that has been anesthetised by chilling (see the section on Standard methods for immobilising, terminating, and storing adult *Apis mellifera* in the *BEEBOOK* miscellaneous paper (Human *et al.*, 2013), into a small glass container with a convenient amount of an apolar solvent (e.g. 1 ml of hexane HPLC grade).

Use clean forceps to avoid any possible contamination. Glass containers that are used need to be carefully washed, rinsed with hexane and kept overnight at high temperature before use; other materials should not be used to avoid contamination.

2. Leave the specimens for extraction at room temperature. The duration of the extraction can vary and preliminary tests are advisable to determine the best duration according to the amount of compounds obtained. In general, 2-10 min can be regarded as a good compromise between extraction efficiency and the need to keep the working time reasonably short.

3. Transfer the extract into a vial and store at -20°C until use.
4. In case an absolute quantification of CHC has to be carried out, add a convenient internal standard to the sample before extraction (see section 2.2.4.2 above for more details on this subject).

Such extracts are normally suitable for analysis without further processing apart from solvent evaporation under a stream of nitrogen to concentrate the sample for the analysis.

Other extraction methods can be used to sample CHC; in particular Nazzi *et al.* (2002 and 2004) used solid phase microextraction (SPME) to study short chain hydrocarbons released by honey bee pupae infested by the parasitic mite *Varroa destructor*; Nazzi *et al.* (2002) rinsed the gelatin capsules used for rearing the bees from larval stage to emergence, to sample the hydrocarbons released during pupation. The use of SPME for extracting CHC from living specimens is increasingly being used in other insects and details of this method are available (Bland *et al.*, 2001). For more details on other extraction methods see section 2.2.2 above.

4.2.2. Sample preparation

Crude hexane honey bee extracts are normally already suitable for subsequent analysis without further purification. In some cases CHC are separated for the purpose of behavioural bioassays; in particular, HC can be separated from oxygenated compounds using column chromatography on silica gel.

To do so the following method can be used:

1. Prepare a column, this is done by packing a Pasteur pipette (clogged at its tip with glass wool to prevent the gel running down the pipette) with 100-500 mg of silica gel (200–400 mesh, 60 Å).
2. Condition the column by passing hexane through it 2-3 times.
3. Add the sample to the column using a small volume of solvent.
4. Elute the column with 1-5 ml of hexane to collect the HC. Several elutions can be done to ensure collection of all HCs.
5. Elute the column with 1-5 ml of ether or acetone if interested in more polar compounds.

To further fractionate CHCs, saturated and unsaturated components in the apolar fraction can be separated.

6. Prepare a column packed with 100-500 mg of 10% silver nitrate on silica gel (200–400 mesh, 60 Å).
7. Elute the column with 1-5 ml of hexane to collect saturated hydrocarbons.
8. Elute the column with 1-5 ml of ether to collect unsaturated HC.

In order to remove silver ions,

9. Reduce the ether fraction under nitrogen.
10. Pass the eluate through an identical silica column.
11. Elute with hexane.

4.2.3. Identification

4.2.3.1. Analysis

Nowadays, the most used method for HC identification is mass spectrometry coupled to gas-chromatography (see section 2.2.4 for more details on GC-MS analysis), although satisfactory results can be obtained also using stand-alone gas-chromatography using the retention index method. In this case, peaks in the chromatogram are identified according to their retention index that is calculated from the retention time of the peak and that of adjacent reference alkanes. For the analysis, 0.1 insect equivalent (i.e. 1/10 of the material that can be extracted from a single insect) in 1 µl of hexane can be injected in the chromatograph (injector temperature can be set to 300°C) in splitless mode. A capillary column is normally used; many authors found DB-5 or DB-1 columns particularly suited for the purpose. The temperature of the GC is ramped from 40-50°C to 300-320°C and maintained at that temperature for 20 minutes or more.

The compounds in the extracts are identified by comparison of retention times and spectra (if GC-MS is used) with those of authentic standards. Quantification is based on the peak area with reference of that of the internal standard; a standard solution of hydrocarbons should also be analysed to calculate calibration factors. For more details on quantitative analysis see section 2.2.4.2.

4.2.3.2. Double bond position in unsaturated hydrocarbons

The double-bond position of unsaturated hydrocarbons detected in crude extracts can be assessed by GC-MS after derivatisation with dimethyl disulfide (DMDS) (Carlson *et al.*, 1989). Given its importance, the method will be described in detail.

To derivatise a few µg of HC in 200 µl of hexane, use the following protocol:

1. Add 200 µl of DMDS and 100 µl of iodine solution (60 mg in 1 ml of ether) to catalyse the reaction.
2. Keep at 40°C for 4 hours.
3. Dilute with hexane with 5% sodium thiosulfate to neutralize the iodine.
4. Collect the organic phase which contains the DMDS adducts of the unsaturated hydrocarbons.
5. Remove moisture from the solution by adding a small amount of anhydrous Na_2SO_4 to it.
6. Reduce the solvent under a stream of nitrogen to the desired concentration.

Derivatised alkenes give a single chromatographic peak with a characteristic spectrum composed of two prominent fragments. These are related to the cleavage of the bond between the carbon atoms carrying the methyl sulphide substituent, originally the location of the double bond. The m/z values of these fragments are identified by the series 61+n14; for example, the alkene 9-heptadecene ($CH_3(CH_2)_7$=CH $(CH_2)_6CH_3$) will give a peak with two prominent fragments at 159 ($CH_3(CH_2)_6CHSCH_3$) and 173 m/z ($CH_3(CH_2)_7CHSCH_3$); the molecular ion is normally seen (in the case of 9-heptadecene, this has m/z 332, that is the sum of the two fragments described above).

4.2.3.3. Stereochemistry of alkenes (determining the identities of geometric isomers of unsaturated hydrocarbons)

Carbon-carbon double bonds of alkenes can exist in two alternative forms, known as isomers, using the named (*E*)-(*Z*) notation. To determine the stereochemistry of the alkenes that are present in a sample, they can be co-chromatographed with authentic standards on silver nitrate impregnated thin-layer chromatography plates using hexane as a solvent. This allows separation of the stereoisomers so that the sections of the plate corresponding to the (*Z*) and (*E*) alkenes can then be extracted with ether and analysed by GC-MS (Nazzi *et al.*, 2002).

4.2.3.4. Branching position

The point where the chain of carbon atoms is branched in non-linear hydrocarbons can be determined from the mass spectrum taking into account both the mass spectrum and the Kovats index (Carlson *et al.*, 1998).

4.2.3.5. Synthesis for the purpose of identification

Although the study of retention time and mass spectrum normally provides reliable data for the identification, injection and coinjection represent the definitive proof of the identity. If standard compounds are not readily available they have to be synthesized using organic chemistry techniques that are outside the purpose of this review. Some of these techniques have been used extensively; in particular the Wittig reaction for the synthesis of unsaturated HC. In a few cases suitable alternatives (e.g. the partial hydrogenation of the corresponding alkyne obtained by alkylation of a convenient terminal alkyne with 1-bromoalkanes of suitable chain length according to Sonnet, 1984) have been used giving excellent results (Nazzi *et al.*, 2002).

4.2.4. Data analysis

For the analysis of differences between cuticular compounds or profiles, data can be arranged in a matrix with as many rows as the number of the studied hydrocarbons and one column for each analysed honey bee. In many cases the percentage composition (e.g. the proportion of a compound relative to the whole set of hydrocarbons is used for the analysis, in other cases the absolute amount of each hydrocarbon is used; sometimes this is expressed as the absolute quantity of a compound relative to the weight of the insect. If the percentage composition is used, data transformation according to Reyment (1989) is common using the following formula:

$$Z_{i,j}=\log[X_{i,j}/g(X_j)]$$

where:

$Z_{i,j}$ the transformed area of peak i for specimen j;

$X_{i,j}$ represents the area of peak i for specimen j;

$g(X_j)$ the geometric mean of the areas of all peaks for specimen j.

Different methods for data analysis are applied according to the purpose of the study. Given the distribution of data, possible differences between experimental groups can be tested using parametric methods such as ANOVA if experimental groups are three or more, or Student's t test if only two groups are considered. In this case, the number of tests to be carried out corresponds to the number of CHC considered, that can be rather high causing possible errors related to multiple comparisons; therefore probabilities from the test should be adjusted to allow for possible false positives using convenient formulas (e.g. Bonferroni correction that is very common and conservative, see the *BEEBOOK* paper on statistics by Pirk *et al.* (2013)).

In many cases the whole set of CHC is considered using multivariate techniques such as principal components or discriminant analysis. This can be carried out with most commercial statistical packages and this allows for a plot of the specimens on the plane formed by the derived functions accounting for most of the variability. Possible differences between groups are denoted by isolated clouds of points grouped around the centroids, whose distance from other centroids can be tested for its significance with standard methods.

Discriminant analysis is carried out when samples belong to predefined groups; in this case, to account for multicollinearity, a preliminary principal component analysis is carried out and the discriminant analysis is applied on the extracted factors.

5. Bee attraction bioassay

5.1. Introduction

Chemical communication in the honey bee is very complex with 15 exocrine glands known to produce various pheromonal chemicals. Among these pheromones, the queen retinue pheromone (QRP) is considered the most important as it is essential to maintaining social cohesion in the honey bee colony. This pheromone is composed of 9 components derived from various glands in the queen and elicits 'aggregation', antennating, licking and grooming of the queen by worker bees (Slessor *et al.*, 1988; Wossler and Crewe, 1999; Keeling *et al.*, 2003; Katzav-Gozansky *et al.*, 2001; Slessor *et al.*, 2005). The attractiveness of the queen and her pheromonal cues to worker bees is not stable but varies with the age and reproductive status of the queen, worker sensitivity, seasonal change and genetics (De Hazan *et al.*, 1989; Pankiw *et al.*, 1994, 2000; Kocher *et al.*, 2009; Wossler *et al.*, 2006).

Over the years, several compounds have been tested for their role as queen attractants using various bioassay setups. This section focuses on solvent extraction of pheromones from glands and the evaluation of both crude extract and its components in bioassays. Although we present these methods based on queen pheromones, worker and drone secretions can be analysed in the same manner.

5.2. Stimuli preparation

The prior knowledge regarding the chemical cues (source, nature and etc.) determines the procedure for stimuli preparation. If the glandular origin of the secretion is known, then a chemical extract of the gland/s can be used following these steps:

1. Dissect the glands of interest under double distilled water.
2. Clean unwanted tissues attached to the gland.
3. Wash once or twice with double distilled water.
4. Transfer the gland into a glass vial
5. Extract with solvent.

As described by Millar and Sims (1998) the glands can often be extracted by soaking for a few minutes to hours. Furthermore, the extraction is solvent dependent with solvents such as alcohols penetrating membranes more effectively and yielding greater extraction of cellular content than less polar solvents (e.g. hexane). Although this approach provides relatively large quantities of compounds in the extracts, it might not correctly represent the composition of the emitted odour from the gland. The latter can be better assessed by *in situ* or *ex situ* head space collections (Millar and Sims, 1998) (see sections 2 and 3). The crude extracts can be tested directly or after fractionation using a hand-made column (Katzav-Gozansky *et al.*, 2001) (see section 4.2.2, steps 1-5) or commercially available columns.

Analysis and quantification of the extracted components can be achieved by GC-MS with an internal standard added during the preparation of the extract as previously described in section 4.2.3.1. The amount of the glandular secretion is calculated based on the average amount of several analysed queens, all components included. This amount is considered as queen equivalent (Qeq). Where the chemical nature of the suspected mixture is known, a synthetic blend of it prepared in the natural ratio of the individual components can be used. For example in the case of Dufour's gland, esters can be synthesized from commercially available alcohols and acid chlorides following standard procedures as described by Francke *et al.*, 2000 and Katzav-Gozansky *et al.* (2001), detailed below. The blend of the esters is prepared based on the relative proportions of the esters present in the queens' total glandular extracts. The doses used are calculated as queen equivalents (Qeq).

5.2.1. Preparation of synthetic esters

Esters can be synthesized according to the following standard procedure (Francke *et al.*, 2000):

1. Dissolve 10 mmol of the alcohol in 10 ml absolute pyridine.
2. Add a catalytic amount (ca. 10 mg) of 4-dimethylamino pyridine to facilitate the planned reaction.
3. Cool the solution in an ice bath.

4. Stir the solution and add 1.1 equivalent of the acid chloride, dissolved in 10 ml hexane.
5. Continue to stir the resulting mixture for one hour at room temperature.
6. Quench the reaction by slowly adding 100 ml of aqueous sodium hydrogen carbonate while stirring the mixture.
7. Extract the aqueous phase with two subsequent 20 ml portions of hexane.
8. Wash the combined hexane solution with 30 ml portions of each diluted hydrochloric acid, aqueous sodium hydrogen carbonate, and brine.
9. Dry the solution with anhydrous magnesium sulfate and remove the white material by filtration.
10. Concentrate the clear solution under reduced pressure of Nitrogen.
11. Chromatograph the resulting crude product on silica gel (Merck 60–200 mesh) using hexane/ethyl acetate 30:1 as the eluent.
12. Check the purity of the final product and provide its structural proof, by GC/MS (see section 2.2.4.1.1) and other spectroscopic methods (e.g. nuclear magnetic resonance (NMR)) in consultation with a chemist.

5.3. Stimuli presentation techniques
5.3.1. Use of surrogates
To study the impact of queen pheromone "bouquet" and its components on workers attraction, a surrogate queen is used. The surrogate material needs to be as chemically neutral as possible. Glass slides/ micropipettes are commonly used as surrogates. Apply the cues to be tested, for example glandular extracts, and its suspected chemical component(s) or corresponding solvent (control) on the surrogate and allow the solvent to evaporate for about 5 min. The other possibility is to use workers as a substrate to test a chemical. Presenting a secretion on a bee vs. an inanimate object such as a glass slide provides a more natural situation. When using this procedure, apply the treatment slowly on the worker thorax and allow the solvent to evaporate before the bioassay. The disadvantage of this procedure is that worker behaviour and chemical signals may interfere with the assay.

5.3.2. Stimuli preparation
To prepare a stimulus for testing in assays, potential solvent effects on the living honey bee used as a surrogate queen can be a challenge. Different solvents such as methanol (Kaminsky *et al.*, 1990), ethanol (Katzav-Gozansky *et al.*, 2003), dichloromethane (Wossler and Crewe, 1999) and isopropanol (Kocher *et al.*, 2009) can be used for stimuli presentations on the glass, but for living workers, ethanol is recommended (Katzav-Gozansky *et al.*, 2003). Irrespective of the solvent used, best results can be obtained by preparing the lure just prior to the bioassay.

5.4. Bioassays:
Using bioassays to assess the effect of a stimulus can pose various challenges. To identify behaviour modifying cues, a simple and repeatable bioassay, in which all parameters are well controlled, with the experimental conditions kept as close as possible to the natural ones is recommended. The bioassay should also be a discriminative and quantitative test that measures the induction of the tested behaviour. In studies of workers attraction to the queen, commonly used bioassays can be carried in or out of the colony. Each has its own advantages and shortcomings detailed below.

5.4.1. In colony assays
These can be performed in observation hives of various sizes, established for at least 2 weeks prior to experimentation (see the *BEEBOOK* paper on behavioural studies (Scheiner *et al.*, 2013)). The surrogates can be introduced and removed through portals in the walls situated on either side of the frame.

Pros: This setup almost completely mimics the natural situation. The presence of a large variety of worker bees creates a good chance to record a good response. It is suitable for evaluating changes in queen attractiveness.

Cons: The system is complex, includes a large number of uncontrollable cues changing with time. Moreover, the complexity of such bioassays and its operational cost is high, not only due to the price of the hive itself but all the necessary preparations. The colonies need to be established days in advance, thus making replication difficult and problematic for a routine screening of compounds in question. Behaviour of the bees can be affected by colony and weather conditions.

5.4.2. Choice assays on groups in semi natural conditions (micro-hives)
In this assay, caged bees can be used (see the *BEEBOOK* paper on maintaining adult *Apis mellifera* in cages under *in vitro* laboratory conditions (Williams *et al.*, 2013)). The cage must be fitted with transparent walls allowing for observation, with a small comb glued in the middle and two side openings for lure introduction. Briefly,

1. Place freshly emerged (about 1 day old) bees into the cage (Dor *et al.*, 2005; Malka *et al.*, 2007)
2. Feed them with sugar solution and pollen cakes ad libitum.
3. Keep the cages in a temperature controlled chamber simulating hive conditions (of 30-33°C and 50-70% relative humidity).
4. After a few (1-3) days, of acclimatization in darkness, present workers with the choice of lures.

5. Count the number of workers on the lure licking and antennating at specific time intervals determined in preliminary studies.

Pros: This type of assay offers greater control over environmental conditions; remotely resembles the natural situation of bees as they are kept on a comb; allows for maintenance of bees for about three weeks and enables the comparing of responses of the same group of bees to a series of doses or different compounds. It also enables the conduction of age-specific tests.

Cons: Such bioassays demand construction of special cages and possession of a temperature controlled chamber.

5.4.3. Arena tests

A disposable plastic Petri dish (15 x 2 cm) or 9 cm in diameter lined with a filter paper, can be used as a bioassay arena (Kaminsky *et al.*, 1990; Wossler and Crewe, 1999). The system is especially suitable for a two choice bioasay. For stimulus introduction, side slits for insertion of the tested material or openings for introduction of bees on both sides of the cover can be easily prepared (Katzav-Gozansky *et al.*, 2003). Individual behaviour of 10-20 worker bees can be observed at a time in such an arena. The best approach is to use workers aged 1-12 days which normally attend to the queen (Seeley, 1982). If the exact age of the bees is unknown, workers can be collected from the open brood area of a queenright colony and used for the screening bioassay. Freshly collected bees are best for this study. However, if it is intended to test the responsiveness of workers from a special age and task, these need to be prepared in advance by marking bees at emergence (see the section on obtaining brood and workers of known age in the *BEEBOOK* paper on miscellaneous methods (Human *et al.*, 2013)).

1. Introduce the glass slides (treatment and control) or marked workers (Fig. 7) simultaneously to the arena through the side openings.
2. Place the arena in a temperature controlled room (25-27°C). Either red or day light will do for this assay.
3. Record behaviour towards each one at pre-determined time intervals.

Since the volatility of the tested component is expected to affect the active space of the chemical and thus the detection time, the length of the observation need to be determined in preliminary studies. This set up enables simultaneous evaluation of multiple arenas.

Pros: It is fast and highly reproducible. It is also a sensitive bioassay. Especially well suited for evaluation of chemicals of low volatility. It requires minimal equipment investment. Positive response can be obtained with low amount of material i.e. 10^{-7} Qeq of synthetic 4 component queen mandibular pheromone

(Kaminski *et al.*, 1990) and 1/3 Qeq of Dufours gland extract (Katzav-Gozansky *et al.*, 2003). This bioassay enables the testing of worker responses based on their specific ages and tasks.

Cons: This type of assay examines honey bee behaviour out of the natural context; the assay conducted in closed atmosphere with mostly still air can be problematic for evaluation of highly volatile compounds as the arenas could become saturated with volatiles rather quickly, thus eliminating the gradient to which the test organism could respond (Hare, 1998).

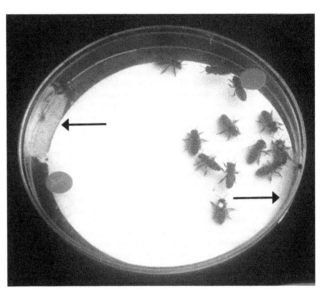

Fig. 7. The arena bioassay in the Petri dish. The red marked bee is pheromone treated while the yellow one is solvent treated control. The blue marks indicate openings for bees' introduction. Arrows indicate slots, for the glass slide introduction, when the latter are used as the surrogates.

5.4.3.1 An example of arena choice bioassay using live workers

Evaluation of the attractiveness of queen Dufour's gland pheromone extract on workers (Katzav-Gozansky *et al.*, 2003).

1. Apply the test material and also the solvent on two individually marked nurse bees in each assay.
2. The amount of secretion applied onto a glass or live bee should be calculated relative to 1 queen equivalent.

To illustrate, let's assume that 1 queen equivalent (Qeq) of Dufour's gland secretion is equal to 20 µg (all constituents included). If the total ester content amounts to 14 µg, then for 1 Qeq of a blend of the esters, prepare a synthetic mixture of the gland esters in ethanol based on the relative proportion of each ester present in the total glandular extract (see Table 2) (Katzav-Gozansky *et al.*, 1997).

3. Apply the gland extract (2 µl), its synthetic constituents (main esters) dissolved in ethanol using a micropipette.

4. Allow the solvent to evaporate for about 5 min.

5. Colour-mark the treated bees on the thorax after applying the extracts on the abdomen and allow the paint to dry for another 5-10 mins.

6. Introduce simultaneously extract and solvent treated bees into the arena.

7. Observe the activity of the bees for 5 min and record the number of workers attracted to the test material (contacting the worker) at 30-sec time intervals (total of 10 times).

For each replicate the sum of contacts for treatment vs. control over 5 min is used as a measure of attraction. The level of preference of the treatment vs. control can be presented as the percentage of time intervals in which there is preference towards the treated bee. Statistical analyses can be performed by any commonly used program that performs non-parametric statistical tests such as Wilcoxon Signed Rank test, Kruskal-Wallis ANOVA or repeated measures ANOVA after appropriate transformation (see also the *BEEBOOK* paper on statistics by Pirk *et al.* (2013)).

Table 2. Ester composition of Dufour's gland secretion of *Apis mellifera* queens and the amount used for preparation of ester synthetic blend (1 Qeq = 14 μg/gland). *- main compounds.

Esters	Amount of synthetic ester compound (μg)
Tetradecyl dodecanoate	0.14
Tetradecyl (Z)-9-tetradecenoate	Trace
Tetradecyl tetradecanoate*	2.88
Tetradecyl (Z)-9-hexadecenoate Tetradecyl (Z)-11-hexadecenoate*	Trace 3.46
Tetradecyl hexadecanoate Hexadecyl tetradecanoate*	1 2.78
Hexadecenyl hexadecenoate	0.88
Tetradecyl (Z)-9-octadecenoate*	1
Hexadecyl (Z)-9-hexadecenoate	0.94
Hexadecyl hexadecanoate	1.02
Octadecyl hexadecanoate	0.28

5.5. Summary

We have described both field and laboratory bioassays with their cons and pros for evaluating queen chemical cues inducing and sustaining the retinue of workers around the queen. This behaviour seems to be mediated by signals that are derived from a number of sources. The discussed bioassay systems offer a means for evaluation of the short range queen attractive compounds from various sources. These bioassays are useful for testing any suspected attractive source as well as for

evaluating changes in worker responses to queen-retinue-inducing-signals and changes in the queen's abilities to produce such cues. Still, retinue behaviour does not necessarily mirror workers' response to the whole queen's chemical bouquet. Isolation of cues indicating queen quality, long range queen attractiveness etc., need to be guided by separate and specific bioassays. Moreover, in order to clearly distinguish the roles of volatile versus contact cues, the responding workers should be separated from the queen by double mesh screen to prevent any contact (Katzav-Gozansky *et al.*, 2004).

6. *In vitro* bioassay for studying mechanisms regulating pheromonal gland activity in honey bees

6.1. Introduction

The pheromonal system of the honey bee, *Apis mellifera* L. is complex and composed of several caste specific chemical signals mediating various colonial activities in an integrative manner. These caste specific signals are known to be produced by over 15 exocrine glands in workers and queens (Free, 1987; Blum, 1992). Pheromone production by these glands shows age, task and caste-based variations (Blum, 1992; Winston, 1987; Pankiw *et al.*, 1998; Robinson and Huang, 1998), which suggest plasticity in pheromone biosynthesis. Such biosynthetic plasticity has been demonstrated for the mandibular glands (Plettner *et al.*, 1996), Dufour's glands (Katzav-Gozansky *et al.*, 1997; Martin and Jones, 2004) and may also be true for some of the other exocrine glands.

Regulation of the biosynthesis of these pheromones within glands is mediated by hormones. For instance juvenile hormone (JH) is known to regulate the activity of the mandibular and Koshevnikov glands (Robinson, 1985) while pheromone production in the Dufour glands is regulated by unidentified brain factors (Katzav-Gozansky *et al.*, 2007). It is probable that pheromone biosynthesis in other exocrine glands are similarly regulated by various hormones.

To study the mechanisms behind the control of pheromone production, it is important to isolate the gland of interest and evaluate its performance *in vitro* via manipulation with potential regulatory factors. This is especially crucial when studying organs of a social organism whose function is affected at any moment by multiple external cues followed by activation of internal cues. The *in vitro* isolation of the gland enables the separation of the target tissue from the impact of suspected factors on its performance both at the organismal and molecular levels. At the organismal level we consider pheromone production, whilst at the molecular level we refer to gene expression evaluation via genomic and proteomic tools. So far, proteomic tools (see the *BEEBOOK* paper on physiology and biochemistry methods (Hartfelder *et al.*, 2013)) have been used in studies of mandibular glandular performance *in vivo* (Hasegawa *et al.*, 2009; Malka *et al.*, 2009). The effect of each factor, their interaction and regulation mechanisms can be delineated using classical pharmacological methods but also with innovative molecular tools (e.g. RNAi-RNA interference, see the *BEEBOOK* paper on molecular methods by Evans *et al.* (2013)) without the interference of any social factors.

The idea of isolating pheromone producing glands was first successfully explored studying the control mechanism of sex-pheromone production in female moths (Soroker and Rafaeli, 1987), and thereafter to study ant postpharyngeal gland secretion (Soroker and Hefetz, 2000) and to discover the source of the locust pheromone phenylacetonitrile (Seidelmann *et al.*, 2003). This approach has been used largely to study moth sex-pheromone production and control (Rafaeli, 2009). In the case of honey bee exocrine glands it has only been briefly explored to study pheromone production by Dufour's and Mandibular glands (Katzav-Gozansky *et al.*, 2000 and Soroker and Katzav-Gozansky, unpublished data).

6.2. Methods

While developing a specific method for target tissue studies, several issues should be taken into consideration, in particular related to tissue isolation (section 6.2.1.), incubation conditions (section 6.2.6), appropriate precursors (section 6.2.4) and extraction procedures (section 6.2.8):

6.2.1. Isolation of organs/tissues

Incomplete isolation of the gland from other organs and tissues may negatively affect the glandular performance. For example, residues of fat body or other destroyed tissues can contribute degradation factors such as proteases, RNAases, etc. These can severely reduce glandular activity and thus obscure the studied effects. For example, Dufour's glands should be completely separated and cleared from the sting and the poison gland whilst mandibular glands need to be separated and cleared from labial, hypopharyngial glands and muscles. While dissecting the labial glands, the head glands need to be carefully separated from the hypopharyngeal gland and the thoracic glands for precise separation from the thoracic muscles.

6.2.2. Maintaining the normal performance of the gland in an artificial medium

In the body, the gland is surrounded by haemolymph characterized by specific composition (pH, inorganic salts, sugars, amino acids). Bee medium (Kaatz *et al.*, 1985, modified, as described in section 6.2.8.2.) has successfully been used in the *in vitro* incubation of honey bee Dufour's glands (Katzav -Gozansky *et al.*, 2000) to support glandular function.

6.2.3. Media contamination by microorganisms

Media contamination by microorganisms may severely affect the results directly interfering with glandular performance or by producing substances that may bias the results. Thus, only sterilised media should be used. Although the bees are themselves not sterile, the incubation should be maintained with care to reduce contamination especially when long term incubation is required. In some cases, surface sterilization of the bee by brief dipping in 70% ethanol prior to tissue dissection is recommended.

6.2.4. Selection and labelling of an appropriate precursor

Pheromonal signals, for example those produced by the mandibular gland often consist of multiple components (Blum, 1992). In addition, not all pheromonal components are necessarily derived from the same biosynthetic pathway. Thus precursors need to be selected according to the suspected biosynthetic pathway. For example, acetate is an appropriate precursor for evaluation of *de novo* biosynthesis in the case of pheromones that are derived from the fatty acid biosynthesis pathway such as Dufour's gland esters, hydrocarbons, 11- eicosanol and the acids of mandibular gland pheromone. However, for evaluation of other stages of synthesis, fatty acids or other compounds of appropriate length, saturated and unsaturated, should be used following the corresponding biosynthetic pathway (Blomquist *et al.*, 1987; Martin and Jones, 2004; Plettner *et al.*, 1995, 1996; Stanley-Samuelson, *et al.*, 1988).

Precursors can be labelled with stable or radioactive isotopes. Selection of a type of labelling will determine the methods for analysis of the biosynthetic product and which may in turn affect the detection sensitivity. Using radioactive labelled precursors containing $^{14}C/^{3}H$ has advantages by providing fast evaluation and high sensitivity, but these demand authorization from appropriate agencies as they may pose environmental risks. Labelling precursors with stable isotopes (^{13}C) is less hazardous but demands GC-MS analysis (see section 2.2.4). Precursors such as acetates/propionates/fatty acids are commercially available.

6.2.5. Incubation conditions

Incubation conditions (temperature and humidity) should be defined based on preliminary studies reflecting normal physiological and environmental conditions. Humidity should be maintained to prevent water evaporation and media concentration during the incubation process. Temperature can be maintained by incubating the samples in temperature controlled conditions (water bath or temperature controlled dry blocks). To prevent evaporation, the incubation vials must be kept sealed with protective cover or Parafilm. The number of glands (1 or 2 glands) per incubation may vary and depend mainly on the sensitivity of the detection method and on their secretion rate.

6.2.6. Determining optimum incubation time

Optimum incubation time should be adjusted by preliminary time course experiments, adjusted to the rate of biosynthesis and sensitivity of detection and depending on incubation temperature. However, care should be taken not to extend it over the period of 24 hours due to potential problems of tissue viability and bacterial contamination (unless sterile conditions prevail) and tissue deterioration.

6.2.7. Extraction of pheromone

Extraction of biosynthesized pheromonal components is the final step of an *in vitro* incubation. It involves extraction of the product from the incubation medium by an appropriate solvent and this depends on the chemical properties of the product. Recommended solvents for extracting pheromonal components are pentane/hexane/heptane or dichloromethane. The first group of solvents is especially suited for non-polar compounds such as hydrocarbons while dicholoromethane, a moderately polar solvent, efficiently extracts components such as aldehydes and ketones.

6.2.8. Product analysis

The various newly synthesized components can be separated and analysed using methods as determined by the nature of the precursor label. When radioactive labelling is involved, the major compounds or classes of compounds can be separated by a number of techniques depending on the nature of products, for example HPLC separation with radioactive detection. Other options include thin-layer chromatography (TLC) separation followed by phosphor imager detection (Katzav-Gozansky *et al.*, 2000) as described in section 6.2.8.3. For stable isotopes, GC-MS or LC-MS using single ion monitoring can be used (Tsfadia *et al.*, 2008). Below, we take the example of Dufour's and Mandibular glands for the description of the analysis of exocrine glands products (after Katzav-Gozansky *et al.*, 1997).

6.2.8.1. Dissection and sample preparation

1. Chill freshly collected bees on ice prior to dissection.
2. Dissect the exocrine glands under a stereo microscope (X20) using the medium specified below (see section 6.2.8.2.).
3. Wash the glands twice in fresh medium.
4. Transfer 1 or 2 glands to 80 - 100 μl medium supplemented with a labelled precursor (1 μCi [1-^{14}C] sodium acetate. An aliquot of ethanolic solution of sodium acetate , 1 μCi/gland, is dispensed to a vial, ethanol is evaporated to dryness under N$_2$ prior to addition of the incubation medium (56 mCi/mmol, Perkin Elmer).

For best results, worker glands should be incubated in pairs whilst the queen gland can be individually incubated (see section 6.2.8.2. for incubation medium preparation).

5. Incubate glands at 39°C, for 4-20 hours.

6.2.8.2 Bee incubation medium preparation based on Kaatz (1985)

Amino Acids (AA) (50 ml)

1. Dissolve L-Cysteine (5 mg) in 2 ml 1M HCl,
2. Add L-Tyrosine (10 mg) to 40 ml of double distilled water.
3. Dissolve the remainder Amino Acids one after the other: 60 mg L-alanine, 51 mg L-arginine, 20 mg L-aspargine, 10 mg L-aspartic acid, 25 mg L-glutamic acid, 100 mg L-glutamine, 20 mg L-glycine, 20 mg L-histidine, 10 mg L-isoleucine, 55 mg L-lysine, 5 mg L-methionine, 20 mg L-phenylalanine, 330 mg L-proline, 20 mg DL-serine, 15 mg L-threonine, 5 mg L-tryptophan, 15 mg L-valine.
4. Add double-distilled water to make up the volume to 50 ml.

Inorganic Salts (IS) (20 ml):

Dissolve 179 mg KCl, 40.6 mg MgCl$_2$*6H$_2$0, 49.3 mg MgSO$_4$*7H$_2$O, 69 mg NaH$_2$PO$_4$*H$_2$O, 33.6 mg NaHCO$_3$, in 20 ml DDW.

Sugars (15ml):

Dissolve 400 mg glucose, 250 mg fructose and 6700 mg sucrose in 15ml DDW.

Mixture of all components

1. Mix the three groups of components (AA, IS and sugars) with piperazine-N,N'-bis(2-ethanesulfonic acid buffer (PIPES) (756 mg).
2. Titrate with NaOH until PIPES is completely dissolved.
3. When all components are dissolved, add NaCl (100 mg).
4. Adjust the pH with NaOH to 6.7.
5. Add 2ml of CaCl$_2$*2H$_2$O (56.45 mg/ml in DDW) to the medium.
6. Bring the medium to a final volume of 100 ml.
7. Sterilize the medium by filtration via Millipore 0.22 μm.

6.2.8.3. Isolation and identification of Dufour's biosynthesis products.

The glands and incubation media are extracted in 350 μl dichloromethane for 24 h after incubation and then subjected to TLC. To isolate Dufour's gland biosynthesis products, carry out TLC of the extract using silica gel coated plates (polygram Sil G). The radioactivity of the various TLC fractions can be determined by a phosphor imager (IP Autoradiography System). Quantification of the radioactive fractions (Rfs) is achieved by comparison of its radiation to a standard radiation curve generated using different doses of radioactive acetate. TLC separation is performed in two successive steps:

1. Run the TLC plate using hexane to separate the components.
2. Air-dry the plate in a fume hood.
3. Subject the dried TLC plate to a second separation in a mixture of hexane: diethyl ether: acetic acid (70:30:1) (v/v), as the running solvent.
4. Identify the various lipid classes by comparing their Rf values with those of co-chromatographed standards (unsaturated).
5. Visualize by iodine vapour (in pre-prepared saturated tank).

In case of Dufours gland, suitable standards are: cis-9-tricosane (hydrocarbons); oleic acid (fatty acids); 11-eicosenol (alcohol); palmitic acid myristhyl ester and palmitoleic acid stearyl ester (esters); trinervonin and triolen (triglycerides). This procedure can be validated by using unlabelled products followed by GC/MS analysis as described below.

1. Incubate Dufour's glands as described in section 6.2.8.2. using cold acetate as a precursor (0.3 mg/ml).
2. Apply the glandular extracts on TLC along with commercial TLC standards (as above).
3. Perform TLC separation as previously described and allow the plate to dry.
4. Cut off the part of plates containing the commercial standards and expose it to iodine vapour to prevent modifications to compounds in the sample lanes designated for GC-MS analysis.
5. Mark the position of the standards.
6. In the sample lanes scrape the silica gel of the areas corresponding to standards on the TLC, into a glass vial and extract with chloroform, evaporate to dryness, reconstitute in dichloromethane.
7. Concentrate using N_2 and subject the product to GC-MS to identify relevant peaks as described in section 2.2.4.

Pros: The obvious advantage of the above described system is in its ability to study the activity of the exocrine gland detached from its original controlled environment. By manipulating the environment of the gland, factors and mechanisms regulating glandular activity can be isolated. This technique enables the researcher to separate between the effects of social regulation and possible physiological constraints of any studied gland.

Cons: However, the isolated gland function may not necessarily represent the full range of glandular functions as some of these may be dependent on unknown precursors received from elsewhere in the body of the bee. Thus, it is advisable to compare in vitro function of the gland with in vivo labelling studies, followed by glandular extraction and determination of de novo synthesized products.

6.3. Summary

In this section we have provided a step-by-step approach method for conducting an *in vitro* bioassay. We have also provided the pros and cons associated with using certain solvents, reagents and conditions for isolating gland biosynthetic products. Further reading using the references provide is recommended.

7. Acknowledgements

The COLOSS (Prevention of honey bee COlony LOSSes) network aims to explain and prevent massive honey bee colony losses. It was funded through the COST Action FA0803. COST (European Cooperation in Science and Technology) is a unique means for European researchers to jointly develop their own ideas and new initiatives across all scientific disciplines through trans-European networking of nationally funded research activities. Based on a pan-European intergovernmental framework for cooperation in science and technology, COST has contributed since its creation more than 40 years ago to closing the gap between science, policy makers and society throughout Europe and beyond. COST is supported by the EU Seventh Framework Programme for research, technological development and demonstration activities (Official Journal L 412, 30 December 2006). The European Science Foundation as implementing agent of COST provides the COST Office through an EC Grant Agreement. The Council of the European Union provides the COST Secretariat. The COLOSS network is now supported by the Ricola Foundation - Nature & Culture.

8. References

AGELOPOULOS, N; PICKETT, J A (1998) Headspace analysis in chemical ecology: effects of different sampling methods on ratios of volatile compounds present in headspace samples. *Journal of Chemical Ecology* 24(7): 1161-1172.
http://dx.doi.org/10.1023/A:1022442818196

ALAUX, C; ROBINSON, G E (2007) Alarm pheromone induces immediate-early gene expression and slow behavioural response in honey bees. *Journal of Chemical Ecology* 33(7): 1346-1350.
http://dx.doi.org/10.1007/s10886-007-9301-6

ARBOGAST, R T; TORTO, B; VANENGELSDORP, D; TEAL, P E A (2007) An effective trap and bait combination for monitoring the small hive beetle, *Aethina tumida* (Coleoptera: Nitidulidae). *Florida Entomologist* 90(2): 404-406.

AUGUSTO, F; VALENTE, A L P (2002) Applications of solid-phase microextraction to chemical analysis of live biological samples. *Trends in Analytical Chemistry* 21(6–7): 428-438.
http://dx.doi.org/10.1016/S0165-9936(02)00602-7

BAKER, T C; WILLIS, M A; HAYNES, K F; PHELAN, P L (1985) A pulsed cloud of sex pheromone elicits upwind flight in male moths. *Physiological Entomology* 10: 257-265.

BLAND, J M; OSBRINK, W L A; CORNELIUS, M L; LAX, A R; VIGO, C B (2001) Solid-phase microextraction for the detection of termite cuticular hydrocarbons. *Journal of Chromatography* A932: 119-127.
http://dx.doi.org/10.1016/S0021-9673(01)01239-0

BLOMQUIST, G (2010) Structure and analysis of insect hydrocarbons. In *Blomquist G J; Bagneres A G (Eds) Insect hydrocarbons: biology, biochemistry and chemical ecology.* Cambridge University Press, UK. 491 pp.

BLOMQUIST, G J; HOWARD, R W; MCDANIEL, C A; REMALEY, S; DWYER, L A; NELSON, D R (1980) Application of methoxymercuration-demercuration followed by mass spectrometry as a convenient microanalytical technique for double-bond location in insect-derived alkenes. *Journal of Chemical Ecology* 6(1): 257-269. http://dx.doi.org/10.1007/BF00987544

BLOMQUIST, G J; NELSON, D R; RENOBLES DE, M (1987) Chemistry, biochemistry, and physiology of insect cuticular lipids. *Archives of Insect Biochemistry and Physiology* 6(4): 227-265. http://dx.doi.org/10.1002/arch.940060404

BLUM, M S (1992) Honey bee pheromones. In *Graham, J M (Ed.). The hive and the honey bee.* Dadant; Hamilton, IL, USA. pp 373-400.

BLUM, M S; FALES, H M (1988) Eclectic chemisociality of the honey bee: a wealth of behaviors, pheromones, and exocrine glands. *Journal of Chemical Ecology* 14(11): 2099-2107. http://dx.doi.org/10.1007/BF01014252

BOCH, R; SHEARER, D A; STONE, B C (1962) Identification of iso-amyl acetate as an active component in the sting pheromone of the honey bee. *Nature* 195(4845): 1018-1020. http://dx.doi.org/10.1038/1951018b0

BOCH, R; SHEARER, D A (1964) Identification of nerolic and geranic aicds in the Nassanoff pheromone of the honey bee. *Nature* 202 (4930): 320-321. http://dx.doi.org/10.1038/202320a0

BREED, M D (1998) Recognition pheromones of the honey bee. *BioScience* 48(6): 463-470. http://dx.doi.org/10.2307/1313244

BREED, M D; WILLIAMS, K R; FEWELL, J H (1988) Comb wax mediates the acquisition of nest-mate recognition cues in honey bees. *Proceedings of the National Academy of Sciences* 85(22): 8766-8769.

CARLSON, D A; BOLTON, A B (1984) Identification of Africanized and European bees using extracted hydrocarbons. *Bulletin of the Entomological Society of America*, 30(1): 32-35.

CARLSON, D A; ROAN, C; YOST, R A; HECTOR, J (1989) Dimethyl disulfide derivatives of long chain alkenes, alkadienes, and alkatrienes for gas chromatography / mass spectrometry. *Analytical Chemistry* 61(14): 1564-1571. http://dx.doi.org/10.1021/ac00189a019

CARLSON, D A; BERNIER, U R; SUTTON, B D (1998) Elution patterns from capillary GC for methyl-branched alkanes. *Journal of Chemical Ecology* 24(11): 1845-1865. http://dx.doi.org/10.1023/A:1022311701355

CARROLL, M J; DUEHL, A J (2012) Collection of volatiles from honey bee larvae and adults enclosed on brood frames. *Apidologie* 43 (6): 715-730. http://dx.doi.org/10.1007/s13592-012-0153-x

CHRISTENSEN, T A (2004) *Methods in insect sensory neuroscience.* CRC Press; Florida, USA. 435 pp.

D'ALESSANDRO, M; TURLINGS, T C J (2006) Advances and challenges in the identification of volatiles that mediate interactions among plants and arthopods. *Analyst* 131(1): 24-32. http://dx.doi.org/10.1039/B507589K

DANI, F R; JONES, G R; CORSI, S; BEARD, R; PRADELLA, D; TURILLAZZI, S (2005) Nest mate recognition cues in the honey bee: differential importance of cuticular alkanes and alkenes. *Chemical Senses* 30(1): 1-13. http://dx.doi.org/10.1093/chemse/bji040

DE HAZAN, M; LENSKY, Y; CASSIER, P (1989) Effect of queen honey bee (*Apis mellifera* L.) aging on her attractiveness to workers. *Comparative Biochemistry and Physiology* A93(4): 777-783. http://dx.doi.org/10.1016/0300-9629(89)90501-X

DEL PICCOLO, F; NAZZI, F; DELLA VEDOVA, G; MILANI, N (2010) Selection of *Apis mellifera* workers by the parasitic mite *Varroa destructor* using host cuticular hydrocarbons. *Parasitology* 137(6): 967-973. http://dx.doi.org/10.1017/S0031182009991867

DIETEMANN, V; NAZZI, F; MARTIN, S J; ANDERSON, D; LOCKE, B; DELAPLANE, K S; WAUQUIEZ, Q; TANNAHILL, C; FREY, E; ZIEGELMANN, B; ROSENKRANZ, P; ELLIS, J D (2013) Standard methods for varroa research. In *V Dietemann; J D Ellis; P Neumann (Eds) The COLOSS BEEBOOK, Volume II: standard methods for* Apis mellifera *pest and pathogen research. Journal of Apicultural Research* 52(1): http://dx.doi.org/10.3896/IBRA.1.52.1.09

DOR, R; KATZAV-GOZANSKY, T; HEFETZ, A (2005) Dufour's gland pheromone as a reliable signal for reproductive dominance among workers honey bee (*Apis mellifera*). *Behavioural Ecology and Sociobiology* 58(3): 270-276. http://dx.doi.org/10.1007/s00265-005-0923-9

DUCRET, A; TRANI, M; LORTIE, R (2008) General guidelines for authors for submission of manuscripts that contain identifications and shntheses of compounds. *Journal of Chemical Ecology* 34(8): 984-986.

EVANS, J D; SCHWARZ, R S; CHEN, Y P; BUDGE, G; CORNMAN, R S; DE LA RUA, P; DE MIRANDA, J R; FORET, S; FOSTER, L; GAUTHIER, L; GENERSCH, E; GISDER, S; JAROSCH, A; KUCHARSKI, R; LOPEZ, D; LUN, C M; MORITZ, R F A; MALESZKA, R; MUÑOZ, I; PINTO, M A (2013) Standard methodologies for molecular research in *Apis mellifera*. In *V Dietemann; J D Ellis; P Neumann (Eds) The COLOSS BEEBOOK, Volume I: standard methods for* Apis mellifera *research. Journal of Apicultural Research* 52(4): http://dx.doi.org/10.3896/IBRA.1.52.4.11

FRANCKE, W; LÜBKE, G; SCHRÖDER, W; RECKZIEGEL, A; IMPERATRIZ-FONSECA, V; KLEINERT, A; ENGELS, E; HARTFELDER, K; RADKE, R; ENGELS, W (2000). Identification of oxygen containing volatiles in cephalic secretions of workers of Brazilian stingless bees. *Journal of Brazilian Chemical Society* 11 (6): 562-571.

FREE, J B (1987) *Pheromones of social bees*. Chapman and Hall; London, UK. 218 pp.

FRANCIS, B R; BLANTON, W E; NUNAMAKER, R A (1985) Extractable surface hydrocarbons of workers and drones of the genus *Apis*. *Journal of Apicultural Research* 24(1): 13-26.

FRANCIS, B R; BLANTON, W E; LITTLEFIELD, J L; NUNAMAKER, R A (1989) Hydrocarbons of the cuticle and haemolymph of the adult honey bee (Hymenoptera: Apidae). *Annals of the Entomological Society of America* 82(4): 486-493.

GENERSCH, E; GISDER, S; HEDTKE, K; HUNTER, W B; MÖCKEL, N; MÜLLER, U (2013) Standard methods for cell cultures in *Apis mellifera* research. In *V Dietemann; J D Ellis; P Neumann (Eds) The COLOSS BEEBOOK, Volume I: standard methods for* Apis mellifera *research. Journal of Apicultural Research* 52(1): http://dx.doi.org/10.3896/IBRA.1.52.1.02

GILLEY, D C; DEGRANDI-HOFFMAN, G; HOOPER, J E (2006) Volatile compounds emitted by live European honey bee (*Apis mellifera* L.) queens. *Journal of Insect Physiology* 52(5): 520-527. http://dx.doi.org/10.1016/j.jinsphys.2006.01.014

GRAMACHO, K P; SPIVAK, M (2003) Differences in olfactory sensitivity and behavioral responses among honey bees bred for hygienic behavior. *Behavioral Ecology and Sociobiology* 54(5): 472-479. http://dx.doi.org/10.1007/s00265-003-0643-y

HANDLEY, A J; ADLARD, E R (2005) *Gas chromatographic techniques and applications*. CRC Press; Boca Raton, FL, USA.

HARE, J D (1998) Bioassay methods in terrrestrial invertebrates. In *Millar, J G; Haynes, K F (Eds). Methods in chemical ecology: bioassay methods*. Kluwer Academic Publishers; Boston, USA. pp 212-271.

HASEGAWA, M; ASANUMA, S; FUJIYUKI, T; KIYA T; SASAKI, T; ENDO D; MARIOKA, M; KUBO, T (2009) Differential gene expression in the mandibular glands of queen and worker honey bees, *Apis mellifera. Insect Biochemistry and Molecular Biology* 39(10): 661-667. http://dx.doi.org/10.1016/j.ibmb.2009.08.001

HARTFELDER, K; GENTILE BITONDI, M M; BRENT, C; GUIDUGLI-LAZZARINI, K R; SIMÕES, Z L P; STABENTHEINER, A; DONATO TANAKA, É; WANG, Y (2013) Standard methods for physiology and biochemistry research in *Apis mellifera*. In *V Dietemann; J D Ellis; P Neumann (Eds) The COLOSS BEEBOOK, Volume I: standard methods for* Apis mellifera *research. Journal of Apicultural Research* 52(1): http://dx.doi.org/10.3896/IBRA.1.52.1.06

HEATH, R R; MANUKIAN, A (1994) An automated system for use in collecting volatile chemicals released from plants. *Journal of Chemical Ecology* 20(3): 593-608. http://dx.doi.org/10.1007/BF02059600

HUMAN, H; BRODSCHNEIDER, R; DIETEMANN, V; DIVELY, G; ELLIS, J; FORSGREN, E; FRIES, I; HATJINA, F; HU, F-L; JAFFÉ, R; KÖHLER, A; PIRK, C W W; ROSE, R; STRAUSS, U; TANNER, G; TARPY, D R; VAN DER STEEN, J J M; VEJSNÆS, F; WILLIAMS, G R; ZHENG, H-Q (2013) Miscellaneous standard methods for *Apis mellifera* research. In *V Dietemann; J D Ellis; P Neumann (Eds) The COLOSS BEEBOOK, Volume I: standard methods for* Apis mellifera *research. Journal of Apicultural Research* 52(4): http://dx.doi.org/10.3896/IBRA.1.52.4.10

HUNT, G J; WOOD, K V; GUZMAN-NOVOA, E; LEE, H D; ROTHWELL, A P; BONHAM, C C (2003) Discovery of 3-methyl-3-buten-1-yl acetate, a new alarm component in the sting apparatus of Africanized honey bees. *Journal of Chemical Ecology* 29(2): 453-463. http://dx.doi.org/10.1023/A:1022694330868

KAATZ, H H; HAGEDORN, H H; ENGELS, W (1985) Culture of honey bee organs: development of a new medium and the importance of tracheation *in vitro. Cellular and Developmental Biology* 21: 347-352.

KAMINSKI, L A; SLESSOR, K N; WINSTON, M L; HAY, N W; BORDEN, J H (1990) Honey bee response to queen mandibular pheromone in laboratory bioassays. *Journal of Chemical Ecology* 16(3): 841-850. http://dx.doi.org/10.1007/BF01016494

KATZAV-GOZANSKY, T; SOROKER, V; HEFETZ, A; COJOCARU, M; ERDMANN, D H; FRANCKE, W (1997) Plasticity of caste-specific Dufour's gland secretion in the honey bee (*Apis mellifera* L.). *Naturwissenschaften* 84(4): 238-241. http://dx.doi.org/10.1007/s001140050386

KATZAV-GOZANSKY, T; SOROKER, V; HEFETZ, A (2000) Plasticity in caste-related exocrine secretion biosynthesis in the honey bee (*Apis mellifera*). *Journal of Insect Physiology* 46(6): 993-998. http://dx.doi.org/10.1016/S0022-1910(99)00209-7

KATZAV-GOZANSKY, T; SOROKER, V; IBARRA, F; FRANCKE, W; HEFETZ, A (2001) Dufour's gland secretion of the queen honey bee (*Apis mellifera*): an egg discriminator pheromone or a queen signal? *Behavioral Ecology and Sociobiology* 51(1): 76-86. http://dx.doi.org/10.1007/s002650100406

KATZAV-GOZANSKY, T; SOROKER, V; HEFETZ; A (2003) Honey bee egg laying workers mimics the queen signal. *Insectes Sociaux* 50 (1): 20-23. http://dx.doi.org/10.1007/s000400300003

KATZAV-GOZANSKY, T; BOULAY, R; SOROKER, V; HEFETZ, A (2004) Queen-signal modulation of worker pheromonal composition in honey bees. *Proceedings of the Royal Society of London: Biological Sciences* 271(1552): 2065-2069. http://dx.doi.org/10.1098/rspb.2004.2839

KATZAV-GOZANSKY, T; HEFETZ, A; SOROKER, V; (2007) Brain modulation of Dufour's gland ester biosynthesis *in vitro* in the honey bee (*Apis mellifera*).*Naturwissenschaften* 94(5): 407-411. http://dx.doi.org/10.1007/s00114-006-0206-y

KEELING, C I; SLESSOR, K N; HIGO, H A; WINSTON, M L (2003) New components of the honey bee (*Apis mellifera* L.) queen retinue pheromone. *Proceedings of the National Academy of Sciences* 100 (8): 4486-4491.
http://dx.doi.org/ 10.1073/pnas.0836984100

KOCHER, S D; RICHARD, F-J; TARPY, D R;. GROZINGER, C M (2009) Queen reproductive state modulates pheromone production and queen worker interactions in honey bees. *Behavioral Ecology* 20 (5): 1007-1014. http://dx.doi.org/10.1093/beheco/arp090

KRAUS, B; VELTHUIS, H H W (1997) High humidity in the honey bee (*Apis mellifera* L.) brood nest limits reproduction of the parasitic mite *Varroa jacobsoni* Oud. *Naturwissenschaften* 84(5): 217-218.
http://dx.doi.org/10.1007/s001140050382

MAISONNASSE, A; LENOIR, J C; COSTAGLIOLA, G; BESLAY, D; CHOTEAU, F; CRAUSER, D; BECARD, J M; PLETTNER, E; LECONTE, Y (2009) A scientific note on E-*β*-ocimene, a new volatile primer pheromone that inhibits worker ovary development in honey bees. *Apidologie* 40(5): 562-564.
http://dx.doi.org/10.1051/apido/2009024

MAISONNASSE, A; LENOIR, J C; BESLAY, D; CRAUSER, D; LECONTE, Y (2010) E-β-Ocimene, a volatile brood pheromone involved in social regulation in the honey bee colony (*Apis mellifera*). *PLoS ONE* 5(10): e13531.
http://dx.doi.org/10.1371/journal.pone.0013531

MALKA, O; KRUNGER, I; YEHESKEL, A; MORIN, S; HEFETZ, A (2009) The road to royalty-differential expression of hydroxylating genes in the mandibular glands of the honey bee. *Federation of European Biochemical Societies Journal* 276(19): 5481-5490.
http://dx.doi.org/10.1111/j.1742-4658.2009.07232.x

MALKA, O; SHNIEOR, S; HEFETZ, A; KATZAV-GOZANSKY, T (2007) Reversible royalty in worker honey bees under the queen influence. *Behavioural Ecology and Sociobiology* 61(3): 465-473.
http://dx.doi.org/10.1007/s00265-006-0274-1

MARTIN, S J; JONES, G R (2004) Conservation of bio synthetic pheromone pathways in honey bees *Apis*. *Naturwissenschaften* 91(5): 232-236. http://dx.doi.org/10.1007/s00114-004-0517-9

MASTERMAN, R; ROSS, R; MESCE, K; SPIVAK, M (2001) Olfactory and behavioralresponse thresholds to odors of diseased brood differ between hygienic and non-hygienic honey bees (*Apis mellifera* L.). *Journal of Comparative Physiology* A187(6): 441-452.
http://dx.doi.org/10.1007/s003590100216

MILLAR, J; HAYNES, K F (1998) *Methods in chemical ecology: chemical methods.* Chapman & Hall; USA. 420 pp.

MILLAR, J G; SIMS, J J (1998) Preparation, cleanup and preliminary fractionation of extracts. In *Millar, J G; Haynes, K F (Eds). Methods in chemical ecology: bioassay methods.* Kluwer Academic Publishers; Boston, USA. pp 2-37.

MORITZ, R; CREWE, R M (1991) The volatile emission of honey bee queens (*Apis mellifera* L). *Apidologie* 22(3): 205-212.
http://dx.doi.org/10.1051/apido:19910304

NAUMANN, K; WINSTON, M L; SLESSOR, K N; PRESTWICH, G D; WEBSTER, F X (1991) Production and transmission of queen (*Apis mellifera* L.) mandibular gland pheromone. *Behavioral Ecology and Sociobiology* 29(5): 321-332.
http://dx.doi.org/ 10.1007/BF00165956

NEUMANN, P; PIRK, C W W; SCHÄFER, M O; EVANS, J D; PETTIS, J S; TANNER, G; WILLIAMS, G R; ELLIS, J D (2013) Standard methods for small hive beetle research. In *V Dietemann; J D Ellis, P Neumann (Eds) The COLOSS BEEBOOK: Volume II: Standard methods for* Apis mellifera *pest and pathogen research. Journal of Apicultural Research* 52(4):
http://dx.doi.org/10.3896/IBRA.1.52.4.19

NAZZI, F; MILANI N; DELLA VEDOVA, G (2002) (Z)-8-Heptadecene from infested cells reduces the reproduction of *Varroa destructor* under laboratory conditions. *Journal of Chemical Ecology* 28(11): 2181-2190. http://dx.doi.org/10.1023/A:1021041130593

NAZZI, F; MILANI, N; DELLA VEDOVA, G (2004) A semiochemical from larval food influences the entrance of *Varroa destructor* into brood cells. *Apidologie* 35(4): 403-410.
http://dx.doi.org/10.1051/apido:2004023

NAZZI, F; BORTOLOMEAZZI, R; VEDOVA, G D; DEL PICCOLO, F; ANNOSCIA, D; MILANI, N (2009) Octanoic acid confers to royal jelly varroa-repellent properties. *Naturwissenschaften* 96(2): 309-314. http:dx.doi.org/10.1007/s00114-008-0470-0

NÚÑEZ, A J; GONZALEZ, L F; JANAK, J (1984) Pre-concentration of headspace volatiles for trace organic analysis by gas chromatography. *Journal of Chromatography* 300(1): 127-162.
http://dx.doi.org/10.1016/S0021-9673(01)87583-X

PANKIW, T (2004) Cued in: honey bee pheromones as information flow and collective decision-making. *Apidologie* 35(2): 217-226.
http://dx.doi.org/10.1051/apido:2004009

PANKIW, T; HUANG, Z-Y; WINSTON, M L; ROBINSON, G E (1998) Queen mandibular gland pheromone influences worker honey bee (*Apis mellifera* L.) foraging ontogeny and juvenile hormone titers. *Journal of Insect Physiology* 44(7-8): 685-282.
http://dx.doi.org/10.1016/S0022-1910(98)00040-7

PANKIW, T; WINSTON, M L; SLESSOR, K N (1994) Variation in worker response to honey bee (*Apis mellifera L.*) queen mandibular pheromone (Hymenoptera: Apidae). *Journal of Insect Behaviour* 7 (1): 1-15. http://dx.doi.org/10.1007/BF01989823

PANKIW, T; WINSTON, M; FONDRK, M K; SLESSOR, K N (2000) Selection on worker honey bee responses to queen pheromone (*Apis mellifera* L.). *Naturwissenschaften* 87(11): 487-490.
http://dx.doi.org/10.1007/s001140050764

PICKETT, J A; WILLIAMS, I H; MARTIN, A P; SMITH, M C (1980) Nasonov pheromone of the honey bee *Apis mellifera* L. (Hymenoptera: Apidae) Part I. Chemical characterization. *Journal of Chemical Ecology* 6(2): 425-432. http://dx.doi.org/10.1007/BF01402919

PLETTNER, E; SUTHERLAND, G R J; SLESSOR, K N; WINSTON, M L (1995) Why not be a queen? Regioselectivity in mandibular secretions of honey bee castes. *Journal of Chemical Ecology* 21 (6): 1017-1029. http://dx.doi.org/10.1007/BF02033805

PLETTNER, E; SLESSOR, K N; WINSTON, M L; OLIVER, J E (1996) Caste-selective pheromone biosynthesis in honey bees. *Science* 271(5257): 1851-1853. http://10.1126/science.271.5257.1851

RAGUSO, R A; PELLMYR, O (1998) Dynamic headspace analysis of floral volatiles: a comparison of methods. *Oikos* 81(2): 238-254.

REYMENT, R A (1989) Compositional data analysis. *Terra Nova* 1: 29-34. http://dx.doi.org/10.1111/j.1365-3121.1989.tb00322.x

RICKLI, M; DIEHL, P A; GUERIN, P M. (1994) Cuticle alkanes of honey bee larvae mediate arrestment of bee parasite *Varroa jacobsoni*. *Journal of Chemical Ecology* 20(9): 2437-2453. http://dx.doi.org/10.1007/BF02033212

RAFAELI, A (2009) Pheromone biosynthesis activating neuropeptide (PBAN): Regulatory role and mode of action. *General Comparative Endocrinology* 162(1): 69-78. http://dx.doi.org/10.1016/j.ygcen.2008.04.004

ROBINSON, G E (1985) Effects of a juvenile hormone analogue on honey bee foraging behaviour and alarm pheromone production. *Journal of Insect Physiology* 31(4): 277-282. http://dx.doi.org/10.1016/0022-1910(85)90003-4

ROBINSON, G E A; HUANG, Z-H (1998) Colony integration in honey bees: genetic, endocrine and social control of division of labour. *Apidologie* 29(1-2): 159-170. http://dx.doi.org/10.1051/apido:19980109

ROMEO, J T (2009) New SPME guidelines. *Journal of Chemical Ecology* 35(12): 1383. http://dx.doi.org/10.1007/s10886-009-9733-2.

SALVY, M; MARTIN, C; BAGNÈRES, A G; PROVOST, E; ROUX, M; LECONTE, Y; CLÉMENT, J L (2001) Modifications of the cuticular hydrocarbon profile of *Apis mellifera* worker bees in the presence of the ectoparasitic mite *Varroa jacobsoni* in brood cells. *Parasitology* 122(2): 145-159. http://dx.doi.org/10.1017/S0031182001007181

SCHEINER, R; ABRAMSON, C I; BRODSCHNEIDER, R; CRAILSHEIM, K; FARINA, W; FUCHS, S; GRÜNEWALD, B; HAHSHOLD, S; KARRER, M; KOENIGER, G; KOENIGER, N; MENZEL, R; MUJAGIC, S; RADSPIELER, G; SCHMICKLI, T; SCHNEIDER, C; SIEGEL, A J; SZOPEK, M; THENIUS, R (2013) Standard methods for behavioural studies of *Apis mellifera*. In *V Dietemann; J D Ellis; P Neumann (Eds) The COLOSS BEEBOOK, Volume I: standard methods for* Apis mellifera *research. Journal of Apicultural Research* 52(4): http://dx.doi.org/10.3896/IBRA.1.52.4.04

SCHMITT, T; HERZNER, G; WECKERLE, B; SCHREIER, P; STROHM, E (2007) Volatiles of foraging honey bees *Apis mellifera* (Hymenoptera: Apidae) and their potential role as semiochemicals. *Apidologie* 38(2): 164-170. http://dx.doi.org/10.1051/apido:2006067

SCHNEIDER, D (1957). Elektrophysiologische untersuchungen von chemorezeptoren und mechanorezeptoren der antenne des seidenspinners *Bombyx mori* L. *Zeitschrift fur Vergleichende Physiologie* 40: 8-41. http://dx.doi.org/10.1007/BF00298148

SCHÖNING, C S; GISDER, S; GEISELHARDT, S; KRETSCHMANN, I; BIENEFELD, K; HILKER, M; GENERSCH, E (2012) Evidence for damage-dependent hygienic behaviour towards *Varroa destructor*-parasitised brood in the western honey bee, *Apis mellifera*. *The Journal of Experimental Biology* 215(2): 264-271. http://dx.doi.org/10.1242/jeb.062562

SEELEY, T D (1982) Adaptive significance of the age polytheism schedule in honey bee colonies. *Behavioural Ecology and Sociobiology* 11(4): 287-293. http://dx.doi.org/10.1007/BF00299306

SEIDELMANN, K; WEINERT, H; FERENZ, H-J (2003) Wings and legs are production sites for desert locust courtship-inhibition pheromone, phenyl acetonitrile *Journal of Insect Physiology* 49 (12): 1125-1133. http://dx.doi.org/10.1016/j.jinsphys.2003.08.005

SKOOG, D A; HOLLER, F J; CROUCH, S R (2007) *Principles of instrumental analysis, (6th Ed.)*. Brooks / Cole, Thomson Learning; Belmont, CA, USA.

SLESSOR, K N; KAMINSKI, L A; KING, G G S; BORDEN, H J; WINSTON, M L (1988) Semiochemical basis of the retinue response to queen honey bees. *Nature* 332(6162): 354-356. http://dx.doi.org/10.1038/332354a0

SLESSOR, K N; WINSTON, M; LE CONTE, Y (2005) Pheromone communication in the honey bee (*Apis mellifera* L.). *Journal of Chemical Ecology* 31(11): 2731-2745. http://dx.doi.org/10.1007/s10886-005-7623-9

SONNET, P E (1984) Tabulation of selected methods of syntheses that are frequently employed for insect sex pheromones, emphasizing the literature of 1977–1982. In *Hummel, H E; Miller, T A (Eds). Techniques in pheromone research*. Springer-Verlag; New York, USA.

SOROKER, V; HEFETZ, A (2000) Hydrocarbon site of synthesis and circulation in the desert ant *Cataglyphis niger*. *Journal of Insect Physiology* 46(7): 1097-1102. http://dx.doi.org/10.1016/S0022-1910(99)00219-X

SOROKER, V; RAFAELI, A (1989) *In vitro* hormonal stimulation of acetate incorporation by *Heliothis armigera* pheromone glands. *Journal of Insect Biochemistry* 19(1): 1-5. http://dx.doi.org/10.1016/0020-1790(89)90002-4

STANLEY-SAMUELSON, D W; JURENKA, R A; CRIPPS, C; BLOMQUIST, G J; de RENOBLES, M (1988) Fatty acids in insects: composition, metabolism, and biological significance. *Archives of Insect Biochemistry and Physiology* 9(1): 1-33. http://dx.doi.org/10.1002/arch.940090102

SWANSON, J A; TORTO, B; KELLS, S A; MESCE, K A; TUMLINSON, J H, SPIVAK, M (2009) Odorants that induce hygienic behavior in honey bees: identification of volatile compounds in chalkbrood-infected honey bee larvae. *Journal of Chemical Ecology* 35(9): 1108-1116. http://dx.doi.org/10.1007/s10886-009-9683-8

SYNTECH (2004) *Electroantennography: a practical introduction.* Hilversum; Netherlands.

TEAL, P E A; TORTO, B; BOUCIAS, D; TUMLINSON, J H (2006) In-hive trap and attractant composition for the control of the small hive beetle, *Aethina tumida. US Patent, 20060141904.* 29 June 2006.

THOLL, D; BOLAND, W; HANSEL, A; LORETO, F; RÖSE, U S R; SCHNITZLER, J-P (2006) Practical approaches to plant volatile analysis. *The Plant Journal* 45(4): 540-560. http://dx.doi.org/10.1111/j.1365-313X.2005.02612.x

THOM, C; GILLEY, D C; HOOPER, J; ESCH, H E (2007) The scent of the waggle dance. *PLoS Biology* 5(9): 1862-1867. http://dx.doi.org/10.1371/journal.pbio.0050228

TORTO, B; BOUCIAS, D G; ARBOGAST, R T; TUMLINSON, J H; TEAL, P E A (2007a) Multi-trophic interaction facilitates parasite-host relationship between an invasive beetle and the honey bee. *Proceedings of the National Academy of Science, USA* 104(20): 8374-8378. http://dx.doi.org/10.1073/pnas.0702813104

TORTO, B; ARBOGAST, R T; VAN ENGELSDORP, D; WILLMS, S; PURCELL, D; BOUCIAS, D; TUMLINSON, J H; TEAL, P E A (2007b) Trapping of *Aethina tumida* Murray (Coleoptera: Nitidulidae) from *Apis mellifera* L. (Hymenoptera: Apidae) colonies with an in-hive baited trap. *Environmental Entomology* 36(5): 1018-1024. http://dx.doi.org/10.1603/0046-225X(2007)36[1018:TOATMC]2.0.CO;2

TORTO, B; SUAZO, A; ALBORN, H; TUMLINSON, J H; TEAL, P E A (2005) Response of the small hive beetle (*Aethina tumida*) to a blend of chemicals identified from honey bee (*Apis mellifera*) volatiles. *Apidologie* 36(4): 523-532. http://dx.doi.org/10.1051/apido:2005038

TSFADIA, O; AZRIELI, A; FALECH, L; ZADA A; ROELOFS W; RAFAELI, A (2008) Pheromone biosynthesis pathway: PBAN-regulated rate-limiting steps anddeferential expression of desaturase genes in moth species. *Insect Biochemistry and Molecular Biology* 38(5): 552-567. http://dx.doi.org/10.1016/j.ibmb.2008.01.005

WILLIAMS, G R; ALAUX, C; COSTA, C; CSÁKI, T; DOUBLET, V; EISENHARDT, D; FRIES, I; KUHN, R; MCMAHON, D P; MEDRZYCKI, P; MURRAY, T E; NATSOPOULOU, M E; NEUMANN, P; OLIVER, R; PAXTON, R J; PERNAL, S F; SHUTLER, D; TANNER, G; VAN DER STEEN, J J M; BRODSCHNEIDER, R (2013) Standard methods for maintaining adult *Apis mellifera* in cages under *in vitro* laboratory conditions. In *V Dietemann; J D Ellis; P Neumann (Eds) The COLOSS BEEBOOK, Volume I: standard methods for* Apis mellifera *research. Journal of Apicultural Research* 52(1): http://dx.doi.org/10.3896/IBRA.1.52.1.04

WINSTON, M L (1987) *The biology of the honey bee.* Harvard University Press; Cambridge, Massachusetts, USA. 281 pp.

WOSSLER, T C; CREWE, R M (1999) The releaser effects of the tergal gland secretion of queen honey bees (*Apis mellifera*). *Journal of Insect Behavior* 12(3): 343-351. http://dx.doi.org/10.1023/A:1020839505622

WOSSLER, T C; JONES, G E; ALLSOP, M H; HEPBURN, R (2006) Virgin queen mandibular gland signals of *Apis mellifera capensis* change in age and affect honey bee worker response *Journal of Chemical. Ecology* 32(5): 1043-1056. http://dx.doi.org/10.1007/s10886-006-9053-8

ZABARAS, D; WYLLIE, S G (2002) Rearrangement of p-menthane terpenes by Carboxen during HS-SPME. *Journal of Separation Science* 25(10-11): 685-690. http://dx.doi.org/10.1002/1615-9314(20020701)25:10/11<685::AID-JSSC685>3.0.CO;2-9

Journal of Apicultural Research 52(1): (2013)
DOI 10.3896/IBRA.1.52.1.03

REVIEW ARTICLE

Standard methods for estimating strength parameters of *Apis mellifera* colonies

Keith S Delaplane[1*], Jozef van der Steen[2] and Ernesto Guzman-Novoa[3]

[1]Department of Entomology, University of Georgia, Athens, GA 30602, USA.
[2]Plant Research International, bees@wur, Wageningen University and Research, Wageningen, The Netherlands.
[3]School of Environmental Sciences, University of Guelph, Guelph, Ontario, N1G 2W1, Canada.

Received 30 April 2012, accepted subject to revision 17 July 2012, accepted for publication 22 October 2012.

*Corresponding author: Email: ksd@uga.edu

Summary

This paper covers measures of field colony strength, by which we mean population measures of adult bees and brood. There are generally two contexts in which an investigator wishes to measure colony strength: 1. at the beginning of a study as part of manipulations to produce uniform colonies and reduce experimental error and; 2. as response variables during or at the end of an experiment. Moreover, there are two general modes for measuring colony strength: 1. an objective mode which uses empirical measures and; 2. a subjective mode that relies on visual estimates by one or more observers. There is a third emerging mode for measuring colony strength; 3. computer-assisted digital image analysis. A final section deals with parameters that do not directly measure colony strength yet give important indicators of colony state: flight activity at the hive entrance; comb construction; and two proxy measures of colony fitness: production of queen cells and drone brood.

Métodos estándar para estimar parámetros sobre la fortaleza de las colonias de *Apis mellifera*

Resumen

Este trabajo trata sobre mediciones de campo de la fortaleza de colonias de abejas, es decir, mediciones de la población de abejas adultas y de cría. En general hay dos contextos en los que un investigador desearía medir la fortaleza de una colonia: 1. al comienzo de un estudio como parte de las manipulaciones para producir colonias uniformes y reducir el error experimental y; 2. como variables de respuesta durante o al final de un experimento. Además, en general hay dos maneras de medir la fortaleza de las colonias: 1. una manera objetiva usando mediciones empíricas y; 2. una manera subjetiva que se basa en estimaciones visuales de uno o mas observadores. Hay una tercer forma más reciente de medir la fortaleza de una colonia; 3. análisis computacional de imágenes digitales. La sección final trata sobre parámetros que no miden directamente la fortaleza de la colonia pero que son importantes indicadores del estado de ésta: actividad de vuelo en la piquera de la colmena, construcción de panales y dos mediciones equivalentes de aptitud de la colonia: producción de celdas reales y de cría de zánganos.

评估西方蜜蜂蜂群群势的标准方法

本文综述了多种测量蜂群群势的方法，主要包括成年蜂和幼虫的群势。通常研究者在两种情况下需要测量蜂群群势：1 在实验开始时，作为处理的一部分，希望获得统一的群势以减少实验误差；2 群势是实验中或实验结束时的反应变量。此外，测量蜂群群势有两种基本的模式：1 客观模式，应用实验技术测量；2 主观模式，依赖于一个或多个观察者的视觉估计。还有第三种新兴的模式，3 计算机辅助的数字图像分析。最后一部

Footnote: Please cite this paper as: DELAPLANE, K S; VAN DER STEEN, J; GUZMAN, E (2013) Standard methods for estimating strength parameters of *Apis mellifera* colonies. *In* V Dietemann; J D Ellis; P Neumann (Eds) The COLOSS BEEBOOK, *Volume I: standard methods for* Apis mellifera *research. Journal of Apicultural Research* 52(1): http://dx.doi.org/10.3896/IBRA.1.52.1.03

分是关于利用重要的指示参数，开展非直接测量评估，如：巢门口的飞行频率、群内巢脾的建造数度，以及王台和雄蜂的产生和发展，这两个指示蜂群适应度指标的测定。

Keywords: bee population, brood area, brood pattern, honey stores, disease symptoms, parasite symptoms, flight activity, queen cells, drone brood, digital image analysis, honey bee, *BEEBOOK*, COLOSS

1. Introduction

Herein, we cover measures of honey bee field colony strength, by which we primarily mean population measures of adult bees and brood. We will also talk about secondary measures such as: quantity of stored honey and pollen; "brood pattern" by which is meant the degree of worker brood solidity or contiguity; and the expression of visible disease or parasite symptoms. Strictly speaking these secondary measures are not so much indicators of a colony's immediate state as they are legacy effects or predictors of future condition.

For our purposes there are two contexts in which an investigator wishes to measure colony strength: 1. at the beginning of a study as part of manipulations to produce uniform colonies and reduce experimental error and; 2. as response variables during or at the end of an experiment. Moreover, there are two general modes of measuring colony strength: 1. an objective mode which uses empirical measures such as weight (mg, g, or kg) or area (cm^2), covered in sections 3 and 4.1. and; 2. a subjective mode that relies on visual estimates by one or more observers, covered in sections 4.2. and 5.1. The objective mode is the more accurate of the two, but it is also invasive and disruptive to the bees, constituting in some cases the complete deconstruction and reassembly of colonies with disruption to any social cohesion formerly intact. For this reason we consider the objective mode best suited to the beginning and end of experiments. In contrast, the subjective mode is less accurate, but far less disruptive to the bees and therefore appropriate for collecting response variables during the experiment when the investigator has an interest in preserving the social cohesion and health of experimental colonies. One exception to this would be if the sampling intervals are sufficiently distanced (2-3 times per year) to justify the objective mode throughout. Nevertheless, with safeguards in place such as we describe below, the subjective mode is an acceptably robust technique.

There is a third emerging mode for measuring colony strength; 3. computer-assisted digital image analysis, covered in sections 5.2. – 5.5. This method is minimally invasive, automatically generates archival images for data traceability and verification, provides objective empirical data, and can be done moderately quickly. Its chief disadvantages are cost and dependence on technology. Moreover, it is the opinion of some that the speed and ease of visual estimates surpass the advantages of objectivity and archival properties of digital methods. Nevertheless, we will probably see technical improvements and increasing use of this mode in the near future. Computer-assisted digital image analysis is useful for experiments that call for measures of bee health or development but fall short of field-scale colony strength assessment, chief examples being laboratory studies in environmental toxicology or nutrition.

The sections 5.6. – 5.9. cover methods that do not fit neatly into the other sections. These include: measuring flight activity at the entrance; comb construction; and two proxy measures of colony fitness: production of queen cells; and drone brood.

A note is warranted here on a couple omissions from this chapter; gross colony weight and X-ray tomography. Gross colony weight is a useful metric in the context of seasonal changes in forage availability. Hive-scale data have long interested beekeepers for their usefulness in tracking local nectar flows, and more recently these kinds of data have been used to monitor flowering phenology in the context of climate change (Nightingale *et al.,* 2008). As a measure of colony strength *per se*, however, gross colony weight is ambiguous and unreliable, owing to the fact that workers from health-compromised colonies may express precocious foraging with the result that weights of colonies may increase, not decrease, in response to disease or other disorders (Mayack and Naug, 2009). X-ray tomography offers what is probably the most empirically quantifiable, thorough, and non-invasive means of monitoring colony strength of colonies of honey bees or other social insects (Greco, 2010). Although it sets a gold standard, its formidable technical requirements keep this method out of reach of most honey bee researchers.

2. An optimal colony configuration

In establishing colonies for experiments, it is useful to have some guidance on how colony population size can be expected to affect colony growth, behaviour, and survivorship. The best guidance in this matter comes from John Harbo (1986) who compared brood production, worker survival, and honey gain in colonies begun with no brood and 2,300, 4,500, 9,000, 17,000, or 35,000 bees while fixing bee density at *ca.* 230 per 1000 cm^3 hive space. The experiment was repeated in each of the months of Feb, Apr, Jun, Aug, and Oct and terminated before brood emerged. Worker survival (22 days) was significantly higher in colonies with 2,300-9,000 bees than in colonies with 35,000 bees. Larger populations tended to store more nectar per bee during times of nectar flow and consume less during times of nectar dearth. However, smaller populations produced more brood per bee. Harbo concluded that colonies established with 9,000 bees are a good optimum between the honey hoarding efficiency of large populations

Table 1. Pros and cons of two variations of an objective mode for starting up colonies of uniform initial strength.

Method	Pros	Cons
Classical 3.1.	1. Results in colonies with initial populations of adult bees normalized for genetics and pathogen load.	1. Disease and parasite legacy effect in brood, although normalized, is nevertheless sustained into the experimental period.
	2. Results in colonies with brood of all stages, accelerating colony growth.	2. Because brood of all stages is present, progeny turn-over from new experimental queens is correspondingly delayed.
	3. Results in maximized colony uniformity in regards to initial adult bee populations.	
Shook swarm 3.2.	1. Does not require use of a customized cage to house common pool of bees.	1. Because adult bees are not normalized, between-colony variation in genetics and pathogen load remains at pre-experiment levels.
	2. Sustains colony-specific identity from pre-experimental to experimental period	2. Because colonies begin broodless, colony growth is delayed.
	3. Because of #2, drift is not a concern and it is not necessary to move experimental colonies from the source apiary.	
	4. Because all brood is removed and replaced with frames of foundation, disease and parasite legacy effect is minimized.	
	5. Because of #4, if varroa control is an element of experimental design, the initial broodless period provides an ideal opportunity to treat for mites.	

and the brood rearing efficiency of small colonies. Colonies that are significantly larger than this are costly and labour-intensive to set up and less suitable for measures of population growth because they are already near their maximum. Colonies significantly smaller than this may do well at the height of the season, but they are more vulnerable to winter and summer stresses. Bees are normalized for colony-source genetics and parasite loads if colonies are set up with the common pool of bees as described in section 3.1. If colony growth is a measure of interest, the investigator can invite a greater range of expansion if colonies are begun with no brood. But if one prefers to provide colonies with brood, it is reasonable to stock colonies of 9,000 bees with no more than two combs of brood of various ages, allowing plenty of open cells to accommodate growth.

3. Setting up experimental colonies of uniform strength

This section describes two variations of an objective mode for setting up uniform colonies for experiments. The first (3.1.), which we call the classical objective mode and a variation (3.2.), the so-called "shook swarm" method. Table 1 explains the pros and cons.

3.1. Classical objective mode

One of the recurring pitfalls of honey bee field research is large experimental error which handicaps the investigator's attempts to discriminate statistical differences among effects of interest. One of the best ways to minimize this problem is to begin experiments with colonies as uniform as possible in regard to comb space, food

resources, and populations of adult bees and brood. It is the job of the investigator to distribute these resources equitably among experimental colonies.

The following synthesis draws from methods pioneered by John Harbo who was interested chiefly in reducing environmental variation in honey bee breeding programs (Harbo, 1983, 1986, 1988, 1993; Delaplane and Harbo, 1987) and adapted later by workers who recognized the utility of these methods for field research on *Varroa destructor* (Delaplane and Hood, 1997, 1999; Ellis *et al.*, 2001; Strange and Sheppard, 2001; Berry *et al.*, 2010) and colony growth (Berry and Delaplane, 2001).

1. The goal is field colonies equalized with regard to bees, brood, mites, and food resources within units of higher-order experimental replication, i.e. blocks or whole plots, usually based on geography.

2. Empty hives are pre-stocked with brood, empty combs, syrup feeders, and a caged queen in advance of receiving worker bees (Fig. 1). Bottom boards and hive bodies are stapled together to prepare them for moving. Hive entrances are screened to temporarily trap bees; this is done for two reasons: 1. experimental colonies often need to be moved to a permanent site and away from the source colonies from which workers are collected and; 2. a period of in-hive confinement, usually overnight, seems to help bees orient to their new hive and queen.

3. Brood for incipient experimental colonies can be collected from the same source colonies used to collect adults. A near-equal quantity of brood is then assigned to experimental colonies without regard to source. We do not prescribe

Fig. 1. Investigators pre-stocking experimental hives with equal numbers of brood combs and honey combs and caged queens in preparation for receiving worker bees from a common cage.

"random" brood assignment because the investigator should place a higher priority on equalizing quantity of brood over concerns of non-random assignment of brood. Efforts should be made to equalize the relative quantity of sealed versus open brood. A measure of beginning quantity of brood is done by overlaying on each side of every brood comb a grid pre-marked in cm^2 and visually summing the area of brood (Fig. 2) The area (cm^2) of brood can be converted to cells of brood by multiplying cm^2 by the average cells per cm^2. This value varies by geography and bee genetics (Table 2), or the investigator can determine local average cell density per cm^2 by counting the number of cells directly in a square equalling one cm^2 and using the mean of at least ten measurements.

4. In a similar fashion the investigator can derive and equalize the beginning number of cells of honey or pollen or even cells that are empty. Depending on one's standards for strict uniformity, it may be simpler to provide nothing but brood or empty cells and to provide uniform nutrition across the experiment by use of sugar syrup and protein supplements.

5. Variation due to bee genetics is minimized by providing each colony a sister queen reared from the same mother and open-mated in the same vicinity. A more robust option is to instrumentally inseminate sister queens (see the *BEEBOOK* paper on instrumental insemination of queens (Cobey *et al.,* 2013)) with the same pool of mixed drone semen.

6. Adult bees are collected for experimental set-up by shaking workers from a diversity of source colonies into one large, common, ventilated cage, allowing workers (and diseases and parasites) to freely mix. With African subspecies, it helps minimize loss from flight to first spray bees on the comb with water mist. The cage is maintained in cool conditions to prevent bee death from over-heating for at least 24 hours to allow thorough admixing of bees, resulting in a uniformly heterogeneous mixture. The weight of bees collected (kg) should exceed the target weight of bees needed for the study by at least 2 kg, or at least by a third in the case of African subspecies, to account for bee loss through death or flight. Bee survival in the cage is greatly improved if the investigator designs it to accommodate 5-6 Langstroth sized brood combs to provide clustering surface (Fig. 3).

Fig. 3. A ventilated cage made to hold a large common heterogeneous mixture of bees for starting experiments.

7. In order to equalize initial colony populations, it is preferable to make colony-specific caged cohorts. Empty screened cages, ideally made to fit on top of an empty hive, are each pre-weighed or tared with a balance in the field. The large common cage is opened, the bees sprayed with water to reduce flight, then bees transferred from the common cage into the smaller colony-specific cages with the aid of cups or scoops (Fig. 4). Bees are added or removed from each colony-cage until the target weight is achieved and recorded, preferably ≥ 2 kg.

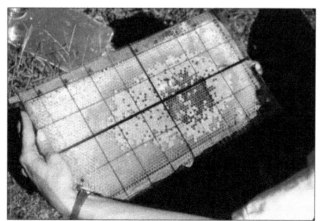

Fig. 2. A transparent plastic grid etched in square centimetres is used to visually sum the surface area of brood.

Table 2. Surface area of some regionally common frame types and expected bee density when frame is fully occupied by worker bees.

Region	Local frame type	Number bees per fully-occupied side	Surface (cm²) per side of frame	Bees / cm²	Ref	Worker cells / cm²
North America	Deep Langstroth	1215	880	1.38	a	3.7[c]-3.9[d]
North America	3/4s	910	655	1.39	a	
North America	Western	785	565	1.39	a	
North America	Shallow	640	461	1.39	a	
Europe	Swiss	1200	930	1.29	b	4.0[e]
Europe	Dadant	1400	1130	1.24	b	4.0[e]
Europe	German normal	900	720	1.25	b	4.0[e]
Europe	Langstroth	1100	880	1.25	b	4.0[e]
Europe	Zander	1000	810	1.23	b	4.0[e]
South and Central America	Jumbo for brood chamber (modified Dadant)	1980	1130	1.75	f	4.1-4.7[g]
South and Central America	Jumbo for super (modified Dadant)	920	520	1.77	f	4.1-4.7[g]
Africa	We are not aware of published methods for determining bee numbers and cell density with *A. mellifera* in Africa. However, these bees are ca. 3% smaller than African bees in South America[h], so it is reasonable to apply this conversion to the values given above for South and Central America.					

[a]Burgett and Burikam (1985), [b]Imdorf and Gerig (2001), [c]Harbo (1993), [d]Harbo (1988), [e]Imdorf *et al.* (1987), [f]Gris (2002), [g]Guzman-Novoa *et al.* (2011), [h]Buco et al. (1987).

8. A sample of *ca.* 300 workers is collected from each incipient colony into a pre-weighed or -tared screw-top container, weighed fresh, then the number of bees counted in the lab to derive a colony-specific measure of average fresh weight of individuals (mg per bee). To count bees it is necessary to first immobilize them, either by freezing them or non-sacrificially with CO_2 narcosis. Dividing initial colony cohort size (kg, from step 7) by average fresh weight of individuals (mg) gives initial bee population for the colony.

9. A variation of steps 7 and 8 is available if the investigator is using nucleus hives small enough to weigh entire in the field (Fig. 5). In these cases, the intermediate step of a colony-specific cage is not necessary and the investigator can scoop bees from the common cage directly into the pre-weighed or tared hive. The net weight (kg) of bees is recorded, then initial population determined the same way as given in step 8.

10. If initial measures reveal outliers in the amount of bees, brood, honey, pollen, and empty cells, corrective action should be taken. In general, corrections aimed at minimizing experimental error are permissible until the point at which treatments are begun.

11. After bees and all resources are placed inside hives (with entrances screened, but without sugar syrup which can spill in transit), hives are then moved to their permanent apiary site. Over-heating is a risk, and hives must be kept as cool as possible. There is a special advantage to setting up colonies late in the day and moving hives to the experimental apiary at night. Not only is it cooler, but once hives are unloaded and entrances opened, the bees do not fly because of the darkness and this protracted period inside the hive seems to

Fig. 4. Bees are transferred from the common cage to hive-specific cohort cages by use of cups or scoops.

Fig. 5. Nucleus colonies are small enough to be weighed directly in the field, bypassing the need for intermediate hive-specific cohort cages.

help them orientate to the new queen and reduces drifting. Colonies can be given sugar syrup after they are unloaded or 24 hours later after bees have settled down.

12. Apiaries should be arranged to limit worker drift between colonies. This can be done by "complicating" the visual field of bees with orientating landmarks near their nest entrances. This can be as simple as using rocks or trees or more deliberate such as painting varying geometric shapes on hive fronts. Arranging hives in a strongly linear arrangement is not good because hives at the ends tend to accumulate bees. For this reason some investigators place hives in a circle.

13. Colony maintenance should include control of non-target diseases and disorders, queen conservation, swarm prevention, and feeding as necessary. Of these, queen loss and swarming tend to be the most disruptive to colony populations. Cutting out queen cells, adding honey supers, marking queens, and regular inspections can reduce these problems. If honey supers are added, it is best to add them above a queen excluder to limit the range of the queen's egg-laying activity. The goal of these manipulations is to decrease experimental residual error.

3.2. Shook swarm objective mode

1. With this method it is assumed that investigators will use a pre-existing apiary and modify it for the experiment's purposes. It is important that the experiment start at a time of year that the bees can draw out foundation into comb.

2. In the days leading up to set-up, queens in each colony are caged and returned to the colony to save time on set-up day. Colonies are removed from the apiary if they are expressing disease symptoms, significantly under-performing, or otherwise causing excessive between-colony variation. New colonies are imported if needed to reach the target colony number and treated similarly.

3. A number of empty hives equal to the target number of colonies is brought to the apiary, each stocked with brood chamber frames of new foundation, including honey supers with frames of foundation if the nectar flow warrants, and sugar syrup feeders. If affordable, it is good to start new colonies on factory-new woodenware to avoid confounding issues of any disease legacy effects.

4. Each hive in the apiary is moved aside and an empty hive set in its place. Roughly half of the frames of foundation are momentarily removed to create space, then the caged queen is suspended between two centre-most frames of foundation.

5. Combs of bees from the original colony are then sequentially removed and the adult bees shaken off the combs into the new box. Bees are bounced or brushed out of the supers and the bee-free combs returned to them and covered to discourage robbing behaviour.

6. Once all bees are shaken into the new hive, the frames of foundation initially removed are now returned to the new boxes, gently to avoid injuring bees which may be heaped on the floor. Unless there is a strong nectar flow in progress, it is advisable to feed experimental colonies sugar syrup to encourage drawing out the new foundation.

7. The old bee-free boxes of combs are then removed from the experimental apiary and the combs used as needed elsewhere as supplemental brood or feed.

8. After one day, the caged queens in experimental colonies are released. Colonies are subsequently monitored for queen performance and normal colony development. Poor-performing queens are replaced as needed to minimize within-apiary experimental error. Once colonies reach a development state consistent with the experiment's objectives, treatments may be applied and the experiment begin.

9. The expected outcome of this manoeuvre is a high degree of within-apiary consistency in colony developmental state.

4. Measuring colony strength at end of experiment
4.1. Objective mode

This section is derivative of the references cited in section 3.1.

1. The day before the experiment is ended, each queen is found, caged with attendants, and returned to her colony. This will save a great deal of time the next day. Additionally, any hive cracks or gaps are sealed with duct tape to prevent bee loss.

2. The night or early morning before colonies are dismantled, the entrance of each colony is securely closed with a ventilated screen to trap workers inside.

3. Ending colony adult bee population is derived from net colony bee weight (kg) and average fresh bee weight (mg). Each screened whole hive is weighed in the field, then opened, all bees brushed off every comb and surface (usually into a temporary holding hive), and the hive re-weighed without bees. The difference in weight is the net weight of bees. A sample of *ca.* 300 live bees is collected into a pre-weighed or -tared container, weighed, the bees frozen or narcotized with cold or CO_2, and counted to determine average fresh weight (mg) per bee. Net colony bee weight is divided by average fresh weight per bee to derive colony bee population. If the fresh bee sample is frozen or stored in alcohol, it can be used to later determine adult loads of diseases, varroa mites, or other parasites of the investigator's choice.

4. Combs are labelled to preserve colony-specific identity and moved to the laboratory for further measures.

5. Number of brood cells is derived as described in section 3, using a grid pre-marked in cm^2, visually summing the area of

Fig. 6. A piece of cardboard with a square equal in size to 10 x 10 cells is laid over a patch of brood. Percentage brood solidness is measured directly as (100 - no. empty cells).

brood (Fig. 2), and converting area of brood (cm²) to cells of brood by multiplying cm² by the average cell density per cm² appropriate to one's locality (Table 2). This same method can be used to derive cell number of any comb resource of interest to the investigator; honey, pollen, or empty cells.

6. Brood solidness is determined by placing a grid that delimits 100 cells over a section of sealed brood and subtracting empty cells to estimate percentage brood solidness (Fig. 6).This measure is repeated on different patches of brood to derive a mean of at least ten observations.

7. Alternatively to reporting comb resources as cells, many investigators report these resources empirically as total are (cm²).

8. In the case of honey, it is traditional to report this variable by weight (kg). In these cases, the investigator is aided with the use of queen excluders that restrict brood to the lower hive bodies. If supers are pre-weighed before adding to hives, the investigator can determine honey yield by simply weighing bee-free honey supers at the end of the experiment.

9. Visible brood disorders can be quantified by first selecting a relatively contiguous patch of brood in the late larval / capped

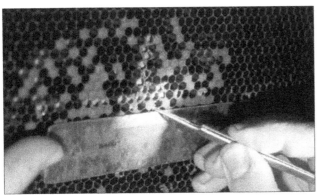

Fig. 7. A cross-shaped 10 x 10-cm transect intersects in the middle of a patch of contiguous brood, and every cell along the transect is opened and assessed for visible disorders.

stage (stage more likely to express visible symptoms), and overlaying on the patch a 10-cm horizontal transect and a 10 cm vertical transect intersecting at the centre (Fig. 7). Along each transect every cell of brood is examined under strong light and magnification for visible disorders, i.e., symptoms typical of American foulbrood, European foulbrood, sacbrood, or chalkbrood. The parameter is reported as percentage of brood expressing visible disorders.

4.2. Subjective mode

This section and the next describe the subjective mode of measuring colony strength and are thus best suited for collecting response variables while the field experiment is in progress. The gist of the method is the use of human observers who visually estimate the surface area of a comb covered by a target; bees, brood, honey, pollen, etc., and if necessary convert comb surface to target-appropriate units, i.e. bees, cm², or cells. The syntheses in sections 4 and 5 draw from the work of Burgett and Burikam (1985) and derivative papers from North America (Skinner *et al.,* 2001, Delaplane *et al.,* 2005, 2010), Imdorf *et al.* (1987) and Imdorf and Gerig (2001) from Europe, and Gris (2002) and Guzman-Novoa *et al.* (2011) from Central America.

1. Visual estimates of bees on combs will vary according to time of day and bee foraging activity. For this reason it is important to control for this effect – either by limiting observations to a narrow time window on successive days, randomly assigning time of inspection such that day effect is equitably and randomly distributed across treatments, or closing hive entrances in the early morning until bees are counted.

2. Estimates should be carried out by no fewer than two human observers, preferably each with a dedicated secretary who writes down numbers, or each fitted with an audio recorder.

3. A colony is opened and combs of bees sequentially removed. Each observer looks at one side of a comb, visually estimates the percentage of the comb surface covered by bees, and records the number with the secretary or audio recorder. It is convenient to label frames 1-X, with each side indicated A or B. For beginners it is advisable to "calibrate the individual" with estimates made by an experienced observer. Observers describe the process as a kind of mental "resorting" the bees, such that the bees are imaginatively moved into a contiguous mass on the comb surface, at which point the reader estimates the percentage surface of the comb they cover. It is important to visually sort the bees into a contiguous mass that approximates their density if the frame were fully covered because the bee densities given in Table 2 (1.23 – 1.77 bees per cm²) apply to combs at full carrying capacity.

4. Investigators can use the values in Table 2 or calculate the comb side surface area unique to their equipment. Fig. 8 is a

G2   fx   =(F2*880)

	A	B	C	D	E	F	G	H
1	colony	frame	side	observer 1	observer 2	mean of observers	cm2 covered by bees	bees per side
2	1	1	A	0.2	0.3	0.25	220	303.6
3	1	1	B	0.3	0.5	0.4	352	485.76
4	1	2	A	0.4	0.4	0.4	352	485.76
5	1	2	B	0.5	0.6	0.55	484	667.92
6	1	3	A	0.7	0.8	0.75	660	910.8
7	1	3	B	0.7	0.7	0.7	616	850.08
8	1	4	A	0.5	0.4	0.45	396	546.48
9	1	4	B	0.3	0.3	0.3	264	364.32
10	1	5	A	0.4	0.2	0.3	264	364.32
11	1	5	B	0.1	0	0.05	44	60.72
12							colony bee population	5039.76
13	2	1	A	0	0	0	0	0
14	2	1	B	0.2	0.3	0.25	220	303.6
15	2	2	A	0.4	0.4	0.4	352	485.76
16	2	2	B	0.5	0.6	0.55	484	667.92
17	2	3	A	1	0.9	0.95	836	1153.68
18	2	3	B	0.9	0.8	0.85	748	1032.24
19	2	4	A	0.7	0.8	0.75	660	910.8
20	2	4	B	0.5	0.5	0.5	440	607.2
21	2	5	A	0.3	0.4	0.35	308	425.04
22	2	5	B	0	0	0	0	0
23							colony bee population	5586.24

Fig. 8. Example Excel worksheet for converting raw observer data into colony bee population. See 4.2., paragraph 4.

screenshot of an Excel datasheet demonstrating the conversion of raw data from two observers into colony bee population. There are two fictional colonies, each with 5 North American deep frames, each with two sides. Columns D and E show the respective visual estimates of two observers for percentage comb surface covered by bees, and column F is the mean of the two. Column G converts the mean percentage surface covered by bees into area (cm^2) covered by bees, using the surface area for one side of a North American deep frame from Table 2 (880 cm^2). Column H converts cm^2 bees to number of bees with the appropriate bee density (1.38 bees / cm^2). Finally, rows 12 and 23 sum the bees of each frame and side to yield colony bee population.

5. If investigators use colonies with different sized supers and frames it will be necessary to adjust calculations for the one-side surface area unique to each comb type. Bee density at full carrying capacity is consistent within North America or Europe, so it should be adequate to pick a density value from Table 2 that best fits one's local situation.

4.3. Special considerations for measuring adult bee populations in African bees

When it comes to African bees, the methods described in 4.2. must be modified to account for the fact that these bees immediately fly when disturbed.

1. Two observers plus a dedicated data recorder are recommended as before.
2. The key difference is that each observer makes a visual estimate of the percentage surface of the comb occupied by bees immediately as one of them withdraws it from the hive. Each observer is responsible for only one side of a comb.
3. To minimize loss of bees it is necessary to keep the hive intact as much as possible, working downward, removing the lid

then first measuring bees in honey supers, then bees in the brood chambers.

4. Raw data are converted into colony bee population using Table 2 and the methods given in Fig. 8.
5. An alternative to visually estimating bees on combs is to rapidly remove each frame and immediately shake the bees into a large plastic bag. The bag is weighed to determine total net weight (kg) of bees, then a fresh sample of *ca.* 20 g bees collected. The bees in the bag are returned to the colony and the sample taken to the lab where it is weighed, frozen, and the bees counted to determine g per bee. Dividing total net weight by g per bee gives colony bee population.

5. Measuring brood, honey, pollen, and other colony strength parameters during the experiment

5.1. Subjective mode

This section describes a subjective mode for reporting quantity of any kind of colony resource stored in cells: open brood, sealed brood, honey, or pollen. The methods are similar to those described for measuring colony bee populations subjectively in section 4.2. The only difference concerns whether the investigator wants to report the resource in units of area (cm^2) or number of cells. Authors have also reported resources in units of "frames," but this is unnecessarily ambiguous and makes it harder to compare data to other studies. As mentioned before, honey is traditionally reported as weight (kg), and it is best to use queen excluders and pre-weighed honey supers as described in paragraph 8, section 4.1. However, if the investigator wants to report honey occurring in combs alongside brood it may be necessary to report it in units of cm^2 or cells as described in this section. These methods for measuring brood, honey, or pollen are fundamentally the same for African bees, given that the investigator uses the region-specific multipliers in Table 2.

1. Estimates are carried out by no fewer than two observers, preferably each with a dedicated secretary who writes down numbers, or each fitted with an audio recorder.
2. A colony is opened and combs of bees sequentially removed. Each observer looks at one side of a comb, visually estimates the percentage of the comb surface occupied by the target resource, and records the number with the secretary or audio recorder. It is convenient to label frames 1-X, with each side indicated A or B. As described in the previous section, the observer is imaginatively sorting the resource into one contiguous mass and making a decision on the percentage surface area of the comb the contiguous resource occupies. This can be difficult in cases of spotty brood where widely

	G12				fx	=SUM(G2:G11)	
	A	B	C	D	E	F	G
1	colony	frame	side	observer 1	observer 2	mean of observers	cm2 covered by open brood
2	1	1	A	0.1	0.2	0.15	132
3	1	1	B	0.3	0.4	0.35	308
4	1	2	A	0.6	0.4	0.5	440
5	1	2	B	0.7	0.6	0.65	572
6	1	3	A	0.8	0.8	0.8	704
7	1	3	B	1	1	1	880
8	1	4	A	0.9	1	0.95	836
9	1	4	B	0.8	1	0.9	792
10	1	5	A	0.5	0.4	0.45	396
11	1	5	B	0.1	0	0.05	44
12						sum cm2 of colony open brood	5104
13	2	1	A	0	0	0	0
14	2	1	B	0.1	0.3	0.2	176
15	2	2	A	0.5	0.4	0.45	396
16	2	2	B	0.5	0.6	0.55	484
17	2	3	A	0.9	0.9	0.9	792
18	2	3	B	1	0.8	0.9	792
19	2	4	A	0.7	0.7	0.7	616
20	2	4	B	0.3	0.4	0.35	308
21	2	5	A	0	0.1	0.05	44
22	2	5	B	0	0	0	0
23						sum cm2 of colony open brood	3608

Fig. 9. Example Excel worksheet for converting raw observer data into cm² open brood. See 5.1., paragraph 3.

separated cells must be imaginatively grouped together. It is to be expected that the accuracy of this mode is best when target resources are massed together in convenient contiguous patches.

3. Fig. 9 is a screenshot of an Excel datasheet demonstrating the conversion of raw data from two observers into cm² of target resource, in this example open cells of brood. There are two fictional colonies, each with 5 North American deep frames, each with two sides. Columns D and E show the respective visual estimates of two observers for percentage comb surface occupied by brood, and column F is the mean of the two. Column G converts the mean percentage surface occupied by open brood into area (cm²), using the surface area for one side of a North American deep frame from Table 2 (880 cm²). Rows 12 and 23 sum the area of open brood for each colony.

4. If investigators use colonies with different sized supers and frames it will be necessary to adjust calculations for the one-side surface area unique to each comb type. This would affect the area conversion factor used in Fig. 9, column G.

	H2					fx	=(G2*3.7)	
	A	B	C	D	E	F	G	H
1	colony	frame	side	observer 1	observer 2	mean of observers	cm2 covered by open brood	cells open brood
2	1	1	A	0.1	0.2	0.15	132	488.4
3	1	1	B	0.3	0.4	0.35	308	1139.6
4	1	2	A	0.6	0.4	0.5	440	1628
5	1	2	B	0.7	0.6	0.65	572	2116.4
6	1	3	A	0.8	0.8	0.8	704	2604.8
7	1	3	B	1	1	1	880	3256
8	1	4	A	0.9	1	0.95	836	3093.2
9	1	4	B	0.8	1	0.9	792	2930.4
10	1	5	A	0.5	0.4	0.45	396	1465.2
11	1	5	B	0.1	0	0.05	44	162.8
12							sum cells of colony open brood	18884.8
13	2	1	A	0	0	0	0	0
14	2	1	B	0.1	0.3	0.2	176	651.2
15	2	2	A	0.5	0.4	0.45	396	1465.2
16	2	2	B	0.5	0.6	0.55	484	1790.8
17	2	3	A	0.9	0.9	0.9	792	2930.4
18	2	3	B	1	0.8	0.9	792	2930.4
19	2	4	A	0.7	0.7	0.7	616	2279.2
20	2	4	B	0.3	0.4	0.35	308	1139.6
21	2	5	A	0	0.1	0.05	44	162.8
22	2	5	B	0	0	0	0	0
23							sum cells of colony open brood	13349.6

Fig. 10. Example Excel worksheet for converting raw observer data into cells open brood. See 5.1., paragraph 5.

5. To report a resource in units of cells, it is necessary to multiply the cm² of resource by the average cell density per cm². This value varies by geography; conversion factors range from 3.7 - 4.7 (Table 2). It is advisable for investigators to determine this value for their local conditions. Figure 10 shows a modification of Fig. 9 taking the data from cm² open brood to cells of open brood, using a conversion factor of 3.7.

5.2. Computer-assisted digital image analysis

Computer-assisted digital image analysis can be used to directly measure surface area of comb occupied by bees or other colony resources such as open brood, sealed brood, or pollen. There are two kinds of data output: 1. direct surface measurements (cm² or dm²) of target (Cornelissen *et al.*, 2009) (section 5.4.) and; 2. ratio of surface target relative to total comb surface (Yoshiyama *et al.*, 2011) (section 5.5.). In the case of 1., it is possible to convert surface to units of bees or cells using conversion values in Table 2.

5.3. Technology and photographic considerations

1. A high-resolution camera (3648 x 2736, 10 megapixels, or higher) is preferred. We recommend a DSLR or similar camera. Compact cameras will work fine too, but it is unlikely that eggs and young brood will be visible.

2. Use image formats with the least amount of compression (resulting in the larger file size). For DSLR cameras this will be either RAW or TIFF format, and for compact cameras JPEG. As Image J software doesn't support the use of RAW images, conversion to either TIFF or JPEG files (uncompressed) is required. This can be done using free-ware such as Irfan-View (http://www.irfanview.com/).

3. Use of a tripod with a fixed distance to the frames is recommended. This makes image analysis easier and pictures more comparable.

4. Make sure the object (comb frame) completely covers the picture. This will result in the highest resolution and optimal lighting conditions.

5. It is advisable to use Shutter speed priority (indicated with an "S" on camera) with a setting of 1/125. Lower shutter speeds can result in blurred bees as their movement is caught on camera. In low light conditions use a flash or adjust ISO values.

6. Aperture settings are dependent on the type of lens used. For DSLR we suggest using a fixed 50 mm lens. These generally are affordable, sharp, fast (low f-value, e.g. f1:1.4), and have little distortion. Using these lenses the f-value should be above f4.5 for sharp pictures. For zoom lenses aim at an f-value of between f6.7 and f13.

7. Cloudy conditions can create low light levels; likewise the sun can obscure details due to high contrasts. Optimal results are

possible with a shaded location and a flash, but in our experience this is not practical due to terrain difficulties and limited battery life. Cloudy weather is no problem when using a fast lens. When it is sunny, it is best to take pictures with one's back to the sun.

5.4. Direct surface measurements of target

1. Photographic records of bees on combs will vary according to time of day and bee foraging activity. For this reason, it is important to control for this effect, either by limiting observations to a narrow time window on successive days, randomly assigning time of inspection such that day effect is equitably and randomly distributed across treatments, or closing hive entrances in the early morning until bees are counted. This constraint does not apply to cell-based resources such as brood, honey, or pollen.

2. Hives are lightly smoked, opened, and frames permanently labelled: frame 1 side A or B, frame 2 side A or B, and so forth.

3. Each frame is removed and photographed on each side in such a way that colony and frame labelling are recorded. It is preferable to use a custom-built holding mount where each comb is placed in a holder and the distance between the comb and camera fixed.

4. Combs are first photographed with bees. If additional comb resources are of interest, then the bees are brushed into a holding box and the comb photographed again to expose brood, honey, or pollen. It is important to avoid brushing bees back into the hive because this will affect the photographic bee record of subsequent frames. Eggs and 1 – 3 day old larvae may be hard to see and if these brood stages are the objective of the study it is preferable to apply digital cell / location recognition software.

Table 3. Colour- and number-coding of cell contents according to OECD (2007).

Brood stages / stores	Colour	Number
empty cells	Brown	0
eggs	Blue	1
young larvae (1-3 days)	Green	2
older larvae (4-6 days)	Red	3
pupae (capped brood)	Yellow	4
nectar	orange	5
pollen / bee bread	deep pink	6
dead larvae / pupae	dark salmon	7
not characterized (nc)	White	8

5. The digital photos are analysed using a computer program such as ImageJ, available free at http://rsbweb.nih.gov/ij/. Post-hoc, the photos are uploaded in the computer and analysed as diagrammed in Fig. 11.

6. The results of the calculated area is in dm^2 or cm^2 depending on the scale that has been set. To finish the analysis, the number of bees or cells are derived with Excel or a similar spreadsheet program using the expected density of bees per cm^2 or cells per cm^2 given in Table 2. Surface data from this digital analysis could be inserted into Column G in Figs 8-10.

5.5. Ratio of target surface relative to total comb surface area

This method, also applying ImageJ, yields the ratio between selected area and the total comb area. Prior to computer analysis, each digital image of a comb side is digitally edited by the investigator to delineate comb areas of target resource, e.g. capped brood area, open brood, pollen, or honey with unique identifying colours. In Fig. 12, this application is outlined.

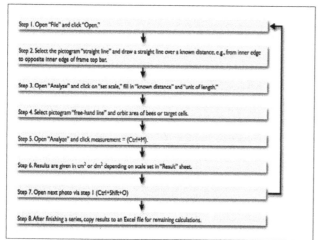

Fig. 11. Outline of the method of Cornelissen *et al.* (2009). Flow chart of computer assisted image analysis applying ImageJ software. Step 2 can be skipped by making the photos in a fixed position at which the distance between camera and frame is constant.

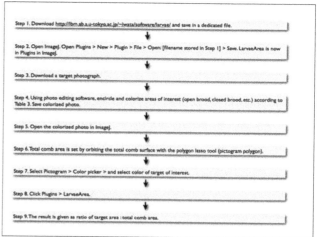

Fig. 12. Outline of method of Yoshiyama *et al.* (2011) for determining ratio of target surface : total comb surface. It is used with the OECD colour codes in Table 3.

5.6. Flight activity

1. Bee flight activity can be monitored visually at hive entrances to gain a relative measure of colony foraging effort. To control for between-colony variation due to time of day, the investigator should: 1. limit observations to days and time of day with good flight conditions; 2. randomize the numeric order in which colonies are measured; 3. measure all colonies within a relatively narrow window of hours, and; 4. limit colony observations to the same time window over successive days.

2. Two observers sit to the side of a colony, each positioned well enough to the side to avoid obstructing the flight of the bees. Each observer has a hand-held counting device and one keeps time.

3. For one 15 min counting episode, each observer counts and records the number of bees exiting the colony (but see below). Exiting bees are simpler to count because returning foragers land with less predictability; some directly into the entrance, others onto the front of the hive.

4. The mean of the two observers is derived and the data reported as exiting foragers per min.

5. Investigators may want to focus on returning, instead of exiting foragers, especially if pollen foraging is a parameter of interest. In these cases observers need to count foragers returning with, and without, corbicular pollen loads in order to derive proportion of foragers collecting pollen.

5.7. Comb construction

This section draws upon methods of Matilla and Seeley (2007).

1. This metric can be collected only during times of rich nectar flow when bees can draw out new comb.

2. Colonies are each provided a hive body provisioned with ten new frames; 5 combless and 5 with wax foundation, alternating. The use of alternating frames of foundation encourages bees to build combs in compliance with the removable-frame parallel orientation of Langstroth equipment.

3. Measures of area of comb constructed (both natural and on the foundation) by each colony can begin two days after establishment and weekly thereafter until all comb is finished or the nectar season is over. Comb area on both sides of every frame is determined and summed by colony, either with the Objective mode (4.1.) or Subjective mode (5.1.). Inexperienced observers will need to be trained to discriminate differences between natural comb and the imprinted beeswax foundation.

5.8. Queen cell production

This measure can be determined while the colony is being opened and measured for other strength metrics. It should be done after bee population measures have been taken. Every brood comb is shaken free of bees and examined carefully for the presence of queen cells provisioned with royal jelly and a larva. The cells are counted and then cut out for two reasons: 1. to prevent swarming (unless swarming is a variable of interest) and; 2. to prevent redundant duplicate observations on subsequent samples. For each block of the experiment this variable can be reported as sum of queen cells constructed per colony.

5.9. Drone brood production

This measure is best taken in spring or early summer when drones are being actively reared. It is nothing more than an extension of the Objective mode (4.1.) or Subjective mode (5.1.) limiting observations to drone cells filled with larvae or capped with pupae. Values for drone cells per cm^2 for European bees range from 2.3 (J A Berry, University of Georgia, USA, pers. obs.) and 2.6 (Dadant, 1963); a good estimate for African races is 3.0 (Buco *et al.*, 1987; Hepburn, 1983).

References

BERRY, J A; DELAPLANE, K S (2001) Effects of comb age on honey bee colony growth, brood survivorship, and adult mortality. *Journal of Apicultural Research* 40: 3-8.

BERRY, J A; OWENS, W B; DELAPLANE, K S (2010) Small-cell comb foundation does not impede varroa mite population growth in honey bee colonies. *Apidologie* 41: 41-44. http://dx.doi.org/10.1051/apido/2009049

BUCO, S M; RINDERER, T E; SYLVESTER, H A; COLLINS, A M; LANCASTER, V A; CREWE, R M (1987) Morphometric differences between South American Africanized and South African (*Apis mellifera scutellata*) honey bees. *Apidologie* 18(3): 217-222.

BURGETT, M; BURIKAM, I (1985) Number of adult honey bees (Hymenoptera: Apidae) occupying a comb: a standard for estimating colony populations. *Journal of Economic Entomology* 78: 1154-1156.

COBEY, S W; TARPY, D R; WOYKE, J (2013) Standard methods for instrumental insemination of *Apis mellifera* queens. In *V Dietemann; J D Ellis; P Neumann (Eds) The COLOSS* BEEBOOK, *Volume I: standard methods for* Apis mellifera *research. Journal of Apicultural Research* 52(4): http://dx.doi.org/10.3896/IBRA.1.52.4.09

CORNELISSEN, B; BLACQUIÈRE, T; VAN DER STEEN, J (2009) Estimating honey bee colony size using digital photography. In *Proceedings of 41st International Apicultural Congress, Montpellier, France.* p. 48.

DADANT, H C (1963) Beekeeping equipment. In *R A Grout (Ed.) The hive and the honey bee.* Dadant & Sons; Hamilton, Illinois, USA. pp 219-240.

DELAPLANE, K S; HARBO, J R (1987) Effect of queenlessness on worker survival, honey gain and defence behaviour in honey bees. *Journal of Apicultural Research* 26: 37-42.

DELAPLANE, K S; HOOD, W M (1997) Effects of delayed acaricide treatment in honey bee colonies parasitized by *Varroa jacobsoni* and a late season treatment threshold for the south-eastern United States. *Journal of Apicultural Research* 36: 125-132.

DELAPLANE, K S; HOOD, W M (1999) Economic threshold for *Varroa jacobsoni* Oud. in the south-eastern USA. *Apidologie* 30: 383-395.

DELAPLANE, K S; BERRY, J A; SKINNER, J A; PARKMAN, J P; HOOD, W M (2005) Integrated pest management against *Varroa destructor* reduces colony mite levels and delays economic threshold. *Journal of Apicultural Research* 44: 117- 122.

DELAPLANE, K S; ELLIS, J D; HOOD, W M (2010) A test for interactions between *Varroa destructor* (Acari: Varroidae) and *Aethina tumida* (Coleoptera: Nitidulidae) in colonies of honey bees (Hymenoptera: Apidae). *Annals of the Entomological Society of America* 103(5): 711-715. http://dx.doi.org/10.1603/AN09169

ELLIS, J D; DELAPLANE, K S; HOOD, W M (2001) Efficacy of a bottom screen device, Apistan, and Apilife VAR in controlling *Varroa destructor. American Bee Journal* 141: 813-816.

GRECO, M K (2010) *Imaging techniques for improved bee management.* ALP Technical-Scientific Information Series 534. 38 pp.

GRIS, A (2002) Efecto de las diferencias en población entre colonias de abejas (*Apis mellifera* L.) con una y dos reinas en el comportamiento productivo y de pecoreo de sus obreras. MSc thesis, National University of Mexico, Mexico D.F. 47 pp.

GUZMÁN-NOVOA, E; CORREA BENÍTEZ, A; MONTAÑO, L G E; GUZMÁN NOVOA, G (2011) Colonization, impact, and control of Africanized honey bees in Mexico. *Revista Veterinaria México* 42 (2): 149-178.

HARBO, J R (1983) Effect of population size on worker survival and honey loss in broodless colonies of honey bees, *Apis mellifera* L. (Hymenoptera: Apidae) *Environmental Entomology* 12: 1559-1563.

HARBO, J R (1986) Effect of population size on brood production, worker survival and honey gain in colonies of honey bees. *Journal of Apicultural Research* 25: 22-29.

HARBO, J R (1988) Effect of comb size on population growth of honey bee (Hymenoptera: Apidae) colonies. *Journal of Economic Entomology* 81: 1606-1610.

HARBO, J R (1993) Worker-bee crowding affects brood production, honey production, and longevity of honey bees (Hymenoptera: Apidae). *Journal of Economic Entomology* 86: 1672-1678.

HEPBURN, H R (1983) Comb construction by the African honey bee, *Apis mellifera adansonii. Journal of the Entomological Society of South Africa* 46: 87-101.

IMDORF, A; BUEHLMANN, G; GERIG, L; KILCHENMANN, V; WILLE, H (1987) Überprüfung der Schätzmethode zur ermittlung der Brutfläche und der anzahl Arbeiterinnen in freifliegenden Bienenvölkern. *Apidologie* 18: 137-146.

IMDORF, A; GERIG, L (2001) Course in determination of colony strength. Swiss Federal Dairy Research Institute, Liebefeld CH-3003 Bern Switzerland (after L Gerig, 1983. Lehrgang zur Erfassung der Volksstärke). *Schweiz Bienen-Zeitung* 106: 199-204.

MATTILA, H R; SEELEY, T D (2007) Genetic diversity in honey bee colonies enhances   productivity and fitness. *Science* 317: 362-364. http://dx.doi.org/10.1126/science.1143046

MAYACK, C; NAUG, D (2009) Energetic stress in the honey bee *Apis mellifera* from *Nosema ceranae* infection. *Journal of Invertebrate Pathology* 100: 185-188.

NIGHTINGALE, J M; ESAIAS, W E; WOLFE, R E; NICKESON, J E; MA, P L A (2008) Assessing honey bee equilibrium range and forage supply using satellite-derived phenology. Geoscience and Remote Sensing Symposium, IGARSS 2008. *IEEE International* (3): 763-766. http://dx.doi.org/10.1109/IGARSS.2008.4779460

OECD (2007) Guidance document on the honey bee (*Apis mellifera* L.) brood test under semi-field conditions. *Series on testing and assessment, Guidance Document* 75, ENV/JM/MONO(2002)22.

SKINNER, J J; PARKMAN, J P; STUDER, M D (2001) Evaluation of honey bee miticides, including temporal and thermal effects on formic acid gel vapours, in the central south-eastern USA. *Journal of Apicultural Research* 40: 81-89.

STRANGE, J P; SHEPPARD, W S (2001) Optimum timing of miticide applications for control of *Varroa destructor* (Acari: Varroidae) in *Apis mellifera* (Hymenoptera: Apidae) in Washington state, USA. *Journal of Economic Entomology* 94(6): 1324-1331. http://dx.doi.org/10.1603/0022-0493-94.6.1324

YOSHIYAMA, M; KIMURA, K; SAITOH, K; IWATA, H (2011) Measuring colony development in honey bees by simple digital analysis. *Journal of Apicultural Research* 50: 170-172. http://dx.doi.org/10.3896/IBRA.1.50.2.109

Journal of Apicultural Research 52(4): (2013)
DOI 10.3896/IBRA.1.52.4.07

REVIEW ARTICLE

Standard methods for research on *Apis mellifera* gut symbionts

Philipp Engel[1†]**, Rosalind R James**[2*]**, Ryuichi Koga**[1,4]**, Waldan K Kwong**[1]**, Quinn S McFrederick**[3] **and Nancy A Moran**[1]

[1]Ecology and Evolutionary Biology, Yale University, New Haven, CT, USA.
[2]USDA-ARS Pollinating Insects Research Unit, Logan, UT, USA.
[3]Integrative Biology, University of Texas, Austin, TX, USA.
[4]Bioproduction Research Institute, National Institute of Advanced Industrial Science and Technology, Tsukuba, Ibaraki 305-8566, Japan.

Received 1 March 2013, accepted subject to revision 11 April 2013, accepted for publication 20 May 2013.

†All authors are in alphabetical order.

*Corresponding author: Email: rosalind.james@ars.usda.gov

Summary

Gut microbes can play an important role in digestion, disease resistance, and the general health of animals, but little is known about the biology of gut symbionts in *Apis mellifera*. As part of the *BEEBOOK* series describing honey bee research methods, we provide standard protocols for studying gut symbionts. We describe non-culture-based approaches based on Next Generation Sequencing (NGS), methodology that has greatly improved our ability to identify the microbial communities associated with honey bees. We also describe Fluorescent *In Situ* Hybridization (FISH) microscopy, which allows a visual examination of the microenvironments where particular microbes occur. Culturing methods are also described, as they allow the researcher to isolate particular bacteria of interest for further study or gene identification, and enable the assignment of particular functions to particular gut community members. We hope these methods will help others advance the state of knowledge regarding bee gut symbionts and the role they play in honey bee health.

Métodos estandar para investigar simbiontes intestinales de *Apis mellifera*

Resumen

Los microbios intestinales pueden jugar un papel importante en la digestión, la resistencia a las enfermedades, y la salud general de los animales, pero se conoce poco sobre la biología de los simbiontes intestinales en *Apis mellifera*. Como parte de la serie BEEBOOK que describe los métodos de investigación en la abeja, ofrecemos protocolos estándar para el estudio de simbiontes intestinales. Se describen métodos no basados en cultivos sino sobre la base de la secuenciación de nueva generación (NGS según sus siglas en inglés), metodología que ha mejorado en gran medida nuestra capacidad para identificar las comunidades microbianas asociadas con la abeja de la miel. También describimos la microscopía de hibridación in situ fluorescente (FISH), la cual permite un examen visual de los microambientes donde viven microbios particulares. También se describen métodos de cultivo, que permiten al investigador aislar bacterias de interés particular para posteriores estudios o para la identificación de genes, y permitir asignar funciones particulares a determinados miembros de la comunidad intestinal. Esperamos que estos métodos ayudarán a otros a avanzar en el estado del conocimiento sobre simbiontes intestinales de abejas y el papel que desempeñan en la salud de las abejas de la miel.

Footnote: Please cite this paper as: ENGEL, P; JAMES, R R; KOGA, R; KWONG, W K; MCFREDERICK, Q S; MORAN, N A (2013) Standard methods for research on *Apis mellifera* gut symbionts. In *V Dietemann; J D Ellis; P Neumann (Eds) The COLOSS* BEEBOOK, *Volume I: standard methods for* Apis mellifera *research.* *Journal of Apicultural Research* 52(4): http://dx.doi.org/10.3896/IBRA.1.52.4.07

西方蜜蜂肠道共生体研究的标准方法

摘要

肠道微生物在动物的消化、抗病性和常规健康中起重要作用。然而，目前对于西方蜜蜂肠道共生体的生物学知之甚少。作为*BEEBOOK*蜜蜂研究方法系列的一部分，我们为研究肠道共生体提供了标准方案。介绍了基于新一代测序（NGS）而免于生物培养的研究方法，该方法极大地提高了我们鉴定蜜蜂相关微生物群落的能力。描述了荧光原位杂交（FISH）显微技术，该技术实现了对一些特殊微生物出现的微环境进行目视检测。同时，也描述了一些培养方法，这些方法使得研究人员可分离出特定感兴趣的细菌，以供进一步研究或进行基因鉴定，并使得将特定的功能与特定肠道菌群相联系成为可能。我们希望这些方法能帮助大家提升对于蜜蜂肠道共生体及其在蜜蜂健康方面所起作用的了解。

Keywords: COLOSS, *BEEBOOK*, fluorescent *in situ* hybridization, gut symbionts, methods, microbial culture, next generation sequencing, honey bee

Table of Contents

1. Introduction

In honey bees, the gut is the primary location for digestion and food processing, as well as the site of infection for a variety of pathogens, including *Paenibacillus larvae* (de Graaf *et al.*, 2013), *Ascosphaera apis* (Jensen *et al.*, 2013) *Nosema ceranae* (Fries *et al.*, 2013), and probably many of the honey bee viruses (de Miranda *et al.*, 2013). Biologists have increasingly recognized that gut microorganisms play a beneficial role in many aspects of the health of animals, animals as

diverse as mammals and insects. The role of symbiotic gut microbes in digestion, resistance to infectious disease, and the general health of honey bees, both individual bees and the colony at large, is an intriguing area of research where much is still to be learned. In this chapter we present a number of basic protocols for investigating the microbial communities found in honey bee guts.

Traditional microbiological studies, from the time of Pasteur, have relied on the axenic (pure) isolation of individual microbes, which were then characterized based on their metabolic, biochemical, and morphological characteristics. These culture-based studies remain the foundation of microbiology. However, for most environmental samples, estimates of microbial density based on microscopic counts of cell numbers tend to be far higher than estimates based on colony-forming units on culture plates (Staley and Konopka, 1985). Furthermore, it has long been known that many microbes cannot be cultured in the laboratory, or require specialized conditions yet to be discovered, but no real solution to this problem existed for environmental microbiology until molecular sequencing technology became available. Using sequencing techniques, Rappé and Giovannoni (2003) showed that the microorganisms readily cultured from a given environment are only a tiny subset of the species actually living there. Typically, only about 1% of bacteria from a given habitat will grow in culture (Staley and Konopka, 1985). With attention to specific aspects of the culture media and atmospheric conditions, more organisms might grow, but consistently, many organisms sampled from most environments do not appear in lab cultures (Stevenson *et al.*, 2004). Therefore, good estimates of "what is really there" in microbiology depend on non-culture-based studies. Non-culture-based studies may be followed by culturing efforts so that the microbes of interest can be isolated, and thus better described chemically and morphologically, and to conduct experiments with them. In addition, molecular methods such as fluorescent *in situ* hybridization (FISH) can be used to locate specific microorganisms in their precise locations within a sample; for example, FISH can be used to identify where certain microbes occur in the honey bee gut, or where infection takes place for a pathogen.

Numerous researchers have performed studies on organisms cultured from bee guts and the hive, documenting a variety of metabolic and functional activities (Gilliam and Prest, 1972, 1987; Gilliam and Valentine, 1974; Gilliam *et al.*, 1974; Gilliam, 1978; Evans and Armstrong, 2006). However, non-culture-based surveys have presented a contrasting view of the dominant members of the honey bee gut microbiota, revealing that readily cultured, fully aerobic organisms comprise only a small portion of the diversity of microbes present. A set of eight major taxa dominate the honey bee gut environment, and these fall within the Gammaproteobacteria, Betaproteobacteria, Alphaproteobacteria, Lactobacillales, and Actinomycetes. These eight bacterial taxa, which correspond to the typical definition of bacterial species, have been found in *A. mellifera* worldwide (Jeyaprakash *et al.*, 2003; Mohr and Tebbe, 2006; Babendreier *et al.*, 2007; Cox-Foster *et al.*, 2007; Martinson *et al.*, 2011; Cornman *et al.*, 2012; Engel *et al.*, 2012; Li *et al.*, 2012; Martinson *et al.*, 2012; Moran *et al.*, 2012; Sabree *et al.*, 2012; Tian *et al.*, 2012). Close relatives of some of these taxa have been found in other *Apis* species in Asia (Ahn *et al.*, 2012; Li *et al.*, 2012) and in many species of bumble bees (*Bombus*) (Koch and Schmid-Hempel, 2011, 2012; Koch *et al.*, 2012, 2013). These surveys used a variety of sequencing methodologies, yet consistently retrieved a similar set of organisms.

The primary molecule currently used for surveying microbial diversity and verifying taxonomic identities is the small subunit ribosomal RNA, which is referred to as the 16S rRNA in Bacteria and Archaea, and the 18S rRNA in Eukaryota. This molecule is present in all cells and provides a molecular label for a particular species or taxon, and can be compared against public databases to determine whether a sampled sequence corresponds to previously studied organisms (McDonald *et al.*, 2012). The microbial community in a bee gut can be determined by extracting the DNA, using targeted PCR to amplify the bacterial rRNA genes present, followed by high throughput sequencing technologies such as Illumina, 454 or others (Sogin *et al.*, 2006; Tringe and Hugenholtz, 2008), often called next-generation sequencing (NGS) technologies. Bioinformatic searches are then used to compare the resulting sequences to those previously identified and stored in publically available databases. We present protocols for these methods.

Sequencing technologies are becoming increasingly cost-effective while also yielding improved data quality, primarily through an increase in the length of the sequencing read. Longer and more accurate sequences give higher quality information for identifying microbial taxa. Because these technologies are evolving rapidly, we give a generalized overview for the methods, recognizing that details of techniques will change as these technologies, and the methods for analysing the results, evolve. Our recommendations are for specific techniques that are likely to remain static for some time, such as DNA extraction, PCR, and a general approach for analysing microbial communities.

The 16S rRNA can also be used to design *in situ* hybridization probes. Many ribosomes are present in the cytoplasm of each cell, and probes corresponding to diagnostic regions of the rRNA gene can be used to selectively label and visualize specific cells containing the corresponding RNA sequences. In addition, the 16S rRNA can be used to identify bacteria isolated using culture-based methods. DNA can be extracted from isolated colonies of the bacteria, amplified using selective PCR, followed by sequencing. Again, a bioinformatic search can then be used to compare the unknown sequence to those previously identified and stored in publically available databases.

Metagenomics includes several approaches useful for studying functions of gut biota, but is not included among our listed protocols. The original use of the word "metagenomics" referred to cloning relatively large fragments of DNA from community DNA samples, and then attempting to screen these cloned fragments for functional activities (Handelsman, 2004). A primary limitation of this approach is that genes underlying many functional activities maybe present but not expressed, although metagenomics can be an effective method for detecting certain functions, such as antibiotic resistance (Tian *et al.*, 2012). Since major members of the bee microbiota can be cultured, the study of cultured isolates may offer a more robust approach for finding functional capabilities. Currently, metagenomics generally refers to the use of deep sequencing of total genomic contents of a microbial community to identify and compare the prevalence of functional genes, such as those for enzymes associated with cellulose or pectin degradation (e.g. Warnecke *et al.,* 2007; Brulc *et al.,* 2009; Engel *et al.,* 2012). Typically, metagenomic sequencing is combined with a 16S rRNA gene sequencing approach, with the former being used for inferring functional capabilities, and the latter used to infer community membership and diversity. The study by Engel *et al.* (2012) is the only example to date of using deep sequencing metagenomics to understand functions of the bee gut microbiota. Metagenomic methods encompass rapidly evolving approaches to environmental microbiology. Since standard methodologies are still under development at this time, these were not included here.

We provide here standard protocols for studying bee gut bacteria, including NGS methods for surveying bacterial diversity, FISH microscopy to precisely locate where these microorganisms occur in the insect gut, and culturing methods for known gut symbionts. Although we focus on bacteria, our methods can be extended to other microorganisms and viruses, with appropriate changes in the oligonucleotide primers used for diversity surveys and FISH.

2. Bacterial community analysis using next-generation sequencing (NGS)

Currently, one of the easiest and more cost effective tools available for characterizing the microbial communities associated with honey bee guts is sequencing the diversity of the 16S rRNA gene in honey bee guts using NGS technology. NGS surveys have provided insights into the composition of bee gut-associated bacterial communities, symbiont host-specificity, and conditions conducive to the co-evolutionary dynamics of bees and their associated microbes (Ahn *et al.,* 2012; Martinson *et al.,* 2012; McFrederick *et al.,* 2012; Moran *et al.,* 2012; Newton and Roeselers, 2012).

2.1. Extraction, PCR, and sequencing

2.1.1. Extraction of bacterial community DNA
Extraction is the first step towards identifying the microbial communities present in a sample (Fig. 1). Extraction methods for diverse bacterial communities must include a method to disrupt the cell walls of spores and the more recalcitrant bacteria (Marmur, 1961). We recommend using a method that includes bead-beating because that is one of the best methods for cell disruption, and it has been the method of choice for recent bee-associated NGS surveys (Mattila *et al.,* 2012; McFrederick *et al.,* 2012; Moran *et al.,* 2012).

1. First determine what you consider to be your sample: the entire gut or part of the gut. It is very important to work in a clean, sterile environment using sterile materials and tools. Unless you intend to include the microbial community occurring on the insect cuticle, it is imperative to disinfect and clean the surface of the bees before dissections. Microbes and their nucleic acids can be removed from bee cuticle using a 1% aqueous solution of chlorine (using either sodium dichloro -s-triazinetrione-dihydrate (swimming pool or spa chlorine) or sodium hyperchlorite (bleach)). Soak each bee for at least 2 min but not more than 7 min., then rinse three times in sterile, purified water. This can be done in a 24 well plate on a shaker table, or in a series of 1.5 ml disposable centrifuge tubes with gentle mixing on a vortex mixer. It is important that all of the chlorine be removed from the bees prior to dissection because chlorine degrades DNA and thus can inhibit the PCR reactions.

2. Once you obtain your sample, place it in a sterile 2 ml microcentrifuge tube with a sterile 5 mm stainless steel bead (QIAGEN, Valencia, CA), 500 µl of 0.1 mm glass beads (Scientific Industries, Inc.;Bohemia, NY, USA), and 500 µl RLT buffer (QIAGEN; Valencia, CA, USA) with 10 µl of β-mercaptoethanol per ml buffer. Note that β-mercaptoethanol (including the used buffer) must be handled as hazardous waste.

3. Run in tissue lyser or bead beater at 30 Hz for 5 min.

4. Centrifuge samples briefly to separate the beads and the buffer. If a foam layer has developed, it can be eliminated with a longer centrifugation period (several minutes at 5000 rpm at 5°C).

5. In a new tube add 100 µl of 100% ethanol and 100 µl of supernatant (200 µl of 100% ethanol and 200 µl of supernatant may be used for small samples, but it is important to treat all samples within a study the same so that the data are comparable). Gently mix using a vortex mixer. Centrifuge briefly, if needed, to consolidate the sample, which may get spread around in the tube.

6. For DNA recovery, apply the sample from step 5 to a QIAamp mini spin column and follow the tissue protocol of QIAamp DNA minikit (QIAGEN, Valencia, CA) (starting at the step where the supernatant is applied to the spin column, Step 7 on page 35 in the handbook published 4/2010).

7. Elute DNA in 30 µl of sterilized nanopure water, or in the provided QIAGEN buffer AE if long-term storage at -20 is desired.

8. Quantify DNA and adjust to a standard concentration across all samples. We have used 20 ng/µl, but higher concentrations may be desirable with small samples.

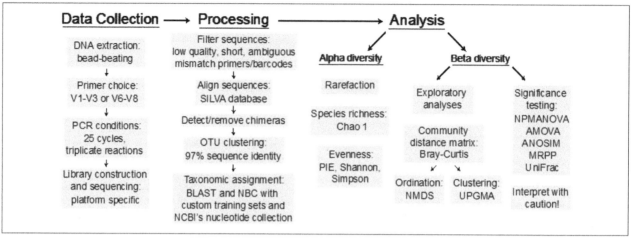

Fig. 1. Suggested protocol for data collection, processing, and analysis of bacterial community data using 16S rRNA next generation sequencing (NGS).

2.1.2. Primer choice and 16S rRNA regions.

Several factors influence primer choice. Universality of the primer across the Bacteria is the primary concern, and for this we recommend using the 16S rRNA gene as the target. The region of the 16S rRNA gene that the primers amplify also deserves careful consideration. Different regions can return different results, and therefore, studies that use different regions of the 16S rRNA gene are not directly comparable (Engelbrektson *et al.*, 2010). We recommend two regions that have recently been used successfully for identifying the bee-associated microbiome: V1-V3 (Ahn *et al.*, 2012; Mattila *et al.*, 2012; McFrederick *et al.*, 2012) and V6-V8 (Martinson *et al.*, 2012; Moran *et al.*, 2012). V1 is particularly useful for discriminating *Lactobacillus* species (McFrederick *et al.*, 2013). Both of these regions provide good taxonomic resolution of bee-associated bacteria, but universal primers associated with these regions can also amplify eukaryotic DNA (McFrederick, unpublished data; Moran *et al.*, 2012). As NGS read lengths increase, obtaining overlapping paired-end reads (i.e. sequencing 250 bases from each end of a 400 bp amplicon) will become feasible, and additional variables will factor into primer choice.

1. Suggested V1-V3 region primer pair.
   Forward primer, 28F: 5'– GAGTTTGATCNTGGCTCAG –3'
   Reverse primer, 519R: 5'–GTNTTACNGCGGCKGCTG –3'.
2. Suggested V6-V8 region primer pair.
   Forward primer, 926F: 5'– AAACTYAAAKGAATTGACGG –3'
   Reverse primer, 1392R 5'– ACGGGCGGTGTGTRC –3'

2.1.3. PCR conditions

PCR bias is known to occur in multiple template PCR amplifications, leading to a situation where the relative concentrations of templates are not equal to the relative concentrations of products. To minimize PCR bias we recommend using no more than 25 PCR cycles,

performing reactions in triplicate, and then combining triplicate reactions before sequencing.

1. Use HotStarTaq Plus Master Mix Kit (QIAGEN, Valencia, CA) for 50 µl reaction.
   a. 25 µl HotStarTaq plus master mix (2x).
   b. 2 µl each of 10 µM forward and reverse primer (0.4 µM final concentration).
   c. 5 µl DNA template (100 ng).
   d. 16 µl sterilized nanopure water.
2. PCR conditions.
   a. 95°C for 5 minutes.
   b. 25 cycles of: 94°C for 30 seconds, 60°C for 40 seconds, 72°C for 1 minute.
   c. Final elongation of 72°C for 10 minutes.
3. PCR purification
   a. Remove unincorporated primers, dNTPs, etc., using a PCR purification kit, such as Agencourt AMPure XP (Agencourt Bioscience Corporation, MA) or MinElute PCR columns (QIAGEN, Valencia, CA).

2.1.4. Library construction

Library construction is the process of preparing DNA for sequencing on the NGS platform of choice. This process typically consists of ligating platform-specific adapters to the flanking ends of DNA fragments. Kits are typically available for this step. For 16S rRNA gene sequence (henceforth 16S amplicon) bacterial community analyses, sequencing adapters and barcodes for sample identification are designed into 16S amplicon universal primers and added in the PCR step above. As using paired-end reads to analyse bacterial communities becomes more commonplace, primer design and library construction will continue to evolve (Degnan and Ochman, 2012).

2.1.5. Sequencing

Only a few laboratories maintain their own next generation sequencers, and subsequently, most studies outsource their sequencing to core facilities or off site laboratories specializing in NGS. Potentially this will change as smaller and cheaper version of sequencers become available. However, sequencing protocols are usually carried out by service facilities, and so are not detailed here.

2.2. Quality filtering and data analysis

Entry into the analysis of large data sets produced by NGS may appear intimidating at first, but several software pipelines are available to assist you in these analyses. The most commonly used pipelines (based on the number of citations) are QIIME (Caporaso *et al.*, 2010b) and mothur (Schloss *et al.*, 2009), but other useful software, such as the Joint Genome Institute's PyroTagger (Kunin and Hugenholtz, 2010), also exist. QIIME and mothur, with a couple of exceptions that we note below, have all of the following analyses available in their respective pipelines, and both pipelines are freely available. Both of these packages take NGS files as input and additionally take user-defined files (mapping files in QIIME or oligonucleotide files in mothur) that contain:

1. primer sequences.
2. sample names.
3. barcode sequences specific to each sample.
4. optional metadata for QIIME mapping file.

These files allow the multiplexed reads to be assigned to the sample from which they came and allow for alpha and beta diversity analyses downstream in the pipeline. Some popular analyses, such as UniFrac (Lozupone and Knight, 2005; Hamady *et al.*, 2009), are newer methods designed to explicitly deal with 16S amplicon community surveys, while other commonly used analyses have a long history of use in community ecology.

2.2.1. Quality filtering

Quality control of NGS data will coevolve with sequencing technology. Currently, the following steps are recommended:

1. Check for sequence errors. Quality filtering steps are conducted in any of the above pipelines and begin with the analysis of raw sequence to remove sequences with:
    a. low quality scores (sequences with average quality scores less than 25),
    b. short length (sequences less than 50% of the expected sequence length, e.g. for 454 titanium data we recommend removing reads < 200 bases),
    c. ambiguous base calls (for 454 data, reads with ambiguous base calls are correlated with other errors and should be discarded (Huse *et al.*, 2007), for other platforms, sequences can be trimmed at the first ambiguous base

call, then discarded if they no longer meet the specified length criteria),
    d. mismatches to the primer sites or barcodes, i.e. reads where there is a sequencing error in the primer sequence or that do not match any of the barcodes used to label which sample each read came from,
    e. for 454 data, sequences that have runs of 6 or more of the same nucleotide (homopolymers), which is a characteristic error for 454 sequencing (Schloss *et al.*, 2011).

2. Check alignment. As an additional quality control step and to prepare data for downstream analyses, align the sequences against a 16S rRNA database such as SILVA (Quast *et al.*, 2013). Sequences that fall out of the alignment window are discarded. We recommend NAST based aligners (Schloss, 2009; Caporaso *et al.*, 2010a) as they create high quality alignments when used with the SILVA database.

3. Check for chimeric sequences. As long as microbial community analyses continue to rely on PCR methods, chimeric sequences will remain a problem for which to monitor. Chimeras occur when sequences from two separate templates are combined during the reaction, and chimeras thereby artificially inflate diversity estimates. Several chimera checkers have been developed (e.g. Edgar *et al.*, 2011; Haas *et al.*, 2011), and although some chimeras will remain undetectable, the rate can be greatly reduced with these tools. As honey bee gut symbionts are still not well represented in the curated reference databases, we currently recommend using UCHIME in *de novo* mode, which does not require a reference database (Edgar *et al.*, 2011).

2.2.2. Identifying operational taxonomic units (OTUs)

The next step is to cluster sequences into OTUs. No single sequence identity threshold exists for matching bacterial species, as defined by phenotype (Schloss and Westcott, 2011). Therefore, the standard practice is to use ≥ 97% sequence identity to cluster sequences into OTUs. Several programs are available to cluster sequences into OTUs including PyroTagger (Kunin and Hugenholtz, 2010), CD-HIT (Huang *et al.*, 2010), UCLUST (Edgar, 2010), and mothur (Schloss and Westcott, 2011).

Undetected chimeras can persist as spurious OTUs, and because the chimera rate and sequencing errors are compounded with sequencing depth in NGS, it is important to randomly subsample each community using a standardized number of sequences (Schloss *et al.*, 2011). We recommend subsampling a minimum of 1000 sequences, or the highest number that will avoid discarding too many samples. Additionally, many studies exclude rare and/or singleton OTUs from further analysis. One approach is to analyse the data twice, once with singleton OTUs included and then again with them excluded, and report the results of both analyses in publication supplemental files (e.g. Martinson *et al.*, 2012).

2.2.3. Taxonomic assignment of OTUs

The most commonly used software programs for taxonomic assignment of OTUs are the Ribosomal Database Project's Naïve Bayesian Classifier (NBC, (Wang *et al.*, 2007)) and the Basic Local Alignment Search Tool (BLAST, (Altschul *et al.*, 1990)). The database against which sequences are compared, however, appears to be more important than the tool used to assign sequences to taxonomy (Newton and Roeselers, 2012; Sabree *et al.*, 2012). For example, many of the 16S rRNA gene sequences from uncultured and undescribed honey bee gut symbionts have not been incorporated into highly curated databases such as SILVA and greengenes, and use of these databases may result in inconsistent assignment of *A. mellifera* gut symbionts (Newton and Roeselers, 2012). NBC classifications using custom training sets that explicitly include representatives of *A. mellifera* gut symbionts will obviate such inconsistencies (Newton and Roeselers, 2012) as will searches against the complete nucleotide collection at NCBI (Sabree *et al.*, 2012). Two of the *A. mellifera* core gut symbionts, *Gillamella apicola* and *Snodgrasella alvi*, have recently been cultured and formally described (Kwong and Moran, 2013), and are presently represented in the greengenes, SILVA, and ribosomal database project alignments. As formal description of the *A. mellifera* microbiota expands, the highly curated alignments will more accurately classify these sequences. Searches against NCBI's nucleotide collection, however, will continue to provide insight into uncultured sequences that may be closely related to the query sequence and are recommended.

2.2.4. Alpha diversity

Estimators of within-community (alpha) diversity have been proposed and refined for decades (Whittaker, 1972; Magurran, 2004). For NGS surveys of bacterial symbionts, three measurements of alpha diversity are commonly used: rarefaction curves, species richness estimators (often in conjunction with rarefaction curves), and community diversity indices. Bee-associated bacterial surveys commonly report all three measures, but it should be noted that the abundance of 16S amplicon sequences can be a poor predictor of relative bacterial abundances (Amend *et al.*, 2010). Estimates of within and between community diversity that rely on 16S amplicon sequence abundance should therefore be interpreted with caution. Recently, a method to account for 16S gene copy number in estimating bacterial abundance was developed (Kembel *et al.*, 2012), which may help improve the accuracy of bacterial diversity measurements based on 16S amplicons.

1. Species richness estimators estimate the total number of species present in a community. The *Chao 1* index is commonly used, and is based upon the number of rare classes (i.e. OTUs) found in a sample (Chao, 1984):

$$S_{est} = S_{obs} + \left(\frac{(f_1)^2}{2f_2}\right)$$

where S_{est} is the estimated number of species, S_{obs} is the observed number of species, f_1 is the number of singleton taxa (taxa represented by a single read in that community), and f_2 is the number of doubleton taxa. If a sample contains many singletons, it is likely that more undetected OTUs exist, and the *Chao 1* index will estimate greater species richness than it would for a sample without rare OTUs. Besides the Chao1 estimator, mothur includes several other species richness estimators and a wrapped version of CatchAll, which calculates 12 different estimators and proposes a best estimate of species richness (Bunge *et al.*, 2012). QIIME also includes the Chao1 estimator along with several other species richness estimators.

2. Rarefaction curves are used to determine whether sampling depth was sufficient to accurately characterize the bacterial community being studied. To build rarefaction curves, each community is randomly subsampled without replacement at different intervals, and the average number of OTUs at each interval is plotted against the size of the subsample (Gotelli and Colwell, 2001). The point at which the number of OTUs does not increase with further sampling is the point at which enough samples have been taken to accurately characterize the community. Mothur and QIIME will both calculate rarefaction for observed and estimated species richness. QIIME will additionally create graphs of rarefaction curves, while mothur outputs results that can be imported into graphing software.

3. Community diversity indices combine species richness and abundance into a single value of evenness. Communities that are numerically dominated by one or a few species exhibit low evenness while communities where abundance is distributed equally amongst species exhibit high evenness (Gotelli, 2008). Two of the most widely used indices are the Shannon (or Shannon-Wiener) index (Shannon, 1948) and Simpson's index (Simpson, 1949). A recommended index that is not sensitive to sample size is the Probability of an Interspecific Encounter (PIE [Hurlbert, 1971]):

$$PIE = \left(\frac{N}{N-1}\right)\left(1.0 - \sum_{i=1}^{s}(p_i)^2\right)$$

where N is the sample size, p_i is the proportion of the sample that is made up of individuals of species i, and S is the number of species in the sample. PIE is bounded between 0 (a community comprised of a single species), and 1 (a community comprised of an infinite number of equally abundant species), but is not currently included in either mothur or QIIME. Both mothur and QIIME include multiple community diversity indices.

2.2.5. Exploratory techniques: beta diversity

The main goal of most bacterial community studies is to compare the composition of different communities (beta diversity). The communities being compared differ in some trait or treatment, such as which section of the gut the samples are from. There are numerous ways to visualize and analyse beta diversity, and a thorough review of multivariate techniques that are commonly used by microbial ecologists is presented by Ramette (2007). The beta diversity analyses that have been used in studies of bee-associated bacteria fall into two categories: exploratory techniques and tests of significance. We recommend the following steps for ordination and hierarchical clustering (exploratory techniques):

1. Determine the distance/dissimilarity matrix. The goal of ordination and clustering is to visually compare community composition. Both approaches utilize community distance matrices as input, and these matrices are commonly computed using two methods.

    a. Bray-Curtis dissimilarity (Bray and Curtis, 1957):

    $$\bar{C} = 1 - \frac{2w}{a + b}$$

    where w is the sum of the of the lesser scores for only those species which are present in both communities, a is the sum of the measures of taxa in one community and b is the sum of the measures of taxa in the other community. When proportional abundance is used, a and b equal 1 and the index collapses to 1-w.

    b. UniFrac distances (Lozupone and Knight, 2005). UniFrac distances are based on branches in a phylogenetic tree that are either shared or unique amongst samples. UniFrac distance matrices therefore depend on the quality of the input tree, which can be problematic for short NGS data (Ochman *et al.*, 2010). Given that caveat, UniFrac distances are commonly used, and can be calculated in QIIME given an OTU table that lists the abundance of each OTU in a sample and a phylogenetic tree.

2. Evaluate ordination patterns. The Bray-Curtis dissimilarity matrix or UniFrac distance matrix is used as input for ordination and clustering analyses. The two most common methods for ordination of NGS bacterial community data are principal coordinates (PCoA) and nonmetric multidimensional scaling (NMDS). NMDS is recommended, as NMDS is non-parametric, free of assumptions, and can reduce the data into fewer axes than PCoA (Quinn and Keough, 2002; Ramette, 2007). The number of axes for the NMDS ordination is determined beforehand, and will likely be a tradeoff between interpretability and goodness of fit (Quinn and Keough, 2002). When Kruskal's stress formula is used, it is recommended to use as few dimensions as possible, while achieving stress values of less than at least 0.20 and preferably less than 0.10 (Quinn and Keough, 2002). Although currently not implemented in mothur or QIIME, analyses such as canonical correspondence analysis (CCA) relate environmental variables to ordination patterns (Ramette, 2007). CCA can also be used to determine which OTUs correspond with specific environmental variables.

3. Hierarchical community clustering. To visualize community relatedness in the same format as a phylogenetic tree, we recommend UPGMA, or the Unweighted Pair Group Method with Arithmetic mean (Sokal and Michener 1958). Jackknife support for the branching patterns in the resulting dendrogram can be calculated in QIIME (Kuczynski *et al.*, 2011), providing an estimate of confidence in the clustering patterns.

2.2.6. Testing for significant differences in communities

Several analyses have been used to test for differences in microbial community composition (Schloss, 2008). It is likely, however, that new methods specifically developed for analysis of deeply sequenced microbial communities will complement or even replace the existing methods.

1. Non-parametric MANOVA (NPMANOVA, (Anderson, 2001), also called Adonis in the vegan R package and QIIME) is a non-parametric analyses of variance that has been used to test for differences in microbial community composition. In NPMANOVA and the following analyses the response variable is the dissimilarity or distance matrix calculated above and the independent variable is the group of samples being tested, (e.g. samples grouped by different regions of the bee gut or by an experimental treatment). Multiple response permutation procedure (MRPP) is a related test for differences between communities. NPMANOVA (Adonis) and MRPP are available in QIIME and require:

    a. a distance or dissimilarity matrix.
    b. a mapping file.
    c. a categories list that details the category in the mapping file that will be tested (i.e. the independent variable).

2. AMOVA (Excoffier *et al.*, 1992) is another non-parametric analyses of variance that is available in mothur and tests whether the variation in each sample differs from the variation of the pooled samples. To run AMOVA mothur requires:

    a. a distance or dissimilarity matrix.
    b. a design file that lists the sample name and corresponding group (i.e. the independent variable).

Although these tests are non-parametric, false positive results may be problematic and results should be interpreted with caution.

3. Fluorescent *in situ* hybridization (FISH) for image analysis of specific microorganisms in the digestive tract

FISH with fluorescently labelled oligonucleotides targeting bacterial rRNAs can be a very powerful tool for detecting specific bacteria in various environmental samples such as active sludge, stool, clinical specimens, animal tissues and so on (Wagner *et al.* 2003, Amann and Fuchs 2008). In this chapter, we describe a basic FISH protocol suitable for the detection of specific bacteria in the honey bee gut (Fig. 2) and probably also applicable for other insect samples. We provide useful tips and mention important points to be considered when performing these experiments.

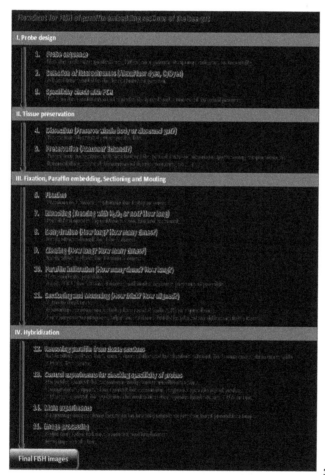

Fig. 2. Flowchart of the steps involved in fluorescent *in situ* hybridization (FISH) of the honey bee gut using paraffin embedded sectioning.

3.1. Designing Probes

3.1.1. Probe sequence

Requirements for oligonucleotide probes used in FISH are partially the same as those for oligonucleotide primers used in conventional PCR, and as such, general primer design software and web-based services can be used as a starting point for designing probes (e.g. Primer 3 (Rozen and Skaletsky, 2000), SILVA (Quast *et al.*, 2013) and probeBase (Loy *et al.*, 2007)). The probeBase database represents a good resource to analyse potential probe sequences (http://www.microbialecology.net/probebase/default.asp). At this web site, one can also find published oligonucleotides that have been shown to work in a given target bacterium.

Probes should be designed using the following guidelines:

1. 15-25 bases in length
2. 50-70% GC content
3. 50-60°C Tm
4. Specificity of a FISH probe is increased by including unique bases to a given target bacterium (or bacterial group) either in the middle part of the primer or evenly distributed over the entire sequence.
5. Choose a suitable region of the 16S rRNA molecule to target for hybridization. The brightness of a FISH signal is largely affected by the accessibility of a given region for probe hybridization (Fuchs *et al.*, 1998; Behrens *et al.*, 2003, Yilmaz *et al.* 2006). This factor was experimentally examined for a few bacterial species. Because secondary structures of rRNAs can slightly differ between bacterial species, regions known to result in good hybridization efficiency in these bacterial species might not be the most suitable ones in another target bacterium.
6. Examine the specificity of a designed probe by conducting diagnostic PCR using an unlabelled oligonucleotide with the same sequence as the probe and a universal forward primer for the bacterial 16S rRNA gene such as 16SA1 (5′-AGAGTTTGATCMTGGCTCAG-3′) or 16SA2 (5′-GTGCCAGCAGCCGCGGTAATAC-3′).
7. The unlabelled oligonucleotide is also useful for competitive suppression control experiments, which are later described in this chapter. You will need at least 500 µM of the unlabelled oligonucleotide.

3.1.2. Selecting fluorochromes

Recently, various types of fluorochromes have become available which emit fluorescent light at different wavelengths. This allows the simultaneous detection of distinct bacteria using specific probes labelled with different fluorochromes. A high-performance laser scanning microscope (LSM) in combination with multiple types of probes dually-labelled with such fluorochromes increases the number of bacteria that can be detected simultaneously (Valm *et al.*, 2011). However, most of epifluorescence microscopes used routinely in conventional laboratories may be equipped with only a few fluorescence filters. This limits the number of fluorochromes that can be detected unambiguously. Moreover, because insect tissues generally have stronger autofluorescence in shorter (blue to green) wavelengths, signals of fluorochromes emitting fluorescence in this wavelength range tend to be obscured by the autofluorescence of the insect tissue (Fukatsu *et al.*, 1998; Koga *et al.*, 2009). Accordingly, only fluorochromes emitting longer wavelength light are generally suitable as reporter dyes of FISH probes for insect materials (Table 1).

Green fluorochromes are sometimes allocated for counter staining, especially when using LSM, because these microscopes are typically not equipped with a laser generator to detect blue fluorescence dyes such as DAPI. Different companies offer a broad variety of fluorochromes for oligonucleotide labelling. Alexa Fluor® dyes seem to outperform traditional fluorochromes, such as FITC, TRITC or Cy™ Dyes, in fluorescence intensity and photo stability. However, these patented dyes are relatively expensive, and for most applications, other dyes will be sufficient to generate good results.

Table 1. Fluorochromes suitable for experiments in *Apis mellifera*. [1]If autofluorescence is low in tissues.

Use	Recommended Fluorochromes	Emission colour
Counter staining DNA	DAPI	Blue
	Hoechst33342	Blue
	SYTOX Green®	Green
Labeling oligonucleotide probes	Alexa Fluor® 555	Orange
	Alexa Fluor® 647	Far-red
	Cy™3	Orange
	Cy™5	Far-red
	TYE™563	Orange
	TYE™665	Far-red
	Alexa Fluor® 488[1]	Green

3.2. Preserving insect materials for FISH

Specimen quality can substantially influence the success of FISH experiments. Immediately processed fresh insect specimens represent the optimal material. The tissue should be dissected (see Carreck *et al.*, 2013) in an isotonic solution, such as phosphate buffered saline (PBS) or Insect Ringer's solution, before being fixed, as this will facilitate the infiltration of reagents into tissue later. Dissection prior to fixation also aids in orienting the specimen for later sectioning.

However, insects are often collected in the field, and thus they cannot immediately be processed. If the insects must be preserved before FISH experiments can be performed, acetone and ethanol can be used as preservatives. Acetone is an amphiphilic organic solvent that deprives water from fresh tissue samples very quickly. This property makes acetone an excellent preservative agent for insects with cuticle exoskeletons that prevent efficient penetration of many aqueous fixative agents (Fukatsu, 1999). However, acetone is not always the best choice for FISH experiments, especially in case of bee gut specimens. The major flaw of acetone is that tissues soaked in it become brittle. This makes the dissection of an intact gut from whole body preservations in acetone extremely challenging. Hence, absolute ethanol is recommended if entire insects need to be preserved for later dissection. However, if the desired tissue can be dissected prior to preservation then acetone is the primary choice.

Use excessive amounts of preservative agent. The total volume of insect tissue should be roughly 10% of the preservative volume. These preservatives are highly volatile and their volume can decline fairly quickly over time. Thus, be sure to check the volume periodically in preserved specimens.

3.3. Fixation, paraffin embedding, sectioning and mounting

In histological analyses, paraffin sectioning is one of the most widely used techniques, and many good protocols are available (e.g. Barbosa (1974) or Presnell *et al.* (1997)). One can refer to these protocols to learn the principles and details of different techniques. Here, we describe the conditions useful for preparing paraffin sections of the honey bee gut.

3.3.1. Fixing honey bee gut samples from fresh bees

1. Dissect the gut from the bee in PBS (0.8% NaCl [w/v], 0.02% KCl [w/v], 0.115% Na$_2$HPO$_4$ [w/v], 0.02% KH$_2$PO$_4$ [w/v], pH 7.5) or Insect Ringer's solution.
2. Freshly prepare formaldehyde fixative mixing 9 vol. of PBS and 1 vol. of 40% formaldehyde solution (i.e. 4% formaldehyde in PBS). It would be better to use more than 10 volumes of the fixative to that of a specimen.
3. Fix the tissue by soaking the gut samples in the freshly prepared formaldehyde fixative, overnight at 4°C.
4. After fixation, wash the tissue in PBS at least three times for 10 min at room temperature.
5. Follow with at least three washes in 75% ethanol for 30 min at room temperature.
6. Transfer the gut samples to absolute ethanol for storage until used. You can keep the samples in absolute ethanol at least a week, and perhaps longer.

3.3.2. Fixing honey bee gut samples from preserved bees

1. If the honey bees have been preserved in absolute ethanol or acetone, the tissue has to be softened before dissection. Achieve this by soaking and dissecting the specimen in aqueous alcohol (usually 70-80% ethanol), then wash the gut sample once more with absolute ethanol just prior to fixation. (See 3.3.1 for working with fresh samples.)

2. Fix the samples by soaking them in Carnoy's fixative (ethanol: chloroform:acetic acid = 6: 3: 1 [v/v]) at room temperature, overnight. Tissues can be exposed to Carnoy's fixative for longer time periods if autofluorescence from tissue needs to be reduced further. However, this is not typically necessary for honey bee gut tissues.

3. Then wash with absolute ethanol at least three times for 20 min at room temperature.

3.3.3. Bleaching samples to reduce autofluorescence in the gut tissues

Autofluorescence can be reduced in tissues using hydrogen peroxide (H_2O_2) (Koga *et al.*, 2009), as described here. However, autofluorescence is not very strong in honey bee gut tissue and H_2O_2 treatments produce oxygen bubbles in the lumen causing insufficient penetration of the paraffin. Thus, bleaching with H_2O_2 is not recommended unless autofluorescence is problematic in your tissues for some reason.

1. Prepare a solution of H_2O_2-ethanol using 1 volume of 30% H_2O_2 and 4 volume of absolute ethanol.

2. Soak the gut samples in the H_2O_2-ethanol for several days to several weeks, changing the solution every two to three days. During soaking, the colour of the tissue should become lighter.

3. Wash with absolute ethanol at least three times for 20 min at room temperature.

4. The specimen can be kept at -20°C until use.

3.3.4. Dehydration, clearing, paraffin infiltration and embedding

Complete replacement of water in tissues with paraffin is crucial for obtaining high-quality sections. Firstly, water is replaced with an amphiphilic solvent (e.g. ethanol). Then, this solvent will be replaced with a hydrophobic solvent (e.g. xylene). Finally, the hydrophobic solvent is replaced with paraffin. Longer exposure to each of the different reagents will ensure complete replacement, but may also adversely affect the tissues. Therefore, it is important to optimize conditions, such as the exposure time for each solvent, depending on the type of specimen. Here, we describe conditions for paraffin embedding of the midgut and hindgut of the honey bee.

1. Dehydrate the gut tissues by soaking in absolute ethanol for 1 hr at room temperature, repeat twice or more to ensure the replacement of water with ethanol in the tissue.

2. Clear dehydrated tissues by soaking in absolute xylene for 20 min at room temperature; repeat twice more for a total of 3 washes.

3. Infiltrate paraffin into the tissues by applying liquid paraffin at 65-70°C, 3 times for 1 hr.

4. Put the tissues in melted paraffin in a desired orientation and 'snap cool' it in ice-cold water. Do this in a mould (plastic or metal), which has been pre-warmed to the same temperature as the paraffin. Use a Pasteur pipette (also pre-warmed) to pipette melted paraffin into the mould. Add the specimen into the mould filled with paraffin, then 'snap-cool' it immediately in the ice-cold water. After leaving it for 5 min in the ice-cold water, the paraffin should have hardened and the block including the embedded tissue can be removed from the mould.

Determine the direction of your sectioning (cross or sagital) before embedding it in the paraffin block, and then remember the orientation of the sample. Once embedded, the tissue is difficult to see in solid paraffin. Thus, the orientation of the tissue should be noted down for proper orientation of the block when preparing sectioning later.

3.3.5. Sectioning and mounting

Use the following procedure to section paraffin-embedded tissues with a microtome. The resulting sections should then be mounted onto glass slides. In general, cutting tissue sections needs training and practice. Therefore, we recommend consulting a person who is experienced in histological techniques when using a microtome the first time.

1. Remove (trim) excess parts of the paraffin block containing the tissue to a few millimetres around the tissue. Use a scalpel for trimming.

2. Fix the trimmed paraffin block onto the specimen holder of the microtome for preparing tissue sections. The procedure of fixing the paraffin block depends on the model of the microtome. Please refer to the manual of the microtome.

3. Trim the paraffin block on the specimen holder again to make the upper and lower sides of the block parallel to each other.

4. Mount the specimen holder onto the microtome.

5. Cut the paraffin block into 5 µm thick sections. You can adjust the thickness of the sections to your purpose and type of tissue. For example, you may have to use thicker sections when preparing hindgut tissue that is filled with solid waste, such as pollen shells.

6. Mount the sections on a commercially available glass slide coated with aminosilane (APS) or equivalent. Place each section on a separate slide. It is important to have several slides with the same tissue sample section, so that different stains can be used and controls can be included.

3.4. Hybridization

The protocol for conducting the hybridization is essentially the same as described by Fukatsu *et al.* (1998) and Koga *et al.* (2009). Its use in honey bee research is discussed further in Evans *et al.* (2013).

3.4.1. Hybridization procedures

1. Freshly prepare a hybridization buffer containing the probe(s) (20 mM Tris-HCl [pH 8.0], 0.9M NaCl, 0.01% sodium dodecyl sulphate, 30% formamide, 100 pmol/ml each of the probes, 10 µg/ml DAPI or equivalents). Prepare 150 µl of this solution per slide.

2. Remove paraffin from the sections using absolute xylene (soak for 3 min., and repeat this 3 times), followed by absolute ethanol (soak for 3 min, twice), then rinse with RNase-free water.

3. Apply 150 µl of the hybridization buffer onto the sections and cover with a coverslip, taking care not to trap air bubbles.

4. Put the glass slides in a humidified chamber and leave at room temperature, overnight, in the dark.

5. The next day, carefully remove the coverslip by applying freshly prepared PBSTx (0.8% NaCl [w/v], 0.02% KCl [w/v], 0.115% Na$_2$HPO$_4$ [w/v], 0.02% KH$_2$PO$_4$ [w/v], 0.3% Triton X-100 [v/v]) between the glass slide and coverslip using a wash bottle.

6. Wash the slides with PBSTx with gentle agitation (~50 rounds per minute) in the dark, 3 times for 10 min at room temperature. You can use conventional staining trays or containers.

7. Apply 100 µl of a commercially available antifade solution such as SlowFade (Invitrogen) or DABCO-glycerol (prepared by mixing 1.25 g of 1,4-Diazabicyclo[2.2.2]octane [DABCO], 10 mL of PBS and 90 mL of glycerol) directly onto the specimen and cover the slide with a coverslip; avoid trapping air bubbles.

8. Seal all sides of the coverslip with clear fingernail polish.

9. The samples are now ready for microscopy and should be immediately analysed. These slides cannot be stored, because the rRNA probes and fluorochromes are unstable, especially when exposed to light.

3.4.2. FISH controls

Control experiments are necessary to validate the specificity of the probes and to discriminate between the fluorescent signals of probes and autofluorescence of insect tissues. All three of the following control experiments should be conducted, because they are crucial for verifying any FISH results obtained.

1. No probe control. Conduct hybridization without the fluorescent probe(s) to examine the levels of autofluorescence from the tissues.

2. Competitive suppression control. Conduct hybridization with an excess amount (usually 50x) of an unlabelled oligonucleotide using the same sequence as a fluorescent probe to observe sequence-specific, competitive suppression of the hybridization signals.

3. RNase A control. Prior to hybridization, treat the tissue sections with 10 µg/mL of RNase A in PBSTx at 37°C for 30 min. Wash the section with PBSTx and then conduct hybridization as usual. The purpose of this control is to verify that the probe specifically binds to RNA molecules.

3.4.3. Combining probes

Probes that do not overlap in their emission wavelengths can be combined in the same sample to detect different bacteria simultaneously (Fig. 3). Combining a universal eubacterial probe with a more specific probe is also useful to detect the presence of additional bacteria not identifiable with a specific probe.

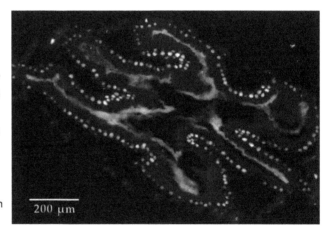

Fig. 3. Micrograph using fluorescent *in situ* hybridization (FISH) of *Snodgrassella alvi* (green) and Firm-5 (blue) in an oblique section of the honey bee ileum. The fluorescent tags used here were linked to probes hybridizing with 16S rRNA specific to these bacterial types. In addition, a DNA stain (white) was used to stain the honey bee nuclei.

Image prepared by W Kwong.

3.4.4. Image acquisition and adjustments

FISH signals can be detected by using an epifluorescence microscope using standard fluorescence microscopy procedure for your instrument. Here is some basic information that might be helpful for beginners.

1. Frequently, each channel (i.e. each signal from a given probe or DNA counter-stain) is acquired as a black & white image. Each of these images will later be coloured with a specific colour (arbitrarily chosen) and merged, resulting in a single pseudo-coloured image. Therefore, the actual fluorescent colour seen can differ from that in the final images.

2. The probe will dissociate from the target RNA molecules over time in the mounting media. Thus, after mounting in antifade medium, observation should be conducted as soon as possible.

3. Minimize exposure of your slides to the excitation light to prevent photobleaching of fluorochromes (i.e. reduction of signal intensity).

4. Irradiation at higher magnification causes localized photobleaching. This photobleached region will be observed as a dark spot in a lower-magnified image. Therefore, images should be acquired from lower to higher magnifications.

3.5. Concluding remarks about FISH

Preparing fine paraffin sections is the key to a successful FISH experiment. However, this step requires technical proficiency and practice. Therefore, whole-mount FISH (wFISH) (Koga *et al.*, 2009) might be an attractive alternative. This technique does not require tissue sectioning, but a laser-scanning microscope is essential for analysis. In addition, wFISH has several other limitations, such as poor performance on large and thick specimens and vulnerability to autofluorescence. A combination of both FISH techniques would be optimal to obtain a comprehensive picture of the structural organization of microbial communities in the gut environment of bees and other insects.

4. Culture conditions for the dominant members of the bee gut microbiota

Culture-independent studies show that the gut microbiota of the honey bee is dominated by only 8 or so groups of bacteria (Martinson *et al.*, 2011; Ahn *et al.*, 2012; Moran *et al.*, 2012). Given the difficulty of drawing delineations between species when it comes to bacteria, each "group" may actually represent one or more species. Nonetheless, these groups are valid in that they comprise distinct phylogenetic clades, and the bacteria within each clade are found in association with bees and not with other animals or environmental sources (Martinson *et al.*, 2011; Ahn *et al.*, 2012; McFrederick *et al.,* 2012).

Bacteria from each of these groups can be cultured on standard microbiological media if supplied with the correct growth conditions. Recent culture-based studies have begun to support the findings of culture-independent surveys, giving due recognition to the dominant members of the microbiota (Olofsson and Vásquez, 2008; Engel *et al.,* 2012; Tian *et al.*, 2012; Vásquez *et al.*, 2012; Kwong and Moran, 2013). By combining the low-bias culture-independent analysis of bacterial communities and the powerful molecular tools available in culture-based approaches, a rigorous study of the microbial associates of bees can be realized.

The following is a brief overview of the current state of knowledge regarding the cultivation of the bee gut microbiota for each of the dominant groups. All bacteria described here can be cultured by plating homogenized guts (aseptically removed from the bee using methods described in Section 2.1.1.) onto a suitable medium (often a standard nutrient-enriched agar that can be purchased from commercial suppliers). The plates are then incubated for several days, typically at 35-37°C, to allow growth of bacterial colonies.

Most of the bacteria associated with honey bee guts require a low-O_2 atmosphere for optimal growth. To supply these conditions, plates may be incubated in dedicated CO_2 incubators or sealed in pouches or jars with CO_2-generating packets. Bacteria that require anaerobic conditions can be grown in dedicated anaerobic chambers by displacing air with nitrogen gas, or in sealed pouches or jars with the appropriate packets for generating anaerobic atmospheres. Packets for both CO_2 and anaerobic atmospheres are available from commercial suppliers such as BD Biosciences's GasPak system (Franklin Lakes, NJ, USA) and Oxoid's CO_2Gen and AneroGen (Basingstoke, Hampshire, UK). Identification of bacterial isolates should be carried out by DNA sequencing (as described in section 2.1.2. and 2.1.3.) rather than by phenotypic observations, as traits such as colony morphology and biochemical activities may be heterogeneous among members of the same species.

A word of caution, culturing methods will always lead to bias in what is recovered, and this bias is affected by the culture conditions you select. What is presented here is a general guide to methods that have successfully cultured some members of each dominant group, but it is important to recognize that these techniques may not always capture the full strain diversity within each group.

4.1. Genus *Snodgrassella*

4.1.1. Optimal growth conditions

1. Atmosphere of 5% CO_2
2. Temperature of 35-37°C
3. Culture media: trypticase soy agar, trypticase soy agar + 5% defibrinated sheep blood, heart infusion agar, brain heart infusion agar, and LB agar; grows weakly in trypticase soy broth (Kwong and Moran, 2013).

4.1.2. Microbe characteristics

Colonies smooth, white, round, ~1mm diameter, and form within 2 days (Fig. 4). Cells are Gram negative, rod-shaped, and non-motile. *Snodgrassella* can use citric acid or malic acid as the main carbon source; are positive for catalase and nitrate reductase activity, negative for oxidase.

The sole species described for this genus is *Snodgrassella alvi*, a member of the family *Neisseriaceae* and class *Betaproteobacteria* (Martinson *et al.*, 2012; Kwong and Moran, 2013). In early publications (Babendreier *et al.*, 2007; Martinson *et al.*, 2011; Moran *et al.*, 2012), *S. alvi* is referred to as the "Beta" or "Betaproteobacteria". *Snodgrassella alvi* strains have been isolated from honey bees and bumble bees, and the type strain is *S. alvi* wkB2[T] (Kwong and Moran, 2013). The type strain can be procured from bacterial culture collections (accession BAA-2449[T] at the American Type Culture Collection in Manassas, VA, USA [ATTC] or 14803[T] at the National Collection of Industrial, Food, and Marine Bacteria in Aberdeen, UK [NCIMB]). *Snodgrassella* has been estimated to comprise 0.6-39% of bacteria in the guts of individual workers (Moran *et al.*, 2012).

4.2. Genus *Gilliamella*

4.2.1. Optimal growth conditions

1. Atmosphere of 5% CO_2
2. Temperature of 35-37°C
3. Culture media: trypticase soy agar or broth, trypticase soy agar + 5% defibrinated sheep blood, heart infusion agar, and LB agar (Kwong and Moran, 2013).

4.2.2. *Gilliamella* characteristics

Smooth, white, round colonies ~2.5mm diameter that form within 2 days incubation; however, colony morphology can vary between strains (Fig. 4). Cells are Gram negative, rod-shaped, non-motile, and may form filament chains. *Gilliamella* is negative for catalase, nitrate reductase, and oxidase (Kwong and Moran, 2013).

The sole species described for this genus is *Gilliamella apicola*, a member of the family *Orbaceae* and class *Gammaproteobacteria* (Martinson *et al.*, 2012; Kwong and Moran, 2013). In early publications (Babendreier *et al.*, 2007; Martinson *et al.*, 2011; Moran *et al.*, 2012), *G. apicola* is referred to as the "Gamma-1" or "Gammaproteobacteria-1". *G. apicola* strains have been isolated from honey bees and bumble bees, and the type strain is *G. apicola* wkB1[T] (Kwong and Moran, 2013). The type strain can be procured from bacterial culture collections (accession BAA-2448[T] at ATCC, or 14804[T] at NCIMB). *Gilliamella* has been estimated to comprise 0.6-30% of bacteria in the guts of individual workers (Moran *et al.*, 2012).

4.3. Genus *Frischella*

4.3.1. Optimal growth conditions

1. Atmosphere of 5% CO_2, or anaerobic
2. Temperature of 35-37°C
3. Culture media: trypticase soy agar + 5% defibrinated sheep blood, heart infusion agar, brain heart infusion agar, and trypticase soy broth (Engel *et al.*, 2013).

4.3.2. *Frischella* characteristics

Frischella are facultative anaerobic, but will not grow in fully aerobic conditions (Engel *et al.*, 2013). These bacteria produce smooth, round, flat, semi-transparent colonies of ~1mm diameter in about 3 days (Fig. 4). Cells are rod-shaped and may form filaments or chains. *Frischella* can obtain carbon through fermentation of glucose, fructose, and mannose. They are positive for catalase and negative for nitrate reductase and oxidase.

The species strain described for this group is *Frischella perrara* (Strain PEB0191), a member of the family *Orbaceae* and class *Gammaproteobacteria*. Although in the same family as *Gilliamella*, its closest relatives are in the *Orbus* genus. *Frischella perrara* was referred to as "Gamma-2" or "Gammaproteobacteria 2" (Babendreier *et al.*, 2007; Martinson *et al.*, 2011; Moran *et al.*, 2012), and has recently been formally named (Engel *et al.*, 2013). *Frischella* has been isolated from *A. mellifera* but has not been detected in bumble bees (Koch and Schmid-Hempel, 2011). The type strain is PEB0191[T] and can be procured from bacterial culture collections (accession BAA-2450[T] at ATCC, or 14821[T] at NCIMB). *Frischella* is present in most adult workers and has been estimated to comprise up to 13% of the gut bacteria (Moran *et al.*, 2012).

4.4. Genus *Lactobacillus*

4.4.1. Optimal growth conditions

1. Atmosphere can be aerobic or anaerobic
2. Temperature of 35-37°C
3. Culture media: tryptone soy broth agar (Oxoid; Basingstoke, Hampshire, UK), tomato juice agar (Oxoid; Basingstoke, Hampshire, UK), Rogosa agar (Merck), and MRS (de Man-Rogosa-Sharpe) agar (Oxoid; Basingstoke, Hampshire, UK), Lactobacillus Carrying Media (Efthymiou and Hansen, 1962) and MRS broth supplemented with 0.5% L-cysteine or 20% fructose (Olofsson & Vásquez, 2008; Forsgren *et al.* 2010).

4.4.2. *Lactobacillus* characteristics

Cells of the *Lactobacillus* genus can vary from long and slender rods to short coccobacilli rods (Hammes and Hertel, 2009). Colony morphologies also vary, but are typically convex, smooth, and opaque without pigment (Hammes and Hertel, 2009). The bee-associated *Lactobacillus* are negative for catalase and sporulation, positively Gram-staining, and produce lactic acid by homofermentation (Olofsson and Vásquez, 2008). The *Lactobacillus kunkeei* clade is fructophilic, preferentially utilizing fructose over glucose as the carbon source (Neveling *et al.*, 2012).

Lactobacillus species are ubiquitous in nature and are commonly found in association with many animals, plants, and foodstuffs. Lactobacilli are widely considered probiotic, meaning their presence is beneficial to the health of the host organism (Kleerebezem and Vaughan, 2009). The bee-associated *Lactobacillus* fall into two main clades, "Firm-4" and "Firm-5" (Babriendier *et al.*, 2007; Martinson *et al.*, 2011; Moran *et al.*, 2012). These clades are only distantly related to other *Lactobacillus*, with 16S rRNA identities ~90% (Olofsson and Vásquez, 2008; Vásquez *et al.*, 2012), and thus may eventually be classified as novel species.

Another species, *L. kunkeei*, may be the most frequently recovered member of *Lactobacillus* in culturing experiments (Tajabadi *et al.*, 2011; Vásquez *et al.*, 2012; Neveling *et al.*, 2012). However, culture-independent studies show "Firm-4" and "Firm-5" are the dominant bee gut lactobacilli, not *L. kunkeei* (Moran *et al.*, 2012; Ahn *et al.*, 2012). *L. kunkeei* has also been found on flowers (Neveling *et al.*, 2012) and wine (Edwards *et al.*, 1998), suggesting they may have other naturally occurring habitats outside the bee gut (McFrederick *et al.*, 2012). *Lactobacillus* is the most abundant group in the bee gut and has been estimated to comprise 20-99% of bacteria in individual workers (Moran *et al.*, 2012).

4.5. Genus *Bifidobacterium*

4.5.1. Optimal growth conditions

1. Atmosphere that is aerobic or anaerobic
2. Temperature of 37°C
3. Culture media: blood agar, peptone-yeast extract-glucose broth (Biavati *et al.*, 1982), MRS medium (Olofsson & Vásquez, 2008; Bottacini *et al.* 2012).

4.5.2. *Bifidobacterium* characteristics

Bifidobacterium are typically anaerobic or microaerophilic; however, members of the bee gut *Bifidobacterium* have been found to grow aerobically (Bottacini *et al.*, 2012). Colonies appear within 2 days and are punctiform, convex, smooth, and greyish-white (Biavati *et al.*, 1982) (Fig. 4). The bee gut *Bifidobacterium* are negative for catalase and sporulation, Gram positive, and produce lactic and acetic acid (Olofsson and Vásquez, 2008).

Like Lactobacilli, Bifidobacteria are common members of animal gut microbial communities and have been used as probiotics (Kleerebezem and Vaughan, 2009). Three species of *Bifidobacterium* specific to honey bees have been described: *B. asteroides*, *B. coryneforme*, and *B. indicum* (Scardovi and Trovatelli, 1969; Biavati *et al.*, 1982; Felis and Dellaglio, 2007). Other species have been found in bumble bees (Killer *et al.*, 2011). The type strains are available from ATCC: *B. asteroides* 25910[T], *B. coryneforme* 25911[T], and *B. indicum* 25912[T]. The genome of *B. asteroides* strain PRL2011 has been sequenced (Bottacini *et al.*, 2012). *Bifidobacterium* is present in most adult workers and has been estimated to comprise up to 15% of the gut bacteria (Moran *et al.*, 2012).

4.6. Alpha-1 bacteria

4.6.1. Optimal growth conditions

1. Atmosphere of 5% CO_2
2. Temperature of 35-37°C
3. Culture media: trypticase soy agar, trypticase soy agar + 5% defibrinated sheep blood, and heart infusion agar (PE, unpublished).

4.6.2. Alpha-1 characteristics

This group of bacteria has not been formally described, but in our experience they form smooth, round, white colonies after 1 day of growth (Fig. 4; PE, unpublished). The Alpha-1 group belongs to the class *Alphaproteobacteria* and is related to several ant-associated bacteria and to *Bartonella*, a genus that includes opportunistic intracellular parasites (Jeyaprakash *et al.*, 2003; Babriendier *et al.*, 2007; Martinson *et al.*, 2011). Alpha-1 is typically found at low frequency (< 4 %) in the honey bee gut and are probably not present in all individuals (Moran *et al.*, 2012).

4.7. Other bacteria

Alpha-2.1 and Alpha-2.2 (Martinson *et al.*, 2011) are two other groups within the *Alphaproteobacteria* commonly found in the bee gut. Alpha-2.1 is related to *Gluconobacter* and *Acetobacteraceae*, while Alpha-2.2 is related to *Saccharibacter floricola*, a bacterium associated with flowers (Martinson *et al.*, 2011). Members of both of these groups have been cultured (Mohr and Tebbe, 2007), although they remain poorly studied. These two groups are typically found at low frequency in the bee gut (< 6%), although they appear to be present in most individuals (Moran *et al.*, 2012).

Other bacterial genera may also be recovered from culturing efforts, such as *Pantoea* (Loncaric *et al.*, 2009) and *Bacillus* (Evans and Armstrong, 2006). However, these likely represent transient members of the bee microbiota, as they are not among the dominant groups found in culture-independent studies.

4.8. Preservation of bacterial cultures

Cryopreservation is a standard practice for the long-term storage of bacterial cultures. Although a wide range of preservation methods are available, the protocol described here is sufficient for general strain archival purposes.

1. Streak out the strain on a plate and incubate for 1 day or until it is dense enough to harvest. Liquid media may also be used, if the strain is able to grow in it.
2. Prepare the cryoprotectant: Sterilize a 30% (v/v) glycerol solution by autoclaving at 121°C for 15-20 min.
3. Harvest the culture and resuspend in 500 μl of suitable liquid media (ex. for *Snodgrassella*, use trypticase soy broth).
4. Add the resuspended culture (500 μl) and 30% glycerol (500 μl) to a 2 ml microfuge tube and mix thoroughly. Cryotubes designed for storage at low temperatures, such as Nunc Cryotubes (Sigma-Aldrich; St. Louis, MO, USA), are recommended.
5. Store frozen at -80°C.
6. To reactivate frozen bacteria, remove tube from freezer. Do not thaw; keep tube on ice or dry ice, or work quickly. Scrape a small amount of frozen culture out of the tube using a sterile tool (an inoculation loop or toothpick works well), and streak out on a fresh agar plate. Return tube to -80°C, and incubate the plate in the optimal conditions for your strain.

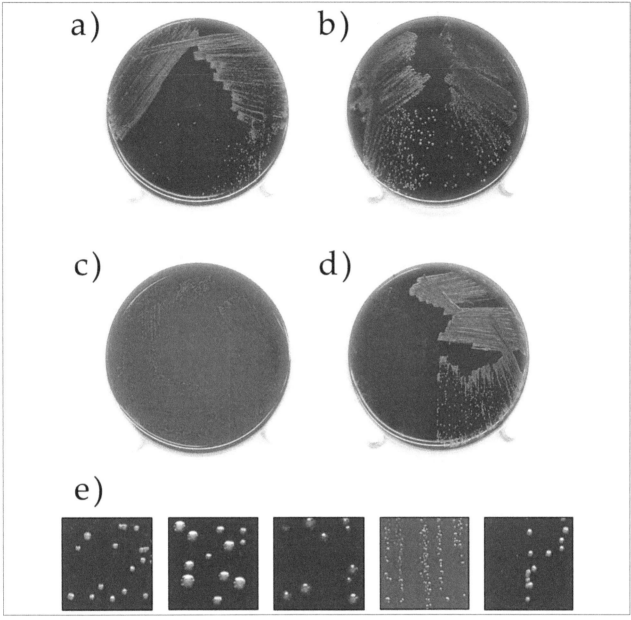

Fig. 4. Examples of dominant gut microbiota isolated from *Apis mellifera:* **a)** *Snodgrassella alvi* wkB2[T]; **b)** *Gilliamella apicola* wkB7 (left) and *G. apicola* wkB1[T] (right); **c)** *Frischella perrara* PEB0191[T], grown anaerobically; **d)** *Bifidobacterium* strain wkB3 (left) and Alpha-1 strain PEB0122 (right); **e)** enlargements of 1x1 cm sections for (left to right) *S. alvi* wkB2[T], *G. apicola* wkB7, *G. apicola* wkB1[T], *F. perrara* PEB0191[T], and *Bifidobacterium* wkB3/Alpha-1 PEB0122. Strains grown on heart infusion agar + 5% sheep blood at 37ºC and 5% CO_2 for 48 h, or as indicated.

Images by W Kwong and P Engel.

5. Conclusions and outlook

We have described key methods for studying symbiotic microorganisms in honey bee guts. Non-culture-based approaches in general, and NGS sequencing in particular, have greatly improved our capabilities when it comes to identifying the microbial communities associated with honey bees. FISH microscopy provides a means to visually examine the microenvironments where particular microbes occur. Culturing methods are also important, allowing researchers to isolate particular bacteria of interest for further study or gene identification, and enabling the assignment of particular functions to particular gut community members. The culture conditions required for bacterial species found in the honey bee gut, as described here, clearly show that these gut symbionts tend to require low oxygen conditions, as is the case for many microbes isolated from the intestines of mammals, but not always true for insect gut symbionts. Also, the bacterial symbionts in the digestive tracts of honey bees require fairly high culturing temperatures, but temperatures consistent with honey bee colonies. The techniques described in this chapter focus on bacterial communities, but can easily be adapted for other microbial groups, for the study of pathogens, or to elucidate interactions between potentially beneficial gut microorganisms and pathogens. In addition, many of these methods can readily be used to study the gut microbiota of other bee species, allowing for comparative studies across hosts. We hope the methods we describe will help others advance the state of knowledge regarding bee gut symbionts, an intriguing area of research.

6. Acknowledgements

We thank J Ellis for inviting us to prepare this chapter and providing editorial support, and K Hammond for assisting with the references and formatting. QSM was supported by the National Science Foundation Grant PRFB-1003133 during the preparation of this work. The COLOSS (Prevention of honey bee COlony LOSSes) network aims to explain and prevent massive honey bee colony losses. It was funded through the COST Action FA0803. COST (European Cooperation in Science and Technology) is a unique means for European researchers to jointly develop their own ideas and new initiatives across all scientific disciplines through trans-European networking of nationally funded research activities. Based on a pan-European intergovernmental framework for cooperation in science and technology, COST has contributed since its creation more than 40 years ago to closing the gap between science, policy makers and society throughout Europe and beyond. COST is supported by the EU Seventh Framework Programme for research, technological development and demonstration activities (*Official Journal L 412, 30 December 2006*). The European Science Foundation as implementing agent of COST provides the COST Office through an EC Grant Agreement. The Council of the European Union provides the COST Secretariat. The COLOSS network is now supported by the Ricola Foundation - Nature & Culture.

7. References

AHN, J H; HONG, I P; BOK, J I; KIM, B Y; SONG, J; WEON, H Y (2012) Pyrosequencing analysis of the bacterial communities in the guts of honey bees *Apis cerana* and *Apis mellifera* in Korea. *Journal of Microbiology* 50(5): 735-45. http://dx.doi.org/10.1007/s12275-012-2188-0

ALTSCHUL, S F; GISH, W; MILLER, W; MYERS, E W; LIPMAN, D J (1990) Basic local alignment search tool. *The Journal of Molecular Biology* 215(3): 403–410. http://dx.doi.org/10.1016/S0022-2836(05)80360-2

AMANN, R; FUCHS, B M (2008) Single-cell identification in microbial communities by improved fluorescence *in situ* hybridization techniques. *Nature Reviews Microbiology* 6(5): 339–348. http://dx.doi.org/10.1038/nrmicro1888

AMEND, A S; SEIFERT, K A; BRUNS, T D (2010) Quantifying microbial communities with 454 pyrosequencing: does read abundance count? *Molecular Ecology* 19(24): 5555–5565. http://dx.doi.org/10.1111/j.1365-294X.2010.04898.x

ANDERSON, M J (2001) A new method for non-parametric multivariate analysis of variance. *Austral Ecology* 26(1): 32–46. http://dx.doi.org/10.1111/j.1442-9993.2001.01070.pp.x

BABENDREIER, D; JOLLER, D; ROMEIS, J; BIGLER, F; WIDMER, F (2007) Bacterial community structures in honey bee intestines and their response to two insecticidal proteins. *FEMS Microbiology Ecology* 59(3): 600–610. http://dx.doi.org/10.1111/j.1574-6941.2006.00249.x

BARBOSA, P (1974) *Manual of basic techniques in insect histology.* Autumn Publishers; Amherst, Massachusetts, USA. 245 pp.

BEHRENS, S; FUCHS, B M; MUELLER, F; AMANN, R (2003) Is the *in situ* accessibility of the 16S rRNA of *Escherichia coli* for Cy3-labeled oligonucleotide probes predicted by a three-dimensional structure model of the 30S ribosomal subunit? *Applied and Environmental Microbiology* 69(8): 4935–4941. http://dx.doi.org/10.1128/AEM.69.8.4935-4941.2003

BIAVATI, B; SCARDOVI, V; MOORE, W E C (1982) Electrophoretic patterns of proteins in the genus *Bifidobacterium* and proposal of four new species. *International Journal of Systematic and Evolutionary Microbiology* 32(3): 358-373. http://dx.doi.org/10.1099/00207713-32-3-358

BOTTACINI, F; MILANI, C; TURRON, I F; SÁNCHEZ, B; FORONI, E; DURANTI, S; SERAFINI, F; VIAPPIANI, A; STRATI, F; FERRARINI, A; DELLEDONNE, M; HENRISSAT, B; COUTINHO, P; FITZGERALD, G F; MARGOLLES, A; VAN SINDEREN, D; VENTURA, M (2012) *Bifidobacterium asteroides* PRL2011 genome analysis reveals clues for colonization of the insect gut. *PLoS One.* 7(9): e44229. http://dx.doi.org/10.1371/journal.pone.0044229

BRAY, J R; CURTIS, J T (1957) An ordination of the upland forest communities of southern Wisconsin. *Ecological Monographs* 27(4): 325–349. http://dx.doi.org/10.2307/1942268

BRULC, J M; ANTONOPOULOS, D A; BERG MILLER, M E; WILSON, M K; YANNARELL, A C (2009) Gene-centric metagenomics of the fiber-adherent bovine rumen microbiome reveals forage specific glycoside hydrolases. *Proceedings of the National Academy of Science* 106(6): 1948-1953.

BUNGE, J; WOODARD, L; BÖHNING, D; FOSTER, J A; CONNOLLY, S; ALLEN, H K (2012) Estimating population diversity with CatchAll. *Bioinformatics* 28(7): 1045–1047. http://dx.doi.org/10.1093/bioinformatics/bts075

CAPORASO, J G; BITTINGER, K; BUSHMAN, F D; DESANTIS, T Z; ANDERSEN, G L; KNIGHT, R (2010a) PyNAST: a flexible tool for aligning sequences to a template alignment. *Bioinformatics* 26(2): 266–267. http://dx.doi.org/10.1093/bioinformatics/btp636

CAPORASO, J G; KUCZYNSKI, J; STOMBAUGH, J; BITTINGER, K; BUSHMAN, F D; COSTELLO, E K; FIERER, N; PEÑA, A G; GOODRICH, J K; GORDON, J I; HUTTLEY, G A; KELLEY, S T; KNIGHTS, D; KOENIG, J E; LEY, R E; LOZUPONE, C A; MCDONALD, D; MUEGGE, B D; PIRRUNG, M; REEDER, J; SEVINSKY, J R; TURNBAUGH, P J; WALTERS, W A; WIDMANN, J; YATSUNENKO, T; ZANEVELD, J; KNIGHT, R (2010b) QIIME allows analysis of high-throughput community sequencing data. *Nature Methods* 7(5): 335–336. http://dx.doi.org/10.1038/nmeth.f.303

CARRECK, N L; ANDREE, M; BRENT, C S; COX-FOSTER, D; DADE, H A; ELLIS, J D; HATJINA, F; VANENGELSDORP, D (2013) Standard methods for *Apis mellifera* anatomy and dissection. In *V Dietemann; J D Ellis; P Neumann (Eds) The COLOSS* BEEBOOK, *Volume I: standard methods for* Apis mellifera *research. Journal of Apicultural Research* 52(4): http://dx.doi.org/10.3896/IBRA.1.52.4.03

CHAO, A (1984) Nonparametric estimation of the number of classes in a population. *Scandinavian Journal of Statistics* 11(4): 265–270. http://dx.doi.org/10.2307/4615964

CORNMAN, R S; TARPY, D R; CHEN, Y; JEFFREYS, L; LOPEZ, D; PETTIS, J S; VANENGELSDORP, D; EVANS, J D (2012) Pathogen webs in collapsing honey bee colonies. *PloS One* 7(8):e43562. http://dx.doi.org/10.1371/journal.pone.0043562

COX-FOSTER, D L; CONLAN, S; HOLMES, E C; PALACIOS, G; EVANS, J D; MORAN, N A QUAN, P L; BRIESE, T; HORNIG, M; GEISER, D M; MARTINSON, V; VANENGELSDORP, D; KALKSTEIN, A L; DRYSDALE, A; HUI, J; ZHAI, J; CUI, L; HUTCHISON, S K; SIMONS, J F; EGHOLM, M; PETTIS, J S; LIPKIN, W I (2007) A metagenomic survey of microbes in honey bee colony collapse disorder. *Science.* 318(5848): 283–287. http://dx.doi.org/10.1126/science.1146498

DEGNAN, P H; OCHMAN H (2012) Illumina-based analysis of microbial community diversity. *The ISME Journal* 6(1): 183–194. http://dx.doi.org/10.1038/ismej.2011.74

DE GRAAF, D C; ALIPPI, A M; ANTÚNEZ, K; ARONSTEIN, K A; BUDGE, G; DE KOKER, D; DE SMET, L; DINGMAN, D W; EVANS, J D; FOSTER, L J; FÜNFHAUS, A; GARCIA-GONZALEZ, E; GREGORC, A; HUMAN, H; MURRAY, K D; NGUYEN, B K; POPPINGA, L; SPIVAK, M; VANENGELSDORP, D; WILKINS, S; GENERSCH, E (2013) Standard methods for American foulbrood research. In *V Dietemann; J D Ellis; P Neumann (Eds) The COLOSS* BEEBOOK, *Volume II: standard methods for* Apis mellifera *pest and pathogen research. Journal of Apicultural Research* 52(1): http://dx.doi.org/10.3896/IBRA.1.52.1.11

DE MIRANDA, J R; BAILEY, L; BALL, B V; BLANCHARD, P; BUDGE, G; CHEJANOVSKY, N; CHEN, Y-P; GAUTHIER, L; GENERSCH, E; DE GRAAF, D; RIBIÈRE, M; RYABOV, E; DE SMET, L; VAN DER STEEN, J J M (2013) Standard methods for virus research in *Apis mellifera.* In *V Dietemann; J D Ellis; P Neumann (Eds) The COLOSS* BEEBOOK, *Volume II: standard methods for* Apis mellifera *pest and pathogen research. Journal of Apicultural Research* 52(4): http://dx.doi.org/10.3896/IBRA.1.52.4.22

EDGAR, R C (2010) Search and clustering orders of magnitude faster than BLAST. *Bioinformatics* 26(19): 2460–2461. http://dx.doi.org/10.1093/bioinformatics/btq461

EDGAR, R C; HAAS, B J; CLEMENTE, J C; QUINCE, C; KNIGHT, R (2011) UCHIME improves sensitivity and speed of chimera detection. *Bioinformatics* 27(16): 2194–2200. http://dx.doi.org/10.1093/bioinformatics/btr381

EDWARDS, C G; HAAG, K M; COLLINS, M D; HUTSON, R A; HUANG, Y C (1998) *Lactobacillus kunkeei* sp. nov.: a spoilage organism associated with grape juice fermentations. *Journal of Applied Microbiology* 84(5): 698-702. http://dx.doi.org/10.1046/j.1365-2672.1998.00399.x

EFTHYMIOU, C; HANSEN, P A (1962) An antigenic analysis of *Lactobacillus acidophilus. The Journal of Infectious Diseases* 110 (3): 258–267. http://dx.doi.org/10.1093/infdis/110.3.258

ENGEL, P E; KWONG, W; MORAN, N A (2013) *Frischella perrara,* gen. nov., sp. nov. a gammaproteobacterium isolated from the gut of *Apis mellifera. International Journal of Systematic and Evolutionary Microbiology* (in press).

ENGEL, P; MARTINSON, V G; MORAN, N A (2012) Functional diversity within the simple gut microbiota of the honey bee. *Proceedings of the National Academy of Sciences of the United States of America* 109(27): 11002-11007. http://dx.doi.org/10.1073/pnas.1202970109

ENGELBREKTSON, A; KUNIN, V; WRIGHTON, K C; ZVENIGORODSKY, N; CHEN, F; OCHMAN, H; HUGENHOLTZ, P (2010) Experimental factors affecting PCR-based estimates of microbial species richness and evenness. *The ISME Journal* 4(5): 642–647. http://dx.doi.org/10.1038/ismej.2009.153

EVANS, J D; ARMSTRONG, T N (2006) Antagonistic interactions between honey bee bacterial symbionts and implications for disease. *BMC Ecology* 6: 4. http://dx.doi.org/10.1186/1472-6785-6-4

EVANS, J D; SCHWARZ, R S; CHEN, Y P; BUDGE, G; CORNMAN, R S; DE LA RUA, P; DE MIRANDA, J R; FORET, S; FOSTER, L; GAUTHIER, L; GENERSCH, E; GISDER, S; JAROSCH, A; KUCHARSKI, R; LOPEZ, D; LUN, C M; MORITZ, R F A; MALESZKA, R; MUÑOZ, I; PINTO, M A (2013) Standard methodologies for molecular research in *Apis mellifera.* In *V Dietemann; J D Ellis; P Neumann (Eds) The COLOSS* BEEBOOK, *Volume I: standard methods for* Apis mellifera *research. Journal of Apicultural Research* 52(4):http://dx.doi.org/10.3896/IBRA.1.52.4.11

EXCOFFIER, L; SMOUSE, P E; QUATTRO, J M (1992) Analysis of molecular variance inferred from metric distances among DNA haplotypes: application to human mitochondrial DNA restriction data. *Genetics* 131(2): 479–491.

FELIS, G E; DELLAGLIO, F (2007) Taxonomy of *Lactobacilli* and *Bifidobacteria*. *Current Issues in Intestinal Microbiology* 8(2): 44–61.

FORSGREN, E; OLOFSSON, T C; VÁSQUEZ, A; FRIES, I (2010) Novel lactic acid bacteria inhibiting *Paenibacillus* larvae in honey bee larvae. *Apidologie* 41(1): 99-108.
http://dx.doi.org/10.1051/apido/2009065

FRIES, I; CHAUZAT, M-P; CHEN, Y-P; DOUBLET, V; GENERSCH, E; GISDER, S; HIGES, M; MCMAHON, D P; MARTÍN-HERNÁNDEZ, R; NATSOPOULOU, M; PAXTON, R J; TANNER, G; WEBSTER, T C; WILLIAMS, G R (2013) Standard methods for *Nosema* research. In *V Dietemann; J D Ellis; P Neumann (Eds) The COLOSS BEEBOOK: Volume II: Standard methods for* Apis mellifera *pest and pathogen research. Journal of Apicultural Research* 52(1):
http://dx.doi.org/10.3896/IBRA.1.52.1.14

FUCHS, B M; WALLNER, G; BEISKER, W; SCHWIPPL, I; LUDWIG, W; AMANN, R (1998) Flow cytometric analysis of the *in situ* accessibility of *Escherichia coli* 16S rRNA for fluorescently labelled oligonucleotide probes. *Applied and Environmental Microbiology* 64(12): 4973–4982.
http://dx.doi.org/10.1128/AEM.67.2.961-968.2001

FUKATSU, T (1999) Acetone preservation: a practical technique for molecular analysis. *Molecular Ecology* 8(11): 1935–1945.
http://dx.doi.org/10.1046/j.1365-294x.1999.00795.x

FUKATSU, T; WATANABE, K; SEKIGUCHI, Y (1998) Specific detection of intracellular symbiotic bacteria of aphids by oligonucleotide-probed *in situ* hybridization. *Applied Entomology and Zoology.* 33 (3): 461–472.

GILLIAM, M (1978) Bacteria belonging to the genus *Bacillus* isolated from selected organs of queen honey bees, *Apis mellifera*. *Journal of Invertebrate Pathology* 31(3): 389–391.
http://dx.doi.org/10.1016/0022-2011(78)90235-5

GILLIAM, M; PREST, D B (1972) Fungi isolated from the intestinal contents of foraging worker honey bees, *Apis mellifera*. *Journal of Invertebrate Pathology* 20(1): 101–103.
http://dx.doi.org/10.1016/0022-2011(72)90087-0

GILLIAM, M; PREST, D B (1987) Microbiology of faeces of the larval honey bee, *Apis mellifera*. *Journal of Invertebrate Pathology* 49 (1): 70–75. http://dx.doi.org/10.1016/0022-2011(87)90127-3

GILLIAM, M; VALENTINE, D K (1974) Enterobacteriaceae isolated from foraging worker honey bees, *Apis mellifera*. *Journal of Invertebrate Pathology* 23(1): 38–41.
http://dx.doi.org/10.1016/0022-2011(74)90069-X

GILLIAM, M; WICKERHAM, L J; MORTON, H L; MARTIN, R D (1974) Yeasts isolated from honey bees, *Apis mellifera*, fed 2,4-D and antibiotics. *Journal of Invertebrate Pathology* 24(3): 349–356.
http://dx.doi.org/10.1016/0022-2011(74)90143-8

GOTELLI, N J (2008) *A primer of ecology (4th Ed.).* Sinauer Associates, Inc.; Sunderland, Massachusetts, USA. 290 pp.

GOTELLI, N J; COLWELL, R K (2001) Quantifying biodiversity: procedures and pitfalls in the measurement and comparison of species richness. *Ecology Letters* 4(4): 379–391.
http://dx.doi.org/10.1046/j.1461-0248.2001.00230.x

HAAS, B J; GEVERS, D; EARL, A M; FELDGARDEN, M; WARD, D V; GIANNOUKOS, G; CIULLA, D; TABBAA, D; HIGHLANDER, S K; SODERGREN, E; METHÉ, B; DESANTIS, T Z; HUMAN MICROBIOME CONSORTIUM; PETROSINO, J F; KNIGHT, R; BIRREN, B W (2011) Chimeric 16S rRNA sequence formation and detection in Sanger and 454-pyrosequenced PCR amplicons. *Genome Research* 21(3): 494–504.
http://dx.doi.org/10.1101/gr.112730.110

HAMADY, M; LOZUPONE, C; KNIGHT, R (2009) Fast UniFrac: facilitating high-throughput phylogenetic analyses of microbial communities including analysis of pyrosequencing and PhyloChip data. *The ISME Journal* 4(1): 17–27. http://dx.doi.org/10.1038/ismej.2009.97

HAMMES, W P; HERTEL, C (2009) Genus Lactobacillus Beijerinck 1901, 212[AL]. In *P De Vos; G M Garrity; D Jones; N R Krieg; W Ludwig; F A Rainey; K Schleifer; W B Whitman (Eds). Bergey's manual of systematic bacteriology (2nd Ed.). Vol. 3,* The Firmicutes. Springer; New York, USA. pp 465–490.

HANDELSMAN, J (2004) Metagenomics: application of genomics to uncultured microorganisms. *Microbiology and Molecular Biology Reviews* 68(4): 669-685.
http://dx.doi.org/10.1128/MMBR.68.4.669-685.2004

HUANG, Y; NIU, B; GAO, Y; FU, L; LI, W (2010) CD-HIT Suite: a web server for clustering and comparing biological sequences. *Bioinformatics* 26(5): 680–682.
http://dx.doi.org/10.1093/bioinformatics/btq003

HURLBERT, S H (1971) The nonconcept of species diversity: a critique and alternative parameters. *Ecology* 52(4): 577–586.
http://dx.doi.org/10.2307/1934145

HUSE, S; HUBER, J; MORRISON, H (2007) Accuracy and quality of massively parallel DNA pyrosequencing. *Genome Biology* 8(7): R143.
http://dx.doi.org/10.1186/genomebiology.com/2007/8/7/R143

JENSEN, A B; ARONSTEIN, K; FLORES, J M; VOJVODIC, S; PALACIO, M A; SPIVAK, M (2013) Standard methods for fungal brood disease research. In *V Dietemann; J D Ellis, P Neumann (Eds) The COLOSS BEEBOOK: Volume II: Standard methods for* Apis mellifera *pest and pathogen research. Journal of Apicultural Research* 52(1):
http://dx.doi.org/10.3896/IBRA.1.52.1.13

JEYAPRAKASH, A; HOY, M A; ALLSOPP, M H (2003) Bacterial diversity in worker adults of *Apis mellifera capensis* and *Apis mellifera scutellata* (Insecta: Hymenoptera) assessed using 16S rRNA sequences. *Journal of Invertebrate Pathology* 84(2): 96–103.
http://dx.doi.org/10.1016/j.jip.2003.08.007

KEMBEL, S W; WU, M; EISEN, J A; GREEN, J L (2012) Incorporating 16S gene copy number information improves estimates of microbial diversity and abundance. *PLoS Computational Biology* 8 (10): e1002743. http://dx.doi.org/10.1371/jounnal.pcbi.1002743

KILLER, J; KOPEČNÝ, J; MRÁZEK, J; KOPPOVÁ, I; HAVLÍK, J; BENADA, O; KOTT, T (2011) *Bifidobacterium actinocoloniiforme* sp. nov. and *Bifidobacterium bohemicum* sp. nov., from the bumble bee digestive tract. *International Journal of Systematic and Evolutionary Microbiology* 61(6): 1315–1321. http://dx.doi.org/10.1099/ijs.0.022525-0

KLEEREBEZEM, M; VAUGHAN, E E (2009) Probiotic and gut lactobacilli and bifidobacteria: molecular approaches to study diversity and activity. *Annual Review of Microbiology* 63: 269-290. http://dx.doi.org/10.1146/annurev.micro.091208.073341

KOCH, H; ABROL, D P; LI, J; SCHMID-HEMPEL, P (2013) Diversity and evolutionary patterns of bacterial gut associates of corbiculate bees. *Molecular Ecology* 22(7): 2028-2044. http://dx.doi.org/10.1111/mec.12209

KOCH, H; CISAROVSKY, G; SCHMID-HEMPEL, P (2012) Ecological effects on gut bacterial communities in wild bumble bee colonies. *Journal of Animal Ecology* 81(6): 1202–1210. http://dx.doi.org/10.1111/j.1365-2656.2012.02004.x

KOCH, H; SCHMID-HEMPEL, P (2011) Bacterial communities in central European bumble bees: low diversity and high specificity. *Molecular Ecology* 62(1): 121–133. http://dx.doi.org/10.1007/s00248-011-9854-3

KOCH, H; SCHMID-HEMPEL, P (2012) Gut microbiota instead of host genotype drive the specificity in the interaction of a natural host-parasite system. *Ecology Letters* 15(10): 1095–1103. http://dx.doi.org/10.1111/j.1461-0248.2012.01831.x

KOGA, R; TSUCHIDA, T; FUKATSU, T (2009) Quenching autofluorescence of insect tissues for *in situ* detection of endosymbionts. *Applied Entomology and Zoology* 44(2): 281-291. http://dx.doi.org/10.1303/aez.2009.281

KUCZYNSKI, J; STOMBAUGH, J; WALTERS, W A; GONZÁLEZ, A; CAPORASO, J G; KNIGHT, R (2011) Using QIIME to analyse 16S rRNA gene sequences from microbial communities. *Current Protocols in Bioinformatics* Unit 10.7: 1-20. http://dx.doi.org/10.1002/0471250953.bi1007s36

KUNIN, V; HUGENHOLTZ, P (2010) PyroTagger: A fast, accurate pipeline for analysis of rRNA amplicon pyrosequence data. *The Open Journal* 1: 1–8.

KWONG, W K; MORAN, N A (2013) Cultivation and characterization of the gut symbionts of honey bees and bumble bees: *Snodgrassella alvi* gen. nov., sp. nov., a member of the *Neisseriaceae* family of the *Betaproteobacteria*; and *Gilliamella apicola* gen. nov., sp. nov., a member of *Orbaceae* fam. nov., *Orbales* ord. nov., a sister taxon to the *Enterobacteriales* order of the *Gammaproteobacteria*. *International Journal of Systematic and Evolutionary Microbiology*, 62: 2008-2018. http://dx.doi.org/10.1099/ijs.0.044875-0

LI, J; QIN, H; WU, J; SADD, B M; WANG, X; EVANS, J D; PENG, W; CHEN, Y (2012) The prevalence of parasites and pathogens in Asian honey bees *Apis cerana* in China. *PloS one* 7(11):e47955. http://dx.doi.org/10.1371/journal.pone.0047955

LONCARIC, I; HEIGL, H; LICEK, E; MOOSBECKHOFER, R; BUSSE, H J; ROSENGARTEN, R (2009) Typing of *Pantoea agglomerans* isolated from colonies of honey bees (*Apis mellifera*) and culturability of selected strains from honey. *Apidologie* 40(1): 40-54. http://dx.doi.org/10.1051/apido/2008062

LOY, A; MAIXNER, F; WAGNER, M; HORN, M (2007) probeBase--an online resource for rRNA-targeted oligonucleotide probes: new features 2007. *Nucleic Acids Research* 35(suppl 1): D800–4. http://dx.doi.org/10.1093/nar/gkl856

LOZUPONE, C; KNIGHT, R (2005) UniFrac: a new phylogenetic method for comparing microbial communities. *Applied and Environmental Microbiology* 71(12): 8228–8235. http://dx.doi.org/10.1128/AEM.71.12.8228-8235.2005

MAGURRAN, A E (2004) *Measuring biological diversity*. Blackwell Publishing; Malden, Massachusetts, USA. 264 pp.

MARMUR, J (1961) A procedure for the isolation of deoxyribonucleic acid from micro-organisms. *Journal of Molecular Biology* 3(2): 208 –218. http://dx.doi.org/10.1016/S0022-2836(61)80047-8

MARTINSON, V G; DANFORTH, B N; MINCKLEY, R L; RUEPPELL, O; TINGEK, S; MORAN, N A (2011) A simple and distinctive microbiota associated with honey bees and bumble bees. *Molecular Ecology* 20(3): 619-628. http://dx.doi.org/10.1111/j.1365-294X.2010.04959.x

MARTINSON, V G; MOY, J; MORAN, N A (2012) Establishment of characteristic gut bacteria during development of the honey bee worker. *Applied and Environmental Microbiology* 78(8): 2830– 2840. http://dx.doi.org/10.1128/AEM.07810-11

MATTILA, H R; RIOS, D; WALKER-SPERLING, V E; ROESELERS, G; NEWTON, I L (2012) Characterization of the active microbiotas associated with honey bees reveals healthier and broader communities when colonies are genetically diverse. *PLoS One* 7 (3): e32962. http://dx.doi.org/10.1371/journal.pone.0032962

MCDONALD, D; PRICE, M N; GOODRICH, J; NAWROCKI, E P; DESANTIS, T Z; PROBST, A; ANDERSEN, G L; KNIGHT, R; HUGENHOLTZ, P (2012) An improved Greengenes taxonomy with explicit ranks for ecological and evolutionary analyses of Bacteria and Archaea. *The ISME Journal* 6(3): 610–618. http://dx.doi.org/10.1038/ismej.2011.139

MCFREDERICK, Q S; WCISLO, W T; TAYLOR, D R; ISHAK, H D; DOWD, S E; MUELLER, U G (2012) Environment or kin: whence do bees obtain acidophilic bacteria? *Molecular Ecology* 21(7): 1754-1768. http://dx.doi.org/10.1111/j.1365-294X.2012.05496.x

MCFREDERICK, Q S; CANNONE, J J; GUTELL, R R; KELLNER, K; PLOWES, R M; MUELLER, U G (2013) Host specificity between Hymenoptera and Lactobacilli is the exception rather than the rule. *Applied and Environmental Microbiology* 79(6): 1803-1812.

MOHR, K I; TEBBE, C C (2006) Diversity and phylotype consistency of bacteria in the guts of three bee species (*Apoidea*) at an oilseed rape field. *Environmental Microbiology* 8(2): 258–272. http://dx.doi.org/10.1111/j.1462-2920.2005.00893.x

MOHR, K I; TEBBE, C C (2007) Field study results on the probability and risk of a horizontal gene transfer from transgenic herbicide-resistant oilseed rape pollen to gut bacteria of bees. *Applied Microbiology and Biotechnology* 75(3): 573-82. http://dx.doi.org/10.1007/s00253-007-0846-7

MORAN, N A; HANSEN, A K; POWELL, J E; SABREE, Z L (2012) Distinctive gut microbiota of honey bees assessed using deep sampling from individual worker bees. *PLoS One* 7(4): e36393. http://dx.doi.org/10.1371/journal.pone.0036393

NEVELING, D P; ENDO, A; DICKS, L M (2012) Fructophilic *Lactobacillus kunkeei* and *Lactobacillus brevis* isolated from fresh flowers, bees and bee-hives. *Current Microbiology* 65(5): 507-15. http://dx.doi.org/10.1007/s00284-012-0186-4

NEWTON, I L; ROESELERS, G (2012) The effect of training set on the classification of honey bee gut microbiota using the Naïve Bayesian Classifier. *BMC Microbiology* 12: 221. http://dx.doi.org/10.1186/1471-2180-12-221

OCHMAN, H; WOROBEY, M; KUO, C H; NDJANGO, J B N; PEETERS, M; HAHN, B H; HUGENHOLTZ, P (2010) Evolutionary relationships of wild hominids recapitulated by gut microbial communities. *PLoS Biology* 8(11): e1000546. http://dx.doi.org/10.1371/journal.pbio.1000546

OLOFSSON, T C; VÁSQUEZ, A (2008) Detection and identification of a novel lactic acid bacterial flora within the honey stomach of the honey bee *Apis mellifera*. *Current Microbiology* 57(4): 356-363. http://dx.doi.org/10.1007/s00284-008-9202-0

PRESNELL, J K; SCHREIBMAN, M P; HUMASON, G L (1997) *Humason's animal tissue techniques (5th Ed.)*. Johns Hopkins University Press; Baltimore, USA. 572 pp

QUAST, C; PRUESSE, E; YILMAZ, P; GERKEN, J; SCHWEER, T; YARZA, P; PEPLIES, J; GLÖCKNER, F O (2013) The SILVA ribosomal RNA gene database project: improved data processing and web-based tools. *Nucleic Acids Research* 41(D1): D590–D596. http://dx.doi.org/10.1093/nar/gks1219

QUINN, G P; KEOUGH, M J (2002) *Experimental design and data analysis for biologists*. Cambridge University Press; Cambridge, UK. 556 pp.

RAMETTE, A (2007) Multivariate analyses in microbial ecology. *FEMS Microbiology Ecology* 62(2): 142–160. http://dx.doi.org/10.1111/j.1574-6941.2007.00375.x

RAPPÉ, M S; GIOVANNONI, S J (2003) The uncultured microbial majority. *Annual Review of Microbiology* 57: 369–394. http://dx.doi.org/10.1146/annurev.micro.57.030502.090759

ROZEN, S; SKALETSKY, H (2000) Primer3 on the WWW for general users and for biologist programmers. *Methods in Molecular Biology* 132: 365–386. http://dx.doi.org/10.1385/1-59259-192-2:365

SABREE, Z L; HANSEN, A K; MORAN, N A (2012) Independent studies using deep sequencing resolve the same set of core bacterial species dominating gut communities of honey bees. *PloS One* 7 (7): e41250. http://dx.doi.org/10.1371/journal.pone.0041250

SCARDOVI, V; TROVATELLI, L D (1969) New species of bifid bacteria from *Apis mellifera* L. and *Apis indica* F. A contribution to the taxonomy and biochemistry of the genus *Bifidobacterium*. *Zentralblatt fur Bakteriologie, Parasitenkunde, Infektionskrankheiten und Hygiene*. 123: 64-88.

SCHLOSS, P D (2008) Evaluating different approaches that test whether microbial communities have the same structure. *The ISME Journal* 2(3): 265–275. http://dx.doi.org/10.1038/ismej.2008.5

SCHLOSS, P D (2009) A high-throughput DNA sequence aligner for microbial ecology studies. *PLoS One* 4(12): e8230. http://dx.doi.org/10.1371/journal.pone.0008230

SCHLOSS, P D; GEVERS, D; WESTCOTT, S L (2011) Reducing the effects of PCR amplification and sequencing artifacts on 16S rRNA -based studies. *PLoS One* 6(12): e27310. http://dx.doi.org/10.1371/journal.pone.0027310

SCHLOSS, P D; WESTCOTT, S L (2011) Assessing and improving methods used in operational taxonomic unit-based approaches for 16S rRNA gene sequence analysis. *Applied and Environmental Microbiology* 77(10): 3219–3226. http://dx.doi.org/10.1128/AEM.02810-10

SCHLOSS, P D; WESTCOTT, S L; RYABIN, T; HALL, J R; HARTMANN, M; HOLLISTER, E B; LESNIEWSKI, R A; OAKLEY, B B; PARKS, D H; ROBINSON, C J; SAHL, J W; STRES, B; THALLINGER, G G; VAN HORN, D J; WEBER, C F (2009) Introducing mothur: open-source, platform-independent, community-supported software for describing and comparing microbial communities. *Applied and Environmental Microbiology* 75(23): 7537–7541. http://dx.doi.org/10.1128/AEM.01541-09

SHANNON, C E (1948) A mathematical theory of communication. *The Bell System Technical Journal*, 27: 379–423, 623–656.

SIMPSON, E H (1949) Measurement of diversity. *Nature* 163: 688. http://dx.doi.org/10.1038/163688a0

SOGIN, M L; MORRISON, H G; HUBER, J A; MARK WELCH, D; HUSE, S M; NEAL, P R; ARRIETA, J M; HERNDL, G J (2006) Microbial diversity in the deep sea and the underexplored "rare biosphere". *Proceedings of the National Academy of Sciences of the United States of America* 103(32): 12115–12120. http://dx.doi.org/10.1073/pnas.0605127103

SOKAL, R R; MICHENER, C (1958) A statistical method for evaluating systematic relationships. *The University of Kansas Science Bulletin* 38(2): 1409–1438.

STALEY, J T; KONOPKA, A (1985) Measurement of *in situ* activities of nonphotosynthetic microorganisms in aquatic and terrestrial habitats. *Annual Review of Microbiology* 39: 321–346. http://dx.doi.org/10.1146/annurev.mi.39.100185.001541

STEVENSON, B S; EICHORST, S A; WERTZ, J T; SCHMIDT, T M; BREZNAK, J A (2004) New strategies for cultivation and detection of previously uncultured microbes. *Applied and Environmental Microbiology* 70(4): 4748-4755. http://dx.doi.org/10.1128/AEM.70.8.4748-4755.2004

TAJABADI, N; MARDAN, M; ABDUL MANAP, M Y; SHUHAIMI, M; MEIMANDIPOUR, A; NATEGHI, L (2011) Detection and identification of *Lactobacillus* bacteria found in the honey stomach of the giant honey bee *Apis dorsata*. *Apidologie* 42(5): 642-649. http://dx.doi.org/10.1007/s13592-011-0069-x

TIAN, B; FADHIL, N H; POWELL, J E; KWONG, W K; MORAN, N A. (2012) Long-term exposure to antibiotics has caused accumulation of resistance determinants in the gut microbiota of honey bees. *mBio* 3(6): e00377-12. http://dx.doi.org/10.1128/mBio.00377-12

TRINGE, S G; HUGENHOLTZ, P (2008) A renaissance for the pioneering 16S rRNA gene. *Current Opinions in Microbiology* 11 (5): 442–446. http://dx.doi.org/10.1016/j.mib.2008.09.011

VALM, A M; MARK WELCH, J L; RIEKEN, C W; HASEGAWA, Y; SOGIN, M L; OLDENBOURG, R; DEWHIRST, F E; BORISY, G G (2011) Systems-level analysis of microbial community organization through combinatorial labelling and spectral imaging. *Proceedings of the National Academy of Sciences of the United States of America* 108(10): 4152–4157. http://dx.doi.org/10.1073/pnas.1101134108

VÁSQUEZ, A; FORSGREN, E; FRIES, I; PAXTON, R J; FLABERG, E; SZEKELY, L; OLOFSSON, T C (2012) Symbionts as major modulators of insect health: lactic acid bacteria and honey bees. *PLoS One*. 7(3):e33188. http://dx.doi.org/10.1371/journal.pone.0033188

WAGNER, M; HORN, M; DAIMS, H (2003) Fluorescence *in situ* hybridisation for the identification and characterisation of prokaryotes. *Current Opinion in Microbiology* 6(3): 302–309. http://dx.doi.org/10.1016/S1369-5274(03)00054-7

WANG, Q; GARRITY, G M; TIEDJE, J M; COLE, J R (2007) Naïve Bayesian Classifier for rapid assignment of rRNA sequences into the new bacterial taxonomy. *Applied and Environmental Microbiology* 73(16): 5261–5267. http://dx.doi.org/10.1128/AEM.00062-07

WARNECKE, F; LUGINBUHL, P; IVANOVA, N; GHASSEMIAN, M; RICHARDSON, T H (2007) Metagenomic and functional analysis of hindgut microbiota of a wood-feeding higher termite. *Nature* 450 (7169): 560-565.

WHITTAKER, R H (1972) Evolution and measurement of species diversity. *Taxon* 21(2/3): 213–251.

YILMAZ, L S; ÖKTEN, H E; NOGUERA D R (2006) Making all parts of the 16S rRNA of *Escherichia coli* accessible *in situ* to single DNA oligonucleotides. *Applied and Environmental Microbiology* 72(1): 733-744. http://dx.doi.org/10.1128/AEM.72.1.733-744.2006

Journal of Apicultural Research 52(4): (2013)
DOI 10.3896/IBRA.1.52.4.08

REVIEW ARTICLE

Standard use of Geographic Information System (GIS) techniques in honey bee research

Stephanie R Rogers[1]* and Benno Staub[1]

[1]Department of Geosciences, Geography, University of Fribourg, Fribourg, Switzerland.

Received 4 January 2013, accepted subject to revision 11 February 2013, accepted for publication 11 March 2013.

*Corresponding author: Email: stephrogers5@gmail.com

Summary

Geographic Information Systems (GIS) have been used in various fields and disciplines to summarize and analyse spatial patterns and distributions, for the purpose of understanding how geographic and non-geographic entities interact with each other over space and time. Although honey bees are directly related to and influenced by their local environment, few studies have incorporated honey bee data into GIS for the purposes of gauging these spatial relationships. This paper will briefly discuss some of the types of spatial analyses and GIS methods that have been used for bees, and also, how some methodologies developed in the non-*Apis* bee domain could be applied to honey bee research. With this paper, we aim to stimulate spatial thinking processes and thus the future use of GIS analyses to better understand the relationships between environmental characteristics and honey bee health and abundance. We will introduce the framework and some important basic concepts of GIS, as well as provide detailed instructions for becoming familiar and comfortable in using the GIS softwares ArcGIS and Quantum GIS (QGIS) (a commercial and free GIS package) for the basics of geospatial research.

Uso estándar de las técnicas de Sistemas de Información Geográfica (SIG) en la investigación de la abeja de la miel

Resumen

Los Sistemas de Información Geográfica (SIG) se han utilizado en diversos campos y disciplinas para resumir y analizar los patrones y distribuciones espaciales, con el fin de entender cómo las entidades geográficas y no geográficas interactúan entre sí en el espacio y en el tiempo. Aunque las abejas melíferas están directamente relacionadas e influidas por su ambiente local, pocos estudios han incluido estos datos en los SIG para analizar estas relaciones espaciales. Este artículo discutirá brevemente algunos de los tipos de análisis espaciales y métodos de SIG que se han utilizado para las abejas, así como también, algunas metodologías desarrolladas en dominios no-*Apis* que podrían ser aplicadas a la investigación de la abeja de la miel. El objetivo de este trabajo es fomentar los procesos de pensamiento espacial y por lo tanto el futuro uso de los análisis SIG para entender mejor las relaciones entre las características ambientales con la salud de las abejas melíferas y con la abundancia de estas. Vamos a introducir el marco y algunos conceptos básicos importantes de los SIG, así como proporcionar instrucciones detalladas para familiarizar y facilitar el uso de los programas SIG, ArcGIS y Quantum GIS (QGIS) (un paquete SIG comercial y libre) para los fundamentos de la investigación geoespacial.

Footnote: Please cite this paper as: ROGERS, S R; STAUB, B (2013) Standard use of Geographic Information System (GIS) techniques in honey bee research. In V Dietemann; J D Ellis; P Neumann (Eds) *The COLOSS BEEBOOK, Volume I: standard methods for* Apis mellifera *research. Journal of Apicultural Research* 52(4): http://dx.doi.org/10.3896/IBRA.1.52.4.08

NOTE: Figures prefixed by "S" are contained in the Online Supplementary Material to be found at http://www.ibra.org.uk/downloads/20130812_23/download

地理信息系统技术在蜜蜂研究中的标准应用

摘要

为了理解地理和非地理的实体之间如何在空间和时间上互作，地理信息系统（GIS）已被用于多种领域和学科以概括和分析空间模式和分布。尽管蜜蜂与它们的生活区域环境直接相关并受其影响，但很少有研究将蜜蜂的数据与GIS结合起来测定这些空间关系。本文将简要讨论已用于蜜蜂的一些类型的空间分析和GIS方法，以及如何将一些在非蜜蜂领域发展起来的方法应用于蜜蜂研究。通过本文，我们意图刺激空间思考的进程，从而实现未来通过GIS分析来更好的理解环境特征与蜜蜂健康和丰富度的关系。我们将介绍GIS的框架和一些重要的基本概念，同时提供GIS软件

ArcGIS和Quantum GIS（两者分别是商业化和免费的GIS包）详细的使用说明，使人们熟悉并轻松地使用以作为地理空间研究的基础。

Keywords: COLOSS *BEEBOOK*, GIS, spatial analysis, honey bee, ArcGIS, Quantum GIS

地理信息系统技术在蜜蜂研究中的标准应用

Table of Contents

Table of Contents Cont'd

1. Introduction

Due to the close relationship between honey bees and their surroundings, and because of recent increases in colony mortality in many regions of the world (Carreck and Neumann, 2010; Williams *et al.*, 2010), there is an urgent need to better understand how environmental changes affect habitat, life patterns, and the overall health of honey bees. This type of research can be facilitated through the use of Geographic Information Systems (GIS), which can briefly be described as a software system for handling and analysing map

Table 1. Publication names, authors, and websites of free GIS Books and information sources.

Name of publication	Author(s), year	Website
Map Projections: A Working Manual	Snyder, 1987	http://pubs.er.usgs.gov/publication/pp1395
Geospatial Analysis - A comprehensive guide	De Smith, Goodchild, and Longley, 2007	http://www.spatialanalysisonline.com
A Practical Guide to Geostatistical Mapping	Hengl, 2009	http://www.lulu.com/shop/tomislav-hengl/a-practical-guide-to-geostatistical-mapping/ebook/product-17389015.html
GIS Commons: An Introductory Textbook on Geographic Information Systems	Schmandt, 2009	http://giscommons.org/
Various ESRI E-books	Many	http://www.esri.com/industries/ebooks

data through the use of coordinate systems and georeferenced information to aid in the analysis of spatial patterns that may exist between entities on the Earth's surface (Burrough and McDonnell, 1998). Because of its broad application, GIS methods are used in a diverse array of disciplines, including geoscience (Bonham-Carter, 1994), economics (Pogodzinski and Kos, 2012), and biology (Kozak *et al.,* 2008). For example, GIS has been used extensively to undertake spatial analyses of the effects of land-use and climate change on declines of native bee abundance and health (Arthur *et al.,* 2010; Biesmeijer *et al.,* 2006; Chifflet *et al.,* 2011; Choi *et al.,* 2012; Fitzpatrick *et al.,* 2007; Giannini *et al.,* 2012; Kremen, 2002; Kremen *et al.,* 2004; Watson *et al.,* 2011; Williams *et al.,* 2012).

Through the use of spatial analyses, researchers using GIS were able to determine that native bees were more efficient crop pollinators when access to nearby forests was provided (Arthur *et al.,* 2010; Kremen *et al.,* 2004; Watson *et al.,* 2011). Additionally, relationships between climate and native bee species richness and abundance have been identified using GIS modelling procedures (Giannini *et al.,* 2012). For honey bees, GIS analyses have been incorporated into relatively few research initiatives (Berardinelli and Vedova, 2004; Naug, 2009; Henry *et al.,* 2012). With this paper, we aim to stimulate spatial thinking processes and thus the future use of GIS analyses to better understand the relationships between environmental characteristics and honey bee health and abundance for a broader geographic overview of the mechanisms affecting their plight (Naug, 2009). In this paper, we introduce the framework and some important basic concepts of GIS, and provide detailed instructions about using the GIS softwares ArcGIS and Quantum GIS (QGIS) for the geospatial study of honey bees. We hope that the knowledge gained from this paper will further the field of honey bee science, but also promote the use of GIS in other fields of study.

2. What is GIS?

2.1. Definition

Geographical Information Systems (GIS) are computer systems based on hardware, software, and georeferenced data that can be used to collect, store, manage, process, analyse, and visualize both spatial and non-spatial information representing real-world geographic phenomena (Burrough and McDonnell, 1998; Neteler and Mitasova, 2008). Georeferenced data refers to any data that are linked to a location on the Earth's surface through the use of a geographic or projected coordinate system. GIS data are digital objects which represent real-world entities (Longley *et al.,* 1999) and are defined by: their geometric properties (spatial location), their attributes (characteristics associated with each object), and their topology (definition of how entities are related to others in space) (Burrough and McDonnell, 1998). In other words, data provide means to locate them in space and can be overlaid, calculated, manipulated, visualised and analysed along with other data layers that use the same coordinate system. Each entity in the real world is represented by a data layer with geometric and topologic properties and an associated set of attributes, in the form of a table, which define the characteristics of that entity. GIS facilitates the analysis of spatial relationships within datasets based on the topological properties within the data. Topology refers to the interconnectivity and interrelated properties between data and defines and describes how spatial objects relate to their neighbours in space. This geographic descriptor is what sets GIS data apart from other data types (Burrough and McDonnell, 1998) (for a list of free GIS books and other information sources see Table 1).

2.2. GIS framework

The main principles of GIS (geometric properties, attributes, and topology) are built into two data structures: vector (e.g. points, lines, and polygons) and raster (e.g. climatic, altitude, or satellite imagery). The terms "layer" or "dataset" are used to denote a raster or a vector file type that contains a similar theme, for example, varying elevations across a terrain or locations of honey bee colonies, respectively. It is possible to convert a raster layer to a vector layer and vice versa, depending on what type of analysis will be conducted with the data. Each layer contains information about its geometric properties and a table of attributes associated to and linked with, its respective geometric properties (Fig. 1). You can view each layer's properties in GIS to see the resolution, file type, size of the file, and other information of importance.

2.2.1. Vector data structure

Vector layers represent discrete features in space in the forms of points, lines, and polygons. They represent unchanging static entities

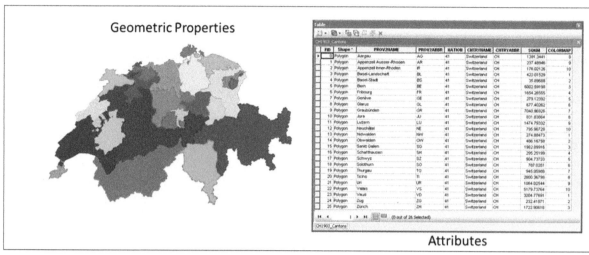

Fig. 1. The GIS data framework is made up of both geometric properties and attributes associated with the geometry. A row in the attribute table exists for every polygon on the map. Some of the attributes for this Swiss layer include the names of cantons and their abbreviations, the country code, the country name and abbreviation, and the area. Reproduced with permission of Swisstopo (BA13016).

and do not contain spatial or temporal information (Burrough and McDonnell, 1998). For example, a polygon layer could represent the extent of an apiary, and a point layer could represent a honey bee colony on a map. Line layers can represent rivers, roads, tracks, and any other linear features. In Fig. 2 a polygon, point, and line layer are shown as examples.

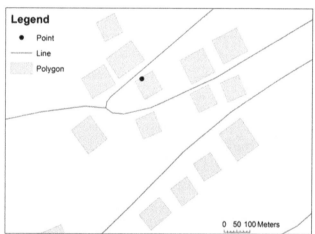

Fig. 2. A map displaying a point, lines, and polygons. Reproduced with permission of Swisstopo (BA13016).

2.2.2. Raster data structure

Raster layers represent continuous features such as aerial and satellite imagery, as well as Digital Elevation Models (DEM), which store elevation data across a surface (see section 3.2.4. for more information). The value of each pixel of a thematic raster layer represents the attribute value. For satellite imagery, pixel values represent physical values of surface reflectances and for terrain-model

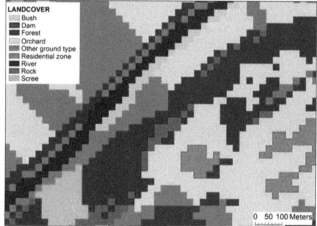

Fig. 3. A map displaying the raster data structure of a landcover layer which was converted from vector format. The legend denotes the landcover properties which are displayed. Reproduced with permission of Swisstopo (BA13016).

data pixel values represent absolute topographic heights (Campbell, 2002). For example, if the resolution of the image is 25 m, then each 25 m pixel in the image will be representative of the average value of that 25 m on the ground. In Fig. 3, a land cover vector layer was transformed to a raster to visualize the pixel differences and value types.

2.3. GIS functionalities

The main functionalities of GIS include interfaces for database management, geoprocessing and spatial analysis, as well as visualisation and map creation. Here, these functionalities are introduced and the specific tools will be discussed more in-depth in further sections.

2.3.1. Database management

A GIS database is used to store, create, organize, manipulate, and query spatial and their linked non-spatial datasets (Burrough and McDonnell, 1998). GIS facilitates the link between geographic and non-geographic data and enables the comparison and assessment of various data types (Rigaux *et al.*, 2002). Compared to traditional (non-spatial) databases (e.g. those without geographic coordinates), geographic databases make explicit locational distinctions for each dataset stored in the database, that is, all geographic information is linked to a location on the earth's surface through coordinates (Arctur and Zeiler, 2004).

2.3.2. Geoprocessing and spatial analysis

GIS can be used to calculate, analyse, and manipulate spatial data to examine spatial relationships and to create new data (De Smith *et al.*, 2007). Basic manipulations of geographic datasets such as extracting new information from existing layers or clipping specific geographic extents can be referred to as geoprocessing. Geoprocessing refers to the process of creating a new geographic data layer after the calculation of an input layer(s) (Wade and Sommer, 2006).

Spatial analysis is a central concept in GIS and refers to more complex calculations which have been developed from various quantitative methodologies outside of GIS and incorporated into GIS over time (Longley *et al.*, 1999; Conolly and Lake, 2006; De Smith *et al.*, 2007). Spatial analyses can be used to summarize and analyse the spatial properties of geographic distributions, to solve spatial problems through modelling, and finally, to aid in spatial decision making (Longley *et al.*, 1999). Because this is a basic introduction to GIS, we will focus more on the geoprocessing aspects of GIS and touch briefly on more complex spatial analyses.

2.3.3. Data visualisation and map creation

One of the main benefits of GIS is that it gives researchers the ability to visualize their data in a variety of formats, which enables the creation of new links and relationships between spatial entities while discovering spatial interactions that were perhaps unknown before. GIS is also a common cartographic platform which facilitates the production of maps, either to present a geographic area or the results of data analysis using spatial data. The results of GIS analysis can be displayed as maps using a variety of symbolisations and annotation features as the final step in a GIS investigation.

3. Geographic data

GIS data represent real-world phenomena based on their geometric properties, attributes, and topology (Burrough and McDonnell, 1998).

3.1. File types and extensions

Often, GIS software has its own file types for vector and raster data. For example, the Environmental Systems Research Institute's (ESRI) commercial ArcGIS software most commonly uses shapefiles for vector data and grids for raster data. Shapefiles and grids have come to be some of the most commonly known GIS formats and can often be read by and incorporated into other GIS programs without the need for conversion, as is the case for Quantum GIS (QGIS). It is important to note that most layers are made up of multiple files with different file extensions. For example, a shapefile may look like a single file in the GIS, but actually, it is composed of multiple file extensions. It is detrimental to the data layer if one of these extensions gets lost or deleted. Some of the extensions often associated with shapefiles are: *.shp (required file that stores the feature geometry), *.shx (required file that stores the index of the feature geometry), *.dbf (database table that stores the attribute information) or *.prj (file that stores the coordinate system information). The entire list of extensions can be found by searching "shapefile file extensions" in the ArcGIS help menu (http://resources.arcgis.com/en/help/main/10.1/).

Multiple raster dataset file formats including ASCII (*.asc), GeoTiff (*.tif), JPEG (*.jpg), and GIFs (*.gif), to name only a few, can be supported in most GIS software. Again it is important to keep all files when copying and moving data. For example, a Geotiff file often comes with two files, *.tif and *.tfw. The *.tif holds the image, and the *.tfw holds the information about the georeference.

3.2. Common GIS data types

The geometric aspect associated with GIS data is what separates geographic databases from other databases (Burrough and McDonnell, 1998). Geographic databases are sometimes referred to as geodatabases, which refer to databases specific to geographic data. Until recently, most GIS data were generated and distributed by government or private industries and had to be purchased. Now, there are many sources for downloading free data (Table 2). The amount of available data largely depends on the area of interest, as urban areas tend to have more data available than rural or remote areas. Another important aspect of geographic data is the metadata. Metadata are the information provided about the data, thus, data about data (Burrough and McDonnell, 1998). When downloading data online or receiving data from an outside source, metadata are important for obtaining knowledge about when the data were created, for what purpose, by whom, and using which coordinate system, among other things. Without this information, data would be of limited use only. In the following section we will introduce some of the common file formats and extensions, as well as some geographic

Table 2. Names, websites, and types of data available from various free data sources.

Name	Website	Types of data
Aster GDEM	http://asterweb.jpl.nasa.gov/gdem.asp	30 m DEM of the world
CloudMade	downloads.cloudmade.com	Layers created from Open Street Maps for various places in the world
Corine Land Cover 2006 (European Environment Agency (EEA))	http://www.eea.europa.eu/data-and-maps/data/corine-land-cover-2006-raster-2	Land cover raster data for Europe 100 m and 250 m resolutions
European Environment Agency (EEA)	http://www.eea.europa.eu/data-and-maps/data#c5=all&b_start=0&c9=CLC2006	Land use, economics, natural resources, water, etc.
FOSSGIS (Free and Open Source Software for GIS)	http://freegis.org/	Many from multiple sources
Global Land Cover Facility (GLCF)	Glcf.umd.edu/data/	Satellite imagery, products derived from satellite imagery, vector products
Open Street Map	www.openstreetmap.org	Global or individual country or state map data
Swisstopo Landsat Image	http://www.swisstopo.admin.ch/internet/swisstopo/en/home/products/images/ortho/landsat.html	Digital satellite mosaic
United Nations Geographic Information Working Group (UNGIWG) Second Administrative Level Boundaries (SALB)	http://www.unsalb.org/	International borders
United States Geological Survey (USGS) Earth Resources Observation and Science (EROS) Center	http://eros.usgs.gov/#/Find_Data	Aerial Photography Satellite Imagery Elevation Land Cover Digitized Maps Image Gallery Collections

data examples. The possibilities of types of data in GIS are becoming broader as the technology develops. The following is a non-exhaustive list that shows some of the data types you may encounter when using GIS.

3.2.1. Maps

Topographic maps (Fig. 4) are representations of the terrain which include both natural and man-made features, including, relief (in the form of contour lines), towns, villages, roads, and other geographic characteristics (Burrough and McDonnell, 1998). They are those maps which people are most familiar with. In GIS, topographic maps are stored in raster format in their entirety, but can also be broken down into their respective categories (rivers, buildings, roads, etc.) in vector layers.

3.2.2. Aerial imagery

Photos taken from the air, usually from an aircraft, are referred to as aerial imagery or photographs (Fig. 5). These images provide valuable information about landscape changes by comparing photos over time. When aerial images are georeferenced, i.e. located with respect to a geographic or projected coordinate system, they can be used as raster layers in GIS (Burrough and McDonnell, 1998). The different types of aerial images include black and white, true colour, panchromatic, and infrared and can each be used for different purposes (Campbell, 2002).

Fig. 4. A sample topographic map (scale 1: 200 000) covering the study area of Fribourg, Switzerland. This map comes from Swisstopo (2013) and must only be used for visualization purposes in this paper. Reproduced with permission of Swisstopo (BA13016).

3.2.3. Satellite imagery

Satellite images are those that have been taken from satellites. Landsat (Fig. 6) is one of the longest running platforms for image acquisition using the Thematic Mapper (TM) or the Enhanced

Fig. 5. A sample orthophoto, with a resolution of 50 cm, showing the study area of Fribourg, Switzerland. This photo comes from Swisstopo (2013) and must only be used for visualization purposes in this paper. Reproduced with permission of Swisstopo (BA13016).

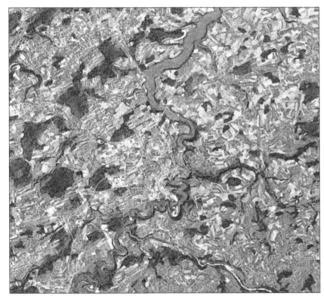

Fig. 6. A section of the Swisstopo (2013) Landsat image (freely available from at http://www.swisstopo.admin.ch/internet/swisstopo/en/home/products/images/ortho/landsat.html) with a resolution of 25m, covering the study area of Fribourg, Switzerland. Reproduced with permission of Swisstopo (BA13016).

Thematic Mapper (ETM) satellite image acquisition systems (http://landsat.gsfc.nasa.gov/) which provide multispectral imagery for the whole world (Campbell, 2002). Satellite imagery is often expensive to buy as the equipment is difficult to operate and maintain, but conveniently, Landsat and other satellite images can be added as raster layers directly into the GIS interface.

3.2.4. Digital Elevation Model (DEM)

A DEM (Fig. 7) is a raster dataset that represents the altitude of the terrain. Depending on the resolution of the image, each pixel represents the average altitude of the terrain covered by that pixel. For example, if the image resolution is 30 m, the altitude of the terrain on the ground that is covered by that pixel within that 30 m square is averaged and the value is assigned to that pixel.

With a DEM, numerous calculations can be performed including the calculation of slope (in degrees or percent) of the terrain, the direction each slope is facing (aspect), and where shadows exist when the sun is hitting the terrain from a certain angle (hillshade). A DEM is also the base dataset for calculations of solar radiation, hydrographic, and various modelling calculations.

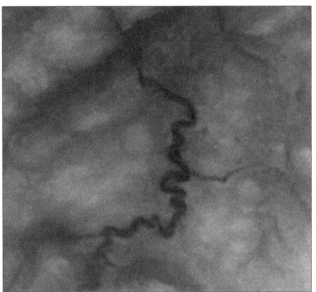

Fig. 7. A sample Digital Elevation Model (DEM) (30 m resolution) covering the study area of Fribourg, Switzerland. This map was a free download from ASTER GDEM which was projected to the Swiss projected coordinate system (CH1903 LV03).

3.2.5. Thematic vector data

Vector (Fig. 2) data represent a broad range of information, as they include all layers represented by points, lines, or polygons. In a GIS database, common vector layers might represent administrative boundaries (the locations of countries, provinces, states, cantons, etc.), hydrography (rivers, dams, lakes, etc.), built-up areas (buildings, stations, airports, etc.), and land cover.

4. Coordinate systems

This section will briefly introduce the concept of coordinate systems in GIS. For any analysis, it is important to know the coordinate system of the data which represent the study area (region of interest) and to ensure that all of your data layers are already in, or can be converted

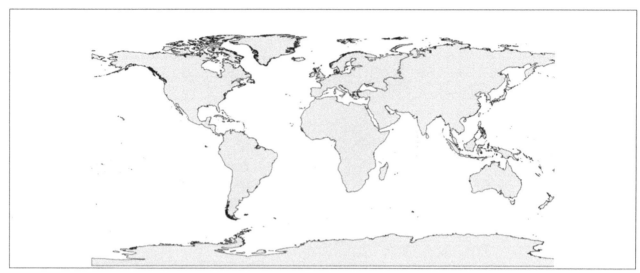

Fig. 8. A map of the world shown in the WGS 84 geographic coordinate system.

to, that coordinate system. Basic knowledge about coordinate systems is recommended before any type of spatial analysis is conducted as they can affect the resulting outputs. It is also important to know the capabilities and weaknesses of each specific coordinate system used in analysis. For more in-depth information about coordinate systems in GIS see Seeger (1999).

4.1. Geographic coordinate systems

A geographic coordinate system is defined by the ellipsoid and datum that are used to calculate longitudes and latitudes which represent locations on the Earth's surface (Snyder, 1987). It is represented in degrees, minutes, and seconds north or south of the equator (latitude) or east and west of the equator (longitude) (i.e. Fribourg, Switzerland is at 46° 48' 00" N, 7° 09' 00" E), and is the simplest solution for representing places on the globe (Burrough and McDonnell, 1998). Decimal degrees are also used to represent latitude/longitude coordinates and can be calculated with the following equation: Decimal Degrees = Degrees + Minutes/60 + Seconds/3,600 (http://support.esri.com/en/knowledgebase/techarticles/detail/27215) (i.e. same location above would be represented as 46.8 N, 7.15 E). There are various online sources for calculating conversions between latitude/longitude and decimal degrees (e.g. http://transition.fcc.gov/mb/audio/bickel/DDDMMSS-decimal.html).The most commonly used geographic coordinate system is the World Geodetic System (WGS) 1984 (Fig. 8). However, geographic coordinate systems do not transfer nicely into two dimensions, causing major distortions in distance, area, shape, or direction (Snyder 1987; Seeger, 1999) on a map or computer screen. To solve these issues, projected coordinate systems are used.

4.2. Projected coordinate systems

Projected coordinate systems are created to allow a two-dimensional representation on a screen or map sheet (Snyder, 1987). Latitudes

and longitudes are converted to meters with respect to the centre of projection by applying calculations to geographic coordinate systems to counteract and offset the distortions on the map. Map projections have been developed for both local and regional scales. There is a large body of research based on the calculations of map projections (see Snyder, 1987 or Seeger, 1999 for more information). One of the most important things to remember is that all map projections create distortion of a least one parameter of the following: distance, direction, scale, conformity (shape), and area (Snyder, 1987; Seeger, 1999). The goal of map projections is to show specific areas with the least amount of distortion. For specific regions, take Switzerland as an example, a map projection has been created (CH 1903 LV03 - the coordinates for Fribourg in the CH1903 LV03 projection are 577965.97 m E, 183244.73 m N) to display the least amount of distortion for the country, but very high levels of distortion elsewhere (Fig. 9).

Fig. 9. A map of the world shown in the CH 1903 projected coordinate system.

5. Types of GIS software

Various types of GIS software exist, each with their own strengths and weaknesses. For the purpose of this paper, we will focus on and discuss one commercial off-the-shelf GIS (COTS GIS) software, ArcGIS, and one Free and Open Source GIS (FOSS GIS) software, QGIS. These two software suites were chosen due to their popularity and use in the field of GIS. Throughout the next sections, some basic functions of GIS will be introduced and explained.

5.1. COTS GIS - ArcGIS

One of the most well-established and popular GIS programs is ESRI's ArcGIS suite. ArcGIS was established in 1969 originally as a research group focused on landuse planning initiatives (ESRI, 2013). Since then, ArcGIS has grown exponentially and is now the leading commercial GIS software which incorporates mobile, desktop, server, and online platforms. Until recently, the licence levels in ArcGIS were organized by the naming scheme ArcView, ArcEditor, and ArcInfo, which denoted the amount of access to tools and functionalities within the software. Now, the different levels are called Basic, Standard, and Advanced, respectively, to make the licence levels more intuitive. The Basic level allows access to mapping and interactive visualization functionalities, the Standard level offers those as well as multiuser editing and advanced data management, and the Advanced level offers all of the above, plus, advanced analysis, high-end cartography, and extensive database management possibilities (ESRI, 2013). These different levels allow the buyer to choose the package that fits their needs. Also, you can add extensions such as Spatial Analyst, 3D Analyst, and Geostatistical Analyst, which are sets of tools for specific tasks (ESRI, 2013). Extensions allow the user to add more advanced functions without the need for upgrading to a different licence. However, one major problem with ArcGIS and other ESRI products are costs, which can often only be afforded by large corporations, universities, and government agencies. Students or research groups can often acquire versions for free, or at reduced costs, respectively. However, the price is often a deterrent when a company or individual is choosing which GIS software to purchase. Free trials are also available from http://www.ESRI.com/software/arcgis/arcgis-for-desktop/free-trial. Also, various online training courses (http://training.esri.com/), tutorials (http://resources.arcgis.com/en/help/main/10.1/index.html#//00qn0000013t000000), and other resources (http://resources.arcgis.com/en/home/) are available online to help with specific tasks.

5.2. FOSS GIS - QGIS

Open Source software is gaining ground in the GIS software business. The scientific community is coming together to use, create, and enhance open source GIS tools. The development of open source GIS is usually driven by very active communities closely collaborating with different university-level institutions and thus oriented to concrete solutions, e.g. in environmental science or modelling and on technological improvements. Besides that, most open source software is available for free and the users can benefit from the availability and transparency of the open source (program) code, which may be adapted for specific consumer needs and can again be shared within the community. Furthermore, open source software becomes more and more platform-independent and integrative between various projects, libraries, and standards. Some of the software that exists for open source GIS includes the System for Automated Geoscientific Analysis (SAGA, http://www.saga-gis.org), the Geographic Resources Analysis Support System (GRASS, http://grass.osgeo.org), and Quantum GIS (QGIS, http://www.qgis.org/). QGIS is based on an intuitive mapping interface with optional plugins (supplementary program code providing additional functionalities, similar to ESRI ArcGIS extensions) for geoprocessing, analysis, and interoperability with various other software, standards, and data types. GRASS is more advanced in terms of its inherent spatial analytical ability especially for raster data, but some programming knowledge is of advantage for more sophisticated analysis and script automation. The integration of GRASS applications in QGIS through a plugin has a long tradition in the QGIS history, meaning that one can benefit from the strengths of both systems at once. Furthermore, the latest releases of QGIS provide a toolbox called "SEXTANTE" (http://www.sextantegis.com) where various Open Source software and libraries like SAGA can be accessed through the straightforward user interface of QGIS. Chapter 7 provides a tutorial for QGIS with basic applications adapted to honey bee research. For further information about open source GIS software and the different projects mentioned above see the Open Source Geospatial Foundation website http://www.osgeo.org.

5.3. Tutorial data

The following sections contain step by step information on how to complete certain GIS tasks using both ArcGIS and QGIS. To follow these steps, the file "GISBEEBOOK", can be downloaded from (http://www.ibra.org.uk/articles/GISBEEBOOK). Download the data and save them in a location with a few gigabytes of space available. Within the GISBEEBOOK folder, there are two sub-folders, one for ArcGIS and one for QGIS, each containing the same original data. When working through each tutorial, it is recommended that you save your outputs in the respective software's folder to help with file and data organisation. Both folders are located within the GISBEEBOOK folder. The GISBEEBOOK file contains the two folders (ArcGIS and QGIS) with the following data layers in each:

- Point file: This file was created solely for the purpose of this exercise and does not represent the actual locations of honey bee colonies.
  ◊ ArcGIS: An Excel file containing point locations: *colony_locations.xlsx*.

◊ QGIS: A comma-separated text file (*.csv) containing the same point locations: *points-table.csv*.

• A shapefile containing commune information: *commune_boundary_FR.shp* (plus 6 other file extensions which must stay together). This data comes from a clip of the Vector200 Swisstopo (Swisstopo, 2013) data layer which is provided to the readers of this paper for testing purposes only and cannot be shared or used for data analysis purposes outside of this tutorial. Communes are a type of administrative unit in Switzerland.

• 30 m DEM:

◊ ArcGIS: Located within the DEM folder: *aster30_frib_ch.tif* (along with 4 other extensions that should be kept together with this file for georeferencing purposes).This DEM was a free download from the ASTER GDEM website (http://gdem.ersdac.jspacesystems.or.jp/). The projection was previously changed to the Swiss coordinate system (CH1903) and clipped to the area around Fribourg.

◊ QGIS: ASTER30_WGS84.tif. Also downloaded from the ASTER GDEM website but left in the WGS 1984 coordinate system.

6. Using ArcGIS (version 10.1)

6.1. Main components of ArcGIS

The main components of ArcGIS include ArcMap, ArcCatalog, and ArcToolbox, however, the other platforms that are included in ArcGIS for Desktop are ArcScene and ArcGlobe which both enable 3D data viewing and manipulation. ArcGIS has also grown to include mobile applications, map viewers (ArcGIS Explorer: http://www.esri.com/software/arcgis/explorer), and cloud mapping (www.arcgis.com), which will be discussed in section 6.10.10. For more information about ArcGIS in general refer to Ormsby *et al.*, 2010.

6.1.1. ArcMap

ArcMap (Fig. S 10) is the main interface for conducting analyses and creating maps. Here, feature classes and shapefiles can be populated, data can be edited, calculations can be performed, and finally, maps can be created for displaying the results of the GIS analysis. ArcMap projects are stored as map documents (*.mxd) which save all of the layers added to the map as well as the results of any of the geoprocessing tools which were used. It is important to note that the *.mxd does not store the actual data, but only references to them and user-defined layer organization and symbols. Upon first opening ArcMap (*Start menu > All programs > ArcGIS > ArcMap 10.1*) a dialog box will pop-up. Select a recently used map from the list, browse to a map that was previously saved, or just click *Cancel* or close the window to open a blank map.

6.1.2. ArcCatalog

ArcCatalog (Fig. S 11) is where GIS data can be searched for, previewed, and managed. Also, some ArcToolbox tasks can be performed and visualized, but not overlaid with other layers like in ArcMap. Since ArcGIS version 10.0, a more compact version of ArcCatalog can also be accessed directly in ArcMap. ArcCatalog should be used when geographic data are added, moved, or deleted from GIS in order to avoid losing important file extensions (e.g. the geographic reference, see section 3.1 for more information). To open ArcCatalog: *Start menu > All programs > ArcGIS > ArcCatalog 10.1*.

6.1.3. ArcToolbox

ArcToolbox (Fig. S 12) is where all of the geoprocessing and spatial analysis tools are located. There are hundreds of tools available and they range from basic to advanced. The toolbox can be accessed in both ArcMap and ArcCatalog by clicking the 🗔 icon from the standard tools menu or by selecting *Geoprocessing > ArcToolbox* from the main menu. To easily find a tool, type the name of the tool into the *Search* window to save time from manually searching through the toolbox. In the following sections, some of the more basic toolbox tools will be discussed, but this is by no means an exhaustive tutorial.

6.2. Navigating the software

At first, the multiple functionalities and options within ArcGIS may seem overwhelming, but there are a lot of resources that exist to help navigate the software and search for help if there are problems.

6.2.1. Help menus

Help menus can be accessed by choosing *Help* from the main menu, or by pressing *F1*, in both ArcMap and ArcCatalog. The entire ArcGIS help menu can also be found online, which is slightly more intuitive and easier to search through. The online help menu can be found at: http://resources.arcgis.com/en/help/main/10.1/. For previous versions of ArcGIS, the help menus can be found at http://resources.arcgis.com/en/help/previous-help/index.html.

6.2.2. Search window

The *Search window* can be found in the standard toolbar (🗔) or by selecting *Customize > Search* from the main menu. The search window can be used to search for GIS content including data located on the computer and tools within ArcGIS.

6.2.3. Toolbars

In ArcMap there are a variety of toolbars, used for different tasks, to choose from. To display or hide toolbars, from the main menu select *Customize > Toolbars* and check which ones to display and hide. By hovering over the tools in the toolbars you can see the name of each respective tool.

6.2.4. General reminders, shortcuts, timesavers

Here is a list of items that will save time when using ArcGIS:

- Naming schemes for files, folders, and anything else associated with ArcGIS should not have spaces. Instead, underscores ("_") should be used when a space is needed.
- In case data or project files (*.mxd) are moved to a different location, it is useful to save the relative paths of the data so that ArcMap is able to automatically locate the data without having to manually point to the data every time in ArcMap (when ArcMap cannot find the data, a small red exclamation (!) point appears beside the layer in the table of contents (TOC)). To do this:
  ◊ *File > Map Document Properties > Pathnames: check the "Store relative pathnames..." option.*
  ◊ To make this feature permanent, go into the ArcMap options *Customize > ArcMap Options > General tab > and check the "Make relative paths the default for new map documents".*
- Also while in the ArcMap options, have a look at the other options and customize your ArcMap interface to fit your needs.
- When both ArcCatalog and ArcMap are open on the desktop, files from ArcCatalog can be directly dragged and dropped into ArcMap without having to use the *Add Data* button.
- After performing geoprocessing or spatial analyses, the results for all tasks are stored in the results window which can be opened by selecting *Geoprocessing > Results* from the main menu in ArcMap. The *Results* window is useful for keeping track of the tools you already used and easily allows you to reuse them again with alterations to the inputs.
- The naming of raster layers in grid format (ArcGIS native format) cannot be more than 13 characters long.
- Save frequently to avoid losing work!

6.3. Setting the coordinate system and adding data to ArcMap

Upon opening ArcMap, first close the pop-up to open a blank map. It is good practice to set the coordinate system to match the data which will be added, but this step is not required as ArcMap will take on the same coordinate system as the first layer that is added. All subsequent data will be projected "on-the-fly" to the coordinate system specified in the first layer. This will become more apparent later.

6.3.1. Setting a coordinate system and saving a new map document (*.mxd)

Setting the correct map units is important for all types of quantitative analysis and for appropriate cartography. Despite on-the-fly projection capabilities, if layers are in different coordinate systems, this will pose major problems in the display and analysis of data.

To set the coordinate system of the map document:
1. Go to *View.*
2. Select *Data Frame Properties.*
3. Open the *Coordinate System* tab.
4. For the remainder of this paper, we will use data from Switzerland, therefore we will use the projected coordinate system called CH1903 LV03. Double-click *Projected Coordinate Systems.*
5. Double-click *National Grids.*
6. Double-click *Europe.*
7. Select the *CH1903 LV03* option.
8. Click *OK.*

Or:
1. Using the search bar, directly type in the following: CH1903 LV03.
2. Click the *Search* button.
3. Double-click on *Projected coordinates folder >* select the *CH1903 LV03* option from the list
4. Click *OK.*
5. To save these properties in the map document (*.mxd), go to *File.*
6. Click *Save As.*
7. Name the file and click *OK.*

6.3.2. Add data button

Data can be added from *File > Add Data > Add Data* or by selecting the Add Data symbol ⊕ from the standard toolbar. Alternatively, data can be added via drag and drop from ArcCatalog. Since version 10.0, there are also the options of adding base map and other data from ArcGIS online. This will be discussed further in the cloud mapping section (6.10.10.).

6.3.2.1. Connecting to folders

To access the tutorial data, first the GISBEEBOOK folder must be connected to. Browse to the location of the GISBEEBOOK folder using this icon ⊡ in the Add Data dialog box. This path to the GISBEEBOOK folder is now saved for the next time data are added from the same folder.

6.3.3. Importing vector and raster data

Vector data are point, line, and polygon features on the map which represent real world entities while raster data represent continuous features such as altitude or satellite imagery (see section 3.2. for data explanations). Here we will add a polygon layer which represents the administrative boundaries in and around Fribourg, Switzerland, and a DEM from the same area.

To add a vector layer:
1. Click the *Add Data* button.
2. Browse to the commune_boundary_FR shapefile.

Depending on the type of vector added, it will have a point ⬚ , line ⬚ , or polygon ⬚ icon. This one is a polygon layer.

3. Single-click the commune_boundary_FR shapefile.
4. Click *Add*.

Now the vector layer can be seen in the TOC and in the map.

5. Change the symbology to make the polygons hollow (no fill, thicker line) so the DEM is visible (Fig. S 13).
6. In the TOC the name of any layer can be changed by slowly double-clicking the layer name or by going into the *Properties > General* and change the layer name there to make it more intuitive, if needed.
7. Now would also be a good time to save the map document (*.mxd). Go to *File > Save* (or use the shortcut Ctrl + s).

To add a raster layer:

1. Click the *Add Data* button.
2. Open the DEM sub-folder in the GISBEEBOOK folder.
3. Double-click the aster30_frib_ch.tif file to add it to the map.

In ArcGIS rasters are represented by this icon ⬚ . Now the DEM can be seen in the TOC and on the map (Fig. S 14).

6.3.3.1. Importing XY data from text files

Many people work with spatially related data even without recognizing it, for example when dealing with observation or address data from multiple locations. It is easy to make use of spatial information also from data stored in text files, calculation spreadsheets or any database. Often, point data representing X, Y, and sometimes Z (altitude) coordinates in the real world are contained within text files and can be imported into ArcMap as an Excel spreadsheet (*.xls or *.xlsx), a text file (*.txt), or comma separated value (*.csv) file. To ensure that this is done without problems or annoyances, ensure the data are properly formatted. For example, the field names in the header line of the text file should not use spaces or special characters, nor should the filename.

To add points from a text file (Fig. S 15):

1. Go to *File*.
2. Select *Add Data*.
3. Select *Add XY Data*.
4. Browse to the GISBEEBOOK folder.
5. Double-click the colony_locations.xlsx file.
6. Double-click Sheet1$.
7. Choose the respective fields that hold the X and Y data (or longitude and latitude, depending on which coordinate system is used). If the file contains altitude information, add that field in the Z field. In this case:
    a. X Field: POINT_X
    b. Y Field: POINT_Y

8. The coordinate system has already been defined for this layer.
9. Click *OK*.

It is important to note that this layer is loaded into ArcGIS and available for any display and map creation purpose and also some basic analysis, but not physically stored as geodata in the file system. A warning dialog box (Fig. S 16) may appear which indicates that the created layer will have to be exported to a shapefile if all functionalities are needed.

10. Click *OK*.

Now the points can be seen in the map and the layer can be seen in the TOC.

11. Right-click the "Sheet1$ Events" layer in the TOC.
12. Select *Data*.
13. Select *Export Data* to export these temporary points and make a permanent shapefile.
14. Choose to use the layer's source data as the correct coordinate system should have already been set in the previous steps.
15. Browse to the GISBEEBOOK/ArcGIS folder to name the files intuitively, for example, we will name this layer colony_locations.
16. Save as type: *Shapefile*.
17. Click *OK*.
18. Click *Yes* when asked to display the new layer on the map.
19. Remove the Sheet1$ Events layer by right-clicking on *Sheet1$ Events layer*.
20. Select *Remove*. We now have the permanent shapefile, so this temporary layer is no longer needed (Fig. S 17).
21. To change the symbology of the point layer to make the points more visible, either:
    a. Single-click on the symbol in the TOC and choose the new symbol and size.
    b. Go into the properties of the layer by right-clicking *Layer*.
    c. Select *Properties*.
    d. Click the *Symbology* tab.
    e. Change the symbol to a yellow circle for example.
    f. Click OK.

Also to note, there are multiple symbol libraries that can be added by clicking the *Style References* button when in the *Symbol Selector* dialog box. The symbol libraries provide more options for symbol styles.

6.3.3.2. Importing GPS data (*.gpx files)

Geographic coordinates, often collected via a handheld Global Positioning Systems (GPS) receiver can be imported and displayed in ArcGIS (since version 10.1). GPS data are in the GPS eXchange Format (*.gpx) and can be converted to shapefiles using the *GPX To Features (Conversion)* tool, which can be found in the *Conversion Tools* toolbox > *From GPS* toolset.

6.3.4. Layer organisation and navigation

It is essential to have some basic knowledge about layer organisation and navigation in order to optimize the map display (e.g. to change the order and how layers are drawn in the *Data View*) and to preserve a good organisation among all the different data, intermediate calculation results and map elements, especially for a growing GIS project with many layers. In ArcMap, the layers will be displayed on the map in the same order as in the TOC. In Fig. S 13 for example, the colony_locations layer is above commune_boundary_FR and aster30_frib_ch.tif layers therefore everything can be seen clearly, since we made the commune boundaries hollow. If the DEM was at the top of the layer list in the TOC then the other two layers would be hidden. At the top of the TOC there are five icons which allow different views of the data in the TOC. To rearrange the layers in the TOC, select the first icon (*List by Drawing Order*) and click and drag the layers to their new positions. The layers can be turned on and off by checking or unchecking the box beside the layer.

6.3.5. Layer properties

The properties of one specific layer can be accessed by right-clicking the layer and selecting *Properties* from the context menu, or simply by double-clicking the layer name. The *Layer Properties* might vary between different file formats, but every layer will at least have tabs for *Symbology* options, *General* settings and provide metadata (see section 3.) in the *Source* tab. Look at the metadata of the layer commune_boundary_FR.shp by opening the *Source* tab in the *Layer Properties*. Apart from the extent (maximum and minimum values for X and Y), data source, and geometry type, there is also information about the coordinate system and projection used. From the *Display* tab, the transparency of the layer can be set and the *Labels* tab enables the labelling of features in the layer.

6.4. Database investigations and editing

Some databases have hundreds, if not thousands or millions, of entries. This makes manual searching almost impossible. Like any other form of database management system, ArcMap offers a number of options to efficiently browse, select, export and edit these records.

6.4.1. Queries

One of the characteristics of GIS is the possibility to combine information about several layers and datasets by using a geographic reference, e.g. a given point, a set of polygons or specific grid cells in a raster image. Queries are used to locate specific records from tables or places on the map based on the shared properties of layers. In GIS, a query can refer to two different functions. The first is a very basic GIS function that enables the user to interactively obtain information about a specific location in the GIS, using the *Identify* tool. The *Identify* tool is used to get information about specific pixel values within a raster image or to look at the attribute values of vector feature. In the example (Fig. S 18), it can be seen that the location on the map which was clicked is somewhere in the commune of Düdingen and has an altitude of about 612 m. This information comes from the commune layer and the DEM that are currently in the ArcMap *.mxd.

The second, more advanced, type of query in GIS refers to the database query. This type of query uses a structured query language (SQL), adopted from database management systems, which allows the retrieval of information from the database (Longley *et al.*, 1999). In ArcMap, the tool that incorporates SQL is the *Select by Attributes* (6.4.1.1.) tool. To go beyond the database search, the *Select by Location* (6.4.1.2.) tool can be used for geographic or locational queries of data. Selections can also be conducted manually using the *Select Features* tool (6.4.1.3.). These three selection methods can also be used in combination by changing the method of selection (*Create new selection, Add to current selection, Remove from current selection, Select from current selection*) in each respective dialog box. Any selection can be removed by clicking the *Clear Selected Features* button.

6.4.1.1. Select by attributes

The *Select by Attributes* tool is located on the main menu in *Selection > Select by Attributes*, or also within the attribute table of each layer: *Right-click layer > Open Attribute Table > Table Options* ⋮ ▾ > *Select by Attributes*. This tool allows the user to select a set of data from a layer based on the attribute properties of the layer. In the following example, all the communes within the canton of Fribourg, Switzerland, will be selected because all of the honey bee colonies are within that canton.

1. From the main menu, go to *Selection*.
2. Click *Select by Attributes*.
3. The input layer is the layer from which data will be selected. In this case, choose the commune_boundary_FR layer as input.
4. There are multiple methods available:
   a. *Create new selection*,
   b. *Add to current selection*,
   c. *Remove from current selection*,
   d. *Select from current selection*.
5. Choose *Create new selection*.

In the next section there is a list of attributes corresponding to those of the input layer.

6. Select the *KANTONSNR* attribute by double-clicking on the name.
7. Click the equals = button.
8. Click the *Get Unique Values* button to see all the values in that attribute field. In this case, the values represent two cantons in Switzerland.
9. Double-click the *10* to add it to the equation. In this example, 10 was clicked because that is the canton number for Fribourg.
10. Now the SQL equation looks like this: SELECT * FROM commune_boundary_FR WHERE: "KANTONSNR" = 10 (Fig. S 19).

11. Click *OK*
12. Open the attribute table of the layer from which data were selected (*Right-click > Open attribute table*).
13. Click the *Show selected records* button to view the records corresponding to the selection 🔲.
14. The selected records (49 out of 51) in the table match those selected on the map (Fig. S 20).
15. Now the selection can be cleared using the icon 🔲 to begin a new selection in the next section.

6.4.1.2. Select by location

Now two layers will be used to select the communes which have colonies inside them. To do this (Fig. S 21):

1. Go to *Selection*.
2. Choose *Select by Location*.
3. Choose the Selection method: select features from the target layer: commune_boundary_FR.
4. Select the Source layer: colony_locations.
5. Select the Spatial selection method: intersect the source layer feature (this tool has various options for spatially querying data and applying search distances to discover geographic relationships).
6. The selected results: 11 records out of 51 (Fig. S 22), meaning 11 communes had colonies inside their boundaries.

6.4.1.3. Manual selection

Data can also be selected manually with the *Select Features* tool 🔲 ▾ in the *Tools* toolbar. With this tool, you can select the data by various shapes (rectangle, polygon, circle, etc.) and export the selected items to their own shapefile (see section 6.4.2.).

6.4.2. Creating new layers from selected data

If a selection of data has to be exported, to send to someone who needs only a part of a data layer, for instance, a new data layer can be created with the selected features. We will create a new layer with the data selected in section 6.4.1.2.

1. Perform step 6.4.1.2.
2. Close the *Select by Location* dialog box.
3. Right-click on the commune_boundary_FR layer in the TOC.
4. Go to *Data*.
5. Select *Export Data*.
6. Select Export: *Selected features*.
7. Use the same coordinate system as the source data.
8. Save as a new shapefile in the ArcGIS BEEBOOK folder. Call it colony_communes.shp.
9. Click *Save*.
10. Click *OK*.
11. Click Yes to add the new layer to the map.
12. Clear the selected features, go to *Selection*.
13. Click *Clear selected features*.

14. Turn off the commune_boundary_FR and aster30_frib_ch.tif (DEM) layers in the TOC.
15. Right-click on the colony_communes layer.
16. Click *Zoom to layer* (Fig. S 23) to focus the map on that layer.

6.4.3. Attribute table data editing

Not only can the attributes of layers be displayed and used for selection, but they may also be changed or removed. The data in tables can be edited and new fields can be added and populated. The attribute table for the colony_locations layer has two fields with no data in them. To populate information about the site location and the apiary number (as examples), the table can be edited.

To edit attribute tables:

1. Right-click the colony_locations shapefile.
2. Go to *Open Attribute Table*.
3. Open the *Editor* toolbar.
4. Go to *Customize*.
5. Select *Toolbars*.
6. Select *Editor*.
7. From the *Editor* toolbar, select *Editor*.
8. Select *Start Editing*.
9. Choose the colony_locations layer to be edited.
10. Click *OK*.

Previously the column names in the attribute table were grey; when in editing mode they will turn white which means they are editable (Fig. S 24).

11. Click in the cell to edit and type the information. Information can be manually typed or if the information exists in a text file, copy and paste them into the table while in editing mode.
12. Periodically save your edits: Click *Editor*.
13. Click *Save Edits*.
14. Click *Editor* again.
15. Click *Stop Editing* to end the editing session.

6.4.4. Add new field

To add a new field to the table:

1. Open the table options 🔲 ▾ .
2. Select *Add field*. Note: Fields can only be added outside of the editing mode!!
3. Name the new field you would like to add.
4. Select the type of data which will populate it (see Table 3 to learn about data types).
5. Click *OK*.
6. A message might warn you that you are performing calculations outside of an editing session. Check the *Don't warn me again* option.
7. Click *Yes*.
8. To manually add information to the new attribute field, go into the editing mode and type them into the new field.

Table 3. Description of data types which are found in attribute tables, as well as the range of numbers or text they store and their uses. Adapted from the ArcGIS help menu (ESRI, 2013).

Data type	Storable range	Applications
Short integer	-32,768 to 32,767	Numeric values without fractional values within specific range; coded values
Long integer	-2,147,483,648 to 2,147,483,647	Numeric values without fractional values within specific range
Single-precision floating-point number (float)	approximately -3.4E38 to 1.2E38	Numeric values with fractional values within specific range
Double-precision floating-point number (double)	approximately -2.2E308 to 1.8E308	Numeric values with fractional values within specific range
Text	---	Store alphanumeric symbols
Date	---	Store dates, times, or both

6.4.4.1. Advanced table calculations

1. For more advanced table management:
   a. Use the field calculator to calculate the inputs in the field:
      i. Right-click new field name.
      ii. Select *Field Calculator*.
      iii. Use computer programming code to write the equation for the calculation.
   b. Use the calculate geometry option to calculate the x and y coordinates of point layers, or areas, and centroids of polygon layers:
      i. Right-click on *New Field Name*.
      ii. Select *Calculate Geometry*.
      iii. Choose one of the calculation options.

6.5. Basic vector tools

The following section shows how to perform selected vector data calculations. The name of the tool will be given and the name of the toolbox in which the tool is located in ArcToolbox will be given in brackets (tool (toolbox)).

6.5.1. Clip

Using the *Clip tool*, two layers can be overlaid with each other. The output layer (which has to be specified as a new vector file) will contain the content of the first layer (Input vector layer) reduced to the geometry and extent of the second layer (Clip feature) (Fig. 25). A similar process is also followed when clipping raster data, but the *Clip (Data Management)* tool is used instead. This tool is also useful if you require all data layers to have the same spatial extent, for example if you are working with a specific region and only need a small piece of

a bigger layer. Note: A specific example is not shown here, but this is how the tool would be used. To use the Clip tool:

1. Open the *Search* window .
2. Type "clip" (or browse to *Analysis Tools > Extract > Clip* in the ArcToolbox).
3. Choose the *Clip (Analysis)* tool. This one is for vector data.
4. The Input features layer is the layer that will be clipped.
5. The Clip Feature will be the new extent of the clipped layer.
6. Name and save the new clipped layer.
7. Click *OK*.

6.5.2. Intersect

The *Intersect (Analysis)* tool computes a geometric intersection of the layers. Only those input features which overlap will be included in the output layer. This tool is useful for connecting multiple layers and joining their attributes together, by discarding the portions which do not overlap. In the next example, we will intersect the colony_locations layer (colony locations) with the colony_commune layer so that the intersected layer will have the point locations but also the commune information for each point. To use the intersect tool (Fig. S 26):

1. Open the *Search* window .
2. Type "intersect" (or browse to *Analysis Tools > Overlay > Intersect* in the ArcToolbox).
3. Choose the *Intersect (Analysis)* tool.
4. The Input Features will be all of those to intersect, in this case the colony_locations and colony_communes layers.
5. Save the output as intersect.shp.
6. Leave the defaults for the rest of the options.
7. Click *OK*.

In this case, the resulting geometry of the point file has not changed, but the attribute table of the intersected layer now includes the colony locations and commune information (Fig. S 27).

6.5.3. Union

The *Union (Analysis)* tool is used to create a geometric connection between polygon layers where all features and geometries will be added to the resulting layer. The union function is useful when the

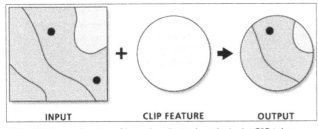

| INPUT | CLIP FEATURE | OUTPUT |

Fig. 25. A visualisation of how the clip tool works in ArcGIS taken from the help menu (ESRI, 2013).

geometries and attributes of different layers should be merged. In ArcGIS, all input polygons will be transferred to the output layer, regardless as to whether they spatially overlap or not, leaving the resulting dataset with three feature types: those found only in the first input layer, those found only in the second input layer, and those found in both the first and second input layers (Ormsby *et al.*, 2010).

6.5.4. Spatial join

In section 6.4.1. spatial queries were performed. In a similar manner, multiple layers can be spatially linked to each other using the *Spatial Join (Analysis)* tool. Spatial joins allow the attributes of features in separate layers to be linked together based on their shared spatial locations. This is useful for amalgamating different types of information into the same layer. The *Spatial join (Analysis)* tool can be used to perform this task. The *Target Features* can be any spatial data source supported by ArcGIS and will be transferred to the *Output Feature Class* along with the *Join Features*. The *Join Operation* determines how joins between target and join featured will be handled if multiple join features have the same spatial relationship with a single target feature; one-to-one or one-to-many (search spatial join in the ArcGIS online help menu for more information).

6.5.5. Buffer(s)

Buffers are one of the most important geoprocessing operations and frequently used for data analysis and cartography. A buffer increases the features area by a given radius, exactly following its original geometry (point, line or polygon). Buffers are also useful for determining proximities. For honey bees, this tool could be useful for determining the extent of the foraging radius with regards to various land covers, for example.

6.5.5.1. Single buffer

To create a single buffer (Fig. S 28):

1. Type "buffer" into the *Search* window (or browse to *Analysis Tools > Proximity > Buffer* in the ArcToolbox).
2. Select the *Buffer (Analysis)* tool.
3. Input features will be the colony locations: colony_locations.
4. Create a new output feature class: buffers.shp.
5. Choose "2000 m" as the search radius.
6. Leave the default options for the remaining.
7. Click *OK.*

The results show the buffered area surrounding each point (Fig. S 29).

6.5.5.2. Multiple ring buffers

To create multiple ring buffers (Fig. S 30):

1. Type "buffer" into the *Search* window (or browse to *Analysis Tools > Proximity > Multiple Ring Buffer* in the ArcToolbox).
2. Select the *Multiple Ring Buffer (Analysis)* tool.

3. Input features will be the colony locations: colony_locations.
4. Create a new output feature class: multi_buffers.shp.
5. Type "100".
6. Click the plus sign (+).
7. Do the same for 200, 300, 400, 500, and 1000 m.
8. Choose the buffer unit (meters in this case).
9. Keep the field name of "distance" for the new attribute table of the multiple buffer layer.
10. Keep the default for the dissolve option, otherwise all buffers will be aggregated to one single feature without the different attributes (see section 6.5.7. for more information about Dissolve).
11. Click *OK.*

The results show the multiple buffers created surrounding the points (Fig. S 31).

6.5.6. Summary statistics

A *Summary Statistics (Analysis)* tool exists in ArcGIS and is useful for performing basic calculations on fields in an attribute table. This can be valuable for summarizing large quantities of data within GIS (similar to basic statistics that can be calculated in Excel). For example, the sum, mean, min, max, range, standard deviation, and count can be calculated for each selected attribute. The results are saved in an output table and can be viewed directly in ArcMap. Also, to access the statistics of single fields in an attribute table, *Right-click* the layer in the TOC > *Open attribute table* and then *Right-click field name* and choose the *Statistics* option.

6.5.7. Dissolve

Dissolve is another very useful function that merges several features together, based on their attributes. All features within a given vector layer containing the same attribute value will be combined to one single feature (one entry in the *Attribute table*). Shared borders between these geometries will disappear and in the case of spatially divided (geometrically non-adjacent) input features, the result will be a so-called "multipart" feature. The Dissolve (Data Management) tool is useful for portraying or visualizing specific information from a shapefile. For example, here we used the commune_boundary layer for Switzerland (basically the same as the commune_boundary_FR layer, but for all of Switzerland). The geometry of the layer currently represents the communes of Switzerland, but we want to visualize countries instead of communes. In the attribute table, there is a column indicating which country each commune belongs to, thus we can "dissolve" the administrative boundaries based on that field. In the next example, the attributes in the commune layer will be dissolved to create a new layer based on the country geometry. NOTE: Here we will use a layer that has not been provided in the tutorial. This is to illustrate what might be done using the dissolve tool.

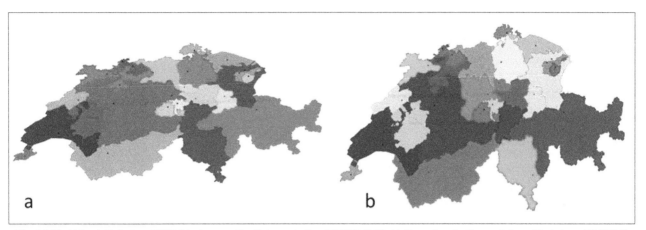

Fig. 36. World Geodetic System (WGS 84, a) versus the Swiss projection (CH1903 LV03, b). Reproduced with permission of Swisstopo (BA13016).

To use the dissolve tool (Fig. S 32):

1. Type "dissolve" into the *Search* window (or browse to *Data Management Tools > Generalization > Dissolve* in the ArcToolbox).
2. Select the *Dissolve (Data Management)* tool.
3. The Input features will be the commune_boundary (for all of Switzerland) in this case.
4. Name and save the output layer as a shapefile.
5. In this case we will dissolve based on the COUNTRY attribute.
6. Leave the default options for the remaining.
7. Click *OK*.

The result (Fig. S 33) shows that the geometry and the attributes represent only the countries.

6.5.8. Project

Data layers often need to be "projected" to get them into the same coordinate system as other layers. For example, if data comes from different places (e.g. online, governmental agencies) or represents different areas on the map, there is a good chance that the coordinate systems between layers will differ, therefore a projection of the data is required. If layers are not in the same coordinate system and not overlapping correctly, there will be errors in the resulting layers created after analyses. The Project (Data Management) tool is used to change a vector layer from one coordinate system to another. The same process can also be used for raster layers, but the Project Raster (Data Management) tool is used instead. If the layer's coordinate system is unknown, it must first be defined, either in the layer properties in ArcCatalog or by using the Define Projection (Data Management) tool in ArcMap. In this example, we will project the colony_communes layer from the Swiss projection to WGS 1984 and import the results into Google Earth (KML file) (*6.5.9.*).

To use the project tool (Fig. S 34):

1. Type "project" in the Search window (or browse to *Data Management Tools > Projections and Transformations > Feature > Project* in the ArcToolbox).

2. Select the *Project (Data Management)* tool.
3. Input layer: colony_communes.
4. The coordinate system of the layer will automatically be added.
5. Save the output as: commune_wgs84.shp.

Indicating the new coordinate system in the name is useful for layer organisation.

6. Choose the output coordinate system.
7. Click the *Browse* button.
8. Expand the Geographic coordinate systems option.
9. Click *World*.
10. Select the WGS 1984 coordinate system.
11. Select a transformation from the predefined list. In this case, the first is sufficient.
12. Click *OK*.
13. Click *OK* again.

The resulting layer will possess the new projection and is automatically added to the map document (Fig. S 35). The green commune_wgs84 polygon layer is slightly shifted to the south compared to the blue colony_communes layer underneath. As another example of the differences in coordinate systems, Fig. 36 shows an example from Switzerland between the WGS 1984 coordinate system (a) and the CH1903 projection (b).

6.5.9. Layer to KML and import to Google Earth

A KML (Keyhole Markup Language) file is a geographic data layer which was developed for use with Google Earth. Using these steps, any layer (shapefile or raster) from GIS can be imported to Google Earth (conversely, the *KML to Layer* tool can be used to bring data from Google Earth into ArcGIS). This conversion is useful for sharing geographic information with people who do not use GIS. In previous ArcGIS versions, a shapefile needs to be saved first as a "layer" file before converting it to KML. To do that, the *Save to Layer File* tool can be used. A layer file is basically a pointer to the shapefile which stores specific symbolization properties. In version 10.1, the user is no longer required to save as a layer file before converting to a KML.

To export a layer from ArcMap for import into Google Earth (Fig. S 37):

1. Type "kml" in the *Search* window.
2. Select the *Layer to KML (Conversion)* tool.
3. Input layer: commune_wgs84.
4. Output: GE_commune_wgs84.kmz.
5. Leave the other options with their default values.
6. Click *OK*.
7. Open Google Earth (if not already installed, it can be downloaded from: http://www.google.com/earth/index.html)
8. In Google Earth go to *File*.
9. Click *Open*.
10. Select the GE_commune_wgs84.kmz layer.
11. Click *OK*.

The layer created in ArcMap should now be displayed and correctly georeferenced in Google Earth (Fig. S 38).

6.6. Basic raster tools

Some of the previously mentioned vector data tools also exist specifically for rasters, for example, *Clip* and *Project*. However, most raster tools have different names and functionalities which are unrelated to vector analyses. Below, we will briefly touch on some of the basic raster calculations that can be performed, specifically some of the most commonly used tools for terrain analysis and raster file management.

6.6.1. Mosaic

The *Mosaic (Data Management)* tool allows multiple raster layers to be combined with an existing raster dataset, while the *Mosaic to New Raster (Data Management)* tool allows multiple raster layers to be combined together to create a new raster dataset. The input raster images can vary in resolution and extent, but the defined output resolution may cause some input images to increase or decrease in resolution, sometimes leading to a loss in information. Also, the input rasters must have the same predefined coordinate system and the same number of bands.

6.6.2. Surface Analysis

The Surface Toolset in ArcGIS allows you to calculate and visualize different properties of the terrain from an input DEM (see section 3.2.1.4.). The surface toolset is part of the Spatial Analyst toolbox in ArcGIS. In the following section we will perform calculations based on the same 30 m DEM introduced earlier. For the next sections, one of either Spatial or 3D Analyst extensions is required. To turn on extensions select *Customize > Extensions* from the main menu. If you do not own a license, you will get a message when you try to click an extension.

6.6.2.1. Slope

The Slope (Spatial Analyst or *3D Analyst)* tool calculates the slope of the terrain, either in degrees or percentage rise.

To calculate the slope of the terrain (Fig. S 39):

1. Turn on *Spatial and 3D Analyst extensions*, go to *Customize* (from the main menu).
2. Select *Extensions*.
3. Check the *Spatial Analyst* (and/or *3D Analyst*) extension.
4. Click *Close*.
5. Type "slope" into the *Search* window (or browse to *Spatial Analyst Tools > Surface > Slope* in the ArcToolbox).
6. Select the *Slope (Spatial Analyst or 3D Analyst)* tool.
7. Choose the DEM as the input raster (aster30_frib_ch.tif).
8. Save the output raster as "slope".
9. Choose the output measurement to be used, here we will use degrees.
10. If the X, Y, and Z (altitude) coordinates are all in meters, leave the Z factor as 1. Otherwise click the *Show Help* button in the tool to read about how to convert the Z factor.
11. Click *OK*.
12. Turn off all other layers in the TOC, right-click *Layer* at the top of TOC.
13. Select *Turn all layers off*.
14. Turn on the new slope raster in the TOC.

The results show the default classification scheme in degrees ranging from 0 to about 58 degrees in eight categories (Fig. S 40).

Now the results will be classified in a different way by opening the layer properties (Fig. S 41).

15. Right-click *slope* layer in the TOC.
16. Select *Properties*.
17. Select the *Symbology* tab.
18. Choose the "stretched" option in the Show dialog box.
19. Select a colour ramp.
20. Choose the stretch type of Minimum-Maximum.
21. Click *OK*.

The results (Fig. S 42) show that the region is not so diverse in terms of slope.

6.6.2.2. Aspect

The Aspect (*Spatial Analyst or 3D Analyst*) tool calculates the direction in which slopes are facing. Values are given in degrees in regards to the degrees on a compass (with north as 0 and 360 degrees).

To calculate the aspect of the terrain (Fig. S 43):

1. Type "aspect" into the *Search* window (or browse to *Spatial Analyst Tools > Surface > Aspect* in the ArcToolbox).
2. Select the *Aspect (Spatial Analyst or 3D Analyst)* tool.
3. Choose the DEM as the input raster.
4. Save the output raster as aspect.
5. Click *OK*.

In the TOC the results (Fig. S 44) can be seen in degrees.

6.6.2.3. Hillshade

Shaded relief rasters (also called hillshade) are frequently used for visualization and cartographic purposes and are calculated from a DEM and predefined values for the azimuth (horizontal deviation from north, clockwise) and the solar elevation angle (vertical inclination between the sun and the horizon). The *Hillshade (Spatial Analyst or 3D Analyst)* tool calculates the shaded relief of the terrain in ArcGIS. This tool is useful for determining which areas of the terrain are shaded and which are not, during certain hours of the day, month, or year.

To calculate the hillshade of the terrain (Fig. S 45):

1. Type "hillshade" into the *Search* window (or browse to *Spatial Analyst Tools > Surface > Hillshade* in the ArcToolbox).
2. Select the *Hillshade (Spatial Analyst or 3D Analyst)* tool.
3. Choose the DEM as the input raster.
4. Save the output raster as aspect.
5. Click *OK*.

The results will be shown in a greyscale with values ranging from 0 to 254 (Fig. S 46), where 0 is black, representing completely shaded areas, 254 is white, representing illuminated areas, and all values in between representing an increasingly lighter shade of grey, thus different levels of illumination from the sun.

Hillshade rasters are good for visualizing terrain under different layers. To display the terrain properties under the DEM:

1. Drag the hillshade layer to the bottom of the layer list in the TOC.
2. Turn off all of layers except for the hillshade and the DEM.
3. Open the properties of the DEM, right-click *DEM* layer.
4. Select *Display*.
5. Set the transparency to 50%.
6. Go to the *Symbology* tab.
7. Select a different colour ramp.
8. Click *OK*.

The resulting image (Fig. S 47) enables a more intuitive view of the terrain with high elevations in red, low elevations in green, and the ability to see shadows from the hillshade.

6.6.2.4. Contour

The *Contour (Spatial or 3D Analyst)* tool generates an output vector file containing elevation levels (contour lines) from an input raster (in most cases a DEM). Contour lines are frequently used in cartography as they can easily be combined with other raster images (like aerial photos or pixel maps).

To generate contour lines from a DEM:

1. Open the *Search* window.
2. Type "contour" (or browse to *Spatial Analyst Tools > Surface > Contour* in the ArcToolbox).
3. Use a DEM as the *Input raster*.

4. Choose a location to save the *Output polyline features* (vector format).
5. Define the *Contour interval* which will be the distance between the contours (if you are using the Swiss projection this will be in meters).
6. Set the optional *Base contour* value, if desired (contours will start after the base value).
7. If using a different unit than meters, click the *Z factor* input box.
    a. Click *Show Help* to learn about Z factor conversions.
8. Click *OK*.

The new vector line file will be generated and added to the document. To achieve better results for visualisation, a smoothing algorithm could be performed on the line feature, e.g. with the tool *Simplify* or *Smooth Line* (*Cartography Tools > Generalization > Simplify* or *Smooth Line* from the ArcToolbox).

6.6.3. Zonal statistics

The *Zonal statistics (Spatial Analyst)* tool calculates the statistics on values of a raster within the zones of another dataset. For example, buffers could be used as input zones to calculate the statistics from any raster layer (DEM, slope, aspect, etc.) to determine the mean, majority, maximum, median, minimum, minority, range, standard deviation, sum, or variety of the raster pixels which are located within the buffer confines. The results are contained in a new raster layer and this should be conducted for each statistic required. To obtain multiple statistics at one time, the *Zonal Statistics as Table (Spatial Analyst)* tool can be used to generate the statistical results in table form for either all or selected statistic types. For an example of the use of the zonal tools, see the case study in section 6.9.7.

6.6.4. Raster calculator

The raster calculator is used for more advanced raster calculations, including raster math and building conditional and trigonometric expressions. It can be found using the search window or in the *Spatial Analyst > Map Algebra* toolset. For example, the formula "aster30_frib_ch.tif" * 3.28084" would convert the elevation data in the DEM from meters into feet by creating a new raster image. For more information search "raster calculator" in the help menu (http://resources.arcgis.com/en/help/main/10.1/).

6.7. Format conversions

ArcGIS provides a variety of different conversion functions between vector and raster data formats similar to other GIS software. As both of these formats have their advantages and disadvantages (see section 2.2.) in terms of data management, analysis, and display options, format conversions should be clearly linked to a precise purpose (e.g. when an elevation model is only available as raster data, but a map containing contour curves has to be created). There are various tools available to switch between data formats. See the *Conversion Tools* toolbox to view all of the options. The most common conversions are between raster and vector formats. Conversion operations can lead to data loss or decreased precision – so apply these functions with care.

6.7.1. Vector to raster

This function rasterizes vector geometries (Fig. 2) into the band(s) of a raster image (Fig. 3).

To convert to a raster from a vector:

1. Expand the *Conversion Tools* in the ArcToolbox.
2. Select *To Raster*.
3. Depending on the type of vector data you want to convert, select either *Point to Raster*, *Polygon to Raster*, or *Polyline to Raster*.
4. Indicate where the output will be saved.
5. Click *OK.*

6.7.2. Raster to vector

To convert to a vector from a raster:

1. Expand the *Conversion Tools* in the ArcToolbox.
2. Select *From Raster*.
3. Depending on the type of vector data you want to convert to, select either *Raster to Point*, *Raster to Polygon*, or *Raster to Polyline*.
4. Indicate where the output will be saved.
5. Click *OK.*

6.8. Spatial Statistics

The purpose of this section is to introduce the *Spatial Statistics* toolset in ArcGIS so that researchers are aware of its existence. Since most of the same tools can be found in other statistical packages, they will not be explained here in-depth, but their geographic links will be discussed. For more information on these tools see the overview in the ArcGIS online help menu by typing "An overview of the Spatial Statistics toolbox" into the search window. Additionally, see Lee and Wong (2001) or Wong and Lee (2005) for further tool descriptions. Be aware that these texts used previous versions of ArcGIS so the tutorials are not completely compatible. Unfortunately, these are the most updated versions available.

6.8.1. Analysing patterns

This toolset includes: Average Nearest Neighbour, High/Low Clustering (Getis-Ord General G), Incremental Spatial Autocorrelation, Multi-Distance Spatial Cluster Analysis (Ripley's K Function), and Spatial Autocorrelation (Morans I). These tools can be used for analysing spatial patterns based on features, or the values associated with features.

6.8.2. Mapping clusters

Cluster and Outlier Analysis (Anselin Local Morans I), Grouping Analysis, and Hot Spot Analysis (Getis-Ord Gi*) are located in this toolbox and can be used to identify statistically significant hot spots, cold spots, and outliers.

6.8.3. Measuring geographic distributions

These tools enable the researcher to ask spatially defined questions about their data by using the Central Feature, Directional Distribution (Standard Deviational Ellipse), Linear Directional Mean, Mean Center, Median Center, and Standard Distance tools.

6.8.4. Modelling spatial relationships

This toolset can be used to conduct regression analyses or creating spatial weights matrices. The tools include Exploratory Regression, Generate Network Spatial Weights, Generate Spatial Weights Matrix, Geographically Weighted Regression, and Ordinary Least Squares.

6.9. Case study – zonal statistics with land cover properties

Until now, several concepts and tools in ArcGIS have been introduced. Now, to combine some of the different tasks discussed, a case study will be presented. Within this short case study, geostatistical relationships between honey bee colonies and land cover properties will be calculated. The content of this case study is equal to the one for QGIS (see section 7.9.), but adapted to the tools of ArcGIS. The prerequisite for this case study is an ArcMap document (see section 6.3.) with a point vector layer (e.g. the honey bee colony locations colony_locations.shp, created in section 6.3.3.1.) and a buffer polygon layer (e.g. buffer.shp, created in section 6.5.5.1.). In this case study, the percentages of land cover types within 2 km radii of honey bee colony locations will be calculated. This will allow the user to obtain information about forage availability near colony locations. First, you will go through the process of downloading the Corine (which means "coordination of information on the environment") Land Cover 2006 (CLC06) dataset from an online source. The CLC06 layer is a free 100 m resolution raster representing 50 different land cover classifications for the continent of Europe including, agricultural areas, forests, wetlands, and water bodies (for more information see http://sia.eionet.europa.eu/CLC2006). This layer will then be projected to the Swiss coordinate system and clipped to the study region around Fribourg, Switzerland in order to make analysis and map rendering more time efficient. You will then learn how to assign land cover codes and a predefined colour scheme to the CLC06 layer and how to optimise the layer for the study region. After, raster calculator will be used to calculate conditional rasters which will be used as the basis of the statistical analysis for determining land cover types within the 2 km radii surrounding each honey bee colony location.

6.9.1. Downloading data and changing projection

First, land cover data for the region surrounding Fribourg will be downloaded and extracted.

1. *Download* the dataset *g100_06.zip* (CLC06 – 100 m) from the website http://www.eea.europa.eu/data-and-maps/data/ corine-land-cover-2006-raster-2 to a local directory (e.g. on the Desktop).

2. *Unzip* the file *g100_06.zip* with a file archiving program of your choice (if you need additional software for this task, the program "7-zip" is a good choice: www.7-zip.org) into your ArcGIS working directory (e.g. the GISBEEBOOK folder). Note: data in zipped folders cannot be seen in ArcGIS applications so you must first unzip.

3. Open ArcCatalog.

4. Navigate to folder containing the land cover raster *g100_06.tif* and select the *Preview* tab to get a visual impression of the data. As you can see, the data layer covers all of Europe.

5. Now open the *Properties* of the raster layer by double-clicking the layer name.

From the *General* tab in the layer properties, you see that CLC06 is a 1 band raster dataset with 8 bit unsigned integer values (that correspond to different land cover categories) at 100 m horizontal resolution. Because it covers all Europe, the CLC06 raster is quite large and the rendering on the screen may take some time. We will improve the performance by clipping to the extent of interest. The coordinate system is different (ETRS89_ETRS_LAEA) than the one used in Switzerland therefore it is best to project it into the same coordinate reference system as the other layers (CH1903 LV03).

6. Now, add the CLC06 layer into a new blank *.mxd by dragging the layer from ArcCatalog into ArcMap directly, or using the *Add Data* button in ArcMap.

7. To reproject the raster to the Swiss coordinate system, open the *Project Raster (Data Management)* tool (browse to *Data Management Tools > Projections and Transformations > Raster > Project Raster*).

8. Select *g100_06.tif* as the *Input Raster*. The *Input Coordinate System* will already be filled in.

9. Save the *Output Raster Dataset* to your working folder as *g100_06_ch1903.tif*.

10. Choose CH1903 LV03 as the Output Coordinate System *(Projected Coordinate Systems > National Grids > Europe > CH1903 LV03)*.

11. Choose the *NEAREST Resampling Technique*.

12. Leave the default *Output Cell Size* (100 m) (Fig. S 48).

13. Click *OK*.

14. Click *OK* again to run the tool.

Be patient, this is a large file and it will take some time to complete (~ 10 minutes but depends on your processing system). You will get a pop-up at the bottom right-hand side of your screen when the process has finished.

The resulting projected raster (g100_06_ch1903.tif) will be automatically added to the *.mxd. The original CLC06 layer can now be removed from the document as it is no longer needed:

15. Right-click the g100_06.tif layer in the TOC.

16. Select *remove*.

6.9.2. Adding vector layer and changing document properties

Since the CLC06 layer covers all of Europe, it should first be clipped to the extent of the required study area using the *Clip (Data Management)* tool so that map rendering is less time consuming in the next steps. This process can also be conducted using the *Extract by Rectangle (Spatial Analyst)* tool. We will use the commune_boundary_FR.shp to define the output extent of the clipped land cover raster layer so add the layer to the map now. Note: because the first map layer which was added to the document (g100_06.tif) was in the ETRS89_ETRS_LAEA coordinate system, the *.mxd properties automatically take on the properties of that coordinate system. For that reason, you will get a pop-up warning when you try to add layers from the Swiss coordinate system. The warning will state that there may be a problem because a layer with a different coordinate system is being added. Close the warning and change the document properties:

17. Go to menu *View*.

18. Select *Data Frame Properties*.

19. Open the *Coordinate System* tab.

20. Manually change the coordinate system to *CH1903 LV03* (same path as step 10).

21. Click *OK*.

22. Save the *.mxd to keep these changes (*File > Save As...*).

Now the map document coordinate system corresponds to the data being used in the Swiss projection. Next we will clip to the extent of the study area.

6.9.3. Clipping to study area extent

23. Open the *Clip (Data Management)* tool *(Data Management Tools > Raster > Raster Processing > Clip)*.

24. Input Raster: g100_06_ch1903.tif.

25. Output Extent: commune_boundary_FR.shp. The X and Y maximums and minimums will be automatically populated based on the output extent layer.

26. Check the *Use Input Features for Clipping Geometry* option.

27. Save the *Output Raster Dataset* as g100_06_ch_FR.tif.

28. Change the NoData Value to 9999 (Fig. S 49).

29. Click *OK*.

30. Turn off the g100_06_ch1903.tif and commune_boundary_FR.shp layers (uncheck in TOC).

31. Right-click the g100_06_ch_FR.tif layer.

32. Select *Zoom to layer* to view the result (Fig. S 50).

6.9.4. Applying a colour map

The layer g100_06_ch_FR.tif is now an efficient size for our analysis, but we still do not know what the numbers (raster values) in the raster represent (see Fig. S 50). This is due to the fact that the attribute table of the g100_06.tif layer does not include the official CLC06 land cover codes because of a purely technical reason: the codes (50 in total) range from 111 to 999 and cannot be displayed in a single band 8-bit integer raster dataset because the range of values is limited to 0-255. Of course a 32-bit floating-point raster image could be used to display the whole range of values from 111 to 999. However, this would almost quadruple the size of the dataset which would cause problems for download, storage, and analysis. For this reason, the European Environment Agency (EEA) provides a layer file along with the raster image to indicate the code and colour map information which can be easily integrated into ArcGIS.

To add the code and colour map information:

33. Browse to the unzipped land cover folder (g100_06.tif).
34. Add the clc_colortable_92.lyr file to ArcMap (section 6.3.3.).
35. In the TOC, you should now see a layer called: lceugr250_00_pct.tif (which corresponds to the clc_colortable_92.lyr file) with a red exclamation point (!) beside it. That means that the layer does not know where the data are located. The codes (e.g. 111, 112, etc.) in this file represent the land covers type from the CLC06 dataset.
36. Right-click the lceugr250_00_pct.tif layer.
37. Click *Data*.
38. Now click *Repair Data Source* to tell the layer where the data are located.
39. Browse to the *g100_06_ch_FR.tif* layer.
40. Click *Add*. The red exclamation point beside *lceugr250_00_pct.tif* should have disappeared.

6.9.5. Selecting and labelling land cover values

Now we will remove the land cover values which are not within our study area in order to further optimise the data to fit our needs.

41. Open the properties of the *lceugr250_00_pct.tif* layer.
42. Open the *Symbology* tab.
43. Manually select and remove all values which have a "0" *Count* value (use the Ctrl button on the keyboard to select more than one at a time) (Fig. S 51).
44. Click *OK*. This leaves 14 values in that layer.

Now open the excel file from the previously unzipped folder (*clc_legend.xlsx*) so the land cover names can be added as labels to the codes. This step is not completely necessary but it adds another element to the data. For each code, find the corresponding name ("Label 3") and add it to the TOC by slowly right-clicking the code and pasting the name beside (Fig. S 52). In the following tasks we will make use of the code numbers and names which were just added manually as labels.

6.9.6. Conditional expressions using the raster calculator

To statistically analyse the area covered by three land cover types of interest (arable land, coniferous, and mixed forest) within the 2 km buffers (created in section 6.5.5.1.), we will first create three new raster layers using raster calculator.

45. Open the *Raster calculator (Spatial Analyst Tools > Map Algebra > Raster calculator)*.
46. Combine *Layers and Variables* and functions until the Raster calculator expression equals "Con("g100_06_ch_FR.tif" == 12, 1, 0)" (Fig. S 53) (12 is the raster value for "Non-irrigated arable land" in the CLC06 dataset). The expression is a conditional statement (Con) which literally means: if the value of a pixel from the g100_06_ch_FR.tif layer equals 12, assign that pixel a new value of 1. If not, assign a value of 0. This will create a binary raster.
47. Specify an output raster, e.g. lc_class211.tif, which denotes that the new layer will represent the 211 (Non-irrigated arable land) land cover code.
48. Click *OK*.
49. Repeat steps 45 to 48 with the following raster values: 24 (replaces 12 in the equation in step 46.) (class 312 as the output, Coniferous forest) and 25 (class 313, Mixed forest).

The newly created raster images contain zones with the value 1 (where the *Raster calculator expression* was true, i.e. the land cover represents the non-irrigated arable land category) and 0, where the expression was false (i.e. the land cover does not represent the non-irrigated arable land category).

6.9.7. Zonal statistics of land cover values within buffers

With this binary classification, we can now do a simple statistical analysis of the land cover within each buffer polygon layer. Below, the *Zonal Statistics as Table* tool will be used to summarise the values of the land cover layer within the buffers created in section 6.5.5.1. and output the results into a table. Specifically, we will calculate the mean percentages of non-irrigated arable land, coniferous forest, and mixed forest within a 2 km radius of the honey bee colony locations.

50. Open the *Zonal Statistics as Table* Tool (*Spatial Analyst Tools > Zonal > Zonal Statistics as Table*).
51. Input raster or feature zone data: buffer2000.shp (browse to or add to *.mxd).
52. Zone field: FID (Unique identifier).
53. Input value raster: one of the newly created binary raster images (e.g. *lc_class211.tif*).
54. Output table: stats211.
55. Check the *Ignore NoData in calculations* option.
56. Choose the Statistics type *MEAN* (Fig.54).
57. Click *OK*.
58. Open the *Results* window (*Geoprocessing > Results*) and double click the *Zonal Statistics as Table* option (since that was the last process completed).

59. Repeat steps 50 to 57 for the other input rasters by altering the input dialog for the classes 312 and 313 (choosing appropriate filenames for the results).

Now look at the table results in the TOC (Fig. S 55). If you cannot see the tables in the TOC, make sure you are in *List by Source* view instead of *List by Drawing Order* (section 6.3.4.). In each table you will see columns for Rowid, FID (unique identifier corresponding to buffer ID), COUNT (number of pixels of particular land cover class inside the buffers), AREA (calculation of square meters of land cover class within the buffers), and MEAN (in percentage). In Fig. S 55, the mean values directly correspond to the percentage of non-irrigated arable land within each buffer (uniquely identified by its FID value).

6.9.8. Joining statistical information to attribute table

Now we will do a table join to connect the calculated statistical information directly to the colony_locations.shp shapefile based on the unique identifier in both the statistics table and the attribute table of the colony locations layer. By doing this, statistical information will be directly linked to the colony locations attribute table. This allows for the symbolisation and visualisation of statistical information on the map based on the geographic location of the colonies. For example, in section 6.10.4. we will symbolize each colony based on its statistical properties.

60. Right-click on the colony locations layer in the TOC.
61. Select *Joins and Relates*.
62. Click *Join*.
63. Select the option to *Join attributes from a table* (Fig. S 56).
64. Choose the FID as the fields on which to base the join with the stats211 table.
65. *Keep all records*.
66. Click *OK*.
67. Open the *Attribute table* for buffer2000.shp and scroll over to the right. You will see the values from the stats211 table have been added to the end. Since we are only interested in the *MEAN*, turn off the other fields (*right-click column name > Turn field off*). For table organisation purposes, other fields can also be turned off (this is not permanent – to undo go to *Table Options* (top left corner of table) and *Turn all fields on*).
68. To distinguish this *MEAN* column (211) from the other 2 other mean value columns (312 and 313), which we will add in the next steps, let's rename the column (*Right-click column name > properties*).
69. In the *Alias* field, change the name to "211MEAN".
70. Click *OK*.
71. Repeat 60-70 for the 312 and 313 statistics tables.

6.9.9. Adding and populating a new field in attribute table

Now, to obtain a value for all of the other land cover classes besides 211, 312, and 313, we will create a new column in the table and then calculate the values of "other" land cover values inside each buffer.

72. Open the *Table Options* for the colony_locations layer.
73. Click *Add Field*.
74. *Name*: LC_other.
75. Choose the *Type* Float.
76. Use the field calculator to populate this column. *Right-click* on LC_other field name.
77. Select *Field calculator*. Fill out the dialog box according to Fig. S 57 with the equation: 1 - ([stats211:MEAN] + [stats312:MEAN] + [stats313:MEAN]) which calculates the percentage of "other" land cover types in relation to the three we have already used.
78. You will end up with a table similar to the one in Fig. S 58. We will use these four new fields later in the Map creation section (6.10.).

Note: as you can see in Fig. S 58, there are "<NULL>" values for FID 16. This seems to be a glitch in the *Zonal Statistics as Table* tool. There are ways this can be fixed in the table but we will not deal with them now.

6.9.10. Visualisation of attributes using graphs and charts

To visualise the statistical data another way, graphs and charts can also be created in ArcMap. To create a graph similar to that in Fig. S 59., go to *View > Graphs > Create Graph*. Use the mean values for 211, 312, 313, and LC_other. In the next section, a final map will be created showing the case study results.

6.10. Map creation

Map creation is one of the main functionalities in ArcMap. Here we will give a brief tutorial about how to create a map as the final product of the GIS analysis. This task assumes that all layers created in the previous steps are already in the map document (including the results of the case study).

6.10.1. Data view *vs.* layout view

In ArcMap there are two different views: one for viewing, manipulating, and analysing data (Data View, Fig. S 60), and one for creating the final map product (Layout View, Fig. S 61). To switch between the two, toggle the buttons at the bottom left hand corner of the data frame (map display, see Fig. S 60).The toolbars used in each of these views can be confusing as there are multiple tools for zooming. The *Tools* toolbar is used for zooming in and out of the data frame, thus the actual data, and the *Layout* tools toolbar is used to zoom in and out on the final map document. See Figs. S 60 and S 61 to help understand the differences between these two views and also the different toolbars.

6.10.2. Symbology

The layer symbology can be changed within the properties of each layer under the *Symbology* tab. Depending on the type of data you want to symbolize, there are various options. For example, layers can be symbolized based on one set of features (Features), categories of attributes based on unique values or unique values using many fields (Categories), or quantities of attributes (Quantities). As an example, we will use the *commune_boundary_FR* layer and change the symbology so that each commune is represented by a different colour.

To change symbology to represent unique features:

1. Right-click commune_boundary_FR layer.
2. Select *Properties*.
3. Select *Symbology*.
4. Select *Categories*.
5. Select *Unique values*.
6. For the *Value Field* choose the attribute to be symbolized. In this case it is GEMNAME.
7. Click the *Add All Values* button.
8. Click *OK*.

The results show each commune represented by a different colour (Fig. S 62).

6.10.3. Labels

Labels can be automatically added for any layer.

To add labels to a layer:

1. Right-click the commune_boundary_FR layer.
2. Click the *Labels* tab.
3. Check the *Label features in this layer* option.
4. Label field: GEMNAME.
5. Click *OK*.

Now the name of each commune can be seen (Fig. S 63).

6. To make the label larger or change the font Properties > Labels.

6.10.4. Advanced symbolization

In this step, the statistical results of the case study will be symbolized using pie charts.

1. To begin, open the layer properties of colony_locations.shp.
2. Open the *Symbology* tab.
3. Choose *Charts* and make sure *Pie* is selected.
4. Add the *MEAN* attributes (including LC_other) to the Symbol/ Field box. Change the colours to correspond with the original land cover values (Fig. S 64).
5. Click *OK*.
6. Now label the colony locations to denote the FID identifier number (see above).

The map should now look something like Fig. S 65, which is the data in *Data View*. Maybe you will notice that FID 16 is missing. Because of the error we encountered earlier, it did not get symbolized or labelled.

6.10.5. Final map preparation

Now we will put the finishing touches on the map in *Layout View*. For the final map the pie charts and communes will be displayed with the background of the DEM and the hillshade for terrain visualization.

To finalize the map:

1. Turn on the commune_boundary_FR, colony_locations (symbolized as pie charts), hillshade, and DEM layers in the TOC.
2. Go into *Layout View* (bottom left hand corner of the map window).
3. The page size is automatically set to normal printing page size. To change the page size, go to *File*.
4. Select *Page and Print Set-up*.
5. Uncheck *Use Printer Paper Settings*.
6. Insert your own values for the width and height of the page (22 x 24 cm were used here).
7. Click *OK*.
8. Rescale the data frame (dotted green line) to fit inside the page border (solid black). Leave some white space between the data frame and the page border to insert a grid later.
9. Right-click the colony locations layer.
10. Select *Zoom to layer* to focus the map on this layer.

Remember to use the standard tools to zoom in and out on the data, and the layout tools to zoom in and out on the page.

11. Play around with the transparency values of the layers to make a nice end product. Go to *Properties*.
12. Select the *Display* tab.
13. Insert a value between 0 (no transparency) and 100 (completely transparent).

6.10.6. North arrow, scale bar, legend, title, text

To add a north arrow:

1. From *Layout view*, click *Insert* (main menu).
2. Click *North arrow*.
3. Select the style of the north arrow.
4. Click *OK*.

To add a scale bar:

1. Click *Insert*.
2. Select *Scale bar*.
3. Choose the style.
4. Click *Properties*.
5. Select the number of divisions and units.
6. Click *OK*.
7. Click *OK* again.
8. Double-click *Scale* to change the properties.

To add a legend:

1. Click *Insert*.

2. Select *Legend*.

3. Choose the layers to add to the legend.

4. Click *Next*.

5. Choose the legend title.

6. Click *Next*.

7. Choose a border, background, or shadow.

8. Click *Next*.

9. Click *Next* again.

10. Click *Finish*.

To add a title or other text:

1. Click *Insert*.

2. Click *Text*.

3. Type the text.

4. Hit *Enter*.

5. Double-click on the text box to open the properties.

6. Change the size and font.

7. Click *OK*.

Once added to the map, you can double-click on any feature in the layout view to open its properties. Also, the scale bar and legend are linked directly to the map, therefore they change as the map changes. For example, the scale bar will adjust to the scale of the map, and the legend items will adjust if the symbologies or names of layers change in the TOC. To adjust these features manually they must first be converted into graphics and ungrouped.

To manually change the legend:

1. Right-click the legend.

2. Select *Convert to graphics* (this unlinks the feature from the map).

3. Right-click the legend again.

4. Select *Ungroup* to have the ability to select individual features and edit them (some features required a second "ungrouping" to be editable, e.g. the symbols and labels).

5. Select the item to edit.

Note: once the item has been converted to graphics, there is no going back (you would have to insert the item again from scratch) and, the item is no longer dynamically linked to the map contents. Therefore changing layers and map scales will no longer change legend items or scales. It is recommended that conversion to graphics is the final step in the map creation process.

6.10.7. Drawing toolbar

The *Drawing* toolbar is useful for adding graphics such as shapes, text, and lines to the map.

To open the toolbar:

1. Go to *Customize*.

2. Select *Toolbars*.

3. Select *Draw*.

6.10.8. Adding a grid

To add a numbered grid around the map:

1. Go to *View* (main menu).

2. Select *Data Frame Properties*.

3. Select *Grids*.

4. Select *New Grid*.

5. Select the *Measured Grid* option (this is for projected coordinate systems (meters)), *Graticule* is for Geographic coordinate systems (latitude/longitude)).

6. Choose the *Labels only* option.

7. Click *Next*.

8. Here you could change the axes and labels, click *Next*.

9. Click *Finish*.

10. Click *OK*.

To adjust the properties of the grid, go back into the *View > Data Frame Properties > Grids* menu. Here, there is also the option to convert the grid to graphics in order to manually adjust it. See Fig. S 66 for the final map in ArcMap.

6.10.9. Exporting a map

To export the final map product:

1. Go to *File*.

2. Select *Export Map*.

3. Name it and choose the format (JPEG, PDF, etc.).

4. Choose the resolution (high quality resolution is 300 dpi (dots per inch) and draft quality is normally 96 dpi; also line art (contour lines, labels, ...) should be at 600 dpi and raster graphics at 300.).

5. Click *Save*.

6. Browse to the exported map.

7. Open it to view the results and see if any changes are needed (see Fig. 67 to view the exported map).

6.10.10. Cloud mapping and ArcGIS Explorer

Recently, ArcGIS has created a cloud mapping platform online which allows users to share their maps and data with others, access their maps online, on their desktop, and on mobile devices through the use of the cloud. On ArcGIS online, users can create maps by adding their own data, choose from different base layers, and use other people's shared data that has been uploaded to the site. This site is useful for map creation when you do not have access to the ArcGIS software, but want to make a basic map. All you need is an ESRI Global Account which is free to sign up for. For more information about ArcGIS online visit https://www.arcgis.com/home/.

ArcGIS Explorer is an application that can be downloaded for the purposes of exploring, visualising, and sharing geographic information. Users can also access ArcGIS' basemaps and layers,

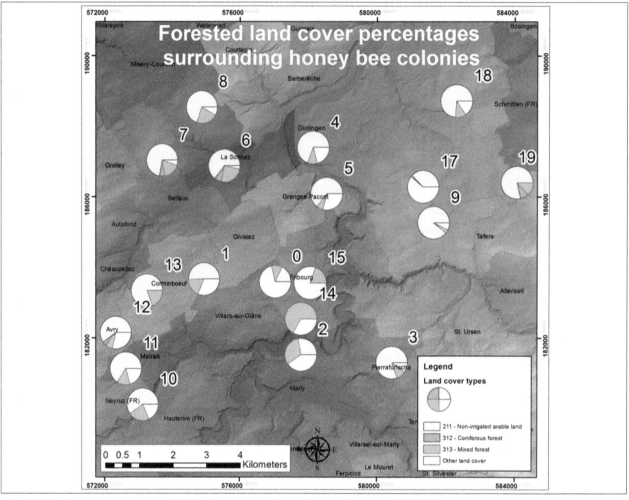

Fig. 67. The final map exported as a JPEG file. Reproduced with permission of Swisstopo (BA13016).

customise maps by adding data, photos, reports, and videos. Basic spatial analyses can also be performed. ArcGIS Explorer can be downloaded from http://www.esri.com/software/arcgis/explorer.

7. Using Quantum GIS (version 1.8.0.)

This section provides a brief hands-on tutorial as an introduction to Quantum GIS (QGIS) and shows examples of how open source GIS can be used instead of proprietary software for the same research applications as demonstrated in the previous chapter. QGIS is licensed under the GNU General Public License (http://www.gnu.org) and is an official project of the Open Source Geospatial Foundation (http://www.osgeo.org). This tutorial is based on QGIS version 1.8.0. "Lisboa" that was the latest stable release at the end of 2012 and runs on the 64-bit Microsoft Windows 7 operating system. Detailed information about QGIS, its functionalities, and the installation procedure on different platforms is well documented on the project website (http://www.qgis.org) and in the manual to version 1.8.0 (QGIS Development Team, 2013).

7.1. Main components of QGIS

QGIS includes the main program QGIS Desktop, QGIS Browser for data management, and a variety of optional plugins that provide specific functionalities.

7.1.1. QGIS Desktop

QGIS Desktop is the main interface for data collection, analysis, and presentation on the client-side and is similar to ArcMap (see section 6.1.1.). The language of the GUI is chosen automatically by the system defaults, but it can be changed within the menu *Settings > Options > Locale*. Most of the menus and symbols (Fig. S 68) are self-evident and should be easy to use for anyone with basic knowledge in GIS techniques, but in particular the menu "Plugins" is different than other software and will be discussed more in detail (see 7.1.3.). Projects (equivalent to map documents within ArcGIS, see section 6.3.1.) saved within QGIS are stored using the XML-Standard (Extensible Markup Language, machine and human-readable) and with the extension *.qgs.

Table 4. QGIS Plugins and Repositories list used or mentioned in this tutorial.

Plugin name	Repository
Add Delimited Text Layer (see section 7.3.3.1.)	Core plugin, automatically installed with QGIS (can be enabled and disabled in the Plugin Manager)
fTools (see section 7.6.)	Core plugin, automatically installed with QGIS (can be enabled and disabled in the Plugin Manager)
Spatial Query (see section 7.4.1.)	Core plugin, automatically installed with QGIS (can be enabled and disabled in the Plugin Manager)
Raster Terrain Analysis Plugin (see section 7.7.2.)	Core plugin, automatically installed with QGIS (can be enabled and disabled in the Plugin Manager)
GDAL Tools (see section 7.5. and 7.7.)	Faunalia Repository (http://www.faunalia.it/qgis/plugins.xml)
qgSurf (see section 7.8.3.)	QGIS Official Repository (http://plugins.qgis.org/plugins/plugins.xml)
Chart Maker Plugin (see section 7.9.)	QGIS Contributed Repository (http://pyqgis.org/repo/contributed)
SEXTANTE (se section 7.8.2.)	QGIS Official Repository (http://plugins.qgis.org/plugins/plugins.xml)
QGIS Cloud Plugin (see section 7.9.5.)	QGIS Official Repository (http://plugins.qgis.org/plugins/plugins.xml)
XyTools (see section 7.3.3.1.)	QGIS Official Repository (http://plugins.qgis.org/plugins/plugins.xml)

7.1.2. QGIS Browser

Starting in version 1.7., a new feature has been added to QGIS: The QGIS Browser (Fig. S 69). This browser is a data management program similar to ArcCatalog (see section 6.1.2.), simplifies file navigation within the file system, and provides preview options for various file types. Items can be easily added to the Map View of QGIS Desktop by drag and drop.

7.1.3. Plugins

In addition to its core functionalities, QGIS can be extended modularly with additional features and functionalities, downloadable from so-called repositories through the internet. By the end of 2012, about 175 plugins were listed on the project website provided by various authors (http://plugins.qgis.org/plugins). Because the application is open source, everyone is authorized to change or add features for special purposes by programming in the languages C++ or Python. Often, popular external plugins that have a big impact on the user community are implemented into the core functionalities of QGIS in the following release.

Plugins can be loaded and unloaded from the main menu *Plugins > Fetch Python Plugins* (Fig. S 70) from various internet repositories directly into the QGIS Desktop. Some of the plugins might require additional software packages or libraries, or are still marked as "experimental", indicating an early stage of development.

Within this tutorial the plugins listed in Table 4 will be used, so it is recommended to install them now:

1. Open QGIS, go to the *Start Menu*.
2. Go to *All Programs*.
3. Open the *Quantum GIS Lisboa* folder.
4. Select *Quantum GIS Desktop 1.8.0*.
5. From the main menu go to *Plugins*.
6. Select *Fetch Python Plugins*.

7. Open the *Repositories* tab.
8. Click *Add...* and type in the name and URL for the "Faunalia Repository" corresponding to Table 4.
9. Click the checkbox to enable the repository.
10. Click *OK*.
11. Verify if the "Faunalia Repository" is listed and if the connection works correctly (indicated with a green icon).
12. Repeat steps 8 to 11 also for the Repository called QGIS "Contributed Repository" (see Table 4).
13. Switch to the tab *Options*.
14. Activate the checkbox *Check for updates on startup*.
15. Set the *allowed plugins* to *Show all plugins, even those marked as experimental*.
16. Switch back to the tab *Plugins* within the QGIS Python Plugin Installer.
17. Filter for a plugin name (e.g. "GdalTools") and select it in the list.
18. Click *Install/upgrade plugin*.
19. Repeat the steps 17 and 18 for all these Plugin names: "GdalTools", "qgSurf" and "Chart Maker".
20. Close the QGIS Python Plugin Installer.

To remain focused on the most important functionalities for your daily work, it is recommended to make use of the Plugin Manager (Fig. S 71):

1. Open *Plugins*.
2. Select *Manage Plugins*.
3. Click *Select All*.
4. Click *OK*.

For more detailed information about managing and troubleshooting QGIS plugins on different platforms, see the QGIS user guide (QGIS Development Team, 2013).

7.2. Navigating the software

One motivation for the development of QGIS was to provide a user friendly GIS software with a simple and straightforward graphical user interface (GUI, QGIS Development Team, 2013). This was also one major motivation for why we chose QGIS as the representative for open source GIS within this tutorial.

7.2.1. Help menu and tooltips

The *Help* menu is placed at the right of the main menu bar in *QGIS Desktop* (Fig. S 68) or can be opened with the shortcut *F1* (on Windows). The majority of the functions and tools are well documented in the user guide (QGIS Development Team, 2013) and additional information is accessible at the QGIS project website (http://www.qgis.org). Useful information is provided by the tooltips (that usually appear when the mouse cursor is placed over an icon for a few seconds) indicating the name or functionality of the buttons and plugins. In addition, several functions and plugins have their own help button like the example in Fig. S 70. Thanks to a huge effort from the community, the GUI and large parts of the help menu have already been translated into many different languages.

7.2.2. User guide and documentation

All documentation for QGIS can be found on the project website http://www.qgis.org, divided into the two categories "Manuals" (containing user guides and tutorials) and "Developer" (providing programming help for developers). The official manual, the QGIS user guide (QGIS Development Team, 2013) cited in this tutorial, was released on the 15th of January 2013 and might have changed since then.

7.3. Adding data and working with different file formats

Adding geodata to QGIS Desktop is as easy as with ArcMap (see section 6.3.): a variety of file formats can simply be dragged and dropped, for example, from a folder within the file system directly into the *Map Legend* panel on the left of QGIS Desktop or loaded from the *Manage Layers* toolbar. The following hints and remarks will provide a start-up aid for users new to GIS and for some specific data types.

7.3.1. Setting a coordinate system and saving the new QGIS project (*.qgs)

Similar to the ArcGIS section (6.3.1.), it is important and recommended to first set the coordinate system according to the data used for the analysis or the target projection of the final map product, respectively (Fig. S 72):

1. Go to *Settings*.
2. Select *Project Properties*.

3. Select *Coordinate Reference Systems (CRS)*.
4. Activate the checkbox on top called *Enable "on-the-fly" CRS transformation*.
5. Select the projection called *CH1903 / LV03* with the EPSG-code *21781* within the list of the *Coordinate reference systems of the world* (Group *Swiss. Obl. Mercator*, see Fig. S 72) with a double-click.
6. Click *OK* to apply these settings and close the Project Properties dialog.
7. Verify by checking if the EPSG-code 21781 is written in the Status Bar (see Fig. S 68).

Note: This step is very important for the rest of this tutorial, because choosing the wrong CRS may lead to various problems in geoprocessing.

Advanced users who deal with a variety of different projections might appreciate the possibility to filter the list of all coordinate systems (Fig. S 72) by names or the identification number of the EPSG Geodetic Parameter Dataset (http://www.epsg-registry.org). Switching back to the *General* tab of the *Project Properties*, the *Layer units* will be now set to meters, because the Swiss oblique Mercator projection uses metric units.

8. Save the new project and settings, go to *File*.
9. Select *Save Project As,* or click the button 🔲 to the desired destination folder (A QGIS project file with the extension *.qgs will then be created).
10. It can be reopened directly by a double-click on the file or within the QGIS main menu *File > Open Project* or the *open project* button 🔲 from the *File* toolbar.

7.3.2. Add data to the project

Most of the functions to add and remove data or to change layer properties and appearance can be accessed through the main menu *Layer*. Of particular interest are the functions (listed in Table 5) contained in the *Manage Layers* toolbar:

Only data of the appropriate type (see sections 2.2.1. and 2.2.2.) will be selectable as a source for adding vector or raster data, respectively. In case the data type or the file extension is unknown, try to drag and drop the files directly from the source folder into the *Map Legend* in QGIS Desktop (Fig. S 68). Take into consideration that some common file formats consist of multiple files and that the main file has to be chosen (e.g. the file with the extension *.shp for a shapefile or the *.tif for a raster in GeoTIFF format, respectively, see also section 3.1.). Due to the Geospatial Data Abstraction Library for raster data translation GDAL (http://www.gdal.org) and the OGR Simple Feature Library (http://www.gdal.org/ogr) QGIS has excellent data interoperability capabilities and almost every geodata file format from any source can be loaded into the *Map View*.

Table 5. The most important functions of the *Manage Layers toolbar*. This toolbar can be enabled and disabled by the main *menu View > Toolbars > Manage Layers.*

Menu Entry Layers	Icon	Purpose
New		Creates a new layer without content (shapefile or SpatiaLite database)
Add Vector Layer		Add new vector layer(s) (see section 7.3.3.)
Add Raster Layer		Add new raster layer(s) (see section 7.3.3.)
Add PostGIS Layers		Manage connections to PostGIS databases, load PostGIS layers and tables (http://www.postgis.org/)
Add WMS Layer		Add and manage layers from Web Map Services (http://www.opengeospatial.org/standards/wms)
Add WFS Layer		Add and manage layers from Web Feature Services (http://www.opengeospatial.org/standards/wfs)
Remove Layer		Remove layer(s) (see section 7.3.3.1.)
Add Delimited Text Layer		Import delimited text files (see section 7.3.3.)
Create new GPX Layer		Create a GPX file for the use in Handheld GPS devices (see section 7.3.3.2.)

7.3.3. Importing vector and raster data

Now it is time to add some data to the new empty QGIS project:

1. Choose *Layers.*
2. Select *Add Vector Layer.*
3. Browse to the source folder of your Geodata (for example select commune_boundary_FR.shp, see section 5.3.).
4. Press *Open* to load the shapefile.

As an intermediate result, a rectangle full of polygons (see section 2.2.1.) representing different administrative districts, called communes, within the canton of Fribourg (western Switzerland) appears (Fig. S 73).

In the same way, but with the *Add Raster Layer* function, a DEM (see section 3.2.4.) can be added to the map:

1. Open *Layers.*
2. Select *Add Raster Layer.*
3. Browse to the source folder of your Geodata (for example select ASTER30_WGS84.tif).
4. Click *Open* to load the DEM into QGIS.

A rectangular greyscale raster image appears around the city of Fribourg in western Switzerland (Fig. S 73). If the raster is not visible, please check that the *"on-the-fly" CRS transformation* is enabled within the dialog accessible through the menu *Settings > Project Properties > Coordinate Reference Systems (CRS)* (see also Fig. S 72).

If the raster image is displayed in a uniform grey, the display options must be adapted:

5. Open *Layer.*
6. Click *Properties.*
7. Select the *Style* tab.
8. Click *Load* in the *Load min/max values from band* option.
9. In the *Contrast enhancement* option, choose *Stretch to MinMax.* The raster values range from 407 (black) to 883 (white), that represents the altitude above sea level in meters.
10. Click *OK.*

The new data will be added to the top of the map, thus probably hiding some other content behind it (the display order is defined by the hierarchy in the *Map Legend*). Considering the current map display, we see that the raster image has a smaller extent than the vector layer and only few a details are visible at this scale. To zoom to the region of interest around Fribourg, suitable operations are:

- Using the *Zoom* and *Pan* icons in the *Map Navigation* toolbar or the same functions in the *menu View*

- Right-click on the raster image in the *Map Legend* (in our example called ASTER30_WGS84) and choose the option *Zoom to layer extent.*
- Zoom with the mouse wheel placing the cursor at the centre of the area of interest.

Now the topography and the Sarine River crossing the city of Fribourg from south to north are visible (Fig. S 74). First of all, from this step we learn that QGIS has several tools for similar applications, and every user will find and choose the most efficient manner to achieve their goals. Second, it is beneficial to use (or even try) the right mouse button, for example to set the *Map View* exactly to the extent of one specific layer or to the access the *Layer Properties.* Advanced users might be comfortable using the mouse wheel to change the zoom level while other tools stay activated (e.g. during digitisation).

7.3.3.1. Add delimited text layer (*.txt, *.csv)

In this section, a simple text file will be added to QGIS to visualize and analyse it spatially (for more information about this refer to section 6.3.3.1.):

1. Open *Layer.*
2. Choose *Add Delimited Text* or click the appropriate button from the *Manage Layers* toolbar (Table 5).

3. Choose the source text file (in our case the comma delimited textfile points_table.csv).

4. Give a name to the output layer (e.g. bees_colony_locations_tab) .

5. Select the appropriate delimiters (e.g. semicolon or comma) and the columns containing the coordinates in East (X) and North (Y) according to Fig. S 75.

6. Click *OK*.

7. Choose the CH1903/LV03 projection from the CRS selector list.

8. Click *OK*.

9. Go to *File*.

10. Click *Save Project*.

As a result, the honey bee colonies are shown as point information on top of the *Map View* (Fig. S 76). It is important to note, that this layer is loaded into QGIS and available for any display and map creation purpose and also some basic analysis, but not physically stored as geodata in the file system. To get the full functionality of a point vector file, save this layer e.g. as a shapefile by performing the following steps:

1. Right-click the layer in the *Map Legend* (Fig. S 68).

2. Choose the *Save as* option in the context menu.

3. Select a file format (in our case it will be an ESRI Shapefile).

4. Select a target destination folder within the file system (e.g. with the filename bees_colony_locations.shp).

5. If special text characters are included, choose an encoding that is suitable (e.g. for Western Europe utf-8 or latin1 are good options).

6. Select the correct CRS (in our case it is CH1903 / LV03).

7. Enable the option *Add saved file to map* (Fig. S 77).

8. Click *OK*.

A message ("Export to vector file has been completed") will appear and the layer will be added on top of the *Map Legend*.

9. Select the old text file from the *Map Legend* using a left mouse click.

10. Press the button *remove layer(s)* (see Table 5) to remove it from the project – we will only need our newly created shapefile for all future steps.

Using the *plugin* called "XyTools" (Tab. 4) it is also possible to import Microsoft Excel or OpenOffice spreadsheets, similar to the procedure presented in section 6.3.3.1.

7.3.3.2. GPS and KML data, working with different coordinate systems

Other important functionalities of modern GIS are those to import and export GPS data (e.g. those acquired by handheld GPS devices), and the transformation between different coordinate systems. Because the GPS eXchange Format *.gpx is common and integrated within the OGR library (see section 7.3.2.), the import of existing GPS data is as simple as adding other vector data (see section 7.3.3.). Assuming that

our point vector layer added in the previous step represents colony locations of honey bees we would like to locate in the field, this vector layer needs to be exported to a new *.gpx-file to be uploaded onto a handheld GPS device. This can be done with the same function used to save a layer to a new shapefile (Fig. S 77), but modified settings:

1. Right-click on the layer in the *Map Legend*.

2. Select *Save as*.

3. Choose the GPS eXchange Format (supported by a majority of GPS devices).

4. Set the target coordinate system to "WGS84" (because GPS devices are natively working in World Geodetic System 1984).

5. Enable the option *Skip attribute creation* (because we only need the coordinates for a stakeout of the colony locations in the field, without any attributes).

6. Click *OK* and verify if the new file (e.g. with the file name bees_colony_locations_WGS84.gpx) has been created within the chosen destination folder.

Behind this two very important technical GIS functionalities are working: first, the translation from a source file format (*.shp) to a target file (*.gpx) format and second, a precise transformation between two different coordinate systems, even with different map units. If you have a software like Google Earth for instance, you can easily verify if the coordinate transformation to WGS84 worked correctly (if you do not have Google Earth installed, you can skip this paragraph, or download the software from http://www.google.com/earth/index.html).

1. Open Google Earth.

2. To open the *.gpx-file in GoogleEarth, go to *File*.

3. Click *Open*.

4. Browse to the location of *.gpx file.

5. Beside the *File name* box, choose the *Gps* option from the dropdown menu.

6. Click *Open*.

7. The GPS data Import dialog window will appear asking if KML tracks should be created from the GPS data. Check the options *Create KML Tracks* and *Adjust altitudes to ground height* (with this function, all elements are attached on top of the elevation model behind GoogleEarth and you are sure that all features will be visible) and confirm with *OK*.

8. Google Earth will do a file format conversion that results in a new temporary layer named "GPS device" in the panel *Places > Temporary Places*. The new created placemarks in Google Earth should be at the same location as in the QGIS project.

9. Right-click on the layer *GPS device*.

10. Click *Save Place As* to save this temporary layer to the file system in the KML format (Fig. S 78, make sure to choose KML instead of the zipped KMZ format).

The Keyhole Markup Language (KML) is another file format supported by the OGR library. It provides great possibilities to share spatial information of vector based geodata with people who are not familiar with GIS software, since almost everybody can handle software like Google Earth. Of course it is also possible to import the KML file back to QGIS. We will see if the coordinate transformation from WGS84 back to the Cartesian CRS of Switzerland works fine:

1. Just drag the KML file from its data source to the *Map Legend* in your QGIS project.
2. By activating and deactivating the checkbox to the left of the layers you can check if the geometries are congruent.

If there is a lateral shift between the point layers or an error in scale, the option *"on-the-fly" CRS transformation* was probably not enabled – you can fix this by navigating to the main menu *Settings > Project Properties* where you choose the second tab *Coordinate Reference Systems (CRS)*.

7.3.4. Layer organisation and navigation

Managing layers with QGIS is very similar to ArcGIS, thus the following hints are complementary to sections 6.3.4. and 6.3.5:

- The drawing order of one specific layer relative to the others can be changed in the *Map Legend* (Fig. S 68): click on the layer and hold the left mouse button to move layers upwards or downwards in the hierarchy.
- Within a large project it is recommended to make groups (right-click on the *Map Legend > Add New Group*), to be more flexible to turn a whole group of layers on and off together, for instance, or to get more space in the layer hierarchy by collapsing multiple legends.
- To avoid losing the most important buttons and menus, have a look at the main menu *View > Panels and View > Toolbars*. Here you can enable additional panels to the QGIS GUI (e.g. the *Overview* or the *QGIS Browser* window) and show or hide several toolbars and buttons (the most essential toolbars are *Attributes, File, Manage Layers* and *Map Navigation*).

To achieve a good visualization result, changes in the layer *Layer Properties* are often required – the topic of the following section.

7.3.5. Layer properties

The properties of one specific layer can be accessed by the main menu *Layer > Properties*, a right-click on the layer name and choosing the entry *Properties* in the context menu or simply by double-clicking the layer name. The *Layer Properties* might vary between different file formats, but every layer will at least have tabs for *Style* options, *General settings* and provide *Metadata* information. Let's first have a look at the *Metadata*, the information about the data: opening the *Metadata* for the layer commune_boundary_FR.shp, the following information appears (Fig. S 79):

Apart from the file format, data source, and geometry type, we can also learn something about the number of features (n = 51), the

coordinate reference system (CRS) used for projection and the spatial extent (minimal and maximal values for X and Y). The current administrative districts are digitized as polygons, displayed with an opaque colour fill. To combine them with our elevation model, the fill colour could be removed and the border set to a bright colour:

1. Open the *Layer Properties* window by a method of your choice (see above).
2. Switch to the tab *Style*.
3. Click the button *Change* to the left and a further window with *Symbol properties* will open.
4. Select an appropriate *Symbol layer type* in the *Symbol properties* window (dropdown-menu on top) (Fig. S 80). By default, polygon layers are visualized by a "simple fill" with a uniform colour for the fill area and black lines representing the polygons boarders. Change the layer type e.g. to *Outline: simple line*.
5. Fill with a colour of your choice.
6. Click to close the *Symbol properties*.
7. Click *Apply* to implement the changes.
8. Close the *Layer Properties* with *OK*.
9. Move the polygon layer one level higher in the *Map Legend* on top of the elevation model.

7.4. Data selection and editing

This section presents the most important selection queries and editing tasks using different techniques in QGIS.

7.4.1. Spatial query and select by location

Similar to ArcGIS (section 6.4.1.), different functions exist to obtain information for defined locations: The *Identify Features* button is the same as the *Identify* tool of ArcGIS (see section 6.4.1.) and is used to obtain information about specific pixel values within a raster image or to look at the attribute values of a vector feature. Before applying the *Identify Features tool*, it is necessary to activate and select the appropriate layer in the *Map Legend*. In addition, various spatial operations are provided within the main menu *Vector*. Of particular interest is the function *Spatial Query* that allows more advanced queries by using vector geometries. To demonstrate this functionality with QGIS we will ask for all communes that contain at least one honey bee colony.

1. Choose *Vector* from the main menu.
2. Select *Spatial Query* (a dialog window will appear).
3. Select the layer commune_boundary_FR.shp as *Source Features* and the point vector file with the honey bee colonies as *Reference Features*.
4. Choose the operator *Contains* to be applied between these two features.
5. Choose to create a *New Selection* (alternatively, it could be added to or removed from an existing selection).

6. Click *Apply* and close the dialog, if the result looks similar to Fig. S 81.

Intersection operations can also be applied with the help of the function *Select by location*:

1. Go to *Vector*.
2. Choose *Research Tools*.
3. Select *Select by location*.
4. Choose the settings according to Fig. S 82 to achieve similar results as displayed in Fig. S 81.

In practice, spatial selection operations are frequently used to assign attributes of one layer to another layer and for raster calculation tasks or modelling tasks.

7.4.2. Select by attributes

Let's have a look at the result of the spatial query from the last section: In total eleven features were highlighted (that means selected) within the layer bees_colonies_locations.shp. So far so good, but what if the research question was to get the names of all communes containing honey bee colonies? To achieve this goal:

1. Right-click the commune_boundary_FR.shp layer.
2. Choose *Open Attribute Table* or use the button 🖹 within the *Attributes* toolbar.

It is written in the headline of the *Attribute table* that the selected layer contains 51 entries in total (representing communes). Because they are topologically correct (see section 2.1.), none of these communities intersects with another and all the borders are congruent (that means that there are no blank "islands" in between).

3. To focus on the selected features within a large dataset, use the checkbox named *Show selected only* within the *Attributes table* (Fig. S 83). Browsing to the right within the different attributes, within the column called "GEMNAME", the name of each community is given.
4. The *Attribute table* can stay open while other functions are performed, e.g. during digitizing or editing tasks.

Assuming that the goal to produce a map that shows only the municipalities of the canton Fribourg, the following workflow can be applied:

1. Right-click the commune_boundary_FR.shp layer.
2. Select *Query*.

The *Query Builder* dialog opens (Fig. S 84). The attribute "KANTONSNR" contains an ID that represents the canton in Switzerland and in the case for Fribourg it is 10. So it is now our goal to restrict the features by means of a formula like "KANTONSNR" = 10. QGIS will then perform a restriction on the polygons displayed similar to an SQL Selection-Query by the use of a "where clause" (http://www.w3schools.com/sql/sql_where.asp).

3. Try to rebuild the formula mentioned above by activating the options *Fields*, *Operators* and *Values* (otherwise it can also be written manually).
4. Confirm the query with *OK*.
5. Right-click on layer and *Zoom to the layer extent*.

Two polygons in the north-west outside of the canton of Fribourg will have disappeared from the *Map View* and will also no longer be selectable within the *Attributes table* (there are only 49 features left, Fig. S 85.). However, they are not deleted from the file and can be restored by removing any SQL "where clause" in the *Query Builder*.

For other applications where a section of a vector file should be selected without blanking out all the other features, the same *Query Builder* dialog can be accessed by the *Attribute table > Advanced search*. Of particular use is the combination of different "where clauses", e.g. selecting attributes above or below a certain threshold or containing a given text string.

7.4.3. Manual selection

Sometimes the easiest way to select geometries or entries in the *Attribute table* is by hand, especially for digitizing or editing tasks. To create new selections use the *selection* button 🖑 within the *Attributes* toolbar. Clicking on the little black arrow to the right, a drop-down menu opens, containing some more options for the selection using rectangles, polygons, freehand forms, or a radius. To select multiple features or to remove some from a given selection, hold the CTRL key (Windows) down while selecting or deselecting the features with the left mouse button (see the manual (QGIS Development Team, 2013) for the same function on Mac and Linux).

Any selection can be removed using the main menu *View > Select > Deselect Features from All Layers* or the button 🖱 within the *Attributes* toolbar.

7.4.4. Creating new layers from selected data

If a given selection has to be exported into a new layer, e.g. to share the file with business colleagues, this can be done by saving a vector layer (see section 7.3.3.): open the layers context menu by a right-click within the *Map Legend > Save Selection As*. Choose all the polygons in the layer commune_boundary_FR.shp that intersect with at least one of the honey bee locations with a selection tool of your choice and load the new created layer into the QGIS project (Fig. S 86.).

The newly created polygon layer will only contain 11 features and will again be symbolized by a non-transparent simple fill pattern.

7.4.5. Attribute table data editing

QGIS provides a set of editing tools and conventions similar to ArcGIS (see section 6.4.3.). To edit a vector layer, it has to be set into "edit

Table 6. The most important functions and the corresponding buttons for editing tasks.

Function	Icon	Purpose
Toggle editing mode		Start and End editing sessions. Changes have to be confirmed (or rejected) at the end of an editing session.
Save Edits		Saves the edits permanently.
Delete Selected Features		Deletes only the selected features
New column		Inserts a new column (field) at the end of the attribute table. A field name and a data type must be defined.
Delete column		Deletes the selected column(s)
Open field calculator		Opens the field calculator

mode" by the *Toggle Editing* function. As usual, this function is accessible by several buttons and menu entries:

- Main menu *Layer > Toggle Editing.*
- The button *Toggle Editing* within the *Digitizing* toolbar or the *Attribute table.*
- The context menu within the *Map Legend > Toggle Editing.*

If the *Toggle Editing* function is greyed out, the selected layer is probably not in an editable vector file format (not all the file formats that can be imported and displayed with QGIS may also be changed, so that in some cases a file conversion is necessary first) or accessed by another program or other users.

1. Set the newly created layer containing all municipalities where colonies of honey bees were sighted into the *Edit* mode.
2. In the *Map Legend* a pencil will appear to the left of the layers name and in the *Map View* the different vertices of the polygons are marked by transparent circles.
3. Open the *Attribute table* and look at the buttons to the left (see Table 6):
4. Now add a new field (column in the *Attribute table*) with the button , where additional values (e.g. the number of honey bee colonies) can be filled in.
5. Within the *Add column* dialog, choose appropriate settings for *name* (avoiding special characters and blanks, e.g. 'bees_colonies'), *data type* (e.g. integer, only whole numbers will be needed) and *width* (2 digits are sufficient as there will not be more than 99 colonies per entry).
6. Now, fill in some numbers into the newly created fields as a test (Fig. S 87).

Because of the restriction to integer values of 2 digits maximum width, all types of characters and also numbers higher than 99 will be refused. This is one of the ways GISystems deal with aspects of data integrity.

7. Click the *Save Edits* button to make the changes permanent, i.e. written to the file system.

Save often to avoid data loss and finish successful edit sessions by turning the edit mode off (*Toggle Edits* button again > *save all changes*).

The number of honey bee colonies could be counted and written to the table manually, but of course QGIS provides several tools for this task that run automatically, even when the *editing mode* is not active or no empty field has been created before:

1. Go to *Vector.*
2. Choose *Analysis Tools.*
3. Select *Points in Polygon.*
4. Choose the appropriate *input layers* and a *name* for the *new attribute field* (e.g. "bee_colony")
5. Specify a path and a name for the new shapefile that will be created (Fig. S 88).
6. Click *OK.*
7. Click *Yes* when asked if the newly created shapefile should be added to the table of contents (TOC).

In which commune are most honey bee colonies located?

1. Switch back to the *Attribute table* of the new shapefile.
2. Sort the attributes by descending in the field "bee_colony" (click the arrow beside of the column name).

Within the commune of "Düdingen", with the "BFSNR" = 2293, 5 colonies of honey bees were found.

7.4.5.1. Using the field calculator and adding geometry columns

As the communes in this example have different areal extents, one could also ask for the commune area in relation to the number of honey bee colonies. With GIS software, this kind of problem is easy to solve. Also the extraction of statistical information about geometries like *point coordinates, line length, perimeter* or *area* from the features of a vector layer can be efficiently processed.

Let's demonstrate this with the polygon layer already containing the attribute "bee_colony" created in the previous section:

1. Open the *Attribute table.*
2. Again toggle on the *Edit mode.*
3. Open the *Field calculator.*
4. Activate the checkbox to *create a New field.*
5. Set the *output field name* to something like "area_bee" and the other parameters as follows: *field type* (decimal number), *width* (10) and *precision* (3).

6. Search now for "area" within the *Function List > geometry* and select the function with the name *$area* by a double click.

7. Click the appropriate button for a division operation and select the field containing the number of honey bee colonies.

8. Confirm with *OK*, if the resulting *Expression* corresponds to "$area / bee_colony" (Fig. S 89, "bee_colony" represents the column name, selectable by *Function List > Fields and Values*).

The values in the field "bee_area" (to the right within the *Attribute table*, ratio area per bee colonies) could be a first indicator about the occurrence of honey bees (of course with strong limitations, as the polygon layer only represents administrative boundaries): The minimum value belongs to the community of "Granges-Paccot" with the "BFSNR" 2198.

We learned in this section to calculate the area of a polygon layer and write these values into an *attribute field*. Another possibility to add standard geometry information (e.g. point coordinates, line length, perimeter or area) is the *function Export/Add geometry columns*. This function is of particular use in combination with different coordinate reference systems (CRS), which can easily be demonstrated with the KML file created in section 7.3.3.2. (e.g. bees_colony_locations_WGS84.kml):

1. Go to *Vector*.

2. Select *Geometry Tools (Fig. S 90)*.

3. Choose *Export/Add geometry columns*.

4. Select bees_colony_locations_WGS84 as Input vector layer.

5. Choose to calculate the geometry columns using the *Layer CRS* (e.g. WGS84).

6. Specify the *path* and *filename* for the destination file that will contain the geometry columns (it is the best to convert it back to a shapefile again).

7. Click *OK* and wait until the progress bar is at 100%.

8. Confirm to import this new shapefile by clicking *Yes* on the *next* dialog and close the *Export/Add geometry columns* window.

9. Open the *Attribute table* of the new shapefile.

Like its origin (the KML file), the shapefile is referenced in WGS84 but because of the "on-the-fly" CRS transformation, the points representing the honey bee colonies are still congruent with those referenced in the projected coordinate system (which is defined as CH1903 / LV03). Within the *Attribute table* two new fields have been created that show the latitude and longitude in degrees (according to WGS84).

7.4.5.2. Table manager plugin

New attribute fields can be added or deleted directly within the *Attribute table* or the *Field calculator* dialog. But at this stage there is no option to rename any given field. For all changes within the table structure (adding, renaming and removing operations), the *Table manager plugin* is the perfect tool. As this is an optional plugin, it has to be downloaded first from the Official QGIS Repository (see section

7.1.3.). In this example, we rename the geometry columns:

1. Go to *Plugins*.

2. Select *Table*.

3. Select *Table Manager* (the *edit mode* has to be turned off).

4. Locate the *field* in the list called "XCOORD".

5. Click the *Rename* button 📝 and rename it as "longitude".

6. Click *OK*.

7. Do the same with "YCOORD" renaming it as "latitude".

8. Confirm the changes by the button *Save*.

If changes are made for fields used for layer styling or labelling, some features may become invisible. This should not be the case in this example.

9. Click *Yes* to keep the current layer style.

The *Table Manager* is very strict in terms of the length of field names and the characters allowed. Due to this, it may be difficult to find suitable field descriptions, but this is the price to pay for preserving interoperability options between different file formats.

10. Save the layer to a new shapefile into the CRS CH1903/LV03 (see section 7.4.4., e.g. with the name bees_colony_locations_coordinates.shp)

11. Import this new shapefile (e.g. and add geometry columns again, see section 7.4.5.1.)

12. Open its *Attribute table*.

13. Compare the different columns.

Now the coordinates of the points are listed as attributes in the two CRS, one in degrees (WGS84) and one in meters (CH1903/LV03).

7.5. Format conversions and projections

QGIS provides a variety of different conversion functions between vector and raster data formats similar to other GIS software like ArcGIS (sections 2.2 and 6.7.). The most important plugin for data conversion functions within QGIS is called *GDAL Tools*, so make sure it is installed and activated (see section 7.1.3.).

7.5.1. Vector to raster

The *Vector to raster* function rasterizes vector geometries into the band(s) of an existing or a new raster image.

1. Go to *Raster*.

2. Select *Conversion*.

3. Choose *Rasterize (Vector to raster)*.

4. Choose the input shapefile (e.g. the layer commune_boundary_FR.shp) and a suitable *attribute field* (e.g. "Shape_Area").

Be aware that any layer intersection (different features overlaying each other) cannot be transformed to raster data without data loss (in the intersecting areas).

5. Specify an output raster file and a resolution in pixel value (Fig. S 91).

6. Launch the conversion by clicking *OK*.

The newly created raster image will probably be displayed in uniform grey colour. To apply different colours for the corresponding values (generated by the shape area of each polygon) proceed as follows:

1. Open the *Layer Properties.*
2. Select *Style.*
3. Choose *Single band properties.*
4. Select *Colormap* in the dropdown menu *Color map.*
5. Switch to the tab *Colormap* and increase the number of entries, e.g. to 5.
6. Click *Apply.*
7. Navigate to the tab *Transparency* and make 0 values transparent (otherwise regions outside of the original polygon remain filled).
8. Have a look at the changes on the *Map View* by clicking *Apply.*
9. Close the *Layer Properties* with *OK,* if the result looks similar to Fig. S 92.

Notice that the resolution of the polygon boundaries is coarse in the raster and not as smooth as in the original vector layer (zoom to the boundaries).

7.5.2. Raster to vector

As the name of this function suggests, a polygon layer will be created according to the different raster values (contrary to the function *Contour* demonstrated in section 7.7.2.4., which creates a line vector layer based on raster grid values).

1. Go to *Raster.*
2. Select *Conversion.*
3. Choose *Polygonize (Raster to vector).*
4. Choose the newly created raster from the previous section (7.5.1.) and convert it back to a vector file (so specify an output shapefile, see Fig. S 93).
5. Start the conversion by clicking *OK.*
6. Open the newly created shapefile.

An attribute field containing the former raster values will be created.

Beside the fact that the geometry became coarser (the boundaries are no longer smooth), we see that many of the original attributes have been lost (e.g. as seen with the *Identify* tool or within the *Attribute table*).

7.5.3. Reproject data to a new coordinate reference system (CRS)

Sometimes, not only one raster file needs to be converted to another coordinate reference system, but a whole folder with many raster images. In this example, the ASTER30_WGS84 DEM will be projected to the CRS CH1903/LV03 of Switzerland.

1. Go to *Raster.*
2. Select *Projections.*
3. Choose *Warp (Reproject).*

4. Select *single input* and *output* files (e.g. ASTER30_dem_CH1903) or source and destination folders (for processing in batch mode).
5. Specify both the source and *target reference system (SRS).*
6. Choose *bilinear* or *cubic* as *Resampling method.*
7. Activate the checkbox *Load into canvas when finished* (Fig. S 94).
8. Start the projection by clicking OK.

Depending on the type of the raster image values different *Resampling methods* should be used (e.g. nearest neighbour for categorized thematic data and bilinear or cubic for continuous data such as raster images like elevation models). In addition, it might be useful to define a value for pixels without data (so that they can be made transparent afterwards).

7.6. Basic vector tools

The standard tools for data processing and analysis provided by different GIS software have a similar appearance. All tools described in the following sections can be accessed by the main menu *Vector > Geoprocessing Tools* (if the plugin *fTools* has been successfully installed and activated before, see section 7.1.3.).

7.6.1. Buffer(s)

In QGIS, the *Buffer* tool can be accessed in the following way (see section 6.5.5. for a description of the tool):

1. Go to *Vector.*
2. Select *Geoprocessing Tools.*
3. Choose *Buffer.*
4. Select a realistic *buffer value* for honey bees (e.g. 2,000 meters).
5. Leave the *dissolve option* blank (otherwise all buffers will be aggregated to one single feature without the different attributes, see Fig. S 95).
6. Specify the *path* and *name* for the *destination file.*
7. Click *OK.*
8. Click *YES* to confirm loading the newly created vector file to the QGIS project.

One should consider the use of the buffer function may lead to topologically incorrect vector files, e.g. when the distances between the features of origin are closer than the double buffer radius, leading to overlaps between different features. The circle polygons created in this step are visible in Fig. 99.

7.6.2. Clip

See section 6.5.1. for a description of the *Clip* tool. As an example, the clip operation could be performed between the honey bee locations and the commune of Fribourg.

1. Activate the layer commune_boundary_FR.shp in the *Map Legend.*
2. First select the polygon that represents the city of Fribourg ("BFSNR" = 2196) with an appropriate selection tool (see section 7.4.).

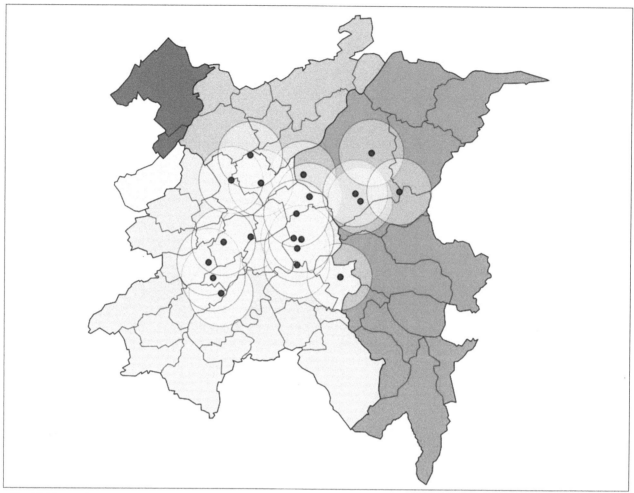

Fig. 99. After applying the Dissolve tool, the polygon layer is divided into 5 different districts. The thin black lines on top represent the former state (51 different municipalities). The yellow circles were created using a radial buffer of 2000 m around of all the honey bee colony locations. Reproduced with permission of Swisstopo (BA13016).

3. Open the Clip tool, go to *Vector*.
4. Select *Geoprocessing Tools*.
5. Choose *Clip*.
6. Choose the point layer containing the honey bee colony locations as *Input vector layer* and the polygon layer as *Clip layer*.
7. Make sure that the checkbox the Input Vector Layer *Use only selected features* (with this the operation could be performed on a selection of features only) remains blank (there is no selection), and that the one for the *Use only selected features* Clip Layer (beneath) is checked (Fig. S 96).
8. Launch the Clip operation by clicking *OK*.
9. Click *YES* to confirm loading the newly created vector layer (containing only 3 points) to the project.

7.6.3. Intersect

Identically to ArcMap, the intersect operation in QGIS computes a geometric overlap of two layers. This can be best demonstrated by the use of two polygon layers with different geometries:

1. Go to *Vector*.
2. Select *Geoprocessing Tools*.
3. Choose *Intersect*.
4. Choose the newly created buffer layer as *Input vector layer*.
5. Leave the checkbox beneath *Use only selected features* blank.
6. Select the layer commune_boundary_FR.shp (where the city of Fribourg should still be selected).
7. Check *Use only the selected feature* of this layer of the city of Fribourg (Fig. S 97).
8. Click *OK*.

The new layer contains only the common area of the two polygons (the geometry of the buffer within the city of Fribourg).

7.6.4. Dissolve

In section 7.4.2., a definition query was set from the polygon layer representing different municipalities within the canton of Fribourg (Fig. S 85). To aggregate all features that contain the same attribute "KANTONSNR" by producing a vector layer of all districts, the *Dissolve* function is a good option (see section 6.5.7. for further explanation about the dissolve tool):

1. Open the *Query Builder* for the polygon layer you want to use by right-clicking.
2. Make sure that no restriction is made within the SQL "where clause" (If you use the layer commune_boundary_FR.shp, delete the entry "KANTONSNR" = 10, if this still exists).
3. Confirm with *OK*.
4. Go to *Vector*.
5. Select *Geoprocessing Tools*.
6. Choose *Dissolve*.
7. Locate the layer commune_boundary_FR.shp as *Input vector layer*.
8. Leave the checkbox *Use only selected features* blank (with this the operation could be performed on a selection of features only, but in our case no selection has to be applied).
9. Select an attribute (e.g. "BEZIRKSNR") as *Dissolve field*.
10. Choose a *path* and *name* for the *destination file* to be created (Fig. S 98).
11. Finish by clicking *OK*.

Now the whole canton of Fribourg is selectable at once (representing only one line within the *Attribute table*), and the "island polygons" in the north-western part demonstrate what is meant by a "multipart polygon". What happened to the attributes? – Of course some of the attributes are no longer thematically correct: the dissolve function allocates the attributes from the first line of each value that was aggregated. The display of the new layer can be improved:

12. Open the *Layer Properties*.
13. Select *Style*.
14. Set a categorized symbol style with different fill colours for all 5 districts (by using the column "BEZIRKSNR").
15. Choose a colour ramp.
16. Click *Classify*.
17. Apply the changes.
18. Close the *Layer Properties* dialog by clicking *OK*.
19. The result will look similar to Fig. 99.

7.6.5. Union

In QGIS the *Union* tool can be found within *Vector > Geoprocessing Tools > Union* (see section 6.5.3. for an explanation of the *union* function).

7.6.6. Join attributes by location

In section 7.4.1. spatial queries were performed. In a similar manner, different layers and their features can also be spatially linked to each other, e.g. to relate thematic data to known coordinates or geometries:

1. Go to *Vector*.
2. Select *Data Management Tools*.
3. Choose *Join attributes by location*.

4. Select a *target* (e.g. *bee_colony_locations.shp*) and a *join vector layer* (where the attributes are taken from, e.g. commune_boundary_FR.shp).
5. Choose how the attributes should be attached to the target layer (by the first matching feature is usually the best choice if text fields are joined, but in the case of numerical attributes different aggregation modes may also be suitable).
6. Set the *path* and *name* for the destination shapefile (Fig. S 100).
7. Click *OK*
8. Load the new shapefile into the project by clicking *Layer > Add Vector Layer*.

Within the *Attribute table* of the new shapefile, the attributes of the commune layer will be added.

7.6.7. Basic statistics, list unique values

Especially for large vector data files containing a variety of attribute fields and features, it can be useful to display some statistical information before performing calculations with these attribute values. Two very useful tools are provided by the *fTools plugin* (see section 7.1.3.) and explained in the following examples:
To display basic statistics based on attributes,

1. Go to *Vector*.
2. Select *Analysis Tools*.
3. Choose *Basic statistics*.
4. Any vector layer can be chosen within the QGIS project as *Input Vector Layer*.
5. Select a *target field*, where the statistics (mean value, standard deviation, sum, minimum, maximum etc.) should be performed.
6. Click *OK* (Fig. S 101).

Sometimes the total number of attribute values within a given field are not important, but the different values themselves are:

1. Go to *Vector*.
2. Choose *Analysis Tools*.
3. Select *List unique values*.
4. Select a *target vector layer* and an *attribute field*.
5. Click *OK* to perform the calculation.

It is possible to export the values to the clipboard if they are used in another program (e.g. calculation spread sheet or text editor).

7.7. Basic raster tools

In the following section, some basic raster functionalities in QGIS will be discussed. Many of these tools rely on the *GDAL Tools plugin* (see section 7.1.3.), so install or activate this first.

7.7.1. Merge

The *Merge* function is similar to the mosaic tool of ArcMap, where several raster layers and files can be combined to a new raster dataset (see section 6.6.1. for more information).

1. Go to *Raster*.
2. Select *Miscellaneous*.
3. Choose *Merge*.
4. Select the *input raster images* (or even a whole *directory*).
5. Create one single *output raster file* in a *destination folder*.
6. Enable the lowermost checkbox so that the new raster image will be added directly to the project.

7.7.2. Terrain analysis

Beneath the elevation information itself, DEMs can serve as a source of data for various terrain analysis operations to calculate geometrical properties (e.g. slope or aspect) and other derivatives (e.g. clear sky factor). All functions for terrain analysis are accessible by the menu *Raster > Terrain analysis*. In some cases these functions are also useful for analysis or visualisation of thematic raster data.

7.7.2.1. Slope

The *Slope (Terrain analysis)* tool calculates the slope angle of each raster cell in degrees.

To calculate the slope of the terrain (Fig. S 102):
1. Go to *Raster*.
2. Select *Terrain analysis*.
3. Choose *Slope*.
4. Choose the DEM as the *input raster*.
5. Browse for a destination path and save the *output raster* as "slope".
6. Choose the *output format* to be used from the list (various formats within GDAL).
7. Click *OK*.
8. Adapt the raster symbolization if necessary (*Layer Properties > Symbol*).

Besides the *Terrain analysis* tools there is a set of *Terrain analysis* functions (with even more options) provided by the *GDALTools Plugin* (Fig. S 103): these tools are accessible by menu *Raster > Analysis > DEM (Terrain models)*.

7.7.2.2. Aspect

The *Aspect (Terrain analysis)* tool uses a DEM to calculate the direction in which slopes are facing (see section 6.6.2.2. for more information). To calculate the aspect based on a DEM:
1. Go to *Raster*.
2. Select *Analysis*.
3. Choose *DEM (Terrain models)*.
4. Choose the mode *Aspect* (Fig. S 103) (the function *Raster > Terrain analysis > Aspect* is just another option to achieve the same result).
5. Select the DEM as the *input raster*.
6. Browse for a destination path to save the *output raster* as ASTER30_aspect_CH1903.tif.

7. Some *optional settings* can be activated or deactivated in the checkboxes, to make flat areas (that do not have a valid aspect value by default) zero values, for instance.
8. Click *OK*.
9. Adapt the raster symbolization if necessary (*Layer Properties > Symbol*).

7.7.2.3. Hillshade

Hillshade rasters are useful for determining which areas on the terrain are shaded from the sun and which are not and are calculated from a DEM (see section 6.6.2.3) for further information about the tool. To calculate the hillshade based on a DEM:
1. Go to *Raster*.
2. Select *Terrain analysis*.
3. Choose *Hillshade*.
4. Select the DEM as the *input raster*.
5. Specify an *output (raster) layer*.
6. Choose reasonable values for the horizontal and vertical angle of the illumination (in degrees), the Z factor can usually stay at 1 (otherwise the terrain will be super-elevated when working in a CRS).
7. Click *OK*.
8. Adapt the raster symbolization if necessary (*Layer Properties > Symbol*, notice that the results are shown in degrees (Fig. S 104)).

7.7.2.4. Contour

This function generates an output vector file containing elevation levels (contour lines) from an input raster (see section 6.6.2.4. for more information). To create contours from a DEM:
1. Go to *Raster*.
2. Select *Extraction*.
3. Choose *Contour*.
4. Specify the *input DEM* and the *output vector file*.
5. Choose an appropriate interval between the different contours (e.g. 50 meters).
6. Activate the first checkbox to store the elevation information within an *attribute* and specify a *field name* (Fig. S 105).
7. Click *OK*.
8. Load the resulting vector layer into QGIS.

To achieve better results for visualisation, a smoothing algorithm could be performed on the line feature, e.g. with the tool *Simplify geometries* (*Vector > Geometry Tools > Simplify geometries*) or the *Generalizer plugin* (*Plugins > Generalizer > Generalizer*).

Contour lines can also be created by the *Contour plugin* that provides some more options (e.g. minimal and maximal height). After the installation (see section 7.1.3.) it can be started in the menu *Plugins > Contour > Contour*.

7.7.3. Raster calculator

The raster calculator is a very flexible and useful tool to combine raster images with conditional and trigonometric expressions and any other task that requires raster math operations. It can be found within the menu *Raster > Raster calculator*. The formula "ASTER30_dem_CH1903@1 * 3.28084" would convert the elevation data in the DEM from meters into feet by creating a new raster image (with an elevation range from 1,332 to 2,982 feet).

7.8. Spatial statistics and scientific modelling

QGIS provides several tools for spatial statistics and interfaces to statistical software like R (http://www.r-project.org). This section demonstrates some introductory examples of how statistics of spatially related data can be processed, as well as suggestions for useful plugins and software to perform more sophisticated spatial analysis. If you search for a specific application linked to spatial statistics, browse through the *QGIS Plugins* (see section 7.1.3.) and have also a look at the projects and their features listed on the OSGEO website (http://www.osgeo.org). A more general introduction to statistics in honey bee research is provided in the *BEEBOOK* paper on statistics (Pirk *et al.*, 2013).

7.8.1. Example 1: Computing mean coordinates

A very basic analysis could be to compute the mean coordinates of the honey bee locations. Of course this could be done by computing the X and Y coordinates into an attribute first, but now we will use the tool mean coordinate(s):

1. Go to *Vector*.
2. Select *Analysis Tools*.
3. Choose *Mean coordinate(s)*.
4. Select the *input vector layer* (e.g. bees_colony_locations.shp).
5. Specify the *output shapefile* (Fig. S 106).
6. Start the calculation with *OK*.

The new shapefile will contain one point at the coordinates 577289.7 / 184622.7.

7.8.2. Example 2: Nearest Neighbour Analysis and Distance matrix

Is the spatial distribution of our honey bee colonies disperse or clustered? – The level of clustering in a point vector layer can be assessed by *Nearest Neighbour Statistics (Analysis tools)*:

1. Go to *Vector*.
2. Select *Analysis Tools*.
3. Choose *Nearest Neighbour Analysis*.
4. Select the *point vector layer* representing the bee's locations.
5. Click *OK* to view the result.

In this case, the observed mean distance between nearest neighbours is 1356 meters. If the distances between different points (or groups of points) are of interest, the *Distance matrix* function can be useful:

1. Go to *Vector*.
2. Select *Analysis Tools*.
3. Choose *Distance Matrix*.
4. Select the bees_colony_locations.shp as both *input* and *target vector layer*.
5. Select the field ID as *unique field* (Fig. S 107).

This function could also be performed using two different point layers, or using the same layer but different ID fields.

6. Click *OK* to perform the calculation and view the result.

The resulting delimited text file contains a list of all point combinations and the distance (computed in map units, thus meters in our example) in-between these points. In case of our honey bee locations, the maximal distance (12.5 kilometres) is between the honey bee colony Nr. 12 and 20, the minimal distance is only 484.5 meters (between ID 1 and 16).

7.8.3. Example 3: Computing intersections between a DEM and planes

GIS is very useful for applications in natural sciences and a variety of tools are provided, e.g. for visibility analysis or to delimit watersheds, calculate flow paths, or even simulate natural hazards. This example demonstrates a very specific function, based on the *qgSurf plugin* for QGIS:

1. Load and activate the *plugin qgSurf* (see section 7.1.3.).
2. Go to *Plugins*.
3. Select *qgSurf*.
4. Select the DEM ASTER30_dem_CH1903 and choose *source point* for the plane (e.g. around Fribourg 578000/183000/627).
5. Play around with the values for *orientation* and *inclination* of the plane and *calculate the intersections* (Fig. S 108).
6. Save the output as point or line shapefile.

To analyse spatially related processes in more detail or for modelling, functions within GRASS GIS (http://grass.osgeo.org, Neteler and Mitasova, 2008) or SAGA (http://www.saga-gis.org) can be useful. To access these, GRASS and/or SAGA have to be installed on your computer and within QGIS the *SEXTANTE Toolbox plugin* (http://plugins.qgis.org/plugins/sextante) needs to be installed and configured correctly (see QGIS Development Team, 2013). If you work with R, the plugin *Spatial Data Analysis for Point Pattern (SDA4PP)* might also be a good option (see section 7.1.3.).

7.9. Case study – zonal statistics with land cover properties

Until now, several concepts and tools in QGIS have been introduced. Now, to combine some of the different tasks discussed, a case study will be presented. Within this short case study, geostatistical relationships between honey bee colonies and land cover properties will be calculated. The content of this case study is equal to the one for ArcGIS (see section 6.9.), but adapted to the tools of QGIS. The prerequisites for this case study include an opened QGIS project (see sections 7.3.1. and 7.3.3.) with a point vector layer (e.g. the honey bee colony locations bees_colony_locations.shp, created in section 7.3.3.1.) and a buffer polygon layer (e.g. bees_colony_locations_buffer.shp, created in section 7.6.1.). For more information about this case study see section 6.9.

7.9.1. Downloading and extracting

First we need land cover information for the region around Fribourg. If you have already downloaded the CLC06 dataset for the ArcGIS case study (see section 6.9.), skip the following paragraph and continue with step 4.

1. Download the dataset *g100_06.zip* (CLC06 – 100 m) from the website http://www.eea.europa.eu/data-and-maps/data/corine-land-cover-2006-raster-2 to a local directory (e.g. on the Desktop).

2. Unzip the file g100_06.zip with a file archiving program of your choice (if you need additional software for this task, the program "7-zip" is a good choice: www.7-zip.org) into your QGIS working directory (e.g. "BeeBook_data").

3. Navigate to the raster g100_06.tif in QGIS Browser to open the tab *Metadata*. For a first visual impression of the data, you can also check out the *Preview* tab.

From step 3 you see that CLC06 is a 1 band raster dataset with 8 bit unsigned integer values (that correspond to different land cover categories) at 100 m horizontal resolution. Because it covers all Europe, the CLC06 raster is quite big and the rendering on the screen may take some time. We will improve the performance by clipping on the extent of interest (see steps 15-22).

7.9.2. Projecting data into another coordinate reference system (CRS)

The coordinate system is different (ETRS89 / ETRS-LAEA, EPSG-code 3035) from that one of Switzerland, and for our purpose its best to bring it into the same coordinate reference system as the other layers (CH1903 / LV03).

4. Switch to QGIS Desktop.
5. Open the menu *Raster*.
6. Choose *Projections*.
7. Click *Warp (Reproject)*.
8. Select the raster image g100_06.tif as *Input file*.
9. Choose a name and destination for the *Output file* (e.g. g100_06_ch1903.tif).
10. The Source SRS should be automatically detected (ETRS89 / ETRS-LAEA), but the *Target SRS* has to be defined: click *Select >* filter for "CH1903 / LV03" and select the remaining CRS listed.
11. Confirm with *OK*. The EPSG-codes should be set corresponding to Fig. S 109.

12. Choose the Resampling method Near because we have categorized data and want to keep the original integer values (see section 7.5.3.).
13. Activate the checkbox *Load into canvas when finished*.
14. Click *OK*. The projection process of large raster datasets can take several minutes depending on the computer processing power, so please be patient.

7.9.3. Clipping to study area extent

The resulting raster g100_06_ch1903.tif still covers all Europe, so we will now clip it to the extent of our polygon layer commune_boundary_FR.shp.

15. Open *Raster*.
16. Select *Extraction*.
17. Open *Clipper*.
18. Select g100_06_ch1903.tif as *Input file*.
19. Choose CLC06_Fribourg.tif as *Output file* name.
20. Choose *Extent* within the *Clipping mode* and enter the minimum and maximum values for x and y according to Fig. S 110 (xMin: 564835, yMin:168930, xMax: 593200, yMax: 196915). To look up these values, double-click the layer commune_boundary_FR.shp in the *Map Legend > Layer Properties > Metadata*. In GIS, minimum and maximum x and y values can also be obtained manually by hovering over a location on the map and reading the x and y values from the bottom right-hand corner of the screen (Fig. S 110).
21. Activate the checkbox *Load into canvas when finished*.
22. Click *OK* to start the clipping process.
23. After the clipping has finished, remove the big layer g100_06_ch1903.tif from the *Map Legend*.
24. The result should then look similar to Fig. S 111.

7.9.4. Raster layer styling and predefined colour maps

The layer CLC06_Fribourg.tif is now quite handy for our purpose, but so far we do not know for what the numbers (raster values) stand for (see Fig. S 111). Luckily, a colour map is provided by the European Environment Agency (EEA) and can be loaded into QGIS (similar to the ArcGIS layer file, see section 6.9.4. for more information):

25. Double-click on the layer CLC06_Fribourg in the *Map Legend* to open its *Layer Properties*.
26. Navigate to the Tab *Colormap* and Click the Symbol [⬚] *Load color map from file*.
27. Navigate to the folder that you have unzipped in step 2 (e.g. CORINE_2006_100m), select the text file clc_legend_qgis.txt and *click Open*.
28. Click *Apply* (see Fig. S 112) and close the Layer Properties with *OK*.

Note that the raster values do not correspond with the numbers (categories) indicated in the labels (Fig. S 112). You can verify this by using the *Identify Features* button [icon] and comparing this with the entries in the Map Legend. This might be a little confusing, but the reason is of technical nature: the official CLC06 land cover category numbers (50 in total) ranging from 111 to 999 cannot be displayed with a single band 8-bit integer raster dataset, where the range of raster values is limited to 0-255. Of course a 32-bit floating-point raster image could be used to display the whole range from 111 to 999, but then the dataset would become almost 4 times larger and unhandy for download, storage and analysis. In the following tasks we will make use of numbers written in the labels of CLC06_Fribourg.tif.

7.9.5. Conditional expressions using the raster calculator

Let's now analyse the area covered by some land use types of interest (arable land, forests) within the buffer zones of our honey bee colonies (see section 6.9.6. for more information).

29. Open the menu *Raster > Raster calculator*.
30. Combine *Raster bands* and *Operators* until the *Raster calculator expression* equals "CLC06_Fribourg@12 = 211" (211 is the raster value for "non-irrigated arable land in the CLC06 dataset).
31. Specify an *Output layer*, e.g. CLC06_class211.tif.
32. Activate the Raster band CLC06_Fribourg@1 again and click *Current layer extent* to make sure that the output layer has the same extent as the source.
33. Verify that the parameters are set according to Fig. S 113 and click *OK* to start the calculation.
34. Double-click on the new greyscale Layer CLC06_class211 to open the *Layer Properties > Tab Style*.
35. Check *Estimate (faster)* within *Load min/max values from band* and click *Load*, select *Stretch to MinMax in the Contrast enhancement (see section 7.3.3.)*.
36. Click *Apply* to make the layer styling effective and close the *Layer Properties* dialog with *OK*.
37. Repeat steps 29 to 36 with the following raster values: 24 (class 312, Coniferous forest) and 25 (class 313, Mixed forest).

7.9.6. Zonal statistics of land cover values within buffers

The newly created raster images contain zones with the value 1 (where the *Raster calculator expression* was true) and 0, where the expression was false. This binary classification can now be used to calculate the percentage of each land cover class within any polygon layer, e.g. the honey bee colony buffers (also see section 6.9.7.):

38. Open the menu *Raster > Zonal statistics > Zonal statistics*.
39. Choose one of the newly created binary raster images (e.g. CLC06_class211.tif) as input *Raster layer*.
40. Select bees_colony_locations_buffer.shp as input Polygon layer containing the zones.

41. Choose an appropriate column prefix (e.g. the land cover class "211") and Click *OK*.
42. Repeat steps 38 to 41 for the other input rasters (choosing appropriate filenames for the results, see Fig. S 114).

Opening the attribute table of the layer bees_colony_locations_buffer.shp, you see new fields at the end: The field "211mean" represents the average percentage of the land cover category "non-irrigated arable land" within the respective buffer (honey bee colony).

7.9.7. Visualisation of attribute values

The proportion of the three land cover categories can be used for any symbolisation and vector calculation, but also visualized within a simple chart:

43. Open *Plugins > Chart Maker > Chart Maker*. If you don't see this entry, you probably have to install or enable first the *Chart Maker plugin* (see section 7.1.3.).
44. Select the *Vector layer* bees_colony_locations_buffer.shp.
45. Set the *attribute fields* for the X and Y axis you want to display (e.g. the ID for X and a land cover percentage for Y).
46. Click *Plot chart* and look at your chart.
47. Feel free to display all land cover categories you have processed by incrementing the *number of series* (see Fig. S 115).
48. If you want, you can save the chart as picture file.

7.9.8. Mathematical operations using the field calculator

Within most of the honey bee colony buffers there are also other land cover categories that can be calculated using the following procedure:

49. Right-click on bees_colony_locations_buffer.shp in the *Map Legend > Open Attribute Table*.
50. Toggle on the *editing mode* with the button [icon].
51. Open the *Field calculator* [icon].
52. Make sure the checkbox Create a new field is activated and specify an *Output field name* (e.g. "other").
53. Change the *Output field type* to *decimal*.
54. Increase the *Precision* to 3.
55. Combine the *Attribute Fields and Values* and different operators so that the resulting *Expression* equals *1-("211mean"+"312mean"+"313mean")* (see Fig. S 116.).
56. Click *OK* to perform the calculation and have a look at the new field created on the right within the attribute table..
57. Repeat steps 51 to 56 and add a new *field (type decimal)* called "sum" that summarizes all 3 land use categories 211, 312 and 313 (the *Expression* should equal *211mean"+"312mean"+"313mean")*.
58. Toggle off the *editing mode* with the button [icon].

The resulting attribute field "other" represents the remaining area not covered by the CLC06 categories 211,312, and 313 relative to the total area of the polygon layer. The field "sum" summarizes the relative area covered by these three land cover categories of interest.

7.9.9. Joining attributes from other vector layers

For visualisation purposes (see map creation in section 7.10.) we must join the land cover percentage per honey bee colony buffer (see sections 7.9.6. and 7.9.8.) to the original point layer bees_colony_locations.shp (see 6.9.8. for more information). To do this:

59. Double-click on bees_colony_locations.shp in the *Map Legend* to open its *Layer Properties*.
60. Switch to the *Tab Joins*.
61. Click the button ⊕ to *add a join to another vector layer*.
62. Select bees_colony_locations_buffer.shp as *Join layer* (Fig. S 177).
63. Make sure the *Join and Target fields* are set to their identifier "ID".
64. Leave unchecked both of the checkboxes and click *OK*.
65. Click *OK* to close the *Layer Properties*.

Congratulations, you have finished your first geo-statistical analysis with QGIS. The next section is about the creation of maps.

7.10. Map creation

Creating maps with QGIS is similar to the procedure with other GIS software like ArcMap. This section collects the most important steps from managing geographic data to its visualisation (see section 2.3.3.) and requires the results of the case study (see section 7.9.).

7.10.1. Layer styling and labelling

Some basic settings in the *Layer Properties* for styling were already discussed in this tutorial (see sections 7.3.4. and 7.3.5.). Based on this experience, we will look at some more techniques to visualise different geometry types and attribute information together:

1. Add the vector layer commune_boundary_FR.shp to the *QGIS project* (see section 7.3.3.).
2. Double-click this layer in the *Map Legend* to open the *Layer properties*.
3. Select *Style*.
4. Select *Graduated* instead of *Single Symbol*.
5. Choose an *attribute* (e.g. "Shape_Area").
6. Choose a number of *Classes* (e.g. 7) and the *Mode Natural Breaks (Jenks)*.
7. Choose colour ramp (e.g. "BuPu").
8. Click *Classify*.
9. Click *Apply* (the *Layer Properties* can stay open).

As an intermediate result, all polygons (representing different municipalities) are filled with a colour from the chosen spectrum corresponding to the chosen attribute (e.g. "Shape_Area"). This is already a good start, but perhaps the map should also provide the names of the communes?

10. Switch to the *Label* tab (Fig. S 118).
11. Activate the *checkbox* to display labels.
12. Choose an *attribute field* that should be displayed (e.g. "GEMNAME").

13. Feel free to adjust font size and colour or the placement options below.
14. Click *OK*.
15. Activate and select the layer bees_colony_locations_buffer.shp created in section 7.6.1.
16. Place it at least one level higher than commune_boundary_FR.shp.
17. Open again the *Layer Properties*.
18. Select the *Style* tab.
19. Click the button *Change*.
20. Choose *Outline: Simple line* as *Symbol layer type*.
21. Select a dark *colour* (e.g. black), a *Pen width* of about 0.25 and *Dot Line* as *Pen style*. Confirm with *OK*.
22. Set the *transparency* to about 30%.
23. Close the *Layer Properties* of bees_colony_locations_buffer.shp.
24. Move the layer bees_colony_locations.shp on top of the *Map Legend* and open its *Layer Properties*.
25. Not only do colours and attributes matter for the visualisation of spatial data, but also size. To demonstrate this, select the size 5, click *Advanced > Size scale field* and choose an *attribute field* of your choice (e.g. "sum").
26. Adjust the *Size* to reach a good result on the *Map View*. The honey bee colonies close to the city of Fribourg have now the smallest symbols, the ones closer to arable land and forests are bigger.
27. Switch to tab *Diagrams* (Fig. S 119).
28. Activate the checkbox *Display diagrams* and choose *Pie chart* as *Diagram type*.
29. Set the *Size* to about 8 mm, the *Placement AroundPoint* with a *Distance* of 6.
30. Now add some content for the pie charts: Use the Button to add the *fields 211mean, 312mean, 313mean* and *other* ⊕ (see section 7.9.). The colours can be adjusted by double-clicking on the coloured bar.
31. Make the changes permanent by clicking *Apply* and close the *Layer Properties* with *OK*.
32. Save the QGIS project (and all the layer styling you have done so far) by the menu *File > Save*.

The result (Fig. S 120) is already a good start for a first visualization of geodata and it will serve as an example for the next paragraphs describing map creation functions.

In some cases it makes sense to save the layer symbology not only within a project, but also for different layers. This can be done layer by layer at the bottom of the *Layer Properties* dialog (see Fig. S 118) using the button *Save Style*. This will safeguard all styling information in a *.qml-file. You can import these styles later also into other projects and apply the styles to more layers (*Layer Properties > Load Style*) – so this is probably the most important time-saver for larger mapping projects within QGIS.

7.10.2. Print composer

The *Print Composer* is the main mapping program within QGIS, equivalent to the layout view of ArcMap (see section 6.10.1.).

1. Click *File.*
2. Select *New Print Composer* and a new window opens.
3. Select *Layout.*
4. Choose *Add Map* in the menu on top.
5. Draw a rectangle to the map canvas below to define the map position.

Within the *Item Properties* to the right you can now see the dimensions and the scale of your map. To get familiar with the *Print Composer* (Fig. S 121), look at all the buttons on top and try to:

- Move or change the map extent using the *Select/Move Item* button
- Move the map content within the frame with the *Move Item* button
- Save the current *map as template* using the button .

A map like this may be suitable for certain tasks, but it is not complete. In the next section, it will be shown how additional map elements can be added.

7.10.3. North arrow, scale bar, legend, title, labels

The final map should provide information about the map scale and orientation and explain the map content. Let's now add the most common map elements using the *QGIS Print Composer*:

1. Open the menu *Layout.*
2. Choose *Add Image* and click somewhere on the map canvas (all items can be moved afterwards).
3. Select a north arrow of your choice (within the set of preloaded images).
4. Now add a scale bar (*Layout > Add Scalebar*) and click somewhere on the map. A basic scale bar will show up and within the *Item Properties* the unit label must be defined (e.g. m or meters). There are a lot of other optional settings.
5. Up to now we have no information about the different map elements shown, so we add a map legend. Go to *Layout.*
6. Select *Add Legend*, click somewhere on the map canvas and customize the items listed.
7. The names of the different layers can be changed in QGIS desktop by a right click to the *Map Legend.*
8. Select *Rename.*
9. A map needs a title, or sometimes additional labels to explain some features. Go to *Layout.*
10. Click *Add Label* and place it on top of the map. Labels and legend items can be formatted and changed within the *Item Properties.*
11. Last but not least, try to also add a grid according to the CRS used (in this case it is metric, CH1903/LV03). Click with the *Item selection tool* on the map frame and locate within the *Item Properties* the section *Grid.*
12. Activate the checkbox and experiment with the settings below. At the end, the map should look somehow similar to Fig. S 122.
13. Save the template again, and be proud of your first map made with QGIS!

7.10.4. Exporting a map

Maps can be exported from the QGIS Print Composer in different ways:

- In most cases the aim will be to create a printable PDF file. Go to *File* and select *Export as PDF*. By default the settings should be already in the range for good printing results.
- For special map dimensions or to change the resolution settings, go to the *Composition* tab beside the *Item Properties.*
- Of course a map can also directly be sent to a printer device (*File > Print*).
- For a special purpose it might make sense to export a map to a vector or raster image. The two corresponding functions are *File > Export as SVG* and *File > Export as Image*, respectively.

7.10.5. Cloud mapping and sharing maps online

By the means of the QGIS *Cloud plugin* (see section 7.1.3. and Table 4) developed by Sourcepole (http://www.qgiscloud.com), users can easily publish their own maps directly from QGIS desktop to a PostGIS 2.0 database within a cloud online storage. There are also many other ways to share geodata and maps produced by open source GIS using Web Map (WMS) and Feature Services (WFS), for example by the means of the open source projects OpenLayers, MapServer or degree. The official website of the Open Source Geospatial Foundation (OSGEO, http://www.osgeo.org) provides a good overview about the different projects and their technical capabilities.

8. Future perspectives

The main purpose of the paper was to stimulate spatial thinking in honey bee research by introducing the basic principles and methods of GIS. With an introduction to the basics of GIS, we hope that more honey bee researchers will build on techniques already developed in the native bee domain, and discover new geospatial methods for exploring the relationships between honey bees and their environment. As of now, relatively few honey bee researchers have studied the effects of how environmental characteristics and mechanisms have affected honey bee health, abundance, and honey production (Berardinelli and Vedova, 2004; Naug, 2009; Henry *et al.*, 2012). Because of the recent colony losses in many parts of the world (Carreck and Neumann, 2010; Williams *et al.*, 2010), it is important to mechanistically study and attempt to discover new factors which directly and indirectly affect honey bee populations (Naug, 2009) from both broad and local scales.

GIS and spatial analyses have been used to make great progress in regards to studying the effects of landscape characteristics in research about native bees (Arthur *et al.*, 2010; Biesmeijer *et al.*, 2006; Chifflet *et al.*, 2011; Choi *et al.*, 2012; Fitzpatrick *et al.*, 2007; Giannini *et al.*, 2012; Kremen, 2002; Kremen *et al.*, 2004; Watson *et al.*, 2011; Williams *et al.*, 2012). Various researchers have looked at the links between land cover properties and their effects on bees using satellite imagery to perform land cover classifications (Arthur *et al.*, 2010; Kremen *et al.*, 2004) and aerial imagery to quantify land cover properties (Watson *et al.*, 2011) using GIS. Positive correlations were established between native bees and their surrounding landscapes, perhaps most notably the link between wooded areas and the increase in bee populations and abundance. In terms of climate, Giannini *et al.* (2012) have used GIS to help reveal that native bees populations are decreasing because of a changing climate and are projected to continue to decline in the future due to the loss of suitable habitats.

Because of the obvious similarities between native bees and honey bees, it is assumed that similar relationships will be found between honey bees and their surrounding landscapes and changing climate. Naug (2009) was one of the first to use GIS to test the effects of land cover changes on honey bee colony health and productivity. By looking at the effects of nutritional stress due to habitat loss in the U.S.A., Naug (2009) established that the relative extent of open land area is an important predictor of colony losses, and that states with the largest open area proportions had higher honey yields, meaning that their honey bees were more productive. It is necessary to study these types of relationships from a broad scale in order to get a better general understanding of the overall processes linking honey bees to the surrounding environments. Research of this type can and should be expanded upon to explore more specific geographic regions as well as local scales. Climate research, similar to the work by Giannini *et al.* (2012), can also be conducted in GIS to study not only its effects on honey bees directly, but how changes in climate affect honey bee pathogens.

In conclusion, there are numerous opportunities for expanding geospatial analyses in the honey bee domain and we hope that this paper will motivate and stimulate spatial thinking for researchers.

9. Acknowledgements

We would like to acknowledge the Swiss National Science Foundation (SNF) for providing funds to the authors and Swisstopo for the use of their data through a contract with the University of Fribourg for research purposes. We would like to thank Geoffrey Williams, Jamie Ellis, and Vincent Dietemann for suggesting the idea for this paper, and Christina Zweifel for stepping into the world of GIS to test the tutorials. The COLOSS (Prevention of honey bee COlony LOSSes) network aims to explain and prevent massive honey bee colony losses. It was funded through the COST Action FA0803. COST (European Cooperation in Science and Technology) is a unique means for European researchers to jointly develop their own ideas and new initiatives across all scientific disciplines through trans-European networking of nationally funded research activities. Based on a pan-European intergovernmental framework for cooperation in science and technology, COST has contributed since its creation more than 40 years ago to closing the gap between science, policy makers and society throughout Europe and beyond. COST is supported by the EU Seventh Framework Programme for research, technological development and demonstration activities (*Official Journal L 412, 30 December 2006*). The European Science Foundation as implementing agent of COST provides the COST Office through an EC Grant Agreement. The Council of the European Union provides the COST Secretariat. The COLOSS network is now supported by the Ricola Foundation - Nature & Culture.

10. References

ARCTUR, D K; ZEILER, M (2004) *Designing geodatabases: case studies in GIS data modeling*. ESRI Press; Redlands, USA. 411 pp.

ARTHUR, A D; LI, J; HENRY S; CUNNINGHAM, S A (2010) Influence of woody vegetation on pollinator densities in oilseed *Brassica* fields in an Australian temperate landscape. *Basic and Applied Ecology* 11(5): 406-414.
http://dx.doi.org/10.1016/j.baae.2010.05.001

BERARDINELLI, I; VEDOVA, G D (2004) Use of GIS in the management of apiculture: preliminary note. *APOidea - Rivista Italiana di Apicoltura*. v. 1(1): 31-36. Available at: http://agris.fao.org/agris-search/search/display.do?f=2007/IT/IT0639.xml;IT2006601932 [accessed 10/01/13].

BIESMEIJER, J C; ROGERS, S P M; REEMER, M; OHLEMÜLLER, R; EDWARDS, M; PEETERS, T; SCHAFFERS, A P; POTTS, S G; KLEUKERS, R; THOMAS, C D (2006) Parallel declines in pollinators and insect-pollinated plants in Britain and the Netherlands. *Science* 313(5785): 351–354.
http://dx.doi.org/10.1126/science.1127863

BONHAM-CARTER, J M (1994) *Geographic Information Systems for geoscientists: modelling with GIS*. Elsevier; Oxford, UK. 420 pp.

BURROUGH, P A; MCDONNELL, R A (1998) *Principles of Geographical Information Systems*. Oxford University Press; Oxford, UK. 333 pp.

CAMPBELL, J B (2002) *Introduction to remote sensing (3rd Edition)*. Guilford Press; London, UK. 621 pp.

CARRECK, N L; NEUMANN, P (2010) Honey bee colony losses. *Journal of Apicultural Research* 49(1): 1-6.
http://dx.doi.org/10.3896/IBRA.1.49.1.01

CHIFFLET, R; KLEIN, E K; LAVIGNE, C; LE FÉON, V; RICROCH, A E; LECOMTE, J; VAISSIERE, B E (2011) Spatial scale of insect-mediated pollen dispersal in oilseed rape in an open agricultural landscape. *Journal of Applied Ecology* 48 (3): 689-696.
http://dx.doi.org/10.1111/j.1365-2664.2010.01904.x

CHOI, M-B; KIM, J-K; LEE, J-W (2012) Increase trend of social hymenoptera (wasps and honey bees) in urban areas, inferred from moving-out case by 119 rescue services in Seoul of South Korea. *Entomological Research* 42(6): 308-319. http://dx.doi.org/10.1111/j.1748-5967.2012.00472.x

CONOLLY, J; LAKE, M (2006) *Geographical Information Systems in archaeology*. Cambridge University Press; Cambridge, UK. 338 pp.

DE SMITH, M J; GOODCHILD, M F; LONGLEY, P A (2007) *Geospatial analysis: a comprehensive guide to principles, techniques and software tools*. Troubador Publishing Ltd; Leicester, UK. 418 pp.

ESRI (Environmental Systems Research Institute) (2013). Available at: www.esri.com.

FITZPATRICK, U; MURRAY, T E; PAXTON, R J; BREEN, J; COTTON, D; SANTORUM, V; BROWN, M J F (2007) Rarity and decline in bumble bees - a test of causes and correlates in the Irish fauna. *Biological Conservation* 136(2): 185-194. http://dx.doi.org/10.1016/j.biocon.2006.11.012

GIANNINI, T C; ACOSTA, A L; GARÓFALO, C A; SARAIVA, A M; ALVES-DOS-SANTOS, I; IMPERATRIZ-FONSECA, V L (2012) Pollination services at risk: bee habitats will decrease owing to climate change in Brazil. *Ecological Modelling* 244(0): 127–131. http://dx.doi.org/10.1016/j.ecolmodel.2012.06.035

HENRY, M; FRÖCHEN, M; MAILLET-MEZERAY, J; BREYNE, E; ALLIER, F; ODOUX, J-F; DECOURTYE, A (2012) Spatial autocorrelation in honey bee foraging activity reveals optimal focus scale for predicting agro-environmental scheme efficiency. *Ecological Modelling* 225(0): 103-114. http://dx.doi.org/10.1016/j.ecolmodel.2011.11.015

KOZAK, K H; GRAHAM, C H; WIENS, J J (2008) Integrating GIS-based environmental data into evolutionary biology. *Trends in Ecology & Evolution* 23(3): 141–148. http://dx.doi.org/10.1016/j.tree.2008.02.001

KREMEN, C (2002) Crop pollination from native bees at risk from agricultural intensification. *Proceedings of the National Academy of Sciences* 99(26): 16812-16816. http://dx.doi.org/10.1073/pnas.262413599

KREMEN, C; WILLIAMS, N M; BUGG, R L; FAY, J P; THORP, R W (2004) The area requirements of an ecosystem service: crop pollination by native bee communities in California. *Ecology Letters* 7(11): 1109-1119. http://dx.doi.org/10.1111/j.1461-0248.2004.00662.x

LEE, J; WONG, D W S (2001) *Statistical analysis with ArcView GIS*. John Wiley & Sons; New York, USA. 192 pp.

LONGLEY, P; GOODCHILD, M F; MAGUIRE, D J; RHIND, D W (Eds) (1999) *Geographical Information Systems: principles, techniques, applications and management (2nd Ed.), Volume 1*. Wiley; New York, USA. 580 pp.

NAUG, D (2009) Nutritional stress due to habitat loss may explain recent honey bee colony collapses. *Biological Conservation* 142 (10): 2369-2372. http://dx.doi.org/10.1016/j.biocon.2009.04.007

NETELER, M; MITASOVA, H (2008) *Open source GIS: a GRASS GIS approach (3rd Ed.). The International Series in Engineering and Computer Science, Volume 773*. Springer; New York, USA. 406 pp.

ORMSBY, T; NAPOLEON, E J; BURKE, R; GROESSL, C; BOWDEN, L (2010) *Getting to know ArcGIS: desktop*. ESRI Press; Redlands, USA. 584 pp.

PIRK, C W W; DE MIRANDA, J R; FRIES, I; KRAMER, M; PAXTON, R; MURRAY, T; NAZZI, F; SHUTLER, D; VAN DER STEEN, J J M; VAN DOOREMALEN, C (2013) Statistical guidelines for *Apis mellifera* research. In *V Dietemann; J D Ellis; P Neumann (Eds) The COLOSS BEEBOOK, Volume I: standard methods for* Apis mellifera research. *Journal of Apicultural Research* 52(4): http://dx.doi.org/10.3896/IBRA.1.52.4.13

POGODZINSKI, J M; KOS, R M (2012) *Economic development and GIS*. ESRI Press; Redlands, USA. 229 pp.

QGIS Development Team (2013) *QGIS 1.8.0 Geographic Information System User Guide. Open Source Geospatial Foundation Project*. Electronic document: http://docs.qgis.org/1.8/pdf/QGIS-1.8-UserGuide-en.pdf

RIGAUX, P; SCHOLL, M O; VOISARD, A (2002) *Spatial databases: with application to GIS*. Morgan Kaufmann; San Francisco, USA. 411 pp.

SCHMANDT, M (2009) *GIS Commons: an introductory textbook on Geographic Information Systems*. Available at: http://giscommons.org/

SEEGER, H (1999) Spatial referencing and coordinate systems. In *Longley, P; Goodchild, M F; Maguire, D J; Rhind, D W (Eds) Geographical Information Systems: principles, techniques, applications and management (2nd Ed.), Volume 1*. Wiley; New York, USA. pp 427-436.

SNYDER, J P (1987) *Map projections - a working manual*. United States Government Printing Office; Washington, USA. 410 pp.

SWISSTOPO (2013) *Federal Office of Topography*. Available at: www.swisstopo.ch.

WADE, T; SOMMER, S (2006) *A to Z GIS: an illustrated dictionary of geographic information systems*. ESRI Press; Redlands, USA. 268 pp.

WATSON, J C; WOLF, A T; ASCHER, J S (2011) Forested landscapes promote richness and abundance of native bees (Hymenoptera: Apoidea: Anthophila) in Wisconsin apple orchards. *Environmental Entomology* 40(3): 621-632. http://dx.doi.org/10.1603/EN10231

WILLIAMS, G R; TARPY, D R; VANENGELSDORP, D; CHAUZAT, M-P; COX-FOSTER, D L; DELAPLANE, K S; NEUMANN, P; PETTIS, J S; ROGERS, R E L; SHUTLER, D (2010) Colony Collapse Disorder in context. *BioEssays* 32(10): 845-846. http://dx.doi.org/10.1002/bies.201000075

WILLIAMS, N M; REGETZ, J; KREMEN, C (2012) Landscape-scale resources promote colony growth but not reproductive performance of bumble bees. *Ecology* 93(5): 1049-1058. http://dx.doi.org/10.1890/11-1006.1

WONG, D W S; LEE, J (2005) *Statistical analysis of geographic information with ArcView GIS and ArcGIS*. Wiley; New Jersey, USA. 446 pp.

Journal of Apicultural Research 52(1): (2013)
DOI 10.3896/IBRA.1.52.1.04

© IBRA 2013

REVIEW ARTICLE

Standard methods for maintaining adult *Apis mellifera* in cages under *in vitro* laboratory conditions

Geoffrey R Williams[1,2*†], Cédric Alaux[3], Cecilia Costa[4], Támas Csáki[5], Vincent Doublet[6], Dorothea Eisenhardt[7], Ingemar Fries[8], Rolf Kuhn[1], Dino P McMahon[6,9], Piotr Medrzycki[4], Tomás E Murray[6], Myrsini E Natsopoulou[6], Peter Neumann[1,10], Randy Oliver[11], Robert J Paxton[6,9], Stephen F Pernal[12], Dave Shutler[13], Gina Tanner[1], Jozef J M van der Steen[14] and Robert Brodschneider[15†]

[1]Swiss Bee Research Centre, Agroscope Liebefeld-Posieux Research Station ALP-HARAS, Bern, Switzerland.
[2]Department of Biology, Dalhousie University, Halifax, Nova Scotia, Canada.
[3]INRA, UR 406 Abeilles et Environnement, Avignon, France.
[4]Agricultural Research Council - Honey bee and Silkworm Research Unit (CRA-API), Bologna, Italy.
[5]Institute for Wildlife Conservation, Szent István University, Gödöllő, Hungary.
[6]Institute for Biology, Martin-Luther-University Halle-Wittenberg, Halle (Saale), Germany.
[7]Department of Biology / Chemistry / Pharmacy, Institute for Biology-Neurobiology, Free University of Berlin, Berlin, Germany.
[8]Department of Ecology, Swedish University of Agricultural Sciences, Uppsala, Sweden.
[9]School of Biological Sciences, Queen's University Belfast, Belfast, UK.
[10]Department of Zoology & Entomology, Rhodes University, Grahamstown, South Africa.
[11]ScientificBeekeeping.com, Grass Valley, California, USA.
[12]Agriculture and Agri-Food Canada, Beaverlodge, Alberta, Canada.
[13]Department of Biology, Acadia University, Wolfville, Nova Scotia, Canada.
[14]Bee Unit, Plant Research International, Wageningen University and Research Centre, Wageningen, The Netherlands.
[15]Department of Zoology, Karl-Franzens-University, Graz, Austria.

[†]All authors, except those listed first and last, are alphabetical.

Received 15 May 2012, accepted subject to revision 9 July 2012, accepted for publication 18 November 2012.

*Corresponding author: Email: geoffrey.r.williams@gmail.com

Summary

Adult honey bees are maintained *in vitro* in laboratory cages for a variety of purposes. For example, researchers may wish to perform experiments on honey bees caged individually or in groups to study aspects of parasitology, toxicology, or physiology under highly controlled conditions, or they may cage whole frames to obtain newly emerged workers of known age cohorts. Regardless of purpose, researchers must manage a number of variables, ranging from selection of study subjects (e.g. honey bee subspecies) to experimental environment (e.g. temperature and relative humidity). Although decisions made by researchers may not necessarily jeopardize the scientific rigour of an experiment, they may profoundly affect results, and may make comparisons with similar, but independent, studies difficult. Focusing primarily on workers, we provide recommendations for maintaining adults under *in vitro* laboratory conditions, whilst acknowledging gaps in our understanding that require further attention. We specifically describe how to properly obtain honey bees, and how to choose appropriate cages, incubator conditions, and food to obtain biologically relevant and comparable experimental results. Additionally, we provide broad recommendations for experimental design and statistical analyses of data that arises from experiments using caged honey bees. The ultimate goal of this, and of all COLOSS *BEEBOOK* papers, is not to stifle science with restrictions, but rather to provide researchers with the appropriate tools to generate comparable data that will build upon our current understanding of honey bees.

Footnote: Please cite this paper as: WILLIAMS, G R; ALAUX, C; COSTA, C; CSÁKI, T; DOUBLET, V; EISENHARDT, D; FRIES, I; KUHN, R; MCMAHON, D P; MEDRZYCKI, P; MURRAY, T E; NATSOPOULOU, M E; NEUMANN, P; OLIVER, R; PAXTON, R J; PERNAL, S F; SHUTLER, D; TANNER, G; VAN DER STEEN, J J M; BRODSCHNEIDER, R (2013) Standard methods for maintaining adult *Apis mellifera* in cages under *in vitro* laboratory conditions. In *V Dietemann; J D Ellis; P Neumann* (Eds) *The COLOSS BEEBOOK, Volume I: standard methods for* Apis mellifera *research. Journal of Apicultural Research* 52(1):
http://dx.doi.org/10.3896/IBRA.1.52.1.04

Métodos estándar para el mantenimiento de adultos de *Apis mellifera* en cajas bajo condiciones de laboratorio *in vitro*

Resumen

Las abejas adultas se mantienen *in vitro* en cajas de laboratorio para una variedad de propósitos. Por ejemplo, los investigadores pueden realizar experimentos con las abejas de miel enjauladas individualmente o en grupos para estudiar aspectos de la parasitología, toxicología y fisiología en condiciones muy controladas, o pueden meter en las cajas panales completos para obtener obreras recién emergidas de cohortes de edad conocida. Independientemente del propósito, los investigadores deben manejar una serie de variables, que van desde la selección de los sujetos a estudiar (por ejemplo, la subspecies de abeja), al ambiente experimental (por ejemplo, temperatura y humedad relativa). Aunque las decisiones tomadas por los investigadores no tienen por qué poner en peligro el rigor científico de un experimento, si que pueden afectar profundamente a los resultados, y pueden dificultar las comparaciones con estudios similares pero independientes. Centrándonos principalmente en obreras, ofrecemos recomendaciones para mantener adultos en condiciones de laboratorio *in vitro*, si bien reconocemos algunas lagunas en nuestro conocimiento que requieren una mayor atención. En especial, se describe cómo obtener correctamente abejas, y cómo elegir cajas adecuadas, las condiciones de incubación, y los alimentos para obtener resultados experimentales biológicamente relevantes y comparables. Además, ofrecemos recomendaciones generales para el diseño experimental y el análisis estadístico de los datos que surgen de experimentos con abejas enjauladas. El objetivo final de éste, y de todos los artículos de *BEEBOOK* y COLOSS, no es limitar la ciencia con restricciones, sino más bien proporcionar a los investigadores las herramientas necesarias para obtener datos comparables que se basen en el conocimiento actual de las abejas melíferas.

实验室条件下笼中饲养成年西方蜜蜂的标准方法

很多研究都需要在实验室内应用蜂笼饲养成年蜜蜂，比如，研究者可能应用单个蜂笼或多个蜂笼开展严格控制条件下的寄生虫学、毒理学或生理学研究。也可能把整个巢脾关入笼中来得到日龄明确的刚羽化出房的蜜蜂。不管目的如何，研究者必须控制多个变量：从研究对象（不同的蜜蜂亚种）到实验环境（温度和相对湿度等）。虽然研究者的选择可能不一定会损害实验的科学性，但可能会显著影响实验结果，使独立实验成为相关实验。围绕饲养工蜂，我们推荐了在实验室条件下饲养成年蜂的方法。特别描述了如何恰当的饲养蜜蜂以及如何选择饲养蜂笼、温箱和饲料以得到具生物学意义并具可比性的实验结果。此外，针对实验设计和数据的统计分析还给出了大量建议。本文以及本书中所有文章所涉及的研究方法，其最终目的是给研究者提供合适的研究工具，得到具有可比性的数据，推进我们对蜜蜂的认识，而不是设立技术障碍，限制科学发展。

Keywords: *Apis mellifera*, honey bee, colony losses, hoarding, cage, *in vitro*, laboratory, COLOSS, *Nosema*, toxicology, *BEEBOOK*

Table of Contents

Table of Contents cont'd

Table of Contents cont'd

1. General introduction

Recent dramatic losses of honey bee (*Apis mellifera*) colonies in many regions of the world are primarily attributed to introduced and native parasites and diseases, environmental toxins, genetic constraints, beekeeper management issues, and socio-economic factors, acting singly or in combination (Neumann and Carreck, 2010; vanEngelsdorp and Meixner, 2010; Williams *et al.*, 2010). We can study potential effects of many of these factors at the colony-level under field or semi-field (e.g. in tunnel tents) conditions, or at the individual or small group level in a laboratory under relatively controlled settings using honey bees isolated from the outdoors.

Regardless of purpose, maintaining adult honey bees *in vitro* in the laboratory prior to or during experiments is often required, and in many cases can provide better control of extraneous variables. For example, host-parasite interactions (e.g. Forsgren and Fries, 2010), parasite management products (e.g. Maistrello *et al.*, 2008), toxicology (e.g. Johnson *et al.*, 2009) and physiology (e.g. Alaux *et al.*, 2010) can be studied. Honey bees can also be caged individually for the evaluation of learning and memory using techniques such as the proboscis extension reflex (e.g. Frost *et al.*, 2011, 2012; Giurfa and Sandoz, 2012).

Here we discuss important factors that researchers must consider when maintaining adult worker honey bees under *in vitro* conditions in the laboratory using cages that restrict movement to the surrounding outdoor environment. We also briefly describe the maintenance of queens and drones. Because an individual's condition can have profound effects on experimental results, it is vital that adults be maintained under appropriate, controlled conditions that enhance repeatability of experiments. Ultimately, our discussions and recommendations presented here are aimed at facilitating and standardising general care of workers in the laboratory for use in scientific investigations. Additional and more specific information on laboratory methods and settings best suited for the purpose of one's study can be found in greater detail in other parts of the COLOSS *BEEBOOK* (Williams *et al.* 2012), such as in the nosema (Fries *et al.*, 2013), toxicology (Medrzycki *et al.*, 2013), larval rearing (Crailsheim *et al.*, 2013), and behaviour (Scheiner *et al.*, 2013) papers of the COLOSS *BEEBOOK*.

2. Experimental design

2.1. Important experimental design considerations before caging adult workers in the laboratory

Although this paper discusses maintenance of adult worker honey bees in a laboratory outside of a colony, regardless of study type or purpose, it is important to highlight that careful consideration needs to be given to experimental design and statistical analysis of the ensuing data before any practical work should commence. Importantly, one must determine if sufficient resources are available to perform rigorous research with an appropriate level of reproducibility; if constraints preclude good science, it may not be worth conducting experiments in the first place.

General recommendations for design of experiments and analysis of data can be found in the *BEEBOOK* paper on statistical methods (Pirk *et al.,* 2013).

2.2. Independence of observations for laboratory cage experiments involving adult workers

A fundamental aspect of good experimental design is independence of observations; what happens to one experimental unit should be independent of what happens to other experimental units before results of statistical analyses can be trusted.

Until shown otherwise, workers within the same cage are not independent, so each cage becomes the minimum unit to analyse statistically (i.e. the experimental unit). Caging workers individually is therefore extremely desirable because each honey bee can be considered to be an independent experimental unit. Although a method for maintaining workers individually for one week exists (section 5.2.3), one that enables individual workers to be maintained in isolation in the laboratory for even longer periods would be beneficial for certain experiments (so long as social interaction is not the focus of investigation or necessary to the phenomenon(a) investigated).

Additionally, careful consideration is required when performing experiments on which volatiles emitted by workers can influence measured parameters. This might require using separate incubators.

2.3. Appropriate worker and cage replicates for laboratory experiments involving adult workers

A minimum sample of 30 independent observations per treatment is relatively robust for conventional statistical analyses (e.g. Crawley, 2005); however, financial constraints and large effect sizes (e.g. difference among treatments for the variable (s) of interest; see statistics paper (Pirk *et al.* (2013)) will no doubt lower this limit, especially for experiments using groups of caged workers. Larger sample sizes (i.e. number of cages and workers per cage) reduce the probability of uncontrolled factors producing spurious insignificance or significance, and help to tease apart treatments with low effect size. Repeated sampling of individuals over time to observe development of parasite infection, for example, will also require larger samples.

Furthermore, it is important to consider biological relevance of the numbers of individuals in each cage. Unsurprisingly, isolated workers die much quicker than those maintained in groups, possibly due to timing of food consumption (Sitbon, 1967; Arnold, 1978), so experimenters must be aware of expected duration of survival. Possible individual and social behaviours that are of interest should also be considered (e.g. Beshers *et al.,* 2001). For example, > 75 workers were needed to consistently elicit clustering behaviour (Lecomte, 1950), whereas 50 workers and a queen were needed for the initiation of wax production (Hepburn, 1986).

A Monte Carlo simulation model incorporating average lifespan (and standard deviation) for treatments and controls has been created to determine percentage of cases where a significant difference is obtained between groups. Without preliminary trials to determine the magnitude of an effect elicited by an experimental treatment as well as the variation between cages in that effect, statistical power may be impossible to know in advance. In such cases, it is advisable to maintain as many cages per treatment ($\geq$ 3) and individuals per cage ($\geq$ 30) as possible. Examination of the literature for similar studies may also help choose sample size; however, caution should be exercised due to differences in experimental conditions. Refer to the *BEEBOOK* paper on statistical methods (Pirk *et al.,* 2013) for further details on the Monte Carlo simulation and on selecting appropriate sample sizes.

2.4. Appropriate randomisation of study organisms for laboratory cage experiments involving adult workers

When designing studies it is crucial that experimenters avoid bias when choosing study subjects. Workers, for example, can exhibit significant genetic variation for expression of mechanical, physiological, immunological, and behavioural responses used in disease resistance (Evans and Spivak, 2010). This diversity can occur among workers in the same colony or among honey bees from different geographic regions. Additionally, timing and method of collection, as described in section 4, can also have a significant influence on results. Because of this, careful consideration is needed when choosing colonies from which to collect experimental honey bees. To promote a repeatable investigation that is representative of a honey bee population in question, workers should be collected from as many, and as diverse a set of, colonies as possible. It is generally recommended to randomly mix workers from all source colonies among all cages during a study to minimize potential colony-level effects on experimental results. Refer to the *BEEBOOK* statistics paper by Pirk *et al.* (2013) for details on determining number of colonies from which to source individuals and for how to properly randomize individuals and cages for experiments, as well as section 4 for choosing and obtaining workers for experiments.

3. Statistical analyses

Specific details on statistical analyses of honey bee experimental data can be found in the statistics paper of the *BEEBOOK* (Pirk *et al.*, 2013) or in statistical texts.

3.1. Where the response variable is not mortality during laboratory experiments involving adult workers

If a response variable to be measured (e.g. a phenotype of interest that may change with treatment) is quantitative or qualitative (i.e. diseased versus not diseased), then a generalised linear mixed model (GLMM) can be used to analyse data in which 'cage' is a 'random effect' parameter and treatment is a 'fixed effect' parameter (Crawley, 2005; Bolker *et al.*, 2009). Several fixed and random effect parameters can be analysed in the same statistical model. If individuals in two or more experimental cages used in the same treatment group are drawn from the same colony, then a GLMM with 'source colony' as a random effect parameter should also be used to analyse data. This random effect accounts for the fact that, within the same treatment, variation between two cages of honey bees drawn from the same colony may not be the same as variation between two cages drawn from two separate colonies. This statistical approach accounts for the problem of pseudoreplication in experimental design. If the factor 'cage' and 'source colony' are non-significant, an experimenter may be tempted to treat individual honey bees from the same cage as independent samples (i.e. ignore 'cage'). Logically, however, workers drawn from the same cage are not truly independent samples and therefore it would inflate the degrees of freedom to treat individual workers as individual replicates. This point requires further attention by statisticians. In lieu of an immediate solution to this statistical issue, an experimenter can consider using a nested experimental design in which 'individual honey bee' is nested within 'cage', as presented above.

3.2. Where the response variable is mortality during laboratory experiments involving adult workers

If survival of workers is the response variable of interest, a typical survival analysis can be undertaken, such as the parametric Kaplan-Meier survival analysis for 'censored' data (so-called right-censored data in which bees are sampled from a cage during an experiment) or the non-parametric Cox proportional hazards model (Cox model) for analysing effects of two or more 'covariates', or predictor variables, such as spore intensity of the microsporidian *Nosema ceranae* or black queen cell virus titres (Zuur *et al.*, 2009; Hendriksma *et al.*, 2011).

3.3. Statistical software for laboratory experiments involving adult workers

Numerous statistical computing programmes are available to handle analyses mentioned in sections 3.1 and 3.2, such as the freeware R (R Development Core Team; Vienna, Austria), as well as other packages, including Minitab® (Minitab Inc.; State College, USA), SPSS® (SPSS Inc.; Chicago, USA), and SAS® (SAS Institute Inc.; Cary, USA). See the statistics paper of the *BEEBOOK* (Pirk *et al.*, 2013) for details.

4. Obtaining adult workers for laboratory experiments

4.1. Considerations for choosing and obtaining adult workers for laboratory experiments

Consideration of honey bee material to be used for experiments must be made prior to practical work because environmental and genetic factors can profoundly influence results (e.g. Fluri, 1977; Evans and Spivak, 2010). Here we discuss a number of factors that may influence worker collection for experiments.

4.1.1. Seasonal timing of adult worker collection for laboratory experiments

In temperate climates workers can be classified either as short-lived "summer" or long-lived "winter" individuals. Physiological differences, such as in juvenile hormone and vitellogenin levels (Fluri *et al.*, 1977; Crailsheim, 1990; Seehus *et al.*, 2006; Corona *et al.*, 2007; Strand, 2008), are mainly driven by quantity of protein consumption and level of brood rearing by the colony (Maurizio, 1950; Amdam *et al.*, 2004; 2005b). "Summer" individuals can be collected beginning in late spring, after colonies have replaced old "winter" honey bees, and up until late summer, when colonies start to prepare for winter. For specific experiments in which the susceptibility of winter bees is the object of study, one can cage the queen within the broodnest for greater than 21 days so that the queenright colony contains no brood (Maurizio, 1954; Fluri *et al.*, 1982). This mimics the broodless period experienced by honey bees in temperate climates.

4.1.2. Subspecies of adult workers used for laboratory experiments

Honey bees subspecies can exhibit great morphological, behavioural, physiological, and genetic variation (Ruttner, 1987), with subsequent differences in productive traits and in disease susceptibility (Evans and Spivak, 2010; DeGrandi-Hoffman *et al.*, 2012). The same subspecies of honey bees should be used for an experiment.

If one wants to further limit influence of genetics on experimental results, individuals from a single colony or multiple colonies that are headed by sister queens can be collected. This will, however, limit the ability of experimental findings to be more broadly generalized across the study population compared to studies that obtained experimental individuals from multiple, genetically diverse colonies of the same subspecies. Refer to section 2 on experimental design in this paper, as well as *BEEBOOK* papers by Meixner *et al.* (2013) for characterizing honey bee subspecies and Delaplane *et al.* (2013) for discussions on preparing colonies for experiments.

4.1.3. Age of adult workers used for laboratory experiments

Adult workers differ greatly in their physiology depending on their age. For example, changes in host immune response (Amdam *et al.*, 2005a) and morphology (Rutrecht *et al.*, 2007) over time can result in differences in disease resistance and susceptibility to parasites (Villa, 2007). Choice of age of experimental workers will reside solely on the purpose of the experiment, and is largely related to collection method (see sections 4.2, 4.3, 4.4). Researchers must ensure that experimental individuals are of a homogeneous age. If they are not, then heterogeneously aged individuals, or those of undefined age, should be evenly distributed among all cages.

4.1.4. Queen status of source colonies used to obtain adult workers for laboratory experiments

A queen is the typical reproductive phenotype in honey bee colonies. Not only is she responsible for egg production, but also for producing pheromones that can greatly influence worker behaviour (e.g. queen rearing) and physiology (e.g. worker ovary development) (Winston, 1987; Winston and Slessor, 1992; Slessor *et al.*, 2005). Health and age of queens are critical, as Milne (1982) observed that progeny of some queens exhibited early death in laboratory cages; this likely had a genetic component, and could be avoided when young laying queens were used. Experimental honey bees should be obtained from colonies that possess a young, mated, laying queen.

4.1.5. Strength of source colonies used to obtain adult workers for laboratory experiments

Source colonies for experimental honey bees should contain appropriate adult brood : food (i.e. honey and bee bread) ratios to ensure that workers are properly nourished, as well as adult and developing individuals of all ages, and food stores from poly-floral sources. Colonies should also be of approximately equal strength because size can influence colony defensive behaviour which can subsequently effect honey bee collection (Winston, 1987). Refer to the *BEEBOOK* papers by Delaplane *et al.* (2013) for how to estimate colony strength, Human *et al.* (2013) for estimating age of developing honey bees, and Delaplane *et al.* (2013) for estimating floral sources, as well as section 4 in this paper for obtaining workers from colonies for experiments.

4.1.6. Health of source colonies used to obtain adult workers for laboratory experiments

Multiple environmental pressures, such as pests, pathogens, and agricultural practices, acting singly or in combination, can influence honey bee health (Neumann and Carreck, 2010; vanEngelsdorp and Meixner, 2010; Williams *et al.*, 2010), and therefore potentially their response to experimental treatments. Ideally, workers used for experiments, as well as the colonies they are sourced from, should be free of pathogens, parasites, pests, and contaminants. In most cases this may not be possible, so at the very least factors potentially confounding results should be stated. Colonies with clinical symptoms of disease (e.g. chalkbrood mummies, foulbrood scales, dysentery, and individuals with deformed wings) should not be used, and infestation levels of the parasitic mite *Varroa destructor* on adults should be below economic and treatment thresholds for the particular region and time of year.

The purpose of the experiment will determine if presence/absence of certain pathogens, parasites, and pests of honey bees need to be considered. Refer to respective *BEEBOOK* papers for pathogen-specific diagnostic methods (Anderson *et al.* (2013) for *Tropilaelaps* spp., de Graaf *et al.* (2013) for American foulbrood, de Miranda *et al.* (2013) for viruses, Dietemann *et al.* (2013) for *Varroa* spp., Ellis *et al.* (2013) for wax moth, Forsgren *et al.* (2013) for European foulbrood, Fries *et al.* (2013) for *Nosema* spp., Jensen *et al.* (2013) for fungi, Neumann *et al.* (2013) for small hive beetle, and Sammataro *et al.* (2013) for tracheal mites).

4.1.7. Beekeeper management of source colonies used to obtain adult workers for laboratory experiments

Beekeeper management practices can greatly influence a honey bee colony. For example, miticides used to control *V. destructor* can be found at high levels in honey bee products (Mullin *et al.*, 2010), and could potentially be responsible for sub-lethal or synergistic effects on individuals (Alaux *et al.*, 2009; Wu *et al.*, 2011). Additionally, pathogens can occur in bee products (Gilliam, 1979), and be a local source of infection (Fries, 1993). Both chemicals residues and pathogens can accumulate on comb over time. It is important to fully understand beekeeper management of source colonies in the months, and even years, preceding collection of honey bees for laboratory tests. This includes gathering information on timing and type of medications, addition or removal of honey supers, condition of comb (e.g. old versus new), timing of previous comb replacement, queen age, requeening events, and origin of honey bee materials (e.g. wax foundation sourced locally or not, organic versus non-organic, etc.). Workers should not be collected during, or within 8 weeks of, the application of any honey bee pest or parasite control treatment. This will ensure that newly emerging workers and most "summer" individuals performing tasks inside the hive were not exposed to treatments (Winston, 1987). Researchers should acknowledge that residues from some treatments may persist in honey bee products and colonies for an extended period (Lodesani *et al.*, 2008; Mullin *et al.* 2010).

4.1.8. Environment surrounding source colonies used to obtain adult workers for laboratory experiments

Source colonies should not be located in intensive agricultural areas with high agricultural chemical use or low bee-plant diversity because of potential sub-lethal or synergistic effects of residues (Alaux *et al.*, 2009; Wu *et al.*, 2011) and the importance of nutrition to honey bee vitality (Brodschneider and Crailsheim, 2010), respectively. Additionally, knowledge of neighbouring apiaries is useful because of the potential for disease transmission. Note that honey bee poisoning can also occur in non-agricultural areas (e.g. natural or urban areas), normally because of misuse of pesticides on attractive flowering garden plants. These toxic pesticides used during blooming may cause important honey bee loss, although their residues may not necessarily will be found in hive matrices as individuals may die before returning to the colony. These deaths can alter the age profile of workers available for collection for experiments. Therefore, one should not collect workers from colonies that experience unexpected depopulation or abnormal honey bee mortality in front of the hive. Although costly, analyses of honey bees and their products (especially bee bread) can be used to quantify chemical residues within colonies. Local information on pesticide applications may also be gleaned from agricultural pesticide-use databases when they are available.

Vegetation surveys can be performed within normal worker foraging distances from the colony – within a 2 km radius of the hive (Winston, 1987) – to identify major nectar and pollen producing plants. Careful inspection of bee bread will also determine diversity of floral sources. This can be performed by visualizing pollen grain morphology using microscopy, or more crudely by colour differentiation (see Delaplane *et al.* (2013) in the pollination paper of the *BEEBOOK* for details on identifying plant species using pollen grains).

4.1.9. Weather before and during collection of adult workers for laboratory experiments

Weather events prior to honey bee collection can have a dramatic influence on colony strength and health. Periods of dearth or drought can greatly reduce food reserves within colonies (Schmickl and Crailsheim, 2001); whereas, prolonged periods of unfavourable flying conditions (e.g. rain, snow, wind) can confine workers to colonies for extended periods, and may promote overall colony stress (Schmickl and Crailsheim, 2007) and intra-colony disease transmission (Fries, 1993).

Current weather can also greatly affect flying patterns, and therefore potentially influence worker collection. Age polyethism observed in honey bees typically dictates that older individuals perform tasks outside of the colony, such as ventilating and guarding the colony, as well as collecting food (Winston, 1987). Therefore during unfavourable conditions a high number of older individuals will be present in the colony.

Both temperature and solar radiation influence foraging patterns (Burrill and Dietz, 1981). For example, foraging activity is positively related to temperature between 12 - 20°C (below 12°C honey bees typically do not search for food). Similarly, a positive relationship between foraging and solar radiation exists at low radiation intensities (i.e. < 0.66 langley (common unit of energy distribution for measuring solar radiation); the opposite occurs at high intensities). Expectedly, higher winds and rainfall also results in decreasing foraging activity, and therefore a greater number of older individuals in the colony (Winston, 1987). Sunny, warm weather conditions are optimal for collecting workers for experiments because fewer constraints are likely to limit the ability of workers to perform their required tasks. Regardless of weather, current conditions during collection, or unusual weather events prior to collection that may influence the nature of worker collection, should always be noted.

4.1.10. Diurnal timing of collection of adult workers for laboratory experiments

Flight patterns can also be influenced by time of day, possibly because of variations in flower nectar production (Winston, 1987). Foraging peaks typically late in both the morning and the afternoon, but lulls during the early afternoon (i.e. during the high sun period), and is infrequent between dusk and dawn (i.e. during the night) (Burrill and Dietz, 1981). Periods of high foraging activity are typically suitable for collecting workers for experiments because workers are more likely to be performing their tasks normally.

4.2. Collecting newly emerged workers for laboratory experiments

4.2.1. Considerations for choosing to use newly emerged workers for laboratory experiments

Collecting newly emerged workers, or "tenerals" as described by Winston (1987), is an easy and accurate method for obtaining large quantities of adults of a homogenous age. Newly emerged adults can be an important source of relatively 'clean' individuals because they are exposed to hive and environmental conditions less than older ones. It should be noted that it is virtually impossible to prevent, with 100% certainty, horizontal residue or pathogen contamination because of conditions in which workers develop within the colony (i.e. developing individuals are fed bee products in a wax cell) and because newly emerged workers, even caged on a frame in the laboratory, will feed on frame food stores, manipulate wax, and interact with previously emerged individuals. Newly emerged workers are also appropriate to use when examining possible treatment effects on honey bee longevity, or intra-host parasite development because individuals can be maintained in the laboratory for a number of weeks.

4.2.2. Obtaining newly emerged workers for laboratory experiments without caging queens

Here is the most practical way to obtain newly emerged workers with relatively low chemical residue or pathogen exposure:

1. Choose appropriate colonies from which to collect workers from based on health, environmental, genetic, and experimental design considerations discussed in sections 2 and 4.1.

2. Select frames containing enough capped brood that will emerge in one to three days (i.e. pupae with dark eyes and cuticle) to ensure that the required number of adults can be obtained. Consult the *BEEBOOK* paper on miscellaneous methods by (Human *et al.*, 2013) for information on how to obtain brood and adults of known age. Frames should be relatively new, not appear dark in colour or be soiled with faecal material or fungi, and should have few food stores.

3. Remove all adult honey bees from the frame using a bee brush or by gently shaking the frame over the colony.

4. Place the frame in an appropriate frame cage (see sections 5.2.1 and 5.3.1) that is outfitted with food (see section 7). Frame food stores and emerging honey bees can be segregated by cutting away honey and bee bread, or by installing 0.3-cm diameter aluminium hardware cloth screen around the stored food to keep workers from feeding.

5. Transfer the frame cage to a laboratory incubator maintained at conditions discussed in section 6.

6. Monitor the frame frequently to limit exposure of newly emerged workers to the frame. Individuals should be removed from the brood frame at least every 12-24 hours to obtain age homogeneity; however, frequency of worker removal from the frame can be adjusted according to the needs of the study and to reduce contamination by pathogens and chemical residues.

7. Gently brush newly emerged individuals into appropriate hoarding cages containing appropriate food (see sections 5 and 7). Newly emerged adults can also be removed gently from cells using a forceps before full emergence to further reduce potential for contamination. These individuals can be identified by small perforations in the wax capping of the brood cells. Care must be taken because the cuticle may not be fully hardened, and individuals can be easily damaged.

8. Immediately place the hoarding cage containing newly emerged adults in a laboratory incubator maintained at conditions discussed in section 6.

4.2.3. Obtaining newly emerged workers for laboratory experiments by caging queens

Newly emerged workers can also be collected from pre-selected brood frames that queens were previously restricted onto.

To obtain newly emerged workers from a frame that the queen was caged onto:

1. Identify suitable source colonies, as discussed in section 4.1, and brood frames, as discussed by Crailsheim *et al.* (2013) in the *in vitro* larval rearing paper of the *BEEBOOK*. A frame previously used for brood production that is relatively new (i.e. not containing dark, soiled comb) and has adequate empty cells is most suitable, and will likely contain fewer pathogens and environmental contaminants. A frame from the source colony will likely be most successful for rearing known age cohorts of workers; however, one from a different colony can also be used. Number of empty cells available for egg laying will be determined by the number of individuals needed for experiments. Brood mortality of approximately 20% should be expected (Fukuda and Sakagami, 1986).

2. Locate the queen in the source colony and gently place her on the chosen brood frame by grasping her wings. A clip queen catcher cage can also be used to move her. Refer to the *BEEBOOK* paper by Human *et al.* (2013) for handling honey bees. Ensure that at least a few hundred workers are on the frame before the queen is moved. These workers can either be ones that were on the frame originally or ones brushed from another brood frame in the same colony that contains open brood. This will serve to calm her and will lessen the chances that she runs or flies, or is crushed during caging.

3. Carefully place the frame, containing the queen and workers, in a queen excluder cage (Fig. 1), and seal it, ensuring the queen is not crushed. See section 5.2.1 for discussions on minimizing pathogen and environmental contaminant exposure when using cages.

Fig. 1. A brood frame containing workers, the queen, and many empty cells is being inserted into a queen excluder cage. Slits between 4.3 and 4.4 mm wide allow worker movement to and from the frame, but restrict queen passage.

Fig. 2. A frame caged in a queen excluder placed in the middle of the brood nest, between frames containing eggs and larvae.

4. Place the caged frame in the broodnest, preferably between two brood frames containing eggs and larvae (Fig. 2). This will improve chances that the newly-laid eggs are accepted by the colony. Refer to Human *et al.* (2013) in the miscellaneous methods paper of the *BEEBOOK* for estimating developing worker bee age.

5. After a defined period of time, remove the frame from the queen excluder cage and place it, with brood and the queen, back into the colony in its previous position. Mark the frame with a permanent marker or a coloured drawing pin to help locate it in the future. The number of honey bees required for experiments will determine the length of time the queen is confined to the frame. Queens typically lay between 5-35 eggs per hour (Allen, 1960), and frames can be checked every 24 hours to determine if enough eggs have been laid by inspecting cells through the queen excluder cage with the aid of a flashlight. Refer to the miscellaneous methods paper of the *BEEBOOK* by Human *et al.* (2013) for identifying eggs. It is possible that the queen will not begin egg laying until a few hours after initial isolation. Queens should not be confined to the frame for more than 72 hours, or when the availability of cells for egg laying is low, to avoid significant disruption of brood rearing in the colony. Homogeneity of age of newly emerged bees will also determine the length the queen is restricted to the frame, although this can also be controlled for during regular removal of newly emerged adults from the frame.

6. Remove the frame 19-20 days after initial queen restriction, just prior to adult emergence (Winston, 1987). The frames can be removed later if egg laying was significantly delayed, but care must be taken to prevent workers from emerging in the colony. Although a worker will usually emerge from a cell 21 days after an egg was laid, development time can vary

between 20-28 days depending on environmental conditions such as temperature and nutrition (Winston, 1987).

7. The frame and newly emerged adults can be subsequently handled according to #5, 6, and 7 of section 4.2.2.

4.2.4. Obtaining newly emerged workers for laboratory experiments by *in vitro* rearing

Newly emerged workers can also be obtained for experiments using *in vitro* rearing techniques described by Crailsheim *et al.* (2013) in the *in vitro* rearing paper of the *BEEBOOK*. This option is particularly useful to study experimental treatment effects in adults exposed during development.

4.3. Collecting adult workers of an undefined age for laboratory experiments

4.3.1. Considerations for choosing to use adult workers of an undefined age for laboratory experiments

Under certain circumstances it is not necessary to collect individuals of a known age. Although there is a tendency due to age polyethism (i.e. temporal division of labour) for young and old workers to be found in the centre or periphery of the broodnest, respectively (Seeley, 1982), or for older workers to perform jobs outside of the hive (Winston, 1987), distribution of age cohorts throughout the colony is dynamic and can be influenced by colonial needs (Calderone, 1995; van der Steen *et al.,* 2012). See Human *et al.* (2013) for a summary of worker development. We describe here how to sample workers of an undefined age. Under the appropriate conditions (see sections 4.1.9 and 4.1.10) broad functional groups of workers can be collected (e.g. individuals performing tasks in the hive versus those performing tasks outside the hive).

4.3.2. Challenges associated with collecting adult workers of an undefined age for laboratory experiments

Obtaining workers of an undefined age for an experiment usually requires the collector to physically open the colony or stand immediately in front of it to retrieve individuals. Collecting flying workers at the colony entrance can particularly agitate colonies, and may initiate a defensive response that will result in a mass exodus of guards from the hive (Breed *et al.,* 2004). Thus, agitation of colonies should be minimized because it can influence worker collection.

4.3.3. Collecting flying adult workers of an undefined age for laboratory experiments

Workers performing tasks outside of the hive are generally older than individuals working within (Winston, 1987), but as discussed in section 4.3.1., collecting workers of a particular age, or even performing a specific task, may not be straightforward. Returning pollen foragers can easily be observed by presence of corbicular pollen on their hind legs (Fig. 3).

Fig. 3. A foraging worker honey bee with corbicular pollen (black arrow) on its hind leg.

Fig. 4. Collecting exiting worker honey bees using a clear container with mesh bottom from a colony with a reduced entrance size.

It may be helpful to reduce the size of the hive entrance when performing certain collection methods to limit the area individuals may pass in or out of the colony. Completely sealing the hive for short periods (i.e. < 30 minutes) can also be used to collect returning flying individuals as they accumulate on the landing board. Time required to collect an appropriate number of flying workers can be estimated by observing the hive entrance for 2 - 3 minutes. Most foragers perform approximately 10 - 15 trips per day (Winston, 1987); however, length of collection time will be influenced by time of day and weather (as discussed in sections 4.1.9 and 4.1.10), as well as size of colony.

4.3.3.1. Collecting flying adult workers of an undefined age for laboratory experiments using a forceps

Exiting workers can be collected individually using forceps.

1. Stand beside, and not in front of, the colony.
2. During normal flight activity, grasp appropriate individuals by a leg or wing using forceps. Care must be taken that individuals are not damaged during collection. Refer to Human *et al.* (2013) in the miscellaneous methods paper of the *BEEBOOK* for details on handling honey bees using forceps.
3. Place collected workers in a ventilated hoarding cage with appropriate food (see sections 5 and 7).
4. Immediately transfer the hoarding cage to a laboratory incubator maintained at conditions discussed in section 6.

4.3.3.2. Collecting flying adult workers of an undefined age for laboratory experiments using a container

Workers leaving the hive can also be collected using a clear, wide-mouthed, well ventilated transparent container (with associated lid) as they depart the hive entrance (Fig. 4). Ventilation can be provided by perforating the container with numerous 2 mm-sized holes or by replacing a large portion of the base of the container with a mesh screen. Efficiency of this method depends on flying patterns of the colony, the ease of attaching the lid to the container, and the reflexes of the collector. Alternatively, a UV light-permeable plexiglass pyramid

(height = 30 cm, apex 3,5 x 3, 5 cm, base 18 x 18 cm) that is closable at the apex and the base can be placed tightly around the hive entrance to prevent exiting foragers from escaping (e.g. Felsenberg, 2011; Matsumoto *et al.*, 2012).

1. Stand beside the colony and hold a wide-mouthed clear container immediately against the front of the colony so that exiting individuals will fly or walk into the container. It may be helpful to reduce the size of the hive entrance to funnel greater numbers of exiting workers directly into the container and to use a container with a rectangular shaped opening that fits better to the flight board and hive entrance.
2. Seal the container when an appropriate quantity of workers is collected.
3. Shake the collected individuals gently into a ventilated hoarding cage containing food (described in sections 5 and 7).
4. Transfer the hoarding cage to a laboratory incubator maintained at conditions discussed in section 6.

4.3.3.3. Collecting flying adult workers of an undefined age for laboratory experiments using an entrance trap

Entrance traps allow for a large number of exiting workers to be collected from colonies with minimal disturbance because workers will eventually not view the trap as a foreign object. The Bologna Trap has a particularly effective design (Medrzycki, 2013).

4.3.3.3.1. Bologna Trap description for collecting adult workers for laboratory experiments

The Bologna Trap acts as a funnel that can be placed over the lower front portion of a hive. Because the trap can remain on the colony for an indefinite period of time in an open position, workers will pass in and out of the colony normally (Fig. 5). The bottom of the funnel acts as an extension of the flight board, sealing tightly to it and to the front of the hive so that exiting individuals leave the hive and enter the trap by walking (Fig. 6). The funnel is curved upwards, reaching an

Fig. 5. Bologna Traps, without collection containers, attached to the entrance of honey bee colonies.

Fig. 6. A detached Bologna Trap. Arrow points to proximal portion of the trap that can be attached to the lower front portion of the hive to completely and securely surround the hive entrance.

Fig. 7. Terminal end of the Bologna Trap. The ring (*i.e.*, a lid with a large hole cut away) accommodates a collection container that can be attached to obtaining flying honey bees exiting the colony.

inclination of approximately 30°; any greater inclination may result in dead honey bees accumulating in the trap. The funnel ends with an adaptor (i.e. a lid with a large hole cut out) for where a collection container to be attached (Fig. 7). See section 4.3.3.2 for a description of a collection container.

4.3.3.3.2. Collecting flying adult workers of an undefined age for laboratory experiments using the Bologna Trap

1. Fix the trap, without the collection container, to the hive for at least 5 days before collecting workers to accustom the colony to the device. Acceptance of the trap can be verified when undertaker workers remove dead individuals from the trap. The trap can remain installed on the hive for the entire season, apart from when cleaning and repairs are required.
2. When experimental workers are needed, observe the hive entrance for 2-3 minutes, noting the number of exiting workers, to estimate approximate length of time collection is needed.
3. Install the collection container to the distal end of the funnel (Fig. 8).
4. When the appropriate number of flying workers are collected (Fig. 9), remove the collection container quickly and seal it (Fig.10).
5. Transfer collected workers by gently shaking the collection container over an open hoarding cage containing food (as discussed in sections 5 and 7)
6. Transfer the hoarding cage to a laboratory incubator maintained at conditions discussed in section 6.

4.3.4. Collecting intra-hive adult workers of an undefined age for laboratory experiments

Workers can be easily collected from frames within the colony. Because of the dynamic nature of honey bee age polyethism (Calderone, 1995; van der Steen *et al.*, 2012), it is not possible to accurately collect individuals of known ages based on location within the colony. For example, van der Steen *et al.* (2012) observed no difference in worker age classes among frames in a colony, and that approximately 60% of workers on frames were one or two weeks old.

1. Inspect the frame from which workers are to be collected from for the queen. If present, gently move her to an adjacent frame.
2. Gently brush individuals into a suitable hoarding cage (see section 5) placed below the frame using a beekeeping brush or similar tool with soft bristles. Alternatively, the frame can be gently shaken over a suitably sized open-mouthed container prior to transferring collected workers to a suitable hoarding cage.

Fig. 8. A Bologna Trap with a ventilated collection container installed on the foreground colony to obtain exiting honey bees.

Fig. 9. A ventilated collection container obtaining flying workers exiting the hive. Note that the bottom of the container is replaced with a fine mesh that is held in place using an elastic.

Fig. 10. Removing the collection container filled with exiting honey bees from the Bologna Trap.

3. Gently shake the opened hoarding cage or container for ~ 1 minute to prevent young workers from escaping by walking and to allow older flying workers to exit.
4. Close hoarding cage, or transfer remaining workers into a suitable hoarding cage with food (sections 5 and 7).
5. Immediately transfer the hoarding cage to a laboratory incubator maintained at conditions discussed in section 6.

4.4. Recommendations for choosing and collecting adult workers for laboratory experiments

The choice of type of honey bees to use during experiments, as well as when and how to collect them, is intimately tied to the hypothesis being tested. At a minimum, all possible characteristics of the experimental individuals (e.g. age), source colonies, (e.g. strength, health, subspecies), surroundings (e.g. availability of multiple nectar and pollen sources), as well as conditions during collection (e.g. time of day and year, weather conditions) and collection method (e.g. brushing from a brood frame versus collecting exiting flying workers using a hive entrance trap), should be described in detail in the methods section of each publication. Importantly, researchers must ensure that all treatments contain experimental honey bees were handled identically. The easiest approach to guarantee this is to mix honey bees from all sources evenly among all experimental cages, as suggested in this paper in section 2.4. Additional information on choosing source colonies is provided by Pirk *et al.* (2013) in the statistics paper of the *BEEBOOK*.

5. Cages in which to maintain adult workers in the laboratory

5.1. Types of cages in which to maintain adult workers in the laboratory

Generally, three types of cage design exist for maintaining adult worker honey bees outside of a colony in a laboratory:

a) caged on a frame (i.e. using a frame cage)
b) caged off a frame in a group (i.e. using a hoarding cage)
c) caged off a frame individually (i.e. using an isolation cage)

Even within these types numerous variants exist (Fig.11). Yet, despite the diversity of cage designs, very little work has investigated the influence of these differences on results of experiments using honey bees.

5.2. Choosing a suitable cage to maintain adult workers in the laboratory

5.2.1. Minimum criteria for frame and hoarding cages in which to maintain adult workers in the laboratory

Generally, frame and hoarding cages of all types should meet the following minimum criteria; however, discretion may be used

Fig. 11. The diverse assemblage of cages used for honey bee research brought by those attending a COLOSS workshop in November 2011 in Bologna, Italy.

depending on the purpose of containing honey bees (e.g. for caging newly-emerged adults in a brood frame or for performing experiments using hoarding cages).

- Cages should be used once and discarded, or sterilised and cleaned if used multiple times, to minimise contamination by pathogens and chemical residues.
- Single-use cages are recommended for studies involving pesticide toxicology because of the difficulty in removing chemical residues.
- Multiple-use cages can be used for honey-bee pathogen studies and should be made from materials that are easily sterilised (e.g. autoclaved or irradiated), such as stainless steel and glass. Type of sterilisation required will depend on the nature of the study. For example, exposure to 121°C for 30 minutes will destroy *N. ceranae* spores (Fenoy *et al.*, 2009). Metal and plastic cages can be further decontaminated using acetone[*]:

1. Wash cages using a standard laboratory dish washer
2. Apply a sparse quantity of technical grade 100% acetone (the preferred solvent in toxicology laboratories) to a cloth and wipe cage clean. Attention should be paid to effects of acetone on plastic cages.
3. Soak a new cloth in warm soapy water and wash/rinse cage.
4. Rinse cage with water.
5. Dry cage using a new cloth, and air-dry until all liquid evaporates.
   [*]Refer to your own laboratory safety manual to learn how to properly work with acetone.

- Materials used to make cages should be inexpensive, and easily accessible and manipulated. Plastic and wood allow for easy

modification of cages when, for example, an additional feeding device is needed.
- Cages should have a sufficient quantity of air holes to provide ventilation.
- To reduce risk of contamination by pathogens and chemical residues among cages maintained in the same incubator, ventilation holes should be covered by filter paper or similar breathable material. If vents are unfiltered, cages should face in opposite directions and should be placed sufficiently far apart to prevent inter-cage trophallaxis or frass movement.
- Cages should allow both living and dead honey bees to be easily removed during the experiment, and should prevent live bees from accidentally escaping.
- At least a portion of the cage should be transparent to allow honey bees to be observed.
- Cage size will depend on the number of honey bees to be detained. For example, 500 cm^3 (i.e. 500 ml) can easily accommodate several hundred workers, whereas cages of 100 cm^3 are suitable for maintaining 30 workers. Generally, a ratio of ~3:1 (cm^3/bee) is appropriate for maintaining less than a few hundred workers.

5.2.2. Supplementary frame and hoarding cage materials to be used when maintaining adult workers in the laboratory

Additional materials, such as comb or wax foundation (e.g. Czekońska, 2007) and plastic devices for releasing queen mandibular pheromone (QMP) (e.g. Alaux *et al.*, 2010), are sometimes used to provide more realistic conditions to honey bees. For the former, comb and wax foundation should be used with caution because both can contain chemical residues (Mullin *et al.*, 2010) and pathogens (Melathopoulos *et al.*, 2004); however, organic wax foundation is available. For the latter, QMP, composed of 5 compounds ((E)-9-oxodec-2-enoic acid (9-ODA), both enantiomers of 9-hydroxydec-2-enoic acid (9-HDA), methyl p-hydroxybenzoate (HOB) and 4-hydroxy-3-methoxyphenylethanol (HVA)) (Slessor *et al.*, 1988), likely promotes honey bee health and reduces stress, as well as influences brain development (Morgan *et al.*, 1998), resistance to starvation (Fischer and Grozinger, 2008), age-related division of labour (Pankiw *et al.*, 1998), and worker ovary activation (Hoover *et al.*, 2003). More studies are needed to fully understand effects of QMP on caged honey bees before it can be recommended as a regular requirement for maintaining adults in the laboratory.

5.2.3. Minimum criteria for isolation cages in which to maintain adult workers in the laboratory

In contrast to frame and hoarding cages, isolation cages are rarely used outside of studies investigating behaviour or learning. Many of the principles discussed above for frame and hoarding cages also apply to isolation cages, such as the importance of providing a sterile, well-ventilated cage.

5.3. Suitable cages in which to maintain adult workers in the laboratory

The following cage descriptions are provided by the authors to give examples of those generally meeting minimal criteria listed above. There are no doubt other cages described in detail elsewhere that are equally suitable (e.g. hoarding cages: Pernal and Currie, 2000; Evans *et al.*, 2009).

5.3.1. Example of a frame cage in which to maintain adult workers in the laboratory

Generally, a frame cage allows for a single frame to be suspended within it, and contains one or two ventilated sides that can be slid away to allow access to the frame (Fig. 12).

Fig. 12. A frame cage containing a Zander-sized frame and composed of a wooden casing, a metal screen, a glass removable sliding side, and two feeding devices. Cage courtesy of the Swiss Bee Research Centre.

5.3.2. Examples of hoarding cages in which to maintain adult workers in the laboratory

Classic hoarding cages are shaped similar to frame cages, and also contain one or two sides that may be removed (Fig. 13), although other designs exist that are cup-shaped (Fig. 14) or are modifications of the classic design with the cage rested on its side so that the top is removable (Figs. 15 and 16).

5.3.3. Examples of isolation cages in which to maintain adult workers in the laboratory

For isolation cages, modified straws with pins placed at either end, 1.5 ml microcentrifuge tubes with breathing holes drilled through the tip (Fig. 17), or 0.8 cm wide plastic Eppendorf tubes cut in half longitudinally with sticky tape restraining harnesses (Fig. 18), can be used. To our knowledge, researchers do not maintain individuals in these types of cages for more than one week. Future studies should investigate effects of isolation cages on survival and health of caged

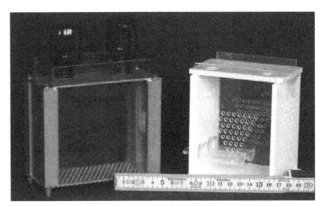

Fig. 13. Examples of 'classic' hoarding cages equipped with transparent and removable sides, ventilation holes, and multiple inputs for feeding devices. Cages courtesy of the Swiss Bee Research Centre (left) and INRA (right).

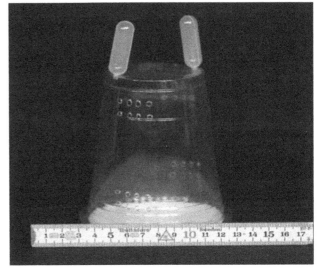

Fig. 14. Cup-shaped hoarding cage with removable base, multiple ventilation holes, and two feeding devices. Modified from Evans *et al.* (2009). Cage courtesy of ScientificBeekeeping.com

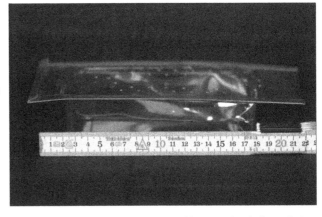

Fig. 15. Hoarding cage containing removable top, and multiple ventilation holes and feeding device inputs. Cage courtesy of Szent István University.

honey bees, as well as work to develop an appropriate method for maintaining individuals in isolation cages for an extended period of

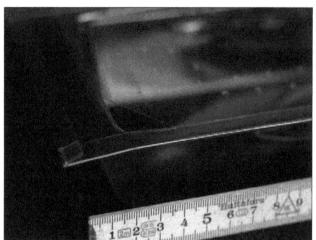

Fig. 16. Magnification of the sliding, removable top of the cage presented in *Fig. 15.* Note the removed corner to facilitate addition or removal of honey bees. Cage courtesy of Szent István University.

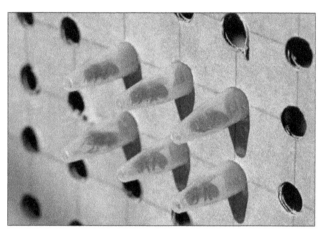

Fig. 17. Isolation cages created by drilling 2-3 mm ventilation holed in the tip of 1.5-ml microcentrifuge tubes. Cages and storing device courtesy of Ulrike Hartmann, Swiss Bee Research Centre.

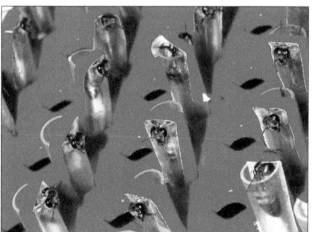

Fig. 18. Isolation cage constructed using a plastic Eppendorf tube cut in half longitudinally and sticky tape harnesses. Tube height, and outer and inner diameters = 3, 1, and 0.8 cm, respectively. Cages courtesy of CWW Pirk.   Photo: V Dietemann

time. This could potentially greatly increase experiment sample size compared to hoarding cages that must include cages, rather than individuals, as number of replicates.

6. Incubator conditions

6.1. Regulation of biophysical properties within colonies

Honey bees are renowned for cooperatively maintaining nest homeostasis by regulating biophysical properties such as temperature, humidity, and respiratory gases within a colony. In doing so, they create a suitable environment that moderates adverse conditions (Danks, 2002). When maintained in an incubator, however, appropriate conditions must be provided, regardless of if honey bees are kept individually or in small groups. Because biophysical properties, whether in a colony or an incubator, are intimately connected to water loss, researchers must consider not only chamber conditions, but also water availability (i.e. both drinking and vapour water), when maintaining adults in the laboratory. Although honey bees are relatively tolerant to changes in thermal and moisture conditions, it is recommended that adult honey bees be maintained in conditions as close to their natural environment as possible.

6.2. Temperature

6.2.1. Honey bee intra-hive temperature requirements

Despite considerable changes in ambient air temperature, honey bees typically maintain their brood nest between 32 and 36°C by adjusting their metabolism and by using a number of behavioural methods (Stabentheiner *et al.*, 2010) to ensure optimal brood development. Nevertheless, outer edges of honey bee clusters can drop to as low as 10°C in winter when no brood is present (Seeley, 2010). Most laboratory studies maintained caged honey bees between 25-34°C (e.g. Webster, 1994; Higes *et al.,* 2007; Paxton *et al.*, 2007; Alaux *et al.*, 2009), and 25 ± 2°C is recommended for testing acute oral toxicity of chemicals (OECD, 1998).

6.2.2. Recommendations for incubator temperature for maintaining adult workers in the laboratory

Frames of brood should be maintained at 34.5°C for optimal brood development (Heran, 1952; Crailsheim *et al.*, 2012); whereas, we recommend keeping adults at 30°C, based on optimal respiration at 32°C (Allen, 1959) and honey bee thermal preference of 28°C (Schmolz *et al.*, 2002). Generally these recommendations are also appropriate when performing experiments; however, further adjustments to temperature may be required. For example, a recent study investigating acute oral toxicity of chemicals to honey bees under laboratory conditions suggested that these evaluations should be performed at both 25 and 35°C to account for the wide range of temperatures to which honey bees are exposed (Medrzycki and Tosi, 2012). When obtaining newly-emerged honey bees from a brood frame maintained at 34.5°C in an incubator, young honey bees (i.e. individuals 0 to 24 hours old) should be transferred from a caged frame containing brood to one or more cages that are maintained at 30°C. More details on obtaining newly emerged honey bees from brood comb are provided in section 4.2.

6.3. Relative humidity

6.3.1. Honey bee intra-hive relative humidity requirements

Humidity within a colony can also be influenced by honey bees, albeit to a lesser extent than temperature (Human *et al.*, 2006). Similar to temperature, relative humidity can differ among areas of a colony (Human *et al.*, 2006), but also fluctuate substantially because of breathing events that exchange stale air at optimal humidity with air at ambient humidity (Southwick and Moritz, 1987). Relative humidity within honey bee colonies (among frames and not within capped brood cells) is typically between 50 and 80% (Human *et al.*, 2006; V. Dietemann, pers. comm.), and when given a choice between a range of relative humidities (i.e. 24, 40, 55, 75, and 90%), honey bees showed a preference for 75% (Ellis *et al.*, 2008). The OECD (1998) recommends relative humidity to be between 50-70% for laboratory testing of acute oral toxicity of chemicals.

6.3.2. Regulating incubator relative humidity for maintaining adult workers in the laboratory

If the laboratory is not equipped with an incubator capable of automatically regulating a desired relative humidity, then it can be attained easily using two methods. One can also refer to methods discussed in the *in vitro* rearing paper of the *BEEBOOK* Crailsheim *et al.* (2013) for appropriate relative humidity conditions for maintaining brood in the laboratory.

6.3.2.1. Regulating incubator relative humidity for maintaining adult workers in the laboratory using an open water basin

Relative humidity can be regulated by placing open containers filled with water at the bottom of the incubator (Fig. 19). In some cases, a suitably hung cloth wick can be used to promote evaporation.

6.3.2.2. Regulating incubator relative humidity for maintaining adult workers in the laboratory using a saturated salt solution

If an open basin of water cannot maintain the incubator at the desired condition, then further regulation can be provided using saturated salt solutions. Use of these salts is summarized here, but discussed in further detail by Wexler and Brombacker (1951) and Winston and Bates (1960).

Fig. 19. Regulation of incubator relative humidity using an open water basin.

6.3.2.2.1. Criteria for using saturated salts to regulate incubator relative humidity for maintaining adult workers in the laboratory

The following criteria are discussed by Winston and Bates (1960). Expected relative humidity values produced using saturated salt solutions may vary when experimental conditions do not permit all criteria to be met.

1. Container (i.e. incubator) must be a closed system.
2. A fan to distribute air should be provided when incubator volume is >1 litre.
3. Surface area of the solution should be as large as possible.
4. Reagent grade chemicals should be used to allow for reproducibility.

6.3.2.2.2. Choosing appropriate saturated salts for regulating incubator relative humidity for maintaining adult workers in the laboratory

A variety of salts can produce a wide range of relative humidities at many defined temperatures (see Table 1 in Winston and Bates (1960)). Choice of these salts should be determined by desired relative humidity and temperature conditions. Sodium chloride (NaCl) is easily available and can maintain relative humidity at ~75% over various temperatures when certain criteria are met (section 6.3.2.2.1). Sodium chloride can still be used despite circumstances when all criteria are not met, especially when an open basin of water alone cannot regulate the desired conditions; however, constant vigilance of relative humidity is required (section 6.3.3).

6.3.2.2.3. Preparing a saturated salt solution for regulating incubator relative humidity for maintaining adult workers in the laboratory

The following is an example of how to create approximately one litre of sodium chloride saturated salt solution:

1. Heat one litre water slowly in a two litre glass beaker.
2. Place beaker on standard laboratory magnetic stirrer.
3. During heating, gradually add ~400 g sodium chloride to water until crystals do not dissolve any further; this will slightly increase the volume of the solution.
4. Mix solution using stirrer.
5. Continue adding sodium chloride until a gentle boil is reached and no further salt will dissolve.
6. Remove solution from heat, pour in appropriate, open-mouthed basin, and let cool before transferring to the incubator. Solution should contain a mixture of crystals and liquid.
7. Use salt solution for multiple weeks; replace when no water is present or when fungi or bacterial growth occurs.

6.3.3. Monitoring and recording incubator relative humidity when maintaining adult workers in the laboratory

Small changes in ambient weather, as well as the opening of the incubator door, can significantly affect incubator relative humidity, especially when the total volume of the chamber is greater than one litre (Rockland, 1960; Winston and Bates, 1960). Because of this, an accurate, reliable data recorder or a digital measuring device should be used to document relative humidity, as well as temperature, over time. Numerous types of equipment are available, such as the iButton (Maxim Integrated Products, San Jose, United States) or HOBO (Onset Computer Corporation, Cape Cod, United States).

6.3.4. Recommendations for incubator relative humidity for maintaining adult workers in the laboratory

Considering natural colony conditions and worker preference, we recommend that adult workers of all ages should be maintained at 60-70% relative humidity in the laboratory.

Pre-trials will be needed to determine water surface area, frequency of water replacement, and choice of salt needed to sustain appropriate levels because incubator size and air exchange with the ambient surroundings will greatly influence relative humidity.

6.4. Light

6.4.1. Natural honey bee light conditions

Honey bees typically spend a considerable amount of their lives in mostly dark conditions within the hive, although late in life, light-dark cycles play a crucial role in determining foraging rhythm of workers (Moore, 2001). An exception includes some *Apis mellifera adansonii*

that nest in the open (Fletcher, 1978). Despite phototaxis (i.e. movement toward or away from a light stimulus) varying relative to bee age, light intensity, and light wavelength (Menzel and Greggers, 1985; Ben-Shahari *et al.*, 2003; Erber *et al.*, 2006), permanent exposure to honey bee-visible light can affect hoarding behaviour (i.e. the collection and storage of food in the honey stomach) of caged honey bees (Free and Williams, 1972). To our knowledge, honey bees in the laboratory are always maintained in complete darkness (e.g. Malone and Stefanovic, 1999; Maistrello *et al.,* 2008; Alaux *et al.*, 2009); however, many studies fail to report light conditions.

6.4.2. Recommendations for incubator light conditions for maintaining adult workers in the laboratory

Caged workers should be maintained in an incubator under dark conditions. Workers and cages should be examined and manipulated under dim light conditions, preferably using red light that emits 660-670 nm wavelengths that are not visible to honey bees (Menzel and Backhaus, 1991). To produce light of this wavelength, special bulbs can be purchased or standard incandescent bulbs emitting human-visible light can be covered with a red lens so that light produced is of the appropriate wavelength.

6.5. Ventilation

6.5.1. Honey bee ventilation requirements

Honey bees rely on a permanent supply of oxygen to survive. Because carbon dioxide within colonies can reach levels much higher than normal atmospheric levels (0.04%) (Nicolas and Sillans, 1989), honey bees use fanning and gas exchange events to expel carbon dioxide rich air (Southwick and Moritz, 1987; Nicolas and Sillans, 1989) to maintain levels between 0.1 - 4.3% (Seeley, 1974). Carbon dioxide can also reach high levels within cages and incubators that do not provide adequate air exchange and ventilation with ambient fresh air.

6.5.2. Recommendations for incubator ventilation with ambient air for maintaining adult workers in the laboratory

It is extremely important that cages allow for appropriate ventilation, and that incubators are equipped with air exchangers or passive vents at a minimum. Although air exchange occurs every time an incubator is opened, this technique should not be relied upon because air exchange should be permanent and opening the chamber regularly will disturb caged workers. To minimise effects of potential differences in gas composition within an incubator on experimental honey bees, cages of each treatment group should be homogenously distributed in the useable space of an incubator and a small fan should be used to promote air homogenisation. More information on effects of carbon dioxide on honey bees can be found in the *BEEBOOK* paper on miscellaneous methods (Human *et al.*, 2013).

7. Nutrition

7.1. Nutritional requirements of worker honey bees

Diet can affect honey bees in numerous ways including, for example, longevity (Schmidt *et al.*, 1987) and physiology (Alaux *et al.*, 2010). Under natural conditions, honey bees receive carbohydrates and proteins they require by consuming nectar and pollen stored in a colony as honey and bee bread, respectively. Carbohydrates are the source of energy for workers; whereas, proteins are crucial for building and maintaining tissues (e.g. Hersch *et al.*, 1978; Pernal and Currie, 2000). Additional nutrients, such as vitamins, minerals, and lipids, are also obtained from pollen, although their importance are not well understood (Brodschneider and Crailsheim, 2010).

For proper growth and maintenance, each worker larva requires 59.4 mg of carbohydrates and 5.4 mg of pollen during their development (Rortais *et al.*, 2005); whereas adult workers require ~4 mg of utilizable sugars (Barker and Lehner, 1974) and consume ~5 mg pollen (Pernal & Currie 2000) per day. Interestingly, under laboratory conditions caged workers self regulated their intake at approximately 10% proteins and 90% carbohydrates (Altaye *et al.*, 2010). Although providing laboratory workers with these natural food types may not always be practical, or even ideal, it is necessary that they receive in some form appropriate quantities of essential nutrients that provide energy and promote proper growth and development (e.g. Pernal and Currie, 2000; Brodschneider and Crailsheim, 2010).

7.2. Carbohydrates

7.2.1. Types of carbohydrates to provide to caged adult workers in the laboratory

Honey bees are capable of surviving long periods on carbohydrates alone, although median lethal time (LT_{50}) can vary significantly by substrate (i.e. LT_{50} = 56.3, 37.7, and 31.3 days, respectively, for sucrose, high-fructose corn syrup, and honey) (Barker and Lehner, 1978). Additionally, recent data suggest type of carbohydrate can influence detoxification in honey bees (Johnson *et al.*, 2012), further underlining the importance of carefully choosing source of carbohydrate to feed to workers.

7.2.1.1. Providing honey to caged adult workers in the laboratory

Honey is the natural carbohydrate source of honey bees, and can be easily collected from a colony; however, it is difficult to standardize given variation in composition due to floral diversity (e.g. White and Doner, 1980). Additionally, it may contain chemical residues (Chauzat *et al.,* 2009) and microflora (Gilliam, 1997), including pathogens (Bakonyi *et al.,* 2003), despite its antibacterial properties (Kwakman *et al.,* 2010). Honey can be collected from honey supers and provided pure, diluted 1:1 (volume/volume) with tap water, or as a paste consisting of 70% (volume/volume) powdered sucrose and 30% pure honey (e.g. Alaux *et al.,* 2011a). Refer to section 7.5 in this paper for a discussion on providing water to caged honey bees in the laboratory.

7.2.1.2. Providing sucrose solution to caged adult workers in the laboratory

Sucrose solutions can sustain workers for long periods of time in the laboratory (Barker and Lehner, 1978), and they are frequently used (e.g. Malone and Stefanovic, 1999; Paxton *et al.*, 2007; Forsgren and Fries, 2010). Solutions can be made simply by dissolving sucrose sugar in water. The sucrose should be white refined table sugar intended for human consumption that can be purchased in a supermarket.

To make a 100 ml volume 50% (weight/volume) solution, for example:

1. Add 50 g table sugar (sucrose) to a 200 ml glass beaker.
2. Add tap water until total volume reaches 100 ml.
3. Stir until all sugar is dissolved (i.e. < 5 mins.). If needed, water can be briefly warmed to < 50°C to help dissolve the sugar, but it should be cooled to room temperature before it is provided to caged workers.
4. Provide immediately to caged workers.
5. Store surplus solution for no more than 2-3 days at 4°C. Prior to feeding, remove solution from fridge a few hours before providing it to caged workers in order to prevent feeder leakage caused by the solution warming.

7.2.1.3. Providing sucrose paste to caged adult workers in the laboratory

Although it is used less frequently during laboratory assays compared to sucrose solutions (e.g. Maistrello *et al.,* 2008; Alaux *et al.,* 2009), sucrose paste is often provided to queens and accompanying nurses that are maintained in cages in a laboratory. Because it is a solid, the paste should be provided using devices designed for protein distribution, as explained in section 7.3.2. Water should also be given in a separate feeder when sucrose paste is the sole source of carbohydrates; refer to section 7.5. for details on providing water to caged workers.

To make 100 g of 95% (weight/weight) sucrose candy, for example:

1. Add 95 g powdered sucrose sugar to a 200 ml glass beaker.
2. Add 5 g tap water to the beaker.
3. Stir until a paste is created. Consistency should be similar to soft dough, and it should not ooze.

7.2.2. Feeding devices for providing carbohydrates to caged adult workers in the laboratory

Numerous types of devices can be used to provide liquid carbohydrates to caged honey bees. Feeding devices must fulfil the following minimum criteria:

- Allows workers to drink safely, without drowning.
- Holds the respective volume securely, minimises evaporation, and prevents leakage; a small piece of paper tissue can be inserted in the feeder over top of the feeding site to prevent leakage.

Fig. 20. A disposable 5 ml plastic syringe with Luer connection fitting removed to create a 2 mm hole revealing the black plunger.

- Ensures feeding sites are not easily blocked by crystallisation; size of feeding site hole that is dispensing food, as well as water concentration of carbohydrate, will influence crystallisation. Since no data are currently available on the subject, pre-trials will determine an appropriate size of feeding site.
- Allows for quick and easy replenishment of the solution, as well as measurement of consumption, that minimises accidental escape of experimental individuals and preferably does not require opening cages.

For workers in frame or hoarding cages (refer to section 5), a simple disposable feeding device can be made using a microcentrifuge tube (< 2 ml) with two to three small holes 1-2 mm wide drilled into the bottom or by using a syringe with the needle removed and adaptor cut away to reveal a 2-5 mm wide hole (Fig. 20). Alternatively, a feeding device can be created by drilling a single 2-5 mm wide hole in the base, as well as two 2-5 ml sized holes on the sides ~5 mm from the tip to prevent air bubbles from forming at the bottom; a small piece of tissue paper can be inserted into the tip to prevent leakage. Gravity feeders, created by inverting a jar with a lid containing a single large hole (i.e. 5 mm) screened with multiple layers of cheese cloth or a lid with three to five 1-mm holes without cheese cloth, can also be used; however, one must be careful of leakage and crystallisation. Quantity and size of feeders should be adapted to the number of workers requiring food and to the interval between food replenishment. At least 2 devices should be used to reduce the risk of workers starving if one feeder becomes defective, especially if it leaks. Leaky feeders can result in workers starving or drowning; use of ventilation holes or absorptive material on the bottom of cages can prevent the latter. Workers in isolation cages can be individually fed using a micropipette (section 7.8.2).

Refer to section 7.3.2 for a description of providing solid food to caged workers in the laboratory.

7.2.3. Measuring carbohydrate consumption by caged adult workers in the laboratory

Consumption by caged workers can be measured by determining the change in weight or volume of carbohydrate over a given period of time, although most experiments measure the former (Barker and Lehner, 1974). Regardless of method used, consumption should be adjusted for length of feeding period and number of caged individuals to calculate food consumed per honey bee per 24 hours. An easy approach is to simply record consumption every 24 hours, but when this is not possible, recording within 36 hours will suffice, depending upon the size of the feeder and number of caged workers.

To measure average daily carbohydrate consumption per worker for each cage when feeders are not checked every 24 hours:

1. Fill feeder with food.
2. Record mass of food-filled feeder ($MASS_{INITIAL}$).
3. Provide feeder to caged workers; record date and time (hours and minutes) of insertion ($TIME_{INITIAL}$) and number of living caged workers ($WORKERS_{INITIAL}$).
4. Remove feeder after given interval (see section 7.2.4 for frequency of feeder replenishment).
5. Record date and time of removal ($TIME_{FINAL}$), and number of living caged workers ($WORKERS_{FINAL}$).
6. Record mass of feeder ($MASS_{FINAL}$)
7. Determine mass of food consumed (CONSUMED) by subtracting $MASS_{FINAL}$ from $MASS_{INITIAL}$
8. Calculate number of hours (HOURS) the feeder was provided to caged workers using $TIME_{INITIAL}$ and $TIME_{FINAL}$.
9. Calculate hourly cage consumption ($CONSUMED_{HOURLYCAGE}$) by dividing CONSUMED by HOURS.
10. Calculate hourly worker consumption ($CONSUMED_{HOURLYWORKER}$) by dividing $CONSUMED_{HOURLYCAGE}$ by $WORKERS_{FINAL}$; note that consumption is measured for the final living workers, rather than the initial number of living workers or an average of the number of initial and final living workers.
11. Calculate daily worker consumption ($CONSUMED_{DAILYWORKER}$) by multiplying $CONSUMED_{HOURLYWORKER}$ by 24.

Consult section 7.7 to correct for mass of food stuff lost through evaporation.

7.2.4. Replenishing carbohydrates provided to caged adult workers in the laboratory

Care must be taken when renewing carbohydrates because workers are at a higher risk of escaping or being damaged during this time. In theory, 1 ml of 50% (weight/volume) sucrose solution should be adequate for approximately 100 individuals during a 24-hour period because adult workers require 4 mg useable sugar per day to survive (Barker and Lehner, 1974). As worker consumption may vary according to treatment, at least 5 ml of 50% (weight/volume) sucrose solution for 100 workers should be provided daily to ensure that they do not run out of food.

Carbohydrates should be replenished frequently to ensure they are provided *ad libitum* (i.e. caged workers are never without carbohydrates), or at least every three days to prevent microbial growth or drying when sucrose pastes are provided.

If carbohydrates cannot be provided *ad libitum* to honey bees in isolation cages, individuals can be fed to satiation immediately upon caging, and 16 µl (four 4 µl droplets) of approximately 30% (weight/volume) sucrose solution every 24 hours; this should maintain them for at least one week (Felsenberg *et al.*, 2011).

7.2.5. Recommendations for providing carbohydrates to caged adult workers in the laboratory
The use of a self made sucrose solution is easy, reduces chances of contamination, and depending on type of sugar used, can sustain honey bees for several weeks. Therefore, a good option for providing workers maintained in the laboratory with carbohydrates is to feed 50% (weight/volume) sucrose-tap water solution *ad libitum* (Barker and Lehner, 1978) using a feeder that meets the minimum criteria described previously (section 7.2.2.). Refer to section 7.5 for providing water to caged honey bees.

7.3. Proteins

7.3.1. Types of proteins to provide to caged adult workers in the laboratory
Similar to carbohydrates, source and type of protein (i.e. protein content and amino acid composition) can significantly influence honey bee development, longevity, and immunity (e.g. Haydak, 1970; Pernal and Currie, 2000; Brodschneider and Crailsheim, 2010; DeGrandi-Hoffman *et al.*, 2010; Alaux *et al.*, 2011a). Proteins can be fed to laboratory workers in a variety of forms, although nutritive value, palatability, and digestibility will vary. For example, individuals survived longer (Beutler and Opfinger, 1948) and had higher protein titre levels (Cremonez *et al.*, 1998) when fed pollen collected from the comb (i.e. bee bread) versus pollen traps (i.e. corbicular pollen). Additionally, Peng *et al.* (2012) found that head weight (a surrogate for hypopharyngeal gland size) was larger in young workers fed pollen substitutes compared to various pollen diets.

7.3.1.1. Providing bee bread to caged adult workers in the laboratory
Bee bread, a mixture of fermented pollen, regurgitated nectar, honey, and glandular secretions (Herbert and Shimanuki, 1978), is the natural and most nutritious protein source for young workers. However, it can contain pathogens (Gilliam, 1979) and chemical residues (Genersch *et al.*, 2010; Mullin *et al.*, 2010), and harvesting it is difficult and takes considerable time. A small, metal micro-spatula with a concave blade that is 3-4 mm wide can be used to collect multi-floral bee bread (see

section 4.1.8 and the *BEEBOOK* pollination paper by Delaplane *et al.* (2013) for details on identifying multi-floral bee bread). Alternatively, an entire area of cells containing bee bread can be removed from the frame by cutting cross-sections of all cells near their bases. This allows bee bread to be 'popped' out of each cell. Refer to Human *et al.* (2013) in the miscellaneous techniques paper of the *BEEBOOK* for specific instructions on collecting bee bread from colonies. Bee bread can be provided to workers as a 50% (weight/weight) homogeneous paste mixture with sucrose paste (e.g. Cremonez *et al.*, 1998). Refer to section 7.2.1.3 for creating sucrose paste. Quantities may vary, depending upon the nature of the bee bread.

7.3.1.2. Providing corbicular pollen to caged adult workers in the laboratory
Corbicular pollen pellets are units of worker-collected pollen that can be harvested before they are stored in a colony. They provide a common and simple way to provide workers with proteins, and can be collected by outfitting colonies with pollen traps, such as those attached to the hive entrance or those placed under the brood box but above the original colony entrance, as described by Human *et al.* (2013) in the *BEEBOOK* paper on miscellaneous methods. Similar to honey and bee bread, however, corbicular pollen can contain chemical residues and pathogens (e.g. Higes *et al.*, 2008; Mullin *et al.*, 2010), and typically provides relatively fewer proteins than bee bread, possibly because of its reduced digestibility or degradation during storage (*e.g.* Hagedorn and Moeller, 1968; Herbert and Shimanuki, 1978; Dietz and Stevenson, 1980; Cremonez *et al.*, 1998).

To make a 100 g paste containing 90% (weight/weight) fresh corbicular pollen with water (Alaux *et al.* 2010), for example:
1. Add 90 g fresh corbicular pollen to suitable sized glass beaker.
2. Add 10 g tap water to the beaker.
3. Knead using gloved fingers or a spatula until a thick paste is created. Consistency should be similar to soft dough, and it should not ooze.
4. Feed to caged workers, or wrap it in aluminium foil within an air-tight container and store for a few days at -20°C until it is needed.

To make a 100 g paste containing 50% (weight/weight) fresh corbicular pollen with 95% (weight/weight) sucrose candy, for example:
1. Create 50 g of 95% (weight/weight) sucrose candy as described in section 7.2.1.3. in a suitably sized glass beaker.
2. Add 50 g fresh corbicular pollen to the beaker.
3. Knead using gloved fingers or a spatula until a thick paste is created. Consistency should be similar to soft dough, and it should not ooze.
4. Feed to caged workers, or wrap it in aluminium foil and store for a few days at -20°C until it is needed.

7.3.1.2.1. Collecting and storing corbicular pollen to feed to caged adult workers in the laboratory

Based on storage methods described by Pernal and Currie (2000) that successfully maintained honey bee-collected pollen pellets for up to one year without decreasing its nutritional value, the following procedure allows for proper collection and storage of fresh pollen for at least a single field season.

1. Identify a suitable colony to collect pollen from. Refer to section 4.1 for a brief discussion on choosing source colonies for worker collection because pollen should also be collected from healthy colonies.
2. Install a thoroughly cleaned trap (see section 5.2.1. for cleaning equipment using acetone) to collect pollen from incoming foragers sporadically over the course of a few weeks, rather than continuously for more than two days at a time, to ensure colony pollen supplies remains sufficient (see the *BEEBOOK* paper on miscellaneous methods by Human *et al.* (2013)).
3. Carefully separate pollen from other trap debris (i.e. plant material, honey bee body parts) using sterile forceps or a small fine-tipped paint brushe.
4. Separate a subsample of each pollen species based on colour (e.g. Moore and Webb, 1983), and store at −18°C or colder to allow for possible future identification of plant species if needed (see the *BEEBOOK* paper on pollination methods by Delaplane *et al.* (2013)).
5. Homogenise collected pollen to ensure uniform distribution of colony-specific pollen, and store it fresh in air-tight containers at −18°C or colder. Minimize or evacuate air in storage containers.
6. Remove from cold storage only when needed and prepare for feeding as discussed in section 7.3.1.2.

7.3.1.3. Providing pollen substitutes to caged adult workers in the laboratory

Pollen substitutes are artificial diets that do not contain pollen, but rather protein from, for example, soybean, brewer's yeast, milk, or algae (Brodschneider and Crailsheim, 2010). Much like sucrose solution as an artificial source of carbohydrates, these substitutes should contain no honey bee-related pathogens, few chemical residues, and can be more easily standardised among laboratories, especially when purchased from a commercial manufacturer that has strict quality assurance practices.

Both self-made, such as soybean and corn meal patties (e.g. van der Steen, 2007; Ellis and Hayes, 2009), as well as commercially produced substitutes containing essential amino acids, such as Bee-Pro® and Ultra Bee® (Mann Lake Ltd.; Hackensack, USA), Feed-Bee® (Bee Processing Enterprises Ltd; Scarborough, Canada), and MegaBee® (S.A.F.E. R&D; USA) can provide proteins, and possibly other nutrients and vitamins, required by honey bees (e.g. Cremonez *et al.*, 1998; De Jong *et al.*, 2009; Brodschneider and Crailsheim, 2010). Care must

Table 1. Pollen substitute composition from van der Steen (2007).

Component	Proportion of total mass
Soya flower (degreased)	0.143
Beer yeast flour	0.095
Calcium caseinate flour (milk protein 90%)	0.152
Whey protein flour (milk protein 80%)	0.038
Sucrose solution (50% (weight/volume) in tap water)	0.476
Linseed oil	0.095

be taken because, for example, even soybean flour formulations can vary widely, and ingredients may not be ubiquitously available (Cremonez *et al.*, 1998). Although various homemade recipes exist, the following soy-based pollen substitute was readily consumed by colony honey bees and promoted individual longevity (van der Steen, 2007) (Table 1); however, nutrition tests on caged workers are required.

7.3.2. Feeding devices for providing proteins to caged adult workers in the laboratory

Similar to sugar solution feeding devices, multiple methods exist for providing protein to workers, and the minimum criteria for protein feeding devices are similar to those required for carbohydrate feeding devices. Disposable plastic trays provide the easiest route for providing protein, and can be created by cutting plastic tubes in half to resemble a trough used for feeding livestock that can simply be inserted into cages from the exterior (Fig. 21). Alternatively, a feeder can be created by removing the lower 8 mm tip of a 1.5 ml microcentrifuge tube to reveal a 6-7 mm diameter hole (Fig. 22). This allows workers to enter the feeder and eat the protein upwards. Care must be taken that the protein paste does not leak out the bottom when exposed to incubator conditions (section 6).

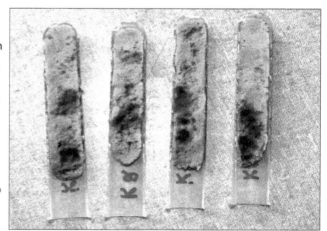

Fig. 21. Protein paste provided to honey bees in 10-ml plastic test tubes cut in half longitudinally. The dark orange-brown areas were moistened by workers during 24 hours in a hoarding cage.

Fig. 22. Protein paste provided in a 1.5 ml micro-centrifuge tube with its base removed to reveal a 6-7 mm diameter hole that allows workers to enter the feeder to consume protein.

7.3.3. Measuring protein consumption by caged adult workers in the laboratory

Consumption can be measured by weighing remaining food, and similar to carbohydrates, should be adjusted to calculate amount consumed per honey bee per 24 hours as detailed in section 7.2.3. It may also be appropriate to calculate quantity of protein consumed, rather than total mass of food stuff providing the protein. This can be determined when the proportion of protein in the food stuff is known. Consult section 7.7 to correct for mass of food stuff lost through evaporation.

7.3.4. Replenishing proteins provided to caged adult workers in the laboratory

Similar to carbohydrates, care must be taken when replenishing proteins to avoid harming caged workers. Feeding pre-trials should be performed to determine quantity needed to ensure workers are fed *ad libitum*. Daily worker consumption should not exceed 3 mg protein; therefore, 3 g of protein paste, at least made from corbicular pollen pellets, should be sufficient to meet daily needs of 100 caged workers. Protein should be replaced at least every three days to prevent drying and microbial growth.

7.3.5. Recommendations for providing proteins to caged adult workers in the laboratory

Under natural conditions, adult workers meet the majority of their protein needs by consuming bee bread within 10 days of emergence (Crailsheim *et al.*, 1992). This protein is vital for proper gland and tissue development, such as the hypopharyngeal and wax glands, flight muscles, and fat bodies (Maurizio, 1959), and consuming it can extend worker longevity beyond that of individuals which only receive carbohydrates (Schmidt *et al.*, 1987). Although caged workers can survive extended intervals on carbohydrates alone, providing proteins

is recommended when newly emerged or intra-hive workers of an undefined age are caged (see sections 4.2 and 4.3.4 for instructions on how to collect newly emerged and intra-hive workers for laboratory experiments). Protein is not required when flying workers are collected and maintained in the laboratory because they are likely greater than 10 days old and have therefore met their protein consumption demands (Winston, 1987).

Currently we cannot recommend one specific source of protein to provide to caged workers due to lack of data. Multi-floral beebread and corbicular pollen as described previously (sections 7.3.1.1 and 7.3.1.2, respectively) is sufficient for providing proteins as long as it contains minimal pathogens or environmental contaminants. This can be accomplished by sterilising bee products (section 7.6) and collecting from multiple colonies located in non-intensive agricultural areas or from those certified as organic. These multiple colonies ensure that the same, florally diverse pollen is provided to all workers during an entire experiment. Section 4.1 discusses how to select appropriate colonies to collect workers from; similar insights can be used towards the collection of pollen. Alternatively, inexpensive and nutritious pollen substitutes (section 7.3.1.2) that are subject to rigid quality control are ubiquitously available, and may provide a more standardised, sterile protein source to caged workers. Future studies should explore their use, especially those that are fermented by micro-organisms like bee bread to aid their preservation (Ellis and Hayes, 2009).

When used, protein can be provided *ad libitum* using feeders as discussed previously (section 7.3.2), and replaced at least every three days (section 7.3.4). Quality of protein (e.g. nutrition, contamination) should always be considered (see section 7.6 for food sterilisation).

7.4. Lipids, minerals, and vitamins

The importance of lipids, minerals, and vitamins for brood-rearing in a colony is well-known, whereas, in adults it is not (Haydak, 1970; Brodschneider and Crailsheim, 2010). It is likely that reserves stored in the body during development may be used during adulthood (Maurizio, 1959; Haydak, 1970). Honey bees typically receive these nutrients when consuming bee bread (Brodschneider and Crailsheim, 2010), although many protein substitutes can also contain lipids, minerals, and vitamins. Additionally, soluble vitamins of known concentrations can be added to sugar solution, and protein patties or other formulations can be supplemented with lipids, vitamins, and minerals (Herbert *et al.*, 1980; 1985; Herbert and Shimanuki, 1978). Little information is available on this subject regarding caged honey bees. More research is needed to better understand effects of lipids, minerals, and vitamins on caged workers, and to determine if they should be provided to individuals as a standard to promote honey bee health in the laboratory. Currently, we recommend to not provide lipid, mineral, and vitamin supplements to caged individuals.

7.5. Water

Water is needed for metabolism and cooling, and is generally obtained by caged workers during ingestion of sugar solutions. In nature, water can also act as an important source of minerals (Brodschneider and Crailsheim, 2010), which can be highly variable depending upon source (WHO, 2005). In North America, for example, tap water provides important sources of calcium, magnesium, and sodium, at least for humans (Azoulay *et al.*, 2001). It is not known how these differences may affect caged workers. Water is essential for maintaining worker honey bees in the laboratory. Carbohydrate solutions containing ≥ 50% (weight/volume) water are sufficient for hydration; if any less is provided, or if only sucrose paste is given, then a separate feeder containing tap water must be offered. Pre-trials for testing feeder leakage may be necessary due to the lower viscosity of water than sucrose solution. Tap water can be boiled to kill harmful micro-organisms, but it should be allowed to return to room temperature before it is given to caged workers.

7.6. Food sterilisation and detoxification

7.6.1. Pathogens and environmental contaminants found in bee products

All bee products, including honey, corbicular pollen, and bee bread, can contain pathogens, environmental contaminants, and agro-chemical residues (e.g. Bromenshenk *et al.*, 1985; Higes *et al.*, 2008; Chauzat *et al.*, 2009; Mullin *et al.*, 2010). A number of methods are available for sanitation of bee products.

7.6.2. Sterilising bee products to destroy pathogens

Bee products can be sterilised to kill pathogens using radiation and temperature treatments.

7.6.2.1. Sterilising bee products to destroy pathogens using radiation

Radiation generally does not alter physiochemical properties of nutrients (Yook *et al.*, 1998) when the appropriate dosage (i.e. treatment intensity and length) is provided (Undeen and Vander Meer, 1990). Greater than 2 kGy of gamma radiation from cobalt[60] destroyed *N. apis* spores (Katznelson and Robb, 1962), 500 Gy gamma radiation from a caesium[137] irradiator damaged developmental stages of *N. apis* (Liu *et al.*, 1990), and 10 kGy of high velocity electron-beam radiation effectively sterilised spores of the bacteria *Paenibacillus larvae* and the fungus *Ascophaera apis*, responsible for American foulbrood and chalkbrood disease, respectively (Melathopoulos *et al.*, 2004). Although 3.8 J/cm^2 of 254 nm ultraviolet radiation can reduce viability of *Nosema algerae* spores from moths (Undeen and Vander Meer, 1990), it can also degrade nutrients such as fatty acids (Yook *et al.*, 1998) and may not kill all organisms because the entire food stuff was not penetrated.

For pathogen control, the United States Department of Agriculture currently permits a number of fresh or frozen foods destined for human consumption to be irradiated up to a maximum of 5.5 kGy; dried food may be irradiated up to 30 kGy (USDA, 2008).

7.6.2.2. Sterilising bee products to destroy pathogens using temperature

Temperature treatment can be used to sterilise food stuffs; however, nutrient degradation may occur (Barajas *et al.*, 2012). For example, heat treating *N. apis* spores at 49°C for 24 hours will result in their destruction; whereas, freezing *N. ceranae* at -18°C for one week significantly reduces numbers of infective spores (Fries, 2010). Heating honey greater than 49°C should be performed with caution due to the possible production of dangerous levels of toxic hydroxymethylfurfural (HMF) (Brodschneider and Crailsheim, 2010).

7.6.3. Detoxifying bee products to destroy chemicals

Chemicals can be degraded by various methods, such as radiation and temperature treatments; however, rates of degradation vary tremendously depending on compound chemistry, and break down products produced during degradation can also be dangerous to honey bees. Currently, little is known about degradation of chemicals relevant to honey bee health, particularly those in food stuffs.

7.6.4. Recommendations for sterilising and detoxifying bee products fed to caged adult workers in the laboratory

Development of specific protocols to sterilise and detoxify food made from bee products against a broad range of pathogens and environmental contaminants is urgently required. Until then, use of non-honey bee products (sections 7.2.1.2, 7.2.1.3, 7.3.1.3) provide a relatively effective, safe, and standardised approach to supplying food to honey bees. If bee products are fed to caged workers, those products collected from colonies in non-intensive agricultural areas, or from colonies certified as organic, provides a good alternative because they will contain limited chemicals residues and can be sterilised using radiation to kill pathogens.

7.7. Controlling for water evaporation from food provided to caged adult workers in the laboratory

Food consumption is determined by calculating the difference between food provided and food remaining (sections 7.2.3 and 7.3.3). In most cases, evaporation does not need to be considered because all experimental variables should be conserved among treatment groups except for the variable of interest, thereby creating systematic conservative errors among cages.

If water loss from both carbohydrate or protein diets needs to be measured during the course of a study, it can be calculated:

1. Prepare three 'mock' cages (MOCK$_{CAGE}$) cages in the same incubator used to hold experimental cages for food of interest (i.e.

carbohydrates or protein). All conditions for these evaporative control cages should be identical to experimental cages (e.g. type of food provided, frequency of food replacement, type of cage used, incubator maintained in, etc.).

2. Within each MOCK$_{CAGE}$, one 'mock' feeder (MOCK$_{FEEDER}$) should be protected by a breathable mesh; whereas, a second feeder is only provided to allow workers to feed.

3. For each MOCK$_{CAGE}$, calculate average daily mass reduction of MOCK$_{FEEDER}$ feeding device per worker (MOCK$_{BEE}$) according to methods described in sections 7.2.3 and 7.3.3 for measuring carbohydrate consumption.

4. Determine average daily mass reduction per worker among all three cages (MOCK$_{TOTAL}$) using the three MOCK$_{BEE}$ values.

5. Determine loss via evaporation by subtracting MOCK$_{TOTAL}$ from daily per worker food consumed per cage (DAILYWORKER) as determined according to sections 7.2.3 and 7.3.3 for all experi mental cages of interest; negligible negative adjusted con sumptions should be set to zero.

Interestingly, licking or moistening of protein patties by honey bees can adulterate consumption (Fig. 21). At this time, we do not know how to include this behaviour into calculations of food consumption, although it likely has little influence.

7.8. Feeding tests using caged adult workers in the laboratory

Some investigations (e.g. nutrition, toxicology, virology and nosema studies) require that workers receive experimental treatments orally. Typically, the test substance is mixed with food, such as 50% (weight/volume) sucrose solution (section 7.2.1.2). For workers, typical quantities of sucrose solution consumed in nature in a short interval is 50 µl (Seeley, 1994), whereas the honey stomach of drones usually can contain approximately 30 µl (Hoffmann, 1966).

7.8.1. Starving caged adult workers in the laboratory prior to performing a feeding test

Workers are usually starved to ensure that the entire oral treatment is consumed within a short time. So far, no commonly accepted method for starving individuals prior to oral application of a treatment exists; however, within fields of study there are some consistencies. For example, most experimental laboratory investigations of *Nosema* starve groups of young workers for two to four hours (e.g. Fries *et al.*, 1992; Malone and Stefanovic, 1999; Higes *et al.*, 2007; Maistrello *et al.*, 2008); starvation for this length is also recommended by Fries *et al.* (2013) in the *BEEBOOK* paper describing methods used to study nosema in honey bees. Similarly, up to two hours of starvation is recommended for acute, oral toxicity experiments (OECD, 1998). Amount of food in the honey stomach will no doubt influence length of required starvation time, and resilience to starvation will likely depend on type of collected

workers (e.g. age, health, etc.), and if they were starved individually or in a group.

Future studies should investigate effects of both short and long-term starvation on honey bees, in addition to the influence of honey bee condition, age, and subspecies. Generally, groups of adult workers should be starved for no more than four hours to ensure rapid consumption of a test substance. Workers starved in isolation should be without food for less time – no longer than two hours. Individuals starved for any longer are more likely to be injured or to die. Starved honey bees that do become impaired (e.g. behaviourally) or that exhibit other unusual signs should be discarded from experimental studies. Pre-trials will determine the minimum length of starvation period needed to consistently induce feeding of entire food treatment quickly.

7.8.2. Feeding a liquid test substance to individual adult workers in the laboratory

Individual feeding is used when specific, known quantities of test substance are required to be ingested by individual workers. Although precise, individual feeding can be extremely time-consuming and may inadvertently limit sample size.

The easiest way to orally feed workers liquid test substances individually is to provide a micropipette filled with a known quantity of test substance to an individual as detailed below. A specific quantity (i.e. the same volume for each experimental worker) between 3-10 µl should be provided. This will ensure that all workers can easily consume the same volume of homogeneously mixed test substance.

1. Because some workers may not feed, it is appropriate to starve more individuals than will be required for the experiments. Pre-trials testing starvation times and test substance consumption will help determine how many workers will be needed.

2. Remove a starved individual worker from its cage using a forceps by gently grasping a leg. Refer to Human *et al.*, (2013) in the *BEEBOOK* paper discussing miscellaneous methods for details on how to handle adult honey bees.

3. Gently grasp the wings together at their base using the thumb and index finger so that her mouthparts are exposed (i.e. wings facing down) and her stinger is pointing away from your body (Fig. 23).

4. Vortex the food test substance for 5 seconds.

5. Feed a specific volume (i.e. a volume between 3-10 µl) of liquid test substance to the worker using a micropipette, which allows for a precise volume to be administered. Place the end of the loaded pipette tip in front of the individual's mouthparts or beneath the mandibles in front of the maxillae and create a small droplet at the open end of the pipette tip to promote feeding (Fig. 23). Additionally, the pipette tip can be gently placed against an antenna when the honey bee is reluctant to feed.

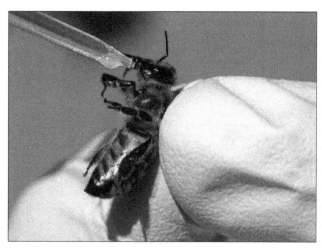

Fig. 23. A worker honey bee being individually fed using a micropipette. Note that the individual is held by gently squeezing its wings between the index finger and thumb, and that the distal part of the abdomen is pointed in such a way that the honey bee cannot sting the handler.

6. Provide the remaining test substance by depressing the pipette plunger gently to ensure that nothing spills when the individual begins to feed. Discard the individual and start over using a new worker if she does not consume all of the test substance within one minute.

7. Place the fed worker in an appropriate cage (section 5) with food (section 7) maintained under proper incubator conditions (section 6).

8. If needed, isolate the worker for 20-30 minutes to observe regurgitation or to ensure that none of the test substance is passed to another worker via trophallaxis. Isolation is not required when individually feeding queens and drones because they are not food providers (Crailsheim, 1998), and therefore will not discard the test substance to another individual. Orally transmitted pathogens take fewer than 15 minutes to enter the ventriculus after ingestion (Kellner and Jacobs, 1978; Verbeke *et al.,* 1984).

Although typically less efficient, individuals withheld in isolation cages, such those shown in Figs 17 and 18, can also be fed a test substance using a micropipette, and may minimize the handling of honey bees.

7.8.3. Feeding a liquid test substance to groups of caged adult workers in the laboratory

In contrast to feeding a liquid test substance to individuals, group-feeding has fewer logistic and time constraints. It mimics consumption and transfer of food among honey bees in a colony via trophallaxis because food is typically consumed by only a small proportion of workers but ultimately shared among nearly all worker nest-mates within 24 hours (Nixon and Ribbands, 1952; Crailsheim, 1998).

Although not well studied, the primary disadvantage of group

feeding a test substance is its potential unequal distribution among individuals over time (Furgala and Maunder, 1961). Although many factors may influence food consumption, such as parasitism (Mayack and Naug, 2009), quantity of honey bees and level of starvation will most importantly dictate volume of test substance to provide. Generally, ten workers can consume 100 - 200 µl of 50% sucrose solution in 3-4 hours (OECD, 1998), or at an individual rate of 2.5 - 6.6 µl per hour. To group feed workers a known quantity of inoculum, a single feeding device containing a minimal amount of test substance should be used to ensure all contents are consumed in a timely fashion. Generally, the entire volume should be consumed in less than 24 hours, but exact duration of consumption should be determined by the specific experiment. The test substance should be replaced with standard food when the total volume of the test substance is ingested; constant vigilance is required to ensure that workers do not go without food. One should assume equal consumption by all caged workers when determining ingestion of test substances. For example, 1,000,000 *N. ceranae* spores are required to inoculate 30 workers with 33,333 spores each.

7.8.4. Feeding a solid test substance to groups of adult workers caged in the laboratory

Test substances can also be provided orally to a group of caged workers in a solid form, such as in 95% (weight/weight) sucrose paste (section 7.2.1.3). See section 7.8.3. for a discussion on feeding a group of workers in a single cage. Ensure the solid test substance is well mixed and homogeneous, and perform a pre-trial to determine how much sucrose paste is required.

7.8.5. Recommendations and considerations for oral exposure of a test substance to caged adult workers in the laboratory

Choice of whether to inoculate workers individually or as a group will mainly depend upon the particular experiment. Few investigations have compared results from individual versus group feeding of a test substance, although it is clear that both can be effective. For example, Tanner *et al.* (2012a) demonstrated no difference in individual versus group feeding of intra-host *N. ceranae* spore development 14 days post-inoculation. Research should also examine homogeneity of test substances, especially suspensions containing particles such as *Nosema* spores that may settle in liquid. In these cases, a test substance fed as a solid may ensure a more even distribution of particles.

8. Queens and drones

So far we have discussed how to properly maintain worker honey bees under *in vitro* laboratory conditions, mainly ignoring queens and drones. Because workers are generally required to provide food via trophallaxis to both of these groups, many of the methods described above, such as incubator conditions, cage designs, and nutrition, are also valid for maintaining adult queens and drones in the laboratory. When choosing worker attendants, researchers must also consider that workers can horizontally transmit pathogens to both drones and queens (e.g. Higes *et al.*, 2009).

8.1. Maintaining queens under *in vitro* laboratory conditions

Adult queens can be maintained safely in the laboratory when kept in cages with workers collected from brood frames from the same source colony as the queen. For up to five days, a queen can be placed in a standard queen cage provided with sucrose candy *ad libitum* and four to seven workers (Fig. 24); however, for longer intervals a queen should be maintained with at least 10 workers in a standard worker hoarding cage as discussed in section 5. To obtain and maintain virgin queens in the laboratory, a cell from which a queen is expected to emerge from within two to four days can be placed in a hoarding cage with workers (Alaux *et al.*, 2011b) under appropriate incubation conditions described in section 6. When performing experiments, it is important for researchers to consider nutrients that should be provided to caged queens and workers because of the importance of protein to tissue

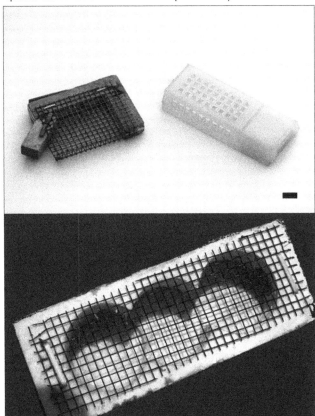

Fig. 24. Standard queen cages suitable for maintaining a queen and approximately five nurse worker honey bees safely in a growth chamber for up to five days when sucrose candy is provided *ad libitum*. Top and bottom images not equal in scale; black lines denote ~0.9 cm.

and organ development (*e.g.* Hersch *et al.*, 1978; Pernal and Currie, 2000). More detailed instructions on rearing and maintaining queens can be found in the *BEEBOOK* paper on queen rearing and selection (Büchler *et al.*, 2013).

8.2. Maintaining drones under *in vitro* laboratory conditions

Similar to queens, drones should be maintained in the laboratory with workers collected off brood frames. Preferably, these workers should come from the same colony as the drones to facilitate the latter's acceptance. Additionally, attention must be paid to the type of nutrients provided to caged drones and workers because of the potential importance of protein to development of tissues, including gonads (Jaycox, 1961). Unlike queens, multiple drones can be kept in the same cage, and at a 2:1 drone:worker ratio (Jaycox, 1961; Huang *et al.*, 2012). This will ensure that drones survive at least until they reach maturity, approximately 8-9 days post-emergence (Jaycox, 1961). If caged individuals die during the experiment, one should consider maintaining this drone:worker ratio by adding or removing workers.

Because of the affinity of the parasitic mite *Varroa destructor* to drones, researchers must also consider the influence of parasitism during development when designing experiments. Drones should be maintained in conditions previously recommended for adult workers because they exhibit a similar thermo-preference (Kovac *et al.*, 2009). However, future studies should evaluate alternative temperature and feeding regimes when evaluating drone reproductive traits because of the sensitivity of sperm production. For example, Jaycox (1961) recommended that drones be kept between 31 and 34°C, and suggested that drones can be caged without workers when appropriate feeding devices provide honey rather than sucrose because of drones' difficulty to invert sugars. General methods for maintaining drones more appropriately in the laboratory urgently need development because of their greater sensitivity to *in vitro* conditions (Tanner *et al.*, 2012b).

9. Conclusions and future directions

In this paper we have primarily discussed methods for maintaining adult worker honey bees *in vitro* in the laboratory. The main purpose for providing these recommendations is to promote standardisation of research methods that will facilitate comparison of data generated by different laboratories. Although methods for maintaining adult workers *in vitro* are typically capable of sustaining workers for many weeks, the real issue lies in creating an experimental environment that can produce biologically relevant data. Honey bees are highly social organisms; no doubt placing 30 workers in a cage without a queen will have consequences for their behaviour and physiology. Additionally, proper nutrition is paramount to immune responses of all animals, including honey bees (Alaux *et al.*, 2010), yet the majority of laboratory experiments feed carbohydrates only, ignoring key nutrients honey bees normally consume in the natural environment (Brodschneider and Crailsheim, 2010). Researchers should also assume that laboratory settings provide a relatively stressed environment compared to the colony arena, which has many buffering mechanisms to fend off external threats. Therefore, most studies performed in the laboratory should represent the first step in performing hypothesis-driven research, with further studies carried out under natural "field" conditions.

Ultimately, a laboratory setting can provide an environment in which one can perform controlled investigations using honey bees to test falsifiable hypotheses using appropriate experimental designs. Given the potential influence of the myriad variants in a laboratory, researchers must maintain honey bees under appropriate and repeatable conditions, and should always provide sufficient details about their experiments so that data can be more easily interpreted and compared.

10. Acknowledgements

We acknowledge: the Ricola Foundation - Nature and Culture; COST (Action FA0803); the UK Insect Pollinators Initiative (joint BBSRC, NERC, DEFRA, Scottish Government, Wellcome Trust) grant BB/I000100/1; the EU (FP7 project Bee Doc, FP7-KBBE-2009-03-244956); and the German BLM project FitBee for providing financial support. Additionally, we thank: members of the Swiss Bee Research Centre, especially Jean-Daniel Charrière and Verena Kilchenmann; Marie-Pierre Chauzat (Anses); Alex English (Iotron); and editor Vincent Dietemann for providing fruitful discussions. Two anonymous reviewers provided helpful comments

11. References

ALAUX, C; BRUNET, J-L; DUSSAUBAT, C; MONDET, F; TCHAMITCHAN, S; COUSIN, M; BRILLARD, J; BALDY, A; BELZUNCES, L P; LE CONTE, Y (2009) Interactions between *Nosema* microspores and a neonicotinoid weaken honey bees (*Apis mellifera*). *Environmental Microbiology* 12(3): 774-782. http://dx.doi.org/10.1111/j.1462-2920.2009.02123.x

ALAUX, C; DUCLOZ, F; CRAUSER, D; LE CONTE, Y (2010) Diet effects on honey bee immunocompetence. *Biology Letters* 6(4): 562-565. http://dx.doi.org/10.1098/rsbl.2009.0986

ALAUX, C; DANTEC, C; PARRINELLO, H; LE CONTE, Y (2011a) Nutrigenomics in honey bees: digital gene expression analysis of pollen's nutritive effects on healthy and varroa-parasitized bees. *BMC Genomics* 12: 496. http://dx.doi.org/10.1186/1471-2164-12-496

ALAUX, C; FOLSCHWEILLER, M; MCDONNELL, C; BESLAYA, D; COUSIN, M; DUSSAUBAT, C; BRUNET, J-L; LE CONTE, Y (2011b) Pathological effects of the microsporidium *Nosema ceranae* on honey bee queen physiology (*Apis mellifera*). *Journal of Invertebrate Pathology* 106(3): 380-385. http://dx.doi.org/10.1016/j.jip.2010.12.005

ALLEN, M D (1959) Respiration rates of worker honey bees of different ages and at different temperatures. *Journal of Experimental Biology* 36(1): 92-101.

ALLEN, M D (1960) The honey bee queen and her attendants. *Animal Behaviour* 8(3/4): 201-208. http://dx.doi.org/10.1016/0003-3472(60)90028-2

ALTAYE, S Z; PIRK, C W W; CREWE, R M; NICOLSON, S W (2010) Convergence of carbohydrate-biased intake targets in caged worker honey bees fed different protein sources. *The Journal of Experimental Biology* 213(19): 3311-3318. http://dx.doi.org/10.1242/jeb.046953

AMDAM, F V; HARTFELDER, K; NORBERG, K; HAGEN, A; OMHOLT, S W (2004) Altered physiology in worker honey bees (Hymenoptera: Apidae) infested with the mite *Varroa destructor* (Acari: Varroidae): a factor in colony loss during overwintering? *Journal of Economic Entomology* 97(3): 741-747. http://dx.doi.org/10.1603/0022-0493(2004)097[0741:APIWHB]2.0.CO;2

AMDAM, G V; AASE, A L T O; SEEHUUS, S-C; FONDRK, M K; NORBERG, K; HARTFELDER, K (2005a) Social reversal of immunosenescence in honey bee workers. *Experimental Gerontology* 40(12): 939-947. http://dx.doi.org/10.1016/j.exger.2005.08.004

AMDAM, G V; NORBERG, K; OMHOLT, S W; KRYGER, P; LOURENÇO, A P; BITONDI, M M G; SIMÕES, Z L P (2005b). Higher vitellogenin concentrations in honey bee workers may be an adaptation to life in temperate climates. *Insectes Sociaux* 52(4): 316-319. http://dx.doi.org/10.1007/s00040-005-0812-2

ANDERSON, D (2013) Standard methods for *Tropilaelaps* mites research. In *V Dietemann; J D Ellis; P Neumann (Eds) The COLOSS BEEBOOK, Volume II: standard methods for* Apis mellifera *pest and pathogen research. Journal of Apicultural Research* 52(4): http://dx.doi.org/10.3896/IBRA.1.52.4.21

ARNOLD, P G (1978) Les variations annuelles dans l'effet groupe chez l'abeille et l'origine de la mort précoce des isolées. *Insectes Sociaux* 25(1): 39-51.

AZOULAY, A; GARZON, P; EISENBERG, M J (2001) Comparison of the mineral content of tap water and bottled waters. *Journal of General Internal Medicine* 16(3): 168-175. http://dx.doi.org/10.1111/j.1525-1497.2001.04189.x

BAKONYI, T; DERAKHSHIFAR, I; GRABENSTEINER, E; NOWOTNY, N (2003) Development and evaluation of PCR assays for the detection of *Paenibacillus larvae* in honey samples: comparison with isolation and biochemical characterization. *Applied and Environmental Microbiology* 69(3): 1504-1510. http://dx.doi.org/10.1128/AEM.69.3.1504–1510.2003

BARAJAS, J; CORTES-RODRIGUEZ, M; RODRÍGUEZ-SANDOVAL, E (2012) Effect of temperature on the drying process of bee pollen from two zones of Colombia. *Journal of Food Process Engineering* 35(1): 134-148. http://dx.doi.org/10.1111/j.1745-4530.2010.00577.x

BARKER, R J; LEHNER, Y (1974) Acceptance and sustenance value of naturally occurring sugars fed to newly emerged adult workers of honey bees (*Apis mellifera* L.). *Journal of Experimental Zoology* 187(2): 277-285. http://dx.doi.org/10.1002/jez.1401870211

BARKER, R J; LEHNER, Y (1978) Laboratory comparison of high fructose corn syrup, grape syrup, honey, and sucrose syrup as maintenance food for caged honey bees. *Apidologie* 9(2): 111-116. http://dx.doi.org/10.1051/apido:19780203

BEN-SHAHARI, Y; LEUNG, H T; PAK, W L; SOKOLOWSKI, M B; ROBINSON, G E (2003) cGMP-dependent changes in phototaxis: a possible role for the *foraging* gene in honey bee division of labor. *Journal of Experimental Biology* 206(14): 2507-2515. http://dx.doi.org/10.1242/jeb.00442

BESHERS, S N; HUANG, Z Y; OONO, Y; ROBINSON, G E (2001) Social inhibition and the regulation of temporal polyethism in honey bees. *Journal of Theoretical Biology* 213(3): 461-479. http://dx.doi.org/10.1006/jtbi.2001.2427

BEUTLER, R; OPFINGER, E (1948) Pollenernahrung und Nosemabefall der Honigbiene. *Naturwissenshaften* 35(9): 288-288.

BOLKER, B M; BROOKS, M E; CLARK, C J; GEANGE, S W; POULSEN, J R; STEVENS, M H H; WHITE, J S S (2009) Generalized linear mixed models: a practical guide for ecology and evolution. *Trends in Ecology and Evolution* 24(3): 127-135. http://dx.doi.org/10.1016/j.tree.2008.10.008

BREED, M D; GUZMÁN-NOVOA, E; HUNT, G J (2004) Defensive behaviour of honey bees: organization, genetics, and comparisons with other bees. *Annual Review of Entomology* 49: 271-298. http://dx.doi.org/10.1146/annurev.ento.49.061802.123155

BRODSCHNEIDER, R; CRAILSHEIM, K (2010) Nutrition and health in honey bees. *Apidologie* 41(3): 278-294. http://dx.doi.org/10.1051/apido/2010012

BROMENSHENK, J J; CARLSON, S R; SIMPSON, J C; THOMAS, J M (1985) Pollution monitoring of Puget Sound with honey bees. *Science* 227 (4687): 632-634. http://dx.doi.org/10.1126/science.227.4687.632

BÜCHLER, R; ANDONOV, S; BIENEFELD, K; COSTA, C; HATJINA, F; KEZIC, N; KRYGER, P; SPIVAK, M; UZUNOV, A; WILDE, J (2013) Standard methods for rearing and selection of *Apis mellifera* queens. In *V Dietemann; J D Ellis; P Neumann (Eds) The COLOSS BEEBOOK, Volume I: standard methods for* Apis mellifera *research. Journal of Apicultural Research* 52(1): http://dx.doi.org/10.3896/IBRA.1.52.1.07

BURRILL, R M; DIETZ, A (1981) The response of honey bees to variations in solar radiation and temperature. *Apidologie* 12(4): 319-328. http://dx.doi.org/10.1051/apido:19810402

CALDERONE, N W (1995) Temporal division-of-labor in the honey-bee, *Apis mellifera*: a developmental process or the result of environmental influences? *Canadian Journal of Zoology* 73(8): 1410-1416. http://dx.doi.org/10.1139/z95-166

CHAUZAT, M-P; CARPENTIER, P; MARTEL, A-C; BOUGEARD, S; COUGOULE, N; PORTA, P; LACHAIZE, J; MADEC, F; AUBERT, M; FAUCON, J-P (2009) Influence of pesticide residues on honey bee (Hymenoptera: Apidae) colony health in France. *Environmental Entomology* 38(3): 514-523.

CORONA, M; VELARDE, R A; REMOLINA, S; MORAN-LAUTER, A; WANG, Y; HUGHES, K A; ROBINSON, G E (2007) Vitellogenin, juvenile hormone, insulin signaling, and queen honey bee longevity. *Proceedings of the National Academy of Sciences* 104(17): 7128-7133. http://dx.doi.org/10.1073/pnas.0701909104

CRAILSHEIM, K (1990) The protein balance of the honey bee worker. *Apidologie* 21(5): 417-429. http://dx.doi.org/10.1051/apido:19900504

CRAILSHEIM, K (1998) Trophallactic interactions in the adult honey bee (*Apis mellifera* L.). *Apidologie* 29(1-2): 97-112. http://dx.doi.org/10.1051/apido:19980106

CRAILSHEIM, K; SCHNEIDER, L H W; HRASSNIGG, H; BUHLMANN, G; BROSCH, U; GMEINBAUER, R; SCHOFFMANN, B (1992) Pollen consumption and utilization in worker honey bees (*Apis mellifera carnica*) - dependence on individual age and function. *Journal of Insect Physiology* 38(6): 409-419. http://dx.doi.org/10.1016/0022-1910(92)90117-V

CRAILSHEIM, K; BRODSCHNEIDER, R; AUPINEL, P; BEHRENS, D; GENERSCH, E; VOLLMANN, J; RIESSBERGER-GALLÉ, U (2013) Standard methods for artificial rearing of *Apis mellifera* larvae. In *V Dietemann; J D Ellis; P Neumann (Eds) The COLOSS BEEBOOK, Volume I: standard methods for* Apis mellifera *research. Journal of Apicultural Research* 52(1): http://dx.doi.org/10.3896/IBRA.1.52.1.05

CRAWLEY, M J (2005) *The R book*. John Wiley & Sons; Chichester, UK. 942 pp.

CREMONEZ, T M; DE JONG, D; BITONDI, M M G (1998) Quantification of haemolymph proteins as a fast method for testing protein diets for honey bees (Hymenoptera: Apidae). *Journal of Economic Entomology* 91(6): 1284-1289.

CZEKOŃSKA, K (2007) Influence of carbon dioxide on *Nosema apis* infection of honey bees (*Apis mellifera*). *Journal of Invertebrate Pathology* 95(2): 84-86. http://dx.doi.org/10.1016/j.jip.2007.02.001

DANKS, H V (2002) Modification of adverse conditions by insects. *Oikos* 99(1): 10-24. http://dx.doi.org/10.1034/j.1600-0706.2002.990102.x

DE GRAAF, D C; ALIPPI, A M; ANTÚNEZ, K; ARONSTEIN, K A; BUDGE, G; DE KOKER, D; DE SMET, L; DINGMAN, D W; EVANS, J D; FOSTER, L J; FÜNFHAUS, A; GARCIA-GONZALEZ, E; GREGORC, A; HUMAN, H; MURRAY, K D; NGUYEN, B K; POPPINGA, L; SPIVAK, M; VANENGELSDORP, D; WILKINS, S; GENERSCH, E (2013) Standard methods for American foulbrood research. In *V Dietemann; J D Ellis; P Neumann (Eds) The COLOSS BEEBOOK, Volume II: standard methods for* Apis mellifera *pest and pathogen research. Journal of Apicultural Research* 52(1): http://dx.doi.org/10.3896/IBRA.1.52.1.11

DEGRANDI-HOFFMAN, G; CHEN, Y; HUANG, E; HUANG, M H (2010) The effect of diet on protein concentration, hypopharyngeal gland development and virus load in worker honey bees (*Apis mellifera* L.). *Journal of Insect Physiology* 56(9): 1184-1191. http://dx.doi.org/10.1016/j.jinsphys.2010.03.017

DEGRANDI-HOFFMAN, G; ECKHOLM, B J; HUANG, M H (2012) A comparison of bee bread made by Africanized and European honey bees (*Apis mellifera*) and its effects on hemolymph protein titres. *Apidologie* (in press). http://dx.doi.org/10.1007/s13592-012-0154-9

DE JONG, D; DA SILVA, E J; KEVAN, P G; ATKINSON, J L (2009) Pollen substitutes increase honey bee haemolymph protein levels as much or as more than does pollen. *Journal of Apicultural Research* 48(1): 34-37. http://dx.doi.org/10.3896/IBRA.1.48.1.08

DELAPLANE, K S; VAN DER STEEN, J; GUZMAN, E (2013) Standard methods for estimating strength parameters of *Apis mellifera* colonies. In *V Dietemann; J D Ellis; P Neumann (Eds) The COLOSS* BEEBOOK, *Volume I: standard methods for* Apis mellifera *research. Journal of Apicultural Research* 52(1): http://dx.doi.org/10.3896/IBRA.1.52.1.03

DELAPLANE, K S; DAG, A; DANKA, R G; FREITAS, B M; GOODWIN, M (2013) Standard methods for pollination research in *Apis mellifera*. In *V Dietemann; J D Ellis; P Neumann (Eds) The COLOSS* BEEBOOK, *Volume I: standard methods for* Apis mellifera *research. Journal of Apicultural Research* 52(4):http://dx.doi.org/10.3896/IBRA.1.52.4.12

DE MIRANDA, J R; BAILEY, L; BALL, B V; BLANCHARD, P; BUDGE, G; CHEJANOVSKY, N; CHEN, Y-P; VAN DOOREMALEN, C; GAUTHIER, L; GENERSCH, E; DE GRAAF, D; KRAMER, M; RIBIÈRE, M; RYABOV, E; DE SMET, L VAN DER STEEN, J J M (2013) Standard methods for virus research in *Apis mellifera*. In *V Dietemann; J D Ellis; P Neumann (Eds) The COLOSS* BEEBOOK, *Volume II: standard methods for* Apis mellifera *pest and pathogen research. Journal of Apicultural Research* 52(4): http://dx.doi.org/10.3896/IBRA.1.52.4.22

DIETEMANN, V; NAZZI, F; MARTIN, S J; ANDERSON, D; LOCKE, B; DELAPLANE, K S; WAUQUIEZ, Q; TANNAHILL, C; FREY, E; ZIEGELMANN, B; ROSENKRANZ, P; ELLIS, J D (2013) Standard methods for varroa research. In *V Dietemann; J D Ellis; P Neumann (Eds) The COLOSS* BEEBOOK, *Volume II: standard methods for* Apis mellifera *pest and pathogen research. Journal of Apicultural Research* 52(1): http://dx.doi.org/10.3896/IBRA.1.52.1.09

DIETZ, A; STEVENSON, H R (1980) Influence of long term storage on the nutritional value of frozen pollen for brood rearing of honey bees. *Apidologie* 11(2): 143-151. http://dx.doi.org/10.1051/apido:19800204

ELLIS, A M; HAYES, JR, G W (2009) An evaluation of fresh versus fermented diets for honey bees (*Apis mellifera*). *Journal of Apicultural Research* 48(3): 215-216. http://dx.doi.org/10.3896/IBRA.1.48.3.11

ELLIS, J D; GRAHAM, J R; MORTENSEN, A (2013) Standard methods for wax moth research. In *V Dietemann; J D Ellis; P Neumann (Eds) The COLOSS* BEEBOOK, *Volume II: standard methods for* Apis mellifera *pest and pathogen research. Journal of Apicultural Research* 52(1): http://dx.doi.org/10.3896/IBRA.1.52.1.10

ELLIS, M B; NICOLSON, S W; CREWE, R M; DIETEMANN, V (2008) Hygropreference and brood care in the honey bee (*Apis mellifera*). *Journal of Insect Physiology*, 54(12): 1516-1521. http://dx.doi.org/10.1016/j.jinsphys.2008.08.011

ERBER, J; HOORMANN, J; SCHEINER, R (2006) Phototactic behaviour correlates with gustatory responsiveness in honey bees (*Apis mellifera* L.) *Behavioural Brain Research* 174(1): 174-180. http://dx.doi.org/10.1016/j.bbr.2006.07.023

EVANS, J D; CHEN, Y P; DI PRISCO, G; PETTIS, J; WILLIAMS, V (2009) Bee cups: single-use cages for honey bee experiments. *Journal of Apicultural Research* 48(4): 300-302. http://dx.doi.org/10.3896/IBRA.1.48.4.13

EVANS, J D; SPIVAK, M (2010) Socialized medicine: individual and communal disease barriers in honey bees. *Journal of Invertebrate Pathology* 103(S): S62-S72. http://dx.doi.org/10.1016/j.jip.2009.06.019

FELSENBERG, J; GEHRING, K B; ANTEMANN, V; EISENHARDT, D (2011) Behavioural pharmacology in classical conditioning of the proboscis extension response in honey bees (*Apis mellifera*). *Journal of Visualized Experiments* 47: e2282. http://dx.doi.org/10.3791/2282.

FENOY, S; RUEDA, C; HIGES, M; MARTÍN-HERNÁNDEZ, R; DEL AGUILA, C (2009) High-level resistance of *Nosema ceranae*, a parasite of the honey bee, to temperature and desiccation. *Applied and Environmental Microbiology* 75(21): 6886-6889. http://dx.doi.org/10.1128/AEM.01025-09

FISCHER, P; GROZINGER, C M (2008) Pheromonal regulation of starvation resistance in honey bee workers (*Apis mellifera*). *Naturwissenschaften* 95(8): 723-729. http://dx.doi.org/10.1007/s00114-008-0378-8

FLETCHER, D J C (1978) The African bee, *Apis mellifera adansonii*, in Africa. *Annual Review of Entomology* 23: 151-171.

FORSGREN, E; FRIES, I (2010) Comparative virulence of *Nosema ceranae* and *Nosema apis* in individual European honey bees. *Veterinary Parasitology* 170(3-4): 212-217. http://dx.doi.org/10.1016/j.vetpar.2010.02.010

FORSGREN, E; BUDGE, G E; CHARRIÈRE, J-D; HORNITZKY, M A Z (2013) Standard methods for European foulbrood research. In *V Dietemann; J D Ellis, P Neumann (Eds) The COLOSS BEEBOOK: Volume II: Standard methods for* Apis mellifera *pest and pathogen research. Journal of Apicultural Research* 52(1): http://dx.doi.org/10.3896/IBRA.1.52.1.12

FLURI, P (1977) Juvenile hormone, vitellogenin and haemocyte composition in winter worker honey bees (*Apis mellifera*). *Experientia* 33(9): 1240-1241.

FLURI, P; LÜSCHER, M; WILLE, H; GERIG, L (1982) Changes in weight of the pharyngeal gland and haemolymph titres of juvenile hormone, protein and vitellogenin in worker honey bees. *Journal of Insect Physiology* 28(1): 61-68. http://dx.doi.org/10.1016/0022-1910(82)90023-3

FREE, J B; Williams, I H (1972) Hoarding by honey bees (*Apis mellifera* L.). *Animal Behaviour* 20(2): 327-334. http://dx.doi.org/10.1016/S0003-3472(72)80054-x

FRIES, I (1993) *Nosema apis* - a parasite in the honey bee colony. *Bee World* 74(1): 5-19.

FRIES, I (2010) *Nosema ceranae* in European honey bees (*Apis mellifera*). *Journal of Invertebrate Pathology*, 103(S): S73-S79. http://dx.doi.org/10.1016/j.jip.2009.06.017

FRIES, I; GRANADOS, R R; MORSE, R A (1992) Intracellular germination of spores of *Nosema apis* Z. *Apidologie* 23(1): 61-70. http://dx.doi.org/10.1051/apido:19920107

FRIES, I; CHAUZAT, M-P; CHEN, Y-P; DOUBLET, V; GENERSCH, E; GISDER, S; HIGES, M; MCMAHON, D P; MARTÍN-HERNÁNDEZ, R; NATSOPOULOU, M; PAXTON, R J; TANNER, G; WEBSTER, T C; WILLIAMS, G R (2013) Standard methods for nosema research. In *V Dietemann; J D Ellis, P Neumann (Eds) The COLOSS BEEBOOK: Volume II: Standard methods for* Apis mellifera *pest and pathogen research. Journal of Apicultural Research* 52(1): http://dx.doi.org/10.3896/IBRA.1.52.1.14

FROST, E H; SHUTLER, D; HILLIER, N K (2011) Effects of cold immobilization and recovery period on honey bee learning, memory, and responsiveness to sucrose. *Journal of Insect Physiology* 57(10): 1385-1390. http://dx.doi.org/10.1016/j.jinsphys.2011.07.001

FROST, E H; SHUTLER, D; HILLIER N K (2012) The proboscis extension reflex to evaluate learning and memory in honey bees (*Apis mellifera*): some caveats. *Naturwissenschaften* 99(9): 677-686. http://dx.doi.org/10.1007/s00114-012-0955-8

FUKADA, H; SAKAGAMI, S F (1968) Worker brood survival in honey bees. *Researches on Population Ecology* 10(1): 31-39.

FURGALA, B; MAUNDER, M J (1961) A simple method of feeding *Nosema apis* inoculum to individual honey bees. *Bee World* 42(10): 249-252.

GENERSCH, E; VON DER OHE, W; KAATZ, H H; SCHROEDER, A; OTTEN, C; BÜCHLER, R; BERG, S; RITTER, W; MÜHLEN, W; GISDER, S; MEIXNER, M; LIEBIG, G; ROSENKRANZ, P (2010) The German bee monitoring project: a long term study to understand periodically high winter losses of honey bee colonies. *Apidologie* 41(3): 332-352. http://dx.doi.org/10.1051/apido/2010014

GILLIAM, M (1979) Microbiology of pollen and bee bread: the genus *Bacillus*. *Apidologie* 10(3): 269-274. http://dx.doi.org/10.1051/apido:19790304

GILLIAM, M (1997) Identification and roles of non-pathogenic microflora associated with honey bees. *FEMS Microbiology Letters*, 155(1): 1-10.

GIURFA, M; SANDOZ, J C (2012) Invertebrate learning and memory: fifty years of olfactory conditioning of the proboscis extension response in honey bees. *Learning and Memory* 19(2): 54-66. http://dx.doi.org/10.1101/lm.024711.111

HAGEDORN, H H; MOELLER, F E (1968) Effect of the age of pollen used in pollen supplements on their nutritive value for the honey bee. I. Effect on thoracic weight, development of hypopharyngeal glands, and brood rearing. *Journal of Apicultural Research* 7(2): 89-95.

HAYDAK, M H (1970) Honey bee nutrition. *Annual Review of Entomology* 15: 143-156. http://dx.doi.org/10.1146/annurev.en.15.010170.001043

HENDRIKSMA, H P; HÄRTEL S; STEFFAN-DEWENTER, I (2011) Honey bee risk assessment: new approaches for *in vitro* larvae rearing and data analyses. *Methods in Ecology and Evolution*, 2(5): 509-517. http://dx.doi.org/10.1111/j.2041-210X.2011.00099.x

HEPBURN, H R (1986) *Honey bees and wax: an experimental natural history*. Springer-Verlag; Berlin, Germany. 205 pp.

HERAN, H (1952) Untersuchungen über den Temperatursinn der Honigbiene unter besonderer Berücksichtigung der Wahrnehmung strahlender Wärme. *Zeitschrift für Vergleichende Physiologie* 34(2): 179-206.

HERBERT, E W; SHIMANUKI, H (1978) Chemical composition and nutritive value of bee-collected and bee-stored pollen. *Apidologie* 9(1): 33-40. http://dx.doi.org/10.1051/apido:19780103

HERBERT, E W; SHIMANUKI, H; SHASHA, B S (1980) Brood rearing and food consumption by honey bee colonies fed pollen substitutes supplemented with starch encapsulated pollen extracts. *Journal of Apicultural Research* 19(2): 115-118.

HERBERT, E W; VANDERSLICE, J T; HIGGS, D J (1985) Effect of dietary vitamin C levels on the rate of brood production of free-flying and confined colonies of honey bees. *Apidologie* 16(4): 385-394. http://dx.doi.org/10.1051/apido:19850403

HERSCH, M I; CREWE, R M; HEPBURN, H R; THOMPSON, P R; SAVAGE, N (1978) Sequential development of glycolytic competence in muscles of worker honey bees. Comparative Biochemistry and Physiology B61(3): 427-431. http://dx.doi.org/10.1016/0305-0491(78)90149-9

HIGES, M; GARCÍA-PALENCIA, P; MARTÍN-HERNÁNDEZ, R; MEANA, A (2007) Experimental infection of *Apis mellifera* honey bees with *Nosema ceranae* (Microsporidia). *Journal of Invertebrate Pathology* 94(3): 211-217. http://dx.doi.org/10.1016/j.jip.2006.11.001

HIGES, M; MARTÍN-HERNÁNDEZ, R; GARRIDO-BAILÓN, E; GARCÍA-PALENCIA, P; MEANA, A (2008) Detection of infective *Nosema ceranae* (Microsporidia) spores in corbicular pollen of forager honey bees. *Journal of Invertebrate Pathology* 97(1): 76-78. http://dx.doi.org/10.1016/j.jip.2007.06.002

HIGES, M; MARTÍN-HERNÁNDEZ, R; GARCÍA-PALENCIA, P; MARÍN, P; MEANA, A (2009) Horizontal transmission of *Nosema ceranae* (Microsporidia) from worker honey bees to queens (*Apis mellifera*). *Environmental Microbiology Reports* 1(6): 495-498. http://dx.doi.org/10.1111/j.1758-2229.2009.00052.x

HOFFMANN, I (1966) Gibt es bei Drohnen von *Apis mellifica* L. ein echtes Füttern oder nur eine Futterabgabe? *Zeitschrift für Bienenforschung* 8: 249-255.

HOOVER, S E R; KEELING, C I; WINSTON M L; SLESSOR, K N (2003) The effect of queen pheromones on worker honey bee ovary development. *Naturwissenschaften* 90(10): 477-480. http://dx.doi.org/10.1007/s00114-003-0462-z

HUANG, Z Y; ROBINSON, G E (1992) Honey bee colony integration: worker-worker interactions mediate hormonally regulated plasticity in division of labor. *Proceedings of the National Academy of Sciences* 89(24): 11726-11729. http://dx.doi.org/10.1073/pnas.89.24.11726

HUANG, Q; KRYGER, P; LE CONTE, Y; MORITZ, R F A (2012) Survival and immune response of drones of a nosemosis tolerant honey bee strain towards *N. ceranae* infections. *Journal of Invertebrate Pathology* 109(3): 297-302. http://dx.doi.org/doi:10.1016/j.jip.2012.01.004

HUMAN, H; NICOLSON, S W; DIETEMANN, V (2006) Do honey bees, *Apis mellifera scutellata*, regulate humidity in their nest? *Naturwissenschaften* 93(8): 397-401. http://dx.doi.org/10.1007/s00114-006-0117-y

HUMAN, H; BRODSCHNEIDER, R; DIETEMANN, V; DIVELY, G; ELLIS, J; FORSGREN, E; FRIES, I; HATJINA, F; HU, F-L; JAFFÉ, R; KÖHLER, A; PIRK, C W W; ROSE, R; STRAUSS, U; TANNER, G; TARPY, D R; VAN DER STEEN, J J M; VEJSNÆS, F; WILLIAMS, G R; ZHENG, H-Q (2013) Miscellaneous standard methods for *Apis mellifera* research. In *V Dietemann; J D Ellis; P Neumann (Eds) The COLOSS BEEBOOK, Volume I: standard methods for* Apis mellifera *research. Journal of Apicultural Research* 52(4): http://dx.doi.org/10.3896/IBRA.1.52.4.10

JAYCOX, E R (1961) The effects of various foods and temperatures on sexual maturity of the drone honey bee (*Apis mellifera*). *Annals of the Entomological Society of America* 54(4): 519-523.

JENSEN, A B; ARONSTEIN, K; FLORES, J M; VOJVODIC, S; PALACIO, M A; SPIVAK, M (2013) Standard methods for fungal brood disease research. In *V Dietemann; J D Ellis, P Neumann (Eds) The COLOSS BEEBOOK: Volume II: Standard methods for* Apis mellifera *pest and pathogen research. Journal of Apicultural Research* 52(1): http://dx.doi.org/10.3896/IBRA.1.52.1.13

JOHNSON, R M; POLLOCK, H S; BERENBAUM, M R (2009) Synergistic interactions between in-hive miticides in *Apis mellifera*. *Journal of Economic Entomology* 102(2): 474-479.

JOHNSON, R M; MAO, W; POLLOCK, H S; NIU, G; SCHULER, M A; BERENBAUM, M R (2012) Ecologically appropriate xenobiotics induce cytochrome p450 in *Apis mellifera*. *PLoS ONE* 7(2): e31051. http://dx.doi.org/10.1371/journal.pone.0031051

KATZNELSON, H; ROBB, J A (1962) The use of gamma radiation from cobalt-60 in the control of diseases of the honey bee and the sterilization of honey. *Canadian Journal of Microbiology* 8(2): 175-179.

KELLNER, N; JACOBS, F J (1978) In hoeveel tijd bereiken de sporen van *Nosema apis* Zander de ventriculus van de honingbij (*Apis mellifera* L.)? *Vlaams Diergeneeskundig Tijdschrift* 47(3): 252-259.

KOVAC, H; STABENTHEINER, A; BRODSCHNEIDER, R (2009) Contribution of honey bee drones of different ages to colonial thermoregulation. *Apidologie* 40(1): 82-95. http://dx.doi.org/10.1051/apido/2008069

KWAKMAN, P H S; TE VELDE, A A; DE BOER, L; SPEIJER, D; VANDENBROUCKE-GRAULS, C M J E; ZAAT, S A J (2010) How honey kills bacteria. *FASEB Journal* 24(7): 2576-2582. http://dx.doi.org/10.1096/fj.09-150789

LECOMTE, J (1950) Sur le determinisme de la formation de la grappe chez les abeilles. *Zeitshfrift fur Vergleichende Physiologie* 32(5): 499-506. http://dx.doi.org/10.1007/BF00339925

LIU, T P; BATCHELOR, T A; MUNN, R J; MARSTON, J M; JUDSON, C L (1990) The effects of low-doses of gamma irradiation on the ultrastructure of *Nosema apis in vitro*. *Journal of Apicultural Research* 29(3): 165-171.

LODESANI, M; COSTA, C; SERRA, G; COLOMBO, R; SABATINI, A G (2008) Acaricide residues in beeswax after conversion to organic beekeeping methods. *Apidologie* 39(3): 324-333. http://dx.doi.org/10.1051/apido:2008012

MAISTRELLO, L; LODESANI, M; COSTA, C; LEONARDI, F; MARANI, G; CALDON, M; MUTINELLI, F; GRANATO A (2008) Screening of natural compounds for the control of nosema disease in honey bees. *Apidologie* 39(4): 436-445. http://dx.doi.org/10.1051/apido:2008022

MALONE, L A; STEFANOVIC, D (1999) Comparison of the responses of two races of honey bees to infection with *Nosema apis* Z. *Apidologie* 30(5): 375-382. http://dx.doi.org/10.1051/apido:19990503

MATSUMOTO, Y; MENZEL, R; SANDOZ, J –C; GIURFA, M (2012) Revisiting olfactory classical conditioning of the proboscis extension response in honey bees: a step toward standardized procedures. *Journal of Neuroscience Methods* 211(1): 159-167. http://dx.doi.org/10.1016/j.jneumeth.2012.08.018

MAURIZIO, A (1950) The influence of pollen feeding and brood rearing on the length of life and physiological condition of the honey bee. *Bee World* 31(2): 9-12.

MAURIZIO, A (1954) Pollenernäahrung und Lebensvorgöange bei der Honigbiene (*Apis mellifera* L.). *Sonderdruck Landwirtschaft Jahrbuch Schweiz* 68(2): 115-182.

MAURIZIO, A (1959) Factors influencing the lifespan of bees. In *G E W WOLSTENHOLME; M O'CONNER (Eds) Ciba Foundation Symposium – The lifespan of animals* 5: 231-243. http://dx.doi.org/10.1002/9780470715253.ch13

MAYACK, C; NAUG, D (2009) Energetic stress in the honey bee *Apis mellifera* from *Nosema ceranae* infection. *Journal of Invertebrate Pathology* 100(3): 185-188. http://dx.doi.org/10.1016/j.jip.2008.12.001

MEDRZYCKI, P (2013) Funnel trap – a tool for selective collection of exiting forager bees for tests. *Journal of Apicultural Research* (in press).

MEDRZYCKI, P; TOSI, S (2012) Nutritional status and other parameters influence results of toxicological tests on bees. In *Proceedings of the COLOSS WG3 Workshop 'Honey bee nutrition', Bled, Slovenia, 22- 23 October 2012*. p 19.

MEDRZYCKI, P; GIFFARD, H; AUPINEL, P; BELZUNCES, L P; CHAUZAT, M-P; CLAßEN, C; COLIN, M E; DUPONT, T; GIROLAMI, V; JOHNSON, R; LECONTE, Y; LÜCKMANN, J; MARZARO, M; PISTORIUS, J; PORRINI, C; SCHUR, A; SGOLASTRA, F; SIMON DELSO, N; STEEN VAN DER, J; WALLNER, K; ALAUX, C; BIRON, D G; BLOT, N; BOGO, G; BRUNET, J-L; DELBAC, F; DIOGON, M; EL ALAOUI, H; TOSI, S; VIDAU, C (2013) Standard methods for toxicology research in *Apis mellifera*. In *V Dietemann; J D Ellis; P Neumann (Eds) The COLOSS BEEBOOK, Volume I: standard methods for Apis mellifera research. Journal of Apicultural Research* 52(4): http://dx.doi.org/10.3896/IBRA.1.52.4.14

MEIXNER, M D; PINTO, M A; BOUGA, M; KRYGER, P; IVANOVA, E; FUCHS, S (2013) Standard methods for characterising subspecies and ecotypes of *Apis mellifera*. In *V Dietemann; J D Ellis; P Neumann (Eds) The COLOSS BEEBOOK, Volume I: standard methods for Apis mellifera research. Journal of Apicultural Research* 52(4): http://dx.doi.org/10.3896/IBRA.1.52.4.05

MELATHOPOULOS, A P; NELSON, D; CLARK, K (2004) High velocity electron-beam radiation of pollen and comb for the control of *Paenibacillus larvae* subspecies *larvae* and *Ascosphaera apis*. *American Bee Journal* 144(9): 714-720.

MENZEL, R; BACKHAUS, W (1991) Colour vision in insects. In *E Gouras (Ed.). Vision and visual dysfunction: the perception of colour*. MacMillan Press; London, UK. pp. 262-288.

MENZEL, R; GREGGERS, U (1985) Natural phototaxis and its relationship to colour vision in honey bees. *Journal of Comparative Physiology* A157(3): 311-321. http://dx.doi.org/10.1007/BF00618121

MILNE, C P (1982) Early death of newly emerged worker honey bees in laboratory test cages. *Journal of Apicultural Research* 21(2): 107-110.

MOORE, D (2001) Honey bee circadian clocks: behavioural control from individual workers to whole-colony rhythms. *Journal of Insect Physiology* 47(8): 843-857. http://dx.doi.org/10.1016/S0022-1910(01)00057-9

MOORE, P D; WEBB, J A (1983) *An illustrated guide to pollen analysis*. Hodder and Stoughton; London, UK. 131 pp.

MORGAN, S M; HURYN, V M B; DOWNES, S R; MERCER, A R (1998) The effects of queenlessness on the maturation of the honey bee olfactory system. *Behavioural Brain Research* 91(1-2): 115-126. http://dx.doi.org/10.1016/S0166-4328(97)00118-6

MULLIN, C A; FRAZIER, M; FRAZIER, J L; ASHCRAFT, S; SIMONDS, R; VANENGELSDORP, D; PETTIS, J S (2010) High levels of miticides and agrochemcials in North American apiaries: implications for honey bee health. *PLoS ONE* 5(3): e9754. http://dx.doi.org/10.1371/journal.pone.0009754

NEUMANN, P; CARRECK, N L (2010) Honey bee colony losses. *Journal of Apicultural Research* 49(1): 1-6. http://dx.doi.org/10.3896/IBRA.1.49.1.01

NEUMANN, P; PIRK, C W W; SCHÄFER, M O; ELLIS, J D (2013) Standard methods for small hive beetle research. In *V Dietemann; J D Ellis, P Neumann (Eds) The COLOSS BEEBOOK: Volume II: Standard methods for Apis mellifera pest and pathogen research. Journal of Apicultural Research* 52(4): http://dx.doi.org/10.3896/IBRA.1.52.4.19

NICOLAS, G; SILLANS, D (1989) Immediate and latent effects of carbon dioxide on insects. *Annual Review of Entomology* 34: 97-116. http://dx.doi.org/10.1146/annurev.en.34.010189.000525

NIXON, H L; RIBBANDS, C R (1952) Food transmission within the honey bee community. *Proceedings of the Royal Society* B140 (898): 43-50. http://dx.doi.org/10.1098/rspb.1952.0042

OECD (1998) Test no. 213: honey bees, acute oral toxicity. In *OECD Guidelines for Testing of Chemical Section 2: effects on biotic systems*. OECD Publishing. http://dx.doi.org/10.1787/9789264070165-en

PANKIW, T; HUANG, Z –Y; WINSTON, M L; ROBINSON, G E (1998) Queen mandibular gland pheromone influences worker honey bee (*Apis mellifera* L.) foraging ontogeny and juvenile hormone titers. *Journal of Insect Physiology*, 44(7-8): 685-692. http://dx.doi.org/10.1016/S0022-1910(98)00040-7

PAXTON, R J; KLEE, J; KORPELA, S; FRIES, I (2007) *Nosema ceranae* has infected *Apis mellifera* in Europe since at least 1998 and may be more virulent than *Nosema apis*. *Apidologie* 38(6): 558-565. http://dx.doi.org/10.1051/apido:2007037

PENG, Y; D'ANTUONO, M; MANNING, R (2012) Effects of pollen and artificial diets on the hypopharyngeal glands of newly hatched bees (*Apis mellifera* L.). *Journal of Apicultural Research* 51(1): 53-61. http://dx.doi.org/10.3896/IBRA.1.51.1.07

PERNAL, S F; CURRIE, R W (2000) Pollen quality of fresh and 1-year-old single pollen diets for worker honey bees (*Apis mellifera* L.). *Apidologie* 31(3): 387-409. http://dx.doi.org/10.1051/apido:2000130

PIRK, C W W; DE MIRANDA, J R; FRIES, I; KRAMER, M; PAXTON, R; MURRAY, T; NAZZI, F; SHUTLER, D; VAN DER STEEN, J J M; VAN DOOREMALEN, C (2013) Statistical guidelines for *Apis mellifera* research. In *V Dietemann; J D Ellis; P Neumann (Eds) The COLOSS BEEBOOK, Volume I: standard methods for* Apis mellifera *research. Journal of Apicultural Research* 52(4): http://dx.doi.org/10.3896/IBRA.1.52.4.13

ROCKLAND, L B (1960) Saturated salt solutions for static control of relative humidity between 5° and 40°C. *Analytical Chemistry* 32 (10): 1375-1376. http://dx.doi.org/10.1021/ac60166a055

RORTAIS, A; ARNOLD, G; HALM, M-P; TOUFFET-BRIENS, F (2005) Modes of honey bees exposure to systemic insecticides: estimated amounts of contaminated pollen and nectar consumed by different categories of bees. *Apidologie* 36(1): 71-83. http://dx.doi.org/10.1051/apido:2004071

RUTRECHT, S T; KLEE, J; BROWN, M J F (2007) Horizontal transmission success of *Nosema bombi* to its adult bumble bee hosts: effects of dosage, spore source and host age. *Parasitology* 134: 1719-1726. http://dx.doi.org/10.1017/S0031182007003162

RUTTNER, F (1987) *Biogeography and taxonomy of honey bees.* Springer; Berlin, Germany. 290 pp.

SAMMATARO, D; DE GUZMAN, L; GEORGE, S; OCHOA, R (2013) Standard methods for tracheal mites research. In *V Dietemann; J D Ellis, P Neumann (Eds) The COLOSS* BEEBOOK*: Volume II: Standard methods for* Apis mellifera *pest and pathogen research. Journal of Apicultural Research* 52(4): http://dx.doi.org/10.3896/IBRA.1.52.4.20

SCHEINER, R; ABRAMSON, C I; BRODSCHNEIDER, R; CRAILSHEIM, K; FARINA, W; FUCHS, S; GRÜNEWALD, B; HAHSHOLD, S; KARRER, M; KOENIGER, G; KOENIGER, N; MENZEL, R; MUJAGIC, S; RADSPIELER, G; SCHMICKLI, T; SCHNEIDER, C; SIEGEL, A J; SZOPEK, M; THENIUS, R (2013) Standard methods for behavioural studies of *Apis mellifera*. In *V Dietemann; J D Ellis; P Neumann (Eds) The COLOSS* BEEBOOK*, Volume I: standard methods for* Apis mellifera *research. Journal of Apicultural Research* 52(4): http://dx.doi.org/10.3896/IBRA.1.52.4.04

SCHMICKL, T; CRAILSHEIM, K (2001) Cannibalism and early capping: strategy of honey bee colonies in times of experimental pollen shortages. *Journal of Comparative Physiology* A187(7): 541-547. http://dx.doi.org/10.1007/s003590100226

SCHMICKL, T; CRAILSHEIM, K (2007) HoPoMo: A model of honey bee intracolonial population dynamics and resource management. *Ecological Modelling* 204(1-2): 219-245. http://dx.doi.org/10.1016/j.ecolmodel.2007.01.001

SCHMIDT, J O; THOENES, S C; LEVIN, M D (1987) Survival of honey bees, *Apis mellifera* (Hymenoptera: Apidae), fed various pollen sources. *Annals of the Entomological Society of America* 80(2): 176-183.

SCHMOLZ, E; HOFFMEISTER, D; LAMPRECHT, I (2002) Calorimetric investigations on metabolic rates and thermoregulation of sleeping honey bees (*Apis mellifera carnica*). *Thermochimica Acta* 382(1-2): 221-227. http://dx.doi.org/10.1016/S0040-6031(01)00740-7

SEEHUUS, S-C; NORBERG, K; GIMSA, U; KREKLING, T; AMDAM, G V (2006) Reproductive protein protects functionally sterile honey bee workers from oxidative stress. *Proceedings of the National Academy of Science* 103(4): 962-967. http://dx.doi.org/10.1073/pnas.0502681103

SEELEY, T D (1974) Atmospheric carbon dioxide regulation in honey bee (*Apis mellifera*) colonies. *Journal of Insect Physiology* 20(11): 2301-2305. http://dx.doi.org/10.1016/0022-1910(74)90052-3

SEELEY, T D (1982) Adaptive significance of the age polyethism schedule in honey bee colonies. *Behavioural Ecology and Sociobiology* 11(4): 287-293. http://dx.doi.org/10.1007/BF00299306

SEELEY, T D (1994) Honey bee foragers as sensory units of their colonies. *Behavioural Ecology and Sociobiology* 34(1): 51-62. http://dx.doi.org/10.1007/BF00175458

SEELEY, T D (2010) *Honey bee democracy*. Princeton University Press; Princeton, USA. 273 pp.

SITBON, G (1967) L'effet de group chez l'abeille: L'abeille d'hiver; survie et consummation de candi des abeilles isolees ou groupees. *Ann Abeille* 10(2): 67-82. http://dx.doi.org/10.1051/apido:19670201

SLESSOR, K N; KAMINSKI, L –A; KING, G G S; BORDEN, J H; WINSTON, M L (1988) Semiochemical basis of the retinue response to queen honey bees. *Nature* 332(6162): 354-356. http://dx.doi.org/10.1038/332354a0

SLESSOR, K N; WINSTON, M L; LE CONTE, Y (2005) Pheromone communication in the honey bee (*Apis mellifera* L.). *Journal of Chemical Ecology* 31(11): 2731-2745. http://dx.doi.org/10.1007/s10886-005-7623-9

SOUTHWICK, E E; MORITZ, R F A (1987) Social control of air ventilation in colonies of honey bees, *Apis mellifera*. *Journal of Insect Physiology* 33(9): 623-626. http://dx.doi.org/10.1016/0022-1910(87)90130-2

STABENTHEINER, A; KOVAC, H; BRODSCHNEIDER, R (2010) Honey bee colony thermoregulation - regulatory mechanisms and contribution of individuals in dependence on age, location and thermal stress. *PLoS ONE* 5(1): e8967. http://dx.doi.org/10.1371/journal.pone.0008967

STRAND, M R (2008) The insect cellular immune response. *Insect Science* 15(1): 1-14.
http://dx.doi.org/10.1111/j.1744-7917.2008.00183.x

TANNER, G; WILLIAMS, G R; MEHMANN, M; NEUMANN, P (2012a) Comparison of group versus individual inoculation of worker honey bees with *Nosema ceranae*. In *Proceedings of the 8th COLOSS Conference / MC meeting FA0803, Halle-Saale, Germany, 1-3 September 2012*. p 59.

TANNER, G; WILLIAMS, G R; MEHMANN, M; NEUMANN, P (2012b) Differential susceptibility of drone versus worker honey bees towards infections with *Nosema ceranae* and black queen cell virus? In *Proceedings of the 5th European Conference of Apidology (EurBee 5), Halle-Saale, Germany, 3-7 September 2012*. p 239.

UNDEEN, A H; VANDER MEER, R K (1990) The effect of ultraviolet radiation on the germination of *Nosema algerae* Vavra and Undeen (Microsporida: Nosematidae) spores. *Journal of Protozoology* 37(3): 194-199.
http://dx.doi.org/10.1111/j.1550-7408.1990.tb01127.x

USDA (2008) Foods permitted to be irradiated under FDA Regulations, Title 21 CFR Part 179.26 of the Code of Federal Regulations.

VAN DER STEEN, J (2007) Effect of a home-made pollen substitute on honey bee colony development. *Journal of Apicultural Research* 46(2): 114-119. http://dx.doi.org/10.3896/IBRA.1.46.2.09

VAN DER STEEN, J J M; CORNELISSEN, B; DONDERS, J; BLACQUIÈRE, T; VAN DOOREMALEN, C (2012) How honey bees of successive age classes are distributed over a one storey, ten frames hive. *Journal of Apicultural Research* 51(2): 174-178.
http://dx.doi.org/10.3896/IBRA.1.51.2.05

VANENGELSDORP, D; MEIXNER, M D (2010) A historical review of managed honey bee populations in Europe and the United States and the factors that may affect them. *Journal of Invertebrate Pathology* 103(S): s80-s95.
http://dx.doi.org/10.1016/j.jip.2009.06.011

VERBEKE, M; JACOBS, F J; DE RYCKE, P H (1984) Passage of various particles through the ventriculus in the honey bee (*Apis mellifera* L.). *American Bee Journal* 124(6): 468-470.

VILLA, J D (2007) Influence of worker age on the infestation of resistant and susceptible honey bees (*Apis mellifera*) with tracheal mites (*Acarapis woodi*). *Apidologie* 38(6): 573-578.
http://dx.doi.org/10.1051/apido:2007050

WEBSTER, T C (1994) Fumagillin affects *Nosema apis* and honey bees (Hymenoptera: Apidae). *Journal of Economic Entomology* 87(3): 601-604.

WEXLER, A; BROMBACHER, W G (1951) Methods of measuring humidity and testing hygrometers. *National Bureau of Standards (USA)* 512.

WHITE, J W; DONER, L W (1980) Honey composition and properties. In *Beekeeping in the United States*. US Department of Agriculture, Agricultural Handbook. pp. 82-91.

WHO (2005) *Nutrients in drinking water*. WHO Press; Geneva, Switzerland. 186 pp.

WILLIAMS, G R; TARPY, D R; VANENGELSDORP, D; CHAUZAT, M P; COX-FOSTER, D L; DELAPLANE, K S; NEUMANN, P; PETTIS, J S; ROGERS, R E L; SHUTLER, D (2010) Colony Collapse Disorder in context. *BioEssays* 32(10): 845-846.
http://dx.doi.org/10.1002/bies.201000075

WILLIAMS, G R; DIETEMANN, V; ELLIS, J D; NEUMANN P (2012) An update on the COLOSS network and the "*BEEBOOK*: standard methodologies for *Apis mellifera* research". *Journal of Apicultural Research* 51(2): 151-153.
http://dx.doi.org/10.3896/IBRA.1.51.2.01

WINSTON, M L (1987) *The biology of the honey bee*. Harvard University Press; Cambridge, USA. 281 pp.

WINSTON, M L; SLESSOR, K N (1992) The essence of royalty – honey bee queen pheromone. *American Scientist* 80(4): 374-385.

WINSTON, P W; BATES, D H (1960) Saturated solutions for the control of humidity in biological research. *Ecology* 41(1): 232-237.
http://dx.doi.org/10.2307/1931961

WU, J Y; ANELLI, C M; SHEPPARD, W S (2011) Sub-lethal effects of pesticide residues in brood comb on worker honey bee (*Apis mellifera*) development and longevity. *PLoS ONE* 6(2): e14720. http://dx.doi.org/doi:10.1371/journal.pone.0014720

YOOK, H-S; LIM, S-I; BYUN, M-W (1998) Changes in microbiological and physicochemical properties of bee pollen by application of gamma irradiation and ozone treatment. *Journal of Food Protection* 61(2): 217-220.

ZUUR, A F; GENDE, L B; IENO, E N; FERNANDEZ, N J; EGUARES, M J; FRITZ, R; WALKER, N J; SAVELIEV, A A; SMITH, G M (2009) Mixed effects modelling applied on American Foulbrood affecting honey bee larvae. In *A F Zuur; E N Ieno; N J Walker; A A Saveliev; G M Smith (Eds). Mixed effects models and extensions in ecology with R*. Springer Science & Business Media; New York, USA. pp. 447-458.

Journal of Apicultural Research 52(1): (2013)
DOI 10.3896/IBRA.1.52.1.05

IBRA
INTERNATIONAL BEE
RESEARCH ASSOCIATION

REVIEW ARTICLE

Standard methods for artificial rearing of

Apis mellifera larvae

Karl Crailsheim[1*†], **Robert Brodschneider**[1†], **Pierrick Aupinel**[2], **Dieter Behrens**[3], **Elke Genersch**[4], **Jutta Vollmann**[1] **and Ulrike Riessberger-Gallé**[1†]

[1]Department of Zoology, Karl-Franzens-University Graz, Universitätsplatz 2, 8010 Graz, Austria.
[2]INRA, UE1255, UE Entomologie, F-17700 Surgères, France.
[3]Institute of Biology, Martin-Luther-University Halle-Wittenberg, Hoher Weg 4, 06099 Halle (Saale), Germany.
[4]Institute for Bee Research Hohen Neuendorf, Friedrich-Engels-Str. 32, 16540 Hohen Neuendorf, Germany.

Received 26 March 2012, accepted subject to revision 13 August 2012, accepted for publication 25 October 2012.

*Corresponding author: Email: karl.crailsheim@uni-graz.at

[†]The first two authors and the last author contributed equally to this paper.

Summary

Originally, a method to rear worker honey bee larvae *in vitro* was introduced into the field of bee biology to analyse honey bee physiology and caste development. Recently, it has become an increasingly important method in bee pathology and toxicology. The *in vitro* method of rearing larvae is complex and can be developed as an art by itself, especially if the aim is to obtain queens or worker bees which, for example, can be re-introduced into the colony as able members. However, a more pragmatic approach to *in vitro* rearing of larvae is also possible and justified if the aim is to focus on certain pathogens or compounds to be tested. It is up to the researcher(s) to decide on the appropriate experimental establishment and design. This paper will help with this decision and provide guidelines on how to adjust the method of *in vitro* rearing according to the specific needs of the scientific project.

Métodos estándar para la cría *in vitro* de larvas de *Apis mellifera*

Resumen

Originalmente, el método para la cría *in vitro* de larvas de obreras de abejas melíferas se introdujo en el campo de la biología de las abejas para analizar la fisiología y el desarrollo de las castas. Recientemente, se ha convertido en un método cada vez más importante para la patología y la toxicología de la abeja. El método de cría *in vitro* de larvas es complejo y constituye un arte en sí mismo, especialmente si el objetivo es obtener reinas o abejas obreras que, por ejemplo, puedan ser re-introducidas en la colonia como miembros activos. Sin embargo, un enfoque más pragmático de la cría de larvas *in vitro* también es posible y justificado si el objetivo es centrarse en ensayos con ciertos patógenos o compuestos. Corresponde al investigador (es) decidir sobre el adecuado establecimiento experimental y el diseño. Este artículo ayudará con esta decisión y proporcionará directrices sobre cómo ajustar el método de cría *in vitro* en función de las necesidades específicas del proyecto científico.

人工饲养西方蜜蜂幼虫的标准方法

人工饲养工蜂幼虫的方法最早始于蜜蜂生物学研究，用来分析蜜蜂的生理机能和级型发育。近来，其在蜜蜂病理学和毒理学领域的应用也日趋重要。人工饲养幼虫方法十分复杂但也极富技巧，特别是当培养目的是获得可以重新介绍入蜂群中的蜂王或工蜂，并还可成为群体的有效成员时将更加困难。目前一个经过改进并更为实用的饲养技术方案已经形成，以此技术方案为基础，略加调整就可成为用于某单一病原体或混合病原体

Footnote: Please cite this paper as: CRAILSHEIM, K; BRODSCHNEIDER, R; AUPINEL, P; BEHRENS, D; GENERSCH, E; VOLLMANN, J; RIESSBERGER-GALLÉ, U (2012) Standard methods for artificial rearing of *Apis mellifera* larvae. In *V Dietemann; J D Ellis; P Neumann (Eds) The COLOSS BEEBOOK, Volume I: standard methods for Apis mellifera research. Journal of Apicultural Research* 52(1): http://dx.doi.org/10.3896/IBRA.1.52.1.05

测试实验所需的饲养方案。由此类推，研究者可根据自己的研究目的，以本方案为基础，经再加工就可设计或调整出可满足自己实验需要的饲养方案。本文针对各类科研项目的需求，就如何调整人工饲养蜜蜂幼虫方案给出了指导方针。

Keywords: honey bee, *Apis mellifera*, larval rearing, standardization, risk assessment, *BEEBOOK*, COLOSS

1. Introduction

Honey bees are important pollinators and are responsible for much of the world's agricultural production and the conservation of biodiversity (Klein *et al.*, 2007; Gallai *et al.*, 2009). In many regions of the world, the number of honey bee colonies is declining, thus possibly endangering pollination (Aizen *et al.*, 2009; VanEngelsdorp and Meixner, 2010; van der Zee *et al.*, 2012). Parasites, pathogens and pesticides are three of the major threats to honey bees and are believed to be partially responsible for the abovementioned declines (Neumann and Carreck, 2010). Therefore, the effects of these handicaps and the combination of two or more sublethal effects are extensively investigated (Mullin *et al.*, 2010; Moritz *et al.*, 2010; Genersch *et al.*, 2010). An important tool for this research is the rearing of honey bee larvae *in vitro* (i.e. in the laboratory and in the absence of nurse bees) because it allows more controllable conditions compared to *in vivo* (i.e. in the hive by nurse bees). However, artificial larval rearing can also be regarded as an *in vivo* method conducted in an *in vitro* system. In particular, the testing of the toxicity of plant protection products on brood can be conducted in a reproducible and standardized way only in the laboratory, because a defined uptake of food containing the testing compound is not feasible using in-hive methods (Wittmann and Engels, 1981; Oomen *et al.*, 1992; Schuur *et al.*, 2003; Becker *et al.*, 2009; Aupinel *et al.*, 2007a).

The first attempt at mass application and standardization of *in vitro* rearing of honey bee larvae and testing plant protection products was a ring test, in which seven different laboratories assessed the LD_{50} for dimethoate 48 hours after acute larval exposure (Aupinel *et al.*, 2009). This experiment underlined the variability in results and the importance of further investigation of factors such as colony origin of larvae, effect of season and larval heterogeneity at grafting.

In this paper, we give an overview of existing *in vitro* protocols, present the crucial points of rearing honey bee larvae in the laboratory, and discuss where further research or standardization of methods might be useful or necessary. The aim of this paper is not to present a rigorous recipe for rearing larvae *in vitro* (as these are mostly well documented in individual papers), but rather to discuss advantages and disadvantages of different protocols and give recommendations where necessary and meaningful.

2. Larval nourishment

Honey bee larvae do not feed on a readily available vegetal food; rather, they are fed by adult bees. In a colony, worker larvae are fed worker jelly according to their age by their adult sisters. This fact prevented artificial rearing for a long time. A larva is frequently inspected and progressively fed 135 - 143 times with worker jelly during its larval development (Lindauer, 1952; Brouwers *et al.*, 1987). According to a rough estimation, one worker larva is fed a total of 59.4 mg of carbohydrates and 25-37.5 mg of protein (Hrassnigg and Crailsheim, 2005).

The composition of jelly fed to workers has been analysed and compared to royal jelly (food fed to immature queens, see Section 5.10. Royal jelly). Findings suggest that royal jelly and the jelly of worker larvae up to the age of three days are similar, at least regarding protein, sugar, and lipid content (Rhein, 1933; Brouwers *et al.*, 1987). Pantothenic acid, for example, was found to be five times higher in royal jelly than in worker jelly; no other differences in vitamin content were reported (Rembold and Dietz, 1965).

Worker jelly is produced by the hypopharyngeal and mandibular glands of nurse bees (Jung-Hoffmann, 1966; Moritz and Crailsheim, 1987). Biochemical analyses show a broad variation among season, age of fed larvae, investigation methods and colony supply (Jung-Hoffmann, 1966; Brouwers, 1984). According to Kunert and Crailsheim (1987), worker jelly is composed of 39.2-53.3% protein dry weight for larvae younger than four days. After that age, the protein content decreases to 15.7-26.1% dry weight. Correspondingly, the sugar content increases from 7.8-19.6% to 32.2-64.6% dry weight after day four (Fig. 1). These findings were confirmed by Brouwers *et al.* (1987), who moreover described the decrease of lipids in the food of worker larvae with increasing age. Asencot and Lensky (1988) measured the fructose / glucose ratio of worker jelly and also confirmed the increase of sugar content with age.

Four- and five-day-old worker larvae are additionally fed an increasing, although still small, amount of unprocessed pollen (e.g. Rhein, 1933; Jung-Hoffmann, 1966). After day six, larvae cease feeding and larval cells are sealed by adult workers. During pupation, larvae make use of the material gained during larval nourishment for anabolism and catabolism. The metabolism of glycogen and lipids in particular plays an important role during metamorphosis (Hrassnigg and Crailsheim, 2005).

3. History of rearing honey bee larvae in the laboratory

Isolating worker larvae from adult bees in a hive and rearing them to the adult stage has challenged many early researchers who were mainly interested in honey bee physiology and caste determination.

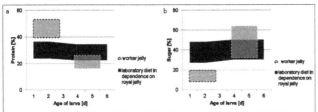

Fig. 1. Comparison of: **a.** protein and; **b.** sugar content of brood food of young (1-3 d) and old (4-6 d) worker larvae (Data summarized from Kunert and Crailsheim, 1987) and laboratory diet, both percentages of dry weight. Protein decrease and sugar increase are due to the increase in sugar added to diets (Aupinel *et al.*, 2005). Range of content is determined by variation in royal jelly protein (Sabatini *et al.*, 2009) and sugar (Brouwers, 1984) content. Protein and sugar in yeast were excluded from calculations.

Bertholf (1927) kept three-day-old larvae in the laboratory alive for more than two days by feeding them different concentrations of sucrose solution, and Velich (1930) reared them to adults. The first important report of hand-feeding honey bee larvae in the laboratory was published by Rhein (1933). He found differences in the jelly fed to workers and queens (royal jelly) and successfully used royal jelly to rear two- to three-day-old worker larvae to adults. He also recognized the problem of worker-queen intermediates. Due to the focus of research, the consistent production of females belonging to the worker caste was one of the main challenges during these early days of *in vitro* rearing of larvae. Presently, the *in vitro* rearing of larvae is being used as a routine method in pesticide testing and for many applications in honey bee physiology and pathology (Table 1). Hence, the reduction of individual feedings and, thus the work load for the experimenter, is appreciated and presents a challenge to those further developing these methods. In contrast to the numerous progressive feedings of larvae in a colony, larvae in the laboratory are only fed daily or even less frequently, preferably without influencing survival rate (Aupinel *et al.*, 2005; Kaftanoglu *et al.*, 2010).

After initial attempts, many researchers developed methods to rear honey bee larvae in the laboratory by modifying or optimizing the basic diet, which consisted of royal jelly diluted with an aqueous solution of glucose and fructose. Fraenkel and Blewett (1943) discovered the importance of yeast in artificial insect diets for *Drosophila* larvae, but this was later replaced with B vitamins and polyunsaturated fatty acids (Vanderzant, 1974). The use of yeast was also applied to honey bee larval diets, and yeast extract is still in use (Michael and Abramovitz, 1955; Peng *et al.*, 1992; Aupinel *et al.*, 2005) but has been omitted by others without a reduction in individual survival rates (Genersch *et al.*, 2005; Genersch *et al.*, 2006). Yeast has been demonstrated not to be a differentiating agent for female castes, but it may be beneficial because of phagostimulatory or nourishing effects (Rembold and Lackner, 1981; Vandenberg and Shimanuki, 1987). Since *in vitro* rearing can also be completed successfully without the addition of yeast extract, the role of yeast needs to be questioned critically. In addition to using yeast, other attempts have been made to enrich the larval diet, for example with a readily available vitamin formulation (Multibionta; Merck; Herrmann *et al.*, 2008) which did not affect larval weight or mortality significantly.

Michael and Abramovitz (1955), amongst others, developed self-feeding dishes and fed an aqueous honey solution containing 10% dehydrated yeast extract. They were the first to inoculate larvae in the laboratory with European foulbrood. Rearing larvae on diluted royal jelly is "an art", as stated by Weaver (1974), and yielded only small numbers of individuals and also queen-worker intermediates. More information on early *in vitro* rearing of honey bee larvae and the success of different researchers can be found in Jay (1964). These early experiments formed the basis for later protocols, produced valuable scientific information and permitted biological testing of different qualities of royal jelly (Weaver, 1955; Smith, 1959; Mitsui *et al.*, 1964; Asencot and Lensky, 1984). The use of worker jelly (which is also disproportionately labour intensive to harvest compared to royal jelly) was doomed, because survival is low and larvae do not pupate

Table 1. Research topics using laboratory-reared honey bee larvae and selected references.

Research topic	Authors
Caste differentiation	Rhein, 1933; Weaver, 1955; Smith, 1959; Shuel and Dixon, 1960, 1986; Mitsui *et al.*, 1964; Rembold *et al.*, 1974; Weaver, 1974; Asencot and Lensky, 1976, 1984, 1988; Shuel *et al.*, 1978; Rembold and Lackner, 1981; Vandenberg and Shimanuki, 1987; Wittmann and Engels, 1987;
Diploid drones	Woyke, 1963; Herrmann *et al.*, 2005
Larval pathogens	Michael and Abramovitz, 1955; Peng *et al.*, 1992, 1996; Brødsgaard *et al.*, 1998, 2000; Genersch *et al.*, 2005, 2006; Behrens *et al.*, 2007, 2010; Jensen *et al.*, 2009; Forsgren *et al.*, 2010; Vojvodic *et al.*, 2011a, 2011b, 2012; Vasquez *et al.*, 2012; Foley *et al.*, 2012
Toxicity	Wittmann and Engels, 1981; Davis *et al.*, 1988; Czoppelt and Rembolt, 1988; Aupinel *et al.*, 2005, 2007a, 2007b, 2009; Medrzycki *et al.*, 2010; Da Silva Cruz *et al.*, 2010; Gregorc and Ellis, 2011; Hendriksma *et al.*, 2011a; Gregorc *et al.*, 2012
Transgenic plants	Malone *et al.*, 2002; Brødsgaard *et al.*, 2003; Hendriksma *et al.*, 2011b, 2012

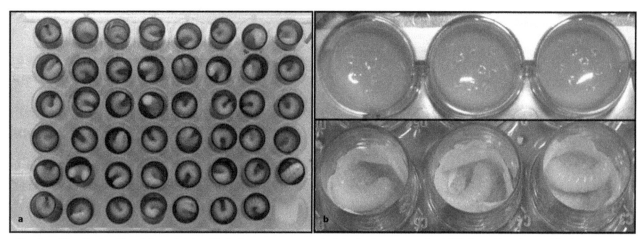

Fig. 2. **a.** 48-well microtiter plate containing 47 plastic queen cups on dental rolls. Larvae are about 6-7 days old and reared in the same cup from first instar on (Photo: Pierrick Aupinel). **b.** 3 groups of 10 each first instars (upper panel) which will be progressively isolated in new wells lined with filter paper, in the end resembling the 3 engorged larvae in one cup each (lower panel, Photo: Elke Genersch).

(Rhein, 1933; Herrmann *et al.*, 2008). Survival on worker jelly can be increased when sugars are added, and Asencot and Lensky (1976; 1984; 1988) believed that sugars together with juvenile hormone could be the queen determinant. More recent results regarding the importance of sugars for pupation are available in Kaftanoglu *et al.* (2011). In turn, to find the caste-determining factor, Rembold *et al.* (1974) reared young worker larvae in thimbles fed a basic food derived from royal jelly via a very extensive alcohol extraction. This diet produced more adult workers, but also more queens and intermediate forms, than did diluted royal jelly. Shuel and Dixon (1986) prepared a complex diet composed of soluble and insoluble protein extracts from royal jelly enriched with solutions containing vitamins, minerals and other compounds.

Wittmann and Engels (1981) suggested investigating the effects of plant protection products on *in vitro*-reared honey bee larvae, and Davis *et al.* (1988) were among the first to administer carbofuran and dimethoate dissolved in royal jelly to honey bee larvae. At the same time, Czoppelt and Rembold (1988) assessed the toxicity of parathion to larvae. Aupinel *et al.* (2005) imitated the age-dependent increase of sugars and dry matter in the larval food in the colony (Brouwers *et al.*, 1987) by gradually increasing glucose, fructose and yeast extract in the diet of larvae (Fig. 1b). They also replaced beeswax cups or thimbles with readily available plastic cups (cupula - used mainly by beekeepers for queen rearing purposes) placed in 48-well microtiter plates (Fig. 2a). This method allows for adoption in testing insecticides in different laboratories (Aupinel *et al.*, 2009).

Two principal *in vitro* rearing methods are employed which we will discuss in this paper. In the first method, one larva is reared per cup and the exact amount of diet a larva consumes daily is administered, following the protocol of Rembold and Lackner (1981) with modifications by Vandenberg and Shimanuki (1987). This amount of diet is estimated to be 160µl in total during larval development (Aupinel *et al.*, 2005) or 164µl for Africanized honey bees (Silva *et al.*, 2009). The second

approach administers excess diet, and consequently the larvae have to be transferred to new dishes regularly. At the start of this process several worker (Peng *et al.*, 1992; Genersch *et al.*, 2005; 2006) or drone (Behrens *et al.*, 2007, 2010) larvae are reared in one culture plate well (Fig. 2b). With increasing age, the larvae are either progressively isolated by daily grafting on fresh food (Genersch *et al.*, 2005, 2006) or still grafted in groups to new petri dishes (Kaftanoglu *et al.*, 2010). During *ad libitum* feeding, the individual uptake of diet and compounds / pathogens is not restricted to a certain volume (and hence, dosage) as in single-cup protocols. Therefore, it is only possible to determine the LC_{50} (median lethal concentration) for a given substance or pathogen; however, the LD_{50} (median lethal dose) can be roughly estimated (when necessary) by assuming the average consumption to be the abovementioned 160µl per larvae. In many cases it will be sufficient to determine the lethal concentration of a substance or pathogen, because under natural conditions bees and larvae will be exposed to matrices containing a certain concentration of a substance or pathogen. Administering a certain dosage, as is done in medical treatments of humans and other vertebrates, will not be the issue with honey bees, because drugs are administered to bees not in a given dosage per bee but rather dissolved in sugar solution fed to the bee colony. However, it has to be taken into account that all calculated dosage estimates are derived from the assumed consumption, and larvae will consume more if more food is provided. Therefore only individual larval rearing with limited feedings and control of complete consumption allows accurate dosage calculations (Aupinel *et al.,* 2005).

Many of the methods previously published resulted in high mortalities and were labour intensive. However, progress has been made and some of the protocols are sufficient to consistently rear worker honey bee larvae with little or no mortality in the laboratory and can therefore be used for mass rearing and the application of routine testing of compounds (Aupinel *et al.*, 2005; Hendriksma *et al.*, 2011a). This might suggest that no more research efforts to develop new diets would be necessary. However, researchers need to adapt existing diets adequately for their studies or develop new diets to understand the nutritional requirements of honey bee larvae in detail. There is still no chemically defined diet available, as there are for other insects, and it is unlikely that such a diet will be produced for honey bee larvae (Vanderzant, 1974; Shuel and Dixon, 1986). So far, all published diets have been composed of crude materials (royal jelly), imitating the natural food to a great degree and thus are categorized as a third type of artificial diets as described by Vanderzant (1974).

4. Crucial points for rearing honey bee larvae in the laboratory

4.1. Study design

When testing the effect of plant protection products or insect growth regulators on immature honey bees, the design of the study (i.e. the number of replicates and larvae) is crucial. First, clarity regarding the minimum number of replicates is needed. A replicate is a repetition within a treatment group, commonly made up of a series of microtiter plates or petri dishes, where each larva can be regarded as a single statistical event. We recommend a minimum of three replicates with a minimum of 30 larvae per replicate. In toxicity testing, a replicate typically consists of a control plate without solvent, one control with solvent (if necessary) and several plates with the doses or concentrations of pesticides to be tested. As a reference, one treatment with a substance of known toxicity (e.g. dimethoate, Aupinel *et al.*, 2007a,

2007b) must be conducted (see the *BEEBOOK* paper on toxicology methods (Medrzycki *et al.*, 2013)).

Decisions about study design should carefully consider the aim of the investigation. Depending on the study design, it is possible to study the influence of the genetic background of the bees (Fig. 3a) or of the season (Fig. 3b) on the susceptibility of larvae to a certain substance or pathogen. Mixing larvae from different colonies (Fig. 3c) and testing them even over the entire season (Fig. 3d) will reveal the effect of a substance / pathogen that is not influenced by the genetic background of the bees or by the season. To fully understand the effect of a substance / pathogen, ideally all four approaches should be carried out. If the variance due to an effect of the genetic background needs to be reduced or controlled for, the usage of haploid drone larvae instead of diploid worker offspring of different patrilines may be desirable. This approach however requires that the phenotypes of the drones resemble those of the workers or that differences have been investigated. If usage of diploid workers is inevitable the genetic variance of the study population can be reduced by taking the worker offspring of a single drone inseminated queen. In any case, the minimum required sample size must be calculated and for this we recommend consulting a statistician.

4.2. Caging of queen

- Cage queen on an empty brood comb
- Confirm oviposition by visual inspection after few hours
- 72 hours after confirmation of eggs - Section 5.5. Grafting of larvae

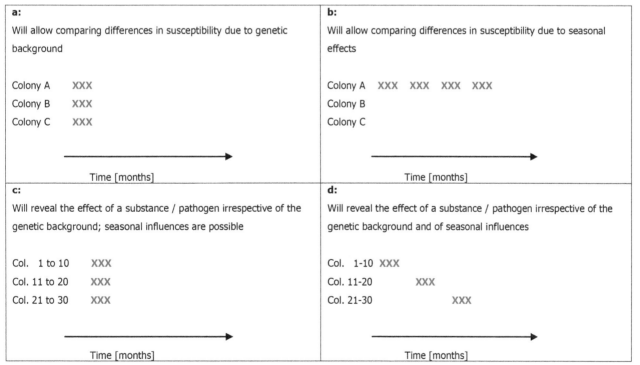

Fig. 3. **a-d** Examples of four different study designs to conduct three replicates of an experiment with different levels of genetic and seasonal variability. X indicates date of grafting and origin of larvae used for experiments.

The developmental stages of honey bee larvae exhibit different susceptibilities towards pesticides and pathogens, and larvae should be exposed to pathogens or toxins in experiments at biologically relevant stages (Davis *et al.*, 1988). If honey bees are to be reared from the very early larval stages, first instar larvae are needed. Two definitions of age are known: the biological age (larval stage) and the chronological age. To differentiate these two morphological parameters need to be investigated. During their development, honey bee larvae undergo five instars, and each instar lasts a different length of time (Bertholf, 1925). Head diameter (Rembold *et al.*, 1980) and the developmental stages of the mouthpart, wing buds, leg buds and the gonapophyses (see drawings in Myser, 1954 and the *BEEBOOK* paper on miscellaneous research methods (Human *et al.*, 2013)) are proper parameters to characterize each of the five instars. As the first instar stage lasts 14-20 hours, one should be aware that a larva that has reached the age of one day (24 hours post hatching from the egg) already has reached the second instar (Bertholf, 1925; Rembold *et al.*, 1980). Larval age can also be determined by weighing (Rembold and Lackner, 1981; Vandenberg and Shimanuki, 1987; Davis *et al.*, 1988). Weight of larvae can be confirmed using data of Wang (1965) and the *BEEBOOK* paper on miscellaneous research methods (Human *et al.*, 2013): 12 hour old larvae have on average a weight of 0.36 ± 0.024 mg but this may depend on honey bee race. However, weighing will delay grafting and increase the risk of contamination and mortality.

Age of first instar larvae for grafting (for the grafting method, see the *BEEBOOK* paper on queen rearing (Büchler *et al.*, 2013) is usually determined chronologically, and larvae younger than 12 hours after hatching are safely first instars. Chronological age can be controlled by caging the queen on a brood comb with a queen excluder cage (for information on obtaining brood and adult bees of known ages, see the *BEEBOOK* paper on miscellaneous research methods (Human *et al.*, 2013)). Queens do not start oviposition immediately after caging, so onset of egg-laying has to be confirmed. It has to be taken into account that larvae hatch from eggs 66 to 93 hours after oviposition (Collins, 2004), hence general time schedules based on 72 hours need to be confirmed for every trial. We recommend that the queen lays eggs in large areas of the comb; larvae of the same age are usually found in ring-like areas on the comb. Releasing the queen from the excluder cage afterwards is important only for colony development and a fixed time cannot be given, as this strongly depends on the queen.

Instead of grafting, larvae can also be obtained using plastic queen cups mounted on commercially available artificial combs (Cupularve Nicotplast, France). On day four after caging the queen and depending on larval hatching, plastic cups containing first instar larvae can quickly be collected (Hendriksma *et al.*, 2011a).

4.3. Sterile environment

- Clean incubator and desiccators
- Sterilize equipment and water for diet preparation

- Sterilize rearing equipment (cups)
- Put plates with cups under UV

When rearing honey bee larvae in the laboratory, strict sanitation is required. A sterile environment with sterile materials and chemicals is crucial; otherwise infections from bacteria or fungi will impede experiments. We recommend laboratories to strive to achieve optimum levels of sterility using their preferred method. All equipment, such as glassware, tools, consumables and water should be autoclaved. Work areas and material for control and exposure / infection treatments should be separated. Autoclaving sugar solution should be avoided because of heat-derived formation of hydroxymethylfurfural (HMF). Two possibilities for the treatment of sugar-yeast-water solutions are recommended: first, to filter with 0.22µm pore membrane filters, and second, to use autoclaved water for the preparation.

Mask and hand disinfection are important to reduce infections. It is crucial to use incubators only for larval rearing, as any other usage needlessly increases the risk of contamination. Disinfect incubators between experiments using ethanol (> 70%) or heat (e.g. more than one hour at more than 100°C) but check operators manual of equipment. As a regular test of sterility, or for troubleshooting, we recommend incubating open agar petri dishes and analysing them to detect any possible sources of contamination. Tools such as paintbrushes (for grafting) must be washed in ethanol and rinsed in autoclaved water before and regularly throughout usage. Dental rolls can be inserted in wells of rearing plates and 500µl of 15.5% glycerol solution added (see 5.8. Incubation conditions). Wet dental rolls and plastic cups can be put under UV-light for sterilization for at least one hour. Methyl benzethonium chloride (MBC) was proposed for the disinfection of plastic cups and also on wetted filter paper or dental roll to prevent microbial growth during incubation (Vandenberg and Shimanuki, 1987; Aupinel *et al.*, 2005). Due to its high price, MBC is no longer in use. In early protocols, methyl-4-hydroxybenzoate was used as a fungicide in the diet (Rembold *et al.*, 1974; Rembold and Lackner, 1981) or Nyastin was added to royal jelly by Herrmann *et al.* (2008). Plastic queen cups can be disinfected in a chlorine solution such as Milton® Sterilising Tablets (Brodschneider *et al.*, 2009). To reduce bacterial infections, Penicillin G (e.g. 375 ppm) can be added to the diet on day one, most of which will degrade within a few days (Riessberger-Gallé *et al.*, 2011).

4.4. Preparation and storage of diets

- Estimate the needed amount of diet
- Prepare or thaw sugar solutions (w/v)
- Mix sugar solutions with royal jelly 1:1 (w/w)
- Warm diet to 34.5°C in the incubator - Section 5.9. Feeding of diets
- Dump remaining diet or store diet maximum 2 days at 4°C

Most artificial diets for honey bee larvae contain royal jelly plus 3-9% (w/v) each of glucose and fructose and 0-2% (w/v) yeast extract (Rembold and Lackner, 1981; Vandenberg and Shimanuki, 1987; Peng *et al.*, 1992; Aupinel *et al.*, 2005; Genersch *et al.*, 2005).

Table 2. Daily amounts and composition for limited feeding according to the age of worker larvae in percentage of weight (Aupinel *et al.*, 2005). For details on preparation of diets (w/v and w/w in particular) see section 5.4.

Day of feeding	Amount [µl]	% D glucose	% D fructose	% yeast extract	% royal jelly
1	10	6	6	1	50
2	10	6	6	1	50
3	20	7.5	7.5	1.5	50
4	30	9	9	2	50
5	40	9	9	2	50
6	50	9	9	2	50

As an example, we present a detailed preparation description for 10 g diet with 6% (w/v) glucose, 6% (w/v) fructose and 1% (w/v) yeast extract (diet A which is fed to larvae on day one according to Aupinel *et al.*, 2005): 1.2 g of glucose, 1.2 g of fructose and 0.2 g of yeast extract are weighed (w) and solved in about 5 ml autoclaved water. This clear solution is filled up to exactly 10 ml (v). For preparing the final diet the aqueous solution must be mixed with fresh royal jelly 1:1 (w/w), for the given example 5 g of each are needed. If frozen royal jelly is used it can be recommended to freeze it in aliquots (e.g. amounts used to feed one plate), ready to be filled up with the same weight (w) of aqueous solution. To optimize worker rearing and avoid intercastes composition of diets and amounts of food must be changed in later larval stages (Table 2).

In another recipe (Genersch *et al.*, 2005; 2006), the larval diet consists of 3% (w/v) glucose, 3% (w/v) fructose, and 66.6% (v/v) royal jelly in sterile, double-distilled water. It is recommended to have a glucose/fructose stock solution containing 9% (w/v) glucose and 9% (w/v) fructose solved in sterile double-distilled water and stored at 4°C until use. Aliquots of royal jelly are stored frozen at -20°C. Every day fresh larval diet is prepared by thawing royal jelly shortly before use and mixing two volume parts of royal jelly with one volume part of glucose/fructose stock solution to obtain the above given concentrations. If an aqueous solution of yeast extract is also added, the concentration of the glucose/fructose stock solution needs to be adjusted to ensure that the larval diet does not contain less than 66.6% (v/v) royal jelly and not more than 33.3% (v/v) aqueous solutions. Otherwise the larval diet will contain too much water and young larvae will tend to drown in the diet or will have problems with digestion.

If investigations start in early spring, no fresh royal jelly will be available; therefore frozen royal jelly has to be used. Also, freeze-dried royal jelly stored at -70°C can be used (Peng *et al.*, 1992). In general, we suggest preparing new diet every day or using prepared and refrigerated diets within three days. Re-freezing should be avoided.

4.5. Grafting of larvae

- Prepare plates
- Apply diets into cups

- Pre-incubate plates and thermoblocks for temperature regulation during grafting
- Take combs to the lab and start grafting
- Section 5.8. Incubation conditions

According to the chosen method, rearing plates or petri dishes have to be prepared in advance, because larvae are already grafted into the first day's diet. Grafting is the collection of larvae from a comb in the same way as beekeepers graft larvae for queen breeding (see the *BEEBOOK* paper on queen rearing for a detailed explanation of grafting (Büchler *et al.*, 2013). A grafting tool or a paintbrush and a binocular loupe are recommended. The first priority in grafting is not to injure larvae. If the first grafting try fails (i.e. one suspects that the larva is injured), another larva should be taken. Larvae lie on their side in the bottom of the comb cell. Consequently, they breathe from the spiracles located on the upper side of their body. Upside-down grafting (i.e. larvae flipped over during the grafting process) should be avoided, because larvae should remain in the same position in the queen cup as originally on the comb due to their use of the open spiracles. Grafting should not take longer than 20 minutes per 48 larvae and a warm (> 20°C, place pre-incubated thermoblocks below rearing plates, see 5.9. Feeding of diets) and humid environment should be maintained throughout. While grafting from one comb, additional brood combs should be stored in an incubator at 34.5°C and more than 60% RH.

4.6. Exclusion of grafting effect

A commonly used option, depending on and if justified by the research aim, may be the exclusion of grafting effects by excluding dead larvae from the experiment 24 hours after grafting. However, this is impossible if susceptibility to a certain substance or pathogen decreases with increasing age, as is the case with *Paenibacillus larvae*, the etiological agent of American Foulbrood. If experiments can be performed equally well with second or older instar larvae, then it might be more appropriate to graft the more robust second or older instar larvae right away instead of replacing larvae in the course of the experiment. In acute toxicity testing, sometimes a surplus of larvae is used at day four to replace larvae which died prior to treatment administration. Of course, this procedure increases the quality of the resulting honey bees and thus biases results. Whether or not to replace larvae during the experiment must be considered carefully prior to use, explicitly documented, and critically discussed when evaluating the results.

4.7. Randomization

Random application of treatments to test larvae may be used to improve dose-effect curves in toxicity studies (Pierrick Aupinel, unpublished data). We recommend randomizing larvae within a study in order to avoid an eventual bias due to the minor differences of age and then a difference of susceptibility to the test pathogen or compound. Randomization can be conducted at different moments according to the applied protocol. Randomization is easily accomplished when determining the acute toxicity of a test compound. For example, one can randomly exchange queen cups on microtiter plates on day four, just before one ordinarily provides larvae with contaminated diet in an acute toxicity test. If testing in chronic conditions, cup randomization must be done at the grafting stage because larval exposure to test compounds begins at this moment. Larvae can be randomly dispatched directly from the comb to the plastic cups.

4.8. Incubation conditions

- Clean environment is needed (Section 5.3. Sterile environment)
- Prepare saturated solutions of K_2SO_4 and NaCl, place in open dishes in desiccator
- Incubate larvae at 34.5°C and 95% RH for the first 6 days
- On day 7 change humidity to 80% RH

Honey bee larvae may be kept in hermetic plexiglass desiccators or other air-tight plastic containers (to facilitate humidity control, e.g. Tupperware) placed inside incubators. Incubators alone are also adequate when the necessary humidity can be maintained in the chamber. It is important to maintain the desired rearing temperature of 34.5°C with maximum precision (± 0.5°C), as suboptimal larval temperature affects adult bee longevity and adult bee resistance to dimethoate and induces malformed wings (Medrzycki *et al.*, 2010). Furthermore, pathogens may increase or lower in pathogenicity according to temperature (Vojvodic *et al.*, 2011a). The deviation of incubation temperature must be kept as low as possible and temperature should be verified with data loggers to help explain any possible problems with mortality. Beside temperature, honey bee larvae also require constant and high humidity (Human *et al.*, 2006). Most researchers propose a humidity of 95-96% RH during the first 6 days followed by a reduction to 70-80% RH, which proved to be appropriate (Rembold and Lackner, 1981; Vandenberg and Shimanuki, 1987; Peng *et al.*, 1992; Aupinel *et al.*, 2005). The humidity adjustment is accomplished by first placing a dish with a saturated solution of K_2SO_4 (to achieve 95% RH) and later a saturated solution of NaCl (to achieve 80% RH) on the bottom of the desiccator. More on preparation of these solutions can be found in section 6.3 "Relative Humidity" of the *BEEBOOK* paper on maintaining adult bees in the laboratory (Williams *et al.*, 2013). Humidity should also be measured with data loggers regularly to verify accuracy. If glycerol soaked dental rolls are used (see 3. Sterile environment) they have to be removed at day 7 (Aupinel *et al.*, 2005). Alternatively, good humidity results have been obtained with a much easier approach by just filling up to one fourth of the wells of the plate with water, closing the plate with the accompanying lid, and placing the plates in bacterial incubators, i.e. incubators without CO_2 and humidity adjustment (Genersch *et al.*, 2005, 2006).

We recommend protecting larvae and reducing the handling of their rearing dishes to reduce mortality. The exposure of larvae to

temperature and humidity conditions other than those in the incubator increase bacterial or fungal infections and mortality. Larvae should be weighed only if necessary for experiments. Transferring to new dishes or carefully cleaning larvae at the beginning of or after defecation (Smith, 1959; Shuel and Dixon, 1986, Genersch *et al.*, 2005, 2006) is not performed in many methods, although it is reported to increase successful pupation and reduce pupal mortality (Vandenberg and Shimanuki, 1987).

4.9. Feeding of diets

- Warm diet to 34.5°C in the incubator
- Place rearing plates on pre-incubated thermoblocks
- Check mortality - Section 5.12. Assessing survival
- Remove dead larvae
- Place food by using a pipette next to the head of each larvae

Feedings can either be administered every day (for 6 days) in different amounts or, if allowed by the study design, the first two portions (10 + 10µl) can be administered the first day. This option, recommended by Aupinel *et al.* (2005), gives one day off from the lab and it is assumed that this does not affect mortality or the quality of the test. The total amount of diet given to each larva is 160µl and should be reported if altered. When feeding 160µl of diets in total, honey bee larvae will consume all the food administered: therefore the cleaning of plastic cups or transfer to new cups is not necessary, as each manipulation increases mortality. Prior to feeding, the diet is carefully heated on a magnetic stirrer or in incubators to 34.5°C. During the feeding procedure, the pipette should be placed on the inner side of the plastic cup and the food drop should be placed next to the mouth of the larvae. Drowning of the larvae should be avoided. If drowning occurs occasionally, lift the larvae with a sterile paintbrush; if it occurs systematically, check method (e.g. humidity inside the desiccator or water content of diet).

When larvae are reared following a protocol implying feeding *ad libitum*, each well is filled with an appropriate volume of diet depending on the volume of the well (e.g. 500µl per well of a 24-well plate), and larvae are then added to the well by placing them carefully on top of the larval diet. Larvae are transferred to newly filled wells every day and the number of larvae per well is adjusted to accommodate the size of the growing larvae. While ten first instar larvae can easily be placed into one well of a 24-well plate, each engorged larva needs a separate well (Fig. 2b). When using this rearing protocol, it is essential to remove any dead larva from the wells and to not transfer dead larvae into newly filled wells. Shortly before pupation, i.e. shortly before or after defecation, larvae are gently cleaned from adhering food and faeces by carefully rolling them over tissue paper (Kimwipes) and then transferred into pupation wells lined with filter paper (Peng, 1992; Genersch *et al.*, 2006). If defecation continues in the pupation-well, developing larvae are placed in new wells lined with clean filter paper.

It is important to pay attention to temperature during feeding. Thermoblocks (pre-warmed in incubators at 34.5°C) underneath the rearing plates can be used to stabilize the temperature of the larvae.

4.10. Royal jelly

- Check quality of royal jelly (preliminary experiments)
- Store royal jelly in aliquots at -20°C until usage
- Thaw royal jelly briefly before (Section 5.4. Preparation and storage of diets)

Royal jelly is the brood food produced by nurse bees to specifically feed queen larvae. The immature queen is superabundantly fed, and older queen larvae are fed a different formation than worker larvae (Brouwers *et al.*, 1987). This jelly is harvested by beekeeping operations and commercially available. Due to its manifold use, including for humans, much research on quality standards of this undefined hive product has been conducted. Royal jelly makes up a great part of the larvae's laboratory diet; thus its quality strongly influences *in vitro* rearing success. The exact composition of royal jelly and the importance of several of its components for larval development are not clear. Fresh royal jelly contains roughly 9-18% protein, 7-18% sugars and 3-8% lipids (Sabatini *et al.*, 2009). The water content varies between 50 and 70% (Rembold and Dietz, 1965; Sabatini *et al.*, 2009; Zheng *et al.*, 2011). Royal jelly contains 10-hydroxy-2-decenoic acid (10-HDA), an antibacterial substance that is analysed as a freshness parameter in routine testing.

The biochemical composition of royal jelly depends on harvesting season and regional origin (Sabatini *et al.*, 2009). Moisture and protein content also depend on the harvesting day (i.e. first, second, third) after grafting of larvae (Zheng *et al.*, 2011). The latter fact has long been demonstrated to influence the development of ovaries of *in vitro* -reared larvae (Mitsui *et al.*, 1964). Accordingly, differences even from several batches of royal jelly may significantly alter the quality (e.g. protein content, see Fig. 1) of *in vitro* diets and influence rearing success or susceptibility against pathogens. The quality of royal jelly is presumably also altered by shipping and storage condition and duration. Differences in freeze-dried, frozen or fresh royal jelly (harvested during investigation year) can be expected. For experiments involving the testing of pathogens, the presence of unwanted antibiotics in royal jelly must be excluded, hence we recommend the use of organic royal jelly or own production. Royal jelly may also be irradiated at 20 kGy for purposes of sterilization. Though it is not known if this will impact its structural integrity or affect its developmental impact on larvae, Gregorc and Ellis (2011) and Gregorc *et al.* (2012) reared larvae successfully on irradiated royal jelly. It is advisable to test every new batch of royal jelly to make sure that the results obtained with the 'old' batch can be reproduced with the 'new' batch. Sometimes it might be necessary to test several batches before a batch suitable to replace the consumed charge can be identified.

4.11. Application of test substances

Substances (toxins or pathogens) can be tested in chronic or acute conditions. In chronic testing, a substance is mixed into the diets at a constant concentration and provided to the larvae at each feeding day. Acute exposure can occur at every feeding day by mixing the tested substance into the diet once. The testing compound is dissolved in water. If it is not soluble in water at the experimental concentrations, other solvents such as acetone can be used. In that case, it is required to have a second control group fed with diet containing the solvent at the same concentration as the treated samples. The proportion of the solvent must not exceed 10% of the final volume. A different rate has to be justified. In all cases, a constant volume for the different dilutions must be used in order to maintain a constant concentration between the control diet and the test substance diet. In order to assess the adequate LD_{50} or LC_{50} range, it is recommended to run a preliminary experiment where doses of the test substance may vary according to a geometrical ratio from 5 to 10 (for examples, see the *BEEBOOK* paper on toxicology methods (Medrzycki *et al.*, 2013)).

4.12. Assessing survival

Larvae can be classified as dead when respiration ends, when they lose body elasticity, or when they develop oedema and change colour to greyish or brownish (Genersch *et al.*, 2005, 2006). Dead larvae should be removed from the incubator to prevent decomposition by bacterial or fungal saprophytes and subsequent contamination of other larvae; in pathology studies dead larvae need to be sampled anyway. Non-emerged individuals on day 22 after grafting are counted as dead during pupal stage. Live adults and dead adults that have left their cell and show a regular development are both counted upon emergence or at the latest at day 22.

4.13. Control mortality

If *in vitro* larval rearing methods are supposed to become a commonly accepted method to test plant protection products, the maximum tolerated control mortality must be considered for validation. Techniques have improved over the years, and control mortality has progressively reached low levels. Nonetheless, control mortality has to be reported regularly, to demonstrate successful trials. Fukuda and Sakagami (1968) found in a normal colony setting, that 85% of adult workers emerge from the eggs laid by a queen, with mortality occurring in all premature stages. This data suggest a natural mortality of about 15% during larval and pupal development. Therefore, in control samples of *in vitro*-reared larvae, total mortality until late pupal development ($\sim$ day 14 after grafting of first instar larvae) should be lower or equal to 15%. This is especially important for the assessment of the median lethal dose (LD_{50}) or the median lethal concentration (LC_{50}), while 20% control mortality can be accepted for the assessment of a NOAEL (no observable adverse effect level) or a NOAEC (no observable adverse effect concentration). In case of higher mortality in control samples, the replicate is invalidated or convincing reasons, like adverse weather conditions before grafting of larvae or late season, must be given. However, when *in vitro* rearing of larvae is just one method among others to answer a scientific question, we assume a control mortality of 5% from grafting to day 7 to be very good, and 10% tolerable; from grafting to day 22, 20% is very good, while 25% can be tolerated. Care should be taken in presenting these parameters unambiguously and the way they were calculated. These levels should be achieved without exclusion of the grafting effect (see 5.6. Exclusion of grafting effect). Several research institutes have already accomplished control mortalities below these suggested ones (Aupinel *et al.,* 2005; Aupinel *et al.,* 2009; Brodschneider *et al.,* 2009; Hendriksma *et al.,* 2011a; Kaftanoglu *et al.,* 2010, 2011).

4.14. Capping

In order to investigate parameters of adult honey bees, capping of the rearing plates with perforated bees wax and turning the plates to a vertical position on day eleven has been proven as a useful add-on (Riessberger-Gallé *et al.,* 2008; Brodschneider *et al.,* 2009). A wax layer big enough to cover one 48 well rearing plate is obtained by squeezing about 20 cm^2 of soft (warm) but not liquid comb foundation between two sheets of paper. The bees are able to bite through the thin, almost transparent, wax layer, which allows for the use of the eclosion rate as an additional parameter of survival. Rotating the rearing plates not only reduces malformations due to the unnatural vertical position of larvae in horizontal rearing plates, but capping may also improve microclimate (humidity) in cells and act as an additional barrier against fungi.

4.15. Evaluation of rearing success

The control mortality for evaluation of rearing success was already discussed (see 5.13. Control mortality). Researchers also measured physiological parameters such as body weight, body size, ovary development and flight performance to evaluate rearing methods. A comparison with sisters naturally reared in a colony provides information about the quality of the artificial diet and rearing condition. This comparison reveals that larvae reared in the laboratory are close to those reared in a colony. They have slightly reduced wing areas and thorax dry weights, but reach the age of foragers when introduced into a colony and engage in long persisting flights similar to those of naturally reared control bees (Brodschneider *et al.,* 2009).

Another problem is the predictable production of individuals unambiguously belonging to the worker caste, which might be necessary for certain scientific questions. Early attempts to rear worker larvae in the laboratory frequently resulted in inter-castes (Rhein, 1933; Weaver, 1974; Rembold *et al.,* 1974; Rembold and Lackner, 1981; Shuel and Dixon, 1986). Therefore, and also because larval nutrition affects ovary development (Hoover *et al.,* 2006), the developmental stages of

ovaries and number of ovarioles of laboratory-reared honey bees have been investigated (Weaver, 1955; Mitsui *et al.,* 1964; Kaftanoglu *et al.,* 2010). Also other characteristics such as developmental time, tongue length, spermatheca, chaetotaxy of metathoric legs, sting lancet, mandibular notch, weight at adult emergence and juvenile hormone titer are used for differentiation (Weaver, 1974; Asencot and Lensky, 1976, 1984; Shuel and Dixon, 1986; Kucharski *et al.,* 2008; Kamakura, 2011).

4.16. Rearing reproductives in the laboratory

According to investigations on honey bee workers, there are many reports on rearing queen bees in the laboratory (Hanser, 1983; Patel *et al.,* 2007). Most recently, Kamakura (2011) produced queens feeding just one basic diet with 6% glucose, 6% fructose and 1% yeast extract, and no increase of matters in the following diets. Amounts of diets and reason for queen development are not given. The same plastic cups can be used, as they are also applied in queen breeding in the colony.

In vitro rearing of drone larvae is also possible feeding previously established worker diets at least up to the prepupal phase (Herrmann *et al.,* 2005; Behrens *et al.,* 2007, 2010) and even up to adult emergence (Woyke, 1963; Takeuchi *et al.,* 1972). However, in some cases the control mortality in drones during in vitro rearing was higher than in workers. This could be due the fact that drone larvae might in fact need a slightly different food composition than workers (Hrassnigg and Crailsheim, 2005) which has not been studied in detail yet. But also the longer developmental time (especially during the very critical pupal phase) or lethal factors that become visible in the haploid genome might play a role. Clearly, more work is needed to obtain results in drones that are comparable to those of workers and queens. Especially the timing of changing diets would need to be adapted to the development of drone larvae.

5. Discussion

Rearing honey bee larvae *in vitro* is of great importance for research on pathogens and risk assessment. Recent studies have developed protocols that gained success in producing unambiguous workers with low mortalities and facilitated the use of readily available materials (Vandenberg and Shimanuki, 1987; Aupinel *et al.,* 2005). This enables laboratories to rear high numbers of workers compared to the significant efforts from pioneer studies (Rhein, 1933). However, the quality of food, incubation conditions and occurrence of infections can also result in high mortality, and we therefore present some recommendations to reduce undesirable effects.

The production of honey bee workers in the laboratory is still a challenge, in contrast to queen breeding. Queen breeding (see the *BEEBOOK* paper on rearing queens (Büchler *et al.,* 2013) has been employed by beekeepers for decades. It is a process in which larvae grafted into queen cups are raised to queens in a colony, and has also been employed in the laboratory, where larvae are grafted on royal jelly *in vitro*. In all described methods royal jelly is diluted to reduce its queen-potency and produce workers. Alternatively, the recent findings regarding 'Royalactin' may provide another possibility. According to Kamakura (2011), 'Royalactin', a 57 kDa protein, is the queen-determining factor in royal jelly. He demonstrated that not only worker larvae, but even *Drosophila* larvae fed on the synthesized protein exhibit features normally associated with the honey bee queen: shortened developmental time, enhanced longevity and increased egg production. Even more intriguing is the fact that this protein seems to degrade with storage, and honey bee larvae reared on royal jelly stored at 40°C for 30 days all develop with a full worker morphotype. The potential of these findings to allow mass rearing of workers on degraded royal jelly needs further attention and investigation.

We gave much detailed information on what must or can be taken into account to make *in vitro* rearing of honey bee larvae comparable among different laboratories. Researchers also need the freedom to adopt the given recommendations as needed but in accordance with their research purpose and time management. In the end, researches must carefully consider a method (e.g. larval grafting stage, single or group rearing, *ad libitum* or limited feeding of diets or quantity and quality of manipulations) that is justified by the research topic.

In conclusion, it is absolutely essential for the successful rearing of honey bee larvae to use fresh materials of high quality and to gain experience in the chosen method before applying it to any scientific question or to routine testing of compounds and pathogens. However, even with the best of all methods, *in vitro*-reared larvae will always be *in vitro*-reared larvae and never become indistinguishable from larvae reared in a healthy colony. Therefore, researchers have to remain critical regarding results obtained with *in vitro*-reared larvae. Researchers need to decide how close to 'natural' (meaning reared in a colony, *in vivo*) bees they want to come with their *in vitro* efforts and adjust efforts accordingly. In general, the similarity of bees produced by current protocols to natural bees has been proven. This is a strong argument for the use of these methods in research and as the official testing method for plant protection products. Like with all other *in vitro* methods, these larvae are, although a valuable research tool, just a model for reality rather than reality itself.

Acknowledgement

The establishment of the larval rearing method at Karl-Franzens-University-Graz was financed by the Land Steiermark (A3.16Z1/2006-5).

References

AIZEN, M A; L D HARDER (2009) The global stock of domesticated honey bees is growing slower than agricultural demand for pollination. *Current Biology* 19: 915-918. http://dx.doi.org/10.1016/j.cub.2009.03.071

ASENCOT, M; LENSKY, Y (1976) The effect of sugars and juvenile hormone on the differentiation of the female honey bee larvae (*Apis mellifera* L.) to queens. *Life Sciences* 18: 693-699.

ASENCOT, M; LENSKY, Y (1984) Juvenile hormone induction of 'queenliness' on female honey bee (*Apis mellifera* L.) larvae reared on worker jelly and on stored royal jelly. *Comparative Biochemistry and Physiology* B:78: 109-117.

ASENCOT, M; LENSKY, Y (1988) The effect of soluble sugars in stored royal jelly on the differentiation of female honey bee (*Apis mellifera* L.) larvae to queens. *Insect Biochemistry* 18: 127-133.

AUPINEL, P; FORTINI, D; DUFOUR, H; TASEI, J N; MICHAUD, B; ODOUX, J-F; PHAM-DELÈGUE, M H (2005) Improvement of artificial feeding in a standard *in vitro* method for rearing *Apis mellifera* larvae. *Bulletin of Insectology* 58: 107-111.

AUPINEL, P; FORTINI, D; MICHAUD, B; MAROLLEAU, F; TASEI, J-N; ODOUX, J-F (2007a) Toxicity of dimethoate and fenoxycarb to honey bee brood (*Apis mellifera*), using a new *in vitro* standardized feeding method. *Pest Management Science* 63: 1090-1094. http://dx.doi.org/10.1002/PS.1446

AUPINEL, P; MEDRZYCKI, P; FORTINI, D; MICHAUD, B; TASEI, J N; ODOUX, J-F (2007b) A new larval *in vitro* rearing method to test effects of pesticides on honey bee brood. *Redia* 90: 91-94.

AUPINEL, P; FORTINI, D; MICHAUD, B; MEDRZYCKI, P; PADOVANI, E; PRZYGODA, D; MAUS, C; CHARRIERE, J-D; KILCHENMANN, V; RIESSBERGER-GALLE, U; VOLLMANN, J J; JEKER, L; JANKE, M; ODOUX, J-F; TASEI J-N (2009) Honey bee brood ring-test: method for testing pesticide toxicity on honey bee brood in laboratory conditions. *Julius-Kühn-Archiv* 423: 96-102.

BECKER, R; VERGNET, C; MAUS, C; PISTORIUS, J; TORNIER, I; WILKINS, S (2009) Proposal of the ICPBR Bee Brood Group for testing and assessing potential side effects from the use of plant protection products on honey bee brood. *Julius-Kühn-Archiv* 423: 43-44.

BEHRENS, D; FORSGREN, E; FRIES, I; MORITZ, R F A (2007) Infection of drone larvae (*Apis mellifera*) with American foulbrood. *Apidologie* 38: 281-288. http://dx.doi.org/10.1051/apido:2007009

BEHRENS, D; FORSGREN, E; FRIES, I; MORITZ, R F A (2010) Lethal infection thresholds of *Paenibacillus larvae* for honey bee drone and worker larvae (*Apis mellifera*). *Environmental Microbiology* 12: 2838–2845. http://dx.doi.org/10.1111/j.1462-2920.2010.02257.x

BERTHOLF, L M (1925): The moults of the honey bee. *Journal of Economic Entomology* 18: 380-384.

BERTHOLF, L M (1927) The utilization of carbohydrates as food by honey bee larvae. *Journal of Agricultural Research* 35: 429-52.

BRODSCHNEIDER, R; RIESSBERGER-GALLÉ, U; CRAILSHEIM, K (2009) Flight performance of artificially reared honey bees (*Apis mellifera*). *Apidologie* 40: 441-449. http://dx.doi.org/10.1051/apido/2009006

BRØDSGAARD, C J; RITTER, W; HANSEN, H (1998) Response of *in vitro* reared honey bee larvae to various doses of *Paenibacillus larvae larvae* spores. *Apidologie* 29: 569-578.

BRØDSGAARD, C J; RITTER, W; HANSEN, H; BRØDSGAARD, H F (2000) Interactions among *Varroa jacobsoni* mites, acute paralysis virus, and *Paenibacillus larvae larvae* and their influence on mortality of larval honey bees *in vitro*. *Apidologie* 31: 543-554. http://dx.doi.org/10.1051/apido:2000145

BRØDSGAARD, H F; BRØDSGAARD C J; HANSEN, H; LÖVEI, G L (2003) Environmental risk assessment of transgene products using honey bee (*Apis mellifera*) larvae. *Apidologie* 34: 139-145. http://dx.doi.org/10.1051/apido:2003003

BROUWERS, E V M (1984) Glucose / fructose ratio in the food of honey bee larvae during caste differentiation. *Journal of Apicultural Research* 23: 94-101.

BROUWERS, E V M; EBERT, R; BEETSMA, J (1987) Behavioural and physiological aspects of nurse bees in relation to the composition of larval food during caste differentiation in the honey bee. Journal of Apicultural Research 23: 94-101.

BÜCHLER, R; ANDONOV, S; BIENEFELD, K; COSTA, C; HATJINA, F; KEZIC, N; KRYGER, P; SPIVAK, M; UZUNOV, A; WILDE, J (2013) Standard methods for rearing and selection of *Apis mellifera* queens. In *V Dietemann; J D Ellis; P Neumann (Eds) The COLOSS BEEBOOK, Volume I: standard methods for* Apis mellifera *research*. *Journal of Apicultural Research* 52(1): http://dx.doi.org/10.3896/IBRA.1.52.1.07

COLLINS, A M (2004) Variation in time of egg hatch by the honey bee, *Apis mellifera* (Hymenoptera: Apidae). *Annals of the Entomological Society of America* 97: 140-146. http://dx.doi.org/10.1603/0013-8746(2004)097[0140:VITOEH]2.0.CO;2

CZOPPELT, C; REMBOLD, H (1988) Effect of parathion on honey bee larvae reared *in vitro*. *Anzeiger für Schädlingskunde und Pflanzenschutz* 61: 95-100.

DA SILVA CRUZ, A; DA SILVA-ZACARIN, E C; BUENO, O C; MALASPINA, O (2010) Morphological alterations induced by boric acid and fipronil in the midgut of worker honey bee (*Apis mellifera* L.) larvae: Morphological alterations in the midgut of *A. mellifera*. *Cell Biology and Toxicology* 26: 165-76. http://dx.doi.org/10.1007/s10565-009-9126-x

DAVIS, A R; SOLOMON, K R; SHUEL, R W (1988) Laboratory studies of honey bee larval growth and development as affected by systemic insecticides at adult-sublethal levels. *Journal of Apicultural Research* 27: 146-161.

FOLEY, K; FAZIO, G; JENSEN, A B; HUGHES, W O H (2012) Nutritional limitation and resistance to opportunistic *Aspergillus* parasites in honey bee larvae. *Journal of Invertebrate Pathology* 111: 68-73. http://dx.doi.org/10.1016/j.jip.2012.06.006

FORSGREN, E; OLOFSSON, T C; VÁSQUEZ, A; FRIES, I (2010) Novel lactic acid bacteria inhibiting *Paenibacillus larvae* in honey bee larvae. *Apidologie* 41: 99-108. http://dx.doi.org/10.1051/apido/2009065

FRAENKEL, G; BLEWETT, M (1943) The basic food requirements of several insects. *Journal of Experimental Biology* 20: 28-34.

FUKUDA, H; SAKAGAMI, S F (1968) Worker brood survival in honey bees. *Researches on Population Ecology* 10: 31-39. http://dx.doi.org/10.1007/BF02514731

GALLAI, N; SALLES, J-M; SETTELE, J; VAISSIÈRE, B E (2009) Economic valuation of the vulnerability of world agriculture confronted with pollinator decline. *Ecological Economics* 68: 810-821. http://dx.doi.org/10.1016/j.ecolecon.2008.06.014

GENERSCH, E; ASHIRALIEVA, A; FRIES, I (2005) Strain- and genotype-specific differences in virulence of *Paenibacillus larvae* subsp. *larvae*, a bacterial pathogen causing American foulbrood disease in honey bees. *Applied and Environmental Microbiology*, 7551-7555. http://dx.doi.org/10.1128/AEM.71.11.7551-7555.2005

GENERSCH, E; FORSGREN, E; PENTIKÄINEN, J; ASHIRALIEVA, A; RAUCH, S; KILWINSKI, J; FRIES, I (2006) Reclassification of *Paenibacillus larvae subsp. pulvifaciens* and *Paenibacillus larvae* subsp. *larvae* as *Paenibacillus larvae* without subspecies differentiation. *International Journal of Systematic and Evolutionary Microbiology* 56: 501-511. http://dx.doi.org/10.1099/ijs.0.63928-0

GENERSCH, E; EVANS, J D; FRIES, I (2010) Honey bee disease overview. *Journal of Invertebrate Pathology* 103: S2-S4. http://dx.doi.org/10.1016/j.jip.2009.07.015

GREGORC, A; ELLIS, J D (2011) Cell death localization *in situ* in laboratory reared honey bee (*Apis mellifera* L.) larvae treated with pesticides. *Pesticide Biochemistry and Physiology* 99: 200-207. http://dx.doi.org/10.1016/j.pestbp.2010.12.005

GREGORC, A; EVANS, J D; SCHARF, M; ELLIS, J D (2012) Gene expression in honey bee (*Apis mellifera*) larvae exposed to pesticides and varroa mites (*Varroa destructor*). *Journal of Insect Physiology* 58: 1042-1049. http://dx.doi.org/10.1016/j.jinsphys.2012.03.015

HANSER, G (1983) Rearing queen bees in the laboratory. In *F Ruttner (Ed.). Queen rearing: Biological basis and technical instruction*. Apimondia Publishing House; Bucharest, Romania. pp: 63-81.

HENDRIKSMA, H P; HÄRTEL, S; STEFFAN-DEWENTER, I (2011a) Honey bee risk assessment: new approaches for *in vitro* larvae rearing and data analyses. *Methods in Ecology and Evolution* 2: http://dx.doi.org/10.1111/j.2041-210X.2011.00099.x

HENDRIKSMA, H P; HÄRTEL, S; STEFFAN-DEWENTER, I (2011b) Testing pollen of single and stacked insect-resistant bt-maize on *in vitro* reared honey bee larvae. *PLoS ONE* 6: e28174. http://dx.doi.org/10.1371/journal.pone.0028174

HENDRIKSMA, H P; HÄRTEL, S; BABENDREIER, D; OHE, W V D; STEFFAN-DEWENTER, I (2012) Effects of multiple Bt proteins and GNA lectin on *in vitro*-reared honey bee larvae. *Apidologie*, online first, http://dx.doi.org/10.1007/s13592-012-0123-3

HERRMANN, M; TRENZCEK, T; FAHRENHORST, H; ENGELS, W (2005) Characters that differ between diploid and haploid honey bee (*Apis mellifera*) drones. *Genetics and Molecular Research* 4: 624-641.

HERRMANN, M; ENGELS, W D; ENGELS, W (2008) Optimierte Futtermischungen zur Laboraufzucht von Bienenlarven (Hymenoptera: Apidae: Apis mellifera). *Entomologia Generalis* 31: 13-20.

HOOVER, S E; HIGO, H A; WINSTON, M L (2006) Worker honey bee ovary development: seasonal variation and the influence of larval and adult nutrition. *Journal of Comparative Physiology B* 176: 55-63. http://dx.doi.org/10.1007/s00360-005-0032-0

HRASSNIGG, N; CRAILSHEIM, K (2005) Differences in drone and worker physiology in honey bees (*Apis mellifera* L.). *Apidologie* 36: 255–277. http://dx.doi.org/10.1051/apido:2005015

HUMAN, H; BRODSCHNEIDER, R; DIETEMANN, V; DIVELY, G; ELLIS, J; FORSGREN, E; FRIES, I; HATJINA, F; HU, F-L; JAFFÉ, R; KÖHLER, A; PIRK, C W W; ROSE, R; STRAUSS, U; TANNER, G; VAN DER STEEN, J J M; VEJSNÆS, F; WILLIAMS, G R; ZHENG, H-Q (2013) Miscellaneous standard methods for Apis mellifera research. In *V Dietemann; J D Ellis; P Neumann (Eds) The COLOSS* BEEBOOK, *Volume I: standard methods for* Apis mellifera *research. Journal of Apicultural Research* 52(4): http://dx.doi.org/10.3896/IBRA.1.52.4.10

HUMAN, H; NICOLSON, S W; DIETEMANN, V (2006) Do honey bees, *Apis mellifera scutellata*, regulate humidity in their nest? *Naturwissenschaften* 93: 397-401. http://dx.doi.org/10.1007/s00114-006-0117-y

JAY, S C (1964) Rearing honey bee brood outside the hive. *Journal of Apicultural Research* 3: 51-60.

JENSEN, A B; PEDERSEN, B V; EILENBERG, J (2009) Differential susceptibility across honey bee colonies in larval chalkbrood resistance. *Apidologie* 40: 524-534. http://dx.doi.org/10.1051/apido/2009029

JUNG-HOFFMANN, I (1966) Die Determination von Königin und Arbeiterin der Honigbiene. *Zeitschrift für Bienenforschung* 8: 296-322.

KAFTANOGLU, O; LINKSVAYER, T A; PAGE JR, R E (2010) Rearing honey bees (*Apis mellifera* L.) *in vitro*: effects of feeding intervals on survival and development. *Journal of Apicultural Research* 49: 311-317. http://dx.doi.org/10.3896/IBRA.1.49.4.03

KAFTANOGLU, O; LINKSVAYER, T A; PAGE, R E (2011) Rearing honey bees, *Apis mellifera*, in vitro 1: Effects of sugar concentrations on survival and development. *Journal of Insect Science* 11: 96. http://dx.doi.org/10.1673/031.011.9601

KAMAKURA, M (2011) Royalactin induces queen differentiation in honey bees. *Nature* 473: 478-483. http://dx.doi.org/10.1038/nature10093

KLEIN, A; VAISSIÈRE, B E; CANE, J H; STEFFAN-DEWENTER, I; CUNNINGHAM, S A; KREMEN, C; TSCHARNTKE, T (2007) Importance of pollinators in changing landscapes for world crops. *Proceedings of the Royal Society B* 274: 303-313. http://dx.doi.org/10.1098/rspb.2006.3721

KUCHARSKI, R; MALESZKA, J; FORET, S; MALESZKA, R (2008) Nutritional control of reproductive status in honey bees via DNA Methylation. *Science* 319: 1827-1830. http://dx.doi.org/10.1126/science.1153069

KUNERT, K; CRAILSHEIM, K (1987) Sugar and protein in the food for honey bee worker larvae. In *J Eder and H Rembold (Eds). Chemistry and Biology of Social Insects.* Peperny; München, Germany. pp 164-165.

LINDAUER, M (1952) Ein Beitrag zur Frage der Arbeitsteilung im Bienenstaat. *Zeitschrift für Vergleichende Physiologie* 34: 299-345.

MALONE, L A; TREGIDGA, E L; TODD, J H; BURGESS, E P J; PHILIP, B A; MARKWICK, N P; POULTON, J; CHRISTELLER, J T; LESTER, M T; GATEHOUSE H S (2002) Effects of ingestion of a biotin-binding protein on adult and larval honey bees. *Apidologie* 33: 447-458. http://dx.doi.org/10.1051/apido:2002030

MEDRZYCKI, P; GIFFARD, H; AUPINEL, P; BELZUNCES, L P; CHAUZAT, M-P; CLAßEN, C; COLIN, M E; DUPONT, T; GIROLAMI, V; JOHNSON, R; LECONTE, Y; LÜCKMANN, J; MARZARO, M; PISTORIUS, J; PORRINI, C; SCHUR, A; SGOLASTRA, F; SIMON DELSO, N; STEEN VAN DER, J; WALLNER, K; ALAUX, C; BIRON, D G; BLOT, N; BOGO, G; BRUNET, J-L; DELBAC, F; DIOGON, M; EL ALAOUI, H; TOSI, S; VIDAU, C (2013) Standard methods for toxicology research in *Apis mellifera*. In *V Dietemann; J D Ellis; P Neumann (Eds) The COLOSS* BEEBOOK, *Volume I: standard methods for* Apis mellifera *research. Journal of Apicultural Research* 52(4): http://dx.doi.org/10.3896/IBRA.1.52.4.14

MEDRZYCKI, P; SGOLASTRA, F; BORTOLOTTI, L; BOGO, G; TOSI, S; PADOVANI, E; PORRINI, C; SABATINI, A G (2010) Influence of brood rearing temperature on honey bee development and susceptibility to poisoning by pesticides. *Journal of Apicultural Research* 49: 52 -59. http://dx.doi.org/10.3896/IBRA.1.49.1.07

MICHAEL, A S; ABRAMOVITZ, M (1955) A new method of rearing honey bee larvae *in vitro. Journal of Economic Entomology* 48: 43-44.

MITSUI, T; SAGAWA, T; SANO, H (1964) Studies on rearing honey bee larvae in the laboratory. I. The effects of royal jelly taken from different ages of queen cells on queen differentiation. *Journal of Economic Entomology* 57: 518-521.

MORITZ, B; CRAILSHEIM, K (1987) Physiology of protein digestion in the midgut of the honey bee (*Apis mellifera* L.). *Journal of Insect Physiology* 33: 923-931.

MORITZ, R F A; DE MIRANDA, J; FRIES, I; LE CONTE, Y; NEUMANN, P; PAXTON, R J (2010) Research strategies to improve honey bee health in Europe. *Apidologie* 41: 227-242. http://dx.doi.org/10.1051/apido/2010010

MULLIN, C A; FRAZIER, M; FRAZIER, J L; ASHCRAFT, S; SIMONDS, R; VANENGELSDORP, D; PETTIS, J S (2010) High levels of miticides and agrochemicals in North American apiaries: implications for honey bee health. *PLoS ONE* 5: e9754. http://dx.doi.org/10.1371/journal.pone.0009754

MYSER, W C (1954) The larval and the pupal development of the honey bee *Apis mellifera* Linnaeus. *Annals of the Entomological Society of America* 47: 683-711.

NEUMANN, P; CARRECK, N L (2010) Honey bee colony losses. *Journal of Apicultural Research* 49: 1-6. http://dx.doi.org/10.3896/IBRA.1.49.1.01

OOMEN, P A; DE RUIJTER, A; VAN DER STEEN, J (1992) Method for honey bee brood feeding tests with insect growth-regulating insecticides. *EPPO Bulletin* 22: 613-616.

PATEL, A; FONDRK, M K; KAFTANOGLU, O; EMORE, C; HUNT, G; FREDERICK, K; AMDAM, G V (2007) The making of a queen: TOR pathway is a key player in diphenic caste development. *PLoS ONE* 2: e509. http://dx.doi.org/10.1371/journal.pone.0000509

PENG, Y S C; MUSSEN, E; FONG, A; MONTAGUE, M A; TYLER, T (1992) Effects of chlortetracycline on honey-bee worker larvae reared *in vitro*. *Journal of Invertebrate Pathology* 60: 127-133. http://dx.doi.org/10.1016/0022-2011(92)90085-I

PENG, Y S C; MUSSEN, E; FONG, A; CHENG, P; WONG, G; MONTAGUE, M A (1996) Laboratory and field studies on the effects of the antibiotic tylosin on honey bee *Apis mellifera* L. (Hymenoptera: Apidae) development and prevention of American foulbrood disease. *Journal of Invertebrate Pathology* 67: 65-71.

REMBOLD, H; DIETZ, A (1965) Biologically active substances in royal jelly. *Vitamins and Hormones* 23: 359-383.

REMBOLD, H; LACKNER, B; GEISTBECK, I (1974) The chemical basis of honey bee, *Apis mellifera*, caste formation. Partial purification of queen bee determinator from royal jelly. *Journal of Insect Physiology* 20: 307-314.

REMBOLD, H; KREMER J-P; ULRICH, G M (1980): Characterization of postembryonic developmental stages of the female castes of the honey bee, *Apis mellifera* L. *Apidologie* 11: 29-38.

REMBOLD, H; LACKNER, B (1981) Rearing of honey bee larvae *in vitro*: Effect of yeast extract on queen differentiation. *Journal of Apicultural Research* 20: 165-171.

RHEIN, W V (1933) Über die Entstehung des weiblichen Dimorphismus im Bienenstaate. *Wilhelm Roux Archiv für Entwicklungsmechanik der Organismen* 129: 601-665.

RIESSBERGER-GALLÉ, U; VOLLMANN, J; BRODSCHNEIDER, R; AUPINEL, P; CRAILSHEIM, K (2008) Improvement in the pupal development of artificially reared honey bee larvae. *Apidologie* 39: 595. http://dx.doi.org/10.1051/apido:2008041

RIESSBERGER-GALLÉ, U; SCHUEHLY, W; FEIERL, G; HERNANDEZ-LOPEZ, J; CRAILSHEIM, K (2011) Rearing of honey bee larvae under aseptic conditions.*Apidologie* 42: 795. http://dx.doi.org/10.1007/s13592-011-0095-8

SABATINI, A G; MARCAZZAN, G L; CABONI, M F; BOGDANOV, S; DE ALMEIDA-MURADIAN, L B (2009) Quality and standardisation of Royal Jelly. *Journal of ApiProduct and ApiMedical Science* 1: 1-6. http://dx.doi.org/10.3896/IBRA.4.1.01.04

SCHUR, A; TORNIER, I; BRASSE, D (2003) Honey bee brood ring-test in 2002: method for the assessment of side effects of plant protection products on the honey bee brood under semi-field conditions. *Bulletin of Insectology* 56: 91-96.

SHI, Y Y; HUANG, Z Y; ZENG, Z J; WANG, Z L; WU, X B; YAN, W Y (2011) Diet and cell size both affect queen-worker differentiation through DNA methylation in honey bees (*Apis mellifera*, Apidae). *PLoS ONE* 6: e18808. http://dx.doi.org/10.1371/journal.pone.0018808

SHUEL, R W; DIXON, S E (1960) The early establishment of dimorphism in the female honey bee, *Apis mellifera* L. *Insectes Sociaux* 7: 265-282.

SHUEL, R W; DIXON, S E (1986) An artificial diet for laboratory rearing of honey bees. *Journal of Apicultural Research* 25: 35-43.

SHUEL, R W; DIXON, S E; KINOSHITA, G B (1978) Growth and development of honey bees in the laboratory on altered queen and worker diets. *Journal of Apicultural Research* 17: 57-68.

SILVA, I C; MESSAGE, D; CRUZ, C D; CAMPOS; L A O; SOUSA-MAJER, M J (2009) Rearing Africanized honey bee (*Apis mellifera* L.) brood under laboratory conditions. *Genetics and Molecular Research* 8: 623-629.

SMITH, M V (1959) Queen differentiation and the biological testing of royal jelly. *Cornell University Agricultural Experiment Station Memoir* 356: 3-56.

TAKEUCHI, K; WATABE, N; MATSUKA, M (1972) Rearing drone honey bees in an incubator. *Journal of Apicultural Research* 11: 147-151.

VAN DER ZEE, R; PISA, L; ANDONOV, S; BRODSCHNEIDER, R; CHARRIÈRE, J-D; CHLEBO, R; COFFEY, M F; CRAILSHEIM, K; DAHLE, B; GAJDA, A; GRAY, A; DRAZIC, M; HIGES, M; KAUKO, L; KENCE, A; KENCE, M; KEZIC, N; KIPRIJANOVSKA, H; KRALJ, J; KRISTIANSEN, P; MARTIN HERNANDEZ, R; MUTINELLI, F; NGUYEN, B K; OTTEN, C; ÖZKIRIM, A; PERNAL, S F; PETERSON, M; RAMSAY, G; SANTRAC, V; SOROKER, V; TOPOLSKA, G; UZUNOV, A; VEJSNÆS, F; WEI, S; WILKINS, S (2012) Managed honey bee colony losses in Canada, China, Europe, Israel and Turkey, for the winters of 2008-9 and 2009-10. *Journal of Apicultural Research* 51: 100-114. http://dx.doi.org/10.3896/IBRA.1.51.1.12

VANENGELSDORP, D; MEIXNER, M D (2010) A historical review of managed honey bee populations in Europe and the United States and the factors that may affect them. *Journal of Invertebrate Pathology* 103: S80-S95. http://dx.doi.org/10.1016/j.jip.2009.06.011

VANDENBERG, J D; SHIMANUKI, H (1987) Technique for rearing worker honey bees in the laboratory. *Journal of Apicultural Research* 26: 90-97.

VANDERZANT, E S (1974) Development, significance, and application of artificial diets for insects. *Annual Review of Entomology* 19: 139-160.

VASQUEZ, A; FORSGREN, E; FRIES, I; PAXTON, R J; FLABERG, E; SZEKELY, L; OLOFSSON, T C (2012) Symbionts as major modulators of insect health: lactic acid bacteria and honey bees. *PLoS ONE* 7 (3): e33188. http://dx.doi.org/10.1371/journal.pone.0033188

VELICH, A V (1930) Entwicklungsmechanische Studien an Bienenlarven. *Zeitschrift für wissenschaftliche Zoologie* 136: 210-222.

VOJVODIC, S; JENSEN, A B; JAMES, R R; BOOMSMA, J J; EILENBERG, J (2011a) Temperature dependent virulence of obligate and facultative fungal pathogens of honey bee brood. *Veterinary Microbiology*: 149, 200-205. http://dx.doi.org/10.1016/j.vetmic.2010.10.001

VOJVODIC, S; JENSEN, A B; MARKUSSEN, B; EILENBERG, J; BOOMSMA, J J (2011b) Genetic variation in virulence among chalkbrood strains infecting honey bees. *PLoS ONE* 6: e25035. http://dx.doi.org/10.1371/journal.pone.0025035

VOJVODIC, S; BOOMSMA, J J; EILENBERG, J; JENSEN, A B (2012) Virulence of mixed fungal infections in honey bee brood. *Frontiers in Zoology* 2012, 9:5. http://dx.doi.org/10.1186/1742-9994-9-5

WANG, D I (1965) Growth rates of young queen and worker honey bee larvae. *Journal of Apicultural Research* 4: 3-5.

WEAVER, N (1955) Rearing of honey bee larvae on royal jelly in the laboratory. *Science* 121: 509-510.

WEAVER, N (1974) Control of dimorphism in the female honey bee. 3. The balance of nutrients. *Journal of Apicultural Research* 13: 93-101.

WILLIAMS, G R; ALAUX, C; COSTA, C; CSÁKI, T; DOUBLET, V; EISENHARDT, D; FRIES, I; KUHN, R; MCMAHON, D P; MEDRZYCKI, P; MURRAY, T E; NATSOPOULOU, M E; NEUMANN, P; OLIVER, R; PAXTON, R J; PERNAL, S F; SHUTLER, D; TANNER, G; VAN DER STEEN, J J M; BRODSCHNEIDER, R (2013) Standard methods for *in vitro* rearing of *Apis mellifera* larvae. In *V Dietemann; J D Ellis; P Neumann (Eds) The COLOSS* BEEBOOK, *Volume I: standard methods for* Apis mellifera *research. Journal of Apicultural Research* 52(1): http://dx.doi.org/10.3896/IBRA.1.52.1.04

WITTMANN, D; ENGELS, W (1981) Development of test procedures for insecticide-induced brood damage in honey bees. *Mitteilungen der Deutschen Gesellschaft für Allgemeine und Angewandte Entomologie* 3: 187-190.

WITTMANN, D; ENGELS, W (1987) Welche Diät ergibt Arbeiterinnen bei *in vitro*- Aufzucht von Honigbienen? *Apidologie* 18: 279-288.

WOYKE, J (1963) Rearing and viability of diploid drone larvae. *Journal of Apicultural Research* 2: 77-84.

ZHENG, H-Q; HU, F-L; DIETEMANN, V (2011) Changes in composition of royal jelly harvested at different times: consequences for quality standards. *Apidologie* 42: 39-47. http://dx.doi.org/10.1051/apido/2010033

Journal of Apicultural Research 52(1)

Journal of Apicultural Research 52(4): (2013)
DOI 10.3896/IBRA.1.52.4.09

REVIEW ARTICLE

Standard methods for instrumental insemination of *Apis mellifera* queens

Susan W Cobey[1]*, David R Tarpy[2] and Jerzy Woyke[3]

[1]Department of Entomology, Washington State University, 166 FSHN Box 646382 Pullman WA.99164-6382, USA.
[2]Department of Entomology, Campus Box 7613, North Carolina State University, Raleigh NC 27695-7613, USA.
[3]Warsaw University of Life Science, 166 Nowoursynowska 02-787 Warsaw, Poland.

Received 1 March 2012, accepted subject to revision 10 August 2012, accepted for publication 5 June 2013.

*Corresponding author: Email: s.cobey@wsu.edu

Summary

Honey bee queens are highly polyandrous and mate in flight. Instrumental insemination is an essential tool that provides complete control of honey bee mating for research and breeding purposes. The technique requires specialized equipment to anesthetize and immobilize the queen and to collect and deliver semen from the drones. Semen is harvested from mature drones by hand eversion of the endophallus and collected into a syringe. The queen is placed in a chamber and anesthetized during the procedure of insertion of semen into the oviducts. Queens are introduced into colonies and their performance can equal to that of naturally mated queens, given proper technique and care.

Métodos estándar para la inseminación artificial de reinas de *Apis mellifera*

Resumen

Las reinas de las abejas melíferas presentan un elevado grado de poliandria y se aparean durante el vuelo. La inseminación artificial es una herramienta esencial que proporciona un control completo del apareamiento de las abejas con fines de investigación y de cría. La técnica requiere de un equipo especializado para anestesiar e inmovilizar a la reina y para colectar y administrar el semen de los zánganos. El semen se obtiene de zánganos maduros por eversión manual del endofalo y se recoge en una jeringa. La reina se coloca en una cámara y se mantiene anestesiada durante el proceso de inserción del semen en los oviductos. Las reinas se introducen en las colonias y su rendimiento puede ser igual al de las reinas que se aparearon de forma natural, si se realiza la técnica y se da la atención adecuada.

西方蜜蜂蜂王人工授精的标准方法

摘要

蜂王高度多雄交配并在飞行时交尾。人工授精是在科研和育种中完全控制蜜蜂交尾的必要手段。该技术需要专业化设备来麻醉和固定蜂王，以及收集和释放雄蜂精液。精液需通过手工外翻成熟雄蜂的内阴茎采集，并收集到注射器中。蜂王放置于操作室，并在精液注入输卵管过程中处于麻醉状态。若给予适当的技术和护理，人工授精蜂王介绍到蜂群后的表现可与自然交尾蜂王等同。

Keywords: COLOSS, *BEEBOOK*, honey bee, *Apis mellifera*, queen, drone, insemination, valvefold, oviduct, eversion, endophallus, semen, spermatheca

Footnote: Please cite this paper as: COBEY, S W; TARPY, D R; WOYKE, J (2013) Standard methods for instrumental insemination of *Apis mellifera* queens. In *V Dietemann; J D Ellis; P Neumann (Eds) The COLOSS* BEEBOOK, *Volume I: standard methods for* Apis mellifera *research. Journal of Apicultural Research* 52(4): http://dx.doi.org/10.3896/IBRA.1.52.4.09

Table of Contents

1. Introduction

The natural mating behaviour of honey bees presents a unique challenge to control. Queens are highly polyandrous and mate in flight with an average of 10 to 20 drones (Tarpy & Nielsen, 2002) at congregation areas consisting of 10,000 to 30,000 drones from diverse genetic sources (Koeniger 1986). Instrumental or artificial insemination (I.I.) is an essential tool that provides complete control of honey bee mating for research and breeding purposes (Laidlaw, 1977).

1.1. Applications

The technique of I.I. enables controlled mating of honey bees and provides the capability to create crosses beyond what might occur naturally. Novel crosses can be created to advance research and breeding efforts.

- A single drone can inseminate one or even several queens to isolate, enhance, and select a specific trait, which may not be expressed due to the effects of a queen naturally mating with many drones
- Semen from hundreds of drones can be pooled to inseminate a group of queens, which increases the uniformity and effective breeding population size for stock improvement and maintenance purposes

- Varying degrees of inbreeding can be created, including "selfing": the mating of a virgin queen to her own drone sons
- Provides the ability to store honey bee semen. Semen viability can be maintained at room temperature for a few weeks, for convenience in insemination scheduling and the transport of semen

1.2. Current uses

I.I. has been widely used by the scientific community for research and breeding purposes. It has been slow to be adopted by the commercial industry due to the various steps involved to realize the benefits of a breeding programme. Most successful breeding programmes have been limited to cooperative efforts between industry and research institutions with the resources to provide the varied and required expertise.

The general procedure of I.I. is to anesthetize and immobilize a queen bee, manually open her sting chamber, and inject collected drone semen into her vaginal orifice with a syringe. Beyond mastering insemination techniques, the steps required to conduct successful I.I. in honey bees include:

- applying knowledge of breeding principles
- understanding the unique challenges of working with a haploid-diploid super-organism
- practically selecting proper methods and record keeping
- access to and maintaining a large and genetically diverse breeding population (see Meixner *et al.*, 2013)
- acquiring the resources, labour, and a long-term commitment to a breeding programme, and
- learning advanced beekeeping skills, such as queen rearing and drone production (see Büchler *et al.,* 2013)

2. Equipment

The basic technique of I.I. has not changed significantly since its development in the 1950s. Proficiency requires practice, precision, and sanitary conditions. Specialized beekeeping skills and proper care of queens and drones are essential to quality control. Several options of instrumentation are currently available, which offer choice but can vary in quality and lack standardization. The basic instrument consists if a stand, a set of hooks, queen holder assembly, syringe, and syringe tips (Figs. 1, 2, and 3). The microscope stand must be compatible with the instrument and provide sufficient depth of field and instrument clearance (Fig. 1). A cold light source is also recommended to prevent heating and drying. A source of carbon dioxide with flow regulator and flexible tubing to the instrument are also required.

Equipment requirements include:
- Complete insemination instrument, including an instrument stand, manipulators, syringe, and accessories (available through specialty honey bee supply companies)
- Binocular stereozoom microscope, 10x to 20x, and cool light source
- Carbon dioxide source with flow regulator and tubing
- Saline solution (see section 2.2.)
- Sterile vials
- Pipettes and bulb or syringes
- Distilled water
- 95% ethanol
- Sodium hypochlorite
- Sterile tissues and cotton swabs
- Squeeze bottles
- Paper towels or kimwipes
- Autoclave or pressure cooker (for sterilization)
- Queen cages
- Drone holding cages and drone flight box (Fig. 4)

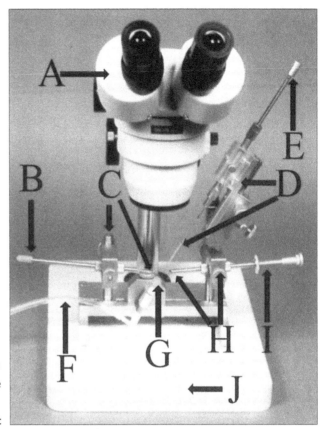

Fig. 1. Standard equipment and arrangement for performing instrumental insemination of honey bee queens. (**A**) a dissecting microscope, (**B**) handle for ventral hook, (**C**) ventral hook base (left **C** arrow) and ventral hook (right **C** arrow), (**D**) syringe base (upper **D** arrow) and syringe (lower **D** arrow), (**E**) syringe plunger, (**F**) plastic tubing leading to CO_2 source, (**G**) chamber in which queen is placed, (**H**) sting hook (left **H** arrow) and sting hook base (right **H** arrow), (**I**) handle for sting hook, and (**J**) microscope base.

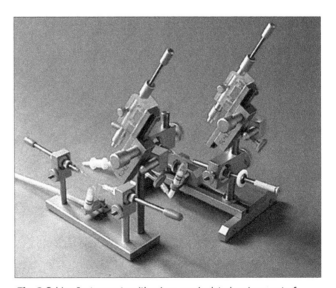

Fig. 2. Schley Instruments with micro-manipulated syringe, set of dorsal and ventral hooks, queen holder assembly with CO_2 attachment (for labelled parts, see Fig. 1).

Drone holding cages are made of queen excluder material to allow worker bees access to care for them when held in nursery colonies. Cages can vary in size, although they should be sized to fit in a frame space in nursery colonies and also fit easily in a flight box. Drones are released into the flight box for easy access during semen collection (Fig. 4).

2.1. Instrument options and choices

Instrument quality, precision, and accuracy will determine the ease and repeatability of the technique. Most instruments offer micro-manipulators that provide precision in movement and fine adjustment. Various designs of sting manipulation tools offer personal choice in techniques. Large capacity syringes provide efficiency in semen collection and a practical method of semen storage and shipment. While the protocol describing the procedure is general for all instruments, the pictures depict herein are the Schley Instrument (Fig. 2) and the Harbo large-capacity syringe (Fig. 3).

2.2. Saline diluent formulas

Two saline diluent formulas are recommended. The simple formula (section 2.2.1.) is for insemination with fresh collected semen used for insemination the same day. The second formula (section 2.2.2.) is for storage and mixing of semen (Hopkins *et al.,* 2012).

- Use double distilled water to make all solutions. Add all of the components to a volumetric flask then add distilled water to make up a final volume
- To sterilize the final product, use bacteriological filtering (pore size 0.2 μm)
  - Solutions can also be heat sterilized at ~177ºC for 30 minutes
  - Add amino acids and antibiotic <u>only after</u> heating
- Adjust the pH of the final product to 8.6
  - To increase the pH, use NaOH, sodium hydroxide
  - To decrease the pH, use HCl, hydrochloric acid

2.2.1. Simple saline formula

The simple saline diluent formula (Table 1) can be used for same-day semen collection and insemination (Williams and Harbo, 1982). An alternative, very basic physiological saline solution is adequate (0.9% NaCl, 0.1% glucose and antibiotic).

2.2.2. Formula for semen storage and mixing

A more complex saline diluent formula, HHBSE Saline Formula (Tables 2 and 3), is recommended when mixing and storing semen, including storage at temperatures above freezing and in liquid nitrogen (Hopkins *et al.,* 2012).

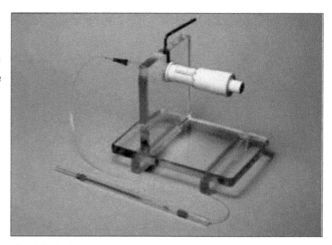

Fig. 3. Harbo large capacity syringe designed for semen collection and storage.

Fig. 4. Drone holding cage and flight box. Drones are collected in holding cages made of queen excluder material (left photograph). Cages should be sized to fit in a frame space in a nursery colony and also fit into a flight box (photograph on right: holding cage is left of white/screened flight box). Drones are released into the flight box for easy access during semen collection.

As a few of the components required (EDTA, glycine, and tylosin) are at very low concentrations and are difficult to measure on conventional scales. Consequently, it is necessary to make a 100x solution that can be added to the solution produced in Table 2 to make up the finished product.

1. Add the reagents (Table 2) (excluding EDTA, glycine, and tylosin) to the volumetric flask and fill to 100 ml (final volume) with distilled water to make the base solution.
2. The minor components, EDTA, glycine and tylosin, are made up as a 100X solution (Table 3) to be added to the solution in Table 2. Using 10 ml of the base solution (Table 2), mix in the components from Table 3. This results in 10 ml of a 100X solution of the three minor components. Then add 0.9 mL of the 100X solution back to the now 90 ml remaining of the base solution. Sterilize using bacteriological filters (pore size 0.2 μm).

Table 1. Simple saline formula based on Williams and Harbo, 1982. Add all of the ingredients to a volumetric flask then add distilled water to make up a final volume of 100 ml.

Ingredient	g ingredient / 100 ml final volume
Dihyrdostreptomycin	0.25
Glucose	0.10
L-lysine	0.01
L-arginine	0.01
L-glutamic acid	0.01
Trizma HCL	0.35
Trizma base	0.35
NaCl	1.11

Table 2. Recipe for base solution that is used in HHBSE saline solution for semen storage and mixing. Add all of the ingredients (excluding EDTA, glycine, and tylosin) to a volumetric flask, then add distilled water to make up a final volume of 100 ml. Use 10 ml of this solution to make the solution in Table 3. Add 0.9 ml of the resulting solution from Table 3 back to the 90 ml of the remaining base solution (presented here) to make the final product. *Tylosin, EDTA, and glycine compose only a small amount of the finished solution. Consequently, it is necessary to make a 100x solution of these ingredients to add back to the rest of the ingredients listed here. See Table 3 and section 2.2.2. for details.

Ingredients	g ingredient / 100 ml final voume
Penicillin	0.05
Streptomysin	0.044
Kanomyosin	0.06
Tylosin*	0.0032
EDTA (ultra pure)*	0.0002923
TES (acid)	0.6879
Tris (base)	0.3635
Sodium Phosphate Dibasic	0.0142
Sodium citrate	0.02942
Arginine	0.01
Glycine*	0.00075
Proline	0.05
Catalase	0.002
BSA (lipid rich)	0.002
KCl	0.61131
NaCl	0.4847
NaHCO3	0.042

Table 3. Recipe for 100x solution of ingredients that compose a small amount of the HHBSE saline solution (in Table 2). Pipette 10 ml of the solution from Table 2 to mix with the components listed below. Add 0.9 ml of this solution to 90 ml of the base solution (in Table 2) to create the final HHBSE saline solution.

Ingredients	g ingredient /10 ml final volume
Tylosin	0.032
EDTA (ultra pure)	0.003
Glycine	0.0075

3. Insemination Techniques

3.1. Eversion of the endophallus

Semen is collected directly from mature drones, 14 days post-emergence or older. For identification purposes, drones can be collected immediately after emergence (i.e. capturing "fuzzy" drones that are newly enclosed) and stored in cages placed in a bank colony (another honey bee colony that will tend the drones; see Büchler *et al.,* 2013 for a discussion of "bank" colonies). Mature drones can be captured the day prior to or the day of insemination by capturing drones returning from failed mating flights or collecting them from the outside combs within the colony. To expose semen, the endophallus is readily everted by hand in two-steps: the partial eversion, and the full eversion.

Maintain sanitary conditions, as drones often defecate during the procedure. Hold the drone to avoid the endophallus touching the drone body or your fingers and keep a towel soaked in alcohol to clean up. The eversion of the endophallus is preformed within a few seconds. Evaluation of drone maturity and semen quality must be determined instantly; any drone that does not evert properly or does not present sufficient (~1 µl) semen on the bulb (see section 3.2) should be discarded. Semen collection is tedious, therefore proper techniques and practice will greatly increase efficiency. Plan to have a plentiful supply of mature drones, more than is needed, as not all will yield semen. Keep drones warm and well fed until they are used. A light above the flight box provides warmth and bee candy or a piece of honey comb will extend their activity.

Procedure for everting drones:
1. Assemble and prepare the syringe
   a. All parts should be sterilized, either by heat or an alcohol wash, and rinsed with distilled water.
2. To obtain partial eversion, grasp the head and thorax of the drone between the thumb and forefinger, ventro-dorsally, with the abdomen facing upward. It is helpful, if the individual is right-handed, to hold the drone's head with the right hand and squeeze the abdomen with the left so that the drone remains held in the left in position for sperm collection.
3. Roll or crush the thorax between your fingers.
   a. If mature, the abdomen will contract and a pair of yellow-orange cornua emerge (Fig. 5).
   b. If the abdomen remains soft or the cornua lacks colour, the drone is immature and will not yield semen (Fig. 6).
4. To obtain full eversion, grasp the base of the abdomen near the thorax with the thumb and forefinger and apply pressure along the sides of the abdomen, starting at the anterior base and working toward the posterior tip.
   a. Squeeze and roll your fingers together in one steady forward motion, forcing the eversion to complete.

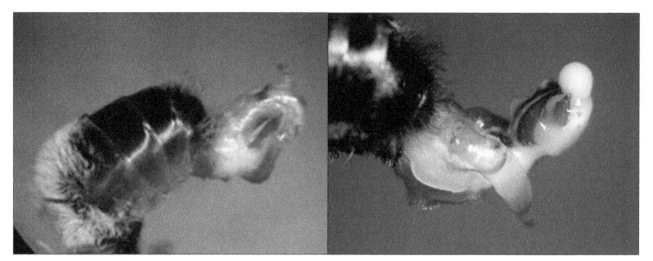

Fig. 5. Partial eversion of the endophallus of a mature drone (left) and full eversion (right). At the stage of the partial eversion, the abdomen will contract and a pair of yellow-orange cornua appear.

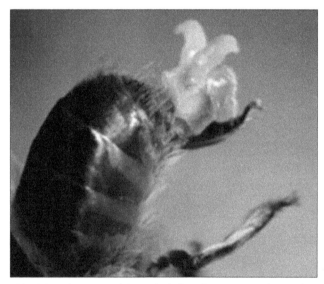

Fig. 6. Partial eversion of the endophallus of an immature drone. The abdomen is soft and the cornua lacks colour.

Fig. 7. Semen contaminated. Avoid contamination of semen during the eversion process. Position the drone to stop the endophallus from falling back onto your fingers or the drone's abdomen.

b. Hold the drone with his abdomen pointing downward to keep the endophallus from falling back onto your fingers and contaminating the semen as in Fig. 7. This positioning also provides ready placement under the microscope.

c. The exposed semen is a creamy, marbled tan colour, with an underlying layer of white mucus (Fig. 8).

3.2. Semen collection

Semen is collected directly from the endophallus of many drones into a syringe and stored in glass capillary tubes. The amount and consistency of semen obtained from each drone varies and depends on skill and experience. Generally, each drone will yield approximately 1 µl of semen. The standard volume of semen to inseminate one queen is ~8 to 12 µl. Maintain sanitary conditions, as drones often defecate during eversion. It is recommended to have a paper towel readily available to wipe drone faeces. Although less common, queens may also defecate. Discard the queen if this happens during the procedure.

Procedure for collecting semen:

1. After assembly of the syringe, collect an air space (~5 µl) to separate the saline and semen column in the syringe. Then collect a small drop of saline into the glass tip (~2 µl). This will be the last fluid to be injected into the last queen inseminated, which will help wash any remaining semen adhering to the capillary walls out of the tip and into the queen.

2. Collect another small air space and collect a small drop (0.5 µl) of saline in the syringe tip (Fig. 9). Use the drop of saline to make contact with the semen on the endophallus of the first drone.

3. Skim the semen off the mucus layer and draw it into the syringe (Fig. 10).

Fig. 8. Full eversion of the endophallus with semen exposed. The exposed semen is a creamy, marbled tan colour, with an underlying layer of white mucus.

a. Avoid collecting the viscous mucus layer at all costs. If resistance is felt, back off or expel any mucus in the tip. Excess mucus in the tip can leave it clogged (Fig. 11).

4. Repeat Step 3 until the total desired amount of semen is collected.

   a. Expel a small drop of semen from the syringe tip on to the semen load of the next drone and draw semen into the syringe (Fig. 12)

   b. Avoid collection of air bubbles and additional saline in the semen column. The column of semen should be uniform in colour and density.

   c. Between semen loads, keep the tip moist with saline to prevent drying, taking care not to excessively dilute the semen. Drone semen quickly dries and the sperm die when exposed to air.

Fig. 9. Collecting semen into the syringe. Collect an air space to separate the saline and semen column in the syringe. After the air space, collect a small drop of saline in the syringe tip to make contact with the semen on the endophallus of the first drone.

Fig. 11. Accidental collection of mucus into the syringe. Avoid collecting the viscous mucus layer as pictured here. If resistance is felt, back off or expel any mucus in the tip to avoid clogging the tip.

Fig. 10. Collecting semen. Skim the semen, a marbled tan colour, off the underlying layer of viscous white mucus and draw this into the syringe.

Fig. 12. Collecting semen from subsequent drones. To repeat the process of semen collection from the next drone, expel a small drop of semen from the syringe tip onto the semen load of the next drone and draw this into the syringe.

3.3. Insemination of the queen

Inseminate queens between 5 and 12 days post-emergence. Carbon dioxide is used to anesthetize the queen during the procedure and also stimulates oviposition. Queens can be emerged in a queenless bank or, preferably, in their own colonies (typically small hives each with several hundred adult workers with a single virgin queen, called mating nuclei). If mating nuclei are used, cage the queen cells (so queens emerge into a cage) or be sure that the hive entrances are covered with queen excluder material to prevent unwanted natural mating flights. Queenless banks, mating nuclei, and caging queens cells are all discussed in detail in Büchler *et al.*, 2013.

Procedure for inseminating queens:

1. Two CO_2 treatments are usually required. Give the first CO_2 treatment, a 1 to 4 minute exposure, one or two days before the insemination procedure. The dose can be applied by individually caging queens and placing them in a jar or plastic bag filled with CO_2. The second treatment is administered during the procedure.

2. Align the syringe and queen holder on the instrument stand at a 30º to 45º angle (dependent upon the instrument used) to facilitate bypassing the valvefold (Figs 13 and 18).

3. Place the queen in the holding tube abdomen first, ventral side up, with her abdomen protruding several segments (Fig. 13), and administer a slow continuous flow of CO_2.

4. Separate the abdominal plates to expose the vaginal orifice using a pair of hooks or forceps (Fig. 14).

5. Lift the sting structure dorsally, to expose the vaginal cavity (Figs 15-17).

    a. During this manipulation, position the ventral hook only to stabilize the queen.

6. Position the syringe tip dorsally above the "V", defining the vaginal orifice. Insert the tip into the vaginal orifice 0.5 to 1.0 mm, slightly forward of the apex of the "V" (Fig. 18).

7. Insert the tip further, another 0.5 to 1.0 mm, while using the tip to lift the valvefold ventrally (Fig. 18). Use a slight "zigzag" movement to bypass the valvefold.

    a. The valvefold, a stretchy flap of tissue covering the median oviduct, must be bypassed or semen will backflow out of the vaginal orifice.

    b. Correctly inserted, the tip slips easily past the valvefold without resistance.

    c. As outlined in section 3.2., it is useful to leave a tiny air bubble between the saline and semen and release some of the saline for lubrication before inserting the tip in the queen.

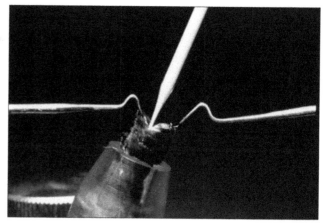

Fig. 13. Virgin queen positioned in the holding tube. The syringe and queen holder are aligned at a 45º angle on the device to facilitate bypassing the valvefold.

Fig. 14. Separating the abdominal plates of the queen to expose the sting structure using the perforated sting hook. The ventral hook is on the left while the perforated sting hook (seen with small hole to accommodate the sting) is on the right.

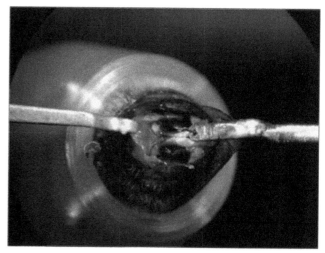

Fig. 15. Threading the sting through the perforated sting hook (on right). The ventral hook is on the left and is used only to stabilize the queen.

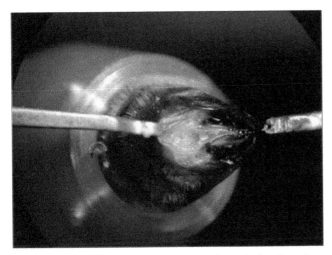

Fig. 16. Lifting the sting structure to expose the vaginal cavity, using a perforated sting hook (on right).

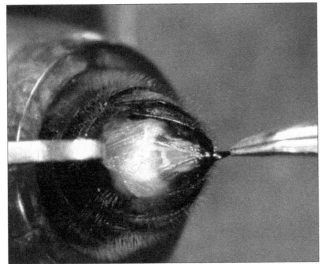

Fig. 17. Lifting the sting structure, using Schley pressure grip forceps (on right).

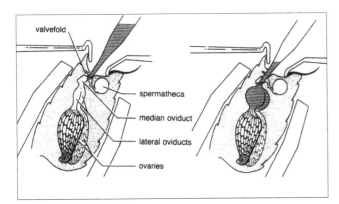

Fig. 18. Bypassing the valvefold. To bypass the valvefold, position the syringe tip dorsally above the "V", defining the vaginal orifice. Insert the tip about 0.5 to 1.0 mm, slightly forward of the apex of the "V". Insert the tip another 0.5 to 1.0 mm lifting the valvefold ventrally, using a slight "zigzag" movement to manoeuvre around the valvefold.

8. Deliver a measured amount of semen directly into the median oviduct (Fig. 19).
   a. The standard dosage is 8 to 12 µl per queen. When giving a 12 µl semen dose, release the queen directly into her mating nucleus colony to promote sperm migration or give two 6 µl semen doses 48 hours apart.
   b. With practice, the insertion of semen is preformed quickly and precisely, requiring only seconds per queen.
9. After insemination, remove the syringe tip, collect a small air space and small drop of saline, (~0.5 µl) to precede the next insemination.
   a. Keep a drop of saline in the tip to prevent any residual semen from drying and to initiate subsequent semen collection.
   b. If inseminating queens with different drones where precise genetic crossings are paramount, rinse the insemination tip with distilled water then saline to completely cleanse the syringe of semen from the previous drone.
10. Release the queen from the holder, place her in a cage, and return her to her nucleus colony.

3.4. Field dissection of the spermatheca

When learning the procedure of insemination, it is helpful to check the spermatheca to determine the degree of insemination success. Sperm migration requires about 40 hours post insemination. After insemination, a subset of queens can be held in a nursery colony until tested.

Procedure for field dissection of spermatheca:
1. Sacrifice the queen, by crushing her head and thorax.
2. Grasp the queen's terminal abdominal segments, dorsally and ventrally.
3. Pull and separate the terminal segments from the rest of the queen's body, with your fingernails or forceps (Fig. 20).
   a. The spermatheca is a white, spherical structure about 1 mm in diameter, and appears rough in texture due to the trachea net covering (Fig. 21).
4. Tease the spermatheca out of the body cavity with your thumbnail or forceps.
5. To remove the tracheal net, gently roll the spermatheca between your fingers. The net will collapse in a small white mass.
   a. The colour shade and density of the spermatheca indicates the relative insemination success of the queen.
   b. The spermatheca of a virgin queen is clear (Fig. 22).
   c. A cloudy or milky appearance of the spermatheca indicates an inadequate insemination or a failing queen (Fig. 23).
   d. For a fully inseminated queen, the spermatheca is a creamy tan colour with a pattern of marbled swirls (Fig. 21).

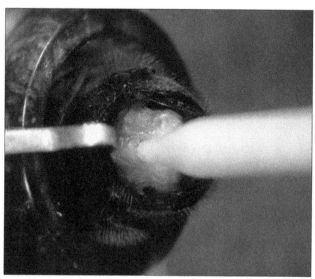

Fig. 19. Insertion of semen in to the median oviduct. Positioned correctly, the tip slips easily past the valvefold without resistance.

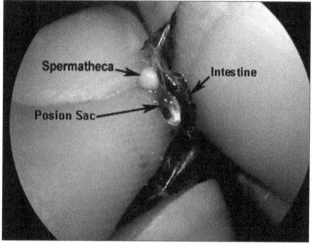

Fig. 20. Exposing the spermatheca. To expose the spermatheca, grasp the queen's terminal abdominal segments, dorsally and ventrally and pull to separate these segments from the rest of the queen's body, with your fingernails.

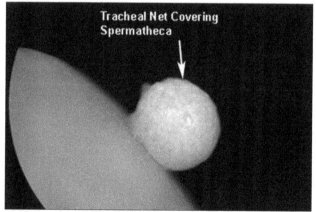

Fig. 21. The spermatheca is a white, spherical structure about 1 mm in diameter and appears rough in texture due to the trachea net covering.

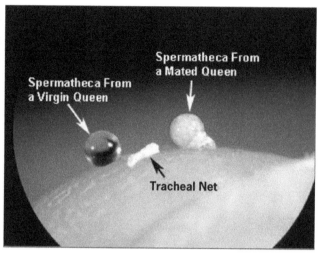

Fig. 22. Comparison of spermathecas of a virgin queen (clear) and mated queen (tan with a pattern of marbled swirls).

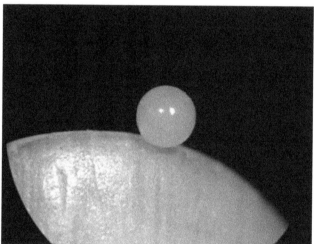

Fig. 23. The spermatheca of a failing or poorly mated queen is a milky colour.

4. Maintenance of queens and drones and factors affecting queen performance

Given proper care, instrumentally inseminated queens are capable of heading productive colonies and enduring the rigors of performance testing in the field (Cobey, 2007). Many factors that influence queen performance can be optimized through proper beekeeping management practices. The quality of the insemination, in terms of technique and sanitation, are critical. The treatment of queens before and after the insemination will influence the amount of semen stored and queen performance. Natural conditions should be maintained as much as possible.

4.1. Maintenance of drones for instrumental insemination

Producing a large number of mature drones from select sources can be more challenging then queen rearing, especially if seasonal conditions are not optimal. Drones have a high rate of attrition and drift heavily among colonies. Free flying drones have better survival and have a prior opportunity to void faeces.

1. Use strong, healthy, well-fed colonies for drone production.
   a. Colonies headed by older queens tend to rear and maintain more drones.
   b. Minimize colony stressors, especially exposure to pathogens, parasites and chemical residues (see the *BEEBOOK* papers on pests and diseases: Dietemann *et al.,* 2013; Forsgren *et al.,* 2013; Fries *et al.,* 2013; de Graaf *et al.,* 2013; Jensen *et al.,* 2013; de Miranda *et al.,* 2013; Sammataro *et al.,* 2013).
2. Drones require 12 -14 days post-emergence to sexually mature (Woyke and Jasinski, 1978).
3. Maintain the identity of drone sources.
   a. For free-flying drones, newly emerged drones can be emerged from field-collected combs placed in an incubator set at broodnest conditions (34°C, ~50% RH), marked, and returned to colonies to reach sexual maturity. Mark several times the number needed to ensure an adequate supply.
   b. To confine drones, emerge drone brood above a queen excluder in a healthy, strong colony. Screening the top box (i.e. replacing the colony lid with screen mesh) may be used to collect the drones as mature drones are attracted to light.
4. Collect drones in holding cages made of queen excluder material (Fig. 4).
   a. Drone flight times vary with the season and weather conditions. Watch the colonies in the local area to determine drone flight time.
   b. Caged drones can be banked in queenless nursery colonies or queen cell builders (see Büchler *et al.,* 2013 for more information on queen cell builders).
      i. Minimize banking time as drones are perishable, over-night to a few days.
      ii. Banked drones accumulate faeces that need to be voided.
5. Release mature drones into a flight box for semen collection (Fig. 4).
   a. Bring in only the number of drones that can be collected in about 30-40 minutes (100 to 150), as inactive drones are difficult to evert.
   b. Providing heat (a light above the flight box) and food (bee candy) will extend the active period of drones.

4.2. Maintenance of queens before and after instrumental insemination

The pre- and post-insemination treatment of queens will influence their performance. Maintain queens with a high proportion of nurse bees in well-fed nursery and/or nucleus colonies (see Büchler *et al.,* 2013). Direct release of queens into colonies after insemination enhances sperm migration (Woyke, 1983).

1. Optimize queen rearing conditions.
   a. Rearing conditions influence queen size, the number of ovarioles, and spermatheca capacity (see Büchler *et al.,* 2013).
2. Place mature queen cells in individual cages before emergence.
   a. Emerge virgins in a nursery colony or incubator.
   b. After emergence, remove their wax cells to prevent queens crawling into them and dying.
3. Cage each newly emerged virgin queen individually in a nucleus colony for several days (5-14) before insemination.
   a. Queens banked in a nursery colony can be subject to injury by workers resulting in damaged legs and tarsi (Woyke, 1988).
   b. If banking is necessary, use a well-fed queenless colony maintained with open brood (eggs and larvae) and a high proportion of young nurse bees.
4. After insemination, release each queen directly into her established nucleus colony. Use a spray of sugar water (preferably scented with anise oil some other fragrant) or a small candy plug in a queen cage to facilitate introduction.
   a. Note the behaviour of the queen upon release. If aggressive worker behaviour is observed, cage the queen and provide a slow release (i.e. release her 3-4 days later and observe worker behaviour; repeat if necessary).
   b. If queens have been banked, use a screened push in cage (Fig. 24) or another similar slow-release method (see Büchler *et al.,* 2013).
   c. To prevent unwanted natural mating flights of I.I. queens (Woyke and Janinski, 1992), place queen excluder material on the colony entrance until eggs are observed.
5. Allow queens to build their own populations naturally.
   a. I.I. queens are slower than naturally mated queens to develop their full pheromone blends and tend to supersede when placed in large colonies initially.

Fig. 24. A queen "push-in" cage. These cages can be used to introduce newly-inseminated queens into colonies. To use, place the queen on a section of capped/emerging brood. Push the cage around the queen and brood, firmly into the wax comb. Leave queen in cage with emerging brood for 3-4 days. Following this time, the cage can be removed and the queen released into the colony. If eggs are not observed at release, place a queen excluder over the colony entrance until the queen begins laying.

4.3. Factors influencing results of instrumental insemination

1. The optimal age for insemination of queens is 5 to 14 days post emergence.
   a. Queens inseminated older than 2 weeks tend to store less sperm in their spermathecae. Queens inseminated less than 4 days old have high mortality (Woyke and Jasinski, 1976).
2. The standard semen dosage given to each queen is 8 to 12 µl.
   a. An insufficient semen dose can result in premature queen supersedure or premature queen failure.
3. Post-insemination care of queens influences sperm storage (Woyke 1979).
   a. Active movement of queen, appropriate brood nest temperatures, and attendance by worker bees promote sperm migration into the queen's spermatheca.
   b. Queens confined in cages after insemination tend to store less sperm and retain semen in their oviducts.

5. Specialized techniques

5.1. Homogenizing honey bee semen

To homogenize or mix honey bee semen from numerous drones requires dilution, mechanical movement and reconstitution of semen. Current techniques using centrifugation result in a high percentage of damaged spermatozoa, although 50% viability of spermatozoa is sufficient to produce normal brood patterns (Collins, 2000). Semen is very dense, tends to clump, and the long, fragile tails of spermatozoa are subject to damage during processing and some components of the seminal fluid are removed.

Migration of sperm from the oviducts into the spermatheca is a complex process involving contraction of muscles mediated by the specialized composition of fluids in the semen and the oviduct as well as active sperm movement (Koeniger, 1986). Queens are very active after natural mating which also promotes sperm migration; therefore, use a direct queen introduction release method (Büchler *et al.*, 2013).

Procedure by centrifugation:

1. Collect semen into glass capillary tubes (section 3.2).
   a. Collect 10 µl for each queen to be inseminated.
2. Add diluent (the recipe can be found in Table 4) to Eppendorf tube.
   a. Ratio should be 10 parts diluent to 1 part semen by volume.
   b. Up to 700 µl of semen can be added per Eppendorf tube.
3. Expel semen into an Eppendorf tube.
4. Mix semen by inversion, gently shaking, until suspension is uniform.
   a. Strong mixing with a Pasteur pipette may damage spermatozoa.
5. Centrifuge at a 45° angle until a semen pellet is formed.
   a. Use speeds of 82 or 250 g at 20–30 or 10–20 min, respectively.
   b. Higher speeds can damage tails (see Collins, 2003).
6. Carefully remove supernatant.
7. Draw semen into syringe and use immediately.
8. Inseminate queens with the required dosage.

Table 4. Recipe for making desired volumes of diluent for counting sperm. Mix the ingredients together and add distilled water to achieve the final volume.

Reagents	Desired Volume			
	2 ml	5 ml	10 ml	20 ml
1M HEPES	20 µl	50 µl	100 µl	200 µl
NaCl	0.017 g	0.0425 g	0.085 g	0.17 g
BSA	0.20 g	0.50 g	1.0 g	2.0 g
add water to bring total volume to:	2.0 ml	5.0 ml	10.0 ml	20.0 ml

5.2. Short term semen storage at above freezing temperature

Honey bee semen can be held at room temperature for several weeks without significant loss of sperm viability. The Harbo syringe, with detachable capillary tubes, is designed for semen storage (Fig. 25). For step by step pictorial instructions, see:
honey beeinsemination.com/uploads/HarboSyringeAssembly.pdf
honey beeinsemination.com/uploads/HarboSemen_Storage.pdf

5.2.1. Sealing semen-filled capillary tubes

1. Collect semen into glass capillary tubes of the syringe.
2. Remove glass tip and detach the filled capillary tube.
3. Force petrolatum into one end of the tube (~7 mm).
   a. Insert the capillary tube into the petroleum jelly perpendicularly several times until a sufficient plug is formed.
   b. The petroleum can be touching the semen or a small airspace can be collected between the semen and petrolatum seal.
4. Reconnect the sealed end of the tube to the syringe and push the column of semen forward to allow space to seal the other end.
5. Detach and place a petrolatum seal in the other end of the filled capillary tube.
6. An alternative method, a glass bead connected with a small piece of silicone tubing, can also be used to seal the capillary tube. The glass bead is made by heating a small piece of capillary tube to seal both ends.
7. Store in the dark at a temperature of 20ºC. Avoid sunlight and temperature fluctuations.

5.2.2. To remove the petroleum seal:

1. Assemble the syringe and load the capillary tube without the glass tip.
2. Push out the petrolatum seal with the action of the syringe.
3. Attach the glass tip and pick up a small drop of saline to precede the first insemination.
   a. Leave the seal at the terminal of the capillary tube.
   b. After inseminations are complete, collect saline to move the seal out of the tip and up into the capillary. Discard the capillary tube.

5.3. Cryopreservation of semen

The maintenance of honey bee stocks currently requires costly and labour intensive annual propagation. Current threats to the biodiversity of honey bees, and the need to select lines tolerant to pests and diseases, creates a need to develop techniques for the cryopreservation of honey bee germplasm. Repositories

Fig. 25. Sealed capillary tubes of semen ready for transport.

would provide a resource for breeding purposes and the preservation and recovery of selected stocks and endangered populations.

High viability and motility of honey bee semen cryopreserved in liquid nitrogen and thawed has been demonstrated, although fertility is greatly reduced in the spermatheca of queens. Current techniques demonstrate that fertility is adequate to produce sequential generations of queens inseminated with frozen-thawed semen for breeding purposes, although they are not sufficient to head productive colonies (Hopkins *et al.*, 2012). Further research is being conducted to perfect these techniques.

Current recommendations for cryopreservation:

- Cryoprotectants: dimethyl sulfoxide (DMSO) and ethylene glycol (see Wegener & Bienfeld, 2010 and Hopkins and Herr, 2010)
- Programmable freezing rate: 3ºC / min, from 4ºC to -40ºC, then place samples in liquid nitrogen (see Hopkins *et al.*, 2012).
- Thawing rate, 40ºC for 10 seconds (see Hopkins *et al.*, 2012)

5.4. Techniques for counting sperm

5.4.1. Queen spermathecae

1. Place the test queen in a freezer (-20ºC is sufficient) 4-6 mins or until immobilized.
2. Remove the queen from freezer and weigh to the nearest 0.1 mg on digital scale. Measure the queen's thorax and head using micro-calipers to the nearest 0.1 mm and record.
   *NOTE: Non-destructive morphometric measures such as these may be helpful and potentially important correlates of other measures of queen reproductive potential (see Delaney *et al.*, 2011).
3. Dissect out the spermatheca:
   a. Euthanize the queen by removing her head and pin her body to a dissection tray.

b. Cut her abdomen along both sides.

c. Grasp stinger with forceps and gently pull out until the ovaries are exposed.

d. Gently push hindgut aside to reveal the spermatheca (off-white, semi-hard sphere; see Fig. 21).

4. Carefully pull the spermatheca out and place on a watchglass. Remove the tracheal netting covering it if still attached.

5. Set dissection microscope to maximum magnification, and use graduations on ocular to measure spermatheca diameter. Measure two diameters and record the average. This is an optional but potentially useful measure for spermatheca volume, which can be used to calculate the theoretical maximum storage capacity and therefore the percentage filled (see Tarpy *et al.*, 2011).

6. Place spermatheca in 0.5 ml sperm diluent (e.g. Table 4) in a small glass beaker, then use forceps to burst it to release sperm.

7. Immediately place remainder of queen in a labelled 1.5 ml microcentrifuge tube and place in -80°C freezer for any further analyses (e.g. PCR, GC-MS, etc.).

5.4.2. Drone seminal vesicles

1. Cut off the drone abdomen and immediately freeze head/ thorax at -80°C for any further analyses.

2. Pin abdomen down in petri dish and dissect out the seminal vesicle:

   a. Cover with 0.9% saline solution, then make ventral incisions.

   b. Pull the two cut sides apart and grab the bursa, then gently pull out until mucous glands and seminal vesicles come with it.

   c. Cut seminal vesicles away from mucous glands.

3. Place seminal vesicle into a 50 ml beaker containing 1 ml sperm diluent (see Table 4 for recipe).

5.4.3. Sperm count protocol (unstained sperm)

This protocol is covered in some detail in Human *et al.*, 2013. Despite that, we include an expanded version of the technique here as we list details specific to counting sperm.

1. Dissect out the spermatheca of a mated queen or obtain the sperm sample from a desired drone (queen: section 5.4.1., drone: section 5.4.2.). *NOTE: Always use glassware when dealing with sperm, because they can stick to plastic.

2. Add 10.0 ml of diluent or water to a glass petri dish. *NOTE: water causes live sperm to contract and contort their shape, so only use water with frozen or dead sperm.

3. Break open the spermatheca using forceps, taking care to remove all sperm, and remove the remaining tissue.

4. Mix well with a clean glass pipette about 40 times until all sperm are dispersed. Use caution to prevent air bubbles or excess splattering.

5. Immediately add a drop of the diluent/sperm mixture to both sides of a haemocytometer on which a cover slip has been placed. Capillary action will fill each chamber (the area between the cover slip and slide) with solution. View under 250× magnification. Start counts on the gridded section after the sperm have settled (~ 20 seconds).

6. The sperm appear headless, translucent, and filamentous. They are usually about 0.25 mm long, but are often coiled or looped. It is best to bring the focus slightly upwards from the grid and to keep the light somewhat dim in order to best see them. Rapidly change the fine-tuning focus on the microscope to observe those that are not laying on the bottom of the haemocytometer.

7. Count the number of sperm in five of the large 1.00 mm^2 squares in the grid, preferably the large squares in the four corners and the one in the centre (5 bold, black squares seen in Fig. 26). The centre square will contain a smaller grid (used to count red blood cells at a greater magnification), while the remaining squares will be divided into 16 squares (4 × 4 grids). *NOTE: There are different types of haemocytometers so it is important to follow manufacturer's directions when using one as calculations and haemocytometer volume may differ between types. Our calculations are done using a Bright-Line haemocytometer (Hausser Scientific).

8. Since sperm often overlap the boundaries of the squares, only count those sperm which are entirely within a square or are *only* on the top and left boundaries (or bottom and right, if you prefer). This procedure will prevent double counting of sperm and give a more accurate count.

Calculate the total number of sperm in 5 large squares in *both* chambers of the haemocytometer. Divide the total number of sperm in the 10 large squares (5 large squares per 2 chambers) by 100 to estimate the number of millions of sperm. (The volume of each large square is 1.0 mm × 1.0 mm × 0.1 mm = 100 nl, making the total volume counted for ten squares = 1000 nl, Table 5. Thus the number of sperm, for an initial dilution of 10.0 ml, is to the order of 10^4 which is equivalent to 10^{-2} million).

Alternative method for counting sperm:

In this method, the sperm are induced to coil in the solution, which facilitate their counting (Woyke 1979).

1. Add a drop of saline solution into a small porcelain evaporator, preferably one having blue bottom.

2. Break open the spermatheca, remove all sperm, and remove the remaining tissue.

3. Stir with the dissecting needle and add more solution up to a total of 1 ml.

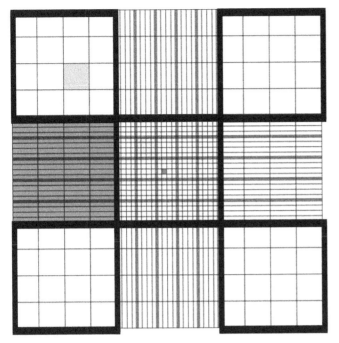

7. Add a drop of the diluent with sperm on both grids and cover with a cover slip.
8. Wait for the sperm to settle.
9. Use 250× magnification for counting the sperm.
10. Count the number of sperm in 5 large 1 mm² squares on both grids (totalling 10 squares). The sperm will be easy to count because they will be coiled. Some sperm will be located over the square boundary. To avoid double counting, count only those that are over the top and left boundaries.
11. Calculate the total number of sperm as follows: the total volume of the solution over the 10 large squares (5 squares in both grids) in which the sperm are counted is: $1 \times 1 \times 0.2 \times 10 = 2$ mm³. However, the total volume of the solution in which the sperm were dispersed is 10 cm³ = 10,000 mm³. Thus, the volume of the dispersion solution is 10,000:2 = 5,000 times higher. To get the total number of spermatozoa, multiply the number of spermatozoa counted over the 10 large squares by 5,000.

Fig. 26. Haemocytometer grid. Red square (and each of the 5 large squares with bold, black lines added) = 1 mm² (100.00 nl); green square = 0.0625 mm² (6.250 nl); yellow square = 0.040 mm² (4.00 nl); blue square = 0.0025 mm² (0.25 nl). All squares are at a depth of 0.1 mm. For area and volume calculations per certain grid dimensions on the haemocytometer, see Table 5. Information is for Bright-Line Haemocytometer (Hausser Scientific). Figure from Wikipedia.

Table 5. Area and volume calculations for Haemocytometer grids seen in Fig. 26. The volume is calculated as L × W × D with the L and W provided in the "Dimensions" column and the D set as 0.1 mm with the standard Haemocytometer (such as a Bright-Line Haemocytometer, Hausser Scientific).

Dimensions	Area	Volume
1 × 1 mm (1 red and the 5 black bolded squares in Fig. 26)	1 mm²	100 nl
0.25 × 0.25 mm (green square in Fig. 26)	0.0625 mm²	6.25 nl
0.25 × 0.20 mm (square not highlighted in Fig. 26)	0.05 mm²	5 nl
0.20 × 0.20 mm (yellow square in Fig. 26)	0.04 mm²	4 nl
0.05× 0.05 mm (blue square in Fig. 26)	0.0025 mm²	0.25 nl

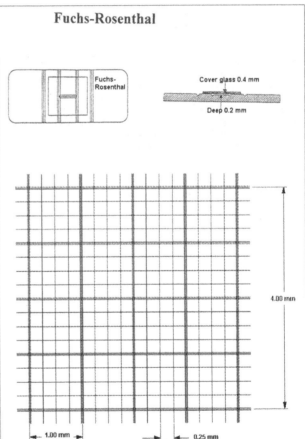

Fuchs-Rosenthal

Fig. 27. Diagram of a Fuchs-Rosenthal grid that can be used for counting sperm.

4. Add 9 ml of tap water, which will result in a total of 10 ml of solution, and mix well. The tap water causes the sperm to coil.
5. Mix well with a clean glass pipette until all sperm are dispersed. Use caution to prevent air bubbles or excess splattering.
6. The Fuchs-Rosenthal count chamber is used to count the sperm (Fig. 27). There are two counting grids and the depth of the counting chamber is 0.2 mm.

5.5. Sperm viability measures

5.5.1. Reagents

- Diluent (10 mM HEPES, 150 mM NaCl, 10% BSA, pH 7.4, see Table 4)
- Invitrogen live/dead sperm staining kit (#L7011)
- Dimethyl sulfoxide (DMSO)

5.5.2. Equipment

- Fluorescent microscope with Rhodamine and FITC filters (for live and dead cells, respectively)
- Dissecting microscope
- Haemocytometer
- Micro-weight balance
- Micro-calipers
- Microdispenser (e.g. Drummond, Fisher Scientific #3-000-225)
- 2 ml glass screw-thread vials and caps (Fisher Scientific #03-391-16)
- Two forceps, dissection tray, and pins
- 1.5 ml microcentrifuge tubes (if you want to save the rest of the queen)
- Water bath set at 36°C

5.5.3. Procedure

1. Make Sybr 14 working solution:
   a. Add 20 µl Sybr 14 (from Invitrogen kit) to 980 µl DMSO
   b. Store at -20°C in the dark.
2. Make sufficient volume of sperm diluent of HEPES, NaCl, and BSA (HNB). See Table 4
3. For spermathecae
   a. Transfer diluent with sperm (from step 5.4.1.) to a labelled 2 ml glass vial.
   b. After all spermathecae have been dissected, add 10 µl Sybr 14 in DMSO to each vial.
   c. Incubate 5-10 min at 36°C.
   d. Add 5 µl propidium iodide.
   e. Incubate 5-10 min at 36°C.
4. For seminal vesicles
   a. Gently disrupt seminal membrane to release sperm in diluent.
   b. Discard seminal membrane and pour sperm solution into a 2 ml glass vial.
   c. Stain seminal vesicles:
   i. Add 10 µl Sybr 14/DMSO.
   ii. Incubate 15 min at 36°C.
   iii. Add 6 µl propidium iodide.
   iv. Incubate another 15 min at 36°C.

5. Turn on fluorescent lamp and camera, and open photo software on computer.
6. Use microdispenser to load haemocytometer with 10 µl sperm solution across both chambers.
7. Place haemocytometer on microscope table and examine under low magnification to centre view on grid (Fig. 26).
8. Switch to high magnification (200x then 400x).
9. Turn off visible light and open fluorescence. Focus using FITC filter, then take an image using the camera. This will be a picture of the <u>dead</u> sperm in the field.
10. Without moving the haemocycometer, switch to Rhodamine filter and take another image. This will be a picture of the <u>live</u> sperm in the same field.
11. Move to a new field of view and repeat; a total of five replicates should be minimum, preferably greater particularly if there is large variation among fields.
12. Save pictures using descriptive names indicating sample number, live or dead sperm, and picture number.
13. Switch back to low-resolution and close fluorescent lamp aperture.
14. Clean haemocytometer and cover slip with a kimwipe.
15. Repeat for each new sample.

6. Acknowledgements

The COLOSS (Prevention of honey bee COlony LOSSes) network aims to explain and prevent massive honey bee colony losses. It was funded through the COST Action FA0803. COST (European Cooperation in Science and Technology) is a unique means for European researchers to jointly develop their own ideas and new initiatives across all scientific disciplines through trans-European networking of nationally funded research activities. Based on a pan-European intergovernmental framework for cooperation in science and technology, COST has contributed since its creation more than 40 years ago to closing the gap between science, policy makers and society throughout Europe and beyond. COST is supported by the EU Seventh Framework Programme for research, technological development and demonstration activities (Official Journal L 412, 30 December 2006). The European Science Foundation as implementing agent of COST provides the COST Office through an EC Grant Agreement. The Council of the European Union provides the COST Secretariat. The COLOSS network is now supported by the Ricola Foundation - Nature & Culture.

7. References

BÜCHLER, R; ANDONOV, S; BIENEFELD, K; COSTA, C; HATJINA, F; KEZIC, N; KRYGER, P; SPIVAK, M; UZUNOV, A; WILDE, J (2013) Standard methods for rearing and selection of *Apis mellifera* queens. In *V Dietemann; J D Ellis; P Neumann (Eds) The COLOSS BEEBOOK, Volume I: standard methods for* Apis mellifera *research. Journal of Apicultural Research* 52(1): http://dx.doi.org/10.3896/IBRA.1.52.1.07

COBEY, S W (2007) Comparison studies of instrumentally inseminated queens and naturally mated queens and factors affecting their performance. *Apidologie* 38: 390-410. http://dx.doi.org/10.1051/apido:2007029

COBEY, S W (2013) Harbo syringe users manual. Syringe assembly, semen storage. honey beeinsemination.com/Equipment.html

COLLINS, A M (2003) A scientific note on the effect of centrifugation of pooled honey bee semen. *Apidologie* 34: 469-470. http://dx.doi.org/10.1051/apido:2003030

COLLINS, A M (2000) Relationship between semen quality and performance on instrumentally inseminated honey bee queens. *Apidologie* 31:421-429. http://dx.doi.org/10.1051/apido:2000132

COLLINS, A M; DONOGHUE A M (1999) Viability assessment of honey bee, *Apis mellifera* sperm using dual fluorescent staining. *Theriogenology* 51: 1513-1523.

DE GRAAF, D C; ALIPPI, A M; ANTÚNEZ, K; ARONSTEIN, K A; BUDGE, G; DE KOKER, D; DE SMET, L; DINGMAN, D W; EVANS, J D; FOSTER, L J; FÜNFHAUS, A; GARCIA-GONZALEZ, E; GREGORC, A; HUMAN, H; MURRAY, K D; NGUYEN, B K; POPPINGA, L; SPIVAK, M; VANENGELSDORP, D; WILKINS, S; GENERSCH, E (2013) Standard methods for American foulbrood research. In *V Dietemann; J D Ellis; P Neumann (Eds) The COLOSS* BEEBOOK, *Volume II: standard methods for* Apis mellifera *pest and pathogen research. Journal of Apicultural Research* 52(1): http://dx.doi.org/10.3896/IBRA.1.52.1.11

DELANEY, D A; KELLER, J J; CAREN, J R; TARPY, D R (2011) The physical, insemination, and reproductive quality of honey bee queens (*Apis mellifera*). *Apidologie* 42: 1-13. http://dx.doi.org/10.1051/apido/2010027

DE MIRANDA, J R; BAILEY, L; BALL, B V; BLANCHARD, P; BUDGE, G; CHEJANOVSKY, N; CHEN, Y-P; GAUTHIER, L; GENERSCH, E; DE GRAAF, D; RIBIÈRE, M; RYABOV, E; DE SMET, L; VAN DER STEEN, J J M (2013) Standard methods for virus research in *Apis mellifera*. In *V Dietemann; J D Ellis; P Neumann (Eds) The COLOSS* BEEBOOK, *Volume II: standard methods for* Apis mellifera *pest and pathogen research. Journal of Apicultural Research* 52(4): http://dx.doi.org/10.3896/IBRA.1.52.4.22

DIETEMANN, V; NAZZI, F; MARTIN, S J; ANDERSON, D; LOCKE, B; DELAPLANE, K S; WAUQUIEZ, Q; TANNAHILL, C; FREY, E; ZIEGELMANN, B; ROSENKRANZ, P; ELLIS, J D (2013) Standard methods for varroa research. In *V Dietemann; J D Ellis; P Neumann (Eds) The COLOSS* BEEBOOK, *Volume II: standard methods for* Apis mellifera *pest and pathogen research. Journal of Apicultural Research* 52(1): http://dx.doi.org/10.3896/IBRA.1.52.1.09

FORSGREN, E; BUDGE, G E; CHARRIÈRE, J-D; HORNITZKY, M A Z (2013) Standard methods for European foulbrood research. In *V Dietemann; J D Ellis; P Neumann (Eds) The COLOSS* BEEBOOK, *Volume II: Standard methods for* Apis mellifera *pest and pathogen research. Journal of Apicultural Research* 52(1): http://dx.doi.org/10.3896/IBRA.1.52.1.12

FRIES, I; CHAUZAT, M-P; CHEN, Y-P; DOUBLET, V; GENERSCH, E; GISDER, S; HIGES, M; MCMAHON, D P; MARTÍN-HERNÁNDEZ, R; NATSOPOULOU, M; PAXTON, R J; TANNER, G; WEBSTER, T C; WILLIAMS, G R (2013) Standard methods for *Nosema* research. In *V Dietemann; J D Ellis; P Neumann (Eds) The COLOSS* BEEBOOK: *Volume II: Standard methods for* Apis mellifera *pest and pathogen research. Journal of Apicultural Research* 52(1): http://dx.doi.org/10.3896/IBRA.1.52.1.14

HOPKINS, B K; HERR, C (2010) Factors affecting the successful cryopreservation of honey bee spermatozoa. *Apidologie* 41: 548-556. http://dx.doi.org/10.1603/0022-0493-93.3.568

HOPKINS, B K; HERR, C; SHEPPARD, W S (2012) Sequential generations of honey bee (*Apis mellifera*) queens produced using cryopreserved semen. *Reproduction, Fertility and Development* 24(8): 1079-1083. http://dx.doi.org/10.1071/RD11088

HUMAN, H; BRODSCHNEIDER, R; DIETEMANN, V; DIVELY, G; ELLIS, J; FORSGREN, E; FRIES, I; HATJINA, F; HU, F-L; JAFFÉ, R; JENSEN, A B; KÖHLER, A; MAGYAR, J; ÖZIKRIM, A; PIRK, C W W; ROSE, R; STRAUSS, U; TANNER, G; TARPY, D R; VAN DER STEEN, J J M; VAUDO, A; VEJSNÆS, F; WILDE, J; WILLIAMS, G R; ZHENG, H-Q (2013) Miscellaneous standard methods for *Apis mellifera* research. In *V Dietemann; J D Ellis; P Neumann (Eds) The COLOSS* BEEBOOK, *Volume I: standard methods for* Apis mellifera *research. Journal of Apicultural Research* 52(4): http://dx.doi.org/10.3896/IBRA.1.52.4.10

JENSEN, A B; ARONSTEIN, K; FLORES, J M; VOJVODIC, S; PALACIO, M A; SPIVAK, M (2013) Standard methods for fungal brood disease research. In *V Dietemann; J D Ellis; P Neumann (Eds) The COLOSS* BEEBOOK, *Volume II: Standard methods for* Apis mellifera *pest and pathogen research. Journal of Apicultural Research* 52(1): http://dx.doi.org/10.3896/IBRA.1.52.1.13

LAIDLAW, H H JR (1977) *Instrumental insemination of honey bee queens.* Dadant and Sons; Hamilton, IL, USA.

KOENIGER, G (1986) Reproduction and mating behaviour. In *Rinderer, T (Ed.). Bee genetics and breeding.* pp 255-275.

KUHNERT, M E; CARRICK, M J; ALLEN, L F (1989) Use of homogenized drone semen in a breeding program in Western Australia. *Journal of Apicultural Research* 20: 371-381.

MEIXNER, M D; PINTO, M A; BOUGA, M; KRYGER, P; IVANOVA, E; FUCHS, S (2013) Standard methods for characterising subspecies and ecotypes of *Apis mellifera.* In *V Dietemann; J D Ellis; P Neumann (Eds) The COLOSS* BEEBOOK, *Volume I: standard methods for* Apis mellifera *research. Journal of Apicultural Research* 52(4): http://dx.doi.org/10.3896/IBRA.1.52.4.05

MORITZ, R F A (1983) Homogeneous mixing of honey bee semen by centrifugation. *Journal of Apicultural Research* 22: 249-255.

RUTTNER, F (1976) *Instrumental insemination of honey bee queens.* Apimondia; Bucharest, Romania.

SAMMATARO, D; DE GUZMAN, L; GEORGE, S; OCHOA, R; OTIS, G (2013) Standard methods for tracheal mites research. In *V Dietemann; J D Ellis; P Neumann (Eds) The COLOSS* BEEBOOK, *Volume II: Standard methods for* Apis mellifera *pest and pathogen research. Journal of Apicultural Research* 52(4): http://dx.doi.org/10.3896/IBRA.1.52.4.20

TARPY, D R; NIELSEN, D I (2002) Sampling error, effective paternity, and estimating the genetic structure of honey bee colonies (Hymenoptera : Apidae). *Annals of the Entomological Society of America* 95: 513–528. http://dx.doi.org/10.1603/0013-8746(2002)095[0513:SEEPAE]2.0.CO;2

WILLIAMS, J; HARBO, J R (1982) Bioassay for diluents of honey bee semen. *Annals of the Entomological Society of America* 75: 457-459.

WATSON, L R (1927) Controlled mating in the honey bee. *Report of State Apiarist, Iowa:* 36-41.

WEGNER, J; BIENEFELD, K (2012) Toxicity of cyroprotectants to honey bee semen and queens. *Theriogenology* 77: 600-607. http://dx.doi.org/10.1016/jtheriogenology.2011.08.036

WOYKE, J (1979) Effect of the access of worker honey bees to the queen on the result of instrumental insemination. *Journal of Apicultural Research* 19(2): 136-143.

WOYKE, J (1983) Dynamics of entry of spermatozoa into the spermatheca of instrumentally inseminated queen honey bees. *Journal of Apicultural Research* 22: 150-154.

WOYKE, J (1988) Problems with queen banks. *American Bee Journal* 124(4): 276-278.

WOYKE, J; JASINSKI, Z (1976) The influence of age on the results of instrumental insemination of honey bee queens. *Apidologie* 7(4): 301-306.

WOYKE, J; JASINSKI, Z (1978) Influence of age of drones on the results of instrumental insemination of honey bee queens. *Apidologie* 9(3): 203-212.

WOYKE, J; JASINSKI, Z (1992) Natural mating of instrumentally inseminated queen bees. *Apidologie* 23(3): 225-230.

Journal of Apicultural Research 52(4): (2013)
DOI 10.3896/IBRA.1.52.4.10

REVIEW ARTICLE

Miscellaneous standard methods for *Apis mellifera* research

Hannelie Human[1]*, Robert Brodschneider[2], Vincent Dietemann[1,3], Galen Dively[4], James D Ellis[5], Eva Forsgren[6], Ingemar Fries[6], Fani Hatjina[7], Fu-Liang Hu[8], Rodolfo Jaffé[9], Annette Bruun Jensen[10], Angela Köhler[1], Josef P Magyar[11], Asli Özkýrým[12], Christian W W Pirk[1], Robyn Rose[13†], Ursula Strauss[1], Gina Tanner[3,14], David R Tarpy[16], Jozef J M van der Steen[15], Anthony Vaudo[16], Fleming Vejsnæs[17], Jerzy Wilde[18], Geoffrey R Williams[3,14] and Huo-Qing Zheng[8]

[1]Department of Zoology & Entomology, University of Pretoria, Pretoria, 0002, South Africa.
[2]Department of Zoology, Karl-Franzens-University, Graz, Austria.
[3]Swiss Bee Research Centre, Agroscope Liebefeld-Posieux Research Station ALP-Haras, Bern, Switzerland.
[4]University of Maryland, College Park, MD 20742-4454, USA.
[5]Honey bee Research and Extension Laboratory, Department of Entomology and Nematology, University of Florida, Gainesville, Florida, USA.
[6]Department of Ecology, Swedish University of Agricultural Sciences, Uppsala, Sweden.
[7]Hellenic Institute of Apiculture (N.AG.RE.F.), N. Moudania, Greece.
[8]College of Animal Sciences, Zhejiang University, Hangzhou 310058, China.
[9]Laboratório de Abelhas, Depto. de Ecologia, Instituto de Biociências, Universidade de São Paulo (USP), Rua do Matão 321, 05508-090 São Paulo-SP, Brazil.
[10]Department of Agriculture and Ecology, University of Copenhagen, Thorvaldsensve, 40, 1817 Frederiksberg C, Denmark.
[11]NEXTREAT, Weltistrasse 11, 5000 Aarau, Switzerland.
[12]Bee Health Laboratory, Department of Biology, Hacettepe University, Beytepe, Ankara, Turkey.
[13]United States Department of Agriculture, Animal and Plant Health Inspection Service, USA.
[14]Institute of Bee Health, Vetsuisse Faculty, University of Bern, Bern, Switzerland.
[15]Plant Research International, Wageningen University and Research Centre, Business Unit Plant Research International, Wageningen, Netherlands.
[16]Department of Entomology, North Carolina State University, Raleigh NC, USA.
[17]Konsulent Danmarks Biavlerforening, Fulbyvej 15, DK-4180 Sorø, Denmark.
[18]Apiculture Division, Faculty of Animal Bioengineering, Warmia and Mazury University, Sloneczna 48, 10-710 Olsztyn, Poland.

[†]The views expressed in section 4.7 are those of the author and do not reflect the views of the United States Department of Agriculture, Animal and Plant Health Inspection Service (USDA APHIS).

Received 22 May 2012, accepted subject to revision 11 July 2012, accepted for publication 9 May 2013.

*Corresponding author: Email: hhuman@zoology.up.ac.za

Summary

A variety of methods are used in honey bee research and differ depending on the level at which the research is conducted. On an individual level, the handling of individual honey bees, including the queen, larvae and pupae are required. There are different methods for the immobilising, killing and storing as well as determining individual weight of bees. The precise timing of developmental stages is also an important aspect of sampling individuals for experiments. In order to investigate and manipulate functional processes in honey bees, e.g. memory formation and retrieval and gene expression, microinjection is often used. A method that is used by both researchers and beekeepers is the marking of queens that serves not only to help to locate her during her life, but also enables the dating of queens. Creating multiple queen colonies allows the beekeeper to maintain spare queens, increase brood production or ask questions related to reproduction. On colony level, very useful techniques are the measurement of intra hive mortality using dead bee traps, weighing of full hives, collecting pollen and nectar, and digital monitoring of brood development via location recognition. At the population level, estimation of population density is essential to evaluate the health status and using beelines help to locate wild colonies. These methods, described in this paper, are especially valuable when investigating the effects of pesticide applications, environmental pollution and diseases on colony survival.

Footnote: Please cite this paper as: HUMAN, H; BRODSCHNEIDER, R; DIETEMANN, V; DIVELY, G; ELLIS, J; FORSGREN, E; FRIES, I; HATJINA, F; HU, F-L; JAFFÉ, R; JENSEN, A B; KÖHLER, A; MAGYAR, J; ÖZKÝRÝM, A; PIRK, C W W; ROSE, R; STRAUSS, U; TANNER, G; TARPY, D R; VAN DER STEEN, J J M; VAUDO, A; VEJSNÆS, F; WILDE, J; WILLIAMS, G R; ZHENG, H-Q (2013) Miscellaneous standard methods for *Apis mellifera* research. In *V Dietemann; J D Ellis; P Neumann* (Eds) *The COLOSS BEEBOOK, Volume I: standard methods for* Apis mellifera *research. Journal of Apicultural Research* 52(4): http://dx.doi.org/10.3896/IBRA.1.52.4.10

Métodos estándar diversos para la investigación en *Apis mellifera*

Resumen

En la investigación de la abeja de la miel, se han usado una variedad de métodos que se diferencian en función del nivel en el que se realiza la investigación. Al nivel individual, el manejo de las abejas individuales es necesario, incluyendo a la reina, las larvas y las pupas. Existen diferentes métodos para la inmovilización, mortandad y almacenamiento, así como para la determinación del peso individual de las abejas. La precisión en la sincronización de las etapas de desarrollo es también un aspecto importante de los experimentos con muestreos individuales. La microinyección se utiliza a menudo con el fin de investigar y manipular los procesos funcionales de las abejas melíferas, como por ejemplo, la formación y recuperación de la memoria y la expresión génica. Un método utilizado tanto por investigadores como apicultores es el marcado de las reinas, que sirve no sólo para ayudar a localizarlas durante su vida, sino que también permite su datación. La creación de varias colmenas a partir de reinas permite al apicultor mantener reinas de repuesto, aumentar la producción de cría o hacer preguntas relacionadas con la reproducción. Al nivel de colmena, la medición de la mortalidad intra colmena utilizando trampas de abejas muertas, el pesaje de las colmenas completas, la recolección de polen y néctar, y el seguimiento digital del desarrollo de la cría a través del reconocimiento de su ubicación, son algunas de las técnicas más útiles. Al nivel poblacional, la estimación de la densidad de población es fundamental para evaluar el estado de salud y el uso de líneas rectas para ayudar a localizar colmenas silvestres. Los métodos descritos en este artículo, son especialmente valiosos en la investigación de los efectos de la aplicación de pesticidas, la contaminación ambiental y las enfermedades sobre la supervivencia de la colmena.

西方蜜蜂研究的杂项标准方法

摘要

由于研究实施所针对的水平不同，用于蜜蜂研究的方法多种多样。在个体水平上，对蜜蜂个体（包括蜂王、幼虫和蛹）的操作是必须的。蜜蜂固定、处死和储存以及个体称重都存在很多不同方法。对发育阶段的精准确定也是实验中个体取样的重要方面。显微注射通常会用于蜜蜂功能过程的研究和操作，比如记忆形成、记忆提取和基因表达。被研究者和蜂农共同使用的一个方法就是蜂王标记，它不仅帮助定位蜂王，还能够标记蜂王年龄。建立多王群不仅让蜂农可保留剩余的蜂王，增加产子量，也为研究繁殖相关的问题提供基础。蜂群水平上，十分有用的技术是测量蜂巢内死亡率，可以利用蜂尸捕集器、整箱称重、采集花粉和花蜜或通过位置识别对子脾发展进行数字监控。种群水平上，种群密度估算是评价健康状况的要素，同时利用蜜蜂飞行直线有助于定位野生群。本章描述的这些方法尤其对研究农药应用、环境污染和病害对蜂群存活的影响具有价值。

Keywords: COLOSS *BEEBOOK*, immobilising bees, killing bees, storing bees, bee weight, microinjection, marking and clipping queens, haemocytometer, colony density, hive weight, dead bee traps, collecting pollen and nectar, digital recognition

Table of Contents

Table of Contents Cont'd

1. Introduction

Honey bees are one of the most studied insects, primarily due to their crucial role in agriculture and the ecosystem and their high economic value. In light of the concern over global honey bee decline experienced in many regions of the world, and with their economic importance in mind, funding has been readily available for research. The honey bee is a fascinating research model, its positive perception in general and its eusociality and importance for the food security and eco-system services makes it a model organism of choice. Therefore it is not surprising that a huge variety of research methods have been employed, evaluating and investigating different aspects of this organism, e.g. their interactions with parasites and pests (Volume 2 of the *BEEBOOK*), the behavioural and chemical ecology of this superorganism as well as aspects of breeding and population dynamics (Volume 1 of the *BEEBOOK*), to name a few. Since the interest in honey bees reaches from applied to fundamental research, numerous basic techniques are used across all disciplines. In this chapter, we will present various methods on recording basic demographic parameters like estimating number of dead bees, the weighing of a colony or of an individual, using a haemocytometer as well as pollen trapping. In addition, we describe ways of marking queens, how to inject, immobilise, kill and store honey bees, and how to obtain brood and adults of known age. Finally we discuss how to locate wild honey bee colonies, estimate honey bee colony density, create multiple queen colonies, and digitally monitor brood development via location recognition.

2. Research methods at the individual level

2.1. Standard methods for immobilising, killing and storing adult *Apis mellifera* in the laboratory

2.1.1. Introduction

Laboratory studies with honey bees usually involve a certain amount of handling of the experimental bees and often the termination and subsequent storage of the bee samples. There are a wide range of potential methods to immobilise, kill and store bees. Standardised methods for these experimental steps enable the comparison within the same trial and between different studies. The following section displays available methods, advantages and disadvantages of the different approaches and recommendations in terms of application.

2.1.2. Immobilising adults

Researchers are often required to immobilise adult honey bees, for example, when inoculating individuals with parasites during experiments (see section 1.3 on microinjection) or when removing live honey bees from hoarding cages or colonies to study intra-host parasite development. It is essential that sensitive body parts of the honey bee, such as the abdomen, antennae, eyes, and mouthparts, are not disturbed or damaged during immobilisation.

2.1.2.1. Physical immobilisation

Fine tip forceps can be used to gently grasp wings and legs; however, butterfly or featherweight forceps are more forgiving and can be used to grasp the thorax, in addition to appendages. The most effective, and sensitive, method for immobilising honey bees is to pinch the wings together gently above their base to ensure that the individual is securely held and cannot sting (Fig. 1).

Fig. 1. A worker honey bee held by gently squeezing its wings between the index finger and thumb. Note that the distal part of the abdomen is pointed in such a way that the honey bee cannot sting the handler.
Photo: G R Williams.

2.1.2.2. Chemical and physical immobilisation

In some cases a general anaesthetic is needed to facilitate the handling or immobilisation of very young or mature adult honey bees because of insufficient exoskeletal development or high activity, respectively. Both chemical (*e.g.*, carbon dioxide, diethyl ether, nitrogen, ethyl acetate) and physical (*e.g.* chilling, freezing) anaesthetics are available. Below we discuss only the two most commonly used methods employed by researchers to immobilise adult honey bees.

2.1.2.2.1. Carbon dioxide

Exposure to carbon dioxide deprives individuals of oxygen, and depending on dose, can lead to anoxia or asphyxiation in various tissues, as well as the accumulation of acid metabolites that can impair physiological processes, especially in the nervous system (Nicolas and Sillans, 1989). Exposure to carbon dioxide can result in premature aging and reduced lifespan of worker honey bees (*e.g.*, Mackensen, 1947; Austin, 1955; Woyciechowski and Moron, 2009), as well as affect behaviour and memory (Erber, 1975; Nicolas and Sillans,

1989). Although, exposure to carbon dioxide can influence intra-host parasite development (Czekońska, 2007), it is uncertain if honey bees exposed to the gas are subsequently more susceptible to parasitic diseases.

Phenotypic response to carbon dioxide is dose-dependent. Whereas large dosages and long exposure of carbon dioxide (*i.e.*, > 95% for 105 min) result in significant mortality and behavioural changes (Rueppel *et al.*, 2010), much shorter exposure duration can still affect workers. For example, pure carbon dioxide treatments greater than 15 seconds influenced sucrose response, foraging behaviour, and survival, although, in some cases certain symptoms may abate over time (Ebadi *et al.*, 1980; Pankiw and Page, 2003). Similar to workers, queens receiving a carbon dioxide anaesthetic can also exhibit symptoms; for example, higher carbon dioxide: nitrogen ratios resulted in significantly earlier oviposition events (Chuda-Mickiewicz *et al.*, 2012). More details on anaesthetising queens can be found in the *BEEBOOK* paper on instrumental insemination (Cobey *et al.*, 2013). To immobilise worker honey bees using carbon dioxide, researchers should provide individuals to pure gas for 10-15 seconds (Ebadi *et al.*, 1980); this should render individuals unconscious for approximately 15-30 seconds.

Protocol to immobilise bees with carbon dioxide:
1. Place honey bees in a well-ventilated cage.
2. Transfer the cage to a sealable plastic container with a small opening in the lid. Place the caged honey bees at the bottom of the sealed container as an added precaution to ensure full carbon dioxide exposure (Ebadi *et al.*, 1980), since carbon dioxide is heavier than air.
3. Connect a tube to the gas source (carbon dioxide bottle).
4. Insert the other end of the tube into the opening of the plastic container lid.
5. Provide constant supply of carbon dioxide (*e.g.*, 100 ml per minute) for 10-15 seconds.

2.1.2.2.2. Chilling

Cold temperatures can temporarily immobilise adult honey bees by reducing the amplitude of neuron action potentials (Wieser, 1973). Similar to carbon dioxide, length of exposure and dose, as well as recovery time (Frost *et al.*, 2011), can greatly influence phenotypic response to chilling exposure. For example, chilling for 3 min at -20°C did not affect worker longevity, orientation, or foraging behaviour (Ebadi *et al.*, 1980); whereas, ice-chilling at 0°C for the minimum amount of time needed to immobilise individuals significantly impaired learning, but not sugar responsiveness, compared to refrigeration at 4-5°C or freezing at -18°C (Frost *et al.*, 2011). Additionally, honey bee age can influence response to chilling, as newly emerged individuals less than 18 h old normally move at 22°C compared to 17°C for older foragers (Allen, 1959), and 85% of one

Table 1. Examples of methods used to kill honey bees depending on purpose of the study.

Method of termination	Termination description	Body part examined and purpose	Reference
Thermal	Exposed to -20°C in freezer	Worker ovarian development and midgut and rectum protein content	Human *et al.* (2007)
	Exposed to -80°C in freezer	Worker abdomen for molecular analyses of *Nosema* infection	Williams *et al.* (in prep.)
	Exposed to -20°C in a freezer	Worker body viral analyses	Yañez *et al.* (2012)
Mechanical	Removed internal organs and decapitated	Queen spermatheca, gut, ovaries, haemolymph, head, eviscerated body virus levels	Chen *et al.* (2006)
	Decapitated	Drone photoreceptor and glial cell intracellular potassium movement	Coles and Orkhand (1983)
	Crushed head and thorax	Queen spermatheca removal for gamete- backcross mating	Gladstone *et al.* (1964)
	Crushed thorax	Worker thorax mass	Heinrich (1979)
	Crushed thorax	Worker hypopharyngeal gland and ovarian development	Pernal and Currie (2000)
Mechanical and chemical	Crushed body and immersion in RNALater®	Worker body virus analyses	Williams *et al.* (in prep.)
Chemical and thermal	Exposed to dry ice in a container	Worker body chemical residue analyses	Mullin *et al.* (2010)
	Exposed to dry ice in a box	Worker gut polystyrene microparticle quantity	Naug and Gibbs (2009)
	Immersed in liquid nitrogen in a container	Adult bee genetic analyses	Zayed *et al.* (2005)
Chemical	Immersed in 95% ethanol	Drone genetic analyses	Jaffé *et al.* (2009b)
	Exposed to potassium cyanide in killing jar	Worker crop load	Visscher *et al.* (1996)

day old workers died when exposed for 3 min to -20°C (Robinson and Visscher, 1984) when no death in older workers receiving the same dose was observed (Ebadi *et al.*, 1980).

An exposure of the bee to -20°C for 3 min is recommended to immobilise mature individuals greater than 1 day old using chilling. At this time no recommendation can be made for chilling time of individuals younger than this due to seemingly adverse effects.

Protocol to immobilise honey bees with chilling:

1. Place required number of honey bees in a cage.
2. Transfer the cage into a freezer (-20°C).
3. Remove the cage with the immobilised bees from the freezer after 3 min.

2.1.2.2.3. Anaesthesia considerations

Anaesthetics should be easy to apply, repeatable, cheap, non-hazardous to humans, and have no or limited long-term effects on honey bees. Regardless of method chosen, and because of dose-dependence, all experimental individuals should receive the same dose, exposure length, and frequency of exposure, and methods should be described in full detail. Additionally, recordings of observations, such as honey bee mortality or responsiveness to sucrose, for example, should be delayed at least 1 h to provide anaesthetised honey bees with a recovery period (Pankiw and Page, 2003). Because honey bee anaesthetising provides a relatively poorly understood sublethal dose of a potentially lethal agent, the benefits of its use for an experiment should be clear. Conflicting data in the scientific literature suggest that carbon dioxide may be a more ideal anaesthetic than chilling, at least until specific methods can be developed for particular experiments that may use differently aged honey bees or need individuals to be sedated for varying lengths of time.

2.1.3. Killing adults

Adult honey bees used for research are often killed during or after experiments to allow for further examination, such as to take measurements of internal organs, to quantify parasite intensity or gene expression (e.g. Pernal and Currie, 2000; Maistrello *et al.*, 2008; Antúnez *et al.*, 2009), or simply to dispose of them safely. Generally, termination methods can be categorised as thermal, mechanical, or chemical; the method chosen will largely depend on the purpose for termination (Table 1).

2.1.3.1. Thermal killing
2.1.3.1.1. Cold

Freezing is a common method for killing adult honey bees because it can be easily and effectively applied, and will preserve genetic material. Freezing can, however, result in damage to cell structures, and therefore it is not recommended for studies that require internal tissues to remain intact, such as for quantifying hypopharyngeal development or midgut parasitism by *Nosema* spp. Exposing individuals to temperatures below -20°C will result in quick death; however, time required will vary depending on temperature and the number of individuals being collectively frozen (i.e. a higher number of honey bees collectively together will take longer to kill because of clustering behaviour). Placing individuals in a -20°C freezer for 2 h usually sufficient. Conversely, honey bees can be placed in a box of dry ice (e.g. Naug and Gibbs, 2009) or immersed in liquid nitrogen (e.g. Zayed *et al.*, 2005) for near instant termination.

2.1.3.1.2. Heat

Heat can also be used to kill honey bees, although its use is much less common than freezing, likely because it results in the denaturation of macromolecules such as nucleic acids (e.g. DNA and RNA) and proteins that in many cases may be studied post-mortem. Honey bees will typically die within one hour of exposure to 46°C (Allen, 1959), but this will depend on crop content and relative humidity.

2.1.3.2. Mechanical killing

Numerous studies kill adult honey bees by physically damaging or removing an essential body section (e.g. head, thorax, or abdomen) using forceps, one's index finger and thumb, or a scalpel. This method is relatively easy to perform, depending upon activity level and quantity of honey bees, and avoids the use of chemicals or other equipment that perhaps are not easily accessible. If there are many bees to kill, this can be a tedious method. Mechanical termination usually leaves the unaffected body part(s) intact; however, it may potentially promote parasite transmission when the exoskeleton is ruptured. The precise method of mechanical termination chosen will largely depend on the purpose of the study, but it can be monotonous (Table 1).

2.1.3.3. Chemical killing

The use of chemicals, including water, to kill honey bees commonly occurred in the 20th century; in recent years fewer studies use this technique. Because of the dangers of cyanide, and the numerous adequate alternatives, the use of this substance is not recommended. Care should be taken when using any chemical in the laboratory or field.

Asphyxiates such as carbon dioxide or ethyl acetate can also effectively kill honey bees, provided the appropriate dose is applied. For ethyl acetate, or alternatively nail polish remover, a sealable glass killing jar <500 ml in volume and lined at the bottom with 1-2 cm of plaster of Paris can be created or purchased from a entomological supply store. Ethyl acetate should be pipetted onto the plaster until satiation, and excess liquid removed, before insects destined to be killed are introduced (Steyskal, 1986). Five minutes within the sealed container should be sufficient to kill honey bees, although this may depend on the volume of the jar, the number of individuals being killed, and the quantity of ethyl acetate provided. Care must be taken to ensure that exposed honey bees are killed, rather than anaesthetised. When maintained in the killing jar for a number of hours, or even days, individuals can still be easily manipulated because of the ability of ethyl acetates to hold moisture, although decomposition may set in. Additionally, asphyxiation by drowning can be performed using pure water, soapy water, or ethanol. The latter, when 95% pure, will also, to some extent, preserve honey bees, as well as organisms and chemical residues present within them; water will promote decomposition. As mentioned earlier (see section 1.3.1.1.), honey bees can also be exposed to dry ice (Naug and Gibbs, 2009) or liquid nitrogen (Zayed *et al.*, 2005) for quick termination.

2.1.4. Storing dead adults

When post-mortem examinations, or necropsies, are to be performed for a particular study it is imperative that honey bees to be examined are maintained under appropriate conditions to ensure degradation does not occur. Ideally, samples should be placed under optimal preservation conditions as soon as possible after death if analyses or examination does not occur immediately. Storage conditions, as well as the materials to be preserved, will largely depend upon the question being asked.

Generally, freezing is the best and most commonly used strategy for maintaining well preserved samples; however, when this is not available certain chemical stabilisers (e.g. RNALater® (Qiagen, Hilden, Germany), and TN, Kiev and TRIS-NaCL buffers) may provide alternative options, at least in the short term (Table 2). Careful attention must be paid during examination of easily degradable material, such as DNA and in particular RNA because of its single stranded architecture and because of endogenous RNases that occur ubiquitous in organisms and the environment (Chen *et al.*, 2007; Winnebeck *et al.*, 2010; Dainat *et al.*, 2011). Additionally, pheromone, pesticide residue, and whole tissue examination also require appropriate preservation (Table 2).

Ideally, samples should be preserved at -80°C; however, freezing at -20°C or less should be sufficient for relatively short-term storage. More in depth discussions on sample preservation can be found in respective papers of the *BEEBOOK*, such as de Miranda *et al.* (2013) for viruses, Fries *et al.*, (2013) *for Nosema*, and Medrzycki *et al.* (2013) for toxicology.

2.2. Determination of individual bee weight

The fresh weight of an *Apis mellifera* worker drastically increases during its 21 days of development from an egg weighing about 0.03-0.1 mg to about 120 mg at adult emergence. In contrast, drones reach 277-290 mg after emergence (Hrassnigg and Crailsheim, 2005). Hence, the weight of larvae is, among others factors, important in determining their age (see section 2.5 Obtaining adult and brood of known age; Wang, 1965). Determining the weight of individual honey bees can also be important when assessing the effect of pathogens, parasites or toxins on their development and health or when assessing their nutritional intake. In this section, we describe procedures to obtain fresh weight of immatures (see section 2.2.2.), adult honey bees (2.2.3.) or their parts (2.2.4.) as well as dry weight of adults (2.2.5.). Larvae or adults collected for later analysis are best stored frozen to prevent desiccation.

2.2.1. Balance required for weighing individual bees or larvae or body parts

A well calibrated and sufficiently sensitive analytical balance should be

Table 2. Examples of methods used to store honey bees and selected bee products depending on purpose of the study.

Method of storage	Storage description	Body part stored and purpose	Reference
Cold	-20°C	Adult worker ventriculi for *Nosema* qPCR quantification	Forsgren and Fries (2010)
	-20°C	Whole adult workers for *Nosema* species identification	Williams *et al.* (2008; 2011)
	-20°C	Adult workers, honey & beeswax for gas chromatography (GC)-tandem mass spectrometry (MS/MS) & liquid chromatography (LC-MS/MS) chemical residue analyses	Nguyen *et al.* (2009)
	-20°C	Beebread, brood, adult workers for LC/MS-MS and GC/MS pesticides residue analyses	Mullin *et al.* (2010)
	-80°C	Mature queen spermathecal fluid protein profiling using gel electrophoresis	Baer *et al.* (2009)
	-80°C	Mature queen ovaries & eviscerated abdomens (cuticle with attached fat bodies) for quantitative real-time PCR of Vitellogenin gene expression	Kocher *et al.* (2008)
	-80°C	Adult drone antennae for microarray and qPCR sex pheromone gene expression quantification	Wanner *et al.* (2007)
	-80°C	Whole adult workers for quantitative real-time PCR of immune gene expression	Antúnez *et al.* (2009)
	-80°C	Extracted RNA from adult workers, eggs, queen faeces & queen tissues for RT-PCR analyses of viruses	Chen *et al.* (2006)
	-80°C	Brood comb (beeswax, beebread and brood) and adult workers for LC/MS-MS and GC/MS pesticides residue analyses	Mullin *et al.* (2010)
Cold & chemical	-20°C & Kiev buffer	Queen spermathecae for sperm counting	Kocher *et al.* (2008)
	-80°C & RNALater®	Worker honey bee RT-PCR virus analyses	Williams *et al.* (2009)

used, automated data transfer to computer, a standard feature, facilitate higher sampling rates. Remember, most analytical balances have a precision or readability of 0.1 mg which is sufficient for larvae and adults, but near the weight of an egg. To precisely determine the latter, we recommend measuring egg length following Henderson (1992) or use a high precision micro balance.

2.2.2. Weighing of larvae
1. Carefully take the larva out of its cell.

   Injuries of larval surfaces result in the loss of haemolymph and injured larvae must be discarded.

   If the larvae are to be weighed when stored frozen, they have to be brought to room temperature to prevent convection during weighing and then weighed quickly before desiccation.
2. Wash larvae with either saline, alcohol or distilled water to remove remainders of the larval food.
3. Quickly dry them on filter paper.

   Due to the low weight and quick drying up of young larvae, time between sampling and weighing should be minimised.
4. Tare the balance with container.
5. Place larvae in container.
6. Weigh larvae.
7. Record weight.

2.2.3. Weighing of adult honey bees
The fresh weight of adult bees is a measure for nutritional and health state and can be measured from live (see section 2.1. on immobilising honey bees) or dead honey bees.

1. Tare the balance with container.
2. Place honey bee in container.
3. Weigh honey bee.

   Note that the weight of newly emerged bees is influenced by the meconium (faeces that is expelled in purging flights after emergence, Jackson and Hart, 2009). Adult weight is also influenced by the consumption of food, and standardised starvation of bees in an incubator corrects for this increase (see the *BEEBOOK* paper on maintaining adults *in vitro* in the laboratory, Williams *et al.*, 2013).
4. Record weight.

2.2.4. Weighing body parts
Before establishing the weight of dead honey bees (see section 2.2.3), we suggest separating the body with small scissors into head, thorax (including legs and wings) and abdomen. This allows to roughly ascribe weight deficiencies to one of these body parts. Total dry weight is the sum of all body parts. For example, the fresh weight of the head correlates with the acini-size of hypopharyngeal glands (Hrassnigg and Crailsheim, 1998) and the fresh and dry weight of the thorax is a measure for the development of flight musculature (Brodschneider *et al.*, 2009). Note that the weight of the abdomen is often determined without the gastrointestinal tract, because of the pollen or meconium in it (Hrassnigg and Crailsheim, 2005; Jackson and Hart, 2009).

2.2.5. Determining dry weight
The dry weight of adult bees is determined by putting whole bees or

their body parts (see section 2.2.4.) in individually labelled and pre-weighed Eppendorf tubes in an incubator at 55-70°C; 60°C can be recommended (Henderson, 1992). The tubes remain open during incubation. Dry weight is reached when the sample show constant weight in successive measures, which usually occurs within 7 days in honey bees, depending on incubation temperature. Samples can also be transferred to weighing dishes, but care must be taken not to lose extremities. Refer to the procedure described in section 2.2.3 to weigh the dried samples. The dry weight of newly emerged bees indicates the nutritional investment in larvae, which may result in different emerging weights during season. In adult bees, dry weight changes depending on nutrition and age, reaching a maximum after five days, and decreases again towards the foraging age (Hrassnigg and Crailsheim, 2005). Finally, parasites like the varroa mite can exert weight differences of more than 10% in emerging honey bee drones (Duay *et al.*, 2003).

2.3. Microinjection

2.3.1. Introduction

Injection is a technique widely employed to manipulate functional processes in honey bees. Injections have been performed at different stages of the honey bee life cycle, from early embryos to adults (Lozano *et al.*, 2001; Aase *et al.*, 2005; Kucharski *et al.*, 2008). In adult honey bees, the injection of receptor antagonists into the brain or antennal lobes provided insights into pathways involved in memory formation and retrieval (Lozano *et al.*, 2001; Farooqui *et al.*, 2003; Wright *et al.*, 2010). Gene expression can be manipulated by injecting double stranded RNA (Schlüns and Crozier, 2007; see also Section VI – RNA interference). Injections of pathogens (Wilson and Rothenbuhler, 1968) and insecticides (Bendahou *et al.*, 1999), as well as injection of labelled markers to trace substance distributions (Crailsheim, 1992) are further applications.

Irrespective of the substance being injected, rupturing the cuticle with the needle is invasive and causes an immune response in honey bees, including increased expression of the immune response gene Defensin2 and antimicrobial peptide production (Richard *et al.*, 2008; Laughton *et al.*, 2011). In addition, researchers should be aware that handling during injection induces a stress response and the tissue damage further poses a risk of secondary infection (Kucharski and Maleszka, 2003). The stress and immune response may even result in death of injected individuals. In adult *A. mellifera*, a 20% mortality rate was observed within 48 h of injection with control buffers (Picard-Nizou *et al.*, 1997). Most studies do not report survival rates following injection, but immune responses and mortality risks should be considered when choosing to inject substances.

2.3.2. Microinjection using a Hamilton syringe

In multiple studies, honey bees have been injected using a Hamilton syringe (Kucharski and Maleszka, 2003; Barron *et al.*, 2007; Schlüns

and Crozier, 2007; Wright *et al.*, 2010; Köhler *et al.*, 2012). The following method is suggested:

1. Select workers to be injected.

   When using newly emerged workers, no anaesthesia is required as they do not sting or fly. Older workers need to be anaesthetised (see section 2.1 standard methods for immobilising, killing and storing adult *Apis mellifera* in the laboratory).

2. Hold the honey bee gently on the side of the thorax between thumb and index finger of one hand.

3. Inject bee with the other hand.

   The most common place of injection is between tergites (the needle can easily be inserted specifically between the 3rd and 4th tergite) at the side of the abdomen. The needle should be inserted parallel to the tergite to avoid puncturing of the gut. Handling time must be kept to a minimum (few seconds per bee) to prevent unnecessary stress. Saline or insect ringer are typically injected as carrier and control (Lozano *et al.*, 2001; Barron *et al.*, 2007; Schlüns and Crozier, 2007; Wright *et al.*, 2010).

2.3.3. Microinjection of small volumes using the Nanoject device and other micro injectors

For very small injection volumes (<1 μl), the Nanoject injector (Drummond) can be used. It consists of a microinjection pipette with an automated microprocessor that can precisely inject a set volume. The injection tips are made from glass capillaries. This automated injection method eliminates vibration and thus minimises tissue injury. It may therefore reduce deleterious effects on honey bees, compared to manual injections. The second advantage over manual injections is the high precision of the injector that will eliminate variations in injection volumes. The Nanoject injector has been successfully used on insects (e.g. Teixeira *et al.*, 2008; Yamane and Miyatake, 2010). Furthermore, embryonic injections have been performed using a microscope with a micromanipulator and a microinjector (Narishige) with glass capillary (Sasaki and Ishikawa, 2000). Beye *et al.* (2002) injected honey bee eggs under a microscope using an Oxford micromanipulator (Singer) and a microinjector with borosilicate capillaries. Lozano *et al.* (2001) injected adult honey bee workers using a custom-made microinjection system consisting of a glass micropipette mounted on a microelectrode puller (Campden Instruments).

1. Prior to injection, adult individuals should be anaesthetised (see section 2.1).

   Anaesthesia with CO_2 is not recommended, given known CO_2 driven physiological and behavioural modifications in honey bees (Ebadi *et al.*, 1980; Koywiwattrakul *et al.*, 2005).

2. Immobilise the individual to be injected.

   This can be done by physical means (see section 2.2.1.) or by chilling them (section 2.2.2). Anaesthesia with CO_2 is not recommended, given known CO_2 driven physiological and behavioural modifications in honey bees (Ebadi *et al.*, 1980; Koywiwattrakul *et al.*, 2005).

3. Inject the individual. The method of injection will depend on the device used (for methods see references in the introduction of this section).

2.3.4. Perspectives

Future studies should compare survival rates following manual and automated injection at different parts of the body, e.g. injection into the thorax vs. the abdomen. The abdomen has been chosen as the injection site in different studies (e.g. Amdam *et al.*, 2006; Schlüns and Crozier, 2007; Richard *et al.*, 2008). The intersegmental membrane is soft between the tergites and can be easily punctured by the needle. Drugs are directly administered into the haemolymph and spread throughout the body with haemolymph circulation. However, if the worker has ingested a large meal prior to injection, one may puncture the full stomach with the needle. This may be avoided by injecting individuals immediately following emergence prior to the first ingestion of honey from the comb.

Survival rates may also differ between honey bees injected with or without anaesthesia. Workers handled without anaesthesia showed lower sucrose responsiveness 30 min after handling than immobilised individuals (Pankiw and Page, 2003), and a delayed onset of feeding may increase mortality. Control individuals should be subjected to the same handling times to control for stress effects.

Lastly, nutritional composition of the diet may also affect survival. Worker survival on sucrose-only solutions was drastically reduced following injection (Köhler *et al.*, 2012), but adding protein to the diet may increase the production of immune system components (e.g. antibacterial peptides), which may help in fighting infections and improve survival (Alaux *et al.*, 2010; DeGrandi-Hoffman *et al.*, 2010); see also the *BEEBOOK* paper on 'maintaining adult honey bees *in vitro* under laboratory conditions' by Williams *et al.* (2013). Mortality rate following injection may depend on multiple factors, including injection technique, types and amounts of injected substances, type of needle, needle thickness and sharpness, age of the honey bees, handling stress, and type of anaesthesia (if any). It may prove valuable to assess the mortality following injection in a particular experimental setup to be able to adjust the sample size for the study (see the *BEEBOOK* paper on statistics (Pirk *et al.*, 2013). The effects of injections should be considered when deciding on a technique for substance application and all parameters or injection methods should be described in detail.

Checklist for injections:

1. Decide on the injection method (Hamilton syringe, Nanoject device, etc.).

2. Decide on a suitable injection site (e.g. abdomen of adults, between tergites).
3. Decide whether anaesthesia is required. Young workers do not sting or fly, older workers may need to be immobilised by cooling prior to injection.
4. Make sure to have defined age-cohorts.
5. Determine the dose and injection volume (ideally < 5 µl for adult workers).
6. Decide on a suitable buffer (saline, insect ringer).
7. Reduce possibilities of unintended secondary infections: use new glass capillaries; disinfect the Hamilton syringe (e.g. ethanol, acetone) before and after use; the needle can be sterilised in a flame to avoid contamination between individuals.
8. Take initial high mortality into account.
9. A test run to practise the injection technique is recommended. This way survival rate following injection can also be determined.

2.4. Marking honey bee queens

2.4.1. Colour-marking queens

Not only does marking the queen helps in finding her in the hive, but a queen which has been marked and recorded can be 'dated' by reference to the hive card or record book. It also makes it possible to ascertain if and when she has been superseded, or if she has attempted to leave with a swarm, in which case she is usually lost. Generally, queens are marked before being introduced, but they can be marked at any time.

A wide variety of markers have been used to assess insect population dynamics, dispersal, trophic-level interactions, and other ecological interactions. The ideal marker should persist without inhibiting the insect's 'normal' biology. Furthermore, the marker should be environmentally safe, cost-effective, and easy to use (Hagler and Jackson, 2001).

2.4.1.1. Marking type

Queens can be marked with a variety of paints or equipped with numbered and coloured Opalith discs on top of the thorax (Fig. 2).

- Queen marking pens are handy and make queen marking easier (no risk of spilling the bottle with paint or glue), but usually wear off very quickly.
- Fast drying nail varnishes are also good markers.
- Model car paint can be used.
- The longest lasting queen marker is made by mixing a pigment with shellac.
- Opalith discs (Fig. 2) are commercially available in a variety of colours.

Fig. 2. A queen marked with Opalithplättchen.   Photo: W Wei.

- A special glue is provided with the discs but partially dried shellac or cyanoacrylate ester glues (e.g. Super Glue®) will serve the same purpose.

2.4.1.2. Procedure for paint marking

1.a. Hold onto the legs or thorax of the queen with one hand (Fig. 3 bottom)

1.b. Alternatively, introduce the queen in a special 'marking tube'. The queen is inserted through the open end of a glass tube and carefully pressed upward with a soft plunger against a net on top of tube. This holds the queen stable during the marking process, thus facilitating it.

2. Dab the marking stick in the paint.

   Only the minimum necessary amount of paint should be transferred onto the stick in order not to smear too much material on the queen's thorax and other appendages.

3. Mark the queen by quickly dabbing the paint on the dorsal side of her thorax.

   The mark should be small, so that it does not cover any other part of the queen and impair her behaviour.

4. Give paint ample time to dry before the queen is released into the colony.

2.4.1.3. Procedure for marking with Opalith discs

1.a. Hold onto the legs or thorax of the queen with one hand.

1.b. Alternatively, introduce the queen in a marking tube (Section 2.4.1.2.).

2. Dab the marking stick in the glue.

   Only the minimum necessary amount of glue or paint should be transferred onto the stick in order not to smear too much material on the queen's thorax and other appendages.

3. Place the glue on the dorsal side of the queen's thorax applying the glue on an area the size of the disc.

4. Moisten the opposite end of the marking stick (where there is no glue).

5. Touch the numbered side of an Opalith disk with this wet end. This allows the disc being picked up.

6. Apply the disc with a slight pressure on the glue.

7. Give glue ample time to dry before the queen is released into the colony.

2.4.1.4. Colour-marking code

An International Colour Code system exists within the beekeeping industry to indicate the year the queen was introduced and facilitates recognition of queen age (Table 3). Since queens do not live more than 5 years, the colour coding starts over in the sixth year (Table 3).

Marking queens with a dot of paint is cost-effective and easy to use and thus practical in beekeeping. While Opalith disks, enabling individual identification of bees (Figs. 2 and 3), have been widely used in research and breeding where it is essential to know the pedigree / history of the queens and colonies.

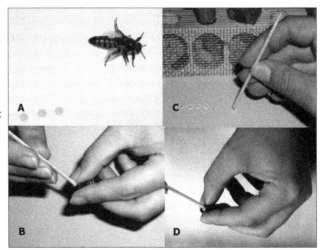

Fig. 3. Step-by-step marking of a queen with numbered plastic disk. **A.** queen with Opalith disks. **B.** the marking stick is dipped in glue and touched to the queen's thorax. **C.** the marking stick (the end opposite of the glue) is moistened and touched to the numbered side of an Opalith disk. **D.** the Opalith disk is affixed to the thorax and held in place by the glue.

Photos: J Wilde.

Table 3. International colour code used for marking queens.

International queen marking colour code:	
colour:	for years ending in:
White	1 or 6
Yellow	2 or 7
Red	3 or 8
Green	4 or 9
Blue	5 or 0

2.4.2. Clipping queens' wings

Queens can be marked by clipping the tip of one forewing. If queens are replaced every two years, the beekeeper can clip the left wing on queens introduced in odd years, and the right wing on queens introduced in even years. The clipping practice may also supplement the paint spot technique as a back-up, should the queen lose her paint mark. Honey bee queens are mated in the air, it should therefore be ascertained that the queen has mated before her wing is clipped.

Another reason to clip the wings of a queen is to prevent her swarming off, besides other methods such as keeping colonies headed by young queens and removing all queen cells. Swarming is the process by which honey bee colonies can reproduce (Seeley, 1986). When swarming occurs, half of the bees will leave the hive. This results in a hive that is unable to rebuild its population before the nectar flow starts, thus decreasing the production of honey. Therefore, swarm control is a very important part of beekeeping management. When conducting experiments during swarming season, swarm control is even more critical since a swarm may take away half the experimental bees.

From beekeeping experience, it is apparent that queens with fully clipped wings are more prone to fall to the bottom of the hive and are often superseded more quickly than those with unclipped wings. It is recommended to clip less than half of one forewing (Fig. 4) to prevent the queen from flying with a swarm, but not to impair other behaviours. With less than half of one forewing clipped, the queen's

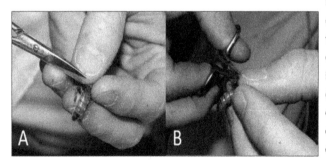

Fig. 4. Clipping the wing of a queen honey bee. The head and thorax of the queen honey bee are lightly grasped between the thumb, index, and middle fingers **(A)**. The wings and abdomen point away from the hand **(A)**. The scissors should be used to tease out the forewing on only one side of the body **(B)**. Using the scissors, clip approximately one fourth, but no more than half of the forewing from the body.

Photos: A Ellis.

ability to function properly inside the colony will not be significantly affected. If the queen tries to fly with the swarm, she will most likely drop in front of the hive. She may then crawl up the leg of the hive stand and re-enter the hive. If not, she can be collected by the beekeeper and put back. The swarming bees will fly away for a short time, but will return to their hive when they are unable to find their queen. Occasionally the clipped queens may fly despite the clipping, but their range is limited, which makes retrieval easier. However, sometimes queens may be lost if they cannot find a way to re-enter the hive after dropping in front of it.

Yet another use of wing clipping is the non-lethal collection of queen DNA. In this case, the purpose is not to prevent flight, since virgin queens that still have to perform a mating flight might be needed for DNA extraction. If a sufficiently small wing piece is clipped (*c.* 1.3 mm^2, 7.5% of each forewing surface is sufficient to genotype them), the mating success of these clipped-wing queens is not affected (Châline *et al.*, 2004).

Wing clipping procedure:
1. Lightly grasp the queen by the thorax between the thumb, index and middle finger of one hand so that the forewing to be clipped points upwards and the abdomen points away from the hand (Fig. 4A).
2. Hold the scissors with the other hand and slide one tip between the fore- and hind wing to separate them (Fig. 4B).
3. Cut approximately one fourth of the forewing without damaging the hind wing.
4. Mark the queen with paint (see section 6.1.2.) when desired.

2.5. Obtaining brood and adults of known age

The development of honey bees is divided between egg, larval and pupal stages. Only at the larval stage do immature bees grow, thanks to the abundant food provided by the nurse bees. Their weight increases by a factor of 1,500 during this stage. There are six moults during their development. The timing of moulting and the growth varies according to caste (worker or queen) and sex (drone or queen) of the individual, and to the lineage to which the honey bees of interest belong. Queen development (16 days) is faster than worker's (21 days) and in turn worker development is faster than that of drones (24 days). Table 4 summarises the major events and developmental times for *A. mellifera*. Variations of a few hours in developmental time of the various stages can occur between individuals in a colony, but also between subspecies (Michelette and Soares, 1993; Allsopp, 2006). For this reason the measurements given in the following methods are valid only for the subspecies or lineage they have been obtained for. Where known, the subspecies is indicated for relevance of use. When another subspecies is investigated and precise timing of developmental stages is needed, it is recommended to verify their timing, for which we give a method (see section 3.5.1.1.).

Table 4. Development time and events for workers, queens and drones of *Apis mellifera* (modified from Bertholf, 1925).

Day	Workers Stages	Workers Events	Queens Stages	Queens Events	Drones Stages	Drones Events
1						
2	egg		egg		egg	
3		hatching		hatching		hatching
4	1st larval instar	*1st moult*	1st larval instar	*1st moult*	1st larval instar	*1st moult*
5	2nd larval	*2nd moult*	2nd larval instar	*2nd moult*	2nd larval instar	*2nd moult*
6	3rd larval instar	*3rd moult*	3rd larval instar	*3rd moult*	3rd larval instar	*3rd moult*
7	4th larval instar	*4th moult*	4th larval instar	*4th moult*	4th larval instar	*4th moult*
8	5th larval instar		5th larval instar		5th larval instar	
9		*cell is capped*		*cell is capped*		
10	prepupa		prepupa	*5th moult*		*cell is capped*
11		*5th moult*				
12			pupa		prepupa	
13						
14						*5th moult*
15				*6th moult*		
16	pupa			*emergence*		
17						
18					pupa	
19						
20		*6th moult*	imago			
21		*emergence*				
22						*6th moult*
23	imago					*emergence*
24					imago	
25 ...		...		...		...

2.5.1. Obtaining brood of known age

To obtain brood of known age, a queen can be caged on a frame on a particular day for a few hours. The duration of the caging is determined by the quantity and age range of the brood needed. For example, if more brood is required, the longer the queen is left in the cage. In this case, the age range of that brood increases. For example, the comb area in which the queen laid eggs during 4 h will hold brood between 20 and 24 h of age, 24 h post caging the queen. Given that queens can lay 2,000 eggs per day during the fastest growing stage of the colony, approximately 100 eggs can be obtained every hour. During less beneficial periods, a lower number of eggs can be obtained. If larger amounts of brood of a narrower age range are needed, queens of several colonies need to be caged. This also allows reducing the disturbance of individual colonies in case the brood needs to be collected at frequent intervals. If large amounts of brood of known age are needed from a single colony, several replicates at different times need to be done.

2.5.1.1. Procedure to obtain worker or drone brood of known age

1. Find the queen in a colony.
2. Place an empty comb with worker or drones cells (depending on the needs of the experiment) in a cage with sides made from queen excluder material or purchase a trap cage which encloses the frame and prevents the queen from leaving the comb, but allows workers to move freely in and out to take care of the brood and queen (see Fig. 1 from the *BEEBOOK* paper on maintaining adult honey bees *in vitro* under laboratory conditions (Williams *et al.*, 2013)).
3. Place the queen in the cage for a predetermined time. Note: the longer the queen is caged, the larger the range of age of the brood becomes.
4. Remove the comb and queen from the cage.
5. Reintroduce the queen into her colony.
6. Mark on a transparent sheet of acetate the area of comb in which the queen oviposited for future localisation (Fig. 5). The sheets should be laid over the surface of the comb and the position of cells can be recorded on the sheet using a permanent marker. Be sure to label the sheet and mark it according to its position on the frame (Fig. 5) to be able to place it accurately when using it later and thus avoid confusion.
7. Remove the sheet.
8. Replace the comb in the colony in middle of the brood area. The comb can be placed in the cage again to prevent further ovipositing by the queen (now on the other side of the excluder) in this comb. If done this way, the brood produced while the queen was caged will not be mixed with younger brood produced later.
9. Collect brood when it reaches the desired age or observe developmental stages at regular intervals, according to the purpose of the experiment.

Fig. 5. Marking the cells freshly oviposited in by a queen on an acetate sheet fixed on the frame. The acetate sheet is fixed to the frame with thumb tacks and its position is marked with lines drawn across the sheet and frame (see left side of the image) for precise repositioning on next use.   Photo: V Dietemann.

It is possible to narrow down the age range of the larvae considered for experiments by considering only those that hatched during a chosen period. If larvae are collected for weighing, they should be rinsed in physiological saline to rid them of adhering food.

2.5.1.2. Procedure to obtain queen brood of known age

Follow steps 1 to 8 as described in section 2.5.1.1.

9. Allow for larval hatching approximately 3 days later.
10. Graft larvae of similar age, but younger than 24 h into queen cups.
11. Follow the instructions given in the *BEEBOOK* paper on queen rearing and selection for queen rearing (Büchler *et al.*, 2013).
12. Collect queen brood when it reaches the desired age or observe developmental stages at regular intervals, according to the purpose of the experiment.

2.5.2. Obtaining pupae of known age

Instead of caging the queen and waiting until pupation to obtain pupae of desired age, freshly capped cells can be identified. This saves time since larval development time can be 'spared' and one need only wait the desired time after capping before obtaining pupae for experiments (see Table 5 for a timeline for worker pupae).

1. Remove frames containing many mature (L5) larvae from the colony.

2. Place an acetate sheet over each frame.
    Be sure to label the sheet and mark it according to its position on each frame to be able to place it accurately when using it later and thus avoid confusion.
3. Mark the position of all sealed brood on the sheets of acetate (Fig. 6).
4. Remove the acetate sheets.
5. Replace the frames in the hives.
6. Remove and re-examine frames at regular intervals (as needed for the experiment, usually a minimum of 2 h).
7. At each interval, mark the position of cells which have been capped since the last check.
    To do this, the acetate sheet is returned to the surface of the frame and aligned with the original point of reference (Fig. 6).
8. Remove the acetate sheet.
9. Replace the comb in the colony.
10. Remove the relevant combs from colonies at pre-determined times and collect pupae of desired age, as indicated by the transparent sheets.

The average duration of the sealed brood stage is 12 days (288 h) for workers and 14-15 days (340-360 h) for drones in *A. mellifera* in

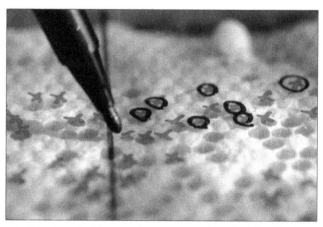

Fig. 6. Marking freshly capped cells on an acetate sheet fixed to the frame.   Photo: V Dietemann.

the U.K. (Martin, 1994, 1995). Relatively high variations are reported for different localities and subspecies (up to 19 h for worker development, Milum, 1930; Le Conte and Cornuet, 1989; 40 h for duration of capped stage in *A. mellifera capensis* workers, Allsopp, 2006). The expected durations should be verified before starting an experiment since these vary from subspecies to subspecies.

The same principle can be used to obtain drone pupae of known age when open drone brood is available. Table 6 gives the timeline for drone pupae development. For queens, follow the procedures described in the *BEEBOOK* paper on queen rearing and selection for queen rearing (Büchler *et al.*, 2013). Table 7 gives the timeline for queen pupae development.

Table 5. Colour changes in worker pupae, modified from Jay (1962). Days are counted from cell capping to correspond to Fig. 8. Body parts mentioned in the table are annotated on Fig. 8.

Number of days from capping	Colour	Body parts
5	slightly marked light pink	eyes, ocelli
6	light pink-purple	eyes
	dark pink	ocelli
7	dark pink-purple	eyes, ocelli
8	slightly marked light brown	head, thorax
	light brown	tibio-tarsal joints, sutures outlining mesonotum, wing bases
9	light yellow	abdomen, legs
	light brown	head, thorax
	light to medium brown	leg joints, claws, mandibles, antennae, sting, spurs, spines, hair
	medium brown	tibio-tarsal joints, wing bases, sutures outlining mesonotum
	dark purple	eyes, ocelli
10	light grey	wing pads
	medium brown	flagellar segments, leg joints, wing bases, mandibles, claws, sting, spines, spurs, hair, sutures outlining mesonotum
	dark yellow	abdomen, scapes, pedicel, tongue, legs
	dark brown	head, thorax
	brownish-purple	ocelli
	black	eyes
11	medium grey	wing pads
	dark yellow to light brown	abdomen, scapes, pedicel, tongue, legs
	dark brown	leg joints, wing bases, claws, sting, spines, spurs, hair, sutures outlining mesonotum
	dark grey	head, thorax
	dark brownish-black	ocelli
	black	eyes, flagellar segments
12		pupal moult complete

Table 6. Colour changes in drone pupae according to Jay (1962). Days are counted from pupation.

Number of days from pupation	Colour	Body parts
2	slightly marked light pink	eyes, ocelli
3	light pink-purple	lower parts of eyes, ocelli
	dark pink	eyes, ocelli
4	dark pink	eyes, ocelli
	dark pink-purple	lower parts of eyes
5	light pink-purple	eyes, ocelli
	dark purple	lower parts of eyes
6	light yellow	wing base
	dark pink-purple	eyes, ocelli
	dark purple	lower parts of eyes
7	slightly marked light yellow	abdomen, tongue, antennae, wing pads, head, thorax, legs, wing bases
	light brown	tibio-tarsal joints, claws, mandibles, sutures outlining mesonotum
	dark purple	eyes, ocelli
8	light yellow	abdomen, tongue, scapes, pedicel, legs
	light brown	head, thorax, spurs, spines, hair, flagellar segments
	light grey	wing pads, tip of abdomen
	medium brown	leg joints, wing bases, claws, mandibles, sutures outlining mesonotum
	dark purple	eyes, ocelli
9	light brown	scapes, pedicel, tongue
	light grey	wing pads
	medium brown	head, thorax, spines, spurs, hair
	dark yellow	abdomen, legs
	dark brown	leg joints, wing bases, claws, mandibles, sutures outlining mesonotum, tip of abdomen
	purple-black	eyes, ocelli, flagellar segments
10	medium to dark grey	wing pads
	dark yellow to light brown	abdomen, scapes, pedicel, tongue, legs
	dark brown	leg joints, wing bases, mandibles, claws, spines, spurs, hair, sutures outlining mesonotum
	dark grey to dark brown	head, thorax
	black	eyes, ocelli, flagellar segments, tip of abdomen
11		pupal moult complete

Table 7. Colour changes in queen pupae according to Jay (1962). Days are counted from pupation.

Number of days from pupation	Colour	Body parts
1	light pink	eyes
2	light pink	ocelli
	medium pink	eyes
3	light pink-purple	eyes
	light yellow	head, thorax, mandibles
	dark pink	ocelli
4	light yellow	abdomen, legs, antennae
	light brown	head, thorax, leg joints, claws, sting, sutures outlining mesonotum
	dark pink-purple	eyes, ocelli
	dark brown	mandibles
5	light grey	wing pads
	medium grey	head, thorax
	dark yellow to light brown	abdomen, legs, frons, clypeus, tongue, scapes, pedicel
	dark brown	leg joints, claws, sting, mandibles, spines, spurs, hair, sutures outlining mesonotum
	black	eyes, ocelli, flagellar segments
6		pupal moult complete

Table 8. Average head diameter and body weight range of workers and queens of *A. mellifera carnica* from Germany and Africanised honey bees from Brazil (after Rembold *et al.*, 1980; Michelette and Soares, 1993).

		Instar	Head diameter (mm ± SD)	Weight (mg, min. – max.)
Apis mellifera carnica	Workers	L1	0.33 ± 0.018	0.10 – 0.45
		L2	0.47 ± 0.030	0.35 – 1.50
		L3	0.70 ± 0.051	1.3 – 6.0
		L4	1.05 ± 0.058	4.2 – 32
		L5	1.58 ± 0.078	27 – 280
	Queens	L1	0.33 ± 0.020	0.10 – 0.45
		L2	0.48 ± 0.026	0.35 – 150
		L3	0.72 ± 0.044	1.3 – 7.0
		L4	1.11 ± 0.072	3.8 – 44
		L5	1.69 ± 0.097	31 – 360
Africanised honey bees	Workers	L1	0.32 ± 0.026	0.11 – 0.30
		L2	0.44 ± 0.032	0.31 – 1.05
		L3	0.65 ± 0.045	1.50 – 4.45
		L4	0.92 ± 0.094	4.80 – 24.8
		L5	1.49 ± 0.048	24.30 – 126.7

2.5.3. Recognising the instar of larvae

Rembold *et al.* (1980) and Michelette and Soares (1993) described the different larval instars based on head diameter for *A. mellifera carnica* from Germany (workers and queens; Table 8) and Africanised honey bees form Brazil (workers; Table 8), respectively. These measures provide a reliable method to identify larval instars, since head size of the various stages grow in a stepwise manner at each ecdysis. These authors also give the weight range of the different instars, which can also help identify them. However, the weight of the heaviest larvae of an instar can overlap with that of the lightest larvae of the next instar.

2.5.4. Recognising the age of larvae

When queen caging is not an option to obtain larvae of known age, the age of worker larvae can be assessed visually or by weighing. Visual recognition can be done based on Fig. 7. This however, only allows for a rough estimate of age. Because the growth is exponential,

visual estimation of age is error prone. A more accurate way is to weigh the larvae after having cleaned them from jelly residues and absorbed the excess water from their surface. Table 5 gives equations that allow the calculation of larva age for workers, queens and drones. Given the exponential growth of larvae, Thrashyvoulou and Benton (1982) divided the larval development of honey bees of Italian origin in several phases that could be described with regression equations for workers, queens and drones (Tables 9 and 10). The high coefficient of correlations obtained (between 92.3 and 99.7) shows that their formulas are reliable for the population measured. An equation was also produced to describe the complete development, but with lower precision and is therefore not given here (coefficient of correlations between 81.7 and 90.6). Despite the good fit of these equations, deviations might occur according to variations between bee populations and subspecies and they should be recalculated for different populations or subspecies.

Table 9. Regression equations for weight categories of honey bee workers and queens. X designate age and Y the measured weight within the category given in the second column (after Thrashyvoulou and Benton, 1965).

Age (h)	Workers		Queens	
	Weight (mg)	Regression equation	Weight (mg)	Regression equation
6 – 30	0.20 – 0.80	X = (Y - 1.41) / 32.60	0.12 – 0.69	X = (Y – 4.79) / 51.40
31 – 54	0.81 – 7.00	X = (Y – 31.90) / 2.71	0.70 – 8.50	X = (Y – 33.50) / 3.29
55 – 90	7.10 – 46.00	X = (Y – 50.60) / 0.87	8.60 – 37.90	X = (Y – 48.80) / 1.12
91 – 120	46.10 – 140.00	X = (Y – 73.30) / 1.69	38.00 – 186.00	X = (Y – 85.10) / 0.16

Table 10. Regression equations for weight categories of honey bee drones. X designate age in hours and Y the measured weight in mg within the category given in the second column (after Thrashyvoulou and Benton, 1965).

Age (h)	weight (mg)	regression equation
9 – 54	0.29 – 3.50	X = (Y – 8.82) / 11.60
55 – 98	3.51 – 42.00	X = (Y – 52.80) / 1.09
99 – 120	42.10 – 129.00	X = (Y – 64.30) / 0.47
121 – 163	129.42 – 311.54	X = (Y – 91.6) / 0.23

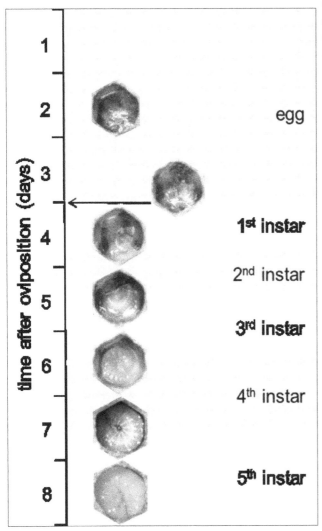

Fig. 7. Development of a worker larva, starting from egg-laying by the queen. A rough assessment of larva age can be obtained by observing the space occupied by the larva in the cell. Larval instars are represented by greyed areas.   Photo: V Dietemann.

2.5.5. Recognising the age of pupae

When queen caging or marking freshly capped cells are not an option, it is possible to recognise the approximate age of pupae based on their morphology and colouration of body parts. Figure 8 can be consulted for identification of worker pupa age as well as Table 5, compiled from the observations of Jay (1962) on pupa appearance. Jay (1962) also describes the appearance of immature drone (Table 6) and queen (Table 7) pupae. The body parts described by Jay (1962) are annotated on Fig. 9. The work of Jay is presented here since it describes the appearance of pupae according to days of development. Others describe colour changes more precisely since they base their description on colour standards, but they only mention the appearance of different stages without linking it to age (Rembold *et al.*, 1980; Michelette and Soares, 1993).

2.5.6. Obtaining workers of known age

Obtaining workers of known age (counted from emergence) can be accomplished by having them emerge in an incubator, marking them and replacing them in their colony for the desired duration. Refer to the *BEEBOOK* paper on 'maintaining adult honey bees *in vitro* under laboratory conditions' by Williams *et al.* (2013) for more details on incubator conditions.

1. Select a brood comb with capped cells.
2. Inspect the comb for emerging workers.
   If none are observed, uncap a few cells to determine the age of the pupae (see section 2.5.5. 'Recognising the age of pupae'). The presence of late stage pupae (dark eyes) indicates that workers will begin emerging within a few days.
3. Place the selected frame in a frame cage (see Fig. 12 of the *BEEBOOK* paper on maintaining adult honey bees *in vitro* under laboratory conditions by Williams *et al.* (2013)).
4. Place in the incubator at 35°C and 60-70% RH.
5. Inspect daily (or when needed) and remove freshly emerged workers.
6. Collect workers when a sufficient amount can be collected at once; discard or reintroduce the workers into colonies if their number is insufficient and wait until enough young workers have emerged.
7. Mark an excess of workers with colour paints (see section 2.3. 'Marking individual bees' of the *BEEBOOK* paper on behavioural methods by Scheiner *et al.*, 2013). Different colours or marking

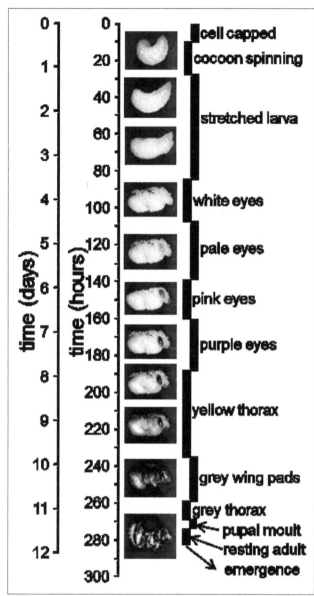

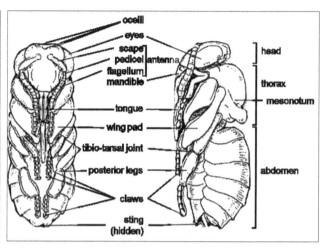

Fig. 9. Anatomy of the worker honey bee pupa with annotations corresponding to Tables 5, 6 and 7. Adapted from Dade (2009).

2.5.7. Conclusion

The possible variation in developmental time between different lineages should be taken into account when designing experiments in which the age of immatures or adult is important. In the literature, development times can be given in hours or days and counted from different starting points (oviposition by the queen, hatching, emergence). This makes the body of work available difficult to rely on and should be considered to avoid mistakes in experimental design. The margin of error when considering development time in days is rather large (24 h) and makes it challenging to set boundaries between developmental stages.

3. Other equipment used in the laboratory

3.1. Using a haemocytometer to estimate the concentration of cells, spores or sperms

In fields of quantitative experimental research e.g. cell culture and microbiology (including bee pathology), it is important to determine the exact concentration or number of bacteria, cells, or spores and even small organisms (hereafter referred to as particles) to guarantee accuracy and reproducibility of experiments (Hefner *et al.*, 2010). The quickest reliable method is direct microscopic or total cell counts of a culture or a suspension through the use of a counting chamber or haemocytometer (Cantwell, 1970; Paul, 1975; Strober, 1997). This method takes into account all cells or spores, cultivable or not, as long as they have a recognisable shape or trait and are not confused with other material in the sample. Further methods can be used to detect culturable (i.e. viable) particles. The plate count method allows for the counting of clonal unicellulars that form colonies and can be cultivated on an appropriate medium (see the European foulbrood paper of the

Fig. 8. Timing and duration of sealed worker brood development. Y-axis starts at capping time. Morphological categories after Martin (1994) for UK honey bees. For simplicity black vertical bars are represented without overlap, but developmental time of each stage can vary.   Photo: S Camazine.

      codes can be used to mark workers of the same colony on different days.

8. Allow some time for the paint to dry.
9. Reintroduce workers into their colonies.
    If workers are attacked by nest mates, spray them with sugar water or reintroduce them in a cage plugged with candy (for a recipe, see the *BEEBOOK* paper on 'maintaining adult honey bees *in vitro* under laboratory conditions' by Williams *et al.* (2013)) so that they can eat their way out. This will increase their acceptance.
10. Inspect colonies and collect marked workers at the desired time.

BEEBOOK (Forsgren *et al.*, 2013)). It is also possible to use spore germination test (see *BEEBOOK* paper on fungi (Jensen *et al.*, 2013)) or fluorescence staining, Fenoy *et al.* (2009) for this purpose.

3.1.1. Total or microscopic count

A haemocytometer (Fig. 10) is used to determine the number of particles found within a demarcated region of a slide haemocytometer containing a known volume. The number of cells counted in this volume is used to extrapolate the number of cells in the total sample. There are several kinds of haemocytometers, but they all consist of a microscope slide with a grid etched into the bottom of a cavity (the counting chamber, Fig. 11). The size of the counting chambers can vary with model and manufacturer (e.g. Helber Z30000, Fuchs-Rosenthal, Neubauer, Neubauer improved, Thoma, Thoma new). A typical chamber depth is 0.1 mm, but to be able to count smaller particles (bacteria) a smaller depth (0.02 mm, e.g. Petroff-Hausser) is required. The grid is divided in squares of different sizes that allow for the counting of particles of different sizes. The number of squares also depends on the model (Neubauer: 3x3; Neubauer improved 5x5; Thoma 4x4) as is the number of lines separating the squares (Figs. 12 and 13). A cover glass closes up the top of the cavity, determining a specific chamber volume. It is possible to obtain disposable counting chambers (e.g. Fastread, UK), which have the advantage of not requiring cleaning between measurements.

Procedure to follow when using a haemocytometer

1. Carefully clean haemocytometer and cover glass with lens paper with sterilised distilled water to avoid contamination or counting errors.
2. Dry with lens paper.
3. Slightly moisten the edges of haemocytometer.
4. Apply cover glass.
   Make sure to use the provided cover glasses - these glasses are thicker than the standard cover glasses so that surface tension will not deform them.
5. Press firmly until the Newton rings appear where slide and cover come into contact.
   This is important for accuracy of the measurement since only a proper placement ensures a correct volume and therefore counting.

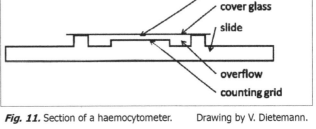

Fig. 11. Section of a haemocytometer.   Drawing by V. Dietemann.

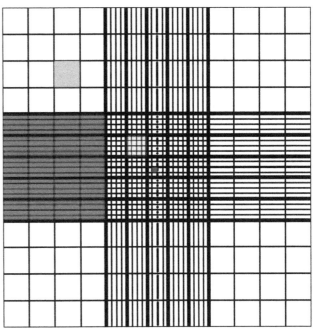

Fig. 12. Haemocytometer grid: red square = 1 mm², 100 nl, green square = 0.0625 mm², 6.25 nl, yellow square = 0.04 mm², 4 nl, blue square = 0.0025 mm², 0.25 nl, at a depth of 0.1 mm. Source: Wikipedia
In an improved Neubauer haemocytometer total number of cells can be determined by number of cells found in grid (red square) x 10⁴ (10,000).

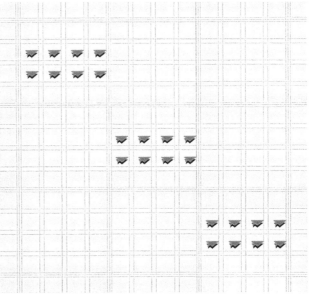

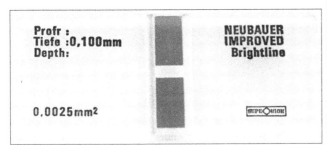

Fig. 10. A haemocytometer.   Photo: V Dietemann.

Fig. 13. Suggested counting of 24 squares in a haemocytometer.

6. Prepare your sample according to description in other papers of the *BEEBOOK* (American foulbrood, De Graaf *et al.*, 2013; European foulbrood, Forsgren *et al.*, 2013; fungi, Jensen *et al.*, 2013; *Nosema*, Fries *et al.*, 2013; queen rearing and selection, Büchler *et al.*, 2013; instrumental insemination, Cobey *et al.*, 2013).

    The samples, especially if they include bees or bee parts, should be carefully ground or dissected and mixed with water. The solution should contain 5-50 particles per square. If stock solution has more particles, it can be diluted until values in this range are obtained. This determines the dilution factor. To facilitate calculations of the dilution factor, it is recommended to use one ml of water per sample or bee that has to be counted or to use 10 times dilution series. In this case, mixing the samples by vortexing during the dilution process is necessary to ensure a homogeneous suspension of the particles. Vortexing is also necessary to homogenise the solution before each counting.

7. Mix sample properly to ensure uniform/ homogenous suspension before introducing the suspension to the periphery of one of the v-shaped wells with pipette. The area under the cover slip fills by capillary action.

8. Place haemocytometer under microscope, adjust to appropriate magnification.

9. Use a weak magnification to facilitate localisation of the grid.

10. Adjust to appropriate magnification for counting (see the *BEEBOOK* papers on *Nosema*, European foulbrood and fungi for more details, Fries *et al.*, 2013; Forsgren *et al.*, 2013; Jensen *et al.*, 2013, respectively).

    Do not crash the objective into the cover glass when focusing! Remember the haemocytometer is much thicker than regular slides.

11. Allow 2 min for the particles to settle in the chamber before counting.

12. Count the particles in the appropriate squares depending on the size of the particles to be counted, making sure that different areas of the chamber are counted (e.g. for *Nosema* spore sized particles, Fig. 13).

    Count at least 300 particles in order to minimise errors. Particles that are only partially inside a particular square must be dealt with in a systematic manner to prevent double counting when the neighbouring square is counted. Count only those particles which are entirely within a square and only those crossing over the top and left boundaries, Fig. 14 (or bottom and right, if you prefer). If squares are separated by several lines, chose one as a boundary.

13. Calculate the number of particles per ml of the original sample from the known volume of the counting chamber.

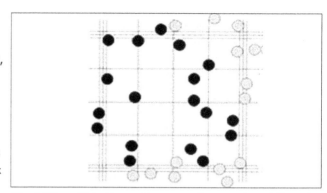

Fig. 14. To avoid double counting, spores that are only partially inside a particular square must be dealt with in a systematic manner. In this example, only the particles which are entirely within a square and only those crossing over the top and left middle lines are counted.

Formula:

$$\frac{\text{total number of counted particles x dilution factor }(\delta)}{\text{area of squares counted }(mm^2) \text{ x chamber depth }(mm)} = \text{items per ml}$$

14. To obtain the total number of particles in the sample, multiply the concentration obtained by the initial sample volume.

Example:

total number of counted particles: 288

area of small squares counted: 24 x 0.04= 0.96 mm^2

chamber depth: 0.1mm

dilution: 1:200 (dilution factor (δ)=200)

$$\frac{288 \times 200}{0.96 \times 0.1} = 600,000$$

Say total sample volume was 0.5 ml, there are 0.5 x 600,000 = 300,000 particles in the samples.

Pros: Haemocytometers are inexpensive and commonly used. They are long-lasting and versatile and a very effective way to count particles.

Cons: Using a haemocytometer requires a phase contrast microscope. Statistical robustness is lacking when counting low concentrations. In addition subjectivity may be a problem among users and it is a tedious and time consuming method (Hefner *et al.*, 2010). It is a monotonous and time consuming task, only reliable for clearly recognisable particles or in samples without structures looking similar to the particles of interest. The viability of the particles counted is unknown.

The automated cell counting method, including flow cytometry, Scepter cell counters and vision based counters, may be a more reliable alternative method to use for particle counting. Not only is it less time consuming, it eliminates subjectivity and it also provides counting algorithms. In future it may even become a necessity in laboratories.

Table 11. Examples of application of hive scale networks: honey meters in different countries.

Name	Country	Webpage	Info
Trachtmeldedienst der Landesverbände Badischer und Württembergischer Imker	Germany	http://lbi.volatus.de/trachtmeldedienst/ Trachtmeldedienst.html	Restricted website. Only for members of the associations.
Nordic/Baltic honey meter	Denmark	www.stadevægt.dk you can also use: http://biavl.volatus.de/bsm0/ BSM.html# if the Danish letter 'æ' is not available on your keyboard	Open website. Access for all beekeepers.
Apistische Beobachtungen – Waagvölker Verein deutschschweizerischer und rätoromanischer Bienenfreunde	Switzerland	http://www.vdrb.ch/service/waagvlker.html	Open website. Access for all beekeepers (scales offline in winter)
US honey beenet	United States of America	http://honey beenet.gsfc.nasa.gov/Sites/ reg_map_button.htm example single scale: http://honey beenet.gsfc.nasa.gov/Sites/ScaleHiveSite.php? SiteID=MD003	Presumably the oldest available hive scale data on the internet. Most of the scales are manual scales.

Table 12. Comparison of four different scales. The design of the electronic scales follows the same basic concept.

Name	Capaz	Penso	BeeWatch® Professional	Apiscale
Webpage	www.capaz.de	www.bienenwaage.net	www.beewatch.de	www.beehive-scales.com
Frame	Stainless steel	Stainless steel (green or black colour)	Stainless steel	Stainless steel
Size in mm (L x W x H)	Fixed frame size (420 x 480 x 86 mm)	Fixed frame size (500 x 420 x 70 mm)	520 x 390 – 520 x 60 (adjustable)	Base set on a base under each bee hive
Max. weight measured	200 kg	200 kg	200 kg	160 kg
Accuracy	100 g	100 g	20g	100 g
Temperature range	-10° up to +45° C.*	-30°C to +65°C.	-30 ° C to 60 ° C	
Electronic transmission	Transmission via wire, contacts.	Wire free transmission, no contacts	Wire free transmission, no contacts	Manual
Battery (build in)	Rechargeable 12 V, 7.2 AH	3x Battery AA	3x Battery AA	No
Battery life	200 days	Scale 2 years, GMS box 1 year	1 year	No
Sensors:				
Temperature	Yes – Standard	Yes - Standard	Yes – Standard	No
Humidity	Yes – Standard	Option	Yes – Standard	No
Rain-gauge	Option	Option	Option	No
Weather station		Option	Option	No
Brood temperature	Option		Option	No
Measurement cycle	Standard 1 or 2 h – during daytime (but optional)	1 h	15 min / 30 min / 1 h	Manual
Software (Standard)	Web and software	Software	Web and software	Manuel web software
Configuration software	Yes	By GMS box	Yes	No
Comment	Under Scandinavien conditions there have been no problems with temperature down to -40°C	Transmission of data from scale in the apiary to external GMS box.		Manual scale. Api-Scale system consists of 2 parts a weighing frame and a beehive base set

4. Research methods at the colony level

4.1. Weighing full hives

4.1.1. Introduction

Over the last few years, new technology has been taken into use in modern beekeeping. For example, electronic hive scales can easily supply the beekeeper/ scientist with important information on several important events from honey bee colonies' life cycles (McLellan, 1977; Buchmann and Thoenes, 1990; Meikle and Holst, 2006). The weight of full colonies (i.e. the summed up weight of the box, combs with food stores and the bees) can be measured to monitor: 1. the occurrence of nectar flow during the foraging season (for examples see Table 11) or daily gain in nectar stores (Meikle *et al.*, 2008; Okada *et al.*, 2012), 2. the reduction of food stores during non-foraging periods (Seeley and Visscher, 1985) and 3. the occurrence of swarming events (Meikle *et al.*, 2008).

All electronic scales are designed following the same basic concept namely to function as a honey meter, much like a weather forecast with which beekeepers can get vital information on the nectar flow, food consumption, but also humidity, temperature, rainfall and brood temperature. Some hive scales also measure wind velocity or sun hours. A state of the art hive scale is designed to automatically transmit these data either directly via internet, to the beekeepers cell phone or to personal computer software.

Since there is an increasing number of scales available on the market, we give a short comparison of three different commercially available electronic and one manual scales in Table 12 and focus on the most widely used Capaz hive scale and its application as a honey meter.

4.1.2. The Capaz hive scale

The first prototype of the Capaz hive scale was developed in 1997. In 2003, it was ready to be put on the commercial market. The scale is a 420 x 480 x 86 mm platform made of aluminium and stainless steel (Fig. 15). The scale can weigh up to 200 kg with a precision of 100 grams. Weight, ambient temperature, humidity are measured by default. The amount of rain collected and brood temperature can be added to the parameters measured. The number of measurements per unit time can be adjusted and depends on the topic of the study (Seeley and Visscher, 1985; Meikle *et al.*, 2008). A very easy way to change the setup of the scale is by connecting the scale to the configuration software that accompanies the scale (Fig. 16). The battery lasts for approximately 200 days, but has a shorter life in the wintertime due to low temperature extremes. So far, no problems due to cold Nordic winter conditions (down to − 40°C) have been reported (Flemming Vejsnæs; pers. comm.).

The scale will send an SMS that is transformed to an e-mail. Using SMS limits the quantity of data that can be sent per unit time (Fig. 17). As standard the scale records data every second or every hour during the daytime. It is possible to change the setup of the scale, sending more SMSs per day, thus increasing the number of daily data. Every day, at a time determined by the user, the scale will send the data.

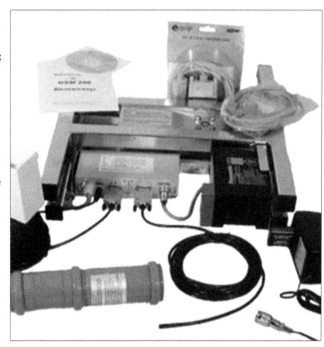

Fig. 15. The Capaz scale is an H-shaped platform made from aluminium and stainless steel, with the dimensions 420 x 480 x 86 mm (long x wide x high). Data are transmitted by cell phone. The rechargeable battery (12 V) lasts for 200 days. Ambient temperature and humidity are measured by default. Additional equipment is the rain collector and brood temperature sensor. Changes of the standard setup of the scale are done via the computer software.   Photo: Capaz.

Fig. 16. Configuration of the Capaz scale directly in the apiary.

Photo: F Vejsnæs.

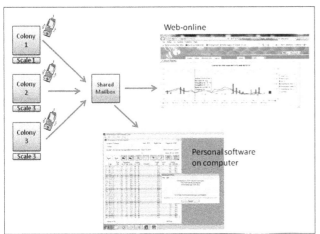

Fig. 17. With the Capaz scale, all data are sent by a cell phone, as an SMS. The SMS is converted to an e-mail sent to an e-mail account. From here, it can be uploaded on the internet and/or downloaded by the software. A sample output received via e-mail follows the format:

!33H74F71301122XXXX#0601+027033891+1970800+025033890+
1961200+029033791+1961400+035033690+1931600+042033683+
1952000+042033682+1962200+040033580+198!

Together with the scale, well developed software (Fig. 18) is provided, where all data can be downloaded directly from the e-mail mailbox. Data can easily be exported to excel spreadsheets. In addition, a web application can upload data directly to the internet. Different scale companies offer different online web applications.

4.1.3. The honey meter

One of the most important and widely used applications of hive scales is the so called 'honey meter', a nectar flow tracking or honey forecasting

system. In other words, the aim of the honey meter is to monitor the timing and potential honey harvest of healthy colonies of a local area, based on the entering amount of nectar. It is of course an open discussion as to how representative data from single colonies are. The best solution is to place all colonies in an apiary on scales, but this is not feasible due to economic reasons. The Capaz Company has therefore designed a pallet scale for existing metal pallet system, having four load cells, thus providing average data for four colonies. The four load cells can measure up to 1.200 kg. Some examples of scale networks functioning as honey meter are given in Table 11. An example dataset for the weight changes of a colony in Jutland, Denmark are shown in Fig. 19.

4.1.4. The use of the data from an electronic scale

Very often it is difficult to judge what is going on in honey bee colonies. The hive scale is an important tool and gives a good assessment if food consumption has been high over a longer period and whether there is a need for feeding. In most countries, it is important to know how big the winter storage is since it will tell if spring feeding of carbohydrates is needed. In addition, it gives a very good assessment of periods without any flow in the summertime and hence can warn of starvation danger. Finally, it gives a very good evaluation of how intense the nectar flow is, in other words, if there is a need to provide the colonies with additional supers. Commercial beekeepers use hive scales to save unnecessary visits to the apiary when they do long-distance migration. Examples are German commercial beekeepers having colonies for pollination of white clover in Denmark. With the scale, such migration has become profitable, since the driven kilometres can be kept to a strict minimum. The German hive scale system (see Table 12) is especially a warning system for the start of honeydew flow. It tends to start suddenly and

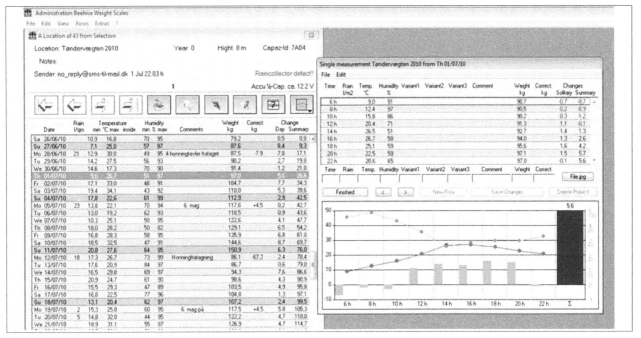

Fig. 18. The well designed software that comes with the Capaz scale allows easy exporting of data to spreadsheets.

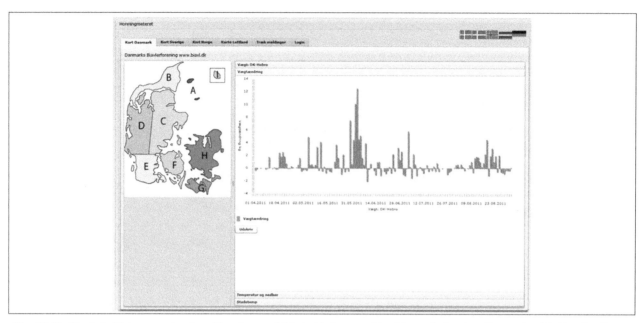

Fig. 19. The Nordic/baltic honey meter. Over 66 scales are distributed in four countries (Denmark, Sweden, Norway, Latvia). Here food/honey consumption/weight gain for the period 1.4.2011-1.9.2011 is shown for a scale located close to the town Hobro in Jutland, Denmark.

can be massive. The scales also give very good information about when the nectar flow stops. In Denmark, information from the honey meter has shown that the main nectar flow stops, in general, earlier than the beekeepers expect. Some Danish beekeepers make all their varroa treatments according to the figures of the honey meter with the positive result of earlier and therefore more efficient varroa summer treatments.

Measuring brood temperature indicates when there is no brood in the colonies. This is the optimal time of the year for varroa-treatment with oxalic acid. But note that since there is only one temperature sensor, one has to ensure correct sensor placement in the centre of the winter cluster. With scales, obtaining an indication of colony swarming through a decrease in its weight (Meikle *et al.*, 2008) is also possible. A necessary requirement for accurate measurements and predictions is to monitor good/ well running colonies on the scales.

Having colonies on hive scales is providing very important complementary data on colonies used for experiments. The disadvantage of the system is that the figure of increasing/ decreasing bees and brood in the colonies are influenced by variation in food stores. A very nice experiment is that of Meikle *et al.* (2008) who used precise bench scales (± 10 grams) that measured every hour. They weighed separately the bees, brood and food, showing that the main part (76%) of the colony weight throughout 2005 was food. However using scales with a precision of 10 grams in the field entails large errors due to the accumulation of rain or even due to wind pressure on the hive body. It is important to use Styrofoam boxes, since wooden boxes absorb moisture and thereby bias weight measurements.

In wintertime, it is important to have the colonies protected from snow and ice in order to have reliable day to day measurements. Procedure to follow when using a hive scale (Capaz scale)

1. Place the hive scale on a levelled platform – to protect against moisture from the ground.
2. Connect all plugs. Test for cell phone connection according to manual.
3. Secure a protection cover for the scale, protecting against debris from the colony, from precipitation, driving rain etc.
4. Check battery charge regularly – recharge at least every spring and fall.
5. Keep all plugs clean and dry – otherwise rust problems will arise.
6. Use Styrofoam boxes, since wooden boxes will absorb moisture.
7. Ensure that water runs of the hive cover, otherwise water can accumulate or be absorbed, biasing results. In countries experiencing snowfall, scales and colonies should be protected in a house, external cover or shed. Otherwise winter measurements will be biased.
8. If using the Capaz brood chamber sensor, ensure that the sensor is placed in the centre of the winter cluster during winter.
9. Refer to scale manuals for data downloading.
10. Download the data regularly and make backups, since it is an enormous amount of data that is collected.

4.2. Using beelines to locate wild honey bee colonies

4.2.1. Introduction

Locating honey bee colonies is obviously essential for any researcher wishing to collect data from naturally occurring or feral populations. This is important for a variety of research interests whether they are determining nest site selection, population densities (see section 4.3 on bee density), collecting samples of bees and other nest constituents, determining parasite loads, studying colony strength, etc. Locating colonies is also important for people who utilise the various nest constituents of honey, pollen, brood, wax, and propolis for food, medicine, or craft. Different cultures throughout history have developed and utilised methods of tracking and 'hunting' honey bee colonies that vary from random searching for colonies to following honeyguide birds (Crane, 1999). Most methods, however, including those used in current academic research, follow the flight paths of honey bees to their colony of origin, known as beelines.

A beeline is defined in this section as the direct flight path taken by foraging honey bees to and from their colony's nest, to and from any particular foraging resource (e.g. flowers, water, propolis, etc.). Beelines are established first by worker honey bees called scouts that locate the resource. Using the waggle dance in the nest, these scouts will communicate the location of the resource to other foragers in the colony (von Frisch, 1967; Seeley, 1983). Once foragers have located and travelled to and from the resource enough times, remembering its location, they fly the most optimal path; and this same path is taken by many foragers. Thus, the beeline is established and can be present for as long as the foraging source is available. Beelines are often quite direct and are essentially a straight line to and from the colony's nest.

There are essentially only three steps associated with locating wild honey bee colonies by beelining. They are 1) establishing a beeline, 2) following the beeline, and 3) locating the honey bee nest. This section details these steps from practiced methods used in Vaudo *et al.* (2012 a, b) and references listed below. This method has been optimised to potentially locate multiple colonies from a single foraging source.

4.2.2. Suggested materials

1. Feeding station.

   Detailed instructions for creating feeding stations are provided in section 4.2.3.1.

2. Mobile feeding station, bee box.

   Detailed descriptions are provided in section 4.2.4.3. and 4.2.6.2.

   Bait - 1:3:3 honey:sugar:water by volume.

   Handheld GPS and/ or map and compass.

   Field proof laptop or note pad.

   Binoculars.

   Camera.

   Personal protective equipment:

   a. epinephrine autoinjector (e.g. EpiPen, Twinject, etc.).

   b. bee suit (veil, gloves, and long clothing).

   c. water and food for a day in the field.

4.2.3. Establishing a beeline

The most essential step of tracking bees using this method is to establish a beeline. The easiest way to establish beelines is to provide a highly attractive foraging source or feeding station, for honey bees and allow the foragers time to locate the source. This requires minimal effort for the observer to attract foraging honey bees. It also allows the observer to place a foraging source in a location that will make it easy for him or her to track the beelines. Once beelines are established, the foraging source can be replenished and beeline maintained. Beelines can also be established in locations where foraging honey bees are already located such as flower patches or water sources (see sections 4.2.6.1 and 4.2.6.2).

4.2.3.1 Setting up a feeding station

The feeding station is the honey bee tracker's means to establishing and maintaining beelines and is composed of two parts, a stand and foraging source. A feeding station consists of a 10 cm^2 iron plate welded to a 2 m tall iron rod that is angled to a point at the bottom (Fig. 20). A 10 cm iron crosspiece is welded 0.5 m from the bottom of the station at 90° from the main rod. The crosspiece and angled bottom facilitate station insertion into hard ground. The iron plate at the top of the station has a 5 mm hole drilled through each corner. A plastic container (~ 11.5 × 17 × 4 cm, L × W × H) is affixed to a 25 cm^2 wooden platform using a nail or screw through the centre of the container. The wooden plate then can be mounted to the iron stand with bolts through each of the holes on the iron plate. Less complex feeding stations can be made and other construction materials used in case sturdiness is not an issue.

1. Construct the stand.

   The stand can be made of any material, such as wood or iron, so long as it will not be knocked over in the field. It is important that the plate of the stand sets high (~1.5-2m) so that is easily found by the foraging honey bees and the observer can easily view the beelines.

2. Place the feeding station in an open field so that the observer can easily see beelines against the sky and they can be tracked without difficulty. It can also be placed in the sunlight, so that bees can be spotted against a dark background.

3. Produce the bait.

   The bait can vary from scented sugar water to honey filled comb. A volume of 0.5L to 1L of a mixture of 1:3:3 honey:sugar:water (by volume) in a plastic container should be sufficient for attracting a large numbers of bees as they tend to be readily attracted to the scent and taste of honey.

Fig. 20. A honey bee feeding station. Arrow **A** points toward the container which is partially filled with bait (see photograph **B** at the right for a close view). Arrow **B** indicates the removable feeding plate. Arrow **C** shows the main iron rod that can be driven into the ground using the crosspiece (Arrow **D**). Photos: A Vaudo.

Scented sugar syrup, using only a few drops of ~50% anise extract per litre of solution (Seeley; pers. comm.) will attract fewer honey bees. This can be useful if one needs to reduce the number of honey bee arrivals to the feeding station in order to obtain accurate round trip times more easily (Wells and Wenner, 1971; see section 4.2.6.4).

4. Place the bait in a container on top of the plate.
5. Place sticks and twigs in the feeding container so the bees do not drown in the liquid bait.
6. Record the location of your feeding station on a handheld GPS (by creating a new waypoint) or map so you may find it easily in the future.

Fig. 21. An example of foraging honey bees feeding from a feeding station used to establish beelines. Note how the bees are easily observable against the blue sky and the bees are lost against the mountains and shrubs. Photo: A Vaudo.

7. Once the feeding station is baited, leave it overnight to allow the honey bees to locate the foraging source and establish their beelines. Usually, scouting foragers will locate the feeding station the morning after it is erected. Foraging bees may not find the feeding station the day it is placed in the field because they have already established foraging sources for the day.

4.2.4. Following the beeline
4.2.4.1. Observing beelines

1. Take position a few meters beyond the feeding station to accurately observe the beeline. The observer should be able to see many bees flying. It is helpful to squat below the feeding station to see the contrast of the dark bees against the sky (Fig. 21).
2. Look at the group of bees above the station.
   At the feeding station, it will appear that the bees are landing and taking flight in random directions from the bait. When forager bees leave the station, they often will circle up in the air to orient themselves then quickly dart off in a straight direction into the beeline and toward their nest. When one observes a strong beeline, it appears as a 'highway' of bees flying both directions.
3. Note the direction of bees leaving the feeding station to return to the colony and not those approaching the station.
4. Circle around the feeding station to determine the directions of all beelines established.
5. Use a GPS or compass to determine and record the direction that each beeline is heading. Stand at the feeding station and record the direction that the beeline is heading from the station.

4.2.4.2. Tracking the beeline

Once beelines are established at your feeding station, it is now time to follow the beeline toward the honey bee nest. This can be time consuming and require some energy. Be prepared to walk through wild vegetation and traverse difficult terrain. The beeline is very direct, so it will transverse over buildings, dense woods, cliffs, marshes, lakes, etc. One should bring ample water and food to spend the day in the field.

1. Refill the feeding station so the bees continue to maintain the original beelines prior to moving in the direction of the beeline.
2. Walk a short distance in the direction determined as that of the bees flying to their colony.
3. Look for the beeline.
   Bees are recognisable from other insects by their direct line of flight.
4. If you are correct in locating the beeline and the direction it is

heading, continue walking that direction in a straight path. One may not see the beeline any longer while moving away from the feeding station. However, one can reasonably trust that the honey bee nest is in that direction.

5. Keep track of one's path with a map or GPS device if there are obstacles that have to be circumvented or scaled. A handheld GPS with a tracking option is useful so that you can visualise your path and return to it if you have to deviate temporarily or return to the feeding station.

6. If the path is followed directly, one should be lead straight to the location of the colony.

   However, there it is difficult to determine the exact distance between your feeding station and the colony (usually less than a 1km but potentially up to 5km). One could estimate the distance by using the techniques outlined in sections 4.2.6.3 and 4.2.6.4.

7. Look for the colony nest entrance as you follow the beeline (section 4.2.5).

8. If you reach an obstacle that prevents travelling further, set up a feeding station and establish a new beeline on the other side of your obstacle. From this point, you can pick up the beeline again and continue your search.

4.2.4.3 Using a mobile feeding station

One may not be successful locating the honey bee nest on the first attempt. You may have lost the beeline, the beeline could have terminated, or you could have reached an obstacle preventing you from continuing on your path. If you get lost while following the path and fail to find the nest, one possible solution is to carry a mobile secondary feeding station with you. Examples of mobile feeding stations include a bee box (see section 4.2.5.2) or another transportable container filled with bait (e.g. a bucket used as a stand and another feeding container like that used for the feeding station).

1. Return to the original feeding station, set up your mobile feeding station beside it and allow the bees to start foraging from the bait. The bees should begin foraging from it quickly.

2. Carry the station with the bees with you in the direction of the beeline once you have many foragers on your mobile station.

3. Stop and let the bees establish a new beeline that you can follow once you reach a considerable distance from the original feeding station. This point can be where you previously lost the beeline.

4. Repeat this process with your mobile feeding station as many times as necessary until you get close enough to locate the nest.

Sometimes you may even travel beyond the colony and see the beeline from your mobile station heading back the way you came. Now you know that the colony is located between the last stations' and your current location. It is advisable to carry a bottle of bait while tracking the bees so the mobile feeding station can be replenished.

4.2.5. Locating the honey bee nest

Honey bee nests can be located in cavities (~40l by volume) at any height, in the ground, or high in a building, tree, or cliff, depending on the environment (Vaudo *et al.*, 2012a) (Fig. 22). Many African subspecies of honey bees also nest in the open, hanging on branches, or overhangs of cliffs and buildings. Generally, wild bees will be located in a wooded or at least covered area. Consequently, their

Fig. 22. Examples of honey bee nest site locations. The white arrows indicate the entrances of the colonies.   Photos: A Vaudo.

nests can be difficult to find. Locating the exact position of the nest requires both your sense of hearing and sight. One must constantly listen and look for the honey bee nest and look at every potential nest site along one's path. This is why it is good to place the feeding station in an open area. It will allow you to determine a definitive direction to head (use landmarks visible from the defined path) prior to entering a wooded or otherwise congested area.

It is advisable to bring personal protective equipment (a bee suit or veil, gloves, and long clothing) when locating a nest in case the honey bee colony is defensive or if one plans on investigating the nest closely. One should keep an epinephrine autoinjector (e.g. EpiPen, Twinject, etc.) at all times in case an allergic reaction is experienced if/ when stung.

1.   Look for the activity of insects flying in, out, and around a specific location.

     One can see almost a 'funnel' or cloud of bees in an open area close to their colony as they fly in and out of the nest (similar to the activity of bees taking off and landing from your feeding station). This activity can be seen against the sky where their black bodies and glistening wings will be apparent. Nest entrances can be quite small, so follow this activity as it narrows to where the nest entrance is located. Active colonies tend to be obvious with many workers flying in and out and a number hanging outside the entrance. Consequently, nests can be easy to find in late or mid-to-late spring when colonies typically are large and actively foraging on available pollen and nectar. Additionally, using a highly attractive bait at your feeding station as suggested can assist in making a colony more active.

2.   Use the sound of the bees.

     If the beeline is strong and the colony is active, you should be able to hear a distinct hum of honey bees (similar to the sound of a swarm) once close to the nest.

3.   Approach the location and confirm that you have located the entrance to the colony.

     Having binoculars could be useful to confirm the colony's location if it is high.

4.   Make sure you have located a nest hosting a live colony.

     The occurrence of pollen foragers shows that there is no ongoing robbing of the nest of a dead colony and that the activity witnessed is not that of scouts looking for a new nest site.

5.   Mark the exact location of the colony with a GPS or on a map once it is found.

6.   Mark the nest to make it easier to locate in the future (Fig. 23).

7.   Take a photograph of the area so you can easily find it again.

4.2.6. Alternative methods
4.2.6.1. Following bees from water sources
Usually in hot conditions, honey bees forage for water to be used for nest temperature regulation. They can be found at fresh water sources,

Fig. 23. Marking a honey bee nest. The nest entrance (not shown) is in the ground nearby.   Photo: A Vaudo.

such as water troughs, small pools of water, and edges of ponds, lakes, and rivers. Colonies tend to nest close to water sources, so following bees from these sources can reduce one's search time. Similar to beelining from a bait station, one can follow the direction of the bees leaving the water source. They will travel directly to the colony.

Another technique is using water to honey foraging conversion as suggested by Wenner *et al.* (1992). Simply, a few drops of undiluted honey can be placed on a stick upslope and close to where individual bees are foraging for water. If the bees switch to imbibing the honey, they will begin to recruit other foragers to the water source and soaked sponges of honey water placed in the area (Wenner *et al.*, 1992). You can now follow the beelines to the colony.

Pros: no installation necessary, water foragers always come to the same place; colonies usually nearby, reducing the search time.

Cons: no easy triangulation done; few water foragers for each colony so finding the water foragers may be difficult.

4.2.6.2. Beelining with a bee box
Several authors described methods to locate nests using a portable device called a bee box (Edgell, 1949; Visscher and Seeley, 1989). Locating a honey bee colony using a bee box (Fig. 24) uses the same basic concepts outlined in section 3.2.3 and 3.2.4. However, the main difference is that the bee box allows you to trap individual honey bees off of flowers rather than allowing bees to find a feeding station. One can trap a number of bees in a section of the box and using glass and trap doors manipulate them into a second section in the box, a section that contains bait material, be it honey, sugar water, or a combination of both, placed in a small sponge or piece of honey bee comb (Fig. 24). When a beeline is established, one can close the lid of the box and carry the bees trapped with the bait along the path to the

colony then stop, open the box, and allow the bees to establish new beelines. Refer to the procedure described below and Fig. 24 for the methodology of using the bee box.

1. Trap individual bees off of flowers in chamber A.
2. Darken chamber A, open divider (D), open window (F), and allow bee into chamber E.
3. Close divider and repeat until enough bees are captured into chamber E.
4. Place bait in chamber A.
5. Close window to chamber E, allow light into window (C) to chamber A, and open divider (D). The bees will eat from the bait. Allow 10-15 min for bees to consume the bait.
6. Open lid (B) to chamber A and allow bees to travel to their colony and back.
7. Replenish bait and wait for enough bees to visit so that a beeline is established.
8. Close foraging bees in chamber A and follow path of bee line.
9. When needed, stop and open chamber A and allow a new beeline to form.
10. Repeat and keep following the beeline to the colony's nest.

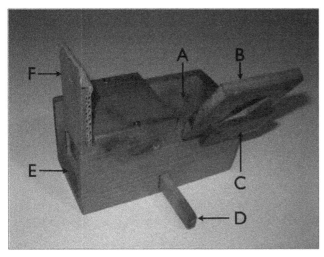

Fig. 24. An example of a bee box. **(A)** Chamber used to trap bees off of flowers and establish bee lines using bait. **(B)** Lid to chamber A to trap bees. **(C)** Window cover to allow light in or darken chamber A. Allowing light in will attract bees toward chamber A while chamber E is darkened. **(D)** Sliding divider between chambers A and E. Opening and closing the dividers allows or blocks bees from moving between chambers. **(E)** Chamber to store bees while trapping individuals in chamber A. **(F)** Sliding window cover to allow light or darken chamber E. Allowing light in will attract bees to chamber E while chamber A is darkened. Refer to Edgell, 1967 and Visscher and Seeley, 1989 for specifications. Photo: A Vaudo.

Pros: bees can be caught directly from foraging sources; the box is transportable and bees can be carried along and new beelines can be established during your search for the nest.

Cons: the major limitation to this technique is the size of the bee box and quantity of bait that can be provided to the bees, limiting the number of bees that will establish a beeline. If there is ample forage in the field, the bees will not readily recruit to the small amount of comb and bait used for the bee box. This technique does work in times prior to or after major blooming periods.

4.2.6.3. Triangulating with feeding stations

Visscher and Seeley (1989) described a method to locate the approximate location of a honey bee colony by triangulation using multiple feeding stations. Refer to Fig. 25 for the following methodology.

1. Place two feeding stations at an arbitrary distance from one another (*c*, baseline) and mark each one's location on a map or GPS and calculate the distance between the two using the map legend or GPS function. If you are placing the feeding stations in the same open area or forest clearing, place them at least a couple hundred meters from each other. It may also be useful to find two different clearings in a forest to set up the feeding stations.
2. Calculate the angles (*A* and *B*) of the beelines from the baseline with a compass or GPS. This can be done easily on a handheld GPS or compass by recording difference in degrees between the direction of the opposite feeding station and the beeline.
3. Calculate the angle from the honey bee nest (*C*) to each feeding station.

$$180° - A - B = C$$

4. Using the 'law of sines', calculate the distances from each feeding station (*a* and *b*) to the nest.

$$\frac{c}{\sin C} = \frac{a}{\sin A} = \frac{b}{\sin B}$$

5. Mark the approximate location of the nest on a map or as a new waypoint on the handheld GPS.
6. Follow the beeline from either feeding station toward the colony for the calculated distance and search the area for the nest.

Pros: potentially reduce searching time by calculating the approximate location of the bee nest, especially in a heavily wooded area.

Cons: beelines may be from different colonies and do not converge on the same location.

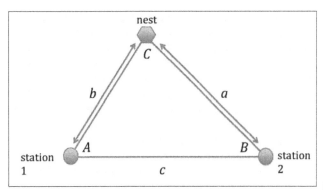

Fig. 25. Triangulating the location of a honey bee nest using two feeding stations. **A, B,** and **C** represent angles and **a, b,** and **c** represent the length of the sides opposite their respective angle. Arrowed lines represent beelines to and from each feeding station.

4.2.6.4. Calculating the distance between a honey bee nest and feeding station by timing a forager's round trip

Visscher and Seeley (1989) calculated the round trip time it takes for a forager to return to its colony and back to a feeding station in order to determine the distance of the colony from the feeding station. This round trip time is calculated as the time from when a forager leaves a feeding station until the time it returns. They found that a 5 min round trip time indicated that the colony was approximately 0.9 km away. A 10 min round trip indicated that the colony was approximately 1.4 km away. Finally, a 15 min round trip indicated that the colony was approximately 1.7 km away. These values, however, can vary based on the environment (e.g. vegetation cover, wind conditions etc.).

Wenner *et al.* (1992) suggested using the following formula to approximate the distance to a colony from a feeding station:

$$x = 150y - 500$$

The distance in yards or meters (x) is approximated by the time between arrivals at the feeding station (y). Note the difference between this measurement of round trip time and that of Visscher and Seeley (1989). The constant (500) represents the approximate amount of time the forager takes to fill at the feeding station and unload in the colony (Wenner *et al.*, 1992).

1. Mark foraging bees (3-6 bees) while they are feeding from the feeding station.
   This can be done by placing a dot of paint on the thorax of the forager bee, between its wings. See section 2.3 of the *BEEBOOK* paper on behavioural methods for marking technique (Scheiner *et al.*, 2013).
2. Record each bee's round trip time (~10 times per individual bee).
   Use the time from when a forager leaves the feeding station until it returns for Visscher and Seeley's (1989) approximation.

Use the time between landings at the feeding station for Wenner *et al.* (1992) formula.

3. Select the third or fourth shortest time for each bee as its representative round trip time.
4. Use the appropriate calculation suggested above to estimate the distance to the nest.
5. Repeat with several foragers marked differently to obtain an average distance.

Pros: optimise search time; distinguish colonies located in the same direction, at different distances.

Cons: the presence of wind can increase flight time, flight times are variable and distance approximation is not exact; marked foragers may not return to feeding station if they have been predated or recruited to another foraging source.

4.3. Honey bee colony density estimations

It can be difficult to determine the number of wild honey bee colonies per unit area (colony density) due to the cryptic nature of honey bee nesting sites. Consequently, research involving data collection on colony density can be complex and time consuming. There are a number of instances where knowing colony density would be beneficial. For example, one can employ GIS technology (Geographic Information System, see the *BEEBOOK* paper on the topic, Rogers and Staub, 2013) to determine how colony density varies over land use patterns or within/ between various ecosystems. Furthermore, one could track population size and health over time, monitor migration patterns, determine disease spread within a population, etc. Yet, these applications seem out-of-reach because of our inability to determine colony density accurately.

Currently, the only way to determine true colony density is to search a landscape thoroughly and locate all of the colonies in a given area by bee lining (identifying the direction of home flight and finding the colony on this line, see section 4.2. (Using beelines to locate wild honey colonies) or extensive search for nests (e.g. Oldroyd *et al.*, 1997). This seems challenging due to the cryptic nature of some nesting sites or in areas where accessing colonies is difficult or dangerous such as on cliff faces or high in trees. As a result, researchers have turned to indirect methods for assessing colony density.

Herein, we present two methods that can be used to assess the density of honey bee colonies in an area. The first method (using feeding stations) assesses the relative density of honey bee colonies in an area through indexing while the second method (using genetic markers) provides a direct estimate of colony density.

4.3.1. Determining a colony density index using feeding stations

It is not known if density indices can be used to provide accurate estimates of the actual number of colonies present, and this should be a subject of future investigation. However, it is believed that the indices are useful for determining relative colony density. The indices rely on approximating the number of forager honey bees that visit established feeding stations spaced throughout a landscape. The indices' reliability rests on the assumption that colony density is positively correlated with the number of bees visiting the feeding stations. The following method is based on Vaudo *et al.* (2012a, b).

4.3.1.1. Material used

See section 4.2.2. 'Setting up a feeding station' in the beelining method for a description of the feeding station and food container required.

4.3.1.2. Procedure

1. Place feeding stations along a transect ~2 km from one another and throughout the study site about 24 h before monitoring.

   The reason for the 2 km recommendation stems from the necessity to minimise the possibility that one colony will be attracted to two stations. The 24 h gives bees time to navigate the unique environmental conditions in an area, find the stations and reach maximum foraging activity prior to station monitoring.

2. Bait feeding stations with ~600 ml of a 1:3:3 mixture of honey:water:sugar respectively and by volume; alternatively, pure honey can be used.

   Pure honey is likely more attractive to bees than a mixture with water and sugar. The choice of bait does not compromise the index when all stations are stocked with the same bait. The honey used should be from a single source to control for possible differences in the attractiveness of honey from varied sources. The honey can be irradiated to kill all pathogens and eliminate the risk of disease spread to wild colonies (see the section 7.6. 'Food sterilisation and detoxification' of the *BEEBOOK* paper on maintaining adult *Apis mellifera* in cages under *in vitro* laboratory conditions (Williams *et al.*, 2013).

3. Place sticks, twigs, or other floatation devices on the bait to provide foraging bees a surface on which to land.

   This minimises the chance that bees will drown in the bait.

4. Monitor feeding stations at similar weather, time and season points.

   For example (1) only during sunny weather, (2) with little or no wind, and (3) between the hours of 09:00 and 15:00.

5. Visit feeding stations in the order and about the same time that they were erected the day before, thus keeping the time

of bee acclamation to stations as close to a standard 24 h time period as possible.

6. Refill the stations prior to data collection in instances where bees removed all of the bait from the feeding stations within 24 h.

   Vaudo *et al.* (2012b) reported that bees are attracted to reprovisioned stations almost immediately.

4.3.1.3. Index data

Three types of index data can be collected from bait stations:

1. Establish the number of bee lines to each station - A beeline is defined as 'the flight path taken to and from a food source and the colony' (see section 4.2. on bee lines).

   They can be determined best when a foraging bee is leaving the food because it takes a more direct flight to return to its colony than when landing on the station. Individual stations should be monitored until all beelines are recorded. The working assumption is that more beelines will be formed to a feeding station when more colonies are nesting in an area, though this assumption needs to be verified.

2. Field rating of bee density on feeding stations – Rate each station on a scale of 0 (no bees foraging) to 3 ('many' bees foraging).

   The rating is based on the intensity of the foraging visits on a station. This is a qualitative and subjective rating but it provides a quick index of visiting intensity, working on the assumption that higher field ratings indicate more colonies nesting in the area.

3. Photograph rating of bee density on feeding stations – Rather than making a subjective rating of foraging intensity at feeding stations, one can take a picture of each feeding station and assign a station rating per the number of bees counted feeding at each station, for example: (0) = zero foraging bees, (1) = 1-50 foraging bees, (2) = 51-200 foraging bees, (3) > 200 foraging bees (Fig. 26). It is assumed that higher ratings indicate more colonies present in the environment.

4.3.1.4. Statistical analyses

The number of bee lines per feeding station can be analysed by an assigned independent variable using a weighted one-way ANOVA. The ANOVA is weighted for the number of feeding stations within an independent variable. For example, Vaudo *et al.* (2012b) looked at land use effects on the number of bee lines, with land use being recognised as stations on (1) game reserves or (2) livestock farms. Since the authors did not place the same number of feeding stations at locations of both types, the ANOVA analyses were weighted for the number of feeding stations used, giving greater weight to sites having more stations. The field and photograph indices of numbers of bees at

Fig. 26. Photograph Field Ratings. **(A)** = Rating 0 (0 foraging bees); **(B)** = Rating 1 (1-50 foraging bees); **(C)** = Rating 2 (51-200 foraging bees); **(D)** = Rating 3 (> 200 foraging bees).   Photos: A Vaudo.

feeders can be analysed by Pearson's χ^2 tests to determine if there is a difference in the distribution of ratings between feedings stations categorised in two or more independent variables.

Pros: this is a relatively inexpensive method.
Cons: time consuming, reliability not established.

4.3.2. Determination of honey bee colony density using genetic markers

The difficulty of locating cryptic honey bee nests for density estimation can be overcome by exploiting their mating behaviour. Drones fly to drone congregation areas (DCAs) to find sexual partners. It is thus possible to locate these DCAs to which colonies in an area contribute drones and queens instead of locating all the nests these come from. DCAs can be located by observing the terrain or transecting it with a pheromone trap, which can then be used to samples drones (Williams, 1987). Using genetic tools, it is then possible to genotype the drones and infer the genotype of their mothers. Because drones are produced parthenogenetically and only carry alleles from their mother, genotyping drones allows for their easy assignment to specific queens. Similarly, by genotyping workers of a single queen, it is also possible to deduce the genotype of the queen and that of her mates (honey bee queens mate with many haploid drones). Since honey bee colonies are headed by a single queen, obtaining the number of queens in an area equals

counting the number of colonies in this area (Baudry *et al.*, 1998; Jaffé *et al.*, 2009a). A recent model verified the validity of using locally mated queens to estimate colony densities based on the genotype of their brood. They conclude that at least 10 mated queens are needed to detect order of magnitude differences in colony density estimates (Arundel *et al.*, 2012).

4.3.3. Sampling

4.3.3.1. Drone sampling

Honey bee drones can be lured by synthetic queen pheromone into a trap kept aloft by a weather balloon. See section on trapping drones in the *BEEBOOK* paper on behaviour (Scheiner *et al.*, 2013).

- Capture drones (ideally between 100-200 individuals) from a previously identified DCA by flying the pheromone trap between 12:00 and 17:00 hours (depending on the region and season), above 17°C, under sunny and windless conditions, during the swarming season. See section 13.4. on locating DCAs in the *BEEBOOK* paper on methods to study behaviour (Scheiner *et al.*, 2013).

4.3.3.2. Worker sampling

- Identify at least 10 colonies headed by locally mated queens.
- Collect freshly emerged workers (ideally between 12-24 workers per colony) directly from brood combs upon opening of the hives in order to avoid sampling workers that drifted from a foreign colony into the sample hive.

  Using this approach, failing to detect some fathers in a colony would be equivalent to failing to sample some drones at a DCA.

4.3.3.3. Genotyping

See section (6.3.1.) on microsatellite markers in the *BEEBOOK* paper on molecular methods (Evans *et al.*, 2013) for the method to determine individual genotypes. The use of independent sets of tightly linked microsatellite markers (Shaibi *et al.*, 2008, Table 15) to reconstruct queen genotypes from a sample of drones has been shown to result in a very high detection power (see section 4.3.3.5. on non-detection errors below), even allowing the identification of closely related queens (Jaffé *et al.*, 2009a). For the details of the linked markers refer to Shaibi *et al.* (2008).

4.3.3.4. Genetic diversity measures and reconstruction of queen genotypes

The drone genotypes are obtained either directly, by genotyping drones caught in a DCA, or indirectly, by inferring their genotype from the worker offspring of a single queen.

1. Construct tables with the genotypes of all drones for each sample set (see Tables 13 and 14).

Table 13. Genotypes obtained from genotyping drones.

Drone ID	Locus 1	Locus 2	Locus 3
1	a	c	a
2	a	b	b
3	b	a	c

Table 14. Genotypes inferred from genotyping workers of a single queen.

Worker ID	Locus 1 (b/b)	Locus 2 (a/b)	Locus 3 (c/c)
1	a/b	c/a	a/c
2	a/b	b/a	b/c
3	**b**/b	a/b	c/c

Table 15. Characteristics of the microsatellite DNA toolkit of Shaibi *et al.* (2008). DNA was Chelex-extracted (Walsh *et al.*, 1991) from one leg of each bee. Multiplex PCR solutions contained 10 µl of 10–100 ng DNA, 1× PCR-Master-Mix (Promega), and 0.2 µm of each primer (5'-label). PCR programme: denaturation for 5 min at 95°C, 35 cycles of 30 s at 95°C, 30 s of annealing at 55°C, extension for 1 min at 72°C, final elongation of 20 min at 72°C.

Locus	PCR reaction	Primers sequence (5'–3')	Repeat motif
HB-SEX-01	2	F: HEX-AGTGCAAAATCCAAATCATC R: ATTCGATCACCCAAAGAA	(A)15
UN351	2	F: FAM-AGCATACTTCTTCACCGAACCAC R: TCCGTTTATGCTTCATTTTCGA	(AT)13
HB-SEX-02	1	F: HEX-ACGCATTGAAGGATATTATGA R: AATTTGAACATTCGATCACC	(A)16
HB-SEX-03	2	F: TET-AACGTGGAAGATAACTTTAACAA R: ACAATGTTATGATTTTTCACGA	(TA)12
HB-THE-01	1	F: FAM-GACGATTTACGAGGTTTCAC R: TCGATTTCGTTTCGTTTTAT	(TA)9
HB-THE-02	2	F: TET-GGGAAAGATATTAGGGAGGA R: CGACGAAAAATTACAAGGAC	(TA)12
HB-THE-03	1	F: FAM-TAACTGGTCGTCGGTGTT R: CACGTAGAGAATCCCATTGT	(TA)11 (TC)12
HB-THE-04	2	F: HEX-GCTGGAAGGGAACTGTAGA R: GGACGCGTTTTAATATCTCA	(GA)9
HB-C16-01	2	F: HEX-AAAATGCGATTCTAATCTGG R: TTGCCTAAAATGCTTGCTAT	(GA)35
AC006	1	F: TET-GATCGTGGAAACCGCGAC R: CACGGCCTCGTAACGGTC	(TCT)5 (TTC)10
HB-C16-02	2	F: TET-TAGTATCGTGCTGTTCATCG R: ACATACATCTCTTGGCGAGT	(TA)23
HB-C16-05	1	F: FAM-ATTTTATGCGCGTTTCGTA R: CATGGCTCCTCCATTAAATC	(TC)23
A079	2	F: HEX-CGAAGGTTGCGGAGTCCTC R: GTCGTCGGACCGATGCG	(CCT)10 (GA)10
AP043	2	F: TET-GGCGTGCACAGCTTATTCC R: CGAAGGTGGTTTCAGGCC	(CT)24
A113	2	F: FAM-CTCGAATCGTGGCGTCC R: CCTGTATTTTGCAACCTCGC	(TC)2,5,8,5
A024	1	F: TET-CACAAGTTCCAACAATGC R: CACATTGAGGATGAGCG	(CT)10
A107	1	F: HEX-CCGTGGGAGGTTTATTGTCG R: CCTTCGTAACGGATGACACC	(CT)23
A007	1	F: FAM-GTTAGTGCCCTCCTCTTGC R: CCCTTCCTCTTTCATCTTCC	(CT)3 (T)7(CT)24

Note: Queen genotypes inferred from worker genotypes are given in parenthesis. Drone genotypes inferred from workers and queens are highlighted in bold.

2. When using unlinked markers, rearrange the tables by grouping all individuals sharing allelic combinations in three or more loci to facilitate the identification and counting of their colonies of origin. The more loci the individuals share, the higher the probability they share a mother queen (see section 4.3.3.5. on non-detection errors below). When using linked markers (Shaibi *et al.*, 2008, Table 15), first group all

individuals sharing the same allelic combination at all loci within each linkage group. The haplotypes found in each linkage group are equivalent to individual alleles.

2. Exclude individuals that showed two or less successfully amplified loci, or that could not be assigned to a specific haplotype in at least one linkage group (because of low polymorphism or misamplifications at some loci).

4. Introduce the alleles/ haplotypes into a sibship reconstruction software (e.g. COLONY, Wang, 2004) to reconstruct the genotype of individual drone-producing queens.

4.3.3.5. Non-detection and non-sampling errors

Two kinds of errors affect estimated number of drone-producing queens:

1. Non-detection errors (the probability of obtaining two identical genotypes in two different individuals by chance). Non-detection errors (NDE) are determined by the number of markers employed and their level of polymorphism and are an indicator of the resolution of these markers. It should always be reported along with the results, but there is no need to correct the results. To calculate NDE the following formula can be used:

$$NDE = \left(\sum q_i^2\right)\left(\sum r_i^2\right)...\left(\sum z_i^2\right)$$

    where

    q_i are the allele/ haplotype frequencies at the first locus,
    r_i are the allele/ haplotype frequencies at the second locus, and
    z_i are the allele/ haplotype frequencies at the last locus.
    This calculation assumes all loci/ linkage groups are unlinked and under Hardy-Weinberg equilibrium.

2. Non-sampling errors (the number of queens remaining undetected because of an insufficient sample). In contrast to NDE, the final number of queens detected should be corrected for non-sampling errors (NSE). In other words, the number of undetected queens should be accounted for. The following procedure describes how to account for NSE.

    2.1. Construct a frequency distribution table with the number of drones found to be assigned to each colony (see Fig. 27).

    2.2. Fit a Poisson distribution to the real data by calculating the expected frequency for each category.
    Expected frequencies of a Poisson distribution can be calculated using most commercial statistical packages (e.g. STATISTICA or SPSS).

    2.3. Obtain the expected frequency for the zero or less than one category.

    2.4. Add the undetected colonies (or colonies with an expected frequency of zero, see Fig. 27) to the detected ones to correct result for non-sampling errors.

4.3.3.6. Density estimation

1. Exclude colonies represented by less than a median number of drones in all density calculations in order to overcome the limitation that distant colonies will contribute fewer drones than colonies located in the vicinity of a DCA.

2. Quantify the number of colonies represented by an equal or higher than median number of drones.

3. Divide this number by the mean mating area of drones (for the drone samples, 2.5 km², Jaffé *et al.*, 2009a) or queens (for the worker samples, 4.5 km², Jaffé *et al.*, 2009a) to obtain an estimate of the local density of colonies at the sampling location (see Fig. 28).

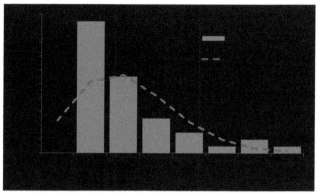

Fig. 27. Estimating the number of non-sampled colonies through a fitted Poisson distribution. While observed frequencies are plotted with blue bars, expected frequencies (fitted Poisson distribution) are shown in a red dashed line. In this example, the number of non-detected colonies is 4.7.

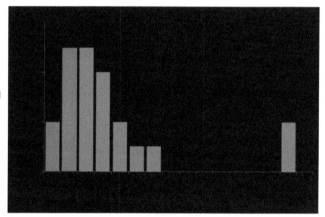

Fig. 28. Schematic representation of the approach to estimate honey bee colony densities based on the frequency distribution of drones among the reconstructed colonies. For a given sample of drones from a specific location, the median number of drones per colony is first calculated. In order to estimate the local density of colonies, those colonies represented by less than a median number of drones (red columns) need to be discarded. The number of remaining colonies (blue columns), are then divided by the mean mating area of drones or queens. This approach aims to avoid the overestimation of colony densities due to the inclusion of low-represented colonies, likely to be located beyond mean flight distances of drones or queens.

Pros: less tedious than finding all nests in an area. Method independent of nest spatial distribution (Arundel *et al.*, 2012).

Cons: Fails to detect colonies that do not produce drones. Season dependence when based on drone trapping, and thus a relevant density figure can only be obtained during mating season when most colonies produce drones. Assumes a similar drone investment by all colonies. Inaccuracy due to variable/ non predictable size of mating areas of drones and queens, which can be different between regions and honey bee populations. High costs involved in genetic analyses, and a suitable lab space and equipment is needed.

4.3.4. Future research needs and perspectives

1. The set of linked markers described in Table 15 might not prove useful for some honey bee populations because of misamplifications or low polymorphism. Additional genetic markers should be identified and tested to create a larger set of tightly linked markers located on different chromosomes.

2. A model accounting for a variable drone production per colony might increase accuracy of the method based on genetic markers.

3. Further studies on the mating area of drones and queens in different regions and populations might also increase accuracy of the method based on genetic markers.

4. The method based on genetic tools should be calibrated against populations of known absolute density.

5. One needs to determine the number of feeding stations that should be deployed per unit area before a site can be considered 'adequately represented'. For example, determining an index for colony density with 10 feeding stations on a 10,000 hectare area hardly seems accurate.

6. Because honey bees can forage 4-6 km from the nest (Winston, 1987), the distance between feeding stations necessary to limit the chances of one colony going to more than one site needs to be determined.

7. The accuracy of the indices should be confirmed by comparing the results from the indices to the actual colony density in an area (determined by methodical search and location of wild colonies in a landscape) and to other published colony density estimation methods (Oldroyd *et al.*, 1997; Baum *et al.*, 2005; Moritz *et al.*, 2008; Jaffé *et al.*, 2009a).

8. Reliability – Vaudo *et al.* (2012b) suggest that the field and photograph ratings provide more reliable indices than counting the number of bee lines, though this assumption needs to be validated.

4.4. Estimating the number of dead honey bees expelled from a honey bee colony with a trap

4.4.1 Aim of using dead bee traps

The assessment of intra hive mortality through dead bee traps is useful for acquiring data on honey bee survival when exposed to pesticides, environmental pollution, or honey bee diseases (Gary, 1960; Atkins *et al.*, 1970; Perez *et al.*, 2001; Porrini *et al.*, 2003).

For determination of bee mortality the removal of dead and sick honey bees (undertaking behaviour) needs to be considered (Gary, 1960; Perez *et al.*, 2001). Heavier objects e.g. bee bodies are usually dropped below the hive opening by bees and dragged away (several metres), while lighter objects are carried by the bees and disposed of at a good distance (several hundred metres) away from the hive (Gary, 1960; Porrini *et al.*, 2002a). Dead bee traps provide an obstacle to this

behaviour and allows for the collection and counting of the majority of the discarded bodies at hive entrance.

4.4.2. Limitations of using dead bee traps

The use of dead bee traps unfortunately does not account for the bees that have died in the field or on their way home (Porrini *et al.*, 2002a). Originally dead bee traps, e.g. the Gary trap, was intended to be used for short periods of time, but ever since bees have become biological indicators, traps are now being used throughout the year (Accorti *et al.*, 1991). These traps can become a problem when bees begin to treat them as an integral part of the hive that also needs to undergo the same cleaning processes as the rest of the hive (Accorti *et al.*, 1991). We therefore recommend, first to clean the trap on a regular basis and second to ensure that the trap is not continuously attached to the colony.

In general, studies tend to report the efficiency of traps, but not the effect on the colonies (Stoner *et al.*, 1979). In their study Stoner *et al.* (1979) reported the negative effect of a modified Todd trap on colonies showing less adult bees were present in colonies with dead bee traps. One should keep in mind when designing experiment using dead honey bee traps that the efficiency and suitability of a trap is not only depending on its design, but also on other factors like season, colony strength and environmental conditions (Porrini *et al.*, 2002a).

4.4.3. Types of dead bee traps

Many dead bee traps have been designed for Langstroth and Dadant type hives but currently the Todd, Gary, Münster and underbasket dead bee traps are the most frequently used (Table 16) (Illies *et al.*, 2002; Porrini *et al.*, 2002a). However, preliminary data on the performance of an experimental dead bee trap called the barrier trap indicates high efficiency (Porrini *et al.*, 2002b). In addition to existing traps, Hendriksma and Härtel (2010) constructed an entrance trap made of plastic ice cream containers that can be used for risk assessment in small hives.

There are several fundamental requirements for the design of a dead bee trap. They are reported in the Table 16 for each trap model and in section 4.3.4. for the general case.

4.4.4. Dead bee traps requirements as gathered from the literature

- Traps have to be well designed to allow for easy sample collection.
- Traps have to be very efficient at trapping only dead bees.
- Dead bee traps should not obstruct the normal behaviour, productivity and flight of bees.
- Predators/ scavengers should not be able to enter the dead bee traps.
- Traps have to be resistant to adverse weather conditions.
- Small, drainage holes for rain water should be present.

- The dead bee trap should allow for straightforward construction and cleaning.
- The attachment and removal of the traps from the hives should be uncomplicated.
- Dead bee traps should be as cost-effective as possible.

4.4.5. Recommended dead bee traps to use

We recommend using the Münster trap (Illies *et al.*, 1999, 2002), the underbasket trap (Accorti *et al.*, 1991) and the trap for small hives (Hendriksma and Härtel, 2010). No negative interference with colony activity was reported in these traps. The recovery rates of dead bees in the Münster trap (Illies *et al.*, 1999, 2002) were somewhat lower than some of the other traps (Table 16), but the artificial honey bee mortality resulting from the use of this trap was lower. As a cheaper alternative, the underbasket trap can be used since it does not interfere with the normal activity of the hive and reportedly has a very good recovery rate of dead bees (Table 16). The small hive trap (Hendriksma and Härtel, 2010) is fairly new in the bee research field, but it has a high potential of being a very successful dead bee trap that is also cost-effective in terms of both construction and maintenance.

4.4.6. Building a dead bee trap

For exact measurements please refer to the articles describing the original traps and their modifications.

4.4.7. Protocol for calibrating dead bees in traps

Before using the selected bee trap, it needs to be calibrated to establish its recovery rate. The following calibration protocol is derived from the work of Gary (1960), Illies *et al.* (2002) and Hendriksma and Härtel (2010).

1. Connect the trap to the hive for several days before the experiment begins, to allow the bees to become accustomed to the new addition.
2. Collect a known number of bees (e.g. 100) from the colony on which the trap is mounted.
3. Kill these workers (see section 2.1.3.).
4. Mark these workers (see the section 2.3. of the *BEEBOOK* paper on behavioural studies (Scheiner *et al.*, 2013).
5. Open the hive on which the trap is mounted.
6. Place the dead workers on top of the frames.
7. Close the hive.
8. Record the number of recovered bees every 15 min during the first hour, then again after 2, 4, 8 and 24 h (e.g., 2, 5, 1, 10, 22, 35, 6, 12).
   The efficiency can then be calculated based on 8 data points.
9. The percentage recovery rate of these marked dead bees is calculated to get an estimate of trapping efficiency.

Recovery rate = (number of recovered bees/ number of dead bees introduced) x 100.

In our example, recovery rate = (93/100) x 100 = 93%

4.4.8. Protocol for using a dead bee trap

1. Equalise colony size or assess colony size (see the *BEEBOOK* paper on colony strength (Delaplane *et al.*, 2013) to obtain a mortality rate. Do regular size assessment if it is a long term experiment.
2. Connect the trap to the hive for several days before the experiment begins, to allow the bees to become used to the new addition.
3. Remove and count the number of dead bees at regular predetermined intervals.
4. Clean the trap if necessary after counting.
5. Calculate the corrected mortality rate based on the recovery rate determined in section 4.4.7.) (Gary, 1960):

$$\text{Corrected mortality} = \frac{1}{\% \text{ recovered controls} \times \text{obs body recovery}}$$

4.4.9. Dead bee trap trade-offs

The most important trade-off among the different trap designs is that of a high recovery of dead bees versus interference with normal colony activity, in particular with undertaker bees and foragers. Another trade-off is the ease to attach and clean the traps versus the exposure of the trap content to the environment and potential predators which could utilise the trap as a feeding ground.

4.5. Creating multiple queen colonies

Recently, a method to create multiple queen honey bee colonies composed of young workers was created by clipping part of the mandibles of queens (Figs. 29 and 30). The crucial part of the method is the clipping of part of their mandibles. This operation does not significantly affect the general activity and mandibular gland profile of queens (Dietemann *et al.*, 2008; Zheng *et al.*, 2012). Queens with their mandibles ablated refrain from lethal fighting, resulting in cohabitation of queens (Dietemann *et al.*, 2008).

This procedure is described in section 4.4.1. In the following sections (4.4.2, 4.4.3.), the preparation and maintenance of multiple queen colonies is described. Multiple queen honey bee colonies (Fig. 31) are of significance both in beekeeping and research. In some areas of China, these colonies are used in beekeeping as supporting colonies to: (1) build up populous colonies faster in spring prior to major nectar flows and to maintain the population year-round when needed; (2) provide the 1-day-old larvae necessary for grafting larvae in queen cells for royal jelly production and (3) provide replacement queens when necessary. Furthermore, they can contribute to package bee

Table 16. Different types of dead bee traps being used in honey bee studies with their main characteristics, their pros and cons.

GARY TRAP (Gary, 1960)	
 Front view of Gary trap, modified from Gary (1960)	According to Gary (1960) the trap can be used for long-term experiments without affecting colony activity and/or the consistency of the information recorded. **Pros:** Efficient collection (84.6%) of dead bees (Gary, 1960). **Cons:** This trap unfortunately detains large numbers of live bees resulting in increased mortality rates and it modifies the behaviour of the undertaker bees (Illies *et al.*, 2002).
TODD TRAP (Atkins *et al.*, 1970; Stoner *et al.*, 1979)	
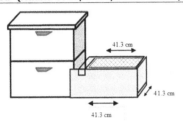 Side view of Todd trap, modified from Atkins *et al.* (1970)	Modifications were made to the trap that permitted the drainage of rain and irrigation water (Atkins *et al.*, 1970). **Pros:** This trap is reported to be efficient (90-95%) at preventing the elimination of dead bees (Atkins *et al.*, 1970; Herbert *et al.*, 1983). **Cons:** Compared to other traps the Todd trap seems to be more difficult to clean from debris by the experimenter.
MÜNSTER DEAD BEE TRAPS (Illies *et al.*, 1999, 2002)	
 Side view of Münster trap, modified from Illies *et al.* (2002)	**Pros:** The entrance of this trap does not interfere with normal flight behaviour and bees adjust quickly to this trap (Illies *et al.*, 1999, 2002). Recovery amounts to 76.4% of dead bees (Illies *et al.*, 2002). The trap also prevented predators from removing dead bees and provided shelter from wind (Illies *et al.*, 1999). **Cons:** The recovery rate is relatively low compared to the other traps mentioned here.
UNDERBASKET (Accorti *et al.*, 1991; Porrini *et al.*, 2002a)	
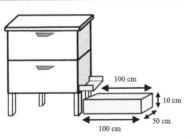 Side view of underbasket modified from Porrini *et al.* (2003)	The trap does not form part of the hive and is located on the ground underneath the hive opening (Accorti *et al.*, 1991; Porrini *et al.*, 2002a). **Pros:** Underbasket traps are easy to attach and clean. They seem to be highly efficient and do not interfere with undertaker bees' activities (Accorti *et al.*, 1991). A dead bee recovery rate of 71-96% was recorded in this trap (see Porrini *et al.*, 2002a). **Cons:** The trap is very exposed to the environment and predators.
TRAP FOR SMALL TEST HIVES (Hendriksma and Härtel, 2010)	
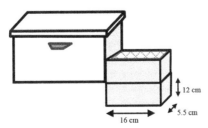 Side view of small trap modified from Hendriksma and Härtel (2010)	**Pros:** This is the first trap developed for small hives. Hendriksma and Härtel (2010) recorded a dead bee recovery rate of 93%. It seems easy to attach and clean, sounds highly efficient and does not interfere with normal hive behaviour. Most of all, it is very cheap to construct (Hendriksma and Härtel, 2010). **Cons:** This hive needs further testing

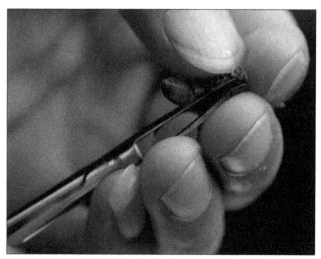

Fig. 29. Clipping mandibles of a queen. The queen's thorax is held between the thumb, index finger and middle finger of one hand while one third to half of mandibles on both sides is cut with small scissors held in the other hand.   Photo: W Wei.

production by providing large numbers of workers. In research, theyare helpful to deepen our understanding of basic questions on the evolution of sociality, such as division of reproductive labour and the evolution of polygyny (Dietemann *et al.*, 2009b).

4.5.1. Mandible clipping procedure

1. Hold the queen lightly by the thorax between the thumb, index- and middle finger of one hand (Fig. 29).
2. Hold the scissors with the other hand.
3. Cut approximately one third to half of both mandibles.
   Take care not to hurt other appendages of the queen.
4. Mark the queen with paint (see section 2.4.1.2.) when desired.

4.5.2. Preparation of colonies destined to host the multiple queens

The method that follows was described by Zheng *et al.* (2009a).

1. Mark the queens (see section 2.4.1) to allow future identification.
2. Select combs of emerging brood for the receiver colony.
3. Slightly shake the combs to trigger flight in the older bees, while young bees tend to remain on the comb.
4. Place the combs in a hive box with the young bees still clinging to them.
   Alternatively, combs with emerging brood can be kept in an incubator, if available, at 34°C for two days to collect young bees. One to three-day-old workers are preferred to freshly hatched individuals, which may not be able to care for the queens efficiently enough. The amount of combs and bees to be used in the multiple-queen colony depends on the number of queens to be introduced. Four to six combs are used for three to six queen colonies.
5. Add combs of honey and pollen beside the brood combs to provide enough food.
   Providing stored food is necessary because the colony is deprived of foragers at the beginning.
6. Place the hive 5-10 m away from their original location to ensure that all remaining foragers (older bees) do not re-enter.
7. Two days after the receiving colonies were prepared, take the queens out of their original colonies.
   To increase the chance for successful introduction, select queens older than six months since younger queens are more aggressive towards each other. The large abdomens of the egg laying queens might further reduce their ability to fight.

a   b

Fig. 30. (A) A queen with intact mandibles; (B) A queen with mandibles clipped.   Photo: H-Q Zheng.

Fig. 31. Five queens on one side of a comb.

Photo: W Wei.

8. Cut off a third to a half of both the queens' mandibles with small scissors.

   A good quality pair of small or micro scissors is necessary. Great care should be taken to avoid hurting other appendages of the queens, specifically their antennae, proboscises and forelegs. It is recommended to practice with workers before clipping queens.

9. Introduce the queens on different frames in the host hives. Observe the queens for a minute after their introduction. If the queens are attacked by workers, take them out and spray some honey water on both the workers and queens and then reintroduce the queens into the hive. If the queens are attacked, which may occasionally happen if some of the workers are too old to accept multiple queens, host colonies should be reorganised, making sure that the majority of the workers are young.

To ensure the multiple queen social structure, great care should be taken to maintain the receiver colonies. The necessary steps are described in the next section.

4.5.3. Steps for maintenance of an artificially established multiple-queen social organisation

The method that follows was described by Zheng *et al.* (2009b).

1. Supply the multiple queen colony with sufficient food at regular intervals.

   The strong egg laying capacity of a multiple queen colony results in most of the combs being occupied by brood, decreasing the space available for food storage and increasing the need for food to rear the brood. Consequently, these colonies must be fed more frequently compared to single queen colonies when there is decreased nectar flow, especially when no supers have been added.

2. Prevent robbing of the multiple queen colonies and drifting by placing food away from other colonies. Regularly monitor the occurrence of robbing.

3. Destroy newly built queen cells.

   This is to ensure that one or more queens are not killed after the occasional production of young queen(s).

4. Abandon foragers before migration.

   The agitation of old bees resulting from the transport during migration may lead to queen elimination. To reduce the possibility of queen losses, these old workers must be removed before migration. For this, the hive hosting the multiple-queen colony should be moved during an active foraging period a short distance away from its original location two days before the migration takes place. A hive with one comb should be placed at the original location to collect the old forager bees that will fly back.

4.6. Digital monitoring of brood development via location recognition

Jeker *et al.* (2011) designed a method to record subsequent development stages in a fixed number of cells selected on a frame at the start of a study following a (pesticide) treatment or other environmental impact. This technique is used as a digital documentation and an automation of the data evaluation according to the OECD guidance document 75 (2007). The method is used for GLP-compliant ecotoxicity tests, focused on subsequent recording of the content of marked cells during brood development. Besides studying the impact of pesticides it can be applied to follow brood development in studies about the impact of pathogens e.g. virus, brood pathogens and in-hive conditions.

4.6.1. Introduction

The development of the bee brood is assessed in individually-marked brood cells of all colonies within a study. At the assessment before the application of a treatment (BFD = Brood Fixing Day), one or more brood combs are taken out of each colony and identified with the study code, treatment group, hive number, comb number and comb site. The frames are photographed with a high-quality digital photo camera (full frame CMOS chip with a resolution of 20 megapixels or more) controlled via a laptop computer. In the laboratory, all photos are transferred to a personal computer and cells to be analysed, are chosen on the screen. Cells with any type of cell content can be selected, although for a typical evaluation according to Oomen *et al.* (1992), only egg-containing cells would be selected. The exact position of the markers and of each cell and its content are stored in a computer file that serves as a template for later assessments. The same cells are assessed on each of the following assessment dates (see Table 17). Thus, the development of each individually marked cell can be determined throughout the duration of the study (pre-imaginal development period of worker honey bees typically averages 21 days). For studies focussing on specific brood development stages e.g. young/ old larvae, the BFD may start at this stage and the assessment days are adjusted automatically to the expected development time of the specific brood stage. Following the OECD guideline 75, the brood development is checked 5 times: start with eggs, five days later these eggs have turned from young to old larvae, sixteen days after the start the brood in the marked cells are in the pupal stage and 22 days later the cells should be empty or contain again eggs. The program can cope with any number of observation days, meaning that frames, if necessary in the scope of the study, can be analysed each day. All data evaluation and files (with results), are adapted automatically. The program will generate additional result files for each of the starting stages (or starting contents). Depending on the study objective it is possible to start with other brood stages and with more frequent check dates. On brood fixing day 0 (BFD 00) cells with any brood stage can be selected. If egg-containing cells are selected and if in addition

Table 17. Assessment of the development of the bee brood starting with the brood stage "egg".

Assessment days	Determination brood stage in marked cells
BFD	egg
Assessment days	**Expected brood stage in marked/ selected cells**
+ 5 days (± 1 day) after BFD	Young to old larvae
+ 10 days (± 1 day) after BFD	Capped brood
+ 16 days (± 1 day) after BFD	Capped brood shortly before hatch
+ 22 days (± 1 day) after BFD	Empty cells or egg containing cells

to the standard data evaluation, images BFD 01 or BFD 02 or BFD 03 are analysed, the presumed age of the egg is calculated accordingly and the time resolution of the study is improved from four days to a maximum of one day.

When photographing the brood containing frames, the bees of the frames to be checked need to be brushed gently from the combs. The combs should not be shaken since too harsh handling might disturb the brood. In order to prevent drying out of the brood and so disturbing the normal development, the frames are taken from the colony, photographed immediately and transferred back to the colony as quickly as possible. Using fixed apparatuses the taking of the photograph requires only minutes.

For the European honey bee, the egg stage varies and is approximately 3 days (70 – 76 h). The larval stage (unsealed stage is considered as the larval stage) can varies between 5 and 6 days, with an average of 5.5 days and the pupal stage (capped cells) is 12 days (Winston, 1987; Jean-Prost and Médori, 1994; refer also to the section 1.5 on obtaining adults and brood of known age in this paper). Working with, for instance African honey bees, the assessment days must be adjusted to the duration of the development stages of the brood (see Fletcher, 1978).

4.6.2. Procedure for data acquisition
4.6.2.1. Software requirements
In order to apply the "Bee Brood Analyser", two programs must be installed on the computer:

- FIJI, a freeware image analysis program.
  (http://pacific.mpi-cbg.de/wiki/index.php/Fiji)
- NEXTREAT Bee Brood Analysis Software Package.
  (NEXTREAT, email: info@nextreat.ch)
  This software is regularly updated for optimal performance. Along with the Bee Brood Analysis Software Package, a User Manual is provided.

The user manual provides detailed instructions. Therefore here, only an outline is presented of the subsequent steps and results.

4.6.2.2. Before starting the project

Put orientation hallmarks (coloured thumbtacks) in the middle of the upper and lower long side of the frame.

4.6.2.3. Image acquisition.

1. Take out a frame.
2. For GLP reasons it is advised to label the frames with an identifier. Preferentially the ID should be used, which corresponds to the ID-System used by the software. The ID pattern is the following "AAAAAA_BB_CCD_EE", whereas "_" is a mandatory separator.
   2.1. AAAAAA: ID of the study in six characters,
   2.2. BB: ID of the hive in two numeric characters, e.g. "05",
   2.3. CC: ID of the frame in two numeric characters, e.g. "03",
   2.4. D: ID of the side of the frame e.g. "a" or "b",
   2.5. EE: ID of the BFD, e.g. 00 for BFD 0 (the day of study start).
   An example of a label on the frame would look like:
   "Study1_01_02a" for a permanent label or
   "Study1_01_02a_00" for a label made specifically for the day of image acquisition.
3. For unequivocal identification of the image, the label is attached on the front side of the frame and must be visible and to be photographed at every recording.
4. Make a picture of the frames using a fixed distance.
   4.1. The minimal photographic distance is calculated in order to allow visibility of at least 75% of the bottom of a cell at the outermost rim of the image. The calculation is based on an average cell diameter of 5.3mm and an average cell depth of 11mm. The photographic distance fulfilling of the above requirement is $11/(5.3 * 0.25)/2 = 4.15$ fold the long axis of the frame.
   4.2. The camera to be used should be connected to and controlled by a computer.
   The control software is necessary for a number of reasons: It enables triggering of the camera without the need to touch it. It enables a magnified live-view of the image allowing the directed focusing on the eggs. The camera's autofocus will always focus on the upper rim of the cell wall.
   4.3. Ideally, a setup should be created, allowing keeping the fixed distance, defined illumination and minimal vibrations.
   4.4. Illumination should be optimised to minimise reflections.
5. After the pictures are made, download the pictures from the camera to the computer.

5.1. The images have to be re-named according to the pattern "AAAAAA_BB_CCD_EE.jpg".
5.2. The image files of the same frame should be stored in the same folder.

4.6.3. Image analysis

4.6.3.1. Analysis of the first image (BFD 00)

1. Use the command keys in the User's Manual
   1.1. Standardise the size of the cell
   1.2. Position the mouse and press on the number-pad
   0 - empty cell
   1 - -an egg
   2 – a young larva
   3 - an old larva
   4 - a pupa
   5 - nectar
   6 - pollen
   7 - a dead larva
   8 - non characterised cell (nc)
   A circular mark will be set at the cell area, generating a circular region of interest (ROI).
   1.3. Do so for the number of cells required for the study.
   1.4. Once the selection of the cells is completed, define the two hallmarks.
   The hallmarks should always be the last ROI.
2. Once the ROI and both hallmarks have been selected, the process is finalised by automated saving the ROI file (AAAAAA_BB_CCD_EE_ROI.zip).
3. Simultaneously a copy of the ROI file is saved in the folder AAAAAA_BB_CCD_Archive with a time-stamped name (AAAAAA_ BB_CCD_EE_ROI yymmdd_hhmmss.zip; yymmdd_hhmmss corresponds to a date and time of the saving).
4. An image file is generated with all selected cells, hallmarks and additional GLP-relevant information is "burned" into the image (AAAAAA_BB_CCD_EE_selections.jpg).

4.6.3.2. For all consecutive images (BFD + 05, 10, 16, 22)

Consecutive images are processed by:

1. Selecting the hallmarks.
2. Letting the program transpose the selections from BFD 0. The program re-classifies the content of all cells to "nc" (not classified), ensuring that previous classifications are not carried forward. The user is forced to a re-classification of the cells. If a cell is classified as "nc" at any of the observation days, than the data of this cell are excluded of all of the subsequent analyses. The event of exclusion is documented, the data are not deleted.

3. Re-classification of the ROI's by the user.

   By presenting one cell after the other, the user has to re-classify the cells with the same keys on the number-pad as used for the selection of the cells on the image from BFD00 (see step 3.1. above), with the difference, that the cells are presented by the program and not chosen by the user.

4.6.4. Finalisation of the analysis.

The data evaluation is based on a developmental described by the following pattern: 1111222333444444444444, with the digits representing the expected developmental stage on consecutive days during larval development, i.e. the first four digits (1) correspond to days 0 to 3 with egg stage, the fifth to the seventh digits (2) correspond to days 4 to 6 with young larva stage, etc. If necessary, the user has the possibility to change this pattern and/ or to assign a maximum of two days of tolerance for either delayed or accelerated development. Once all images of a frame have been processed the analysis is finalised by pressing F6 or choosing the menu "Make gallery". The program will then run the analyses.

1. The program creates a folder with the name AAAAAA_BB_CCD Results yymmdd_hhmmss", where all results files of the evaluation are saved to ("yymmdd_ hhmmss" corresponds to a date and time of the analysis). Copies of all ROI files used for the analysis are saved into this folder.

2. The ROI data from subsequent days of the same frame are pooled into one tab-delimited file and saved as AAAAAA_BB_CCD_RawData.xls.

3. The program populates the classification data of each individual cell from the different observation days as numeric values (data of one cell are in one row; data of the same day are in columns "BFDnn"; e.g. BFD05 for the fifth brood fixing day).

4. The program rates the development as normal or terminated by comparing the set developmental pattern to the developmental stage expected for that cell on that day. These data are populated to the others in columns "BTRnn" (BTR = Brood Termination Rate). Brood termination-rate: Based on the brood termination-rate the failure of individual eggs or larvae to develop is quantitatively assessed. For the calculation of the brood termination-rate the observed cells are split into 2 categories:

   a. The bee brood in the observed cell reached the expected brood stage at the different assessments days or was found empty or containing an egg after hatch of the adult bee on BFD +22 = successful development.

   b. The bee brood in the observed cell did not reach the expected brood stage at one of the assessment days or termination of the bee brood development.

For the final calculation the number of cells, where a termination of the bee brood development was recorded, is summed up for each treatment and colony, is multiplied by 100 and divided by the number of cells observed in order to obtain of the brood termination-rate in %.

5. The program determines the brood index (BI) for each cell and each day and populates the data to the previous ones in the columns "BInn".

   Brood Index:

   The brood-index is an indicator of the bee brood development and facilitates a comparison between different treatments. The brood-index is calculated for each assessment day and colony. Therefore the brood development in each cell will be checked starting from BFD 0 up to BFD +22. The cells are classified from 1 to 5 (1: egg stage, 2: young larvae (L1 – L2), 3: old larvae (L3 – L5), 4: pupal stage (capped cell), 5: empty after hatching or again filled with brood (eggs and small larvae) if the cells contain the *expected* brood stage at the different assessment days. If a cell does not contain the expected brood stage or food is stored in the cell, the cell has to be counted 0 at that assessment day and also on the following days, irrespective whether the cell is filled again with brood. For the final calculation the values of all individual cells in each treatment, assessed at the same day, are summed up and divided by the number of observed cells in order to obtain the average brood-index.

6. The program determines the compensation index (CI) for each cell on each observation day and populates the results to the previous ones in the columns "CInn".

   Compensation index:

   The compensation-index is an indicator for recovery of the colony and will also be calculated for each assessment day and colony. The cells are classified from 1 to 5 (see brood index), solely based on the identified growth stage on the assessment days. By that the compensation of bee brood losses will be included in the calculation of the indices. For the final calculation the values of all individual cells in each treatment, assessed at the same day, are summed up and divided by the number of observed cells in order to obtain the average compensation-index.

7. The program does a frequency analysis for each day and parameter and populates the results below all the other data.

8. The program summarises all data by calculating the brood termination rate (BTR), BI and CI for each observation day and populates the results below the other data.

9. Finally, the developmental pattern and the tolerances used in the analysis are written at the end of this file. This file is saved as a tab-delimited file under the name "AAAAAA_BB_CCD_FinalData.xls".

10. If on BFD00 other than egg-containing cells are selected, then the program separates the cells based on their developmental stage or content and performs the above analyses (steps 5.6. to 5.9.) for each of the developmental stage or content separately and creates additional files with the names according to the following examples:

"AAAAAA_BB_CCD_StartAge_00_03.xls" for cells containing egg on BDF00,"AAAAAA_BB_CCD_StartAge_10_21.xls" for capped cells on FD00,

"AAAAAA_BB_CCD_StartContent_empty.xls" for cells empty on BFD00.

11. The program creates a gallery, where the images of the individual cells on the different observation days are assembled together side-by-side (similarly to a stamp-collection). This allows the user to have a visual verification of the assessment at a glance. This file is saved as a multi-page TIF file under the name "AAAAAA_BB_CCD_gallery.tif".

The temporal resolution of a standard study (observation on BFD00 followed by observation on BFD05, etc.) is four days, because the egg stage is four days-long. Insertion of an additional observation day before the end of the egg-stage allows the refinement of the calculated age of the eggs. This can enhance the temporal resolution of the study by a maximum of four fold. If an additional observation before BFD04 was inserted (e.g. on BFD02), than the program separates the egg containing cells according to their expected age into separate files and performs the above analyses (steps 4 to 9) for each age separately and creates additional tab-delimited files with the names according to the following examples:

"AAAAAA_BB_CCD_StartAge_00_01.xls" for egg containing cells, where the eggs had a calculated age of 0 to 1 days.

4.6.5. Conclusion

This program is a sophisticated tool for further study of stressors on brood development and the impact of stressors on colony level in field situations. The parameter "brood development" provides additional information about the vitality and plasticity of honey bee colonies confronted with stressors. Stressors are not restricted to pesticides but can also be read as the impact of pathogens and the environmental, both in terms of feed and pollution.

4.7. Collecting pollen and nectar from bees and flowers

4.7.1 Introduction

Pollen and nectar (Fig. 32) are produced by flowers as rewards for pollinators in exchange for pollination. Pollen is essential in the reproduction of plants while nectar, a sugary solution, secreted by glands called nectaries, is a product that is not part of the sexual system of plants (Dafni, 1992), but attracts pollinators to ensure the spread of the pollen. Both pollen and nectar are collected for various

Fig. 32. *Aloe greatheadii* var *davyana* flower showing pollen on anthers and a droplet of nectar. Photo: V Dietemann.

reasons in honey bee research, particularly in studies addressing foraging biology, pollination research and exposure risks to environmental pollutes (Sammataro and Avitabile, 2011; see also the *BEEBOOK* paper on toxicology, Medrzycki *et al.*, 2013).

Studies have shown a change in both appearance and nutritional composition of pollen during collection and storage by honey bees (Fig. 33) (Human and Nicolson, 2006). Through the addition of nectar and glandular secretions (Winston, 1987; Roulston and Cane, 2000) and certain bacterial flora associated with stored pollen the digestibility and nutritional value of the beebread/ stored pollen is increased (Herbert and Shimanuki, 1978). The sampling and collection methods depend upon the intended use of the floral source and the specific endpoints of measurements. However, it is important to know that quality of pollen decreases over time and should be stored appropriately and preferably be used within a year of sampling (Pernal and Currie, 2000). Here we describe methods to collect pollen (from the flowers, from the bees and stored in their combs) as well as various methods to collect nectar.

4.7.2. Methods for pollen collection

These methods are mostly used for studies on pesticide residues (Dively and Kamel, 2012, see also the *BEEBOOK* paper on toxicology, Medrzycki *et al.*, 2013) and nutritional content of pollen (e.g., Human and Nicolson, 2006; see references therein). Fresh pollen can be collected directly from flowers where the bees are foraging. There are three basic examples of fresh pollen collection; using bags over the flowers (section 4.7.2.1.1.), by physically shaking the flowers over plastic trays (section 4.7.2.1.2.) or by gently brushing off the pollen from the male anthers with a paint brush (section 4.7.2.1.3.).

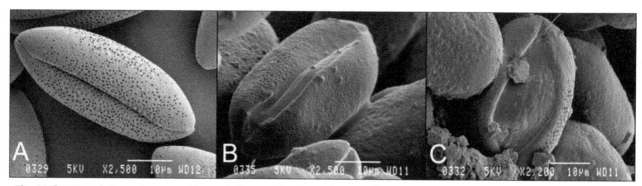

Fig. 33. Scanning electron microscopy pictures of *Aloe greatheadii* var. *davyana* pollen showing physical differences occurring in pollen grains after addition of nectar and glandular secretions; **(A)** Fresh pollen, **(B)** Bee collected pollen and **(C)** Stored pollen.   Photos: H Human.

Whenever fresh pollen is to be collected, flower buds that are open and ready to start shedding pollen, need to be covered with fine gauze or pollination bags the day before collection in order to prevent insect visitation and thus possible contamination. Bee collected pollen can be collected with pollen traps at the hive entrance or manually from the combs in which it has been stored (as bee bread).

4.7.3. Nectar collection

Foraging behaviour of honey bees is closely linked to colony needs and nectar production (volume and quality/ sugar concentration). Plants not only display particular rhythms of nectar secretion, but also nectar reabsorption (Nicolson *et al.*, 2007). In general nectar secretion is influenced by a variety of environmental factors e.g. humidity and temperature (Pacini and Nepi, 2007). Knowledge of these factors is essential for a proper understanding of the relationship between plants and honey bees.

Nectar secretion varies between plants, time of day and is even influenced by age of flowers (Pacini and Nepi, 2007). Nectar volume varies enormously between species; from less than a microlitre to thousands of microlitres (Pacini *et al.*, 2003). Similarly there is an extreme variation in nectar sugar concentration of plants (between and within species); from 7-70%. An example for between species variation is the low sugar concentration of less than 10% in *Aloe*

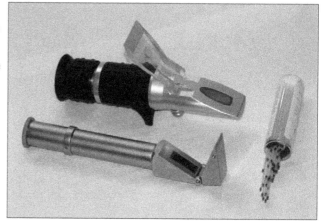

Fig. 35. Calibrated micropipettes/ micro-capillary tubes and refractometers used for measurements of nectar concentration and volume.   Photo: A Switala.

castanea (Aasphodelaceae) (Nicolson and Nepi, 2005, Fig 34) and an average of 66.5% in *Carum carvi* (Apiaceae) (Langenberger and Davis, 2002). It is generally known that the plants producing more concentrated nectar are the ones being visited and pollinated by insects, including bees (Pyke and Waser, 1981; Baker and Baker, 1982).

The method used for nectar collection will be determined by the intended use as well as by flower size, volume and concentration of nectar. Calibrated micropipettes/ micro-capillary tubes (1-20 µl) (Fig. 35) are commonly used to extract nectar with volumes > 0.5 µl and concentrations lower than 70%. Calibrated syringes (Hamilton microsyringes) and filtered paper wicks are other methods for nectar collection (see Kearns and Inouye, 1993) for more detailed descriptions of the various techniques). We here described those most commonly used for collection from honey bees (section 4.7.3.1.) and from flowers (see section 4.7.3.2.). Refractometers (Fig. 35) are normally used for the measurement of sugar concentration (% weight/ weight). In the case of very small amounts of nectar alternative methods are required (Kearns and Inoye, 1993; Dafni *et al.*, 2005). There are various techniques for measurements of nectar volume and concentration is discussed by Dafni (1992) and Kearns and Inoye (1993) and the more common methods used in honey bee research will be discussed here.

Fig. 34. Nectar (arrows) in base of *Aloe castanea* flowers. Photo: M Nepi.

4.7.3.1. Collecting nectar from honey bees

Honey bee foragers collect nectar from flowers. This nectar is stored in their impermeable crops for transfer back to their hives. The crop can greatly expand for storage and it has been shown that workers can carry crop loads close to their own body mass (Nicolson, 2008). By inducing bees to regurgitate, full nectar loads can be collected (Roubik and Buchman, 1984; Roubik *et al.*, 1995; Nicolson and Human, 2008; see the *BEEBOOK* paper on methods for behavioural studies (Scheiner *et al.*, 2013) for the latter method).

1. Capture honey bees visiting flowers on the plant of interest or at the entrance of hives on their way back from nectar gathering.
2. Compress the thorax of individual bees gently dorsoventrally to obtain nectar to induce regurgitation of the content of the honey stomach (Roubik and Buchman, 1984). This should be done within 10 min of capture, to prevent the honey bee using her stomach load as fuel.
3. Collect the liquid nectar from the mouthparts in micro capillary tubes through capillary action.
4. Measure nectar volume.
    Volumes (µl) are determined from the column length in micro-capillary tubes (length 75 mm/75 ml).
5. Measure nectar concentration with a pocket refractometer (e.g. Bellingham and Stanley Ltd, Tunbridge Wells, UK) by placing a drop of nectar onto the prismatic surface of the refractometer (through capillary action). Concentration is measured as % w/w sucrose equivalents.

Pros:
- Bees are not killed.
- Non-invasive method as far as the hive is concerned.

Cons:
- Honey stomachs may contain nectar from the hive used as fuel for flight, which could dilute the nectar collected (Roubik and Buchmann, 1984; Nicolson and Human, 2008).
- It has been shown that nectar concentration can be changed during flight back to the hive (Nicolson and Human, 2008).

4.7.3.2. Nectar collection from flowers

It is necessary to prevent insect visitation to flowers before measuring their nectar production/ secretion since consumption by insects will reduce the volume available. Nectar is collected from flowers in disposable micro capillary/ hematocrit tubes (length 75 mm, capacity 75 µl) through capillary action (e.g. Human and Nicolson, 2008; see references therein) (Fig. 36). It is standard procedure to measure both volume and concentration of nectar (the minimal information required) in any nectar/ foraging studies since this information is crucial.

Fig. 36. Collection of nectar from (*Aloe zebrina*) through capillary action into micro-capillary tubes. The clear nectar is visible in the lower part of the tube.   Photo: A Switala.

1. Cover flowers to be examined with gauze (2mm mesh size) to exclude visitation of any pollinators.
2. Remove flower petals gently to reveal nectar at the base of the flowers.
3. Withdraw/ collect the nectar from the flower in disposable micro-capillary tubes (length 75 mm, capacity 75 µl) by capillary attraction.
4. Determine volumes of nectar from column length in the micro-capillary tubes (75 mm is equivalent to 75 µl).
5. Release the nectar onto the prismatic surface of a pocket refractometer.
6. Measure the nectar concentration as percent (w/w) sucrose equivalents.
7. Depending on the purpose of nectar collection, samples should either be used immediately in the field or transported to the lab on either dry ice or on filter paper (Whatman no 1) (Dafni *et al.*, 2005) after which it should be stored in 15 ml centrifuge tubes at -20°C until ready for composition or residue analysis.

Pros: This is a cheap and easy way of nectar collection.

Cons: These methods are very tedious because of the small quantities of nectar that may be available per flower, and thus several hundred flowers may need to be extracted to collect the required quantities for analysis.

4.7.4. Precautions when sampling pollen and nectar for residue analyses

Pesticide residue levels in pollen and nectar are generally detected in the range of parts per billion (ppb). These extremely low traces of residues can easily occur due to cross-contamination. Therefore, it is essential that all steps in sample collection and processing, be optimised and quality assurance measures be deployed (e.g., use separate tools for each treatment sample, change disposable gloves between samples, etc.).

To quantify pesticide residues at the lowest level of detection, most analytical laboratories require samples of 3g of pollen or 1.5 ml of nectar, so different male flowers (usually 40-50 for pumpkin) may need to be extracted over the flowering period to collect the required quantities for analysis. In this case, detected residues in nectar and pollen represent the cumulative average level during the entire collection period. For more information on toxicology, see the relevant *BEEBOOK* paper by Medrzycki *et al.* (2013).

4.7.4.1. Collection of fresh pollen from flowers
4.7.4.1.1 Using paper bags to collect fresh pollen
Pollen collection with wax coated paper bags can be used for crops such as maize and pumpkin (Fig. 37).

1. Place wax-coated paper bags over maize tassels just prior to anthesis (the time when a flower is fully open and functional, timing of anthesis require observations beforehand) to prevent pollinator visits. The same method can be followed for pumpkins (Stoner and Eitzer, 2012) (Fig. 38).
2. Twist the bag's opening around the stem of the flower, for securing it to the plant.
   It is not necessary to seal tightly.
3. Remove bags from maize plants after one or two days. In the case of pumpkins, bags should be removed the next day when nectar production peaks, because nectar may contaminate the pollen.
4. Clean collected pollen by using sieves (pore sizes 0.119 and 0.0043 cm) to remove anthers, insects, and other debris (Fig. 39).
5. Store collected pollen at -20°C until ready for further testing.

Fig. 38. Pumpkin flowers covered with bags.   Photo: G Dively.

Fig. 39. Cleaning of pollen with sieves.   Photos: G Dively.

4.7.4.1.2. Manual collection of fresh pollen
Fresh pollen can also be collected from e.g. maize by literally shaking the tassels.

1. Shake maize tassels over large plastic trays at peak anthesis (when pollen shedding is at the highest, normally between 09.00h and 10.00h on field of sweet corn).
   Collect early morning after the dew dries, but before pollen shedding is complete.
2. Transfer fallen pollen into containers.
3. Clean collected pollen by using sieves (pore sizes 0.119 and 0.0043 cm) to remove anthers, insects, and other debris (Fig. 39).
4. Store collected pollen at -20°C until ready for further testing.

Fig. 37. Pollen collection with wax coated paper bags can be used for maize.   Photo: G Dively.

4.7.4.1.3. Using a paint brush for collection of fresh pollen

In the case of flowers where pollen is accessible from the outside of flowers e.g. sunflowers and aloes, one can also use a paint brush (Fig. 40; Human and Nicolson, 2006; Nicolson and Human, 2008).

1. Pick flowers.
2. Keep the flowers in containers in the laboratory at room temperature.
3. Use a paint brush to gently brush of pollen from the anthers into a container.
4. Continue collecting pollen this way on a daily basis until pollen shedding is complete.
5. Clean collected pollen using sieves (pore sizes 0.119 and 0.0043 cm) to remove anthers, insects, and other debris.
6. Store collected pollen at -20°C until ready for further testing.

Pros:
- Above mentioned methods allow for the relatively easy collection of a large amount of pollen.
- Allow for the collection of pollen of single and known plant origin.

Cons:
- Methods such as the paint brush collection method is very time consuming and requires a large number of flowers (up to 30,000 in the case of aloes, see Human and Nicolson, 2006) to enable one to collect enough pollen.
- Sieving samples of pollen to clean all debris from collected pollen is time consuming.
- Working with large amounts of fresh pollen can be detrimental to health and increase allergies.

4.7.4.2. Collection of bee collected pollen using pollen traps

A common method of pollen collection is the use of a trapping device placed at the entrance of hives. A variety of specific types of "pollen traps" are commercially available, all designed to force returning foragers entering the hive to crawl through small openings/ a grid

Fig. 40. Using a paint brush to collect pollen.   Photo: A Switala.

4.7.2.1.4. Collection of fresh pollen from smaller flowers such as canola

1. Collect flower clusters in the early morning when plants are 40 -50% flowering.
2. Place the clusters into containers.
3. Allow the clusters to dry at a processing location.
4. Brush flowers over food strainers to separate pollen from anthers.
5. Clean samples of pollen by sifting through multiple sieves of different pore sizes (pore sizes 0.119 and 0.0043 cm).
6. Store collected pollen at -20°C until ready for further testing.

Fig. 41. Example of an Auger-Hole pollen trap with a front and cross sectional view. Source: E R Harp from Sammataro and Avitabile, 2011.

(size of openings depends on the race of bees; African bees are known to be smaller than European races of bees (Johannsmeier, 2001)), which dislodge pollen pellets from their hind legs (see Fig. 41). The pellets then fall into a collection tray. Trap design varies in the size of the openings, installation location on the hive, and mechanism for accessing the collection tray to remove pollen. An effective pollen trap is easy to use, tightly fits the hive box, and can collect at least 60% of the foraged pollen pellets brought to the hive with minimum disturbance and climatic exposure to the colony and trapped pollen. Refer to the to the 'Pollen trapping' section of the *BEEBOOK* paper on pollination (Delaplane *et al.*, 2013) for a method to measure trapping efficiency and how to use pollen traps.

4.7.4.3. Ensuring quality of bee collected pollen

Pollen traps are used in studies to measure foraging activity, identify pollen sources, analyse pollen for toxic residues, and to collect pollen for feeding studies. Dependent upon the intended use, steps should be taken to ensure the quality of trapped pollen. A heap of moist pollen is an ideal breeding place for small hive beetles (where they occur, see also the *BEEBOOK* paper on small hive beetle, Neumann *et al.*, 2013) and wax moths (see the *BEEBOOK* paper on wax moths, Ellis *et al.*, 2013) and is very attractive to ants (Johannsmeier, 2001). Pollen quickly degrades and will start to become mouldy if it gets wet. Pollen should therefore be collected every day, cleaned of larger debris either by hand or by sieving through different sized sieves (see section 4.7.2.1.3., step 5) and be stored immediately as a frozen or dried sample to maintain quality. This is essential for samples collected for pesticide residue analysis, which should be stored on ice in coolers in the field and then frozen immediately to -20°C to prevent pesticide degradation until samples are processed.

Pros:

- Pollen traps are a less invasive technique of collecting bee collected pollen.
- Easy to collect a large quantity of pollen.
- Pollen from certain plants is more suitable for collection because of their abundance and high yield.
- Pollen pellets are usually of single plant origin, but may occasionally be a combination from different species.

Cons:

- Nutritional composition of pollen pellets may already be modified due to addition of nectar and glandular secretions added by bees.
- Pollen traps may reduce water and nectar collection because the congestion at the hive entrance slows the movement of foragers, which could stress the colony.
- Weaker colonies may be more stressed by pollen traps than strong colonies in an experiment, resulting in a confounding factor.

- If traps are left too long on hives there may be a reduction in brood rearing and honey production.

5. Acknowledgements

The COLOSS (Prevention of honey bee COlony LOSSes) network aims to explain and prevent massive honey bee colony losses. It was funded through the COST Action FA0803. COST (European Cooperation in Science and Technology) is a unique means for European researchers to jointly develop their own ideas and new initiatives across all scientific disciplines through trans-European networking of nationally funded research activities. Based on a pan-European intergovernmental framework for cooperation in science and technology, COST has contributed since its creation more than 40 years ago to closing the gap between science, policy makers and society throughout Europe and beyond. COST is supported by the EU Seventh Framework Programme for research, technological development and demonstration activities *(Official Journal L 412, 30 December 2006).* The European Science Foundation as implementing agent of COST provides the COST Office through an EC Grant Agreement. The Council of the European Union provides the COST Secretariat. The COLOSS network is now supported by the Ricola Foundation - Nature & Culture.

6. References

AASE, A L T O; AMDAM, G V; HAGEN, A; OMHOLT, S W (2005) A new method for rearing genetically manipulated honey bee workers. *Apidologie* 36: 293-299. http://dx.doi.org/10.1051/apido:2005003

ACCORTI, M; LUTU, F; TARDUCCI, F (1991) Methods for collecting data on natural mortality in bee. *Ethology, Ecology and Evolution* 1: 123-126.

ALAUX, C; DUCLOZ, F; CRAUSER, D; LE CONTE, Y (2010) Diet effects on honey bee immunocompetence. *Biology Letters* 6: 562-565. http://dx.doi.org/10.1098/rsbl.2009.0986

ALLEN, M D (1959) Respiration rates of worker honey bees of different ages and at different temperatures. *Journal of Experimental Biology* 36: 92-101.

ALLSOPP, M (2006) Analysis of *Varroa destructor* infestation of Southern African honey bee populations. Dissertation, University of Pretoria, South Africa. 285 pp.

AMDAM, G V; NORBERG, K; PAGE, R E; ERBER, J; SCHEINER, R (2006) Downregulation of *vitellogenin* gene activity increases the gustatory responsiveness of honey bee workers (*Apis mellifera*). *Behavioural Brain Research* 169: 201-205. http://dx.doi.org/10.1016/j.bbr.2006.01.006

ANTÚNEZ, K; MARTIN-HERNANDEZ, R; PRIETO, L; MEANA, A; ZUNINO, P; HIGES, M (2009) Immune suppression in the honey bee (*Apis mellifera*) following infection by *Nosema ceranae* (Microsporidia). *Environmental Microbiology* 11(9): 2284-2290. http://dx.doi.org/10.1111/j.1462-2920.2009.01953.x

ARUNDEL, J; OLDROYD, B P; WINTER, S (2012) Modelling honey bee queen mating as a measure of feral colony density. *Ecological Modelling* 247: 48-57. http://dx.doi.org/10.1016/j.ecolmodel.2012.08.001

ATKINS, E L; TODD, F E; ANDERSON, L D (1970) Honey bee field research aided by Todd dead bee hive entrance trap. *California Agriculture* 24(10): 12-13.

AUSTIN, G H (1955) Effect of carbon dioxide anaesthesia on bee behaviour and expectation of life. *Bee World* 36(3): 45-47.

BAER, B; EUBEL, H; TAYLOR, N L; O'TOOLE, N; MILLAR, A H (2009) Insights into female sperm storage from the spermathecal fluid proteome of the honey bee *Apis mellifera*. *Genome Biology* 10(6): R67. http://dx.doi.org/10.1186/gb-2009-10-6-r67

BAKER, H G; BAKER, I (1982) Chemical constituents of nectar in relation to pollination mechanisms and phylogeny. In *M H Nitecki (Ed.). Biochemical aspects of evolutionary biology*. University of Chicago Press, Chicago, USA. pp 131-171.

BARRON, A B; MALESZKA, J; MEER, R K V; ROBINSON, G E; MALESZKA, R (2007) Comparing injection, feeding and topical application methods for treatment of honey bees with octopamine. *Journal of Insect Physiology* 53: 187-194.

BAUDRY, E; SOLIGNAC, M; GARNERY, L; GRIES, M; CORNEUT, J-M; KOENIGER, N (1998) Relatedness among honey bees (*Apis mellifera*) of a drone congregation. *Proceedings of the Royal Society of London* B265: 2009-2014.

BAUM, K A; RUBINK, W L; PINTO, M A; COULSON, R N (2005) Spatial and temporal distribution and nest site characteristics of feral honey bee (Hymenoptera: Apidae) colonies in a coastal prairie landscape. *Environmental Entomology* 34: 610-618. http://dx.doi.org/10.1603/0046-225X-34.3.610

BENDAHOU, N; BOUNIAS, M; FLECHE, C (1999) Toxicity of cypermethrin and fenitrothion on the haemolymph carbohydrates, head acetylcholinesterase, and thoracic muscle Na$^+$, K$^+$-ATPase of emerging honey bees (*Apis mellifera mellifera*. L). *Ecotoxicology and Environmental Safety* 44: 139-146. http://dx.doi.org/10.1006/eesa.1999.1811

BERTHOLF, L M (1925) The moults of the honey bee. *Journal of Economic Entomology* 18: 380-384.

BEYE, M; HÄRTEL, S; HAGEN, A; HASSELMANN, M; OMHOLT, S W (2002) Specific developmental gene silencing in the honey bee using a homeobox motif. *Insect Molecular Biology* 11: 527-532. http://dx.doi.org/10.1046/j.1365-2583.2002.00361.x

BRODSCHNEIDER, R; RIESSBERGER-GALLÉ, U; CRAILSHEIM, K (2009) Flight performance of artificially reared honey bees (*Apis mellifera*). *Apidologie* 40: 441-449. http://dx.doi.org/10.1051/apido/2009006

BÜCHLER, R; ANDONOV, S; BIENEFELD, K; COSTA, C; HATJINA, F; KEZIC, N; KRYGER, P; SPIVAK, M; UZUNOV, A; WILDE, J (2013) Standard methods for rearing and selection of *Apis mellifera* queens. In *V Dietemann; J D Ellis; P Neumann (Eds) The COLOSS BEEBOOK, Volume I: standard methods for* Apis mellifera *research*. *Journal of Apicultural Research* 52(1): http://dx.doi.org/10.3896/IBRA.1.52.1.07

BUCHMANN, S L; THOENES, S C (1990) The electronic scale: honey bee colony as a management and research tool. *Bee Science* 1: 40-47.

CANTWELL, G E (1970) Standard methods for counting nosema spores. *American Bee Journal* 110(6): 222-223.

CHÂLINE, N; RATNIEKS, F L W; RAINE, N E; BADCOCK, N S; BURKE, T (2004) Non-lethal sampling of honey bee, *Apis mellifera*, DNA using wing tips. *Apidologie* 35(3): 311-318. http://dx.doi.org/10.1051/apido:2004015

CHEN, Y P; EVANS, J; HAMILTON, M; FELDLAUFER, M (2007) The influence of RNA integrity on the detection of honey bee viruses: molecular assessment of different sample storage methods. *Journal of Apicultural Research* 46(2): 81-87. http://dx.doi.org/10.3896/IBRA.1.46.2.03

CHEN, Y P; PETTIS, J S; COLLINS, A; FELDLAUFER, M F (2006) Prevalence and transmission of honey bee viruses. *Applied and Environmental Microbiology* 72(1): 606-611. http://dx.doi.org/10.1128/AEM.72.1.606-611.2006

CHUDA-MICKIEWICZ, B; CZEKOŇSKA, K; SAMBORSKI, J; ROSTECKI, P (2012) Success rates for instrumental insemination of carbon dioxide and nitrogen anaesethetised honey bee (*Apis mellifera*) queens. *Journal of Apicultural Research* 51(1): 74-77. http://dx.doi.org/10.3896/IBRA.1.51.1.09

COBEY, S W; TARPY, D R; WOYKE, J (2013) Standard methods for instrumental insemination of *Apis mellifera* queens. In *V Dietemann; J D Ellis; P Neumann (Eds) The COLOSS BEEBOOK, Volume I: standard methods for* Apis mellifera *research. Journal of Apicultural Research* 52(4): http://dx.doi.org/10.3896/IBRA.1.52.4.09

COLES, J A; ORKHAND, R K (1983) Modification of potassium movement through the retina of the drone (*Apis mellifera*) by glial uptake. *Journal of Physiology* 340(1): 157-174.

CRAILSHEIM, K (1992) The flow of jelly within a honey bee colony. *Journal of Comparative Physiology* B162: 681-689. http://dx.doi.org/10.1007/BF00301617

CRANE, E (1999) *The world history of beekeeping and honey hunting*. Duckworth; London, USA. 720 pp.

CZEKONSKA, K (2007) Influence of carbon dioxide on *Nosema apis* infection of honey bees (*Apis mellifera*). *Journal of Invertebrate Pathology* 95(2): 84-86. http://dx.doi.org/10.1016/j.jip.2007.02.001

DADE, H A (2009) *Anatomy and dissection of the honey bee*. International Bee Research Association; Cardiff, UK. 196 pp.

DAFNI, A (1992) *Pollination ecology: a practical approach*. Oxford University Press; Oxford, UK. 250 pp.

DAFNI, A; KEVAN, P G; HUSBAND, B C (2005) *Practical pollination biology.* Enviroquest Ltd; Cambridge, Ontario, Canada. 583 pp.

DAINAT, B; EVANS, J D; CHEN, Y P; NEUMANN, P (2011) Sampling RNA quality for diagnosis of honey bee viruses using quantitative PCR. *Journal of Virological Methods* 174(1-2): 150-152. http://dx.doi.org/10.1016/j.jviromet.2011.03.029

DE GRAAF, D C; ALIPPI, A M; ANTÚNEZ, K; ARONSTEIN, K A; BUDGE, G; DE KOKER, D; DE SMET, L; DINGMAN, D W; EVANS, J D; FOSTER, L J; FÜNFHAUS, A; GARCIA-GONZALEZ, E; GREGORC, A; HUMAN, H; MURRAY, K D; NGUYEN, B K; POPPINGA, L; SPIVAK, M; VANENGELSDORP, D; WILKINS, S; GENERSCH, E (2013) Standard methods for American foulbrood research. In *V Dietemann; J D Ellis; P Neumann (Eds) The COLOSS* BEEBOOK, *Volume II: standard methods for* Apis mellifera *pest and pathogen research. Journal of Apicultural Research* 52(1): http://dx.doi.org/10.3896/IBRA.1.52.1.11

DEGRANDI-HOFFMAN, G; CHEN, Y P; HUANG, E; HUANG, M H (2010) The effect of diet on protein concentration, hypopharyngeal gland development and virus load in worker honey bees (*Apis mellifera* L.). *Journal of Insect Physiology* 56: 1184-1191. http://dx.doi.org/10.1016/j.jinsphys.2010.03.017

DELAPLANE, K S; VAN DER STEEN, J; GUZMAN, E (2013) Standard methods for estimating strength parameters of *Apis mellifera* colonies. In *V Dietemann; J D Ellis; P Neumann (Eds) The COLOSS* BEEBOOK, *Volume I: standard methods for* Apis mellifera *research. Journal of Apicultural Research* 52(1): http://dx.doi.org/10.3896/IBRA.1.52.1.03

DE MIRANDA, J R; BAILEY, L; BALL, B V; BLANCHARD, P; BUDGE, G; CHEJANOVSKY, N; CHEN, Y-P; GAUTHIER, L; GENERSCH, E; DE GRAAF, D; RIBIÈRE, M; RYABOV, E; DE SMET, L VAN DER STEEN, J J M (2013) Standard methods for virus research in *Apis mellifera*. In *V Dietemann; J D Ellis; P Neumann (Eds) The COLOSS* BEEBOOK, *Volume II: standard methods for* Apis mellifera *pest and pathogen research. Journal of Apicultural Research* 52(4): http://dx.doi.org/10.3896/IBRA.1.52.4.22

DIETEMANN, V; ZHENG, H-Q; HEPBURN, C; HEPBURN, H R; JIN, S H; CREWE, R M; RADLOFF, S E; HU, F-L; PIRK, C W W (2008) Self-assessment in insects: honey bee queens know their own strength. *PLoS ONE* 3(1): e1412. http://dx.doi.org/10.1371/journal.pone.0001412

DIVELY, G P; KAMEL, A (2012) Insecticide residues in pollen and nectar of a cucurbit crop and their potential exposure to pollinators. *Journal of Agricultural and Food Chemistry* 60: 4449-4456. http://dx.doi.org/10.1021/jf205393x

DUAY, P; DE JONG, D; ENGELS, W (2003) Weight loss in drone pupae (*Apis mellifera*) multiply infested by *Varroa destructor* mites. *Apidologie* 34: 61-65. http://dx.doi.org/10.1051/apido:2002052

EBADI, R; GARY, N E; LORENZEN, K (1980) Effects of carbon dioxide and low temperature narcosis on honey bees, *Apis mellifera. Environmental Entomology* 9(1): 144-147.

EDGELL, G H (1949) *The bee hunter.* Harvard University Press; Cambridge, MA, USA.

ELLIS, J D; GRAHAM, J R; MORTENSEN, A (2013) Standard methods for wax moth research. In *V Dietemann; J D Ellis; P Neumann (Eds) The COLOSS* BEEBOOK, *Volume II: standard methods for* Apis mellifera *pest and pathogen research. Journal of Apicultural Research* 52(1): http://dx.doi.org/10.3896/IBRA.1.52.1.10

ERBER, J (1975) The dynamics of learning in the honey bee (*Apis mellifica carnica*). *Journal of Comparative Physiology* 99(3): 231-242.

EVANS, J D; SCHWARZ, R S; CHEN, Y P; BUDGE, G; CORNMAN, R S; DE LA RUA, P; DE MIRANDA, J R; FORET, S; FOSTER, L; GAUTHIER, L; GENERSCH, E; GISDER, S; JAROSCH, A; KUCHARSKI, R; LOPEZ, D; LUN, C M; MORITZ, R F A; MALESZKA, R; MUÑOZ, I; PINTO, M A (2013) Standard methodologies for molecular research in *Apis mellifera*. In *V Dietemann; J D Ellis; P Neumann (Eds) The COLOSS* BEEBOOK, *Volume I: standard methods for* Apis mellifera *research. Journal of Apicultural Research* 52(4): http://dx.doi.org/10.3896/IBRA.1.52.4.11

FAROOQUI, T; ROBINSON, K; VAESSIN, H; SMITH, B H (2003) Modulation of early olfactory processing by an octopaminergic reinforcement pathway in the honey bee. *The Journal of Neuroscience* 23(12): 5370-5380.

FENOY, S; RUEDA, C; HIGES, M; MARTÍN-HERNÁNDEZ, R; DEL AGUILA, C (2009) High-level resistance of *Nosema ceranae*, a parasite of the honey bee, to temperature and desiccation. *Applied and Environmental Microbiology* 75(21): 6886-6889. http://dx.doi.org/10.1128/AEM.01025-09

FLETCHER, D J C (1978) The African bee, *Apis mellifera adansonii*, in Africa. *Annual Review of Entomology* 23: 151-171.

FORSGREN, E; FRIES, I (2010) Comparative virulence of *Nosema ceranae* and *Nosema apis* in individual European honey bees. *Veterinary Parasitology* 170(3-4): 212-217. http://dx.doi.org/10.1016/j.vetpar.2010.02.010

FORSGREN, E; BUDGE, G E; CHARRIÈRE, J-D; HORNITZKY, M A Z (2013) Standard methods for European foulbrood research. In *V Dietemann; J D Ellis; P Neumann (Eds) The COLOSS* BEEBOOK, *Volume II: Standard methods for* Apis mellifera *pest and pathogen research. Journal of Apicultural Research* 52(1): http://dx.doi.org/10.3896/IBRA.1.52.1.12

FRIES, I; CHAUZAT, M-P; CHEN, Y-P; DOUBLET, V; GENERSCH, E; GISDER, S; HIGES, M; MCMAHON, D P; MARTÍN-HERNÁNDEZ, R; NATSOPOULOU, M; PAXTON, R J; TANNER, G; WEBSTER, T C; WILLIAMS, G R (2013) Standard methods for nosema research. In *V Dietemann; J D Ellis; P Neumann (Eds) The COLOSS* BEEBOOK, *Volume II: Standard methods for* Apis mellifera *pest and pathogen research. Journal of Apicultural Research* 52(1): http://dx.doi.org/10.3896/IBRA.1.52.1.14

FROST, E H; SHUTLER, D; HILLIER, N K (2011) Effects of cold immobilization and recovery period on honey bee learning, memory, and responsiveness to sucrose. *Journal of Insect Physiology* 57(10): 1385-1390. http://dx.doi.org/10.1016/j.jinsphys.2011.07.001

GARY, N E (1960) A trap to quantitatively recover dead and abnormal honey bees from the hive. *Journal of Economic Entomology* 53(5): 782-785.

GLADSTONE, H C; GOWEN, I; GOWEN, J W (1964) Gamete-backcross matings in the honey bee. *Genetics* 50(6): 1443-1446.

HAGLER, J R; JACKSON, C G (2001) Methods for marking insects: current techniques and future prospects. *Annual Review of Entomology* 46: 511-543.
http://dx.doi.org/10.1146/annurev.ento.46.1.511

HEFNER, E; HSIUNG, F; MCCOLLUM, T; RUBIO, T (2010) Comparison of count reproducibility, accuracy and time to results between a haemocytometer and the TC10™ automated cell counter. *Cell counting.* Tech note 6003, Bio-Rad Laboratories; USA.

HEINRICH, B (1979) Thermoregulation of African and European honey bees during foraging, attack, and hive exits and returns. *Journal of Experimental Biology* 80: 217-229.

HENDERSON, C E (1992) Variability in the size of emerging drones and of drone and worker eggs in honey bee (*Apis mellifera* L.) colonies. *Journal of Apicultural Research* 31: 119-123.

HENDRIKSMA, H P; HÄRTEL, S (2010) A simple trap to measure worker bee mortality in small test colonies. *Journal of Apicultural Research* 49(2): 215-217. http://dx.doi.org/10.3896/IBRA.1.49.2.13

HERBERT, E W; SHIMANUKI, H (1978) Chemical composition and nutritive value of bee-collected and bee-stored pollen. *Apidologie* 9: 33-40. http://dx.doi.org/10.1051/apido:19780103

HERBERT, E W; SHIMANUKI, H; ARGAUER, R J (1983) Effect of feeding pollen substitutes to colonies of honey bees (Hymenoptera: Apidae) exposed to carbaryl. *Environmental Entomology* 12: 758-762.

HRASSNIGG, N; CRAILSHEIM, K (1998) Adaptation of hypopharyngeal gland development to the brood status of honey bee (*Apis mellifera* L.) colonies. *Journal of Insect Physiology* 44: 929-939.
http://dx.doi.org/10.1016/S0022-1910(98)00058-4

HRASSNIGG, N; CRAILSHEIM, K (2005) Differences in drone and worker physiology in honey bees (*Apis mellifera* L.). *Apidologie* 36: 255-277. http://dx.doi.org/10.1051/apido:2005015

HUMAN, H; NICOLSON, S W (2008) Flower structure and nectar availability in *Aloe greatheadii* var. *davyana*: an evaluation of a winter nectar source for honey bees. *International Journal of Plant Sciences* 169: 263-269. http://dx.doi.org/10.1086/524113

HUMAN, H; NICOLSON, S W (2006) Nutritional content of fresh, bee-collected and stored pollen of *Aloe greatheadii* var. *davyana* (Asphodelaceae). *Phytochemistry* 67: 1486-1492.
http://dx.doi.org/10.1016/j.phytochem.2006.05.023

HUMAN, H; NICOLSON, S W; STRAUSS, K; PIRK, C W W; DIETEMANN, V (2007) Influence of pollen quality on ovarian development in honey bee workers (*Apis mellifera scutellata*). *Journal of Insect Physiology* 53(7): 649-655.
http://dx.doi.org/10.1016/j.jinsphys.2007.04.002

ILLIES, I; MÜHLEN, W; DÜCKER, G; SACHSER, N (1999) A study of undertaking behaviour of honey bees (*Apis mellifera* L.) by use of different bee traps. In *L P Belzunces; C Pélissier; G B Lewis (Eds). Hazards of pesticides to bees. Avignon, France, 7-9 September.* pp 237-244.

ILLIES, I; MÜHLEN, W; DÜCKER, G; SACHSER, N (2002) The influence of different bee traps on undertaking behaviour of the honey bee and development of a new trap. *Apidologie* 33: 315-326.
http://dx.doi.org/10.1051/apido:2002014

JACKSON, D E; HART, A G (2009) Does sanitation facilitate sociality? *Animal Behaviour* 77: e1-e5.
http://dx.doi.org/10.1016/j.anbehav.2008.09.013

JAFFÉ, R; DIETEMANN, V; ALLSOPP, M H; COSTA, C; CREWE, R M; DALL'OLIO, R; DE LA RÚA, P; EL-NIWEIRI, M A A; FRIES, I; KEZIC, N; MEUSEL, M; PAXTON, R J; SHAIBI, T; STOLLE, E; MORITZ, R F A (2009a) Estimating the density of honey bee colonies across their natural range to fill the gap in pollinator decline censuses. *Conservation Biology* 24: 583-593.
http://dx.doi.org/10.1111/j.1523-1739.2009.01331.x

JAFFÉ, R; DIETEMANN, V; CREWE, R M; MORITZ, R F A (2009b) Temporal variation in the genetic structure of a drone congregation area: an insight into the population dynamics of wild and African honey bees *(Apis mellifera scutellata). Molecular Ecology* 18: 1511-1522. http://dx.doi.org/10.1111/j.1365-294X.2009.04143.x

JAY, S C (1962) Colour changes in honey bee pupae. *Bee World* 43: 119-122.

JEAN-PROST, P; MÉDORI, P; KAULA, R K (1994) *Apiculture: know the bee, manage the apiary.* Intercept; Andover, UK.

JENSEN, A B; ARONSTEIN, K; FLORES, J M; VOJVODIC, S; PALACIO, M A; SPIVAK, M (2013) Standard methods for fungal brood disease research. In *V Dietemann; J D Ellis; P Neumann (Eds) The COLOSS BEEBOOK, Volume II: Standard methods for* Apis mellifera *pest and pathogen research. Journal of Apicultural Research* 52(1): http://dx.doi.org/10.3896/IBRA.1.52.1.13

JOHANNSMEIER, M F (2001) *Beekeeping in South Africa.* Plant Protection Handbook No. 14, Agricultural Research Council; Pretoria, South Africa. 288 pp.

KEARNS, C A; INOUYE, D W (1993) *Techniques for pollination biologists.* University Press of Colorado; Niwot, Colorado, USA. 583 pp.

KLEINSCHMIDT, G J; KONDOS, A C (1978) The effect of dietary protein on colony performance. *Australian Beekeeper* 80: 251-257.

KOCHER, S D; RICHARD, F J; TARPY, D R; GROZINGER, C M (2008) Genomic analysis of post-mating changes in the honey bee queen (*Apis mellifera*). *BMC Genomics* 9(232): 1-15. http://dx.doi.org/10.1186/1471-2164-9-232

KÖHLER, A; PIRK, C W W; NICOLSON, S W (2012) Simultaneous stressors: additive effects of an immune challenge and dietary toxin are detrimental to honey bees. *Journal of Insect Physiology* 58: 918-923. http://dx.doi.org/10.1016/j.jinsphys.2012.04.007

KOYWIWATTRAKUL, P; THOMPSON, G J; SITTHIPRANEED, S; OLDROYD, B P; MALESZKA, R (2005) Effects of carbon dioxide narcosis on ovary activation and gene expression in worker honey bees, *Apis mellifera*. *Journal of Insect Science* 5: 36.

KUCHARSKI, R; MALESZKA, R (2003) Transcriptional profiling reveals multifunctional roles for transferrin in the honey bee, *Apis mellifera*. *Journal of Insect Science* 3: 1-8.

KUCHARSKI, R; MALESZKA, J; FORET, S; MALESZKA, R (2008) Nutritional control of reproductive status in honey bees via DNA methylation. *Science* 319: 1827-1830. http://dx.doi.org/10.1126/science.1153069

LANGENBERGER, M W; DAVIS, A R (2002) Temporal changes in floral nectar production, reabsorption, and composition associated with dichogamy in annual caraway (*Carum carvi*, Apiaceae). *American Journal of Botany* 89(10): 1588-1598. http://dx.doi.org/10.3732/ajb.89.10.1588

LAUGHTON, A M; BOOTS, M; SIVA-JOTHY, M T (2011) The ontogeny of immunity in the honey bee, *Apis mellifera* L. following an immune challenge. *Journal of Insect Physiology* 57: 1023-1032. http://dx.doi.org/10.1016/j.jinsphys.2011.04.020

LE CONTE, Y; CORNUET, J M (1989) Variability of the post-capping stage duration of the worker brood in three different races of *Apis mellifera*. In *R Cavalloro (Ed.). Proceedings of the Meeting of the EC-Experts' Group, Udine 1988*. CEC; Luxembourg. pp. 171-174.

LOZANO, V C; ARMENGAUD, C; GAUTHIER, M (2001) Memory impairment induced by cholinergic antagonists injected into the mushroom bodies of the honey bee. *Journal of Comparative Physiology* A187: 249-254. http://dx.doi.org/10.1007/s003590100196

MACKENSEN, O (1947) Effect of carbon dioxide on initial oviposition of artificially inseminated and virgin queen bees. *Journal of Economic Entomology* 40(3): 344-349.

MAISTRELLO, L; LODESANI, M; COSTA, C; LEONARDI, F; MARANI, G; CALDON, M; MUTINELLI, F; GRANATO, A (2008) Screening of natural compounds for the control of nosema disease in honey bees. *Apidologie* 39: 436-445. http://dx.doi.org/10.1051/apido:2008022

MARTIN, S J (1994) Ontogenesis of the mite *Varroa jacobsoni* Oud. in worker brood of the honey bee *Apis mellifera* L. under natural conditions. *Experimental and Applied Acarology* 18: 87-100. http://dx.doi.org/10.1007/BF000550033

MARTIN, S J (1995) Ontogenesis of the mite *Varroa jacobsoni* Oud. in the drone brood of the honey bee *Apis mellifera* L. under natural conditions. *Experimental and Applied Acarology* 19(4): 199-210. http://dx.doi.org/10.1007/BF00130823

MCLELLAN, A R (1977) Honey bee colony weight as an index of honey production and nectar flow: a critical evaluation. *Journal of Applied Ecology* 14: 401-408.

MEDRZYCKI, P; GIFFARD, H; AUPINEL, P; BELZUNCES, L P; CHAUZAT, M-P; CLAßEN, C; COLIN, M E; DUPONT, T; GIROLAMI, V; JOHNSON, R; LECONTE, Y; LÜCKMANN, J; MARZARO, M; PISTORIUS, J; PORRINI, C; SCHUR, A; SGOLASTRA, F; SIMON DELSO, N; VAN DER STEEN, J; WALLNER, K; ALAUX, C; BIRON, D G; BLOT, N; BOGO, G; BRUNET, J-L; DELBAC, F; DIOGON, M; EL ALAOUI, H; TOSI, S; VIDAU, C (2013) Standard methods for toxicology research in *Apis mellifera*. In *V Dietemann; J D Ellis; P Neumann (Eds) The COLOSS BEEBOOK, Volume I: standard methods for Apis mellifera research. Journal of Apicultural Research* 52(4): http://dx.doi.org/10.3896/IBRA.1.52.4.14

MEIKLE, W G; HOLST, H (2006) Using balances linked to data loggers to monitor honey bee colonies. *Journal of Apicultural Research* 45: 39-41. http://dx.doi.org/10.1051/apido:2008055

MEIKLE, W G; RECTOR, B G; MERCADIER, G; HOLST, N (2008) Within-day variation in continuous hive weight data as a measure of honey bee colony activity. *Apidologie* 39: 694-707. http://dx.doi.org/10.1051/apido:2008055

MICHELETTE, E R; SOARES, A E E (1993) Characterization of preimaginal developmental stages in Africanized honey bee workers (*Apis mellifera* L). *Apidologie* 24: 431-440. http://dx.doi.org/10.1051/apido:19930410

MILUM, V G (1930) Variations in time of development of the honey bee. *Journal of Economic Entomology* 23: 441-447.

MORITZ, R F A; DIETEMANN, V; CREWE, R M (2008) Determining colony densities in wild honey bee populations (*Apis mellifera*) with linked microsatellite DNA markers. *Journal of Insect Conservation* 12: 455-459. http://dx.doi.org/10.1007/s10841-007-9078-5

MULLIN, C A; FRAZIER, M; FRAZIER, J L; ASHCRAFT, S; SIMONDS, R; VANENGELSDORP, D; PETTIS, J S (2010) High levels of miticides and agrochemicals in North American apiaries: implications for honey bee health. *PLoS ONE* 5(3): e9754. http://dx.doi.org/10.1371/journal.pone.0009754

NAUG, D; GIBBS, A (2009) Behavioural changes mediated by hunger in honey bees infected with *Nosema ceranae*. *Apidologie* 40(6): 595-599. http://dx.doi.org/10.1051/apido/2009039

NEUMANN, P; PIRK, C W W; SCHÄFER, M O; EVANS, J D; PETTIS, J S; TANNER, G; WILLIAMS, G R; ELLIS, J D (2013) Standard methods for small hive beetle research. In *V Dietemann; J D Ellis; P Neumann (Eds) The COLOSS BEEBOOK, Volume II: Standard methods for Apis mellifera pest and pathogen research. Journal of Apicultural Research* 52(4): http://dx.doi.org/10.3896/IBRA.1.52.4.19

NGUYEN, B K; SAEGERMAN, C; PIRARD, C; MIGNON, J; WIDART, J; THIRIONET, B; VERHEGGEN, F J; BERKVENS, D; DE PAUW, E; HAUBRUGE, E (2009) Does imidacloprid seed-treated maize have an impact on honey bee mortality? *Journal of Economic Entomology* 102(2): 616-623. http://dx.doi.org/10.1603/029.102.0220

NICOLAS, G; SILLANS, D (1989) Immediate and latent effects of carbon dioxide on insects. *Annual Review of Entomology* 34: 97-116. http://dx.doi.org/10.1146/annurev.ento.34.1.97

NICOLSON, S W (2008) Water homeostasis in bees, with the emphasis on sociality. *Journal of Experimental Biology* 212: 429-434. http://dx.doi.org/10.1242/jeb.022343

NICOLSON, S W; HUMAN, H (2008) Bees get a head start on honey production. *Biology Letters* 4: 299-301. http://dx.doi.org/10.1098/rsbl.2008.0034

NICOLSON, S W; NEPI, M (2005) Dilute nectar in dry atmospheres: nectar secretion patterns in *Aloe castanea* (Asphodelaceae). *International Journal of Plant Sciences* 166: 227-233. http://dx.doi.org/10.1086/427616

NICOLSON, S W; NEPI, M; PACINI, E (2007) *Nectaries and Nectar.* Springer; Dordrecht, The Netherlands. 396 pp.

ORGANISATION FOR ECONOMIC CO-OPERATION AND DEVELOPMENT (OECD) (2007) Guidance document on the honey bee (*Apis mellifera* l.) brood test under semi-field conditions Number 75. ENV/JM/MONO22.

OKADA, R; AKAMATSU, T; IWATA, K; IKENO, H; KIMURA, T; OHASHI, M; AONUMA, H; ITO, E (2012) Waggle dance effect: dancing in autumn reduces the mass loss of a honey bee colony. *The Journal of Experimental Biology* 215: 1633-1641. http://dx.doi.org/10.1242/jeb.068650

OLDROYD, B P; THEXTON, E G; LAWLER, S H; CROZIER, R H (1997) Population demography of Australian feral bees (*Apis mellifera*). *Oecologia* 111: 381-387. http://dx.doi.org/10.1007/s004420050249

OOMEN, P; DE RUIJTER, A; VAN DER STEEN, J (1992) Method for honey bee brood feeding tests with insect growth-regulating insecticides. *Bulletin OEPP/EPPO* 22: 613-616.

PACINI, E; NEPI, M (2007) Nectar production and presentation. In *S W Nicolson; M Nepi; E Pacini (Eds). Nectaries and nectar.* Springer; Dordrecht, The Netherlands. pp. 177-182.

PACINI, E; NEPI, M; VESPRINI, J L (2003) Nectar biodiversity: a short review. *Plant Systematics and Evolution* 238: 7-21. http://dx.doi.org/10.1007/s00606-002-0277-y

PANKIW, T; PAGE, R E (2003) Effect of pheromones, hormones, and handling on sucrose response thresholds of honey bees (*Apis mellifera* L.). *Journal of Comparative Physiology* A189(9): 675-684. http://dx.doi.org/10.1007/s00359-003-0442-y

PAUL, J (1975) *Cell and tissue Culture.* Churchill Livingstone; London, UK. 102 pp.

PEREZ, J L; HIGES, M; SUAREZ, M; LLORENTE, J; MEANA, A (2001) Easy ways to determine honey bee mortality using dead-bee traps. *Journal of Apicultural Research* 40(1): 25-28.

PERNAL, S F; CURRIE, R W (2000) Pollen quality of fresh and 1-year-old single pollen diets for worker honey bees (*Apis mellifera* L.). *Apidologie* 31: 387-409. http://dx.doi.org/10.1051/apido:2000130

PICARD-NIZOU, A L; GRISON, R; OLSEN, L; PIOCHE, C; ARNOLD, G; PHAM-DELÈGUE, M H (1997) Impact of proteins used in plant genetic engineering: toxicity and behavioral study in the honey bee. *Journal of Economic Entomology* 90: 1710-1716.

PIRK, C W W; DE MIRANDA, J R; FRIES, I; KRAMER, M; PAXTON, R; MURRAY, T; NAZZI, F; SHUTLER, D; VAN DER STEEN, J J M; VAN DOOREMALEN, C (2013) Statistical guidelines for *Apis mellifera* research. In *V Dietemann; J D Ellis; P Neumann (Eds) The COLOSS BEEBOOK, Volume I: standard methods for* Apis mellifera *research. Journal of Apicultural Research* 52(4): http://dx.doi.org/10.3896/IBRA.1.52.4.13

PORRINI, C; GHINI, S; GIROTTI, S; SABATINI, A G; GATTAVECCHIA, E; CELLI, G (2002a) Use of honey bees as bioindicators of environmental pollution in Italy. In *J Devillers; M-H Pham-Delegue (Eds). Honey bees: estimating the environmental impact of chemicals.* Taylor and Francis; Florence, Italy. pp. 186-247.

PORRINI, C; MEDRZYCKI, P; BENTIVOGLI, L; CELLI, G (2002b) Studies to improve the performance of dead honey bee collection traps for monitoring bee mortality. In *VIII Simposio Internazionale ICPBR on hazards of pesticides to bees, Bologna, Italy, 4-6 September 2002.* pp 29.

PORRINI, C; SABATINI, A G; GIROTTI, S; GHINI, S; MEDRZYCKI, P; GRILLENZONI, F; BORTOLOTTI, L; GATTAVECCHIA, E; CELLI, G (2003) Honey bees and bee products as monitors of the environmental contamination. *Apiacta* 38: 63-70.

PYKE, G H; WASER, N M (1981) The production of dilute nectars by hummingbird and honeyeater flowers. *Biotropica* 13: 260-270. http://dx.doi.org/10.2307/2387804

REMBOLD, H; KREMER, J-P; ULRICH, G M (1980) Characterisation of postembryonic developmental stages of the females castes of the honey bee, *Apis mellifera* L. *Apidologie* 11: 29-38. http://dx.doi.org/10.1051/apido:19800104

RICHARD, F-J; AUBERT, A; GROZINGER, C M (2008) Modulation of social interactions by immune stimulation in honey bee, *Apis mellifera*, workers. *BMC Biology* 6: 50.

ROBINSON, G E; VISSCHER, P K (1984) Effect of low temperature narcosis on honey bee (Hymenoptera: Apidae) foraging behaviour. *Florida Entomologist* 67(4): 568-570. http://dx.doi.org/10.2307/3494466

ROGERS, S R; STAUB, B (2013) Standard use of Geographic Information System (GIS) techniques in honey bee research. In *V Dietemann; J D Ellis; P Neumann (Eds) The COLOSS BEEBOOK, Volume I: standard methods for* Apis mellifera *research. Journal of Apicultural Research* 52(4): http://dx.doi.org/10.3896/IBRA.1.52.4.08

ROUBIK, D W; BUCHMANN, S L (1984) Nectar selection by *Melipona* and *Apis mellifera* (Hymenoptera: Apidae) and the ecology of nectar intake by bee colonies in a tropical forest. *Oecologia* 61: 1-10. http://dx.doi.org/10.1007/BF00379082

ROUBIK, D W; YANEGA, D; ALUJA, M; BUCHMANN, S L; INOUYE, D W (1995) On optimal nectar foraging by some tropical bees (Hymenoptera, Apidae). *Apidologie* 26: 197-211. http://dx.doi.org/10.1051/apido:19950303

ROULSTON, T H; CANE, J H (2000) Pollen nutritional content and digestibility for animals. *Plant Systematics and Evolution* 222: 187-209. http://dx.doi.org/10.1007/BF00984102

RUEPPEL, O; HAYWORTH, M K; ROSS, N P (2010) Altruistic self-removal of health-compromised honey bee workers from their hive. *Journal of Evolutionary Biology* 23: 1538-1546. http://dx.doi.org/10.1111/j.1420-9101.2010.02022.x

SAMMATARO, D; AVITABILE, A (2011) *The beekeeper's handbook.* Cornell University Press; Ithaca, New York.

SASAKI, T; ISHIKAWA, H (2000) Transinfection of *Wolbachia* in the Mediterranean flour moth, *Ephestia kuehniella*, by embryonic microinjection. *Heredity* 85: 130-135.

SCHEINER, R; ABRAMSON, C I; BRODSCHNEIDER, R; CRAILSHEIM, K; FARINA, W; FUCHS, S; GRÜNEWALD, B; HAHSHOLD, S; KARRER, M; KOENIGER, G; KOENIGER, N; MENZEL, R; MUJAGIC, S; RADSPIELER, G; SCHMICKLI, T; SCHNEIDER, C; SIEGEL, A J; SZOPEK, M; THENIUS, R (2013) Standard methods for behavioural studies of *Apis mellifera*. In *V Dietemann; J D Ellis; P Neumann (Eds) The COLOSS BEEBOOK, Volume I: standard methods for* Apis mellifera *research. Journal of Apicultural Research* 52(4): http://dx.doi.org/10.3896/IBRA.1.52.4.04

SCHLÜNS, H; CROZIER, R H (2007) Relish regulates expression of antimicrobial peptide genes in the honey bee, *Apis mellifera*, shown by RNA interference. *Insect Molecular Biology* 16: 753-759.

SEELEY, T D (1983) Division of labour between scouts and recruits in honey bee foraging. *Behavioural Ecology and Sociobiology* 12: 253-259. http://dx.doi.org/10.1007/BF00290778

SEELEY, T D (1986) *Honey bee ecology: a study of adaptation in social life.* Princeton University Press; New Jersey, USA. 49 pp.

SEELEY T D; VISSCHER, P K (1985) Survival of honey bees in cold climates: the critical timing of colony growth and reproduction. *Ecological Entomology* 10: 81-88. http://dx.doi.org/10.1111/j.1365-2311.1985.tb00537.x

SHAIBI, T; LATTORF, H M G; MORITZ, R F A. (2008) A microsatellite DNA toolkit for studying population structure in *Apis mellifera*. *Molecular Ecology Resources* 8: 1034-1036. http://dx.doi.org/10.1111/j.1755-0998.2008.02146.x

STEYSKAL, G C; MURHPY, W L; HOOVER, E M (1986) Collecting and preserving insects and mites: techniques and tools. USDA Miscellaneous Publication no. 1443.

STONER, A; MOFFETT, J O; WARDECKER, A L (1979) The effect of the Todd dead bee trap, used alone and in combination with the Wardecker waterer, on colonies of honey bees (Hymenoptera: Apidae). *Journal of the Kansas Entomological Society* 2(3): 556-560.

STONER, K A; EITZER, B D (2012) Movement of soil-applied Imidacloprid and Thiamethoxam into nectar and pollen of squash (*Cucurbita pepo*). *PLoS ONE* 7(6): e39114. http://dx.doi.org/10.1371/journal.pone.0039114

STROBER, W (1997) Monitoring cell growth. In *J E Coligan; A M Kruisbeek; D H Margulies; E M Shevach; W Strober (Eds). Current protocols in immunology.* John Wiley & Sons; Hoboken, New Jersey, USA.

TEIXEIRA, L; FERREIRA, A; ASHBURNER, M (2008) The bacterial symbiont *Wolbachia* induces resistance to RNA viral infections in *Drosophila melanogaster*. *PLoS Biology* 6: e1000002. http://dx.doi.org/10.1371/journal.pbio.1000002

THRASYVOULOU, A T; BENTON, A W (1965) Rates of growth of honey bee larvae. *Journal of Apicultural Research* 21: 189-192.

VAUDO, A D; ELLIS, J D; CAMBRAY, G A; HILL, M (2012a) Honey bee (*Apis mellifera capensis|A.m. scutellata* hybrid) nesting behaviour in the Eastern Cape, South Africa. *Insectes Sociaux* 59: 323-331. http://dx.doi.org/10.1007/s00040-012-0223-0

VAUDO, A D; ELLIS, J D; CAMBRAY, G A; HILL, M (2012b) The effects of land use on honey bee (*Apis mellifera*) population density and colony strength parameters in the Eastern Cape, South Africa. *Journal of Insect Conservation* 16: 601-611. http://dx.doi.org/10.1007/s10841-011-9445-0

VISSCHER, P K; SEELEY, T D (1989) Bee-lining as a research technique in ecological studies of honey bees. *American Bee Journal* 129: 536-539.

VISSCHER, P K; CRAILSHEIM, K; SHERMAN, G (1996) How do honey bees (*Apis mellifera*) fuel their water foraging flights? *Journal of Insect Physiology* 42(11-12): 1089-1094. http://dx.doi.org/10.1016/S0022-1910(96)00058-3

VON FRISCH, K (1967) *The dance language and orientation of bees.* Harvard University Press; Cambridge, MA, USA.

WALSH, P S; METZGAR, D A; HIGUCHI, R (1991) Chelex-100 as a medium for simple extraction of DNA for PCR-based typing from forensic material. *BioTechniques* 10: 506-513.

WANG, D I (1965) Growth rates of young queen and worker honey bee larvae. *Journal of Apicultural Research* 4: 3-5.

WANG, J (2004) Sibship reconstruction from genetic data with typing errors. *Genetics* 166: 1963-1979. http://dx.doi.org/10.1534/genetics.166.4.1963

WANNER, K W; NICHOLS, A S; WALDEN, K K O; BROCKMANN, A; LUETJE, C W; ROBERTSON, H M (2007) A honey bee odorant receptor for the queen substance 9-oxo-2-decenoic acid. *Proceedings of the National Academy of Sciences* 104(36): 14383-14388. http://dx.doi.org/10.1073/pnas.0705459104

WELLS, P H; WENNER, A M (1971) The influence of food scent on behaviour of foraging honey bees. *Physiological Zoology* 44: 191-209.

WENNER, A M; ALCOCK, J E; MEADE, D E (1992) Efficient hunting of feral colonies. *Bee Science* 2: 64-70.

WIESER, W (1973) Effects of temperature on ectothermic organisms. In *W Wieser (Ed.). Temperature relations of ectotherms. A speculative review.* Springer Verlag; Berlin, Germany. pp. 1-23.

WILLIAMS, J L (1987) Wind-directed trap for drone honey bees. *Journal of Economic Entomology* 80: 532-536.

WILLIAMS, G R; ALAUX, C; COSTA, C; CSÁKI, T; DOUBLET, V; EISENHARDT, D; FRIES, I; KUHN, R; MCMAHON, D P; MEDRZYCKI, P; MURRAY, T E; NATSOPOULOU, M E; NEUMANN, P; OLIVER, R; PAXTON, R J; PERNAL, S F; SHUTLER, D; TANNER, G; VAN DER STEEN, J J M; BRODSCHNEIDER, R (2013) Standard methods for maintaining adult *Apis mellifera* in cages under *in vitro* laboratory conditions. In *V Dietemann; J D Ellis; P Neumann (Eds) The COLOSS* BEEBOOK, *Volume I: standard methods for* Apis mellifera *research. Journal of Apicultural Research* 52(1): http://dx.doi.org/10.3896/IBRA.1.52.1.04

WILLIAMS, G R; ROGERS, R E L; KALKSTEIN, A L; TAYLOR, B A; SHUTLER, D; OSTIGUY, N (2009) Deformed wing virus in western honey bees (*Apis mellifera*) from Atlantic Canada and the first description of an overtly-infected emerging queen. *Journal of Invertebrate Pathology* 101(1): 77-79. http://dx.doi.org/10.1016/j.jip.2009.01.004

WILLIAMS, G R; SHAFER, A B A; ROGERS, R E L; SHUTLER, D; STEWART, D T (2008) First detection of *Nosema ceranae*, a microsporidian parasite of European honey bees (*Apis mellifera*), in Canada and central USA. *Journal of Invertebrate Pathology* 97 (2): 189-192. http://dx.doi.org/10.1016/j.jip.2007.08.005

WILLIAMS, G R; SHUTLER, D; LITTLE, C M; BURGHER-MACLELLAN, K L; ROGERS, R E L (2011) The microsporidian *Nosema ceranae*, the antibiotic Fumagilin-B®, and western honey bee (*Apis mellifera*) colony strength. *Apidologie* 42(1): 15-22. http://dx.doi.org/10.1051/apido/20100230.

WILSON, W T; ROTHENBUHLER, W C (1968) Resistance to American foulbrood in honey bees VIII. Effects of injecting *Bacillus larvae* spores into adults. *Journal of Invertebrate Pathology* 12: 418-424. http://dx.doi.org/10.1016/0022-2011(68)90349-2

WINNEBECK, E C; MILLAR, C D; WARMAN, G R (2010) Why does insect RNA look degraded? *Journal of Insect Science* 10(159): 1-7.

WINSTON, M L (1987) *The biology of the honey bee.* Harvard University Press; Cambridge, Massachusetts, USA, 281 pp.

WOYCIECHOWSKI, M; MORON, D (2009) Life expectancy and onset of foraging in the honey bee (*Apis mellifera*). *Insectes Sociaux* 56 (2): 193-201. http://dx.doi.org/10.1007/s00040-009-0012-6

WRIGHT, G A; MUSTARD, J A; SIMCOCK, N K; ROSS-TAYLOR, A A R; MCNICHOLAS, L D; POPESCU, A; MARION-POLL, F (2010) Parallel reinforcement pathways for conditioned food aversions in the honey bee. *Current Biology* 20: 2234-2240. http://dx.doi.org/10.1016/j.cub.2010.11.040

YAMANE, T; MIYATAKE, T (2010) Reduced female mating receptivity and activation of oviposition in two *Callosobruchus* species due to injection of biogenic amines. *Journal of Insect Physiology* 56: 271-276. http://dx.doi.org/10.1016/j.jinsphys.2009.10.011

YAÑEZ, O; ZHENG, H-Q; HU, F-L; NEUMANN, P; DIETEMANN, V (2012) A scientific note on Israeli acute paralysis virus infection of Eastern honey bee *Apis cerana* and vespine predator *Vespa velutina*. *Apidologie* 43(5): 587-589. http://dx.doi.org/10.1007/s13592-012-0128-y

ZAYED, A; PACKER, L; GRIXTI, J C; RUZ, L; OWEN, R E; TORO, H (2005) Increased genetic differentiation in a specialist versus a generalist bee: implications for conservation. *Conservation Genetics* 6: 1017-1026. http://dx.doi.org/10.1007/s10592-005-9094-5

ZHENG, H-Q; DIETEMANN, V; HU, F-L; CREWE, R M; PIRK, C W W (2012) A scientific note on the lack of effect of mandible ablation on the synthesis of royal scent by honey bee queens. *Apidologie* 43(4): 471-473. http://dx.doi.org/10.1007/s13592-011-0114-9

ZHENG, H-Q; JIN, S H; HU, F-L; PIRK, C W W (2009a) Sustainable multiple queen colonies of honey bees, *Apis mellifera ligustica*. *Journal of Apicultural Research* 48(4): 284-289. http://dx.doi.org/10.3896/IBRA.1.48.4.09

ZHENG, H-Q; JIN, S H; HU, F-L; PIRK, C W W; DIETEMANN, V (2009b) Maintenance and application of multiple queen colonies in commercial beekeeping. *Journal of Apicultural Research* 48(4): 290-295. http://dx.doi.org/10.3896/IBRA.1.48.4.10

Journal of Apicultural Research 52(4): (2013)

Journal of Apicultural Research 52(4): (2013)
DOI 10.3896/IBRA.1.52.4.11

REVIEW ARTICLE

Standard methods for molecular research in *Apis mellifera*

Jay D Evans[1*], **Ryan S Schwarz**[1], **Yan Ping Chen**[1], **Giles Budge**[2], **Robert S Cornman**[1], **Pilar De la Rua**[3], **Joachim R de Miranda**[4], **Sylvain Foret**[5], **Leonard Foster**[6], **Laurent Gauthier**[7], **Elke Genersch**[8], **Sebastian Gisder**[8], **Antje Jarosch**[9], **Robert Kucharski**[5], **Dawn Lopez**[1], **Cheng Man Lun**[10], **Robin F A Moritz**[9], **Ryszard Maleszka**[5], **Irene Muñoz**[3] and **M Alice Pinto**[11]

[1]USDA-ARS, Bee Research Lab, BARC-E Bldg 306, Beltsville MD 20705, USA.
[2]National Bee Unit, Food and Environment Research Agency, Sand Hutton, York, YO41 1LZ, UK.
[3]Department of Zoology and Physical Anthropology, Faculty of Veterinary, University of Murcia, Spain.
[4]Department of Ecology, Swedish University of Agricultural Sciences, Uppsala, 750-07, Sweden.
[5]Research School of Biology, The Australian National University, Australia Australian National Univ. Canberra, ACT 0200, Australia.
[6]Centre for High Throughput Biology, University of British Columbia, Vancouver, BC, Canada.
[7]Swiss Bee Research Centre, Agroscope Liebefeld-Posieux Research Station ALP, Bern, CH-3003, Switzerland.
[8]Institute for Bee Research, Friedrich-Engels-Str. 32, 16540 Hohen Neuendorf, Germany.
[9]Martin-Luther-University Halle-Wittenberg, Dept. of Biology, Molecular Ecology, Hoher Weg 4, 06099 Halle, Germany.
[10]George Washington University, Department of Biological Sciences, Lisner Hall 340, 20234 G St. NW, Washington, DC 20052, USA.
[11]Mountain Research Centre (CIMO), Polytechnic Institute of Bragança, Bragança, Portugal.

Received 19 March 2012, accepted subject to revision 25 September 2012, accepted for publication 11 February 2013.

*Corresponding author: Email: jay.evans@ars.usda.gov

Summary

From studies of behaviour, chemical communication, genomics and developmental biology, among many others, honey bees have long been a key organism for fundamental breakthroughs in biology. With a genome sequence in hand, and much improved genetic tools, honey bees are now an even more appealing target for answering the major questions of evolutionary biology, population structure, and social organization. At the same time, agricultural incentives to understand how honey bees fall prey to disease, or evade and survive their many pests and pathogens, have pushed for a genetic understanding of individual and social immunity in this species. Below we describe and reference tools for using modern molecular-biology techniques to understand bee behaviour, health, and other aspects of their biology. We focus on DNA and RNA techniques, largely because techniques for assessing bee proteins are covered in detail in Hartfelder *et al.* (2013). We cover practical needs for bee sampling, transport, and storage, and then discuss a range of current techniques for genetic analysis. We then provide a roadmap for genomic resources and methods for studying bees, followed by specific statistical protocols for population genetics, quantitative genetics, and phylogenetics. Finally, we end with three important tools for predicting gene regulation and function in honey bees: Fluorescence in situ hybridization (FISH), RNA interference (RNAi), and the estimation of chromosomal methylation and its role in epigenetic gene regulation.

Footnote: Please cite this paper as: EVANS, J D; SCHWARZ, R S; CHEN, Y P; BUDGE, G; CORNMAN, R S; DE LA RUA, P; DE MIRANDA, J R; FORET, S; FOSTER, L; GAUTHIER, L; GENERSCH, E; GISDER, S; JAROSCH, A; KUCHARSKI, R; LOPEZ, D; LUN, C M; MORITZ, R F A; MALESZKA, R; MUÑOZ, I; PINTO, M A (2013) Standard methodologies for molecular research in *Apis mellifera*. In *V Dietemann; J D Ellis; P Neumann (Eds) The COLOSS BEEBOOK, Volume I: standard methods for* Apis mellifera *research. Journal of Apicultural Research* 52(4): http://dx.doi.org/10.3896/IBRA.1.52.4.11

Métodos estándar para la investigación molecular en *Apis mellifera*

Resumen

Las abejas de miel han sido durante mucho tiempo un organismo clave para avances fundamentales en biología a partir de estudios de su comportamiento, comunicación química, genómica y de biología del desarrollo, entre otros muchos. Con la secuencia del genoma en la mano y herramientas genéticas mucho mejores, las abejas son ahora un blanco aún más atractivo para responder a las preguntas más importantes de la biología evolutiva, la estructura de las poblaciones y la organización social. Al mismo tiempo, los incentivos agrícolas para entender cómo las abejas caen enfermas, o evadir y sobrevivir a sus muchas plagas y patógenos, han presionado para comprender genéticamente la inmunidad individual y social en esta especie. A continuación se describen y se hace referencia a herramientas que hacen uso de modernas técnicas de biología molecular para entender el comportamiento de las abejas, su salud y otros aspectos de su biología. Nos centramos en las técnicas de ADN y ARN, en gran parte debido a que las técnicas de evaluación de las proteínas de la abeja se tratan en detalle en Hartfelder *et al.* (2013). Cubrimos las necesidades prácticas de toma de muestras de abejas, su transporte y almacenamiento, y luego se discuten una serie de técnicas actuales de análisis genético. A continuación, se proporciona una hoja de ruta para los recursos genómicos y métodos para estudiar las abejas, seguido de protocolos estadísticos específicos de la genética de poblaciones, la genética cuantitativa y la filogenia. Finalmente, se termina con tres herramientas importantes para predecir la regulación génica y la función en las abejas melíferas: la hibridación in situ fluorescente (FISH), la interferencia de ARN (iARN), y la estimación de la metilación cromosómica y su papel en la regulación epigenética de los genes.

西方蜜蜂分子研究的标准方法

摘要

通过行为、化学通讯、基因组和发育生物学等方面的研究，蜜蜂已经成为用于在生物学基础研究领域取得重大突破的一种重要模式生物。结合已有的基因组序列和多种改进的遗传学工具，蜜蜂已经越加成为回答进化生物学、种群结构和社会性结构等方面重大问题极具吸引力的研究目标。与此同时，农业上为了了解蜜蜂如何因于病害或者避开和幸存于多种害虫和病原菌的危害，也促进了对这一物种个体和社会免疫的遗传学理解。以下我们介绍和引用了一些运用现代分子生物学技术研究蜜蜂行为、健康、以及其它方面生物学的工具。Hartfelder等2013已对研究蜜蜂蛋白做了详细的论述，因此我们将重点放在DNA和RNA技术上。本文也包含了在蜜蜂采样、运输和保存过程中的实际需要，并讨论了当前的一系列遗传分析技术。然后我们提供了研究蜜蜂时所需的基因组资源和方法的路线图，以及群体遗传学、数量遗传学和系统发生学研究中特定的统计学方法。最后，我们以预测蜜蜂基因调控和功能的三个重要工具收尾：荧光原位杂交（FISH）、RNA干扰（RNAi）和染色体甲基化及其在表观遗传基因调控中作用的估算。

Keywords: *Apis mellifera*, pollination, disease, development, genomic, Colony Collapse Disorder, population genetics, methylation, RNA interference, RNAi, Southern Blot, Northern Blot, *In situ* Hybridisation, DNA extraction, Next-generation sequencing, mitochondrial DNA, microsatellite, quantitative PCR, COLOSS, *BEEBOOK*, honey bee

Table of Contents

Table of Contents cont'd

1. Sample management

1.1. Introduction

In order to best reflect honey bee biology, data generated from molecular-genetic studies should reflect as closely as possible the state of honey bee tissues, entire bees, or colonies just prior to sampling. This fact places a premium on collecting and storing samples in a way that retains this state. Although technological developments in molecular biology allow for a great diversity of insights from collected bee samples, it is often forgotten how much these insights are hampered by errors in the collection, storage and processing of samples (Chernesky *et al.,* 2003). These problems are especially evident when data from different studies or laboratories are compared (Birch *et al.,* 2004). The only solution to this is optimization of collection, storage and primary processing protocols, so as to minimize the influence of sample degradation on the molecular analyses and the reliability of the data. As is often the case, cues can be taken from other areas of biology, notably the medical field, where such practices are widely adopted (Valentine-Thon *et al.,* 2001; Verkooyen *et al.,* 2003).

A secondary consideration is that a sample may be used for several different analyses; proteins, nucleic acids, fats and lipids, metabolites *etc.*, requiring a collection and processing protocol suitable for all compounds analysed. Usually this means that the sample management conditions follow the requirements for the least stable of the compounds, which for bee research is usually the RNA. RNA is highly sensitive to degradation by robust RNAse enzymes found in all cells, unless the sample is stabilized with RNAse-inhibiting additives and/or frozen as soon as possible. Given the necessity of RNA analyses for many questions related to bees and their parasites and pathogens (e.g., de Miranda *et al.* 2013), field-appropriate methods for stabilizing RNA are required.

1.2. Sample collection

The optimum strategy for collection and transport of bee samples depends partly on what type of sample is collected. Bees, pupae, larvae and eggs can be sampled whole or as field-dissected components, such as heads, thoraxes, abdomens, guts, endophalli, semen, ovaries *etc.* Many bee viruses are shed in large amounts in the guts, as are many bacterial and protozoan pathogens (Shimanuki, 1997; Fries, 1997). Faeces may therefore be a good marker for the infection status of the whole bee, although care has to be taken to distinguish between passively acquired/passaged microbes and true tissue infections. Faeces also allow bees to be sampled repeatedly, and non-destructively. It may therefore be useful for determining the virus status of queens (Hung, 2000), especially if these are a major source of infection of the worker population (Chen *et al.,* 2005b; Fievet *et al.,* 2006), or for following disease progression in individual bees.

Below are suggestions for the collection of different types of bee samples. In all cases a priori decisions are all needed with respect to the use of chemical stabilizers, collection cards and the temperatures during transport and storage.

1.2.1. Adult bees
1.2.1.1. Nurse bees
Inspect each frame in a colony and find a frame with sealed and unsealed brood which is covered by adhering nurse bees and then take the frame out of the colony.

1.2.1.2. Foraging bees
Block the hive entrance where foraging bees are accumulating and collect the returning foraging bees.

1.2.2. Pupae
1. Cut out a section of sealed brood, to be transported whole. Such a brood section can be sent through the post, although with the caveat that such transport away from the hive might affect bee or parasite gene activities.
2. Uncap brood cells, lift pupae by their neck by curling fine curved forceps underneath their heads and transfer to a suitable transport medium, either individual microcentrifuge tubes or collection cards (see section 1.3.5).

1.2.3. Larvae
1. Cut out a section of open brood and transport in a temperature -humidity controlled box, to prevent dehydration.
2. Remove larvae from the comb using either a blunt grafting needle (small larvae) or soft forceps (large larvae) and transfer to individual microcentrifuge tubes or collection cards.

1.2.4. Eggs
1. Cut out a section of comb with eggs and transport in a temperature-humidity controlled box, to prevent dehydration.
2. Remove eggs using a blunt needle and transfer individually or in bulk to microcentrifuge tubes or collection cards.

As an alternative for the rapid collection of massive amounts of eggs and early embryos:
1. Strike soundly a frame containing early-stage bees onto a sterile surface twice.

This releases over half of the eggs and embryos held by that frame,
2. Brush or lift into a new vessel.

While uncapped honey will drip via this method, if done at the right intensity, older uncapped larvae will remain in their cells.

1.2.5. Extracted guts

1. Grab the stinger and last integument of adult worker bees firmly with a pair of fine, straight forceps.
2. Pull backwards gently, removing the whole hindgut and midgut.
3. Transfer guts to individual microcentrifuge tubes or collection cards (see section 1.3.5.).

1.2.6. Drone endophallus and semen

1. Turn drone upside down and grip laterally between thumb and index finger.
2. With the other hand, gently but persistently squeeze the abdomen of the drone dorso-ventrally, exerting pressure backwards, until the endophallus is extruded from the drone.
3. Apply more severe pressure, again backwards, to avert the endophallus and, for mature drones, cause ejaculation of semen.
4. Cut off the entire endophallus with scissors, or collect the exposed semen (brown-red colour) and/or seminal fluid (translucent white) with a sterile micropipette.
5. Collect the material individually in microcentrifuge tubes or on collection cards.

1.2.7. Faeces

If destructive sampling is allowed:

1. Remove the whole gut from individual bees (see above) and expel faecal mass.

If repeated sampling is required:

1. Place adult bees into a Petri dish until defecation has occurred.
2. Collect faeces individually or pooled in microcentrifuge tubes or on collection cards.

1.2.8. Dead bee samples

Many bee disease experiments involve bee death as a parameter. Dead bee samples from such experiments are, of course, valid material for analysis. They should be treated like freshly killed material and frozen as soon as possible to minimize the effects of decay on RNA integrity, using the collection methods appropriate for the sample type, as given above. Dead bee traps attached to hives are suitable for collecting such bees and should be emptied daily to minimize the effects of decomposition.

Passive surveys also involve dead bee samples, in this case those sent in by beekeepers for *post-mortem* analysis of the cause of colony death. These bees will have been dead long enough for decomposition and drying to have severely affected the integrity of the RNA, including viral RNAs. Such degradation can severely affect the accuracy and reliability of detecting and quantifying individual RNAs (Bustin and Nolan, 2004; Fleige and Pfaffl, 2006; Becker *et al.,* 2010). This means

that only positive results from such samples are informative, since negative results can be either due to the absence of virus or the degradation of the RNA.

It is possible to adjust for differential RNA degradation in the different samples with quantitative RT-qPCR techniques, by using host internal reference gene levels for normalizing the virus titers (Dainat *et al.,* 2011) and setting the threshold for template detection with the most degraded sample, so that all samples are evaluated by the same degradation criteria. How to determine the detection thresholds using RT-qPCR assays is covered in section 4.4.

1.3. Sample transport

Sample transport from the collection site to the laboratory is the most critical step in sample management, since this is where the integrity of the sample is most easily compromised (Chen *et al.,* 2007). Sample integrity can be preserved to different degrees with the following methods, given in order of effectiveness. The gold standard for sample collection and transport is to freeze on-site, but this is not always possible. All alternatives are basically aimed at getting the samples as quickly and conveniently as possible into a freezer, with minimum degradation. The most useful tool for transporting frozen material is a liquid nitrogen-based 'dry shipper', which is specifically developed and approved for international shipment of biological samples at ultra-low temperatures (-150°C). The best can hold these temperatures for more than one week. Other options, for more local transport, are (dry) ice-boxes and portable/car freezers. Courier and mail services are less reliable, both with respect to the maintenance of temperature and the duration of transport.

1.3.1. Freezing with dry ice

- Samples: all.
- Conditions: freeze instantly; keep frozen throughout transport using blocks of dry ice in a cooler.
- Transport: restricted transport; dry ice must be replenished ca. every 48 hours.
- Processing: transfer samples to freezer.
- **Pros:** gold standard; fast.
- **Cons:** very expensive; complex field operation.

1.3.2. Freezing with 'wet' ice

Short-term field-storage on 'wet' (frozen water or ice packs) ice is cheap and very practical for many field-studies and surveys. The samples should be frozen as soon as possible, ideally within hours, and kept frozen continuously until RNA processing (a complete frozen transport chain). If a complete frozen transport chain cannot be guaranteed, then a chemical stabilizing agent (see section 1.3.4.) should be used to prevent degradation of the RNA by RNAses, until the samples enter a frozen transport chain. The most important rule for RNA preservation is to keep the samples as cold as possible, as long as possible and to avoid thawing the sample after it has been frozen unless it is to extract RNA.

- Samples: all.
- Conditions: collect in freezer bags, store on wet ice.
- Transport: cold transport; wet-ice; < 12 hours.
- Processing: transfer samples to freezer.
- **Pros:** simple; fast; cheap field operation.
- **Cons:** heavy, expensive mail transport, leaks due to thawing.

1.3.3. Live transport

Bees can also be transported live, which allows them to be sent much more quickly, cheaply and reliably by post than frozen samples. One drawback is that the stress of live transport may affect the expression of host genes, and possibly virus replication, which should be taken into account when planning experiments.

1. Adult bees can be transported live 1) In a well-ventilated bee shipping box containing queen candy and a sponge soaked in water glued to the bottom of the box or 2) in units of 10-15 bees in commercial queen cages with queen candy. Such queen cages are readily available to most beekeepers.
- Samples: adults.
- Conditions: room temperature.
- Transport: < 48 hours.
- Processing: freeze on arrival.
- **Pros:** simple; fast; suitable for beekeepers.
- **Cons:** stress during transport.

2. Pupae can be transported live 1) as a section of capped brood in a well-ventilated bee shipping box, preferably in a warm environment to prevent chilling, 2) as queen cells for queen pupae in a specialized temperature-humidity controlled queen-cell transport container, available from beekeeping suppliers. Such cells should be handled with great care, as developing queen pupae are very sensitive to disturbance, or 3) as a whole frame in a specialized carrier box for frames, available from beekeeping suppliers, or in a swarm box/nucleus hive.
- Samples: pupae.
- Conditions: room temperature.
- Transport: < 48 hours.
- Processing: remove samples from comb and freeze.
- **Pros:** simple, fast.
- **Cons:** pupae may emerge during transport.

3. Larvae and eggs can be transported live 1) as a section of comb, in a temperature and humidity-controlled box or 2) as a whole frame in a specialized carrier box for frames, or in a swarm box/nuc.

- Samples: larvae; eggs.
- Conditions: controlled temperature and humidity.
- Transport: less than 48 hours.
- Processing: remove samples from comb and freeze.
- **Pros:** simple, fast.
- **Cons:** expensive by mail, unsealed larvae are subject to temperature stress and starvation.

1.3.4. Chemical stabilizers

There are a number of chemicals that can be used to help stabilize nucleic acids during transport. Their purpose is to inhibit nucleases, especially the resilient RNAses, and in doing so destroy all enzymatic activity in the sample. So, if the final assays include natural enzymatic activity, these stabilizers should be avoided. For similar reasons, many stabilizers are also incompatible with serological detection methods, such as ELISA.

A large excess (5-fold by weight) of stabilizer should be added to ensure a high enough concentration within the tissues for inhibiting RNAses. It is also essential that the solution penetrates the tissues completely to abolish all RNAse activity. This is a major difficulty for aqueous stabilizers, which cannot penetrate the hydrophobic insect exoskeleton. These are therefore only suitable for extracted tissues, eggs and small larvae, unless bodies are partially disrupted at the start. Organic preservatives, such as 100% ethanol, have much more effective penetration of the exoskeleton and are therefore better for stabilizing whole adult bee samples. Although 100% ethanol is suitable for preserving RNA destined for short-fragment RT-qPCR-based assays, storage in 70% ethanol has been shown to result in strong degradation (Chen *et al.,* 2007). However, recent data using a short amplicon (124 bp) diagnostic for Deformed wing virus (DWV) in a Taqman assay (Chantawannakul et al., 2006), showed no loss of DWV signal after adult bees were stored for 4 weeks in 70% EtOH at room temperature compared to snap frozen controls (G. Budge, unpublished data). RNA can also be stabilized by high concentration sulphate salt solutions (Mutter *et al.,* 2004), of which RNAlater® (Qiagen) is the best known. A generic version can be made as follows:

700 g   di-ammonium sulfate
40 ml   0.5M EDTA (pH 8.0)
25 ml   1M tri-sodium citrate (di-hydrate salt; 29.4 g/100 ml)
1l   sterile water
~1.3 l total volume

Once stabilized, RNase activities will be inhibited and samples can be stored for up to 1 month at 4°C, and long-term at -20°C or -80°C with minimal degradation. The stabilizer should be removed from the bee sample prior to homogenization and RNA extraction.

1. 100% ethanol
- Samples: whole adult bees; pupae; large larvae; tissues.
- Use: 5 volumes by weight.
- Storage: 1 month at room temperature or lower.
- Processing: Remove ethanol and process samples as normal.
- **Pros:** Cheap; effective penetration.
- **Cons:** Evaporation; possible transport restrictions; heavy; incompatible with serological assays .

2. RNAlater® & generic equivalent
- Samples: tissues; eggs; small larvae.
- Use: 5 volumes by weight.
- Storage: 1 month at room temperature, or lower.
- Processing: Remove stabilizer and process samples as normal.
- **Pros:** Non-hazardous; effective penetration.
- **Cons:** Expensive (except generic version); heavy.

It is possible to use RNAlater® for darkened pupae and adult bees, if they are crushed into a paste or cut into 5mm sections (Chen *et al.*, 2007). This is laborious and risks losing virus particles and RNA to the stabilizing solution, but may be required in certain circumstances. In such cases, the crushed bees should be centrifuged at 1,000 rpm for 5 minutes at 4°C before removing the stabilizer and processing the crushed bee tissues.

1.3.5. Sample collection cards

Samples can also be dried on filter paper-based collection cards. In this case the molecules are stabilized primarily through desiccation, rather than low temperature, so thorough drying during sampling and low humidity during storage is essential for this method. The FTA™-cards produced by Whatman are furthermore impregnated with chemicals to prevent bacterial or enzymatic degradation of nucleic acids (Becker *et al.*, 2004; Rensen *et al.*, 2005). The method is ideal for liquid samples (blood, urine, sputum *etc.*) but has also been used for insect samples (Harvey, 2005) including honey bee larvae, pupae, extracted tissues and mites. Such filter-dried samples can be analysed for all manner of compounds (Jansson *et al.*, 2003; Chamoles *et al.*, 2004; Li *et al.*, 2005; Zurfluh *et al.*, 2005) including RNA (Karlson *et al.*, 2003; Prado *et al.*, 2005). The major advantages are the ease and reduced costs of collection, transport, labelling and long-term storage at room temperature, reducing the requirements for freezer space, boxes and tubes (Kiatpathomchai *et al.*, 2004; Harvey, 2005; Karlson *et al.*, 2003; Rensen *et al.*, 2005; Prado *et al.*, 2005). The major disadvantages are the uneven distribution of target across the filter paper and the gradual loss of target during prolonged storage (Chaisomchit *et al.*, 2005). Samples collected on collection cards should therefore also be processed as soon as possible, by cutting out the entire dried sample and soaking this in an appropriate buffer, as recommended by the FTA™-card protocol, for the recovery of nucleic acids.

FTA™ collection cards
- Samples: Tissues; faeces; eggs; larvae; pupae; mites.
- Use: Squash sample on card and air-dry.
- Storage: At room temperature in dry container. Not in freezer.
- Processing: Cut out sample and soak directly in extraction buffer for 15 minutes. Proceed as for fresh samples.
- **Pros:** Excellent preservation; light; easy storage and indexing; versatile; preservation of faeces.
- **Cons:** Expensive; variable processing; uneven distribution across card; not suitable for adult bees, not suitable for bulk samples.

1.4. Long-term sample storage

The critical factors for long-term sample preservation, as with degradation in the weeks after collecting, are minimizing the activity of nucleases. This can be achieved by a combination of:

1.4.1. Freezing

Freezing at -80°C is the gold-standard for long-term storage of bee samples intended for RNA analysis. Freezing at -20°C also provides good storage for preserving the quality of bee samples. However, significant to complete degradation of RNA can occur within days in dead bees kept at 4°C (Chen *et al.*, 2007; Dainat *et al.*, 2011). It is therefore strongly recommended to transfer frozen bees to the -80°C freezer immediately after samples are brought back from the field to the laboratory, if analysis is not initiated immediately.

1.4.2. Drying

Apart from drying soft bee stages and tissues on collection cards, bulk samples of whole bees can also be freeze-dried, or lyophilized. Lyophilization is a convenient way to store samples long-term at room temperature and preserves the chemical integrity of most compounds, although some functional activity may well be lost. Freeze-drying/lyophilization requires a specialized instrument that draws a vacuum while the samples are kept below the point where solid and liquid phases can co-exist (below -50°C), so that the ice sublimates, *i.e.* changes directly to vapour without melting first. Any biological sample can be lyophilized and the instructions for this come with the particular lyophilizing apparatus. It should be noted here that reconstituted dried tissue is fundamentally different from frozen wet tissue, with different and more variable recovery efficiencies for the different biomolecules than for fresh tissues. Lyophilized samples are stored at room temperature in a sealed box with desiccating packages, to prevent re-hydration.

1.4.3. Chemical stabilizers

There are several chemical agents that inhibit RNAses and thus reduce RNA degradation during handling and storage (see Section 4.4.4.). They do not provide any additional benefit to frozen samples, but can be useful for storing samples temporarily at room temperature. Their effectiveness varies and they do not prevent degradation absolutely (Chen *et al.*, 2007) but they are useful if minor degradation can be tolerated and the samples can be processed within a few months of collection.

2. Sample processing

2.1. Introduction

The initial processing of a sample is another key step in ensuring the uniformity and reliability of an assay. Nevertheless, generally little attention is paid to optimizing this part of the protocol for both maximum recovery of the target molecule(s) and for reducing variability. In general, the shorter and faster the protocol the better, since each additional step will contribute to the overall error and reduce the recovery efficiency, both of which compromise results. Here we will describe generalities of sample processing before independent chapters describing RNA and DNA extraction from samples.

2.2. Sample homogenisation

A highly variable step in sample processing is sample homogenization, not only between different homogenization options but also between different samples using the same protocol. The choice of homogenization method depends on the sample type and number of bees per sample. There are numerous options outlined below.

2.2.1. Bead-mill homogenizers

These are the best option for uniform and reproducible homogenization of small (individual) bee samples. The samples are mixed with glass, ceramic or steel 1-3 mm beads and extraction buffer in sturdy disposable plastic tubes and shaken at high velocity in a machine. They also provide consistent cell wall disruption of bacteria and other microbes for parasite/pathogen or microbiome surveys.

- **Pros:** Low-medium volume; rapid; uniform; no cross-contamination.
- **Cons:** Generally only suitable for small bee samples (1-10 bees).

1. Place single bee in a 2 ml screw-cap microcentrifuge tube.
2. Add four 2 mm glass beads.
3. Add 500 µl ice-cold buffer.

For medium-large volume beadmills, increase the number of bees, beads and buffer in proportion to the maximum allowable volume of the disposable container.

4. Make sure that the bead mill is balanced, if this is a requirement.
5. Shake for 5-10 minutes at the highest setting.

2.2.2. Blender

An excellent, cheap alternative to the beadmills, especially for large volumes is homogenisation with a blender.

- **Pros:** Large volume; rapid; uniform.
- **Cons:** Cross-contamination risk due to re-use of blender; incompatible with organic solvents; corrosion of blender due to salts.

1. Add between 30-200 frozen bees to blender.
2. Add 500 µl ice-cold buffer per bee.
3. Homogenise by gradually raising the blender settings, for about 5 minutes total homogenization.

2.2.3. Paint shaker

A paint shaker (e.g. Automix shaker; Merris Engineering ltd) is a surprisingly efficient method of grinding bulk samples which has been used for various matrices including soil , grains, rice, wheat, honey, and bees (Woodhall *et al.*, 2012; Budge, unpublished data). The method is completely scalable ranging using polypropylene wide-mouth environmental bottles (Nalgene) ranging in size from 60 ml to 2000 ml.

- **Pros:** Large volume; easy; cheap; no cross-contamination; high throughput.
- **Cons:** Large piece of equipment required.

1. Place 30-1000 frozen bees in an appropriately sized bottle (Nalgene) containing 5 x 25.4 mm stainless steel ball bearings.
2. Dry grind on the paint shaker for 8 minutes until the sample is sufficiently disrupted.
3. Add the required volume of extraction buffer depending on the protocol.
4. The addition of 1% Antifoam B (Sigma) to GITC, GHCl or CTAB extraction buffers can aide buffer recovery and reduce cross contamination.
5. Wet grind for a further 4 minutes.
6. Spin at 6000 g for 5 mins.
7. Recover supernatant.

2.2.4. Mortar and pestle

Traditional manual method for pulverizing samples.

- **Pros:** Medium volumes; cheap; low maintenance.
- **Cons:** Cross-contamination risk; time consuming; lack of uniformity.

1. Place 1-30 bees in a pre-frozen mortar of appropriate size.
2. Add liquid nitrogen to cool samples to well below freezing.

3. Grind the bees to a powder using an appropriate sized, pre-frozen pestle.
4. Transfer the powder to a plastic tube or bottle.
5. Add 500 µl extraction buffer per bee.
6. Shake tube until the powder has suspended fully in the buffer.

2.2.5. Mesh bags

Mesh bags are sturdy plastic bags with a small pore fine mesh inside. The sample is placed on one side of the mesh, ground from the outside and the homogenate is collected from the other side of the mesh, filtering out large particles.

- **Pros:** Medium-large volume; easy; cheap; no cross-contamination.
- **Cons:** Lack of uniformity; split bags.

1. Place up to 30 frozen bees in a disposable mesh-bag (e.g., www.bioreba.com; #430100).
2. Add 500 µl buffer per bee.
3. Flash-freeze the entire bag in liquid nitrogen.
4. Remove from liquid nitrogen.
5. Wait 30 seconds.
6. Pulverize contents by grinding the bag with a large pestle for 2 minutes on a hard surface, taking care not to damage the bag.
7. Massage the bag until completely thawed.
8. Remove one (or more) 100 µl aliquots of homogenate.
9. As an alternate method, described in section 4.3.2. samples can be crushed in disposable mesh bags using a heavy rolling pin.

2.2.6. Micropestle

You will need individual disposable micro pestles that fit microcentrifuge tubes. These can be bought or made by heating a 1000 µl blue tip in a flame and moulding it into a disposable pestle in a microcentrifuge tube while it cools.

- **Pros:** Single bees; cheap; low maintenance.
- **Cons:** Time consuming; lack of uniformity.

1. Grind a frozen bee tissue or larval sample with the micropestle in a microcentrifuge tube.
2. Discard pestle.
3. Add 500 µl buffer per bee.
4. Mix with a vortex.

2.2.7. Robotic extraction

Companies produce robotic extraction stations to facilitate high-throughput analysis of samples. Comparisons between several such systems, or between automated and manual extraction, generally find little difference in terms of assay sensitivity and reproducibility (Rimmer *et al.*, 2012; Bruun-Rasmussen *et al.*, 2009; Agüero *et al.*, 2007; Petrich *et al.*, 2006; Knepp *et al.*, 2003, but see Schuurman *et al.*, 2005). Such systems are generally only suitable for easily disrupted, soft tissues or samples.

- **Pros:** Single bees; rapid; high throughput; uniform; low cross contamination risk.
- **Cons:** Expensive; inflexible protocols; soft tissues only.

Follow manufacturers' protocol for sample processing.

3. DNA extraction and analysis

3.1. Introduction

Isolating and analysing an organism's DNA is key for developing insights into species or strain identification, for uncovering variants useful in breeding or a more thorough understanding of biology, and for discovering the microbes carried by individuals. DNA extraction methods must be robust for small amounts of starting material even if that material has become degraded. They must deliver extracted DNA of sufficient quality, purity, and quantity for downstream efforts ranging from target identification (e.g., via the Polymerase Chain Reaction, PCR, below in section 6.3.1.), sequence analysis, and cloning, among others. Below are tested protocols for common DNA analyses of diverse bee samples, starting with the isolation and purification of DNA. Isolating DNA from tissues can be accomplished using a variety of commercial kits, or via procedures built on standard disrupting and separating agents as below. Here we describe protocols made from primary ingredients, since this is illustrative of the critical components in these and pre-made extraction protocols.

3.2. Genomic DNA extraction from adult bees

3.2.1. DNA extraction using CTAB

This protocol is for the extraction of DNA from bee abdomens and/or the thorax, using a lysis buffer containing CTAB, a compound that is able to separate polysaccharides from other cell materials. The choice of tissues avoids eye contaminants such as pigments, which can inhibit PCR and other downstream applications. The method can be scaled down for the extraction of *Varroa destructor* mites (see the *BEEBOOK* paper on varroa (Dietemann *et al.*, 2013) for details on sampling) or bee embryos and up for larger larvae and pupae (see section 1.2. for their collection). Volumes should be adjusted accordingly based on sample volume (i.e. initial grinding in 5X sample volume of buffer, ca.25-> 200 µl). The subsequent two extraction protocols are simpler, but the CTAB procedure is excellent for problematic samples and is flexible in terms of tissue disruption, separation, and rescue of nucleic acids.

1. Extract only the abdomen and/or thorax if possible. If a whole animal is extracted, use a Qiagen or similar column following manufacturer's protocol for final purification of extracted DNA in order to reduce pigments that can inhibit genetic assays.

2. Put tissue from a single bee in a 1.5 ml microcentrifuge tube.

3. Add 500 µl of CTAB + 2 µl 2-mercaptoethanol (0.2%).

CTAB buffer:

- 100 mM Tris-HCl, pH 8.0
- 1.4 M NaCl
- 20 mM EDTA
- 2% w/v hexadecyl-trimethyl-ammonium bromide (CTAB)

This buffer both stabilizes nucleic acids and aids in the separation of organic molecules. See MSDS as CTAB is a potential acute hazard.

4. Homogenize with pestle.

5. Add 50 µg proteinase K and 25 µl of RNase cocktail.

While this step is optional, proteinase K improves yields by disrupting cell and organelle boundaries and is critical for extraction of DNA from many microbes.

6. Vortex briefly to mix.

7. Incubate at 55-65°C from several hours to overnight. Invert occasionally during incubation (e.g. once every 30 minutes for the first two hours).

8. Centrifuge for 1 min at maximum speed (~14,000 rpm).

Unwanted tissue debris will form a pellet at the bottom of the microcentrifuge tube.

9. Transfer liquid to fresh tube, leaving tissue debris pellet behind.

10. Add equal volume phenol:chloroform:isoamyl alcohol (25:24:1).

11. Invert several times (10-20 times) to mix then put on ice for 2 min.

12. Spin at full speed (~14,000 rpms) for 15 min at 4°C.

13. Transfer upper phase to fresh tube.

14. Add 500µl cold isopropanol + 50 µl 3 M NaOAc.

15. Vortex to mix, then incubate at 4°C > 30 min.

Samples can be stored at ambient temperature at this point for several days if needed for transport or timing, otherwise 4°C is best.

16. Spin at full speed (~14,000 rpms) for 30 min at 4°C.

17. Carefully decant liquid from DNA pellet.

18. Add 1 ml 4°C 75% EtOH. Tap vortex briefly to loosen pellet.

19. Spin at full speed for 3 min at 4°C.

20. Decant liquid from pellet.

21. Air dry pellet about 10 minutes to evaporate all residual traces of alcohol.

Do not over dry pellet, as it will be hard to resuspend.

22. Resuspend in 50-100 µl nuclease-free water (overnight at 4°C).

23. Check DNA quantity and integrity on an agarose gel.

24. First, prepare TBE gel buffer (an aqueous solution with a final working concentration of 45 mM Tris-borate and 1 mM EDTA). This is often prepared first as a '5x' concentration comprised of 4 g Tris base (FW = 121.14) and 27.5 g boric acid (FW = 61.83) dissolved into approximately 900 ml deionized water. Add 20 ml of 0.5 M EDTA (pH 8.0) to this solution and adjust the solution to a final volume of 1l.

Confusingly, the 'working solution' of this buffer for most uses is as 0.5x = a 1/10 dilution of the stock buffer.

25. For a 1.5% agarose gel on a large-format gel rig, add 3 g of sterile agarose to 200 ml TBE buffer in a 500 ml or larger Erlenmayer flask, microwave at high heat for ca. 45 s (without boiling). For smaller gel rigs the volume of the gel can be from 50 to 100 ml. Take flask out and swirl, then heat in the microwave again until at full boil for 45 seconds, monitoring to avoid spillover. The agarose must fully dissolve so the liquid is perfectly clear

26. Let the solution cool while swirling every minute until the flask can be held for several seconds without unbearable heat

27. While hot, pipette in 10 µl ethidium bromide solution (EtBr, 0.5 mg/ml, used with caution as EtBr is a carcinogen and mutagen) and swirl until mixed

28. Pour into a horizontal gel rig and insert plastic combs holding ca. 10 µl of sample each

29. Let the gel solidify fully; gels can be wrapped in plastic wrap for longterm storage (overnight in place or for days at 4°C).

30. Mix 5 µl of the extraction solution with 2 µl of a 40% weight/volume sucrose load buffer (made as 4 g sucrose and 25 mg bromophenol blue in 10 ml distilled water)

31. Submerge gel in a rig containing 0.5 x TBE, remove gel comb and load the 7 µl of sample/dye mix in separate wells using DNA molecular weight standards (e.g., 500 bp molecular ruler, www.biorad.com)

32. Draw the DNA across the gel toward the anode/positive charge at ca. 100 V depending on the gel rig size and specifications.

33. Monitor via the blue bromophenol blue stain movement (which tracks a DNA size fragment of ca. 300 bp in a 1.5% gel), stopping the gel and visualizing the DNA using ultraviolet light when it has progressed enough.

34. DNA can also be quantified via a spectrophotometer such as the Nanodrop (www.nanodrop.com), following manufacturer's protocol: Briefly, after calibration 1 µl of nucleic acid solution is placed onto a cleaned pedestal, the lid is closed and a reading is taken prior to cleaning by wiping the pedestal in preparation for the next sample. The machine will estimate concentration using the equation dsDNA: A^{260} 1.0 = 50 ng/µl.

35. Store at -20°C or below.

3.2.2 DNA extraction using Qiagen Blood and Tissue DNA kits

This is a reliable extraction method using a commercial kit sold by Qiagen (www.quiagen.com), it is suitable for honey bee guts, small larvae or tissues from larger larvae or adults (avoid using the compound eyes).

1. Place 50 mg honey bee material in a centrifuge tube and mince thoroughly on ice with a mini pestle
2. Add 180 µl Buffer ATL and 20 µl Proteinase K at the provided concentration
3. Vortex 30 seconds and incubate at 56°C for 1 hour, vortexing for 30 seconds after 30 min
4. Premix equal volumes of Buffer AL and ethanol (96-100%), mixing enough to provide 400 µl per sample plus 10% extra
5. Vortex samples 30 seconds and add 400 µl AL/EtOH mix each, vortex again 30 seconds
6. Pipette all into DNeasy Mini spin column nested in a 2 ml collection tube.
7. Centrifuge at ≥ 8000 rpm in a microcentrifuge (6k g). Discard flow-through and collection tube
8. Place spin column in new 2 ml collection tube, add 500 µl Buffer AW1, centrifuge 1 min at ≥ 8000 rpm. Discard flow-through and collection tube
9. Place spin column in new 2 ml collection tube, add 500 µl Buffer AW2, centrifuge 3 min at ≥ 14000 rpm. Discard flow-through and collection tube
10. Remove spin column, checking to be sure ethanol is gone and place into a clean 1.7 ml centrifuge tube
11. Add 200 µl Buffer AE to the centre of the membrane, incubate at room temperature and then centrifuge for 1 min at ≥ 8000 rpm. Eluted DNA will be in tube. Check quantity by Nanodrop or agarose gel as in section 3.2.1 above.

3.2.3. DNA extraction using Chelex

The Chelex method (Walsh *et al.*, 1991) provides a very rapid way to protect DNA from degradative enzymes and from some of the potential contaminants that might inhibit experiments. In principle, the Chelex resin will trap salts needed by degradative enzymes, leaving DNA in solution. In practice, Chelex extractions can be prone to degradation, and should be kept in the freezer when not in use, or these extractions should be used within 24 hours of extraction. If the extracted tissues contain pigments and other inhibitors for downstream experiments, it is often successful to dilute the Chelex extraction 1:9 with distilled water before use. Finally, when drawing from these extractions it is important to pipette from the top of the aqueous layer, avoiding the resin itself. Below is a recipe that works well for legs from adult bees or beetles, for whole varroa mites, or for other tissues of about that size.

1. Add two posterior legs into Eppendorf tubes.
2. Allow them to dry until the EtOH evaporates.
3. Transfer to each tube:
- 100 µl of Chelex® (5% solution in water),
- 5 µl of proteinase K (10 mg/ml).
4. Incubate the samples in a thermocycler with the following program:
- 1 h at 55°C,
- 15 min at 99°C,
- 1 min at 37°C,
- 15 min at 99°C,
- Pause at 15°C.

3.3. DNA detection using southern blots with DIG labelling

Southern blotting was invented by Edward M Southern as a means for detecting specific nucleotide sequences in a complex mixture and determining the size of the restriction fragments, which are complementary to a probe. Southern blotting combines transfer of restriction-enzyme-digested and then electrophoresis-separated DNA fragments from a gel to a membrane and subsequent detection by probe hybridization. A variety of non-radioactive methods have been developed to label probes for detection of specific nucleic acids. The Roche Applied Science DIG system is a simple adaptation of enzymatic labelling and offers a non-radioactive approach for the safe and efficient labelling of probes for hybridization reactions.

3.3.1. Restriction enzyme digestion and agarose gel electrophoresis

This step is carried out in order to array chromosomal sections across a one-dimensional space so that unique sections can be probed for matches to a query sequence. In principle, the targeted gene will be embedded in a single chromosomal segment flanked by specific sequences that match the restriction enzyme used.

1. Digest 5-10 µg of genomic DNA in a volume of 30 µl with an appropriate restriction enzyme by setting up reaction as follows:
- 3 µl 10X buffer,
- 0.3 µl of BSA if needed (this will be on the restriction enzyme label),
- 3 µl enzyme (10U/µl),
- 5-10 µg genomic DNA,

Add sterile water to reach a total volume of 30 µl.

Generally, enzymes that cut frequently in the target genome are used here (e.g., 'four-cutters' that cut at a specific four-base-pair sequence, an event expected to occur ca. once every several hundred base-pairs).

2. Allow the digestive reaction to go for overnight at 37°C (or temperature appropriate to your specific enzyme).

3. Run the full 30 μl of reaction mixture with 3 μl 6X loading dye on 1% agarose gel (see section 3.2.1) containing ethidium bromide (1 μg/ml) for 2 hours at 100 volts. Include one lane of a DIG-labelled DNA Molecular Weight Marker at the appropriate level.

4. Take a picture of the digestion.

5. Depurinate the agarose gel for exactly 10 min in 0.25 M HCl if DNA fragment > 4 kb.

6. Denature the gel in freshly made denaturing solution (0.5M NaOH, 1.5 M NaCl) for 2 x 15 min at RT, slowly shaking on rotating shaker.

Denaturation of the DNA into single strands allows hybridization with a probe possible.

7. Rinse the gel with sterile water.

8. Neutralize the gel in neutralizing solution (0.5 M Tris-HCl, pH 7.4, 1.5 M NaCl) for 2 x 15 min, slowly shaking on rotating shaker.

9. Equilibrate gel in 20X SSC for 10 min.

3.3.2. Assembly of the transfer setup and transfer of DNA from gel to membrane

DNA is here pulled from the gel into a nylon membrane by capillary action pulled by the positive charge of the membrane. Once in contact with the membrane, DNA is attached using high-voltage cross-linking.

1. Set up capillary transfer using 20X SSC as a transfer agent: Inside a baking glass dish filled with 20X SSC, place a glass plate that is elevated by four rubber stoppers that is slightly larger than the gel.

2. Cover the glass plate with a piece of wick-blotting paper that has to be long enough so that it is in contact with the 20X SSC transfer solution.

The buffer flows up the wick-blotting paper by capillary action, then through the gel to the membrane.

3. Smooth out the air bubbles between the glass and the blotting paper by gently rolling with a glass pipette.

4. Place the gel facing down on the wet blotting paper.

5. Cut a small triangular piece from the top left-hand corner to simplify orientation.

6. Smooth out the air bubbles.

7. Cut one piece of positively charged nylon membrane to match size of the gel.

8. Soak the membrane in water for 2-3 min to wet and then float in 20X SSC.

9. Gently place the membrane on the top of the gel.

10. Mark well positions on the membrane.

11. Smooth out the air bubbles.

12. Cut 4-5 sheets of Whatman 3MM paper to the same size as the gel and place on top of the membrane.

13. Place a stack of paper towels on top of the Whatman 3MM papers.

14. Add a 200-400 g weight to hold everything in place.

15. Allow the DNA to transfer for 10-16 hours.

16. After transfer, rinse the membrane briefly in 6X SSC.

17. Immobilize DNA to the membrane by UV cross-linking (120,000 microjoules per cm^2). Membrane is now ready for labelling (section 3.3.3.).

3.3.3. Synthesis of DIG-labelled DNA probe

DIG-labelled probes offer a method to identify where probes have attached on the membrane (e.g.., the location of their targeted DNA match). These DIG probes are an alternative to highly regulated and more dangerous radio isotopic probes.

1. Mix the DIG-labelled PCR reaction components from the Roche Applied Science PCR DIG Labelling Mix with the probe template as follows:

- 5 μl PCR Buffer (10X),
- 5 μl PCR DIG Labelling Mix,
- 0.5 μl Upstream Primer (25 μM),
- 0.5 μl Downstream Primer (25 μM),
- 0.5-1 μl Template (plasmid DNA, 10-100 pg, or genomic DNA, 1-50 ng),
- 0.75 μl Enzyme Mix,
- Add sterile water until total reaction volume is equal to 50 μl.

2. Set the annealing temperature of PCR reaction to reflect the predicted annealing temperature of the primers, also reported at the time of purchase.

3. The kit contains a post-hoc check for probe labelling efficiency that is recommended.

3.3.4. Hybridizing the DIG-labelled DNA Probe to DNA on the Blot

This procedure relies on the Roche Applied Science DIG Easy® Hyb, DIG Wash and Block Buffer Set, (with the fluorescent reporter CSPD®).

1. Pre-warm an appropriate volume of DIG Easy® Hyb solution® to the hybridization temperature.

2. Pre-hybridize membrane in a small volume of pre-warmed DIG Easy® Hyb solution (20 ml if in a 200 ml hybridization tube).

3. While the membrane is pre-hybridizing, denature 10 μl of DIG -labelled DNA by boiling for 5 min.

4. Rapidly cool on ice.

5. Add appropriate amount of denatured probe to give you (25 ng/ml) into DIG Easy® Hyb solution.

6. Incubate with agitation in a hybrid oven at 55-58°C for overnight.

7. Wash membrane in 25-50 ml Washing Solution-1 (2X SSC, 0.1% SDS) 2X for 5 min at room temperature under constant agitation.

8. Wash membrane in 25-50 ml Washing Solution-2 (0.1% SSC, 0.1% SDS) 2X for 5 min at 68°C under constant agitation.

9. Wash membranes briefly (1-5 min) in 25 ml of 1X Washing Buffer provided in DIG Wash kit.

10. Incubate membranes for 30 min in 1X Blocking Solution diluted in maleic acid buffer (supplied in the kit).

11. Incubate membrane in Anti-body solution for 30 min.

To make anti-body solution, add 1 µl anti-body to 20 ml 1X blocking solution.

12. Wash membrane in 1X Washing buffer 2X for 15 min.

Make sure membrane is immersed in the Washing buffer.

13. Prepare 20 ml 1X Detection Buffer.

14. Equilibrate membrane in 20 ml 1X Detection Buffer for 2-5 min.

15. Transfer the membrane with DNA side facing up to a Plastic wrap that is at least twice the size of the membrane.

16. Apply 1 ml of CSPD®, ready-to-use (about 20-25 drops) to the membrane.

17. Quickly cover the membrane with the plastic wrap.

18. Incubate for 5 min at RT.

19. Drain off excess buffer by gently brushing across the top of the membrane covered by plastic wrap with a paper towel.

20. Tape the membrane into a film cassette.

21. Close the cassette and incubate at 37°C for 10 min to enhance the luminescent reaction.

22. Remove the film for development using a standard x-ray film developer.

4. RNA methods

4.1. Introduction

Analyses based on RNA have two major advantages over DNA analyses. First, they are by definition restricted to a step in the expression of proteins from an organism's genome. This means that RNA pools are generally far less complex than are pools of DNA representative of the organism's entire genome, and that a quantitative estimate of different RNA's can provide a useful surrogate for the proteins produced at that time point for a specific organism or tissue within an organism. Second, since nearly all of the recognized viral threats to honey bee exist without a DNA stage, these threats are only visible via RNA analyses. These arguments make RNA the resource of choice for many honey bee analyses; despite greater concerns over storage and preservation of tissues.

A common strategy is to extract total nucleic acids directly in strongly denaturing buffers, so as to inactivate RNAses immediately during homogenisation. RNAses have numerous disulphide bridges. This makes them very stable in a very wide range of conditions, such that strong denaturants are required to permanently inactivate them. Heat, detergents (sodium dodecyl sulphate), organic solvents (phenol), proteinases, chaotropic salts (guanidine isothiocyanate), reducing agents (β-mercaptoethanol; dithiothreitol) and nucleic acid protecting compounds (CTAB; cetyl trimethylammonium bromide) are some of the more common methods used to inactivate RNAses. The nucleic acids can be purified from other compounds with affinity columns, magnetic bead-linked nucleic acid binding agents or by precipitation with alcohol or lithium chloride. The most common, quickest and most reliable combination is a chaotropic salt/β-mercaptoethanol extraction buffer, followed by purification on disposable affinity columns (Verheyden *et al.*, 2003). The main disadvantage of RNA precipitation (with 2 volumes ethanol, 1 volume isopropanol or with 6M LiCl) is that many undesirable compounds often co-precipitate with the nucleic acid, requiring further precipitations or washes to clean the sample. There are many excellent commercial RNA extraction kits available, based on one or more of these principles. However, their performance in comparative tests varies greatly, depending on the organism, tissue type and nucleic acid extracted (Konomi *et al.*, 2002; Knepp *et al.*, 2003; Wilson *et al.*, 2004; Schuurman *et al.*, 2005; Labayru *et al.*, 2005). Below are two protocols, representing the most common approaches to RNA extraction.

4.2. Affinity column purification

The processing consists of making a primary homogenate from 1-30 bees, purifying RNA from one (or more) aliquots of the homogenate using affinity columns, and measuring the RNA concentration. β-mercaptoethanol is toxic so processing should be done in a fume hood. Prepare all the buffers and tubes before starting.

The protocol below is based on the columns marketed by Qiagen or generic equivalents. The maximum recommended amount of tissue per column is 20 mg. More than 20 mg tissue causes the column to bind too much protein, reducing the yield and quality of the nucleic acid. Bees, pupae and large larvae weigh between 100-180 mg each, and so need to be homogenised first in a primary extract, from which a volume equivalent to 20 mg tissue is then processed on the affinity columns. A suitable denaturing buffer for this primary extract is a Guanidine Iso-Thio Cyanate (GITC) buffer, which has similar properties to the Qiagen RLT buffer:

1. Mix the GITC buffer:
- 5.25 M guanidinium thiocyanate (guanidine isothiocyanate),
- 50 mM TRIS.Cl(pH 6.4),
- 20 mM EDTA,

- 1.3% Triton X-100,
- 1% β-mercaptoethanol.
2. Place exact, pre-determined number of frozen bees in the homogenizer of choice.
3. Per bee, add the following amount of GITC buffer:

Bee	Weight	Buffer	Total volume
Worker bee	120 mg	500 µl	600 µl
Drone	180 mg	700 µl	900 µl
Worker pupa	160 mg	650 µl	800 µl
Drone pupa	240 mg	1000 µl	1200 µl

With these extract volumes, 100 µl extract is approximately 20 mg bee tissue

4. Mix:
- 100 µl bee extract,
- 350 µl RLT buffer + 1% β-mercaptoethanol.
5. Proceed according to the Plant RNA extraction protocol (see Qiagen instructions booklet).

Inclusion of the Qia-shredder column step is not required, but significantly increases yield and purity of nucleic acid.

6. Elute in 100 µl nuclease-free water.
7. Determine nucleic acid concentration and purity (see section 8.4.; "Nucleic Acid Quality Assessment").
8. Store as two separate 50 µl aliquots at -80°C, one for working with and one for storage.
9. Include a 'blank' extraction (*i.e.* an extraction of purified water) after every 24 bee samples, to make sure none of the extraction reagents have become contaminated.

4.3. Acid phenol RNA extraction from adult bees

The below recipes use an acid-phenol phase separation for isolating RNA from DNA and other tissue components. The TRIzol® (Invitrogen™) protocol is the most commonly used, and widely available, method of acid-phenol extraction of RNAs. However, using a generic lysis and acid-phenol buffer (e.g. section 4.3.2) provides a cost effective alternative than TRIzol®, and allows a great reduction in the use of the caustic chemical phenol for pooled samples. We use honey bee abdomens because they provide representation of the microbes and immune components of the honey bee, while avoiding pigments in the eye which can inhibit downstream enzymatic reactions. The procedure is also appropriate for larvae, whole adult bees and pupal RNA extractions, if volumes are scaled upward, i.e. doubled, to reflect the volume of the sample, for the latter two.

4.3.1 TRIzol® extraction

Advanced preparation: You will need RNase-free bench, pipettes, barrier tips, pestles and 1.5 ml microcentrifuge tubes. Bench tops and other glass and plastic surfaces can be treated to remove RNAse contamination by application of RNAse Zap (Ambion), following manufacturer's protocol. Disposable tips, pestles, and microcentrifuge tubes should be purchased RNase–free. You will need cold 75-80% ethanol and 100% isopropanol, both nuclease-free; and a pre-chilled centrifuge (at 4°C for 30 min) for Step 9. Have ready at room temperature, the TRIzol® and other reagents needed. It is recommended to use a vented fume hood for safety when working with TRIzol® and chloroform. It is also very important to work quickly with bee tissue, as it is possible that RNA will degrade if bees thaw for ten or more minutes (Dainat *et al.*, 2011).

In a very sterile (RNAase-free) environment:
1. Add 500 µl of TRIzol® to frozen bee abdomens in 1.5 ml tubes.
2. Mash the tissue until completely homogenized with a pestle and shaking.

All soft tissues should be disrupted completely. Remove pestle and scrape it off along the rim of the microcentrifuge tube. Sample should be viscous.

3. Centrifuge at 5,000 rpm for 1 min to pellet debris.
4. Transfer the TRIzol® suspension to a fresh tube, leaving out the chitinous debris pellet.
5. Add another 500 µl TRIzol® and invert several times to mix.
6. Add 200 µl chloroform.
7. Shake vigorously for 15 sec.

Do not vortex! This may increase DNA contamination in your sample.

8. Incubate at RT for 2-3 min.
9. Spin at 4°C for 15 min at ~14,000 rpm.

NOTE: Be especially diligent about avoiding RNases from this point on!

10. Label a fresh set of RNase-free microcentrifuge tubes.
11. Carefully remove tubes from centrifuge.
12. Use a 1 ml pipette tip with pipettor set at 550 µl to draw off the upper phase and transfer it to a fresh tube.

Carefully avoid the interface (one product that ensures a clean break between phases is the Phaselock gel (5 Prime Inc.) and could be used here).

13. Add 500 µl 100% Isopropanol.
14. Invert 3-5 times to mix gently.
15. Incubate at RT for 10 min.
16. Centrifuge at 4°C for 10 min at full speed (~12,500rpm), placing all tubes in the same rotation (e.g., hinge facing away from arc) so pellet location will be consistent.

17. Carefully siphon off liquid using a 1 ml pipette tip.

Observe the pellet (white) so you do not inadvertently aspirate it into the tip! Be cautious as it may dislodge and float.

18. Add 1 ml of cold 75-80% nuclease-free EtOH.
19. Invert several times to mix.
20. Spin at 4°C for 5 min at full speed.
21. Carefully decant liquid using a 1 ml pipette tip, avoiding the pellet and tilting the tube so no alcohol remains at the bottom of the tube covering the pellet.
22. Let tubes air dry in a clean area just until the EtOH has evaporated (~20-30 min).
23. Resuspend RNA pellet in 100 µl of RNase-free water.
24. Incubate at 55°-60°C for 10 min in water bath, ideally with shaking or flicking tubes for 10 seconds once during this time.
25. Quantify and validate RNA integrity using spectrophotometer (Nanodrop, section 3.2.1), following manufacturer's protocols, or run a small amount on 1-2% agarose gel (see section 3.2.1) to verify RNA quality. This can be accomplished by looking for degradation products migrating as a diffuse smear below the sharp 28S and 18S ribosomal RNA bands, which migrate at an analogous rate to ca. 1.75 and 2 kb double-stranded DNA markers. Alternatively, an Agilent 2100 RNA chip will provide both an accurate quantification and a measure of RNA integrity.
26. Freeze for storage at -80°C for long term storage, -20°C if you plan to use the RNA within 24 hrs.
27. Yields should be at least 100 µg (1 µg/µl) total RNA.

4.3.2. Bulk extraction of RNA from 50-100 whole bees using the acid-phenol method

For colony-level surveys of bee microbes, including pathogens, it is often important to analyse a bulk sample of bees in order to ensure a more accurate view of colony loads (most parasites and pathogens are not found uniformly across all bees in the hive, see section 4. 'Obtaining adult workers for laboratory experiments' of the *BEEBOOK* paper on maintaining adult workers in vitro laboratory conditions (Williams *et al.*, 2013) and the *BEEBOOK* paper on statistics (Pirk *et al.*, 2013) for details on how to sample bees). Similarly, if a colony-level genetic or phenotypic (gene-expression) trait is desired it is often better to generate an estimate that is the average across many colony members rather than a few selected bees. Extractions from a sample of tens of bees can be costly since volumes of reagents must be scaled up. The below protocol greatly reduces the most costly (and hazardous) ingredient used in RNA extractions, acid-phenol, and otherwise generates equivalent yields and purity to the TRIzol® extraction described above.

1. Put whole frozen bees (stored at -80°C since death) into a disposable extraction bag (e.g. www.Bioreba.ch) and add 500 µl lysis/stabilization solution (section 4.3.3) per bee (i.e. for 50 bees add 25 ml solution).
2. Mash until homogenized using a rolling pin, leaving the bag partly open initially to allow air to escape.
3. Allow to settle ~10 min so bubbles go down.

You can mash 10 or so bags consecutively at a time. By the time #10 is finished, the bubbles in #1 have subsided. Keep pending bags on ice in bucket.

4. Transfer 620 µl of extraction liquid into a pre-labelled 1.5 ml micro tube.

Note: It is advisable to save subsamples of the lysed tissues as a reserve (Store at-80°C).

5. Add 380 µl acid phenol.
6. Vortex 30 sec to mix well.
7. Incubate 10 min in a 95°C hot block.

Place weight on top of tubes to prevent lids from popping open.

8. Wearing goggles and a lab coat carefully remove weight and then transfer the tubes from hot block to pre-chilled rack in ice.

It is best to keep hot block in hood to contain the phenol.

9. Incubate on ice for 20 min.
10. Bring to RT.
11. Add 200 µl chloroform.
12. Shake vigorously for 1 sec.
13. Incubate at RT 3 min.
14. Centrifuge at 14,000 rpm for 15 min at 4°C.
15. Transfer 500 µl upper phase to fresh tube.
16. Add equal volume of isopropanol (100%).
17. Invert ten times to mix.
18. Incubate at RT 15 min.
19. Centrifuge at 10,000 rpm for 10 min at 4°C.
20. Carefully decant liquid from pellet.
21. Wash w/ 1 ml of cold 75% EtOH.
22. Centrifuge at 10,000 rpms for 2 min at 4°C.
23. Carefully decant liquid from pellet.
24. Spin 1 min.
25. Remove excess alcohol with pipette tip.
26. Air dry completely.
27. Resuspend in 200 µl nuclease-free H_2O.
28. Solubilize for 10 min at 55°C.
29. Store at -80°C.
30. Yields should be higher than 200 µg (1 µg/µl) total RNA, and extractions should be stable for > 5 years. RNA degradation can be checked using an Agilent Bioanalyzer or by 2% agarose gels looking for the co-migrating large ribosomal RNA's as a sign of largely intact RNA. If extractions are to be shipped or kept at temperatures above -50°C for more than 48 hours, RNA should first be precipitated in an equal volume of isopropyl alcohol, shipped in that state, then suspended starting at step 22 above.

4.3.3. RNA lysis/stabilization buffer

1. Fill a 1l beaker with 300 ml of nuclease-free water and insert a large magnetic stir bar.

2. Following safety procedures (http://www.sciencelab.com/ msds.php?msdsId=9927539) add:
   94.53 g guanidine thiocyanate ($CH_5N_3 \cdot CHNS$; MW = 118.16) (Sigma #50981), 30.45 g ammonium thiocyanate (CH_4N_2S; MW = 76.12) (Sigma #43135), 33.4 ml of 3M sodium acetate (NaOAc), pH 5.5 ml ultrapure molecular biology-grade (USB # 75897 or Sigma #71196).

3. Stir until completely dissolved.

4. Pour into 1l graduated cylinder and bring up to 550 ml with nuclease-free water.

5. Pour from graduated cylinder into autoclave-safe desired 1l bottle.

6. Add: 50 ml glycerol ($C_3H_8O_3$; MW=92.09 g/mol) (Sigma # G6279) and 20 ml Triton-X 100 (Sigma #T8787).

7. Autoclave on liquid cycle for 15 min with slow exhaust.

8. Remove from autoclave immediately, cool and store at 4°C.

This makes a total volume of 620 ml.

4.4. RNA quality assessment

The next step is to determine the condition of the RNA sample, prior to any assay. The three critical parameters are quantity, quality and integrity (i.e. absence of degradation). Quantity and quality are usually assessed by spectrophotometry (Green and Sambrook, 2012), by comparing the peak absorbance at 260 nm (nucleic acids), 280 nm (proteins) and 230 nm (phenolic metabolites). A number of companies now market spectrophotometers and fluorometers that provide a complete UV absorbance profile from 1 µl of sample, from which the concentration of the nucleic acid can be determined, as well as its purity with respect to protein and metabolite contaminants. However, nucleic acid integrity can only be determined by running an electrophoretic trace profile, and assessing the degree of degradation by comparison of different nucleic acid size classes. The most comprehensive RNA quality analysis is through a chip-based microelectrophoresis system that provides a complete electrophoretic trace of the RNA sample which is used to quantify the integrity of the RNA, as well as the amount and purity (Bustin, 2000). Agilent, Qiagen, Invitrogen and BioRad market such systems. However, for fresh samples or those preserved with stabilizers or in a frozen transport chain, with little expected degradation, a simple UV absorbance spectrum is usually sufficient.

- Read the absorbance of an RNA sample at 230 nm, 260 nm and 280 nm.
- A^{260} of 1.0   = 40 ng/µl ssRNA
                         = 37 ng/µl ssDNA
                         = 50 ng/µl dsDNA

- A^{260}/A^{280} < 2.0 indicates contamination with proteins.
- A^{260}/A^{230} < 2.0 indicates contamination with phenolics.

4.5. cDNA synthesis from total RNA

Most downstream measurements of RNA traits rely on the complementary DNA (cDNA) generated by back-transcribing RNA using a commercially available reverse transcriptase such as 'Superscript' (Invitrogen). Reverse transcription is the most delicate step in RT-PCR. This step is very sensitive to inhibitors and contaminants in the sample (Ståhlberg *et al.,* 2004b) such that the efficiency can vary between 0.5% and 95%. This efficiency is furthermore also strongly affected by both the absolute and relative amounts of target RNA in a sample, especially at very low levels of target (Ståhlberg *et al.,* 2004a; 2004b), and by a variety of reaction conditions (Singh *et al.,* 2000).

To minimize this variability, the RNA concentrations should be measured accurately by spectrophotometry (Qubit; Invitrogen), and a constant amount added to the cDNA reactions. If the RNA concentration is very low (< 10 ng/µl), 100 ng neutral carrier tRNA can be added to the reaction prior to addition for cDNA synthesis stabilize reverse transcription and detection reliability. The final major parameter to optimise is the cDNA primer. Different target-specific cDNA primers (such as used in One-step RT-qPCR reactions), can have significantly different reverse transcription reaction efficiencies, which will affect the quantitative estimation of the targets in the sample (Bustin, 2000). A useful, practical approach is therefore to first prepare a fully representative cDNA 'copy' of the entire RNA population, using random 'hexamer' (6-nucleotide) primers. Such a complete cDNA population will have much less quantitative biases between different targets due to variable reverse transcriptase reaction efficiencies, allowing for more accurate quantitative comparison and normalisation between different targets. However, cDNA prepared with random primers can sometimes overestimate the original amount of target RNA (Zhang and Byrne, 1999). Another commonly used technique for sampling RNA pools is to use poly-dT primers targeting the polyadenylated stretch found at the 3' end of most messenger RNAs and also on most of the honey bee viruses.

4.5.1 Reverse Transcription of RNA

The following is a robust reverse-transcription protocol for generating cDNA that is fully representative of the original RNA population:

1. Mix:
- 0.5 µg sample RNA template,
- 1 ng exogenous reference RNA (*e.g.* Ambion RNA250),
- 1 µl 50 ng/µl random hexamers,
- 1 µl 10 mM dNTP,
- up to 12 µl RNAse free water.

2. Heat the mixture to 65°C for 5 min and chill quickly on ice.
3. Add:
    - 4 µl 5X First-Strand Buffer,
    - 2 µl 0.1 M DTT,
    - 1 µl (200 units) of M-MLV RT.
4. Mix by pipetting gently up and down.
5. Centrifuge briefly to collect the contents at the bottom of the tube.
6. Incubate 10 min at 25°C.
7. Incubate 50 min at 37°C.
8. Inactivate the reaction by heating 15 min at 70°C.
9. Dilute the cDNA solution ten-fold with nuclease-free water before using in PCR assays.

4.6. Qualitative RT-PCR for honey bee and pathogen targets

Detection by PCR can be "qualitative", *i.e.* recording only the presence or absence of the target cDNA, by analysing the accumulated "end-point" PCR products after the PCR is completed. The sensitivity of the assay can be raised or lowered as desired by, respectively, increasing or decreasing the number of amplification cycles. Usually PCR does not exceed 40 cycles, which is theoretically sufficient to detect a single molecule of the target DNA in the original template, when analysing the end products by agarose gel electrophoresis. Consider the following rough calculation:

- Assuming perfect doubling with each amplification cycle.
- 2^0 molecules (*i.e.* 1 molecule) prior to PCR = 2^{40} molecules after 40 cycles of PCR.
- 2^{40} molecules of a 100 bp DNA fragment (mw ~ 61,700 g/mol)

= 1.1×10^{12} molecules   x 1 mol/6.0221415×10^{23} molecules

= 1.8×10^{-12} mol   x 61,700 g/mol

= 1.1×10^{-7} g   = 110 ng DNA

Normally, 20 ng DNA is easily visible as a single band on an ethidium bromide-stained agarose gel. Even when allowing for imperfections in the amplification, 40 cycles are therefore theoretically more than sufficient to detect a single molecule in a reaction.

However, such extreme sensitivity is rarely required in practical or even most experimental settings. Furthermore, by aiming for absolute detection at the level of a single molecule of target DNA, the detection system becomes axiomatically susceptible to high rates of detection error: both false positives (accidental amplification of contaminating molecules) and false negatives (non-detection of a single molecule due to amplification insufficiencies).

By raising the detection threshold a few orders of magnitude, to around 1,000 molecules per reaction (~2^{10} molecules prior to PCR) it is possible to produce detectable amounts of target DNA (~2^{40}

molecules) with 30 cycles of amplification (2^{10+30} molecules), again assuming perfect doubling each cycle. This avoids most of the risk of both types of detection errors, since chance contamination events of singular molecules (false positive results) are now below the detection threshold and there is sufficient initial target DNA in the reaction to avoid accidental non-detection (false negative results). A few more cycles beyond 30 can be added to compensate for the imperfections in the PCR efficiency. This means that 35 amplification cycles should be the upper limit for most practical applications. Beyond 35 cycles, the rapidly increasing risk of detection errors outweighs the marginal gains in sensitivity.

4.7. Quantitative RT-PCR for honey bee and pathogen targets

Detection of specific PCR products can also be made continuously as the PCR proceeds (*i.e.* in 'real time'). In this case the cycle number at which the accumulated PCR products reach a fluorescence detection threshold, read after each cycle by laser optics, can be very accurately related to the initial amount of target in the reaction, through the use of exponential algorithms and internal and external quantitation standards (Bustin *et al.*, 2009; 2010). This is the basis for real-time quantitative PCR (qPCR). The great advantage of real-time qPCR, apart from the accurate quantitation of the initial amount of target DNA in the reaction, is that the diagnostic threshold for qualitative detection can be set after the reactions have taken place, or at a number of different levels, from the same data set. This is useful if different diagnostic sensitivities are required for different experimental or reporting purposes, or for quality control management purposes.

There are numerous methods for qPCR in the literature, and this approach has been used for measuring gene activity in honey bees and all of their major parasites and microbial associates. The primary difference in those cases will come in the specific primers used for amplification and in some cases in changes to the chemistry or thermal conditions. One main decision point is between using SYBR green or another non-specific fluorescent marker that measures (amplified) DNA non-discriminately versus reporters that target specific amplified products directly such as TaqMan probes (Applied Biosystems; e.g. Chen *et al.*, 2004). There is considerable debate over the merits of each approach. Assays using Taqman® chemistry and other internal probe methodologies are inherently more specific than those using Sybr chemistry, due to the additional match required in the probe sequence. Therefore, Taqman® assays are more prone to Type II errors (false negative), where a negative result is returned despite the sample being positive (perhaps due to slight modification in the probe region within the sample). Sybr-based assays are more likely to return a Type I error (false positive), due to difficulties in distinguishing between low positive signal at the threshold of detection and non-specific binding. The errors for both methods can be minimized after careful preparatory work.

4.7.1. One-Step versus Two-Step RT-PCR

The buffer conditions for reverse transcription and PCR are largely compatible, which means that the two steps can be coupled in a single tube reaction, with the incubation conditions favouring first the reverse transcription, and then the PCR. Such 'One-Step' RT-PCR kits reduce the number of manipulations and associated errors, both qualitative and quantitative. The disadvantage is that they use up the sample RNA at a much higher rate than 'Two-Step' RT-PCR, where the cDNA is produced independently in a separate reaction. One-Step RT-PCR is also generally less sensitive than Two-Step RT-PCR, since the reaction conditions are not optimised exclusively for reverse transcription, and cannot account easily for variable reverse transcription efficiencies between different assays/primers (Bustin, 2000; Bustin *et al.,* 2009). The main disadvantage of Two-Step RT-PCR is that the additives included in the reverse transcription buffer to enhance primer binding and reaction efficiency, can also encourage the production of non-specific PCR products during PCR, which affects the quantitation accuracy. To minimize such effects, cDNA should be diluted ten-fold with water before being used for Two-Step RT-PCR.

Commercial One-Step or Two-Step RT-qPCR kits have proprietary reagent mixtures that are optimised for the corresponding recommended cycling profiles. Different kits therefore perform differently with particular primers and cycling profiles (Grabensteiner *et al.,* 2001), and the choice of RT-PCR kit is therefore also part of the optimization procedure. To take maximum advantage of such pre-optimized systems, the most practical approach is to design the assays and primers to fit these optimized recommendations, whenever this is possible.

4.7.2. One-Step RT-qPCR

The following is a robust, standard One-Step RT-qPCR protocol for amplifying and quantifying targets <400bp in length, using SYBR-green detection chemistry, and starting with an RNA template:

1. Mix:
   - 3 µl 5 ng/ µl RNA,
   - 0.4 µl 10 µM Forward primer (0.2 µM final concentration),
   - 0.4 µl 10 µM Forward primer (0.2 µM final concentration),
   - 0.4 µl* 10 mM dNTP* (0.2 mM final concentration*),
   - x µl OneStep Buffer + SYBR-green (as per manufacturer),
   - y µl nuclease-free water,
   - r µl reverse transcriptase (as per manufacturer),
   - z µl Taq polymerase (as per manufacturer),
   - 20 µl total volume.

(* dNTPs are often included in the optimized buffer)

2. Incubate in real-time thermocycler:
   - 95°C 5 min,
   - 35 cycles of:
     95°C 10 sec,
     58°C 30 sec *read for qPCR.

3. For Melting Curve analysis of the products, incubate:
   - 95°C 1 min,
   - 55°C 1 min,
   - +0.5°C increments for 5 sec, with reads from 55°C to 95°C.

In addition, DNA sequencing of the amplified products is recommended.

4.7.3. Two-Step RT-qPCR

The following is a robust, standard qPCR protocol for amplifying and quantifying targets <400bp in length, using SYBR-green detection chemistry, and starting with a cDNA template:

1. Mix:
   - 3 µl cDNA (pre-diluted 1/10, in nuclease-free water),
   - 0.4 µl 10 µM Forward primer (0.2 µM final concentration),
   - 0.4 µl 10 µM Forward primer (0.2 µM final concentration),
   - 0.4 µl* 10 mM dNTP* (0.2 mM final concentration*),
   - x µl Buffer + SYBR-green (as per manufacturer),
   - y µl nuclease-free water,
   - z µlTaq polymerase (as per manufacturer),
   - 20 µl total volume.

(* dNTPs are often included in the optimized buffer)

2. Incubate in real-time thermocycler:
   - 95°C for 5 min,
   - 35 cycles of:
     95°C for 10 sec,
     58°C for 30 sec* read (qPCR),

3. For Melting Curve analysis of the products, incubate:
   - 95°C for 1 min,
   - 55°C for 1 min,
   - +0.5°C increments for 5 sec, with reads from 55°C to 95°C.

4.7.4. Two-step Quantitative PCR for high-throughput assays

The below variant of qPCR is for a 96-well plate format on the CFX96 real time system (BioRad) or related machines, and works for both bee transcripts and pathogen targets. The primary difference over the prior protocol is that this one is initiated with cDNA generated in a non-specific way, rather than from *de novo* reverse-transcription for each viral and/or host test and control (as shown in the previous section).

1. Mix 1x SsoFast EvaGreen® supermix (BioRad) with 3 mM of each forward and reverse primer for a given target (final volume 4 µl).
2. Add 1 µl (~8 ng) of cDNA template to specific wells.
3. Use the following cycling conditions:
   - 97°C for 1 min,
   - 45 (maximum 50) cycles of:
     95°C for 2 sec,
     60°C for 5 sec,

Melt curve from 65-95°C at +0.5°C/5 sec increments.

4. Verify amplicon melting points for every positive sample. Amplicons from positive controls and initial samples should be cloned into pGEM-T Easy vector (Promega) to verify sequence.

5. Run four distinct no-template controls on the plate to monitor for contamination and non-specific amplification.

6. Standard curves should be run using a recombinant plasmid dilution series of the primer targets from 10^1 to 10^8 copies, providing a linear equation to calculate the copy number in each sample using $10^{Cq-b/m}$, where Cq = quantification cycle, b = y-intercept, and m = slope.

4.7.5. Multiplex RT-(q)PCR

Often there is a need to amplify several target RNAs from a single sample. This can be done in several parallel 'uniplex' reactions, or in a single 'multiplex' reaction containing the primer pairs for all different targets (Williams *et al.*, 1999; Wetzl *et al.*, 2002; Syrmis *et al.*, 2004; Szemes *et al.*, 2002). Detection of the different amplicons is usually by size difference and electrophoresis for qualitative PCR, or by target -specific labelled probes in real-time quantitative PCR (Mackay *et al.*, 2003). A number of such qualitative multiplex PCR protocols have been designed for honey bee viruses as well (Chen *et al.*, 2004b; Topley *et al.*, 2005; Grabensteiner *et al.*, 2007; Weinstein-Texiera *et al.*, 2008; Meeus *et al.*, 2010). Real-time qPCR can also be multiplexed, by using a range of different fluorophores and excitation-reading laser channels. This is useful for minimizing between-reaction variability, if both target and internal reference standards can be amplified simultaneously, in the same reaction. Other uses are to distinguish between variants of the same gene or pathogen. Currently, up to four different targets can be detected and quantified simultaneously in qPCR.

However, there are many serious disadvantages of multiplexed PCR methods that may ultimately outweigh the advantages of consolidation and efficiency:

- Multiplex RT-PCR is considerably less sensitive than uniplex RT-PCR (the reagents will be exhausted by several targets instead of just one), as much as several orders of magnitude depending on the number of targets (Herrmann *et al.*, 2004).
- Optimization of multiplex (q)PCR assays is considerably more complicated than uniplex (q)PCR, due to the large number of primers and probes that need to be optimized simultaneously for absence of undesired interactions. An alternative to multiplex RT-(q)PCR that avoids many of the assay optimization problems due to the complex primer mixes is the Multiplex Ligation Probe Amplification method. .
- The PCR products need to be resolved on size or by fluorophore, before they can be quantified, nullifying many of the gains in efficiency and cost.

- Amplification (and thus quantification) of one target can be strongly affected by the prior amplification of more abundant targets in the reaction, either through competition for a limited pool of reagents, or through inhibition of the PCR reaction at later stages by the PCR products produced during earlier cycles, which sequester most of the polymerase (Santa Lucia, 2007).

For these reasons, it is often much more practical and simple to use uniplex RT-PCR, even for large volume and throughput projects.

4.8. Primer and probe design

There are numerous primer design software packages around to help design primers and, if appropriate, TaqMan® probes for the amplified regions (Yuryev, 2007). Such software generally recommends using very short amplicons (< 100 nucleotides), which shortens the cycling times, avoids incomplete amplicons and saves reagents, avoiding competition even at late cycles. However, longer amplicons (up to 500 base pairs) provide much greater flexibility in designing an internal (e.g., TaqMan®) probe for the target. The probe should as much as possible be devoid of secondary structures (stem-loops) and have a T^m slightly higher than that of the amplification primers, so that it anneals to the denatured target molecules before any primer-driven polymerisation takes place. 'G' bases should be avoided at the 5' end, where the fluorophore usually resides, since they quench fluorescence, even after cleavage (Bustin, 2000).

4.8.1. Primer length, melting temperature and composition

Both amplification primers should ideally be the same length (around 20 nucleotides) with similar melting temperature (T^m) between 55°C-60°C, giving enough room for experimental annealing temperature optimization and long enough to avoid non-specific amplifications. It is useful to design all assays and primers around the same annealing temperature, so that a single cycling program can be used for all assays, and that different assays can be run concurrently with the same program, on the same plate. 56°C is a good, standard, robust target for the *in silico* estimated T^m for primers. The primer sequences should be evenly balanced between A/T and G/C nucleotides and avoid long homopolymeric stretches (*i.e.* runs of more than 4 of the same nucleotide).

4.8.2. Annealing temperature

The annealing temperature for the assay should be optimized experimentally, with a temperature gradient, which can be generated by most modern thermocyclers in a single run. Set the annealing temperature at 1-2°C below the highest temperature that still generates a signal/band, to make sure the assay is both specific and robust. For primers with a T^m of 56°C, the optimized assay annealing temperature is usually around 58°C and the maximum annealing temperature still generating a (weak) signal around 60°C.

4.8.3. Cycling parameters

The default incubation times recommended for particular kits have been optimized for the reaction components and should be followed unless there are compelling reasons not to. Typical for PCR products < 400 bp is 10 seconds denaturation at 95°C; 15 seconds annealing-extension at 58-60°C. Longer products require an additional incubation of 60 seconds per 1,000 bp at 72°C.

4.9. Assay optimization

Each assay should be optimized experimentally since the various components can significantly affect the reaction dynamics (Caetano-Anolles, 1998). The criteria for optimization can be sensitivity, specificity or reproducibility, and for qPCR also reaction efficiency. Higher primer concentrations and lower annealing temperatures increase sensitivity, but reduce specificity. Optimising for reproducibility usually means identifying the highest annealing temperature, the lowest primer concentrations and the shortest incubation times that consistently generate the right product, without secondary products, at a consistent amplification cycle.

4.9.1. Primer-dimers and other PCR artefacts

PCR is susceptible to qualitative and quantitative errors caused by the accidental, and highly efficient, amplification of short non-target PCR templates, especially when there is little target template available. Such artefactual amplifications arise from fleeting, partial complementarity of the primers with non-target templates, or among the primers themselves (SantaLucia, 2007). The latter version is called 'primer-dimer' and is formed through (self)-complementarity at the 3' end of the amplification primers. For example, if one primer ends in $N_{16}AC$-3' and another primer in $N_{16}GT$-3', the two primers can form a short template through the pairing of these two 3' base-pairs (Fig. 1A). If a primer ends in complementary bases ($N_{16}GC$-3' or $N_{16}AT$-3') it could even create a 2-bp overlap with itself (Fig. 1B), generating a short amplifiable fragment. The risk of primer-dimer increases with the number of unique primers in a reaction, such as in multiplex PCR (see section 4.7.5.; "Multiplex RT-(q)PCR"). Primer-dimer is identified if a product is produced in a template-free reaction. If PCR artefacts are only produced in samples, but not template-free controls, then the cause is less clear, involving most likely other nucleic acids molecules present in the samples. In both cases, the easiest solution is to design new primers and test these experimentally (SantaLucia, 2007).

4.9.2. Primer concentration

Primer concentration can be conveniently optimised at the same time as annealing temperature (Topley *et al.,* 2005; Todd *et al.,* 2007). A useful starting point is 0.2 μM reaction concentration for each primer. Higher concentrations tend to increase sensitivity but also non-specific products, which interfere with accurate quantification. Lower concentrations reduce sensitivity and accurate quantification.

Fig. 1. Formation of primer-dimer through complementarity between the 3' ends of two primers (A) or self-complementarity of the 3' end of a single primer (B).

4.9.3. Magnesium concentration

Magnesium is an essential ion required for most nucleic-acid processing enzymes, particularly polymerases. The use of ion-chelating agents, such as EDTA, to stop or inhibit polymerases and nucleases attests to the importance of Mg^{+2} in nucleic acid reactions (Green and Sambrook, 2012). Most commercial buffers contain optimized concentrations of Mg^{+2}, so that currently there is little need for further Mg^{+2} optimization. Above a certain minimum concentration, magnesium has only marginal influence on reaction efficiency and almost none on reaction specificity.

4.10. Quantitation Controls

In order to accurately quantify the amounts of target in individual samples a number of different controls are used. These can be broadly divided into external reference standards, which are used to quantify the targets, and internal reference standards, which are used to correct the quantitative data for differences between individual samples in overall RNA quality and quantity.

4.10.1. External reference standards

The classic way to relate indirect measurements to absolute amounts of target is through external reference standards. These are established by running the RT-qPCR assay on a dilution series of a known amount of target (external standard), using the resulting data to calculate the relationship between the absolute amount of target and cycle number, and then using this equation to convert the sample data to absolute amounts (Bustin, 2000). All modern real-time PCR thermocyclers have this function automatically included in their software, requiring as only input the absolute concentrations of the external standards. Such curves are also extremely useful during optimization of the RT-qPCR reaction conditions, particularly for determining the reaction efficiency (Bustin *et al.,* 2009).

External reference dilution standards should be prepared for all targets, including the internal reference standards. This is done, in short, by amplifying the appropriate fragment with PCR, purifying and cloning this fragment in a plasmid and preparing purified, well quantified plasmid DNA. This plasmid DNA can be either used directly to prepare DNA-based external standard series, or be used to synthesize RNA transcripts which in turn are quantified accurately and used to prepare RNA-based external standard series. DNA-based standards tend to be more sensitive and reproducible but RNA-based standards are more realistic and also take the cDNA reaction efficiency into account. The professional literature is divided on the issue, with good arguments for both approaches (Pfaffl and Hageleit, 2001). Both curves still require several positive control RNA samples per run, to normalize between runs for differences in reagent mixtures and, in the case of the DNA curve, to account for the reverse transcription step as well.

Reference standards from PCR products:
1. Amplify the target fragment by RT-PCR.
2. Confirm the amplification and absence of secondary products with electrophoresis.
3. If there are secondary products, excise the correct fragment under low-intensity UV light.
4. Purify the fragment using a commercial DNA affinity purification column.

The purified PCR fragments can be used directly to prepare an external reference standard, as follows:
1. Estimate the DNA concentration of the fragment in ng/µl, using spectrophotometry (e.g. Nanodrop, section 3.2.1) or fluorimetry (e.g. Qubit®; www.inVitrogen.com).
2. Estimate the molecular weight of your fragment.

This can be done exactly, based either on actual sequence or on fragment length, in the tools tab at www.currentprotocols.com. An approximate estimate for fragments within the 100-1000 bp range is:

$$MW^{dsDNA} = bp \times 617 \text{ ng/nmol}$$

3. Convert the DNA concentration to copies/µl as follows:
copies/µl = [ng/µl]/[MW^{dsDNA}] x [6.0221415 x 10^{14} copies/nmol]
4. Store the undiluted DNA fragment in aliquots at -80°C.
5. Prepare a working quantification standards series by serial ten-fold dilution of the DNA fragment, ranging from 10^{12} – 10^{0} copies/µl, in 10 ng/µl yeast or *E. coli* tRNA (Bustin *et al.*, 2009), to minimize loss of standard DNA due to adsorption to the microcentrifuge tube walls.

Whether or not the PCR products are used directly for preparing external reference standards, they should also be cloned: for confirmation of the fragment by sequencing, for long-term preservation of a positive DNA control and for the synthesis of RNA-based external reference standards. The fragment should be cloned into a T/A plasmid cloning vector that has T7 and T3 RNA promoters either side of the cloning site. Many molecular supply companies market such T/A cloning vectors, which are specially prepared for cloning PCR fragments.

1. Clone PCR fragments.

Protocols for cloning PCR fragments are beyond the scope of this paper. For this, the reader is referred to the product manuals provided by commercial suppliers of T/A cloning kits, and specialist manuals, such as the outstanding and long-established "Molecular cloning: a laboratory manual", by Green and Sambrook (2012).

2. Confirm candidate bacterial clones by colony PCR. This is a conventional 20 µl PCR reaction using the primers and amplification profile appropriate for the target, containing a small smudge of primary bacterial colony as template.
3. Run the colony-PCR products on an agarose gel (Green and Sambrook, 2012, see section 3.2.1).
4. Identify those colonies containing a plasmid with a cloned target.
5. Prepare small-scale liquid cultures of positive bacterial clones (Green and Sambrook, 2012).
6. Mix 0.5 ml of liquid bacterial culture with 0.5 ml 50% sterile glycerol and store this at -20°C (glycerol stocks).
7. Prepare plasmid DNA from the remaining liquid bacterial culture, using either a commercial plasmid purification kit or home-made reagents recommended in a molecular laboratory manual (Green and Sambrook, 2012).

Make sure that the protocol includes an RNAse step, to digest any bacterial RNA.

8. Purify the plasmid DNA on a commercial DNA affinity purification column.
9. Sequence the plasmid, using universal plasmid-based primers.

This is best done at specialist commercial facilities.

10. Confirm the presence of the insert in the plasmid from the sequence data, and the orientation of the insert in the plasmid.
11. Estimate the DNA concentration of the plasmid in ng/µl, using spectrophotometry (e.g. Nanodrop, section 3.2.1) or fluorimetry (e.g. Qubit®; www.InVitrogen.com). dsDNA A^{260} 1,0 = 50 ng/µl
12. Estimate the molecular weight of the plasmid + insert, by combining their lengths in bp and converting either exactly at www.currentprotocols.com or approximately as follows:

$$MW^{dsDNA} = (bp^{plasmid} + bp^{insert}) \times 607.4 + 157.9 \text{ ng/nmol}$$

13. Convert the DNA concentration to copies/ul as follows:
copies/µl = [ng/µl]/[MW^{dsDNA}] x [6.0221415 x 10^{14} copies/nmol]
14. Store the undiluted plasmid in aliquots at -80°C.
15. Prepare a working quantification standards series by serial ten-fold dilution of the plasmid, ranging from 10^{12} – 10^{0} copies/µl, in 10 ng/µl yeast or *E. coli* tRNA (Bustin *et al.*, 2009).

RNA-based external reference standards

1. Transcribe RNA from purified plasmid DNA, using either the T7 or the T3 promoter, depending on the orientation of the insert and the desired strand polarity of the RNA.
2. Linearize the plasmid with a restriction enzyme that digests right after the cloned fragment, in the desired orientation.

This ensures that the RNA transcripts have a defined length.

3. Transcribe the digested plasmid with a specific commercial T3/T7 RNA transcription kit.

Follow the corresponding instructions. Alternatively, detailed protocols with home-made reagents can be found in Green and Sambrook (2012).

4. Digest the synthetic, transcribed RNA with DNAse, as recommended by the kit manufacturer.

This is to remove contaminating plasmid DNA which may co-amplify and thus interfere with correct quantification.

5. Purify the DNAse-treated RNA on RNA affinity purification columns.
6. Estimate the RNA concentration in ng/µl, using spectrophotometry (e.g. Nanodrop, section 3.2.1) or fluorimetry (e.g. Qubit®; InVitrogen). ssRNA A^{260} 1,0 = 40 ng/µl
7. Calculate the insert size (number of bases from the T3/T7 promoter site to the restriction enzyme site on the other side of the insert used for digesting the plasmid).
8. Estimate the molecular weight of the RNA transcript either exactly at www.currentprotocols.com or approximately as follows:

$$MW^{ssRNA} = nt \times 320.5 + 159.0 \text{ ng/nmol}$$

9. Convert the concentration of the synthetic RNA to copies/µl as follows:

copies/µl = [ng/µl]/[MW^{ssRNA}] x [6.0221415 x 10^{14} copies/nmol]

10. Store the undiluted RNA in aliquots at -80°C.
11. Prepare a working quantification standards series by serial ten-fold dilution of the RNA, ranging from $10^{12} - 10^{0}$ copies/µl.

Do not use an RNA carrier for preparing the dilution series, since this carrier RNA will participate in the reverse transcriptase reaction and thereby significantly affect quantification!! Instead, dilute either in nuclease-free water or in 10 ng/µl of a neutral DNA carrier, obtained from a commercial source.

4.10.2. Internal reference standards

Unfortunately, external standards cannot correct for factors unique to each sample that affect the RT and/or PCR reactions, such as RNA quality and quantity, enzyme inhibitors, sample degradation, internal fluorescence *etc.* To correct for these factors, internal reference standards are used. These come in two forms:

4.10.2.1. Exogenous internal reference standards

Exogenously added internal reference standards are a pure, unrelated RNA of known concentration and integrity that is added to each sample prior to analysis. Such RNAs can be bought commercially (for example, Ambion's RNA250) and can be used to correct the data for the presence of enzyme inhibitors in individual RNA samples (Tentcheva *et al.,* 2006). The amount added per reaction should be low; < 1% of the amount of sample RNA, so as not to affect the RT-qPCR reaction efficiencies.

4.10.2.2. Internal reference standards

Endogenous internal reference standards (commonly called 'housekeeping genes') are relatively invariant host mRNA targets present in every sample that can be used to normalize quantitative data for minor variations between samples in RNA quality and quantity (Bustin *et al.,* 2009; Radonić *et al.,* 2004). The problem is that it is impossible to prove categorically that the expression of any candidate 'invariant' gene is not affected by the expression of the target gene (Radonić *et al.,* 2004). For this reason it is currently recommended to use a battery of 3 or 4 internal controls, from different classes of genes (metabolic enzymes, structural proteins, transcription factors, ribosomal proteins *etc.*) and construct a control-gene index, with which to normalise between samples (Bustin, 2000). Common internal reference standards for honey bee research are β-actin (Chen *et al.,* 2005a; Shen *et al.,* 2005a; 2005b; Locke *et al.,* 2012), rRNA (Chantawannakul *et al.,* 2006), microsomal glutathione-S transferase (Evans and Wheeler, 2000; Gregory *et al.,* 2005); ribosomal proteins RP-S5 (Evans, 2004; 2006; Wheeler *et al.,* 2006), RP49 (Corona *et al.,* 2005; Yañez *et al.,* 2012), RP-S8 (Kucharski and Maleszka, 2002), and transcription factors eIF3-S8 (Grozinger *et al.,* 2003) and eF1α (Toma *et al.,* 2000; Yamazaki *et al.,* 2006).

One technical difficulty with endogenous internal reference standards is the presence of contaminating genomic DNA in a sample, which could be amplified instead of the mRNA. There are two solutions to this:

- Digest the RNA sample with DNAse prior to RT-PCR. Many RNA purification kits come with this option.
- Design the RT-PCR assay such that the primers are separated by a (large) intron in the genomic copy of the gene.

Only the spliced mRNA will be amplified by the assay (Bustin, 2000). Such intron-spanning primers have been designed for the honey bee RP49 mRNA and B-actin-isoform-2 mRNA (de Miranda and Fries, 2008; Yañez *et al.,* 2012; Locke *et al.,* 2012).

As stated above, all internal reference standards also require their own external standards for accurate quantification. The inclusion of internal reference standards obviously greatly increases the cost of a project. The inclusion of internal controls is therefore one of several parameters to be decided on when starting a project, based on the projects' objectives, requirement for quantitative precision and available finances. Generally, the need for internal controls is greater for fully quantitative experiments with highly detailed analysis of relatively few samples. The need is much less for semi-quantitative survey-type studies, with fewer specific analyses and large numbers of samples.

4.10.2.3. External standard for viral target quantification

1. Extract RNA (Qiagen RNeasy® Mini Kit and QiaShredder®, according to manufacturer´s protocol) of bees with an RNA target (in this example DWV).
2. Generate an external standard by amplifying a DWV genomic fragment of 1520 bp via RT-PCR, using the primers Fstd (5´-GGACCATCCTTCCAGTCTACGAT-3´) and Bstd (5´-CTGTAGGTTGTGCTCCTGATGAAGA-3´) and the one-step RT-PCR kit from Qiagen.
3. This fragment contains the 354 bp fragment, which can be amplified by the primer pair F1/B1 (Genersch, 2005), for quantification.
4. Quantify the number of PCR-fragments via photometric analysis at 260 nm wavelength (Nanodrop, section 3.2.1).
5. Prepare a dilution series from the initial concentration through three orders (10-fold dilutions) of concentrated solutions.

This set of fixed dilutions will be used to ensure that PCR efficiency is maintained and to identify the precise predicted copy number for a particular C_q threshold.

4.11. Microarrays

A microarray is a powerful multiplex detection technology consisting of an ordered array of hundreds of molecular probes specific for different target RNAs bound to a solid support, usually a slide. The target sequences in an RNA sample are hybridized to these probes and these hybridization events are detected by a variety of, usually optical, detection chemistries (de Miranda, 2008). The power of the technology lies in the massive multiplexing potential where the relative and absolute amounts of hundreds of different targets can be determined simultaneously (Cheadle *et al.*, 2003; Gentry *et al.*, 2006). As molecular biology, pathology and diagnostics moves away from single organism/gene effects to surveying interactions among pathogens and (host) genes, microarray-based diagnostics will become increasingly relevant. Microarray printing technology is becoming cheaper and more reliable, and single-use disposable microarrays for specific multi-target diagnosis are increasingly available (Yuen *et al.*, 2003; Lieberfarb *et al.*, 2003; Noerholm *et al.*, 2004; Lin *et al.*, 2004; Perreten *et al.*, 2005; Fiorini *et al.*, 2005). Uniformity of hybridisation across the microarray, important for reliability in quantitation, is maximized with a range of nano-technological innovations (Yuen *et al.*, 2003; Noerholm *et al.*, 2004; Fiorini and Chiu, 2005), improved

oligonucleotide design (Rouillard *et al.*, 2003) and with replication of the spots or even whole arrays (an array of arrays) across the slide. The probe-target hybridisation can be detected through FRET-based probes, SYBR-green-I dye, or labelling of the nucleic acid sample containing the target sequences. Microarray technology can also be combined with quantitative RT-PCR, multiplex (pyro)sequencing and label-free electronic or optical detection technologies to increase the speed, accuracy, specificity or information content of the diagnosis (Weidenhammer *et al.*, 2002; Erali *et al.*, 2003; Gharizadeh *et al.*, 2003; Fixe *et al.*, 2004).

Numerous honey bee arrays have already been designed for different research purposes (Whitfield *et al.*, 2002; Evans and Wheeler, 2000; 2001; Robinson *et al.*, 2006). A microarray has also been developed for the semi-quantitative detection of honey bee viruses (Table 5 in Glover *et al.*, 2011) which will be developed further for diagnostic purposes.

Microarrays can also be developed for serology-based detection of proteins (Sage, 2004), using a similar approach as the sandwich ELISA (Enzyme-Linked ImmunoSorbent Assay: see de Miranda *et al.*, 2013). The probe-target recognition events are visualized and detected using similar detection chemistries as for nucleic acid-based microarrays.

4.12. Northern blots using DIG labelling

The primary advantage of using Northern blot analyses for identifying specific predicted RNA's, versus a PCR-based method, comes in the ability to predict the size of the entire transcript that is targeted using standard gel size markers. This is key especially when transcripts are subjected to editing (splice variants or enzymatic cutting as for small RNAs) and editing must be validated using a technique other than PCR. In addition, since probe binding is more permissive of nucleotide changes, Northern blots can be used to verify transcripts that might have mutations at primer sites used for PCR. In addition, this approach has somewhat lower vulnerability to point mutations that might cause a specific primer pair to fail to amplify a predicted target. The disadvantage to using Northern blots versus a PCR method as above is in time and expense and in a somewhat reduced ability to quantify transcript abundance. The below protocol avoids the use of radio-isotopic nucleotides as probes.

4.12.1. Agarose /formaldehyde gel electrophoresis

What follows is a standard protocol for denaturing gels suitable for linear separation of RNA strands:

1. Be RNase free!! Use gel apparatus designated for RNA. Wipe apparatus with "RNaseAway" and rinse thoroughly with RNAse -free water.
2. Prepare 100 ml of 1% agarose/formaldehyde gel:
   1. Dissolve 1 g agarose in 72 ml DEPC-treated water in a 250 ml glass flask.

2. Cool to 60°C in a water bath.

3. Add 10 ml of 5X MOPS running buffer (200 mM MOPS buffer, 50 mM Sodium acetate, 20 mM EDTA, pH 7.0) and 18 ml of 37% formaldehyde.

Precautions: Formaldehyde vapours are toxic. Prepare the gel in a fume hood.

3. Pour the gel to the gel tank and allow it to set.

4. Add sufficient 1X MOP running buffer to fill the tank in order to cover the gel and remove the comb carefully.

5. To prepare samples for gel electrophoresis, mix:
   - 11 µl of each RNA sample (0.5-1 µg/µl),
   - 5 µl 5X MOPS running buffer,
   - 9 µl 37% formaldehyde,
   - 25 µl of 50% formamide.

6. Heat the sample at 65°C for 15 min.

7. Cool on ice for 2 min.

8. Add 3 µl loading dye mix and 2 µl ethidium bromide (0.5 mg/ml).

9. Run the gel immediately after loading samples.

10. When the gel dye bands have separated and migrated at least 2-3 cm into the gel, or as far as 2/3 the length of the gel, visualize under UV light and take picture.

The 28s and 18sribosomal RNA (rRNA) should appear as sharp bands on the gel with no apparent smearing from degradation. The 28S rRNA band should be approximately twice as intense as the 18S.

4.12.2. Assembly of the transfer setup and transfer of RNA from gel to membrane

1. To prepare a gel for transfer, rinse the gel in DEPC-treated water twice for 20 min to remove the formaldehyde, which will otherwise interfere with transfer of RNA from gel to the membrane.

2. Soak the gel in RNase-free 20X SSC (3.0 M NaCl and 0.3 M sodium citrate, pH 7.0) for 45 min before proceed to setting up the transfer.

3. Cut uncharged nitrocellulose membrane to size of gel.

4. Soak the membrane in water for 2-3 min to wet .

5. Float in 20X SSC.

The transfer is conducted by the capillary method (Fig. 2).

1. Place a piece of thick blotting paper on the top of a glass plate that is elevated by four rubber stoppers placed near each corner of a baking glass dish.

2. Drape the ends of the wick blotting paper over the edges of the plate.

3. Fill the glass dish with RNase-free 20X SSC until the wick blotting paper on the top of glass plate is completely wet.

4. Squeeze out all air bubbles by rolling with a glass rod or pipette.

5. Place the gel facing down on the wet blotting paper.

6. Squeeze out air bubbles by rolling a glass pipette.

7. Cut a small triangular piece from the top left-hand corner to simplify orientation.

8. Place the wetted membrane on the surface of the gel by aligning the cut corners.

9. Get rid of any air bubbles under the membrane by rolling a glass pipette.

10. Cut 4-5 sheets of Whatman 3MM paper to the same size as membrane.

11. Place on top of the membrane.

12. Place a stack of paper towels on top of the Whatman 3MM papers.

13. Add a 200-500 g weight to hold everything in place.

14. Allow the transfer of RNA to proceed by capillary action overnight.

15. Disassemble the transfer stack at the next day.

16. Rinse the membrane briefly in 6X SSC.

17. Immobilize RNA to the blot by UV cross linking while the membrane is still damp.

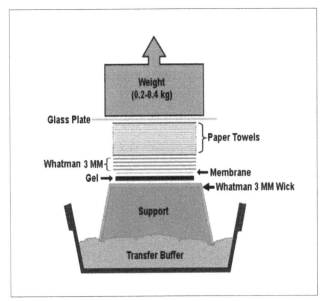

Fig. 2. Schematic diagram of the process used to transfer nucleic acids from a gel onto a binding membrane.

4.12.3. Preparation of DIG labelling (non-radioactive) probe

While DNA probes can also be used to detect RNA targets, a DIG-labeled RNA probe is ideal for detecting RNA on a Northern blot because RNA probes (riboprobes) that are transcribed in vitro are able to withstand more rigorous washing steps preventing some of the background noise. RNA probes give better sensitivity for detecting low amounts of RNA target than DNA probes. The following protocol is based on the Roche DIG RNA Labelling Kit, SP6/T7.

1. Linearize the recombinant plasmid DNAs with the target insert by cutting a restriction enzyme cleavage site downstream from the cloned insert using a restriction enzyme that creates 5′ overhangs (the choice of this enzyme will depend on the sequence of both the plasmid and insert).

2. After restriction digestion, purify the DNA by spin column purification or via phenol/chloroform/isoamyl alcohol extraction and ethanol.

This is commonly referred to as the 'plasmid mini-prep' and there are numerous commercial and home-made recipes for doing so that all work well.

3. Add the 2 µg purified template DNA to the following transcription reaction mixture to make 26 µl probe as follows:

- 4 µl 10X NTP labelling mixture,
- 4 µl 10X Transcription buffer,
- 2 µl Protector RNase Inhibitor,
- 4 µl RNA Polymerase SP6/or T7.

Adjust the volume with additional water until a final volume of 26 µl.

4. Place transcription reaction in the 37°C incubator for 2 hours. Longer incubations do not increase the yield of labeled RNA.

5. Stop reaction with 2µl 0.2M EDTA (pH 8.0).

Labeled probes are stable for at least one year at -15 to -25°C.

4.12.4. Hybridization analysis

(Roche Applied Science DIG Easy® Hyb, DIG Wash and Block Buffer Set, CSPD®Ready-to-use protocol)

1. Pre-hybridize the blot with pre-warmed DIG Easy® Hyb (10– 15 ml per 100 cm^2) in a specialized hybridization bag or any sealable container:
   1. Incubate the blot for 30 min at 65°C.
   2. Agitate gently during the pre-hybridization step.

2. Pipette the desired volume of probe (50-100 ng probe per ml hybridization buffer) into the hybridization bag.

3. Continue to incubate with rotation at 65°C for 10-16 hours.

4. After the hybridization is complete, wash the blot in a tray containing Low Stringency Buffer (2x SSC containing 0.1% SDS) twice by incubating the tray at RT for 5 min with gentle agitation.

5. Transfer the blot to preheated High Stringency Buffer (0.1x SSC containing 0.1% SDS).

6. Incubate the blot twice (2 x 15 min, with shaking) in High Stringency Buffer at 65°C.

7. After last wash, pull out the blot out of the hybridization container.

8. Place it between two Whatman paper sheets.

Do not allow the membrane get too dry so the membrane can be stripped and reused for hybridization.

9. Place blot onto a piece of Plastic wrap that is at least twice the size of the membrane.

10. Add 1 ml detection reagent (anti-digoxigenin-AP conjugate and the premixed stock solution of CSPD® ready-to-use) to stain the membrane and leave for 5 min.

11. Completely wrap up the blot with the plastic wrap.

12. Put it in a film cassette for chemiluminescent detection of hybridization signals.

4.13. *In situ* hybridization

4.13.1. Tissue fixation

1. Dissect out individual tissues.
2. Wash tissues with cold PBS 2-3 times.
3. Fix tissues in freshly made 4% paraformaldehyde in 100mM PBS (pH 7.0) overnight at 4°C.
4. Rinse in nuclease-free water three times.
5. Store tissues in 70% ethanol at 4°C until further use.
6. For tissue dehydration, carry out successive incubation in ethanol (70%, 95% and 100%) and xylol (2 x 5 min each).
7. Embed in paraffin.
8. Cut Paraffin sections into 2-5 micron thick segments.
9. Mount on poly-L-lysinated slides which are to be stored at 4°C overnight.
10. To rehydrate the sections prior to hybridization:
    1. Carry out descending concentration of ethanol (100%, 95% and 70%).
    2. Dewax in xylol.
    3. Treat with proteinase K (10 ug/ml) for 30 min.
    4. Acetylate with 0.33% (v/v) acetic anhydride in 0.1 M triethanolamine-HCl (pH 8.0) for 10 min.

4.13.2. Preparation of DIG labelling (non-radioactive) probe

Using Roche DIG RNA Labelling Kit, SP6/T7. The same procedures as for Northern blot (see section 4.12.).

4.13.3. Hybridization Analysis

1. Pre-hybridize the sections in pre-hybridization solution (50% formamide, 5X SSC, 40 µg/ml salmon sperm) at 58°C for two hours.
2. Incubate in hybridization buffer with Dig-labeled TARGET probe solution to a concentration of 100-200 ng/ml of probe in pre-hybridization solution at 58°C overnight.
3. After hybridization, wash the sections twice in low stringency wash solution (2X SSC, 0.1% SDS) at room temperature for five minutes.
4. Wash twice in high stringency wash solution (0.1 × SSC, 0.1% SDS) at 52°C for 15 min.

Note: The hybridization signals are detected with Alkaline phosphatase (AP)-labeled sheep anti-DIG antibody conjugate (Roche Applied Science).

5. Add the conjugate solution to the dry sections.

6. Incubate at 4°C for two hours in a chamber in which humidity is maintained at > 70% relative humidity.

7. Rinse the slides three times with washing buffers.

8. Perform the colour development by adding the buffer solution containing nitroblue tetrazolium (NBT) and 5-bromo-4-chloro-3-indoyl phosphate (BCIP) on the tissue sections.

9. Incubate for three to six hours at RT protected from bright light.

10. Stop the colour reaction by a 5 min wash in Tris/EDTA (0.1 mM, pH 8.0).

11. Remove the non-specific staining in 95% EtOH overnight.

12. Rehydrate the sections through:

    1. Successive incubation in ethanol (70%, 95%, and 100%).
    2. Incubation in xylol (2 X15 min each).

13. Mount in Eukitt® resin (Sigma).

Note: Negative control reactions include regular dUTP instead of DIG-labeled TARGET probe.

14. Observe and photograph *In Situ* hybridization slides under a light microscope.

The hybridization signals are shown by dark blue sites where the DIG-labeled probe bound directly to the viral RNA. The section of negative control will stain pink only with the nuclear fast red.

5. Proteomic methods

5.1. Introduction

Proteins are the ultimate functional product of most gene expression so optimally one would prefer to look at proteins when trying to understand the mechanisms an organism uses to respond to a given condition. As with any other biomolecule, tools for identifying and quantifying proteins are a prerequisite to their successful study. Single proteins are typically detected using antibodies but very few antibodies have been generated against bee proteins and none have been commercialized. Of all the analytical methods available for studying proteins, mass spectrometry is the most sensitive, most accurate and least biased. Proteins can be identified by mass spectrometry by first hydrolyzing them with a specific protease such as trypsin. The masses and fragmentation patterns of the resulting peptides can then be determined and used to identify the peptides individually and the protein(s) they came from (i.e. proteomics). This process works best when all possible proteins that might be present are known and is only really successful when an organism's genome has been sequenced. To this end, in recent years proteomics has begun to be applied in bees towards understanding a range of paradigms.

Where is the future of proteomics research in bees heading? Mapping protein expression across all tissues and castes in adult bees is the logical next step after sequencing the bee genome. The genome helps to determine which proteins may be present but where are those proteins expressed? The protein expression atlas in bees will tell that and will mark a significant step forward for bees as a model system as this would be the first such comprehensive atlas in any multicellular organism. Additional protein-based methods (protein extraction and immunochemical assays for protein abundance) are covered in further detail in the *BEEBOOK* paper on physiology and biochemistry (Hartfelder *et al.*, 2013).

6. Population genetics

6.1. Introduction

Measuring the current variation in genetic traits within and across populations can give insights into past movement of individuals, population size, and the association of specific genetic histories with honey bee biological traits such as behaviour, disease resistance, and colony life histories. Honey bees, thanks to human transport and breeding, are made up of numerous and often entwined genetic lineages and one aim of population-genetic analyses are to resolve these connections.

6.2. Mitochondrial DNA analysis

In principle, one honey bee (worker or drone) per colony is enough to determine the maternal origin of the whole colony given the maternal inheritance of mitochondrial DNA (mtDNA), i.e. all the daughter workers and son drones from one honey bee queen share the same molecule. Due to the risk of drifting between colonies, it is ideal, and in some cases essential, to make this one collected individual of a life stage where host colony is unambiguous (e.g., a developing bee or one observed exiting from a brood cell). In cases where this is not possible, pre-flight worker bees could be substituted, although it is arguably worth sampling more than one individual to avoid mistakes in assigning colony heritage.

The most widely used mitochondrial region for population genetic studies is the intergenic region located between the tRNA[leu] and cox2 (subunit 2 of the cytochrome oxidase) genes. This region shows length and sequence variation that allows discrimination of honey bee evolutionary lineages (Garnery *et al.*, 1993). It is composed of two types of sequences: P and Q. The sequence P can be absent (lineage C from east Europe) or present in four different forms: P (lineage M from west Europe), P_0 (lineages A from Africa and O from Near East), P_1 (Atlantic African sub-lineage) and P_2 (restricted to the Y lineage from Ethiopia; De la Rúa *et al.*, 2009). The number of Q sequences and the sequence variation developed through a RFLP test with the restriction enzyme *Dra*I (Garnery *et al.*, 1993) can be used to determine the haplotype within each lineage. Below is the protocol to determine the mitochondrial haplotype, modified from Garnery *et al.* (1993) by including a new thermal regime for PCR and optimizing the chemistry of PCR reactions. A full description of how this locus can discriminate among honey bee ecotypes is presented in the *BEEBOOK* paper on characterizing subspecies and ecotypes by Meixner *et al.* (2013).

1. Honey bee samples are immediately transferred into tubes with absolute EtOH and preserved at -20°C until DNA extraction. A single or two legs from one individual are enough to extract total DNA following the Chelex®-based (Biorad, Inc.) protocol (Walsh *et al.*, 1991; see section 3.2.3.).

2. PCR amplify the intergenic region with the primers E2 (5'-GGCAGAATAAGTGACATTG-3') located at the 5' end of the gene tRNA[leu] and H2 (5'-CAATATCATTGATGAACC-3') located close to the 5' end of the gene cox2 (Garnery *et al.*, 1993).

This amplification can be performed by using Ready-To-Go TM PCR Beads (product code 27-9557-01, GE Healthcare), that are pre-mixed and pre-dispensed reactions for PCR featuring, therefore reducing the pipetting steps and the chances to handling error. They contain all the necessary reagents for a 25 µl reaction volume.

3. Add 20.2 µl of PCR-quality water to each tube.

4. Mix by gently flicking it with the fingers.

5. Add 0.4 µl of each primer (10 mM) and vortex and centrifuge the mix to get all the components at the bottom of the tubes.

6. Add 4 µl of DNA extraction solution and mix.

7. Place the reaction mixtures in a thermo cycler with the following amplification program:

Denaturation at 94 °C for 5 min,

Followed by 35 cycles of:

94 °C for 45 sec,

48 °C for 45 sec,

62 °C for 2 min,

Final elongation step of 20 min at 65 °C.

8. To identify successful amplicons, 2 µl of the PCR product of each sample are electrophoresed in a 1.5% agarose gel (see section 3.2.1) with ethidium bromide included and photographed over a UV light screen.

9. Aliquots of the PCR product are then digested with the endonuclease *Dra*I (recognition site 5'-TTTAAA-3') by adding:

10X endonuclease buffer to a final concentration of 1X,

0.06U of *Dra*I/,

10 µL of PCR product,

Incubate at 37°C overnight.

10. To determine RFLP patterns, the digested products of each sample are electrophoresed in a 4% agarose Nusieve® or Metaphor® (Lonza Biosciences) gel at 100 volts for ca. 1 hour 30 min and photographed over a UV light screen.

11. At least one sample with a characteristic RFLP pattern should be directly sequenced using the same primers as for the amplification.

12. Prior to sequencing, purify PCR products:

Either with QIAquick® PCR Purification Kit (Qiagen). Alternatively, PCR products can be purified with isopropanol and 5 M ammonium acetate as follows:

1. To 10 µl of PCR product add:

7 µl of 5 M ammonium acetate,

17 µl of isopropanol.

2. Leave 10 min at room temperature.

3. Centrifuge 30 min to 13,500 rpm.

4. Discard the supernatant.

5. Add 500 µl of cold 70% EtOH.

6. Centrifuge 5 min at 13,500 rpm.

7. Remove supernatant and allow to dry overnight.

8. Re-suspend in 30 µl of water.

6.3. Nuclear DNA analysis

Nuclear markers are biparentally inherited and allow genotyping of workers to obtain information from both the mother queen and the drones she has mated with. Nuclear analyses of *A. mellifera* involve widely used microsatellites and more recently, single nuclear polymorphisms or SNPs.

6.3.1. Microsatellites

Microsatellites consist of short motifs (one to six nucleotides) that are repeated from four to 20+ times at points scattered across all eukaryotic genomes. They are useful as markers for genetic structure since the number of repeats at any given locus is unstable and new repeat variants are constantly arising by mutation and being lost by drift and other population-level events. Strategies to screen a total of 550 polymorphic microsatellite loci have been described in *A. mellifera* (Solignac *et al.*, 2003), and many thousands more are found in the complete honey bee genome (Honey Bee Genome Sequencing Consortium, 2006). The protocol described here has been used to analyse the temporal genetic variation of island honey bee populations (Muñoz *et al.*, 2012), the mating frequency of the Iberian honey bee (Hernández-García *et al.*, 2009) and the population genetic structure of European honey bees (Muñoz *et al.*, 2009; see also the *BEEBOOK* paper on characterizing subspecies and ecotypes by Meixner *et al.* (2013) for a full review of the use of microsatellites in determining honey bee ecotypes). It takes advantage of multiplexing, whereby multiple loci are screened in a single PCR reaction and size assay. These loci are widely used, and it is subsequently possible to compare allelic counts and genotypes across different studies.

6.3.1.1 Microsatellite reaction mix

To prepare the reaction mix, add:

- 1X reaction buffer (provided as a 10x solution with Taq polymerase),
- 1.2 mM MgCl$_2$,

- 0.3 mM of each dNTPs,
- 0.4 µM of each primer,
- 1.5 U Taq polymerase,
- > 5 ng DNA (provided in 2µl DNA solution).

6.3.1.2 Primers for multiplexed honey bee microsatellite loci

A113-F-(FAM)	CTC GAA TCG TGG CGT CC
A113-R	CCT GTA TTT TGC AAC CTC GC
A007-F-(NED)	GTT AGT GCC CTC CTC TTG C
A007-R	CCC TTC CTC TTT CAT CTT CC
AP043-F-(VIC)	GGC GTG CAC AGC TTA TTC C
AP043-R	CGA AGG TGG TTT CAG GCC
AP055-F-(PET)	GAT CAC TTC GTT TCA ACC GT
AP055-R	CAT CCG GTA TGG TAC GAC CT
B124-F-(FAM)	GCA ACA GGT CGG GTT AGA G
B124-R	CAG GAT AGG GTA GGT AAG CAG

6.3.1.3 Thermal cycling conditions for multiplex PCR

Incubate the samples as follows:

5 min at 95°C,

Followed by 30 cycles:

95°C for 30 sec,

54°C for 30 sec,

72°C for 30 sec,

Final extension is 30 min at 72°C.

6.3.1.4 Size estimation of PCR products

PCR products are visualized by capillary electrophoresis and sized with an internal size-standard (e.g., using the Applied Biosystems or MegaBACE systems, both of which have extensive use for microsatellite scoring). Alleles are subsequently scored using GeneMapper v3.7 software (Applied Biosystems™). It is also possible to measure microsatellite size variants using large denaturing polyacrylamide gels (e.g. Evans, 1993) although this method has fallen from favour do to the hazards of polyacrylamide and difficulties in manually scoring allele sizes.

Once genotypes of samples have been established, microsatellite data are well suited for assessing parentage of nestmates (Evans, 1993), for standard population-genetic statistics including ecotype determination (Estoup *et al.*, 1993; reviewed by Meixner *et al.*, 2013), and for genome mapping (Solignac *et al.,* 2003), among other uses.

6.3.2. Single-nucleotide polymorphisms (SNPs)

In the honey bee and other species for which extensive data have been gathered on genomic sequence variants, it is possible to use SNPs to reconstruct past migration events, and to separate races and populations. A SNP is any validated nucleotide change between the genomes of two or more samples, and SNP's can occur both within the coding regions (exons) of genes and in the vast regions that

separate genes or lie in non-coding parts of the genome. SNP analyses are standard in human, veterinary, and agricultural systems, and this approach will continue to increase as a viable option for the study of honey bees. Unfortunately, high-throughput SNP genotyping remains an expensive endeavour that requires cutting edge technologies and the expertise often only available in a core laboratory facility or at larger institutions. In addition, prior to genotyping the honey bee sample of interest, a SNP assay must be developed (or purchased, if commercially available) from sequence data relevant for the study population. At present, there are only two SNP assays developed and published for honey bees. The first one (Whitfield *et al.*, 2006), which consisted of 1536 SNP loci that were selected mainly based on spacing criteria, was developed for genotyping using the Illumina GoldenGate™ assay and is not commercially available, although a honey bee SNP database (over 1 million SNPs) is available at NCBI (http://www.ncbi.nlm.nih.gov/snp/) and this resource could be exploited to establish a system for genotyping. More recently, Spötter *et al.* (2011), published a 44,000 SNP assay designed for analysis of varroa-specific defence behaviour in honey bees. This assay uses Affymetrix™ technology, and it is now commercially available via AROS Applied Biotechnology AS.

As illustrated in Spötter *et al.* (2011), development of a SNP assay is a time and resource intensive undertaking, yet it can be designed to address a specific objective (e.g., to investigate varroa-specific defence behaviour). Once the design stage is accomplished, the assay can then be used to genotype honey bee samples at hundreds to thousands of loci via high-throughput technologies. Illumina® technologies, for example, offer a number of options for high throughput genotyping depending on the number of SNPs to be interrogated. The GoldenGate assay, employed by Whitfield *et al.* (2006), interrogates 96, or from 384 to 1,536 SNP loci simultaneously (plex levels can be 384, 768, or 1,536). For genotyping a number of SNPs larger than 6,000 up to 2,500,000 the Infinium assay (also a product from Illumina) is required.

Both the GoldenGate assay and the Infinium assay take three days for completion and require reasonable quality and accurately quantified genomic DNA. DNA concentrations should be 50 ng/µl, quantified a fluorometric assay (e.g., Picogreen) or spectrophotometry (e.g., Nanodrop, section 3.2.1). DNA can be extracted from the thoraces of honey bees that had been stored at -80°C or in absolute EtOH. The GoldenGate assay involves several steps including DNA activation for binding to paramagnetic particles, hybridization of activated DNA with assay oligonucleotides, washing, extension, ligation, PCR, hybridization onto the BeadChip, and finally analysis of the fluorescence signal on BeadChip by the iScan System.

Unlike in the GoldenGate assay, where universal primers are used to amplify SNP-reactive DNA fragments, in the Infinium assay genomic targets hybridize directly to array-bound sequences. Following hybridization onto the BeadChip, samples are extended and fluorescently stained. As for the GoldenGate assay, the last step consists of analysis of scanned BeadChips using the iScan System. Genotype data generated by both assays using the iScan System (and other systems), are then analysed using the GenomeStudio Genotyping (GT) Module. The calls are automated but can be manually verified and edited if necessary (e.g., if there are signs of unequal proportions of an expected biallellic marker). Finally, summary statistics and results are exported for further analyses using standard population genetics software packages such as STRUCTURE (http://pritch.bsd.uchicago.edu/structure.html).

With increasingly affordable sequencing costs allowed by next generation technologies (e.g., 100 bp or shorter), it will be feasible to carry out population-genetic and strain-identifying projects via whole-genome sequencing. This technique involves a scan (usually 3-fold sequencing depth or more, i.e., > 750 million sequenced bases for the honey bee) of a genome or population of interest followed by an alignment of those short reads to a reference genome (for the honey bee this would be the genome assembly from HGSC, 2006). It is relatively straightforward, using free programs available for download (e.g., http://bioinformatics.igm.jhmi.edu/salzberg/Salzberg/Software.html) to identify and in some cases quantify SNPs that differ among samples. There are also public sites at which one can import data and benefit from a maintained supercomputer dedicated to such genomic analyses (e.g. https://main.g2.bx.psu.edu/). SNP analyses derived from sequencing data have yet to make an impact on honey bee science but they are expected to in the next few years.

7. Phylogenetic analysis of sequence data

7.1. Introduction

The goal of this protocol is to provide the reader with an easy to use, reliable, and technically appropriate method to choose, align and analyse sequence data for phylogenetic analyses of taxa or genes of interest. Analysis of highly conserved loci (i.e. rRNA, cytochrome oxidase I) or population genetic studies from one species, require nucleotide level data to achieve necessary resolution in tree topology. Amino acid sequences are typically used when reconstructing phylogenies from an encoded protein across a large evolutionary distance, which can make alignment at the nucleotide level difficult. Over time, one develops their preferred approach and programs to use in this process, of which there are many. While the following protocol reflects preferences of the authors, it is appropriate for a wide variety of applications, user skill levels, and relies on freely available programs with graphical user interface (GUI)-based options. Detailed information on use is available from each of the program sites, given below. As a disclaimer, concatenation of sequence data,

while appropriate and employed for taxonomy classification, is a more specific approach some users may wish to use but will not be discussed here. Additionally, though PAUP is also widely used in phylogenetic analyses, it requires a small fee and therefore is not discussed here, though labs with frequent phylogenetic needs may wish to purchase this program. MEGA and other software free to the public can invoke many or all of the same phylogenetic analyses as PAUP.

The steps to perform a phylogenetic analysis are:

1. Obtain and format sequences of interest.
2. Format sequence data in FASTA format.
3. Align sequence data.
4. Trim aligned sequence data to equal length.
5. Perform phylogenetic analyses.

Each step is described below in detail.

7.2. Obtaining and formatting sequences of interest for phylogenetics

Once you have obtained DNA sequence data for your study, you may wish to add accessioned sequence data to your analyses. This will be particularly important if you want to root your phylogeny and provide an outgroup (sequence(s) to which all of your sequences are distantly related) to strengthen comparative interpretation of your data. Accessions from nucleotide sequence data banks can be searched, using a keyword(s) or via a BLAST search algorithm (i.e. blastn, megablast, etc.), to identify homologues to your sequence of interest. These include GenBank (via NCBI; (http://www.ncbi.nlm.nih.gov/), EMBL-Bank (via EBI; http://www.ebi.ac.uk/embl/), and DNA Data Bank of Japan (DDBJ; http://ddbj.sakura.ne.jp/). If using rRNA sequence data, SILVA rRNA database (http://www.arb-silva.de/) can be used to retrieve reference sequences that are quality-scored (Pruesse *et al.*, 2007).

7.3. Sequence data in FASTA format

For compatibility in downstream analyses, sequence data should be in a single file and FASTA formatted. Sequence databases include FASTA as an option for output format. An example of FASTA formatted sequences retrieved from GenBank (abbreviated in length for the sake of space):

>gi|21747902|gb|AY114459.1| *Apis mellifera mellifera* isolate melli4 cytochrome oxidase subunit I (COI) gene, partial cds; mitochondrial gene for mitochondrial product
    CCCCGAATAAATAATGTTAGATTTTGATTACTTCCTCCCTCATTAT
    TAATACTTTTATTAAGAAATTTATTTTACCCAAGACCAGGAACTG
    GATGAACAGTATATCC

>gi|14193071|gb|AF153104.1| *Apis cerana* haplotype 4
cytochrome oxidase subunit 1 (COI) gene, partial cds;
mitochondrial gene for mitochondrial product
TTTCTAATTGGAGGTTTTGGAAATTGATTAATTCCTTTAATATTA
GGATCTCCAGATATAGCATTTCCTCGAATAAATAATATTAGATTC
TGATTACTCCCTCCTTC
>gi|67626085|gb|DQ016070.1| *Apis dorsata* haplotype 7
cytochrome c oxidase subunit 1 (CO1) gene, partial cds;
mitochondrial
TTTTTAATTGGAGGATTTGGAAATTGATTAATCCCTTTAATATTA
GGGTCTCCAGATATAGCATTTCCTCGAATAAATAATATTAGATTT
TGATTATTACCTCCTT

The sequence identifier (e.g. accession number) and title for each
entry is preceded by a carrot symbol ">" and ends with a hard return.
The immediate next line below this is the sequence information and
should contain no spaces. The end of the sequence is determined by
a hard return. You will want to abbreviate the title of your sequence
entries now, prior to alignment, using the all-important accession
number or perhaps just the species name. The number of characters
allowed in the sequence title is limited, to varying degrees, by
alignment programs but are typically 30 characters or less. Only
letters, numbers, underscores "_", and pipes "|" are typically allowed.
The above sequence entries are prepared for alignment like this:

>AY114459_A_mellifera
CCCCGAATAAATAATGTTAGATTTTGATTACTTCCTCCCTCATTAT
TAATACTTTTATTAAGAAATTTATTTTACCCAAGACCAGGAACTG
GATGAACAGTATATCC
>AF153104_A_cerana
TTTCTAATTGGAGGTTTTGGAAATTGATTAATTCCTTTAATATTA
GGATCTCCAGATATAGCATTTCCTCGAATAAATAATATTAGATTC
TGATTACTCCCTCCTTC
>DQ016070_A_dorsata
TTTTTAATTGGAGGATTTGGAAATTGATTAATCCCTTTAATATTA
GGGTCTCCAGATATAGCATTTCCTCGAATAAATAATATTAGATTT
TGATTATTACCTCCTT

Note that you may want to keep two copies of your sequence
data files: one with all the original information pertaining to the
sequences and a second with just the abbreviated titles prepared for
alignment analysis.

7.4. Alignment of sequence data

The alignment quality of sequences is critically important to achieving
a strong phylogenetic reconstruction. There are a variety of multiple
sequence alignment programs available, with varying capacity for the
number of sequences input and user-determined parameter
adjustments. Additionally, some aligners may be specific for protein
sequence data vs. nucleotide data. Two alignment programs that are

available for all computing platforms (Mac OSX, Windows, and Unix/
Linux), accessible at an off-site server via the web if installing locally
is not desired, and known for robust alignment algorithms are
discussed here.

7.4.1. Clustal

Clustal (Thompson *et al.*, 1994) is a commonly used alignment
program that will handle protein, DNA, or RNA sequence data and is
actively maintained (http://www.clustal.org/). The version ClustalW,
currently in version 2.1 (Larkin *et al.*, 2007), can be installed locally
and run in command-line or it can be run remotely at an off-site
server where it is already installed (i.e. at EMBL-EBI; http://
www.ebi.ac.uk/Tools/msa/clustalw2/ or at GenomeNet; http://
www.genome.jp/tools/clustalw/). ClustalW allows the user to specify
certain parameters of the alignment algorithm. Users who do not have
the knowledge to make parameter specifications may choose general
purpose, default settings (as at the EMBL-EBI site). ClustalX, a
graphical version and Clustal Omega, specifically for large sets of
protein sequence data are also available.

7.4.2. MUSCLE

Though less commonly used than Clustal, MUltiple SequenCe
aLignmEnt (MUSCLE; http://www.drive5.com/muscle/) is another easy
to use, good option for sequence alignment (currently limited to 500
sequences/1MB of data).

7.5. Trimming aligned sequence data to equal length

To properly compute phylogenetic analyses on a sequence data set,
the number of positions in each sequence should be equal. This
includes gaps and insertions/deletions (indels) in the aligned data set,
not the actual number of nucleotides or amino acids. Use your
sequence alignment editor to trim the aligned files to equal size or to
the size of the region you are interested in analysing (i.e. a specific
domain encoded within your gene).

7.6. Performing phylogenetic analyses

Again, there are a number of options for users to perform
phylogenetic analyses, but only two will be discussed here: MEGA and
SATé.

7.6.1. Using MEGA

The program MEGA (Molecular Evolutionary Genetics Analysis; http://
megasoftware.net) (Tamura *et al.*, 2011), currently available as
version 5.05, is continually being updated and improved. Note: use of
MEGA will require the user to download and install the freely available
software to their computer. MEGA 5 can be used as a platform to
complete all of the above steps (building your sequence data file,

alignment using Clustal or MUSCLE, and trimming sequence data). It gives the user the ability to construct phylogenies using distance based (i.e. Neighbour Joining) and character based (i.e. Maximum Likelihood) methods (see Table 1) and test them using bootstrapping. It also includes a tree viewer that allows for some editing of the final output tree and many additional features not covered here.

7.6.1.1. Converting data to MEGA format

Before phylogenetic tests can be run on your sequence data file, it must be converted into a format that MEGA can read, the *.meg file format.

1. From the 'File' menu in MEGA, select 'Convert File Format to MEGA…', browse to your alignment file, select 'FASTA format' from the 'Data Format' pull down menu, then select 'OK'.
2. A window will open asking for you to specify a name and location for the newly created *.meg file.

The new file will be created and opened under a tab in the same window next to your open FASTA format file.

3. MEGA will warn you to check the file for any errors and adjustments to your *.meg file can be made in this window and saved.

Details about the *.meg format are available in MEGA.

7.6.1.2. Constructing and testing phylogenetic trees

Described below is a basic, distance based Neighbour-Joining analysis using bootstrap statistical tests for robustness (Felsenstein, 1985).

1. From the main MEGA window, open the 'Phylogeny' pull down tab and select 'Construct/Test Neighbour-Joining Tree'.
2. Browse to and open the .meg file you just created. Select the appropriate data type (nucleotide or protein sequence data).

Note the defaults for missing data, alignment gaps and identity and make any changes if necessary then select 'OK'.

3. A window will open asking you to identify your sequence data as protein-coding or not.
4. Another window will open asking for genetic code selection.

For the example, provided here, using the cytochrome oxidase I (COI) gene, select 'Invertebrate Mitochondrial'.

5. A third window opens and allows you to select a number of parameters for your analysis.

A minimum of 100 and typically 1,000 iterations of bootstrapping are used to test the robustness of your phylogeny. For now, we will accept the default parameters for our simple analysis.

6. A progress window will open for you until the test is completed.
7. When complete, a window opens with two tabs showing the 'Original tree' generated, as well as a 'Bootstrap consensus tree', which is the tree you should refer to.

Bootstrap support values show the percentage of iterations supporting the shown topology.

8. The tree image can be saved as a pdf for good resolution for presentation (Fig. 3A), can be saved as a mts Tree Session File for future viewing in MEGA, or exported and saved as a more general Newick (.nwk) format file that is readable by a variety of other tree viewing programs (e.g. FigTree http://tree.bio.ed.ac.uk/software/figtree/ ; TreeDyn http://www.treedyn.org/).

For comparison, a Maximum Likelihood (ML) analysis of the same alignment was performed in a similar manner and is shown in Fig. 3B.

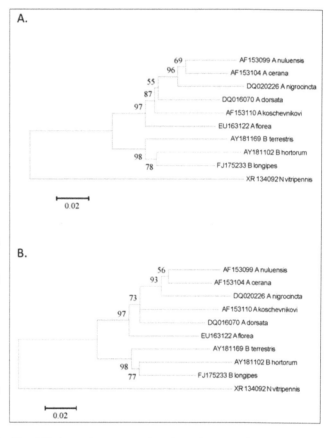

Fig. 3. Phylogenetic reconstruction of cytochrome oxidase I gene from *Apis* and *Bombus* species (Hymenoptera; Apidae) using Neighbour-Joining method (A) and Maximum Likelihood (B). Topology of each was tested with 1,000 bootstrap iterations (consensus tree is shown) using *Nasonia vitripennis* (Hymenoptera; Pteromalidae) as outgroup. Scale represents the substitution rate per site from a total of 981 positions. A) was computed using Maximum Composite Likelihood (Tamura *et al.*, 2004) with uniform rates among sites and pairwise gap deletion. B) was computed using Tamura-Nei model (Tamura and Nei, 1993) with uniform rates at all sites.

7.6.2. Using SATé

An alternative software package to MEGA for ML analyses is SATé (Simultaneous Alignment and Tree estimation; Liu *et al.*, 2012), which infers sequence alignment and tree building concurrently as an iterative process using the ML method. This program must also be downloaded for use, and is currently freely available as SATé-II at http://phylo.bio.ku.edu/software/state/sate.html. The user experience for SATé is still being improved, including recommendations for how to parse phylogenetic runs. A strength of SATé is that it accepts up to 1,000 sequences in the FASTA format as described in section 7.3., and claims speed and precision in phylogenetic analyses. Through changes in subproblem size parameters (below) it is possible to run SATé on desktop machines, but using this software on high-memory and high-CPU clusters will be simpler since those parameters will be less likely to affect performance. Several alignment programs are bundled with the download, including Clustal and MUSCLE. If an alignment is already prepared, SATé will use Randomized a(x)-elerated Maximum Likelihood (RaxML) (Satmatakis, 2006) to infer an initial tree for phylogeny reconstruction.

From the main SATé window, select the desired analysis criteria in the following sections:

7.6.2.1. External tools

SATé breaks the tree topology down into subproblems during each round of analysis and realigns the data for each subset, merges the alignments into a full alignment and re-estimates the tree for full alignment.

1. 'Aligner' is used to select the multiple sequence alignment tool to produce the initial full alignment.
2. 'Merger' is used to select the multiple sequence alignment tool to merge the alignments of subproblems into a bigger and final multiple sequence alignment.
3. 'Tree estimator' uses RAxML for tree estimation with the chosen evolutionary 'Model'.

7.6.2.2. Sequence import and tree building

1. Click 'Sequence file...' to upload your sequence alignment file in FASTA format (Note: the file MUST have the extension *.fas or *.fasta to be read by SATé; see section 7.3.).
2. Select the appropriate 'Data Type' (nucleotide or amino acid).
3. If you have previously generated a 'Tree file', you can upload it as the initial guide for SATé, if appropriate.

7.6.2.3. Job Settings

1. Specify the 'Job Name' for identifying output files created by SATé.
2. Select the folder/directory for storing the created outfile files using 'Output Dir.'

7.6.2.4. SATé Settings

This section allows users to control the details of the algorithm. In each iteration, the dataset will be breaking down into non-overlapping sequence subproblems and these subproblems are given to the chosen alignment tool.

1. There are options under 'Quick Set' to allow a more or less intensive search during the SATé iterative process.
2. 'Max. Subproblem' is used to control the largest dataset that are aligned during the iterative process.
3. Use the 'Fraction' option to express the maximum problem size as a percentage of the total number of taxa in the full dataset.

This value will be limited by available computational power.

4. Use the 'Size' option for size cutoff in absolute number of sequences.
5. Select 'Decomposition' to choose how the process should be broken to create subproblems.
6. 'Apply Stop Rule' is used to control how SATé should be finished.

The decision to stop can be done based on number of iterations (one may be sufficient), the amount of time in hours or 'Blind Mode Enabled' meaning that SATé will terminate if it ever completes one iteration without improving the ML score.

7. Click 'Start' to run the SATé analysis.

There will be five files created in the selected directory after SATé is completed. An alignment file (*.aln), tree file (*.tre), best ML score file (*.score), error file (*.err) and history file (*.out). Unlike MEGA, SATé does not have bundled tree viewing or alignment viewing programs, so the user will need to open the tree file using one of the tree viewing programs described above and the alignment file using Clustal or a similar alignment viewing program. The other files can be opened with a text editor (i.e. Notepad, TextEdit). In addition, SATé does not utilize bootstrap testing to support inferred tree topology. Rather, a similarity score from 0-1 (0 is most similar and 1 is least similar) is placed on the branches to aid in topology interpretation. Hence, other reconstruction methods should be compared to confirm the output.

7.6.3. Building trees using distance and character based methods

To assess the reliability of the tree topology, users should be aware of the phylogenetic tree construction and tree analysis methods according to the data and algorithmic strategy used. Each method has different assumptions that may or may not be valid for the evolutionary process of the given sequence data. For example, the distance based method UPGMA (Unweighted Pair Group Method with Arithmetic mean) assumes a neutral mutation rate proportional to time (a molecular clock). Therefore, it is important to be aware of this fact when evaluating tree topology generated by each method. It is encouraged that one run a variety of distance based methods (Neighbour-Joining, UPGMA, Minimum Evolution) that calculate evolutionary distance between sequences, and character based methods (Maximum Parsimony, Maximum Likelihood, Bayesian) that determine the most probable evolutionary event history between sequences (Table 1).

Statistical testing of topology should also be performed where possible, i.e. bootstrapping analysis. Altering the substitution model, rates and patterns, and treatment of gaps/missing data may also be warranted, though the varying justifications for each of these tests is beyond the detail provided here. Low branch support for any topology shown in the final tree or conflicts in topology determined by multiple testing should be addressed when presenting any phylogenetic data.

Table 1. Classification of phylogenetic analysis methods and strategies.

Tree building strategy	Method	
	Distance based	Character based
Clustering algorithm	UPGMA Neighbour Joining	n/a
Optimality criterion	Minimum Evolution	Maximum Parsimony Maximum Likelihood Bayesian Analysis

8. Genomic resources and tools

8.1. Introduction

Genomic analyses take several forms. At one level, any study that draws inferences for protein-coding genes or other genetic traits in the context of their neighbors on chromosomes is 'genomic'. More recently, genomic studies are those that use massive DNA sequencing strategies to describe and piece together entire sections of the targeted genome, without using PCR or other selective techniques to target specific short regions. Ultimately, genomic studies hope to assemble chromosome-length stretches of an organism's genetic blueprint, and then annotate or describe the functionality of specific regions within chromosomes. The field of genomics is driven by technological advances, including huge cost reductions for the sequencing of samples and advances in both statistical methods and computational resources for analysing the obligatory large datasets. Current estimates indicate that entire pipelines (sets of routines needed for an output analysis) are viable for only six months before becoming obsolete. One helpful review of modern techniques is given by Desai *et al.* (2012) and there are numerous advances and tutorials available via the forums Seqanswers.com and the GALAXY wiki

(http://wiki.galaxyproject.org/FrontPage). Researchers are advised to consult these sources while planning genomic, transcriptomic, and metagenomic projects, as the standards and possible analyses are improving constantly.

8.2. Honey bee genome project

After a multi-year international project, the honey bee genome was described in fall, 2006, in a main overview paper (Honey bee Genome Sequencing Consortium, 2006) and > 30 satellite genome-enabled companion papers (primarily in the journals *Insect Molecular Biology* and *Genome Research*). Sequence data, generated using dideoxy sequencing was assembled into ca. 10,000 contigs (blocks of overlapping sequence reads) spanning *ca.* 238 million base pairs. These contigs are in many cases linked together by scaffolding (a strategy whereby long strands of DNA are sequenced from each end and linked via informatics) leading to an assembly that was > 95% complete for the non-repetitive genome. Honey bee genes and various genomic features are predicted based on homology to other organisms, evidence from RNA expression studies, and evidence for open reading frames. The current genome assembly along with a consensus ("GLEAN") gene set and other resources are available at "Beebase" (www.beebase.org/, Christine Elsik, Univ. Missouri) and at the U.S. National Institutes of Health National Center for Biotechnology Information (http://www.ncbi.nlm.nih.gov/genome?term=apis%20mellifera). Both sites allow for downloading sequences as well as searching the genome via the BLAST family of search algorithms, while the Beebase site also provides the chance to 'browse' the genome visually. Efforts are continuing to improve the primary *Apis mellifera* genome data while adding sequence data from different honey bee strains.

8.3. Honey bee parasite and pathogen genomes

Most of the named RNA viruses for honey bees have been sequenced and published. These genomes are relatively small and tend to be placed into the NCBI databases upon publication (http://www.ncbi.nlm.nih.gov/genomes/GenomesGroup.cgi?taxid=439488). Genome sequences for several parasites and pathogens with larger genomes (e.g., *Paenibacillus larvae*, *Ascosphaera apis*, *Nosema ceranae*, and the mite *Varroa destructor*; (Qin *et al.*, 2006; Cornman *et al.*, 2009; Cornman *et al.*, 2010; Chan *et al.*, 2011) are held at the NCBI as well as at Beebase and can be queried there alongside the above honey bee genome data.

8.4. Comparative genomics

Currently, assembled genome sequences exist for over 30 insects and other arthropods, and that number is soon to increase dramatically (http://arthropodgenomes.org/wiki/i5K). Resources that have proven useful for comparing honey bee genes to those found in other insects (e.g., to *Drosophila* and other insects for which gene function is firmly decided) include the NIH-NCBI and the OrthoDB database run from the University of Geneva (http://cegg.unige.ch/orthodb5). In addition, as each incarnation of the honey bee genome is published, annotations based on presence/absence and functional similarity to other insects is simultaneously added. With projects on four other *Apis* species underway, along with social and solitary apoid bees, strategies for comparative genomics will be in flux for some time. The orthoDB database, along with flybase (www.flybase.org) are good places to start for insights from comparative genomics, and each site has hosts the requisite tools (gene searching/alignment/retrieval) for carrying out comparative analyses.

8.5. Second-generation sequencing

Initial genome projects, from small viruses through the human and honey bee project, all relied on 'Sanger' dideoxy sequencing, a relatively expensive but accurate protocol developed in the 1980's that generates sequences (often drawn from random cloned fragments for the popular 'shotgun' sequence method) of several hundred to 1000 base pairs. Since 2000, there has been a great economization of sequencing, such that current technologies are more than ten-fold less expensive than Sanger sequencing. Nevertheless, Sanger sequencing persists and is often the right strategy when compared to newer technologies (which currently give either quite short or quite inaccurate sequences). Readers should consider ILLUMINA/SOLEXA sequencing (summarized in the methylation section 11. below), 454 pyrosequencing, SOLiD sequencing or the Ion Torrent platform (all platforms are reviewed by Metzger, 2010), and the final decision might rest on local availability along with different strengths of each platform. As of 2012, most DNA and RNA sequencing efforts include at least some component of ILLUMINA sequencing, as that technology is viewed as being most cost-effective.

8.6. Genomic sequence assembly

Standards and tools for genome sequence assembly and analysis are constantly improving and the best strategy for carrying out a genome project is often through collaboration or through mimicking the protocols and computational strategies used by a recent genome project of similar scope (genome size, budget for sequencing and informatics, etc.). Accordingly, we will not list specific pipelines for these processes, but can direct researchers to sites such as http://www.broadinstitute.org/scientific-community/science/programs/genome-sequencing-and-analysis/computational-rd/computational-, http://bioinformatics.igm.jhmi.edu/salzberg/Salzberg/Software.html, and http://soap.genomics.org.cn/soapdenovo.html. Of these options, whole-genome assemblies using the ALLPATHS-LG method (the first link above for the Broad Institute) have been highly successful for both microbes and higher organisms, and this method is arguably the tool of choice currently.

8.7. Transcriptomic analyses ("RNASeq")

Transcriptomic analyses are helpful for seeing trends in honey bee gene expression as well as changes due to experimental conditions, and bee researchers have carried out such studies for many years, adapting as new methods arise. Two recent papers have used the ILLUMINA platform for studying gene regulation in response to nutrition (Alaux *et al.*, 2011) and responsiveness to varroa mites and viruses (Nazzi *et al.*, 2012), respectively. Analyzing RNASeq data will depend on the sequencing platform as well as developments in software and public or personal computational resources, all of which are under constant renewal. Generally, RNASeq experiments rely on differential gene expression (DGE) between categories of one factor (e.g., bees exposed to mites versus controls) and the statistical analysis identifies which regions are up- or down-regulated in the context of this factor. Nazzi *et al.*, 2012 used a technique prescribed by Mortazavi *et al.*, (2008) that, like all current methods, first develops a model for how often a particular expressed region *should* be seen in a sequencing effort, and then uses the number of times that sequence *was* sampled to determine whether it was up-or down regulated compared to an expected level. There are now numerous such methods and both methods and strategies to trim the computational resources for their use are being improved monthly. Public platform with video tutorials for RNASeq analysis that promises to remain current is described at the Galaxy site (https://main.g2.bx.psu.edu/).

8.8. Metagenomics

Metagenomic approaches began as an attempt to study the functional significance of all organisms in a habitat, and the term was first used to describe soil microbes and their collective proteins (Handelsman *et al.*, 1998). In honey bees, usage has so far focused on identifying pathogen taxa (Cox-Foster *et al.*, 2007; Runckel *et al.*, 2011, Cornman *et al.*, 2012) and, more recently, on targeted surveys of bacterial associates in honey bee guts (Martinson *et al.*, 2011; Mattila *et al.*, 2012). There are six key decision points in metagenomic surveys and, rather than choose any specific recipe, we will instead list those decisions and their outcomes:

8.8.1. RNA versus DNA sampling

RNA pools contain those genomes that are actively producing proteins AND genomes of the key RNA viruses in bees. RNA sampling is often used for assessing bee pathogens. DNA sampling is preferable when samples are poorly preserved (given the higher stability of DNA) and as a means of reducing the often overwhelmingly high frequency of ribosomal RNA's (75-80%) in most sequenced RNA samples.

8.8.2. Sample preparation

Both RNA and DNA destined for metagenomic surveys can be extracted using the means described in sections 3 and 4 (e.g. Trizol® extractions for RNA and CTAB extractions for DNA). For more recalcitrant samples (e.g. spore stages, or samples of organisms with impermeable coats, common to bacterial species and fungi) it is important to use mechanical or enzymatic rupturing of the cell coat via proteinase K, as described in the above CTAB DNA extraction protocol (section 3.2.1.), or prolonged shaking with a suspension of low-affinity silica particles or other inert solids.

8.8.3. Amplicon-based or shotgun sequencing

Given the low costs of sequencing, it is feasible now to simply survey all nucleic acids in a sample and then assign them to taxa in various kingdoms via searches of local or online databases. Nevertheless, targeted deep sequencing of specific taxonomic groups can benefit from a selection of specific regions via PCR-based amplification prior to generating the sequencing libraries. This has been done most frequently with the 454 sequencing platform since relatively long read lengths on this platform (> 400 bp) enable the capture of sequence data for a substantial section of the targeted species. Several studies have now used amplicon-based sequencing to describe bacterial populations carried by honey bees. As with any PCR protocol, this approach will under sample taxa with mismatches to the initial primer sequences since no PCR primers are truly 'universal' to a targeted group. Nevertheless, there are many examples of primers that amplify broadly across all of the major bacterial taxonomic groups, and amplicon-based 454 sequencing has appears to provide a consistent and accurate view of bacterial communities in bees. The software environment Qiime (http://qiime.org/) is widely used to match amplicon-based sequences to microbial databases in order to identify and quantify taxa.

8.8.4. Assembly of shotgun sequences vs. read mapping

For sequences generated by shotgun sequencing, it is generally desirable to assemble all sequencing reads into contigs (aggregates of nearly identical sequences from the same region and species) prior to statistical analysis, since this can reduce computational needs greatly while retaining vital statistics including the number of reads per contig. Once the computationally intensive assembly of contigs has taken place (using for example the Metavelvet routine, http://metavelvet.dna.bio.keio.ac.jp/) datasets can be reduced by many orders of magnitude. This is critical if online or 'cloud' databases are searched for microbial matches since the data transfer speeds alone for such searches can be measured in days when using raw sequence reads. In addition, contigs are by definition longer than any individual read and therefore also can provide a more secure match to distant taxa. The count data for sequenced reads per contig provides the measure of depth that, once scaled to contig length, allows estimates of microbial frequency. Once metagenomic sequences have been assembled, moderate experiments can often be enacted without cost to the user at public resources such as GALAXY (https://main.g2.bx.psu.edu/). As with any complicated statistical procedure it is highly possible to get erroneous matches and statistical results, and researchers are advised to enlist the help of colleagues with current expertise here.

In practice, metagenomic analyses are also carried out by mapping (aligning with high probability) individual sequence reads to members of a reference database, and algorithms (including Tophat, http://genomics.jhu.edu/software.html) have been developed that are extremely efficient at doing so. For diagnostic regions with highly conserved sequences (e.g., parts of the rRNA operons) both assembling and mapping are problematic and query sequences often cannot be placed securely to even family-level matches. In this case, it is best to bin sequences at a higher taxonomic level (even Order) rather than force matches into a possibly erroneous taxon. Nevertheless, as genome sequencing of microbial species is increasing exponentially, even rare and distant taxa tend to have a fully sequenced family member in the public databases, as described below in section 8.8.5.

8.8.5. Databases for metagenomics

Several sites have emerged for mapping metagenomic sequence reads and amplicons, including the longstanding Ribosomal Database Project for bacterial and archael 16S alignments (http://rdp.cme.msu.edu/), the SILVA databases for rRNA's generally (http://www.arb-silva.de/), and MEGAN (ab.inf.uni-tuebingen.de/software/megan/), which aspires to map to targets across the tree of life. Each site allows for limited web searching, and for downloading relevant databases for more efficient local searches.

8.8.6. Post-assignment statistics

Quantifying differences between two or more samples in the taxa to which reads or contigs map is the ultimate goal for many metagenomic experiments. While not the only option, MG-RAST (metagenomics.anl.gov/) provides an example of statistical comparisons using read mapping. Assuming read accounts are normalized by size of their target (various methods have been used for this), and then the count frequencies themselves can be used with a variety of standard statistics. Similarly, Qiime, mentioned above in section 8.8.3, provides an effective way for mapping reads to microbial taxa.

9. Fluorescence *In Situ* Hybridization (FISH) analysis of tissues and cultured cells

9.1. Introduction

Fluorescence *in situ* Hybridization (FISH) is a sensitive and specific method for localizing expressed genes or microbes within tissues of the honey bee. In general, a probe matching a specific DNA or RNA sequence is exposed to prepared tissues. This probe can then be localized using fluorescent tags, pointing the researcher to the precise location of a desired target. To date this method has been used successfully to show the locations of bacterial associates of bees (Martinson *et al.*, 2012, Yue *et al.*, 2008, and below).

9.2. Tissue fixation and tissue sectioning exemplified with gut tissue

1. Immobilize about 20 bees with CO_2.
2. Cut of the head.
3. Fix the abdomen on a separation plate with micro pins.
4. Remove carefully the alimentary tract of each bee with forceps.
5. Transfer the hindgut and the midgut into one well of a 24-well microtiter plate.
6. Fix tissues in 4% formalin (Roth) for 24 hours at 4°C by shaking.

The further embedding and blocking procedure using e.g. Technovit 8100 and Technovit 3040 kits (Heraeus-Kulzer) should be performed as given in the manufacturer's protocols (Heraeus-Kulzer, T8100 embedding kit).

7. Wash the alimentary tracts with 6.8% sucrose in 1xphosphate buffered saline (1xPBS, pH 7.0) for 24 hours at 4°C.
8. For dehydration transfer the tissue in 100% acetone for one hour.
9. Pre-infiltrate the organs with T8100 basic-solution and 100% acetone (mixed 1:1) for two hours.
10. Prepare the infiltration solution (0.6 g hardener I in 100 ml T8100 basic solution).
11. Transfer the organs into the infiltration solution.
12. Incubate at 4°C for at least 24 hours and up to one week, depending on tissue size.
13. Apply careful shaking for better infiltration results.
14. Prepare the embedding solution (mix 0.5 ml hardener II with 15 ml infiltrating solution).
15. Fill the mould of a Teflon-embedding form (pre-cooled at -20°C) with the embedding solution.
16. Transfer the tissue and orientate it in the mould.
17. Close the well immediately with a plastic strip.
18. Incubate at 4°C for 3 hours to allow polymerization.
19. Finally, block the polymerized probes with histoblocs and with the Technovit 3040 kit (both from Heraeus-Kulzer) using the manufacturer's protocol.
20. Prepare semi thin-sections (2-4 μm) with a rotation microtome (Leica).
    Use a knife with a hard metal edge (Tungsten).
21. For fluorescence *in situ*-hybridization transfer the tissue sections on Polysine™-covered glass slides (Fisher Scientific, Menzel-Gläser).

9.3. Fixation of cultured cells grown in suspension

1. Transfer 100 μl of cultured cells into a cell funnel-chamber of a cell spin (Tharmac).
2. Centrifuge the cells with 600 rpm (54xg) gently for 5 minutes on a glass slide (VWR).
3. Remove the cell funnel-chamber.
4. Let the medium air dry.
5. Fix the cells in 4% formalin (Roth) at 4°C for 24 hours.

9.4. FISH-analysis of tissue sections and fixed insect cells

1. Wash the slides (tissue sections and fixed insect cells) twice in 1xPBS.
2. For further processing, transfer each slide into a 10 ml dish.
3. Add 10 ml of 1 μg ml^{-1} Proteinase K in Proteinase K-buffer (0.2 M Tris-HCl, pH 7.5).
4. Transfer the dish into a humid chamber (Eppendorf, Thermostat Plus) for 5 min at 37°C.
5. Remove the Prot K and wash each slide with 10 ml of 1xPBS.
6. For post-fixation add 10 ml of 4% formalin (Roth).
7. Incubate at RT for 20 min.
8. Aspirate the formalin.
9. Remove remaining solution by washing the slides three times with 1xPBS-buffer.
10. Prepare hybridization buffer:
    *   200 μl of 100% formamide,
    *   180 μl 5 M NaCl,
    *   20 μl 1 M Tris/HCl,
    *   1 μl 10% SDS,
    *   599 μl DEPC-H_2O, pre-warm to 46°C in a heating block.
11. Add the probes to 37.5 μl pre-warmed (46°C) hybridization buffer:
    7.5 μl species-specific probe annealing to a region of the 16S rRNA or another species-specific genomic region of the pathogen to be detected labelled with fluorescein isothiocyanate-FITC with a final concentration of 15 ng μl^{-1}.

Sequence of *Nosema* spp.-probes: Gisder *et al.* (2011); sequence of DWV-probe: Möckel *et al.* (2011); sequence of *P. larvae*-probe: Yue *et al.* (2008)

> 5 µl Euk516-probe (5'-ACCAGACTTGCCCTCC-3', universally detecting eukaryotic ribosomes by hybridizing to a universal conserved sequence of the eukaryotic 18S rRNA) labelled with sulforhodamine 101 acid chloride-Texas Red® with a final concentration of 10 ng µl⁻¹.

12. Continue incubation in the heating block which is now covered with a lid (i.e. incubation at 46°C in the dark).
13. Cover the slides with LifterSlips (VWR).
14. Pipette 50 µl of hybridization buffer to each slide (tissue sections or fixed cells).
15. Transfer the slide into a hybridization chamber (Corning, Corning chamber), drop 15 µl H$_2$O into the given wells to preserve the humidity.
16. Close the chamber tightly.
17. Put the corning chamber in a 46°C water bath for overnight hybridization.
18. Open the hybridization chamber and remove the cover slips in 1xPBS.
19. Wash the slides three times with 1xPBS.
20. Let them air dry.
21. Stain the nuclei with 50 µl 4′, 6-Diamidin-2-phenylindol- (DAPI, 1 µg ml⁻¹ in 99% methanol) solution for 10 min in the dark.
22. Wash the slides again three times with 1xPBS.
23. Let them air dry.
24. Cover the slides with the ProLong Gold antifade reagent (Invitrogen) and a cover slip (Roth).
25. Analyse the tissue sections and the cells under an inverse fluorescence microscope (e.g. Nikon, Ti-Eclipse) at 100-fold and 600-fold magnification using consecutively a FITC-, TexasRed- and DAPI-filter.

10. RNA interference

10.1. Introduction

RNA interference (RNAi) is a cellular mechanism leading to a knock-down of gene expression mediated by target specific double-stranded RNA (dsRNA) molecules (Fire *et al.*, 1998). Understanding the mechanism of mRNA destruction by these dsRNA molecules dramatically increased the possibilities of functional genomics studies during the last decade especially in organisms where the recovery of mutants is not feasible. Thus RNAi has become a dominant reverse genetic method for the study of gene functions and furthermore, plays an increasing role in therapeutics and in pest control (Maori *et al.*, 2009; Liu *et al.*, 2010; Hunter *et al.*, 2010).

Up to now a dozen studies report on the successful usage of RNAi in honey bees. But the application methods and also the choice of RNAi effective molecules are very diverse. Several studies report on the application of dsRNA to eggs and larvae whether by injection (Aronstein and Salivar, 2005; Beye *et al.*, 2002; Maleszka *et al.*, 2007) or ingestion (Aronstein *et al.*, 2006; Patel *et al.*, 2007; Kucharski *et al.*, 2008; Nunes and Simoes, 2009; Liu *et al.*, 2010). Others report on a successful manipulation of adult bees (Amdam *et al.*, 2003; Farooqui *et al.*, 2004; Seehuus *et al.*, 2006; Schlüns and Crozier, 2007; Maori *et al.* 2009; Paldi *et al.*, 2010; Mustard *et al.*, 2010; Jarosch and Moritz, 2011; Jarosch *et al.*, 2011).

This section aims at a collection of RNAi protocols successfully applied in honey bees beforehand. The well-established protocols for producing dsRNA as well as siRNA (short interfering RNAs, the products of dsRNA once the enzyme Dicer and its partners have processed them) molecules are presented. Moreover, the two application methods feeding and injection are presented and compared to each other. In conclusion, we summarize five important factors that may decrease the effectiveness of target gene expression knock-down.

10.2. Production of RNA interfering molecules

10.2.1. siRNA design and synthesis

So far most bee scientists have used dsRNA rather than siRNA for RNAi experiments. Although dsRNA molecules have advantages in handling, off-target effects (Jarosch *et al.*, 2012) have been reported in honey bees. Therefore the usage of siRNAs is recommended where feasible. This allows the selection of one or a few short sequences to initiate RNAi, rather than the many tens of possible permutations generated by a typical dsRNA construct, any of which might cause effects away from the desired target.

1. Design 3-6 siRNAs for your target gene in order to find an optimal siRNA.

   General guidelines for siRNA design:
   - siRNA targeted sequence is usually 21 nt in length.
   - Avoid regions within 50-100 bp of the start codon and the termination codon.
   - Avoid intron regions.
   - Avoid stretches of 4 or more bases such as AAAA, CCCC.
   - Avoid regions with GC content < 30% or > 60%.
   - Avoid repeats and low complex sequence.
   - Avoid single nucleotide polymorphism (SNP) sites.
   - Perform BLAST homology search to avoid off-target effects on other genes or sequences (16- to 18-nt–long stretches of homology are suggested as the maximum acceptable length in RNAi studies per Ambion siRNA design guidelines).

- Design negative controls by scrambling the target siRNA sequence. This control RNA has the same length and nucleotide composition as the target specific siRNA but in a different order. Make sure that the scrambled siRNA does not show homologies for any known bee gene.

Several web based programs for appropriate siRNA design, which implement the actual siRNA design algorithms, are available (e.g. siRNA target designer version 1.6 (Promega); siDesign center (Dharmacon, Inc); Block-iT™ RNAi Designer (Invitrogen).

2. Use T7 Ribomax™ Express RNAi System (Promega) for siRNA production.
    1. Follow the manufacturers´ instructions.
    2. Incubation time may be increased in order to increase the siRNA yield (A time-course experiment has to be performed beforehand in order to find the optimal incubation time).
3. Assess the quality and quantity by photometric measurements (OD260) and by capillary gel electrophoresis (alternatively agarose gel electrophoresis, see section 3.2.1).

10.2.2. Production of dsRNA

Since dsRNAs can cause off-target effects, you need to be careful in designing them. Nevertheless, RNAi efforts using dsRNA constructs have proven effective in honey bees. To avoid targets that might interfere with other honey bee genes, you need to compare your sequence with the honey bee genome during the design process using the Basic Local Alignment Tool (www.ncbi.nlm.nih.gov). Make sure none of the designed dsRNAs has 20-bp segments identical to any known bee sequence. As dsRNAs are processed by the dicer complex into a cocktail of siRNAs 19–21 nt in length, the absence of 20-nt stretches of homology minimizes the possibility of off-target effects.

1. Use the E-RNAi web application (Horn and Boutros, 2010) for optimal dsRNA design.

Design of dsRNA sequences has to be stringent in order to avoid/minimize off-target effects.

2. Set up appropriate negative controls.

Note: be careful using GFP; this sequence might cause off-target effects in some cases (GenBank ID: U17997, Clontech; Jarosch and Moritz, 2012).

Other possible negative controls: e.g., Q-marker (Beye *et al.*, 2002).

3. Amplify the chosen target fragment by using target specific T7 (TAA TAC GAC TCA CTA TAG GGC GAT) added primer in optimized PCRs using approximately 100-ng genomic DNA obtained by chloroform– phenol extraction (e.g. Maniatis *et al.*, 1982).

4. Clone the amplified fragments into pGem-T easy vectors (Promega) according to the manufacturer's instructions. (Cloning eliminates the possibility of a dsRNA mixture due to a polymorphism of the PCR product).
5. Transform your plasmids into JM109 competent cells (Promega) following the instructions from the manufacturer.
6. Prepare the plasmids according to Del Sal (1988).
7. Analyse the identity of the cloned sequence by Sanger sequencing.
8. Once the right clone has been identified its insert needs to be amplified to serve as a template for dsRNA production by standard PCR using again T7 tailed primers.
    8.1. E.g. use Biotherm DNA Polymerase (Genecraft); chemicals:
    - 0.2 mM dNTPs,
    - 0.3 µM of T7-promotor added primer,
    - 5 U Taq Polymerase,
    - in a total reaction volume of 100 µl.
    8.2. PCR protocol:
    - 5 min DNA denaturation, and Taq activation, at 95°C,
    - 40 cycles of:
      95°C for 30 sec,
      x°C (primer specific annealing temperature) 30 sec,
      72°C for 1 min.
    - A final extension of 20 min at 72°C completes the protocol.
9. Purify the PCR-products with the QIAquick® PCR Purification Kit (Qiagen).
10. Use the T7 RibomaxTM Express RNAi System (Promega) for dsRNA production.

Time course experiments and experiments for optimizing the incubation temperature have to be conducted beforehand (e.g. Jarosch *et al.*, 2011 used an extended transcription time of 5 h at 32°C).

11. Purify the dsRNA by a Trizol® (Invitrogen) - chloroform-treatment following the manufacturers´ instructions.
12. Resolve the pellet in nuclease free water.
13. Assess the dsRNA quality and quantity photometrically and by agarose gels or capillary gel electrophoresis.

The photometric measurement of the OD260/OD280 ratio should be between 1.8 and 2. A lower ratio indicates contamination with proteins. As a contamination with DNA or dsRNA degradation cannot be detected by photometry, visualization of the dsRNA product is necessary. For this 1.5% agarose gels can be used, see section 3.2.1). A single distinct band should be visible.

14. Adjust dsRNA concentrations to 5 µg/µl by diluting with insect ringer (54 mM NaCl; 24 mM KCl; 7 mM $CaCl_2$ x $2H_2O$) right before the injection.

10.3. RNAi Applications

10.3.1. RNAi in adult honey bees via feeding

1. Take newly emerged bees (1-2 d old) from one colony from one brood frame.
2. Set up at least two controls:
   1. Bees fed with 50% sugar water alone.
   2. Bees fed with scrambled siRNA (siRNA with exactly the same nucleotides as the target siRNA but in an altered order lacking any similarity with other known bee genes).
3. Take a mixture of two siRNAs specific for the target gene.

Note: in previous experiments a mixture of two siRNAs was more effective than single siRNAs (Jarosch *et al.*, 2011).

4. Put 35-40 newly emerged bees in wooden cages (see the *BEEBOOK* paper on maintaining adult honey bees in vitro under laboratory conditions (Williams *et al.*, 2013)) supplied with a small comb and pollen *ad libitum*.
5. Put cages in temperature controlled incubators (see the *BEEBOOK* paper on maintaining adult honey bees in vitro under laboratory conditions (Williams *et al.*, 2013)) and feed with 1.5 ml 50% sugar water containing approximately 1 µg siRNA per insect every 24 hours.
6. Dependent on the actual experiments bees can be held for several weeks.
7. Once the experiment is finished, bees should be shock-frozen in liquid nitrogen in order to maintain the RNA status.

10.3.2. RNAi in honey bee larvae via feeding

1. Take a comb with second instar larvae out of the colony.
2. Transfer it to the lab.

The whole treatment is conducted at room temperature.

3. Draw a map of the different treatment groups on the very same comb for future identification of the treated individuals.
4. Apply 1 µl of sugar solution containing the respective amount of dsRNA directly into the cells. Deposit it at the bottom of the worker brood cell that contains a drop of food. Avoid touching the larvae. Successful experiments used dsRNA concentrations between 0.5 µg (Nunes and Simões, 2009) and up to 1.26 µg (Aronstein *et al.*, 2006).

In addition to the first dsRNA feeding you may feed another µg of your dsRNA after 12 hours. This feeding cycle can be repeated for several days (Liu *et al.*, 2010) until the life stage of interest is reached.

5. Place the comb back to its host colony two hours after treatment and take samples at the life stage you are interested in.

10.3.3. Gene knock-down by abdominal injection of target-specific dsRNA/siRNA

RNA interfering molecules injected by intra-abdominal injection do not reach every tissue (Jarosch and Moritz, 2011). But especially the fat body can be easily reached by this user friendly method (Amdam *et al.*, 2003; Jarosch and Moritz, 2011).

1. Take age-defined workers (see the *BEEBOOK* paper on miscellaneous methods (Human *et al.*, 2013)).

Note: newly emerged workers are a little bit more difficult to inject as their abdomen is quite flexible.

2. Immobilise bees by cooling down in at 4°C.
3. Fix the bees on wax plates using small fixing pins.
4. Inject 5 µg of freshly diluted dsRNA or alternatively 3 µg of freshly diluted siRNA (treatment and control dsRNA/siRNA) between the 5th and 6th abdominal segment using a 10 µl microsyringe (e.g. Hamilton).
5. Inject negative controls with insect ringer (54 mM NaCl; 24 mM KCl; 7 mM $CaCl_2$ x $2H_2O$).
6. Keep the injected workers on wax plates until they recover and keep bees not showing haemolymph leakage (visible on their substrate or as a droplet on the cuticle) together with about 25 nurse bees (1-10 days) in cages (see the *BEEBOOK* paper on maintaining adult honey bees in vitro under laboratory conditions (Williams *et al.*, 2013)).
7. Sacrifice the bees by shock-freezing in liquid nitrogen.
8. Store them at -80°C until tissue preparation.
9. Prepare the worker tissues on cooled wax plates using an RNA Stabilization Reagent (e.g. RNAlater®) in order to avoid RNA degradation.

10.4. Concluding remarks

Based on the literature (see Huvenne and Smagghe, 2012 for review) five aspects seem to be most important to conduct successful RNAi knockdown experiments in honey bees.

Concentration of dsRNA: For every target gene the most effective concentration of RNAi molecules has to be determined. It does not follow the general rule: The more the better. Nunes and Simões (2009) for example report on the removal of 2nd instar larvae, which were fed with 3 and 5 µg dsRNA. In contrast, larvae fed with just 0.5 µg dsRNA did not show a significant higher removal rate than the control group, and moreover exhibited an mRNA silencing effect of about 90%.

Nucleotide sequence: Sequences of the RNAi effective molecules have to be carefully designed and tested in order to avoid off-target effects.

Length of the dsRNA fragment: When not using siRNAs the length of the dsRNA fragments may be crucial for uptake and efficient silencing. Most experiments used dsRNA ranging from 300 to 520 bp (see Huvenne and Smagghe, 2012 for review). Moreover, a minimal length of 211 bp is suggested in S2 cells (Saleh *et al.*, 2006).

Honey bee life stage: Although adults are easier to handle, literature of other insects suggest a larger silencing effect in younger life stages. E.g. in fall armyworms (*Spodoptera frugiperda*) the silencing effects after RNAi treatment were reported to be less effective than in *S. frugiperda* larvae (Griebler *et al.*, 2006). Thus the usage of larvae rather than adults where feasible may be advisable in honey bees as well.

Application method: The two application methods presented here both have pros and cons. The feeding regimes lead to an individually different consumption of food and therefore to the ingestion of different dosages of dsRNA. But in contrast to injections protocols, feeding is much easier in handling and moreover, it causes less stress in the target animals. Moreover, studies suggest, that the composition of the target tissue may have some influence on the accessibility of dsRNA when choosing injection as application method (Jarosch and Moritz, 2011).

11. DNA methylation in honey bees

11.1. Introduction

Methylation of chromosomal DNA is a flexible epigenetic mechanism that plays a critical role in gene regulation, and patterns of methylation across the genome are often surrogates for interesting sets of proteins that are regulated in concert with each other and with biological traits. In order to detect methylated bases in genomic DNA (essentially only cytosines are methylated), DNA has to be treated with bisulfite to convert non-methylated cytosines to uracil and subsequently to thymine during the PCR amplification step.

11.2. DNA methylation in honey bees

So far, four full methylomes (genome-wide methylation patterns) have been generated for *Apis mellifera* using the following tissues: adult brains of queens and workers and 96 hrs-old queen and worker larval heads (Lyko *et al.*, 2010; Foret *et al.*, 2012). The below protocol describes methylation analyses of DNAs extracted from the dissected brains of 50 age matched active queens and 50 8-day old workers (dissection of clean, gland-free brains is shown at: http://dl.dropbox.com/u/59152790/Brain%20dissection%20Maleszka%20lab.wmv).

A similar protocol was used to generate larval methylomes (Foret *et al.*, 2012) and in principle this procedure should work for any bee tissue and/or life stage from which intact RNA's can be extracted as below (section 11.3.).

11.3. DNA extraction from various tissues for methylation analysis

Methylation analyses do not depend on a particular DNA extraction method. Nevertheless, the below extraction has been validated in a variety of honey bee tissues.

1. Homogenize tissues in a 1.5 ml microcentrifuge tube in a small volume of NTE buffer:
   - 100 mM NaCl,
   - 50 mM Tris pH 8.2,
   - 10 mM EDTA,
   - 1% SDS,
   - Proteinase K (500 µg per ml, freshly dissolved).
2. Add a small amount (0.01%) of a non-ionic detergent such as Triton X100.

The detergent is beneficial (increases the efficiency of proteinase K digestion), but not necessary.

3. Add more buffer (roughly 500 µl per 20-50 mg of tissue).
4. Incubate at 55°C for 1-3 hrs.
5. Extract with 1 volume of phenol: chloroform.
6. Spin for 10 min at 10,000rpm.
7. Collect the upper phase (repeat the extraction if the upper phase looks cloudy).
8. Add 1 µl of RNase A (10 mg/ml).
9. Incubate for 10 min at 37°C to digest RNA.

This step is not necessary for bisulfite conversion, but the presence of RNA interferes with measuring DNA yield.

10. Precipitate DNA with 1 volume of isopropanol or 2 volumes of EtOH.
11. Spin gently (5,000 rpm for 2 min).
12. Discard the supernatant.
13. Wash the pellet once with 70% EtOH.
14. Remove ethanol, but DO NOT DRY THE PELLET!
15. Dissolve the pellet in TE buffer by heating to 65°C.
16. Store at 4°C (or at -80°C for long term storage).

Clean DNA preps are stable at 4°C for at least 5 years. The majority of DNA strands from the above prep are 200-250 kb in length with the smallest molecules around 70 kb.

Note: DNA preps from larvae might appear milky after this procedure, but such preps are suitable for bisulfite conversions. Alternatively use the MasterPure DNA Purification kit from AMRESCO (Cat. No. MCD85201) yields cleaner larval preps.

11.4. High-throughput sequencing of targeted regions

11.4.1. Fragmentation of DNA

1. Fragment 5µg of high molecular weight DNA using the Covaris S2 AFA System in a total volume of 100µl.
   Fragmentation-run parameters:
   - Duty cycle 10%,
   - Intensity: 5,
   - Cycles/burst: 200,
   - Time: 3min,
   - Number of cycles: 3,
   - This results in a total fragmentation-time of 180s.
2. Confirm fragmentation with a 2100 Bioanalyzer (Agilent Technologies) using a DNA1000 chip, aiming for fragment sizes of 140 bp on average for both queen and worker DNAs.

11.4.2. End-repair of sheared DNA

Having a blunt (neither strand longer than the other) end to each double-stranded DNA section is needed to attach the below adaptors ('handles' that help connect DNA during library formation), and this is achieved as follows:

1. Concentrate fragmented DNA to a final volume of 75 µl using a DNA Speed Vac.
2. End-repair fragmented DNA in a total volume of 100 µl using the Paired End DNA Sample Prep Kit (Illumina, PE-102-1001) following manufacturer's protocol.

11.4.3. Adaptor ligation

Ligate adaptors using the Illumina Early Access Methylation Adaptor Oligo Kit (P/N: 1006132) and the Paired End DNA Sample Prep Kit (Illumina, PE-102-1001), as recommended by the manufacturer.

11.4.4. Size selection of adapter-ligated fragments

For the size selection of the adaptor-ligated fragments use the E-Gel Electrophoresis System (Invitrogen) and a Size Select 2% precast agarose gel (Invitrogen) as below.

1. Load each fragmented DNA on two lanes of an E-gel.
2. Electrophorese using the "Size Select" program for 16 min.
3. Using a size standard (50 bp DNA Ladder, Invitrogen, Cat no. 104 16-014), extract 240 bp fragments from the gel.
4. Pool samples and directly transfer to bisulfite treatment without further purification.

11.4.5. Bisulfite conversion and amplification of the final library

1. Bisulfite treatment can be carried out with the EZ-DNA Methylation Kit (Zymo) as recommended by the manufacturer, with the exception of a modified thermal profile for the bisulfite conversion reaction. Alternatively the QIAGEN EpiTect Bisulfite Kit can be used.
   The conversion is carried out in a thermal cycler using the following thermal profile:
   15 cycles of:
   95° C for 15 sec,
   50° C for 1hr,
   Incubate at 4° C for at least 10 min.
2. Amplify the resulting libraries using the Fast Start High Fidelity PCR System (Roche, 03 553 400 001) with buffer 2, the Illumina PE1.1 and PE2.1 amplification primers, and the below protocol.
   PCR thermal profile:
   95°C- 2min,
   11 cycles of:
   95°C for 30 sec,
   65°C for 20 sec,
   72°C for 30 sec,
   72°C for 7 min,
   20°C hold.
3. Purify products on PCR purification columns (MinElute, Qiagen), eluting in 20 µl elution buffer (Qiagen).

11.4.6. Validation of the libraries

1. Analyse 1 µl of the libraries on a 2100 Bioanalyzer (Agilent Technologies) using a DNA1000 chip.
2. Confirm product size of ca. 240 base pairs and adequate quantity using the DNA1000 size standards.

11.4.7. Sequencing and data analysis

1. Use a Solexa Genoma Analyzer GAIIx with a v2 Paired End Cluster Generation Kit - GA II (Illumina, PE-203-2001) and v3 36 bp Cycle Sequencing Kits (Illumina, FC-104-3002) following manufacturer's protocols, for sequencing.
2. Extract sequences using Illumina Pipeline v1.4 software.
3. Perform image analysis and base calling using Illumina SCS v2.5 software.

8 pM of material is used per sequence lane, generating between 10 and 16M sequence reads.

11.5. Mapping and methylation assessment

1. Trim the above sequence data using the Illumina Data Analysis Pipeline.
2. Align bisulfite-converted sequencing reads to the honey bee genome using the BSSeeker software (http://pellegrini.mcdb.ucla.edu/BS_Seeker/BS_Seeker.html) as described in Foret *et al.* (2012); http://www.pnas.org/content/suppl/2012/03/12/1202392109.DCSupplemental/pnas.201202392SI.pdf#nameddest=STXT.

3. Reads containing consecutive CHN nucleotides are the product of incomplete bisulfite conversion and must first be discarded.

4. To increase the accuracy of methylation calls, only those cytosines fulfilling neighbourhood quality standards are counted (bases of quality 20 or more, flanked by at least three perfectly matching bases with a PHRAP quality score of 15 or more).

5. The methylation status of each cytosine base can be modelled by a binomial distribution with the number of trial equal to the number of mapping reads and the probability equal to the conversion rate.

6. A base is called methylated if the number of reads supporting a methylated status departed from this null model significantly at the 5% level after correcting for multiple testing.

7. Differentially methylated genes are identified using generalized linear models of the binomial family; the response vector *CpGmeth* (number of methylated and non-methylated reads for each CpG in a gene) was modelled as a function of two discrete categorical variables, the caste and the CpG position: CpGmeth = caste * CpGi.

8. P-values are corrected for multiple testing using the Benjamini and Hochberg method. These tests are carried out using the R statistical environment (http://www.r-project.org).

9. Honey bee ESTs and predicted genes are loaded into a Mysql database and visualized with Gbrowse, where CpG methylation levels in queens and workers are added as separate tracks.

11.6. Methylation dynamics and expression of individual genes

Targeted analyses of selected genes can be conducted using 454 sequencing of amplified gene fragments from bisulfite-converted DNAs. For a small-scale testing the amplicons can be cloned into a plasmid, cloned and sequenced using via Sanger dideoxy sequencing (Foret *et al.*, 2009). Both approaches can be used to validate genome-wide methylation data, but 454 sequencing allows for a much higher coverage, as shown in section 8.5.

11.6.1. Amplicon sequence selection

1. Illumina sequencing and BSMAP mapping results can be confirmed by 454 sequencing of a set of bisulfite amplicons.

2. Specific amplicon sequences are selected using raw methylome data and the following arbitrary criteria:
   1. Minimum coverage - 5 mapped reads for each queen and worker sample.

2. Minimum 2 methylated CpGs within a region of ~400-600 bp of sequence showing at least 50% difference in methylation levels between the two samples. This selection is very stringent, but assures that amplicons with high probability of differential methylation are selected.

3. In addition a few regions of mtDNA that is not methylated are selected as controls (optional).

11.6.2. Bisulfite DNA conversion

1. The Qiagen EpiTect Bisulfite Kit and the manufacture's protocol is widely accepted as the most efficient and reliable kit for DNA conversion. The amount of starting materials can range from 0.1 to 2 µg.

2. Because DNA conversion with bisulfite is only ~98% efficient it is highly recommendable to repeat this protocol twice.

3. $1/10^{th}$ of the second conversion reaction is sufficient for subsequent amplification.

11.6.3. Bisulfite PCR

Bisulfite amplicons are amplified using a nested PCR protocol (Wang *et al.*, 2006; Foret *et al.*, 2009). Nested primers contained an additional 9 nucleotide-long linkers with EcoRI or HindIII recognition sequences allowing directional cloning of the amplicons.

PCR reactions are performed in 25 µl volume containing:
- 1x PCR buffer,
- mM $MgCl_2$,
- mM dNTP,
- 50 pmol each forward and reverse primer,
- 5 units Taq polymerase.

Reaction efficiencies are optimized via annealing temperature gradients (Mastercycler gradient PCR machine, Eppendorf) and testing multiple Taq polymerases such as GoTaq (Promega) or FastStart Taq (Roche).

Cycling profile is as follows:
- Initial denaturation at 94^0C for 2 min,
- Followed by 40 cycles of:
    15 sec denaturation at 94^0C,
    15 sec annealing at primer-specific optimal temperature,
    60 sec extension at 72^0C,
- A final extension cycle at 72^0C for 5 min.

When using FastStart Taq polymerase, the denaturation temperature was increased to 95^0C, initial denaturation time to 5 min and cycling denaturation and annealing times to 30 sec.

11.7. RNA extraction

1. RNA for analysis can be extracted using the TRIzol®/QIAGEN RNeasy combination method followed by the QIAGEN Mini or Rneasy Minelute Cleanup kit, or by the detailed protocol in the RNA methods section 4.3. above.

2. RNA concentrations are then evaluated via Nanodrop (section 3.2.1) analyses and integrity assessed by gel electrophoresis (see section 4.4.).

11.8. cDNA synthesis and template quantification

1. Typical first-strand reactions consist of a 20µ volume containing:
   - 0.5-2 mg of total RNA, or 50-100 ng of poly(A)RNA,
   - 100 pmol of anchored d(T)20VN primer,
   - 200 units of Superscript III (Invitrogen),
   - 1X concentration of proscribed buffer.

2. The tube is incubated for 1 h at 50°C.

3. Terminate by adding 30 µl of TE buffer and freezing.

4. Resulting cDNAs can be screened for levels of specific genes via quantitative-PCR as described in the RNA methods in section 4.7.

12. Acknowledgements

We thank four careful reviewers and Editors Peter Neumann and Vincent Dietemann for their insights and advice. The COLOSS (Prevention of honey bee COlony LOSSes) network aims to explain and prevent massive honey bee colony losses. It was funded through the COST Action FA0803. COST (European Cooperation in Science and Technology) is a unique means for European researchers to jointly develop their own ideas and new initiatives across all scientific disciplines through trans-European networking of nationally funded research activities. Based on a pan-European intergovernmental framework for cooperation in science and technology, COST has contributed since its creation more than 40 years ago to closing the gap between science, policy makers and society throughout Europe and beyond. COST is supported by the EU Seventh Framework Programme for research, technological development and demonstration activities (*Official Journal L 412, 30 December 2006*). The European Science Foundation as implementing agent of COST provides the COST Office through an EC Grant Agreement. The Council of the European Union provides the COST Secretariat. The COLOSS network is now supported by the Ricola Foundation - Nature & Culture.

13. References

AGÜERO, M; SAN MIGUEL, E; SÁNCHEZ, A; GÓMEZ-TEJEDOR, C; JIMÉNEZ-CLAVERO, M A (2007) A fully automated procedure for the high-throughput detection of avian influenza virus by real-time reverse transcription–polymerase chain reaction. *Avian Diseases* 51: 235–241. http://dx.doi.org/10.1637/7634-042806R1.1

ALAUX, C; DANTEC, C; PARRINELLO, H; LE CONTE, Y (2011) Nutrigenomics in honey bees: Digital gene expression analysis of pollen's nutritive effects on healthy and varroa-parasitized bees. *BMC Genomics* 12: 496. http://dx.doi.org/doi:10.1186/1471-2164-12-496

AMDAM, G; SIMOES, Z; GUIDUGLI, K; NORBERG, K; OMHOLT, S (2003) Disruption of vitellogenin gene function in adult honey bees by intra-abdominal injection of double-stranded RNA. *BMC Biotechnology* 3: 1. http://dx.doi.org/doi:10.1186/1472-6750-3-1.

ARONSTEIN, K; PANKIW, T; SALDIVAR, E (2006) SID-1 is implicated in systemic gene silencing in the honey bee. *Journal of Apicultural Research* 45(1): 20-24.

ARONSTEIN, K; SALDIVAR, E (2005) Characterization of a honey bee toll related receptor gene am18w and its potential involvement in antimicrobial immune defense. *Apidologie* 36(1): 3-14. http://dx.doi.org/10.1051/apido:2004062

BECKER, C; HAMMERLE-FICKINGER, A; RIEDMAIER, I; PFAFFL, M W (2010) mRNA and microRNA quality control for RT-qPCR analysis. *Methods* 50: 237-243. http://dx.doi.org/10.1016/j.ymeth.2010.01.010

BECKER, S; FRANCO, J R; SIMARRO, P P; STICH, A; ABEL, P M; STEVERDING, D (2004) Real-time PCR for detection of *Trypanosoma brucei* in human blood samples. *Diagnostic Microbiology and Infectious Disease* 50(3): 193-199. http://dx.doi.org/ 10.1016/j.diagmicrobio.2004.07.001

BEYE, M; HÄRTEL, S; HAGEN, A; HASSELMANN, M; OMHOLT, S W (2002) Specific developmental gene silencing in the honey bee using a homeobox motif. *Insect Molecular Biology* 11(6): 527-532.

BIRCH, L; ENGLISH, C A; BURNS, M; KEER, J T (2004) Generic scheme for independent performance assessment in the molecular biology laboratory. *Clinical Chemistry* 50(9): 1553-1559. http://dx.doi.org/10.1373/clinchem.2003.029454

BRUUN-RASMUSSEN, T; UTTENTHAL, Å; HAKHVERDYAN, M; BELÁK, S; WAKELEY, P R; REID; S M; EBERT, K; KING, D P (2009) Evaluation of automated nucleic acid extraction methods for virus detection in a multicenter comparative trial. *Journal of Virological Methods* 155: 87–90. http://dx.doi.org/DOI:10.1016/j.jviromet.2008.09.021

BUSTIN, S A (2000) Absolute quantification of mRNA using real-time reverse transcription polymerase chain reaction assays. *Journal of Molecular Endocrinology* 25(2): 169-193. http://dx.doi.org/10.1677/jme.0.0250169

BUSTIN, S A; BEAULIEU, J F; HUGGETT, J; JAGGI, J; KIBENGE, R; FSBOLSVIK, P A; PENNING, L C; TOEGEL, S (2010) MIQE précis: Practical implementation of minimum standard guidelines for fluorescence-based quantitative real-time PCR experiments. *BMC Molecular Biology* 11: e74. http://dx.doi.org/10.1186/1471-2199-11-74

BUSTIN, S A; BENES, V; GARSON, J A; HELLEMANS, J; HUGGETT, J; KUBISTA, M; MUELLER, R; NOLAN, T; PFAFFL, M W; SHIPLEY, G L; VANDESOMPELE, J; WITTWER, C T (2009) The MIQE Guidelines: minimum information for publication of quantitative Real-Time PCR experiments. *Clinical Chemistry* 55(4): 611-622. http://dx.doi.org/10.1373/clinchem.2008.112797

BUSTIN, S A; NOLAN, T (2004) Pitfalls of quantitative real-time reverse-transcription polymerase chain reaction. *Journal of Biomolecular Techniques* 15(3): 155-166.

CAETANO-ANOLLES, G (1998) DAF optimization using Taguchi methods and the effect of thermal cycling parameters on DNA amplification. *Biotechniques* 25(3): 472-6, 478-80.

CANDOTTI, D; TEMPLE, J; OWUSU-OFORI, S; ALLAIN, J P (2004) Multiplex real-time quantitative RT-PCR assay for hepatitis B virus, hepatitis C virus, and human immunodeficiency virus type 1. *Journal of Virological Methods* 118(1): 39-47. http://dx.doi.org/10.1016/j.jviromet.2004.01.017

CHAISOMCHIT, S; WICHAJARN, R; JANEJAI, N; CHAREONSIRIWATANA, W (2005) Stability of genomic DNA in dried blood spots stored on filter paper. *Southeast Asian Journal of Tropical Medicine and Public Health* 36(1): 270-273

CHAMOLES, N A; NIIZAWA, G; BLANCO, M; GAGGIOLI, D; CASENTINI, C (2004) Glycogen storage disease type II: enzymatic screening in dried blood spots on filter paper. *Clinica Chimica Acta* 347(1-2): 97-102. http://dx.doi.org/10.1016/j.cccn.2004.04.009

CHAN, Q; CORNMAN, R S; BIROL, I; LIAO, N; CHAN, S; DOCKING, T R; JACKMAN, S; TAYLOR, G; JONES, S; DE GRAAF, D; EVANS, J; FOSTER, L (2011) Updated genome assembly and annotation of *Paenibacillus larvae*, the agent of American foulbrood disease of honey bees. *BMC Genomics* 12(1): 450 http://dx.doi.org/10.1186/1471-2164-12-450.

CHANTAWANNAKUL, P; WARD, L; BOONHAM, N; BROWN, M (2006) A scientific note on the detection of honey bee viruses using real-time PCR (TaqMan) in varroa mites collected from a Thai honey bee (*Apis mellifera*) apiary. *Journal of Invertebrate Pathology* 91 (1): 69-73. http://dx.doi.org/10.1016/j.jip.2005.11.001

CHEADLE, C; VAWTER, M P; FREED, W J; BECKER, K G (2003) Analysis of microarray data using Z score transformation. *Journal of Molecular Diagnostics* 5(2): 73-81. http://dx.doi.org/10.1016/S1525-1578(10)60455-2

CHEN, Y P; EVANS, J; HAMILTON, M; FELDLAUFER, M (2007) The influence of RNA integrity on the detection of honey bee viruses: molecular assessment of different sample storage methods. *Journal of Apicultural Research* 46(2): 81-87. http://dx.doi.org/10.3896/IBRA.1.46.2.03

CHEN, Y P; HIGGINS, J A; FELDLAUFER, M F (2005a) Quantitative real-time reverse transcription-PCR analysis of deformed wing virus infection in the honey bee (*Apis mellifera* L.). *Applied and Environmental Microbiology* 71(1): 436-441. http://dx.doi.org/10.1128/AEM.71.1.436-441.2005

CHEN, Y P; PETTIS, J S; FELDLAUFER, M F (2005b) Detection of multiple viruses in queens of the honey bee *Apis mellifera* L. *Journal of Invertebrate Pathology* 90(2): 118-121. http://dx.doi.org/10.1016/j.jip.2005.08.005

CHEN, Y P; SMITH, I B; COLLINS, A M; PETTIS, J S; FELDLAUFER, M F (2004b) Detection of deformed wing virus infection in honey bees, *Apis mellifera* L., in the United States. *American Bee Journal* 144: 557-559.

CHEN, Y; ZHAO, Y; HAMMOND, J; HSU, H T; EVANS, J D; FELDLAUFER, M F (2004) Multiple virus infections in the honey bee and genome divergence of honey bee viruses. *Journal of Invertebrate Pathology* 87: 84-93. http://dx.doi.org/10.1016/j.jip.2004.07.005

CHERNESKY, M; JANG, D; CHONG, S; SELLORS, J; MAHONY, J (2003) Impact of urine collection order on the ability of assays to identify *Chlamydia trachomatis* infections in men. *Sexually Transmitted Diseases* 30(4): 345-347. http://dx.doi.org/10.1097/00007435-200304000-00014

CORNMAN, R S; CHEN, Y P; SCHATZ, M C; STREET, C; ZHAO, Y; DESANY, B; EGHOLM, M; HUTCHISON, S; PETTIS, J S; LIPKIN, W I; EVANS, J D (2009) Genomic analyses of the microsporidian *Nosema ceranae*, an emergent pathogen of honey bees. *PLoS Pathogens* 5(6): e1000466, http://dx.doi.org/10.1371/journal.ppat.1000466

CORNMAN, R S; SCHATZ, M C; JOHNSTON, J S; CHEN, Y P; PETTIS, J; HUNT, G; BOURGEOIS, L; ELSIK, C; ANDERSON, D; GROZINGER, C M; EVANS, J D (2010) Genomic survey of the ectoparasitic mite *Varroa destructor*, a major pest of the honey bee *Apis mellifera*. *BMC Genomics* 11: 602. http://dx.doi.org/10.1186/1471-2164-11-602

CORNMAN, R S; TARPY, D R; CHEN, Y-P; JEFFREYS, L; LOPEZ, D; PETTIS, J S; VANENGELSDORP, D; EVANS, J D (2012) Pathogen webs in collapsing honey bee colonies. *PloS One* e43562. http://dx.doi.org/10.1371/journal.pone.0043562

CORONA, M; HUGHES, K A; WEAVER, D B; ROBINSON, G E (2005) Gene expression patterns associated with queen honey bee longevity. *Mechanisms of Ageing and Development* 126(11): 1230 -1238. http://dx.doi.org/10.1016/j.mad.2005.07.004

COX-FOSTER, D L; CONLAN, S; HOLMES, E C; PALACIOS, G; EVANS, J D; MORAN, N A; QUAN, P L; BRIESE, T; HORNIG, M; GEISER, D M; MARTINSON, V; VANENGELSDORP, D; KALKSTEIN, A L; DRYSDALE, A; HUI, J; ZHAI, J; CUI, L; HUTCHISON, S K; SIMONS, J F; EGHOLM, M; PETTIS, J S; LIPKIN, W I (2007) A metagenomic survey of microbes in honey bee Colony Collapse Disorder. *Science* 318(5848): 283-287. http://dx.doi.org/10.1126/science.1146498

DAINAT, B; EVANS, J D; CHEN, Y P; NEUMANN, P (2011) Sampling and RNA quality for diagnosis of honey bee viruses using quantitative PCR. *Journal of Virological Methods* 174(1-2): 150-152. http://dx.doi.org/10.1016/j.jviromet.2011.03.029

DE LA RÚA, P; JAFFÉ, R; DALL´OLIO, R; MUÑOZ, I; SERRANO, J (2009) Biodiversity, conservation and current threats to European honey bees. *Apidologie* 40: 263–284. http://dx.doi.org/10.1051/apido/2009027

DE MIRANDA, J R; BAILEY, L; BALL, B V; BLANCHARD, P; BUDGE, G; CHEJANOVSKY, N; CHEN, Y-P; GAUTHIER, L; GENERSCH, E; DE GRAAF, D; RIBIÈRE, M; RYABOV, E; DE SMET, L VAN DER STEEN, J J M (2013) Standard methods for virus research in *Apis mellifera*. In *V Dietemann; J D Ellis; P Neumann (Eds) The COLOSS BEEBOOK, Volume II: standard methods for* Apis mellifera *pest and pathogen research. Journal of Apicultural Research* 52(4): http://dx.doi.org/10.3896/IBRA.1.52.4.22

DE MIRANDA, J R (2008) Diagnostic techniques for virus detection in honey bees. In *Aubert, M F A; Ball, B V; Fries, I; Moritz, R F A; Milani, N; Bernardinelli, I (Eds). Virology and the honey bee.* EEC Publications; Brussels, Belgium. pp 121-232.

DE MIRANDA, J R; FRIES, I (2008) Venereal and vertical transmission of deformed wing virus in honey bees (*Apis mellifera* L.). *Journal of Invertebrate Pathology* 98: 184-189. http://dx.doi.org/10.1016/j.jip.2008.02.004

DEL SAL, G; MANFIOLETTI, G; SCHNEIDER, C (1988) A one-tube plasmid DNA mini-preparation suitable for sequencing. *Nucleic Acids Research* 16: 20.

DESAI, N; ANTONOPOULOS, D; GILBERT, J A; GLASS, E M; FOLKER, F (2012) From genomics to metagenomics. *Current Opinion in Biotechnology.* 23(1): 72-76.

DIETEMANN, V; NAZZI, F; MARTIN, S J; ANDERSON, D; LOCKE, B; DELAPLANE, K S; WAUQUIEZ, Q; TANNAHILL, C; FREY, E; ZIEGELMANN, B; ROSENKRANZ, P; ELLIS, J D (2013) Standard methods for varroa research. In *V Dietemann; J D Ellis; P Neumann (Eds) The COLOSS BEEBOOK, Volume II: standard methods for* Apis mellifera *pest and pathogen research. Journal of Apicultural Research* 52(1): http://dx.doi.org/10.3896/IBRA.1.52.1.09

EDGAR, R C (2004) MUSCLE: multiple sequence alignment with high accuracy and high throughput. *Nucleic Acids Research* 32: 1792-1797. http://dx.doi.org/10.1093/nar/gkh340

ERALI, M; SCHMIDT, B; LYON, E; WITTWER, C (2003) Evaluation of electronic microarrays for genotyping factor-V, factor-II and MTHFR. *Clinical Chemistry* 49(5): 732-739. http://dx.doi.org/10.1373/49.5.732

ESTOUP, A; SOLIGNAC, M; HARRY, M; CORNUET, J M (1993) Characterization of (GT)n and (CT)n microsatellites in two insect species *Apis mellifera* and *Bombus terrestris. Nucleic Acids Research* 21: 1427-1431.

EVANS, J D (1993) Parentage analyses in ant colonies using simple sequence repeat loci. *Molecular Ecology* 2: 393-397.

EVANS, J D (2004) Transcriptional immune responses by honey bee larvae during invasion by the bacterial pathogen, *Paenibacillus larvae. Journal of Invertebrate Pathology* 85(2): 105-111. http://dx.doi.org/10.1016/j.jip.2004.02.004

EVANS, J D (2006) Beepath: An ordered quantitative-PCR array for exploring honey bee immunity and disease. *Journal of Invertebrate Pathology* 93(2): 135-139. http://dx.doi.org/10.1016/j.jip.2006.04.004

EVANS, J D; WHEELER, D E (2000) Expression profiles during honey bee caste determination. *Genome Biology* 2(1): http://dx.doi.org/10.1186/gb-2000-2-1-research0001

EVANS, J D; WHEELER, D E (2001) Gene expression and the evolution of insect polyphenisms. *Bioessays* 23(1): 62-68. http://dx.doi.org/10.1002/1521-1878(200101)

FAROOQUI, T (2004) Octopamine receptors in the honey bee (*Apis mellifera*) brain and their disruption by RNA-mediated interference. *Journal of Insect Physiology* 50(8): 701-713. http://dx.doi.org/10.1016/j.jinsphys.2004.04.014

FELSENSTEIN, J D (1985) Confidence limits on phylogenies: An approach using the bootstrap. *Evolution* 39: 783-791.

FIEVET, J; TENTCHEVA, D; GAUTHIER, L; DE MIRANDA, J R; COUSSERANS, F; COLIN, M E; BERGOIN, M (2006) Localization of deformed wing virus infection in queen and drone *Apis mellifera* L. *Virology Journal* 3: e16. http://dx.doi.org/10.1186/1743-422X-3-16

FIORINI, G S; CHIU, D T (2005) Disposable microfluidic devices: fabrication, function, and application. *Biotechniques* 38(3): 429-446. http://dx.doi.org/10.2144/05383RV02

FIRE, A; XU, S; MONTGOMERY, M K; KOSTAS, S A; DRIVER, S E; MELLO, C C (1998) Potent and specific genetic interference by double-stranded RNA in *Caenorhabditis elegans. Nature* 391 (6669): 806-811. http://dx.doi.org/URLhttp://dx.doi.org/10.1038/35888

FIXE, F; CHU, V; PRAZERES, D M; CONDE, J P (2004) An on-chip thin film photodetector for the quantification of DNA probes and targets in microarrays. *Nucleic Acids Research* 32(9): e70. http://dx.doi.org/10.1093/nar/gnh066

FLEIGE, S; PFAFFL, M W (2006) RNA integrity and the effect on the real-time qRT-PCR performance. *Molecular Aspects of Medicine* 27: 126-139. http://dx.doi.org/10.1016/j.mam.2005.12.003

FORET, S; KUCHARSKI, R; PELLEGRINI, M; FENG, S; JACOBSEN, S E; ROBINSON, G E; MALESZKA, R (2012). DNA methylation dynamics, metabolic fluxes, gene splicing, and alternative phenotypes in honey bees. *Proceedings of the National Academy of Sciences of the United States of America* 109(13): 4968- 4973. http://dx.doi.org/10.1073/pnas.1202392109

FRIES, I (1997) Protozoa, In *Morse, R A; Flottum, K (Eds). Honey bee pests, predators, & diseases (3rd edition).* A I Root: Medina, Ohio, USA. pp 57-76.

GARNERY, L; SOLIGNAC, M; CELEBRANO, G; CORNUET, J-M (1993) A simple test using restricted PCR-amplified mitochondrial DNA to study the genetic structure of *Apis mellifera* L. *Experientia* 49: 1016–1021. http://dx.doi.org/10.1007/BF02125651

GENERSCH, E (2005) Development of a rapid and sensitive RT-PCR method for the detection of deformed wing virus, a pathogen of the honey bee (*Apis mellifera*). *Veterinary Journal* 169: 121-123. http://dx.doi.org/10.1016/j.tvjl.2004.01.004

GENTRY, T J; WICKHAM, G S; SCHADT, C W; HE, Z; ZHOU, J (2006) Microarray applications in microbial ecology research. *Microbial Ecology* 52(2): 159-175.
http://dx.doi.org/10.1007/s00248-006-9072-6

GHARIZADEH, B; KÄLLER, M; NYRÉN, P; ANDERSSON, A; UHLÉN, M; LUNDEBERG, J; AHMADIAN, A (2003) Viral and microbial genotyping by a combination of multiplex competitive hybridization and specific extension followed by hybridization to generic tag arrays. *Nucleic Acids Research* 31(22): e146. http://dx.doi.org/10.1093/nar/gng147

GISDER, S; AUMEIER, P; GENERSCH, E (2009) Deformed wing virus: Replication and viral load in mites (*Varroa destructor*). *Journal of General Virology* 90: 463-467.

GISDER, S; MÖCKEL, N; LINDE, A; GENERSCH, E (2011) A cell culture model for *Nosema ceranae* and *Nosema apis* allows new insights into the life cycle of these important honey bee pathogenic microsporidia. *Environmental Microbiology* 13: 404-413. http://dx.doi.org/10.1111/j.1462-2920.2010.02346.x

GLOVER, R H; ADAMS, I P; BUDGE, G; WILKINS, S; BOONHAM, N (2011) Detection of honey bee (*Apis mellifera*) viruses with an oligonucleotide microarray. *Journal of Invertebrate Pathology* 107 (3): 216-219.
http://dx.doi.org/10.1016/j.jip.2011.03.004

GRABENSTEINER, E; BAKONYI, T; RITTER, W; PECHHACKER, H; NOWOTNY, N (2007) Development of a multiplex RT-PCR for the simultaneous detection of three viruses of the honey bee (*Apis mellifera* L.): Acute bee paralysis virus, Black queen cell virus and Sacbrood virus. *Journal of Invertebrate Pathology* 94(3): 222-225. http://dx.doi.org/DOI: 10.1016/j.jip.2006.11.006

GRABENSTEINER, E; RITTER, W; CARTER, M J; DAVISON, S; PECHHACKER, H; KOLODZIEJEK, J; BOECKING, O; DERAKHSHIFAR, I; MOOSBECKHOFER, R; LICEK, E; NOWOTNY, N (2001) Sacbrood virus of the honey bee (*Apis mellifera*): rapid identification and phylogenetic analysis using reverse transcription -PCR. *Clinical and Diagnostic Laboratory Immunology* 8(1): 93-104. http://dx.doi.org/10.1128/CDLI.8.1.93-104.2001

GREEN, M R; SAMBROOK, J (2012) *Molecular cloning: a laboratory manual (4th Ed.).* Cold Spring Harbor Laboratory Press; Cold Spring Harbor, USA. 2028 pp.

GREGORY, P G; EVANS, J D; RINDERER, T; DE GUZMAN, L (2005) Conditional immune-gene suppression of honey bees parasitized by varroa mites. *Journal of Insect Science* 5(7): 1-5. http://dx.doi.org/10.1672/1536-2442(2005)005[0001:CISOHP] 2.0.CO;2

GRIEBLER, M; WESTERLUND, S A; HOFFMANN, K H; MEYERING-VOS, M (2008) RNA interference with the allatoregulating neuropeptide genes from the fall armyworm *Spodoptera frugiperda* and its effects on the JH titer in the hemolymph. *Journal of Insect Physiology* 54(6): 997-1007.
http://dx.doi.org/10.1016/j.jinsphys.2008.04.019

GROZINGER, C M; SHARABASH, N M; WHITFIELD, C W; ROBINSON, G E (2003) Pheromone-mediated gene expression in the honey bee brain. *Proceedings of the National Academy of Sciences of the United States of America* 100(2): 14519-14525. http://dx.doi.org/10.1073/pnas.2335884100

HANDELSMAN, J; RONDON, M; BRADY, S; CLARDY, J; GOODMAN, R (1998) Molecular biological access to the chemistry of unknown soil microbes: a new frontier for natural products. *Chemistry and Biology* 5: 245-249.
http://dx.doi.org/10.1016/S1074-5521(98)90108-9

HARTFELDER, K; GENTILE BITONDI, M M; BRENT, C; GUIDUGLI-LAZZARINI, K R; SIMÕES, Z L P; STABENTHEINER, A; DONATO TANAKA, É; WANG, Y (2013) Standard methods for physiology and biochemistry research in *Apis mellifera*. In *V Dietemann; J D Ellis; P Neumann (Eds) The COLOSS* BEEBOOK, *Volume I: standard methods for* Apis mellifera *research. Journal of Apicultural Research* 52(1):
http://dx.doi.org/10.3896/IBRA.1.52.1.06

HARVEY, M L (2005) An alternative for the extraction and storage of DNA from insects in forensic entomology. *Journal of Forensic Sciences* 50(3): 627-629.

HERNÁNDEZ-GARCÍA, R; DE LA RÚA, P; SERRANO, J (2009) Mating frequency of *Apis mellifera iberiensis* queens. *Journal of Apicultural Research* 48(2): 121–125.
http://dx.doi.org/10.3896/IBRA.1.48.2.06

HERRMANN, B; LARSSON, V C; RUBIN, C J; SUND, F; ERIKSSON, B M; ARVIDSON, J; YUN, Z; BONDESON, K; BLOMBERG, J (2004) Comparison of a duplex quantitative real-time PCR assay and the COBAS amplicor CMV monitor test for detection of cytmomegalovirus. *Journal of Clinical Microbiology* 42(5): 1909-1914. http://dx.doi.org/10.1128/JCM.42.5.1909-1914.2004

HONEY BEE GENOME SEQUENCING CONSORTIUM (2006). Insights into social insects from the genome of the honey bee *Apis mellifera*. *Nature* 443(7114): 931-949. http://dx.doi.org/10.1038/nature05260

HORN, T; BOUTROS, M (2010) E-RNAi: a web application for the multi -species design of RNAi reagents-2010 update. *Nucleic Acids Research* 38(suppl 2): W332-W339. http://dx.doi.org/10.1093/nar/gkq317

HUMAN, H; BRODSCHNEIDER, R; DIETEMANN, V; DIVELY, G; ELLIS, J; FORSGREN, E; FRIES, I; HATJINA, F; HU, F-L; JAFFÉ, R; KÖHLER, A; PIRK, C W W; ROSE, R; STRAUSS, U; TANNER, G; TARPY, D R; VAN DER STEEN, J J M; VEJSNÆS, F; WILLIAMS, G R; ZHENG, H-Q (2013) Miscellaneous standard methods for *Apis mellifera* research. In *V Dietemann; J D Ellis; P Neumann (Eds) The COLOSS BEEBOOK, Volume I: standard methods for* Apis mellifera *research. Journal of Apicultural Research* 52(4): http://dx.doi.org/10.3896/IBRA.1.52.4.10

HUNG, A C F (2000) PCR detection of Kashmir bee virus in honey bee excreta. *Journal of Apicultural Research* 39(3-4): 103-106.

HUNTER, W; ELLIS, J; VANENGELSDORP, D; HAYES, J; WESTERVELT, D; GLICK, E; WILLIAMS, M; SELA, I; MAORI, E; PETTIS, J; COX-FOSTER, D; PALDI, N (2010) Large-Scale field application of RNAi technology reducing Israeli acuteparalysis virus disease in honey bees (*Apis mellifera*, hymenoptera: Apidae). *PLoS Pathology* 6(12): e1001160. http://dx.doi.org/10.1371/journal.ppat.1001160

HUVENNE, H; SMAGGHE, G (2010) Mechanisms of dsRNA uptake in insects and potential of RNAi for pest control: A review. *Journal of Insect Physiology* 56 (3): 227-235. http://dx.doi.org/10.1016/j.jinsphys.2009.10.004

JANSSON, A; GUSTAFSSON, L L; MIRGHANI, R A (2003) High-performance liquid chromatographic method for the determination of quinine and 3-hydroxyquinine in blood samples dried on filter paper. *Journal of Chromatography B Analytical Technologies in the Biomedical and Life Sciences* 795(1): 151-156. http://dx.doi.org/10.1016/S1570-0232(03)00554-3

JAROSCH, A; MORITZ, R F A (2011) Systemic RNA-interference in the honey bee *Apis mellifera*: Tissue dependent uptake of fluorescent siRNA after intra-abdominal application observed by laser-scanning microscopy. *Journal of Insect Physiology* 57(7): 851-857. http://dx.doi.org/10.1016/j.jinsphys.2011.03.013

JAROSCH, A; MORITZ, R F A (2012) RNA interference in honey bees: off-target effects caused by dsRNA. *Apidologie* 43(2): 128-138. http://dx.doi.org/10.1007/s13592-011-0092-y

JAROSCH, A; STOLLE, E; CREWE, R M; MORITZ, R F A (2011) Alternative splicing of a single transcription factor drives selfish reproductive behaviour in honey bee workers (*Apis mellifera*). *Proceedings of the National Academy of Sciences* 108(37): 15282-15287. http://dx.doi.org/10.1073/pnas.1109343108

KARLSON, H; GUTHENBERG, C; VON DÖBELN, U; KRISTENSSON, K (2003) Extraction of RNA from dried blood on filter papers after long-term storage. *Clinical Chemistry* 49(6): 979-981. http://dx.doi.org/10.1373/49.6.979

KIATPATHOMCHAI, W M; JITRAPAKDEE, S; PANYIM, S; BOONSAENG, V (2004) RT-PCR detection of yellow head virus (YHV) infection in *Penaeus monodon* using dried haemolymph spots. *Journal of Virological Methods* 119(1): 1-5. http://dx.doi.org/10.1016/j.jviromet.2004.02.008

KNEPP, J H; GEAHR, M A; FORMAN, M S; VALSAMAKIS, A (2003) Comparison of automated and manual nucleic acid extraction methods for detection of enterovirus RNA. *Journal of Clinical Microbiology* 41(8): 3532-3536. http://dx.doi.org/10.1128/JCM.41.8.3532-3536.2003

KONOMI, N; LEBWOHL, E; ZHANG, D (2002) Comparison of DNA and RNA extraction methods for mummified tissues. *Molecular and Cellular Probes* 16(6): 445-451. http://dx.doi.org/10.1006/mcpr.2002.0441

KUCHARSKI, R; MALESZKA, R (2002) Evaluation of differential gene expression during behavioural development in the honey bee using microarrays and northern blots. *Genome Biology* 3(2): Epub 2002. http://dx.doi.org/10.1186/gb-2002-3-2-research0007

KUCHARSKI, R; MALESZKA, J; FORET, S; MALESZKA, R (2008) Nutritional control of reproductive status in honey bees via DNA methylation. *Science* 319(5871): 1827-1830. http://dx.doi.org/10.1126/science.1153069

LABAYRU, C; EIROS, J M; HERNANDEZ, B; DE LEJARAZU, R O; TORRES, A R (2005) RNA extraction prior to HIV-1 resistance detection using Line Probe Assay (LiPA): comparison of three methods. *Journal of Clinical Virology* 32(4): 265-271. http://dx.doi.org/10.1016/j.jcv.2004.08.007

LARKIN, M A; BLACKSHIELDS, G; BROWN, N P; CHENNA, R; MCGETTIGAN, P A; MCWILLIAM, H; VALENTIN, F; WALLACE, I M; WILM, A; LOPEZ, R; THOMPSON, J D; GIBSON, T J; HIGGINS, D G (2007) Clustal W and Clustal X version 2.0 *Bioinformatics* 23: 2947-2948. http://dx.doi.org/10.1093/bioinformatics/btm404

LI, C C; SEIDEL, K D; COOMBS, R W; FRENKEL, L M (2005) Detection and quantification of human immunodeficiency virus type 1 p24 antigen in dried whole blood and plasma on filter paper stored under various conditions. *Journal of Clinical Microbiology* 43(8): 3901-3905. http://dx.doi.org/10.1128/JCM.43.8.3901-3905.2005

LIEBERFARB, M E; LIN, M; LECHPAMMER, M; LI, C; TANENBAUM, D M; FEBBO, P G; WRIGHT, R L; SHIM, J; KANTOFF, P W; LODA, M; MEYERSON, M; SELLERS, W R (2003) Genome-wide loss of heterozygosity analysis from laser-capture microdissected prostate cancer using single nucleotide polymorphic allele (SNP) arrays and a novel bioinformatics platform dChipSNP. *Cancer Research* 63 (16): 4781-4785.

LIN, M; WEI, L J; SELLERS, W R; LIEBERFARB, M; WONG, W H; LI, C (2004) dChipSNP: significance curve and clustering of SNP-array-based loss-of-heterozygosity data. *Bioinformatics* 20(8): 1233-1240. http://dx.doi.org/10/1093/bioinformatics/bth069

LIU, K; WARNOW ,T J; HOLDER, M T; NELESEN, S; YU, J; STAMATAKIS, A; LINDER, R C (2012) SATé-II: Very fast and accurate simultaneous estimation of multiple sequence alignments and phylogenetic trees. *Systematic Biology* 61(1): 90-106. http://dx.doi.org/10.1093/sysbio/syr095

LIU, X; ZHANG, Y; YAN, X; HAN, R (2010) Prevention of Chinese sacbrood virus infection in *Apis cerana* using RNA interference. *Current Microbiology*. http://dx.doi.org/10.1007/s00284-010-9633-2

LOCKE, B; FORSGREN, E; FRIES, I; DE MIRANDA, J R (2012) Acaricide treatment affects viral dynamics in *Varroa destructor*-infested honey bee colonies via both host physiology and mite control. *Applied and Environmental Microbiology* 78: 227-235. http://dx.doi.org/10.1128/AEM.06094-11

LYKO, F; FORET, S; KUCHARSKI, R; WOLF, S; FALCKENHAYN, C; MALESZKA, R (2010) The honey bee epigenomes: differential methylation of brain DNA in queens and workers. *PLoS Biology* 8 (11): e1000506. http://dx.doi.org/10.1371/journal.pbio.1000506

MACKAY, I M; GARDAM, T; ARDEN, K E; MCHARDY, S; WHILEY, D M; CRISANTE, E; SLOOTS, T P (2003) Co-detection and discrimination of six human herpes viruses by multiplex PCR-ELAHA. *Journal of Clinical Virology* 28(3): 291-302. http://dx.doi.org/10.1016/S1386-6532(03)00072-6

MALESZKA, J; FORÊT, S; SAINT, R; MALESZKA, R (2007) RNAi-induced phenotypes suggest a novel role for a chemosensory protein CSP5 in the development of embryonic integument in the honey bee (*Apis mellifera*). *Development Genes and Evolution* 217 (3): 189-196. http://dx.doi.org/10.1007/s00427-006-0127-y

MANIATIS, T; FRITSCH, E F; SAMBROOK, J (1982) *Molecular cloning: A laboratory manual (2nd Ed.)*. Cold Spring Harbor Laboratory; Cold Spring Harbor, USA. pp 559, 458–460.

MAORI, E; PALDI, N; SHAFIR, S; KALEV, H; TSUR, E; GLICK, E; SELA, I (2009) IAPV, a bee-affecting virus associated with colony collapse disorder can be silenced by dsRNA ingestion. *Insect Molecular Biology* 18(1): 55-60. http://dx.doi.org/10.1111/j.1365-2583.2009.00847.x

MARTINSON, V G; MOY, J; MORAN, N A (2012) Establishment of characteristic gut bacteria during development of the honey bee worker. *Applied Environmental Microbiology* 78: 2830-40. http://dx.doi.org/10.1128/AEM.07810-11

MARTINSON, V G; DANFORTH, B N; MINCKLEY, R L; RUEPPELL, O; TINGEK, S; MORAN, N A (2011) A simple and distinctive microbiota associated with honey bees and bumble bees. *Molecular Ecology* 20(3): 619-628. http://dx.doi.org/10.1111/j.1365-294X.2010.04959.x

MATTILA, H R; RIOS, D; WALKER-SPERLING, V E; ROESELERS, G; NEWTON, I L G (2012) Characterization of the active microbiotas associated with honey bees reveals healthier and broader communities when colonies are genetically diverse. *PLoS ONE* 7 (3): e32962. http://dx.doi.org/10.1371/journal.pone.0032962

MEEUS, I; SMAGGHE, G; SIEDE, R; JANS, K; DE GRAAF, D C (2010) Multiplex RT-PCR with broad-range primers and an exogenous internal amplification control for the detection of honey bee viruses in bumblebees. *Journal of Invertebrate Pathology* 105: 200-2003. http://dx.doi.org/10.1016/j.jip.2010.06.012

MEIXNER, M D; PINTO, M A; BOUGA, M; KRYGER, P; IVANOVA, E; FUCHS, S (2013) Standard methods for characterising subspecies and ecotypes of *Apis mellifera*. In *V Dietemann; J D Ellis; P Neumann (Eds) The COLOSS BEEBOOK, Volume I: standard methods for* Apis mellifera *research. Journal of Apicultural Research* 52(4): http://dx.doi.org/10.3896/IBRA.1.52.4.05

METZKER, M (2010) Sequencing technologies - the next generation. *Nature Reviews Genetics* 11: 31–46. http://dx.doi.org/10.1038/nrg2626

MÖCKEL, N; GISDER, S; GENERSCH, E (2011) Horizontal transmission of deformed wing virus (DWV): Pathological consequences in adult bees (*Apis mellifera*) depend on the transmission route. *Journal of General Virology* 92: 370-377. http://dx.doi.org/10.1099/vir.0.025940-0

MORTAZAVI, A; WILLIAMS, B A; MCCUE, K; SCHAEFFER, L; WOLD, B (2008) Mapping and quantifying mammalian transcriptomes by RNA-Seq. *Nature Methods*. http://dx.doi.org/10.1038/nmeth.1226

MUÑOZ, I; DALL´OLIO, R; LODESANI, M; DE LA RÚA, P (2009) Population genetic structure of coastal Croatian honey bees (*Apis mellifera carnica*). *Apidologie* 40: 617–626. http://dx.doi.org/10.1051/apido/2009041

MUÑOZ, I; MADRID-JIMÉNEZ, M J; DE LA RÚA, P (2012) Temporal genetic analysis of an introgressed island honey bee population (Tenerife, Canary Islands, Spain). *Journal of Apicultural Research* 51(1): 144–146. http://dx.doi.org/10.3896/IBRA.1.51.1.20

MUSTARD, J A; PHAM, P M; SMITH, B H (2010) Modulation of motor behaviour by dopamine and the d1-like dopamine receptor AmDOP2 in the honey bee. *Journal of Insect Physiology* 56(4): 422-430. http://dx.doi.org/URLhttp://dx.doi.org/10.1016/j.jinsphys.2009.11.018

MUTTER, G L; ZAHRIEH, D; LIU, C-M; NEUBERG, D; FINKELSTEIN, D; BAKER, H E; WARRINGTON, J A (2004) Comparison of frozen and RNALater solid tissue storage methods for use in RNA expression microarrays. *BMC Genomics* 5: e88. http://dx.doi.org/10.1186/1471-2164-5-88

NAZZI, F; BROWN, S P; ANNOSCIA, D; DEL PICCOLO, F; DI PRISCO, G; VARRICCHIO, P; VEDOVA, G D; CATTONARO, F; CAPRIO, E; PENNACCHIO, F (2012) Synergistic parasite-pathogen interactions mediated by host immunity can drive the collapse of honey bee colonies. *PLoS Pathogens* 8(6): http://dx.doi.org/10.1371/journal.ppat.1002735

NOERHOLM, M; BRUUS, H; JAKOBSEN, M H; TELLEMAN, P; RAMSING, N B (2004) Polymer microfluidic chip for online monitoring of microarray hybridizations. *Lab on a Chip* 4(1): 28-37. http://dx.doi.org/10.1039/B311991B

NUNES, F M; SIMÕES, Z L (2009) A non-invasive method for silencing gene transcription in honey bees maintained under natural conditions. *Insect Biochemistry and Molecular Biology* 39(2): 157-160. http://dx.doi.org/10.1016/j.ibmb.2008.10.011

OLIVIER, V; BLANCHARD, P; CHAOUCH, S; LALLEMAND, P; SCHURR, F; CELLE, O; DUBOIS, E; TORDO, N; THIÉRY, R; HOULGATTE, R; RIBIÈRE, M (2008) Molecular characterisation and phylogenetic analysis of chronic bee paralysis virus, a honey bee virus. *Virus Research* 132(1-2): 59-68. http://dx.doi.org/10.1016/j.virusres.2007.10.014

PALDI, N; GLICK, E; OLIVA, M; ZILBERBERG, Y; AUBIN, L; PETTIS, J; CHEN, Y; EVANS, J D (2010) Effective gene silencing in a microsporidian parasite associated with honey bee (*Apis mellifera*) colony declines. *Applied and Environmental Microbiology* 76(17): 5960-5964. http://dx.doi.org/10.1128/AEM.01067-10

PATEL, A; FONDRK, M K; KAFTANOGLU, O; EMORE, C; HUNT, G; FREDERICK, K; AMDAM, G V (2007) The making of a queen: TOR pathway is a key player in diphenic caste development. *PloS One* 2(6): http://dx.doi.org/10.1371/journal.pone.0000509

PERRETEN, V; VORLET-FAWER, L; SLICKERS, P; EHRICHT, R; KUHNERT, P; FREY, J (2005) Microarray-based detection of 90 antibiotic resistance genes of gram-positive bacteria. *Journal of Clinical Microbiology* 43(5): 2291-2302. http://dx.doi.org/10.1128/JCM.43.5.2291-2302.2005

PETRICH, A; MAHONY, J; CHONG, S; BROUKHANSKI, G; GHARABAGHI, F; JOHNSON, G; LOUIE, L; LUINSTRA, K; WILLEY, B; AKHAVEN, P; CHUI, L; JAMIESON, F; LOUIE, M; MAZZULLI, T; TELLIER, R; SMIEJA, M; CAI, W; CHERNESKY, M; RICHARDSON, S E (2006) Multicenter comparison of nucleic acid extraction methods for detection of severe acute respiratory syndrome coronavirus RNA in stool specimens. *Journal of Clinical Microbiology* 44: 2681–2688. http://dx.doi.org/10.1128/JCM.02460-05

PFAFFL MW (2001) A new mathematical model for relative quantification in real-time RT-PCR. *Nucleic Acids Research* 29(9): e45

PFAFFL, M W; HAGELEIT, M (2001) Validities of mRNA quantification using recombinant RNA and recombinant DNA external calibration curves in real-time RT-PCR. *Biotechnology Letters* 23(4): 275-282. http://dx.doi.org/10.1023/A:1005658330108

PIRK, C W W; DE MIRANDA, J R; FRIES, I; KRAMER, M; PAXTON, R; MURRAY, T; NAZZI, F; SHUTLER, D; VAN DER STEEN, J J M; VAN DOOREMALEN, C (2013) Statistical guidelines for *Apis mellifera* research. In *V Dietemann; J D Ellis; P Neumann (Eds) The COLOSS BEEBOOK, Volume I: standard methods for* Apis mellifera *research. Journal of Apicultural Research* 52(4): http://dx.doi.org/10.3896/IBRA.1.52.4.13

PRADO, I; ROSARIO, D; BERNARDO, L; ALVAREZ, M; RODRIGUEZ, R; VAZQUEZ, S; GUZMAN, M G (2005) PCR detection of dengue virus using dried whole blood spotted on filter paper. *Journal of Virological Methods* 125(1): 75-81. http://dx.doi.org/10.1016/j.jviromet.2005.01.001

PRUESSE, E; QUAST, C; KNITTEL, K; FUCHS, B; LUDWIG, W; PEPLIES, J; GLÖCKNER, F O (2007) SILVA: a comprehensive online resource for quality checked and aligned ribosomal RNA sequence data compatible with ARB. *Nucleic Acids Research* 35: 7188-7196. http://dx.doi.org/10.1093/nar/gkm864

PUGNALE, P; LATORRE, P; ROSSI, C; CROVATTO, K; PAZIENZA, V; DE GOTTARDI, A; NEGRO, F (2006) Real-time multiplex PCR assay to quantify hepatitis C virus RNA in peripheral blood mononuclear cells. *Journal of Virological Methods* 133(2): 195-204. http://dx.doi.org/10.1016/j.jviromet.2005.11.007

QIN, X; EVANS, J; ARONSTEIN, K; MURRAY, K; WEINSTOCK, G (2006) Genome sequences of the honey bee pathogens *Paenibacillus larvae* and *Ascosphaera apis*. *Insect Molecular Biology* 15(5): 715-8. http://dx.doi.org/10.1111/j.1365-2583.2006.00694.x

RADONIĆ, A; THULKE, S; MACKAY, I M; LANDT, O; SIEGERT, W; NITSCHE, A (2004) Guideline to reference gene selection for quantitative real-time PCR. *Biochemical and Biophysical Research Communications* 313(4): 856-862. http://dx.doi.org/10.1016/j.bbrc.2003.11.177

RENSEN, G; SMITH, W; RUZANTE, J; SAWYER, M; OSBURN, B; CULLOR, J (2005) Development and evaluation of a real-time fluorescent polymerase chain reaction assay for the detection of bovine contaminates in cattle feed. *Food-borne Pathogens and Disease* 2(2): 152-159. http://dx.doi.org/10.1089/fpd.2005.2.152

RIMMER, A E; BECKER, J A; TWEEDIE, A; WHITTINGTON, R J (2012) Validation of high throughput methods for tissue disruption and nucleic acid extraction for ranaviruses (family Iridoviridae). *Aquaculture* 338-341: 23–28. http://dx.doi.org/10.1016/j.aquaculture.2012.01.012

ROBINSON, G E; EVANS, J D; MALESZKA, R; ROBERTSON, H M; WEAVER, D B; WORLEY, K; GIBBS, R A; WEINSTOCK, G M (2006) Sweetness and light: illuminating the honey bee genome. *Insect Molecular Biology* 15(5): 535-539. http://dx.doi.org/10.1111/j.1365-2583.2006.00698.x

ROUILLARD, J M; ZUKER, M; GULARI, E (2003) OligoArray 2.0: design of oligonucleotide probes for DNA microarrays using a thermodynamic approach. *Nucleic Acids Research* 31(12): 3057-3062. http://dx.doi.org/10.1093/nar/gkg426

RUNCKEL, C; FLENNIKEN, M L; ENGEL, J C; RUBY, J G; GANEM, D; ANDINO, R; DERISI, J L (2011) Temporal analysis of the honey bee microbiome reveals four novel viruses and seasonal prevalence of known viruses, Nosema, and Crithidia. *PLoS ONE* 6 (6): e20656. http://dx.doi.org/10.1371/journal.pone.0020656

SAGE, L (2004) Protein biochips go high tech. *Analytical Chemistry* 76 (7): 137A-142A. http://dx.doi.org/10.1021/ac0415408

SALEH, M-C; VAN RIJ, R P; HEKELE, A; GILLIS, A; FOLEY, E; O'FARRELL, P H; ANDINO, R (2006) The endocytic pathway mediates cell entry of dsRNA to induce RNAi silencing. *Nature Cell Biology* 8 (8): 793-802. http://dx.doi.org/10.1038/ncb1439

SANTALUCIA, J (2007) Physical principles and visual-OMP software for optimal PCR design. In *Yuryev, A (Ed.). PCR primer design. Methods in Molecular Biology™ series #402*. Humana Press; Totowa, New Jersey, USA. pp 3-33.

SATMATAKIS, A (2006) RAxML-VI-HPC: maximum likelihood-based phylogenetic analyses with thousands of taxa and mixed models. *Bioinformatics* 22: 2688-2690.

SCHLÜNS, H; CROZIER, R H (2007) Relish regulates expression of antimicrobial peptide genes in the honey bee, (*Apis mellifera*), shown by RNA interference. *Insect Molecular Biology* 16(6): 753-759. http://dx.doi.org/10.1111/j.1365-2583.2007.00768.x

SCHUURMAN, T; VAN BREDA, A; DE BOER, R; KOOISTRA-SMID, M; BELD, M; SAVELKOUL, P; BOOM, R (2005) Reduced PCR sensitivity due to impaired DNA recovery with the MagNA Pure LC total nucleic acid isolation kit. *Journal of Clinical Microbiology* 43(9): 4616-4622. http://dx.doi.org/10.1128/JCM.43.9.4616-4622.2005

SEEHUUS, S-C C; NORBERG, K; GIMSA, U; KREKLING, T; AMDAM, G V (2006) Reproductive protein protects functionally sterile honey bee workers from oxidative stress. *Proceedings of the National Academy of Sciences of the United States of America* 103(4): 962-967. http://dx.doi.org/10.1073/pnas.0502681103

SHEN, M Q; CUI, L W; OSTIGUY, N; COX-FOSTER, D (2005a) Intricate transmission routes and interactions between picorna-like viruses (Kashmir bee virus and sacbrood virus) with the honey bee host and the parasitic varroa mite. *Journal of General Virology* 86(8): 2281-2289. http://dx.doi.org/10.1099/vir.0.80824-0

SHEN, M Q; YANG, X L; COX-FOSTER, D; CUI, L W (2005b) The role of varroa mites in infections of Kashmir bee virus (KBV) and deformed wing virus (DWV) in honey bees. *Virology* 342: 141-149. http://dx.doi.org/10.1016/j.virol.2005.07.012

SHIMANUKI, H (1997) Bacteria. In *Morse, R A; Flottum, K (Eds). Honey bee pests, predators, and diseases (3rd Ed.)*. A I Root; Medina, Ohio, USA. pp 33-54.

SINGH, R P; NIE, X; SINGH, M (2000) Duplex RT-PCR: reagent concentrations at reverse transcription stage affect the PCR performance. *Journal of Virological Methods* 86(2): 121-129. http://dx.doi.org/10.1016/S0166-0934(00)00138-5

SOLIGNAC, M; VAUTRIN, D; LOISEAU, A; MOUGEL, F; BAUDRY, E; ESTOUP, A; GARNERY, L; HABERL, M; CORNUET, J-M (2003) Five hundred and fifty microsatellite markers for the study of the honey bee (*Apis mellifera* L.) genome. *Molecular Ecology Notes* 3: 307–311. http://dx.doi.org/10.1046/j.1471-8286.2003.00436.x

SPÖTTER, A; GUPTA, P; NÜRNBERG, G; REINSCH, N; BIENEFELD, K (2011) Development of a 44K SNP assay focusing on the analysis of a varroa-specific defense behaviour in honey bees (*Apis mellifera carnica*). *Molecular Ecology Resources* 12: 323–332. http://dx.doi.org/10.1111/j.1755-0998.2011.03106.x

STÅHLBERG, A; HÅKANSSON, J; XIAN, X; SEMB, H; KUBISTA, M (2004a) Properties of the reverse transcription reaction in mRNA quantification. *Clinical Chemistry* 50(3): 509-515. http://dx.doi.org/10.1373/clinchem.2003.026161

STÅHLBERG, A; KUBISTA, M; PFAFFL, M (2004b) Comparison of reverse transcriptases in gene expression analysis. *Clinical Chemistry* 50(9): 1678-1680. http://dx.doi.org/10.1373/clinchem.2004.035469

STRAM, Y; KUZNETZOVA, L; GUINI, M; ROGEL, A; MEIROM, R; CHAI, D; YADIN, H; BRENNER, J (2004) Detection and quantitation of Akabane and Aino viruses by multiplex real-time reverse-transcriptase PCR. *Journal of Virological Methods* 116(2): 147-154. http://dx.doi.org/10.1016/j.jviromet.2003.11.010

SYRMIS, M W; WHILEY, D M; THOMAS, M; MACKAY, I M; WILLIAMSON, J; SIEBERT, D J; NISSEN, M D; SLOOTS, T P (2004) A sensitive, specific, and cost-effective multiplex reverse transcriptase-PCR assay for the detection of seven common respiratory viruses in respiratory samples. *Journal of Molecular Diagnostics* 6(2): 125-131. http://dx.doi.org/10.1016/S1525-1578(10)60500-4

SZEMES, M; KLERKS, M M; VAN DEN HEUVEL, J F J M; SCHOEN, C D (2002) Development of a multiplex AmpliDet RNA assay for simultaneous detection and typing of potato virus Y isolates. *Journal of Virological Methods* 100(1-2): 83-96. http://dx.doi.org/10.1016/S0166-0934(01)00402-5

TAMURA, K; NEI, M (1993) Estimation of the number of nucleotide substitutions in the control region of mitochondrial DNA in humans and chimpanzees. *Molecular Biology and Evolution* 10: 512-526.

TAMURA, K; NEI, M; KUMAR, S (2004) Prospects for inferring very large phylogenies by using the neighbour-joining method. *Proceedings of the National Academy of Sciences (USA)* 101: 11030-11035.

TAMURA, K; PETERSON, D; PETERSON, N; STECHER, G; NEI, M; KUMAR, S (2011) MEGA5: Molecular Evolutionary Genetics Analysis using maximum likelihood, evolutionary distance, and maximum parsimony methods. *Molecular Biology and Evolution* 28: 2731-2739.

TENTCHEVA, D; GAUTHIER, L; BAGNY, L; FIEVET, J; DAINAT, B; COUSSERANS, F; COLIN, M E; BERGOIN, M (2006) Comparative analysis of deformed wing virus (DWV) RNA in *Apis mellifera* and *Varroa destructor*. *Apidologie* 37: 41-50. http://dx.doi.org/10.1051/apido:2005057

THOMPSON, J D; HIGGINS, D G; GIBSON, T J (1994) CLUSTAL W: improving the sensitivity of progressive multiple sequence alignment through sequence weighting, position-specific gap penalties and weight matrix choice. *Nucleic Acids Research* 22: 4673-4680.

TODD, J H; DE MIRANDA, J R; BALL, B V (2007) Incidence and molecular characterization of viruses found in dying New Zealand honey bee (*Apis mellifera*) colonies infested with *Varroa destructor*. *Apidologie* 38: 354-367. http://dx.doi.org/10.1051/apido: 2007021

TOMA, D P; BLOCH, G; MOORE, D; ROBINSON, G E (2000) Changes in period mRNA levels in the brain and division of labour in honey bee colonies. *Proceedings of the National Academy of Sciences of the United States of America* 97(12): 6914-6919. http://dx.doi.org/10.1073/pnas.97.12.6914

TOPLEY, E; DAVISON, S; LEAT, N; BENJEDDOU, M (2005) Detection of three honey bee viruses simultaneously by a single Multiplex Reverse Transcriptase PCR. *African Journal of Biotechnology* 4(8): 763-767.

VALENTINE-THON, E; VAN LOON, A M; SCHIRM, J; REID, J; KLAPPER, P E; CLEATOR, G M (2001) European proficiency testing program for molecular detection and quantitation of hepatitis B virus DNA. *Journal of Clinical Microbiology* 39(12): 4407-4412. http://dx.doi.org/10.1128/JCM.39.12.4407-4412.2001

VERHEYDEN, B; THIELEMANS, A; ROMBAUT, B; KRONENBERGER, P (2003) RNA extraction for quantitative enterovirus RT-PCR: comparison of three methods. *Journal of Pharmaceutifcal and Biomedical Analysis* 33(4): 819-823. http://dx.doi.org/10.1016/S0731-7085(03)00312-1

VERKOOYEN, R P; NOORDHOEK, G T; KLAPPER, P E; REID, J; SCHIRM, J; CLEATOR, G M; IEVEN, M; HODDEVIK, G (2003) Reliability of nucleic acid amplification methods for detection of *Chlamydia trachomatis* in urine: results of the first international collaborative quality control study among 96 laboratories. *Journal of Clinical Microbiology* 41(7): 3013-3016. http://dx.doi.org/10.1128/JCM.41.7.3013-3016.2003

WALSH, P S; METZQER, D A; HIGUCHI, R (1991) Chelex 100 as a medium for simple extraction of DNA for PCR-based typing from forensic material. *Biotechniques* 10: 506–512.

WANG, Y; ZHENG, W; LUO, J; ZHANG, D; ZUHONG, L (2006) *In situ* bisulfite modification of membrane-immobilized DNA for multiple methylation analysis. *Analytical Biochemistry* 359(2): 183-8.

WEIDENHAMMER, E M; KAHL, B F; WANG, L; WANG, L; DUHON, M; JACKSON, J A; SLATER, M; XU, X (2002) Multiplexed, targeted gene expression profiling and genetic analysis on electronic microarrays. *Clinical Chemistry* 48(11): 1873-1882. http://www.clinchem.org/content/48/11/1873.full

WEINSTEIN-TEXEIRA, E; CHEN, Y P; MESSAGE, D; PETTIS, J; EVANS, J D (2008) Virus infections in Brazilian honey bees. *Journal of Invertebrate Pathology* 99(1): 117-119. http://dx.doi.org/10.1016/j.jip.2008.03.014

WETZEL, T; JARDAK, R; MEUNIER, L; GHORBEL, A; REUSTLE, G M; KRCZAL, G (2002) Simultaneous RT/PCR detection and differentiation of arabis mosaic and grapevine fanleaf nepoviruses in grapevines with a single pair of primers. *Journal of Virological Methods* 101(1-2): 63-69. http://dx.doi.org/10.1016/S0166-0934(01)00422-0

WHEELER, D E; BUCK, N; EVANS, J D (2006) Expression of insulin pathway genes during the period of caste determination in the honey bee, *Apis mellifera*. *Insect Molecular Biology* 15(5): 597-602. http://dx.doi.org/10.1111/j.1365-2583.2006.00681.x

WHITFIELD, C W; BEHURA, S K; BERLOCHER, S H; CLARK, A G; JOHNSTON, J S; SHEPPARD, W S; SMITH, D R; SUAREZ, A V; WEAVER, D; TSUTSUI, N D (2006) Thrice out of Africa: Ancient and recent expansions of the honey bee, *Apis mellifera*. *Science* 314: 642–645. http://dx.doi.org/10.1126/science.1132772

WHITFIELD, C W; BAND, M R; BONALDO, M F; KUMAR, C G; LIU, L; PARDINAS, J R; ROBERTSON, H M; SOARES, M B; ROBINSON, G E (2002) Annotated expressed sequence tags and cDNA microarrays for studies of brain and behaviour in the honey bee. *Genome Research* 12(4): 555-566. http://dx.doi.org/10.1101/gr.5302

WILLIAMS, K; BLAKE, S; SWEENEY, A; SINGER, J T; NICHOLSON, B L (1999) Multiplex reverse transcriptase PCR assay for simultaneous detection of three fish viruses. *Journal of Clinical Microbiology* 37 (12): 4139-4141.

WILLIAMS, G R; ALAUX, C; COSTA, C; CSÁKI, T; DOUBLET, V;
EISENHARDT, D; FRIES, I; KUHN, R; MCMAHON, D P;
MEDRZYCKI, P; MURRAY, T E; NATSOPOULOU, M E; NEUMANN,
P; OLIVER, R; PAXTON, R J; PERNAL, S F; SHUTLER, D; TANNER, G;
VAN DER STEEN, J J M; BRODSCHNEIDER, R (2013) Standard
methods for maintaining adult *Apis mellifera* in cages under *in
vitro* laboratory conditions. In *V Dietemann; J D Ellis; P Neumann
(Eds) The COLOSS BEEBOOK, Volume I: standard methods for
Apis mellifera research. Journal of Apicultural Research* 52(1):
http://dx.doi.org/10.3896/IBRA.1.52.1.04

WILSON, D; YEN-LIEBERMAN, B; REISCHL, U; WARSHAWSKY, I;
PROCOP, G W (2004) Comparison of five methods for extraction
of *Legionella pneumophila* from respiratory specimens. *Journal of
Clinical Microbiology* 42(12): 5913-5916.
http://dx.doi.org/10.1128/JCM.42.12.5913-5916.2004

WOODHALL, J W; WEBB, K M; GILTRAP, P M; ADAMS, I P; PETERS, J
C; BUDGE, G E; BOONHAM, N (2012) A new large scale soil DNA
extraction procedure and real-time PCR assay for the detection of
Sclerotium cepivorum in soil. *European Journal of Plant Pathology.*
134: 467-473.
http://dx.doi.org/10.1007/s10658-012-0025-2]

YAMAZAKI, Y; SHIRAI, K; PAUL, R K; FUJIYUKI, T; WAKAMOTO, A;
TAKEUCHI, H; KUBO, T (2006) Differential expression of HR38 in
the mushroom bodies of the honey bee brain depends on the
caste and division of labour. *FEBS Letters* 580(11): 2667-2670.
http://dx.doi.org/10.1016/j.febslet.2006.04.016

YAÑEZ, O; JAFFÉ, R; JAROSCH, A; FRIES, I; MORITZ, R F A; PAXTON, R J;
DE MIRANDA J R (2012) Deformed wing virus and drone mating
in the honey bee (*Apis mellifera*): implications for sexual
transmission of a major honey bee virus. *Apidologie* 43: 17-30.
http://dx.doi.org/10.1007/s13592-011-0088-7

YUE, D; NORDHOFF, M; WIELER, L H; GENERSCH, E (2008)
Fluorescence-*in situ*-hybridization (FISH) analysis of the
interactions between honey bee larvae and *Paenibacillus larvae*,
the causative agent of American foulbrood of honey bees (*Apis
mellifera*). *Environmental Microbiology* 10: 1612-1620.
http://dx.doi.org/doi: 10.1111/j.1462-2920.2008.01579.x.

YUEN, P K; LI, G; BAO, Y; MULLER, U R (2003) Microfluidic devices
for fluidic circulation and mixing improve hybridization signal
intensity on DNA arrays. *Lab on a Chip* 3(1): 46-50.
http://dx.doi.org/10.1039/B210274A

YURYEV, A (2007) PCR primer design. *Methods in Molecular Biology™
series #402.* Humana Press; Totowa, New Jersey, USA. 431 pp

ZHANG, J; BYRNE, C D (1999) Differential priming of RNA templates
during cDNA synthesis markedly affects both accuracy and
reproducibility of quantitative competitive reverse transcriptase
PCR. *Biochemical Journal* 337(2): 231-241.
http://dx.doi.org/10.1042/0264-6021:3370231

ZURFLUH, M R; GIOVANNINI, M; FIORI, L; FIEGE, B; GOKDEMIR, Y;
BAYKAL, T; KIERAT, L; GARTNER, K H; THONY, B; BLAU, N
(2005) Screening for tetrahydrobiopterin deficiencies using dried
blood spots on filter paper. *Molecular Genetics and Metabolism*
86: S96-S103. http://dx.doi.org/10.1016/j.ymgme.2005.09.011

Journal of Apicultural Research 52(1): (2013)
DOI 10.3896/IBRA.1.52.1.06

REVIEW ARTICLE

Standard methods for physiology and biochemistry research in *Apis mellifera*

Klaus Hartfelder[1]*, Márcia M G Bitondi[2], Colin S Brent[3], Karina R Guidugli-Lazzarini[2], Zilá L P Simões[2], Anton Stabentheiner[4], Érica D Tanaka[2], and Ying Wang[5]

[1]Faculdade de Medicina de Ribeirão Preto, Universidade de São Paulo, Avenida Bandeirantes 3900, 14049-900 Ribeirão Preto, SP, Brazil.
[2]Depto Biologia, Faculdade de Filosofia, Ciências e Letras de Ribeirão Preto, Universidade de São Paulo, Ribeirão Preto, Brazil.
[3]US Department of Agriculture, US Arid Land Agricultural Research Center, Maricopa, AZ, USA.
[4]Institut für Zoologie, Universität Graz, Graz, Austria.
[5]School of Life Sciences, Arizona State University, Tempe, AZ, USA.

Received 20 June 2012, accepted subject to revision 31 July 2012, accepted for publication 6 November 2012.

*Corresponding author: Email: klaus@fmrp.usp.br

Summary

Despite their tremendous economic importance, and apart from certain topics in the field of neurophysiology such as vision, olfaction, learning and memory, honey bees are not a typical model system for studying general questions of insect physiology. The reason is their social lifestyle, which sets them apart from a "typical insect" and, during social evolution, has resulted in the restructuring of certain physiological pathways and biochemical characteristics in this insect. Not surprisingly, the questions that have attracted most attention by researchers working on honey bee physiology and biochemistry in general are core topics specifically related to social organization, such as caste development, reproductive division of labour and polyethism within the worker caste. With certain proteins playing key roles in these processes, such as the major royal jelly proteins (MRJPs), including royalactin and hexamerins in caste development, and vitellogenin in reproductive division of labour and age polyethism, a major section herein will present and discuss basic laboratory protocols for protein analyses established and standardized to address such questions in bees. A second major topic concerns endocrine mechanisms underlying processes of queen and worker development, as well as reproduction and polyethism, especially the roles of juvenile hormone and ecdysteroids. Sensitive techniques for the quantification of juvenile hormone levels circulating in haemolymph, as well as its synthesis by the *corpora allata* are described. Although these require certain instrumentation and a considerable degree of sophistication in the analysis procedures, we considered that presenting these techniques would be of interest to laboratories planning to specialize in such analyses. Since biogenic amines are both neurotransmitters and regulators of endocrine glands, we also present a standard method for the detection and analysis of certain biogenic amines of interest. Further questions that cross borders between individual and social physiology are related to energy metabolism and thermoregulation. Thus a further three sections are dedicated to protocols on carbohydrate quantification in body fluid, body temperature measurement and respirometry.

Métodos estándar para la investigación de la fisiología y bioquímica de *Apis mellifera*

Resumen

A pesar de su enorme importancia económica, y aparte de ciertos temas en el campo de la neurofisiología, tales como la visión, el olfato, el aprendizaje y la memoria, las abejas no son un sistema modelo típico para el estudio de cuestiones generales sobre la fisiología de los insectos. La razón de ello es su forma de vida social, lo que las diferencia de un "insecto típico" y que durante la evolución social, se ha traducido en la reestructuración de ciertas vías fisiológicas y bioquímicas propias de este insecto. Como era de esperar, las preguntas que han atraído mayor atención por parte de los investigadores que trabajan en la fisiología y la bioquímica de la abeja melífera, son en general temas relacionados específicamente con la organización social, tales como el desarrollo de las castas, la división reproductiva del trabajo y el

Footnote: Please cite this paper as: HARTFELDER, K; BITONDI, M M G; BRENT, C; GUIDUGLI-LAZZARINI, K R; SIMÕES, Z L P; STABENTHEINER, A; TANAKA, D E; WANG, Y (2013) Standard methods for physiology and biochemistry research in *Apis mellifera*. In *V Dietemann; J D Ellis; P Neumann (Eds) The COLOSS BEEBOOK, Volume I: standard methods for* Apis mellifera *research. Journal of Apicultural Research* 52(1): http://dx.doi.org/10.3896/IBRA.1.52.1.06

polietismo de la casta obrera. Dadas aquellas proteínas que juegan un papel clave en estos procesos, como la proteína principal de la jalea real (MRJPs), incluyendo a la royalactina y las hexamerinas en el desarrollo de las castas, y la vitelogenina en la división reproductiva del trabajo, el polietismo vital, una importante sección de este documento presentará y discutirá protocolos básicos de laboratorio establecidos y estandarizados para el análisis de dichas proteínas para abordar esas cuestiones en las abejas. Un segundo tema importante se refiere a los mecanismos endocrinos subyacentes en los procesos de desarrollo de la reina y de las obreras, así como la reproducción y el polietismo, especialmente el papel de la hormona juvenil y los ecdisteroides. Se describen algunas técnicas para la cuantificación de los niveles de hormona juvenil circulantes en la hemolinfa, así como su síntesis por el *allata corpora*. Aunque éstos requieren cierta instrumentación y un grado considerable de sofisticación en los procedimientos de análisis, se consideró que la presentación de estas técnicas podría ser de interés para los laboratorios que planifiquen especializarse en este tipo de análisis. Dado que las aminas biogénicas son neurotransmisoras y reguladoras de las glándulas endocrinas, también presentamos un método estándar para la detección y el análisis de ciertas aminas biogénicas de interés. Otras preguntas entre la fisiología individual y la social están relacionadas con el metabolismo energético y la termorregulación. Así, una sección final está dedicada a los protocolos para cuantificar hidratos de carbono en el fluido corporal, la medición de la temperatura corporal y la respirometría.

西方蜜蜂生理学和生物化学研究的标准方法

蜜蜂不是研究昆虫生理学一般问题的典型模式，虽然蜜蜂具有重要的经济价值，并在诸如视觉、味觉、学习及记忆等神经生理学领域有特殊性，原因在于它们的社会性生活方式，在社会性进化过程中蜜蜂重建了某些生理学通路和生物化学特性，使其有别于"典型的昆虫"。毫无疑问，蜜蜂生理学和生物化学方面的研究多集中在蜜蜂的社会性组织结构，如：级型发育、劳动分工以及工蜂行为多态性。在这些过程中，某些蛋白质发挥着关键作用，如王浆主蛋白（MRJPs），包括"成王"蛋白（royalactin）和储存蛋白（hexamerins）影响级型发育，卵黄蛋白原影响生殖和日龄相关的劳动分工。所以本文将主要阐述和讨论用于建立和标准化蜜蜂蛋白质分析实验的基本实验指南。其次，还阐述了与工蜂和蜂王发育相关的内分泌调控机制，包括繁殖和行为多态性，特别是保幼激素和蜕皮激素的作用。还介绍了定量淋巴液中保幼激素水平以及咽侧体合成量化的灵敏技术。尽管这需要特殊的仪器设备，并且分析过程相当复杂，但我们认为对计划从事这些分析的实验室来说提供这些技术很有价值。由于生物胺既是神经传导物质又是内分泌腺的调节器，所以我们还提供了一些生物胺的测定和分析的标准方法。由个体水平的生理学研究跨越到群体生理学研究，关键在于探明能量代谢和温度调节的规律。所以本文最后一部分给出了蜜蜂体液中碳水化合物的量化、体温测量和呼吸测量的实验方案。

Keywords: haemolymph, protein, electrophoresis, SDS-PAGE, western blotting, immunofluorescence, sucrose, trehalose, juvenile hormone, radioimmunoassay, gas chromatography, *corpora allata*, temperature, thermosensors, thermography, radiation, humidity, operative temperature, respiration, energetics, gas exchange, respirometry, oxygen consumption, calorimetry , COLOSS, *BEEBOOK*, honey bee

1. Protein analysis for honey bee samples

1.1 Introduction

Obtaining protein profiles of an organism is the basis for assessing several aspects of biological processes. The protein content and protein composition in haemolymph, whole body or specific tissue extracts can provide valuable information on developmental stage, reproductive potential, aging processes, health status and correlated processes. Furthermore, quantitative analysis of protein content could be the starting point for standardizing or normalizing measures on other physiological, biochemical, or morphological parameters.

The first step in any protein analysis is usually the assessment of total protein content in a given sample, so as to guide further studies, especially comparative ones. As such, accurate measurement of protein concentration is critical for any further calculations such as, representation of specific proteins in a sample and, even more so, when determining enzyme activity. Errors in the calculation of protein concentration will tend to amplify overall errors in any such further estimates.

We selected a series of classical protocols currently used for an accurate measure of protein content of samples with different natures. The simplest method for quantifying protein content is spectrophotometry at 280 nm. However, this approach is not very reliable or sensitive compared to the two principle approaches that are detailed. The first approach is the Bradford Assay, which is based on the differential binding of a staining compound (Coomassie) through ionic interactions between sulfonic acid groups and positive amine groups on proteins (Bradford, 1976). The second is the bicinchoninic acid (BCA) method, which gained importance as a means for quantification of detergent extracted protein samples (Smith *et al.*, 1985). The Coomassie method is cheaper and very well suited for quantifying haemolymph proteins, but it is sensitive to higher detergent concentrations, as typically used for extracting proteins from tissue. In this case the BCA method is preferable. Other frequently used methods, such as that using Biuret-Folin-Ciocalteu reagents (Lowry *et al.*, 1951) are equally sensitive as the Bradford or BCA methods, but are more laborious, and it is only for the latter reason that we do not describe the Lowry method here.

In contrast to the determination of total protein content, the analysis of protein composition can be done by a plethora of methods and their respective variants. For this reason we decided to focus on just a few which can easily be established in any laboratory with basic equipment for analytical biochemistry and using low cost reagents. These are an electrophoretic separation of proteins according to their molecular mass (actually the Stoke radius of denatured proteins), based on the original method by Laemmli (1970), and two immunological methods (Western blot analysis and rocket immunoelectrophoresis) for the detection of specific proteins in complex mixtures. Each of these methods have been frequently utilized in research on honey bees. Western blot analysis has now become a gold-standard method for identifying specific proteins, but when emphasis is on more precise quantification, rocket immunoelectrophoresis is more precise.

Notwithstanding, the methods outlined here are ones that can fairly easily be implemented in any laboratory, as they do not require sophisticated equipment. Obviously, more advanced methods are available, starting from two-dimensional electrophoresis (2DE) to ever more sophisticated and high throughput proteomics analyses. In 2DE, proteins are usually first separated by isoelectric focusing and then by SDS-PAGE in the second dimension. Such gels have much higher resolution than one-dimensional gels, and spots detected in such gels can be retrieved for amino acid sequencing by Matrix-assisted laser desorption/ionization (MALDI) time of flight (TOF) analysis followed by comparison of amino acid sequences to proteome databases, e.g. MASCOT. 2DE methods have, for instance, been applied to study the honey bee haemolymph proteome (Chan and Foster, 2008; Boegaerts *et al.*, 2009), and MALDI-TOF proteomics analyses have also been applied to a variety of questions in honey bee biology (Santos *et al.*, 2005; Collins *et al.*, 2006; Li *et al.*, 2008).

1.2. Quantification of total protein content in samples

A crucial aspect when assaying protein concentration is the selection of an assay compatible with the sample. As such, whilst simple and well suited for analysing haemolymph protein content, the main disadvantage of Coomassie-based protein assays is the interference of certain detergents at concentrations routinely used to solubilise membrane proteins.

1.2.1. The Bradford assay

The Bradford assay for protein quantification is a popular protein assay because it is simple, rapid, inexpensive, and sensitive. It is based on the direct binding of Coomassie Brilliant Blue G-250 dye (CBBG) to proteins at arginine, tryptophan, tyrosine, histidine, and phenylalanine residues.

1. Prepare the protein reagent (Bradford reagent) by dissolving 100 mg of Coomassie Brilliant Blue G-250 in 50 ml of 95% ethanol.
2. Add 100 ml of 85% (w/v) phosphoric acid.
3. Dilute to 1l with distilled water
4. Let the solution stir overnight to assure maximal dissolution.
5. Filter (e.g. through Whatman #1 paper) and store in a dark bottle.
6. Prepare a standard curve from a 1 mg/ml of bovine serum albumin (BSA fraction V) stock solution by pipetting 1, 2, 5, 7, 10, 15 and 20 μl of this solution into glass test tubes. It is of importance not to extend the range of the standard curve beyond 20 μg/μl when using BSA to guarantee that measurements are within a linear range.
7. Complete each tube with distilled water to a final volume of 20 μl.
8. Prepare these in triplicates.
9. Prepare blank samples containing 20 μl of distilled water.
10. Also prepare triplicates from 20 μl of each unknown sample, that must be adequately diluted (this must be done empirically) to give measurements within the range of the standard curve.
11. Add 1 ml of Bradford reagent to blanks, standards and samples, vortex, leave for 2 min at room temperature and transfer the solution to disposable plastic cuvettes.
12. Set spectrophotometer to a wavelength of 595 nm.
13. Absorbance should be measured after 2 min and before 1h from the moment that Bradford reagent was added.
14. Average the absorbance readings of each of the triplicates and subtract blanks from standards and samples.
15. Plot absorbance values of the standard curve samples against their protein concentration (μg/μl).
16. Determine the concentration of the unknown samples by linear regression.

1.2.2. The bicinchoninic acid (BCA) assay

The BCA assay (Smith *et al.*, 1985) uses bicinchoninic acid (BCA) in a reaction forming Cu^+ from Cu^{2+} by the Biuret complex in alkaline solutions of protein. The advantage of this assay is its high tolerance towards the presence of detergents in protein extracts.

1. Prepare Reagent A (aqueous solution containing 1% of 2,2'-Biquinoline-4,4-dicarboxylic acid disodium salt, 0.16% of sodium tartrate, 0.4% sodium hydroxide, 2% of $Na_2CO_3 \cdot H_2O$, 0.95% $NaHCO_3$) .
2. Adjust pH to 11.25.
3. Prepare Reagent B (4% $CuSO_4 \cdot 5H_2O$ in deionized water). These reagents are stable indefinitely when kept in dark bottles at room temperature.
4. Standard working reagent (S-WR) should be prepared weekly or as needed by mixing 50 volumes of Reagent A with 1 volume of Reagent B.
5. Prepare a convenient standard curve using a solution 1 mg/ml of bovine serum albumin (BSA fraction V) in either isotonic saline or, in the case of any possibly interfering substance (e.g. sodium dodecyl sulphate, SDS), in a solution containing this particular substance.
6. Prepare blanks and triplicates of standards and samples (as described above in step 6 of the Bradford assay).
7. Add 20 volumes of S-WR per volume of sample (e.g. add 950 μl S-WR to a 50 μl sample).
8. Add 1 ml of S-WR to each.
9. Vortex well for a few seconds.
10. Incubate the samples for 30 min at 37ºC.
11. Cool samples to room temperature.
12. Transfer to disposable plastic cuvettes.
13. Set spectrophotometer to a wavelength of 562 nm.
14. Read standard curve and unknown samples.
15. Average the absorbance readings of each of the triplicates and subtract blanks from standards and samples.
16. Plot absorbance values of the standard curve samples against their protein concentration (μg/μl).
17. Determine the concentration of the unknown samples by linear regression.
18. Make sure that the standard curve is linear and that unknown samples are within range.

1.3. One-dimensional SDS gel electrophoresis of proteins

Electrophoresis is used for investigating complex mixtures of proteins by separating these according to their mobility in an electric field. It can be used to analyse subunit composition of certain proteins, to verify homogeneity of protein samples, and to purify proteins for use in further applications.

Polyacrylamide gels are the most commonly used matrices in electrophoretic separations being less costly than agar or agarose gels and providing a very broad range of options for defining matrix pore size (the sieving properties of the gel) through selecting appropriate

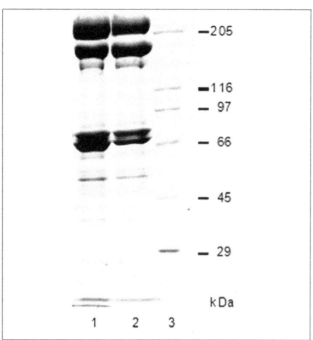

Fig. 1. Haemolymph protein patterns of *Apis mellifera* workers separated by SDS-PAGE (7.5 %), lane 1: 4-day-old worker; lane 2: 6-day-old worker; lane 3: molecular mass marker. Gel stained with Coomassie Brilliant Blue. Modified from Bitondi and Simões (1996). Copyright *Journal of Apicultural Research.*

proportions of polyacrylamide/bisacrylamide and water in the gel mix. In polyacrylamide electrophoresis (PAGE), proteins are separated according to charge and molecular mass/molecule structure. The use of sodium dodecyl sulphate (SDS) as detergent for eliminating differences in charge, the reduction of disulfide bonds by treatment with a reducing reagent, such as ß-mercaptoethanol or dithiothreitol (DTT) and heat to denature protein structure, were innovations by Laemmli (1970). This created the widely used SDS-PAGE protocols for separating proteins according to their Stoke's radius, commonly referred to as molecular mass (not molecular weight, as weight is dependent on gravity).

Unlabelled proteins separated by PAGE are typically detected by staining either with Coomassie Brilliant Blue or with silver salts. Coomassie Brilliant Blue binds nonspecifically to proteins but not the gel, thereby allowing visualization of the proteins as discrete blue bands within a translucent gel matrix. Observe that Coomassie Brilliant Blue G-250 is used in the Bradford assay, but it is Coomassie Brilliant Blue R-250 which is used for staining gels. These are different reagents, so be careful to use the correct one for each application.

Silver staining, although more laborious, is significantly more sensitive, but it may present problems when quantification of protein bands in gel documentation systems is the aim. The coloration of silver-stained bands is not uniform and bands of high protein content may in fact invert colour intensity and appear transparent. Although there are a plethora of variations of the original SDS-PAGE protocol (Laemmli, 1970) adapting the method to specific problems,

we describe a protocol commonly used to assess the haemolymph composition (Fig. 1) of honey bee larvae and adults (Pinto *et al.,* 2000; Barchuk *et al.,* 2002; Guidugli *et al.,* 2005; Bitondi *et al.,* 2006). A major variation in this protocol compared to the original Laemmli protocol is that there is no SDS in the gel, but only in the buffers. This avoids the precipitation of SDS in the gel matrix when running thin gels at low temperatures, conditions used to improve separation. After protein staining with Coomassie Brilliant Blue and scanning the gels on a gel documentation system, it is further possible to quantify specific proteins by Image J software (http://rsbweb.nih.gov/ij/index.html) or by commercial software implemented in gel documentation systems. Gels stained with silver salts are not appropriate for such quantitative analyses as silver staining does not follow linear characteristics.

1.3.1. Preparing samples

Ideally, samples should contain about 1 - 10 µg of total protein to give optimal results, thus protein content of the samples should be assessed by one of the methods described above (Bradford or BCA, section 1.2). Haemolymph proteins are usually best separated in gels with a 7.5% acrylamide concentration.

1. Prepare the sample buffer by dissolving 1.51 g Tris, 20 ml glycerol in 35 ml of double distilled water (ddH$_2$O).
2. Adjust pH to 6.75 with 1 N HCl.
3. Then add:
   3.1.   4 g SDS,
   3.2.   10 ml 2-mercaptoethanol,
   3.3.   0.002 g bromophenol blue,
   3.4.   ddH$_2$O to a final volume of 100 ml.
4. Prepare protein samples (haemolymph) by adding the sample to the sample buffer in a 1:1 (v/v) ratio.
   Very diluted samples may require a 2X concentrated sample buffer to make an adequate volume of 10 – 15 ml.
5. Prepare a sample containing the molecular mass markers, following the manufacturer's instructions.
   Use the same sample buffer (step 1) as that used for your samples.
6. Heat samples and the molecular mass marker sample in a boiling water bath for 1 – 3 min to denature protein structure.
   Perforate tops of Eppendorf tubes to avoid popping of the lid and spilling of the sample as internal pressure increases with heating.
7. Cool on ice for a few minutes.
8. Spin in a refrigerated tabletop centrifuge at maximum speed for 5 min.
9. Use supernatant only for application to the gel, as the precipitate may contain protein and nucleic acid aggregates which may cause streaking along the separation path.

User safety: Mercaptoethanol is an irritant and a foul smelling compound. It can be substituted with dithiotreitol.

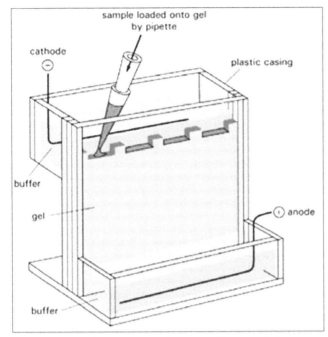

Fig. 2. Setup of a vertical SDS-PAGE system. Observe polarity settings.

1.3.2. Preparing and running vertical slab gels

This protocol is designed for a vertical slab gel with dimensions 100 x 120 x 0.9 mm (Fig. 2). For thicker gels or other gel sizes, the volumes of separating and stacking gels and the electric current must be adjusted accordingly.

1. Prepare 100 ml of an acrylamide stock solution containing
   1.1. 30% (w/v) acrylamide
   1.2. 0.8% (w/v) *N,N* ´-methylene bisacrylamide in ddH$_2$O.
   This solution can be stored refrigerated for 2-3 weeks.
2. Prepare a 1.5 M Tris buffer with a pH of 8.8, for this add:
   2.1. 18.15 g of Tris-base to
   2.2. 80 ml of ddH20.
   2.3. Use 1N HCl to adjust pH to 8.8.
   2.4. Complete volume to 100 ml.
3. Prepare a 0.25 M Tris buffer with a pH of 6.8, for this add:
   3.1. 3 g of Tris-base to
   3.2. 80 ml of ddH20
   3.3. Use 1N HCl to adjust pH to 8.8
   3.4. Complete volume to 100 ml.
   These buffers for the preparation of the separation and stacking gels, respectively, can be stored refrigerated.
4. Prepare 1 l of electophoresis buffer: dilute in 1 l of distilled water
   4.1. 3.03 g Tris.
   4.2. 14.4 g glycine.
   4.3. 1 g SDS.
   A polyacrylamide gel with a 7.5% separating gel is then prepared and run in the following sequence:
5. Mix
   5.1. 2.5 ml of the acrylamide/bis-acrylamide stock solution,
   5.2. 5 ml of the 1.5 M Tris buffer (pH 8.8),
   5.3. 2.3 ml ddH$_2$O.
6. Stir gently so as not to introduce air bubbles, as oxygenation may impede polymerization.
7. Add 190 μl of 1% ammonium persulphate (APS) solution and 40 μl of *N,N,N´,N´*-Tetramethylethylenediamine (TEMED).
   These are the starters for the polymerization process.
8. Quickly mix the reagents and immediately pour the solution into the cassette formed by the two glass plates sandwiched over sealing spacers.
   Leave sufficient space for later pouring the stacking gel on top.
9. Carefully overlay the gel with water to guarantee a smooth and straight surface.
10. Wait for about 30 min until the gel is completely polymerized. If it does not polymerize in due time, your APS solution is probably too old.
11. After polymerization is completed, pour off the water and carefully remove any remaining water with filter paper, but avoid touching the gel surface.
12. Prepare a stacking gel (4.26%) by mixing
    12.1. 0.375 ml of the acrylamide/bisacrylamide stock solution.
    12.2. 1.38 ml of the 0.25 M Tris buffer (pH 6.8) .
    12.3. 0.825 ml ddH$_2$O.
13. Stir gently so as not to introduce air bubbles.
14. Add 51 μl of a 5% ammonium persulphate (APS) solution.
15. Add 10 μl TEMED.
16. Pour the stacking gel on top of the separating gel.
17. Insert a Teflon comb with blunt-ended teeth for creating the sample application wells.
    Be careful to avoid introducing air bubbles.
18. Allow the gel to polymerize completely at room temperature. Once polymerized, gels can be stored in the refrigerator for up to a few days, but make sure to wrap the glass plate-gel sandwich with household PVC film.
19. Mount the gel sandwich in your vertical electrophoresis apparatus.
20. Fill the tanks with electrophoresis buffer.
21. Carefully remove the Teflon comb pulling up evenly by a small amount on each side.
22. Using a micropipette with long tips or a Hamilton-type syringe load the samples into the wells.
    The sample solution will be more dense than the running buffer, and will displace the latter when pipetted into the comb wells; the added bromophenol blue allows this to be visualized.
23. Connect the electrophoresis apparatus to a power supply, make sure that polarity is correct (see Fig. 2).
24. Carry out the electrophoresis run at a constant current of 15 mA, preferably in a cold room or refrigerator, until the bromophenol blue front reaches the desired position, usually 0.5 cm above the gel bottom.

25. Switch off power supply and remove cables.

26. Dismount the cassette, remove the stacking gel and carefully slide the gel off the glass plate into the dish containing the fixation/staining solution.

General advice: Ammonium persulphate (APS) decomposes within a short time, so a fresh working solution should be prepared weekly. Both APS and TEMED are starters for the polymerization process, so make sure to pour gels quickly after these compounds are added.

User safety: Acrylamide and bisacrylamide are neurotoxic compounds, so use protective equipment (fume hood, safety glasses and gloves) when preparing and handling any acrylamide solution. Even though after polymerization, polyacrylamide is no longer toxic, some unpolymerized acrylamide residue is still present, so gels should always be handled with gloves.

1.3.3. Staining gels with Coomassie Brilliant Blue

For preparing the staining solution, which at the same time is used for fixing the proteins within the gel matrix:

1. Dissolve 0.25% (w/v) Coomassie Brilliant Blue R-250 in ethanol, ddH$_2$O and glacial acetic acid [5:5:1 (v/v)].

2. Let the solution stir overnight in the dark.

3. Filter the solution

4. Store in a dark bottle

The staining solution can be reused several times, but do not leave it for prolonged time in the staining dish, as this will cause the evaporation of the ethanol and thus reduce the fixation properties of the solution.

Staining of the gel can be done in a glass dish covered with plastic foil or in a household plastic dish with cover:

1. Immerse the gel in the staining solution.

2. Agitate slowly for 16 h (overnight) at room temperature on a slowly rocking platform or orbital shaker.

3. Remove the staining solution and save it for future use. Destain the gel in the same solution [5:5:1 (v/v) of ethanol, ddH$_2$O and glacial acetic acid] without the dye.

4. Change destaining solution two or three times and continue destaining until blue bands and a clear background are obtained. Inserting a paper tissue into the destaining dish helps to absorb unbound Coomassie Brilliant Blue.

5. Destained gels can be dried in a vacuum gel-drying system; alternatively they can be stored in wet conditions within a sealed plastic container; add a few drops of glycerol to keep the gel soft.

1.3.4. Staining gels with silver salts

Various methods have been developed for staining polypeptides with silver salts after separation by SDS-PAGE, including more or less laborious and more or less sensitive methods. The procedure for silver staining of proteins in polyacrylamide gels described below is based on that described by Blum *et al.* (1987). A polyacrylamide gel with

dimensions 100 x 120 x 0.9 mm requires a volume of at least 200 ml of all solutions. The plastic container used should be adapted to the size of the gels to allow complete immersion. Solutions containing thiosulfate have to be prepared freshly to obtain sensitive and reproducible staining results. Powder-free disposable gloves should be worn when handling/transferring gels, as silver stain will detect proteins and oil transferred from fingers.

1. Prepare a fixative solution containing
   1.1. 50% (v/v) methanol.
   1.2. 12% (v/v) acetic acid.
   1.3. 0.5 ml of 37% formaldehyde.
   1.4. Dilute with ddH$_2$O to 1 l.

2. Incubate gel in the fixing solution for 30 min at room temperature with gentle shaking.

3. Wash for 10 min at room temperature in aqueous solution containing 50% (v/v) ethanol, with gentle shaking.

4. Repeat step 3.

5. Incubate gel for 5 min in a solution containing Na$_2$S$_2$O$_3$·5H$_2$O (0.2 g/l).

6. Rinse gel in water for 20 s.

7. Repeat step 6 twice.

8. Incubate for 10 min at room temperature with gentle shaking in a solution containing:
   8.1. AgNO$_3$ (2 g/l),
   8.2. 37% formaldehyde (0.75 ml/l).

9. Rinse the gel in water for 20 s.

10. Repeat step 9.

11. Add developer solution, made up from:
    11.1. Na$_2$CO$_3$ (60 g/l),
    11.2. 37% formaldehyde (0.5 ml/l),
    11.3. Na$_2$S$_2$O$_3$·5H$_2$O (4 mg/l).

12. Incubate gel at room temperature with gentle agitation and carefully watch the developing process.

13. Stop developing process once the desired band intensity and contrast are obtained by adding a solution containing
    13.1. 50% (v/v) methanol.
    13.2. 12% (v/v) acetic acid.
    13.3. 38% ddH$_2$O.

14. Wait for a few minutes, then wash the gel with ddH$_2$O.

15. Preserve the gel by drying or in wet condition within a sealed plastic container; add a few drops of glycerol to keep the gel soft.

1.4. Western blotting and immunodetection of proteins separated by SDS-PAGE

Western blotting, also known as immunoblotting, refers to the transfer of proteins from a polyacrylamide gel onto a solid support, such as a nitrocellulose, polyvinylidene difluoride (PVDF), or cationic nylon membrane. This membrane is then used in an immunodetection procedure to reveal specific protein(s).

The transfer of proteins from polyacrylamide gels to membranes was originally described by Towbin *et al.* (1979). The original method uses a tank containing a large volume of transfer buffer and is referred to as tank blotting. Subsequently, special western blotting systems were developed for semi-dry transfer. These systems are equal in performance, thus preference will depend on already available laboratory equipment.

After transfer, unspecific binding sites of the membrane are first blocked by excess protein added to the incubation buffer to suppress nonspecific adsorption of antibodies. Subsequently, the immobilized proteins are reacted with a specific polyclonal or monoclonal antibody. Antigen-antibody complexes are finally revealed through a secondary antibody and chromogenic or chemiluminescent reactions.

The following protocol uses a tank blotting system and a horseradish peroxidase-conjugated secondary antibody of an enhanced chemoluminescence (ECL) detection system (GE Healthcare) for revealing antigen-antibody complexes:

1. Prepare the western blotting transfer buffer:
    1.1. 25 mM Tris,
    1.2. 192 mM glycine,
    1.3. 20% (v/v) methanol.
2. Cut the PVDF membrane and filter paper sheets to fit the size of the separating gel.
    Be sure to handle the PVDF membrane using powder free gloves and forceps.
3. Activate the membrane in methanol for 1–2 min.
4. Immerse in water for 1 min to remove the activating solvent.
5. Incubate the PVDF membrane in transfer buffer for 2 min.
6. Assemble the blotting sandwich (Fig. 3) in the following order in a tray containing western transfer buffer (make sure to keep all items submersed and avoid including air bubbles, especially so between the gel and the membrane):
    6.1. A stiff plastic supporting grid.
    6.2. A foam sponge or Scotch-Brite pad (3M).
    6.3. Two sheets of thick filter paper.
    6.4. The polyacrylamide gel.
    6.5. The PVDF membrane.
    6.6. Two sheets of thick filter paper.
    6.7. A foam sponge or Scotch-Brite pad (3M).
    6.8. The second stiff plastic supporting grid.
7. Fill the transfer tank with western transfer buffer.
8. Insert the sandwich into the support holder of the blotting apparatus. Make sure the orientation is correct, as transfer is from the cathode (-) to the anode (+), thus, the gel should face towards the cathode and the membrane face the anode.
9. Connect tank to a high voltage power supply.
    This is not your usual electrophoresis power supply, but one that can go up to 200 V, 2000 mA and 200W).
10. Run transfer at a setting of 30 V for about 2.5 h at room temperature.
11. Shut down power supply, disconnect cables.

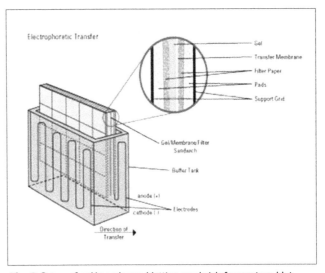

Fig. 3. Setup of gel/membrane blotting sandwich for western blot transfer of proteins. Observe gel/membrane position in the polarity setting [anode (+), cathode (-)].

12. Dismount the blotting sandwich.
13. Mark with a pencil the side of the membrane that faced the gel.

To verify transfer efficiency, the gel can be stained with Coomassie Brilliant Blue after blotting (see section 1.3.3); alternatively, the membrane can be stained with Ponceau S solution [0.5 g of Ponceau S in 100 mL of 1% (v/v) acetic acid aqueous solution] for 2 min, washed two to three times in water and then further destained in water. The PVDF membrane can either be used directly for immunodetection, as described below, or air-dried for later detection (this will require reactivation by immersion in methanol, as described in step 3 of the above list).

For immunodetection:

1. Prepare 1l of a 10 x PBS stock solution:
    1.1. 80 g NaCl,
    1.2. 2 g KCl,
    1.3. 14.4 g Na_2HPO_4,
    1.4. 2.4 g KH_2PO_4,
    1.5. in 1l ddH_2O.
2. Prepare a blocking solution containing 250 ml of buffer A:
    2.1. 3.027 g Tris,
    2.2. 0.147 g $CaCl_2$,
    2.3. 2.33 g NaCl,
    2.4. Adjust pH to 8.5,
    2.5. 50 g non-fat dried milk,
    2.6. Complete the volume to 500 ml with ddH_2O.
3. Block unspecific binding sites by immersing the membrane in this solution for 1 h at room temperature on an orbital shaker. Alternatively, membranes may be left in the blocking solution overnight in a refrigerator.
4. Briefly rinse the membrane with two changes of wash buffer (500 µl Tween 20 in 1l of 1 x PBS (make up from the 10 x stock, described in step 1, and adjust the pH to 7.2).

5. Appropriately dilute the primary antibody in blocking solution. The dilution factor must be determined empirically for each antibody, e.g. through dot blotting of a serial dilution of the antibody.

6. Incubate the membrane in diluted primary antibody for 1 h at room temperature on an orbital shaker.

7. Briefly rinse the membrane with two changes of wash buffer.

8. Keep the membrane in wash buffer for 15 minutes at room temperature.

   Use >4 ml of wash buffer per cm^2 of membrane.

9. Wash the membrane a further three times for 5 min each, in changes of wash buffer.

10. Dilute the horseradish peroxidase (HRP)-conjugated secondary antibody of the ECL kit in wash buffer.

    Again, the dilution factor must be determined empirically for each antibody - a 1:12,000 (v/v) dilution may usually be appropriate.

11. Incubate the membrane in the diluted secondary antibody for 1 h at room temperature on an orbital shaker.

12. Briefly rinse the membrane with two changes of wash buffer..

13. Wash the membrane in > 4 ml/cm^2 of wash buffer for 15 min at room temperature.

14. Wash the membrane a further three times for 5 min each, in changes of wash buffer.

15. Proceed with the detection reaction to obtain the chemoluminescent signal following the manufacturer's instructions. This procedure is specific for each commercial ECL kit.

16. Wrap the blots wetted with ECL solution in household PVC foil and place, protein side up, in an X-ray film cassette.

17. In a dark room place a sheet of autoradiography film on top of the membrane previously wrapped in foil.

18. Close the cassette and expose for a short time, usually 5 min.

19. Immediately develop this first film using commercial X-ray film developer or, if available, an automatic developer system.

20. Based on the obtained band intensity, estimate an optimal exposure time for a second (or third) film.

1.5. Rocket immunoelectrophoresis

Rocket immunoelectrophoresis is a simple, quick and reproducible method for determining the concentration of a single protein in a protein mixture. Like immunodetection following western blotting it is a method based on the affinity of a specific antiserum (which can be mono- or polyclonal) with a specific protein. As it does not use a secondary antibody conjugated with a moiety for high sensitivity detection, but is based on the formation of an antigen-antibody precipitate in a gel matrix. It does not have the sensitivity of the immunodetection method described in section 1.4. It does, however, have the advantages that the presence of a specific protein can be analysed fairly quickly, both qualitatively and quantitatively, in a relatively large number of samples.

In this procedure, appropriately diluted samples are applied to small circular wells cut into an agarose gel which has a specific

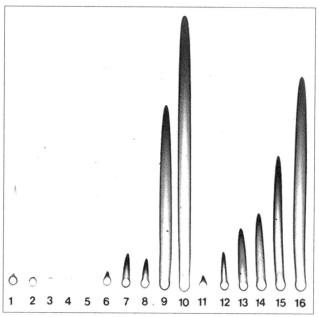

Fig. 4. Rocket immunoeletrophoresis for quantification of haemolymph vitellogenin in 1 to 6 day-old *Apis mellifera* workers reared on different diets. Haemolymph was from bees fed a 0% pollen diet (wells 1-4), a 15% pollen diet (wells 5-7), a 50% pollen diet (wells 8-10), a pollen-free sugar diet supplemented with soybean and yeast (wells 11-13), and naturally fed workers (wells 14-16). Reproduced from Bitondi and Simões (1996).   Copyright *Journal of Apicultural Research*

antibody already incorporated in its matrix. When migrating in an electric field, the protein of interest will eventually reach a critical point of antigen-antibody concentrations resulting in the local formation large precipitating complexes. The agarose gel can then be stained and the rocket-shaped precipitate becomes apparent (Fig. 4). The position of this peak is directly related to the concentration of the protein of interest.

A typical rocket immunoelectrophoresis assay of honey bee haemolymph proteins is done as follows:

1. Prepare a 1% (w/v) agarose solution in 0.06 M Tris-HCl buffer, pH 8.6.

2. Completely dissolve the agarose by boiling for 2-3 min.

3. When the agarose has dissolved, place the flask in a 52ºC water bath.

4. Once the agarose solution has cooled to 52°C (use a thermometer to check temperature), add an appropriate amount of antiserum (this amount has to be determined empirically by serial dilution assays for each antiserum).

   Do not add the antiserum earlier as a higher temperature will cause its denaturation, also do not add it much later, as agarose will soon start to solidify; briefly mix to ensure even dispersal of the antiserum.

5. Pour the agarose solution (15 ml) onto a clean glass plate (usually 10x10 cm) so as to guarantee an even gel thickness (the final gel will be approximately 1.5 mm thick):

5.1.   Place the glass plate on a levelled surface.

5.2.   Start pouring the agarose solution in the middle of the plate. The liquid will spread to the edges of the glass plate, but surface tension will prevent it from running off the edge of the plate (alternatively, tape can be used to seal the edges).

6. Wait for 5–10 min for the agarose gel to harden.

7. Using a steel puncher connected to a suction device punch holes of 1 mm diameter at minimal spacing of 0.5 cm forming a line across the plate at approximately 1 cm from one edge. The holes can also be made using a glass Pasteur pipette with a suction bulb.

8. Place the glass plate with the gel on a horizontal electrophoresis system.

9. Fill the troughs with electrode buffer (0.3 M Tris-HCl buffer, pH 8.6).

10. Apply wicks (made from filter paper) immersed in buffer to form a bridge between the gel edges and the buffer troughs. Make sure that the gel is in correct orientation with respect to polarity of the electrophoresis system [the cathode (-) should be next to the sample wells].

11. Add gel buffer (0.06 M Tris-HCl buffer, pH 8.6) at a 1:1 ratio (v/v) to the appropriately diluted samples.
    It is important to ensure that all samples are prepared to an approximately equal volume and, if possible, contain similar amounts of total protein.

12. Apply samples to the well.
    This should be done as quickly as possible to minimize diffusion from the wells into the gel (alternatively a small current of 1-2 mA may be applied during loading to overcome diffusion problems).

13. Immunoelectrophoresis is then carried out at 20°C for 16 h, at a setting of 0.08 V/cm gel length (make sure to check polarity).

14. For staining after electrophoresis:

    14.1.   First, cover the gel with two layers of filter paper soaked in saline 0.9%.

    14.2.   Add a 2–3 cm thick layer of dry soft paper tissue.

    14.3.   Cover with a thick glass plate to guarantee application of slight and even pressure (about 10 g/cm^2).

15. After 20-30 min the gel layer should have now been reduced to a thin film covering the glass plate.

16. Immerse the gel for 24 h in saline solution.

17. Wash in distilled water for 30 min.

18. Cover the gel with two layers of filter paper soaked in distilled water.

19. Cover with a 2–3 cm thick layer of dry soft tissue paper.

20. Cover with a thick glass plate to apply slight and even pressure (about 10 g/cm^2).

21. Air dry the gel on the glass plate with a hair dryer.
    Hot or cold air can be used.

22. Place gel for 20-30 min in staining solution made up with 0.25% (w/v) Coomassie Brilliant Blue R-250 dissolved in a solution 5:5:1 (v/v) of ethanol, ddH$_2$O and glacial acetic acid (see section 1.3.3).

23. Remove excess dye by washing the gel in the same solution prepared without the dye.

24. After drying at room temperature the gel can be kept as a permanent record.

25. Measure the peak height of the rocket-like precipitates for each sample.

For absolute quantification compare this to a standard sample for the protein of interest run on the same gel; for relative quantification set the sample with the highest peak as 100%.

1.6. Immunofluorescence detection of proteins in tissue: tubulin localization in ovariole whole mounts as an example of a working protocol

One of the most widely used techniques to study the function and/or localization of proteins is known as immunolabelling or immunolocalization. This is a general technical term that defines the use of specific antibodies to identify the location of molecules or structures within cells or tissues, both in whole mounts and histological sections. Depending on the method of antibody detection, these techniques are divided into two major categories: immunofluorescence, which employs a fluorescence-conjugated secondary antibody, and immunocytochemistry, which uses an enzyme-conjugated secondary antibody and a precipitable enzyme substrate for detection. The choice of an immunolabelling protocol will take into account several factors to obtain reliable staining results, such as the specificity of the antibodies and the general conservation of cell and tissue structure. There are currently no commercially available primary antibodies generated against honey bee proteins. For honey bee research, all antibodies are by definition heterologous ones, having been produced against a protein of another species. Depending on the honey bee protein(s) of interest, these heterologous antibodies can have good cross reactivity, as certain immunoreactive protein domains (epitopes) may be conserved. If available, it is of course preferable to use antibodies produced against specific honey bee proteins (e.g. vitellogenin). When choosing a primary antibody, it is furthermore of interest to note whether it is a polyclonal or monoclonal one. Polyclonal ones were generated by conventional immunization of a laboratory animal, and thus are reactive with several domains (epitopes) of a certain single protein, whereas monoclonal ones were generated by hybridization and subsequent selection for a single epitope.

As far as secondary antibodies are concerned, they must, of course, be directed against the immunoglobulin type of the species in which the primary antibody was produced. Furthermore, secondary antibodies can be whole serum, purified immunoglobulins, or antibody fragments (Fab) corresponding to the antigen binding domain. All such choices will eventually depend on the question to be answered, time investment, prior laboratory experience, and availability of antibodies from commercial or non-commercial suppliers (e.g. the Developmental Studies Hybridoma Bank at the University of Iowa, http://dshb.biology.uiowa.edu, or colleagues).

As it is impractical to provide a comprehensive listing and description of all different immunolocalization techniques and their numerous variants herein, we instead describe in this section a specific protocol as an example for general guidance. This is an immunofluorescence protocol for detecting tubulins in whole-mount ovary preparations of adult honey bees. We considered this as a topic of broader interest, considering the importance of reproductive division of labour between queens and workers of *A. mellifera* as a colony trait. In terms of physiology and biochemistry, this involves, on the one hand, vitellogenin produced and secreted by the fat body, and on the other, the structural organization of the ovariole undergoing oogenesis. It is in the latter process where cytoskeleton proteins play a major role in the differentiation and development of the oocytes and nurse cells, both in adult bees and during postembryonic development (Schmidt-Capella and Hartfelder, 2002; Tanaka and Hartfelder, 2004; Florecki and Hartfelder, 2011).

One of the main steps in the developmental determination of the oocyte and its differentiation within a cluster of cystocytes is the presence of a microtubule-organizing centre or centrosome, which is a structure containing, amongst other proteins, all three members of the tubulin family, α- β- and γ-tubulin. Furthermore, α- and β-tubulin also form heterodimers that make up microtubules, e.g. present in the mitotic or meiotic spindle apparatus. Immunofluorescence detection of α-, β- or γ-tubulin can thus reveal several structures of interest, providing relevant information about cytoskeleton organization and organelle distribution in honey bee oogenesis.

1.6.1. Buffers

1. prepare 100 ml of a 0.02M phosphate-buffered saline (PBS) from the following stock solutions:
    1.1. 1.9 ml of 0.2 M $NaH_2PO_4xH_2O$,
    1.2. 8.1 ml of 0.2 M $Na_2HPO4xH_2O$,
    1.3. 0.9 g NaCl,
    1.4. 80 ml ddH_2O,
    1.5. Adjust pH to 7.4,
    1.6. Complete volume to 100 ml.
2. Prepare a microtubule Stabilization Buffer containing:
    2.1. HEPES buffer (25 mM, pH 6.8),
    2.2. 25% glycerol,
    2.3. 0.5 mM $MgCl_2$,
    2.4. 25 µM phenylmethyl sulfonyl fluoride (PMSF),
    2.5. 1% Triton X-100,
    2.6. 0.01% sodium azide.

All molarities and % given as final concentrations.

HEPES (4-(2-hydroxyethyl)-1-piperazineethanesulfonic acid) is a zwitterionic organic chemical buffering agent. It is widely used in cell culture media for maintaining a physiological pH level. PMSF is a serine protease inhibitor which is best prepared as a stock solution of 25mM PMSF in ethanol. Triton X-100 is a detergent, so it should be added to the buffer only after adjusting the pH. Sodium azide is a common preservative of samples and stock solutions, acting as a bacteriostatic.

1.6.2. Dissection and fixation of ovarioles

1. Dissect the ovaries from adult queens or workers (see the *BEEBOOK* paper on anatomy and dissection (Carreck *et al.*, 2013)).
2. Transfer these to a dish containing honey bee culture medium (recipe and preparation steps as described in Rachinsky and Hartfelder, 1998) or a commercial insect culture medium (e.g. TC-100).
3. Individualize the ovarioles and, with the aid of fine watchmakers forceps, manually remove the tracheae and the peritoneal sheath covering each ovariole.
4. Transfer batches of individualized clean ovarioles into a 4-well plate containing sufficient (around 0.5 ml) Stabilization Buffer (see 1.6.1., step 2.) to cover the ovarioles.
5. Keep for 20 min at room temperature (RT) under shaking movement.
6. Fix the ovarioles in cold (-20°C) pure methanol for 10 min at 4°C. Note, the most common fixative used in immunolabelling protocols is 4% paraformaldehyde (PFA) in PBS, but for microtubules, a better result is obtained with methanol fixation following treatment with Stabilization Buffer.
7. Rinse in PBS for 30 min at RT with shaking.
8. Permeabilize in PBS-T (PBS 10 mM, Triton X-100, 0.1%) for 30 min at RT with shaking.
9. Block unspecific binding sites by incubating the ovarioles in PBS-T supplemented with 10% non-immune serum for 20 min at RT with shaking.

The non-immune serum must correspond to a serum of the organism used as source for generating the secondary antibody, e.g., if using a FITC-conjugated goat-anti-mouse-IgG antibody, use a non-immune goat serum for blocking.

10. Rinse in PBS-T for 2x 15 min at RT with shaking.
11. Incubate for 3 h at RT or at 4°C overnight in a monoclonal antibody raised against α-, β- or γ-tubulin, diluted in PBS-T under shaking.

When using the monoclonal antibody raised against $β_{I+II}$-tubulin (Sigma-Aldrich, T8538) a dilution of 1:200 in PBS-T may give good results, but such dilutions must be determined empirically by serial dilution experiments for each antibody and tissue.

12. Rinse in PBS-T for 2 x 30 min at RT with shaking.
13. Remove PBS-T.
14. Leave ovarioles for 1 h in a fluorescence-conjugated secondary antibody diluted in PBS-T in darkness at RT under shaking.

In the case of a FITC-conjugated goat-anti-mouse IgG (Sigma-Aldrich), a dilution of 1:400 in PBS-T is sufficient to obtain a good signal to background ratio.

15. Rinse the ovarioles in PBS-T for 2 x 30 min in darkness at RT with shaking.

16. After this wash step one can add TRITC-conjugated phalloidin to label filamentous actin, and/or DAPI (4',6-diamidino-2-phenylindole) or another reagent intercalating with dsDNA to counterstain nuclei.

17. Rinse the ovarioles in PBS-T for 2 x 30 min in darkness at RT with shaking.

18. Rinse briefly in ddH$_2$O.

19. Transfer ovarioles to microscope slides and embed in glycerol/n-propyl gallate mounting medium:

    19.1.   Glycerol 90%,

    19.2.   N-propyl gallate 3% in PBS,

    19.3.   Supplemented with sodium azide 0.01%.

    N-propyl gallate is an anti-fade reagent used in fluorescence microscopy to reduce photobleaching; alternatively, you may use commercial antifade reagents (e.g. Vectashield).

20. Coverslip.

21. Seal coverslip edges with nail polish.

21. Store slides in dark.

The whole-mount ovariole preparations can be analysed by conventional epifluorescence microscopy or by means of a laser confocal system.

As in any immunolabelling protocol, a negative control staining is mandatory. In the case of commercial antibodies this can be done by substituting the primary antibody for PBS-T. When using an antibody prepared in the proper laboratory or by a colleague, it is even better to use at this step a pre-immune serum, i.e. serum drawn from the respective animal (usually rabbit) prior to immunization.

User safety: Sodium azide is highly toxic. Appropriate ventilation (laboratory chemical hood) and personal protective equipment (such as gloves) must be used to minimize exposure.

2. Measurement of glucose and trehalose in honey bee haemolymph

This protocol describes a sensitive enzymatic method for quantifying glucose and trehalose in honey bee haemolymph. Haemolymph sugar levels do not only need to be regulated, they also provide important information on the physiological state of carbohydrate metabolism, which is associated with gustatory sucrose responsiveness (Amdam et al., 2006; Wang et al., 2012). In general, the glucose titre in honey bee haemolymph is below 20 µg/µl and the trehalose titer is below 30 µg/µl (Wang et al., 2012).

The method allows measuring of glucose concentrations from 0.5 to 100 µg/µl, and trehalose from 0.4 to 94 µg/µl. Such sensitivity makes the method appropriate for estimation of glucose and trehalose in most insect body fluids, without prior concentration or extraction. Furthermore, these methods are specific, reproducible, sensitive, high throughput and do not require extensive sample preparation. Measuring other carbohydrates of the honey bee colony, such as sugar concentration in honey or the crop of foragers, does not require such a sensitive method and can be easily done by means of a refractometer.

The glucose method described herein is based on the reaction between glucose and adenosine triphosphate (ATP). Glucose can be phosphorylated by adenosine triphosphate (ATP) to form glucose-6-phosphate (G6P) in a reaction catalysed by hexokinase. G6P is then oxidized to 6-phosphogluconate in the presence of oxidized nicotinamide adenine dinucleotide (NAD$^+$) in a reaction catalysed by glucose-6-phosphate dehydrogenase (G6PDH). During this oxidation, an equimolar amount of NAD$^+$ is reduced to NADH, which consequently can be spectophotometrically detected as an increase in absorbance at 340 nm (Peterson and Young, 1968; Bondar et al., 1974). The amount of NAD$^+$ consumed in the reaction is directly proportional to the amount of glucose present in the sample. For quantifying trehalose, this disaccharide must first be hydrolysed to two molecules of D-glucose in a reaction catalysed by trehalase (Broughton et al., 2005; Flatt and Kawecki, 2007).

2.1 Sample Collection

1. Anesthetize bees by keeping them at 4°C for a few minutes.

2. Mount them on a wax plate by a pair of insect pins crossing over the waist.

3. Incise dorsally between the 5th and 6th abdominal segment by means of a G 30 needle (BD).

4. Collect 1 µl of haemolymph using a microcapillary.
   Care needs to be taken to avoid contamination from intestine and other surrounding organs.

5. Transfer each haemolymph sample into a 1.5 ml Eppendorf tube.

6. Immediately snap freeze the samples in liquid nitrogen.

7. Store them at -80°C until testing.

2.2. Preparation of the reagents
2.2.1. Benzoic acid solution

Dissolve 0.1 g of benzoic acid in 100 ml ddH$_2$O to make a 0.1% solution.

This solution is stable at room temperature.

2.2.2. Glucose standard stock solution (1 mg/ml)

1. Dry the glucose at 60-80°C for 4 h.

2. Allow to cool in a desiccator.

3. Dissolve 10 g of glucose in benzoic acid solution.

4. Make up to 10 ml in a volumetric flask.

This stock solution is stable for at least six months at 4°C; do not freeze.

2.2.3. Enzyme mix solution

The enzyme mix solution contains 1.5 mM NAD$^+$, 1.0 mM ATP, 1.0 unit/ml of hexokinase, 1.0 unit/ml of glucose-6-phosphate dehydrogenase.

This enzyme mix is commercially available (Sigma-Aldrich) and dissolves readily in ddH$_2$O. When kept at 4°C, this solution should be stable for up to one month.

2.3. Glucose assay

1. To make the glucose standard curve, prepare eight clean glass test tubes (5 ml) and label them S1 through S8.
2. Aliquot 0.5, 1, 5, 10, 30, 50 and 100 µl glucose standard solution (1 mg/ml) into tubes S2-S8.

Do not add any standard to tube S1 as this tube will be the blank.

3. Label appropriate number of clean glass test tubes (5 ml) for samples.
4. Keep the tubes on ice.
5. Take the haemolymph samples from the -80°C freezer and keep them on ice.
6. Add 1 ml enzyme mix solution to each standard tube and sample tube.
7. Invert tubes 4-6 times.
8. Centrifuge tubes for 30 s to spin down contents.
9. Equilibrate reaction mix at room temperature for 15 min.
10. Transfer 200 µl aliquots of each standard (S1-S8) and sample into cuvettes or into a 96-well microplate for spectrophotometry readings.
11. Set spectrophotometer (for cuvettes) or ELISA reader (for microplates) for reading absorbance at 340 nm.

Standards and samples should be read in replicates or triplicates.

12. To calculate glucose concentrations, plot absorbance at 340 nm as a function of glucose concentration of the standard curve samples.

The concentrations of glucose standards (0.5, 1, 5, 10, 30, 50 and 100 µg/ml) are plotted on the X-axis; respective absorbance is on the Y-axis.

13. Determine the equation of the line by linear regression.
14. Each glucose concentration is calculated as:

$$\text{Glucose concentration (ug/ul)} = \frac{[(A_{sample} - A_{blank}) - (\text{y-intercept})]}{\text{Slope}}$$

15. If a sample had to be diluted during preparation, the result must be multiplied by the dilution factor, F.

2.4. Trehalose assay

1. After taking the glucose readings, pipette 0.5 µl of the trehalase enzyme into each cuvette or well of the microplate including the wells containing the glucose standards (Flatt *et al.*, 2008; Ishikawa *et al.*, 2009).

2. Slowly shake the plate on a rocking platform for 1 min.
3. Centrifuge for 2 min.
4. After centrifugation, use a piece of Parafilm to seal the cuvettes or microwell plate.
5. Incubate overnight at 37°C overnight.
6. Centrifuge again.
7. Read absorbance as described above for glucose (step 11 in 2.3)
8. Calculate the trehalose concentration as:

$$\text{Trehalose concentration (ug/ul)} = \frac{\left(\frac{[(A_{sample} - A_{blank}) \cdot (\text{y-intercept})]}{\text{Slope}} - \text{Glucose concentration}\right) \cdot 342.3}{180.2 \cdot 2}$$

The trehalose concentration is calculated as the reading of the proper reaction minus the prior determined glucose concentration. The term is then multiplied by the molecular weight of trehalose (342.3) and, as trehalose is split by trehalase into two glucose molecules, the entire term is finally divided by 2 x the molecular weight of glucose (180.2).

3. Analysis of juvenile hormone and ecdysteroid levels in honey bees

Juvenile hormone (JH) and ecdysteroids are lipid signalling molecules playing fundamental roles in postembryonic development and the reproductive physiology of insects (Nijhout, 1994). In social insects, where these hormones are also involved in regulating caste development, reproductive dominance and division of labour, the role of these hormones has been extensively reviewed (de Wilde and Beetsma, 1976; Nijhout and Wheeler, 1982; Robinson, 1992; Robinson and Vargo, 1997; Hartfelder and Engels, 1998; Hartfelder and Emlen, 2012) ever since their biochemical characterization.

While many of these insights into the roles played by these hormones in social insects have been gained through application of synthetic hormones or hormone analogs, such experiments require confirmation through analyses of circulating hormone titres or rates of hormone synthesis by the respective endocrine glands or peripheral tissues. Conclusions based solely on hormone treatment experiments are frequently and justifiably subject to critique (Zera, 2007) as the applied doses usually exceed endogenous hormone levels by three to six orders of magnitude, thus introducing the possibility of pharmacological effects masking their truly physiological ones. Employing sensitive detection methods is thus paramount to fully understand the role of these lipid signalling molecules.

As a key regulator of insect development, reproduction, and behaviour (Goodman and Cusson, 2012), juvenile hormone (JH) also plays a major role in the social organization of bees, wasps, ants and termites (Hartfelder and Emlen, 2012). In honey bees, JH has been shown to drive caste development in the larval stages (Hartfelder and Engels, 1998), and in adult workers it plays an important role in division of labour (Robinson and Vargo, 1997; Amdam *et al.*, 2007) and sensory modulation (Pankiw and Page, 2003). There are several isoforms of JH in different insects (e.g. JH I, JH II, JH III, bis-epoxy JH III). These differ slightly in their side chains and unsaturated bonds, but in honey bees, as in most insects, JH III is the only isoform produced by the corpora allata (Hagenguth and Rembold, 1978).

Whereas insects can synthesize JH *de novo* from relatively simple compounds (acetyl CoA or proprionyl CoA) they cannot do this for ecdysteroids. Rather, they require dietary steroids for conversion to physiologically active hormone. This conversion occurs in the prothoracic glands of larvae and pupae and in the gonads of the adults. In honey bees, the predominant ecdysteroid moiety in haemolymph of larvae and pupae and ovaries of adult females is makisterone A, quantitatively followed by 20-hydroxyecdysone and ecdysone (Feldlaufer *et al.*, 1985, 1986; Rachinsky *et al.*, 1990). These are the physiologically active ecdysteroids, whereas others, especially a series of different conjugates, are either metabolites or storage forms (Lafont *et al.*, 2012). In honey bee caste development, haemolymph ecdysteroid titres differ between queen and worker larvae and pupae (Rachinsky *et al.*, 1990; Pinto *et al.*, 2002), but they do not seem to play a major role in reproduction or division of labour (Hartfelder *et al.*, 2002).

With the importance of these hormones in honey bee biology in mind, we will focus in this section primarily on currently used and firmly established analytical methods. We detail radioimmunoassay (RIA) and physicochemical detection methods, such as gas chromatography coupled with mass spectroscopy (GC-MS), for hormone titration, as well as a radiochemical *in vitro* assay for determining the JH-synthetic activity of the *corpora allata* (CA). It is important to note that while radioimmunoassays have frequently been substituted by enzyme-linked immunosorbant assays (ELISAs), due to restrictive regulations for the use of radioisotopes, there are no ELISAs of sufficient sensitivity available for the quantification of insect ecdysteroids and JH.

Older methods, such as the *Galleria* bioassay, will not be described herein. While important tools in the early days of JH quantification, including in honey bees (Fluri *et al.*, 1982), these older methods are extremely laborious, and provide only relative measures (e.g. *Galleria* units) rather than absolute quantities (ng JH per ml haemolymph). We also do not present recently developed analysis methods employing liquid chromatography mass spectrometry (LC-MS) that have been developed (Westerlund and Hoffmann, 2004; Li *et al.*, 2006). Although these have been validated for use in honey bees (Zhou *et al.*, 2011) and are comparable in terms of sensitivity to radioimmunoassays and GC-MS (Chen *et al.*, 2007), they are not yet in common use. A recently developed, very elegant and highly sensitive method for quantifying JH based on tagging the epoxy group

of JH with a fluorescent tags, with subsequent analysis by reverse phase high performance liquid chromatography coupled to a fluorescent detector (HPLC-FD) (Rivera-Perez *et al.*, 2012) may, however, eventually become an option. Ultimately, the method of choice for a laboratory will, of course, essentially depend on available equipment and expertise.

3.1. General instructions on haemolymph sample collection and glassware preparation for juvenile hormone assays

3.1.1. Haemolymph collection

The most frequently analysed samples are haemolymph obtained from larvae or adult honey bees. Pupae typically have very low JH titres which are physiologically irrelevant. It is only the pharate adults (pupae undergoing pigmentation of the thorax and abdomen) that may be of interest, as in these stages JH becomes relevant for inducing vitellogenin expression.

1. For collecting haemolymph from feeding-stage larvae:
   1.1. Place the insect on a piece of Parafilm.
   1.2. Identify the position of the dorsal vessel, which is the transparent and easily visible vessel running all along the entire dorsal side of the larva.
   1.3. Puncture this vessel with a pair of forceps.
   1.4. Extruding haemolymph should be clear and can be collect with a microcapillary.
2. For collecting haemolymph from spinning stage larvae and prepupae:
   2.1. Puncture and collect extruding body fluid.

As spinning stage larvae and prepupae are undergoing metamorphosis, the extruding fluid this is not clear haemolymph, but a whitish fluid that contains a lot of tissue debris.

For obtaining the haemolymph fraction:
   2.2. Transfer to a centrifuge tube.
   2.3. Centrifuge at 10,000 x g for 5-10 min.

A thin white sheet containing lipids will have now formed a top layer.
   2.4. Carefully insert a collection capillary tube through this sheet to collect the small volume of clear fluid directly below.

Avoid aspirating the voluminous white bottom layer, which mainly contains cell debris.

3. For collecting haemolymph from pharate adults or adults:
   3.1. Immobilize the bee with two insect pins crosswise inserted into a wax-filled Petri dish and pressing the pins down over the waist.
   3.2. Haemolymph should preferably be collected with a microcapillary from an incision in the neck membrane of adult bees or from the thorax.

This minimizes lipid content in the sample.

When collecting haemolymph, a graduated microcapillary tube should be used to ensure accurate assessment of volume. For both GC-MS and RIA, 2 - 4 µl are normally sufficient for a sample. Depending on the stage being sampled, it may be necessary to pool haemolymph from multiple individuals to get readable results. After registering the exact volume of haemolymph, this is expelled into a glass collection vial already containing appropriate solvent.

The vials should be glass, fitted with a screw cap containing a Teflon-lined rubber septum. The Teflon cover should face the sample. These are low-cost glass vials customarily used for GC analysis. For JH analysis by RIA, vials of 1.25-2 ml volume containing 0.5 ml acetonitrile are recommended. For JH analysis by GC-MS, the samples should be collected into 8 ml vials containing 1.5 ml of 50% acetonitrile in water. Samples can then be stored for long periods of time at -20°C. Special refrigeration on transport for short periods (up to two days) is not required.

Do not use plastic vials, such as Eppendorf tubes, to store samples, as JH binds to plastic. It is also not recommended to store haemolymph in microcapillaries in the freezer, as JH degrading enzymes may retain activity under such conditions.

3.1.2. Glassware preparation

JH is a "sticky" lipophilic molecule which makes the use of clean glassware an imperative component throughout all steps of sample preparation. Furthermore, organic solvents can extract compounds from plastic vials, thus preventing accurate assessment of JH, especially when using a GC-MS protocol. Thus:

- Do not use plastic vials at any step and use disposable glassware whenever possible.
- Do not mix glassware used in the preparation of stock solutions from commercial JH with those used in the preparation of biological samples.
- It is strongly recommended that glassware be extensively cleaned and heated to remove contaminants prior to use, especially if the glassware is being reused.

The following procedures for treating glassware are recommended.

3.1.2.1. For GC-MS analysis

1. Clean all glassware by washing in acetone 3 times.
2. Wash in hexane 3 times.
3. Bake overnight at 150°C.

3.1.2.2. For JH-RIA

1. Wash non-disposable glassware for re-use extensively with ethanol, preferably in a sonicating bath.
2. Wash in soapy water.
3. Rinse thoroughly with distilled water.
4. Incubate overnight in 1 M HCl.
5. Rinse in water.

6. Neutralize in 1 M NaOH for 1 h.
7. Rinse thoroughly in distilled water.
8. Dry and heat for 2 days at 150-250°C.

3.2. Juvenile hormone extraction, purification and quantification by GC-MS

JH has principally been quantified using one of two procedures, radioimmunoassay (RIA) or gas chromatography-mass spectrometry (GC-MS), where mass spectrometer is the quadrupole mass spectrometer, herein referred to by the trade name "Mass Selective Detector" (MSD). Generally, the GC-MSD approach purifies extracted JH III, then converts it to a d3-methoxyhydrin derivative prior to quantification. This technique was initially developed by Bergot *et al.* (1981), modified by Shu *et al.* (1997), and adapted for use in honey bees by Amdam *et al.* (2010). For further protocol details on GC-MS applications in honey bee research see the *BEEBOOK* paper on chemical ecology (Torto *et al.,* 2013).

The major advantage of this protocol is its high resolution, providing the capacity to quantify significant differences between relatively small quantities of the hormone. Its major limitations are the time necessary to process samples, its relatively high cost, and having to maintain sensitive equipment.

3.2.1. JH sample purification and quantification by GC-MSD

1. Clean all glassware by washing in acetone and hexane (3 times each) then baking overnight at 150°C (see section 3.1.1) prior to use.
2. Prepare solutions of:
   2.1. Ethyl ether: hexane 10:90 (v/v).
   2.2. Ethyl ether: hexane 30:70 (v/v).
   2.3. Ethyl acetate: hexane 50:50 (v/v).
   All solvents should be HPLC grade.
   2.4. Store at -20°C until use.
3. Prepare a labelled 8 ml borosilicate glass vial with a Teflon lined cap for each sample.
4. Add 1.5 ml of 50% aqueous acetonitrile to each vial.
5. Dilute 200 pg of farnesol (Sigma-Aldrich) in 10 µl hexane. Farnesol will serve as an internal standard.
6. Add farnesol to the vials prepared at step 4.
7. After collecting a 2-4 µl sample of haemolymph (see section 3.1.1), expel it into the prepared sample vial.
8. Prepare a positive control vial, containing 200 pg JH III (Sigma-Aldrich) in 0.5 ml of 50% aqueous acetonitrile.
9. Prepare a negative control vial with just 50% aqueous acetonitrile.
10. Add 2.5 ml hexane to each sample, using a graduated glass pipette.
11. Mix thoroughly with a vortexer.
12. The JH should partition into the upper hexane layer.
    If the layers fail to separate well, centrifuge the vials at 1000 rpm for 1 min.

13. Using a flint glass Pasteur pipette, remove the hexane layer and transfer it into a fresh 8-ml vial.

14. Repeat the partitioning process twice, adding 2.5 ml hexane each time, for a total volume of 7.5 ml hexane extract.

15. Discard the bottom layer after the third extraction step.

16. Dry the hexane extracts completely by vacuum centrifugation or under a nitrogen stream (UHP grade).

17. While the samples are drying, prepare glass columns under a laboratory hood:

    17.1. Insert a small plug of glass wool at the narrow end of a Pasteur pipette that is sufficient to plug the column, but not so tightly packed as to impede the solvent flow rate.

    17.2. Place the pipettes in the holes of a column holder with a drip tray underneath.

    17.3. Add water [6% (v/w)] to Al_2O_3 powder (activated, neutral, Brockmann I; Sigma-Aldrich).

    17.4. Mix until completely dry.

    17.5. Add 2 ml of the activated Al_2O_3 to the columns.

    17.6. Add 750 µl of hexane to the columns

If the columns drip, add activated Al_2O_3 until they hold the volume.

    17.7. Wash the columns twice with 900 µl of hexane, allowing the hexane to drip into the tray.

18. Add 300 µl of hexane to each sample vial.

19. Cap.

20. Mix thoroughly to dissolve JH from the vessel walls.

21. Transfer each JH sample to a column, using clean glass pipettes.

22. Wash each sample vial twice more, adding 300 µl of hexane each time, thus transferring a total volume of 900 µl per sample to each column.

23. Add another 900 µl of hexane to each column to remove any remaining nonpolar compounds.

24. Wash each column twice with 900 µl of ethyl ether: hexane (10:90), allowing the flow-through to pass into the drip tray.

25. Wash each column with 750 µl of ethyl ether: hexane (30:70), again allowing the flow-through to drip into the tray.

26. Elute each column with 900 µl of ethyl ether: hexane (30:70) and collect the flow-through in new 8 ml vials.

27. Repeat step 26 once and pool both eluates in the same vial, for a total volume of 1.8 ml.

28. Discard the columns.

29. Concentrate the samples to dryness by vacuum centrifugation (~15 min) or under a nitrogen stream.

30. While drying the samples, equilibrate the ampoules containing the methanol-d_4 to room temperature.

Methanol-d_4 can absorb water from air, which may quench the reaction; raising the ampoules' temperature reduces the likelihood of condensation forming.

31. With a micropipettor slowly add 75 µl of methanol-d_4 to each sample vial.

32. Using a micropipettor, add 53 µl of trifluoroacetic acid (TFA) (spectrophotometric grade) to a 1 ml methanol-d_4 ampoule.

33. Mix gently to make enough 5% trifluoroacetic acid:methanol-d_4 solution for 12 samples.

When doing so, evacuate air from the TFA container with nitrogen before storage.

34. Slowly add 75 µl of 5% trifluoroacetic acid:methanol-d_4 to each sample.

35. Tightly cap the vials and gently mix the contents.

36. Heat the sample vials for 20 min at 60°C to produce d3-methoxyhydrin derivatives.

37. Prepare new columns as described above (step 17).

38. Remove the samples from the oven (step 36).

39. Add 500 µl of hexane to each vial.

40. Mix well to remove residue from the walls.

41. Concentrate the samples to dryness by vacuum centrifugation (~15 min) or under a nitrogen stream.

42. Add 300 µl of hexane to each vial containing dried extract.

43. Mix thoroughly.

44. Transfer each sample to a column using glass pipettes.

45. Rinse sample vial twice, adding 300 µl of hexane each time, thus transferring a total volume of 900 µl to each column.

46. Wash each column twice with 900 µl of ethyl ether: hexane (30:70), allowing the flow-through to drip into the tray.

47. Wash each column twice with 750 µl of ethyl acetate: hexane (50:50), allowing the flow-through to drip into the tray.

48. Add 900 µl of ethyl ether: hexane (50:50) to each column and collect the flow-through in new 8-ml vials.

49. Repeat step 48 once, pooling both eluates in the same vial, for a total volume of 1.8 ml.

50. Discard the columns.

51. Concentrate the samples to dryness by vacuum centrifugation (~15 min) or under a nitrogen stream.

52. Add 300 µl of hexane to each sample and mix well to resuspend JH.

53. Evaporate the hexane under nitrogen gas (UHP grade) to ~25 µl of liquid.

54. Use a glass pipette to transfer the fluid into a tapered 250 µl glass vial insert (e.g. Agilent #5181-1270).

55. Repeat the steps 50-52 for obtaining a final concentrate volume of ~50 µl.

56. Using nitrogen, dry the liquid in the vial insert down to a final volume of 3 µl, measured with a graduated 10 µl syringe.

57. Using the syringe, repeatedly wash down the sides of the insert during the drying process to ensure all JH is within the remaining concentrate.

Repeated flushing through the syringe will also help evaporate down the last few microliters.

58. Syringes should be washed after each use by drawing up acetone then hexane (at least 5 times each).

59. Prepare the GC-MSD system for analysing samples:

    59.1.   Install a Zebron ZB-Wax 30 m x 0.25 mm x 0.25 μm GC capillary column.

    59.2.   Set initial temperature of GC to hold for 1 min at 60°C.

    59.3.   Ramp the GC temperature up at 20°C / min to a final temperature of 240°C; hold for 20 min (i.e. a total run time of 30 min).

    59.4.   Set the GC inlet protocol to pulsed splitless injection at 250°C.

    59.5.   Set the pressure at 9.98 psi with a 23.8 ml/min flow rate for the carrier gas (UHP grade helium).

    59.6.   Set the injection quantity to 1μl.

    59.7.   Create a 5-min solvent delay.

This prevents the recording of highly volatile solvents, which evaporate early on and can damage the detector.

60. Manually load the 1 μl of the concentrated sample using an injection syringe approved by the GC-MSD manufacturer.

61. To ensure accuracy, first draw up 1-2 μl of air, then the sample then another 1-2 μl of air.

62. To ensure specificity, monitor MSD results at m/x 76 and 225 for the JH-III derivative, and at 69, 84 and 136 for farnesol.

63. Measure the peak area for JH and adjust the value to account for any changes indicated by the results for farnesol, which was added in step 6 to serve as internal standard

64. Calculate the final titre based on the haemolymph volume and a standard curve.

The latter can be prepared with 5, 25, 125, 250 μg or more JH III, and starting the process at step 30.

3.3. Juvenile hormone quantification by radioimmunoassay

Functionally, radioimmunoassays are competition assays, whereby a radiolabelled ligand competes with equivalent nonradioactive moieties from a sample for antigen binding sites of a highly specific antibody. The limiting factor in such assays is always the concentration of the antibody for which the ligands compete. Thus, to be efficient and sensitive, antibody concentrations must be chosen so as to allow a maximal binding ratio of less than 50% for the radioactive ligand, in a standard solution that is free of unlabelled ligand. Once this competition for antibody binding sites has reached an equilibrium, antibody-complexed antigen is separated from the remaining unbound antigen, either chemically, generally by ammonium sulphate precipitation, or immunologically, by means of a secondary antibody or Protein A. Radioactivity in the resultant precipitate is then counted by scintillation spectrometry. The isotopes most frequently used for labelling antigens are ^{3}H or ^{125}J. Since the latter is suitable only for labelling proteins or peptides, all juvenile hormone and ecdysteroid radioimmunoassays use tritiated (^{3}H) compounds.

An in-depth discussion on the radioimmunoassay for juvenile hormones has been provided by Granger and Goodman (1988),

including a very detailed description on antiserum production, which will not be addressed here, as this is a rather complicated process and, once a suitable antibody has been generated, it is usually shared within the community. Sharing suitable antibodies has the further advantage that assays are easily comparable among laboratories that run such assays.

Hence, the protocol described here uses a specific antiserum produced by conjugation of JH III to thyroglobulin (Goodman, 1990) and injection into rabbits. The general radioimmunoassay protocol for the detection of JH by means of this antiserum has originally been given by Goodman *et al.* (1990) and subsequently, in a slightly modified version, by Goodman *et al.* (1995). We have used this protocol and adapted it for use with honey bees (Guidugli *et al.*, 2005; Amdam *et al.*, 2007; Marco Antonio *et al.*, 2008). As JH III is the only juvenile hormone homolog present in honey bees (Trautmann *et al.*, 1974), just as in most other insect orders, this greatly facilitates data analysis.

The method described here is certainly not the first serum and RIA protocol used for the quantification of JH titres in honey bees. A highly sensitive but slightly more laborious protocol using a different antiserum and an assay based on equilibrium dialysis for detection of insect JHs has been developed by Strambi *et al.* (1981) and applied to honey bee larvae (Rachinsky *et al.*, 1990) and adults (Robinson *et al.*, 1987), and a direct comparison using the two radioimmunoassay on honey bee samples has validated both assays as equally sensitive (Goodman *et al.*, 1993). A third RIA, with an enantiomere-specific antiserum developed by Hunnicut *et al.* (1989) has also been used on honey bee samples, primarily in the context of division of labour in workers (Huang *et al.*, 1994; Jassim *et al.*, 2000). Finally, it is worthy of note that there are no commercially available JH antibodies.

The following protocol, which is the currently most frequently run, uses an antibody developed by Walter G Goodman (Univ. Wisconsin, Madison). It is divided into four parts: sample preparation and JH extraction, preparation of radioimmunoassay solutions and assay standardization, running the assay, and data analysis.

3.3.1. JH extraction

1. Transfer the entire acetonitrile sample into a 12 x 75 mm disposable glass test tube, but avoid transferring debris that may have accumulated during storage.

2. Add 1 ml of 0.9% NaCl.

3. Add 1 ml of hexane.

4. Vortex thoroughly for 10 s.

5. Keep vials on ice for 30 min.

6. Centrifuge for 5 min at low speed for complete phase separation.

7. Transfer the hexane phase (upper layer) to a new 12 x 75 mm glass test tube.

8. Repeat steps 2 to 7 twice, each time adding 1 ml of hexane to the acetonitrile/saline phase.

9. Pool the hexane phases of each extraction step.

10. Evaporate hexane to dryness by vacuum centrifugation or under an N_2-stream.

11. Re-dissolve JH in 40 µl of toluene containing 0.5% propanediol.

12. Transfer liquid to a RIA glass vial (disposable 6-8 x 40-50 mm test tubes).

13. Wash the extraction vial with another 40 µl of toluene/propanediol.

14. Add to RIA vial (step 12).

3.3.2. Preparation of RIA solutions

3.3.2.1. JH-III standard

As commercial JH III (Sigma or other supplier) is usually not sufficiently pure and there is degradation over time, it is recommended that an aliquot of commercial JH III is first purified by thin layer chromatography (TLC). To do so:

1. "Wash" a glass backed TLC plate (5x20 cm or 20x20 cm, Merck silica gel 60 F_{254}) in ethanol. This is done by placing it in a TLC glass tank with 1 cm of ethanol in the bottom; when the solvent reaches the top of the plate, remove and dry it; it may then be stored.

2. Prior to the actual separation, put a washed TLC a plate into a tank containing a 1 cm high volume of toluene: ethylacetate (95:5) solvent mix and run it in the tank until it reaches the top of the plate. This will clean and activate the plate.

3. Mark the top to indicate plate orientation.

4. Dry the plate.

5. Activate it by heat treatment at 60°C for 10 min.

6. On the bottom side, using a soft pencil, draw a line across at 1.5 cm.

7. At the sides, use the edge of a fine spatula to scratch a line at 0.5 cm from each edge to prevent edge migration distortion.

8. Streak an aliquot of the commercial JH III on the 1.5 cm bottom line.

9. Completely evaporate solvent with an air stream.

10. Run the plate in toluene: ethylacetate (95:5) solvent mix until the solvent front has risen approximately 5 cm into the plate.

11. Remove, air dry and return the plate to the solvent to complete the run to about 17 cm of the total plate length.

12. Air dry the plate.

13. Visualize the JH band under UV light (254 nm).

14. Mark its position with a pencil.

15. Scratch the silica layer corresponding to the JH band from the plate using a spatula, collecting the scraps on a piece of aluminium foil.

16. Transfer scraps to a screw cap glass vial containing 5 ml of hexane/ethyl acetate 85:15.

17. Agitate on an orbital shaker overnight.

18. Filter solvent through a glass syringe fitted with a glass filter.

19. Dry the solvent by vacuum centrifugation.

20. Re-dissolve in toluene containing 0.5% propanediole (final conc.). This will be the purified JH solution for storage at -20°C.

For preparing the JH-III standard for preparation of standard curves:

1. Take a 5 µl aliquot of this purified JH.

2. Dilute 1:100 in methanol.

3. Quantify JH-III by UV spectroscopy at $\lambda = 217$ nm. At this wavelength an absorption of 0.502 corresponds to a JH III concentration of 10 µg/ml.

4. Correcting for dilution (step 2), dilute an appropriate aliquot of the purified JH in toluene/propanediol to a final concentration of 50 pg/µl. This will be the stock solution for preparing standard curves.

3.3.2.2. Phosphate buffer

Prepare a 0.1 M phosphate buffer (pH 7.2-7.4) from 0.2 M mono- and dibasic potassium phosphate stock solutions and add sodium azide to a 0.02% final concentration.

3.3.2.3. Saturated ammonium sulphate

1. Add 80 g of ammonium sulphate to 100 ml warm distilled water.

2. Bring to a boil.

3. Make sure that nearly all ammonium sulphate is dissolved.

4. Let cool to room temperature and observe the formation of crystals.

5. Transfer to screw top bottle and store in refrigerator, more crystals will form at the bottom.

6. For use, draw from supernatant without disturbing the crystal bed.

3.3.2.4. Solution of radioactive JH (RIA tracer solution)

The stability of JH-III in aqueous solution is limited. As this solution should be used within 4-6 weeks, it is necessary to estimate the volume needed for each assay series. 100 µl of this solution is required per sample, and the radioactivity should be 6,000-6,500 cpm/100 µl tritiated juvenile hormone III ([10-^{3}H(N)] Juvenile hormone III, Perkin Elmer NET586050UC).

1. Tritiated juvenile hormone III comes as 50 µCi in 0.5 ml toluene/hexane and an appropriate volume to be used in an assay series must first be evaporated to dryness in a glass test tube.

2. Re-dissolve the dry radioactive JH in the appropriate volume of phosphate buffer (see section 3.3.2.2).

3. Transfer to a screw cap glass vial.

4. Keep vial on orbital shaker for 24 h at 4°C.

5. Test radioactivity by liquid scintillation counting of 100 µl aliquots.

6. The level of radioactivity should be 6,000-6,500 cpm/100 µl. If necessary adjust by adding phosphate buffer or tracer.

3.3.2.5. Antibody solution

Depending on the number of samples to be processed, prepare an adequate volume of RIA serum, Typically, a minimum of 10 ml solution should be prepared, which would be sufficient for 100 samples (100 µl of this solution are required per sample). The solution should be stored at 4°C and should also be used within 4-6 weeks.

For making 10 ml of antibody solution:

1. Dissolve 10 mg of bovine serum albumin (BSA, Fraction V)) and 10 mg of IgG from rabbit serum (Sigma I5006) in 10 ml of 0.1 M phosphate buffer (3.3.2.2).
2. Remove a 0.5-1 ml aliquot of this solution to later be used with the non-specific binding (background) tubes in the assay.
3. Add an appropriate volume of the JH antibody to the remaining RIA serum solution.

Note: The appropriate concentration of JH antibody will have to be established in prior tests evaluating sensitivity through a dilution series of the respective antibody, remembering that an RIA is an isotope dilution assay whereby the specific antibody is the limiting factor. Ideal antibody concentrations should give maximal-binding cpm values corresponding to 35-40% of the total radioactivity added to each sample (around 2000-2500 cpm under the above described conditions). Such tests also need to be run periodically to ascertain binding and, if necessary, adjust the antibody concentration.

3.3.3. Running a JH RIA

1. To set up a standard curve (in duplicates), prepare 10 disposable borosilicate glass culture tubes (6 x 50 mm).
    Eight of each series will receive aliquots of 0.5, 1, 2, 5, 10, 20, 50 and 100 µl (these corresponding to 25, 50, 100, 250, 500, 1000, 2500 and 5000 pg of JH- III) of the stock solution of unlabelled JH-III (see section 3.3.2.1).
    Note that pipetting of the standard curve samples must be done with Hamilton-type syringes that are exclusively dedicated to use in this JH RIA.
2. The additional four tubes will not receive any unlabelled JH. One of each series will receive antibody-free background solution (to assess non-specific binding) and the other receives radiolabelled JH only (maximum binding values).
3. Evaporate the solvent from all RIA vials, both the standard curve and the sample tubes.
4. Add 100 µl of the RIA tracer solution (3.3.2.4) to each vial.
5. Vortex vigorously for 10 s.
6. Add 100 µl of the antibody solution (3.3.2.5) to all tubes, except the two for background counting, which will receive an equal amount of the background protein solution (3.3.2.5. step 2).
7. Gently mix the tracer with the serum components.
    Do not vortex as this will cause bubbles to form.
8. Quickly (30 s) spin the tubes in a centrifuge (2000 x g) to remove any solution adhering to the walls of the vials.
9. Cover RIA test tubes with an adhesive sheet to avoid loss by evaporation.
10. Incubate samples overnight at 4°C to attain a binding equilibrium. Do not incubate for more than 24 h.
11. Add 200 µl of saturated ammonium sulphate solution (see section 3.3.2.3) to the test tubes to precipitate proteins, including the antibody-bound JH.

12. Vortex.
13. Store samples at 4°C for 30 min.
14. Centrifuge at 7,500 x g for 15 min at 4°C.
15. Remove supernatant, taking care to leave the pellet undisturbed.
16. Add 400 µL of a 50% ammonium sulphate solution [1:1 saturated ammonium sulphate/water (v:v)] to the test tubes to dissolve the pellet.
17. Vortex.
18. Store samples at 4°C for 30 min.
19. Centrifuge at 7,500 x g for 15 min at 4°C.
20. Remove supernatant, taking care to leave the pellet undisturbed.
21. Add 50 µl of water to the pellet.
22. Vortex to dissolve completely.
23. Transfer the dissolved precipitate to a 20 ml vial appropriate for liquid scintillation counting (LSC) vial.
24. Wash the RIA test tube with another 50 µl of water and add to the same respective LSC vial.
25. Add 5 ml of a biodegradable LSC cocktail (e.g. Optiphase Hisafe3, Perkin-Elmer).
26. Leave the counting vials for 1 h at room temperature before starting the counting process in a β-counter.
27. Counting times should be set to allow a 5 sigma level of confidence.
    This period will vary depending on the number of counts in the sample, counting efficiency and instrument variables. Most LSC instrumentation includes variable counting to provide 5 sigma margins of error in their programs. As a rule of thumb for counting the RIA samples, counting times of 2 min should be sufficient, and the entire series should be counted at least twice

3.3.4. Data analysis

The use of the ImmunoAssay Calculations spreadsheet freely available from Bachem (http://www.bachem.com/service-support/immunoassay -calculator) is highly recommended. This program is designed to run a four parameter non-linear regression analysis of RIA data.

In this spreadsheet the mean of the non-specific binding counts (NSB) is first subtracted from the means of all other samples. Next, each of the standard curve means (B-NSB) is divided by the maximum count (B-NSB/B_0-NSB) and these values are then fitted to a nonlinear regression curve given as:

$$y = ((a-d)/(1 + (x/c)^b))+d$$
$$x = c ((y-a)/(d-y))^{(1/b)}$$

where *a* represents the maximum binding value (set to 1), *b* the slope, *c* the inflection point of the curve (Ic_{50}) and *d* the minimum value. Through a graphic conversion of the minimization procedure, the spreadsheet makes this fitting a user friendly task. Finally, sample values are entered and converted to amounts of hormone in each sample.

Such non-linear regression is far superior over a linear log/logit regression analysis for calculating hormone concentrations in samples close to the detection limit of the RIA.

An important quality check for all radioimmunoassays is the calculation of intra- and inter-assay variation. Whereas intra-assay variation usually reflects the level of pipetting errors, inter-assay variation is of relevance when large sample series are processed or when results are to be compared to other studies. This is done by including in each assay a sample prepared from an aliquot of a large haemolymph-pool sample. As this sample will be included in any of the other assays done on the same species, it allows for quality control of the RIA and to assess inter-assay variation.

3.3.5. User safety

The use of radioactive compounds (([10-^{3}H(N)] Juvenile hormone III) requires authorization for individual user and laboratory according to national regulations and guidelines. Working with gloves is required to avoid direct contact with skin, but no further protection from radiation is necessary due to the low energy of tritiated compounds. Disposable material (glass) used during handling of radioactive compounds and liquid residues, as well as scintillation cocktail containing radioactivity must all be properly disposed (follow national guidelines). Toluene and ethyl acetate are flammable solvents. UV-light (254 nm) causes DNA damage and is especially harmful to the eyes; use protective glasses or a UV shield.

3.4. Quantification of ecdysteroids by radioimmunoassay

As far as ecdysteroid quantification is concerned, the same general principles described for the juvenile hormone RIA also apply to the detection of ecdysteroids, except that not a single but several ecdysteroids may naturally be present in haemolymph or tissues. For this reason, when ecdysteroids are not fractionated and separated by specific chromatography methods prior to radioimmunoassaying, results are given as immunoreactive ecdysteroids in relation to the compound used to set up the standard curve, which is usually 20-hydroxyecdysone.

Basically two ecdysteroid RIAs have been used in honey bees, one developed by de Reggi *et al.* (1975), in a protocol using equilibrium dialysis to quantify larval and pupal and adult ecdysteroids (Rachinsky *et al.*, 1990; Robinson *et al.*, 1991), and the other based on an antiserum and protocol developed by Bollenbacher *et al.* (1983). The latter was used more frequently for studies on honey bee ecdysteroids (Feldlaufer and Hartfelder, 1997; Feldlaufer *et al.*, 1985, 1986; Hartfelder *et al.*, 2002; Pinto *et al.*, 2002; Amdam *et al.*, 2004; Nascimento *et al.*, 2004). It is worthy of note that there are no commercially available ecdysone antibodies.

The following protocol uses an antibody developed in the Gilbert laboratory (Univ. North Carolina at Chapel Hill) against a hemisuccinate derivative of ecdysone (Bollenbacher *et al.*, 1983; Warren and Gilbert, 1986). Tritiated ecdysone serves as the labelled ligand and standard curves are established with 20-hydroxyecdysone

(20E) as the non-radioactive ligand. Results are, therefore, expressed as 20E equivalents. The protocol is essentially divided into four parts: sample preparation, preparation of radioimmunoassay solutions and assay standardization, running the assay, and data analysis.

3.4.1. Sample preparation
3.4.1.1. for haemolymph
1. Collect the haemolymph into a calibrated microcapillary as described for the JH radioimmunoassay (see section 3.1.1).
2. Note its exact volume.
3. Transfer the haemolymph into an Eppendorf tube containing cold methanol in a 1:100 volume ratio (100 μl methanol per μl haemolymph).
4. Keep at 4°C for a couple of hours (best overnight).
5. Samples can be stored at -20°C in these Eppendorf tubes, as ecdysteroids will not adhere to plastic the way JH does.
6. Centrifuge the tubes to pelletize any remaining impurities.

Most haemolymph samples can be assayed directly by RIA. If there are problems, try purifying free ecdysteroids by chromatographic separation (see section 3.4.1.3., below).

3.4.1.2. for tissue
1. Weigh tissue.
2. Homogenize in at least 500x (w:v) cold methanol.
3. Keep at 4°C for a couple of hours (best overnight).
4. Centrifuge to pelletize and remove impurities.
5. Store supernatant at -20°C.

3.4.1.3. in case of lipid-rich samples (e.g. whole larva)
An additional purification step is usually required as lipids may interfere in the ligand-binding and the ammonium sulphate precipitation of the antigen-antibody complex. A rapid and efficient purification is the separation of free ecdysteroids from polar or apolar compounds by means of reverse-phase liquid chromatography on disposable reverse-phase columns (SepPak C$_{18}$, Waters, Waters, WAT051910).
1. Evaporate methanol from sample (after centrifugation to remove impurities) by vacuum centrifugation.
2. Resuspend sample in 30% methanol.
3. Activate SepPak C$_{18}$ cartridge mounted on a 1 ml syringe by slowly passing pure methanol.
4. Equilibrate column by passing 1 ml 30% methanol.
5. Load sample on column
6. Pass 2 x 1 ml 30% methanol.

This eluate contains polar ecdysteroid conjugates that can be discarded, or if of interest, analysed after enzymatic digestion (*Helix pomatia* juice).
7. Pass 2 x 1 ml 60% methanol through column and collect eluate in a 5 ml disposable glass test tube.

This eluate contains the free ecdysteroids

8. Discard column.
9. Evaporate solvent from the free ecdysteroid fraction samples by vacuum centrifugation.
10. Re-dissolve in 100% methanol for the RIA procedure.

3.4.2. Preparation of RIA solutions

3.4.2.1. 20-hydroxyecdysone (20E) standard

Prepare a stock solution of unlabelled 20-hydroxyecdysone (20E) from a commercial source (Sigma) diluted to a concentration of 50 pg/µl in methanol. As commercial ecdysteroids usually are not sufficiently pure and there is degradation over time, it is recommended that an aliquot of commercial 20E is first purified by thin layer chromatography (TLC) or, if available by HPLC. The steps for TLC purification are essentially the same ones as described for cleaning up commercial JH (see section 3.3.2.1), the only differences are:

- the solvent mix used for the chromatography run, which is dichloromethane : methanol (85:15) in steps 2 and 10.
- the solvent used to elute TLC-separated 20E from the silica scraps is methanol (step 16).

For preparing the 20E standard for preparation of standard curves:

1. Take a 5 µl aliquot of this purified 20E and dilute 1:100 in methanol for quantification of 20E by spectroscopy at λ = 245 nm.
2. Calculate the 20E concentration in mg/ml as: (0.48 x absorption x dilution factor)/12.6.
3. Correcting for dilution, dilute an appropriate aliquot of the purified 20E in methanol to a final concentration of 50 pg/µl.

3.4.2.2. Phosphate buffer (this is the same as in the JH RIA, see section 3.3.2.2.)

3.4.2.3. Saturated ammonium sulphate (this is the same as in the JH RIA, see section 3.3.2.3.)

3.4.2.4. Solution of radioactive ecdysone (RIA tracer solution)

The stability of ecdysone in aqueous solution is limited. As this solution should be used within 4-6 weeks, it is necessary to estimate the volume needed for each assay series. 100 µl of this solution is required per sample, and the radioactivity should be 5,000-6,000 cpm/100 µl tritiated ecdysone ([23,24-3H(N)]ecdysone, Perkin Elmer NEN; NET621050UC; 1.85 MBq; 1.85-4.07 TBq/mmol). The solution should be stored at 4°C.

1. As tritiated ecdysone comes as 50 µCi in 0.5 ml ethanol, calculate the volume to be used in an assay series.
2. Transfer to 5 ml disposable glass test tube and evaporated to dryness by vacuum centrifugation or under a N_2-stream.
3. Dissolve the radioactive ecdysone in the appropriate volume of 0.1 M phosphate buffer.
4. Transfer to a screw cap glass vial.
5. Keep this vial on an orbital shaker for 24 h at 4°C.

6. Test radioactivity level by liquid scintillation counting of 100 µl aliquots.

This should give 5,000-6,000 cpm for 100 µl.

7. If necessary, adjust radioactivity level by adding phosphate buffer or tracer.

3.4.2.5. Antibody solution (RIA serum)

Prepare an adequate volume of RIA serum for the number of samples to be run. Typically, a minimum of 10 ml solution should be prepared, which would be sufficient for 100 samples (100 µl of this solution are required per sample). The solution should be stored at 4°C and should also be used within 4-6 weeks. For making 10 ml of antibody solution:

1. Dissolve 10 mg of bovine serum albumin (BSA, Fraction V) and 10 mg of IgG from rabbit serum (Sigma I5006) in 10 ml of 0.1 M phosphate buffer (3.3.2.2.).
2. Remove a 0.5-1 ml aliquot of this solution to later be used with the non-specific binding (background) tubes in the assay.
3. Add an appropriate volume of the JH antibody to the remaining RIA serum solution.

Note: The appropriate concentration of ecdysone antibody will have to be established in prior tests evaluating sensitivity through a dilution series of the respective antibody, remembering that an RIA is an isotope dilution assay whereby the specific antibody is the limiting factor. Ideal antibody concentrations should give maximal-binding cpm values corresponding to 35-40% of the total radioactivity added to each sample (around 1800-2200 cpm under the above described conditions). Such tests also need to be run periodically to ascertain binding and, if necessary, adjust the antibody concentration.

3.4.3. Running an ecdysteroid RIA

1. To set up a standard curve (in duplicates), prepare two series of 10 disposable borosilicate glass culture tubes (6 x 50 mm); to eight tubes each, add serial aliquots of the stock solution of unlabelled 20E, this being 0.5, 1, 2, 7.5, 10, 15, 20, 40 µl (corresponding to 25, 50, 100, 375, 500, 750, 1000 and 2000 pg of 20E).

Note that pipetting of the standard curve samples must be done with Hamilton-type syringes that are exclusively dedicated to use in this ecdysteroid RIA.

2. The remaining two tubes in each series do not receive any unlabelled JH; one will receive antibody-free background solution (to assess non-specific binding) and one will receive radiolabelled JH only (maximum binding values).

All further steps (3-27) are the same as those described for the JH RIA in section 3.3.3.

3.4.4. Data analysis

For data analysis and calculation of intra and inter-assay variation proceed as described for the JH RIA (see section 3.3.4.).

3.4.5. User Safety

The use of radioactive compounds ([23,24-3H(N)]ecdysone) requires authorization for individual user and laboratory according to national regulations and guidelines. Working with gloves is required to avoid direct contact with skin, but no further protection from radiation is necessary due to the low energy of tritiated compounds. Disposable material (glass) used during handling of radioactive compounds and liquid residues, as well as scintillation cocktail containing radioactivity must be properly disposed of following national guidelines. Dichloromethane and methanol are solvents harmful to health if used improperly. UV-light (254 nm) causes DNA damage and is especially harmful to the eyes; use protective glasses or a UV shield.

3.5. Measuring the rate of juvenile hormone biosynthesis by the paired *corpora allata*

This protocol describes a rapid partition radiochemical assays for quantifying rates of juvenile hormone III biosynthesis *in vitro* from honey bee *corpora allata* (CA), the glands responsible for JH production (Goodman and Cusson, 2012). Excised glands are incubated with methionine, which is a methyl donor to the ester function of the JH molecule (Metzler *et al.*, 1972; Judy *et al.*, 1973). Using radiolabelled methionine results in the production of radiolabelled JH III, which can then be quantified after partitioning into an isooctane phase. This technique was initially developed by Pratt and Tobe (1974) and Tobe and Pratt (1974), modified by Feyereisen and Tobe (1981) and adapted for use in honey bees (Rachinsky and Hartfelder, 1990, 1998; Huang *et al.*, 1991). This approach is most useful when quantification of the activity of the glands, rather than circulating JH levels, is desired. Its difficulties lie primarily in the ability to surgically remove and handle the glands without damaging them prior to incubation.

3.5.1. Purifying and preparing radioactive methionine for use in assay

Just prior to use, the labelled methionine will have to be washed to remove any unbound tritium, a source of contaminating "noise".

1. Dilute 100 µCi of L-[methyl-3H]-methionine (>97%, 1 mCi/ml, 70-85 Ci/mmol; PerkinElmer NET061X250UC) in insect saline to a final volume of 100 µl in a 1.5-ml microcentrifuge tube.
2. Add 500 µl of isooctane (2,2,4-trimethylpentane, anhydrous, 99.8%; Sigma-Aldrich 360066) to the methionine.
3. Mix well by vortexing for 1 min.
4. Centrifuge at 10,000 x g for 10 min.
5. Using a pipettor, remove the isooctane (i.e. upper) layer, avoiding to touch the lower methionine/saline layer.
6. Repeat the wash steps 2-5 two more times.
7. After the third wash, transfer 100 µl of isooctane to a scintillation vial to determine how much contaminate remains.
   7.1. Evaporate the isooctane in the vial to near-dryness under nitrogen.

7.2. Add 3 ml of scintillation fluid with a high efficiency for tritium (e.g. ScintiSafe Econo F, Fisher Scientific) to the vial.
7.3. Vortex for 1 min.
7.4. Determine the radioactivity of the isooctane in a scintillation counter.
7.5. If the background count is sufficiently low (i.e., <1000 dpm), the methionine is ready for use.
7.6. If the background count is >1000 dpm, repeat the three wash cycle (steps 2-5).
7.7. If the cleaning process does not reduce the background dpm under 2,000 within two wash cycles, incubate the methionine in 500 µl of isooctane overnight (or longer) to remove more of the degraded isotope.

8. Evaporate any of the isooctane remaining in the vial with nitrogen gas.
9. For ten samples and a negative control, dilute 65 µl of the methionine solution in 1235 µl of Grace's insect medium (from commercial supplier, e.g. Sigma-Aldrich) that is free of L-methionine.

This should provide a final (recommended) concentration of 100 µM methionine per incubation, with a specific activity of 5 µCi/100 µl.

Higher concentrations will ensure that radioactive methionine is incorporated into any JH synthesized, but will also increase the background counts.

3.5.2. Measuring the rate of JH biosynthesis

1. Remove the paired corpora allata (CA) from the base of the bee brain (see the section on isolation of the retrocerebral complex in the *BEEBOOK* paper on anatomy and dissection (Carreck *et al.*, 2013). While minimizing the amount of brain tissue attached to the glands, leave any attached tracheal elements as these enhance buoyancy. This will help ensure the glands stay near the surface of the medium during incubation, enhancing oxygenation and tissue activity (Kaatz *et al.*, 1985; Holbrook *et al.*, 1997).
2. Separately preincubate each dissected pair of CA in 100 µl of nonradiolabelled Grace's medium for 15 min in borosilicate glass culture tubes (6 x 50 mm) at room temperature.
3. Aliquot 100 µl of radiolabelled medium into a duplicate set of labelled glass culture tubes.
4. Using a small copper wire hoop, pick up a droplet of medium containing the preincubated CA and transfer it to the corresponding tube containing the radiolabelled medium. Ensure that the tissue stays close to the surface to enhance oxygenation.
5. Prepare a negative control tube containing radiolabelled media with no added tissue.
6. Loosely cover the tubes with plastic wrap to prevent desiccation, but ensure an adequate supply of oxygen.

7. Incubate the samples on a variable-plane mixer set at a 15°-17° angle at 90 rpm at 27°C for 3h.

The functional capacity of the *A. mellifera* CA begins to degrade within a few hours after dissection, so the 3h incubation allows for an accurate assessment of the rate of release and maximizes the amount of JH available for detection purposes.

8. At the end of the incubation period, remove the CA.
9. Add 250 μl of chilled isooctane to each tube.
10. Incubate on the variable-plane mixer for an additional 15 min.
11. Vortex the samples for 1 min.
12. Centrifuge at 10,000 x g for 10 min.
13. Using a pipettor with a gel loading tip, transfer 100 μl of the upper isooctane layer to each of two labelled scintillation vials so that each extract is assayed in duplicate.
14. Reduce the volume of the isooctane to near-dryness under nitrogen or in a vacuum concentrator.
15. Add scintillation fluid (see section 3.5.1., step 7.2) to the vial.
16. Vortex for 1 min.
17. Let rest for 1 h before measuring the radioactivity in a scintillation counter.

3.5.3. Data analysis

1. Subtract the average background count (from the negative control samples) from each reading to correct for background noise.
2. Determine the quantity of JH the CA released into the medium by converting the cpm value obtained from the previous step into dpm counts taking into account the tritium sensitivity of your scintillation counter (counter sensitivity is determined by running a manufacturer supplied scintillation vial containing a specified amount of the isotope of interest; correct for half-life).
3. Convert this dpm value into pmol JH released, taking into account the specific activity (typically 70 Ci/mmol) of your labelled methionine: considering that 2.22 dpm = 1 pCi, a dpm value of 155,400 = 1 pmol JH.
4. Calculate the hourly release rate by dividing the pmol value obtained (step 3) by the duration of the incubation period (divide value by 3 if incubation was 3 h).

3.5.4. User Safety

The use of radioactive compounds (L-[methyl-3H]-methionine) requires authorization for individual user and laboratory according to national regulations and guidelines. Working with gloves is required to avoid direct contact with skin, but no further protection from radiation is necessary due to the low energy of tritiated compounds. Disposable material (glass) used during handling of radioactive compounds and liquid residues, as well as scintillation cocktail containing radioactivity must be properly disposed of, following national guidelines. Isooctane is a flammable solvent.

4. Biogenic amine extraction and quantification by HPLC-ECD

4.1. Introduction

Biogenic amines have a variety of roles in the lives of insects, acting as neurohormones, neuromodulators, and neurotransmitters that can influence both behaviour and physiology (reviewed in Blenau *et al.*, 2001; Scheiner *et al.*, 2006). Four biogenic amines, dopamine (DA), octopamine (OA), serotonin (5-hydroxytryptamine, or 5-HT), and tyramine (TA), have been identified as being significantly associated honey bee behaviour and may be responsible for orchestrating the complex division of labour within their colonies (Fuchs *et al.*, 1989; Brandes *et al.*, 1990; Harris and Woodring, 1992, 1995; Taylor *et al.*, 1992; Bozic and Woodring, 1998; Wagener-Hulme *et al.*, 1999; Scheiner *et al.*, 2002, 2006; Schulz *et al.*, 2002; Barron and Robinson, 2005; Fussnecker *et al.*, 2006; Agarwal *et al.*, 2011). Several approaches have been used to measure amine concentration in insect central nervous system tissues (histofluorimetry, immunohistochemistry, radioenzymatics, gas chromatography-mass spectroscopy), but one that has been used most frequently in honey bees is high performance liquid chromatography coupled to electrochemical detection (HPLC-ECD). This approach utilizes a carrier solution, referred to as the mobile phase, to bring the sample across a filter column and then to the detector. HPLC-ECD analysis can be very sensitive and provide robust and repeatable results. The primary issue with this approach is that HPLC systems can be difficult to maintain over time, necessitating the processing of samples in small batches and running calibrating standards frequently.

An important caution with any approach used to quantify biogenic amines is that their levels can change rapidly in bees subjected to any of a variety of environmental perturbations (Harris and Woodring, 1992; Harris *et al.*, 1996; Chen *et al.*, 2008). For that reason the handling time when collecting samples should be as short as possible, and then technique used should be very uniform. In addition, these amines are highly sensitive to light and heat, therefore every precaution should be made to keep samples cold and covered prior to analysis.

4.2. Dissection of the brain from the head capsule

While it is sometimes possible to take measurements from a single individual, the most robust and reliable results come from pooling multiple individuals. Depending on the size of the brain sections being used, tissue from up to 20 individuals may be needed per sample. Because the stress of being handled can affect neurotransmitter expression, collect the bees as quickly as possible. Frozen bees can be dissected immediately (preferable) or temporarily stored in liquid nitrogen. See the section on dissection of the brain from the head capsule in the *BEEBOOK* paper on anatomy and dissection (Carreck *et al.*, 2013) for the method to extract the brain out of the head capsule.

Place the desired brains or brain sections into a labelled 0.5 ml microcentrifuge tube. Cover and store at -80°C for up to several weeks if the samples are not to be analysed immediately.

4.3. Pre-analysis preparation of the HPLC

1. Prepare the mobile phase (1l) as follow:
    1.1. 700 ml polished water (this is ddH$_2$0 from which all impurities were removed)
    1.2. 150 ml methanol
    1.3. 150 ml acetonitrile
    1.4. 0.433 g sodium dodecyl sulphate (SDS)
    1.5. 10.349 g sodium phosphate monobasic
    1.6. 1.47g sodium citrate
    1.7. Adjust pH to 5.6 using phosphoric acid, delivered a drop at a time while stirring.
2. Degas the mobile phase for the HPLC in an ultrasonic bath for 5 min prior to running any samples to prevent air from entering the column and detector cells.
3. Set the mobile phase flow rate to 1.0 ml/min (normal pressure ~95 bar; set max pressure to 140 bar).
4. Apply current to the ECD cells (e.g. Channel 1, 650mV; Channel 2, 425mV; Channel 3, 175mV; Channel 4, -175V).

The amount of current per channel will need to be adjusted to optimize peak responsiveness as per the manufacturers recommendations; it will take approximately 45 min for channel voltage to equilibrate.

5. During the equilibration process, prepare the external (ES) and internal standards (IS).

4.4. Preparation of internal and external standards

The internal standard for DA and 5-HT is 3,4-dihydroxybenzylamine (DHBA), while that for OA and TA is synephrine. Because of the instability of amines in solution, new standards should be made up daily to ensure accuracy. However, the stock dilutions may be relatively stable for a few days of use if they are kept properly chilled and shielded from light.

4.4.1 Stock dilutions (1 x 10^6 pg/µl)

For each amine:

1. Weigh out the specified amount (these values are adjusted for the weight of associated HCl and HBr molecules) into separate, labelled 12 ml glass vials.
2. Dissolve the standards in:
    2.1. 9.975 ml chilled 0.2M perchloric acid (PCA; 29.6 ml 70% perchloric acid in 1l polished water) and
    2.2. 25µl 0.1M EDTA solution (3.722 g EDTA in 100 ml polished water)
3. Cover the vials in foil to exclude light.
4. Refrigerate (4°C).

4.4.2. External and Internal standard

External Standards (ES)	Internal Standards (IS)
DA: 0.0124 g	DHBA: 0.0174 g
OA: 0.0124 g	Synephrine: 0.0100 g
5-HT: 0.0121 g	
TA: 0.0127 g	

4.4.2.1. Dilution ES-A

to 7.775 ml chilled 0.2 M PCA add:

- 2.0 ml DA (2000000 pg/µl),
- 25 µl OA (2500 pg/µl),
- 25 µl 5HT (2500 pg/µl),
- 125 µl TA (12500 pg/µl)

4.4.2.2. Dilution IS-A (5000 pg/µl):

to 9.925 ml chilled 0.2 M PCA add

- 50 µl DHBA (50000 pg/µl),
- 25 µl Synephrine (25000 pg/µl)

4.4.2.3. Dilution IS-B (500/250 pg/µl)

add 10 µl of IS-A to 990 µl chilled 0.2 M PCA; this will be used for extracting brain amines.

4.4.2.4. Standard Curve:

- Dilution B: Add 10 µl IS-**A** and 15 µl ES-**A** to 945 µl PCA. (3000 pg/µl)
- Dilution C: Add 10 µl IS-**A** and 12 µl ES-**A** to 954 µl PCA. (2400 pg/µl)
- Dilution D: Add 10 µl IS-**A** and 9 µl ES-**A** to 963 µl PCA. (1800 pg/µl)
- Dilution E: Add 10 µl IS-**A** and 6 µl ES-**A** to 972 µl PCA. (1200 pg/µl)
- Dilution F: Add 10 µl IS-**A** and 3 µl ES-**A** to 981 µl PCA. (600 pg/µl)
- Dilution G: Add 10 µl IS-**A** and 1 µl ES-**A** to 987 µl PCA. (200 pg/µl)
- Dilution H: Add 10µl IS-**A** to 990µl PCA.

4.5. HPLC separation of standards

Once the ECD cells equilibrate, standard samples can be run on the HPLC system.

1. Prior to injecting any samples, prime the HPLC by injecting 10 µl of 0.2 M perchloric acid (PCA).
2. After this and every injection, be sure to flush the syringe with polished water several times.

Any residual PCA would crystallize, causing the plunger to become frozen in the barrel. As an added precaution, store the plunger out of the syringe when not in use.

3. If the PCA run looks clear of peaks, sample analysis can begin.

Standards and samples should be processed in this order: Standards G to B (lower to higher concentration), Standard H, 5-7 samples, Standards G/D/H, 5-7 samples, Standards G/D/H, PCA Blank. Standard H and the PCA clear the HPLC of residual amine peaks. Standards should always be run at the start, middle and end of any series to ensure that the sensitivity of the system has not changed. The cycle can be repeated as necessary. Duration of the run will depend on the flow rate, which in turn is determined by the pressure registered within the HPLC. Adjustments to flow rate and run length may need to occur over the course of the day as contaminates injected with the sample cause a gradual pressure increase.

Because of the sensitive nature of the amines, an autoloader cannot be used to inject the samples, as they would degrade while waiting. Instead, use a microsyringe to directly deliver the sample into the injector port.

While standards are being run on the HPLC, tissue samples can be prepared for analysis.

4.6. Sample preparation

1. Remove the stored brains from the freezer or liquid nitrogen and place them in a covered ice bucket.
2. Add 20 µl of chilled IS-B solution to each tube.
3. Homogenize each brain using a separate tissue grinder.
4. Remove the tissue and liquid from the pestle to ensure that all tissue remains in the solution.
5. Suspend the sample tubes in an ultrasonic bath filled with ice water.
6. Cover to reduce light exposure.
7. Sonicate the tubes for 5 min to disrupt tissue.
8. Leave the tubes covered and in ice water for an additional 20 min to allow the amines to be extracted from the brain tissue.
9. Centrifuge the tubes at 12,000 RCF for 10 min at 4°C.
10. Collect the supernatant and pass it through a 0.22-µm nylon membrane filter into a fresh, labelled microcentrifuge tube. This step will help remove any remaining tissue, protecting the HPLC system from clogging.
11. Store the tube containing the supernatant on ice and protected it from light until used.
12. Place the original tube containing the residual brain tissue in a -80°C freezer for later quantification.

4.7. Separation and quantification of biogenic amines by HPLC

For each standard and sample:

1. Load 10 µl of the collected supernatant into the HPLC system.
2. Identify the amines in an HPLC trace by determining the retention time and relative peak sizes for each of the detector channels. The exact retention times for the amines will vary according to analysis conditions, but they do occur in the specific order given in the example below.

3. For each amine, determine the area of the peak for the one channel on which it was largest.

An example of amine characteristics follows. Although the order in which these peaks appear will remain constant, the specific times and channel responses will vary depending on the HPLC system, flow rate, and applied currents:

- Octopamine (OA): 5.2 min, big Channel 1 peak, smaller Channel 2 peak
- DHBA: 5.5 min, big Channel 2 peak, smaller Channels 1 and 3 peaks
- Synephrine: 6.0 min, big Channel 1 peak, smaller Channel 2 peak
- Dopamine (DA): 6.6 min, big Channel 2 peak, smaller Channel 1 and 3 peaks
- Tyramine (TA): 9.6 min, big Channel 1 peak, smaller Channel 2 peak
- Serotonin (5-HT): 10.2 min, big Channel 2 peak, smaller Channels 1 and 3 peaks

4. If the standards maintained comparable values through multiple batches of samples, a single standard curve can be calculated from their combined data. If the standard values changed over time, each batch of samples will need to be compared to a curve calculated from the standards run just prior to and immediately following each batch.

In order to be sure that any differences in amine content are due to true physiological differences and not just differences in the quantity of tissue from which the amines were extracted, it is necessary to standardize the amine content against tissue content. Several approaches are available to determine protein content, such as the Bradford or Lowry assays (see section 1.2), but the accuracy of many of these assays is compromised by the minute volume of tissue involved in an individual brain. An alternative and simple approach is to determine the dry mass of the brain tissue.

To quantify brain mass:

1. Remove the residual tissue pellet from the sample tube.
2. Place it on a pre-weighed and labelled 1cm square section of foil.
3. Fold these squares over to secure the sample.
4. Place them in an oven set to 100°C.
5. Bake the samples for 24 h.
6. Weigh the squares on a scale that can measure microgram differences.

5. Temperature, radiation and humidity measurement in honey bees

Temperature is a decisive parameter in honey bee development, physiology and behaviour, at the individual as well as at the social level (Heinrich, 1993). Larval respiration and growth is strongly dependent on temperature (Petz *et al.*, 2004). A high speed of development is guaranteed via social regulation of brood temperature in the range of 33-36°C (Kleinhenz *et al.*, 2003; Stabentheiner *et al.*, 2010). Low temperatures during pupal development may impair learning and cause significant differences in the behaviour of adult bees (Tautz *et al.*, 2003; Becher *et al.*, 2009). Colony temperature also affects the basal metabolism of hive bees (Stabentheiner *et al.*, 2003a; Kovac *et al.*, 2007). Social thermoregulation allows for overwintering of colonies in cold climates (Stabentheiner *et al.*, 2003b). During the foraging cycle honey bees are always endothermic, which means that they display a high oxygen consumption (and thus energy turnover) to keep their body temperature well above the ambient level (Heinrich, 1993; Stabentheiner *et al.*, 1995; Kovac and Schmaranzer, 1996; Roberts and Harrison, 1999; Kovac *et al.*, 2010; Stabentheiner *et al.*, 2012). Measurement of honey bee body and colony temperature and of environmental parameters requires care to achieve correct results. The following summary is meant as a guide to avoid the main pitfalls of honey bee body temperature and environmental parameter measurement.

5.1. Contact thermosensors

There is a great variety of contact thermosensors available. For the measurement of ambient temperature, sensor size is not so critical, though smaller sensors react much faster to changes in ambient temperature than larger ones. For honey bee body temperature measurement, however, it is *indispensable* that they are not just small but tiny! Otherwise, large measurement errors are to be expected. Furthermore, for measurement of brood nest (comb) temperature, care has to be taken to place the thermosensors in the correct (desired) position, e.g. in the actual centre of the brood comb.

5.1.1. Thermocouples and thermoneedles

Thermocouples are standard sensors for the measurement of ambient temperature. Their small (i.e. thin) size and flexibility allows installation of many sensors in a honey bee colony (Stabentheiner *et al.*, 2010). Due to their usually small sensor tip – and thus heat capacity – their reaction to temperature changes is fast. Sensitivity and reaction speed can be maximized by using small wire diameters (ones close to 12 µm are available; for example see Omega Engineering Inc., http://www.omega.com). Accuracy depends primarily on the electronic thermometer in use, including the accuracy of the reference (cold) junction compensation in the connector (junction) block. During outdoor measurements in sunshine, the electronic thermometer may not be exposed to solar radiation. Thermal gradients in the connector (junction) block may lead to high measurement errors (up to several °C). Similar errors may occur if ambient temperature changes rapidly, for example when stepping outside. Equilibration of the device is always recommended for utmost accuracy.

5.1.1.1. Thermocouple types

There are different types of thermocouples, with different combinations of wire metals, e.g. NiCr/NiAl (Type K), Cu/Constantan (Type T), etc. Thermocouple types may differ considerably in sensitivity (and, irrelevant for biological applications, measurement range). Make sure to always use the correct type of thermocouple, matching the type and size of connector input on your electronic thermometer.

For honey bee body temperature measurement, thermocouple wires have to be thin (diameter <70 µm). For "grab and stab" measurement of body (core) temperature (Stone and Willmer, 1989) the thermocouples have to be inserted within hypodermic needles. Most practical are thermocouples inserted in "thermoneedles" by the manufacturer (available ready for use and in different sizes for example at Omega Engeneering, Inc.).

5.1.1.2. Calibration

It is an intrinsic property of thermocouples that their characteristic curve is neither linear nor does it follow a persistent function throughout the measurement range (Fig. 5). Modern thermocouple thermometers, however, usually apply appropriate corrections automatically. If an accuracy of better than 0.3°C has to be guaranteed, individual calibration of each thermocouple is necessary (e.g. in a water bath against an accurate laboratory thermometer). In the temperature range relevant in honey bee life, of about -30 to 100°C, linear correction functions are sufficient. Advanced data loggers provide the possibility to store the corrections permanently in a device or connector memory (e.g. Ahlborn Messtechnik, http://www.ahlborn.com).

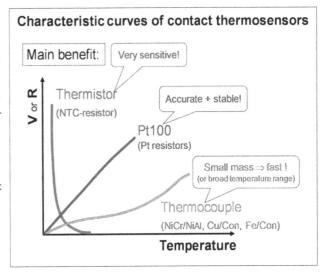

Fig. 5. Characteristic curves of the main types of contact thermosensors (schematic), with main benefits. For main constraints see text. Measured parameters: Voltage (V) or resistance (R, in Ohm).

Graph by Anton Stabentheiner

For air temperature measurements in direct sunlight, heating of the thermocouple junction by solar radiation has to be taken into consideration. To avoid any resulting measurement error, a calibration comparing the readout between two thermocouples, one in sunshine and one in the shade, has to be applied (Stabentheiner *et al.*, 2012). To do this properly, the intensity of solar radiation also has to be measured (see section 5.4.).

5.1.1.3. Use inside colonies

When using probes within colonies, these may require protection from direct contact with the bees, so as to avoid heating by direct bee contact (compare Kleinhenz *et al.*, 2003; Stabentheiner *et al.*, 2010). Logging temperature data at short time intervals helps to uncover such events which will appear as temperature peaks. The actual sensory spot of a thermocouple is the junction between the two wire metals that is closest to the electronic thermometer. This is usually the thermocouple tip. However, hive bees often gnaw at the wire insulation. This is not critical as long as the wires do not touch. But should they touch, location of temperature measurement may change.

5.1.2. Thermoresistors (thermistors, Pt100)

Thermistors and Pt100 thermosensors are not sensitive to thermal gradients within the electronic thermometer. Their size, however, is usually larger. The advantage of thermistors is their high sensitivity (Fig. 5). The strong nonlinearity of their characteristic curve is considered properly by modern electronic thermometers. The strong nonlinearity of the characteristic curve of thermistors, however, implies that their sensitivity decreases with increasing temperature. Arrays of many thermistors may be used to monitor the temperature of many comb cells simultaneously (Becher and Moritz, 2009). If a high accuracy and long-term stability is required, Pt100 sensors (platinum resistors with 100 Ohms at 0°C) are a good choice. In any case, individual calibration is required for utmost accuracy.

5.2. Non-contact temperature measurement

5.2.1. Infrared spot thermometers

Infrared spot thermometers may be used for *surface* temperature measurement of larger targets, such as combs, hive walls, soil, etc. The size of the measurement spot increases with distance to the measured object. In most instruments, spot size is too large for body temperature measurements in honey bees. Some devices offer a close -up lens for spot size reduction. Spot size, however, has to be *considerably* lower than the honey bee thorax diameter of ~4 mm, because accurate measurements in spherical objects are only possible at an angle smaller than ~30° from the normal to the surface. Care has to be taken that a built-in laser indicator for measurement spot identification really hits the target in close vicinity to the measured object. Parallax errors may lead to false measurements. Nowadays,

however, handheld thermographic cameras (see section 5.2.2.1.) have become so cheap that there is no need of using such devices for honey bee body temperature measurement. In fact, they are mostly useless for this purpose.

5.2.2. Infrared thermography

In insect thermoregulation research, infrared (IR) thermography provides several advantages over conventional thermometry. Most importantly, the bees do not have to be killed, as required in measurements with 'thermoneedles'. They neither need to be touched, nor impaired in their natural behaviour. Measurements on the same individual can be repeated several times, and the spatial distribution of the body surface temperature can be evaluated simultaneously (Stabentheiner *et al.*, 1987, 2012). A 'disadvantage' is that only surface temperatures, but not internal body temperatures are obtained.

5.2.2.1. Main thermographic camera types

In earlier times, the infrared (IR) radiation of an object to be measured was scanned by rotating prisms or mirrors and focussed onto a single detector (detector material for example InSb or HgCdTe). Nowadays, many detectors are arranged in focal plane arrays (FPAs). Data readout ('scanning') from the chip is done electronically. Spatial resolution of FPAs may be of 120 x 120, 320 x 240 (see Stabentheiner *et al.*, 2012) or 640 x 480 pixels, or even reach 1280 x 1024 pixels in (rather expensive) top camera models.

There are two main types of thermographic cameras for the practical use in honey bee body temperature measurement. Cameras either use photon-counting detectors, like InSb, PtSi, HgCdTe, or QWIP (quantum well interference photodetectors), or microbolometer arrays. Cameras with photon-counting detectors measure infrared photons directly. To achieve their (usually) high sensitivity (<20 mK in some InSb cameras for example) they require detector cooling. While in earlier times detectors were cooled with liquid N_2 or expanding Ar gas, cooling is nowadays done with miniaturized built-in compressors. InSb and QWIP chip cameras are usually fast in reaction (have a short exposure time), which make them especially convenient for the measurement of flying (or fast moving) honey bees.

Cameras with microbolometer arrays measure infrared radiation indirectly, via (small) temperature changes of the detector pixels. Detector materials in use are for example Vanadium oxide (VOX) or amorphous Silicon (aSi). A disadvantage of (earlier) bolometer cameras is that they may exhibit considerable drift due to internal temperature changes (Stabentheiner *et al.*, 2012). This may require frequent internal and external recalibration. Modern cameras often have a better (though not complete) drift compensation. By now, the sensitivity of the top bolometer cameras is better than 0.03 K. Furthermore, small and relatively cheap handheld bolometer cameras have become available. Many of these allow to set the focus close

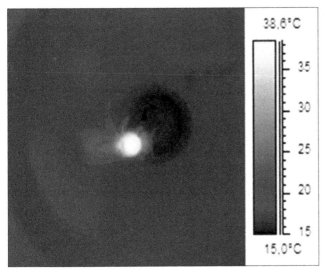

38.6°C

35

30

25

20

15

15.0°C

Fig. 6. Thermogram of a honey bee sucking 0.5 M sucrose solution from an artificial flower. Body surface temperatures: head = 26.1°C, thorax = 38.6°C, abdomen = 23.3°C. Ambient air temperature close to the bee: 16.2°C, ambient radiation: 18.1 W m^{-2}, ambient humidity: 40.6%. Thermographic camera: FLIR i60 (180x180 microbolometer detector). For instrument accuracy see Stabentheiner *et al.* (2012).

Image by Anton Stabentheiner

enough to measure surface temperatures of thorax, abdomen, and head of honey bees (Fig. 6).

The main spectral ranges of thermography cameras in use are the 3-6 μm (short wave) and the 7-14 μm (long wave) wavebands. Short-wave cameras (with e.g. InSb or PtSi FPA) are fine for laboratory use. For field measurements, long wave cameras are recommended (e.g. microbolometer or QWIP FPA), as these are much less sensitive to reflected solar radiation (see next section).

5.2.2.2. Honey bee cuticle infrared emissivity

Thermographic measurement of the surface temperature of an object requires knowledge of its infrared surface emissivity. This quantity is a quotient (ranging from 0-1) describing how much radiation an object emits at a certain temperature in comparison to an ideal black body radiator, which emits the theoretical maximum. The emission and absorption of infrared radiation are directly coupled according to Kirchhoff's law, which means that an opaque surface with an emissivity of 0.97 also has an absorptance of 0.97 (i.e. 97% of incoming infrared radiation is absorbed and 3% are reflected). Therefore, the emissivity value can be used to compensate for ambient infrared radiation reflected into the camera via the object, which adds to the radiation emitted by the object and therefore would produce a measurement error if left uncorrected.

The honey bee cuticle has an infrared emissivity of ~0.97 (Stabentheiner and Schmaranzer, 1987), so approximately 3% of the ambient infrared radiation is reflected via the cuticle into the infrared camera. With standard evaluation software provided with

thermography cameras, this can be compensated by considering ambient 'black body' temperature, and this is usually done by measuring ambient temperature.

During measurements in sunshine, however, there is an additional error caused by solar radiation reflected by the cuticle, which is not corrected for by the above mentioned correction of reflected ambient infrared radiation. This effect can be measured by thermographing insects at different intensities of solar radiation with a real-time infrared camera (frame rate > 30 Hz). The cuticular surface temperature is first measured in sunshine and then the insect is shaded and its temperature measured again within a few milliseconds. The resulting measurement error turns out to be small in the 7-14 μm waveband (long wave cameras). The applied correction factor for reflected solar radiation amounted to 0.2183°C kW^{-2} m^{-2} (Stabentheiner *et al.*, 2012). However, with short wave cameras (3–6 μm waveband, which is closer to the solar radiation maximum in the visible range) the measurement error may be considerably higher (several degrees). Therefore, long-wave cameras should be used for field measurements in sunshine.

5.2.2.3. Thermography camera calibration with reference radiator

Even rather cheap microbolometer cameras have a sensitivity of <0.1°C. This means that they can sense temperature differences between different areas within a picture with this thermal resolution. Measurement accuracy of most cameras, however, is only 2°C or 2% (whatever is smaller). This is true for cameras with both photon counting and microbolometer detector chips. In special cases, for example if macro lenses are used, these specifications may not be valid some time after the camera was turned on or at certain ambient conditions (Stabentheiner *et al.*, 2012). Only a few models offer an accuracy down to 1°C. It has to be kept in mind that this refers to the internal instrument accuracy. Errors resulting from wrong input of surface emissivity (see section 5.2.2.2.) or environmental data (ambient temperature, relative humidity, distance to the measured object) may add to these values. Therefore, calibration with an external reference radiator is indispensable if an accuracy of better than 2°C has to be guaranteed.

A simple precision cavity black body reference radiator of high accuracy (<0.2°C) can be constructed by immersing a hollow metal cylinder (e.g. brass) in a regulated laboratory water bath (Stabentheiner and Schmaranzer, 1987) or, for field use, in a thermal coffee pot filled with warm water (Fig. 7). Its inner surface has to be covered by matt 'black' lacquer (infrared emissivity should be at least 0.95; e.g. Nextel Velvet Coating 811-21). The cavity increases the apparent cavity emissivity to a value >0.995, which is close to an ideal black body radiator (emissivity = 1), if the ratio of cylinder depth to radius is at least 10:1 (Sparrow and Cess, 1970). The accuracy of such a black body depends primarily on the accuracy of the water

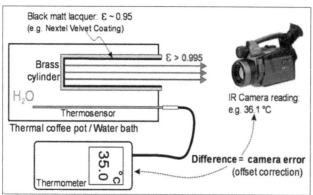

Fig. 7. Infrared camera offset calibration with a cavity black body radiator. Dimensions of the brass cylinder: ratio of cylinder depth to radius ≥ 10:1. ε = infrared emissivity (see section 5.2.2.2.).

Graph by Anton Stabentheiner

bath temperature measurement. An electronic thermometer or a laboratory thermometer with 0.1°C scaling should provide the desired accuracy. Reference temperature should be at least 5°C above ambient temperature for utmost accuracy.

Since thermography cameras may display a drift over time, camera calibration should be performed as often as possible. Correction works best if the reference radiator is permanently visible in the infrared pictures (Stabentheiner *et al.*, 2012).

5.2.2.4. Attenuation of infrared transmissive films

If honey bee body temperature has to be measured inside a hive it is often necessary to do this through infrared transmissive films to prevent the bees from leaving the colony (Stabentheiner *et al.*, 2003a, 2010). Good choices are polypropylene or cellophane films used by florists for wrapping of flowers. They provide good stability and transparency. Polyethylene would be even more transparent to infrared radiation, especially in the 'long-wave' (7–14 μm) spectral band. However, when testing commercially available films, these were mostly of rather poor 'optical' quality, i.e. their surface was not sufficiently smooth. The attenuation of the infrared radiation by the transmissive film can be compensated for by covering part of the reference source with a stripe of the same film (Stabentheiner *et al.*, 2012; and section 6 of this paper). By doing so, camera calibration with the reference radiator compensates for the attenuation of the film, and errors resulting from ambient reflections *via* the film surface can be minimized.

The attenuation of infrared transmissive films may also, with less precision, be corrected by determination of a 'special atmospheric transmission' coefficient (which usually compensates for infrared absorption of the atmosphere at greater measurement distances). This can be done by comparing the direct radiation of a black body radiator (Fig. 7) with radiation passing through the film. An iterative change of the atmospheric transmission coefficient for the film measurement, in the camera or in the evaluation software, until the

temperature reading equals the direct measurement provides the correction coefficient.

Special care has to be taken to avoid different reflection of ambient radiation by room walls, IR camera, operators, etc. from different parts of the film surface. In outdoor measurements, reflection of the sky via the film may produce considerable measurement errors. For more details see Stabentheiner *et al.* (2012). In general, however, it is worthy of note that thermographic measurement through plastic films should be avoided whenever possible because it inevitably adds additional sources of error.

5.3. Operative temperature

In field investigations of honey bee thermoregulation, operative temperature (T_e) provides an integrated measure of the action of several environmental factors on honey bee thermoregulation (Fig. 8). T_e is the body temperature of dead honey bees exposed to the same environmental conditions as the living bees under investigation. It allows separating the endogenous part of thermoregulation (endothermy) from the effect of environmental parameters on body temperature (Kovac *et al.*, 2010). These environmental parameters are ambient air temperature, radiative loss and gain of heat, thermal convection and external air convection (wind), and evaporative heat loss, mainly through the cuticle. In such experiments, dead honey bees are used as thermosensors (Fig. 8). T_e has been determined by using both dried or fresh carcasses. However, using fresh carcasses is recommended, because the use of dried bees strongly reduces honey bee heat capacity (because of their smaller mass), and this way considerably changes the reaction of the T_e-thermometer to environmental factors. For a detailed discussion of this topic see Materials and Methods in Kovac *et al.* (2010).

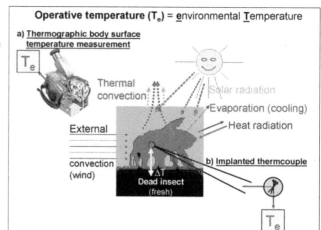

Fig. 8. Operative temperature (T_e) measurement in honey bees. T_e is the body temperature of dead bees exposed to the same environmental conditions as the living individuals under investigation. It can be measured either thermographically as surface temperature **(a)** or by an implanted thermocouple as core temperature **(b)**.

Graph by Anton Stabentheiner

5.4. Radiation sensors

Ambient radiation acting on honey bees is sunlight in the visible and near infrared range (ca. 300-3000 nm wavelength, with the maximum in the visible range) and heat radiation in the middle to far infrared range (ca. 3-20 µm wavelength, with the maximum at about 8.5-11 µm at biologically relevant temperatures). The global radiation is the radiation from the upper hemisphere to a horizontal surface. It is the sum of the direct solar and the diffuse sky radiation (in W m^{-2}). Global radiation sensors measure in the solar spectrum from 0.3 to 3 µm wavelength.

5.4.1. Star pyranometers (according to Dirmhirn)

With star pyranometers, which are widely used in meteorology, the intensity of global radiation (radiant intensity) is measured indirectly, via measurement of the temperature difference between black and white copper plates arranged in the horizontal plane (star shape). The (solar) radiation heats the black plates stronger than the white ones. This temperature difference is measured using a thermopile (i.e. a serial arrangement of thermocouples) attached to the underside of the surfaces. This differential measurement principle minimizes the influence of ambient temperature. The advantage of star pyranometers is their robustness and a broad spectral range. Their main disadvantage is their rather inert reaction to radiation changes.

5.4.2. Photoelectric pyranometers

Photoelectric radiation probes measure the irradiance in the solar spectrum in the visible range and in the short-wave infrared range. Through filtering of the incoming radiation, their spectral range is tuned to the range of wavelengths of interest, e.g. UVB, UVA, visible light. Photoelectric global radiation sensors measure both direct and diffuse solar radiation. They usually cover the spectral range from 400 to 1100 nm. The spectral sensitivity curve should be flat for optimal results. Some companies manufacture miniaturized versions on request (see Stabentheiner *et al.*, 2012)

5.4.3. Measurement of the short-wave radiation balance

For energy balance measurements of foraging honey bees it may be necessary to determine the reflecting ability (albedo) of the ground surface in addition to the direct solar radiation. A combination of a pair of global radiation pyranometers, one directed upwards and the other downwards, allows for measurement of the short-wave radiation balance. The downward directed sensor measures the radiation reflected by the ground. The upward directed sensor measures direct and indirect solar radiation. From the readings of the two pyranometers, the albedo can be calculated.

5.5. Humidity measurement

For humidity measurements, capacitive humidity sensors are in common use. They are easy to use and of sufficient accuracy. They measure the effect of humidity on the dielectric constant of a polymer or metal oxide material. Advanced measurement devices provide calculation of absolute humidity (g m^{-3}), dew point temperature (°C), etc., in addition to relative humidity (%). For outdoor measurements, the probe has to be kept in the shade. The sun will heat the internal temperature sensor, which leads to false calculation of relative humidity.

For application within honey bee colonies miniaturized sensor heads (with a diameter of a few mm; see Kovac *et al.*, 2010; Stabentheiner *et al.*, 2012) are of great advantage. The use within colonies requires protection of the probe from the bees. Otherwise they may cover it over with wax and propolis.

6. Respiration and energetics measurement in honey bees

Respiratory gas exchange is one of the basic physiological mechanisms of insect metabolism. In honey bees, basal or resting metabolism and its temperature dependence (Stabentheiner *et al.*, 2003a; Petz *et al.*, 2004; Kovac *et al.*, 2007), discontinuous gas exchange cycles (Kovac *et al.*, 2007) and energetics of foraging and flying (e.g. Wolf *et al.*, 1989; Harrison and Hall, 1993; Harrison *et al.*, 1996; Woods *et al.*, 2005) have all been investigated so far. Respiratory measurements are also a valuable tool for the determination of (respiratory) critical thermal maxima and minima (CT$_{max}$, CT$_{min}$) and chill coma temperature (Lighton and Lovegrove, 1991; Lighton and Turner, 2004; Kovac *et al.*, 2007; Käfer *et al.*, 2012). Measurements on tiny honey bee larvae (Petz *et al.*, 2004) however require a different setup than measurements on flying (e.g. Roberts and Harrison, 1999) or thermoregulating bees (Goller and Esch, 1991; Blatt and Roces, 2001; Stabentheiner *et al.*, 2003a, 2012), bee groups (Moritz and Southwick, 1986; Southwick and Moritz, 1985, 1987), or whole bee swarms and colonies (Heinrich, 1981a, b; Southwick, 1985, 1988; Van Nerum and Buelens, 1997; see Heinrich, 1993 for more literature). Metabolic differences exist among stages of honey bee development. They are not just a result of differences in mass, but may reflect the bees' development of mitochondrial capacity and enzymatic make-up, as is the case in the first days after emergence (Hersch *et al.*, 1978; Harrison, 1986; Moritz, 1988). Honey bees are also subjected to environmental variation throughout their life cycle, and measurement of whole-body and whole-colony respiration and heat production can provide insights into respiratory and energetic adaptations to the challenges of environmental variation and contribute to a better understanding of the benefits of social cooperation.

This paper is meant as a short guide into the main concepts and setups of whole-animal and colony respiratory and energetic measurements in honey bees. It cannot, however, provide a complete and detailed description of all possible measuring arrangements. A valuable help for a deeper understanding of the possibilities and limitations of insect respiratory and energetic measurement is given by Lighton (2008). For details on the measurement of cellular and sub-cellular respiration and metabolism good references are Suarez *et al.* (1996, 2000) and Suarez (2000).

Any company names mentioned herein are meant as a quick help for the reader to find a faster entry into the field. In no way it means that these companies are the only manufacturers of a certain type of device. Furthermore, since such information may change with time, the reader interested such equipment will eventually have to search for more and updated information.

6.1. Flow-through respirometry

Honey bees are, as honey bee scientists know quite well, extremely social insects. This means that their behaviour may change considerably if they are separated from the community to be put into a respiratory or energetic measurement chamber. In such situations individuals want nothing but find an exit out of the chamber to fly home. This may not matter (much) in experiments where flight energetics (Wolf *et al.*, 1989; Harrison *et al.*, 1996; Woods *et al.*, 2005) or the interrelation of thermoregulation and heat production are under investigation (Stabentheiner *et al.*, 2003a, 2012). Respiratory and energetic measurements also resemble a natural situation quite well if the bees have freely entered the measurement chamber because they expected a reward therein (Balderrama *et al.*, 1992; Moffatt, 2000; Stabentheiner *et al.*, 2012). It is *not* possible – and this has always to be kept in mind –, however, to determine the energy turnover of hive bees directly from respiratory measurements of isolated individuals. Energy turnover may change by more than a 100-fold due to an unpredictable degree of endothermy.

In research on insect respiration and energetics, flow-through respirometry is the state of the art (Lighton, 2008). Its great advantage is that the bees are supplied with fresh air throughout the experiment, which makes possible measurements over extended time periods (Kovac and Stabentheiner, 2007). If necessary, the atmospheric composition may be changed artificially, for example to investigate the impact of tracheal mites on flight capability (Harrison *et al.*, 2001) or of increased hive CO_2 levels on respiration and energy turnover. Modern equipment provides high sensitivity, down to the ppm (O_2 sensors) or sub-ppm range (CO_2 sensors). With appropriate calibration and care measurement accuracy of a few ppm is possible.

6.1.1. Measurement arrangements, measurement chambers and accessories
6.1.1.1. General setup
For respiration and energetics measurements the setup has to be adapted according to the questions under investigation, the developmental stage of the bees and the quantity to be measured. In open-flow investigations of insect respiration, different parallel and serial plumbing setups are in use. For general advice see also Lighton (2008) and Lighton and Halsey (2011). In most cases, scientists will be interested in O_2 consumption or CO_2 production of bees, mostly in comparison to ambient air, rather than in the measurement of absolute concentrations of these gases. Therefore, differential setups, comparing the gas concentration before and after a measurement chamber are the most relevant ones. A differential measurement setup which allows a fast, (semi-) automated switch between different measurement modes (serial or parallel; see sections 6.1.1.3. and 6.1.1.4.), and simultaneous thermographic measurement of body surface temperature is shown in Stabentheiner *et al.* (2012). However, setups with less automation will normally be sufficient to answer a specific question. In general it has to be noted that respiratory equipment is usually not of a 'buy and go' type. Every measurement situation requires its own adaptations. However, many companies provide solutions and help to fit the experimenter's requirements.

1. *Standard conditions:* The measurement output of instruments (volume of O_2 consumption or CO_2 production, or flow rate) usually refers to standard (STPS) conditions (0°C, 101.32 kPa = 760 Torr).

2. *Calculation of O_2 consumption (VO2) or CO_2 production(VCO2):* The difference in concentration measured between the measurement and reference air stream, or before and after a measurement chamber, multiplied by the flow rate, provides the O_2 consumption or CO_2 production (turnover) of the bee in volume units per time interval (Equation 1). Table 1 provides a short reference of how to convert STPS volumes to moles or energy and power units.

VCO2 *or* VO2 = concentration × flow (ppm × ml min^{-1} × 10^{-3} = μl min^{-1})
(Equation 1)

6.1.1.2. Air drying, CO_2 scrubbing and tubing
1. *Chemical scrubbers:* Drying of the air by some means is a prerequisite to obtain accurate results in insect respiratory measurements. Fuel-cell O_2-measurement devices require completely dry air. Chemical scrubbers (desiccants) are inserted in columns, usually before the mass flow controllers. A commonly used desiccant is Drierite (contents: >98% $CaSO_4$ (gypsum), <2% $CoCl_2$). For special purposes, e.g. if CO_2 has also to be absorbed, granular magnesium perchlorate, $Mg(ClO_4)_2$, Ascarite is an excellent CO_2 scrubber.

2. *Cool traps:* For CO_2 measurement with a differential infrared gas analyser (DIRGA) complete air drying is not necessary in all cases. Cool traps (regulated at temperatures of ca. 2-10°C, for example) can bring relative humidity to a low and constant level. This works well because DIRGAs are typically operated at internal temperatures of >50°C. Relative humidity during measurement is therefore low. Since the water content reaching the DIRGA is the

Table 1. Quick reference for result calculations and units conversion. [#] molar volume of an ideal gas at 0°C and 101.32 kPa (760 Torr or 1 atm) of pressure; [##] 21.117/60, [###] caloric equivalent of sucrose (21.117 kJ l[-1], e.g. during sucrose feeding), use different values on demand (Table 2).

Desired results in units of VO2 or VCO2		Calculate
nl/min		ppm × ml/min
nmol/min		ppm × ml/min / 22.414[#]
Units conversion		
From	**To**	**Calculate**
µl/min	nMol/min	µl/min × 44.61497
nMol/min	µl/min	nMol/min × 0.022414
nMol/min	ml/h	nMol/min × 0.0013448
µl O$_2$/min	W (J s[-1])	µl/min × 0.35195[##] × 10^{-3}
µl O$_2$	J	µl × 21.117[###] × 10^{-3}

same during calibration and measurement, and in the measurement and reference gas stream, differential measurements are accurate. However, independent cool traps are needed for the measurement and the reference gas streams (Lighton, 2008; Stabentheiner *et al.*, 2010), which makes the equipment more expensive.

3. *Tubing:* Tubing materials are diverse. A flexible and very durable material well suited for respiratory measurements is Viton®. Specialists of insect respiration also use metal tubes (which of course are less flexible). Inner diameter of tubes should in any case be small, e.g. 2-4 mm, to achieve optimal results with honey bees.

User safety advice: Take care not to inhale Drierite. While fine gypsum particles may already impair respiratory functions, CoCl$_2$, which is used as a humidity indicator, is extremely hazardous. Ascarite is also extremely hazardous. To quote from the Oxzilla 2 (Sable Systems International) manual: "Try not to breath too much of it if you plan to see your grandchildren."

6.1.1.3. Serial measurement arrangement

Serial plumbing has the advantage that during differential measurement of CO$_2$ production or O$_2$ consumption both the measurement and the reference parts of the sensor devices always are operated at exactly the same flow rate. It is also cheaper than a parallel setup (see section 6.1.1.4.) because only one mass flow controller is needed (Lighton, 2008; Stabentheiner *et al.*, 2012).

1. *Air supply:* Fresh air should be taken from outside the laboratory via a tube to avoid disturbance of the measurements by air exhaled by the experimenters. The delay between the reference and measurement tubes or detectors of the measurement devices may otherwise lead to unnecessary baseline fluctuations. To additionally dampen fluctuations in outside air CO$_2$ or O$_2$ content, the air should be passed through a buffer container (5-10 l). An alternative is air from a compressed-air bottle.

2. *Pumps:* A pump feeds the air to a drying unit [e.g. Drierite columns (see safety advice in 6.1.1.2.), or a cool trap regulated at a low temperature of 2–10°C], which brings the water content of the air to zero or a low and constant level. Regulated pumps are of advantage because they help to avoid unnecessary overpressures in the system (offered, for example, by Sable Systems International).

3. *Mass flow regulation:* Subsequently, the air has to pass a mass flow controller (e.g. Brooks 5850S, Brooks Instruments; or Side Trak 840-L, Sierra Instruments; etc.). A range of 0–1000 ml/min is suitable for investigating both resting end endothermic honey bees. A small range mass flow controller of 0–100 ml/min may be recommended to improve accuracy of flow regulation for measurements of tiny honey bee larvae which may weigh considerably less than 1 mg in early stages of development (Petz *et al.*, 2004).

4. *Measurement:* The air then has to pass the reference sensor of the differential gas analyser. Afterwards, the air is put through the measurement (respirometer) chamber containing the honey bee (see section 6.1.1.5.). Before the air is forwarded to the measurement sensor it has to be dried again (and in the same way as the reference air stream).

6.1.1.4. Parallel measurement arrangement

In parallel measurement mode, two sets of pumps and mass flow controllers provide the reference and measurement sensors of the measurement device simultaneously with independent gas streams. This avoids delays between reference and measurement gas stream if the tubing of both sides has the same length. A well-tried sequence of devices and parts for the measurement stream is:

measurement/(reference) chamber(s) – drying columns – pumps – mass flow controllers – measurement device.

Fig. 9. Example of a measurement chamber for use in a temperature controlled water bath. It was milled out of a brass block. Air inlet visible in the picture bottom (front), outlet on top (covered by yellow kitchen cloth). Effective chamber volume is variable by a movable, perforated brass barrier. In this setup, the window in the lid was covered by an infrared transmissive plastic film allowing visual and thermographic behavioural observations and body surface temperature measurements (Kovac *et al.*, 2007; Stabentheiner *et al.*, 2012). Lid tightness was assured by a Viton® O-ring (not visible). Service holes were drilled in one side (left) to accept chromatography septa (11 mm), which allowed tight insertion of thermocouple wires. Tightness has to be proofed by applying an overpressure on the submerged chamber. The rectangle at the right-hand side is a proprietary reference radiator for infrared camera calibration (see Stabentheiner *et al.*, 2012).

Photo by Anton Stabentheiner

Inserting the columns for air drying before the pumps helps in this arrangement to considerably attenuate coupling of pump noise to the measurement chamber at the tube inlet (Stabentheiner *et al.*, 2012).

The parallel mode is preferable with fuel cell devices (for O_2 consumption measurement) because they are very sensitive to pressure differences between the reference and measurement channels (see section 6.1.2.1.).

The parallel mode is also preferable if the respiratory quotient (RQ, quotient VCO2/VO2) is to be determined with open-flow respirometry, with a combination of an infrared gas analyser (VCO2; section 6.1.3.) and a fuel-cell device (VO2; section 6.1.2.) in series. The need for frequent recalibration of fuel-cell instruments necessitates a possibility to bypass the bee measurement chamber. This can be done with manual three-way valves which allow smooth switching. If automated switching is intended magnetic valves are **not** suited because their pressure impulses lead to large unwanted signal offsets (Stabentheiner *et al.*, 2012). Motor stepping valves are be an alternative.

6.1.1.5. Measurement chambers

1. *Resting and active metabolism:* Fig. 9 shows a brass variant of a measurement chamber. For temperature control the chamber can be immersed in a water bath. If the chamber is not completely submerged in the water and the lid is exposed to the laboratory air in order to allow visual or thermographic observation of the bees (Fig. 9; Stabentheiner *et al.*, 2012) the air temperature in the chamber may differ from the water bath temperature. The strength and direction of this deviation depends on the difference between water bath and laboratory air temperature. The resulting temperature gradients inside the chamber require placement of a temperature sensor (see section 5.) inside the chamber. Completely submersing the chamber (and the inlet tubing) avoids this effect.

2. *Outdoor measurement:* A measurement chamber variant for outdoor measurements of foraging honey bees of about 8 ml inner volume was presented in Stabentheiner *et al.* (2012; Fig. 2 and Fig. 5 therein). Its lid could be opened and closed quickly to give the bees fast access to an artificial flower inside. It was also operated in serial mode. These measurement chambers can, of course, also be used in parallel mode. In cases where the sun shine is hits the measurement chamber, cooling via a water bath is indispensable, since the temperature inside can quickly reach critical levels and the bees will no longer enter it (Stabentheiner *et al.*, 2012).

In parallel mode, a different configuration is useful especially for field measurements (Stabentheiner *et al.*, 2012). A set of two identical measurement chambers can be placed at the air inlets of reference and measurement air streams, one for the measuring and the other for the reference air stream. These chambers can be constructed out of a small plastic film cylinder attached to a glass laboratory funnel. Attaching the base of the cylinder to an iron spacer ring as a counterpart of pieces of hard disk magnets fits the chamber to an underlying artificial flower. Air inlets are in the base of the artificial flower.

3. *Compensation of measurement gas loss:* Chamber opening may lead to an unwanted loss of measurement gas. This is especially critical when measuring foraging honey bees where the duration of stay at a food source can be shorter than a minute during unlimited feeding. In such cases as it is not possible to simply cut out a section of the measurement signal and to calculate an average over a certain time interval, special calibration is necessary. Very briefly, such a procedure compares the washout volumes from the chamber containing certain concentrations of CO_2 (or O_2) with and without chamber opening. Stabentheiner *et al.* (2012) provide a detailed description of how to compensate for such losses.

4. *Measurement of larvae and pupae:* For the measurement of tiny honey bee larvae, measurement chambers have to be very small (e.g. 0.5–2 ml). This can be realized by adapting a plastic syringe (Petz *et al.*, 2004), or with plastic photometer cuvets. For very young larvae and for eggs, the closed chamber method has to be used (see section 6.1.8.).

Table 2. Caloric equivalents (oxyjoule equivalents) and calorific values of biological substrates.

Substrate	Caloric equivalent (kJ lO_2^{-1})	Calorific value (kJ g^{-1})
Carbohydrates		
general mean	~21.15	~17.2
sucrose	21.117	16.8
glucose	21.0	15.6
Lipids	~19,6	~38,9
Proteins	~19,65	~17,2

5. *Flight metabolism:* The other extreme of chamber size is required in measurements of flight metabolism. For this purpose, large measurement chambers of about 0.3 to 1 l volume have to be used. Such chambers allow respiration measurements in agitated free flight (Harrison and Hall, 1993; Harrison *et al.*, 1996; Wolf *et al.*, 1996; Roberts and Harrison, 1999; Woods *et al.*, 2005). Measurement of bees in free hovering flight was accomplished in a wind tunnel by Wolf *et al.* (1989), which equals an even larger chamber size of about 4 l volume. Simulating an appropriate virtual environment to stimulate the bees to stay airborne for a longer time period without agitation is, however, a tricky task (Wolf *et al.*, 1989).

6. *Whole colony metabolism:* Even larger chamber volumes will be needed for whole colony measurements (Kronenberg and Heller, 1982; Southwick, 1985).

6.1.2. O_2 consumption

Flow-through oxygen measurement devices are of the fuel cell or paramagnetic type. For individual honey bees, fuel cell devices are in common use.

6.1.2.1. Fuel cell devices

From the different types of fuel cells, electro-galvanic lead-oxygen cells can be operated at room temperature (low temperature cells), while Zirconia cells require internal cell heating to high temperatures (> 500°C). Lead-oxygen cell devices (like the Oxzilla 2, Sable Systems; the S 104, Qubit Systems; etc.) are not extremely fast in reaction but appropriate for honey bee measurement and easy to handle. These cells are very sensitive to pressure differences between reference and measurement channel. Pressure sensitivity is an inevitable consequence of the fact that fuel cell devices do not measure O_2 'concentration' but O_2 'partial pressure'. Therefore, a parallel measurement setup is of great benefit (see section 6.1.1.4.). Their high sensitivity to (partial) pressure changes is, on the other hand, the basis of their high sensitivity. The slower (but usually sufficient) reaction of low temperature fuel cells may (in part) be compensated mathematically (post-hoc) by a *Z* transformation ("instantaneous correction") if restoration of the original shape of a respiratory event is an important parameter in an experiment (Lighton, 2008). Zirconia cells react faster. Appropriate air drying is indispensable in all fuel cells to achieve the best results.

6.1.2.2. Paramagnetic devices

Paramagnetic devices are not so sensitive but are well suited for continuous monitoring of colony oxygen consumption. Their sensitivity to tilting, however, requires installation in a firm position throughout an experiment.

6.1.2.3. Calibration

Calibration of fuel cell and paramagnetic devices is simple, usually just an end point calibration against outside air (20.95% O_2 content). Zero point calibration is usually not necessary. A certain drift, however, is inherent to the nature of fuel cells. Frequent recalibration and baselining during evaluation for drift compensation is, therefore, indispensable (Lighton, 2008).

6.1.2.4. Indirect calorimetry: calculation of energy turnover

Measurement of O_2 consumption (VO2) allows calculation of honey bee energy turnover (P, metabolic power) by multiplication with the caloric equivalent (oxyjoule equivalent, CE). This is called indirect calorimetry.

$$P \text{ (W)} = VO2 \times CE \quad (lO_2 \text{ s}^{-1} \times J \ lO_2^{-1} = J \text{ s}^{-1} = W) \text{ (Equation 2)}$$

The caloric equivalent depends on the substrate combusted in an animal's metabolism. In honey bees this is mainly carbohydrates (Rothe and Nachtigall, 1989). Table 2 summarizes some values of the caloric equivalent and of the calorific value of biological substrates.

6.1.3. CO_2 production

For the measurement of honey bee CO_2 production in open flow respirometry, a differential infrared gas analyser (DIRGA) is recommended. These measure infrared light absorption in the $\lambda = 2.5$ –8 μm wavelength range by the CO_2 present in air (photometric principle according to the Lambert/Beer law). These devices are very sensitive down to the sub-ppm range and usually offer a high baseline stability if their internal temperature is regulated accurately. Internal construction is mostly a set of two tubes (for reference and measurement signal) with an infrared lamp on one side and a detector on the other.

Portable instruments, originally designed for field measurements of plant photosynthesis, can also be used for field measurements of honey bee respiration (e.g. Li-COR LI-6400XT Portable Photosynthesis System). This device also offers the possibility of simultaneous measurement of air H_2O content.

Fig. 10 shows a respiratory CO_2 trace of a honey bee. The bee was kept in the dark during the night, which made her soon entering the resting state. At rest she showed typical patterns of discontinuous respiration. At hive temperatures of 30°C or 34°C a bee breathes on average only once every 37 or 28 seconds, respectively (Kovac *et al.*, 2007). This trace, however, also shows the bees' high proneness to disturbance when caught in a respiratory chamber against their own will. When the experimenter entered the laboratory ('!' in Fig. 10) and afterwards switched on the light, the trace characteristic changed and the bee eventually heated up. Such endothermy increased energy turnover by more than the 100-fold.

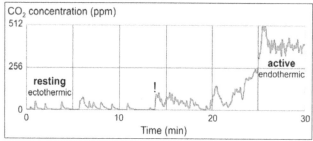

Fig. 10. Respiratory trace of a honey bee (*Apis mellifera carnica*). The bee was kept overnight in a 18 ml respiratory chamber (Fig. 10; 0 min = 07:21 in the morning). Left trace part: discontinuous gas exchange cycles during rest (light off), the bee was ectothermic; exclamation mark: experimenter entering the laboratory (with subsequent light on); right trace part: endothermy (bee prepared for immediate flight) and was actively seeking for an exit out of the chamber. Ambient temperature near the bee: ca. 22°C. For more resting traces see Kovac *et al.* (2007).   Graph by Anton Stabentheiner

6.1.3.1. Measurement range selection

DIRGA measurement range and sensitivity is determined by the tube (or better: infrared beam) length. A frequently offered measurement range is 0–3000 ppm, well suited for the measurement of endothermic or flying bees.

0–250 ppm tubes in the DIRGA provide a high resolution, well suited for resting metabolism measurements of bees and for tiny honey bee larvae. With this measurement range, however, endothermy of adult bees will soon produce a range overflow. With a 0–10,000 or 0–20,000 ppm tube, the respiration of even whole colonies can be measured. A high resolution (small range) and a large range (lower resolution) system in series always provides a calibrated output (ABB, Inc.; Stabentheiner *et al.*, 2012).

It has to be considered, however, that there exists an interrelation between the gas flow and the height of the measurement signal (Gray and Bradley, 2006; see section 6.1.5.). In endothermic foragers (in an 8-18 ml measurement chamber) for example a flow rate of 250 ml

Table 3. Dependence of the respiratory quotient (RQ) of honey bee larvae on age and mass (*A. m. carnica*). * from Petz *et al.* (2004); ** from Melampy and Willis (1939).

age (h)	mean mass (mg)*	RQ**
12	0.36	
36	1.47	
60	10.2	1.42
84	37.73	1.23
108	131.44	1.23
132	159.66	1.13

min^{-1} provides a sufficient signal height with a 0-3000 ppm DIRGA. During investigations of honey bee resting metabolism with a 0–250 ppm tube, a flow of 150 ml min^{-1} turned out to provide a good compromise between sensitivity and temporal resolution of CO_2 production measurements in a wide range of ambient temperatures (2.5–45°C; Kovac *et al.*, 2007). With small honey bee larvae, the flow rate has to be reduced, for example to 20 ml min^{-1} (Petz *et al.*, 2004).

6.1.3.2. DIRGA Calibration

Calibration of a DIRGA is usually performed against outside air at zero point (nowadays about 380 ppm CO_2; e.g. Petz *et al.*, 2004), and against a span gas of known CO_2 concentration in end point (according to the system's measurement range) or with internal calibration cuvettes (e.g. URAS devices of ABB, Inc.). Automatic calibration can be done after switching the air flow with magnetic (or better, stepping motor) valves to bypass the insect measurement chamber (see Stabentheiner *et al.*, 2012). The resolution of sensitive CO_2 measurement equipment is down to <0.2 ppm. With appropriate care, a measurement accuracy of ~2 ppm is feasible.

6.1.3.3. Respiratory quotient (RQ): calculation of energy turnover from CO_2 measurements

Since the respiratory quotient RQ = 1 in adult honey bees (Rothe and Nachtigall, 1989), energy turnover can be calculated directly from measurements of CO_2 production. In honey bee larvae, however, the situation is not just different but, so to say, unusual. Typically, animal RQ values are in the range of 1.0 for carbohydrate combustion, 0.8 for protein and 0.7 for pure lipid metabolism. If carbohydrates are converted to lipids, and during growth, RQ may be >1. In growing honey bee larvae, however, RQ values are not only >1, but also variable (Melampy and Willis, 1939; Table 3).

6.1.4. H_2O balance

In honey bees, there are several pathways of water loss, viz. cuticular transpiration, respiration, the mouthparts (for cooling), and defecation. Except for the latter, net water loss (and this way evaporative heat loss) can be readily measured with a DIRGA, simultaneously with CO_2 production (Roberts and Harrison, 1999).

6.1.5. Impact of flow control and measurement chamber size on sensitivity and temporal resolution

Gas flow has to be accurately regulated by mass flow controllers (compare section 6.1.1.3.). The respiratory output signal, however, depends on several factors (Gray and Bradley, 2006; Lighton, 2008; Terblanche *et al.*, 2010). Always consider that washout phenomena and signal delays influence measurements, depending on the volumes of the measurement chamber and the tubing. If you see respiratory peaks like in Figure 10, for example, keep in mind that signal rise and signal decay characteristics are always distorted. With proper calibrations and mathematical treatment (*Z* transformation; Lighton, 2008), the original signal characteristics may be restored to some extent for individual respiratory peaks or events.

Signal height is also influenced by several factors. In the first place it is influenced by the measurement chamber volume. Smaller volumes lead to a higher CO_2 accumulation and O_2 depletion. A low flow rate (recommended for small individuals) accumulates CO_2 (decreases O_2; Fig. 11) and therefore increases signal height. On the other hand, this reduces temporal resolution. A high flow rate decreases CO_2 concentration (increases O_2) and this way decreases signal height, but improves temporal resolution. A high flow rate may help to avoid signal overflow, for example if larger bee groups or whole colonies are to be measured. One has always to find a compromise between sensitivity and temporal resolution. Taking a measurement chamber as small as possible and increasing flow speed to a not too low value will increase sensitivity, and this way improve detection of small changes in the respiratory signal. A too high flow rate in a small chamber, however, will increase convective heat loss, especially in endothermic or flying bees. As thermoregulating bees probably counteract such additional losses, the measured metabolic rates may be overestimated as compared to natural situations.

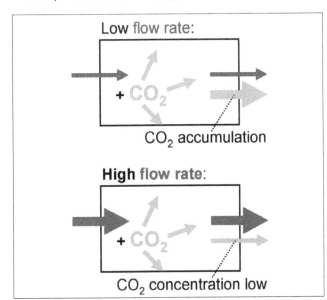

Fig. 11. Effect of flow rate on the respiratory measurement signal.

Graph by Anton Stabentheiner

6.1.6. Controlling relative humidity

Desiccation is usually not a severe issue in endothermic bees because they produce much metabolic water (Roberts and Harrison, 1999). However, it may influence respiratory measurements of larvae or pupae. To avoid desiccation during longer lasting measurements (e.g. overnight), the relative humidity in the measurement chamber can be adjusted by saturating the air with water vapour by passing it through flasks with distilled water immersed in a temperature controlled water bath prior to the respirometer chamber. The temperature of this water bath is adjusted to the dew point temperature ($T_{dewpoint}$, °C) that corresponds to the desired relative humidity (rH_{des}, %) at the desired temperature inside the measurement chamber (T_{des}, °C) (Stabentheiner *et al.*, 2012):

$$T_{dewpoint} = 234.15 \times \ln(VP/6.1078) / [17.08085 - \ln(VP/6.1078)]$$
(Equation 3)
$$VP = rH_{des} \times SVP / 100 \quad \text{(Equation 4)}$$
$$SVP = 6.1078 \times \exp^{(17.08085 \times Tdes) / (234.175 + Tdes)} \quad \text{(Equation 5)}$$

where VP is vapour pressure (mbar), and SVP is the saturation vapour pressure (mbar).

6.1.7. Closed chamber method (CO_2 accumulation)

With honey bee eggs, flow through respirometry is not directly applicable, but this can be done with a closed chamber method. In a protocol described by Mackasmiel and Fell (2000), a measurement chamber (e.g. 2 ml autosampler vials) containing one egg on a piece of cell base wax is closed for some time (e.g. 12–14 h) to accumulate the CO_2 produced by the eggs. After certain periods, air samples of 250 μl are drawn from the vials and injected into an infrared gas analyser (IRGA). Accuracy of egg respiration measurement is ascertained by comparison with the air of empty vials and with the air of vials containing dead eggs (frozen overnight) or containing just a piece of cell base wax (Mackasmiel and Fell, 2000). To achieve a high accuracy of CO_2 measurement, the IRGA should be flushed with air deprived of CO_2 by pumping it through a calcium carbonate container. CO_2 can also be scrubbed with $Mg(ClO_4)_2$, or Ascarite (but see User Safety Advice for these substances in section 6.1.1.2.). It is also possible to consider a setup where the chamber is inserted in the gas path of a DIRGA and CO_2 accumulation is accomplished by closing magnetic valves for appropriate periods.

6.1.8. Thermolimit respirometry

Thermal limits of insects are often determined by visual or video observation (e.g. Hazell *et al.*, 2008; Hazell and Bale, 2011; Terblanche *et al.*, 2011; and citations therein). These protocols test the ability of coordinated movement of the insects. Besides passive observation, such analyses use the "righting response" of insects to regain normal body position after being pushed or nudged by the experimenter.

An alternative is thermolimit respirometry (Lighton and Turner, 2004; Lighton, 2008). With this method, the thermal limits of respiration can be determined. In *A. m. carnica,* chill coma temperature was determined by thermolimit respirometry to be in the range of 9-11°C by the observation that discontinuous respiration ceased (Lighton and Lovegrove, 1990; Kovac *et al.*, 2007; compare Hetz and Bradley, 2005). At these temperatures, muscular and neural functions of bees come to a halt (Esch, 1988). Honey bees can survive chill coma temperatures (<10°C) for a longer time, but are incapable of coordinated movement (Esch, 1988).

Thermolimit respirometry is especially useful in determining the upper critical thermal limit (CT_{max}). By applying a temperature ramp of 0.25°C min^{-1} (Terblanche *et al.*, 2011), the respiratory CT_{max} of *A. m. carnica* was determined as 48.9°C, which equalled the (voluntary) activity CT_{max} of 49.0°C (Käfer *et al.*, 2012). The method uses the fact that cyclic patterns of respiration cease suddenly at the CT_{max} (Lighton and Turner, 2004; Lighton, 2008; Käfer *et al.*, 2012).

In order to better identify the point of respiratory failure, the absolute difference sum (ADS) of the respiratory signal (rADS) can be calculated (Lighton and Turner, 2004). This is done by summing the difference between the absolute values (without considering sign) of successive data points of the respiratory signal. Further improvement can be achieved by calculating the rADS residuals around the estimated temperature of the CT_{max}. The rADS residuals are the difference between the ADS curve and a regression line calculated through the values for a section of, for example, 10 min before and after the estimated thermal limit. A usually sharp inflection point of the rADS residuals accurately indicates the respiratory CT_{max} (Lighton and Turner, 2004; Käfer *et al.*, 2012; Stabentheiner *et al.*, 2012). This method can also be applied for the analysis of other cyclic events like the readout of activity detectors (see section 6.8.2.).

One has to keep in mind, however, that thermolimit respirometry cannot replace conventional methods, which use behavioural cues like voluntary and forced activity to completely determine the CT_{max} and CT_{min}. Especially lethal temperatures have to be determined with appropriate methods of activity monitoring and proper tests of survival (e.g. Ono *et al.*, 1995; Ken *et al.*, 2005; Hazell *et al.*, 2010; Terblanche *et al.*, 2011). Nonetheless, thermolimit respirometry provides a standardized, fast and objective possibility to determine thermal limits because insect respiration depends on active control of spiracle and abdominal respiratory movements to achieve sufficient exchange of respiration gases. Consequently, respiration and muscular and neural activity are closely related. If respiration fails it is likely that other muscular and neural functions in the honey bee body are beyond their limits.

6.2. Chemical-optical oxygen sensors

Due to their small size chemical optical O_2 sensors allow measurement of O_2 concentration in very small volumes with a high spatial and temporal resolution (PreSens - Precision Sensing GmbH). For bees, these can be miniaturized chambers for use in metabolism measurement, including inside of honey bee brood cells. These oxygen microsensors are available with tip sizes of <50 μm. They consist of a silica fibre with a sensor spot at the tip which emits fluorescence when illuminated with the light of a blue LED through the fibre. Through a second (parallel) fibre this fluorescence signal is led back to a photodetector. If the sensor tip encounters an oxygen molecule, the excess light energy is transferred to the O_2 molecule, which decreases (quenches) the fluorescence. The degree of quenching correlates with the partial pressure of oxygen in the sensory material, which is in dynamic equilibrium with oxygen in the sample (e.g. surrounding air). With a pair of these sensors, one before and one after passage through the measurement chamber, it is possible to measure O_2 consumption in a flow-through setup. There are also sensors available for CO_2 measurement. At present, however, they only work in aqueous solutions.

6.3. Manometric and volumetric respirometry

From the variety of manometric and volumetric methods of O_2 consumption and CO_2 production measurement (Lighton, 2008), Warburg manometry is the most important one. It is a closed-system constant-volume method, which means that the measured O_2 concentration in the measurement vessel decreases in the course of an experiment. Therefore, air refreshment in regular intervals is necessary (Balderrama *et al.*, 1992; Stabentheiner *et al.*, 2003; Garedew *et al.*, 2004). This limits experimental time. Air drying is usually done with silica gel. CO_2 scrubbing can for example be done with caustic potash solution (KOH; see Lighton, 2008 for details). Moffatt and Núñez (1997) described a volumetric setup for the measurement of O_2 consumption of honey bees foraging from artificial flowers.

6.4. Titration methods

CO_2 production of foraging honey bees can also be determined by a titration method. By determining the amount of time an air current is needed to titrate 112 nanoequivalents of $Ba(OH)_2$, Moffatt (2000, 2001) determined the energy turnover of honey bees foraging from artificial flowers.

6.5. Isotopic tracer techniques

The doubly-labelled water technique is an appropriate method to measure field metabolic rates of free ranging insects (Wolf *et al.*, 1996; Speakman, 1997, 1998). It uses the differential elimination of ^2H and ^{18}O isotopes from the body tissues to determine metabolic rate. Note, these are stable isotopes and, thus, not radioactive. The principle is explained by Wolf *et al.* (1996) as "the fact that isotope concentrations decrease exponentially with time through the natural 'wash-out' of CO_2 and water. The hydrogen isotope is lost as water only, and the oxygen isotope as both water and CO_2. Therefore, the apparent turnover rate of ^{18}O is higher than that of ^2H, and the difference between the two apparent turnover rates reflects the CO_2 production rate". Wolf *et al.* (1996) validated this method for bumble bees. They injected small volumes (1 µl) of a mixture containing ^2H and ^{18}O. After ~10 minutes, when the mixture had equilibrated with the body water pool, they withdrew 1–2 µl of haemolymph to determine the initial ^{18}O concentration. After 5–7 h, final blood isotope content was analysed by mass spectrometry. This technique actually provides rates of CO_2 production, which can be directly used to calculate energy turnover of honey bees, because their RQ = 1 (Rothe and Nachtigall, 1989).

6.6. Calorimetry

Calorimetry allows the direct measurement of heat production (energy turnover) of individual bees or bee groups (Fahrenholz *et al.*, 1989, 1992). Differential calorimeters, constructed as twin-setups of 'camping box calorimeters' are the most suited variants for honey bee (and insect) calorimetry (e.g. Fahrenholz *et al.*, 1989; Lamprecht and Schmolz, 2000). Depending on construction size, measurements are possible on individual honey bee larvae (Schmolz and Lamprecht, 2000) and adult bees (Schmolz *et al.*, 2002), in whole colonies (Fahrenholz *et al.*, 1992), and even in groups of tiny honey bee parasites (*Varroa destructor*, Garedew *et al.*, 2004). The insects inside these instruments can be continuously supplied with fresh air if necessary, which allows long duration measurements (Schmolz and Lamprecht, 2000; Schmolz *et al.*, 2002).

Construction of such a calorimeter is not difficult ("poor man's calorimeters"; Lamprecht and Schmolz, 2000; see also Lighton, 2008). Peltier units are attached to the bottom of camping boxes to keep the instrument at a desired temperature. It is an inherent property of Peltier elements that heating or cooling of one side can be regulated via the direction of the current applied.

To sense heat flux, such constructions use the fact that Peltier elements can also be used as thermopile heat flux sensors, because a temperature gradient along them produces a current (Seebeck effect). Such boxes are commercially available with variable volumes (~5–30 l). Heat flux sensitivity is about 10±30 mV/W, which is in the range of commercially available instruments (Lamprecht and Schmolz, 2000). To achieve an acceptable baseline stability, a differential (twin) setup is recommended, with one box serving as (empty) reference unit. An additional styrofoam insulation around both boxes compensates for an otherwise poor baseline stability.

A frequent disadvantage of calorimeters is their large time constant (see Lamprecht and Schmolz, 2000; Lighton, 2008). This means that the output signal of heat production may be heavily distorted, hiding short-time dynamic properties of events of heat production or heat release. Appropriate calibration and signal processing may help to compensate for such shortcomings and to restore the dynamic structure of the signal (see pp. 74-75 in Lighton, 2008).

6.7. Energetics derived from measurement of sugar consumption

Honey bees fuel their flight nearly exclusively with sugars (Rothe and Nachtigall, 1989). This provides the opportunity to calculate flight energy consumption from the determination of sugar consumption (Hrassnigg *et al.*, 2005; Brodschneider *et al.*, 2009; see also the *BEEBOOK* paper on behavioural methods (Scheiner *et al.*, 2013)). Usually such experiments are performed in tethered flight in a roundabout or in a wind tunnel. While such measurements are very well suited to answer certain questions of energy metabolism, it has always to be kept in mind that tethered flying bees do not bear their own weight. Calculated absolute values of energy turnover are, therefore, not a full representation of a natural flight situation.

6.8. Activity monitoring:

Energetic considerations and detailed investigation of honey bee respiration often require judgement of activity and behaviour.

6.8.1. Video and thermography:

The most direct method of activity monitoring is direct observation of behaviour and classification according to a (predefined) list of behaviours or stages of activity (Stabentheiner *et al.*, 2003a,b). This can be combined with video analysis. If it is to judge whether or not bees are at rest, infrared video analysis with an active infrared light source allows observation in the dark. Infrared thermography provides the opportunity to simultaneously observe the bees' behaviour in complete darkness and measure their body surface temperature.

6.8.2. Optical activity detectors:

Optical activity detectors, which use an infrared light source (LED) and a photodetector (photodiode, phototransistor, etc.) can monitor changes of the position of bees in the measurement chamber. The information on bee activity is reflected in fluctuations of the signal recorded by the photodetector. According to Lighton (2008), "the one challenge is to create an activity data channel that is easily interpreted." Among other possibilities, a possible approach is to sum the difference between absolute values (not considering sign) of successive data points of the activity signal. The resulting absolute difference sum (aADS; Lighton and Turner, 2004) shows a steep slope in phases of high activity. A clear breakpoint emanates if activity ceases. This method can be enhanced by calculating the aADS residuals (compare section 6.1.9.). It is considered as 'semi-quantitative'. A detector ready for use can be purchased at Sable Systems International.

References

AGARWAL, M; GIANNONI GUZMÁN, M; MORALES-MATOS, C; DEL VALLE DÍAZ, R A; ABRAMSON, C I; GIRAY, T (2011) Dopamine and octopamine influence avoidance learning of honey bees in a place preference assay. *PLoS One* 6: e25371. http://dx.doi.org/10.1371/journal.pone.0025371

AMDAM, G V; OMHOLT, SW (2002) The regulatory anatomy of honey bee lifespan. *Journal of Theoretical Biology* 216: 209–228. http://dx.doi.org/10.1006/jtbi.2002.2545

AMDAM, G V; HARTFELDER, K; NORBERG, K; HAGEN, A; OMHOLT, S (2004) Altered physiology in worker bees infested with *Varroa destructor* as a factor in colony loss during overwintering. *Journal of Economic Entomology* 97: 741-747. http://dx.doi.org/10.1603/0022-0493(2004)097[0741:APIWHB]2.0.CO;2

AMDAM, G V; NORBERG, K; PAGE JR, R E; ERBER, J; SCHEINER, R (2006) Downregulation of vitellogenin gene activity increases the gustatory responsiveness of honey bee workers (*Apis mellifera*). *Behavioural Brain Research* 169: 201-205. http://dx.doi.org/10.1016/j.bbr.2006.01.006

AMDAM, G V; NILSEN, K-A; NORBERG, K; FONDRK, M K; HARTFELDER, K (2007) Variation in endocrine signalling underlies variation in social life history. *American Naturalist* 170: 37-46. http://dx.doi.org/10.1086/518183

AMDAM, G V; PAGE, R E; FONDRK, M K; BRENT, C S (2010) Hormone response to bidirectional selection on social behaviour. *Evolution and Development* 12: 428-436. http://dx.doi.org/10.1111/j.1525-142X.2010.00429.x

BARCHUK, A R; BITONDI, M M G; SIMÕES, Z L P (2002) Effects of juvenile hormone and ecdysone on the timing of vitellogenin appearance in haemolymph of queen and worker pupae of *Apis mellifera*. *Journal of Insect Science* 2: e1.

BALDERRAMA, N M; ALMEIDA DE B., L O; NÚÑEZ, J A (1992) Metabolic rate during foraging in the honey bee. *Journal of Comparative Physiology* B162: 440–447. http://dx.doi.org/10.1007/BF00258967

BARRON, A B; ROBINSON, G E (2005) Selective modulation of task performance by octopamine in honey bee (*Apis mellifera*) division of labour. *Journal of Comparative Physiology* 191A: 659–668. http://dx.doi.org/10.1007/s00359-005-0619-7

BECHER, M A; MORITZ, R F A (2009) A new device for continuous temperature measurement in brood cells of honey bees (*Apis mellifera*). *Apidologie* 40: 577–584. http://dx.doi.org/10.1051/apido/2009031

BECHER, M A; SCHARPENBERG, H; MORITZ, R F A (2009) Pupal developmental temperature and behavioural specialization of honey bee workers (*Apis mellifera* L.). *Journal of Comparative Physiology* A195: 673–679. http://dx.doi.org/10.1007/s00359-009-0442-7

BERGOT, B J; RATCLIFF, M; SCHOOLEY, D A (1981) Method for quantitative determination of the four known juvenile hormones in insect tissue using gas chromatography-mass spectroscopy. *Journal of Chromatography* 204: 231-244. http://dx.doi.org/10.1016/S0021-9673(00)81664-7

BITONDI, M M G; SIMÕES, Z L P (1996) The relationship between level of pollen in the diet, vitellogenin and juvenile hormone titers in Africanized *Apis mellifera* workers. *Journal of Apicultural Research* 35: 27–36.

BITONDI, M M G; NASCIMENTO, A M; CUNHA, A D; GUIDUGLI, K R; NUNES, F M F; SIMÕES, Z L P (2006) Characterization and expression of the Hex 110 gene encoding a glutamine-rich hexamerin in *the honey bee, Apis mellifera*. *Archives of Insect Biochemistry and Physiology* 63: 57–72. http://dx.doi.org/10.1002/arch.20142

BLATT, J; ROCES, F (2001) Haemolymph sugar levels in foraging honey bees (*Apis mellifera carnica*): dependence on metabolic rate and in vivo measurement of maximal rates of trehalose synthesis. *Journal of Experimental Biology* 204: 2709–2716.

BLENAU, W; BAUMANN, A (2001) Molecular and pharmacological properties of insect biogenic amine receptors: lessons from *Drosophila melanogaster* and *Apis mellifera*. *Archives of Insect Biochemistry and Physiology* 48: 13-38. http://dx.doi.org/10.1002/arch.1055

BLUM, H; BEIER, H; GROSS, H J (1987) Improved silver staining of plant proteins, RNA and DNA in polyacrylamide gels. *Electrophoresis* 8: 93–99. http://dx.doi.org/10.1002/elps.1150080203

BOGAERTS, A; BAGGERMAN, G; VIERSTTRAETE, E; SCHOOFS, L; VERLEYEN, P (2009) The haemolymph proteome of the honey bee: gel-based or gel-free? *Proteomics* 9: 3201-3208. http://dx.doi.org/10.1002/pmic.200800604

BONDAR, R J; MEAD, D C; BONDAR, R J; MEAD, D C (1974) Evaluation of glucose-6-phosphate dehydrogenase from *Leuconostoc mesenteroides* in the hexokinase method for determining glucose in serum. *Clinical Chemistry* 20: 586-590

BOLLENBACHER, W E; O'BRIEN, M A; KATAHIRA, E J; GILBERT, L I (1983) A kinetic analysis of the action of insect prothoracicotropic hormone. *Molecular and Cellular Endocrinology* 32: 47–55. http://dx.doi.org/10.1016/0303-7207(83)90096-5

BOZIC, J; WOODRING, J (1998) Variations of brain biogenic amines in mature honey bees and induction of recruitment behaviour. *Comparative Biochemistry and Physiology* 120A: 737–744. http://dx.doi.org/10.1016/S1095-6433(98)10094-6

BRADFORD, M M (1976) A rapid and sensitive method for the quantitation of microgram quantities of protein utilizing the principle of protein-dye binding. *Analytical Biochemistry* 72: 248–254. http://dx.doi.org/10.1016/0003-2697(76)90527-3

BRANDES, C; SUGAWA, M; MENZEL, R (1990) High-performance liquid chromatography (HPLC) measurement of catecholamines in single honey bee brains reveals caste-specific differences between worker bees and queens in *Apis mellifera*. *Comparative Biochemistry and Physiology* 97C: 53–57. http://dx.doi.org/10.1016/0742-8413(90)90171-5

BRODSCHNEIDER, R; RIESSBERGER-GALLÉ, U; CRAILSHEIM, K (2009) Flight performance of artificially reared honey bees (*Apis mellifera*). *Apidologie* 40: 441–449. http://dx.doi.org/10.1051/apido/2009006

BROUGHTON, S J; PIPER, M D W; IKEYA, T; BASS, T M; JACOBSON, J; DRIEGE, Y; MARTINEZ, P; HAFEN, E; WITHERS, D J; LEEVERS, S J; PARTRIDGE, L (2005) Longer lifespan, altered metabolism, and stress resistance in Drosophila from ablation of cells making insulin-like ligands. *Proceedings of the National Academy of Sciences of the United States of America* 102: 3105-3110. http://dx.doi.org/10.1073_pnas.0405775102

CARRECK, N L; ANDREE, M; BRENT, C S; COX-FOSTER, D; DADE, H A; ELLIS, J D; HATJINA, F; VANENGELSDORP, D (2013) Standard methods for *Apis mellifera* anatomy and dissection. In *V Dietemann; J D Ellis; P Neumann (Eds) The COLOSS* BEEBOOK, *Volume I: standard methods for* Apis mellifera *research. Journal of Apicultural Research* 52(4): http://dx.doi.org/10.3896/IBRA.1.52.4.03

CHAN, Q W; FOSTER, L J (2008) Changes in protein expression during honey bee larval development. *Genome Biology* 9: R156. http://dx.doi.org/10.1186/gb-2008-9-10-r156

CHEN, Z; LINSE, K D; TAUB-MONTEMAYOR, T E; RANKIN, M A (2007) Comparison of radioimmunoassay and liquid chromatography tandem mass spectrometry for determination of juvenile hormone titers. *Insect Biochemistry and Molecular Biology* 37: 799-807. http://dx.doi.org/10.1016/j.ibmb.2007.05.019

CHEN, Y-L; HUNG, Y-S; YANG, E-C (2008) Biogenic amine levels change in the brains of stressed honey bees. *Archives of Insect Biochemistry and Physiology* 68: 241–250. http://dx.doi.org/10.1002/arch.20259

COLLINS, A M; CAPERNA, T J; WILLIAMS, V; GARRETT, W M; EVANS, J D (2006) Proteomic analyses of male contributions to honey bee sperm storage and mating. *Insect Molecular Biology* 15: 541-549. http://dx.doi.org/10.1111/j.1365-2583.2006.00674.x

CORONA, M; VELARDE, R A; REMOLINA, S; MORAN-LAUTER, A; WANG, Y; HUGHES, K A (2007) Vitellogenin, juvenile hormone, insulin signalling, and queen honey bee longevity. *Proceedings of the National Academy of Sciences of the USA* 104: 7128-7133. http://dx.doi.org/10.1073/pnas.0701909104

DE REGGI, M L; HIRN, M H; DELAAGE, M A (1975) Radioimmunoassay of ecdysone: An application to *Drosophila* larvae and pupae. *Biochemical and Biophysical Research Communications* 66: 1307-1315. http://dx.doi.org/10.1016/0006-291X(75)90502-1

DE WILDE, J; BEETSMA, J (1976) The physiology of caste development in social insects. *Advances in Insect Physiology* 16: 167-256.

ESCH, H E (1988) The effects of temperature on flight muscle potentials in honey bees and cuculiinid winter moths. *Journal of Experimental Biology* 135: 109–117.

FAHRENHOLZ, L; LAMPRECHT, I; SCHRICKER, B (1989) Thermal investigations of a honey bee colony: thermoregulation of the hive during summer and winter and heat production of members of different bee castes. *Journal of Comparative Physiology B* 159: 551-560. http://dx.doi.org/10.1007/BF00694379

FAHRENHOLZ, L; LAMPRECHT, L; SCHRICKER, B (1992) Calorimetric investigations of the different castes of honey bees, *Apis mellifera carnica. Journal of Comparative Physiology B* 162: 119–130. http://dx.doi.org/10.1007/BF00398337

FELDLAUFER, M F; HARTFELDER, K (1997) Relationship of the neutral sterols and ecdysteroids of the parasitic mite, *Varroa jacobsoni*, to those of the honey bee, *Apis mellifera. Journal of Insect Physiology* 43: 541-545. http://dx.doi.org/10.1016/S0022-1910(97)00005-X

FELDLAUFER, M F; HERBERT JR, E W; SVOBODA, J A (1985) Makisterone A: The major ecdysteroid from the pupae of the honey bee, *Apis mellifera. Insect Biochemistry* 15: 597–600. http://dx.doi.org/10.1016/0020-1790(85)90120-9

FELDLAUFER, M F; SVOBODA, J A; HERBERT JR, E W (1986) Makisterone A and 24-methylenecholesterol from the ovaries of the honey bee, *Apis mellifera. Experientia* 42: 200–201. http://dx.doi.org/10.1007/BF01952468

FEYEREISEN, R; TOBE, S S (1981) A rapid partition assay for routine analysis of juvenile hormone release by insect *corpora allata. Analytical Biochemistry* 111: 372-375. http://dx.doi.org/10.1016/0003-2697(81)90575-3

FLATT, T; KAWECKI, T J (2007) Juvenile hormone as a regulator of the trade-off between reproduction and life span in *Drosophila melanogaster. Evolution* 61: 1980-1991. http://dx.doi.org/10.1111/j.1558-5646.2007.00151.x

FLATT, T; MIN, K J; D'ALTERIO, C; VILLA-CUESTA, E; CUMBERS, J; LEHMANN, R; JONES, D L; TATAR, M (2008) *Drosophila* germ-line modulation of insulin signalling and lifespan. *Proceedings of the National Academy of Sciences of the USA* 105: 6368-6373. http://dx.doi.org/10.1073_pnas.0709128105

FLORECKI, M M; HARTFELDER, K (2011) Cytoplasmic and nuclear localization of cadherin in honey bee (*Apis mellifera* L.) gonads. *Cell Biology International* 35: 45-49. http://dx.doi.org/10.1042/CBI20100333

FLURI, P; LÜSCHER, M; WILLE, H; GERIG, L (1982) Changes in weight of the pharyngeal gland and haemolymph titres of juvenile hormone, protein and vitellogenin in worker honey bees. *Journal of Insect Physiology* 28: 61-68. http://dx.doi.org/10.1016/0022-1910%2882%2990023-3

FUCHS, E; DUSTMANN, J H; STADLER, H; SCHÜRMANN, F W (1989) Neuroactive compounds in the brain of the honey bee during imaginal life. *Comparative Biochemistry and Physiology* 92C: 337–342. http://dx.doi.org/10.1016/0742-8413(89)90065-0

FUSSNECKER, B L; SMITH, B H; MUSTARD, J A (2006) Octopamine and tyramine influence the behavioural profile of locomotor activity in the honey bee (*Apis mellifera*). *Journal of Insect Physiology* 52: 1083-1092. http://dx.doi.org/10.1016/j.jinsphys.2006.07.008

GAREDEW, A; SCHMOLZ, E; LAMPRECHT, I (2004) The energy and nutritional demand of the parasitic life of the mite *Varroa destructor*. *Apidologie* 35: 419– 430. http://dx.doi.org/10.1051/apido:2004032

GOLLER, F; ESCH, H E (1991) Oxygen consumption and flight muscle activity during heating in workers and drones of *Apis mellifera*. *Journal of Comparative Physiology B* 161: 61–67. http://dx.doi.org/10.1007/BF00258747

GOODMAN, W G (1990) A simplified method for synthesizing juvenile hormone-protein conjugates. *Journal of Lipid Research* 31: 354-357.

GOODMAN, W G; CUSSON, M (2012) The juvenile hormones. In *L I Gilbert (Ed.). Insect Endocrinology.* Academic Press; Oxford, UK. pp. 310-365. http://dx.doi.org/10.1016/B978-0-12-384749-2-10008-1

GOODMAN, W G; COY, D C; BAKER, F C; LEI, X; TOONG, Y C (1990) Development and application of a radioimmunoassay for the juvenile hormones. *Insect Biochemistry* 20: 357-364. http://dx.doi.org/10.1016/0020-1790(90)90055-Y

GOODMAN, W G; HUANG, Z-Y; ROBINSON, G E; STRAMBI, C; STRAMBI, A (1993) Comparison of two juvenile hormone radioimmunoassay. *Archives of Insect Biochemistry and Physiology* 23: 147-152. http://dx.doi.org/10.1002/arch.940230306

GOODMAN, W G; ORTH, A P; TOONG, Y C; EBERSOHL, R; HIRUMA, K; GRANGER, N A (1995) Recent advances in radioimmunoassay technology for the juvenile hormones. *Archives of Insect Biochemistry and Physiology* 30: 295-306. http://dx.doi.org/10.1002/arch.940300215

GRANGER, N A; GOODMAN, W G (1988) Radioimmunoassay: juvenile hormones. In *L I Gilbert; T A Miller (Eds). Immunological Techniques in Insect Biology.* Springer; Heidelberg, Germany. pp. 215-251.

GRAY, E M; BRADLEY, T J (2006) Evidence from mosquitoes suggests that cyclic gas exchange and discontinuous gas exchange are two manifestations of a single respiratory pattern. *Journal of Experimental Biology* 209: 1603–1611. http://dx.doi.org/10.1242/jeb.02181

GUIDUGLI, K R; NASCIMENTO, A M; AMDAM, G V; BARCHUK, A R; OMHOLT, S W; SIMÕES, Z L P; HARTFELDER, K (2005) Vitellogenin regulates hormonal dynamics in the worker caste of a eusocial insect. *FEBS Letters 579: 4961– 4965.* http://dx.doi.org/10.1016/j.febslet.2005.07.085

HAGENGUTH, H; REMBOLD, H (1978) Identification of juvenile hormone 3 as the only juvenile hormone homolog in all developmental stages of the honey bee. *Zeitschrift für Naturforschung* 33C: 847–850.

HANAN, B B (1955) Studies of the retrocerebral complex in the honey bee: Part I: Anatomy and histology. *Annals of the Entomological Society of America* 48: 315-320.

HARRIS, J W; WOODRING, J (1992) Effects of stress, age, season and source colony on levels of octopamine, dopamine and serotonin in the honey bee (*Apis mellifera*) brain. *Journal of Insect Physiology* 38: 29–35. http://dx.doi.org/10.1016/0022-1910(92)90019-A

HARRIS, J W; WOODRING, J (1995) Elevated brain dopamine levels associated with ovary development in queenless worker honey bees (*Apis mellifera* L.). *Comparative Biochemistry and Physiology* 111C: 271–279. http://dx.doi.org/10.1016/0742-8413(95)00048-S

HARRIS, J W; WOODRING, J; HARBO, J R (1996) Effects of carbon dioxide on levels of biogenic amines in the brain of queenless worker and virgin queen honey bees (*Apis mellifera*). *Journal of Apicultural Research* 35: 69–78.

HARRISON, J M (1986) Caste-specific changes in honey bee flight capacity. *Physiological Zoology* 59: 175–187.

HARRISON, J F; HALL, H G (1993) African-European honey bee hybrids have low nonintermediate metabolic capacities. *Nature* 363: 258–260. http://dx.doi.org/10.1038/363258a0

HARRISON, J F; FEWELL, J H; ROBERTS, S P; HALL, H G (1996) Achievement of thermal stability by varying metabolic heat production in flying honey bees. *Science* 274: 88–90. http://dx.doi.org/10.1126/science.274.5284.88

HARRISON, J F; CAMAZINE, S; MARDEN, J H; KIRKTON, S D; ROZO, A; YANG, X (2001) Mite not make it home: tracheal mites reduce the safety margin for oxygen delivery of flying honey bees. *Journal of Experimental Biology* 204: 805–814.

HARTFELDER, K; EMLEN, D J (2012) Endocrine control of insect polyphenisms. In *L I Gilbert (Ed.). Insect Endocrinology.* Elsevier; Oxford, UK. pp. 464-522. http://dx.doi.org/10.1016/B978-0-12-384749-2.10011-1

HARTFELDER, K; ENGELS, W (1998) Social insect polymorphism: hormonal regulation of plasticity in development and reproduction in the honey bee. *Current Topics in Developmental Biology* 40: 45-77.

HARTFELDER, K; BITONDI, M M G; SANTANA, W C; SIMÕES, Z L P (2002) Ecdysteroid titer and reproduction in queens and workers of the honey bee and of a stingless bee: Loss of ecdysteroid function at increasing levels of sociality? *Insect Biochemistry and Molecular Biology* 32: 211-216. http://dx.doi.org/10.1016/S0965-1748(01)00100-X

HAZELL, S P; BALE, J S (2011) Low temperature thresholds: are chill coma and CT_{min} synonymous? *Journal of Insect Physiology* 57: 1085–1089. http://dx.doi.org/10.1016/j.jinsphys.2011.04.004

HAZELL, S P; PEDERSEN, B P; WORLAND, M R; BLACKBURN, T M; BALE, J S (2008) A method for the rapid measurement of thermal tolerance traits in studies of small insects. *Physiological Entomology* 33: 389–394. http://dx.doi.org/10.1111/j.1365-3032.2008.00637.x

HEINRICH, B (1981a) Energetics of honey bee swarm thermoregulation. *Science* 212: 565–566. http://dx.doi.org/10.1126/science.212.4494.565

HEINRICH, B (1981b) The mechanisms and energetics of honey bee swarm temperature regulation. *Journal of Experimental Biology* 91: 25–55.

HEINRICH, B (1993) *The hot-blooded insects*. Springer; Berlin, Germany. p. 601.

HERSCH, M I; CREWE, R M; HEPBURN, H R; THOMPSON, P R; SAVAGE, N (1978) Sequential development of glycolytic competence in the muscles of worker honey bees. *Comparative Biochemistry and Physiology* B61: 427–431.

HETZ, S K; BRADLEY, T J (2005) Insects breathe discontinuously to avoid oxygen toxicity. *Nature* 433: 516–519. http://dx.doi.org/10.1038/nature03106

HOLBROOK, G; CHIANG, A-S; SCHAL, C (1997) Improved conditions for culture of biosynthetically active cockroach corpora allata. In vitro *Cellular and Developmental Biology - Animal* 33: 452–458. http://dx.doi.org/10.1007/s11626-997-0063-9

HRASSNIGG, N; BRODSCHNEIDER, R; FLEISCHMANN, P H; CRAILSHEIM, K (2005) Unlike nectar foragers, honey bee drones (*Apis mellifera*) are not able to utilize starch as fuel for flight. *Apidologie* 36: 547–557. http://dx.doi.org/10.1051/apido:2005042

HUANG, Z Y; ROBINSON, G E; TOBE, S S; YAGI, K J; STRAMBI, C; STRAMBI, A; STAY, B (1991) Hormonal regulation of behavioural development in the honey bee is based on changes in the rate of juvenile hormone biosynthesis. *Journal of Insect Physiology* 37: 733-741. http://dx.doi.org/10.1016/0022-1910(91)90107-B

HUANG Z-Y; ROBINSON, G E; BORST, D W (1994) Physiological correlates of division of labour among similarly aged honey bees. *Journal of Comparative Physiology* A174: 731-739. http://dx.doi.org/10.1007/BF00192722

HUNNICUT, D W; TOONG, Y C; BORST, D W (1989) A chiral-specific antiserum for juvenile hormone. *American Zoologist* 29: 48A.

ISHIKAWA, E; ISHIKAWA, T; MORITA, Y S; TOYONAGA, K; YAMADA, H; TAKEUCHI, O; KINOSHITA, T; AKIRA, S; YOSHIKAI, Y; YAMASAKI, S (2009) Direct recognition of the mycobacterial glycolipid, trehalose dimycolate, by C-type lectin Mincle. *Journal of Experimental Medicine* 206: 2879-2888. http://dx.doi.org/10.1084/jem.20091750

JASSIM, O; HUANG, Z-Y; ROBINSON, G E (2000) Juvenile hormone profiles of worker honey bees, *Apis mellifera,* during normal and accelerated behavioural development. *Journal of Insect Physiology* 46: 243-249. http://dx.doi.org/10.1016/S0022-1910(99)00176-6

KAATZ, H-H, HAGEDORN, H H; ENGELS, W (1985) Culture of honey bee organs: development of a new medium and the importance of tracheation. In vitro *Cellular and Developmental Biology - Animal* 21: 347-352. http://dx.doi.org/10.1007/BF02691583

KÄFER, H; KOVAC, H; STABENTHEINER, A (2012) Resting metabolism and critical thermal maxima of vespine wasps (*Vespula* sp.). *Journal of Insect Physiology* 58: 679–689. http://dx.doi.org/10.1016/j.jinsphys.2012.01.015

KAMAKURA, M (2011) Royalactin induces queen differentiation in honey bees. *Nature* 473: 478-483. http://dx.doi.org/10.1038/nature10093

KEN, T; HEPBURN, H R; RADLOFF, S E; YUSHENG, Y; YIQIU, L; DANYIN, Z; NEUMANN, P (2005) Heat-balling wasps by honey bees. *Naturwissenschaften* 92: 492–495. http://dx.doi.org/10.1007/s00114-005-0026-5

KLEINHENZ, M; BUJOK, B; FUCHS, S; TAUTZ, J (2003) Hot bees in empty broodnest cells: Heating from within. *Journal of Experimental Biology* 206: 4217–4231. http://dx.doi.org/10.1242/jeb.00680

KOVAC, H; SCHMARANZER, S (1996) Thermoregulation of honey bees (*Apis mellifera*) foraging in spring and summer at different plants. *Journal of Insect Physiology* 42: 1071–1076. http://dx.doi.org/10.1016/S0022-1910(96)00061-3

KOVAC, H; STABENTHEINER, A; HETZ, S K; PETZ, M; CRAILSHEIM, K (2007) Respiration of resting honey bees. *Journal of Insect Physiology* 53: 1250–1261. http://dx.doi.org/10.1016/j.jinsphys.2007.06.019

KOVAC, H; STABENTHEINER, A; SCHMARANZER, S (2010) Thermoregulation of water foraging honey bees – balancing of endothermic activity with radiative heat gain and functional requirements. *Journal of Insect Physiology* 56: 1834–1845. http://dx.doi.org/10.1016/j.jinsphys.2010.08.002

KRONENBERG, F; HELLER, C (1982) Colonial thermoregulation in honey bees (*Apis mellifera*). *Journal of Comparative Physiology* 148: 65–76.

LAFONT, R; DAUPHIN-VILLEMONT, C; WARREN, J T; REES, H (2012) Ecdysteroid chemistry and biochemistry. In *L I Gilbert (Ed.)*. *Insect Endocrinology*. Academic Press; Oxford, UK. pp.106-176. http://dx.doi.org/10.1016/B978-0-12-384749-2-10004-4

LAEMMLI, U K (1970) Cleavage of structural proteins during assembly of the head of bacteriophage-T4. *Nature* 227: 680–682. http://dx.doi.org/10.1038/227680a0

LAMPRECHT, I; SCHMOLZ, E (2000) Calorimetry goes afield. *Thermochimica Acta* 355: 95–106. http://dx.doi.org/10.1016/S0040-6031(00)00440-8

LI, J K; FENG, M; ZHANG, L; ZHANG, Z H; PAN, Y H (2008) Proteomics analysis of major royal jelly protein changes under different storage conditions. *Journal of Proteome Research* 7: 3339-3353. http://dx.doi.org/10.1021/pr8002276

LI, Y T; WARREN, J T; BOYSEN, G; GILBERT, L I; GOLD, A; SANGAIAH, R; BALL, L M; SWENBERG, J A (2006) Profiling of ecdysteroids in complex biological samples using liquid chromatography/ion trap mass spectrometry. *Rapid Communications in Mass Spectrometry* 20: 185-192. http://dx.doi.org/10.1002/rcm.2294

LIGHTON, J R B (2008) *Measuring metabolic rates*. Oxford University Press; New York, USA.

LIGHTON, J R B; HALSEY, L G (2011) Flow-through respirometry applied to chamber systems: pros and cons, hints and tips. *Comparative Biochemistry and Physiology* A158: 265–275. http://dx.doi.org/10.1016/j.cbpa.2010.11.026

LIGHTON, J R B; LOVEGROVE, B G (1990) A temperature-induced switch from diffusive to convective ventilation in the honey bee. *Journal of Experimental Biology* 154: 509–516.

LIGHTON, J R B; TURNER, R J (2004) Thermolimit respirometry: an objective assessment of critical thermal maxima in two sympatric desert harvester ants, *Pogonomyrmex rugosus* and *P. californicus*. *Journal of Experimental Biology* 207: 1903–1913. http://dx.doi.org/10.1242/jeb.00970

LORENZ, M W; KELLNER, R; WOODRING, J; HOFFMANN, K H; GADE, G (1999) Hypertrehalosaemic peptides in the honey bee (*Apis mellifera*): purification, identification and function. *Journal of Insect Physiology* 45: 647–53. http://dx.doi.org/10.1016/S0022-1910(98)00158-9

LOWRY, O H; ROSEBBROUGH, N J; FARR, A L; RANDALL, R J (1951) Protein measurement with the Folin Phenol Reagent. *Journal of Biological Chemistry* 193: 265–275.

MACKASMIEL, L A M; FELL, R D (2000) Respiration rates in eggs of the honey bee, *Apis mellifera*. *Journal of Apicultural Research* 39: 125–135.

MARCO ANTONIO, D S; GUIDUGLI-LAZZARINI, K R; NASCIMENTO, A; SIMÕES, Z L P; HARTFELDER, K (2008) RNAi-mediated silencing of vitellogenin function turns honey bee (*Apis mellifera*) workers into extremely precocious foragers. *Naturwissenschaften* 95: 953-961. http://dx.doi.org/10.1007/s00114-008-0413-9

MARTINS, J R; NUNES, F M F; CRISTINO, A S; SIMÕES, ZLP; BITONDI, M M G (2010) The four hexamerin genes in the honey bee: Structure, molecular evolution and function deduced from expression patterns in queens, workers and drones. *BMC Molecular Biology* 11: e23. http://dx.doi.org/10.1186/1471-2199-11-23

MELAMPY, R M; WILLIS, E R (1939) Respiratory metabolism during larval and pupal development of the female honey bee (*Apis mellifica* L.). *Physiological Zoology* 12: 302–311.

METZLER, M; MEYER, D; DAHM, K H; RÖLLER, H (1972) Biosynthesis of juvenile hormone from 10-epoxy-7-ethyl-3,1 l-dimethyl-2,6-dodecadenoic acid in the adult cecropia moth. *Zeitschrift für Naturforschung* 27b: 321-322.

MOFFATT, L (2000) Changes in the metabolic rate of the foraging honey bee: effect of the carried weight or of the reward rate? *Journal of Comparative Physiology* A186: 299–306. http://dx.doi.org/10.1007/s003590050430

MOFFATT, L (2001) Metabolic rate and thermal stability during honey bee foraging at different reward rates. *Journal of Experimental Biology* 204: 759–766.

MOFFATT, L; NÚÑEZ, J A (1997) Oxygen consumption in the foraging honey bee depends on the reward rate at the food source. *Journal of Comparative Physiology* B167: 36–42. http://dx.doi.org/10.1007/s003600050045

MORITZ, R F A (1988) Biochemical changes during honey bee flight muscle development. *Biona-report* 6: 51–64.

MORITZ, R F A; SOUTHWICK, E E (1986) Analysis of queen recognition by honey bee workers (*Apis mellifera* L.) in a metabolic bio-assay. *Experimental Biology* 46: 45–49.

NASCIMENTO, A M; CUVILLIER-HOT, V; BARCHUK, A R; SIMÕES, Z L P; HARTFELDER, K (2004) Honey bee (*Apis mellifera*) transferrin - gene structure and the role of ecdysteroids in the developmental regulation of its expression. *Insect Biochemistry and Molecular Biology* 34: 415-424. http://dx.doi.org/10.1016/j.ibmb.2003.12.003

NELSON, C M; IHLE, K E; FONDRK, M K; PAGE JR, R E; AMDAM, G V (2007) The gene vitellogenin has multiple coordinating effects on social organization *PLoS Biology* 5: 673-677. http://dx.doi.org/10.1371/journal.pbio.0050062

NIJHOUT, H F (1994) *Insect hormones*. Princeton University Press; Princeton, NJ, USA.

NIJHOUT, H F; WHEELER, D E (1982) Juvenile hormone and the physiological basis of insect polymorphisms. *Quarterly Reviews of Biology* 57: 109-133. http://dx.doi.org/10.1086/412671

ONO, M; IGARASHI, T; OHNO, E; SASAKI, M (1995) Unusual thermal defence by a honey bee against mass attack by hornets. *Nature* 377: 334–336. http://dx.doi.org/10.1038/377334a0

PANKIW, T; PAGE JR, R E (2003) Effect of pheromones, hormones, and handling on sucrose response thresholds of honey bees (*Apis mellifera* L.). *Journal of Comparative Physiology* A189: 675-684. http://dx.doi.org/10.1007/s00359-003-0442-y

PETERSON, J I; YOUNG, D S (1968) Evaluation of the hexokinase-glucose-6-phosphate dehydrogenase method of determination of glucose in urine. *Analytical Biochemistry* 23: 301-316. http://dx.doi.org/10.1016/0003-2697(68)90361-8

PETZ, M; STABENTHEINER, A; CRAILSHEIM, K (2004) Respiration of individual honey bee larvae in relation to age and ambient temperature. *Journal of Comparative Physiology* B174: 511–518. http://dx.doi.org/10.1007/s00360-004-0439-z

PINTO, L Z; BITONDI, M M G; SIMÕES, Z L P (2000) Inhibition of vitellogenin synthesis in *Apis mellifera* workers by a juvenile hormone analogue, pyriproxyfen. *Journal of Insect Physiology 46: 153–160.* http://dx.doi.org/10.1016/S0022-1910(99)00111-0

PINTO, L Z; HARTFELDER, K; BITONDI, M M G; SIMÕES, Z L P (2002) Ecdysteroid titers in pupae of highly social bees relate to distinct modes of caste development. *Journal of Insect Physiology* 48: 783-790. http://dx.doi.org/10.1016/S0022-1910(02)00103-8

PRATT, G E; TOBE, S S (1974) Juvenile hormone radiobiosynthesized by corpora allata of adult female locusts *in vitro*. *Life Sciences* 14: 575-586. http://dx.doi.org/10.1016/0024-3205(74)90372-5

RACHINSKY, A; HARTFELDER, K (1990) Corpora allata activity, a prime regulating element for caste-specific juvenile hormone titre in honey bee larvae (*Apis mellifera carnica*). *Journal of Insect Physiology* 36: 189-194. http://dx.doi.org/10.1016/0022-1910(90)90121-U

RACHINSKY, A; HARTFELDER K (1998) *In vitro* biosynthesis of juvenile hormone in larval honey bees: comparison of six media. In Vitro *Cellular and Developmental Biology - Animal* 34: 646-648. http://dx.doi.org/10.1007/s11626-996-0014-x

RACHINSKY, A; STRAMBI, C; STRAMBI, A; HARTFELDER, K (1990) Caste and metamorphosis: haemolymph titers of juvenile hormone and ecdysteroids in last instar honey bee larvae. *General and Comparative Endocrinology* 79: 31-38. http://dx.doi.org/10.1016/0016-6480(90)90085-Z

RIVERA-PEREZ, C; NOUZOVA, M; NORIEGA, F G (2012) A quantitative assay for the juvenile hormones and their precursors using fluorescent tags. *PLoS One* 7: e43784. http://dx.doi.org/10.1371/journal.pone.0043784

ROBERTS, S P; HARRISON, J F (1999) Mechanisms of thermal stability during flight in the honey bee, *Apis mellifera*. *Journal of Experimental Biology* 202: 1523–1533.

ROBINSON, G E (1992) Regulation of division of labour in insect societies. *Annual Reviews of Entomology* 37: 637-665. http://dx.doi.org/10.1146/annurev.ento.37.1.637

ROBINSON, G E; VARGO, E L (1997) Juvenile hormone in the social Hymenoptera: gonadotropin and behavioural pacemaker. *Archives of Insect Biochemistry and Physiology* 35: 559–583. http://dx.doi.org/10.1002/(SICI)1520-6327

ROBINSON, G E; STRAMBI, A, STRAMBI, C; PAULINO-SIMÕES, Z L; TOZETTO, S O; BARBOSA, J M N (1987) Juvenile hormone titers in Africanized and European honey bees in Brazil. *General and Comparative Endocrinology* 66: 457-459. http://dx.doi.org/10.1016/0016-6480(87)90258-9

ROBINSON, G E; STRAMBI, C; STRAMBI, A; FELDLAUFER, M F (1991) Comparison of juvenile hormone and ecdysteroid haemolymph titres in adult worker and queen honey bees (*Apis mellifera*). *Journal of Insect Physiology* 37: 929–936. http://dx.doi.org/10.1016/0022-1910(91)90008-N

ROTHE, U; NACHTIGALL, W (1989) Flight of the honey bee. IV. Respiratory quotients and metabolic rates during sitting, walking and flying. *Journal of Comparative Physiology* B158: 739–749. http://dx.doi.org/10.1007/BF00693012

SANTOS, K S; DOS SANTOS, L D; MENDES, M A; DE SOUZA, B M; MALASPINA, O; PALMA, M S (2005) Profiling the proteome complement of the secretion from hypopharyngeal gland of Africanized nurse-honey bees (*Apis mellifera* L.). *Insect Biochemistry and Molecular Biology* 35: 85-91. http://dx.doi.org/10.1016/j.ibmb.2004.10.003

SCHEINER, R; PLÜCKHAHN, S; ÖNEY, B; BLENAU, W; ERBER, J (2002) Behavioural pharmacology of octopamine, tyramine and dopamine in honey bees. *Behavioural Brain Research* 136: 545–553. http://dx.doi.org/10.1016/S0166-4328(02)00205-X

SCHEINER, R; BAUMANN, A; BLENAU, W (2006) Aminergic control and modulation of honey bee behaviour. *Current Neuropharmacology* 4: 259-276.

SCHEINER, R; ABRAMSON, C I; BRODSCHNEIDER, R; CRAILSHEIM, K; FARINA, W; FUCHS, S; GRÜNEWALD, B; HAHSHOLD, S; KARRER, M; KOENIGER, G; KOENIGER, N; MENZEL, R; MUJAGIC, S; RADSPIELER, G; SCHMICKLI, T; SCHNEIDER, C; SIEGEL, A J; SZOPEK, M; THENIUS, R (2013) Standard methods for behavioural studies of *Apis mellifera*. In *V Dietemann; J D Ellis; P Neumann (Eds) The COLOSS* BEEBOOK, *Volume I: standard methods for* Apis mellifera *research. Journal of Apicultural Research* 52(4): http://dx.doi.org/10.3896/IBRA.1.52.4.04

SCHMIDT CAPELLA, I C; HARTFELDER, K (2002) Juvenile-hormone-dependent interaction of actin and spectrin is crucial for polymorphic differentiation of the larval honey bee ovary. *Cell and Tissue Research* 307: 265–272. http://dx.doi.org/10.1007/s00441-001-0490-y

SCHMOLZ, E; LAMPRECHT, I (2000) Calorimetric investigations on activity states and development of holometabolous insects. *Thermochimica Acta* 349: 61–68. http://dx.doi.org/10.1016/S0040-6031(99)00497-9

SCHMOLZ, E; HOFFMEISTER, D; LAMPRECHT, I (2002) Calorimetric investigations on metabolic rates and thermoregulation of sleeping honey bees (*Apis mellifera carnica*). *Thermochimica Acta* 382: 221–227. http://dx.doi.org/10.1016/S0040-6031(01)00740-7

SCHULZ, D J; ELEKONICH, M M; ROBINSON, G E (2002) Biogenic amines in the antennal lobes and the initiation and maintenance of foraging behaviour in honey bee. *Journal of Neurobiology* 54: 406–416. http://dx.doi.org/10.1002/neu.10138

SHU, S; PARK, Y I; RAMASWAMY, S B; SRINIVASAN, A (1997) Haemolymph juvenile hormone titers in pupal and adult stages of southwestern corn borer [*Diatraea grandiosella* (Pyralidae)] and relationship with egg development. *Journal of Insect Physiology* 43: 719-726. http://dx.doi.org/10.1016/S0022-1910(97)00048-6

SMITH, P K; KROHN, R I; HERMANSON, G T; MALLIA, A K; GARTNER, F H; PROVENZANO, M D; FUJIMOTO, E K; GOEKE, N M; OLSON, B J; KLENK, D C (1985) Measurement of protein using bicinchoninic acid. *Analytical Biochemistry* 150: 76–85. http://dx.doi.org/10.1016/0003-2697(85)90442-7

SPARROW, E M; CESS, R D (1970). *Radiation heat transfer.* Brooks/ Cole Publ. Comp.; Belmont, California, USA. 340 pp.

SPEAKMAN, J R (1997) *Doubly labelled water: theory and practice.* Chapman and Hall; London, UK.

SPEAKMAN, J R (1998) The history and theory of the doubly labelled water technique. *American Journal of Clinical Nutrition* 68: 932S–938S.

SOUTHWICK, E E (1985) Allometric relations, metabolism and heat conductance in clusters of honey bees at cool temperatures. *Journal of Comparative Physiology* B156: 143–149. http://dx.doi.org/10.1007/BF00692937

SOUTHWICK, E E (1988) Thermoregulation in honey-bee colonies. In *G R Needham; E R Page Jr.; M Delfinado-Baker; C E Bowman (Eds). African honey bees and bee mites.* Halsted Press; New York, USA. pp. 223-236.

SOUTHWICK, E E; MORITZ, R F A (1985) Metabolic response to alarm pheromone in honey bees. *Journal of Insect Physiology* 31: 389–392. http://dx.doi.org/10.1016/0022-1910(85)90083-6

SOUTHWICK, E E; MORITZ, R F A (1987) Social synchronisation of circadian rhythms of metabolism in honey bees (*Apis mellifera*). *Physiological Entomology* 12: 209–212. http://dx.doi.org/10.1111/j.1365-3032.1987.tb00743.x

STABENTHEINER, A; SCHMARANZER, S (1987) Thermographic determination of body temperatures in honey bees and hornets: Calibration and applications. *Thermology* 2: 563–572.

STABENTHEINER, A; KOVAC, H; HAGMÜLLER, K (1995) Thermal behaviour of round and wagtail dancing honey bees. *Journal of Comparative Physiology* B165: 433–444. http://dx.doi.org/10.1007/BF00261297

STABENTHEINER, A; VOLLMANN, J; KOVAC, H; CRAILSHEIM, K (2003a) Oxygen consumption and body temperature of active and resting honey bees. *Journal of Insect Physiology* 49: 881–889. http://dx.doi.org/10.1016/S0022-1910(03)00148-3

STABENTHEINER, A; PRESSL, H; PAPST, T; HRASSNIGG, N; CRAILSHEIM, K (2003b) Endothermic heat production in honey bee winter clusters. *Journal of Experimental Biology* 206: 353–358. http://dx.doi.org/10.1242/jeb.00082

STABENTHEINER, A; KOVAC, H; BRODSCHNEIDER, R (2010) Honey bee colony thermoregulation – Regulatory mechanisms and contribution of individuals in dependence on age, location and thermal stress. *PLoS One* 5: e8967. http://dx.doi.org/10.1371/journal.pone.0008967

STABENTHEINER, A; KOVAC, H; HETZ, S K; KÄFER, H; STABENTHEINER, G (2012) Assessing honey bee and wasp thermoregulation and energetics - New insights by combination of flow-through respirometry with infrared thermography. *Thermochimica Acta* 53: 77–86. http://dx.doi.org/10.1016/j.tca.2012.02.006

STONE, G N; WILLMER, P G (1989) Endothermy and temperature regulation in bees: a critique of 'grab and stab' measurement of body temperature. *Journal of Experimental Biology* 143: 211–223.

STRAMBI, A; STRAMBI, C; REGGI, M L; HIRN, M; DELAAGE, M A (1981) Radioimmunoassay of insect juvenile hormone and of their diol derivatives. *European Journal of Biochemistry* 118: 401-406. http://dx.doi.org/10.1111/j.1432-1033.1981.tb06416.x

SUAREZ, R K (2000) Energy metabolism during insect flight: biochemical design and physiological performance. *Physiological and Biochemical Zoology* 73: 765–771. http://dx.doi.org/10.1086/318112

SUAREZ, R K; LIGHTON, J R B; JOOS, B; ROBERTS, S P; HARRISON, J F (1996) Energy metabolism, enzymatic flux capacities, and metabolic flux rates in flying honey bees. *Proceedings of the National Academy of Sciences of the USA* 93: 12616–12620.

SUAREZ, R K; STAPLES, J F; LIGHTON, J R B; MATHIEU-COSTELLO, O (2000) Mitochondrial function in flying honey bees (*Apis mellifera*): respiratory chain enzymes and electron flow from complex III to oxygen. *Journal of Experimental Biology* 203: 905–911.

TANAKA, E D; HARTFELDER, K (2004) The initial stages of oogenesis and their relation to differential fertility in honey bee (*Apis mellifera*) castes. *Arthropod Structure and Development* 3: 431-442. http://dx.doi.org/10.1016/j.asd.2004.06.006

TAUTZ, J; MAIER, S; GROH, C; ROESSLER, W; BROCKMANN, A (2003) Behavioural performance in adult honey bees is influenced by the temperature experienced during their pupal development. *Proceedings of the National Academy of Sciences of the USA* 100: 7343–7347. http://dx.doi.org/10.1073/pnas.1232346100

TAYLOR, D J; ROBINSON, G E; LOGAN, B J; LAVERTY, R; MERCER, A R (1992) Changes in brain amine levels associated with the morphological and behavioural development of the worker honey bee. *Journal of Comparative Physiology* 170A: 715–712. http://dx.doi.org/10.1007/BF00198982

TERBLANCHE, J S; CHOWN, S L (2010) Effects of flow rate and temperature on cyclic gas exchange in tsetse flies (Diptera, Glossinidae). *Journal of Insect Physiology* 56: 513–521. http://dx.doi.org/10.1016/j.jinsphys.2009.02.005

TERBLANCHE, J S; HOFFMANN, A A; MITCHELL, K A; RAKO, L; LE ROUX, P C; CHOWN, S L (2011) Ecologically relevant measures of tolerance to potentially lethal temperatures. *Journal of Experimental Biology* 214: 3713–3725. http://dx.doi.org/10.1242/jeb.061283

TOBE, S S; PRATT, G E (1974) The influence of substrate concentrations on the rate of insect juvenile hormone biosynthesis by corpora allata of the desert locust *in vitro. Biochemical Journal* 144: 107-113.

TORTO, B; CARROLL, M J; DUEHL, A; FOMBONG, A T; NAZZI, F; GOZANSKY, K T; SOROKER, V; TEAL, P E A (2013) Standard methods for chemical ecology research in *Apis mellifera*. In *V Dietemann; J D Ellis; P Neumann (Eds) The COLOSS* BEEBOOK, *Volume I: standard methods for* Apis mellifera *research. Journal of Apicultural Research* 52(4): http://dx.doi.org/10.3896/IBRA.1.52.4.06

TOWBIN, H; STAEHELIN, T; GORDON J (1979) Electrophoretic transfer of proteins from polyacrylamide gels to nitrocellulose sheets: Procedure and some applications. *Proceedings of the National Academy of Sciences of the USA* 76: 4350–4354.

TRAUTMANN, K H; MASNER, P; SCHULER, A; SUCHY, M; WIPF, K-H (1974) Evidence of the juvenile hormone methyl(2E,6E)-10,11-epoxy-3,7,11-trimehtyl-2,6 dodecadienoate (JH-3) in insects of four orders. *Zeitschrift für Naturforschung* C29: 757-759.

VAN NERUM, K; BUELENS, H (1997) Hypoxia-controlled winter metabolism in honey bees. *Comparative Biochemistry and Physiology A* 117: 445–455. http://dx.doi.org/10.1016/S0300-9629(96)00082-5

WAGENER-HULME, C; KUEHN, J C; SCHULZ, D J; ROBINSON, G E (1999) Biogenic amines and division of labour in honey bee colonies. *Journal of Comparative Physiology* 184A: 471–479. http://dx.doi.org/10.1007/s003590050347

WANG, Y; BRENT, C S; FENNERN, E; AMDAM, G V (2012) Gustatory perception and fat body energy metabolism are jointly affected by vitellogenin and juvenile hormone in honey bees. *PLoS Genetics* 8: e1002779. http://dx.doi.org/10.1371/journal.pgen.1002779

WARREN, J T; GILBERT, L I (1986) Ecdysone metabolism and distribution during the pupal-adult development of *Manduca sexta. Insect Biochemistry* 16: 65–82. http://dx.doi.org/10.1016/0020-1790(86)90080-6

WESTERLUND, S A; HOFFMANN, K H (2004) Rapid quantification of juvenile hormones and their metabolites in insect haemolymph by liquid chromatography-mass spectrometry (LC-MS). *Analytical and Bioanalytical Chemistry* 379: 540-543. http://dx.doi.org/10.1007/s00216-004-2598-x

WOLF, T J; SCHMID-HEMPEL, P; ELLINGTON, C P; STEVENSON, R D (1989) Physiological correlates of foraging efforts in honey bees. *Functional Ecology* 3: 417–424.

WOLF, T J; ELLINGTON, C P; DAVIS, S; FELTHAM, M J (1996) Validation of the doubly labelled water technique for bumble bees *Bombus terrestris* (L.). *Journal of Experimental Biology* 199: 959–972.

WOODS, W A; HEINRICH, B; STEVENSON, R D (2005) Honey bee flight metabolic rate: does it depend upon air temperature? *Journal of Experimental Biology* 208: 1161–1173. http://dx.doi.org/10.1242/jeb.01510

ZERA, A J (2007). Endocrine analysis in evolutionary-developmental studies of insect polymorphism: hormone manipulation versus direct measurement of hormonal regulators. *Evolution and Development* 9: 499-513. http://dx.doi.org/10.1111/j.1525-142X.2007.00181.x

ZHOU, J; QI, Y; HOU, Y; ZHAO, J; LI, Y; XUE, X; WU, L; ZHANG, J; CHEN, F (2011) Quantitative determination of juvenile hormone III and 20-hydroxyecdysone in queen larvae and drone pupae of *Apis mellifera* by ultrasonic-assisted extraction and liquid chromatography with electrospray ionization tandem mass spectrometry. *Journal of Chromatography B-Analytical Technologies in the Biomedical and Life Sciences* 879: 2533-2541. http://dx.doi.org/10.1016/j.jchromb.2011.07.006

Journal of Apicultural Research 52(1)

Journal of Apicultural Research 52(4): (2013)
DOI 10.3896/IBRA.1.52.4.12

REVIEW ARTICLE

Standard methods for pollination research with *Apis mellifera*

Keith S Delaplane[1*], Arnon Dag[2], Robert G Danka[3], Breno M Freitas[4], Lucas A Garibaldi[5], R Mark Goodwin[6] and Jose I Hormaza[7]

[1]Department of Entomology, University of Georgia, Athens, GA 30602, USA.
[2]Gilat Research Center, Agricultural Research Organization, Ministry of Agriculture, Mobile Post Negev 85280, Israel.
[3]Honey Bee Breeding, Genetics, and Physiology Research, 1157 Ben Hur Road, Baton Rouge, LA 70820, USA.
[4]Departamento de Zootecnia - CCA, Universidade Federal do Ceará, C.P. 12168, Fortaleza – CE, 60.021-970, Brazil.
[5]Sede Andina, Universidad Nacional de Río Negro (UNRN) and Consejo Nacional de Investigaciones Científicas y Técnicas (CONICET), Mitre 630, CP 8400, San Carlos de Bariloche, Río Negro, Argentina.
[6]The New Zealand Institute for Plant and Food Research Limited, Plant and Food Research Ruakura, Private Bag 3123, Hamilton 3240, New Zealand.
[7]Instituto de Hortofruticultura Subtropical y Mediterranea La Mayora (IHSM La Mayora-CSIC-UMA), 29750 Algarrobo-Costa, Málaga, Spain.

Received 30 October 2012, accepted subject to revision 12 February 2013, accepted for publication 20 June 2013.

*Corresponding author: Email: ksd@uga.edu

Summary

In this chapter we present a synthesis of recommendations for conducting field experiments with honey bees in the context of agricultural pollination. We begin with an overview of methods for determining the mating system requirements of plants and the efficacy of specific pollinators. We describe methods for evaluating the pollen-vectoring capacity of bees at the level of individuals or colonies and follow with methods for determining optimum colony field stocking densities. We include sections for determining post-harvest effects of pollination, the effects of colony management (including glasshouse enclosure) on bee pollination performance, and a brief section on considerations about pesticides and their impact on pollinator performance. A final section gives guidance on determining the economic valuation of honey bee colony inputs at the scale of the farm or region.

Métodos estándar para el estudio de polinización con *Apis mellifera*

Resumen

En este capítulo se presenta una síntesis de las recomendaciones para la realización de experimentos de campo con abejas melíferas en el contexto de la polinización agrícola. Comienza con una revisión de los métodos para la determinación de los requisitos del sistema de reproducción de las plantas y de la eficacia de los polinizadores específicos. Se describen métodos para evaluar la capacidad de las abejas como vectores de polen a los niveles de individuos o de colonias, y se continúa con los métodos para la determinación de las densidades óptimas de colonias en campo. Se incluyen secciones para la determinación de los efectos de la polinización en la cosecha, los efectos del manejo de las colonias (incluyendo el cercado en invernaderos) en el rendimiento de polinización de las abejas, y una breve sección sobre consideraciones acerca de los plaguicidas y su impacto en el rendimiento de los polinizadores. Una última sección ofrece una guía para la determinación del valor económico de los gastos de las colonias de abejas melíferas a escala de explotación o de región.

Footnote: Please cite this paper as: DELAPLANE, K S; DAG, A; DANKA, R G; FREITAS, B M; GARIBALDI, L A; GOODWIN, R M; HORMAZA, J I (2013) Standard methods for pollination research with *Apis mellifera*. In *V Dietemann; J D Ellis; P Neumann (Eds) The COLOSS* BEEBOOK, *Volume I: standard methods for* Apis mellifera *research. Journal of Apicultural Research* 52(4): http://dx.doi.org/10.3896/IBRA.1.52.4.12

西方蜜蜂授粉研究的标准方法

摘要

本章给出了蜜蜂授粉田间试验的综合推荐规范。文章开篇概述了测定植物交配系统需求和测定特定授粉者效率的方法。介绍了在个体或群体水平评估蜜蜂的花粉媒介能力的方法，以及确定蜂群田间最适饲养密度的方法。本章节还包括测定授粉的"采摘后"效应和蜂群管理（包括温室环境）对蜜蜂授粉表现的影响，并且简要叙述了对于农药的担忧及农药对授粉表现的影响。最后一节给出了如何在农场或地区层面测定蜂群贡献的经济价值。

Keywords: COLOSS, *BEEBOOK*, honey bee, *Apis mellifera*, pollination, pollen load, pollen deposition, pollination efficacy, pollinator density, crop pollination metrics, pesticides, greenhouse, economic value

Table of Contents

Table of Contents cont'd

1. Introduction

This chapter describes field and lab procedures for doing experiments on honey bee pollination. Most of the methods apply to any insect for which pollen vectoring capacity is the question. What makes honey bee pollination distinctive is its historic emphasis on agricultural applications; hence one finds a preoccupation with matters of bee densities, behaviours, and management with a view to optimizing crop yields and quality. However, the same methods can be modified to address broader questions on plant fitness and ecosystem-level interactions.

2. Plant pollination requirements

The impact of any pollinator, whether in terms of agricultural production or plant fitness, is an interaction between at least two dynamics: the pollen vectoring capacity of the flower visitor and the genetic obligation, or responsiveness, of the plant to pollen deposition on its stigmas (Delaplane, 2011). Most of this chapter is devoted to appraising pollen vectoring capacity, but in this section we begin with the underlying demands of the plant because this is the necessary starting point for understanding and contextualizing any pollination syndrome: the suite of flower characters derived by natural selection in response to pollinating agents, whether biotic or abiotic (see Faegri and Pijl, 1979).

To begin, *pollination* is the transfer of pollen from the anthers to the stigma of flowers of the same species and is essential to the reproduction of most angiosperms (flowering plants). Pollination success is often measured in terms of percentage fruit- or seed-set. Fruit- or seed-set is the ratio of ripe fruit or seeds relative to initial number of available flowers or ovules, respectively. This ratio is rarely 100% owing to such factors as normal levels of fruit abortion, suboptimal pollination conditions, herbivory, or cultural problems.

The degree to which a plant species depends on a particular pollinator is determined in part by the mating and breeding system of the plant (Fig. 1). Some plants can produce seeds or fruits without pollination, and understanding this process is important for understanding when the honey bee can or cannot contribute to fruit- or seed-set and yield enhancement. Asexual reproduction through non-fertilized seeds is called *apomixis* or *agamospermy*. Apomixis happens when an embryo is formed either from an unfertilized egg within a diploid embryo sac that was formed without completing meiosis (blackberries, dandelions) or from the diploid nucleus tissue surrounding the embryo sac (some *Citrus* species, some mango varieties). When fruit forms without fertilization of ovules, either naturally or chemically-induced, this is called *vegetative parthenocarpy* (banana, pineapple, seedless cucumber). In either apomixis or parthenocarpy no fertilization occurs, and pollination is not required. However, in some plant species, pollination or some other stimulation is required to produce parthenocarpic fruits, a chief example being seedless watermelon, a type of stimulative parthenocarpy. Also, in many apomitic plants apomixis does not always occur, or occurs only partially, and sexual reproduction can also take place (*Citrus* and mango).

Most angiosperms, however, need pollination to set seeds and fruits, and with the exception of those whose flowers are capable of autopollinating (ex. many beans, soybean, peach, peanuts), they rely on agents to vector the pollen. Angiosperms have basically two mating systems: outcrossing (*xenogamy*) in which pollination occurs between plants with different genetic constitutions, or selfing (*autogamy*) in which no mixing of different genetic material occurs other than through recombination. Outcrossing is achieved by *cross pollination*, resulting from the transfer of pollen between different flowers of different plants of the same species, while *selfing* is the outcome of pollen transfer within the same flower (*self-pollination*) or between different flowers of the same plant (*geitonogamy*). Some plant species are strictly xenogamous while others are autogamous, but mixed mating systems in which plants use outcrossing and autogamy or even outcrossing, autogamy and agamospermy are not uncommon (Rizzardo *et al.*, 2012).

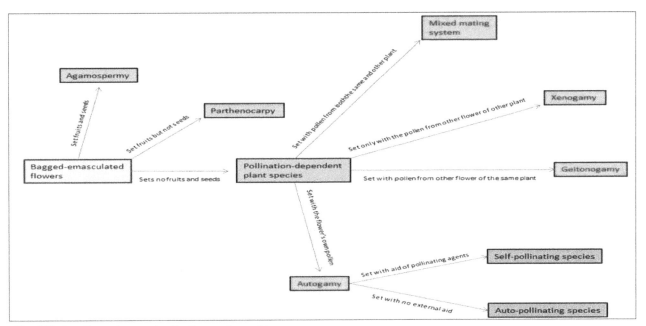

Fig. 1. Plant mating systems and pollination requirements.

The extent to which an angiosperm responds to pollination and the fraction of that pollination that is selfed or out-crossed vary greatly by plant species or variety, and in any particular case a flower visitor must meet specific needs to qualify as a legitimate pollinator. We describe below some field methods for determining the mating system and pollination requirements of plants and the potential pollination role of abiotic and biotic agents, focusing on the level of the individual plant rather than the plant population and drawing heavily upon the following published works (Spears, 1983; Mesquida *et al.*, 1988; Freitas and Paxton, 1996, 1998; Sampson and Cane, 2000; Dafni *et al.*, 2005; Pierre *et al.*, 2010, Vaissière *et al.*, 2011).

2.1. Determining plant mating system

When trying to determine a plant mating system, one can use each of the methods described here as experimental treatments or select only those that appear most relevant to the plant species of interest. In all cases, a positive control in which flowers are marked but otherwise left available for open pollination is necessary to provide a reference for comparison with the manipulative treatments (Fig. 2). In some cases it is also necessary to provide a negative control in which flowers are excluded from all flower visitors for the duration of their dehiscence. It is preferable to reduce background variation by applying distinct treatments to flowers of the same inflorescence, branch, or plant depending on flower abundance and size of the plant.

In the following sections, the performance of a pollinator is implied by the field-scale observation of subsequent fruit- or seed-set. It is also appropriate to measure pollen vectoring capacity at the level of viable pollen on the bee and pollen deposited by the bee onto the stigma. These techniques are covered in sections 3.1. and 3.2.

Fig. 2. Open pollination treatment in soybean plantation: flowers are marked and left open for floral visitors.

2.1.1. Testing for agamospermy (asexual reproduction through non-fertilized seeds)

This test will tell us whether a plant species sets seeds without pollination. If this is so, honey bees cannot contribute to seed- or fruit-set.

1. Choose a given number of flower buds prior to anthesis. The number of buds may vary with availability and ease of access, but larger samples produce more reliable results.
2. Protect half of these buds with pollination bags (Fig. 3) and leave the other half unbagged as a control. Pollination bags are typically made of sheer nylon or similar fine fabric that excludes insects but permits entry of air and light. They are usually semi-transparent nylon and have draw strings to secure the bag around the flower pedicel. The flower should be positioned as much as possible in the centre of the bag so that the mesh does not touch the flower, which could lead to self-pollination. To limit self-pollination further, fix a wire frame around the flower and place the bag over the frame, thus providing structural support to the bag. Identify each treatment with weather-resistant tags. Testing for agamospermy can also be done in a greenhouse without exclusion bags and is thereby easier.

Fig. 3. Restricted pollination treatment: a watermelon flower is bagged throughout its life to prevent honey bee visitation.

3. Before anther dehiscence (depending on the flower species this may happen prior to anthesis), remove the bag (Fig. 4) and emasculate the flower using a fine pair of forceps to minimize injury to floral tissue. After emasculation, replace the pollination bag on the flower to prevent undesired action of pollinating agents. The bags should remain on the flowers while the stigmas are receptive and can be removed afterwards. It is important for the investigator to become familiar with the time of day or floral morphology stage that are conducive to stigma receptivity for a given plant species.

4. After ovule maturation is apparent in the pollinated treatments, check whether fruit has developed from the bagged and emasculated flowers. If none is present, one can conclude that the plant species does not exhibit agamospermy. If fruit does develop, it is necessary to wait until fruit ripening to check for seeds because some plants are parthenocarpic (produce fruits with no seeds and do not depend on pollination). If seeds are set, compare the number of fruits and seeds set per fruit from

Fig. 4. Unbagging watermelon flower for hand pollination.

the emasculated and bagged flowers with those from the control treatment to estimate the proportion of seeds set by agamospermy in that particular plant species.

5. It is important that assessments of seed- or fruit-set occur as early as possible to minimize underestimating set because of losses that occur between set and harvest.

2.1.2. Testing for autogamy (auto- or self-pollination)

This test will tell us whether the flower can set seeds and fruits from its own pollen. In such a situation, the contribution of flower visitors may be little or none, but even in auto-pollinating plants the movements of bees inside the flower can sometimes optimize pollen transfer from anthers to the stigma and increase fruit- or seed-set. Auto-pollination or self-pollination should be distinguished from geitonogamy (see section 2.1.3.). Auto-pollination is associated with hermaphroditic flowers and pollen transfer within that flower that is automatic (soybean) or pollinator-optimized, whereas geitonogamy could apply to monoecious plants in which pollen is self-compatible but the actions of a pollen vector are nevertheless needed.

1. Choose a given number of flower buds prior to anthesis. The investigator must become familiar with the flowering pattern of the model plant because flowers in some species open and close more than once, making anthesis difficult to determine. The number of buds may vary with availability and ease of access, but larger samples produce more reliable results.

2. Protect two thirds of these buds with pollination bags (see section 2.1.1.) and leave the other third unbagged as open controls, or in the case of pollinator shortage pollinate these flowers manually with pollen from another plant of the same species. Identify each treatment with weather-resistant tags.

3. After anther dehiscence and when stigmas are receptive, remove the bags of half of the protected flowers (one third of the total marked buds) and hand-pollinate the stigmas with a soft brush using pollen from the anthers of the same flower. Dehiscence can usually be recognized as anthers with a split in the anther wall, pore, or flap that is exposing the pollen. After hand-pollinating, re-bag the flowers to prevent flower visitors or wind pollination. Leave bags on flowers until they are no longer receptive, then remove the bags.

4. At the end of the season, check whether fruit developed from the flowers that remained bagged throughout the experiment. If all or most of these flowers have developed into fruit, the plant species is autogamous and its flowers are capable of auto-pollinating. Honey bees can contribute little to increasing fruit- or seed-set. If only the hand-pollinated flowers developed into fruits, this means that the plant species is autogamous but flowers need a pollinating agent to transfer the pollen grains from their anthers to the stigmas within the flower. In this case, honey bees may be of great value. The proportion

of fruit- or seed-set obtained from the bagged treatment in comparison to the hand-pollinated treatment will tell the comparative strength of autogamy in this plant species (strictly autogamous, highly autogamous, etc.). If no bagged flowers produce fruit or seeds, this means the species may be self-incompatible and probably needs cross pollen to set fruits and seeds. However, sometimes a few fruits or seeds can set even in self-incompatible plants because self-recognition can be incomplete. But in this case, there is little variation in fruit- or seed-set among the treated plants. One should not confound self-incompatibility with self-sterility resulting from inbreeding depression because in the latter case seed set varies greatly among treated flowers, ranging from low values in more inbred plants to high values in less inbred ones. Confirmation of self-incompatibility must be done by examining pollen tube growth in the pistil, a subject covered in section 3.1.5.

2.1.3. Testing for geitonogamy (selfing within the same plant)

Some flower species do not set fruits/seeds when self-pollinated but do so when receiving pollen from other flowers of the same plant (geitonogamy). This can happen between perfect (hermaphroditic) flowers, but it is obligatory in autogamous plants which are monoecious (unisexual male and female flowers on the same plant). This test will tell us whether the flower can set seeds and fruits when receiving pollen from other flowers of the same plant. This is important information because honey bees tend to explore many flowers per plant before moving to other plants, and this behaviour favours geitonogamy.

1. Repeat the procedures for testing for autogamy (section 2.1.2.), but replace the treatment using the flower's own pollen for a treatment using pollen from another flower on the same plant.
2. Conclusions are similar to those above for testing autogamy (section 2.1.2.), except that if fruits or seeds developed from the geitonogamy treatment it means that the plant sets when pollen is transferred between its own flowers. The proportion of fruit- or seed-set obtained from the geitonogamy treatment in relation to the control treatment will tell the extent to which the plant is responsive to this mode of mating system.

2.1.4. Testing for xenogamy (reliance on out-crossing)

In xenogamy, or cross-pollination, the transfer of pollen to the stigma must occur between plants with different genetic constitutions; the result is offspring with greater genetic diversity than those for species exhibiting self-pollination or geitonogamy. Cross-pollination is also important because some plant varieties, genotypes, and even individuals are entirely self-incompatible and obligated to receive pollen from another variety, genotype, or individual to set fruits. Even self-fertile plants may produce more fruit or seeds of better quality

when cross-pollinated than when self-pollinated, and the extent of this can be determined if outcrossing and selfing (within flower / within plant) are tested at the same time. Crops which grow from highly outcrossed seeds are often more vigorous than ones grown from inbred seeds. Finally, xenogamy is of paramount importance for the production of hybrid varieties and hybrid seed, both of which are of increasing importance.

Knowing the extent to which a plant is obligated to xenogamy helps researchers and growers manage bees optimally and combine compatible cross-pollinating varieties (called *pollinisers*) to promote high rates of pollen transfer (see Jay, 1986; Free, 1993).

1. Repeat the procedures for testing for autogamy (section 2.1.2.), but replace the treatment using the flower's own pollen for one using pollen from a flower of a different plant. In order to prevent using genetically related pollen (parents or siblings), do not collect pollen from plants close to the one whose flowers will be tested.
2. In order to identify compatible pollinisers, the experimental design requires a systematic selection and application of pollen from a number of different varieties of the same plant species. Finding compatible pollinisers is crucial for many commercially important crops such as almond, apple, and plum and is a standard feature of commercial grower guides for planning orchard plantations.
3. Conclusions are similar to those when testing for autogamy (section 2.1.2.), except that if only the xenogamy treatment develops fruit, the plant species is xenogamous and its flowers need a pollinating agent to transfer pollen between flowers of different plants. In this case, honey bees can be of great value. The proportion of fruit- or seed-set obtained from the cross pollination treatment in comparison to the control will tell the extent to which the plant is reliant on a xenogamous mating scheme (strictly xenogamous, highly xenogamous, etc.).

2.1.5. Testing for mixed mating systems

Many plants can set fruit both from self and cross pollen, resulting in a mixed mating system that ensures fruit- or seed-set under autogamy or xenogamy, although one or another may predominate. This test will tell us the extent to which a plant is responsive to either mating scheme.

1. Choose a given number of flower buds prior to anthesis. The number of buds may vary with availability and ease of access, but larger samples produce more reliable results.
2. Protect three fourths of these buds with pollination bags (see section 2.1.1.) and leave the other one fourth unbagged as the control. Identify each treatment with weather-resistant tags.
3. After anther dehiscence and when the stigmas are receptive, remove all bags and hand pollinate one third of the flowers

each with its own pollen (using paint brushes), one third with pollen from another flower of the same plant, and the final third with pollen from multiple plants. After that, bag the flowers again to prevent flower visitors or wind pollination. Leave bags on flowers until they are no longer receptive, then remove the bags.

4. At the end of the season, check whether any fruit developed from the bagged flowers. If all or most bagged flowers have developed into fruits, the plant species has a mixed mating system and the proportion of fruit- or seed-set obtained from the bagged treatments in comparison to the control treatment will tell whether there is a preference for self-pollination, geitonogamy or xenogamy. In the case of mixed breeding systems, honey bees can be highly effective pollinators.

2.2. Testing for pollinating agents and pollination deficit

Once one has learned about the plant mating system, it is of paramount importance to determine the agents capable of pollinating the flowers. Candidate pollinators can be abiotic (wind, water, gravity, electrostatic forces, rain) or biotic (birds, bats, insects and even mammals), but most of the time wind and insects are the major pollinators, and we will concentrate on these. It is useful to know whether a pollinating agent can meet the plant's full potential fruit-set or only a fraction of it. In the latter case, the plant may be under a pollination deficit and its fruit or seed production sub-optimal.

2.2.1. Testing for wind pollination (anemophily)
This test will tell us the extent to which a flower species is wind pollinated (*anemophilous*). It can be exclusively anemophilous in which pollinators do not contribute to fruit- or seed-set or partially anemophilous in which case pollinators can be useful for optimizing yield, examples of which include coconut, canola, olive and castor bean.

1. Choose a given number of flower buds/inflorescences prior to anthesis.
2. Protect half of these buds/inflorescences with muslin bags (mesh large enough to allow pollen grains to pass through but not insects) and leave the other half unbagged as the control. Identify each treatment with weather-resistant tags. In the case of multiple flowers on an inflorescence, a swipe of acrylic paint on the pedicel works well for identifying the treatments. The bags should remain on the flowers/inflorescences while the stigmas are receptive and can be removed afterwards.
3. To control for bag effects on wind transfer of pollen, include inside and outside bags a small sticky surface, such as a microscope slide covered in a thin coat of petroleum jelly, with which one can compare wind-borne pollen deposition in- and outside the bags. Care must be taken in interpreting results as

muslin bags may reduce the level of wind pollination. Observations should be made of the wind direction and the location of the pollen source to determine if a better arrangement of plants might affect the level of wind pollination.

4. A few days later, check whether fruit has developed from the bagged flowers/inflorescences. If not, one can conclude that wind plays little or no role in pollinating that species. In the case of fruit development, the proportion of fruit- or seed-set in relation to the control treatment will tell us the degree of wind dependence by that species.
5. If hand-selfing, geitonogamy, and cross pollination treatments are also performed, one can assess for interactions of these with wind and determine optimum combinations with wind for maximizing fruit- or seed-set. To validate the cross pollination trials it is important to ensure that compatible polliniser varieties are flowering nearby.

2.2.2. Testing for biotic (honey bee) pollination – single visits
With this test, one will be able to check the role of biotic pollinators, in our case the honey bee, in fruit- or seed-setting of a particular plant species. In nature, fruit-set usually happens after repeated flower visits by one or more species of pollinator, but when evaluating different candidate pollinators, it is best to compare fruit-set on the basis of single flower visits; this is the most equitable way to compare innate pollen vectoring capacity among flower visitors. The investigator will bag unopened flowers, un-bag them after they open, observe a single visitor, re-bag the flower, then follow the flower's development for subsequent fruit or seed. The flower now has a history, and the efficacy of the specific agent can be compared with others (Vaissière et al. 1996). It is good practice to have a second flower open at the same time which can be rebagged without being visited to act as a control for bag effects as well as a set of non-manipulated and labelled flowers as open-pollinated controls. Depending on the flower species, the standing stock of nectar or pollen may build up in the bagged flower to the extent that it may influence behaviour of bees visiting newly exposed flowers. To check whether this is affecting forager behaviour, the behaviour of bees visiting previously bagged flowers can be compared to visitors to flowers that have not been bagged.

1. Choose a number of flower buds prior to anthesis.
2. Protect these buds with pollination bags (section 2.1.1.) and identify with weather-resistant tags.
3. After the flower opens remove the bags and watch for the first visit of a honey bee. Rebag the flower after the bee leaves it. The bag should remain on the flower while it is still receptive to avoid undesired visits and should be removed afterwards. Limit observations to the same time each day and to weather conditions that are suitable for insect flight.

4. The following measures may be taken at the time of bee observation and retained for possible use as explanatory covariates: length (sec) of visit, whether the bee is collecting nectar or pollen, ambient temperature, wind speed, and relative humidity.

5. At harvest, check whether fruit has developed from the visited flowers and compare fruit-setting results with those from bagged controls, hand-selfing, geitonogamy, cross pollination, and open-pollinated treatments to know the contribution of a single honey bee visit to the pollination needs of that species.

6. A modification of this method employs a direct measure of Pollinator Effectiveness after Spears (1983):

$$PEi = \frac{(Pi - Z)}{(U - Z)}$$

where Pi = mean number of seeds set per flower resulting from a single visit from pollinator i, Z = mean number of seeds set per flower receiving no visitation, and U = mean number of seeds set per flower resulting from unlimited visitation.

2.2.3. Testing for biotic (honey bee) pollination – multiple visits

Although single-visit fruit-set is a standardized measure of pollination efficiency and independent of pollinator foraging density (Spears, 1983; Sampson and Cane, 2000; Dedej and Delaplane, 2003) in flowers bearing many ovules (ex. apple, pear, melon, pumpkin and kiwi) a single honey bee visit is usually not enough to deposit all the pollen grains needed to set the fruit or to fertilize most of its ovules.

1. Choose a given number of flower buds prior to anthesis.

2. Protect these buds with pollination bags (section 2.1.1.) and identify with weather-resistant tags.

3. Randomly designate each flower as a recipient of 1, 2, 3, or 4 (or more depending on plant species) honey bee visits, remove bags after flowers open, and observe each flower for its assigned number of flower visits.

4. After the assigned number of flower visits is achieved, rebag the flower until it is no longer receptive, after which the bag is removed.

5. A few days later, check whether fruit has developed from the visited flowers and compare fruit-setting results with those from bagged and hand-self, geitonogamy, cross pollination, and open control treatments to know the importance of multiple honey bee flower visits to that particular plant species. The treatment which produces the closest fruit- or seed-set to the best hand-pollinated (or open-pollinated) treatments determines how many honey bee visits are necessary to set acceptable yields.

These can be tiring and time consuming experiments because although one can have many marked flowers within one's visual field, it is usually not possible to observe all flowers at the same time, and bees may take a long time to visit those particular unbagged flowers, especially when there are other flowers around. Some investigators get around this problem by offering freshly-cut female flowers on long extender poles to bees visiting nearby flowers in the patch (Thomson, 1981; Pérez-Balam *et al.*, 2012). This method takes some skill to avoid disturbing the natural foraging behaviour of bees and is obviously only good for destructive measures such as pollen deposition on stigmas (see section 3.2.), but it can greatly speed acquisition of data. In any case, observations should be done at roughly the same period of each day to avoid diurnal variations in flower receptivity. Also, one must not allow a different flower visitor to land on the flower while waiting for specifically honey bee visits; otherwise that flower must be discarded and all work invested on it is lost.

An alternative approach is to use a video camera that follows groups of flowers as they open. A quantitative analysis of the recording will reveal relationships between the number of visits each flower receives and its subsequent seed set. Because flowers are not enclosed, build up of pollen and nectar reflects natural rates. As the relationship between number of bee visits and seed set can only be determined if there is less than full set (once seed set is maximized additional visits are superfluous), it may be necessary to bag flowers (section 2.1.1.) after they have been videoed for an appropriate length of time to prevent full pollination. This method has the advantage that the number of visits required for full pollination can be measured directly rather than estimated as it may be when just measuring the effect of single bee visits.

2.2.4. Fruit-setting experiments at the field level

The methods listed above (sections 2.1. – 2.2.3.) are useful for determining the mating and pollination requirements of a plant and the proportion of a plant's pollen-vectoring needs met by honey bees, other visitors, wind, or self. But honey bees are commonly used as pollinators in high-density agriculture, and when designing fruit-set experiments with crops, one must be aware that cultivated plants can compensate for pollen limitation with longer flowering periods or more flowers. Similarly, fruit- or seed-set can be resource-limited. Therefore, working on the basis of individual flowers or inflorescences may over-estimate yield potentials at the basis of the crop. For these reasons, when working at the scale of agricultural production, the experimental unit should be a plot or a field, and never lower than a whole plant (Vaissière *et al.*, 2011).

Following this argument, at the field level the whole plant or plot (Fig. 5) is to be caged in the exclusion experiments, honey bee colonies are introduced into the areas where their effectiveness as a crop pollinator is to be tested, and fruit or seed production is compared to open fields with no supplemental honey bee introductions. One must also take into account the growth conditions and mating system of the target crop. For example, some crops are negatively affected by shading, others are male sterile and need the presence of male-fertile plants, and others are generally xenogamous and require a compatible cross-variety within the experimental cage.

Fig. 5. Honey bee exclusion experiment: caged plots in soybean plantation.

3. Measuring pollen on bees and pollen deposition on stigmas

Pollinator performance can be thought about with at least three organizing concepts: (1) measuring fruit- or seed-set that results after flower visitation, (2) measuring pollen load on pollinators and their pollen deposition onto stigmas, or (3) measuring plant reproductive success post-pollination, i.e. fertilization efficiency (Gross, 2005; Ne'emen *et al.,* 2010). For our purposes, we are focusing on the first two concepts because reproductive success depends not only on the amount of pollen vectored by pollinators but also on additional factors such as pollen pistil interaction and female choice (Herrero and Hormaza, 1996). In order to evaluate pollinator performance we can study the pollen carried by bees as well as the pollen effectively deposited on the stigmas.

Most of the following methods have been discussed in detail in pollination methodology books (Kearns and Inouye, 1993; Dafni *et al.,* 2005).

3.1. Identifying and evaluating pollen quantity and quality transported by bees

The first step to identify and analyse the pollen transported by bees is to remove the pollen grains from the bees' bodies. Several techniques are available for removing pollen grains from insects, usually mechanically by washing and vortexing the insect body (for example in 50% ethanol), removing the insect, precipitating the pollen grains by centrifugation, and using the pollen grains for further analyses (see Jones, 2012 for a review on pollen extraction from different insects). For studies of pollination success, the pollen packed in the corbiculae should first be removed since it is usually not available for pollination. It is sometimes possible to refine this method by only removing the pollen from the areas of the bees that have been observed to touch the stigma.

3.1.1. Microscopic pollen identification and making archival reference slides

Pollen from different plant species can usually be distinguished based on diagnostic traits such as pollen grain size, exine sculpturing and number and size of the apertures (pores or furrows). It is important to keep in mind when working with fresh pollen that the degree of pollen hydration affects external pollen appearance. Transmitted light microscopy is the most widely used technique for pollen identification using fresh, acetolyzed and stained pollen, but scanning electron microscopy (SEM) is also used to study surface details of the exine. In pollen reference collections, pollen grains are usually subjected to acetolysis that removes the protoplasm and leaves the exine (Erdtman, 1969; Kearns and Inouye, 1993). The acetolysis solution contains glacial acetic acid and concentrated sulphuric acid (9:1). According to Dafni *et al.* (2005) and Kearns and Inouye (1993) the procedure for acetolysis is as follows:

1. Add pollen sample to a solution of glacial acetic acid for 10 min.
2. Centrifuge and discard the supernatant.
3. Add a few ml of acetolysis mixture (glacial acetic acid and concentrated sulphuric acid 9:1).
4. Heat the solution gently to boiling point in a water bath, stirring continuously with a glass rod.
5. Cool the solution for a few minutes, centrifuge and discard the supernatant.
6. Resuspend in distilled water, centrifuge and decant the supernatant. Repeat this step.
7. Pollen is usually stained to increase the contrast. Several stains (such as methyl-green or fuchsin) can be used, but Safranin O is the preferred stain for most uses in palynology, staining the pollen grains pink to red depending on the amount of stain and type of pollen analysed (Jones, 2012).

After acetolysis, pollen can be preserved for further analyses or to make archival reference slides. A common procedure is to use glycerin jelly slides (Erdtman, 1969):

1. Prepare a base stock of jelly by combining 10 g gelatin, 30 ml glycerin, and 35 ml distilled water.
2. On a clean microscope slide add a drop of the prepared jelly and a sample of pollen and stain.
3. Warm the slide gently, stirring to thoroughly homogenize the mixture.
4. Add a cover slip, sealing with nail polish or other varnishes around the edges.

Identification keys and atlases with pollen images are available both in general and for specific taxa (Kearns and Inouye, 1993; http://www.geo.arizona.edu/palynology/polonweb.html).

3.1.2. Pollen identification (palynology) with molecular methods

Alternatively, pollen can be identified using molecular methods. This is straightforward if the pollen grains carried by the bees belong to just a single species since a large amount of pollen grains can be pooled for DNA, but analyses are more difficult if the bees carry a mixture of pollen grains from different species. In this case a possible approach is to use single pollen genotyping strategies (Matsuki *et al.*, 2008; Suyama, 2011) that allow PCR amplification of the genome of single pollen grains. Molecular pollen identification can be useful for assessing bee cross pollination efficacy; if a bee's corbicular pollen load contains a number of variety-specific pollens, this is evidence that the bee is foraging across pollinisers.

3.1.3. Tracking pollen identity

For some purposes in ecological research, target pollen or pollen bearing pollinia can be "tagged" to track pollinator dispersion range and pollination success. Morphological markers such as colour, size and shape polymorphisms may work as long as the work is limited to a few specific taxa. Early attempts to tag non polymorphic pollen were based on the use of radioisotopes or fluorescent dyes as pollen analogues (Dafni, 1992). With progress in genetic engineering, GFP-tagged pollen grains have also been used to track pollen identity in transgenic plants (Hudson *et al.*, 2001).

3.1.4. Pollen quantity

Several methods can be used to evaluate the number of pollen grains attached to bees.

3.1.4.1. Haemocytometers

This is the most common method for counting pollen grains. A drop of a known volume of suspension of collected pollen is placed under the microscope and the number of pollen grains counted allowing the calculation of the total number of pollen grains in the whole volume. Haemocytometers were initially developed to count blood cells, but they can also be used to count the number of pollen grains in a standard volume of liquid containing pollen (see Human *et al.*, 2013 for more information on using haemocytometers in honey bee research). The steps are as follows:

1. Collect pollen as described in section 3.1.
2. Suspend pollen grains in a known volume of 70% ethanol and vortex to assure homogenous mixing.
3. Remove a sample of pollen suspension with a pipette and place in haemocytometer.
4. View and count pollen grains under a microscope. Haemocytometer manufacturers provide the known volume of suspension under the viewing area and provide easy instructions for extrapolating object counts back to absolute counts in the original suspension (sample).
5. When the number of pollen grains is very low, a measured drop of the suspension can be placed on a lined microscope slide and all the grains counted.

3.1.4.2. Alternative methods

More expensive techniques such as electronic particle counting (Kearns and Inouye, 1993) and laser-based counters (Kawashima *et al.*, 2007) can also be used. In some situations it is possible to directly count pollen grains on a bee's body with a stereomicroscope (Nepi and Pacini, 1993).

3.1.5. Pollen viability and quality

There is disparity in results among different methods for appraising pollen viability and quality; for this reason the most robust approach is to use a combination of methods, such as those provided below, that allow a more precise estimation of pollen viability and quality (Dafni and Firmage, 2000).

3.1.5.1. Pollen viability

One approach is to evaluate viability prior to germination. The most common test is the fluorochromatic reaction (FCR test) based in fluorescein diacetate (FDA) (Heslop-Harrison and Heslop-Harrison, 1970; Pinillos and Cuevas, 2008). This test evaluates the integrity of the plasmalemma of the pollen vegetative cell and activity of nonspecific esterases of the cytoplasm, and only viable pollen grains will fluoresce under the microscope (Fig. 6). Another commonly used viability test is Alexander staining (Alexander, 1969); viable pollen stains crimson red while aborted pollen stains green. Additional tests such as the use of tetrazolium dyes, X-Gal, isatin or Baker's reagent can also be used (see Dafni and Firmage, 2000 for a review). However, in most viability tests not all the viable pollen grains are able to germinate, and consequently the percentage of pollen germination is usually lower than the percentage of viable pollen.

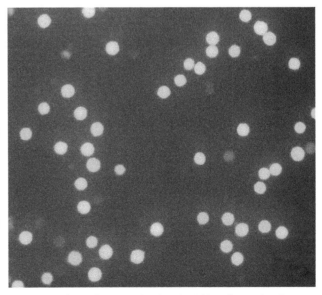

Fig. 6. Pistachio pollen stained with FDA: viable pollen grains show bright fluorescence compared to non-viable pollen grains.

***3.1.5.2. Pollen germination and pollen tube growth* in vitro**

The evaluation of pollen germination *in vitro* takes into account not only pollen viability but also pollen vigour (Shivanna and Johri, 1985; Shivanna *et al.*, 1991). One drawback of the method is that the germination medium and germination conditions (e.g. pollen pre-hydration, temperature conditions) must be optimized for each species to avoid false negatives. Different media for *in vitro* pollen germination have been recommended for several species (Taylor and Hepler, 1997), mainly using the basic medium developed by Brewbaker and Kwack (1963) in a sucrose solution with or without agar, depending on the species. The optimum method has to be tested for each species empirically. A pollen grain is considered as germinated when the length of the pollen tube is at least twice the diameter of the pollen grain.

3.2. Evaluating pollen identity, quantity and quality on stigmas

For some studies, mainly in the field, it is necessary to avoid contaminating stigmas with non-target pollen. This means it is necessary to prevent bees from visiting target flowers. Different strategies have been used (Kearns and Inouye, 1993), including a variety of tubes and capsules for small flowers, plastic pieces to cover just the pistils, or nylon or paper bags to enclose the flowers, the inflorescences, or whole plants before the experiment begins. Errors in flower sampling can be minimized by removing all opened flowers before the experiment begins. The main disadvantage of these enclosures is that the microenvironment in the flower (mainly temperature and humidity) can be altered and depending on the experiment this can have implications for the results. In any case, air-permeable mesh or net bags are likely to have a smaller effect on flower microenvironment than paper bags or plastic enclosures. In some cases, emasculation might be needed to avoid self-pollination, although emasculation can affect subsequent pollinator behaviour. The possibility of emasculating is dependent on the morphology of the flower and should be carried out carefully, especially if dealing with small flowers to avoid accidental self-pollination or damage to the flower (Hedhly *et al.*, 2009).

3.2.1. Identifying pollen on stigmas

Identifying pollen deposited on stigmas can provide evidence of the percentage conspecific pollen deposited and the likelihood of stigmas being clogged by pollen from other species. This constitutes perhaps the most unambiguous and precise measure of pollination success, *sensu stricto*. The stigmas can be collected and washed in 70% ethanol and the pollen grains released can be observed using similar procedures to those described for identifying pollen on bees (see section 3.1.). Molecular markers have also been used to identify pollen deposited on the stigma (Hasegawa *et al.*, 2009).

3.2.2. Quantifying pollen deposited on a receptive stigma per visit or unit time

Usually this parameter is measured by counting the number of pollen grains deposited on the stigma per visit or unit time, regardless of fertilization success. This method implies the microscopic examination of pollen germination and tube growth in the stigma and style. Different stains that stain pollen grains differentially from the surrounding stigmatic tissues can be used. Usually the stigma is gently squashed under a coverslip after staining to better visualize the pollen grains. If needed, stigmas can be fixed in FAA (formaldehyde - acetic acid - 70% ethanol [1:1:18]), 4% paraformaldehyde, glutaraldehyde (2.5% glutaraldehyde in 0.03 mol/L phosphate buffer), 3:1 (v/v) ethanol – acetic acid, or just 70% ethanol and stored at 4ºC for later examination. However, it should be taken into account that before germination, pollen needs to adhere to the stigma and hydrate. Fixing can remove non-adhered pollen grains and consequently the estimate of pollen load may be lower than if fresh stigmas were analysed. In some cases, softening the fixed stigmas should be performed before staining and squashing; this can be done by autoclaving the samples at 1 kg / cm^2 for 10 to 20 min in 5% (w/v) sodium sulphite and rinsing in distilled water or, alternatively, 1M NaOH can be used for 1 h following a rinse in distilled water. Each of the following methods is acceptable for determining number of pollen grains and extent of their germination.

1. Epifluorescence microscopy. This is the most widely used method for visualizing pollen tubes. Stain the pollen grains and pollen tubes with aniline blue (specific for callose, a polysaccharide present in pollen tube walls and plugs produced in pollen tubes of most Angiosperms) (Fig. 7) and observe under fluorescence microscopy. The usual mix is 0.1% (v/v) aniline blue in 0.3 M K3PO4 (Linskens and Esser, 1957). The observer can directly determine the percentage pollen germination on the stigma.

2. Light microscopy. Different methods are available that do not require epifluorescence:

    a. Methyl green and Phloxine B (Dafni *et al.*, 2005). Non-germinating grains stain dark brown-red, whereas in germinating pollen grains, the empty grains stain green and the pollen tubes red.

    b. Stain with 1% basic fuchsin: 1% fast green (4:1) (Kearns and Inouye, 1993). De-stain and soften the tissue in lactic acid for 12 hours and then squash the tissue under a coverslip. Pollen tubes stain maroon and the background remains white.

    c. Acetocarmine/basic fuchsin (Kearns and Inouye, 1993). Add a drop of acetocarmine, followed by a drop of 3% aqueous basic fuchsin and de-stain with a drop of absolute ethanol. Pollen cytoplasm stains red.

3. Scanning electron microscopy can be used to visualize the whole stigmatic surface, but this is a more time-consuming and complicated than light microscopy.

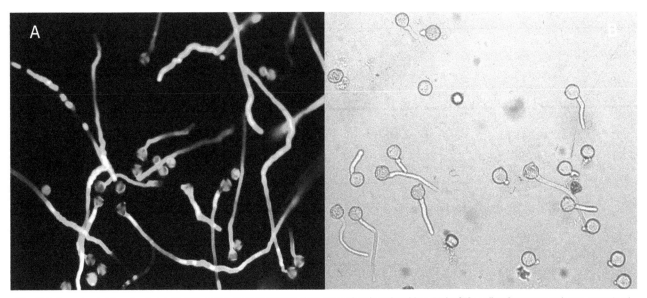

Fig. 7. Pollen germination *in vitro.* Left **(A)**: pollen from Japanese plum stained with aniline blue. Right **(B)**: pollen from sweet cherry, unstained.

3.2.3. Evaluating stigmatic receptivity

Pollination success is dependent on stigmatic receptivity since only insect visits to receptive stigmas can be considered as effective pollination visits. A receptive stigma allows pollen grain adhesion that can be followed by pollen hydration and germination. Stigmatic receptivity can be evaluated directly by studying conspecific pollen germination (see 3.2.4.) or indirectly by studying activity of enzymes (Dafni *et al.,* 2005) such as esterase (with a benzidine solution) or peroxidase (with alpha-naphthyl acetate) or the presence of exudates in wet stigmas stained with Sudan black or auramine O (Kearns and Inouye, 1993). Herein, we discuss determining stigmatic receptivity only by studying conspecific pollen germination.

3.2.4. Evaluating pollen germination and pollen tube growth *in vivo*

The most common procedure is to study pollinated stigmas and styles in squashed preparations stained with aniline blue (specific for callose) and observed under fluorescence microscopy (as described in section 3.1.). The tissues have to be softened, usually in 5% sodium sulfite to allow squashing, and the time of softening is species dependent; it is advisable to start processing the samples overnight and, if needed, the samples can be autoclaved at 1 kg/cm^2 for 10 min in 5% (w/v) sodium sulphite or placed in 1M NaOH for 1 h. Depending on the species, varying concentration of sodium hydroxide (from 1N to 4N) at different temperatures (ambient temperature overnight or 60ºC for an hour) can also be tried for softening. For staining, 0.1% (v/v) aniline blue in 0.3 M K3PO4 (Linskens and Esser, 1957) can be used and the observations made with a fluorescence microscope. Pollen tube walls and the callose plugs produced by growing pollen tubes show a distinct fluorescence signal (Kearns and Inouye, 1993). This test is also useful to determine the presence of gametophytic self-incompatibility in which incompatible pollen tubes get arrested during pollen tube growth in the style.

4. Measures of colony level pollination efficacy

The foraging activity of social insects is in part controlled by their colony and its requirements. Honey bees are usually managed at the colony level for pollination, i.e. colonies may be subject to particular management practices to influence their foragers' pollinating activity.

This section covers measures for assessing the pollination efficiency of honey bees at the colony level. To determine a colony's pollination efficiency, it is usually necessary to have an understanding of the efficiency of individual foragers at pollinating flowers from the crop to determine which assessments should be used. For example, for a crop like kiwi that only produces pollen, a measure of kiwi pollen collected by a colony will indicate the portion of foragers that are visiting the crop. Likewise for avocadoes and hybrid carrots it is only nectar foragers that are carrying out pollination activities and pollen foragers will be of less interest.

The measures of colony performance generally fit into two types. Measures can be made on returning foragers which generally relate to what flowers the foragers are visiting, whether they were foraging for pollen or nectar, and for some crops how effective they were at pollinating the crop. The second approach is to study the behaviour of foragers in the field and then attempt to determine which colony they came from. These types of studies are often carried out to assess the effect of colony level manipulations designed to increase the number of honey bees visiting a crop or their behaviour when they are visiting the crop, e.g. to measure the effect of:

1. the timing of colony introductions,
2. placement of colonies in the crop,
3. colony strength (number of bees and amount of brood),
4. organization of the colony, e.g. location of the brood or proportion of the brood that is uncapped,

5. practices to change foraging behaviour, e.g. feeding pollen or syrup and using pollen traps, and

6. competing floral sources.

This section covers methods for assessing

- pollen foraging
- nectar foraging
- colony foraging rates
- fraction of side working behaviour
- relationship between foragers and their hives

4.1. Proportion of foragers from a colony visiting a crop

4.1.1. Pollen trapping

The number of foragers from a honey bee colony that is collecting pollen can be estimated using pollen traps (Goodwin, 1997; see Human *et al.*, 2013 for more details). This is particularly relevant if the crop of interest only produces pollen (kiwi) or if pollen foragers are more efficient pollinators than nectar foragers (apple, almond) because they have a greater likelihood of contacting the stigma. Pollen traps are devices with grids (Fig. 8) that fit across the entrance of a hive. With some designs, the hive entrance is blocked and the trap forms a new entrance. Returning foragers must walk through the grid to enter their hive. Bees prefer not to walk through pollen traps if they can avoid it and will use any other gaps in a hive body as an entrance once a pollen trap is fitted. These holes need to be blocked to ensure that all bees are using the pollen trap. It is worth checking the hives several days after the trap is fitted to make sure all bees are entering and leaving the hive through the pollen trap.

As returning bees carry pollen through the trap the grid scrapes some of the pollen pellets from their corbicula. The pellets then fall into a tray where they can be collected. The proportion of pollen pellets removed depends on the size and shape of the holes in the grid and the size of the pollen pellets the bees are carrying. Pollen pellet size, and consequently the efficiency of a pollen trap may vary with both the plant species, the time of day the pollen is collected, and meteorological conditions (Synge, 1947). The inside of a pollen trap can become blocked over time. Depending on the design of the trap it may be difficult for hive cleaning bees to carry dead bees through the trap and for drones to move through it. When these accumulate on the inside of the trap it can reduce the ease with which foragers move through the trap and hence the foraging ability of the colony. If traps are to be used for extended periods of time they should be checked regularly for blockages.

4.1.1.1. Determining pollen trap efficiency

Pollen traps are variable in their design, so it is advisable to determine the efficiency of a trap at collecting the pollen. This can be achieved by counting the number of bees entering the trap carrying the pollen

Fig. 8. Grid on a pollen trap.

pellets of interest and determining the percentage that are collected in an empty pollen trap drawer (Levin and Loper, 1984; Goodwin and Perry, 1992).

4.1.1.2. Number of pollen traps

Most studies will require data to be collected from a number of colonies as there can be large differences in the plant species neighbouring colonies are visiting. The level of replication required will depend on the amount of variation between colonies and the size of the difference to be detected. Balancing colonies with regard to the amount of brood and number of bees they contain (see Delaplane *et al.*, 2013) will reduce the amount of variation in the total amount of the pollen collected.

4.1.1.3. Analysing pollen trap contents

Pollen pellets from many plant species can be identified by colour (Fig. 9). To establish the colour of the pellets of interest it is best to catch bees with pollen pellets from the crops of interest and remove their pellets so they can then be used as standards for comparing with pellets in the tray of the pollen trap. It is important to note that the colour of pellets may change depending on the light by which they are viewed and when they dry (Kirk, 2006). Because the colours of pellets from some plant species are sometimes similar, it is often necessary to measure the size of the pollen grains making up the pollen pellet and study the pollen grains' surface features microscopically to establish whether the trap contents can be sorted by eye (see section 3.1.1.).

Because of the large amounts of pollen that may be trapped at times, it may be necessary to subsample the pollen trap contents. The contents of a trap are often layered as bees collect pollen from different plant species at different times of day. It is therefore necessary to mix the contents of the trap thoroughly before subsampling. If determining the weight of different pollens trapped rather than the number of pellets trapped, it is necessary to first dry

Fig. 9. Pollen pellets.

samples to a constant weight as their moisture content may vary between species, time of day, and between days. Samples of pollen pellets can be stored for short periods of time at room temperature. However they may eventually develop mould making them difficult to analyse. It is therefore good practice to freeze samples if they are not going to be analysed at the time of collection.

4.1.1.4. Effect of pollen traps on foraging
Honey bees losing pollen pellets while moving though a trap still go through their normal behavioural repertoire associated with scraping the pollen off their legs into a cell (McDonald, 1968). If using high efficiency pollen traps, it is important to note that they may reduce brood rearing after prolonged use (Eckert, 1942) and cause colonies to increase pollen collection (Levin and Loper, 1984). For this reason, pollen trapping with high efficiency traps should never extend beyond a few days to prevent compromising colony strength.

4.1.2. Nectar collecting
For some crops it is possible to estimate the number of visits that bees from a colony make to collect nectar. This is important for crops were nectar foragers are the most important pollinators (avocadoes, hybrid carrots) This can sometimes be achieved by a chemical analysis of the stored honey if the nectar from the plant of interest has a unique chemical profile e.g. avocados (Dag *et al.*, 2006). It can also sometimes be achieved by an analysis of the pollen contained in the honey; however, care must be taken in interpreting results as pollen grains from some plant species are more likely to be present in honey than other species. Also, the amount of honey produced is not only affected by the amount of nectar collected by a colony but also the amount consumed.

4.1.3. Proportion of colony bees collecting pollen
For many flowers that are visited by both pollen and nectar foragers, pollen foragers are better pollinators (Free, 1966). Because of the

close proximity of anthers and stigmas, pollen foragers are more likely than nectar foragers to touch both structures. Some bees collect both pollen and nectar on a foraging trip. The proportion of pollen and nectar foragers can be determined by observing the behaviour of foraging bees. Nectar gathers will probe the base of the petals while pollen foragers usually scrabble over the anthers.

4.1.4. Colony foraging rate
A colony's foraging rate refers to the number of foraging trips a colony makes during a day. Generally the more foraging trips bees from a colony make to a crop, the more effective the colony will be at pollinating the crop. The number of bees foraging from a colony can be estimated by counting bees entering the hive (Baker and Jay, 1974). This is usually easier than counting bees leaving a hive as returning foragers approach more slowly. When counting the number of returning bees over a set length of time it is important to do this without disturbing the returning bees. The presence of an observer at the front of a hive may confuse bees and delay their return. This can be avoided by using a hide that can be left in front of the hive. Alternatively, a video camera will be less obtrusive and can be left in position for the bees to become accustomed to it. Video has the advantage that allows the action to be observed in slow motion. The data are reported as returning bees per minute.

Depending on the questions being answered the physical counts or video data may need to be backed up with samples of returning bees. Honey bees observed returning with pollen must have been foraging, however honey bees returning without pollen might be nectar foragers or bees going on orientation flights. Returning bees can be captured by blocking the hive entrance and allowing the returning bees to collect on the outside. The bees without pollen can then be captured. Dissecting the bees and measuring their crop weight will differentiate bees that were on orientation flights from bees that were foraging for nectar.

4.1.5. Fraction of bees side-working flowers
On some flower species (almonds, apples) the flower architecture allows bees to approach the nectaries by climbing through the anthers past the stigma (top working bees, Fig. 10) or from the side of the flowers where the push their tongues between the base of the anthers (side working bees, Fig. 11). Side working bees are less likely to touch the stigma than top working bees. The proportion of bees carrying out these behaviours varies with flower architecture and with the experience of the bees. The data are collected by modifying the methods given in section 5.1. to report number of top working, or side working, bees for a given number of flowers or measured area of crop.

Fig. 10. A top-working honey bee visiting an apple flower.

Fig. 11. A side-working honey bee visiting an apple flower.

4.2. Relationship between foragers and their hives

It may be necessary to determine whether the bees in a part of a crop are coming from a particular hive or hives. This might be of interest when manipulations of colonies are carried out to alter honey bee foraging behaviour that cannot be detected by studying foragers as they return to their hive. It might also be used to determine where in a crop bees from particular colonies are foraging. There are several methods of achieving this.

4.2.1. Marking bees in the crop

This can be achieved by catching bees on the crop, marking them, and opening hives in the evening and searching for marked bees. Bees will usually need to be immobilized before marking. This can be done by chilling the foragers, anoxiating them with CO_2 or anesthetizing them with chloroform. Care needs to be taken when choosing to use CO_2 as it has been reported to inhibit pollen collection (Ribbands, 1950, Brito *et al.*, 2010).

Acrylic paints can be used to mark bees if they only need to be marked for a single day as the paint may wear off after this time.

Acetone based paints will last longer. By using a range of colours and positions of spots on the thorax and abdomen it is possible to individually and distinctively mark large numbers of bees. An alternative, more costly method, is to use purpose-made plastic queen tags glued on the thorax of worker honey bees (Fig. 12). As there are often a large number of bee colonies foraging from a crop, it is usually necessary to mark large numbers of bees to obtain adequate recovery rates.

Colour-coded ferrous tags glued to the thoracic dorsum of an individual forager in the field can be retrieved at the hive entrance with magnets (Gary, 1971). This technique enables studies of spatial distribution of bees in an area.

4.2.2. Marking bees according to their hives

Strains of bees with the visible mutation *cordovan* can be used as it is possible to identify these workers in the field (Gary *et al.*, 1981). The bees in a colony can also be fed with radioactive elements (Levin, 1960) or made to walk through a marking block fitted to the entrance of a hive that marks them with coloured dye (Howpage *et al.*, 1998). Workers marked with any of these methods can be searched for in the field.

5. Determining crop-specific recommended pollinator densities

An aim of some research in agricultural pollination is to provide guidelines for stocking honey bee colonies in a crop to maximize pollination in the most economical way. There are two general approaches to designing experiments for determining optimum stocking densities: (1) indirect extrapolations from densities of foraging bees observed in small plots, along transects away from colonies, or in cages; or (2) direct tests of colony densities on whole fields.

Fig. 12. Foragers with queen tags.

5.1. Indirect extrapolations

The stocking rates of colonies required can be estimated indirectly by using the pollination potential of individual foragers (i.e. seed set per flower visit; sections 2.2.2. and 2.2.3.) and extrapolating the number of foraging honey bees and colonies required to pollinate a crop (Goodwin *et al.*, 2011). This presupposes our ability to reliably measure bee densities in the crop.

5.1.1. Bee densities in small field plots

Appropriate bee density measures vary according to plant growth habit, planting arrangement and conformation of flowering. Examples of density measures that may be used include:

- Bees per m^2 or larger area (especially for vine crops)
- Bees per tree (tree fruits and nuts)
- Bees per flower or larger number of flowers (berries, vine crops, cotton, sunflower)

Considerations when counting bees:

- Make several observations through the duration of the flowering period
- Subsample within a day, recording time of samples as local solar time
- Sample during weather conditions that are favourable for foraging, i.e., temperature $\geq 15°C$, wind < 16 km/h, no rain (flowers dry) and preferably sunny
- Also sample for pollinator diversity

Considerations in choosing sampling sites:

- Use multiple sampling sites within a field
- Choose representative sites that are at least 5 m from the field edge
- Use > 1 m of row in row crops
- Use plots of > 1 m^2 in broadcast-seeded plots and non-row crops
- Use individual branches of trees in orchards
- Use individual flowers if large enough (e.g., sunflowers)

Considerations when measuring yield are as in sections 2 and 6.

Vaissière *et al.* (2011) give detailed suggestions about assessing pollination needs of different types of crops. They also provide useful data collection sheets for recording bee count and yield information for crop of different growth habits.

5.1.2. Field-scale transects

Colonies can be placed at one end of a crop field, a linear sampling transect established across the field, and the number of bees visiting flowers counted (visits per min) and seed-set assessed at points along the transect to determine whether bee visitation and yield decline with distance from the colonies (e.g. Manning and Boland, 2000). This may only be useful in large areas of crops. It is good practice to repeat the trial with the colonies at the other end of the crop. For transects,

- Use very long fields (> 400 m)

- Establish a bee density gradient by locating colonies at one end of the field only
- Use multiple sampling sites at intervals of 100-200 m away from the colonies

5.1.3. Cage visitation rates

Observations on visits to flowers of a crop are sometimes made on plants within cages where bee densities can be controlled and replicated more easily than in fields (e.g. Dedej and Delaplane, 2003); cages often are ca. 2 m^3 and constructed of Lumite® (e.g. BioQuip Products; Rancho Dominguez, CA, USA). These tests are useful for gaining insight about the relative impact of bee visitation on the yield response of a crop. The artificial environment within a cage, however, can affect both bee behaviour and plant growth, and this further limits the ability to transfer findings about effects of bee density on crop yield to recommendations about colony stocking rates in normal field situations. Thus cage tests provide only a very general idea of comparative usefulness of different colony densities and are much less useful than studies using small plots or whole fields.

5.2. Direct tests of whole fields to find the required number of colonies per hectare

A less common approach is to stock fields with different numbers of colonies and establish whether the rates used have an effect on pollination (Palmer-Jones and Clinch, 1974; Vaissière, 1991; Brault *et al.*, 1995). Direct comparisons of different colony densities are rare because it is difficult to obtain acceptably large numbers of replicate fields for each treatment. Past research usually involved a few fields (e.g. Eischen and Underwood, 1991), or a few fields repeated over a few years with treatments rotated among fields (e.g. Stern *et al.*, 2004).

If multiple fields are available for testing, they should be:

- As similar as possible regarding cultural practices (e.g. irrigation, drainage, fertilization, pest and weed control), available pollinisers, soil type and surrounding habitat
- Far enough apart (ideally > 3 km) to isolate bee populations

If multiple fields are available, similar fields should be paired and honey bee colonies introduced into half of the fields while the other fields serve as controls without supplemental bees. A recent recommendation (Vaissière *et al.,* 2011) is to use ≥ 5 fields per treatment, with bees introduced at the onset of effective flowering (i.e. at the time of first bloom that would lead to a product).

Considerations when collecting data about pollination outcome include the following:

- Use units of yield per field, plot, plant or flower as appropriate for the crop. Yield may include fruit and seed quantity and quality (see section 6).
- Alternatively, use pollen deposition (see section 3.2.), or fruit- or seed-set (see section 2). It is useful to measure pollination outcomes prior to harvest to prevent losing fruit to events

(e.g. natural herbivory, violent weather) that can confound treatment effects. Note, however, that pre-harvest measures of immature fruits do not reflect outcomes typical of agricultural commerce.

- It is advisable to estimate realized densities of bee foragers in fields resulting from the different numbers of colonies (see section 5.1.1.).

5.3. Appraising risk of competition between plants for pollination

If the target crop blooms in synchrony with neighbouring weeds or crops, there is a risk that it will not be well serviced by a limited pool of pollinators. This underscores the need to monitor forager density on the target crop. It may be useful to gauge the distribution of bees among competing plants species; see the methods outlined in section 4.1. for possible approaches. The most convenient technique may be to use pollen traps to measure the relative proportions of pollen income from different forage sources if bees are collecting pollen from all sources.

5.4. A cautionary note about recommendations

Delaplane and Mayer (2000) list recommended colony densities and their average for many crops is based on information collected from standard pollination references and historical extension bulletins. Recommendations about the optimal number of colonies per hectare ultimately often come from experiences of growers and beekeepers who have adjusted bee densities based on trial and error over time. A commonly recommended starting density is 2.5 colonies of standard strength (often cited as having ≥ 8 combs, two-thirds covered with adult bees or ≥ 6 combs well covered with brood) per hectare. This may be adjusted knowing the relationship of crop yield with factors that affect foraging activity or pollination. Examples of such factors among include:

- Plant reproductive biology, including cultivars that are more difficult to pollinate effectively, e.g. 'Delicious' apples (*Malus domestica*), because of a high frequency of sideworking honey bees; more bees are needed for such plants.
- Field size: larger fields usually need more supplemental pollinators than small fields because small fields often have greater densities of native pollinators.
- Prevailing weather: a region or season with historically poor weather for bee flight may warrant a higher stocking density.
- Competition: the extent to which the target crop is competing with weeds or neighbouring crops for a limited pool of pollinators
- Ambient densities of native pollinators

A case study of how these particular factors have been used to adjust pollination management involves lowbush blueberry (*Vaccinium angustifolium* Aiton and *V. myrtilloides* Michx.) in the northeastern USA. The crop is difficult because it requires much pollen movement, honey bees do not "buzz pollinate" (sonicate) the ericaceous flowers, many commercial fields are large and have insufficient densities of native pollinators, weather during bloom is often poor, and bees seek pollen from sources other than blueberries. Large-scale commercial growers have tested the value of increased stocking densities. Through this experience the largest growers now prefer to rent colonies that are more populous than average and stock them typically at 10-12/hectare and up to 20-25/hectare in historically high-yielding areas (Danka: unpub. obs.).

6. Measuring harvest and post-harvest effects of pollination

The economic impact of pollinators on agricultural output transcends simple yield measures and extends into harvest and post-harvest effects as well (Bommarco *et al.*, 2012; Dag *et al.*, 2007; Gaaliche *et al.*, 2011). These include things like fruit sweetness, shape, weight, texture, and other flavour metrics (Gallai *et al.*, 2009). Many of these quality criteria are affected by seed number which in turn is a result of pollination efficiency (Dag and Mizrahi, 2005; Dag *et al.*, 2007). However, even when fruit have only one seed, pollination can affect fruit-quality parameters since fruit resulting from cross-pollination might differ from those stemming from self-pollination, as has been reported in mango (Dag *et al.*, 1999), avocado (Degani *et al.*, 1990) and other crops. In this section, we describe major quality criteria and provide methods for their quantification. The presented protocols for assessing fruit quality are based on Kader (2002). For each crop, researchers need to define which parameters are relevant in the context of pollination efficiency.

6.1. Visual appearance

- Size: Fruit size can be measured with a sizing ring or calipers (Fig. 13). There is generally a good correlation between size and weight; size can also be expressed as number of units of a commodity per unit weight. Volume can be determined by water displacement or by calculating from measured dimensions.
- Shape: Ratios of dimensions, such as diameter-to-depth ratio, are used as indices of fruit shape (e.g. sweet pepper: Dag *et al.*, 2007) (Fig. 14).
- Colour: The uniformity and intensity of colour are important visual qualities, as is light reflectance which can be measured by any number of dedicated meters. These devices measure colour on the basis of amount of light reflected from the surface of the fruit; examples include Minolta Colorimeter,

Gardner, and Hunter Difference Meters. Internal colour and various internal disorders can be detected with light transmission meters. These devices measure light transmitted through the fruit. Fruit colour can be evaluated on the basis of pigment content, usually a function of quantity of chlorophylls, carotenoids and flavonoids.

- Defects (Fig. 15): Incidence and severity of internal and external defects can be evaluated on a five-point subjective scale (1 = none, 2 = slight, 3 = moderate, 4 = severe, 5 = extreme). To reduce variability among evaluators, detailed descriptions and photographs may be used as guides in scoring a given defect. An objective evaluation of external defects using computer-aided vision techniques appears promising. Internal defects can be evaluated by non-destructive techniques, such as light transmission and absorption characteristics of the fruit, sonic and vibration techniques associated with mass density, and nuclear magnetic resonance imaging.

Fig. 15. Left: regularly and well-shaped fruits resulting from satisfactory pollination. Right: misshapen strawberries due to poor pollination.

6.2. Textural quality

- Yielding quality (firmness/softness): Hand-held testers can be used to determine penetration force. One example is the Magness-Taylor Pressure Tester. The plunger (tip) size used depends on the fruit and varies between 3 and 11 mm. Stand-mounted testers can determine penetration force with more consistent punch speed, one example being the UC Fruit Firmness Tester. Contractual laboratory testing is available for appraising fruit firmness with instruments such as the Instron Universal Testing machine, the Texture Testing system or by measuring fruit deformation using a Deformation Tester.
- Fibrosity and toughness: Shear force can be determined with an Instron or Texture Testing system. Resistance to cutting can be determined using a Fibrometer. Fibre or lignin content can be determined by contractual lab services with various chemical analyses.
- Succulence and juiciness: Any number of commercially available fruit refractometers can be used to measure water content - an indicator of succulence or turgidity.

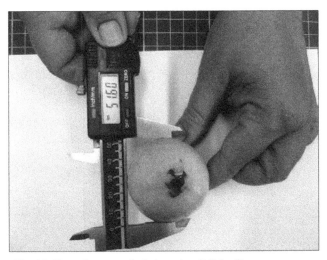

Fig. 13. Measuring guava fruit size using digital calliper.

Fig. 14. Left: large sweet pepper fruit from honey bee-pollinated greenhouse. Right: small fruit from a control, unpollinated greenhouse.

6.3. Flavour

- Sweetness: Sugar content can be determined by chemical analysis for total and reducing sugars or for individual sugars (e.g. yellow pitaya: Dag and Mizrahi, 2005). Total soluble solids can be used as a proxy measure of sugar content because sugars are the predominant component of fruit juice soluble solids. This parameter is measured with a fruit refractometer.
- Sourness (acidity): pH of extracted juice can be quantified with a pH meter or pH indicator paper. Total titratable acidity can be derived by titrating a specific volume of the extracted juice with 0.1 M NaOH to pH 8.1, then calculating titratable acidity as citric, malic, or tartaric acid, depending on which organic acid is dominant in the commodity.
- Astringency is quantified by taste test or by measuring the solubility of tannin or its degree of polymerization.

- Bitterness is quantified by taste test or by measuring alkaloids or the specific glucosides responsible for bitter taste.
- Aroma (odour) is quantified by use of a human sensory panel in combination with identifying the specific volatile components responsible for the aroma of the fruit.
- A comprehensive sensory evaluation can be used to characterize the combined sensory characteristics (sweetness, sourness, astringency, bitterness, overall flavour intensity) of the fruit.

6.4. Nutritional value

Various analytical methods are available to determine total carbohydrates, dietary fibre, proteins and individual amino acids, lipids and individual fatty acids, and vitamins and minerals in fruits and vegetables. For the most part, these kinds of analyses are specialized and require the collaboration of appropriate expertise.

7. Managing bee colonies for optimum pollination

7.1. Bee attractants

Bee attractants are designed to attract bees to crops. There are several approaches to testing their effectiveness, including measuring changes in the numbers of bees visiting flowers and changes in the levels of pollination (Ellis and Delaplane, 2009). The criteria in section 2.2.4. apply, and experimental units should never drop below the level of field plot. The complicating factor in these types of trials is achieving suitable replication and controls because attractants require large areas of a crop to be treated. For this reason, the expedient of using cages is not applicable to attractant studies. Treated and control plots must have large separations between them; otherwise there is risk that if the attractant works it may draw bees away from the control plots, thus artificially enhancing the effect of the attractant. Dependent variables can be collected as described in sections 4.1., 5.2., and 5.3.

7.2. Feeding colonies

7.2.1. Feeding syrup

Colonies can be fed liquid sugar syrup (sucrose) to cause them to increase the amount of pollen they collect (Goodwin, 1997) (Fig. 16). The syrup needs to be fed inside their hives, and the container needs to include flotation to minimize bee death from drowning. Feeding an average of 1 litre of syrup, with between 45 and 65% sucrose concentration, every day has been reported to result in significant increases in pollen collection (Goodwin and TenHouten, 1991). It is important that any syrup that has started to ferment in the feeder is discarded before more syrup is added. Feeding colonies outside their hives is unlikely to cause colonies to collect pollen.

7.2.2. Feeding pollen

Colonies can also be fed pollen or pollen substitutes to promote colony growth. Feeding pollen has, however, been reported to decrease the amount of pollen a colony collects (Free and Williams, 1971). This effect does not happen in all cases (Goodwin *et al.*, 1994).

7.2.3. Testing effects of feeding regimens on pollination performance

Given that colony nutrient state can affect its performance as a pollinator, feeding regimens can be used as experimental treatments in pollination studies. In these cases the criteria in section 2.2.4 apply. Experimental units should be no smaller than a field plot and sufficiently isolated to prevent bees drifting and confounding treatments. If space is limiting, cages can be used to contain sufficient plants with colonies assigned the different treatments. Dependent variables can be collected as described in sections 4.1., 5.2., and 5.3.

7.3. Distribution of colonies within the crop

The distribution of colonies within a crop is often a controversial issue between growers and beekeepers. This is because it is easiest for the beekeeper to drop off hives in single or several large groups. However, many growers want the hives spread evenly throughout the crop in the hope that there will be an equitable distribution of pollinators. Thus, testing for colony distributions that are optimum between these competing interests may be useful. The criteria given in section 2.2.4. apply, and experimental units for different hive distribution scenarios should be no smaller than a field plot and sufficiently isolated to

Fig. 16. Feeding a colony sugar syrup.

prevent bees drifting and confounding treatments. Dependent variables can be collected as described in sections 4.1, 5.2, and 5.3, and reported as treatment means with plot or field as experimental unit. Depending on the hive distribution patterns selected, the use of field transects may be useful as covariates or supplemental information to understand pollinator performance relative to a point source of bees. Section 5.1.2. gives useful guidance on the use of field transects.

8. Conducting pollination research in greenhouses and tunnels

Many high-value cash crops which were once cultivated exclusively in open fields are now grown in greenhouses and net-houses. This shift has been made mainly to protect plants from pests, enable out-of-season production, isolate plants for production of pure seeds, or to limit other environmental hazards (Fig. 17).

Assessing pollination activity in an enclosure is similar to assessing it in the open field (see sections 4 and 5). However, conducting pollination research in this specialized environment requires special considerations, covered below.

8.1. Carbon dioxide (CO_2) level

CO_2 enrichment has been used for many years in greenhouses to increase crop growth and yield. Since nectar production may be related to photosynthesis level, we might expect an increase in floral rewards in greenhouses with enriched CO_2 (Dag and Eisikowitch, 2000). On the other hand, if the greenhouse is completely closed and contains high biomass, intensive photosynthesis may lead to a reduction in CO_2 to lower than ambient (350 ppm) levels. There are different sensors available on the market to assess CO_2 level. Sensors that use non-dispersive infrared (NDIR) technology to measure CO_2 concentration in the greenhouse air are common and generally reliable. The sensor is placed near the leaf, i.e. the organ most affected by CO_2 levels.

Fig. 17. Bee hives placed for greenhouse melon pollination.

8.2. Solar radiation

In recent years, various modifications have been made to the spectral characteristics of greenhouse covers. These alterations are designed to protect greenhouse-grown crops from herbivorous insects and insect-borne viral diseases and to suppress proliferation of foliar diseases. These goals are achieved through partial or complete absorption of solar UV radiation (Raviv and Antigunus, 2004). However, UV radiation is also essential for honey bee and bumble bee navigation (Sakura *et al.,* 2012). Use of UV-absorbing sheets as well as UV-opaque covers (such Perspex and fiberglass) can therefore be damaging to bees' pollination activity in enclosures (Dag, 2008). The degree of UV absorption can be evaluated using UV sensors (10–400 nm), and opacity to UV radiation can be assessed by observing the shadow of an object in the enclosure through a UV filter; if a distinct shadow cannot be seen, then the UV radiation is diffuse.

8.3. Temperature and humidity

Relative humidity tends to be higher in greenhouses than in the open field. This high humidity directly affects floral rewards; the nectar sugar is more dilute since humidity affects rate of evaporative water loss. As a result, nectar sugar concentrations are sometimes below those preferred by honey bees, negatively affecting pollination activity (Dag and Eisikowich, 1999). Furthermore, at high temperatures honey bees lose heat through the evaporative cooling that occurs when they regurgitate nectar from the honey stomach (Heinrich, 1980). Temperature and relative humidity sensors therefore should be placed inside the plant foliage to follow the environmental conditions to which the flower is exposed, and somewhere above the foliage to monitor conditions to which the foragers are exposed.

8.4. Directed air flow

Greenhouses are actively ventilated by forced air ventilation or passively by opening side walls to reduce humidity and overheating that stress plants and promote foliar diseases (Fig. 18). Side walls and screens can also be opened to direct and regulate air flow and velocity within the greenhouse relative to the location of a hive. Air flow direction has been shown to affect honey bee pollination activity and subsequent fruit-set in the greenhouse (Dag and Eisikowitch, 1995), a phenomenon explained by bees' tendency to fly upwind (Friesen, 1973). It is recommended that a wind speed and direction sensor (anemometer) be placed near the hive entrance as well as in the greenhouse in a central location above the plant foliage.

8.5. Limited food resources

The amount of nectar and pollen provided by a crop in an enclosure is generally insufficient for long-term maintenance of honey bee colonies (Free, 1993). Moreover, adverse effects may express in a honey bee colony restricted to a greenhouse monofloral pollen source (Herbert *et al.*, 1970); pollination activity is curtailed, and colonies may deteriorate in the space of a few weeks and eventually collapse (Kalev *et al.*, 2002). There are different solutions for these nutritional deficits (Dag, 2008). One is to allow the honey bees to forage in the open and in the enclosures on alternating days (Butler and Haigh, 1956). Another is to use double entrance hives with one entrance leading into the enclosure and the other leading outside the enclosure to allow bees to forage and feed on the surrounding flora (Free, 1993). A third possibility is to artificially feed the colony (section 7.2.). This was shown to be efficient in a sweet pepper greenhouse (Kalev *et al.,* 2002).

Fig. 18. Greenhouse ventilation system.

9. Pesticides and pollinators

Negative consequences of pesticide interactions with bees pollinating crops are a serious concern. Methods to assess risk to individual bees and colonies from toxic effects of chemicals are established, and methods are expanding to include sublethal behavioural effects such as disorientation of foragers (see Medrzycki *et al.,* 2013). Obviously any environmental toxins which affect the health of a colony may impact the effectiveness of the colony as a pollinating unit by altering (especially diminishing) foraging activity.

Other effects can come from purposeful use of chemical attractants and repellents on a blooming crop. The effects of such chemicals can be measured using techniques to determine bee densities in whole fields or orchards (sections 5.1. or 5.2.) or, more commonly, in small plots. Similar small-plot techniques can be used to gauge any pollination-related effects from GMO crops that potentially arise from altered secretion of nectar and shedding of pollen.

10. Economic valuation of crop pollination by honey bees

Several methods have been proposed for the economic valuation of crop pollination by honey bees and wild insects. This value has been defined as the cost to replace pollination provided by honey bees or wild insects with other sources (e.g. hand pollination) (Allsopp *et al.*, 2008), the income of crop production attributable to pollination (Morse and Calderone, 2000; Gallai *et al.*, 2009), the net income (income - costs) of crop production attributable to pollination (Olschewski *et al.*, 2006; Veddeler *et al.*, 2008; Winfree *et al.*, 2011), or a consumer surplus approach (Southwick and Southwick, 1992). Considering the studies using these methods, some focused on the effects of the *total* depletion of biotic pollination; however, typical management decisions only produce *partial* changes in biotic pollination (Fisher *et al.*, 2008). Therefore, marginal values are most useful when designing management strategies. Here we describe briefly how to quantify the contribution of adding hives of honey bees to the value of crop production at the local scale, using the net income method (Olschewski *et al.*, 2006; Veddeler *et al.*, 2008; Winfree *et al.*, 2011). The critical variables are the increase in yield realized by the addition of X hives (ΔH).

Valuation methods apply differently at different scales such as global *(Gallai et al.*, 2009), national (Southwick and Southwick, 1992; Morse and Calderone, 2000), regional, subregional, or local. In sections 10.1. and 10.2., we focus on the local scale because this is the level at which management decisions are applied, i.e., the optimal density of hives needed for a certain crop species. At regional or larger scales, lower crop yield produced by massive losses of pollinators can generate compensatory increases in cultivated area to maintain crop production (Garibaldi *et al.*, 2011) or increases in market prices of crops (Winfree *et al.*, 2011), and these are treated in sections 10.3.

10.1. Determining yield in response to specific colony density

The reader is directed to sections 4 and 5 that explain these methods in detail, bearing in mind that for our immediate purpose we are interested in the change in net income given a particular increase in hive number. Considering that increased pollinator abundance should augment yield at a decelerating rate to the point that additional individuals do not further increase (e.g. pollen saturation) (Fig. 19) or even decrease (e.g. pollen excess) yield (Chacoff *et al.*, 2008; Morris *et al.*, 2010; Garibaldi *et al.*, 2011), valuation analysis should include a range of hive densities (i.e. from zero to high numbers). At the very least two situations are needed for comparison: fields with hives *vs.* fields without hives (control). Each treatment should be replicated

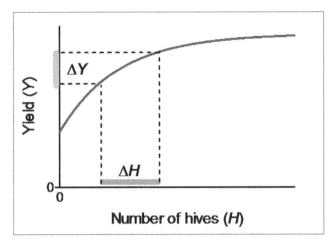

Fig. 19. The marginal benefit to crop yield (e.g. tonnes per ha) of each additional bee colony decreases with colony number. When this marginal benefit equals the cost generated by more yield (variable cost) plus the cost of renting a hive, the net income generated by the addition of the hive is zero and is no longer economically beneficial for the farmer.

with several fields (Prosser, 2010), and the number of necessary fields can be estimated using standard statistical techniques (Anderson *et al.*, 2008). In brief, more fields are necessary if we desire higher statistical power or if we face highly heterogeneous conditions within and between fields. Pollination provided by wild insects should be also measured, as the effect of adding hives of honey bees on crop yield will greatly depend on the "base" level of pollination being provided by wild insects. In addition, the presence of wild insects can enhance honey bees pollination behaviour (Greenleaf and Kremen, 2006; Carvalheiro *et al.*, 2011).

10.2. Response variables and calculations

Net income is the difference between costs from income. We are interested in the change in net income (ΔH) given a particular increase in number of hives (ΔNI):

$$\Delta NI = P * \Delta Y - Cy * \Delta Y - Ch * \Delta H$$

where P is the price that the farmer obtains for each metric tonne of crop, ΔY is the increase in metric tonnes of crop because of the addition of hives (ΔH), Cy is the cost of producing each tonne (i.e. variable costs such as harvest and transportation costs), and Ch is the cost of renting each hive. As mentioned before, at least two treatments are needed to estimate yields without honey bees and subtract them from yields with honey bees ($\Delta Y = Ywith - Ywithout$). In case information for several treatments is available, i.e. several densities of colonies are evaluated, a functional form of yield (Y) with increased number of hives (H) can be estimated. This function should be used to obtain ΔY values for any number of hives within the measured range (Fig. 19).

The data to perform this valuation should be measured at the field scale (e.g. crop yield) or obtained through questionnaires to the farmers and beekeepers (Olschewski *et al.,* 2006). Fruit or seed yield (tonnes per ha) should be measured at ripeness or harvest. The crop price (P) and production costs (including the costs of renting hives) should be obtained from questionnaires.

If honey bees promote yield quality (e.g. bigger and well-formed fruits) in addition to yield quantity (Y), changes in crop prices (**P**) may occur with or without hives. In this case, the price obtained with the desired number of hives should be used in the above formula to estimate *ΔNI* (note that not only ΔY but also all the harvest, **Y**, can show enhanced quality and price). On the other hand, if losses of honey-bee colonies are replaced with other sources of pollination (e.g. hand pollination), this change in production costs should be also accounted in the above formula. Finally, we must account for some management inputs, such as planting hedge rows with flowering species to improve honey-bee colony health, that incur high initial costs but low costs in subsequent years. Therefore, the temporal scale of analyses, as well as equations used to estimate the net income, will depend on the management practices to be evaluated.

By now it should be evident that the quality of the data gathered has a strong influence on the resulting values for the contribution of honey bees. Several of the ideas discussed here (e.g. replication, scale) can also be applied to other objectives such as the evaluation of the impacts of adding hives on the pollination of surrounding wild vegetation or the diversity and abundance of wild pollinators.

10.3. Economic valuation at larger scales

Crop production (supply) is the product of yield (tonnes ha^{-1}) and cultivated area (ha). Lower crop yield (or slower yield growth over years) generated by the lack of adequate pollination can affect production at regional, national or global scales (Garibaldi *et al.*, 2011). In this case, cultivation of more area to compensate for production losses is a likely outcome and should be included in valuation. Another likely outcome is the increase in the market price for the harvested product (Winfree *et al.*, 2011). Valuation at larger than local scales should also account for the welfare of producers as well as consumers (Gallai *et al.*, 2009; Winfree *et al.*, 2011). In economic terms, the welfare of producers can be described as the producer surplus (here we will focus on the net income approach following Winfree *et al.*, 2011), and the welfare of consumers can be described as the consumer surplus (Fig. 20) (Southwick and Southwick, 1992). In addition, it is important for decision making to consider how the value of honey bee pollination to crops changes spatially across the study region (Chaplin-Kramer *et al.*, 2011; Lautenbach *et al.*, 2012).

Here we discuss quantifying the extent to which honey bee numbers affect crop production value loss or gain at the regional scale. Changes in honey-bee abundance within a region will impact

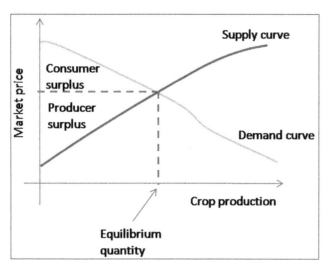

Fig. 20. Consumer surplus occurs because consumers are able to purchase a product at a lower price than the highest price they are willing to pay (presented as area between the demand curve and the market price at the equilibrium quantity). If crop price rises, consumer surplus decreases. Producer surplus occurs because producers sell at a price that is higher than the lowest price they are willing to receive to sell their product (presented as the area between the supply curve and the price at the equilibrium quantity). The functional form of the supply and demand curves are only to exemplify the concept of consumer and producer surplus and need to be estimated from real data for each crop and market.

social welfare (**SW**) in three ways: through the aggregate net income of the crop producers in the affected area (**NIr**), the aggregate net income of producers outside the affected area but sharing the same market (**NIo**), and the consumer surplus (**CS**) (Winfree *et al.*, 2011).

$$SW = NIr + NIo + CS$$

The net income because of a certain variation (suppose losses) in the number of hives (ΔH) for producers within (**NIr**) and outside (**NIo**) an affected area can be estimated similarly to that given above for the local scale:

$$NI = (\Delta P + P) * (\Delta Y + Y) - Cy * (\Delta Y + Y) - (\Delta Ch + Ch) * (\Delta H + H)$$

For producers within the affected area, the loss of honey bees can decrease yield (ΔY) in comparison to the yield previous to the loss of honey bees (Y) and increase crop price (ΔP). Therefore, net income can be reduced because of lower yield on the one hand, but increased because of higher prices on the other. The net outcome will depend on the relative changes in yield and crop prices. In addition, lower yield will reduce the variable costs ($Cy * \Delta Y + Y$) such as costs to harvest and transport crop yield. A significant decrease in the number of honey bee colonies will likely increase the cost of renting each hive ($\Delta Ch + Ch$) and modify the number of hives each producer rents ($\Delta H + H$). For producers outside the affected area,

no change in yield ($\Delta Y = 0$) or variable costs ($Cy * \Delta Y = 0$) may happen, but net income is influenced by changes in crop price and probably by changes in the cost of renting a hive. Finally, it is important to note that ΔP will be a function of the amount of crop production (tonnes) that decreases in the affected area in relation to the total production traded at the market and of the ability of producers to increase crop area to compensate for lower production (i.e. the price elasticity of supply) (Fig. 20) (Garibaldi *et al.*, 2011).

Consumer surplus occurs if consumers are willing to pay a price for the crop product that is higher than market price. Therefore, price increases resulting from honey bee losses will reduce consumer surplus (Fig. 20). Estimating the change in consumer surplus requires estimation of the demand curve using questionnaires or historical market data. The estimations presented here for total welfare effect of a certain variation in honey-bee numbers will also require the estimation of ΔP which can be obtained from the price elasticity of supply and the current crop prices (Fig. 20) (for more details see Winfree *et al.*, 2011). The rest of the data required are the same as discussed in sections 10.2. and 10.3.

We have limited ourselves here to a brief introduction of the key factors for analysing the value of honey bees as pollinators of agricultural crops at a regional scale. The social welfare value obtained with these models depends on the quality of the data gathered. Correspondingly, it is important to study how the resulting values for social welfare change with variation in the assumed functional forms or parameter estimates (i.e. sensitivity analyses; the same is true for the local scale valuation presented before). The functional form of the number of hives on crop yield has not been reliably estimated for most crops at field scales; this is a crucial knowledge deficit in our understanding of the benefits of honey bees to agricultural production.

11. Acknowledgements

The COLOSS (Prevention of honey bee COlony LOSSes) network aims to explain and prevent massive honey bee colony losses. It was funded through the COST Action FA0803. COST (European Cooperation in Science and Technology) is a unique means for European researchers to jointly develop their own ideas and new initiatives across all scientific disciplines through trans-European networking of nationally funded research activities. Based on a pan-European intergovernmental framework for cooperation in science and technology, COST has contributed since its creation more than 40 years ago to closing the gap between science, policy makers and society throughout Europe and beyond. COST is supported by the EU Seventh Framework Programme for research, technological development and demonstration activities (Official Journal L 412, 30 December 2006). The European Science Foundation as implementing agent of COST provides the COST Office through an EC Grant Agreement. The Council of the European Union provides the COST Secretariat. The COLOSS network is now supported by the Ricola Foundation - Nature & Culture.

12. References

ALEXANDER, M P (1969) Differential staining of aborted and non-aborted pollen. *Stain Technology* 44(3): 117-122.

ALLSOPP, M H; DE LANGE, W J; VELDTMAN, R (2008) Valuing insect pollination services with cost of replacement. *PloS ONE* 3: e3128. http://dx.doi.org/10.1371/journal.pone.0003128

ANDERSON, D R; SWEENEY, D J; WILLIAMS, T A (2008) *Statistics for business and economics IIe.* South-Western Cengage Learning; Mason, USA.

AUSTIN, P T; HEWETT, E W; NOITON, D A; PLUMMER, J A (1996) Cross pollination of 'Sundrop' apricot (*Prunus armeniaca* L) by honey bees. *New Zealand Journal of Crop and Horticultural Science* 24(3): 287-294.

BAKER, R G; JAY, S C (1973) A comparison of foraging activity of honey bee colonies with large and small entrances. *The Manitoba Entomologist* 8: 48-54.

BENEDEK, P; KOCSIS-MOLNAR, G; NYÉKI, J (2000) Nectar production of pear (*Pyrus communis* L.) cultivars. *International Journal of Horticultural Science* 6(3): 67-75.

BOMMARCO, R; MARINI, L; VAISSIÈRE, B E (2012) Insect pollination enhances seed yield, quality, and market value in oilseed rape. *Oecologia* 169: 1025-1032.

BRAULT, A M; DE OLIVEIRA, D; MARCEAU, J (1995) Optimization of apple orchard pollination by honey bees in eastern Canada. *Canadian Honey Council Research Symposium Proceedings* 1995 : 55-104.

BREWBAKER, J L; KWACK, B H (1963) Essential role of the calcium ion in pollen germination and pollen tube growth. *American Journal of Botany* 50: 747–858.

BRITO, R M; MCHALE, M; OLDROYD, B P (2010) Expression of genes related to reproduction and pollen foraging in honey bees (*Apis mellifera*) narcotized with carbon dioxide. *Insect Molecular Biology* 19(4): 451-461.

BUTLER, C G; HAIGH, J C (1956) A note on the use of honey bees as pollinating agents in cages. *Journal of Horticultural Science* 31: 295–297.

CARVALHEIRO, L G; VELDTMAN, R ; SHENKUTE, A G; TESFAY, G B; PIRK, C W W; DONALDSON, J S; NICOLSON, S W (2011) Natural and within-farmland biodiversity enhances crop productivity. *Ecology Letters* 14: 251-259.

CHACOFF, N P; AIZEN, M A; ASCHERO, V (2008) Proximity to forest edge does not affect crop production despite pollen limitation. *Proceedings of the Royal Society* B275: 907–913. http://dx.doi.org/10.1098/rspb.2007.1547

CHAPLIN-KRAMER, R; TUXEN-BETTMAN, K; KREMEN, C (2011) Value of wildland habitat for supplying pollination services to Californian agriculture. *Rangelands* 33: 33–41. http://dx.doi.org/10.2111/1551-501X-33.3.33

DAFNI, A (1992) *Pollination ecology: a practical approach.* Oxford University Press; Oxford, UK. 250 pp.

DAFNI, A; FIRMAGE, D (2000) Pollen viability and longevity: practical, ecological and evolutionary implications. *Plant Systematics and Evolution* 222: 113-132.

DAFNI, A; PACINI, E; NEPPI, M (2005) Pollen and stigma biology. In *A Dafni; P G Kevan; B C Husband (Eds). Practical pollination biology.* Envirquest Ltd; Cambridge, Ontario, Canada. pp 83-146.

DAG, A (2008) Bee pollination of crop plants under environmental conditions unique to enclosures. *Journal of Apicultural Research* 47: 162-165.

DAG, A; EISIKOWITCH, D (1995) The influence of hive location on honey bee foraging activity and fruit-set in melon grown in plastic greenhouses. *Apidologie* 26: 511-519.

DAG, A; EISIKOWITCH, D (1999) Ventilation of greenhouses increases honey bee foraging activity in melon, *Cucumis melo. Journal of Apicultural Research* 38: 169-175.

DAG, A; EISIKOWITCH, D (2000) The effect of carbon dioxide enrichment on nectar production in melon under greenhouse conditions. *Journal of Apicultural Research* 39: 88-89.

DAG, A; MIZRAHI, Y (2005) Effect of pollination method on fruit-set and fruit characteristics in the vine cactus *Selenicereus megalanthus* ('Yellow pitaya'). *Journal of Horticultural Science and Biotechnology* 80: 618-622.

DAG, A; GAZIT, S; EISENSTEIN, D; EL-BATSRI, R; DEGANI, C (1999) Effect of the male parent on pericarp and seed weights in several Floridian mango cultivars. *Scientia Horticulturae* 82: 325-329.

DAG, A; AFIK, O; YESELSON, Y; SCHAFFER, A; SHAFIR, S (2006) Physical, chemical and palynological characterization of avocado (*Persea americana* Mill.) honey in Israel. *International Journal of Food Science and Technology* 41(4): 387-394.

DAG, A; ZVIELI, Y; AFIK, O; ELKIND, Y (2007) Honey bee pollination affects fruit characteristics of sweet pepper grown under net-houses. *International Journal of Vegetable Sciences* 13: 45-59.

DEDEJ, S; DELAPLANE, K S (2003) Honey bee (Hymenoptera : Apidae) pollination of rabbiteye blueberry *Vaccinium ashei* var. 'Climax' is pollinator density-dependent. *Journal of Economic Entomology* 96(4): 1215-1220.

DEGANI, H; GOLDRING, A; ADATO, I; EL-BATSRI, R; GAZIT, S (1990) Pollen parent effect on outcrossing rate, yield, and fruit characteristics of Fuerta avocado. *HortScience* 25: 471-473.

DELAPLANE, K S (2011) Understanding the impact of honey bee disorders on crop pollination. In *D Sammataro; J A Yoder (Eds). Honey bee colony health.* CRC Press; Florida, USA. pp 223-228.

DELAPLANE, K S; MAYER, D F (2000) *Crop pollination by bees.* CABI; New York, USA.

DELAPLANE, K S; VAN DER STEEN, J; GUZMAN, E (2013) Standard methods for estimating strength parameters of *Apis mellifera* colonies. In *V Dietemann; J D Ellis; P Neumann (Eds) The COLOSS BEEBOOK, Volume I: standard methods for* Apis mellifera *research. Journal of Apicultural Research* 52(1): http://dx.doi.org/10.3896/IBRA.1.52.1.03

ECKERT, J (1942) The pollen required by a colony of honey bees. *Journal of Economic Entomology* 35: 309-311.

EISCHEN, F A; UNDERWOOD, B A (1991) Cantaloupe pollination trials in the lower Rio Grande Valley. *American Bee Journal* 131: 775.

ELLIS, A; DELAPLANE, K S (2009) An evaluation of Fruit-Boost[TM] as an aid for honey bee pollination under conditions of competing bloom. *Journal of Apicultural Research* 48(1): 15-18.

ERDTMAN, G (1969) *Handbook of palynology.* Munksgaard ; Copenhagen, Denmark.

FAEGRI, K; VAN DER PIJL, L (1979) *The principles of pollination ecology.* Pergamon Press ; Oxford, UK.

FISHER, B; TURNER, K; ZYLSTRA, M; BROUWER, R; DE GROOT, R; FARBER, S; FERRARO, P; GREEN, R; HADLEY, D; HARLOW, J; JEFFERISS, P; KIRKBY, C; MORLING, P; MOWATT, S; NAIDOO, R; PAAVOLA, J; STRASSBURG, B; YU, D; BALMFORD, A (2008) Ecosystem services and economic theory: Integration for policy-relevant research. *Ecological Applications* 18: 2050-2067.

FREE, J B (1966) The pollination of beans *Phaseolus multiflorus* and *Phaseolus vulgaris* by honey bees. *Journal of Apicultural Research* 5: 87-91.

FREE, J B (1993) *Insect pollination of crops (2nd Ed.).* Cardiff University Press; Cardiff, UK.

FREE, J B; WILLIAMS, H I (1971) The effect of giving pollen and pollen supplements to honey bee colonies on the amount of pollen collected. *Journal of Apicultural Research* 10: 87–90.

FREITAS, B M; PAXTON, R J (1996) The role of wind and insects in cashew (*Anacardium occidentale*) pollination in N E Brazil. *Journal of Agricultural Science, Cambridge* 126: 319-326.

FREITAS, B M; PAXTON, R J (1998) A comparison of two pollinators: the introduced honey bee *Apis mellifera* and an indigenous bee *Centris tarsata* on cashew *Anacardium occidentale* in its native range of N E Brazil. *Journal of Applied Ecology* 35: 109-121.

FRIESEN, L J (1973) The search dynamics of recruited honey bees. *The Biological Bulletin* 144: 107-131.

GALLAI, N; SALLES, J M; SETTELE, J; VAISSIÈRE, B E (2009) Economic valuation of the vulnerability of world agriculture confronted with pollinator decline. *Ecological Economics* 68: 810-821.

GAALICHE, B; TRAD, M; MARS, M (2011) Effect of pollination intensity, frequency and pollen source on fig (*Ficus carica* L.) productivity and fruit quality. *Scientia Horticulturae* 130: 737-742.

GARIBALDI, L A; AIZEN, M A; KLEIN, A M; CUNNINGHAM, S A; HARDER, L D (2011) Global growth and stability of agricultural yield decrease with pollinator dependence. *Proceedings of The National Academy of Sciences of the United States of America* 108: 5909–5914. http://dx.doi.org/10.1073/pnas.1012431108

GARY, N E (1971) Magnetic retrieval of ferrous labels in a capture-recapture system for honey bees and other insects. *Journal of Economic Entomology* 64: 961-965.

GARY, N E; WITHERELL, P C; LORENZEN, K (1981) Effect of age on honey bee foraging distance and pollen collection. *Environmental Entomology* 10(6): 950-952.

GOODWIN, R M (1992) Feeding sugar syrup to honey bee colonies to improve pollination: A review. *Bee World* 78: 56-62.

GOODWIN, R M; TEN HOUTEN, A (1991) Feeding sugar syrup to honey bee (*Apis mellifera* L.) colonies to increase kiwifruit pollen collection: Effect of frequency, quantity and time of day. *Journal of Apicultural Research* 30: 41-48.

GOODWIN, R M; PERRY, J H (1992) Use of pollen traps to investigate the foraging behaviour of honey-bee colonies in kiwifruit orchards. *New Zealand Journal of Crop and Horticultural Science* 20(1): 23-26.

GOODWIN, R M; TEN HOUTEN, A; PERRY, J H (1994) Effect of feeding pollen substitutes to honey-bee colonies used for kiwifruit pollination and honey production. *New Zealand Journal of Crop and Horticultural Science* 22: 459-462.

GOODWIN, R M; COX, H M; TAYLOR, M A; EVANS, L J; MCBRYDIE, H M (2011) Number of honey bee visits required to fully pollinate white clover (*Trifolium repens*) seed crops in Canterbury, New Zealand. *New Zealand Journal of Crop and Horticultural Science* 39(1): 7-19.

GREENLEAF, S S; KREMEN, C (2006) Wild bees enhance honey bees' pollination of hybrid sunflower. *Proceedings of the National Academy of Science* 103: 13890-13895.

GROSS, C L (2005) In *A Dafni; P G Kevan; B C Husband (Eds). Practical pollination biology.* Enviroquest Ltd ; Cambridge, Ontario, Canada. pp 354–363.

HASEGAWA, Y; SUYAMA, Y; SEIWA, K (2009) Pollen donor composition during the early phases of reproduction revealed by DNA genotyping of pollen grains and seeds of *Castanea crenata*. *New Phytologist* 182: 994–1002.

HEDHLY, A; HORMAZA, J I; HERRERO, M (2009) Flower emasculation accelerates ovule degeneration and reduces fruit-set in sweet cherry. *Scientia Horticulturae* 119: 455-457.

HEINRICH, B (1980) Mechanisms of body-temperature regulation in honey bees, *Apis mellifera*. I. Regulation of head temperature. *Journal of Experimental Biology* 85: 61-72.

HESLOP-HARRISON, J S; HESLOP-HARRISON, Y (1970) Evaluation of pollen viability by enzymatically induced fluorescence: intracellular hydrolysis of fluorescein diacetate. *Stain Technology* 45: 115–120.

HERBERT, E W; BICKLEY, W; SHIMANUKI, H (1970) The brood-rearing capability of caged honey bees fed dandelion and mixed pollen diet. *Journal of Economic Entomology* 63: 215-218.

HERRERO, M; HORMAZA, J I (1996) Pistil strategies controlling pollen tube growth. *Sexual Plant Reproduction* 9: 343-347.

HOWPAGE, D; SPOONER-HART, R N; SHEEHY, J (1998) A successful method of mass marking honey bees, *Apis mellifera*, at the hive entrance for field experiments. *Journal of Apicultural Research* 37 (2): 91-97.

HUDSON, L C; CHAMBERLAIN, D; STEWART, C N JR (2001) GFP-tagged pollen to monitor pollen flow of transgenic plants. *Molecular Ecology Notes* 1: 321–324.

HUMAN, H; BRODSCHNEIDER, R; DIETEMANN, V; DIVELY, G; ELLIS, J; FORSGREN, E; FRIES, I; HATJINA, F; HU, F-L; JAFFÉ, R; JENSEN, A B; KÖHLER, A; MAGYAR, J; ÖZIKRIM, A; PIRK, C W W; ROSE, R; STRAUSS, U; TANNER, G; TARPY, D R; VAN DER STEEN, J J M; VAUDO, A; VEJSNÆS, F; WILDE, J; WILLIAMS, G R; ZHENG, H-Q (2013) Miscellaneous standard methods for *Apis mellifera* research. In *V Dietemann; J D Ellis; P Neumann (Eds) The COLOSS BEEBOOK, Volume I: standard methods for* Apis mellifera *research. Journal of Apicultural Research* 52(4): http://dx.doi.org/10.3896/IBRA.1.52.4.10

JAY, S C (1986) Spatial management of honey bees on crops. *Annual Review of Entomology* 31: 49-65.

JONES, G (2012) Pollen extraction from insects. *Palynology* 36: 86-109.

KADER, A A (2002) *Postharvest technology of horticultural crops (3rd Ed.)*. Publication 3311, University of California, Agriculture and Natural Resources; Oakland, CA, USA.

KALEV, H; DAG, A; SHAFIR, S (2002) Feeding pollen supplements to honey bee colonies during pollination of sweet pepper in enclosures. *American Bee Journal* 142: 672-678.

KAWASHIMA, S; CLOT, B; FUJITA, T; TAKAHASHI, Y; NAKAMURA, K (2007) An algorithm and a device for counting airborne pollen automatically using laser optics. *Atmospheric Environment* 41: 7987–7993.

KEARNS, C A; INOUYE, D W (1993) *Techniques for pollination biologists*. University Press; Colorado, USA.

KIRK, W D J (2006) *A colour guide to pollen loads of the honey bee (2nd Ed.)*. International Bee Research Association; Cardiff, UK.

LAUTENBACH, S; SEPPELT, R; LIEBSCHER, J; DORMANN, C F (2012) Spatial and temporal trends of global pollination benefit. *PloS ONE* 7: e35954. http://dx.doi.org/10.1371/journal.pone.0035954

LEVIN, M D (1960) A comparison of two methods of mass marking foraging honey bees. *Journal of Economic Entomology* 53: 696-698.

LEVIN, M D; LOPER, G M (1984) Factors affecting pollen trap efficiency. *American Bee Journal* 124(10): 721-723.

LINSKENS, H F; ESSER, K (1957) Uber eine spezifische Anfärbung der Pollen-Shläuche und die Zahl Kallosepfropen nach Selbstung und Fremdung. *Naturwissenschaften* 44: 16.

MANNING, R; BOLAND, J (2000) A preliminary investigation into honey bee (*Apis mellifera*) pollination of canola (*Brassica napus* cv. Karoo) in Western Australia. *Australian Journal of Experimental Agriculture* 40(3): 439-442.

MCDONALD, J (1968) The behaviour of pollen foragers which lose their loads. *Journal of Apicultural Research* 7: 45-46.

MCLAREN, G F; FRASER, J A; GRANT, J E (1992) Pollination of apricots. *Orchardist of New Zealand* 65(8): 20-23.

MATSUKI, Y; TATENO, R; SHIBATA, M; ISAGI, Y (2008) Pollination efficiencies of flower-visiting insects as determined by direct genetic analysis of pollen origin. *American Journal of Botany* 95: 925-930.

MEDRZYCKI, P; GIFFARD, H; AUPINEL, P; BELZUNCES, L P; CHAUZAT, M-P; CLAßEN, C; COLIN, M E; DUPONT, T; GIROLAMI, V; JOHNSON, R; LECONTE, Y; LÜCKMANN, J; MARZARO, M; PISTORIUS, J; PORRINI, C; SCHUR, A; SGOLASTRA, F; SIMON DELSO, N; STEEN VAN DER, J; WALLNER, K; ALAUX, C; BIRON, D G; BLOT, N; BOGO, G; BRUNET, J-L; DELBAC, F; DIOGON, M; EL ALAOUI, H; TOSI, S; VIDAU, C (2013) Standard methods for toxicology research in *Apis mellifera*. In *V Dietemann; J D Ellis; P Neumann (Eds) The COLOSS BEEBOOK, Volume I: standard methods for* Apis mellifera *research. Journal of Apicultural Research* 52(4): http://dx.doi.org/10.3896/IBRA.1.52.4.14

MESQUIDA, J; RENARD, M; PIERRE, J S (1988) Rapeseed (*Brassica napus* L.) productivity: the effect of honey bees (*Apis mellifera* L.) and different pollination conditions in cage and field tests. *Apidologie* 19(1): 51-72.

MORRIS, W F; VÁZQUEZ, D P; CHACOFF, N P (2010) Benefit and cost curves for typical pollination mutualisms. *Ecology* 91: 1276–1285.

MORSE, R A; CALDERONE, N W (2000) The value of honey bees as pollinators of U S Crops in 2000. *Bee Culture* 128.

NE'EMAN, G; JUERGENS, A; NEWSTROM-LLOYD, L; POTTS, S G; DAFNI, A (2010) A Framework for comparing pollinator performance: effectiveness and efficiency. *Biological Reviews* 85: 435-451.

NEPI, M; PACINI, E (1993) Pollination, pollen viability and pistil receptivity in *Cucurbita pepo*. *Annals of Botany* 72: 527-536.

OLSCHEWSKI, R; TSCHARNTKE, T; BENITEZ, P C; SCHWARZE, S; KLEIN, A M (2006) Economic evaluation of pollination services comparing coffee landscapes in Ecuador and Indonesia. *Ecology and Society* 11(1): 7.

PALMER-JONES, T; CLINCH, P (1974) Observations on the pollination of Chinese gooseberries variety 'Hayward'. *New Zealand Journal of Experimental Agriculture* 2: 445-458.

PÉREZ-BALAM, J; QUEZADA-EUÁN, J J G; ALFARO-BATES, R; MEDINA, S; MCKENDRICK, L; SORO, A; PAXTON, R J (2012) The contribution of honey bees, flies and wasps to Avocado (*Persea americana*) pollination in southern Mexico. *Journal of Pollination Ecology* 8(6): 42-47.

PIERRE, J; VAISSIÈRE, B; VALLÉE, P; RENARD, M (2010) Efficiency of airborne pollen released by honey bee foraging on pollination in oilseed rape: a wind insect-assisted pollination. *Apidologie* 41: 109-115

PINILLOS, V; CUEVAS, J (2008) Standardization of the fluorochromatic reaction test to assess pollen viability. *Biotechnic and Histochemistry* 83: 15-21.

PROSSER, J I (2010) Replicate or lie. *Environmental Microbiology* 12: 1806–1810. http://dx.doi.org/10.1111/j.1462-2920.2010.02201.x

RAVIV, M; ANTIGNUS, Y (2004) UV radiation effects on pathogens and insect pests of greenhouse-grown crops. *Photochemistry and Photobiology* 72: 219-229.

RIBBANDS, C (1950) Changes in the behaviour of honey bees following their recovery from anaesthesia. *Journal of Experimental Biology* 27: 302-310.

RIZZARDO, R A G; MILFONT, M O; SILVA, E M S; FREITAS, B M (2012) *Apis mellifera* pollination improves agronomic productivity of anemophilous castor bean (*Ricinus communis*). *Annals of the Brazilian Academy of Sciences* 84: 605-608.

SAKURA, M; OKADA, R; AONUMA, H (2012) Evidence for instantaneous e-vector detection in the honey bee using an associative learning paradigm. *Proceedings of the Royal Society B: Biological Sciences* 1728: 535-542.

SAMPSON, B J ; CANE, J H (2000) Pollination efficiencies of three bee (Hymenoptera: Apoidea) species visiting rabbiteye blueberry. *Journal of Economic Entomology* 93: 1726-1731.

SHIVANNA, K R; JOHRI, B M (1985) *The angiosperm pollen structure and function.* Wiley Eastern Limited; New Delhi, India.

SOUTHWICK, E E; SOUTHWICK, L JR (1992) Estimating the economic value of honey bees (Hymenoptera: Apidae) as agricultural pollinators in the United States. *Journal of Economic Entomology* 85: 621-633.

SHIVANNA, K R; LINSKENS, H F; CRESTI, M (1991) Pollen viability and pollen vigour. *Theoretical and Applied Genetics* 81: 38–42.

SPEARS, E E (1983) A direct measure of pollinator effectiveness. *Oecologia* 57: 196-199.

STERN, R A; GOLDWAY, M; ZISOVICH, A H; SHAFIR, S; DAG, A (2004) Sequential introduction of honey bee colonies increase cross-pollination, fruit-set and yield of Spadona pear (*Pyrus communis*). *Journal of Horticultural Science and Biotechnology* 79: 652-658.

SUYAMA, Y (2011) Procedure for single pollen genotyping. In *Y Isagi; Y Suyama (Eds). Single-pollen genotyping. Ecological Research Monographs* 1.

SYNGE, A D (1947) Pollen collection by honey bees (*Apis mellifera*). *Journal of Animal Ecology* 16: 122-138.

TAYLOR, L P; HEPLER, P K (1997) Pollen germination and tube growth. *Annual Review of Plant Physiology, Plant Molecular Biology* 48: 461–491.

THOMSON, J D (1981) Field measures of flower constancy in bumble bees. *American Midland Naturalist* 105: 377–380.

VAISSIÈRE, B E (1991) Honey bee stocking rate, pollinator visitation and pollination effectiveness in upland cotton grown for hybrid seed production. *Acta Horticulturae* 288: 359-363.

VAISSIÈRE, B E; RODET, G; COUSIN, M; BOTELLA, L; TORRE GROSSA, J-P (1996) Pollination effectiveness of honey bees (Hymenoptera: Apidae) in a kiwifruit orchard. *Journal of Economic Entomology* 89(2): 453-461.

VAISSIÈRE, B E; FREITAS, B M; GEMMILL-HERREN, B (2011) *Protocol to detect and assess pollination deficits in crops: a handbook for its use.* Food and Agriculture Organization of the United Nations (available at http://www.internationalpollinatorsinitiative.org/jsp/documents/documents.jsp).

VEDDELER, D; OLSCHEWSKI, R; TSCHARNTKE, T; KLEIN, A M (2008) The contribution of non-managed social bees to coffee production: New economic insights based on farm-scale yield data. *Agroforestry Systems* 73: 109–114.

WINFREE, R; GROSS, B J; KREMEN, C (2011) Valuing pollination services to agriculture. *Ecological Economics* 71: 80–88. http://dx.doi.org/10.1016/j.ecolecon.2011.08.001.

Journal of Apicultural Research 52(1): (2013)
DOI 10.3896/IBRA.1.52.1.07

REVIEW ARTICLE

Standard methods for rearing and selection of

Apis mellifera queens

Ralph Büchler[1]*, Sreten Andonov[2], Kaspar Bienefeld[3], Cecilia Costa[4], Fani Hatjina[5], Nikola Kezic[6], Per Kryger[7], Marla Spivak[8], Aleksandar Uzunov[2] and Jerzy Wilde[9]

[1]LLH, Bee Institute, Erlenstrasse 9, 35274 Kirchhain, Germany.
[2]Faculty for Agricultural Science and Food, bul. Aleksandar Makedonski b.b., 1000 Skopje, Republic of Macedonia.
[3]Länderinstitut für Bienenkunde Hohen Neuendorf e.V., Friedrich-Engels-Str. 32, 16540 Hohen Neuendorf, Germany.
[4]Consiglio per la Ricerca e la sperimentazione in Agricoltura - Unità di ricerca di apicoltura e bachicoltura, Bee and Silkworm Research Unit, Via di Saliceto 80, 40128 Bologna, Italy.
[5]Hellenic Institute of Apiculture (N.AG.RE.F.), N. Moudania, Greece.
[6]Faculty of Agriculture, University of Zagreb, Svetosimunska 25, 10000 Zagreb, Croatia.
[7]Department of Integrated Pest Management, University of Aarhus, Forögsvej 1, 4200 Slagelse, Denmark.
[8]Department of Entomology, University of Minnesota, 219 Hodson Hall, 1980 Folwell Ave., St. Paul, MN 55108, USA.
[9]Apiculture Division, Faculty of Animal Bioengineering, Warmia and Mazury University, Sloneczna 48, 10-710 Olsztyn, Poland.

Received 28 March 2012, accepted subject to revision 13 June 2012, accepted for publication 5 November 2012.

*Corresponding author: Email: ralph.buechler@llh.hessen.de

Summary

Here we cover a wide range of methods currently in use and recommended in modern queen rearing, selection and breeding. The recommendations are meant to equally serve as standards for both scientific and practical beekeeping purposes. The basic conditions and different management techniques for queen rearing are described, including recommendations for suitable technical equipment. As the success of breeding programmes strongly depends on the selective mating of queens, a subchapter is dedicated to the management and quality control of mating stations. Recommendations for the handling and quality control of queens complete the queen rearing section. The improvement of colony traits usually depends on a comparative testing of colonies. Standardized recommendations for the organization of performance tests and the measurement of the most common selection characters are presented. Statistical methods and data preconditions for the estimation of breeding values which integrate pedigree and performance data from as many colonies as possible are described as the most efficient selection method for large populations. Alternative breeding programmes for small populations or certain scientific questions are briefly mentioned, including also an overview of the young and fast developing field of molecular selection tools. Because the subject of queen rearing and selection is too large to be covered within this paper, plenty of references are given to facilitate comprehensive studies.

Métodos estándar para la cría y selección de reinas de

Apis mellifera

Resumen

Se describe una amplia gama de métodos actualmente en uso y recomendables sobre la cría actual de reinas, su selección y cruzamiento. Las recomendaciones tienen el propósito de servir de igual forma como estándares para fines apícolas tanto científicos como prácticos. Se describen las condiciones básicas y las diferentes técnicas de manejo para la cría de reinas, incluyendo recomendaciones para el equipo técnico adecuado. Dado que el éxito de los programas de mejora depende en gran medida el apareamiento selectivo de reinas, se dedica un subcapítulo a la gestión y control de calidad de las estaciones de apareamiento. Las recomendaciones para el manejo y control de calidad de las reinas completan la sección de cría de reinas. La mejora de las características de colonias por lo general, depende de ensayos comparativos entre colonias. Se presentan recomendaciones normalizadas para la organización de pruebas de rendimiento y la medición de los caracteres de selección más comunes. Aquellos métodos estadísticos y condiciones previas de datos para la estimación de valores de cruzamiento que integren los datos genealógicos y de rendimiento de tantas colonias como sea posible, se describen como los métodos de

Footnote: : Please cite this paper as: BÜCHLER, R; ANDONOV, S; BIENEFELD, K; COSTA, C; HATJINA, F; KEZIC, N; KRYGER, P; SPIVAK, M; UZUNOV, A; WILDE, J (2013) Standard methods for rearing and selection of *Apis mellifera* queens. In *V Dietemann; J D Ellis; P Neumann (Eds) The COLOSS BEEBOOK, Volume I: standard methods for* Apis mellifera *research. Journal of Apicultural Research* 51(5): http://dx.doi.org/10.3896/IBRA.1.52.1.07

selección más eficientes para grandes poblaciones. Se mencionan también pero brevemente, otros programas alternativos de cruzamiento para poblaciones pequeñas, o ciertas preguntas científicas, incluyendo una descripción general del reciente campo de rápido desarrollo de las herramientas de selección molecular. Debido a que el tema de la cría de reinas y la selección es demasiado extenso para ser desarrollado en este trabajo, se proporcionan numerosas referencias para facilitar estudios integrales.

饲养和选择西方蜜蜂蜂王的标准方法

本章列举了当前蜂王的培育、选择和育种中正在使用或值得推荐的方法。这些方法可做为科学研究和实际养蜂操作的标准方法。我们阐述了培育蜂王的基本条件、不同的饲养管理技术、还推荐了对应的育王设备。 由于蜂王是有选择的同雄蜂进行交配，这一行为会极大的影响育种方案的成功性，因此我们专门设立了分章对交配地点的管理和交配质量控制进行了阐述。由此，蜂王的培育和质量控制组成了蜂王培育部分。评价蜂群的整体性状通常应用蜂群间的对比试验，本章介绍了如何组织、评价蜂群开展常规性状测试的方法。对于大群体的评估，选择了最有效选择法，阐述了评估育种值的统计方法和数据预处理方法，育种值的计算整合了尽可能多的蜂群的家谱和蜂群性状指标数据。简述了针对小群体或某些科学问题而开展的特殊育种方法，包括新的快速发展的分子选择技术。由于蜂王的培育和选择是一很大的研究领域，本文不能完全包含，所以给出了大量文献来表述该领域的综合研究现状。

Keywords: Honey bees, selection characters, performance testing, queen production, mating control, molecular selection, breeding values, *BEEBOOK*, COLOSS

1. Introduction

Adaptation through natural selection is the natural response of bee populations to environmental changes and the challenge of pests and diseases. The richness in biodiversity of races and ecotypes of *Apis mellifera* reflects a long lasting, continuous process of adaptation. This diversity represents a highly valuable biological capital that is worth preserving as a basis for future selection and development in response to new ecological and production challenges.

The highly complex reproductive biology of honey bees, including multiple mating of queens, long distance mating flights, male haploidy, excess drone production and drone congregation areas, has evolved as an effective toolbox for the selection of genetically diverse honey bee populations. However, modern beekeeping and breeding techniques may limit or extinguish these natural selection effects (Bouga *et al.*, 2011), which risks lowering the vitality of bee populations.

Responsible breeding activities have to regard the natural reproductive biology of honey bees. Modern techniques of queen rearing, selection and mating control offer very powerful tools to improve the economic, behavioural and adaptive traits of honey bees. Here we describe the available techniques in bee breeding, and recommend scientific and technical standards. Indeed, internationally approved quality standards for queen rearing, mating and testing are needed for the improvement, comparison and exchange of breeding stock, and to fulfil the demands of the market.

The authors share the vision that these recommendations will help preserve the natural diversity in honey bees and to support the production of high quality queens, both in a physiological and in a genetic sense. The use of standard, high-quality queens is a prerequisite for any research on colony development and behaviour as well as for economically successful beekeeping.

2. Queen production
2.1. Queen rearing techniques
2.1.1. Short history of queen rearing
The first queen rearing was practiced in ancient Greece, where bee-keepers put combs with young larvae into queenless colonies in order to raise emergency queen cells. However, at this time very little was known about the biology of honey bee colonies. In 1565 Jacob Nickel was the first in Europe to describe how honey bees can raise queens from worker eggs or very young larvae. In 1861, H Alley, W Carey and E L Pratt, from Massachusetts, USA, began to produce queens for sale. These early producers used narrow strips of comb containing eggs and larvae which they fastened to the top bars of partial combs. Placed in queenless swarms, the bees built queen cells that could be individually distributed to queenless colonies for mating.

The development of modern queen rearing techniques started in the 19th Century. Gilbert Doolittle (1889) in the USA developed a comprehensive system for rearing queen bees which serves as the basis of current production. Essentially, he used wax cups into which he transferred worker bee larvae to start the production of queen cells. His method of queen rearing in queenright colonies with the old queen isolated by a queen excluder (Doolittle, 1915) is still applied. Doolittle emphasized the importance of simulating a swarming or supersedure situation in the cell building colonies and a constant, rich food supply for the production of high quality queens.

Since 1886, queen bees have been delivered by mail with benefits for the beekeepers as well as the breeders (Pellett, 1938). Losses during transit have been reported from time to time, but in general, shipment by mail is satisfactory. Nowadays, about one million queen bees are annually sent by mail, mainly in the USA, Canada, Europe, and Australia (author estimation).

2.1.2. Basic principles of queen rearing

A honey bee colony can produce a new queen without human intervention as long as fertilized eggs are present. Beekeepers have developed techniques to rear large numbers of queen bees to requeen colonies regularly (every year or two), to reduce swarming, to increase brood and honey production, to start new colonies, and to change certain genetic characteristics (Laidlaw and Page, 1997; Ruttner, 1983). Many US beekeepers requeen as often as twice a year.

The key in queen rearing is to take a young (12-24 hours old) larva from a worker cell and place ("graft") it into a queen cell cup suspended vertically in a hive. The larva is fed on a special royal jelly diet by the nurse bees. After 10-11 days, the queen cells, which are ready to emerge, can be transferred to queenless hives or mating nuclei ("nucs") (Woodward, 2007). The success and quality of queen production depends on strong, well fed and healthy nurse colonies and on suitable equipment and colony management.

2.1.3. Equipment for queen rearing

Most systems of queen rearing use standard beekeeping equipment but employ some specialized equipment during the process. Most of the specialized equipment is inexpensive or can be constructed by the beekeeper.

2.1.3.1. Cell cups, bars and frames

- Larvae are placed in artificial queen cell cups (grafted). The cups are placed on bars which, in turn, are placed in frames (Fig. 1). Queen cell cups should measure 8-9 mm in diameter at the rim.
- Cell cups can be produced from beeswax as described by Ruttner (1983) or Laidlaw (1979). Cells should always be rinsed, after removal from the dipping sticks ("cell mandrel"), to eliminate traces of soap. Cups made in advance should be kept free of dust by storing in a sealed box. Most queen producers attach their homemade beeswax cell cups directly to a cell bar with hot wax. Queen producers dip the base of the cell cups in molten beeswax (beeswax melts at 62.3 - 65.2°C) and firmly push the cup base onto the cell bar as the wax cools.
- Alternatively, plastic cell cups can be purchased from beekeeping suppliers. The most popular are JZ-BZ Push In and Base Mount Queen Cell Cups from Mann Lake Ltd (http://www.mannlakeltd.com/) in the USA or Nicot in Europe (http://nicot.fr/).
- Previously used plastic cell cups can be reused after scraping out royal jelly from the base of the cups and washing the cups in warm water with a little detergent (liquid soap, approx. 2 ml for 1000 ml of water). The cups should be left to dry out thoroughly before attaching them to a cell bar. Such cleaning might not prevent an outbreak of black queen cell virus (BQCV), so it is always better to use new ones.

Fig. 1. Different: **a.** wax; and **b-e.** plastic queen cups and ways to attach them to the bars; **f.** frame with bars ready for grafting.

Photos: J Wilde

- Introducing plastic queen cell cups into strong colonies about one day before grafting allows the bees to clean, polish and warm the cells. Plastic cups are attached with molten clean wax as described by Ruttner (1983) or Woodward (2007).
- It is recommended to dip the rim of the outside four cell cups located at each end of the cell bar into wax to increase the acceptance of grafted larvae.
- Special push-in queen cell cups make preparing the cell bars simple. These cells have a raised area on their base that snaps into a groove on the cell bar. The bar then can be inserted into the frame.
- A frame (wooden, plastic or metal) of standard dimensions that will hold 2-4 cell bars can be used.
- Usually, 10-20 cells are attached to each bar with 20-60 cell cups per frame.

2.1.3.2. Grafting tools

An assortment of grafting tools can be used effectively:

- Many different versions of metal grafting needles are produced. Some have a magnifying glass fitted to the stem which can help if one's eyesight is insufficient. Usually both ends are designed for grafting; each offers a different configuration.
- A very small (size no. 000 or 00) artist's paint brush is a suitable tool for grafting. The moistened bristles must stick together to easily slide under a larva.
- A "Chinese" grafting tool is a handy and inexpensive grafting tool that looks like a ball point pen. It consists of a spring loaded bamboo plunger that slides along a thin tongue of flexible plastic. The flexible tongue slips easily under a larva and then a press on the plunger will deposit the larva and any royal jelly that was picked up in the cell to be grafted. A non-slip grip in the middle section gives excellent control. Modern versions of this tool have injection moulded plastic parts, which may help with cleanliness.

In general, grafting is easier from dark wax combs rather than from light wax combs because of the better contrast with the small white larvae. The use of a cool light or an illuminated grafting magnifier will help one see the larvae better. Grafting should be done preferably in a room or in indirect light to ensure the larvae do not dry out or become damaged by UV radiation from direct sunlight.

2.1.3.3. Queen rearing kits

There are several queen rearing kits available (Jenter system, Nicot Queen System, Mann Lake Queen Rearing Kit, Ezi-queen queen rearing system) in which the queen is caged on a plastic comb with removable cell bottoms. The kit systems can be used to transfer larvae without grafting. With a single Karl Jenter kit, about 50 queens can be produced over 50 days. This is suitable for smaller beekeepers producing for their own apiaries. The Ezi-queen system is more effective for a larger production as it uses a cage of 420 cells which can all be transferred in less than 5 minutes. The plastic components used are made of a food grade polycarbonate, which allows for sterilization by autoclaving.

2.1.3.4. Protection of queen cells

In general, the best acceptance and care by nurse bees is achieved when young queens emerge directly into their colony. If possible, ripe queen cells should be transferred from the rearing colony to the mating colony 1-2 days before emergence (Fig. 2).

from chewing down the cells. The most popular are push in cell protectors and top bar cell protectors from Mann Lake Ltd. There are many types of wooden or plastic emergence cages available, which can be used singly or as a block of 10-15 cages, to protect all queen cells on a cell bar.

Fig. 3. Two push-in cell protectors (left) and 2 top bar cell protectors (right) from Mann Lake Ltd. Photo: J Wilde

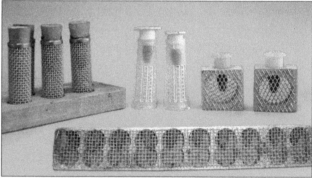

Fig. 4. Queen cells protected by 3 types of cages (from left: iron, plastic and Zander cages) and container for 10 queen cells (below).
Photo: B Chuda-Mickiewicz, J Wilde

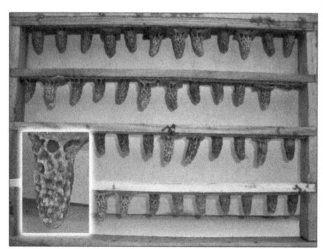

Fig. 2. Sealed queen cells, 1-2 days before emergence, ready to be transferred to mating colonies or an incubator. Photo: J Wilde

However, if queen cells are left to emerge in the nurse colonies or in a brood chamber, they have to be protected against attacks of other queens or workers and to prevent the escape of queens. This can be achieved by cell protectors or emergence cages (Figs. 3 and 4).

Queen cell protectors, made from insulation tape, tin foil or plastic tubing, are placed over the queen cells to prevent the emergence of the queen or to allow the queens emergence but to prevent the workers

2.1.4. Queen rearing methods and management of nurse (or cell builder) colonies

A few queens can be reared very simply by utilizing the natural reproductive impulses of colonies (swarming, supersedure or emergency). For example, in the Alley method (Ruttner, 1983) a strip of cells containing one day old larvae is removed from a comb and placed in a frame with the cells pointing downwards. Every 2nd and 3rd larva is destroyed, leaving adequate spacing for queen cells to be started and finished without having to surgically separate the cells once they are sealed.

Table 1. Methods to stimulate colonies to accept newly grafted queen cells.

Method	Description	Advantages	Disadvantages	Notes
Swarm box	Artificial swarm with plenty of young bees and feed in a 5-6 frame box or a 9-12 frame hive without a queen or open brood, as described by Laidlaw (1979)	Gives perfect starting results independent of the weather conditions The swarm boxes can easily be transferred and used to transport queen cells	Many manipulations Confined bees in the box are stressed and less active compared to free flying colonies	
Free-flying queenless starter colony	Queenless colony without open brood as described by Laidlaw (1979) or by Morse (1979)	No extra hive equipment (like swarm boxes) needed Achieves necessary number of queen cells at any time of the season	Is necessary to cage the queen Works only with very strong colonies Requires extra colonies for queen cells finishing	Need to be supported by the addition of sealed or emerging brood at 7-10 day intervals. Bees should be collected in the morning from open brood of support colonies in other apiaries. The bees should be fed sugar syrup and left caged in a cool dark place until late afternoon before they are added to the starter colonies.
Free-flying queenright colony	Several very popular procedures (Mackensen, Ruttner, Sklenar, Mueller) as described by Ruttner (1983)	Excellent queen quality (Cengiz et al. 2009) Used for starting and finishing the queen cells Possible to graft every day	Swarm prevention necessary	
Queenright starter-finisher	Queenright, two or three story colony as described by Laidlaw & Page (1997)	Achieves optimal cell and queen quality at any time of the season	Needs very strong colony	
Queenless starter-finisher	Queenless two or three story colony, as described by Laidlaw (1979) or one story as described by Morse (1979) or Woodward (2007)	Reliable results widely independent of weather condition and period of season	Needs support of brood and bees from field colonies	Maintained by the addition of about 300-400 g of bees in the evening before each new graft. A frequent addition of this amount of bees is preferable to adding more bees at less frequent intervals. If almost all brood is gone, emerging brood combs are given as well.

However, large scale, systematic production of high quality queens relies on grafting methods and the application of specific colony management schemes. There are several methods available to stimulate colonies to accept newly grafted queen cells and to rear high quality queens. In starter-finisher systems, the queen cells are started in special colonies and transferred to queenright finisher colonies after about two days. In other systems, the queen cells remain in the same colony for the whole rearing period. The most popular methods are listed in Table 1.

If there is no nectar flow available, all nurse colonies or bees in swarm boxes need to be fed with a 50% sugar syrup or candy (powdered sugar with honey, ratio 4:1 by weight) at least three days before grafting during the whole rearing season. The nurse colonies always need to have a good supply of nectar. If necessary, additional pollen combs are put in from other colonies. In any case, the nurse colony needs plenty of young and well fed bees to ensure a rich royal jelly supply for the very young larvae.

2.1.5. Obtaining larvae for grafting

Grafting is easier if the larvae can be removed from dark combs (combs from which 8-10 worker generations have emerged). Before use, dark combs should be placed close (next) to brood combs so the bees will clean and polish the cells for egg laying.

If many larvae from a single queen are to be grafted on certain dates, it is very useful to confine the queen to single combs for 12 - 24 hours four days prior to grafting. After this time, the comb with eggs can be transferred to a queenless nurse colony or can be retained in the brood nest of the source colony. There are several commonly used methods of making queen-confining cages (Morse, 1979):

- A simple method is to use a push-in cage made with wire mesh (with 4 mm spaces) or queen excluder. Push-in cages are usually about 12-15 cm². Worker bees move through the holes in mesh as easily as they do in queen excluders. Sometimes the workers bees will chew the comb around the edge of a push-in cage and may release the queen within two days.
- If a breeder colony is to be used for an extended period, the use of 3-5 comb isolators, made from metal queen excluder, is recommended. The isolators are placed in the centre of the hive. One of the combs should have abundant pollen. The remaining space is filled with one empty comb, sealed and emerging brood and one comb with unsealed honey. Each 24 hours, one comb with eggs is removed and replaced by an empty one. After the four days, larvae on the first comb will be ready for grafting. The system allows for continuous grafting of large cell numbers every day.

One of the best and most convenient methods of obtaining larvae is to use a special full depth hive body insert (Laidlaw and Page, 1997). The breeding queen is confined to three small combs, each about half the size of standard combs, in a compartment with sides made of queen excluder that makes up half of the insert. Three additional half-combs occupy the other half of the insert, which has open sides (see photo in Laidlaw, 1979). A standard comb well filled with pollen is placed next to one side of the insert, such as to the left, and combs with sealed or emerging brood are put in the remaining spaces of the body. Each day a centre comb with eggs is moved from the queenright partition to the non-excluded half of the insert as described by Laidlaw (1979).

2.1.6. Grafting procedure

Respect of the following conditions when transferring the larva from its original cell to the artificial queen cell (Fig. 5) ensures quality queen production:

- Grafting the larvae from the worker comb to the queen cells should be done rapidly and with suitable environmental conditions (24-26°C and RH > 50%).
- The best place to perform the grafting is in a honey house or a laboratory room, as larvae are sensitive to high tempera-tures, direct sun light (UV) and low humidity. Grafting in a room is comfortable for the operator and protects against robbing bees. The location of the grafting room should be just a few steps from the breeder colonies and the nurse colonies that receive the grafted cells.
- Cold lighting must be used to avoid generating too much heat which may damage the larvae.

- Attention must be placed in selecting larvae which are sitting in a pool of royal jelly, as "hungry larvae" will not be readily accepted by the nurse bees nor develop into strong queens.
- The cells and the brood comb should be kept out of the bright sunlight as much as possible. When the weather is hot and dry, a damp cloth may be spread over the cells to prevent them from drying out. A damp cloth also protects the larvae from light and dust.
- With experience and speed, three bars (60 cups) can be completed in 8-10 minutes or less. As soon as one bar is finished, it should be covered with the damp cloth. The grafted cells should be placed into the starters as soon as possible.
- Special carrying boxes for the brood frames and grafted cells exist, which help to protect the larvae from drying from sunshine as well as from chilling on cold days.
- Queen cells can be 'primed' by placing a small drop (about twice the size of a pinhead) of a mixture of half royal jelly and a half warm water before the larva are grafted into the cells. If the cells are primed, it is important that the larvae are not immersed in the royal jelly but are floated off the grafting tool on top of the centre of the drop. Usually it is necessary to prime the queen cells if a standard grafting tool is used while there is no need if a Chinese grafting tool or automatic needle is employed, which tend to transfer royal jelly along with the larva.

Fig. 5. **a.** Larvae that are a few hours old, floating in royal jelly, and ready for grafting; **b.** a larva taken from dark combs is transferred into wax cups using; **c.** a grafting tool. Photos: L Ruottinen

2.1.7. Acceptance of larvae

The number of accepted larvae depends on different factors, as described in detail by Ruttner (1983). The most important factors are: quality, strength and developmental stage of the nurse colonies, age of the workers, age of the grafted larvae, presence or absence of queen in

Table 2. Parameters associated with locating mating apiaries on islands or the mainland.

Mating station type	Accessibility & Applicability	Mating control	Mating risks	Weather conditions	Costs per queen
Mainland	+	o	+	o	+
Island	-	+	o	o	-
+ = optimal, 0 = acceptable, - = suboptimal					

the rearing colony and duration of the queenless stage, presence of open brood in the cell-starting colonies, number of grafted cells, rearing sequence and method of rearing.

Environmental conditions are of major importance for final queen rearing success. Essential factors are: regulation of humidity and temperature by the rearing colony or in the incubator, and vitality of queen cells and the feed supply (nectar flow, supplemental feeding) of the nurse colony. There is also some indirect influence of the weather conditions and of the season. Under well managed conditions at least 80% of the larvae should be accepted even in bad weather conditions.

2.2. Mating control

Honey bee breeding programmes and specific research projects depend on controlling the queen's mating process. In addition to the well-developed instrumental insemination technique (see the *BEEBOOK* paper on instrumental insemination (Cobey *et al.*, 2013)) isolated mating stations can serve as an efficient technique for control of honey bee mating for commercial and scientific purposes.

Because drones completely avoid passing over large stretches of water, islands offer an excellent opportunity to establish a fully controlled genetic composition of drones. On the mainland, mating control depends on the isolation of drone colonies by geographic distance (limited flight range of drones and queens) or barriers (high mountains etc.). A comparison of mating apiaries located in both areas is offered in Table 2.

2.2.1. Criteria for establishment of mating stations

- Absence or minimal presence of managed and unmanaged honey bee colonies and airborne drones in a radius of at least 6 km.
- Favourable pollen and nectar resources.
- Weather conditions with long periods of more than 20°C ambient temperature, and wind speed not more than 24 km/h.
- Undulating landscape and sheltered areas for positioning of mating boxes. Obvious markers, such as stones, trees, bushes or specially installed objects help to minimize queen drifting and losses.
- Sufficient drone colonies to ensure a strong drone population for mating. According to Tiesler and Englert (1989), a minimum of 8 to 10 strong drone colonies, or 1 drone colony per 25 queens, are needed.
- Minimal presence of honey bee predator species.

2.2.2. Maintaining mating boxes and mating stations

- For preventing the presence of alien drones in the mating station, only drone-free mating boxes should be used.
- If possible, mating boxes should not be disturbed during the queen flight period (between 11:00 and 16:00 h).
- Depending on weather conditions, a first inspection of the queens' mating success should happen about 2 weeks after establishing the mating units. Successful mating should occur within 3 weeks after queen emergence. Later mating will result in a reduced fecundity and life expectancy of queens.
- A final evaluation of successful mating should occur upon the appearance of sealed brood in the colony.
- Regular inspections of the storage and supplementary feeding of mating units is needed if they are used over longer periods.

2.2.3. Drone colonies

The main reason for keeping drone colonies is to provide an adequate number of mature drones of selected origin, in the right period, for mating. A single group of sister queens can be used to control the paternal pedigree, or several groups of sister queens each of them derived from a selected breeder colony, can be used for drone production within one mating station, depending on the breeding programme.

- The build up of drone colonies needs to be started in advance of the mating period.
- Drone colonies are managed in standard hives and receive sufficient space to support an optimal population development.
- The drone colonies are established from superior and healthy colonies and special care is taken to provide a continuously rich honey and pollen supply. Regular checks of the health status and the overall development are recommended to achieve a high quality control level.
- Special attention has to be paid to disease treatment. Varroa and other pathogens strongly influence the fitness of drones. Chemical control measures can thus effectively increase the number off fertile drones but at the same time have negative effects on the fertility of drones (De Guzman *et al.*, 1999). On the other hand, reduced treatment can provide a selection pressure that favours colonies with increased varroa resistance. Careful varroa management in drone colonies can thus be an important selection tool within breeding programmes for disease resistance (see Büchler *et al.* (2010) for further details on "tolerance mating stations").
- Up to 2 drone combs are placed within the brood nest of each box to enable a rich production of drones. As the development of drones from egg to maturity takes 40 days and the life expectancy of mature drones last for several weeks, drone production should be started no later than 2 months in advance of the mating period.

Table 3. Meteorological parameters, instruments used to measure the parameters, and units of measure that can be used to characterize mating stations.

Parameter	Instrument	Unit (Abbreviation)
Temperature	Thermometer	Celsius (°C)
Relative humidity	Hygrometer	Percentage (RH)
Wind speed	Anemometer	Meter in second (m/s)
Wind direction	Anemometer	Wind rose (NESW)
Precipitation	Rain gauge	Millimetres on hour (mm/h)
Cloud cover	Campbell-Stokes recorder	Campbell–Stokes recorder card / Subjective cloud coverage in %
Altitude	GPS	meters above sea level (m.a.s)
Position	GPS	Latitude and longitude coordinates
Vegetation	Aerial photography	proportion of different land use, presented as a percentage

- Drone brood combs from selected drone mothers may be removed after capping and placed in nurse colonies, in order to enable production of higher number of drones from the selected queen.
- If the drone colonies are moved to the mating station, queen excluders between the bottom board and the brood box must be used to keep out any other drones. However, those excluders need to be regularly inspected and dead drones removed, which otherwise could block the entrance and ventilation. The queen excluders with all adhering drones should be removed just before moving the drone colonies to the mating station.

2.2.4. Evaluation of a mating station: environmental conditions

In order to better understand and evaluate the requirements and risk factors involved in honey bee mating biology, various research methods have been developed. Consequently, it is useful to characterize mating stations by noting the meteorological phenomena and parameters outlined in Table 3.

2.2.5. Evaluation of a mating station: biological conditions

Mating between the virgin honey bee queen and numerous mature drones occurs in the air, at a certain distance from the hives, in rendezvous sites called "Drone Congregation Areas (DCA) (Koeniger and Koeniger, 2007; Zmarlicki and Morse, 1963). Location of DCAs tends to remain constant over time. When establishing a mating station, it can be useful to assess the presence of surrounding colonies and DCAs. This can be achieved in several ways, as described in the sections below. A comparison of the methods described below can be found in Table 4.

2.2.5.1. Traps to estimate worker presence

- Honey traps, consisting of at least 50 ml of liquid honey on small plate, are positioned in the area surrounding the mating station (see the *BEEBOOK* paper on miscellaneous methods (Human *et al.*, 2013) for more information on using honey traps to estimate worker presence and colony density.
- Alternatively, dark brood combs can be boiled in water in order to attract bees by the intensive and specific smell.
- The traps are regularly checked for the presence of worker honey bees. The total testing time should be not less than 3 h. With regard to common flight distance and speed of honey bee workers (Park, 1923; von Frisch, 1967), the continuous control duration on a single trap should not be less than 15 min.

2.2.5.2. Pheromone traps to estimate drone density

Pheromone traps, prepared from synthesized queen pheromone (9-oxo-2-decenoic acid, abb., 9-ODA) or extracted in acetone ($(CH_3)_2CO$) from honey bee queens can be used to lure airborne drones. Additionally, live or model queens, in which the thorax is fixed or tethered, can serve to attract drones. The details of the technique and necessary equipment are given in the *BEEBOOK* paper on behavioural studies (Scheiner *et al.*, 2013).

2.2.6. Assessment of honey bee queen and drone behaviour

Studying honey bee mating behaviour under local environmental conditions and evaluating the reliability of a mating station are complex tasks and should be organized under specifically controlled circumstances.

- Transparent front extensions and queen excluders can be applied to the mating boxes to accurately observe queen activity (Koeniger and Koeniger, 2007). Thus, the time and duration of each flight attempt as well as the presence of any mating sign on the queen can easily be observed. An experienced person is able to simultaneously follow the queen flight activity of up to 10 mating boxes.
- The starting time of oviposition, the sex of the larvae and the rate of brood mortality can be used as indicators of successful mating.
- The spermathecae of mated queens can be dissected (see the *BEEBOOK* paper on anatomy and dissection of the honey bee (Carreck *et al.*, 2013)); to estimate the number of stored spermatozoa see the *BEEBOOK* paper on miscellaneous research methods (Human *et al.*, 2013).
- For the observation of drone flight activity, the colonies should be equipped with transparent front extensions and entrance reducers to individually follow and count the number of leaving and returning drones in certain intervals as well as to catch and mark individual drones for further observations.

Table 4. A comparison of methods used to determine adult worker and drone honey bee presence in a prospective mating area. + = optimal, 0 = acceptable, - = suboptimal.

Method	Accessibility	Applicability	Efficacy	Price	Notes
Honey traps	+	+	o	+	Attracts worker bees
Wax melting traps	+	+	+	+	Attracts worker bees
Synthesized 9-ODA	-	+	+	-	Attracts drones
Extracted queen pheromone	o	+	+	o	Attracts drones
Fixed live queen	+	o	+	o	Attracts drones
Fixed model queen + pheromone	-	-	+	o	Attracts drones

- Alternatively RFID (Radio Frequency IDentification) technology can be used to individually mark queens and drones and automatically register the exact time of each entrance passage (http://www.microsensys.de).
- Individual drones can be marked with coloured or numbered plates in order to identify them when they return to their colonies or if they are caught again in the field.
- Microsatellite analysis and other molecular methods can be used to identify the individual origin of drones or its semen from certain colonies (see the *BEEBOOK* papers on molecular techniques (Evans *et al.,* 2013), and miscellaneous research methods (Human *et al.*, 2013)). This is a very powerful technique to estimate the number of matings per queen, the realized mating distance of queens and drones, the quantitative contribution of certain drones to the female offspring of a queen etc.

2.3. Handling of adult queens

2.3.1. Marking and clipping queens

See the *BEEBOOK* paper on miscellaneous research methods (Human *et al.*, 2013) for techniques of clipping or marking queens.

2.3.2. Shipment of queens

Queen cages for shipment by mail are usually made from plastic and are offered in a variety of sizes and shapes. The most popular cage has two compartments; the larger one is used to house the queen and 6-12 attendant worker bees, while the smaller one is filled with queen candy to provide food during shipping. If the shipping cages are used to introduce the queen into a colony, a small hole can be created in the end of the candy compartment through which the workers from the hive can slowly reach and free the queen. Several cages can be packed together if care is taken that the queens cannot reach each other through the screened parts. The stack of cages can be placed in an envelope with ventilation holes punched in it and labelled "Live bees" and "Protect against sunshine".

Candy for queen cages should contain little water but nevertheless remain soft. A mixture of powder sugar with about 20% honey (weight:weight) gives suitable results. Whilst is not necessary to give water to queens during transport, it is a good idea to place a drop of water on the screen of a queen cage as soon as it is received. Queens should be introduced to colonies as soon as possible after shipment. As far as possible, caged queens should be kept in a dark place with a medium and stable temperature.

2.3.3. Storage of queens

Large queen breeding operations often have more queens than they can use or ship immediately. They may need to remove mated queens from mating nucs to make space for new emerging queen cells. Mated queens can be caged in regular cages without worker bees or candy and placed together with other similarly caged queens in a "queen bank" colony as described by Morse (1994). It is possible to store up to 60 cages in one frame and up to 120 queens within one colony for 1-2 months with few losses. While queen banking is very popular in the USA, European breeders avoid storing mated queens this way because the queens may become damaged by the workers who may injure the queens' feet, legs, wings and antennae (Woyke, 1988).

Queens lose the ability to fly if the tip of one front wing is clipped (approx. 35 - 40%). Wing clipping has no negative effects on the vitality or longevity of the queens and is therefore a common technique to delay, but not prevent, swarming of the colony. Beekeepers may clip alternate wings in alternate years to keep track of the age of queens.

2.3.4. Requeening colonies

There is no perfectly reliable method to introduce new queens to a colony. The success of queen introduction depends on the attractiveness of the new queen and the previous queen status of the colony. Unmated queens are less attractive than mated queens, and egg laying queens are much more easily accepted than queens that have stopped egg laying due to longer transport or other reasons. The best time for requeening is during a good nectar flow. It is important to make the recipient colony queenless for at least 6-8 hours, sometimes for 1 day. Furthermore, it is essential to destroy queen cells being reared by the colony before releasing the queen (even by hand after several days if the workers are not biting the cage). One should use a push-in cage to introduce queens during a low to marginal nectar flow as this allows the queen to begin oviposition, thus increasing the likelihood of her acceptance.

The most popular method is to replace the previous queen directly with the new one in its shipping cage. The candy compartment on the cage is exposed to allow the bees to slowly release the queen after consuming the candy. The success can be improved if the queen to be replaced is caged for about 7 days before requeening.

Under difficult conditions or for the introduction of highly valuable queens, it is recommended to introduce the queen into a nucleus colony (also known as an "artificial swarm", "split" or "nuc"). Those small units usually accept any kind of queen. The queens can then be safely introduced into strong hives by placing the nucleus with the new queen on top of the strong hives separated by an insert with screens on both sides to avoid direct contact of the bees. Heat from the larger parent colony will pass into the upper unit and support the development of the nucleus colony. As soon as the young queen has built a brood nest and is surrounded by her own young bees, it is ready to be combined with the parent colony. The old queen from the strong colony and the double screen are removed and the young queen in its nuc colony is put on top of the brood box of the strong colony, just separated by a sheet of newspaper containing several slits. In this way, a requeening success of 95-100% can be expected.

2.4. Queen quality control

"Quality" is a subjective term used in relation to queens and drones to describe certain quantitative physical and performance characteristics. It is generally believed that a queen of "high quality" should have the following physical characteristics:

- high body weight (described in section 2.4.1.),
- large number of ovarioles (see the *BEEBOOK* paper on anatomy and dissection (Carreck *et al.*, 2013))
- large size of spermatheca, (see the *BEEBOOK* paper on anatomy and dissection (Carreck *et al.*, 2013))
- high number of spermatozoa (see the *BEEBOOK* paper on miscellaneous research methods (Human *et al.*, 2013).

Once active as the queen of a hive, some of the colony performance traits such as the following can be used as quality criteria:

- high brood production (including number of eggs per day) and large bee population (section 2.4.2. and the *BEEBOOK* paper on measuring colony strength parameters (Delaplane *et al.*, 2013))
- brood solidness (section 2.4.3. and the *BEEBOOK* paper on measuring colony strength parameters (Delaplane *et al.*, 2013))
- disease control (Laidlaw, 1979; Cobey, 2007; see the *BEEBOOK* papers on honey bee diseases: De Graaf *et al.*, 2013; De Miranda *et al.*, 2013; Dietemann *et al.*, 2013; Forsgren *et al.*, 2013; Fries *et al.*, 2013; Jensen *et al.*, 2013).
- increased honey yield (see section 3.3.1.)
- low defensive behaviour (see section 3.3.2.)
- low swarming tendency (see section 3.3.3.)
- intensive hygienic behaviour (see section 3.3.4.)

2.4.1. Body weight

The weight of a fertilized queen can vary considerably due to egg laying intensity, genetic factors (race) and environmental factors that affect egg laying. More uniform conditions can be assured by using very young unfertilised queens and respecting the following conditions:

- Electronic balances with an accuracy of 0.1 mg should be used.
- If unfertilized queens are used, they should be as young as possible. Queens can lose almost 1-2 mg of weight per day after emergence (Skowronek *et al.*, 2004; Kahya *et al.*, 2008).
- Queens can be placed into small cages to facilitate weighing (Fig. 6).
- The genetic origin of the queen influences the weight standards and should thereby be known.
- At least ten queens per line and apiary are collected on the same day when evaluating fertilized queens. Sampling is usually repeated twice during the reproductive season. This parameter can vary considerably due to egg laying intensity and various other factors and mechanisms (genetic, biochemical) that affect egg laying.

2.4.2. Number of eggs per day (fecundity)

- Queen fecundity in a twenty-four-hour period is estimated either once, when the laying of eggs is at its maximum or several times during the productive period.
- The queen should lay more than 2000 eggs in 24 hours period, but this can depend on the bee race.
- A simple way of estimating 24 hours fecundity is with the use of a 5 x 5 cm or 2 x 2 cm grid frame (Fig. 7) or by using the Liebefeld method of estimating brood area (see the *BEEBOOK* paper on estimating colony strength parameters (Delaplane *et al.*, 2013)).

Fig. 6. A queen cage for weighing a queen.   Photo: F Hatjina

Fig. 7. The 2x2 cm grid frame is placed over the surface of the comb and used to estimate the amount of brood (or eggs) in the comb.

Photo: F Hatjina

2.4.3. Brood solidness

- Brood solidness is expressed by the percentage of empty worker cells in a brood patch of a given area. An acceptable level of empty cells is usually less than 10%. To determine brood solidness, see the *BEEBOOK* paper on measuring colony strength parameters (Delaplane *et al.*, 2013).

2.4.4. Disease control

- "High quality" of queens means also that they are free from pests and diseases (Laidlaw, 1979; Cobey, 2007).

Therefore special care has to be taken in order that the productive colonies as well as the mating nuclei show no signs of contaminating diseases such as foulbrood and nosema. Methods for reducing pest/ pathogen loads in colonies can be found in the COLOSS *BEEBOOK* papers on honey bee diseases (De Graaf *et al.*, 2013; De Miranda *et al.*, 2013; Dietemann *et al.*, 2013; Forsgren *et al.*, 2013; Fries *et al.*, 2013; Jensen *et al.*, 2013). One way to ensure that the produced queens are free from nosema spores is to count the number of spores in the alimentary canal on the same sample of queens sacrificed for the other characteristics mentioned above (number of ovarioles, diameter of spermatheca, and number of spermatozoa). According to Rhodes and Somerville (2003), this number should be less than 500,000 spores per queen. However, the queen's attendants in the queen cages can also transmit nosema spores to the queens or to the receiving colony, but the threshold for the accepted limit has still to be evaluated.

3. Performance testing of bee colonies

Performance tests refer to the testing parameters of queen performance across the season, including brood and population production, honey and pollen yield, score of hygienic behaviour, swarming tendency, calmness, overwintering, food consumption etc.

3.1. Preconditions and general recommendations

A breeding programme entails selection of the best individuals for specific traits, and elimination of the worst. To do this, individuals must be assessed in a way that allow genetic effects to be distinguished from environmental influences, and according to a uniform method that allows for comparisons across time and space. The basis of performance testing is that colonies in the test station (apiary) should be placed in similar starting conditions and managed according to a standard protocol. The final result obtained from performance testing is a selection index or breeding value for the chosen traits, which is used to select colonies to reproduce (to use as stock for queen and drone production).

The colonies are started from package bees or uniform nucs (see the *BEEBOOK* paper on estimating colony strength parameters (Delaplane *et al.*, 2013)), into which the queens to be tested are placed. The colonies are normally set up at the beginning of the summer, or so that there is sufficient time for the colony to build up before the winter. The size of the starting package of bees or nuc and the establishment of the test colonies depends on the climatic conditions of the testing station. Methods of equalization (food, space, diseases) of the test colonies are allowed until the last autumn observation, when the first assessment data are taken. This represents the starting point of the test (overwintering).

3.1.1. Location and organization of testing station

Location of the test apiary should ensure a continuous nectar and pollen flow during the testing period for the number of test colonies. The test colonies may be moved to an apiary for the target (main) honey flow. When planning the location of colonies in the apiary, special care must be taken to reduce drifting. Placing hives in straight, long lines or in rows one in front of another is not allowed. In these conditions, colonies are the strongest at the ends of the lines and in the first and last rows due to drifting of the bees when they fly back to the hive.

The following arrangements of hives in the apiary are recommended to reduce drifting among colonies:

- Hives placed on individual stands – recommended
- Hives placed on small group stand (up to 4 hives) (Fig. 8) - acceptable
- Hives distributed irregularly and in smaller groups with their entrances facing to the four coordinates or somehow different directions (U-shaped or circle groupings) - acceptable
- Groups of hives placed in broken lines - acceptable
- Groups of hives separated by hedge or fence (~2 m high) - recommended in test apiaries with more than 30 colonies.

3.1.2. Size of testing station

The number of colonies in the testing station should be at least 10 (representing different sister groups), to allow for statistical calculations (see section 4. selection tools), while the maximum number should be obtained by considering the minimal honey flow potential of the area (sufficient nectar for all colonies throughout the season) and the number of beekeepers involved in the testing. Usually, not more than 30 colonies should be placed in one test apiary.

Fig. 8. Hives placed in small groups and with their entrances facing in different directions.   Photo: N Kezic

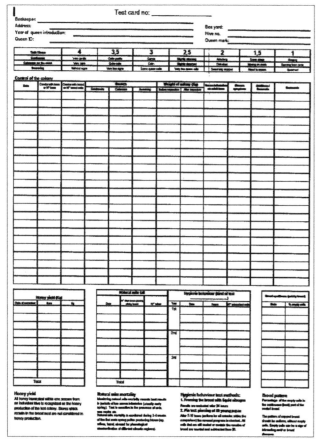

Fig. 9. Recommended protocol to collect all data of repeated performance test controls.

3.1.3. Queens: origin, marking, distribution

Honey bee breeding programmes are based upon evaluation of sister queen groups, in order to estimate the additive genetic component of the chosen traits. At least 12 queens per sister group should be tested, distributed among at least 2 testing apiaries (Ruttner, 1972). Within each test apiary, it is recommended to randomly distribute queens of the same origin. It is not recommended to group and / or isolate sister queens in separate positions within the testing apiary.

The sister queens submitted to performance testing should belong to the same rearing series and be mated at the same mating station (i.e. with the same array of drones). To increase the accuracy of the breeding value calculation, it is important that pedigree information of the queens is known. Each test queen should have an individual code and be unambiguously marked (see section 2.3.1. for details). Hives in the apiary should also be individually numbered and equipped with a test card, on which the performance of the colony is noted. The test card is set up on the basis of the traits chosen for selection. Each control and all specific observations have to be documented in this card. A standard test card that may be used in a testing apiary is shown in Fig. 9.

3.1.4. Timing and duration of test

Performance testing of colonies starts with the last autumn inspection following colony establishment (described in section 3.2.4. establishment of test colonies). Colonies will have been uniformly managed and specific requirements noted. Observations may be made in the first year, taking care that the colony is completely made up of progeny of the test queen (about 40 days after colony establishment). Starting from the spring, qualitative behavioural traits are assessed every time all the hives in the apiary are inspected, with a minimum number of 4 evaluations per trait. Behavioural traits should be evaluated under the same environmental conditions, in other words, tests should be performed on the same day for all colonies present in the testing apiary. Testing continues throughout the season. In the autumn of the queen's second year of life, the test cards are collected and processed for estimation of breeding values.

3.1.5. General recommendations

- The test apiary should be made up of the same kind of material (hives and supers) for uniform management.
- The testing apiary should be run by experienced beekeepers specifically trained to assess production and behavioural traits.
- Assessment of behavioural traits should be performed on all colonies on the same day, preferably by the same tester.
- In migratory beekeeping situations, the apiary should not be split, and the respective colonies should stay together for the whole test period.

3.2. Colony management

Colony management is important and has to be planned and prepared in advance, before the beginning of the test. Colony management has to fulfil specific requirements of the test: standard procedures should be adopted for all colonies in the test to enable comparative results. After the test has started, changes in colony management may significantly influence the results.

During the planning process, decisions should be made on the following issues:

- distribution of queens within the apiary
- type of hives
- kind of wax foundation or comb
- kind of stands for the hives
- water supply
- feeding sources
- nectar and pollen supply / migratory activities.

Large differences exist in different regions regarding colony management. Colony management can significantly influence test results. The main task is to ensure standard conditions for all colonies within each test apiary.

3.2.1. Hives (types, painting, hive components, identification)

3.2.1.1. Type of hive

The type of the hives used must be included in a research report. Common standard hives such as Langstroth or Dadant, are recommended for use, whilst modifications of traditional hives are not recommended.

Use of stands is recommended for the following reasons:

- The hive can be placed on a horizontal level regardless of the terrain configuration.
- It is the most comfortable working position for tester.
- Stands provide protection of the hive from ground moisture.

3.2.1.2. Painting and colouring

Hives should be protected with paint that does not harm bees. If oil dyes are used for hive protection, the overlaying paint has to dry and the polymerisation process has to be finished prior to hive use. Special care should be placed in choice of dyes in order to ascertain that they do not contain insecticides or other components that are long retained in the wood and gradually released. The hive entrances can be painted different colours to help bees in orientation and to reduce drifting between hives.

3.2.1.3. Hive components

Sufficient space for colony development must be provided. Super(s) are added when bees occupy most combs in the brood box (at least ¾). Super(s) should be removed when bees occupy less than two thirds of the combs in lower super.

Fig. 10. Screened bottom boards ensure good hive ventilation and allow for easy control of mite mortality.   Photo: B Binder-Köllhofer

It is recommended that hives in the testing apiary be equipped with screened bottom boards (Fig. 10).They guarantee good ventilation and allow for easy varroa mortality control. The size of the hive entrance has to be adjustable according to colony strength, and time of the year. During winter, a metal mesh / comb should be placed across the entrance as a protection against rodents. The size of the landing board is not important. It is recommended that landing boards should be the same size, but in different colours within the apiary. Regular maintenance of the landing board is important, since it is the place where disturbances to the colony can be noticed and recognized (e.g. to prevent robbing). The use of a queen excluder is not recommended, but if used, it should be placed / removed on all test colonies at the same stage of development. Feeders do not have to be in the hives all the time. If feeding is needed, feeders should be placed in all colonies at the same time and of the same capacity.

3.2.1.4. Hive and colony identification

Multiple types of hive/colony identification are recommended. It is recommended to use an identification number on the bottom board that combines the colony number, hive position in the apiary and number of the queen. Hive identification is complex and can cause problems if the test is long lasting. Clearly identified colonies are the basis for successful test processing. Identification of the queen is not reliable, since queen tags can be removed and an unmarked queen is not easily recognized. Queen identification is, however, useful as an additional ID system.

Frequently used hive identifications:

- An accompanying card under the roof of the hive is good but harsh weather conditions can damage it. Furthermore, during regular work with colonies, cards can be mixed up between neighbour hives.
- Marks on the roof of the hive are good, but roofs are easily switched between hives during regular work.
- Marks regarding hive position within apiary (number on the stand) are a reliable system of identification in the test.
- The best position for hive identification is on the hive bottom board. Usually these hive parts are constant and they need to be changed only in case of damage or for cleaning purposes. Therefore it is recommended to have clean and disinfected bottom boards at the beginning of an experiment.

3.2.2. Water supply

Colonies need to have a sufficient and continuous source of clean water (Figs 11 and 12). Bees can have difficulties in accepting the water source provided by the beekeeper. Therefore, it is important to provide water early in the spring, just after night temperatures are above freezing, or when first establishing the apiary. If there is an interruption of water supply from the designated source, bees may find an alternative water source, and then it is much more difficult to return them to desired water source again. So the water source must be suited to the apiary requirements. Most importantly, the water source has to be protected in such a manner that bees' faeces or dead and dying bees do not end up in the water (Hegić and Bubalo, 2006). It is not recommended to add salt or any other substance in the water. A lack of water may cause problems in digestive tract, especially to young bees intensively feeding on pollen. Also water is needed during hot weather to maintain temperature and humidity in the brood nest.

Fig. 11. Water source in test apiary.   Photo: N Kezic

Fig. 12. A useful water dispenser which can be connected to a water butt in order to provide continuous supply over longer periods. Note that the access to water is covered to reduce the risk of contamination by faeces.   Photo: N Kezic

3.2.3. Wax source

It is recommended that colonies be established on high quality wax foundation, free from pesticides (confirmed with a residue analysis). Residues in wax can significantly influence test results, especially if the wax comes from different suppliers. A part of, or entire supers can contain frames with drawn (built) combs. However, these combs have to be disinfected (acetic acid fumes, gamma rays) (de Ruijter and van der Steen, 1989; Baggio *et al.*, 2005). Frames and supers treated with acetic acid fumes need to be well ventilated prior to use.

3.2.4. Establishment of test colonies

We recommend the use of package bees ("artificial swarms"; Fig. 13) as the healthiest and most uniform start of test colonies. The artificial swarm has to contain at least 2 kg of young and healthy bees. The bees are placed on wax foundation in a disinfected hive. The queen is introduced at the same time as the bees. Bees should have access to sugar solution in feeder. Newly formed colonies are fed for the first few days with small amounts of sugar solution (1:1).

Starting test colonies by requeening existing hives or as nucs with brood is less recommended as it bears a higher risk of contamination with diseases that are not always clearly visible (varroa, nosema, chalkbrood, viruses). However, if this method has to be used for practical reasons, we recommend establishing nucs with at least two frames with brood, two frames with pollen and honey and the rest of the frames with wax foundation. At least 1 kg of bees should be in each nuc (see the COLOSS *BEEBOOK* paper on measuring colony strength parameters (Delaplane *et al.*, 2013)). The source of the bees and combs with brood and honey must be from healthy colonies.

Fig. 13. A uniform and hygienic establishment of test colonies can be achieved by placing artificial swarms placed on wax foundation.

Photo: D Krakar

3.2.5. Feeding

It is not recommended to feed bees with honey in order to avoid the spread of any diseases. During build-up, all colonies in the test apiary should receive the same quantity of sugar solution. Test colonies should always contain of minimum of 10 kg stored honey to support optimal and healthy development. Rescue of weak colonies by adding brood frames or by combining weak colonies is not allowed in test apiaries.

3.3 Testing criteria

At the Apimondia symposium "Controlled mating and selection of the honey bee" held in Lunz in 1972, technical recommendations for methods to evaluate the performance of bee colonies were developed (Ruttner, 1972) which still serve as an international standard for testing and selecting honey bees. However, much technical progress has been achieved since then, and today the beekeeping community is facing new challenges, first of all due to challenges posed by varroa, but also because of rapid environmental and climatic changes (Neumann and Carreck, 2010). Reviews of recent developments in breeding for resistance to *Varroa destructor* in Europe and the USA have been published by Büchler *et al.* (2010) and Rinderer *et al.* (2010) respectively.

The recommendations in the sections below were largely revised and approved by the members of COLOSS Working Group 4 who cooperated in a European-wide experiment with more than 600 test colonies for assessing the impact of genotype-environment interactions on the vitality of honey bee colonies (Costa *et al.*, 2012.).

3.3.1. Honey productivity and feed consumption

- All honey harvested within one season from an individual hive is recognized as the honey production of the test colony. A potential crop of swarms or permanent splits, coming from the test colony, is not regarded.

- Honey stored in the brood nest is not considered toward honey production.

- The supers filled with honey combs are weighed before and after extracting and the difference is noted as the honey harvest. If the extraction procedure does not allow following individual combs, an average net weight of extracted supers can be used instead of weighing individual supers after extraction.

- The result is noted in kg.

- The balance should ensure an accuracy of 100 g.

- Repeated honey harvests during one season are totalled to calculate the total honey production.

- The honey harvest of different periods, however, should be reported separately in order to document the colony's development and adaptability to different crops.

- For more accurate investigations of colony development and food consumption, the total weight of the hives has to be checked in regular intervals. The net weight of all added or replaced equipment has to be noted to calculate the net weight development in defined control intervals, for example during overwintering. See the *BEEBOOK* paper on miscellaneous research methods for techniques associated with weighing full colonies (Human *et al.*, 2013).

- Programmable hive scales are on the market. Some models store the total hive weight in short intervals and can transfer the data via cell phone to central computers. This allows a continuous real-time monitoring of the honey production and food consumption of test colonies.

3.3.2. Gentleness and behaviour on combs

- As a standard protocol in performance testing, defensive behaviour and response of the bees during handling are subjectively classified by an experienced tester (Table 5).

- In accordance with the Apimondia guidelines, the classification of gentleness and calmness are scored on a scale from 1 to 4, where 1 represents the most negative and 4 the most positive phenotype. Intermediate scores (0.5) can be used to better describe slight differences within the population.

- To ensure the comparability of test results colonies should be scored according to the following descriptions. Use intermediate scores (3.5, 2.5, 1.5) if the observed behaviour is somewhere between the given descriptions.

- The evaluation of the behaviour has to be repeated 3-6 times during the season without regard to specific conditions (like weather, honey flow etc.). The arithmetic mean of all evaluations is calculated at the end of season and used as test result.

- All colonies within one test yard need to be evaluated on the same date. As defensive colonies can influence the reaction of neighbouring hives, the order of management should be varied among successive evaluations.

Table 5. Standard scoring criteria for colony defensiveness.

Points	Gentleness	Calmness
4	No use of smoke and no protective clothes are necessary to avoid stings during normal working procedure.	Bees stick to their combs "like fur" without any notable reaction to being handled.
3	Colony can easily be worked without stings, if using some smoke.	Bees are moving, but do not leave their combs during treatment.
2	Single bees attack and sting during working procedure, even if smoke is used intensively.	Bees partly leave their combs and cluster in the edges of frames and supers.
1	In spite of the use of smoke the colony shows a strong defence reaction on being handled, or bees attack without being disturbed.	Bees nervously leave the combs, run out of the supers and cluster inside or outside the hive.

Table 6. Standard scoring criteria for colony propensity to swarm.

Points	Symptoms of swarming behaviour
4	The colony does not show any swarming tendency. There are no swarm cells containing eggs, larvae or pupae.
3	Low swarming tendency: some queen cells with brood are present, but the overall colony condition does not indicate immediate swarming activities. The preparations for swarming may be stopped by destroying the swarm cells and offering additional comb space.
2	Strong swarming tendency as indicated by repeated queen cell construction and advanced symptoms of preparation for swarming (reduction of open brood, emaciated queen, limited comb construction).
1	Active swarming: the test colony swarmed or swarming could be prevented only by extensive intervention (interim nucleus etc.).

Table 7. Methods for determining the level of hygienic behaviour expressed by a colony. *Colonies that are considered hygienic based on the freeze-killed brood assay, i.e. colonies that remove >95% of the freeze-killed brood within 24 hours, will show very high consistency in results between assays, irrespective of strength of colony and nectar flow.

Method	Repeatability	Costs & efforts	Remarks
Freeze killed brood*	High in colonies that remove > 95% of the freeze-killed brood in 24h; variable, in colonies that do not	Moderate	Introduction of freeze killed brood pieces or use of liquid nitrogen
Pin test	Medium	Low	Piercing of 50 young pupae
VSH	Unclear	High	Tests for varroa specific hygiene

- For quantitative research results, black leather balls about the size of tennis balls, marked with alarm pheromone (isopentyl acetate) can be moved in front of the hive entrance to provoke
- stinging by guard bees (Collins and Kubasek, 1982; Free, 1961; Guzman-Novoa *et al.*, 2003; Stort, 1974). The number of stings remaining in the leather after 1 or 5 minutes of exposure can serve to measure differences in defence behaviour.

3.3.3. Swarming behaviour

- As with other behavioural traits (see section 3.3.2.), a 4 point scale is used to classify the swarming behaviour of test colonies (Table 6).
- Note that typical supersedure queen cells are not considered as swarm cells.
- All symptoms of swarming behaviour (score 1-4) are noted on each inspection.
- At the end of the testing season, the lowest registered score, representing the most extreme expression of swarming behaviour, will be assigned as test result.

- All observed (and usually destroyed) queen cells can be counted throughout the season to quantify slight differences between colonies within the same score. Those differences can be expressed be intermediate scores (3.5, 2.5, 1.5).

3.3.4. Hygienic behaviour

Hygienic behaviour is recognized as a natural antiseptic defence against the brood diseases, American foulbrood and chalkbrood, and against varroa (Boecking and Spivak, 1999; Evans and Spivak, 2010; Spivak and Reuter, 2001; Wilson-Rich *et al.*, 2009) and thus may be relevant in breeding programmes for resistance to these pathogens and parasite. Standardized methods for testing hygienic behaviour are based on the removal of freeze killed (Momot and Rothenbuhler, 1971; Spivak and Reuter, 1998) or pin killed brood (Newton and Ostasiewski, 1986). Furthermore, Harbo and Harris (2005) described a method to check for a specific hygiene behaviour induced by reproducing mites in brood cells, called Varroa Sensitive Hygiene (VSH). See Table 7 for more information.

Freezing the brood with liquid nitrogen is more efficient and less destructive to the combs than cutting, freezing, and replacing comb inserts.

3.3.4.1. Freeze-killed brood assay: cutting brood out of comb to freeze

1. Cut a comb section of sealed brood with purple-eyed pupae containing approximately 100 cells on each side (5 x 6 cm) from a frame and freeze it for 24 hours at -20°C.
2. Insert the frozen comb section into a frame of sealed brood in the colony being tested (Fig. 14). Tests have shown that it does not matter if the frozen section comes from the same colony from which it was removed or from a different colony (Spivak and Downey, 1998).
3. Remove the frames no more than 24 hrs later.
4. Record the number of sealed cells. In addition, the number of cells that have been partially or fully uncapped and the dead pupae that have not yet been completely removed from the cells can be recorded.
5. The tests should be repeated on the same colony at least twice
6. A hygienic colony will have uncapped and completely removed over 95% of the frozen brood within 24 hours on both tests. This is the most conservative (strict) assay for hygienic behaviour that should be used for breeding purposes.
7. A less conservative measure of hygienic behaviour calculates the number of frozen pupae completely removed plus those that are in the process of being removed after 24 hours.

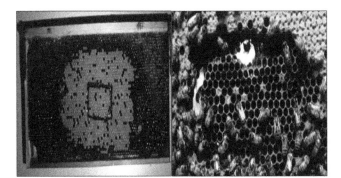

Fig. 14. Freeze-killed brood assay: cutting brood out of comb to freeze. Left: Frozen section of sealed brood is carefully placed into hole cut through comb. Right: Twenty-four hours after being returned to a colony, the amount of freeze-killed brood uncapped and removed is recorded.   Photos: M Spivak

3.3.4.2. Freeze-killed brood assay: freezing brood within comb using liquid N₂

1. Liquid nitrogen must be kept in an appropriate tank (e g. a Dewar tank) and gloves should be used when handling liquid N₂ (Fig. 15)

Fig. 15. Freeze-killed brood assay: freezing brood within comb using liquid N₂. Left: Dewar tank with valve to dispense liquid nitrogen, polystyrene foam cups for pouring liquid N₂ into PVC pipes (black pipes in combs). Right: After 24 hours, this hygienic colony uncapped and removed > 95% of the freeze-killed brood.   Photos: M Spivak

2. Make a 75 mm diameter tube to pour the liquid nitrogen directly on the comb. A metal vent pipe or PVC plumbing pipe can be used. A wider tube will reduce leakage of the nitrogen through empty cells along the perimeter. The tube should be at least 100 mm long.
3. Find a section of sealed brood with purple eyed pupae to freeze.
4. Put the frame horizontally across a support (i.e. an empty super). Press the tube down to the midrib of the comb with a twisting motion until it seals.
5. Record the number of unsealed cells inside the cylinder.
6. Pour 300-400 ml of liquid nitrogen into the tube. Less liquid N₂ may not freeze-kill the brood. Use a 300 ml or larger polystyrene foam (coffee) cup for measuring and pouring. First pour about 5 mm of the liquid nitrogen in the tube. When it evaporates pour the rest.
7. Wait for the liquid nitrogen to evaporate and the tube to thaw before trying to remove it (may take 10 min or more).
8. Return the frames to the colony for 24 hours.
9. The tests should be repeated on the same colony at least twice.
10. A hygienic colony will have uncapped and completely removed over 95% of the frozen brood within 24 hours on both tests. This is the most conservative (strict) assay for hygienic behaviour that should be used for breeding purposes.
11. A less conservative measure of hygienic behaviour calculates the number of frozen pupae completely removed plus those that are in the process of being removed after 24 hours.

Historically, colonies that removed freeze-killed brood within 48 hours were considered hygienic, and if they took more than a week, they were considered non-hygienic (Gilliam *et al.*, 1983). There is, however, a better correlation between the removal of freeze-killed brood and disease resistance when only the removal of freeze-killed brood within 24 hours is considered (Spivak, unpublished data).

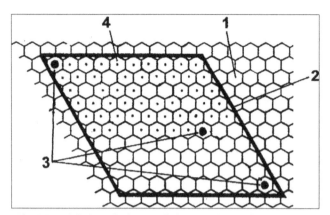

Fig. 16. Pin-killed test for hygienic behaviour. The numbers correspond to text references in Section 3.3.4.3.

Fig. 17. Pin test: **a.** Piercing 50 cells containing young pupae; **b.** Control of brood removal after about 8 hours, many cells are opened but not removed; **c.** Nearly all cells are completely cleaned. Photos: R Büchler

3.3.4.3. Pin-killed brood assay

The pin test method is recommended in Europe as a standard in field selection programmes, because it shows a significant correlation with the removal of varroa infested brood, can be standardized and is easily handled by beekeepers. A statistical tool has been established to include pin test data in the estimation of breeding values for varroa tolerance (see 4.1). For the pin-killed brood assay protocol, see Fig. 16 while following the numbered protocol below. Additionally, Fig. 17 shows images of the protocol being applied in the field.

1. A rhomboid frame of a 10×10 cell wide template (Fig. 16, number 2) is placed on a brood comb containing young pupae (Fig. 16, number 1)
2. The upper left and lower right cells are marked with a colour felt-tip pen (Fig. 16, number 3)
3. 50 capped brood cells are pierced (Fig. 16, number 4) row by row from left to right with a fine insect pin (entomological pin size No 2).
4. Cell 51 is marked to identify the treated brood area (Fig. 16, number 3).
5. The comb is marked on the top bar and placed back to the brood nest in its former position.

6. After 7-15 hours (uniform for all colonies within the comparison) the removal progress is checked. All cells that are still sealed or contain the remains of brood are counted and subtracted from 50. The percentage of completely cleaned cells is noted in the protocol.
7. The highest discriminatory power of the test is reached when all of the test colonies remove an average of 50% of the pupae within the time interval. Therefore, the time interval between piercing the cells and checking should be adapted to the average removal response of the test population. If the average removal rate is much lower than 50%, the time interval should be prolonged to yield higher differences between colonies with high and low hygienic behaviour. If the average removal is much higher than 50%, a shorter time interval should be realized in further test repetitions.
8. The test should be repeated 2-3 times during the main brood season.

3.3.5. Varroa infestation

Regular monitoring of varroa populations is not only a precondition for integrated varroa control, but also an important basis for the selection of mite resistant stock. Several different methods have been developed and tested with regard to systematic field evaluation of varroa densities (Lee *et al.*, 2010). Please also refer to the *BEEBOOK* paper on varroa (Dietemann *et al.*, 2013). We outline in Table 8 the methods commonly used to determine varroa populations in colonies and include information pertinent to the method's uses in stock selection.

As a standard for performance testing, repeated checks of the mite infestation level are recommended. In periods of low infestation (usually early spring), monitoring natural mite mortality reveals best results. Sampling bees is more effective with higher infestation levels that occur later in the season (Büchler, unpublished data). The estimation of breeding values (see 4.1) for varroa resistance is based on mite population growth during the season. For these calculations, natural mite mortality during 3-4 weeks of the first main spring pollen producing bloom (e.g. willow, hazel, almond for phenological standardization of different climatic regions) is combined with the mite infestation of bee samples estimated during summer. Repeated measurements of the bee infestation in intervals of 3-4 weeks improves the accuracy of the test and allows prolongation of the test period without treatment against varroa until defined threshold values (usually 5-10 mites/10 g bees, depending on environmental and beekeeping conditions) are reached.

3.3.6. Other diseases

In general, any disease symptoms of performance test colonies should be carefully registered and documented. Special care should be taken with diseases which can be influenced by the genetics of the bees. These include American foulbrood, chalkbrood and chronic bee paralysis

Table 8. Methods for estimating varroa populations in honey bee colonies (see the *BEEBOOK* paper on varroa for more information on each method, including how to perform the method (Dietemann *et al.*, 2012)).

Method	Repeatability	Effort	Remarks
Natural mite mortality (i.e. mite fall or mite drop)	low	low	Results depend on the amount of emerging brood and colony size; sensitive to the presence of ants, wax moths *et. al.*
Bee samples – washing technique	medium	medium	Doesn´t work with very low infestation rates; independent from colony size; bees are killed
Bee samples - powdered sugar	medium	low	Similar to washing technique, but bees are kept alive; evaluation directly at the bee yard possible; depends on dry weather
Brood samples	low	high	Time consuming; can be combined with investigations on mite reproduction

virus (CBPV or hairless black syndrome). Usually, no prophylactic or acute treatments against those diseases are recommended on test colonies so as to observe potential susceptibility or resistance. However, for a more systematic selection, a uniform initial infection of all colonies should be provided.

A simple, qualitative documentation (symptoms observed: yes/no) may be sufficient for identification and removal of infested colonies from the breeding programme, if the disease prevalence is low among colonies. Furthermore, such data can be used to identify differences among genotypes, if results of related colonies in different test environments and seasons are available. An estimation of breeding values for chalkbrood resistance has recently been developed at the institute in Hohen Neuendorf, Germany (Ehrhardt, pers. communication), based on such a simple data structure. Quantitative protocols may be used for highly prevalent diseases or for more intense selection for resistance to certain diseases. See the respective pest and pathogen *BEEBOOK* papers (De Graaf *et al.*, 2013; De Miranda *et al.*, 2013; Dietemann *et al.*, 2013; Forsgren *et al.*, 2013; Fries *et al.*, 2013; Jensen *et al.*, 2013).

3.3.7. Colony development and wintering

The seasonal development of the bee population and brood activity are important parameters to describe local adaptation, wintering ability and productive potential of test colonies. Therefore, regular notes on the bee and brood status are essential components of each performance test. The strength of the colony (bee population and brood extension) should at least be evaluated before and after wintering (i.e. during the first pollen flow but before plenty of young bees emerge), at the beginning of the honey flow and at the peak of development. An overwintering index, calculated as: bee population at the end of the winter / bee population before winter yields important information on the health of wintered colonies and the wintering ability of the colony. It can be combined with amount of honey consumed during winter (see 3.3.1.) to select for winter hardiness. A high overwintering index and low food consumption indicate healthy colonies that clearly stop rearing brood and have a stable winter cluster. The relation of bees and brood in spring and the overwintering index can be used to classify

the spring development of colonies. Colonies with high brood activity and a quick increase in population are more suitable to exploit a good spring honey flow.

Population estimates measured with high accuracy, as may be needed for scientific investigations, can be achieved by the methods described in the *BEEBOOK* paper on measuring colony strength parameters (Delaplane *et al.*, 2013). When field testing of large numbers of colonies (as in. most honey bee selection programmes), satisfactory results can be achieved using the methods outlined in 3.3.7.1. and 3.3.7.2.).

3.3.7.1. Bee population

- Check each hive box (or super) from the top and bottom (you do not need to take out individual combs) immediately after opening the hive to estimate how many spaces between combs are populated with bees.
- Add up the total number of combs covered with bees. Fully covered spaces between combs count as 1. Partially covered ones are counted proportionately in quarters of a comb (0.25, 0.5, 0.75).
- Seasonal differences in the average density of bees in the cluster do not need to be recorded as the data are mainly used to compare colonies to one another. They are not meant to be an absolute measure of the number of bees.

3.3.7.2. Brood area

- Count the number of combs containing brood. Count the brood as 0.5 if the brood is just on one side of the comb.
- In addition, the brood area on a central brood comb gives useful information on the brood activity of the hive. A 4 point scoring is recommended for the protocol according to following the scheme:
  - ♦ 4 points: brood present on more than 75% of the comb,
  - ♦ 3 points: brood present on 50 – 75% of the comb,
  - ♦ 2 points : brood present on 25-50% of the comb,
  - ♦ 1 point: less than 25% of the whole comb area is covered with brood.

3.3.8. Additional test characters

With regard to specific needs, bees can be tested and selected for further traits. Pollen gathering, length of life and breeding for morphological characters are some examples for successful selection activities (Rinderer, 1986).

Further characters may be included to improve the disease resistance of bees. With regard to varroa resistance, various traits such as the grooming behaviour of bees, the post-capping period duration and others, have been discussed as potential selection criteria but have not been demonstrated to be effective.

However, testing and selection may be more effective if focused on fewer characters. Usually, each additional test parameter needs additional effort and results in additional stress for the colonies. Furthermore, simultaneous selection for several independent characters reduces the selection power for each single trait. Thus, the breeding success depends very much on a clearly defined selection goal and a consequent testing scheme.

4. Selection tools

The goal of beekeeping is to produce many quality products and pollination services with maximum efficiency. An important factor in achieving this goal is genetic improvement in terms of economic, behavioural and adaptive traits of honey bees. Genetic improvement is achieved with selection (Falconer and Mackay, 1996). The rate of improvement is directly linked to accuracy with which queens are ranked based on their breeding value, the intensity with which they are selected, the amount of genetic variation available in the traits and generation interval. All of these issues are part of the breeding programme.

The standardization of performance testing as described in Section 3.3. is a necessary prerequisite for successful breeding. The results will indicate differences between individual colonies that can be utilized for improvement, but these data alone are insufficient. The environment varies greatly between and within apiaries and test stations, and the traits measured are strongly affected by these environmental effects. Only the hereditary disposition is significant in breeding, as only the hereditary disposition (genes) of the animals influence the quality of the offspring. The environmental conditions under which the colonies live unfortunately mask or influence their hereditary properties (breeding value). A breeding programme therefore requires a breeding value or selection index in order to choose which queens to reproduce, according to the aims of the breeding programme.

There are several instruments available for separating the environmental effects of colony performance from genetic disposition. The most sophisticated and accurate method for calculating a selection index is a statistical model called the "BLUP (Best Linear Unbiased Prediction) Animal Model" (Henderson, 1988), which was modified for use in honey bee breeding programmes by Bienefeld *et al.* (2007) (described in section 4.1). However, for small scale breeding programmes, simpler indicators may be used (section 4.2).

4.1. Genetic evaluation with BLUP

The use of the BLUP Animal Model is referred to as "Genetic evaluation" and its outcome, the "breeding values", refers to the probability that the progeny of the selected individuals will be above or below the population average for a certain considered trait.

Genetic evaluation aims at assigning a genetic value to each animal with the goal of ranking animals and selecting animals with the best genetic values. Compared to other livestock which undergo genetic improvement, honey bees have peculiar genetic and reproductive characteristics (haplo-diploid sex determination, arrhenotoky, polyandry) which make simple appliance of the BLUP Animal Model not appropriate (difficulties in calculating the numerator relation matrix, which links information from related colonies (Bienefeld *et al.*, 1989; Fu-Hua and Sandy, 2000)). However, the main methodological problem is that the colony's performance and behaviour result from the interaction between the queen and worker bees. Thus, a trait measured in the honey bee colony is the result of the combined activities of the queen (maternal effect) and workers (direct effect). Bienefeld and Pirchner (1990) found queen and worker effects to be negatively correlated, which strongly hinders selection response (Willham, 1963). Therefore, the BLUP animal model approach was modified to consider worker and queen effects and the negative correlation between them (Bienefeld *et al.*, 2007).

Genetic evaluation via BLUP combines the phenotypic data of the animal itself with data of related animals to rank them according to their (environmentally adjusted) genetic merit. Therefore, this approach needs the individual results of performance tests of all animals and the genetic relationship (pedigree information) between them. All this information must be combined in an appropriate database.

The requirements for the database are the following:

- Controlled (i.e., password-protected) access for data input.
- Software-assisted checking for coherence with existing information, outliers, and logical inconsistencies.
- Clear definition of access rights if several people have written access (e.g. breeder and administrator of a breeding association).
- Data format should fit the requirements of the of the genetic evaluation software.
- Open access for all users regarding the results of the genetic evaluation.

At the moment, just one international database for the honey bee fulfils these requirements (www.beebreed.eu), and so its specifications have been chosen as a standard.

Most breeders use the database not only for efficiently making data of their colonies available for genetic evaluation, but also for running their private studbook. Not all entries of the studbook (e.g. day of birth, tag colour of queen, etc.) are needed for genetic evaluation. To adjust for the environmental effect, information concerning the contemporary group is of central importance. A contemporary group comprises all colonies tested at the same location and management conditions at the same point in time. For genetic evaluation, the contemporary group is formed by combining the following variables: year of birth, login ID of the tester (who is not necessarily the breeder), and a code for the apiary belonging to the tester (one tester may run several apiaries). Ten to 15 colonies per apiary are needed to be able to correctly adjust for the environmental effect of an apiary. However, fewer colonies per apiary are accepted for genetic correlation, but then the colony information at these apiaries is downgraded. Genetic evaluation requires genetic links within the population and is promoted by the simultaneous testing of the different genetic origins (of the same race) at each apiary.

For the reasons explained above (reproductive peculiarities of honey bees) and in contrast to other species, the full pedigree specification in the database used for genetic evaluation consists of the identification number of the (actual) queen, of her mother, of her mating partner, and NOT her father. This model is adapted to the breeding scheme according to which a single drone line is used: a mother queen is selected from whom a group of queen daughters is reared, which will be used for drone production (Ruttner, 1988). The paternal descendent of each queen needed for genetic evaluation is (software-assisted) generated by using pedigree information of her mother. For each drone producing sister group, a dummy father is inserted into the pedigree. The identification number of the mother is a mandatory field in the database, but not for the mating partner, because controlled single-line mating is not adopted by all associations. Pedigree data is combined with performance data for genetic evaluation.

4.1.1. Access to the data input feature

Two options are available:

1. The administrator of a breeding association receives the pedigree and test data from the breeders via lists, copies of their studbooks, etc. to input these data under the breeders login ID.

2. The administrator of a breeding association activates password-protected access to the data input module of database for each breeder: in this case, each breeder inputs all his data alone. However, this stand-alone data input by a breeder requires additional data checking and confirmation by the administrator of the responsible breeding association before these data can be released for genetic evaluation.

4.1.2. Pedigree data

A unique queen identification number is a central requirement for genetic evaluation. The international unique queen identification number (QID) (see www.beebreed.eu for coding) consists of:

- Country code: 2 digits
- Breeder ID (within country)   3 digits
- Queen no. within the studbook of the breeder   5 digits
- Year of birth of the queen   4 digits

The international QID is automatically linked with an alphabetic race code (**C** for *A. m. carnica*, **L** for *A. m. ligustica* and **M** for *A. m. mellifera*) if the authorized breeder enters the corresponding database with his password.

The Statistical model used in the modified BLUP Animal Model is the following:

$$y = Xb + Z_1u_1 + Z_2u_2 + e,$$

where: y = a vector of records/traits of the colonies (e.g., honey production, defence behaviour); b = a vector of fixed year/beekeeper/location effects; u_1 = a vector of random worker (direct) effects; u_2 = a vector of random queen (maternal) effects; e = a vector of random residual effects; X = incidence matrix relating the observations to the corresponding environment (apiary within tester and year effect); Z_1 = incidence matrix relating the observations to corresponding worker effects; Z_2 = incidence matrix relating the observations to the corresponding queen effects.

Solutions are obtained from the following mixed model equations:

$$\begin{pmatrix} X'X & X'Z_1 & X'Z_2 \\ Z_1'X & Z_1'Z_1+A^{-1}\alpha_1 & Z_1'Z_2+A^{-1}\alpha_2 \\ Z_2'X & Z_2'Z_1+A^{-1}\alpha_2 & Z_2'Z_2+A^{-1}\alpha_3 \end{pmatrix} \cdot \begin{pmatrix} b \\ u_1 \\ u_2 \end{pmatrix} = \begin{pmatrix} X'y \\ Z_1'y \\ Z_2'y \end{pmatrix}$$

where

$$\begin{pmatrix} \alpha_1 & \alpha_2 \\ \alpha_2 & \alpha_3 \end{pmatrix} = \begin{pmatrix} \sigma_1^2 & \sigma_{12} \\ \sigma_{12} & \sigma_2^2 \end{pmatrix}^{-1} \cdot \sigma_e^2$$

with: σ_1^2 = additive genetic variance for worker effects; σ_2^2 = additive genetic variance for queen effects; σ_{12} = additive genetic covariance between worker and queen effects; σ_e^2 = residual error variance; A^{-1}= inverse of the additive genetic relationship matrix.

Many production and behavioural traits are correlated genetically (are influenced by some of the same genes). The more traits that are targeted with the breeding programme, the less progress can be made for any single trait. A multi-trait approach, which considers the genetic correlation between traits, is applied so that predicted breeding values for individual traits in the breeding goal are combined according to the demands of the breeders (Ehrhardt and Bienefeld, unpublished).

Phenotypic and genetic parameters (Bienefeld and Pirchner, 1990; Bienefeld and Pirchner, 1991) are re-estimated from time to time. All aspects of estimation procedures for the estimation of variance components (data structure, method and model of estimation, effects included in the model, and so on) should be as similar as possible to the estimation procedures for breeding values.

The accuracy of genetic evaluation depends on the quality of the relationship information and the possibility of the statistical procedures to distinguish the genetic component from the total phenotypic variance. The estimations may even lead to misinterpretation if they are not statistically adequate. Breeding values, inbreeding coefficients, and tools for breeding plans should be published. Breeding values are estimated once a year and are published mid-February of each year.

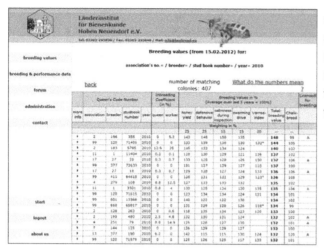

Fig. 18. Screen shot from the breeding value database at www.beebreed.eu.

4.1.3. Outcome of genetic evaluation: breeding values

The breeding value states for a particular characteristic (honey production, varroa tolerance, etc.) the extent to which an animal is genetically different from the average of the population. Breeding values can be expressed as the percentage of a moving genetic average of the population. The moving basis is the last-five-year-genetic-average for each trait. Consequently, breeding values usually depreciate, if genetic response is achieved. Because the traits used for honey bee breeding strongly differ with respect to phenotypic variation (honey 0-150 kg, gentleness 1-4), their breeding values also differ. To ensure their comparability, breeding values of all traits are transformed by fitting to an identical standard deviation of 10.

At www.beebreed.eu, several features are available to select queens meeting the specific demands of breeder or buyers of queens:

- Breeding and inbreeding values of specific queens
- List of queens that meet specific requirements (e.g. breeding value for varroa tolerance > 125% and for other traits ≥ 100%). An example is given in Fig. 18.
- List of queens, including a total breeding value (combination of all traits used for selection) that meets the specific weighting of the traits in which the breeder or buyer is interested.

A breeding plan program is also available at www.beebreed.eu. Entry of the QID of potential parents makes available an estimation of the inbreeding and breeding values of the expected offspring. This allows breeders to visualize the potential results that a specific cross will produce to avoid inbreeding. Inbreeding has been found to be of crucial importance for honey bee breeding programmes. Additionally, a tool is available to search for the mating station that best suits the individual breeding goal.

4.2. Selection indexes and scores

Due to various reasons, there are cases where an organized data collection as described in section 4.1. is not possible or there is an incomplete data structure. In such cases, a direct comparison of the queens based on their performance can be used. However, one

should be aware that this ranking is based on phenotypic value only and does not reflect the genetic potential of the queens. In addition, a lack of pedigree information can lead to inbreeding and it is not reliable in producing the next generation of queens. However, the following approaches can be useful if a breeding programme is not yet established or is in its infancy:

- Regression analyses: In most breeding programmes, several traits are of interest (morphological, behavioural and production level). Evaluation of the colonies is only based on their own performance and additional information gained from ancestors and progeny cannot be linked to them. In most cases, regression analyses can be applied, e.g. linear, logistic or even ordinal, depending of the quantity of information complementing the performance data. The adequate choice is subject to understanding the data structure and statistical methods. Nevertheless, in traits that are described quantitatively, linear regression can be sufficient, with or without previous data transformation for obtaining normality. If the traits are described in categorical values, logistic regression can be used. The estimations will be a compromise between the potential for corrections in environmental factors and the observed individual performance leading to lower accuracy. In some cases, survival analyses are appropriate (Rhodes *et al.*, 2004), particularly in disease tolerance.
- Z-score: a simple way for comparing colonies across apiaries. It assumes that differences between apiary average scores are entirely due to location differences (this is not completely true due to interactions between the genetic origin and the location). Each testing apiary is described in terms of its own mean and standard deviation, then the individual colony performances are transformed into standard deviation units and compared (Rinderer, 1986). The resulting individual score is called z-score: $z = X - M / s$ where: X = colony score; M = apiary average score; s = apiary standard deviation.
- Selection index according to Rinderer (1986): the aim of a selection index is to express the breeding value from the point of view of several traits in a single number. The selection index proposed by Rinderer (1986) considers the colony's individual phenotypic scores, the heritability (h^2) of the traits and the genetic correlations between them, as well as the economic value of the characteristics (based on breeding programme and beekeeper preference). A simple version of the index considers only the z-scores and the relative economic value of the chosen traits: $I = z_a V + z_b$ where: z_a = z-score for trait A; z_b = z-score for trait B; V = relative importance of trait A compared to trait B (e.g. if trait A half as important as trait B then $V = 0.5$).
- The above equation can further incorporate the heritabilities and genetic correlations between traits: $I = z_a V (h^2_a / h^2_b) + z_b (1 - r_g)$ Where: h^2_a = heritability of trait A; h^2_b = heritability of trait B; r_g = genetic correlation between traits (correlation between breeding values).

- Selection index according to Cornuet and Moritz (1987): when groups of sister queens are considered in the testing programme, a selection index J which considers the relationships inside the family (mother-daughter covariance, between sisters covariance and aunt-niece covariance) can be used. Plausible values for covariances result in the following formula, which considers a single trait: $J_{ij} = 0.163 \, (m_{ij} - m_i) + 0.348 \, m_i$ Where: m_{ij} = colony value; m_i = average family value.

4.3. Molecular selection tools

Note: many of the methods mentioned below are outlined in the *BEEBOOK* paper on molecular research techniques (Evans *et al.*, 2013).

The completion of the honey bee genome project held the promise for fast selection of colonies with desirable traits (Weinstock *et al.*, 2006). Knowing the genes coding for any particular trait would, in theory, allow for the selection of queens and drones with desired genotypes for further breeding without evaluation of colony traits. However, at present much knowledge is still needed before delivery on this promise can come through. Complications further arise from the complexity of honey bee genetics. It seems that those colonies that perform best, do so due to a high level of genetic diversity amongst the workers (Seeley and Tarpy, 2007). The colony composition of two generations in form of the queen and her worker offspring and the combinational effects of mostly more than ten chromosome sets due to the multiple matings of the queen. This makes the role that selection for a single trait at individual level can play questionable, especially when transferred into colony performance. In more advanced and complex breeding programmes, genome-wide marker assisted selection may boost accuracy of genetic improvement in honey bees (Meuwissen *et al.*, 2001). The recent developments in sequencing single nucleotide polymorphisms (Harismendy *et al.*, 2009) and bioinformatics' approaches in data evaluation (Pérez-Sato *et al.*, 2010) can make breeding programmes for honey bees more reliable. However, such an approach needs considerable resources and expensive laboratory work.

Even before completion of the honey bee genome, scientists started the search for quantitative trait loci (QTL) in honey bees using different kinds of markers:

- Hunt *et al.* (1995) used bees preselected for variation in their pollen hoarding behaviour to search for the underlying genetic traits. Using genetic markers derived from a technique called random amplification of polymorphic DNA (RAPD), they identified first two and later a third marker (Page *et al.*, 2000). Each marker held predictive power, concerning the preference of a given forager for the collection of either nectar or pollen. The RAPD loci observed are not thought to be directly responsible for the variance in the traits, they are merely closely linked to a genetic region that primes the bees' behaviour in the direction of pollen or nectar collection.

- Using similar RAPD markers with the addition of DNA microsatellites and a sequence tagged site, Lapidge *et al.* (2002) detected seven loci linked to hygienic behaviour in honey bees. This finding conflicts with the only two loci described by Rothenbuhler (1964; see however Moritz, 1988); still it may result from the usage of strains less extremely selected as compared to the earlier studies.

- Today RAPD are all but forgotten, as is their cousin methodology of amplified fragment length polymorphism (AFLP) used by Rüppelt *et al.* (2004) to search for additional markers linked to pollen hoarding behaviour.

A variety of markers with accurate linkage maps today exist for the preliminary screening for QTL:

- At first, the DNA microsatellites carefully mapped by Solignac *et al.* (2004) became the marker of choice.

- Since the genomic information became available (Weinstock *et al.*, 2006), single nucleotide polymorphism (SNPs) also allow cheap and accurate targeting of QTL. Recently a marker set of 44000 has become commercially available (Spötter *et al.*, 2011), providing a robust coverage of the honey bee genome. Using this set of markers in a study of "varroa-specific defence behaviour", it has been shown that it is important to examine several control populations to avoid randomly significant SNPs. In the study at hand, more than 151 SNP differed between the reference sample of "varroa-defence bees" and a set of bees from completely unhygienic colonies, against 7 SNPs differing between varroa-defence bees and related workers not engaging in defensive behaviour, taken at the highest level of significance. Comparing all three groups, merely a single SNP remained. This result demonstrates the value of having appropriate samples available.

The current rapid developments in availability and pricing of DNA sequencing may eventually replace all these linkage bound methods with a direct sequence based search for the underlying genetic variance for each trait.

- A separate methodology to identify marker genes has emerged from the use of microarray techniques. Microarrays consist of a set of known honey bee genes. Using the microarray allows for the detection of mRNA levels in specific workers. The microarrays are built based on expressed sequence tags (EST) results from mRNA of bees, which after cDNA transformation are cloned and can be analysed rather swiftly (Whitfield, 2002). Based on genetic information from *Drosophila melanogaster* many of the gene functions are well known. An example of the application of this technique is the study of honey bee brood reaction to parasitism by varroa mites (Navajas, 2008). The strength of this technique lies in the immediate detection of differential gene activity in bees with variable traits. It is thus feasible to directly identify the action of genes related to specific traits. The currently available microarrays allow for the screening of more than 8000 genes identified from the honey bee brain. Any gene unidentified or not included in the microarray however, will go undetected. This is particularly important for those promoter regions that act as switches for coding genes, as these are likely to go unnoticed from such studies.

- While interactions between coding genes and their regulator genes may go unnoticed by microarray techniques, the use of SNP markers might be particular suitable for the detection of promoter regions. In humans two independent SNPs have been shown to generate lactose tolerance in adults (Tishkoff, 2007).

QTL methods are particularly applicable to honey bees, due to the rather small genome with a high rate of recombination. Furthermore, the haploid stage of the drone allows for direct testing of traits linked to the individual level, but it remains more complex for colony level traits. If workers can be observed to harbour a significant fraction of a colony's traits, like those engaging in hygienic behaviour, these too can be employed for these type of studies. Due to the multiple matings of the queen with haploid drones, a colony will typically consist of more than 10 subfamilies. Each subfamily, often referred to as a "patriline", effectively acts as linkage group sharing the paternal fraction of the genome. Bees with a particular patriline are variable for the remaining queen contributions. This allows for the testing of genotype interactions, both at the individual worker level and at colony level. Finding QTLs or genes affecting complex colony traits, like swarming behaviour, honey production or gentleness will demand thorough testing and considerable skills both at the molecular and computational level. The main problem remains, i.e. to demonstrate, in a considerable set of colonies, that heritable variance exists for the trait of choice. Only once a large sample size is available, representing both variation and similarity between the screened colonies, would it seem worthwhile to conduct a molecular genetic screening.

A caveat in the interpretation of genetic marker data results from the vast number of genes screened, either genetically mapped markers or from microarray studies. Chance differences in marker diversity between tested bees or in the activity of genes unrelated to the trait under study are rather likely given the vast number of comparisons. Hence it is advisable to demand particular strict statistical testing, before accepting a particular marker as involved. One way to reduce this problem is to repeat the study in several independent populations.

While the arrival of molecular markers will allow for rapid selection, some words of caution are needed. It may seem straightforward to select for the identified genotype in a separate population, if this has been found to be associated with particular valuable traits. As a shortcut, it may be equally tempting to inter-cross a set of genes into an unrelated population, and based on marker assisted selection follow their fate in following generations. Organisms resulting from

this technique have been termed cis genetically modified organisms, in contrast to trans genetically modified organisms, as the genetic exchange happens via traditional interbreeding, and genes are not introduced from other completely unrelated species. In theory it could be possible to incorporate a single gene into an unrelated population, however, unless considerable care is taken this will go hand in hand with a significant genetic bottleneck. Whether consumers, be it bee-keepers or honey buyers, will accept such cis techniques as being less problematic than standard trans GM techniques remains an open question. Furthermore, searching for identical genotype variations in unrelated populations hold no warranty for success, as our knowledge of the complex underlying mechanisms are still rather rudimentary. While the future of honey bee breeding may benefit from more advanced molecular methods, it is still an emerging field.

5. Breeding designs

The tools described in section 4 provide an indication on which colonies to use in breeding, i.e. which colonies to use for the production of queens and drones. However, how many colonies should be chosen and how these breeder colonies should be combined depends on the aims, size and resources of the breeding programme.

5.1. Closed-population breeding

In a closed population, there is no introduction of unknown genetic material: this can be achieved by use of completely isolated mating stations (section 2.2.2.) or instrumental insemination (see the *BEEBOOK* paper on instrumental insemination, Cobey *et al.*, 2013). The aim of this kind of design is to rapidly achieve improvement while limiting loss of genetic variability (which would lead to inbreeding depression). Laidlaw and Page (1986) list 3 basic strategies:

- Daughters from all of the breeding queens are each mated (instrumentally inseminated) to 10 drones selected at random from the entire population; replacement breeder queens are selected at random from all the daughters of all the breeder queens, without considering their parentage. To operate this design as a long term plan, about 50 breeder colonies must be selected at each generation, in order to reduce inbreeding.
- Each breeder queen is replaced by one of her daughters, reared as above.
- All queen daughters are inseminated with the same aliquot of mixed semen originating from drones of all breeder queens.

5.2. Open population breeding

In this kind of design, the introduction of foreign genetic material into the population is allowed, thereby reducing the risk of inbreeding. Performance testing with sister queen groups placed in different testing apiaries is particularly useful for the calculation of breeding values.

More simply, significant differences among families, distributed across different apiaries, reveal a heritable effect of the performance. An example of an open breeding scheme is the following (from Cornuet and Chevalet, 1987):

- First generation: selection based on individual values.
- Second generation: colonies ranked according to selection index or breeding value (combination of performance and pedigree data) – the best 10 colonies are used for queen and drone production.
- Mating occurs in mating station where selected and unselected drones are present.

5.3. Special designs for scientific purposes

5.3.1. Bi-directional selection

To understand the physiological or genetic mechanism underlying a specific trait, it can be useful to obtain individuals that manifest extreme values for this trait. A breeding design in which the best and worst individuals are chosen and reproduced is referred to as "bi-directional selection". An example of a bi-directional design is described in detail in Page and Fondrk, 1994. The basic steps are the following:

- The 10 best and 10 worst colonies are selected.
- Five sublines within the best and worst groups are created by inseminating virgin queens with semen from a single drone (from a different colony of the same group). See the *BEEBOOK* paper on instrumental insemination (Cobey *et al.*, 2013).
- At each generation, the best colony of the "best" group and the worst colony of the "worst" group are used for the production of virgin queens and drones.
- The colonies from the 3rd generation queens are used for the experimental observations.

5.3.2. Single drone mating

In some experiments, it is useful to minimize genetic differences among colonies in order to establish the extent of an external factor. For this aim, instrumental insemination (see the *BEEBOOK* paper on instrumental insemination (Cobey *et al.*, 2013)) of one or more queens with semen from a single drone can be used (spermatozoa of a single drone are genetically identical). According to the number of individuals needed for the experiment, the scientist may decide whether to inseminate up to 3 queens with semen from a single drone. However, success in single drone insemination is more likely when a single queen is inseminated. Daughter queens from the single mated queen may then be raised (they will be closely related with degree of relationship = 0.75 i.e. "super-sisters") and according to the required level of homozygosity required in the experiment, may then be inseminated with pooled homogonous semen, or naturally mated in an isolated mating station with selected drones.

References

ALBERT, M; JORDAN, R; RUTTNER, F; RUTTNER, H (1955) Von der Paarung der Honigbiene. *ZeitschriftfürBienenforschung* 3: 1-28.

ARATHI, H S; SPIVAK, M (2001) Influence of colony genotypic composition on the performance of hygienic behaviour in the honey bee, *Apis mellifera* L. *Animal Behaviour* 62(1): 57-66. http://dx.doi.org/10.1006/anbe.2000.1731

BAGGIO, A; GALLINA, A; DAINESE, N; MANZI NELLO, C; MUTINELLI, F; SERRA, G; COLOMBO, R; CARPANA, E; SABATINI, A G; WALLNER, K; PIRO, R; SANGIORGI, E (2005) Gamma radiation: a sanitizing treatment of AFB contaminated beekeeping equipment. *Apiacta* 40: 22-27.

BIENEFELD, K; EHRHARDT, K; REINHARDT, F (2007) Genetic evaluation in the honey bee considering queen and worker effects - a BLUP-animal model approach. *Apidologie* 38: 77-85. http://dx.doi.org/10.1051/apido:2006050

BIENEFELD, K; PIRCHNER, F (1990) Heritabilities for several colony traits in the honey bee (*Apis mellifera carnica*). *Apidologie* 21: 175-183.

BIENEFELD, K; PIRCHNER, F (1991) Genetic correlations among several colony characters in the honey bee (Hymenoptera: Apidae) taking queen and worker effects into account. *Annals of the Entomological Society of America* 84: 324-331.

BIENEFELD, K; REINHARDT, F; PIRCHNER, F (1989) Inbreeding effects of queen and workers on colony traits in the honey bee. *Apidologie* 20: 439-450.

BOECKING, O; SPIVAK, M (1999) Behavioural defences of honey bees against *Varroa jacobsoni* Oud. *Apidologie* 30: 141-158.

BOUGA, M; ALAUX, C; BIENKOWSKA, M; BÜCHLER, R; CARRECK, N L; CAUIA, E; CHLEBO, R; DAHLE, B; DALL'OLIO, R; DE LA RÚA, P; GREGORC, A; IVANOVA, E; KENCE, A; KENCE, M; KEZIC, N; KIPRIJANOVSKA, H; KOZMUS, P; KRYGER, P; LE CONTE, Y; LODESANI, M; MURILHAS, A M; SICEANU, A; SOLAND, G; UZUNOV, A; WILDE, J (2011) A review of methods for discrimination of honey bee populations as applied to European beekeeping. *Journal of Apicultural Research* 50(1): 51-84. http://dx.doi.org/10.3896/IBRA.1.50.1.06

BÜCHLER, R; BERG, S; LE CONTE, Y (2010) Breeding for resistance to *Varroa destructor* in Europe. *Apidologie* 41: 393-408. http://dx.doi.org/10.1051/apido/2010011

CARRECK, N L; ANDREE, M; BRENT, C S; COX-FOSTER, D; DADE, H A; ELLIS, J D; HATJINA, F; VANENGELSDORP, D (2013) Standard methods for *Apis mellifera* anatomy and dissection. In *V Dietemann; J D Ellis; P Neumann (Eds) The COLOSS BEEBOOK, Volume I: standard methods for* Apis mellifera *research. Journal of Apicultural Research* 52(4): http://dx.doi.org/10.3896/IBRA.1.52.4.03

CASAGRANDE-JALORETTO, D C; BUENO, O C; STORT, A C (1984) Numero de ovariolosemrainhas de *Apis mellifera*. *Naturalia* 9: 73-79.

CENGIZ, M; EMSEN, B; DODOLOGLU, A (2009) Some characteristics of queen bees (*Apis mellifera* L.) rearing in queenright and queenless colonies. *Journal of Animal and Veterinary Advances* 8(6): 1083-1085.

COBEY, S W (2007) Comparison studies of instrumental inseminated and naturally mated honey bee queens and factors affecting their performance. *Apidologie* 38: 390-410. http://dx.doi.org/10.1051/apido:2007029

COBEY, S W; TARPY, D R ; WOYKE, J (2013) Standard methods for instrumental insemination of *Apis mellifera* queens. In *V Dietemann; J D Ellis; P Neumann (Eds) The COLOSS BEEBOOK, Volume I: standard methods for* Apis mellifera *research. Journal of Apicultural Research* 52(4): http://dx.doi.org/10.3896/IBRA.1.52.4.09

COLLINS, A M; KUBASEK, K J (1982) Field test of honey bee (Hymenoptera: Apidae) colony defensive behaviour. *Annals of the Entomological Society of America* 75: 383-387.

COLLINS, A M; DONOGHUE, A M (1999) Viability assessment of honey bee, *Apis mellifera* sperm using dual fluorescent staining. *Theriogenology* 51: 1513–1523. http://dx.doi.org/10.1016/S0093-691X(99)00094-1

CORNUET, J-M; MORITZ, R F A (1987) Selection theory and selection programmes. In *Moritz, R F A (Ed.) The instrumental insemination of the queen bee.* Apimondia Publishing House; Romania. pp 125-141.

COSTA, C; BERG, S; BIENKOWSKA, M; BOUGA, M; BUBALO, D; BÜCHLER, R; CHARISTOS, L; LE CONTE, Y; DRAZIC, M; DYRBA, WFILLIPI, J; HATJINA, F; IVANOVA, E; KEZIC, N; KIPRIJANOVSKA, H; KOKINIS, M; KORPELA, S; KRYGER, P; LODESANI, M; MEIXNER, M; PANASIUK, B; PECHHACKER, H; PETROV, P; OLIVERI, E; RUOTTINEN, L; UZUNOV, A; VACCARI, G; WILDE, J (2012) A Europe-wide experiment for assessing the impact of genotype-environment interactions on the vitality of honey bee colonies: methodology. *Journal of Apicultural Science* 56: 147-157. http://dx.doi.org/10.2478/v10289-012-0015-9

DE RUIJTER, A; VAN DER STEEN, J J (1989) Disinfection of combs by means of acetic acid (96%) against nosema. *Apidologie* 21: 503–506.

DE GRAAF, D C; ALIPPI, A M; ANTÚNEZ, K; ARONSTEIN, K A; BUDGE, G; DE KOKER, D; DE SMET, L; DINGMAN, D W; EVANS, J D; FOSTER, L J; FÜNFHAUS, A; GARCIA-GONZALEZ, E; GREGORC, A; HUMAN, H; MURRAY, K D; NGUYEN, B K; POPPINGA, L; SPIVAK, M; VANENGELSDORP, D; WILKINS, S; GENERSCH, E (2013) Standard methods for American foulbrood research. In *V Dietemann; J D Ellis; P Neumann (Eds) The COLOSS BEEBOOK, Volume II: standard methods for* Apis mellifera *pest and pathogen research. Journal of Apicultural Research* 52(1): http://dx.doi.org/10.3896/IBRA.1.52.1.11

DELAPLANE, K S; VAN DER STEEN, J; GUZMAN, E (2013) Standard methods for estimating strength parameters of *Apis mellifera* colonies. In *V Dietemann; J D Ellis; P Neumann (Eds) The COLOSS BEEBOOK, Volume I: standard methods for* Apis mellifera *research. Journal of Apicultural Research* 52(1): http://dx.doi.org/10.3896/IBRA.1.52.1.03

DE MIRANDA, J R; BAILEY, L; BALL, B V; BLANCHARD, P; BUDGE, G; CHEJANOVSKY, N; CHEN, Y-P; VAN DOOREMALEN, C; GAUTHIER, L; GENERSCH, E; DE GRAAF, D; KRAMER, M; RIBIÈRE, M; RYABOV, E; DE SMET, L; VAN DER STEEN, J J M (2013) Standard methods for virus research in *Apis mellifera*. In *V Dietemann; J D Ellis; P Neumann (Eds) The COLOSS BEEBOOK, Volume II: standard methods for* Apis mellifera *pest and pathogen research. Journal of Apicultural Research* 52(4): http://dx.doi.org/10.3896/IBRA.1.52.4.22

DIETEMANN, V; NAZZI, F; MARTIN, S J; ANDERSON, D; LOCKE, B; DELAPLANE, K S; WAUQUIEZ, Q; TANNAHILL, C; ELLIS, J D (2013) Standard methods for varroa research. In *V Dietemann; J D Ellis; P Neumann (Eds) The COLOSS BEEBOOK, Volume II: standard methods for* Apis mellifera *pest and pathogen research. Journal of Apicultural Research* 52(1): http://dx.doi.org/10.3896/IBRA.1.52.1.09

DOOLITTLE, G M (1889) *Scientific queen rearing.* Thomas G Newman & Son; Chicago, USA. 169 pp.

DOOLITTLE, G M (1915) *Scientific queen-rearing as practically applied; being a method by which the best of queen-bees are reared in perfect accord with nature's ways.* American Bee Journal; Hamilton, USA. 126 pp.

EVANS, J D; CHEN, Y P; CORNMAN, R S; DE LA RUA, P; FORET, S; FOSTER, L; GENERSCH, E; GISDER, S; JAROSCH, A; KUCHARSKI, R; LOPEZ, D; LUN, C M; MORITZ, R F A; MALESZKA, R; MUÑOZ, I; PINTO, M A; SCHWARZ, R S (2013) Standard methodologies for molecular research in *Apis mellifera*. In *V Dietemann; J D Ellis; P Neumann (Eds) The COLOSS BEEBOOK, Volume I: standard methods for* Apis mellifera *research. Journal of Apicultural Research* 52(4): http://dx.doi.org/10.3896/IBRA.1.52.4.11

EVANS, J D; SPIVAK, M (2010) Socialized medicine: individual and communal disease barriers in honey bees. *Journal of Invertebrate Pathology* 103: 62-72.

FALCONER, D S; MACKAY; T F C (1996) *Introduction to quantitative genetics (4th Ed.).* Longman; New York, USA.

FORSGREN, E; BUDGE, G E; CHARRIÈRE, J-D; HORNITZKY, M A Z (2013) Standard methods for European foulbrood research. In *V Dietemann; J D Ellis, P Neumann (Eds) The COLOSS BEEBOOK: Volume II: Standard methods for* Apis mellifera *pest and pathogen research. Journal of Apicultural Research* 52(1): http://dx.doi.org/10.3896/IBRA.1.52.1.12

FREE, J B (1961) The stimuli releasing the stinging response of honey bees. *Animal Behaviour* 9: 193-196.

FREE, J B (1987) *Pheromones of social bees.* Chapman and Hall; UK. 218 pp.

FRIES, I; CHAUZAT, M-P; CHEN, Y-P; DOUBLET, V; GENERSCH, E; GISDER, S; HIGES, M; MCMAHON, D P; MARTÍN-HERNÁNDEZ, R; NATSOPOULOU, M; PAXTON, R J; TANNER, G; WEBSTER, T C; WILLIAMS, G R (2013) Standard methods for nosema research. In *V Dietemann; J D Ellis, P Neumann (Eds) The COLOSS* BEEBOOK: *Volume II: Standard methods for* Apis mellifera *pest and pathogen research. Journal of Apicultural Research* 52(1): http://dx.doi.org/10.3896/IBRA.1.52.1.14

VON FRISCH, K (1967) *The dance language and orientation of bees.* Harvard University Press; Cambridge, USA. 566 pp.

FU-HUA, L; SANDY, M S (2000) Estimating quantitative genetic parameters in haplodiploid organisms. *Heredity* 85: 373-382. http://dx.doi.org/10.1046/j.1365-2540.2000.00764.x

GARY, N E (1962) Chemical mating attractants in the honey bee. *Science* 136: 773-774.

GUZMAN-NOVOA, E; MERLOS-PRIETO, D; URIBE-RUBIO, J; HUNT, G J (2003) Relative reliability of four field assays to test defensive behaviour of honey bees (*Apis mellifera*). *Journal of Apicultural Research* 42: 42-46.

HARBO, J R; HARRIS, J W (2005) Suppressed mite reproduction explained by the behaviour of adult bees. *Journal of Apicultural Research* 44: 21-23.

HARISMENDY, O; NG, P C; STRAUSBERG, R L; WANG, X; STOCKWELL, T B; BEESON, K Y; SCHORK, N J; MURRAY, S S; TOPOL, E J; LEVY, S; FRAZER, K A (2009) Evaluation of next generation sequencing platforms for population targeted sequencing studies. *Genome Biology* 10: R32. http://dx.doi.org/10.1186/gb-2009-10-3-r32

HATJINA, F (2012) Greek honey bee queen quality certification. *Bee World* 89: 18-20.

HATJINA, F; BIEŃKOWSKA, M; CHARISTOS, L; CHLEBO, R; COSTA, C; DRAŽIĆ, M; FILIPI, J; GREGORC, A; IVANOVA, E N; KEZIC, N; KOPERNICKY, J; KRYGER, P; LODESANI, M; LOKAR, V; MLADENOVIC, M; PANASIUK, B; PETROV, P P; RAŠIĆ, S; SMODIS-SKERL, M I; VEJSNÆS, F; WILDE, J (2013) Examples of different methodology used to assess the quality characteristics of honey bee queens. *Journal of apicultural Research* (in press).

HEGIĆ, G; BUBALO, D (2006) Hygienic water supply for the bees. *Journal of Central European Agriculture* 7(4): 743-752.

HENDERSON, C R (1988) Theoretical basis and computational methods for a number of different animal models. *Journal of Dairy Science* 71: 1-16.

HUMAN, H; BRODSCHNEIDER, R; DIETEMANN, V; DIVELY, G; ELLIS, J; FORSGREN, E; FRIES, I; HATJINA, F; HU, F-L; JAFFÉ, R; KÖHLER, A; PIRK, C W W; ROSE, R; STRAUSS, U; TANNER, G; VAN DER STEEN, J J M; VEJSNÆS, F; WILLIAMS, G R; ZHENG, H-Q (2013) Miscellaneous standard methods for *Apis mellifera* research. In *V Dietemann; J D Ellis; P Neumann (Eds) The COLOSS* BEEBOOK, *Volume I: standard methods for* Apis mellifera *research. Journal of Apicultural Research* 52(4): http://dx.doi.org/10.3896/IBRA.1.52.4.10

HUNT, G J; PAGE Jr, R E; FONDRK, M K; DULLUM, C J (1995) Major quantitative trait loci affecting honey bee foraging behaviour. *Genetics* 141(4): 1537-1545.

ISHMURATOV, G Y; KHARISOV, R Y; BOTSMAN, O V; ISHMURATOVA, N M; TOLSTIKOV, G A (2002) Synthesis of 9-oxo- and 10-hydroxy-2E-decenoic acids. *Chemistry of Natural Compounds* 38: 1-23.

JACKSON, J T; TARPY, D R; FAHRBACH, S E (2011) Histological estimates of ovariole number in honey bee queens, *Apis mellifera*, reveal lack of correlation with other queen quality measures. *Journal of Insect Science* 11: 82. http://dx.doi.org/10.1016/S0093-691X(99)00094-1

JENSEN, A B; ARONSTEIN, K; FLORES, J M; VOJVODIC, S; PALACIO, M A; SPIVAK, M (2013) Standard methods for fungal brood disease research. In *V Dietemann; J D Ellis, P Neumann (Eds) The COLOSS BEEBOOK: Volume II: Standard methods for* Apis mellifera *pest and pathogen research. Journal of Apicultural Research* 52(1): http://dx.doi.org/10.3896/IBRA.1.52.1.13

KAHYA, Y; GENÇER, Y; WOYKE, J (2008) Weight at emergence of honey bee (*Apis mellifera caucasica*) queens and its effect on live weights at the pre and post mating periods. *Journal of Apicultural Research* 47(2): 118-125. http://dx.doi.org/10.3896/IBRA.1.47.2.06

KOENIGER, N; KOENIGER, G (2007) Mating flight duration of *Apis mellifera* queens: as short as possible, as long as necessary. *Apidologie* 38: 606–611.

LAIDLAW, H H (1979) *Contemporary queen rearing.* Dadant & Sons: Hamilton, USA. 199 pp.

LAIDLAW, H H; PAGE, R E (1986) Mating designs. In *Rinderer, T E (Ed.) Bee genetics and breeding.* Academic Press; Orlando, Florida, USA. pp 323-344.

LAIDLAW, H H; PAGE, R E (1997) *Queen rearing and bee breeding.* Wicwas Press; New York, USA. 224 pp.

LAPIDGE, K E; OLDROYD, B P; SPIVAK, M (2002) Seven suggestive quantitative trait loci influence hygienic behaviour of honey bees. *Naturwissenschaften* 89(12): 565-568. http://dx.doi.org/10.1007/s00114-002-0371-6

LEE, K V; REUTER, G S; SPIVAK, M (2010) Standardized sampling plan to detect varroa densities in colonies and apiaries. *American Bee Journal* 149(12): 1151- 1155.

LEE, K V; MOON, R D; BURKNESS, E C; HUTCHISON, W D; SPIVAK, M (2010) Practical sampling plans for *Varroa destructor* (Acari: Varroidae) in *Apis mellifera* (Hymenoptera: Apidae) colonies and apiaries. *Journal of Economic Entomology* 103(4): 1039-1050. http://dx.doi.org/10.1603/EC10037

MACEDO, P A; WU, J; ELLIS, M D (2002) Using inert dusts to detect and assess varroa infestations in honey bee colonies. *Journal of Apicultural Research* 40: 3-7.

MEUWISSEN, T H; HAYES, B J; GODDARD, M E (2001) Prediction of total genetic value using genome-wide dense marker maps. *Genetics* 157: 1819-1829.

MOMOT, J P; ROTHENBUHLER, W C (1971) Behaviour genetics of nest cleaning in honey bees. *Journal of Apicultural Research* 10: 11-21.

MORITZ, R F A (1988) A re-evaluation of the two-locus model for hygienic behaviour in honey bees (*Apis mellifera* L.). *Journal of Heredity* 79(4): 257-262.

MORSE, R A (1994) *Rearing queen honey bees*. Wicwas Press; Ithaca, New York, USA. 128 pp.

NAVAJAS, M; MIGEON, A; ALAUX, C; MARTIN-MAGNIETTE, M L; ROBINSON, G E; EVANS, J D; CROS-ARTEIL, S; CRAUSER, D; LE CONTE, Y (2008) Differential gene expression of the honey bee *Apis mellifera* associated with *Varroa destructor* infection. *BMC Genomics* 9: 301. http://dx.doi.org/10.1186/1471-2164-9-301

NEUMANN, P; CARRECK, N L (2010) Honey bee colony losses. *Journal of Apicultural Research* 49: 1-6. http://dx.doi.org/10.3896/IBRA.1.49.1.01

NEWTON, D C; OSTASIEWSKI, N J (1986) A simplified bioassay for behavioural resistance to American Foulbrood in honey bees (*Apis mellifera* L.). *American Bee Journal* 126: 278-281.

PAGE, R E Jr; FONDRK, K; HUNT, G J; GUZMÁN-NOVOA, E; HUMPHRIES, M A; NGUYEN, K; GREENE, A S (2000) Genetic dissection of honey bee (*Apis mellifera* L.) foraging behaviour. *Journal of Heredity* 91 (6): 474-479. http://dx.doi.org/10.1093/jhered/91.6.474

PARK, O W (1923) Flight studies of the honey bee. *American Bee Journal* 63:71.

PELLETT, F C (1938) *History of American beekeeping*. Collegiate Press; Ames, Iowa, USA. 393 pp.

PENG, Y-S; LOCKE, S J; NASR, M E; LIU, T P; MONTAGUE, M A (1990) Differential staining for live and dead sperm of honey bees. *Physiological Entomology* 15 (2): 211–217. http://dx.doi.org/10.1111/j.1365-3032.1990.tb00509.x

PÉREZ, P; DE LOS CAMPOS, G; CROSSA, J; GIANOLA, D (2010) Genomic-enabled prediction based on molecular markers and pedigree using the Bayesian Linear Regression Package in R. *The plant genome*. 3(2): 106-116. http://dx.doi.org/10.3835/plantgenome2010.04.0005

PEREZ-SATO, J A; CHALINE, N; MARTIN, S J; HUGHES, W O H; RATNIEKS, F L W (2009) Multi-level selection for hygienic behaviour in honey bees. *Heredity* 102: 609-615. http://dx.doi.org/10.1038/hdy.2009.20

PRATT, E L (1905) *Commercial queen-rearing: cell getting by the Swarthmore labour-saving pressed-cup and interchangeable flange shell plan*. Swarthmore Apiaries, USA. 53 pp.

RHODES, J W; SOMERVILLE, D C (2003) *Rural Industries Research and Development Corporation, Australia. Publication No* 03/049.

RHODES, J W; SOMERVILLE, D C; HARDEN, S (2004) Queen honey bee introduction and early survival - effects of queen age at introduction. *Apidologie* 35: 383–388.

RINDERER, T E (1986) *Bee genetics and breeding*. Academic Press; Orlando, Florida, USA. 426 pp.

RINDERER, T E; HARRIS, J W; HUNT, G J; DE GUZMANN, L I (2010) Breeding for resistance to *Varroa destructor* in North America. *Apidologie* 41: 409-424. http://dx.doi.org/10.1051/apido/2010015

ROTHENBUHLER, W C (1964) Behaviour genetics of nest cleaning in honey bees. IV. Responses of F1 and backcross generations to disease-killed brood. *American Zoologist* 4: 111-123.

RÜPPELL, O; PANKIW, T; PAGE Jr, R E (2004) Pleitropy, epistasis and new QTL: the genetic architecture of honey bee foraging behaviour. *Journal of Heredity* 95(6): 481-491. http://dx.doi.org/10.1093/jhered/esh072

RUTTNER, H (1972) Technical recommendations for methods of evaluating performance of bee colonies. In *F Ruttner. Controlled mating and selection of the honey bee*. Apimondia Publishing House; Bucharest, Romania. pp. 87-92.

RUTTNER, F (ED.) (1983) *Queen rearing: biological basis and technical instruction*. Apimondia Publishing House; Bucharest, Romania. 358 pp.

RUTTNER, F (1988) *Breeding techniques and selection for breeding of the honey bee*. The British Isles Bee Breeders Association by arrangement with Ehrenwirth Verlag; Munich, Germany. 152 pp.

RUTTNER, F (1988) *Biogeography and taxonomy of honey bees*. Springer-Verlag; Berlin, Germany. 284 pp.

SCHEINER, R; ABRAMSON, C I; BRODSCHNEIDER, R; CRAILSHEIM, K; FARINA, W; FUCHS, S; GRÜNEWALD, B; HAHSHOLD, S; KARRER, M; KOENIGER, G; KOENIGER, N; MENZEL, R; MUJAGIC, S; RADSPIELER, G; SCHMICKLI, T; SCHNEIDER, C; SIEGEL, A J; SZOPEK, M; THENIUS, R (2013) Standard methods for behavioural studies of *Apis mellifera*. In *V Dietemann; J D Ellis; P Neumann (Eds) The COLOSS BEEBOOK, Volume I: standard methods for* Apis mellifera *research. Journal of Apicultural Research* 52(4): http://dx.doi.org/10.3896/IBRA.1.52.4.04

SEELEY, T D; TARPY, D R (2007) Queen promiscuity lowers disease within honey bee colonies. *Proceedings of the Royal Society B Biological Sciences* 274(1606): 67-72. http://dx.doi.org/10.1098/rspb.2006.3702

SKOWRONEK, W; BIENKOWSKA, M; KRUK, C (2004) Changes in body weight of honey bee queens during their maturation. *Journal of Apicultural Science* 48(2): 61-68.

SOLIGNAC, M; VAUTRIN, D; BAUDRY, E; MOUGEL, F; LOISEAU, A; CORNUET, J-M (2004) A microsatellite-based linkage map of the honey bee, *Apis mellifera* L. *Genetics* 167: 253-262.

SPIVAK, M; DOWNEY, D L (1998) Field assays for hygienic behaviour in honey bees (Hymenoptera: Apidae). *Journal of Economic Entomology* 91: 64-70.

SPIVAK, M; REUTER, G S (1998) Honey bee hygienic behaviour. *American Bee Journal* 138: 283-286.

SPIVAK, M; REUTER, G S (2001) Resistance to American foulbrood disease by honey bee colonies, *Apis mellifera*, bred for hygienic behaviour. *Apidologie* 32: 555-565.

SPÖTTER, A; GUPTA, P; NÜRNBERG, G; REINSCH, N M; BIENEFELD, K (2012) Development of a 44K SNP assay focussing on the analysis of a varroa-specific defence behaviour in honey bees (*Apis mellifera carnica*). *Molecular Ecology Resources* 12(2): 323-332. http://dx.doi.org/10.1111/j.1755-0998.2011.03106.x

STORT, A C (1974) Genetic study of aggressiveness of two subspecies of *Apis mellifera* in Brazil. 1. Some tests to measure aggressiveness. *Journal of Apicultural Research* 13: 33-38.

TIESLER, F K; ENGLERT, E (1989) *Aufzucht, paarung und verwertung von königinnen.* Ehrenwirth Verlag; München, Germany.

TISHKOFF, S A; REED, F A; RANCIARO, A; VOIGHT, B F; BABBITT, C C; SILVERMAN, J S; POWELL, K; MORTENSEN, H M; HIRBO, J B; OSMAN, M; IBRAHIM, M; OMAR, S A; LEMA, G; NYAMBO, T B; GHORI, J; BUMPSTEAD, S; PRITCHARD, J K; WRAY, G A; DELOUKAS, P (2007) Convergent adaptation of human lactase persistence in Africa and Europe. *Nature genetics* 39: 31-40. http://dx.doi.org/10.1038/ng1946

WEINSTOCK, G M; ROBINSON, G E; *et al.* (2006) Insight into the social insects from the genome of the honey bee *Apis mellifera*. *Nature* 443: 931-949. http://dx.doi.org/10.1038/nature05260

WHITFIELD, C W; BAND, M R; BONALDO, M F; KUMAR, C G; LIU, L; PARDINAS, J R; ROBERTSON, H M; SOARES, M B; ROBINSON, G E (2002) Annotated expressed sequence tags and cDNA microarrays for the studies of brain and behaviour in the honey bee. *Genome Research* 12: 555-566. http://dx.doi.org/10.1101/gr.5302

WILLHAM, R L (1963) The covariance between relatives for characters composed of components contributed by related individuals. *Biometrics* 19: 18-27.

WILLIAMS, J L (1987) Wind-directed pheromone trap for drone honey bees (Hymenoptera: Apidae). *Journal Economical Entomology*. 80: 532–536.

WILSON-RICH, N; SPIVAK, M; FEFFERMAN, N H; STARKS, P T (2009) Genetic, individual, and group facilitation of disease resistance in insect societies. *Annual Review of Entomology* 54: 405-423. http://dx.doi.org/10.1146/annurev.ento.53.103106.093301

WOODWARD, D (2007) *Queen bee: biology, reading and breeding.* Balclutha; New Zealand. 137 pp.

WOYKE, J (1971) Correlations between the age at which honey bee brood was grafted, characteristics of the resultant queens, and results of insemination. *Journal of Apicultural Research* 10(1): 45-55

WOYKE, J (1988) Problems with queen banks. *American Bee Journal* 124(4): 276-278.

ZMARLICKI, C; MORSE, R A (1963) Drone congregation areas. *Journal of Apicultural Research* 2: 64-66.

Journal of Apicultural Research 52(1)

Journal of Apicultural Research 52(4): (2013)
DOI 10.3896/IBRA.1.52.4.13

© IBRA 2013

REVIEW ARTICLE

Statistical guidelines for *Apis mellifera* research

Christian W W Pirk[1*], Joachim R de Miranda[2], Matthew Kramer[3], Tomàs E Murray[4], Francesco Nazzi[5], Dave Shutler[6], Jozef J M van der Steen[7] and Coby van Dooremalen[7]

[1]Social Insect Research Group, Department of Zoology & Entomology, University of Pretoria, Private Bag X20 Hatfield 0028, Pretoria, South Africa.
[2]Department of Ecology, Swedish University of Agricultural Sciences, Box 7044, 750 07, Uppsala, Sweden.
[3]Biometrical Consulting Service, Agricultural Research Service/USDA, Beltsville, MD, 20705, USA.
[4]Institute of Biology, Martin Luther University Halle-Wittenberg, Hoher Weg 8, Halle (Saale) 06120, Germany.
[5]Dipartimento di Scienze Agrarie e Ambientali, Università di Udine, via delle Scienze 206, 33100 Udine, Italia.
[6]Department of Biology, Acadia University, Wolfville, Nova Scotia, B4P 2R6, Canada.
[7]Bees@wur, Bio-Interactions and Plant Health, Plant Research International, Wageningen UR, Droevendaalsesteeg 1, 6708 PB Wageningen, Netherlands.

Received 13 May 2013, accepted subject to revision 15 June 2013, accepted for publication 17 July 2013.

*Corresponding author: Email: cwwpirk@zoology.up.ac.za

Summary

In this article we provide guidelines on statistical design and analysis of data for all kinds of honey bee research. Guidelines and selection of different methods presented are, at least partly, based on experience. This article can be used: to identify the most suitable analysis for the type of data collected; to optimise one's experimental design based on the experimental factors to be investigated, samples to be analysed, and the type of data produced; to determine how, where, and when to sample bees from colonies; or just to inspire. Also included are guidelines on presentation and reporting of data, as well as where to find help and which types of software could be useful.

Guia estadistica para estudios en *Apis mellifera*

Resumen

En este trabajo se proporcionan directrices sobre el diseño estadístico y el análisis de datos para todo tipo de investigación sobre abejas. Tanto las directrices como la selección de los diferentes métodos que se presentan están basadas, al menos en parte, en la experiencia. Este artículo se puede utilizar: para identificar el análisis más adecuado para el tipo de datos recogidos; para optimizar el diseño experimental basado en los factores experimentales a ser investigados, las muestras a analizar, y el tipo de datos que se producen; para determinar cómo, dónde , y cuando muestras abejas de las colonias, o simplemente para inspirar. También se incluyen directrices para la presentación y comunicación de los datos, así como dónde encontrar ayuda y distintos software que puedan ser útiles.

西方蜜蜂研究的统计指南

摘要

在本文中，我们提供了针对蜜蜂所有研究的统计设计和数据分析指南。这些指南和方法的选择至少部分基于我们的经验。本文也可用于：针对收集到的数据类型选择最优分析方法；基于所研究的实验因素、待分析的样本和获得的数据类型优化实验设计；确定从蜂群中采集蜜蜂样本的地点、时间和方式；或者仅为实验提供参考。另外，也包含展示和报告数据时的指南，以及如何寻求帮助和选用何种软件。

Keywords: COLOSS, *BEEBOOK*, honey bees, sampling, sample size, GLMM, robust statistics, resampling, PCA, Power, rule of thumb

Footnote: PIRK, C W W; DE MIRANDA, J R; KRAMER, M; MURRAY, T; NAZZI, F; SHUTLER, D; VAN DER STEEN, J J M; VAN DOOREMALEN, C (2013) Statistical guidelines for *Apis mellifera* research. In *V Dietemann; J D Ellis; P Neumann* (Eds) *The COLOSS BEEBOOK, Volume I: standard methods for* Apis mellifera *research. Journal of Apicultural Research* 52(4): http://dx.doi.org/10.3896/IBRA.1.52.4.13

Table of Contents

1. Introduction

Bees are organisms and, as such, are inherently variable at the molecular, individual, and population levels. This intrinsic variability means that a researcher needs to separate the various sources of variability contained in the measurements, whether obtained by observational or experimental research, into signal and noise. The former may be due to treatments received, bee age, or innate differences in resistance. The latter is largely due to the genetic background (and its phenotypic expression) that characterises individual living organisms. Statistics is the branch of mathematics we use to isolate and quantify the signal and determine its importance, relative to the inherent noise. For the researcher, with an eye toward the statistical analysis to come, and before data collection starts, one should ask:

1) Which variables (VIM, 2008) am I going to measure and what kind of data will those variables generate?

2) What degree of accuracy do I want to achieve and what is the corresponding sample size required?

3) Which statistical analysis will help me to answer my research question? This is related to the question. What kind of underlying process produces data like those I will be collecting?

4) From what population do I want to sample? (What is the statistical population/ statistical universe?) For example, do I want to make inferences about the local, national, continental, or worldwide population?

One function of statistics is to summarise information to make it more usable and easier to grasp. A second is inductive, where one makes generalisations based on a subset of a population or based on repeated observations (through replication or repeated over time). For example, if 50 workers randomly sampled from 20 colonies all produce 10-hydroxydecanoic acid (10-HDAA, one of the major components in the mandibular gland secretion, especially in workers; Crewe, 1982; Pirk *et al.*, 2011), one could infer that all workers produce 10-HDAA. An example of inferring a general pattern from repeated observations would be: If an experiment is repeated 5 times and yields the same result each time, one makes a generalisation based on this limited number of experiments. One should keep in mind that, if one is measuring a quantitative variable, irrespective of how precise measuring instruments are, each experimental unit/replicate produces a unique data value. A third function of statistics is based on deductive reasoning and might involve statistical modelling, in the classical or Bayesian paradigm, to understand the basic processes that produced the measurements, possibly by incorporating prior information (e.g. predicting species distributions or phylogenetic relationships/trees; see Kaeker and Jones, 2003). In this article we will cover, albeit only cursorily, all three functions of statistics. We have largely focused on research with bee pathogens, in part because these are of intense practical and theoretical interest, and in part because of our own backgrounds. However, bee biology rightly includes a much greater spectrum of research, and for much of it there are specialised statistical tools. Some of the ones we discuss are broadly applicable but, by necessity, this section can only provide an uneven treatment of current statistical methods that might be used in bee research. In particular, we do not discuss multivariate methods (other than principal component analysis); Bayesian approaches, and touch only lightly on simulation and resampling methods. All are current fields of investigation in statistics. Molecular, and in particular, genomic research has spawned substantial new statistical methods, also not covered here. These areas of statistics will be included in the next edition of the *BEEBOOK*.

Furthermore, we restrict ourselves here to providing guidelines on statistics for certain kinds of honey bee research, as mentioned above, with reference to more detailed sources of information. Fortunately, there are excellent statistical tools available, the most important of which is a good statistician.

The statistics we describe can be roughly grouped into two main areas, one having to do with sampling to estimate population characteristics (e.g. for pathogen prevalence = proportion of infected bees in an apiary or a colony), and the other having to do with experiments (e.g. comparing treatments, one of which may be a control). Due to the complex social structure of a bee hive, and the peculiar developmental and environmental aspects of bee biology, sampling in this discipline has more components to consider than in most biological fields. Some statistical topics are relevant to both sampling and experimental studies, such as sample size and power. Others are primarily of concern for just one of the areas. For example, when sampling for pathogen prevalence, primary issues include representativeness, and how or when to sample. For experiments, they include hypothesis formulation and development of appropriate statistical models for the processes (which includes testing and assumptions of models). Of course, good experiments require representative samples, and also require a good understanding of sampling. Both areas are important for data acquisition and analysis. We start with statistical issues related to sampling.

1.1. Types of data

There are several points to consider in selecting a statistical analysis including sample size, distribution of the data, and type of data. These points and the statistical analysis in general should be considered **before** conducting an experiment or collecting data. One should know beforehand what kind of measurement and what type of data one is collecting. The dependent variable is the variable that may be affected by which treatment a subject is given (e.g. control *vs.* treated, an ANOVA framework), or as a function of some other measured variable (e.g. age, in a regression framework). Data normally include all measured quantities of an experiment (dependent and independent/predictor/ factor variables). The dependent variable can be one of several types: nominal, ordinal, interval or ratio, or combinations thereof. An example of nominal data is categorical (e.g. bee location A/B/C, where the location of a bee is influenced by some explanatory variables, such as age) or dichotomous responses (yes/no). Ordinal data are also categorical, but which can be ordered sequentially. For example, the five stages of ovarian activation (Hess, 1942; Schäfer *et al.*, 2006; Pirk *et al.*, 2010; Carreck *et al.*, 2013) are ordinal data because undeveloped ovaries are smaller than intermediate ovaries, which are smaller than fully developed. However, one cannot say intermediate is half of fully developed. If one assigned numbers to ranked categories, one could calculate a mean, but it would be most likely a biologically meaningless value. The third and fourth data types are interval and ratio; both carry information about the order of data points and the size of intervals between values. For example, temperature in Celsius is on an interval scale, but temperature in Kelvin is on a ratio scale. The difference is that the former has an arbitrary "zero point" and negative values are used, whereas the latter has an absolute origin of zero. Other examples of data with an absolute zero point are length, mass, angle, and duration.

The type of dependent variable data is important because it will determine the type of statistical analysis that can or cannot be used. For example, a common linear regression analysis would not be appropriate if the dependent variable is categorical. (Note: In such a case a logistic regression, discussed below, may work).

1.2. Confidence level, Type I and Type II errors, and Power

For experiments, once we know what kind of data we have, we should consider the desired confidence level of the statistical test. This confidence is expressed as α; it gives one the probability of making a Type I error (Table 1) which occurs when one rejects a **true** null hypothesis. Typically that level for α is set at 0.05, meaning that we are 95% confident ($1 - \alpha = 0.95$) that we will not make a Type I error, i.e. 95% confident that we will *not* reject a true null hypothesis. For many commonly used statistical tests, the p-value is the probability that the test statistic calculated from the observed data occurred by chance, given that the null hypothesis is true. If $p < \alpha$ we reject the null hypothesis; if $p \geqq \alpha$ we do not reject the null hypothesis.

Table 1. The different types of errors in hypothesis-based statistics.

	The null hypothesis (H$_0$) is	
Statistical result	True	False
Reject null hypothesis	**Type I error,** α value = probability of falsely rejecting H$_0$	Probability of correctly rejecting H$_0$: $(1 - \beta)$ = power
Accept null hypothesis	Probability of correctly accepting H$_0$: $(1 - \alpha)$	**Type II error,** β value = probability of falsely accepting H$_0$

A Type II error, expressed as the probability ß occurs when one fails to reject a **false** null hypothesis. Unlike α, the value of ß is determined by properties of the experimental design and data, as well as how different results need to be from those stipulated under the null hypothesis to make one believe the alternative hypothesis is true. Note that the null hypothesis is, for all intents and purposes, rarely true. By this we mean that, even if a treatment has very little effect, it has some small effect, and given a sufficient sample size, its effect could be detected. However, our interest is more often in biologically important effects and those with practical importance. For example, a treatment for parasites that is only marginally better than no treatment, even if it could be shown to be statistically significant with a sufficiently large sample size, may be of no practical importance to a beekeeper. This should be kept in mind in subsequent discussions of sample size and effect size.

The power or the sensitivity of a test can be used to determine sample size (see section 3.2.) or minimum effect size (see section 3.1.3.). Power is the probability of correctly rejecting the null hypothesis when it is false (power = 1 − ß), i.e. power is the probability of not committing a Type II error (when the null hypothesis is false) and hence the probability that one will identify a significant effect when such an effect exists. As power increases, the chance of a Type II error decreases. A power of 80% (90% in some fields) or higher seems generally acceptable. As a general comment the words "power", "sensitivity", "precision", "probability of detection" are / can be used synonymously.

2. Sampling

2.1. Where and when to sample a colony

Colony heterogeneity in time and space are important aspects to consider when sampling honey bees and brood. For example, the presence and prevalence of pathogens both depend on the age class of bees and brood, physiological status of bees, and/or the presence of brood. Note that pathogens in a colony have their own biology and that their presence and prevalence can also vary over space and time. The relation between a pathogen and particular features of a colony should be taken into account when deciding where and when samples are taken, including the marked seasonality of many pathogen infections.

2.1.1. When to sample?

A honey bee colony is a complex superorganism with changing features in response to (local) seasonal changes in the environment. Average age increases, for example, in colonies in the autumn in temperate regions, because of the transition to winter bees. Age-related tasks are highly plastic (Huang and Robinson, 1996), but after a major change of a colony's organisation it can take some time before the division of labour is restored (Johnson, 2005). Immediately after a colony has produced a swarm, for example, bees remaining in the nest will have a large proportion of individuals younger than 21 days, lowering the average age of bees in the colonies. Over time, these bees will become older and the average age of bees in the colony will increase again. Therefore, it is recommended that if the aim is to have an average / normal / representative sample with respect to age structure, one should only sample established colonies that have not recently swarmed. The same is true for recently caught swarms, because brood will not have had enough time to develop, and one could expect rather an over-aged structure. Age polyethism in honey bees and its implications for the physiology, behaviour, and pheromones is discussed in detail elsewhere (Lindauer, 1952; Ribbands, 1952; Lindauer, 1953; Jassim *et al.*, 2000; Crewe *et al.*, 2004; Moritz *et al.*, 2004).

Furthermore, physiological variables in individual bees (and in pooled samples from a colony) can change over time when these parameters are, for example, related to age of bees or presence of brood. Moreover, build-up of vitellogenin takes place in the first 8-10 days of a bee's adult life and then decreases, but is much faster in summer than in winter when no brood is present and bees are on average older (Amdam and Omholt, 2002), affecting averages of individual bees of the same age, but also averages of pooled samples. *Nosema apis*, *Paenibacillus larvae*, and *Melissococcus plutonius* are examples of organisms with bee age-related prevalence in colonies. *N. apis* infections are not microscopically detectable in young bees; after oral infection it takes three to five days before spores are released from infected cells (Kellner, 1981). *P. larvae* and *M. plutonius* can be detected in and on young bees that clean cells (Bailey and Ball, 1991; Fries *et al.*, 2006). Depending on the disease, higher prevalences can be found in colonies with relatively old and young bees, respectively. Furthermore, seasonal variation in pathogen and parasite loads may also affect when to sample. For example, screening for brood pathogens during brood-less periods (e.g. winter, in temperate climates) is less likely to return positive samples than screening when brood is present.

2.1.2. Where to sample?

To determine proper locations for sampling inside a beehive, one must consider colony heterogeneity in time and space. Feeding brood, and capping and trimming of cells takes place in the brood nest. Other activities such as cleaning, feeding and grooming, honey-storing, and

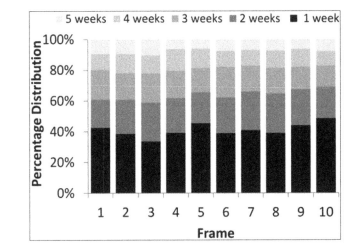

Fig. 1. The percentage distribution of age classes recorded between 24 August and 20 September for pooled colonies of *Apis mellifera* in the Netherlands. The different shades represent different age classes. The distribution of age classes did not differ among frames (p = 0.99). There was also no difference between the mean number of bees per frame (p = 0.94). Adapted from: van der Steen *et al.* (2012).

Table 2. Example of sample sizes needed to detect different infection levels with different levels of probability (from Equation I.).

Proportion of infected bees, **P**	Required probability of detection, **D**	Sample size needed, **N**
0.25	0.95	11
0.25	0.99	16
0.10	0.95	29
0.10	0.99	44
0.05	0.95	59
0.05	0.99	90
0.01	0.95	298
0.01	0.99	459

shaping combs take place all over frames (Seeley, 1985). Free (1960) showed an equal distribution of bees of successive age classes on combs containing eggs, young larvae, and sealed brood, although there were proportionally more young bees (4-5 days old) on brood combs and more old bees (> 24 days) on storage combs. Older bees were overrepresented among returning bees at colony entrances. This was supported by findings of van der Steen *et al.* (2012), who also reported that age classes are distributed in approximately the same ratio over frames containing brood (Fig. 1).

2.2. Probability of pathogen detection in a honey bee colony

For diagnosis or surveys of pathogen prevalence, the more bees that are sampled, the higher the probability of detecting a pathogen, which is particularly important for low levels of infection. An insufficient sample size could lead to a false negative result (apparent absence of a pathogen when it is actually present but at a low prevalence). Historically, 20-30 bees per colony have been suggested as an adequate sample size (Doull and Cellier, 1961) when the experimental unit is a colony. However, based on binomial probability theory, such small sample sizes will only detect a 5% true prevalence in an infected colony with a probability of 65% (20 bees) or 78% (30 bees). If only high infection prevalence is of interest for detection, then small sample sizes may be acceptable, as long as other sampling issues (such as representativeness, see above) have been adequately handled.

In general, sample size should be based on the objectives of the study and a specified level of precision (Fries *et al.*, 1984; Table 2). If the objective is to detect a prevalence of 5% or more (5% of bees

infected) with 95% probability, then a sample of 59 bees per colony is needed. If the objective is to detect prevalence as low as 1% with 99% probability, then 459 bees per colony are required. Above are tabulated sample sizes (number of bees) needed based on such requirements, provided that every infected bee is detected with 100% efficiency. If detection efficiency is less than 100%, this is the equivalent (for sample size determination) of trying to detect a lower prevalence. For example, if only 80% of bees actually carrying a pathogen are detected as positive using the diagnostic test, then the parameter P below needs to be adjusted (by multiplying P by the proportion of true positives that are detected, e.g. use 0.8*P instead of P if the test flags 80% of true positives as positive). Sample size needed for various probability requirements and infection levels can be calculated from Equation I (Equations adapted from Colton, 1974).

Equation I.

$$N = \ln(1-D) / \ln(1-P)$$

where:

N = sample size (number of bees)
ln = the natural logarithm
D = the probability (power) of detection in the colony
P = minimal proportion of infected bees (infection prevalence), which can be detected with the required power D by a random sample of N bees (e.g. detect an infection rate of 5% or more).

Because the prevalence of many pathogens varies over space and time (Bailey *et al.*, 1981; Bailey and Ball, 1991; Higes *et al.*, 2008; Runckel *et al.*, 2011), it is important, prior to sampling, to specify the minimum prevalence (P) that needs to be detected and the power (D). Colony-to-colony (and apiary-to-apiary) heterogeneity exists and needs to be taken into consideration in sampling designs. For example, a large French virus survey in 2002 (Tentcheva *et al.*, 2004; Gauthier *et al.*, 2007) showed that for nearly all virus infections there were considerable differences among colonies in an apiary. This suggests that pooling colonies is a poor strategy for understanding the distribution of disease in an apiary, and that sample size should be sufficient to detect low pathogen prevalence, because the probability of finding no infected bees in a small sample is high if the pathogen prevalence is low, as it may be in some colonies. For a colony with low pathogen prevalence, one might have falsely concluded that the hive is pathogen-free due to low power (D) to detect the pathogen.

For *Nosema* spp. infection in adult bees, the infection intensity (spores per bee) as well as prevalence may change rapidly, particularly in the spring, when young bees rapidly replace older nest mates. To understand such temporal effects on infection intensity or prevalence, sample size must be adequate at each sampling period to detect the desired degree of change (i.e. larger samples are necessary to detect smaller changes). Note that sampling to detect a change in prevalence requires a different mathematical model than simple sampling for prevalence because of the uncertainty associated with each prevalence estimate at different sampling periods. Because, for a binomial distribution, variances are a direct function of sample sizes $n_1, n_2, n_3, ...,$ one can use a rule of thumb which is based on the fact that the variance of a difference of two samples will have twice the variance of each individual sample. Thus, doubling the sample size for each period's sample should roughly offset the increased uncertainty when taking the difference of prevalence estimates of two samples. For determining prevalence, limitations due to laboratory capacity are obviously a concern if only low levels of false negative results can be accepted.

Equation I gives the sample size needed to find a pre-determined infection level (P) with a specified probability level (D) in a sample or, in the case of honey bees, in individual colonies. If we want to monitor a population of colonies and describe their health status, or prevalence in this population, we first have to decide with what precision we want to achieve detection within colonies. For example, for composite samples for *Nosema* spp. or virus detection in which many bees from the same colony are pooled (one yes/no or value per colony), this is not a major concern because we can easily increase the power by simply adding more bees to the pool to be examined. For situations in which individual honey bees from a colony are examined to determine prevalence in that colony, we may not want to increase the power because of the labour involved. But if the objective is to describe prevalence in a population of honey bee colonies, not in the individual colony, we can still have poor precision in the estimates if we do not increase the number of colonies we sample. There could be a trade-off between costs in terms of labour and finance and the precision of estimates of the prevalence in each individual. However, if one decreases the power at the individual level one can compensate by an increase in colonies sampled. The more expensive, or labour intensive, the method for diagnosis of the pathogen is, the more cost effective it

becomes the lower the precision of estimates of prevalence in each individual colony, but increase the number of colonies sampled.

2.2.1. Probability of pathogen detection in a colony based on a known sample size

Instead of focusing on sample size, one can calculate the resulting probability of detection of a disease organism using a specific sample size. This probability can be calculated (for an individual colony) using Equation I, but solving for D, as given in Equation II, below.

Equation II.
$$D = 1-(1-P)^N$$
where

D, P, and N are defined as in Equation I above.

For example, within a colony, if the pathogen prevalence in worker bees is 10% (90% of bees are not infected), then the probability of detecting the pathogen in the colony using a sample size of one bee is 0.10, much lower than that for 30 bees (probability is 0.96). A lower prevalence will lower the probability of detection for the same sample size (Fig. 2).

Based on Equation II, it is also possible to calculate the number of bees that need to be tested (sample size) to detect at least one infected bee as a function of the probability, e.g. at a probability of detection (D), of 95% or 99% (Fig. 3). The number of bees to be tested to detect at least one infected bee is higher if one needs a higher probability of detection (D), i.e. when one needs to be able to detect low prevalence.

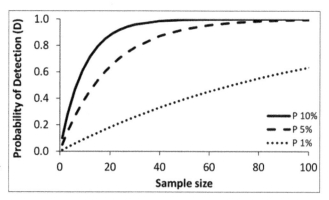

Fig. 2. The probability of detecting a pathogen in a colony (D) as a function of the sample size of bees from that colony, where bees are a completely random sample from the colony. The minimal (true) infection prevalences (P) are 10% (solid line), 5% (dashed line), and 1% (dotted line).

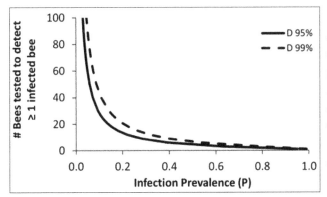

Fig. 3. The number of bees that need to be tested (sample size) to detect at least one infected bee as a function of the prevalence (P), e.g. at a probability of detection (D) of 95% (solid line) or 99% (striped line).

2.2.2. Probability of pathogen detection in a population of colonies

If one wants to calculate the probability of detection of a pathogen in a population of colonies using a known number of colonies, this probability can be calculated (for a population of colonies) according to Equation III.

Equation III.
$$E = 1-(1-P*D)^N$$
where:

E = the probability of detection (in the population)
D and P are defined as above in Equation I and II, N is the sample size, in this case number of colonies

If one wants to determine if a pathogen is present in a population and the probability of pathogen detection in individual colonies is known (see Equation II above), then one can calculate how many colonies need to be sampled in order to detect that pathogen using Equation IV. The computation now calculates the probability of *at least* one positive recording in two-stage sampling situations (the probability of detection in individual colonies and the probability of detection in the population). The probability of pathogen detection in the population can be calculated using Equation III, or it can be set to 0.95 or 0.99, depending on the power required for the investigation.

Equation IV.
$$N = \ln(1-E) / \ln(P * D)$$
where:

N, ln, E, P, and D are defined as above in Equations I, II and III

Equation I, II, III, and IV can easily be entered into a spread sheet for calculation of sample sizes needed for different purposes and desired probabilities of detection.

2.2.3. Extrapolating from sample to colony

A confidence interval of a statistical population parameter, for example, the mean detection rate in brood or the prevalence in the population/colony, can be estimated in a variety of ways (Reiczigel, 2003), most of which can be found in modern statistical software. We do not recommend using the (asymptotic) normal approximation to the binomial method; it gives unreasonable results for low and high prevalence. We show here Wilson's score method (Reiczigel, 2003), defined as:

Equation V.
$$(2N\hat{p} + z^2 \pm z\sqrt{\{z^2 + 4N\hat{p} (1-\hat{p})\}}) / 2(N + z^2),$$

where N is the sample size; $\hat{p}$ is the observed proportion as used by Reiczigel (2003) to indicate that it is an estimated quantity; and z is the $1 - \alpha/2$ quantile, which can be defined as a critical value/threshold, from the standard normal distribution. A shortcoming for all the methods, not only Wilson's method, is that they assume bees in a sample are independent of each other (i.e. there is no over-dispersion, discussed below section 5.2.), which is typically not true, especially given the transmission routes of bee parasites and pathogens (for a detailed discussion of the shortcoming of all methods of confidence interval calculation, see Reiczigel, (2003)).

If the degree of over-dispersion can be estimated, it can be used to adjust confidence limits, most easily by replacing the actual sample size with the effective sample size (if bees are not independent, then the effective sample size is smaller than the actual sample size). One calculates the effective sample size by dividing the actual sample size by the over-dispersion parameter (see section 5.2.3., design effect or *deff* and see Madden and Hughes (1999) for a complete explanation). The latter can be estimated as a parameter assuming the data are beta-binomial distributed, but more easily using software by assuming the distribution is quasi-binomial. The beta-binomial distribution is a true statistical distribution, the quasi-binomial is not, but the theoretical differences are probably of less importance to practitioners than the practical differences using software.

Estimating the parameters of the stochastic model and / or the distribution which will be used to fit the data, based on a beta-binomial distribution (simultaneously estimating the linear predictor, such as regression type effects and treatment type effects, and the other parameters characterising the distribution), is typically difficult in today's software. On the other hand, there are standard algorithms for estimating these quantities if one assumes the data are generated by a quasi-binomial distribution. Essentially, the latter includes a multiplier (not a true parameter) that brings the theoretical variance, as determined by a function of the linear predictor, to the observed variance. This multiplier may be labelled the over-dispersion parameter in software output.

The quasi-binomial distribution is typically in the part of the software that estimates generalised linear models, and requires having bees grouped in logical categories (e.g. based on age or location in a colony), and there must be replication (e.g. two groups that get treatment A, two that get treatment B, etc.). In this kind of analysis, for the dependent variable one gives the number of positive bees and the total number of bees for each category (for some software, e.g. in R, one gives the number of positive bees and the number of negative bees for each category).

Prevalence $\hat{p}$ (estimated proportion positive in the population, as in section 2.2.1. and 2.2.2.) and a 95% confidence interval based on Wilson's score method is given in Fig. 4 for sample sizes (*N*) of 15, 30, and 60 bees. Note that, for the usual sample size of 30, there is still considerable uncertainty about the true infection prevalence (close to 30% if half the bees are estimated to be infected).

3. Experimental design

There are five components to an experiment: hypothesis, experimental design, execution of the experiment, statistical analysis, and interpretation (Hurlbert, 1984). To be able to analyse data in an appropriate manner, it is important to consider one's statistical analyses at the experimental design stage before data collection, a point which cannot be emphasised enough.

Critical features of experimental design include: controls, replication, and randomisation; the latter two components will be dealt with in the next section (3.1.). In terms of a 'control' in an experiment: a negative control group is a standard against which one contrasts treatment effects (untreated or sham-treated control), whereas a positive control group is also often included usually as a "standard" with an established effect (i.e. dimethoate in the case of toxicological studies, see the *BEEBOOK* paper on toxicology by Medrzycki *et al.*, 2013). Additionally, experiments conducted blind or double blind avoid biases from the experimenter or observer. If that is

not possible, one should control for the biases of observers by randomly assigning several different observers to different experimental units or by comparing results from one observer with previous observers to quantify the bias so one can account for it statistically when interpreting results of analyses.

3.1. Factors influencing sample size

A fundamental design element for correct analysis is the choice of the sample size used when obtaining the data of interest. Several factors influencing sample size in an experimental setting are considered below.

3.1.1. Laboratory constraints
Laboratory constraints, such as limitations of space and resources, limit sample size. However, one should not proceed if constraints preclude good science (see the *BEEBOOK* paper on maintaining adult workers in cages, Williams *et al.*, 2013).

3.1.2. Independence of observation and pseudo-replication
A second factor in deciding on sample size, and a fundamental aspect of good experimental design, is independence of observations; what happens to one experimental unit should be independent of what happens to other experimental units before results of statistical analyses can be trusted. The experimental unit is the unit (subject, plant, pot, animal) that is randomly assigned to a treatment. Replication is the repetition of the experimental situation by replicating the experimental unit (Casella, 2008). Where observations are not independent, i.e. there are no true replicates within an experiment, we call this pseudo-replication or technical replication. Pseudo-replication can either be: i) temporal, involving repeated measures over time from the same bee, cage, hive, or apiary; or ii) spatial, involving several measurements from the same vicinity. Pseudo-replication is a problem because one of the most important assumptions of standard statistical analysis is independence.

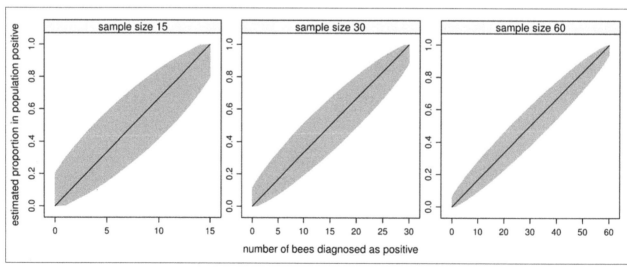

Fig. 4. Estimated proportion of infected bees in a population as a function of the number of bees diagnosed as positive ($\hat{p}$) for various sample sizes (*N* = 15, 30, 60). Lower and upper limits for a 95% confidence interval are based on Wilson's score method.

Repeated measures through time on the same experimental unit will have non-independent (= dependent) errors because peculiarities of individuals will be reflected in all measurements made on them. Similarly, samples taken from the same vicinity may not have non-independent errors because peculiarities of locations will be common to all samples. For example, honey bees within the same cage might not be independent because measurements taken from one individual can be dependent on the state (behaviour, infection status, etc.) of another bee within the same cage (= spatial pseudo-replication), so each cage becomes the minimum unit to analyse statistically (i.e. the experimental unit). An alternative solution is to try estimating the covariance structure of the bees within a cage, i.e. allow for correlation within a cage in the statistical modelling.

But, are honey bees in *different* cages independent? This and similar issues have to be considered and were too often neglected in the past. Potential non-independence can be addressed by including cage, colony, and any other potentially confounding factors as random effects (or fixed effects in certain cases) in a more complex model (i.e. model the covariance structure imposed by cages, colonies, etc.). If pseudo-replication is not desired and is an unavoidable component of the experimental design, then it should be accounted for using the appropriate statistical tools, such as (generalised) linear mixed models ((G)LMM; see section 5.2.).

Some examples may clarify issues about independence of observations.

Example 1: A researcher observes that the average number of *Nosema* spores per bee in a treated cage is significantly higher than in a control cage; one cannot rule out whether the observed effect was caused by the treatment or the cage.

Possible solution 1: Take cage as the experimental unit and pool the observations per cage; including more cages is statistically preferred (yields more power) to including more bees per cage.

Example 2: Relative to cages containing control bees, experimental cages were housed closer to a fan in the lab, resulting in higher levels of desiccation for the experimental cages and, in turn, higher mortality under constant airflow. In this case, the statistical difference between treatments would be confounded by the experimental design.

Possible solution 2: A rotation system could be included to ensure all cages are exposed to the same environmental conditions i.e. placed at identical distances from the fan and for the same periods of time.

Example 3: Honey bees from treated colonies had high levels of a virus and were *A. mellifera mellifera*, whereas control honey bees from untreated colonies that had low levels of a virus were *A. mellifera ligustica*. In such a case the statistical differences could be due to colony differences and/or to subspecies differences and/or due to the treatment and/or due to interactions.

Possible solution 3: Design the experiment using a factorial design with colony as the experimental unit. For half of the colonies in a treatment, use *A. mellifera mellifera* bees and for the other half use *A. mellifera ligustica* bees. Equal numbers of colonies of both subspecies should then be present in the treatment and control groups. Although equal numbers is not a requirement, it is nevertheless preferable to have a completely balanced design (equal numbers in each group or cell) for several reasons (e.g. highest power, efficiency, ease of parameter interpretation, especially interactions). It is, however, also possible to estimate and test with unbalanced designs. In a balanced design the differences between colonies, subspecies, and treatments (and their interactions!) can be properly quantified.

In essence, there are both environmental and genetic factors (which can also interact) that can profoundly affect independence and hence reliability of statistical inference. The preceding examples illustrate, among other things, the importance of randomising experimental units among different treatments. The final solutions of the experimental design are of course highly dependent on the research question and the variables measured.

In summary, randomisation and replication have two separate functions in an experiment. Variables that influence experimental units may be known or unknown, and random assignment of treatments to cages of honey bees is the safest way to avoid pitfalls of extraneous variables biasing results. Larger sample sizes (i.e. replication: number of colonies, cages, or bees per cage) improve the precision of an estimate (e.g. infection rate, mortality, etc.) and reduce the probability of uncontrolled factors producing spurious statistical insignificance or significance. Researchers should use as many honey bee colony sources from unrelated stock as possible if they want their results to be representative, and hence generalisable. One should also not be too cavalier about randomising honey bees to experimental treatments, or about arranging experimental treatments in any setting, including honey bee cage experiments; sound experimental design at this stage is critical to good science; more details are provided below.

3.1.3. Effect size

A third factor affecting decisions about sample size in experimental design is referred to as effect size (Cohen, 1988). As an illustration, if experimental treatments with a pesticide decrease honey bee food intake to 90% that of controls, more replication is needed to achieve statistical significance than if food intake is reduced to 10% that of controls (note that one's objective should be to find biologically meaningful results rather than statistical significance). This is because treatment has a greater effect size in the latter situation. Effect size and statistical significance are substantially intertwined, and there are equations, called power analyses (see section 3.2.1.), for calculating sample sizes needed for statistical significance once effect size is known.

Without preliminary trials, effect size, and also statistical power, may be impossible to know in advance. If one's objective is statistical significance, and one knows effect size, one can continue to sample until significance is achieved. However, this approach is biased in favour of a preferred result. Moreover, it introduces the environmental influence of time; results one achieves in spring may not be replicated in summer e.g. Scheiner *et al.* (2003) reported seasonal variation in proboscis extension responses (previously called proboscis extension reflexes; also see Frost *et al.*, 2012). Removing the influence of time requires that one decides in advance of replication, and accepts results one obtains. Without preliminary trials, it will always be preferable to maintain as many properly randomised cages as possible. A related factor that will influence sample size is mortality rate of honey bees in cages; if control group mortality rates are 20% for individual bees, one will want to increase the number of bees by at least 20%, and even more if variability in mortality rates is high. Alternatively, without knowledge of effect size, one should design an experiment with sufficient replicates such that an effect size of biological relevance can be measured.

3.2. Sample size determination

There are many online sample size calculators available on the internet that differ in the parameters required to calculate sample size for experiments. Some are based on the effect size or minimal detectable difference (see section 3.1.3.); for others input on the estimated mean (μ) and standard deviation (δ) for the different treatment groups is required. Fundamentally, the design of the experiment, the required power, the allowed α and the expected effect size dictate the required sample size. The following two sections (3.2.1. and 3.2.2.) suggest strategies for determining sample size.

3.2.1. Power analyses and rules of thumb

Power (1-β) of a statistical test is its ability to detect an effect of a particular size (see section 3.1.3.), and this is intrinsically linked with sample size (N) and the error probability level (α) at which we accept an effect as being statistically significant (see section 1., Table 1). Once we know two of these values, it is possible to calculate the remaining one; in this case for a given α and β, what is N? Power analyses can incorporate a variety of data distributions (normal, Poisson, binomial, etc.), but the computations are beyond the scope of this paper. Fortunately, there are many freely available computer programs that can conduct these calculations (e.g. G*Power; Faul *et al.*, 2007,

the R-packages "pwr" and "sample size" online programs can be found at www.statpages.org/#Power) and all major commercial packages also have routines for calculating power and required sample sizes.

A variety of 'rules of thumb' exist regarding minimum sample sizes, the most common being that you should have at least 10-15 data points per predictor parameter in a model; e.g. with three predictors such as location, colony and infection intensity, you would need 30 to 45 experimental units (Field *et al.*, 2012). For regression models (ANOVA, GLM, etc.), where you have k predictors, the recommended minimum sample size should be $50 + 8k$ to adequately test the overall model, and $104 + k$ to adequately test each predictor of a model (Green, 1991). Alternatively, with a high level of statistical power (using Cohen's (1988) benchmark of 0.8), and with three predictors in a regression model: i) a large effect size (> 0.5) requires a minimum sample size of 40 experimental units; ii) a medium effect size (of ca. 0.3) requires a sample size of 80; iii) a small effect size (of ca. 0.1) requires a sample size of 600 (Miles and Shevlin, 2001; Field *et al.*, 2012).

These numbers need to be considerably larger when there are random effects in the model (or temporal or spatial correlations due to some kind of repeated measures, which decreases effective sample size). Random effects introduce additional parameters to the model, which need to be estimated, but also inflate standard errors of fixed parameters. The fewer the levels of the random effects (e.g. only three colonies used as blocks in the experiment), the larger the inflation will be. Because random factors are estimated as additional variance parameters, and one needs approximately 30 units to estimate a variance well, increasing the number of levels for each random effect will lessen effects on fixed parameter standard errors. That will also help accomplish the goals set in the first place by including random effects in a designed experiment: increased inference space and a more realistic partitioning of the sources of variation. We recommend increasing the number of blocks (up to 30), with fewer experimental units in each block (i.e. more, smaller blocks), as a general principle to improve the experimental design. Three (or the more common five) blocks is too few. Fortunately, there are open source (R packages "pamm" and "longpower") and a few commercial products (software NCSS PASS, SPSS, STATISTICA) which could be helpful with estimating sample sizes for experiments that include random effects (or temporally or spatially correlated data).

If random effects are considered to be fixed effects and one uses the methods described above for sample size estimation or power, required sample sized will be seriously underestimated and power seriously overestimated. The exemplary data set method (illustrated for GLMMs and in SAS code in Stroup (2013), though easily ported to other software that estimates GLMMs) and use of Monte-Carlo methods (simulation, example explained below, though it is not for a model with random effects) are current recommendations. For count data (binomial, Poisson distributed), one should always assume there will be over-dispersion (see section 5.2.3.).

3.2.2. Simulation approaches

Simulation or 'Monte-Carlo' methods (Manly, 1997) can be used to work out the best combination of "bees/cage" x "number of cages/ group" given expected impact of a certain treatment. Given a certain average life span and standard deviation for bees of a control group and a certain effect of a treatment (in terms of percentage reduction of the life span of bees), one can simulate a population of virtual bees each with a given life span. Then a program can test the difference between the treated group and the control group using increasing numbers of bees (from 5 to 20) and increasing numbers of cages (from 3 to 10). The procedure can be repeated (e.g. 100 times) and a table produced with the percentage of times a significant difference was achieved using any combination of bees/cage x number of cages/ group.

A program using a *t*-test to determine these parameters is given as online supplementary material (http://www.ibra.org.uk/ downloads/20130812/download). It is assumed that the dependent variables in the bee population are normally distributed. The simulation can be run another 100 times simply by moving the mouse from one cell to another. Alternatively, automatic recalculation can be disabled in the excel preferences.

3.2.3. Sample size and individual infection rates

Common topics in honey bee research are pathogens. Prevalence of pathogens can be determined in a colony or at a population level (see section 2.2.). Most likely, the data will be based on whether in the smallest tested unit the pathogen is present or not: a binomial distribution. Hence, sample size will be largely dependent on detection probability of a pathogen. However, with viruses (and possibly other pathogens), concentration of virus particles is measured on a logarithmic scale (Gauthier *et al.*, 2007; Brunetto *et al.*, 2009). This means, for example, that the virus titre of a pooled sample is disproportionately determined by the one bee with the highest individual titre. For the assumption of normality in many parametric analyses we suggest a power-transformation of these data (Box and Cox, 1964; Bickel and Doksum, 1981). For further reading on sample-size determination for log-normal distributed variables, see Wolfe and Carlin (1999).

In summary, a minimum sample of 30 independent observations per treatment (and the lowest level of independence will almost always be cages) may be desirable, but constraints and large effect sizes will lower this quantity, especially for experiments using groups of caged honey bees. Because of this, development of methods for maintaining workers individually in cages for a number of weeks should be investigated. This would be an advantage because depending on the experimental question, each honey bee could be considered to be an independent experimental unit. The same principles of experimental design that apply to the recommended number of cages also apply to other levels of experimental design, such as honey bees

per cage, with smaller effect sizes and more complex questions, recommended sample sizes necessarily increase (in other words the more variables/factors included, the greater the sample size has to be). Researchers must think about, and be able to justify, how many of their replicates are truly independent; 30 replicates is a reasonable starting point to aim for when effect sizes are unknown, but again, this may not be realistic. In the context of wax producing and comb building, colony size and queen status play a role. For example, comb construction only takes place in the presence of a queen and at least 51 workers, and egg-laying occurs only if a mated queen is surrounded by at least 800 workers (reviewed in Hepburn, 1986; page 156). Additionally, novel experiments on new sets of variables means uncertainty in outcomes, but more importantly means uninformed experimental designs that may be less than optimal. Designs should always be scrutinised and constantly improved by including preliminary trials, which could, for example, provide a better idea of prevalence resulting in a better estimate of the required sample size.

4. A worked example

Although a single recommended experimental design, including sample size, may be difficult to find consensus on given the factors mentioned above, we provide below a recommendation for experimental design when using groups of caged honey bees to understand the impact of a certain factor (e.g. parasite or pesticide) on honey bees. For our example, imagine the focus is the impact of the gut parasite *Nosema ceranae* and black queen cell virus on honey bees. One should consider the following:

1. Each cage should contain the same number of honey bees, and be exposed to the same environmental conditions (e.g., temperature, humidity, feeding regime, see the *BEEBOOK* paper on maintaining adult workers in cages, Williams *et al.*, 2013). Each cage of treated honey bees is a single experimental unit, or unit of replication. Because there is no other restriction on randomisation, other than the systematic sampling from different colonies for each replicate (see below), this is a completely randomised design. If one instead put only bees from one colony in a cage, but made sure that all treatments were evenly represented for each colony (e.g. 5 cages from colony A get treatment 1, 5 cages from colony A get treatment 2, etc.), then we would have a randomised complete block design.

2. We recommend 4-9 replicate cages per treatment. Honey bees should be drawn from 6-9 different colonies to constitute each replicate and equal numbers of honey bees from each source colony should be placed in each cage (i.e., in all cages,

including controls and treatments) to eliminate effects of colony; this makes only cage a random factor. For example, if one draws honey bees from 6 source colonies and wants cages to contain 24 honey bees each, then one must randomly select 4 honey bees from each colony for each cage. If one wants to keep colony and cage both as random effects (e.g. to estimate effects of a pathogen on bees from a population of colonies, only some of which were sampled), one should not mix bees from different colonies in the cages. Note that the minimum number of bees also depends on the experimental design. Darchen (1956, 1957) showed that comb construction only started with a minimum number of 51 -75 workers and a queen; in cases of a dead queen, 201-300 workers were needed (summarised in Table 14.1 in Hepburn, 1986). Furthermore, cage design itself can influence behaviour (Köhler *et al.,* 2013) therefore identical cages should be used for all replicates. A group size of 15 workers ensures that the impact of experimentally administered *Nosema* and black queen cell virus in honey bees in general is measured, as opposed to impacts of these parasites on a specific honey bee colony. It also ensures that chance stochastic events, such as all the honey bees dying in a specific treatment cage, do not unduly affect the analysis and interpretation of results. Low numbers of source colonies (i.e. low numbers of replicates) could lead to an over- or under-estimation of the impact of the studied factor(s). A computer simulation based on Monte-Carlo methods (see section 3.2.3.) and parametric statistics supports the appropriateness of the proposed values. Experiments across replicate colonies must be conducted at approximately the same time, because effects such as day length and seasonality can introduce additional sources of error (see section 2.1.1. and section 3. for relationships between model complexity and sources of error).

5. Statistical analyses

5.1. How to choose a simple statistical test

Before addressing the question of how to choose a test, we describe differences between parametric and non-parametric statistics. As stated in the introduction, one has to know what kind of data one has or will obtain. In the discussion below, we use a traditional definition of "parametric" versus "non-parametric tests". In all statistical tests, parameters of one kind or another (means, medians, etc.) are estimated. The distinction has grown murkier over the years as more and more statistical distributions become available for use in contexts where previously only the normal distribution was allowed (e.g. regression, ANOVA). "Parametric" tests assume (1) models where the residuals (the variation that is not explained by the explanatory

variables one is testing, i.e. inherent biological variation of the experimental units), following fitting a linear predictor of some kind, are normally distributed, or that the data follow a (2) Poisson, multinomial, or hypergeometric distribution. This definition holds for simple models only; parametric models are actually a large class of models where all essential attributes of the data can be captured by a finite number of parameters (estimated from the data), so include many distributions and both linear and non-linear models, but the distribution(s) must be specified when analysing the data. The complete definition is quite mathematical. A non-parametric test does not require that the data be samples from any particular distribution (i.e. they are distribution-free). This is the feature that makes them so popular.

For models based on the normal distribution, this *does not* mean that the dependent variable is normally distributed; in fact one hopes it is multimodal, with a different mode for each different treatment. However, if one subtracts (or conditions on) the linear predictor (e.g. subtract each treatment mean from its group of observations), the distribution of each resulting group (and all groups combined) follows the same normal distribution. Also, the discussion below pertains only to "simple" statistical tests and where observations are independent.

Note that chi-square and related tests are often considered "non-parametric" tests. This is incorrect; they are very distribution dependent (data must be drawn from Poisson, multinomial, or hypergeometric distributions), and observations must be independent. Whereas "non-parametric" tests may not require that one samples from a particular distribution, they do require that each set of samples come from the same general distribution. That is, one sample cannot come from a right-skewed distribution and the other from a left-skewed distribution; both must have the same degree of skew and in the same direction. Note that when one has dichotomous (Yes/No) or categorical data, non-parametric tests will be required if we stay in the realm of "simple" statistical tests (Fig. 4). For parametric statistics based on the normal distribution, an important second assumption is that the variance among groups of residuals is similar (homogeneous variances, also called homoscedasticity) (as shown in Fig. 5a) and not heterogeneous variances (heteroscedasticity, Fig. 5b). If only one assumption is violated, a parametric statistic is not applicable. The alternative in such a case would be to either transform the data (see Table 4 and section 5.2.), so that the transformed data no longer violate assumptions, or to conduct non-parametric statistics. The advantage of non-parametric statistics is that they do not assume a specific distribution of the data; the disadvantage is that the power (1-ß, see section 1.) is lower compared to their parametric counterparts (Wasserman, 2006), though the differences may not be great. Power itself is not of such great concern because biologically relevant effects shall be detected with a large enough effect size in a **well-designed experiment**. Table 3 provides a comparison between parametric and non-parametric statistics.

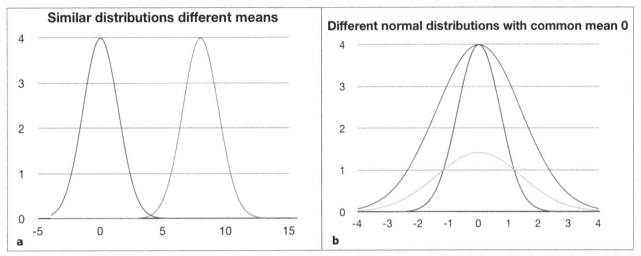

Fig. 5. a. Two similar distributions with different means, where variances of the two groups are homogeneous; **b.** shows three different distributions where the means are the same but the variances of three groups are heterogeneous.

Table 3. Comparison between parametric and non-parametric statistics.

	Parametric	Non-parametric
Distribution	Normal	Any
Variance	Homogenous	Any
General data type	Interval or ratio (continuous)	Interval, ratio, ordinal or nominal
Power	Higher	Lower
Example Tests		
Correlation	Pearson	Spearman
Independent data	t-test for independent samples	Mann-Whitney U test
Independent data more than 2 groups	One way ANOVA	Kruskal Wallis ANOVA
Two repeated measures, 2 groups	Matched pair t-test	Wilcoxon paired test
Two repeated measures, > 2 groups	Repeated measures ANOVA	Friedman ANOVA

5.1.1 Tests for normality and homogeneity of variances

The flow diagram in Fig. 6 gives a simple decision tree to choose the right test; for more examples, see Table 5. Starting at the top, one has to make a decision based on what kind of data one has. If two variables are categorical, then a chi-square test could be applicable. When investigating the relationship between two continuous variables, a correlation will be suitable. In the event one wants to compare two or more groups and test if they are different, one follows the pathway "difference". The next question to answer is how many variables one wants to compare. Is it one variable (for example the effect of a new varroa treatment on brood development in a honey bee colony), or is it the effect of varroa treatment and supplementary feeding on brood development? For the latter, one

could conduct a 2-way ANOVA or an even more complex model depending on the actual data set. For the former, the next question would be "how many treatments?"; sticking with the example, does the experiment consist of two groups (control and treatment) or more (control and different dosages of the treatment)? In both cases, the next decision would be based on if the data sets are independent or dependent. Relating back to the example, one could design the experiment where some of the colonies are in the treatment group and some in the control, in which case one could say that the groups are independent. However, one could as well compare before and after the application of the varroa agent, in which case all colonies would be in the before (control) and after (treatment) group. In this case it is easy to see that the before might affect the after or that the two groups are not independent. A classical example of dependent data is weight loss in humans before and after the start of diet; clearly weight loss depends on starting weight.

To arrive at an informed decision about the extent of non-normality or heterogeneity of variances in your data, a critical first step is to plot your data: i) for correlational analyses as in regression, use a scatterplot ii) for 'groups' (e.g. levels of a treatment factor), use a histogram or box plot; it provides an immediate indication of your data's distribution, especially whether variances are homogeneous. The next step would be to objectively test for departures from normality and homoscedasticity. Shapiro-Wilks W, particularly for sample sizes < 50, or Lilliefors test, can be used to test for normality, and the Anderson-Darling test is of similar if not better value (Stephens, 1974). Similarly, for groups of data, Levene's test tests the null hypothesis that different groups have equal variances. If tests are significant, assumptions that a distribution is normal or its variances are equal **must** be rejected and either the data has to be transformed, a non-parametric test or generalised linear model applied.

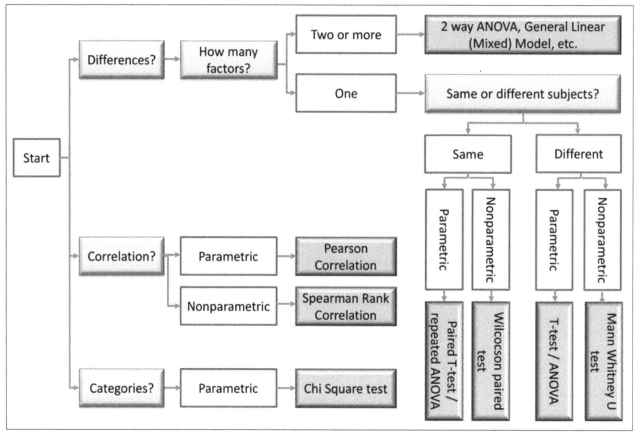

Fig. 6. A basic decision tree on how to select the appropriate statistical test is shown.

Table 4. Common underlying distributions for generalised linear models and their canonical link functions.

Distribution	Canonical Link
Gaussian	identity (no transformation)
Poisson	log
Binomial	logit
Gamma	inverse

5.2. Generalised Linear Mixed Models (GLMM)

A central dogma in statistical analyses is always to apply the simplest statistical test to your data, but ensure it is applied correctly (Zuur *et al.*, 2009). Yes, you could apply an ANOVA or linear regression to your data, but in the **vast majority of cases**, the series of assumptions upon which these techniques are based are violated by 'real world' data and experimental designs, which often include blocking or some kind of repeated measures. The assumptions typically violated are: i) normality; ii) homogeneity; and iii) independence of data.

1. Normality

Although some statistical tests are robust to minor violations of normality (Sokal and Rohlf, 1995; Sokal and Rohlf, 2012), where your dependent variable/data (i.e. the residuals, see section 5.1.) are clearly not normal (positively/negatively skewed, binary data, etc.), a better approach would be to account for this distribution within your model, rather than ignore it and settle for models that poorly fit your data. As an obvious example, a process that produces counts will not generate data values less than zero, but the normal distribution ranges from $-\infty$ to $+\infty$.

2. Homogeneity of variances

As stated above, minor violations of normality can be tolerated in some cases, and the same could be said for heterogeneous dependent variable/data (non-homogenous variance across levels of a predictor in a model, also called heteroscedasticity). However, marked heterogeneity fundamentally violates underlying assumptions for linear regression models, thereby falsely applying the results and conclusions of a parametric model, making results of statistical tests invalid.

3. Independence of data

See section 3.1.2. Simply, if your experimental design is hierarchical (e.g. bees are in cages, cages from colonies, colonies from apiaries) or involves repeated measures of experimental units, your data strongly violate the assumption of independence and invalidate important tests such as the *F*-test and *t*-test; these tests will be too liberal (i.e. true null hypotheses will be rejected too often).

Table 5. Guideline to statistical analyses in honey bee research including examples/ suggestions for tests and graphical representation. Blank fields indicate that a wide variety of options are possible and all have pros and cons.

Subject	Variable	Short description	Fields of research where it is used	Synthetic representation	Measure of dispersion	Statistical test	Graphical representation	Notes
Honey bee	Morphometric variables (e.g. fore-wing angles)	Measures related to body size. Other data can be included here such as, for example, cuticular hydrocarbons	Taxonomic studies	Average	Standard deviation	Parametric tests such as ANOVA. Multivariate analysis such as PCA and DA	Bar charts for single variables, scatterplots for PC, DA	Please note that some morphometric data are ratios; consider possible deviations from normality
	Physiological parameters (e.g. concentration of a certain compound in the haemolymph)	Measures related to the functioning of honey bee systems		Average	Standard deviation		Bar charts or lines	
	Survival			Median	Range	Kaplan Meyer Cox hazard	Bar charts or lines scatterplots	
Pathogens (e.g. DWV, Nosema)	Prevalence	Proportion of infected individuals	Epidemiological studies	Average	Standard deviation can be used but transformation is necessary due to non-normal distribution	Fisher exact solution or Chi square according to sample size	Bar charts, pie charts	
	Infection level	Number of pathogens (e.g. viral particles)	Epidemiological studies, studies on bee-parasite interaction	Average		Parametric tests (e.g. t test/ANOVA) can be used after log transformation otherwise non parametric tests can be used (e.g. Mann-Whitney/Kruskal-Wallis)		
Parasites (e.g. *Varroa destructor*)	Fertility	Proportion of reproducing females	Factors of tolerance, biology of parasites	Average	Range	Fisher exact solution or chi square according to sample size		
	Fecundity	Number of offspring per female	Factors of tolerance, biology of parasites	Average	Standard deviation			

GLMMs are a superset of linear models, they allow for the dependent variable to be samples from non-normal distributions (allowed distributions have to be members of the one and two parameter exponential distribution family; this includes the normal distribution, but also many others). For distributions other than the normal, the statistical model produces heterogeneous variances, which is a desired result if they match the heterogeneous variances seen in the dependent variable. The 'generalised' part of GLMM means that, unlike in linear regression, the experimenter can choose the kind of distribution they believe underlies the process generating their data. The 'mixed' part of GLMM allows for random effects and some degree of non-independence among observations. Ultimately, this level of flexibility within GLMM approaches allows a researcher to apply more rigorous, but biologically more realistic, statistical models to their data.

One pays a price for this advantage. The basic one is that the state of statistical knowledge in this area, especially computational issues, lags behind that for models based on the normal distribution. This translates into software that is buggy, which can result in many kinds of model estimation problems. Also, there are now far more choices to be made, such as which estimation algorithm to use (e.g. the Laplace and quadrature methods do not allow for correlation among observations), and which link function to use. The link function "links" the data scale to the model scale (Table 4). For example, if dependent variable is assumed to be generated by a Poisson process, the typical link function is the log, i.e. log $(E(\mu)) = \mathbf{X}\boldsymbol{\beta} + \mathbf{ZU}$; in words, the natural log of the expected value of the mean is modelled as a sum of fixed and random effects). Tests are based on asymptotic behaviours of various quantities, which can give quite biased results for small samples. One is simultaneously working on two scales: the data scale and the model scale; the two are linked, but model estimates and hypothesis tests are done on the model scale, and so are less easily interpretable (i.e. a change in unit value of a predictor variable has different effects on the data scale depending on whether one is looking at low values or high values). One parameter and two parameter members of the exponential family have to be handled quite differently.

Over-dispersion (see section 5.2.3.) cannot be handled using a quasi-likelihood approach (e.g. using a quasi-binomial distribution); instead, appropriate random effects need to be added (e.g. one for every observation), which can lead to models with many parameters (Note: Over-dispersion means that one has a greater variability than expected based on the theoretical statistical distribution; for example the expected variance of a Poisson distribution is its mean - if the observed variance is larger than the estimated mean, then there is over-dispersion). For some one-parameter members of the exponential distribution (e.g. Poisson, binomial), one can try the analogous two-parameter member (e.g. for a Poisson distribution, it is the negative binomial distribution; for the binomial it is the beta-binomial). Model

diagnosis is in its infancy. While we encourage researchers to explore the use of these models, we also caution that considerable training is necessary for both the understanding of the theoretical underpinnings of these models and for using the software. A recent book using GLMM methodology is Stroup (2013), which developed from experience with researchers in agriculture and covering both analyses and design of experiments. He discusses in detail what we can only allude to superficially; a shortcoming is that the worked examples only use the SAS software.

5.2.1. General advice for using GLMMs

If the response variable to be measured (i.e. the phenotype of interest that may change with treatment) is a quantitative or a qualitative (i.e. yes-diseased/no-not diseased) trait and the experiment is hierarchical (e.g. bees in cages, cages from colonies, colonies from locations), repeated over years, or has some other random effects, then a generalised linear mixed model (GLMM; as provided in the statistical software R, Minitab, or SAS) can be used to analyse the results. The treatment (control, *Nosema*, black queen cell virus) is a 'fixed effect' parameter (Crawley, 2005; Bolker *et al.*, 2009). Several fixed and random effect parameters can be estimated in the same statistical model. The distinction between what is a fixed or a random effect can be difficult to make because it can be highly context-dependent, but in most experiments it should be obvious. To help clarify the distinction between the two, Crawley (2013) suggests that fixed effects influence the mean of your response variable and random effects influence the variance or correlation structure of your response variable, or is a restriction on randomisation (e.g. a block effect). A list of fixed effects would include: treatment, caste, wet *vs.* dry, light *vs.* shade, high *vs.* low, etc. i.e. treatments imposed by the researcher or inherent characteristics of the subjects (e.g. age). A list of random effects would include: cage, colony, apiary, region, genotype (if genotypes were sampled at random, not if the design was to compare two or more specific genotypes), block within a field, plot, subject measured repeatedly.

Example:

The experimenter must consider the structure of the GLMM by addressing two questions, as follows:

- Which underlying distribution?
  Gaussian, useful for data where one expects residuals to follow a 'normal' distribution
  Poisson, useful for count data (e.g. number of mites per bee)
  Binomial, useful for data on proportions based on counts (y out of n) or binary data
  Gamma, useful for data showing a constant coefficient of variation

- What link function to use?

  The link function maps the expected values of the data, conditioned on the random effects, to the linear predictor. Again, this means that the linear predictor and data reside on different scales. Canonical link functions are the most commonly used link functions associated with each 'family' of distributions (Table 4). The term "canonical" refers to the form taken of one of the parameters in the mathematical definition of each distribution.

If two or more experimental cages used in the same treatment group are drawn from the same colony of honey bees (Table 6), then a GLMM with 'source colony' as a random effect parameter should also be included, as described above. This random effect accounts for the hierarchical experimental design whereby, for the same treatment level, variation between two cages of honey bees drawn from the same colony may not be the same as the variation between two cages drawn from two separate colonies. This statistical approach can account for the problem of pseudo-replication in the experimental design.

Finally, if the factor 'cage' and 'source colony' are not significant, the experimenter may be tempted to treat individual bees from the same cage as independent samples; i.e. ignore 'cage'. However, individual bees drawn from the same cage might not truly be independent samples and therefore it would inflate the degrees of freedom to treat individual bees and individual replicates. Because there are currently no good tests to determine if a random effect is 'significant', we suggest retaining any random effects that place restrictions on randomisation - cage and source colony are two such examples - even if variance estimates are small. This point requires further attention by statisticians. The experimenter should consider using a nested experimental design in which 'individual bee' is nested within a random effect, 'cage', as presented above (see section 5.).

5.2.2. GLMM where the response variable is mortality

If survival of honey bees is the response variable of interest, then each cage should contain a minimum of 30 bees so as to provide a more robust estimate of their survival function. A typical survival analysis then needs to be undertaken on the data, e.g. the non-parametric Kaplan-Meier survival analysis for 'censored' data (so-called right-censored data in which bees are sampled from the cage during the experiment) or the semi-parametric Cox proportional hazards model (Cox model) for analysing effects of two or more 'covariates', or predictor variables such as *N. ceranae* or black queen cell virus (Collett, 2003; Zuur *et al.*, 2009; Hendriksma *et al.*, 2011). Note: these models do not only allow for random effects, if the design includes random effects then a GLMM (see section 5.2.) could be an alternative (including some function of time is a predictor variable in the GLMM).

5.2.3. Over-dispersion in GLMM

Over-dispersion is "the polite statistician's version of Murphy's law: if something can go wrong, it will" (Crawley, 2013). It is particularly relevant when working with count or proportion data where variation of a response variable does not strictly conform to the Poisson or binomial distribution, respectively. Fundamentally, over-dispersion causes poor model fitting where the difference between observed and predicted values from the tested model are larger than what would be predicted by the error structure. To identify possible over-dispersion in the data for a given model, divide the deviance (-2 times the log-likelihood ratio of the reduced model, e.g. a model with only a term for the intercept, compared to the full model; see McCullagh and Nelder,1989) by its degrees of freedom: this is called the dispersion parameter. If the deviance is reasonably close to the degrees of freedom (i.e. the dispersion or scale parameter = 1) then evidence of over-dispersion is lacking.

Table 6. Experimental design for studying the impact of *Nosema ceranae* and black queen cell virus (BQCV) on caged honey bees. [†]Notation represents individual cages (Treatment, Colony 1, Cage 1 = T1_1; and Control, Colony 1, Cage 1 = C1_1), each containing equal number of honey bees (e.g. 30) exposed to the same conditions (except experimental treatment differences). Two replicate cages within treatments drawn from the same colony are displayed (T1_1 and T1_2), and more could be used (T1_3, T1_4, etc.). Additional control colonies would then also be required. 'Colony' should be used as a random effect in such cases. But, it is statistically more powerful to maximise inter- as opposed to intra-colony replication; that is, favour the use of replicate cages between colonies, rather than repeated sets of cages per treatment drawn from the same colony. Thus we recommend one set of treatment and control cages per colony of source honey bees rather than repeated sets of cages per treatment and control drawn from a single colony i.e. T1_**1**, T2_**1**, T3_**1** .... T9_**1** and C1_**1**, C2_**1**, C3_**1** .... C9_**1** would be a far superior design compared to T1_**1**, T1_**2**, T1_**3** .... T1_**9** and C1_**1**, C1_**2**, C1_**3** .... C1_**9**.

Treatment	Colony								
	1	2	3	4	5	6	7	8	9
N. ceranae &	T1_1[†],	T2_1,	T3_1,	T4_1,	T5_1,	T6_1,	T7_1,	T8_1,	T9_1,
BQCV	T1_2	T2_2	T3_2	T4_2	T5_2	T6_2	T7_2	T8_2	T9_2
control	C1_1,	C2_1,	C3_1,	C4_1,	C5_1,	C6_1,	C7_1,	C8_1,	C9_1,
	C1_2	C2_2	C3_2	C4_2	C5_2	C6_2	C7_2	C8_2	C9_2

Causes of over-dispersion can be apparent or real. Apparent over-dispersion is due to model misspecification, i.e. missing covariates or interactions, outliers in the response variable, non-linear effects of covariates entered as linear effects, the wrong link function, etc. Real over-dispersion occurs when model misspecifications can be ruled out, and variation in the data is real due to too many zeros, clustering of observations, or correlation between observations (Zuur *et al.*, 2009). Solutions to over-dispersion can include: i) adding covariates or interactions, ii) including individual-level random effects, e.g. using bee as a random effect, where multiple bees are observed per cage, iii) using alternative distributions: if there is no random effect included in the model consider quasi-binomial and quasi-Poisson; if there are, consider replacing Poisson with negative-binomial, and iv) using a zero-inflated GLMM (a model that allows for numerous zeros in your dataset, the frequency of the number zero is inflated) if appropriate. Over-dispersion cannot occur for normally distributed response variables because the variance is estimated independently from the mean. However, residuals often have "heavy tails", i.e. more outlying observations than expected for a normal distribution, which nevertheless can be addressed by some software packages.

5.3. Accounting for multiple comparisons

Thus far, we have assumed that we are investigating two categories of an explanatory variable or experimental treatment (i.e. comparing a treatment group with a control group). However, the objective may instead be to compare multiple levels of an explanatory variable (e.g. different concentrations of a pesticide) or multiple independent kinds of the same sort of explanatory variable (e.g. competing manufacturers of protein substitutes). In addition, one may be interested in testing multiple explanatory variables at the same time (e.g. effects of three different humidity levels and honey bee age on susceptibility to the tracheal mite *Acarapis woodi*). More complex statistical models warrant increased sample sizes for all treatments. Consider the case where one has one control and one treatment group; there is a single comparison possible. Yet if one has one control and 9 treatments groups, there are $9 + 8 + ... + 1 = 55$ possible comparisons. If one rigorously follows the cut-off of $P = 0.05$, one could obtain $0.05 * 55 = 2.8$ significant results by chance or in other words the probability of at least one significant by chance alone is $1 - 0.95^{55} = 0.9405$, so one is likely to incorrectly declare significance at least once (in general, 5% of statistical results will have $p \geq 0.05$ if there are no true differences among treatments, this is what setting $\alpha = 0.05$ represents). *Post hoc* tests or *a posteriori* testing, such as Bonferroni corrections, attempt to account for this excessive testing, but in so doing can become very conservative, and potentially significant results may be overlooked (i.e. correctly control for Type I error, but have inflated Type II errors; Rothman, 1990; Nakagawa, 2004). Less conservative corrections, such as the False Discovery Rate, are now typically favoured as they represent a balance between controlling for

Type I and Type II errors (Benjamini and Hochberg, 1995). Other ways to avoid or minimise this problem include increasing sample size and simplifying experimental design by reducing the number of treatments and variables.

5.4. Principal components to reduce the number of explanatory variables

With an increasing number of explanatory variables (related or not-related, similar or dissimilar units) in one experiment, multivariate statistics may be of interest. Multivariate statistics are widely used in ecology (Leps and Smilauer, 2003), but less often in bee research. Multivariate statistics can be used to reduce the number of response variables without losing information in the response variables (van Dooremalen and Ellers, 2010), or to reduce the number of explanatory variables (especially valuable if they are correlated). A Principle Component Analysis (PCA) can be used to examine, for example, morphometric or physiological variables (such as protein content of different bee body parts or several volatile compounds in the head space of bee brood cells). The PCA is usually used to obtain only the first principal component that forms one new PC variable (the axis explaining most variation in your variables). The correlations between the original variables and the new PC variable will show the relative variation explained by the original variables compared to each other and their reciprocal correlation. The new PC variable can then be used to investigate effects of different treatments (and/or covariates) using statistics as explained above in section 5. For an example in springtails see van Dooremalen *et al.* (2011), or in host-parasite interactions see Nash *et al.* (2008). Note that the new PC variables are uncorrelated with each other, which improves their statistical properties. Unfortunately, it is also easy to lose track of what they represent or how to interpret them. However, by reducing dimensionality and dealing with uncorrelated variables one can transform a data set with a great many explanatory and response variables into one with only a few of each, and ones which capture most of the variability (i.e. the underlying processes) in the data set. Related procedures are factor analysis, partial least squares, non-metric multidimensional scaling (NMDS), and PC regression.

5.5. Robust statistics

Robust statistics were developed because empirical data that considered samples from normal distributions often displayed clearly non-normal characteristics, which invalidates the analyses if one assumes normality. They are usually introduced early on in discussions of measures of central tendency. For example, medians are far more resistant to the influence of outliers (observations that are deemed to deviate for reasons that may include measurement error, mistakes in data entry, etc.) than are means, so the former are considered more robust. Even a small number of outliers (as few as one) may adversely affect a mean, whereas a median can be resistant when up to 50% of observations are outliers. On the other hand, screening for outliers for removal may be subjective and difficult for highly structured data, where a response variable may be functionally related to many independent variables. If "outliers" are removed, resulting variance estimates are often too small, resulting in overly liberal testing (i.e. *p* values are too small).

What are the alternatives when one cannot assume that data are generated by typical parametric models (e.g. normal, Poisson, binomial distributions)? This may be a result of contamination (e.g. most of the data comes from a normal distribution with mean μ and variance σ_1^2 but a small percentage comes from a normal distribution with mean μ and variance σ_2^2, where $\sigma_2^2 >> \sigma_1^2$), a symmetric distribution with heavy tails, such as a *t* distribution with few degrees of freedom, or some highly skewed distribution (especially common when there is a hard limit, such as no negative values, typical of count data and also the results of analytic procedures estimation; e.g. titres). Robust statistics are generally applicable when a sampling distribution from which data are drawn is symmetric. "Non-parametric" statistics are typically based on ordering observations by their magnitude, and are thus more general, but have lower power than either typical parametric models or robust statistical models. However, robust statistics never "caught on" to any great degree in the biological sciences; they should be used far more often (perhaps in most cases where the normal distribution is assumed).

Most statistics packages have some procedures based on robust statistics; R has particularly good representation (e.g. the MASS package). All typical statistical models (e.g. regression, ANOVA, multivariate procedures) have counterparts using robust statistics. Estimating these models used to be considered difficult (involving iterative solutions, maximisation, etc.), but these models are now quickly estimated. The generalised linear class of models (GLM) has some overlap with robust statistics, because one can base models on, e.g. heavy-tailed distributions in some software, but the approach is different. In general, robust statistics try to diminish effects of "influential" observations (i.e. outliers). GLMs, once a sampling distribution is specified (theoretical sampling distributions include highly skewed or heavy-tailed ones, though what is actually available depends on the software package) consider all observations to be legitimate samples from that distribution. We recommend analysing data in several different ways if possible. If they all agree, then one might choose the analysis which best matches the theory (the sampling distribution best reflecting our knowledge of the underlying process) of how the data arose. When methods disagree, one must then determine why they differ and make an educated choice on which to use. For example, if assuming a normal distribution results in different confidence limits around means than those obtained using robust statistics, it is likely that there is serious contamination from outliers that is ignored by assuming a normal distribution. A recent reference on robust statistics is Maronna *et al.* (2006), while the classic one is Huber (1981).

5.6. Resampling techniques

Statistical methodology has benefited enormously from fast and ubiquitous computing power, with the two largest beneficiaries being methods that rely on numerical techniques, such as estimating parameters in GLMMs, and methods that rely on sampling, either from known distributions (such as most Bayesian methods, often called "Monte-Carlo" methods) or from the data (resampling or "bootstrapping"). Resampling techniques are essentially non-parametric, the only assumption is that the data are representative of the population you want to make inferences from. The data set must also be large enough to resample from, following the rules stated earlier for sample sizes for parametric models, (*i.e.* at least 10 observations per "parameter", so a difference between two medians would require at least 20 observations).

As a simple example, if we want to estimate a 95% confidence interval around a median, based on 30 observations, we can draw 100 random resampled data sets (*with replacement*) from the original data set, each of size 30, calculate the median for each of these resampled data sets, and rank those values. The 95% confidence interval is then the interval from the 5th to the 95th calculated median. Even though the original data set and the resampled data sets are the same size (*n* = 30), they are likely not identical because we are sampling with replacement, meaning that there will be duplicates (or even triplicates) of some of the original values in each resampled data set, and others will be missing.

Resampling can be used for statistical testing in a similar way. For example, if we want to know if the difference in medians between two data sets (each of size 30) is significant at α = 0.05, we could use the following approach. Take a random sample (with replacement) of size 30 from data set 1 and calculate its median, do the same for data set 2. Subtract the sample 2 median from the sample 1 median and store the value. Repeat this until you have 1,000 differences. Rank the differences. If the interval between the 50th and 950th difference does not contain zero, the difference in medians is statistically significant.

This general method can be applied to many common statistical problems, and can be shown to have good power (often better than a parametric technique if an underlying assumption of the parametric technique is even slightly violated). It can be used for both quantitative and qualitative (e.g. categorical) data, for example for testing the robustness of phylogenetic trees derived from nucleotide or amino acid sequence alignments, and is also useful as an independent method to check the results of statistical testing using other techniques. It does require either some programming skills or use of a statistical package that implements resampling techniques.

If one writes a program, three parts are required. The first is used for creating a sample by extracting objects from the original data set, based on their position in the data file, using a random number generator. As a simple example, if there are five values, a random number generator (sampling with replacement) might select the values in positions (4, 3, 3, 2, 4). Note that some positions are repeated, others are missing. That is fine because this process will be repeated 10,000 times, and, on average, all data values will have equal representation. The second part is used for calculating the parameters of interest, for example, the median, and is also run 10,000 times. More complicated statistics take longer, and that will affect how long the program takes to complete. The third part stores the results of the second part, and may be a vector of length 10,000 (or a matrix with 10,000 rows, if several statistics are calculated from each resampled data set). Finally, summary statistics or confidence intervals are created, based on the third part. For example, if medians were calculated, one could calculate 90%, 95%, and 99% confidence intervals after ranking the medians and selected appropriate endpoints of the intervals. In general, 10,000 resampled data sets are considered to be a minimum to use for published results, though 500 are usually adequate for preliminary work (and that number is also useful for estimating how long it will take 10,000 to run).

All the major statistical software packages have resampling routines, and some rely almost exclusively on it (e.g. PASS, in the NCSS statistical software). We recommend the **boot** package in the R software, which is very flexible and allows one to estimate many of the quantities of interest for biologists (e.g. differences of means or medians, regression parameters). The classic book is Efron and Tibshirani (1993); Bradley Efron is the developer of the technique. A recent, less technical book is by Good (2013). A related technique is "jack-knifing", where one draws all possible subsamples *without replacement*, typically of size *n* − 1, where *n* is the original sample size.

6. Presentation and reporting of data

Presentation depends on the data collected and what the authors wants to emphasise. For example, to present the mean when one has done a non-parametric test is not meaningful, though a median is (consider using boxplots). The mean is a valid descriptive representation of the location parameter if the distribution is symmetric. The best way to summarise descriptively and represent graphically a given data set depends on both the empirical distribution of the data and the purpose of the statistics and graphs. There are excellent references on this topic such as those by Cleveland (1993) and Tufte (2001), whereas the classic book by Tukey (1977) has a decidedly statistical slant.

Standard error or Standard deviation - the former indicates uncertainty around a calculated mean; the latter is a measurement of the variability of the observed data around the mean. We believe that the standard deviation is the better metric to convey characteristics of the data because the standard error, which is also a function of

sample size, can be made arbitrarily small by including more observations.

Presentation of data might be overlaid with statistics one has applied, such as regression lines or mean separation letters. If data were transformed for the analysis, data on the original scale should be presented, but any means fit from a statistical model back-transformed to the original scale (even though this will create curves in a "straight" line model, like a linear regression). Back-transformed confidence intervals on means should replace standard error bars.

7. Which software to use for statistical analyses?

Statistical programmes, such as the freeware R and its packages, as well as other packages such as Minitab, SPSS, and SAS, can handle the analyses described in this paper. There are several sites comparing the different packages: http://en.wikipedia.org/wiki/Comparison_of_statistical_packages, http://en.wikipedia.org/wiki/List_of_statistical_packages Although spreadsheet software has improved and many statistical tests are available, they often lack good diagnostics on the model fit and checks for the appropriateness of the statistical test.

8. Where to find help with statistics

A statistician, preferably with an understanding of biology, remains the best solution to get one's statistics right. Given the importance of sample size for analyses, it is important to contact one as early as the design stage of an experiment or survey. If your university or institute does not offer the service of a statistician, there are freelance professionals as well as numerous forums on the internet where questions can be posted. Examples of such sites can be found on the support sites for R and commercial programmes. Most maths departments offer some kind of introduction to basic statistics.

9. Conclusion

Guidelines and the selection of the different methods presented are, at least partly, based on experience and we cannot cover all statistical methods available, for example we have not discussed resampling methods like jackknife in detail (for further reading see Good, 2006). More details on designing specific experiments and performing statistical analyses on the ensuing data can be found in respective chapters of the COLOSS *BEEBOOK* (e.g. in the toxicology chapter, Medrzycki *et al.*, 2013).

Experimenters need to use statistical tests to take (or to help take) a decision. A statistical analysis can be conducted only if its assumptions are met, which largely depends on how the experiment was designed, defined during the drafting of the study protocol. Without some effort at the *a priori* conception stage and input from those knowledgeable in statistics and/or experimental design, the resulting analyses are frequently poor and the conclusions can be biased or flat-out wrong. Why spend a year or more collecting data and then realise that, due to poor design, it is not suitable for its original purpose: to test the hypotheses of interest. The most important point to understand about statistics is that one should think about the statistical analysis before collecting data or conducting the experiment.

10. Acknowledgements

We are more than grateful to Ingemar Fries of the Swedish University of Agricultural Sciences for comments on and contributions to earlier versions of the chapter. We thank K L Crous and A A Yusuf for comments on an earlier version of the manuscript and we also thank Werner Luginbühl for a very thoughtful and thorough review of the original submission. The University of Pretoria, the National Research Foundation of South Africa and the Department of Science and Technology of South Africa (CWWP) granted financial support. The COLOSS (Prevention of honey bee COlony LOSSes) network aims to explain and prevent massive honey bee colony losses. It was funded through the COST Action FA0803. COST (European Cooperation in Science and Technology) is a unique means for European researchers to jointly develop their own ideas and new initiatives across all scientific disciplines through trans-European networking of nationally funded research activities. Based on a pan-European intergovernmental framework for cooperation in science and technology, COST has contributed since its creation more than 40 years ago to closing the gap between science, policy makers and society throughout Europe and beyond. COST is supported by the EU Seventh Framework Programme for research, technological development and demonstration activities (Official Journal L 412, 30 December 2006). The European Science Foundation as implementing agent of COST provides the COST Office through an EC Grant Agreement. The Council of the European Union provides the COST Secretariat. The COLOSS network is now supported by the Ricola Foundation - Nature & Culture.

11. References

AMDAM, G V; OMHOLT, S W (2002) The regulatory anatomy of honey bee lifespan. *Journal of Theoretical Biology* 216: 209-228.

BAILEY, L; BALL, B V; PERRY, J N (1981) The prevalence of viruses of honey bees in Britain. *Annals of Applied Biology* 97: 109-118. http://dx.doi.org/10.1111/J.1744-7348.1981.Tb02999.X

BAILEY, L, BALL; B V (1991) *Honey bee pathology*. Academic Press; London, UK.

BENJAMINI, Y; HOCHBERG, Y (1995) Controlling the false discovery rate: a practical and powerful approach to multiple testing. *Journal of the Royal Statistical Society. Series B (Methodological)* 57: 289-300. http://dx.doi.org/10.2307/2346101

BICKEL, P J; DOKSUM, K A (1981) An analysis of transformations revisited. *Journal of the American Statistics Association* 76:

BOLKER, B M; BROOKS, M E; CLARK, C J; GEANGE, S W; POULSEN, J R; STEVENS, M H H; WHITE, J S S (2009) Generalized linear mixed models: a practical guide for ecology and evolution. *Trends in Ecology & Evolution* 24: 127-135. http://dx.doi.org/10.1016/J.Tree.2008.10.008

BOX, G E P; COX, D R (1964) An analysis of transformations. *Journal of the Royal Statistical Society B* 26: 211-252.

BRUNETTO, M R; COLOMBATTO, P; BONINO, F (2009) Bio-mathematical models of viral dynamics to tailor antiviral therapy in chronic viral hepatitis. *World Journal of Gastroenterology* 15: 531-537. http://dx.doi.org/10.3748/Wjg.15.531

CARRECK, N L; ANDREE, M; BRENT, C S; COX-FOSTER, D; DADE, H A; ELLIS, J D; HATJINA, F; VANENGELSDORP, D (2013) Standard methods for *Apis mellifera* anatomy and dissection. In *V Dietemann; J D Ellis; P Neumann (Eds) The COLOSS BEEBOOK, Volume I: standard methods for* Apis mellifera *research. Journal of Apicultural Research* 52(4): http://dx.doi.org/10.3896/IBRA.1.52.4.03

CASELLA, G (2008) *Statistical design*. Springer; Berlin, Germany.

CLEVELAND, W S (1993) *Visualizing data*. Hobart Press; Summit, USA.

COHEN, J (1988) *Statistical Power Analysis for the behavioral sciences (2nd Ed.)*. Lawrence Erlbaum Associates.

COLLETT, D (2003) *Modelling survival data in medical research (2nd Ed.)*. Chapman & Hall/CRC.

COLTON, T (1974) *Statistics in medicine*. Little, Brown and Co.; Boston, USA.

CRAWLEY, M J (2005) *Statistics: an introduction using R*. Wiley & Sons; Chichester, UK.

CRAWLEY, M J (2013) *The R Book (2nd Ed.)*. Wiley & Sons; Chichester, UK.

CREWE, R M (1982) Compositional variability: The key to the social signals produced by honey bee mandibular glands. *The Biology of Social Insects*: 318-322.

CREWE, R M; MORITZ, R F A; LATTORFF, H M (2004) Trapping pheromonal components with silicone rubber tubes: fatty acid secretions in honey bees (*Apis mellifera*). *Chemoecology* 14: 77-79.

DARCHEN, R (1956) La construction sociale chez *Apis mellifica*. *Insectes Sociaux* 3: 293-301. http://dx.doi.org/10.1007/bf02224312

DARCHEN, R (1957) La reine d'*Apis mellifica* les ouvrières pondeuses et les constructions cirières. *Insectes Sociaux* 4: 321-325. http://dx.doi.org/10.1007/bf02224152

DOULL, K M; CELLIER, K M (1961) A survey of incidence of Nosema disease (*Nosema apis* Zander) of the honey bee in South Australia. *Journal of Insect Pathology* 3: 280.

EFRON, B; TIBSHIRANI, R J (1993) *An introduction to the bootstrap.* Chapman & Hall; London, UK.

FAUL, F; ERDFELDER, E; LANG, A-G; BUCHNER, A (2007) G*Power 3: A flexible statistical power analysis program for the social, behavioral, and biomedical sciences. *Behavior Research Methods* 39: 175-191.

FIELD, A; MILES, J; FIELD, Z (2012) *Discovering statistics using R.* SAGE Publications Ltd.; London, UK.

FREE, J B (1960) The distribution of bees in a honey-bee (*Apis mellifera.* L) colony. *Proceedings of the Royal Entomological Society of London* A35: 141-141.

FRIES, I; EKBOHM, G; VILLUMSTAD, E (1984) *Nosema apis*, sampling techniques and honey yield. *Journal of Apicultural Research* 23: 102-105.

FRIES, I; LINDSTROM, A; KORPELA, S (2006) Vertical transmission of American foulbrood (*Paenibacillus larvae*) in honey bees (*Apis mellifera*). *Veterinary Microbiology* 114: 269-274. http://dx.doi.org/10.1016/J.Vetmic.2005.11.068

FROST, E H; SHUTLER, D; HILLIER, N K (2012) The proboscis extension reflex to evaluate learning and memory in honey bees (*Apis mellifera*): some caveats. *Naturwissenschaften* 99: 677-686. http://dx.doi.org/10.1007/S00114-012-0955-8

GAUTHIER, L; TENTCHEVA, D; TOURNAIRE, M; DAINAT, B; COUSSERANS, F; COLIN, M E; BERGOIN, M (2007) Viral load estimation in asymptomatic honey bee colonies using the quantitative RT-PCR technique. *Apidologie* 38: 426-U7. http://dx.doi.org/10.1051/Apido:2007026

GOOD, P I (2006) *Resampling methods (3rd Ed.).* Springer; Berlin, Germany.

GOOD, P I (2013) *Introduction to statistics through resampling methods and R (2nd Ed.).* John Wiley & Sons, Inc.; Hoboken, USA.

GREEN, S B (1991) How many subjects does it take to do a regression -analysis. *Multivariate Behavioral Research* 26: 499-510. ttp://dx.doi.org/10.1207/S15327906mbr2603_7

HENDRIKSMA, H P; HARTEL, S; STEFFAN-DEWENTER, I (2011) Honey bee risk assessment: new approaches for *in vitro* larvae rearing and data analyses. *Methods in Ecology and Evolution* 2: 509-517. http://dx.doi.org/10.1111/J.2041-210x.2011.00099.X

HEPBURN, H R (1986) *Honey bees and wax: an experimental natural history.* Springer Verlag; Berlin, Germany.

HESS, G (1942) Über den Einfluß der Weisellosigkeit und des Fruchtbarkeitsvitamins E auf die Ovarien der Bienenarbeiterin Ein Beitrag zur Frage der Regulationen im Bienenstaat. *Beihefte zur Schweizerischen Bienen-Zeitung* 2: 33-111.

HIGES, M; MARTIN-HERNANDEZ, R; BOTIAS, C; BAILON, E G; GONZALEZ-PORTO, A V; BARRIOS, L; DEL NOZAL, M J; BERNAL, J L; JIMENEZ, J J; PALENCIA, P G; MEANA, A (2008) How natural infection by *Nosema ceranae* causes honey bee colony collapse. *Environmental Microbiology* 10: 2659-2669. http://dx.doi.org/10.1111/J.1462-2920.2008.01687.X

HUANG, Z-Y; ROBINSON, G E (1996) Regulation of honey bee division of labor by colony age demography. *Behavioral Ecology and Sociobiology* 39: 147-158.

HUBER, P J (1981) *Robust statistics.* Wiley; New York, USA.

HURLBERT, S H (1984) Pseudoreplication and the design of ecological field experiments. *Ecological Monographs* 54: 187-211. http://dx.doi.org/10.2307/1942661

JASSIM, O; HUANG, Z Y; ROBINSON, G E (2000) Juvenile hormone profiles of worker honey bees, *Apis mellifera*, during normal and accelerated behavioural development. *Journal of Insect Physiology* 46: 243-249.

JOHNSON, B R (2005) Limited flexibility in the temporal caste system of the honey bee. *Behavioral Ecology and Sociobiology* 58: 219-226. http://dx.doi.org/10.1007/S00265-005-0949-Z

KAEKER, R; JONES, A (2003) On use of Bayesian statistics to make the guide to the expression of uncertainty in measurement consistent. *Metrologia* 40: 235-248.

KELLNER, N (1981) Studie van de levenscyclus van *Nosema apis* Zander in de honingbij (*Apis mellifera*). PhD thesis, Rijksuniversiteit Gent, Belgium.

KÖHLER, A; NICOLSON, S W; PIRK, C W W (2013) A new design for honey bee hoarding cages for laboratory experiments. *Journal of Apicultural Research* 52(2): 12-14. http://dx.doi.org/10.3896/IBRA.1.52.2.03

LEPS, J; SMILAUER, P (2003) *Multivariate analysis of ecological data using CANOCO.* Cambridge University Press; Cambridge, UK.

LINDAUER, M (1952) Ein Beitrag zur Frage der Arbeitsteilung im Bienenstaat. *Zeitschrift für vergleichende Physiologie* 34: 299-345.

LINDAUER, M (1953) Division of labour in the honey bee colony. *Bee World* 34: 63-90.

MADDEN, L V; HUGHES, G (1999) An effective sample size for predicting plant disease incidence in a spatial hierarchy. *Phytopathology* 89: 770-781. http://dx.doi.org/10.1094/Phyto.1999.89.9.770

MANLY, B F J (1997) *Randomization, bootstrap and Monte Carlo methods in biology*. Chapman and Hall; London, UK.

MARONNA, R A; MARTIN, R D; YOHAI, V J (2006) *Robust statistics: theory and Methods*. Wiley; New York, USA.

MCCULLAGH, P; NELDER, J (1989) *Generalized Linear Models*. Chapman & Hall/CRC Press.

MEDRZYCKI, P; GIFFARD, H; AUPINEL, P; BELZUNCES, L P; CHAUZAT, M-P; CLAßEN, C; COLIN, M E; DUPONT, T; GIROLAMI, V; JOHNSON, R; LECONTE, Y; LÜCKMANN, J; MARZARO, M; PISTORIUS, J; PORRINI, C; SCHUR, A; SGOLASTRA, F; SIMON DELSO, N; VAN DER STEEN, J J F; WALLNER, K; ALAUX, C; BIRON, D G; BLOT, N; BOGO, G; BRUNET, J-L; DELBAC, F; DIOGON, M; EL ALAOUI, H; PROVOST, B; TOSI, S; VIDAU, C (2013) Standard methods for toxicology research in *Apis mellifera*. In *V Dietemann; J D Ellis; P Neumann (Eds) The COLOSS BEEBOOK, Volume I: standard methods for* Apis mellifera *research. Journal of Apicultural Research* 52(4) http://dx.doi.org/10.3896/IBRA.1.52.4.14

MILES, J; SHEVLIN, M (2001) *Applying regression and correlation: a guide for students and researchers*. SAGE Publications Ltd.; London, UK.

MORITZ, R; LATTORFF, H; CREWE, R (2004) Honey bee workers (*Apis mellifera capensis*) compete for producing queen-like pheromone signals. *Proceedings of the Royal Society B: Biological Sciences* 271: S98-S100.

NAKAGAWA, S (2004) A farewell to Bonferroni: the problems of low statistical power and publication bias. *Behavioral Ecology* 15: 1044 -1045. http://dx.doi.org/10.1093/Beheco/Arh107

NASH, D R; ALS, T D; MAILE, R; JONES, G R; BOOMSMA, J J (2008) A mosaic of chemical coevolution in a large blue butterfly. *Science* 319: 88-90. http://dx.doi.org/10.1126/Science.1149180

PIRK, C W W; BOODHOO, C; HUMAN, H; NICOLSON, S W (2010) The importance of protein type and protein to carbohydrate ratio for survival and ovarian activation of caged honey bees (*Apis mellifera scutellata*). *Apidologie* 41: 62-72. http://dx.doi.org/10.1051/Apido/2009055

PIRK, C W W; SOLE, C L; CREWE, R M (2011) Pheromones. In *H R HEPBURN; S E RADLOFF (Eds). Honey bees of Asia*. Springer; Berlin Heidelberg, Germany. pp 207-214.

REICZIGEL, J (2003) Confidence intervals for the binomial parameter: some new considerations. *Statistics in Medicine* 22: 611-621. http://dx.doi.org/10.1002/Sim.1320

RIBBANDS, C R (1952) Division of labour in the honey bee community. *Proceedings of the Royal Society B: Biological Sciences* 140: 32-43.

ROTHMAN, K J (1990) No adjustments are needed for multiple comparisons. *Epidemiology* 1: 43-46.

RUNCKEL, C; FLENNIKEN, M L; ENGEL, J C; RUBY, J G; GANEM, D; ANDINO, R; DERISI, J L (2011) Temporal analysis of the honey bee microbiome reveals four novel viruses and seasonal prevalence of known viruses, *Nosema*, and *Crithidia. PLoS ONE* 6: http://dx.doi.org/10.1371/journal.pone.0020656

SCHÄFER, M O; DIETEMANN, V; PIRK, C W W; NEUMANN, P; CREWE, R M; HEPBURN, H R; TAUTZ, J; CRAILSHEIM, K (2006) Individual versus social pathway to honey bee worker reproduction (*Apis mellifera*): pollen or jelly as protein source for oogenesis? *Journal of Comparative Physiology* A192: 761-768.

SCHEINER, R; BARNERT, M; ERBER, J (2003) Variation in water and sucrose responsiveness during the foraging season affects proboscis extension learning in honey bees. *Apidologie* 34: 67-72. http://dx.doi.org/10.1051/Apido:2002050

SEELEY, T D (1985) *Honey bee ecology: a study of adaptation in social life*. Princeton University Press; Princeton, USA.

SOKAL, R R; ROHLF, F J (1995) *Biometry: the principles and practice of statistics in biological research (1st Ed.)*. W H Freeman and Co.; New York, USA.

SOKAL, R R; ROHLF, F J (2012) *Biometry: the principles and practice of statistics in biological research (4th Ed.)*. W H Freeman and Co.; New York, USA.

STEPHENS, M A (1974) EDF statistics for goodness of fit and some comparisons. *Journal of the American Statistical Association* 69: 730-737. http://dx.doi.org/10.1080/01621459.1974.10480196

STROUP, W W (2013) *Generalized Linear Mixed Models: modern concepts, methods and applications*. CRC Press; Boca Raton, Florida, USA.

TENTCHEVA, D; GAUTHIER, L; ZAPPULLA, N; DAINAT, B; COUSSERANS, F; COLIN, M E; BERGOIN, M (2004) Prevalence and seasonal variations of six bee viruses in *Apis mellifera* L. and *Varroa destructor* mite populations in France. *Applied and environmental microbiology* 70: 7185-7191. http://dx.doi.org/10.1128/Aem.70.12.7185-7191.2004

TUFTE, E R (2001) *The visual display of quantitative information*. Graphics Press; Cheshire, USA.

TUKEY, J W (1977) *Exploratory data analysis*. Addison-Wesley Publishing Company; Reading, MA, USA.

VAN DER STEEN, J J M; CORNELISSEN, B; DONDERS, J; BLACQUIÈRE, T; VAN DOOREMALEN, C (2012) How honey bees of successive age classes are distributed over a one storey, ten frame hive. *Journal of Apicultural Research* 51(2): 174-178. http://dx.doi.org/10.3896/IBRA.1.51.2.05

VAN DOOREMALEN, C; ELLERS, J (2010) A moderate change in temperature induces changes in fatty acid composition of storage and membrane lipids in a soil arthropod. *Journal of Insect Physiology* 56: 178-184. http://dx.doi.org/10.1016/J.Jinsphys.2009.10.002

VAN DOOREMALEN, C; SURING, W; ELLERS, J (2011) Fatty acid composition and extreme temperature tolerance following exposure to fluctuating temperatures in a soil arthropod. *Journal of Insect Physiology* 57: 1267-1273. http://dx.doi.org/10.1016/j.jinsphys.2011.05.017

VIM (2008) International vocabulary of metrology - Basic and general concepts and associated terms (VIM). http://www.iso.org/sites/JCGM/VIM/JCGM_200e.html

WASSERMAN, L (2006) *All of nonparametric statistics.* Springer; Berlin, Germany.

WILLIAMS, G R; ALAUX, C; COSTA, C; CSÁKI, T; DOUBLET, V; EISENHARDT, D; FRIES, I; KUHN, R; MCMAHON, D P; MEDRZYCKI, P; MURRAY, T E; NATSOPOULOU, M E; NEUMANN, P; OLIVER, R; PAXTON, R J; PERNAL, S F; SHUTLER, D; TANNER, G; VAN DER STEEN, J J M; BRODSCHNEIDER, R (2013) Standard methods for maintaining adult *Apis mellifera* in cages under *in vitro* laboratory conditions. In *V Dietemann; J D Ellis; P Neumann (Eds) The COLOSS* BEEBOOK, *Volume I: standard methods for* Apis mellifera *research. Journal of Apicultural Research* 52(1): http://dx.doi.org/10.3896/IBRA.1.52.1.04

WOLFE, R; CARLIN, J B (1999) Sample-size calculation for a log-transformed outcome measure. *Controlled Clinical Trials* 20: 547-554. http://dx.doi.org/10.1016/S0197-2456(99)00032-X

ZUUR, A L E N; WALKER, N; SAVELIEV, A A; SMITH, G M (2009) *Mixed effects models and extensions in ecology with R. Vol 1.* Springer; Berlin, Germany.

Journal of Apicultural Research 52(4): (2013)
DOI 10.3896/IBRA.1.52.4.14

© IBRA 2013

REVIEW ARTICLE

Standard methods for toxicology research in

Apis mellifera

INTERNATIONAL BEE
RESEARCH ASSOCIATION

Piotr Medrzycki[1*], Hervé Giffard[2], Pierrick Aupinel[3], Luc P Belzunces[4], Marie-Pierre Chauzat[5], Christian Claßen[6], Marc E Colin[7], Thierry Dupont[8], Vincenzo Girolami[9], Reed Johnson[10], Yves Le Conte[4], Johannes Lückmann[6], Matteo Marzaro[9], Jens Pistorius[11], Claudio Porrini[12], Andrea Schur[13], Fabio Sgolastra[12], Noa Simon Delso[14], Jozef J M van der Steen[15], Klaus Wallner[16], Cédric Alaux[4], David G Biron[17], Nicolas Blot[17], Gherardo Bogo[1], Jean-Luc Brunet[4], Frédéric Delbac[17], Marie Diogon[17], Hicham El Alaoui[17], Bertille Provost[7], Simone Tosi[1] and Cyril Vidau[18]

[1]Agricultural Research Council, Honey Bee and Silkworm Research Unit, Bologna, Italy.
[2]Testapi, Gennes, France.
[3]INRA, UE1255 Entomologie, Surgères, France.
[4]INRA, UR 406 Abeilles & Environnement, Avignon, France.
[5]ANSES, Unit of Honey bee Pathology, Sophia-Antipolis, France.
[6]RIFCON GmbH, Hirschberg, Germany.
[7]Montpellier SupAgro, USAE, Montpellier, France.
[8]Qualilab, Olivet, France.
[9]Department of Agronomy Food Natural Resources Animals and Environment, University of Padua, Legnaro (PD), Italy.
[10]Department of Entomology, The Ohio State University - Ohio Agriculture Research and Development Centre, Wooster, Ohio, USA.
[11]Julius Kühn Institut, Institute for Plant Protection in Field Crops and Grassland, Braunschweig, Germany.
[12]Department of Agroenvironmental Sciences and Technologies – Entomology, University of Bologna, Bologna, Italy.
[13]Bee Department, Eurofins Agroscience Services EcoChem GmbH, Niefern-Öschelbronn, Germany.
[14]Centre Apicole de Recherche et Information (CARI), Louvain La Neuve, Belgium.
[15]Wageningen University, Biointeractions and Plant Health, Wageningen, The Netherlands.
[16]Apicultural State Institute, University Hohenheim, Stuttgart, Germany.
[17]Laboratoire Microorganismes: Génome et Environnement, CNRS UMR 6023, Université Blaise Pascal, Clermont-Ferrand, France.
[18]Laboratoire Venins et Activités Biologiques, EA 4357, PRES-Université de Toulouse, Centre de formation et de recherche Jean-François Champollion, Albi, France.

All authors except those listed first and second are alphabetical.

Received 4 July 2012, accepted subject to revision 7 December 2012, accepted for publication 17 July 2013.

*Corresponding author: Email: piotr.medrzycki@entecra.it

Summary

Modern agriculture often involves the use of pesticides to protect crops. These substances are harmful to target organisms (pests and pathogens). Nevertheless, they can also damage non-target animals, such as pollinators and entomophagous arthropods. It is obvious that the undesirable side effects of pesticides on the environment should be reduced to a minimum. Western honey bees (*Apis mellifera*) are very important organisms from an agricultural perspective and are vulnerable to pesticide-induced impacts. They contribute actively to the pollination of cultivated crops and wild vegetation, making food production possible. Of course, since *Apis mellifera* occupies the same ecological niche as many other species of pollinators, the loss of honey bees caused by environmental pollutants suggests that other insects may experience a similar outcome. Because pesticides can harm honey bees and other pollinators, it is important to register pesticides that are as selective as possible. In this manuscript, we describe a selection of methods used for studying pesticide toxicity/selectiveness towards *Apis mellifera*. These methods may be used in risk assessment schemes and in scientific research aimed to explain acute and chronic effects of any target compound on *Apis mellifera*.

Footnote: Please cite this paper as: MEDRZYCKI, P; GIFFARD, H; AUPINEL, P; BELZUNCES, L P; CHAUZAT, M-P; CLAßEN, C; COLIN, M E; DUPONT, T; GIROLAMI, V; JOHNSON, R; LECONTE, Y; LÜCKMANN, J; MARZARO, M; PISTORIUS, J; PORRINI, C; SCHUR, A; SGOLASTRA, F; SIMON DELSO, N; VAN DER STEEN, J J M; WALLNER, K; ALAUX, C; BIRON, D G; BLOT, N; BOGO, G; BRUNET, J-L; DELBAC, F; DIOGON, M; EL ALAOUI, H; PROVOST, B; TOSI, S; VIDAU, C (2013) Standard methods for toxicology research in *Apis mellifera*. In *V Dietemann; J D Ellis; P Neumann (Eds) The COLOSS BEEBOOK, Volume I: standard methods for* Apis mellifera *research. Journal of Apicultural Research* 52(4): http://dx.doi.org/10.3896/IBRA.1.52.4.14

Métodos estándar para la investigación toxicológica en *Apis mellifera*

Resumen

La agricultura moderna a menudo implica el uso de plaguicidas para proteger los cultivos. Estas sustancias son dañinas para los organismos objetivo (plagas y patógenos). Sin embargo, también pueden dañar a animales que no son objetivo, como artrópodos polinizadores y entomófagos. Obviamente los efectos secundarios indeseables de los plaguicidas sobre el medio ambiente deben ser reducidos al mínimo. Las abejas occidentales (*Apis mellifera*) son organismos muy importantes desde el punto de vista agrícola y son vulnerables a los impactos inducidos por los plaguicidas. Contribuyen activamente a la polinización de los cultivos y de la vegetación silvestre, lo que hace posible la producción de alimentos. Como *Apis mellifera* ocupa el mismo nicho ecológico que muchas otras especies de polinizadores, la pérdida de las abejas melíferas causada por contaminantes ambientales sugiere que otros insectos pueden experimentar un resultado similar. Ya que los plaguicidas pueden dañar a las abejas y a otros polinizadores, es importante registrar los plaguicidas que sean lo más selectivos posible. En este artículo, se describe una selección de los métodos utilizados para el estudio de la toxicidad y el efecto selectivo de los plaguicidas hacia *Apis mellifera*. Estos métodos se pueden utilizar en sistemas de evaluación de riesgo y en la investigación científica para explicar los efectos agudos y crónicos en *Apis mellifera* de cualquier compuesto objetivo.

西方蜜蜂毒理学研究的标准方法

摘要

现代农业经常会使用农药以保护作物。这些物质对害虫和病原菌等靶标生物有害。但是它们也会对诸如授粉昆虫和食虫节肢动物等非靶标动物带来危害。显然，农药对环境的不良副作用应该减少到最低。从农业的角度看，西方蜜蜂是一种重要生物，同时它也极易受到农药的影响。它们对种植的作物和野生植物的授粉发挥了积极的作用，使得粮食生产成为可能。当然，因为西方蜜蜂与很多其它授粉物种处于同一个生态位，由环境污染造成的蜜蜂损失表明其它昆虫可能也在遭遇同样的经历。由于农药会危害蜜蜂和其它授粉昆虫，因此注册登记选择性尽量强的农药显得尤为重要。本文我们选择描述了一些研究针对西方蜜蜂的农药毒性和选择性的方法。这些方法可以应用于风险评估方案和旨在评估某种化合物对于西方蜜蜂的急性和慢性作用的科学研究。

Keywords: COLOSS, *BEEBOOK*, *Apis mellifera*, honey bee, pesticide, exposure, residue, lethal, sublethal, field, semifield, laboratory

Table of Contents

Table of Contents Cont'd

Table of Contents Cont'd

1. Introduction

The presence of toxic substances in the environment may be an important factor contributing to the poor health of honey bee colonies globally. Agrochemicals are of particular interest because they often are accused of causing sublethal effects in individual bees and the bee colony, possibly even leading to the loss of entire colonies and even apiaries (Maini *et al.*, 2010; Desneux *et al.*, 2007).

Honey bees are excellent bioindicators of environmental pollution (Celli and Maccagnani, 2003). Thus, it is easy to imagine that wild pollinators (or other animals occupying the same ecological niche) present in polluted areas will suffer outcomes similar to those experienced by honey bees in the area. For this reason, the research community should work to limit the hazard of toxins to honey bees and, by doing this, will help to protect wild pollinators.

The risk assessment addressing the potential risk for pollinating insects from the use of Plant Protection Products (PPPs) is comprised by oral and contact LD_{50} (Lethal Dose that kills 50% of the population), toxicity exposure ratio (TER) and results of semi-field and field trials (e.g. direct or delayed bee mortality) highlighting the impact on brood development, foraging abilities, etc.

The registration of agrochemicals requires that specific toxicological tests be performed on honey bees, such as those required by the US Environmental Protection Agency (US EPA, 1996) and the European Organisation for Economic Co-operation and Development (OECD, 1998a; OECD, 1998b). These tests must follow specific protocols in order to (1) assess the level of selectiveness of the pesticide to honey bees and (2) satisfy a given country's pesticide regulatory requirements. They must be performed in Good Laboratory Practices (GLP).

The present chapter is not a proposal of guidelines but rather a compendium of methods for testing toxic effects of agrochemicals and other compounds on honey bees. These methods may be used in scientific studies and in official risk assessment schemes where appropriate or where consistent with a given government's requirements. To be used for the latter, the test should undergo regulatory testing and risk assessment systems in order to be properly validated. Nevertheless, both OECD 75 (tunnel test) and acute toxicity standards (OECD 213 and 214) have not been ring-tested despite that they are referenced by all OECD members as standard methodologies.

2. Common terms and abbreviations

Here are some abbreviations and definition of terms used in this manuscript listed in alphabetical order.

Acute oral toxicity: the adverse effects occurring within a maximum period of 96 h of an oral administration of a single dose of test substance.

Acute contact toxicity: the adverse effects occurring within a maximum period of 96 h of a topical application of a single dose of test substance.

AI: active ingredient - the substance composing a commercial formulation of a pesticide which has the desired effects on target organisms.

BFD: Brood area Fixing Day (see sections 5.2.2. and 5.2.3.)

CEB: Biological Tests Commission (Commission des Essais Biologiques), of the French Plant Protection Association (AFPP - Association Française de Protection des Plantes)

Dose (contact): the amount of test substance applied. Dose is expressed as mass (µg) of test substance per test animal (honey bee) or per mg body weight (in non-*Apis* bees).

Table 1. Possible honey bee behavioural effects due to exposure to pesticides in individual tests. Note: "freeze" and "paralysis" bees may be recorded as dead bees at a certain point and later as living bees.

Effect	Looks like	To be recorded as:
Dead	Immobile, no reaction to stimuli such as touching with forceps	Mortality, number of bees
No effect	Bees having normal behaviour	NE, number of bees observed
Freeze	Motionless bees caught in action and looking active such as attached to feeder, standing on the floor but actually completely inactive.	F, number of bees observed
Paralysis	Motionless on the floor of the test cage, responding to stimuli by moving leg, antenna etc.	P, number of bees observed
Spasm	Crawling bees, movement uncoordinated	S, number of bees observed

Dose (oral): the amount of test substance consumed. Dose is expressed as mass (µg) of test substance per test animal (honey bee), or per mg body weight (in non-*Apis* bees). In tests with bulk administration the real dose for each bee cannot be calculated as the bees are fed collectively, but an average dose can be estimated (total test substance consumed/number of test bees in one cage).

EEC: European Economic Community.

ED_{50}: median effective dose - term extending LD_{50} (see below in this section) to the effects other than mortality, e.g. behaviour (see Table 1 and Scheiner *et al.*, 2013)

EFSA: European Food Safety Authority - an agency of European Union (EU) risk assessment regarding food and feed safety. In close collaboration with national authorities and in open consultation with its stakeholders, EFSA provides independent scientific and clear communication on existing and emerging risks. (from: EFSA)

EPPO: European and Mediterranean Plant Protection Organisation - an intergovernmental organisation responsible for European cooperation in plant protection in the European and Mediterranean region. EPPO's objectives are to: (1) protect plants; (2) develop international strategies against the introduction and spread of dangerous pests; and (3) promote safe and effective control methods. EPPO has developed international standards and recommendations on phytosanitary measures, good plant protection practices and on the assessment of PPPs. (from: Wikipedia)

GAP: Good Agricultural Practices - specific methods which, when applied to agriculture, create food for consumers or further processing that is safe and wholesome. The Food and Agricultural Organization of the United Nations (FAO) uses GAP as a collection of principles to apply for on-farm production and post-production processes, resulting in safe and healthy food and non-food agricultural products, while taking into account economic, social and environmental sustainability.

GLP: Good Laboratory Practices - a set of principles that provides a framework within which laboratory studies are planned, performed, monitored, recorded, reported and archived. These studies are undertaken to generate data by which the hazards and risks to users, consumers and third parties, including the environment, can be assessed for pharmaceuticals (only preclinical studies), agrochemicals, cosmetics, food additives, feed additives and contaminants, novel foods, biocides, detergents etc. GLP helps assure regulatory authorities that the data submitted are a true reflection of the results obtained during the study and can therefore be relied upon when making risk/safety assessments. (from: Medicines and Healthcare products Regulatory Agency-UK)

HQ: Hazard Quotient. See section 8.4.2.1.

ICPPR: International Commission for Plant-Pollinator Relationships (formerly ICPBR: International Commission for Plant-Bee Relationships) - an international commission aimed to: (1) promote and coordinate research on the relationships between plants and pollinators of all types. (insect-pollinated plants, bee foraging behaviour, effects of pollinator visits on plants, management and protection of insect pollinators, bee collected materials from plants, products derived from plants and modified by bees); (2) organise meetings, colloquia or symposia related to the above topics and to publish and distribute the proceedings; and (3) collaborate closely with national and international institutions interested in the relationships between plants and bees, particularly those whose objectives are to expand scientific knowledge of animal and plant ecology and fauna protection.

IGR: Insect Growth Regulator - a chemical substance used as an insecticide that inhibits the life cycle of an insect. Normally the IGRs target juvenile harmful insect populations while cause less detrimental effects to beneficial insects.

LD_{50} / LC_{50}: median lethal dose / concentration - a statistically derived single dose /concentration of a substance that can cause death in 50% of animals when administered by the contact or oral route (according to the test), or combined (like in brood test). The LD_{50} value is expressed in μg of test substance per test animal (honey bee), or per mg body weight (in non-*Apis* bees). The LC_{50} value is expressed in concentration units, like mg of test substance / kg or L of the diet (pollen, syrup, honey). For pesticides, the test substance may be either an AI or a formulated product containing one or more than one AI. See section 8.2.1.2.

Moribund bee: a bee is considered moribund when it is not dead (it still moves) but is not able to deambulate actively and in an apparently "normal" way.

Mortality: an animal is recorded as dead when it is completely immobile upon prodding (Ffrench-Constant and Rouch, 1992).

NOAEC: Non Observable Adverse Effect Concentration. See section 8.4.3.

NOAEL/NOAED: Non Observable Adverse Effect Level/Dose (these are two synonyms). See section 8.4.3.

OECD: Organisation for Economic Co-operation and Development - an international economic organisation of 34 countries aimed to stimulate economic progress and world trade.

PER test: Proboscis Extension Reflex (see Scheiner *et al.*, 2013)

PPP: Plant Protection Product - active ingredient of a chemical or biological nature and preparation containing one or more active ingredients, or formulated preparation of microorganisms, put up in the form in which it is supplied to the user, intended to: (1) protect plants or plant products against all harmful organisms or prevent the action of such organisms; (2) influence the life processes of plants, other than as a nutrient, (e.g. growth regulators); (3) preserve plant products; (4) destroy undesired plants; or (5) destroy parts of plants, check or prevent undesired growth of plants. PPPs include: fungicides, bactericides, insecticides, acaricides, nematicides, rodenticides, herbicides, molluscicides, virucides, soil fumigants, insect attractants (e.g. pheromones used in control strategies), repellents (bird, wild life, rodent, insect repellents), stored product protectants, plant growth regulators, products to improve plant resistance to pests, products to inhibit germination, products to eliminate aquatic plants and algae, desiccants and defoliants to destroy parts of plants, products to assist wound healing, products to preserve plants or plant parts after harvest, timber preservatives (for fresh wood), additives to sprays to improve the action of any other PPP, additives to reduce the phytotoxicity of any other PPP. They do not include: fertilizers, timber preservatives (for dried wood). (from: EEC and EPPO)

RQ: Risk Quotient. See section 8.4.2.2.

SSST: Systemic product as Seed and Soil Treatment

Sub-lethal dose/concentration: the dose/concentration inducing no statistically significant mortality.

Sub-lethal effects: the effects of a factor (e.g. intoxication) which was administered at such a low level that the mortality was not significantly higher than in negative reference. These (generally negative) effects can have either behavioural (disorientation, problems with memory, etc.) or physiological nature (pharyngeal gland development impairment, thermoregulation problems, etc.).

TER: Toxicity Exposure Ratio – the ratio between a toxicity index (LD_{50}, LC_{50}, NOAEL...) and the predicted bee exposure in field conditions following a treatment.

3. Effects of toxic substances on adult worker bees: individual assays

This section describes methods for determining the toxicity of test compounds on adult bees in instances where the insects have no possibility of interacting with the hive. The bees are treated individually or within small experimental groups of individuals. The individual adult honey bee is the experimental unit.

3.1. Introduction

3.1.1. Definitions of poisoning and exposure

Poisoning is generally defined as injury or impairment of organ function or death, following exposure to any substance capable of producing adverse effects (Hodgson, 2004). The toxin can have local and/or systemic effects for varying periods of time. Depending on the severity of the effects, poisoning can be considered acute or chronic, both types with varying degrees of intensity. Often, acute poisoning leads to a rapid death.

Exposure is the encounter of the living organism with the poison. It may be characterised by many parameters: duration, number of replications, interval of time, routes of penetration into the body etc. The evaluation of exposure is the key point in experimental toxicology to provide valuable data.

3.1.2. Exploration of acute poisoning using the lethality criterion

Lethality is the most common experimental criterion in bee toxicology. In toxicological tests, an insect usually is considered dead when it exhibits "no movements after prodding" (see section 2). Using this criterion, investigators often use correlation metrics to link the lethality and dose of a toxic substance to a test subject. This assumes that the group of subjects to be tested are randomly selected from a population with a normal distribution (Gaussian) susceptibility to the toxic substance.

The cumulative distribution of the normal probability density is an increasing sigmoidal function (Wesstein, http://mathworld.wolfram.com). In matter of toxicology, the consequence is that the theoretical dose-cumulated lethality (% lethality) relation is a sigmoid ranging from 0% to 100% lethality. To transform the sigmoid into a straight line, Bliss (1934) proposed to use the logarithm of the doses in X axis and the probability units or probits in Y axis, the probit being the percentage of killed individuals converted following a special table. At the present time, a nonlinear regression analysis (Seber and Wild, 1989) can be more relevant and efficient, particularly when using statistical analysis software.

Laboratory experiments to establish the dose-lethality relation involve the administration of increasing doses to groups of selected subjects and the count of the two categories of subjects (dead or alive) after a specified time interval (Robertson *et al.*, 1984). Replications are needed to estimate the variability of each point representing the lethality associated to a particular dose.

From a theoretical point of view, by considering the cumulative distribution function (sigmoid) and its fluctuations due to the experimental replications, the less variable point is the inflection point, in other words the 50% lethality point and its associated dose, the 50% lethal dose or LD_{50} (Finney, 1971). On the contrary, the most variable ones are the extremes of the sigmoid graph. Consequently, when the estimation of the LD_{90} is required, e.g. efficiency of an insecticide against pests, special designs must be used to guarantee its precision (Robertson *et al.*, 1984). From an experimental point of view, the graph of the cumulative distribution function is not necessarily sigmoidal. For instance, after one imidacloprid contact exposure,

Suchail *et al.* (2000) evidenced that mortality rates were positively correlated with doses lower than 7 ng/bee and negatively with doses ranging from 7 to 15 ng/bee. In this situation, the calculation of any lethal dose with the log-probit model is incorrect.

When considering beneficial insect such as bees, the doses which cause slight mortalities (e.g. LD_5, LD_{10}, LD_{25}, etc.) are more pertinent, even if the variability of these LDs due to the toxin is difficult to distinguish from that of the natural mortality deduced from the control groups (Abbott, 1925). This variability is not to be rejected, because its very existence in experimental conditions suggests that the same variability also exists in field conditions.

The variability created by the replications refers mainly to the assumption concerning dealing with the random selection of the subjects and the normal distribution of population from which the subjects are chosen. The variability induced by the replications, meaning that the experiment is identically repeated several times, provides additional information on the reproducibility of the experiment.

For a set of given experimental conditions often recommended by precise guidelines, the LD_{50} should be as reproducible as possible (i.e. with a minimum variability.) Conversely, when the experimental conditions are modified, the LD_{50} correspondingly changes. Zbinden and Flury-Roversi (1981) noted that "every LD_{50} value must thus be regarded as a unique result of one particular biological experiment".

3.1.3. Factors influencing the dose-lethality relation

The scientific literature provides numerous examples of abiotic or biotic factors able to influence the dose-lethality relation.

3.1.3.1. Active ingredient and chemical formulation

An AI is a molecule able to bind on specific receptors of target organisms and produce adverse effects (Hodgson, 2004). Generally the chemical formula is only mentioned, without respect of the spatial arrangements. However, pyrethroids have isomers with varying levels of toxicity (Soderlund and Bloomquist, 1989). The same findings are true for some enantiomers, which have identical physical-chemical properties, but different biological activities (Konwick *et al.*, 2005).

To be used in laboratory conditions, the AI should be formulated as simply as possible, generally with one solvent. The commercial formulation spread in field conditions is more complex because surfactants, stabilizing agents, dispersants, sometimes synergists (Bernard and Philogène, 1993) are added after dilution of the AI. The commercial formulation is targeted at the improvement of the AI activity in time and/or in toxicity. Certain mixtures of AI have synergistic effects, i.e. insecticide and fungicide at sub-lethal doses (Colin and Belzunces, 1992). Some AIs are converted under biological or environmental conditions into products (metabolites) that are often higher in toxicity than the parent compound (Ramade, 1992; Nauen *et al.*, 1998; Suchail *et al.*, 2001; Tingle *et al.*, 2003).

3.1.3.2. Physical formulation

Generally, the higher the concentration of the AI in the formulation, the finer the required dispersion of the formulation in the field. Target application sites can be treated with the same dose of AI in different ways. For instance, the same dose can be sprayed (one method of product delivery) after final dilution in one hundred litres of water for a tractor-drawn device or in three litres of water or oil (ultra-low volumes) by aeroplane. Depending on the spraying method, the concentrations are not identical and the diameter of the droplets ranges between 1 micron to hundreds of microns. Consequently, the delivery method makes the penetration of the AI into the body of living organisms and its toxicological effects different (Luttrell, 1985). In the same order of size as for droplets (1 to 100 μm), plastic micro-capsules are conceived to extend the effective life of AI by releasing slowly through pores of the plastic walls (Stoner *et al.*, 1979). Nanoparticles are patented but their biological and environmental fates are poorly documented (Hodgson, 2004).

3.1.3.3. Temperature and hygrometry

For many substances, a linear relation links ambient temperature and LD_{50}s, negatively for DDT and most of pyrethroids (Ladas, 1972; Faucon *et al.*, 1985), positively for organophosphates and carbamates. Hygrometry is a factor of variation but its true impact on the impact of toxic substances is poorly documented.

3.1.3.4. Exposure features

First, dose and concentration are both to be considered. Local and general consequences on a living organism are quite different if the same dose is concentrated in one microlitre or if diluted in one millilitre. Depending on the toxin, repellent effects could occur at certain concentrations. Inversely, the forced contact with these concentrations would be able to induce local necrosis, with general consequences. Second, the route of administration is important to overall toxicity because it modulates the rapidity and the extension of the toxin in the living organism as well as the triggering of the detoxification pathways. Third, there is a higher probability of poisoning the longer the duration of the exposure to an AI (and/or its toxic metabolites) (Hodgson, 2004). Finally, the temporal features of the exposure often influence the severity of the poisoning. For example, Brunet *et al.* (2009) demonstrated that a dose applied daily for five days can induce higher mortality than a dose five times higher but administered one time.

3.1.3.5. Sex, age and caste

The sex, age, and caste of the insect can influence the impact of the toxin on the individual. For insects, males generally are more susceptible to insecticides than females and newly emerged adults often are more susceptible than older ones (Hodgson, 2004). After emergence, the age-susceptibility relation is variable depending on the target species and toxin. These factors are tightly linked in a social insect colony like the honey bee colony since one female is responsible for egg production while many others perform other activities (some depending on age). To a lesser extent, the same occurs with males; the young male bees remain in the hive while older ones fly outside (Tautz, 2009). The susceptibility to toxins increases with age when bees are nearly inactive gathered in a winter cluster, (Wahl and Ulm, 1983). Thus, it can be more pertinent to consider the social function of the individual than its sex and age when considering toxin impacts on the organism.

3.1.3.6. Weight and diet

The weight of an individual is an important factor influencing the LD_{50} and it is often negatively correlated with toxin susceptibility. Food deprivation can increase the susceptibility of individuals to toxins, with the protein content of the diet being of particular influence (Zbinden and Flury-Roversi, 1981). For honey bees, the amount and quality of pollen ingested in the first days of life can affect the pesticide susceptibility of young and older worker bees independently of their weight (Wahl and Ulm, 1983).

3.1.3.7. Health

The health of the individual or colony can influence the level of poisoning, especially regarding aggravation by or recovery from the toxin. For the honey bee, contact with the toxin can be more frequent during certain activities (for instance, foraging or nursing), thus requiring an acceptable state of health if the impacts of the toxins are to be overcome. The penetration kinetics of the toxin is made easier when injuries are present, for instance broken setae or loss of the epicuticular waxes. The integrity of the intestinal wall and the quantity/quality of the gut flora play an important role in the penetration of the toxin into the body via the digestive route. The fat bodies can trap lypophilic toxins and are important sites of detoxication. Furthermore, the pathogenic action of parasites or microbes influences the severity of poisoning if it modifies the penetration abilities of the toxin, the detoxication capacities, and/or the proteic and energetic metabolisms (Hodgson, 2004). In particular, the interactions between *Nosema* spp. and insecticides have been documented (Ladas, 1972; Alaux *et al.*, 2010; Vidau *et al.*, 2011). Conversely numerous pesticides can have extended general effects, for instance if they inhibit neurosecretion or cellular energy production, impairing the physiology of all the tissues. Bendahou *et al.* (1997), for example, showed that pyrethroids act by decreasing lysozyme concentration and phagocytosis capabilities, thus explaining the observed upsurge of Chronic Bee Paralysis Virus or other diseases in studied honey bees.

3.1.3.8. Genetics and resistance

At the individual level, subspecies and strains of honey bees are not equally susceptible to a given dose of AI (Ladas, 1972; Suchail *et al.*, 2000). Moreover a colony is not genetically homogeneous because of the coexistence of half-sister workers. Part of the tolerance to insecticides is due to genes encoding detoxifying enzymes. However there are significantly fewer genes encoding three major superfamilies of these enzymes in *Apis mellifera* than in other insect groups such as drosophila. Thus the honey bee would have great difficulty to metabolize certain pesticides (Claudianos *et al.*, 2006), making the resistance uncertain and non-uniform across races/subspecies.

3.1.3.9. Density of subjects

The dose-lethality relation typically is determined after submitting small groups of caged subjects to doses of a toxin. Sautet *et al.* (1968) indicated that the susceptibility to DDT increased positively with the number of caged mosquitoes, thus suggesting that individuals within a treatment group are not independent. For honey bees where social interactions occur, Dechaume-Moncharmont *et al.* (2003) concluded that "bees do not die independently of each other" for a continuous chronic exposure.

3.1.3.10. Conclusion

In conclusion, the variation between factors influencing the dose-lethality relation are so numerous, the difference between the lowest and the highest LD_{50} values can be more than a hundred of times (NRCC, 1981). Consequently, the concept of acute toxicity testing must not be restricted to one determination of the LD_{50} but extended to many others, reflecting the biotic and abiotic factors of toxicity variation. In the preliminary evaluation of a compound's toxicity, it is important to establish the dose-lethality relation for the parent molecule and its by-products at three temperatures: internal body temperature for flying (37°C), low wintering bee temperature (12°C, see Stabentheiner *et al.*, 2003), and one intermediate.

Insect death is not always the best determinant of acute toxicity because the moment of insect death often is imprecise, for example when confused with a severe knock-down that fails to result in death (Moréteau, 1991). For insects, the evaluation of acute toxicity would be more accurate if based on the apparition and intensity of severe clinical signs such as intense trembling, paralysis, feeding or warming inabilities, etc. (Vandame and Belzunces, 1998).

3.1.4. Exploration of sub-lethal poisoning

The link between the dose-lethality relation in laboratory conditions and the acute toxicity in field conditions is neither direct nor simple, nor can it be blindly guided by the "useful rule of thumb way of determining the anticipated toxicity hazards of a pesticide to honey bees in the field" (Atkins *et al.*, 1973). For example, this rule stipulates that "since the LD_{50} of parathion is 0.175 µg/bee, we would expect that 0.17 lb/acre of parathion would kill 50% of the bees foraging in a treated field crop at the time of the treatment or shortly afterwards", without mentioning the possibility of sub-lethal toxicity. So the following question remains: can the sublethal toxicology be deduced from the dose-lethality relation?

In the log-probit model itself, the extreme values of the dose-% lethality relation cannot be derived from the LD_{50} and the slope of the regression line (Robertson *et al.*, 1984). Moreover the log-probit model is not necessarily the most adapted model for the dose-lethality relation. For the lowest LD values, the log-probit model is questioned by Calabrese (2005), who mentioned the frequency of the hormesis phenomenon, that is "a modest treatment-related response occur(ing) immediately below the No Observable Effect Level". Consequently, special designs are needed to estimate the low doses effects.

In this complex domain, mortality is not the best criterion for determining toxic effects. During its adult life, the worker bee must be physically able to fly and has to use functional short and long term memories to communicate, care the larvae, form the winter cluster and perform many other social functions. Thus a panel of markers of behavioural, physiological, and molecular origins can provide substantial information in matter of sub-acute poisoning (Desneux *et al.*, 2007). Each sublethal individual assay is important so one can know if the adult bees are capable of accomplishing one of the activities essential for perpetuating the bee colony and maintaining its ecological role (Brittain and Potts, 2011).

3.2. Laboratory methods for testing toxicity of chemical substances on adult bees

3.2.1. Oral application

This method was never ring-tested but was several times reviewed by OECD, EPPO and CEB. It is considered validated.

3.2.1.1. Introduction

The determination of acute oral toxicity on honey bees is required for the assessment and evaluation of chemicals prior to their registration as pesticides (Regulation EC No 1107/2009 of the European Parliament and of the Council of 21 October 2009). In this way, the acute oral toxicity test is conducted to determine the toxicity of all types of compounds to bees (pesticides, specifically, are tested as AIs or as formulated products). The methodology outlined in this section is a general approach of the laboratory test with oral applications and does not present all the details of the referenced guidelines.

Usually an oral exposure study is intended to determine the LD_{50} (see section 8.2.1.2.) and the results are used to define the need for further evaluation. Although the LD_{50} is a common aim of these studies, oral exposure tests can be used to determine NOAEL (see section 8.4.3.). When the LD_{50} cannot be determined because a given compound has a low toxicity, a limit test may be performed in order to demonstrate that the LD_{50} is greater than the standard value of 100 µg of AI/bee.

Data from oral LD_{50} calculations can be used to generate HQ for each compound of interest (see section 8.4.2.1). The LD_{50} calculation provides a raw value only. This result has to be related to the exposure of honey bees in field conditions.

- When the HQ < 50, the product can be considered of low acute risk to adult worker honey bees when ingested. The HQ does not predict product toxicity to brood or the occurrence of any sub-lethal effects on adults or brood.
- When the HQ > 50, more tests are required in semi-field or field conditions for a better evaluation of impact (cf. European scheme for the assessment of impact of PPPs - Guidelines commonly used refer to EPPO (2010b), OECD (1998a) and French CEB (2011). All are similar with main differences occurring on number of the number of replicates.

3.2.1.2. General principle

- Worker honey bees that are all aged or young emerged honey bees that are 1 to 2 days old are kept in laboratory boxes and fed with a sucrose solution for one day.
- Following this, they are exposed to a range of doses of the test substance dispersed in the sucrose solution.
- Usually mortality is recorded up to 48 h and values are used to calculate the LD_{50} with a regression line (see section 8.2.1.2.). Mortality can be recorded after 4 hours to look at an eventual acute effect, and is then recorded at 24 and 48 hours and compared with control values for assessment. When mortality continues to increase, the test can be extended to 72 or 96 hours. In the case of chronic oral toxicity, data are recorded up to 10 days of daily exposure with low doses.

3.2.1.3. Experimental conditions and modalities
3.2.1.3.1. Establishing the hoarding cages

1. Adult honey bees should be collected per Williams *et al.*, 2013. They should be from a single colony in order to provide a similar status regarding origin and health.
2. Upon collection, the adult bees should be kept in hoarding cages that have a syrup feeder. For convenience, plastic containers are recommended as they can be discarded after use in order to avoid contaminations. Glass, wooden or iron boxes that have been used before are not recommended for reuse unless the process of cleaning and sterilization is validated under Good Laboratory Practices. The boxes can be created per Williams *et al.*, 2013.
3. The cages should be individually identified and placed in incubators or in a dedicated controlled room.
4. The cages should be stored at 25 ± 2°C and > 50% rH.
5. Each cage should contain at least 10 bees (EPPO, 2010a; OECD, 1998a). The CEB (2011) recommends 20 bees and up to 50 bees in some specific chronic tests.

3.2.1.3.2. Identifying and replicating the treatment modalities

The number of modalities is defined by the objectives of the study and includes at least the following groups:

1. A control - untreated sugar water, often containing the solvent used to dissolve the test compound in the treatment doses. The control provides the evaluation standard in the assessments.
2. The toxic reference - This reference verifies bee sensitivity to toxic compounds. The toxic standard validates the test. Dimethoate is the main toxic standard used and provides a high subsequent mortality at known doses. It is usually administered at 2-3 doses to cover the expected LD_{50} value. The expected oral LD_{50} for dimethoate ranges from 0.10 to 0.35 µg AI/bee.
3. The test compound at five doses.

Consequently, there are at least 9 "groups" for each study (the control, the toxic standard administered at 3 doses, and the test compound administered at 5 doses). Each group should be replicated three times (i.e. with 3 hoarding cages of 10 to 20 bees) (EPPO 2010a; OECD, 1998a, 1998b). The CEB (2011) guideline requires three "runs" of three replicates/run (3 x 3).

3.2.1.3.3. Substance administration

1. Starve the bees for 1-2 hours before the test so that all bees will feed once the study begins.
2. All bees in a cage are exposed to one of the test substances dispersed in a sucrose solution by being allowed to feed *ad libitum*. The sucrose solution is mixed at 500 g sugar to 1 l distilled water.
3. The number of doses and replicates tested should meet the statistical requirements for determination of LD_{50} with 95% confidence limits. A preliminary test (range finder) is usually conducted with a dose range of factor 10 in order to determine the appropriate doses for the formal test (1, 10, 100, 1000, etc.). Secondly the acute toxicity test is conducted with five doses in a geometric series with a factor 2 in order to cover the range for the LD_{50} (ex; 100, 200, 400, 800, etc.).
4. Bees are provided with 10 µl/per bee of the sucrose solution containing the test substance at the different concentrations. In each test group, the feeder is removed from the box when empty (within 2-4 hours) and replaced with another one containing untreated sucrose solution.
5. In all groups, the eventual remaining treated diet is weighed and replaced with untreated sucrose solution after 6 hours; the amount of treated diet consumed per group is recorded.

Table 2. Example of data sheet: both mortality, number of living bees and abnormal behaviour of living bees are recorded simultaneously. For behavioural effects see Table 1.

Contact/Oral LD$_{50}$ test honeybee (*Apis mellifera* L.)						
Test substance:			Concentration:			
Start date:			Administration: oral / contact (cancel)			
For contact administration:		Start time:				
		Pipet nr:				
For oral administration:		Start feeding time:		Weight feeding device:		
		End feeding time:		Weight feeding device:		
				Amount consumed:		
		Remarks consumption:				
		Remarks trophallaxis:				
General remarks:						

Date	Time	Observation	Test cage / replicate			Initials
			1	2	3	
		Mortality				
		N. living bees				
		Behavioural effects				
		Mortality				
		N. living bees				
		Behavioural effects				
		Mortality				
		N. living bees				
		Behavioural effects				

Note: behavioural effects = NE (no effect), F ("Freeze"), P (paralysis), S (spasms), other

3.2.1.4. Mortality assessment

1. In all treated and control groups, mortality (see section 2) is recorded at 4, 24 and 48 h post exposure. Data should be summarised in tabular form, showing for each treatment group, as well as control and toxic standard groups, the number of bees used, mortality at each observation time, and number of bees with adverse behaviour (Table 2). Any abnormal effects observed during the test are recorded in order to inform about possible subletal effects (Table 1). When mortality continues to increase after 48h, it is appropriate to extend the duration of the test up to 72 or 96 hours.

2. For the validity of the test, mortality in the negative (untreated) reference should be < 10% (OECD, 1998a; CEB, 2011) or 15% (EPPO, 2010b) and the mortality of the toxic standard dimethoate (positive reference) should meet the specified range: almost 50% with the lower dose (0.10 μg AI/bee) to 80-100% for the higher dose (0.35 μg AI/bee). Data from tests failing to meet these standard criteria should not be used.

3. Mortality data are submitted to a statistical analysis. The LD$_{50}$ has to be calculated (see 8.2.1.2.) for each recommended observation time (i.e. 24h, 48h and if relevant, 72h, 96h) based on mortality data corrected for control mortality using Abbott's formula (see 8.4.1.).

3.2.1.5. Extension to other tests

Although the acute oral toxicity test provides an LD$_{50}$ value, this result is not sufficient to appreciate other kinds of pesticide impacts. The oral toxicity test is nevertheless being adapted in other trial protocols related to honey bees. Notably, it is being refined to determine contact toxicity, chronic oral toxicity, seed dust effects, etc. and its evolution is certain to continue.

3.2.2. Topical application

The method outlined in this section (acute contact LD$_{50}$) is based on the OECD guideline 214 (OECD, 1998b) to which later recommendations from EPPO Bulletin 40 (EPPO, 2010b) are added. This method was never ring-tested but was several times reviewed by OECD, EPPO and CEB. It is considered validated.

3.2.2.1. Introduction

Two approaches to determine the contact toxicity of a PPP can be distinguished; a practical approach simulating the contact between a PPP and a honey bee in the field and an academic one assessing the LD$_{50}$. The academic approach is the one presented in this section as it is part of the risk assessment according to the OECD and EPPO guidelines used for legislation of PPP's worldwide.

3.2.2.1.1. Field simulated contact toxicity

To place the contact toxicity of pesticides briefly in a historic framework, two protocols are described briefly. In Stute (1991), the contact toxicity of PPPs applied as a spray, was assessed by exposing the bees to a 150 cm^2 paper, contaminated with twice the recommended field application rate of the target pesticide. The PPPs to be applied in a dusted form were administered using a Lang-Welte-Glocke to cover the surface completely and homogeneously. Johansen (1978) assessed the contact toxicity by placing bees in a bell-jar duster loaded with 200 mg pesticide and administered the pesticide onto the bees via vacuum and subsequent imploding incoming air to disperse the pesticide homogeneously over the bees. Both the Stute (1991) and Johansen (1978) tests provide general information about toxicity. However, in both cases the amount of the PPP actually administered to the bees was unknown. This makes it hard to do further calculations about risk assessment. The other two methods imitating field contact exposure are described in section 3.2.3.

3.2.2.1.2. Contact LD$_{50}$

The acute contact toxicity test is conducted to determine the inherent toxicity of pesticides and other chemicals to bees. The results of this test are used to define the need for further evaluation. The contact LD$_{50}$ is part of the tiered approach; from laboratory to semi-field to field. The tiered approach is implemented in the EU. The contact LD$_{50}$ is assessed for the risk assessment of sprayed PPPs to adult worker bees. The result, a certain dose expressed as μg or ng AI or formulation per bee or per gram of bee is an academic parameter and does not express the hazard of the product in the field. This depends on the concentrations and the field application and is assessed in the HQ

(EPPO, 2010b) or RQ (EPHC, 2009) (see 8.4.2.). When an HQ calculation results in a value lower than 50, the risk to bees is considered to be low. When performing acute contact studies, a toxic standard (positive reference, such as dimethoate) should be used. The results from the test with the toxic standard provide information on potential changes in sensitivity of the test organisms (in time) and consequently the suitability of these populations for further testing. Additionally, information on the precision of the test procedure is generated.

3.2.2.2. Description of the method
3.2.2.2.1. Outline of the test

1. The AI or formulation of a PPP is tested.
2. The PPP is dissolved in acetone if possible. Other solvents should be used only in instances where the compound is insolvable in acetone and these alternative solvents are known to be harmless to bees.
3. When formulations are tested (rather than AIs), they should be water and if needed, an appropriate wetting agent added. If a wetting agent is applied, it should be applied in the positive and negative reference as well.
4. The test substances are administered to anaesthetized bees (Human *et al.*, 2013) in a 1 µl droplet on the dorsal thorax of individual bees.
5. After treatment, the bees are provided *ad libitum* with freshly made sucrose-solution 50% (w/v) and checked daily for mortality and behaviour (see Table 2).

3.2.2.2.2. Collection of bees

Adult worker bees used for this protocol should be collected per Williams *et al.*, 2013. Other special considerations:

1. Adult worker bees of the same race.
2. The bees should be collected in the morning of use or in the evening before the test and kept under test conditions to the next day.
3. Bees collected from frames without brood are suitable.
4. Collection in early spring or late autumn should be avoided, as the bees have an altered physiology during this time.
5. If tests are to be conducted in early spring or late autumn, the bees can be emerged in an incubator and reared for one week with "bee bread" (pollen collected from the comb) and sucrose solution.
6. The bees should not have a treatment history or originate from colonies that have been treated with chemical substances such as antibiotics, anti-*Varroa* agents, etc. Bees can be used from colonies that have been treated with these substance longer than 4 weeks before bee collection.

3.2.2.2.3. Test cages

1. Easy to clean and well-ventilated cages should be used. For recommendations on cage types and maintaining bees in laboratory cages, see Williams *et al.*, 2013.
2. The cages should be lined with filter paper to avoid contamination of the bees from vomit and faeces. Groups of ten bees per cage are preferred.
3. The size of test cages should be appropriate to the number of bees (Williams *et al.*, 2013).

3.2.2.2.4. Handling and feeding conditions

1. Handling procedures, including treatment administration and general observations, may be conducted under daylight conditions.
2. Sucrose solution in water with a final concentration of 50% (w/v) should be used as food for the adult bees and provided *ad libitum* during the test using a feeder device.

3.2.2.2.5. Preparation of bees

1. The collected bees may be anaesthetized with carbon dioxide or nitrogen for application of the test substance (Human *et al.*, 2013). The amount of anaesthetic used and time of exposure should be minimised.
2. Moribund bees, affected by the handling or otherwise, should be rejected and replaced by healthy vital bees before starting the test.

3.2.2.2.6. Preparation of doses

1. The test substance is to be applied as solution in acetone or as a water solution with a wetting agent. As an organic solvent, acetone is preferred but other organic solvents of low toxicity to bees may be used (e.g. dimethylformamide, dimethylsulfoxide). If others are used, they must be administered in the negative reference.
2. For water dispersed formulated products and highly polar organic substances not soluble in organic carrier solvents, solutions may be easier to apply if prepared in a solution of a commercial wetting agent to an extend the product dissolves (e.g. Agral, Citowett, Lubrol, Triton, and Tween).

3.2.2.3. Procedure
3.2.2.3.1. Test and control groups

1. The number of doses and replicates tested should meet the statistical requirements for determination of LD_{50} with 95% confidence limits (OECD, 1998b). Normally, five doses in a geometric series, with a factor not exceeding 2.2, and covering the range for the LD_{50}, are required for the test.

Table 3. Test scheme for the acute contact LD$_{50}$ test. "conc." = concentration.

Test solution	Replicate 1 (colony X)	Replicate 2 (colony Y)	Replicate 3 (colony Z)
Test conc. 1	conc. 1	conc. 1	conc. 1
Test conc. 2	conc. 2	conc. 2	conc. 2
Test conc. 3	conc. 3	conc. 3	conc. 3
Test conc. 4	conc. 4	conc. 4	conc. 4
Test conc. 5	conc. 5	conc. 5	conc. 5
Positive control conc. a	conc. a	conc. a	conc. a
Positive control conc. b	conc. b	conc. b	conc. b
Positive control conc. c	conc. c	conc. c	conc. c
Negative control [solvent: acetone (or other), water, or water with wetting agent]	solvent	solvent	solvent

However, the number of doses has to be determined in relation to the slope of the toxicity curve (dose versus mortality) and with consideration taken to the statistical method which is chosen for analysis of the results.

2. A range-finding test preceding the actual toxicity test enables one to choose the appropriate doses.

3.2.2.3.2. Replicates

1. A minimum of three replicate test groups, each of ten bees, should be dosed with each test concentration. Bees in a single cage (a single replicate group) should be from the same colony, with a different colony being used to populate each cage.

2. The three replicates per dose of the PPP tested are treated with the same preparation of the test solution with a specific concentration, i.e. not a newly prepared test solution for each replicate group (Table 3).

3.2.2.3.3. Toxic reference

1. A toxic (positive) reference should be included in the test series.

2. At least three doses should be selected to cover the expected LD$_{50}$ value.

3. A minimum of three replicate cages, each containing ten bees, should be used with each test dose.

4. The preferred toxic reference is dimethoate. Gough *et al.* (1994) evaluated the use of dimethoate as a reference compound for acute toxicity tests on honey bees. The results of 63 contact tests of technical dimethoate were evaluated, using the 95% confidence linear regression of logit transformation on log10 dose (µg/bee), adjustments using Abbott's correction. The contact LD$_{50}$ assessed with six concentrations in acetone, control acetone and administration on the thorax was 0.16 (min 0.11, max 0.26) µg AI/bee. LD$_{50}$ values ranging from 0.075 to 0.30 µl AI/bee in groups of 10 bees should be considered as valid results of the toxic standard Dimethoate. The LD$_{50}$ (48 h) was similar to 24 h. For the contact LD$_{50}$ tests, the contact LD$_{50}$ (24 h) should be in the range of 0.10-0.30 µg AI/bee.

5. A dose range of 0.075 to 1.0 µg/bee is recommended and results falling in this range validate the test.

6. Other toxic standards would be acceptable where sufficient data can be provided to verify the expected dose response (e.g. parathion).

3.2.2.3.4. Administration of doses

1. Anaesthetized bees (Human *et al.*, 2013) are individually treated by topical application.

2. The bees are randomly assigned to the different test doses and controls.

3. A volume of 1µl of solution containing the test substance at the suitable concentration should be applied with a validated micro applicator to the dorsal side of the thorax of each bee.

4. Other volumes may be used, if justified.

5. After application, the bees are allocated to test cages and supplied with sucrose solutions (50% w/v).

3.2.2.3.5. Test conditions

1. The bees should be held in the dark in an experimental room at a temperature of 25 ± 2°C.

2. The relative humidity, normally around 50-70%, should be recorded throughout the test.

3.2.2.3.6. Duration and observations

1. The number of dead or affected bees (see Table 1) is counted at 4 h after dosing and thereafter at 24 h intervals for up to 48 h or longer if mortality is still increasing (> 15% increase in mortality in the 25-48 h period).

2. Additional assessments at shorter intervals may be useful in specific cases.

3. It is appropriate to extend the duration of the test to a maximum of 96 h.

4. Mortality is recorded daily and compared with values from the positive and negative references.

5. All abnormal behavioural effects observed during the testing period should be recorded.

6. Therefore the total number of bees having yes/no effect should be recorded at each recording. These data allow the calculation of ED_{50}.

3.2.2.4. Calculation of the LD_{50}

The results are analysed in order to calculate the LD_{50} at 24 and 48 h and, in case the study is prolonged, at 72 h and 96 h. The mortality data should be analysed using appropriate statistical methods (LD_{50} calculated based on data corrected for control mortality, see 8.2.1.2. and 8.4.1.)

3.2.2.5. Limit test

In some cases (e.g. when a test substance is expected to be of low toxicity), a limit test may be performed using 100 µg AI/bee in order to demonstrate that the LD_{50} is greater than this value. The same procedure outlined in section 3.2.2.2. should be used, including three replicate test groups for the test dose, the relevant controls, and the toxic reference. If mortality occurs, a full study should be conducted. If sublethal effects are observed, these should be recorded.

3.2.2.6. Validity of the test

The test is valid if:
1. The LD_{50} of the toxic standard meets the specified range (see section 3.2.2.3.3.)
2. Control mortality in 48 h ≤ 15% (EPPO, 2010b).

3.2.2.7. Data and reporting

3.2.2.7.1. Data

- The LD_{50} is expressed in µg AI test substance or µg formulation/bee.
- In case the LD_{50} is applied for the HQ calculation (see 8.4.2.), the LD_{50} of the AI should be used.
- Data should be summarised in tabular form, showing for each treatment group, as well as control and toxic standard groups, the number of bees used, mortality at each observation time, and number of bees with adverse behaviour (see Table 1).

3.2.2.7.2. Test report

The test report must include the following information:

3.2.2.7.2.1. Test substance

- physical nature and relevant physical-chemical properties (e.g. stability in water, vapour pressure);
- chemical identification data, including structural formula, purity (i.e. for pesticides, the identity and concentration of AI).

3.2.2.7.2.2. Test bees

- scientific name, race, approximate age (in weeks), collection method, date of collection;

- all relevant information on colonies used for collection of test bees, including health, any adult disease, any pre-treatment, etc.

3.2.2.7.2.3. Test conditions

- temperature and relative humidity of experimental room;
- housing conditions including type, size and material of cages;
- methods of administration of test substance, e.g. carrier solvent used, volume of test solution applied, anaesthetics used;
- test design, e.g. number and test doses used, number of controls; for each test dose and control, number of replicate cages and number of bees per cage;
- date of test.

3.2.2.7.2.4. Results

- results of preliminary range-finding study if performed;
- raw data: mortality at each concentration tested at each observation time;
- graph of the dose-response curves at the end of the test;
- LD_{50} values, with 95% confidence limits, at each recommended observation time, for test substance and toxic standard;
- statistical procedures used for determining LD_{50};
- mortality in controls;
- other biological effects observed and any abnormal responses of the bees;
- any deviation from the Test Guideline procedures and any other relevant information.

3.2.2.8. Recommendation

It may be useful to have the test solutions analysed to verify the concentrations administered.

3.2.3. Toxicity of residues on foliage

3.2.3.1. Testing toxicity of contaminated dust from pesticide-dressed seed by indirect contact

3.2.3.1.1. Introduction

In some cases, the indirect toxicity tests can be preferred to topical tests because they better simulate the field conditions of the exposure and provide fast and applicable data (see 3.2.2.1.1.). In the indirect or residual toxicity tests, bees enter in contact with the test substance by walking on contaminated substrate in a hording cage (Williams *et al.*, 2013). The "OPPTS 850.3030 Honey bee toxicity of residues on foliage" is the unique official guideline designed to develop data on residual toxicity to honey bees for spray products but no official methods are available to test contaminated dust in laboratory. In fact, individual compounds can show different levels of toxicity depending on formulation (spray *vs.* dust for example) but, specific tests should be adopted to estimate the toxicity of powder products when pesticides are applied as seed treatment.

Table 4. Example of the calculation of dust and AI quantity to distribute in the bottom surface of the hoarding cage.

Quantity of AI deposited during sowing on the ground at 5 m ($\mu g/m^2$) A	Percentage of AI in the dust obtained by Heubach cylinder P	Quantity of the AI-containing dust deposited on the ground at 5 m ($\mu g/m^2$) Q = A*100/P	Surface of the bottom of the hoarding cage (cm^2) S	Quantity of the AI-containing dust (in μg) on the surface of the hoarding cage D = Q*S/10,000	Concentrations	Quantity of the AI-containing dust per cage (μg in 0.01 g of talc)
2.25	33%	6.82	56.72	0.039	x 1000	39
					x 100	3.9
					x 10	0.39
					x 1	0.039

Several bee mortalities in Europe and USA have been linked with contaminated dust dispersed during maize sowing operations (Alix *et al.*, 2009; Bortolotti *et al.*, 2009; Pistorius *et al.*, 2009; Krupke *et al.*, 2012). Pesticides can be dispersed by air during sowing operations when pesticide-dressed seeds are used and contaminated dusts can subsequently deposit on soil and vegetation, posing an exposure risk to foraging bees (Greatti *et al.*, 2003, 2006). In this section, a method to test the impact of contaminated dusts on honey bees is proposed.

3.2.3.1.2. Test procedures
3.2.3.1.2.1. Background
This protocol follows the method of Arzone and Vidano (1980) applied for spray products but adapted to soil/seed treatments. This method has been applied in Italy in order to investigate the effects of pesticides drifted from maize seed dressing on honey bees when bees forage in the edge of the maize field during sowing operation (APENET, 2009, 2010; Sgolastra *et al.*, 2012).

3.2.3.1.2.2. Dust extraction
1. Dust from maize-dressed seeds is obtained by Heubach method. This method is commonly performed to measure the seed dustiness (Heimbach, 2008). In the Heubach method, treated seeds are mechanically stressed inside a rotating drum. A vacuum pump produces an air flow through the rotating drum, the connected glass cylinder and the attached filter. By the air flow, abraded dust particles are transported out of the rotating drum through the glass cylinder and subsequently through the filter unit. Fine dust particle (Ø < 0.5 mm) are deposited onto a filter while coarse non-floating particles are separated and collected in the glass cylinder.
2. The dust retained by the Heubach cylinder filter and the other particles extracted with Ø < 45 μm should be used in the toxicity test. Fine and coarse dust particles are mashed and sieved with a precision 45 μm mesh sieve in order to use only small particles for the test, which are more likely to drift.

3.2.3.1.2.3. Dosages
As a worst case, the quantity of contaminated dusts deposited on the ground during sowing at a maximum of 5 m distance from the edge of the field should be used. The distance was chosen based on the previous results of field studies (APENET, 2010) where the amount of the AIs deposited on the ground during sowing at 5, 10, 20 m distances from the field's edge was measured and a decline in pesticide concentration was observed as distance increased (APENET, 2009, 2010). The dose of AI deposited on the ground was measured in field studies following the indication of the agricultural industry in agribusiness field trials, which in turn were taken over from a methodology designed to study liquid pesticide drift (BBA, 1992; APENET 2009, 2010).

3.2.3.1.2.4. Contaminated dust preparation
1. The AI-containing dust, obtained from dressed seed with Heubach cylinder (see 3.2.3.1.2.2.), is analysed. The percentage of AI content in the dust is used to calculate the quantity of dust to distribute on the surface of the bottom of the hoarding cage (Table 4).
2. To allow homogeneous dispersal of dust on the cage substrate in proportion to the quantity of AI deposited at 5 m, it is necessary to mix the dust with an inert material (talc) through geometric dilutions, starting from a dose that is 1000 times more concentrated. An appropriate quantity of talc is used as a dispersing agent in order to reach the desired concentration (Table 4).

Talc has been suggested as a dispersing agent because it is a common mineral material, not toxic to bees, usually added to seed boxes to reduce friction and stickiness, and to ensure smooth flow of seed during planting. Krupke *et al.* (2012) found that waste talc expelled during and after sowing represents a route of pesticide exposure for bees.

3.2.3.1.2.5. Substrate
1. Leaves collected from a plant that is as far as possible from

possible pollution sources. Other removable substrates (e.g. plastic or Plexiglas surface) may also be used.

2. Before the test, samples of leaves can be analysed for the residues in order to exclude previous contaminations.

3.2.3.1.2.6. Dust application

1. 0.01 g of total dust (the AI-containing dust plus the talc powder) per cage should be distributed on the leaves (Table 4). This quantity was considered adequate for a homogeneous distribution on the surface of approximately 50-70 cm^2. For bigger cages, a proportionally higher amount should be used.

2. A small sieve obtained from a modified Eppendorf tube can be used as shown in Fig. 1.

3.2.3.1.2.7. Control

A negative (untreated) reference is required during the test. The control substrate should be treated with pure talc. Control and treated bees should be kept under the same laboratory conditions (see section 3.2.3.1.2.12.).

3.2.3.1.2.8. Exposure to test substance

1. Forager bees, collected per Williams *et al.* (2013) are exposed to the dust by walking for 3 h on treated apple leaves or other substrate, placed on the bottom of a standard hoarding cage (e.g. 13 x 6 x 11 cm or one from Williams *et al.*, 2013).

2. The leaves are removed from the cage after 3 h.

3.2.3.1.2.9. Number of animals tested

Usually 10 bees per cage should be used.

3.2.3.1.2.10. Number of replicates

3 to 5 cages per treatment (see section 8.4.4.)

3.2.3.1.2.11. Duration of the test

At least 3 days or when the control mortality is >20%.

3.2.3.1.2.12. Test conditions

1. During the trials, the cages containing the bees should be maintained in a darkened incubator at 25 ± 1°C and with 60 - 80% RH.

2. Each cage should be equipped with a dispenser containing sugar solution for the bees (50% w/v). It is important to avoid the dropping of the sugar solution on the treated surface during the exposure period.

3.2.3.1.2.13. Endpoints

1. Cumulative mortality is assessed, then LC$_{50}$ is calculated (see 8.2.1.2.) and any noted sub-lethal effects are registered (see Table 1).

2. The PER assay (Scheiner *et al.*, 2013) can also be performed after bees have been exposed to contaminated dust for 3 h following the above test procedure (APENET, 2010).

3.2.3.2. Testing contact toxicity on bees exposed to pesticide-contaminated leaves

3.2.3.2.1. Introduction

The assessment of the toxicity of residues on foliage to bees can be managed with several methodologies related to the mode of action and the way of application. From 1998 to 2003 the subject of high bee mortalities during spring when sowing of seeds is common became an important topic. On a review of different hypotheses, it was decided to investigate the ability of seeder machines to leave dust residues in the environment, a suspicion identified because of the use of insecticide coated seeds in southwest France. Consequently, it became necessary to determine if increased bee mortalities were related to the dust from coated seeds or alternative routes of exposure.

Crops of maize and sunflower were suspected to trigger such mortalities because of the numerous surfaces and AIs of the insecticide seed protection. As mortalities were mainly located in apiaries of this area, a major link was established with the sowing time of sunflowers.

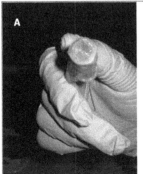

Fig. 1A. Small sieve obtained from a modified Eppendorf tube (the bottom is removed from the tube and replaced with screen mesh); *B.* Dust application on the apple leaves; *C.* Leaves placed in the bottom of the hoarding cage.

The following field-lab protocol was developed after initial tests of dust emission.

1. In indoor conditions, non-moving seeder machines are used to collect dust from different varieties of seeds and dressings. Seeders are equipped with filters that permit one to analyse the source and quantity of dust when working.

2. Coated seeds are classified from a screening with different kinds of varieties as well as different dressings for the same variety.

3. Among all dressing coated seeds, two modalities are selected for comparison of pesticide impact on honey bees. One concerns the low level of dust emission and is expected to have a minor impact when contacting honey bees. The second modality focuses on higher dust emission data and is tested for assessment of an eventual impact to honey bees.

4. The field part of this protocol aims to collect dust from a sowing operation in agricultural conditions. Fields of at least two hectares are separate from one another by about three kilometres in order to avoid a cross-contamination under wind conditions. These fields are bordered by a hedge on the edge of plot so that the wind creates turbulence on site. Dusts are expected to drop to the ground instead of being borne away. Dedicated sentinel plants are arranged on the ground to catch dust. They must have hairy leaves with good hair disposition on the upper leaf surface such as with *Tibouchina* (Order: Myrtales; Family: Melastomataceae) or other ornamental plants. They are placed in fields before sowing starts and they remain in the field for 2 days post sowing.

3.2.3.2.2. Methodology

1. The design includes 4 treatment groups:
   - the 2 sunflower varieties,
   - the untreated control
   - toxic standard (positive reference with dimethoate at 400g AI/ha)

2. The untreated control and the toxic standard are kept in an open space close to the laboratory.

3. The control group receives no treatment. There is no "dusted" toxic reference; thus to ensure bee sensitivity and to validate the design, the toxic reference is treated with a liquid spray of dimethoate (i.e. Dimezyl 1 l/ha = 400 g AI/ha).

4. In this method, the four treatment groups do not have the same route of exposure; the two varieties with coated dressings are tested from dust issued from agricultural practices whereas the toxic standard is a spray and the control is untreated or water treated.

5. Assessments are conducted under controlled conditions where bees are exposed to foliage in hoarding cages similar to LD_{50} tests (see Williams *et al.*, 2013 and section 3.2.3.1.2. of this

manuscript). Sentinel plant foliage is collected 2 hours after seed sowing to look for acute toxicity effects on bees.

6. The surface in each hoarding cage is covered with foliage taken from sentinel plants. The surface of foliage is exactly adapted in number of cm². Twenty honey bees are introduced into all hoarding cages and are allowed to contact the leaves from the sentinel plants. Bees are taken from one single and healthy beehive and dispatched in the 4 groups and containers at random and per Williams *et al.*, 2013.

7. The foliage from the sentinel plants is removed after 24 hours but bees are left in boxes for 2 additional days; thus the test duration is 72 h. Then the laboratory part of this methodology is very similar to standardized LD_{50} test: CEB 230 (CEB, 2011), EPPO 170 (EPPO, 2010), OECD 214 (OECD, 1998b), with mortality assessments at 4 hours, 24, 48 and 72 h after exposure.

8. From the raw data, the average mortalities are calculated in three (3) replicates of each treatment group using usual formulas in statistical analysis (see section 8.4.1.).

9. These results are validated by mortality at 24 hours of 0% in the control and over 90% in the toxic standard.

10. Item modalities induce intermediate mortalities close to the control or higher according to the amount of dust in contact with bees.

11. Assuming no cross contamination is possible, some lethal effects are observed on bees following the use of one treated seed, and absolutely no effect for the other one.

3.3. Field methods for testing toxicity of chemical substances on individual adult bees

3.3.1. In-field exposure to dust during sowing

3.3.1.1. Introduction

It has been shown that bees can be contaminated with potentially lethal doses of insecticide simply by flying in the vicinity of a pneumatic drilling machine using seeds coated with insecticide (Marzaro *et al.*, 2011; Girolami *et al.*, 2012). The fragments of this coating are emitted into the atmosphere and constitute a toxic cloud the size of which may be estimated at some tens of metres in diameter. Only bees in flight were considered when reporting these observations about powdering, not bees possibly exposed to powder that fell to the ground and could contaminate on contact.

The following reported techniques presuppose an evaluation of the contamination, mortality and chemical analysis of a single bee. Once the bees are treated with powder, one must avoid the possibility that the bees in the same cage could contact and exchange contaminants, thus altering the results. For this reason, bees were kept separately one per cage. The test reports the evaluation of the acute toxicity which can cause the death of bees between 24 and 48 h and for maximum practicality should be conducted under normal laboratory conditions (see section 3.3.1.3. below).

3.3.1.2. The management of the bees after exposure

1. In the contamination trials (be it in free flight or in mobile cages as reported below), the bees should be placed singly in small cages with a cubic steel skeleton of 5 cm and all the six sides enclosed entirely in tulle (with mesh of 1.1 mm) (Fig. 2).
2. The bees should be fed with small drops of honey during the period of observation. The honey can be placed on the top of the cage.
3. Additionally, so as to avoid honey dissolving, soiling and to prevent rapid ingestion, parallelepipeds of sponge can be placed on each cage. These can be 2 x 2 x 1 cm and made of normal, non-soluble domestic sponge soaked in 0.5 ml of honey.
4. The cages are ideal for observation when placed in a transparent container (for example, a polystyrene box 24 x 35 x 10 cm) sufficient to contain 12 small cages with a sheet of absorbent paper underneath (Fig. 3). The cages should be kept raised above the base of the cages by means of a net of folded metal. This device was used to prevent accumulations of honey on the base of the cages and to prevent the cage from contacting other liquids.

Fig. 2. Cages employed to expose bees to seed drill emissions and to evaluate survival after exposure.

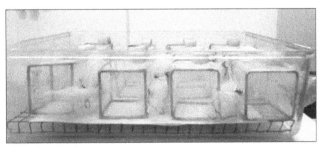

Fig. 3. Transparent polystyrene container with 12 small cages. Feeders placed on the cages are shown.

3.3.1.3. Study conditions

1. The containers with cages are kept at 23 ± 2°C with natural light, or added artificial light, in cloudy condition during the day.
2. The containers should be closed with a sheet of transparent plastic.
3. It is possible to keep the bees at a relative humidity close to saturation simply by wetting an absorbent sheet of paper on the bottom of the container with distilled water.

This system enables the evaluation of the influence of high humidities without wetting the cages. Thus, bees are prevented from contact with the water. Because of the high humidity, the sponges soaked with honey should be replaced every 6-12 h; otherwise, the bees continue to suck at the light with an increase of mortality possible in untreated controls.

3.3.1.4. Capturing the bees
3.3.1.4.1. Inducing the bees to visit the dispenser

In order to apply the trials in free flight, the bees must be conditioned to visit a dispenser simulating normal foraging trips.

1. In order to condition the bees rapidly to take sugar solution (about 50% w/v) from a dispenser placed not less than 30 metres from the apiary, a little flat dispenser with sugar solution is first placed on a running board (the dispenser must be refilled for minimum 2 days).
2. When the bees become accustomed to feed and crowd on the dispenser, it can be placed some metres of distance from the hive. The change of position must be gentle to keep bees from flying away.
3. To achieve visits from a particular hive, the above method can be employed using an isolated hive.
4. Once the bees associate the dispenser with the sugar solution, it is possible to put the dispenser with bees in a cage and transport them even hundreds of metres away.
5. When the bees are freed from the cage, some of them associate with the new position of the dispenser and indicate it to their companions once they re-enter the hive.
6. After the hive is conditioned to the required distance, it is possible to attract hundreds of foraging worker bees by replacing the sugar solution once daily. This is better done at the same hour each day.
7. The solution can be quickly and practically produced by mixing equal quantities of water and sucrose (approximately 50% w/v).

3.3.1.4.2. Collecting bees for use during the study

This topic is reviewed in detail in Williams *et al.*, 2013.

1. The most accurate method of collecting the bees is to put them singly, at the dispenser, into glass test tubes with a diameter greater than 1 cm and 10 cm in height (Falcon vials).

2. The collection can be accelerated by the use of the end section of an "insect vacuum" (Fig. 4).

3. For safety reasons, the vacuum necessary to suck a bee into the tube can be provided by an electric pump. If done manually, a fine, soft mesh should be placed at the mouth of the insect vacuum and a second protective diaphragm over the mouth of the test tube. This should be a thin, fine mesh.

4. It is necessary to limit captured bee exposure to any sort of rubbery material where they could insert their sting and die.

5. If it is not necessary to capture the bees singly at the dispenser (for example in the free flight trials), the bees may be caught en masse in a 20 cm tulle cage (or similar), placing it at the entrance to the hive (Fig. 5) (section 4.3.3.2 in Williams *et al.*, 2013).

6. The dispenser should be withdrawn from the cage, the cage closed and taken to the laboratory.

7. At the laboratory, the bees may be fed with honey placed on the upper part of the cage (Fig. 3).

8. The bees may be transferred from the cage to be kept singly in the laboratory, as described for the capture at the dispenser (step 1 above).

9. It is ideal that the bees not used at the end of the trial be freed to be renewed on successive days of experimentation.

10. Wherever possible, the powdering trials should be conducted using bees collected at the dispenser, avoiding using bees collected with an entomological net in front of the hive. This ensures that no juvenile bees are captured and used during the study.

11. If necessary, in the winter, bees can be caught in front of the hive, taking care to catch those bees returning to the hive (thus, certainly foraging worker bees) and not those exiting the hive who could be solely engaged in orientation flights. Nevertheless, it should be noted that winter bees normally should not be used for standardized ecotoxicological testing.

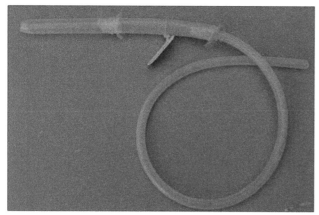

Fig. 4. Insect vacuum (aspirator) used to capture the bees. The two mesh diaphragms safely prevent the bee from being sucked into the mouth of the operator.

Fig. 5. Capture of bees from the hive. Method described in section 4.3.3.2 in Williams *et al.*, 2013.

3.3.1.5. Trials in mobile cages

1. In trial set up to evaluate the presence, consistency, extent and duration of the toxic cloud surrounding pneumatic seed drills during the maize sowing season, and using seed coated with insecticide, the powdering was evaluated by means of an aluminium bar 4 m long, to which cages, each containing a single bee, were attached every 0.3 m (12 in total) (Figs. 2 and 6).

2. The bar was supported at each end by a vertical pole of 2.5 m. The bar was passed by two people at a fast walking pace (6-8 km/h) by the side of the drilling machine, at a variable height according to how the exhaust air was emitted from the machine, taking into account that bees fly predominantly at 1-3 m over ploughed land (unpublished data). The cages may be numbered considering the progressive distances from the drill.

3. The people with the bar followed and passed the tractor on the right hand side (in the first 30 m of the plot) (Fig. 6). The tractor then reduced speed and waited while the people with the bar made a U-turn and again passed the machine, once more at working speed, on the left hand side. In this way, the bees were twice exposed to the cloud in a similar way to foragers in free flight making a round trip over the sowing area.

4. To evaluate the mortality, once the bees had been exposed to the insecticide dust in a cage in the field, they were transferred (inside the same cage) to a room at a controlled temperature (22 ± 1.5°C) and in conditions of high humidity (Girolami *et al.*, 2012b but see section 3.3.1.2.-3.3.1.3.).

3.3.1.6. Trials in free flight

This method is used to evaluate the effect of direct exposure of a bee in flight to the powder emitted by the drill while sowing coated maize

Fig. 6. Exposure of bees using the mobile cage method.

seed. Such a method is needed to test the hypothesis that bees, in
repeated flights to flowering plants, can be expected to fly over plots
being sown with coated maize seed and become lethally poisoned
with powder acquired during the flight.

1. Bees from 4 hives can be conditioned to visit a feeder some
   25 cm in diameter, containing a sucrose solution (50 w/v).
   The feeder can be progressively distanced from the hives up
   to a final distance of 100 m (see section 3.3.1.4.1.).
   Observing the bees, it is possible to count hundreds of bees
   flying, at an average height of 2 m, to and from the hives to
   the food source.
2. From the beginning of the sowing and at succeeding 15 minute
   intervals, bees can be caught in test tubes at the feeder and
   placed singly in small tulle cages (5 x 5 x 5 cm) and fed with a
   drop of honey placed on the mesh of the cage, and periodically
   renewed (every 6-12 h).
3. 24 samples can be captured at each time period, the first when
   the tractor starts and then every 15 minutes thereafter.
4. Each sample of 24 bees can be taken in cages to the laboratory
   and kept at a conditioned temperature of 22 ± 1.5°C (see
   section 3.3.1.2.-3.3.1.3.).
5. For each time interval, 12 cages chosen at random are kept at
   laboratory humidity and the remaining 12 cages placed in a
   box at high humidity close to saturation (>95%). The raised
   relative humidity was obtained by placing the cages in a
   transparent plastic box sealed, but not hermetically, with a
   sheet of Plexiglas, and by placing a sheet of wetted absorbent
   paper at the base. The walls and the cover were sprayed with
   water and the cages were raised with a strip of polystyrene so
   that the bees could not get wet from any water that might
   remain on the base (Girolami *et al.*, 2012a).

3.3.1.7. Collection and analysis of data

1. For both tests (mobile cage and free flight), the comparison
   between bee survival at the beginning of the trial, i.e. before
   the start of drilling and after every 15 minutes is obtained (for
   a maximum of four samples, but are sufficient two samples).

2. To compare different bee samples (treatments, humidity levels
   and collection times), the null hypothesis that the mortality is
   independent on the considered parameters should be tested
   using a chi-squared goodness-of-fit test.
3. To verify the influence of relative humidity, the cages with the
   bees, are randomly divided and held in laboratory or high
   humidity (see section 3.3.1.3.).
4. In the mobile cage test, the distance from the driller, which
   causes no acute bee mortality, also can be estimated.
5. This method of bee mortality evaluation in the field (in particular
   the mobile cage) is an innovative biological test that can be
   applied to verify the efficiency of driller modifications.

4. Effects of toxic substances on bee colonies

This section describes methods of testing effects of toxic substances
on honey bee colonies. The experimental unit consists of the colony
or its different components (brood, stores, bee community etc.). If
the observed subjects are not the colonies but single bees, these are
free to interact with the entire colony. This assures that the bee
behaviour is as natural as possible.

4.1. Introduction

The honey bee colony can be considered as a superorganism including
numerous bees of different castes, ages and sex acting together to
develop the nest. The evolutionary success of honey bee colonies is
based on social organization between the workers and the queen for
colony growth and development. The social organization is based on
division of labour that depends on individual endogenous biotic factors
like hormonal, genetic, immune and neurobiological backgrounds and
on exogenous biotic factors like chemical communications, social
immunity and behavioural interactions, with all of these factors capable
of being modulated by the external environment.

Bearing in mind the complexity of the functioning colony, when
significant variability in the response to toxic substances of bees is
demonstrated using cage experiments, it is reasonable to expect that
the difference in response will be even greater between bees in cages
and in natural conditions. Depending on the questions to be addressed,
it may be necessary to consider working either at individual (cage) or
colony level. Thus, for studying the molecular effects of a toxin on bees,
cage experiments using very controlled environment may be the best
choice. However, in the end, the effect of the toxin in the real life of
the bee, i.e. in natural conditions, should be addressed, even if it is
much more difficult to manage honey bee colonies than cages.

Ideally, studies on the effects of toxic substances at the colony
level require contiguous treated and non-treated areas of a field where

colonies can be placed. Unfortunately, these protocols are not easy to use as the bees will forage in both non-treated and treated areas. Moreover, the sites at which bees can forage in field conditions are not controlled at all, even when colonies are placed close to the observation areas. Thus, it is proposed to observe the behaviour of foragers directly on the target crops, in addition to overall colony development or in semi-field trials (in tunnels), to determine the effect of treated crops on honey bee colonies in semi-controlled conditions. These semi-field trials are informative, but with the bias that usually the colonies do not develop as well as colonies placed in natural conditions. Another approach consists of mimicking the exposure to a substance on the field crop by forced in-hive feeding with syrup or pollen patties and observing the colony development and the impacts on individuals using various investigation methods. This approach can be used to test the effects of acute, chronic, lethal or sub-lethal exposures to different substances. Different parameters can be studied using those methods for testing the toxins on bees at the colony level: individual adult and brood mortality, clinical symptoms or colony development. However, individual observations on behaviour are particularly interesting for gathering information on sublethal effects of the toxins. Different technologies such as honey bee counters, RFID labelling or harmonic radars have been proposed for this purpose.

This section gives information on techniques used to study the effects of toxic substances, including dusts dispersed during sowing and systemic substances distributed in plant matrices, at the colony level. Different field or semi-field protocols are described and in the future could be the basis of procedures used in the risk assessment of pesticides.

4.2. Determining pesticide toxicity on bee colonies in semi-field conditions
4.2.1. Introduction
After the determination of LD_{50}'s on individual honey bees in laboratory conditions, it is necessary to enlarge the assessment of pesticide impacts using outdoor tests at the colony level. These higher-tiered semi-field tests are performed under insect-proof tunnels. A key characteristic of such tunnels, which are similar to those used for the production of some vegetable crops, is that they must be of sufficient size to permit "normal" bee activity (flight and foraging). Tunnels should be at least 120 m²-150 m² (7-8 m x 20 m) and covered with a net that allows wind and rain into the tunnel to duplicate natural climatic conditions. In contrast, small cages of 9 m² (3 x 3 m) typically dedicated to plant selection cannot be considered for semi-field tests for various methodological reasons. The available space is too small and the numerous limited bees cannot fly around the queen -less one-frame hive.

Semi-field studies under insect-proof tunnels are largely based on the existing French CEB protocol n 230 (CEB, 2011). This kind of a test is intended to assess effects from a worst-case exposure scenario, where bees are confined to plants treated with a pesticide. Such studies under insect-proof tunnels are used to determine the following parameters:

- daily mortality,
- foraging activity and repellence effects,
- brood development,
- colony strength,
- behaviour of forager bees,
- residues on apiarist matrices (bees, honey, brood, wax...)

4.2.2. Tunnel description
1. The tunnels (Fig. 7) are placed side by side and separated from each other by a minimum distance of 2 m. All tunnels have the same orientation for common disposal. The tunnel nets are stretched out and embedded alongside the tunnel, thus creating a closed environment limiting foragers' flights. This space appears nevertheless sufficient after adaptation. Rain and wind, though weakened, are able to pass through the net. Temperature is sometimes a little higher in the tunnel than outside, but generally, there is small difference between the two environments (± 1°C).

2. Attractive plants are grown under tunnels in order to trigger foraging activity. These include *Phacelia tanacetifolia,* oilseed rape (*Brassica napus*) or mustard (*Sinapis alba*). When the trial is dedicated to behaviour assessment, sunflowers are convenient for their large flowers where forager bees can be easily observed. In the special case of the use of a pesticide against aphids on cereals, the crop should be winter wheat where bees are attracted by the daily spray of a sugar solution simulating the aphids' honeydew.

3. Inside each tunnel, 4 plots of the same size (2m x 8m) are delimited and separated by areas covered with a film of synthetic material, where vegetation has been removed (Lane 1 to Lane 6, see Fig. 8). The dimensions of these plastic-covered areas are adapted to the tunnel dimensions but the peripheral paths (Lane 1, Lane 3, Lane 4 and Lane 6) are at least 1m wide. The 4 plots (T1 to T4) receive foliar applications. The same relative plot position is adopted in all tunnels.

4. The hives (see section 4.2.5.) are placed in the central parts of the tunnels (Lane 2), as shown in Fig. 8. The entrance of the hive is directed towards the water supply on the central path. After placing the colonies in the tunnels, a water source is provided on the central path. The water source is removed during the foliar application.

5. After a few days of confinement, foraging bees' activity is adapted to the considered area.

Fig. 7. Example of a tunnel used for semi-field toxicity tests.

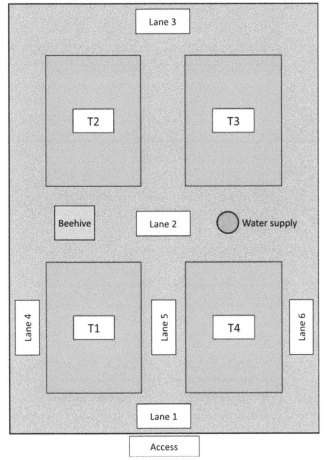

Fig. 8. Tunnel design of 4 plots to be treated and dedicated covered plastic lanes to collect dead bees.

4.2.3. Mortality assessment

1. By agreement, daily mortalities are collected all over the dedicated surfaces of plastic covered lanes. Bees dying among the crops are not collected.

2. Dead bees are collected every day in the morning in order to be accurate, and data express the mortality of the previous day. Additionally, bees can be collected twice on the treatment day (D0 in the morning, and D0+ in the evening in order to look at an eventual acute effect). The total mortality rate recorded in a tunnel for a given day results from adding up mortality rates observed in each of the six plastic lanes in the tunnel (lane 1 to lane 6).

3. During the first days, as well as in the control tunnel, mortality could be considered "normal" without, therefore, being natural. Bees hurt themselves against the net when introduced in the tunnel or when trying to escape. They try to locate themselves above the hive and at both ends of the tunnel. So in all tunnels, part of the recorded mortality during the first days is linked to biological and technical reasons. The impact of substances should be considered over this level and is usually recorded in the control.

4.2.4. Foraging activity assessment

1. Foraging activity is observed on all the crop plots during the trial. It is possible to adapt the time of counting to the environment of the trial and to active foraging periods. All the bees present on the crop plots are considered as forager bees. They are all counted one after the other. Counts can be shifted if activity is not considered satisfying (late activity due to morning mist or disturbed by rainfall, etc.).

2. Assessments are managed at least once a day, except on the day of application where assessments are recorded twice before application with one count just before, and three times after with one count 35 min after application.

4.2.5. Hive description

1. A first selection of the hives is made before experimentation in order to choose appropriate colonies. At least two apiarist visits are needed in the beginning and at the end of experimentation, in order to assess colony development. Parameters taking into account include adult bee population and the quantity of brood the quality of the brood (different stages observed), and amount of reserves (see Delaplane *et al.*, 2013b).

2. The structures of colonies are comparable to each other at the beginning of the test period. Colonies are homogenous regarding population, colony strength, food storage, brood and preparation. Beehives, each with a colony of approx. 15,000 to 20,000 bees (see Delaplane *et al.*, 2013b), are local bred. The colonies have queens of the same maternal origin and the same age, one to two years old. Preparation of the

colonies starts in an appropriate temporal distance to the beginning of the study. The colonies are established in Dadant hives with 6 to 10 frames comprising 4-5 frames for brood of all ages, and at least 1 storage frame and 1 empty frame. Hives are introduced into the tunnels 2 to 5 days before crop plot treatments during flowering. In case of applications before flowering, the hives are established in the tunnels during early flowering.

4.2.6. Treatment methodology

1. After hive settlement under the tunnels, the bees will forage on crop plots and strength parameters can be assessed (Delaplane *et al.*, 2013b) for 2 to 5 days until decreasing mortalities are homogeneous within modalities.

2. The number of semi-field tunnels is defined by the objectives of the study and includes at least 4 tunnels:
   - two tunnels for the pesticide in question
   - control tunnel (negative reference)
   - reference tunnel (positive reference)

3. The tested pesticide has to be applied in two modalities. The first duplicates GAP (i.e. applied according to label) and the second includes "the worst case of exposure". Therefore the first pesticide application occurs during flowering but when bees are not present in order to avoid contact with forager bees (after bee flight generally at night). The second tunnel receives a pesticide application while the bees are foraging on the test crop. To ensure adequate bee exposure for the second modality, there should be at least 5 forager bees/m² crop at the time of the foliar application

4. In the negative reference tunnel, the test crop plots are treated with water in order to determine any physical effect of the spray.

5. The reference tunnel (positive reference) exists to demonstrate bee sensitivity to a pesticide and to validate the trial. Dimethoate (400 g AI/ha) should serve as the toxic standard in the reference tunnel. It provides a high peak in mortality after application. It is, therefore, possible to add replicates of these four initial tunnels in a single study, or to conduct the study again in other conditions.

6. When the semi-field test is used to determine the behaviour of forager bees exposed to no foliar pesticide application (i.e. coated seeds or soil treatment), the test design has no toxic reference (positive reference, one does not exist) and only two modalities are needed (treated and negative reference).

4.2.7. Applications

1. Foliar applications are conducted after the stabilisation of daily mortalities in 2 to 5 days.

2. The four crop plots inside a tunnel receive the treatment, first the water control, then the study item, and the toxic reference at the end.

3. The application is conducted using a 2m long side sprayer boom set with nozzles.

4. The test pesticide and positive reference are applied with an air sprayer.

5. Spraying is performed at a steady speed that guarantees a homogenous deposit level over all sprayed areas. The application is performed with a volume of solution of nominally 200 l/ha at a pressure of 1-2 bar. Walking speed is established during the calibration procedure. The calibration procedure of the equipment used for the application is documented in the raw data.

4.2.8. Comparison of impacts

1. The use of the control and the toxic reference provides predictable impacts to which the impact of the test pesticide can be compared. Mortality is standard and predictable in the control though the foraging activity to the flowering crop may vary with climatic conditions.

2. Colony strength and development (measured per Delaplane *et al.*, 2013b) should be similar at the beginning and end of the experimental phase under the tunnel.

3. On the contrary, the reference dimethoate 400 g AI/ha induces a high mortality the day after application and continues for several days. During the same time, the count of forager bees (see section 4.2.4.) drops to zero because of the pesticide's high repellent effect.

4.2.9. Extension to other topics in semi-field tests

Foliar application on flowering crops is the main classic topic addressed using semi-field tests. However, as previously mentioned, it is possible to perform semi-field tests with other special aims:

- forager behaviour on treated sunflowers: =observe specific parameters associated with individual forager bees (mobile/ immobile, cleaning signs, clinic intoxication signs, etc., see Scheiner *et al.*, 2013)

- brood parameters associated with foliar applications and specific assessment along a 21-day brood cycle (see sections 5.2.2.2.5.4. and 5.2.3.). The OECD (2007) guidance document highlights the problems caused to brood development: assessment of the brood, including an estimate of adults, the area containing cells, eggs, larvae and capped cells (termination of the brood development and eventual compensation).

- residue studies in controlled conditions in pollen, nectar, dead bees, as well as in honey wax, soil and plant (flowers or the whole plant).

4.3. Testing toxicity on bee colonies in field conditions

4.3.1. Problems related to the experimental design

4.3.1.1. Introduction

In the current EPPO guideline (EPPO, 2010a), the field test is designed as the higher tier for the bee risk assessment of PPPs. In fact, according to the EPPO, field tests provide the most reliable risk assessment because it is based on data gathered under conditions which are most similar to agricultural practice. However, field studies are not often repeated because of the complexity of their establishment and their high cost. Only replicates over time can be conducted but, they are subjected to climate variations. Moreover, several methodological limitations, especially related to honey bees' underexposure, make it difficult to assess the realistic risk of a given pesticide to bees using field tests. In this section, the problems related to the experimental design of the field test and how to deal with these problems are discussed. The recent considerations from the EFSA Opinion on the risk assessment of PPPs on bees were taken into account (EFSA, 2012).

4.3.1.2. Replicates

Field studies are more difficult to conduct than semi-field and laboratory studies. One of the main critical points concerns the replicates. In fact, it has always been affirmed that one replicate consists in more colonies located in a single area. Nevertheless this assertion is controversial. In fact, in a field study it is always very difficult to replicate the same environmental conditions in independent trials (it is necessary to have no interference between treated/untreated colonies and replications). For these reasons, in the field every single colony needs to be considered a replicate. In this way, a field experiment using about 10 colonies per apiary can be considered adequate. Furthermore, if it was impossible to find two experimental fields in the same conditions for the comparison of the treatments, then it should be allowed to perform the test on a single plot (before and after the chemical treatment in the same field).

4.3.1.3. External factors

The results of the field studies can be affected by several factors outside the intrinsic toxicity of the substance. This includes the attractiveness of the target crop and the other plants surrounding the test field, the weather conditions during the experimental test, and the modality of the treatments. Honey bees forage an average of 1.5 km radius around their nest (Crane, 1984). However, this can extend to > 9 km under stressed food conditions (Seeley, 1985). For this reason, it is possible that bees from colonies in treated fields could forage in untreated areas and *vice versa*, thus underestimating pesticide exposure. In order to reduce this "dilution factor", the colonies in the test field should be isolated from other important blossoms and the test crop should be very attractive to bees (see section 4.3.1.4.).

Ploughed fields, rivers and highways can be used as natural barriers to isolate the test fields. The negative reference field (if present) should be located at least 4 km from the treated field and in an area with similar climatic and landscape conditions.

4.3.1.4. Application of treatment

EPPO guideline 170 (EPPO, 2010a) suggests to make treatments using the formulated product applied on the blooming crop (e.g. rape, mustard, *Phacelia* or another attractive crop to bees). The product should normally be applied at the highest dose recommended for practical field use. EPPO guideline 170 suggests treating a crop area of about 1 ha. This field range may be sufficient if the crop is very attractive to bees, with high nectar and pollen production, and a high number of flowers per area unit. However, this treated test area is much smaller than the mean foraging area (700 ha) and the level of exposure could be considerably underestimated. An area of at least 2 ha should be used in field tests and it should be isolated from other flowering crops in the bee foraging area. Otherwise, the plot size will be increased proportionally so as to maximise the exposure of foraging bees.

4.3.1.5. Colonies

The colonies should have queens of the same age (1-2 years) and from the same mother origin. Colonies should be homogeneous in size (adult bees and brood – Delaplane *et al.*, 2013b), in brood composition (about same number of young and capped larvae) and in food supply among treatments. The colonies should be visited regularly, at least once or twice a week, for purposes of monitoring the health status and should be free of pathogens before the pesticide application (see Volume II of the *BEEBOOK* for methods to choose colonies that are free of the various pests/diseases). Each colony should have a bee population that covers at least 7 to 10 frames, containing at least: 5 brood frames, 2-3 frames of food, and 1-2 empty frames in order to allow colony growth. The hives should be placed in the edge of the field from 7 to 5 days before the application of the pesticide to the crop to allow the colony to adapt to the surroundings. In order to prevent the bees from foraging in another field, the installation of the hives should be made at the beginning of flowering and a minimum of 7 days before pesticide application. In order to consider the inter-colony variability, at least 10 hives equipped with dead bee traps should be installed in each field.

4.3.1.6. Level of exposure

An important issue in field studies is to demonstrate that all age cohorts of bees (forager and in-hive bees), have been exposed to the test pesticide at the level from which we want to protect them when considering the worst case exposure scenario. For spray products, three exposure routes should be considered: oral, contact and inhalation. Honey bees can be exposed orally through nectar, pollen, and water but also directly during flight or when walking on contaminated substrates. These exposure routes should be considered both for forager and in-hive bees, even if in-hive bees are exposed mainly through residues in the food. The contact and inhalation exposures for in-hive bees should be assessed only in certain cases (e.g. fumigant and liposoluble products with high wax-affinity).

In order to determine if the experimental conditions in the field tests allow one to achieve the target exposure level, several observations and analysis should be performed. For forager bees, the level of exposure can be assessed by observing the number of bees on the test crop, the number of bees entering the nest with pollen loads and the flight activity (e.g. counting the number of bees exiting from the nest in 30 seconds (Porrini, 1995). Confirmed contact with the treated crop can derive from the palynological analysis of the pollen load (see Delaplane *et al.*, 2013a). Pesticide residues should be analysed in honey bees, as well as in the plant matrices (nectar, pollen and guttation droplets) and in the hive (honey, wax, stored pollen and larvae) in order to know the amount of the target pesticide potentially available for forager and in-hive bees following the "destiny" of the compounds from the plant to the hive. For systemic compounds or for pesticides sprayed during bloom, residue analysis should be always carried out in the hive matrices. These analyses can be used to know the potential exposure routes for bees and their duration over time.

4.3.1.7. Mode of assessment and recording

Meteorological data should be recorded at appropriate interval during the whole trial period. These data should include at least: temperature, relative humidity, rainfall and wind speed and direction. All parameters should be assessed at least from 7 days before to 15 days after pesticide application. Post-application assessment should last at least two brood cycles; this evaluation should be extended in case of residues in wax, honey or pollen. In any case, the colonies should be monitored until the following spring, when bees have consumed the food stores.

All parameters should be recorded at least for 7 days after treatments or during the whole exposure period (blooming) for systemic products. After that, assessments should be limited to determining colony size (Delaplane *et al.*, 2013b) until 42 days after treatment (two complete brood cycles). Because time of the day can affect several bee parameters (e.g. flight activity), assessments should be performed approximately at the same time of day.

4.3.1.8. Interpretation of results
4.3.1.8.1. Simultaneous trials

In case the treatment and the control trials were carried out simultaneously, in two different fields, the study could be considered valid if it meets the following conditions:

- before application, the mortality and the foraging activity among the hives of the two treatments are similar and standard (mortality comparable to that detected in the same period in hives located in the same area in good health conditions and without environmental stress);

- in the untreated field, the mortality and the sanitary status of the colonies are comparable before and after application;
- weather conditions during the test allowed normal foraging behaviour.

4.3.1.8.2. Consecutive trials

In case the treatment and the control trials were carried out consecutively (control trial: first week, treatment trial: second week), in the same field, the study could be considered valid if it meets the following conditions:

- before application, the mortality and the foraging activity of the hives are standard (mortality comparable to that detected in the same period in hives located in the same area in good health conditions and without environmental stress);
- weather conditions during the trials are similar;
- the tested crop's attractiveness to bees is higher, compared to the surrounding area, during the trials.

4.3.1.8.3. Data processing

Appropriate statistical analysis should be done for each assessed parameter in order to detect differences between treatments and among days, in particular before and after pesticide application (see Pirk *et al.*, 2013). The magnitude and the duration of the effects should always be detected for following parameters:

- bee mortality and behaviour deviance (see Table 1),
- strength of the colony and honey production (Delaplane *et al.*, 2013b),
- bee activity (Scheiner *et al.*, 2013).

Moreover, an analysis of the statistical power to detect a certain magnitude of effect should be provided in the test (Cresswell, 2011). In fact, the hazard of a pesticide should be defined in terms of magnitude and of temporal scale. For instance, in the treated fields, the bee mortality is increased x times compared with the control for y days. This information can be of use to the risk manager for mitigation actions (see section 8.4.4.).

4.3.2. Forced in-hive nutrition
4.3.2.1. Introduction

Forced in-hive nutrition has been used to investigate the distribution of a xenobiotic within the colony (honey bees) and within the hive (beeswax, pollen, honey) and determine the effects of exposure on honey bee colonies and the development of honey bee colonies.

The selection of the conditions to conduct tests with honey bee colonies is driven by the goal of the experiment. When studying pesticides, the exposure – acute or chronic - is the first parameter to determine. Secondly, experimental conditions have to be chosen for the observation of the targeted parameters such as the mortality of honey bees (adults and larvae), the behaviour of honey bees (Scheiner *et al.*, 2013), the presence/absence of bee pests and diseases (see *BEEBOOK* Volume II, and typical bee disorders (absence of eggs, absence of foraging activity, etc.).

Forced, in-hive nutrition has been used to study veterinary drugs given to colonies (antibiotics and acaricides (Adams *et al.*, 2007)), pesticides used for plant protection (Faucon *et al.*, 2005; Pettis *et al.*, 2012) and the effects of various diets, whether artificial or natural, on colony development (Mattila and Otis, 2006b). The last point does not imply the study of any AI but has generated many publications describing how to artificially feed colonies. These publications also described the parameters observed to assess colony development and some biological traits of honey bees: estimation of the number of populated frames; estimation of the total comb area with sealed brood, open brood (eggs and larvae), stored pollen, or stored honey (see Delaplane *et al.*, 2013b); assessment of worker longevity, monitoring of behaviour –including memory through the use of PER reflex (proboscis-extension response)- and foraging pattern (see Scheiner *et al.*, 2013); measurement of protein content of workers; and the measurement of *Nosema* spore levels in workers (Mattila and Otis, 2006a; Mattila and Otis, 2006b; Mattila and Otis 2007; DeGrandi-Hoffman *et al.*, 2008; Mattila and Smith 2008; Avni *et al.*, 2009, Fries *et al.*, 2013).

4.3.2.2. Methods
4.3.2.2.1. The use of test syrup
There are multiple reasons for using syrups (sugar water) in the study of honey bee colonies. In this section, we will only focus on syrup use to study pesticide effects on colony or pesticide repartition within the colony. The use of syrup to distribute an AI for varroa control such as the trickling method (pouring syrup directly onto the bees between the frame spaces with a syringe) will not be reviewed but can be found in Dietemann *et al.*, 2012.

4.3.2.2.1.1. For pesticide studies
Only a few studies report the use of supplemented syrup to study the influence of pesticide on the colonies maintained in field conditions. Faucon and collaborators (Faucon *et al.*, 2005) studied the effect of imidacloprid exposure on colonies by feeding them with two concentrations of the pesticide diluted into syrup. One litre of syrup was given to each colony twice a week during two months. Bee activity, bee mortality, colony weight, honey production, observation of disease symptoms and pesticide repartition within the colony were assessed.

In 2007, the European Commission indicated that some guidelines related to setting maximum residue limits (MRL) should be produced for pesticides in honey within the EU regulation framework (EC-396/2005) using colonies fed with supplemented syrup. The working group led by French Food Safety Agency - AFSSA (now incorporated in French Agency for Food, Environmental and Occupational Health Safety - ANSES) identified a gap in the regulation when pesticide residues may arise in honey through residues present in feeding stuffs. MRLs established in this case should in principle be set on the basis of appropriate supervised residue trials data. Therefore the group

produced a document including a protocol to study the transfer of pesticide residues from syrup to honey (AFSSA, 2009). The principle of the test is based on spiked sugar syrup placed in a colony feeder. The honey bees collect it and store it in the cells of beehive frames. After transformation, the ripe honey is analysed to determine the "residue" of the tested AI. Control syrup is spiked with the solvent used to dilute the test compound. The quantity of syrup given to each colony depends on the strength of the tested colony. A quantity of 5l for a colony of 10 combs and 20,000 honey bees is considered sufficient. Syrup is distributed in the feeder all at once. In this protocol, only residues in honey are assessed. However, it is possible to adapt other observation concerning the biological traits of honey bees if needed.

4.3.2.2.1.2. For antibiotic studies
When experiments are set to study antibiotics, they usually aim at documenting the repartition of antibiotic residues within the apicultural matrices. Antibiotics are mixed with syrup made usually with sucrose. Syrup can be poured into frames (Adams *et al.*, 2007), or fed to the colony with through feeders. Control colonies are fed with non-supplemented syrup (Martel *et al.*, 2006).

4.3.2.2.2. The use of pollen patties
Patties have been used mainly to document the influence of diet on colony development. In some experiments, they have been use to investigate the effects of chronic pesticide exposure on honey bee health (Pettis *et al.*, 2012). Patties are principally made with some kind of protein (commercial products or pollen collected by honey bees) and sugar (syrup or honey) (Mattila and Otis 2006a; Degrandi-Hoffman *et al.*, 2008). Quantities given to colonies are dependent on the purpose of the experiment and on the size of the colony. When patties are used for pesticide studies, they are spiked with the given AI. In the latter case, it is recommended to sample the fresh patties and analyse it for pesticide levels to insure the proper delivery of the target dose to the colony.

4.3.3. Dust dispersion during sowing
4.3.3.1. Introduction
In contrast to targeted spray applications, where bees are exposed in the treated crop, exposure of bees to dusts is caused by dusts in the seed bag and dusts abraded from the seeds which are emitted into the environment during loading of sowers and during sowing and drift into neighbouring flowering crops. The contamination of nectar and pollen in adjacent field crops and contact exposure to dusts on the treated plants are the most important routes of exposure of bees to dusts. To achieve a realistic pesticide exposure to bees foraging on flowers from bee attractive plants located next to fields sown with pesticide-treated seeds, specific requirements in terms of study design, test item application, and field experiment establishment need to be met.

As no commercial machinery for a targeted dust application on flowering crops is available, it is not possible to administer precisely target doses of AI/ha on flowering crops. Most field trials are conducted by sowing treated seeds and measuring drift into neighbouring areas. To achieve meaningful results, appropriate establishment of trials with sowing and drift of dusts into adjacent crops must be accomplished and one must generate proof of achieving the targeted exposure to bees. While the development of appropriate methods for dust trials continues, experimental designs that allow assessing pesticide effects on bee colonies have been effective and are described in this section.

4.3.3.2. Methods and general requirements for dust exposure field studies

4.3.3.2.1. Requirements for establishment of field trials

4.3.3.2.1.1. Set up and location of bee hives

Field colonies should be set up directly at the field border and sowing activity should be carried out during full bee flight to ensure bees will be exposed by flying through dust clouds during sowing.

4.3.3.2.1.2. Seeds

1. Seed treatment quality data should be obtained before the trial. As the treatment quality may vary between seed treatments and batches, a poor seed treatment quality should be used as a worst case scenario. The total emission from the sowing machine is influenced by the dust abrasiveness (Heubach-value) as well as by the content of AI in dust. The seed quality used for trials needs to be documented for both, amount of dust and content of AI, before the trial starts and given in the report. Since 2008 the Heubach-Dustmeter test method (Heimbach, 2008) was introduced and proposed as a standardized measure of dust abrasion. The Heubach method mainly detects fine dust particles which are most prone for drifting.

2. Residue analysis of the AI in the dust needs to be given in the study as well as information on the AI and the treatment rates.

3. Furthermore, dusts may be present at the bottom of the seed bags. Thus, before the trials, seed bags should be checked to determine if any dusts remain at the bottom. All contents from the bag should be filled into the driller.

4.3.3.2.1.3. Amount of seeds used per hectare

The amount of seeds used per hectare influences the emission for the field sown into neighbouring areas. Therefore the amount of seeds drilled per hectare (amount filled into the drillers minus amount still in the driller after the sowing) needs to be calculated and reported.

4.3.3.2.1.4. Machinery and modifications of sowing machines

1. The machinery used will influence the potential emission. Depending on the crop, mechanical or pneumatic seeders are used for sowing of different crops. Mechanical seeders usually release only small amounts of dusts which is in contrast to precision airplanters with pneumatic vacuum singling of seeds. A number of sowing machines and their accessory kits regarding the potential for dust emission during sowing have been tested for their dust emission potential. Compared to unmodified standard equipment, the drift of these models with deflectors was at least 90% reduced.

2. Depending on the study aim, it should be decided if deflectors should be used. All details on the machinery and deflectors used for sowing need to be documented and given in the report. Preferably tested sowers should be used (e.g.http://www.jki.bund.de/no_cache/en/startseite/institute/ anwendungstechnik/geraetelisten/abdriftmindernde- maissaegeraete.html). For dust drift trials, different machinery types, e.g. pneumatic or mechanic sowing machines, may be used depending on the study aim. Also deflectors may be used depending on study aim. All details about the machinery used need to be given in reports.

4.3.3.2.1.5. Location of fields

An isolated location ensuring exposure of bees in an attractive, exposed crop adjacent to the sowing needs to be chosen. As with all standard field tests, it should be ensured that no other bee attractive crops are present in a range of at least 2 km to ensure maximum exposure.

4.3.3.2.1.6. Soil conditions

Humid soil surface is more likely to retain dust particles on the field sown. As a worst case situation, a dry soil surface is recommended which will allow dust particles to travel and drift even after having touched the soil surface. Soil condition and soil humidity for the time of the sowing have to be reported.

4.3.3.2.1.7. Wind conditions, direction, weather conditions

1. The field site needs to be carefully chosen as it should be determined that sufficient drift directed into the exposed flowering crop occurs.

2. Wind speed and wind direction especially during sowing needs to be documented and reported. For achieving the worst case exposure, fields should be established to ensure that all dusts drift into the flowering crop. Since it is not possible to predict the wind direction several days before start of the experiment, it is recommended to have flowering neighbouring crops on two sides, representing two main wind directions. The trial set up and the availability of uncontaminated forage needs to be carefully considered in the interpretation of the results.

3. Other weather conditions before, during and after sowing have to be reported in the same way as for experiments with spray applications.

4.3.3.2.1.8. Sowing

The sowing area should be sufficiently large. Dust drift may travel far wider than spray drift. Therefore the sowing width should be sufficiently wide (about 50 m or more). The start and end of the sowing area has to be reported.

4.3.3.2.1.9. Foraging conditions during full bee flight

To ensure the exposure of flying and foraging bees to the pesticide, sowing should be done during full bee flight activity when bees are actively foraging on the crop neighbouring the sowing area to ensure the worst case exposure to contaminated plant surfaces, nectar, pollen, and to dusts present in the air during the sowing process.

4.3.3.2.1.10. Crop for sowing

As the seed treatment quality and the potential of crop exposure may vary greatly between different crops, the crop needs to be selected according to the study aim.

4.3.3.2.1.11. Flowering adjacent crops

Adjacent to the sowing area, a bee attractive crop (e.g. Winter Oilseed Rape, *Phacelia* or Mustard) is needed. The crop should be at full flowering (BBCH 65-67).

4.3.3.2.1.12. Residue samples (plants, bees, bee matrices) proof of exposure

1. To demonstrate the exposure achieved in the contaminated adjacent crop, Petri dishes with wet filter paper should be placed at least at 1, 3, 5, 10 and 20 m in free cut areas (on at least 30 m length) in the neighbouring crop.
2. Also, flower samples may be taken very carefully to avoid a loss of dust particles.
3. Foraging bees returning to the hive should be collected for residue analyses of nectar and pollen.
4. Additionally, samples of fresh nectar in combs, freshly stored pollen, honey and bee bread or other matrices (e.g. Royal Jelly) may be obtained.
5. Because soil particles may drift during sowing, a residue analysis of the upper soil layer is recommended.

4.3.3.2.2. Setup of field trials using other devices for a direct dust application

A few testing facilities have developed machinery for a direct application of dusts in field trials. As only small amounts of contaminated dust containing insecticides are emitted during sowing operations, only very small amounts of these dusts have to be applied homogenously. To ensure a good dispersion of small amounts of insecticidal dusts during application in the field, an inert filling material may be necessary. Different materials may be used for filling purposes. Small dust particles of soil seem to represent real field situations best and are recommended.

A good mixing of the contaminated dust and the filling material needs to be ensured. It is important to ensure that appropriate particle sizes of dusts and of the filling material are used. In semi-field trials with manual application of dusts on flowering crops, it has been demonstrated that smaller particles, e.g. below 160 µm, result in higher effects. Small particles are also more likely to drift into adjacent crops. See section 3.2.3.1.2.4. of the present manuscript for the method.

4.3.4. Foraging on a treated crop

4.3.4.1. Returning foragers as a tool to measure the pesticide confrontation and the transport into the bee colony

After the application of a pesticide in blooming cultivations or orchards, forager bees might be contaminated during their flight (Schur and Wallner, 1998). Also systemic pesticides may reach nectar and pollen of seed treated plants or after spray applications before the blooming stage (Wallner, 2009). The bee body itself and the collected goods contain residues of the applied ingredients.

Residue analysis with honey showed that this bee product is inadequate to measure the realistic level with which single bees are confronted. During honey preparation, honey bees have a remarkable influence on the residue level in honey. Reduction factors up to 1000 times have been shown between the nectar contamination and prepared honey. Based on the lypophilic character of the pesticide, colonies are more or less successful at reducing the contamination level. As a general rule, harvested honey is less contaminated than harvested nectar (Wallner, 2009). Therefore honey cannot be used to access the pesticide levels that bees have to handle on their flights. A much better tool, even to demonstrate that there was a contact to sprayed fields, is the analysis of returning foragers and their loads (Reetz *et al.*, 2012). This can be done in field experiments as well as in tent tests with reasonable plot sizes.

Besides the analysis of returning foragers at the hive entrance, it is also possible to collect bees directly from plants or flowers. In this case, a 12 Volt vacuum, which can be run with a car battery, is useful (Wallner, 1997). Residue analysis is performed on the basis of single bees (pollen loads or honey stomach content) or pooled groups of one sampling date.

4.3.4.1.1. Reasons for collection of forager bees

- Residues at worst case level (no dilution, nectar present in the crop)
- In combination with sampling plants/flowers and matrices from the bee hive (honey, pollen, bee bread), the route of transfer of residues from a pesticide in the bee hive can be demonstrated
- Determination of realistic residue values for the risk assessment and further evaluations/studies (e.g. bee brood study in lab)
- Assessment of exposure in the field via pollen source determination

- Assessment of exposure to contaminated water sources, e.g. guttation (Reetz *et al.*, 2011).

4.3.4.1.2. Collection of forager bees in tunnel tents or in the field

On each sampling day, one sample of approximately 300-600 forager bees will be taken per hive. At each sampling, the hive entrances will be sealed before the sampling and the forager bees will be subsequently collected as they return to the hive e.g. by suction with a vacuum, by brushing them into a box filled with dry ice, or by using a pair of tweezers. After each sampling interval, the hive will be re-opened allowing honey bees to return to and leave the hive.

Directly after sampling, each sample will be divided into two sub-samples (A and B). Each sub-sample should approximately 150 bees, one for preparation (A) and one as a retained sample (B). To avoid squeezing during storage and shipment, the bees will be transferred into containers. If <300 bees are collected per hive and sampling day, then sub-sample A will be composed of up to 150 bees with any remainder being allotted to sub-sample B. Details of the approximate numbers of bees collected for each sub-sample will be recorded in the raw data. Each sub-sample will be labelled uniquely.

All samples will be chilled during transport to the freezer and subsequently will be stored deep frozen at ≤-18 °C. Storage conditions will be recorded by use of a data logger or a min/max thermometer and will be documented in the raw data.

4.3.4.1.2.1 Preparation of the honey stomachs

The forager bees collected as described above will be stored deep frozen (≤-18°C) in separate containers for each treatment group until preparation in house of the honey stomachs. In principle, it is possible to determine the nectar source of single bees with pollen analysis of the honey stomach content. Successful foragers could be identified by their body weight before the preparation process.

The preparation of the honey stomachs from forager bees will be done as follows (see Carreck *et al.*, 2013 for more information):

1. All bees of one sample will be allowed to thaw for a few minutes.
2. Bees will be fixed at their thorax and their abdomens will be stretched flat with a pair of tweezers.
3. The abdomens or the tergite plates will be removed, so that the honey stomachs will be free.
4. The honey stomach will be held at the lowest part of the oesophagus (see Carreck *et al.*, 2013).
5. The main front part of the oesophagus should be removed.
6. The honey stomach will be held with a pair of tweezers at the small remaining part of the oesophagus.
7. The total weight of the honey stomachs will be determined.
8. The honey stomach contents from one sampling time, treatment and replicate hive will be pooled to get at least 0.2g per sample. The number of prepared bees per sampling time, treatment

and replicate, will be recorded. The nectar sample will be transferred into the freezer immediately after the preparation of one forager bee sample.

9. Bees from the control sampling will be processed first. Once this task has been completed, the process will be started with the last sampling.
10. After preparation, the contents of the honey stomachs will be stored separately for each sample at ≤ -18°C.

4.3.4.1.2.2. Preparation of the pollen loads

The preparation of the pollen loads will be carried out as follows (see Delaplane *et al.*, 2013a and Carreck *et al.*, 2013 for more information):

1. All bees from sub-sample A are kept on a deep frozen metal plate (≤ -18°C).
2. The pollen loads will be detached from the legs of the forager bees and placed into a vial.
3. All pollen loads from sub-sample A will be collected and pooled in order to get at least 100 mg of pollen for residue analysis. If < 100 mg is obtained from sub-sample A then sub-sample B will be prepared. If this is the case, all bees of sub-sample B will be prepared in the same way as sub-sample A and added to sub-sample A. The total number of prepared bees and the sub-samples used will be recorded.

The pollen samples will be unfrozen during the preparation of one sub-sample. The bees and pollen will be transferred back to the freezer immediately after the preparation of one sub-sample. Each sub-sample will be labelled 3 times and will include at least the information given below. All samples will be frozen at ≤ -18°C outside of the sample preparation time.

4.3.5. Systemic toxins expressed in plant matrices
4.3.5.1. Introduction

Systemic products have the capacity to enter into the plants independently of their application pattern. Commercial products containing these AIs exist for treatments of seeds, soils, for applications as spray or directly to the roots or bulbs. Other application patterns may render systemic any AI, as is the case of stem injections. Pesticide formulations may contain other AIs or co-formulants that increase the systemicity of the AI under study (Dieckmann *et al.*, 2010).

This section focuses on the proposal of a protocol evaluating the impact on honey bees exposed to the pollen and nectar coming from a crop that has received a treatment different from spraying with systemic products in field conditions. Exposure to guttation water or honeydew would require specific modifications of the methodology. Therefore, it should be dealt with separately.

Different methodologies for different application patterns: a different section should deal with the study of the impact of pesticides with systemic properties applied on spray.

4.3.5.2. Application of systemic products as seed and soil treatment (SSST), bulbs or root bathing

4.3.5.2.1. Introduction

The methodology presented here focuses on the exposure of bees to contaminated flowers resulting from treated plants (as seed and soil treatments, bulbs or roots bathing). Observations are done at the level of the colony and only individual observations on bees are included insofar as they may affect colony development. In principle, guttation water would not be a major source of exposure given that normally these droplets occur mainly in early plant developmental stages (Girolami *et al.*, 2009; Tapparo *et al.*, 2012). However, the individual geographical and meteorological conditions of each area should be considered to exclude this potential exposure route.

The EFSA has published an extensive review about the risk assessment of pesticides on bees (EFSA, 2012). In this document, a thorough analysis has been conducted concerning the adequacy of the international standards (EPPO, 2010a) recommended for field-testing to the exposure of bees to systemic pesticides. The following recommendations are based on the limitations identified on the EFSA document.

4.3.5.2.2. Principle of the trial

Beehives come from a similar background, the same apiary or constituted in the same way. Their health status and strength are evaluated before the beginning of the trial. Then they are placed on the test fields as soon as the crop presents a number of flowers enough to allow the visit of foragers (5 to 15% of the flowers are flowering). The crop must have been treated at the time of seeding/planted when it starts to bloom. After the flowering period, the colonies are returned to a common area where they will remain until the following season.

The observation of effects continues during and after bloom. The monitoring can be extended until the spring of the following year. Especially when the tests is run during the period of production of winter bees, this monitoring until the spring becomes more relevant. Ideally, the generic observations on the full colonies should be complemented with individual tests studying the impact of sublethal doses on bees, e.g. homing flight tests or with more specific observations (fecundity, growth and development of individual honey bees), though many of the sublethal effects may be captured in the full colony assessments.

4.3.5.2.3. Preliminary steps

Seeding/planting/pesticide application should follow GAP. Bee colonies should be conducted following Good Beekeeping Practices. A flight entrance observation system (e.g. Floriade), which includes a climate control station as well as bee tracking system, could be placed in the area of testing. It should collect the meteorological data (temperature, relative humidity and rainfall) and provide information about the bees' activity all along the duration of the trial. Should such a system not be available, alternatives should be found to collect the mentioned data (meteorological data, foraging activity, etc.).

4.3.5.2.4. Environment of the trial

The aim of the information collected from the environment of the colonies under study is identifying potential interferences of the exposure of bees to the AI or potential synergies in their action on bee colonies.

It is well known that bees cover wide surfaces when foraging, mean distances being around 1.5-3 km, extreme distances being around 10 km (Vischer and Seeley, 1982; Winston, 1987; Seeley, 1995; Steffan-Dewenter and Kuhn, 2003), average surface ranging from 7 to over 100 km^2. International standards, however, normally recommend a treated area of 2,500 m^2 or 1 ha.

With the help of satellite imaging or similar, the environment of 3 km around the placement of the colonies could be audited and noted. All software should be up-to-date. Whenever possible, any chemical treatments happening in this area should be registered and considered for the study.

4.3.5.2.5. Trial plots: experimental and control

4.3.5.2.5.1. Crops planted in the trial plots

In order to increase the likelihood that bees will forage in treated plots, crops attractive to bees should be used. Special attention should be put on the nutritional value of the pollen of the chosen crop. Rich pollens as that of oilseed rape or *Phacelia* may mask the effects of the exposure to the pesticide. Ideally, an attractive crop with pollen of lower nutritional value would better evidence any toxicological problems (e.g. sunflower). For regulatory purposes, the crop for which the authorisation is to be requested should be used.

4.3.5.2.5.2. Size of the trial plots

Trial plots should be a minimum of 5 ha. Should this not be the case, testers should make sure that the treated crop represents a major nutritional source for the colonies of the test during the crop flowering period. Treated seeds or granules with the formulated product can be used as well. It should contain the highest dose recommended for field application. Should less attractive crops be used, specific attention should be put on assuring that exposure occurs.

4.3.5.2.5.3. Location of the colonies at the trial plots

One can possibly increase exposure by placing the colonies on the edge of the field. Studies have shown that pesticides affect the navigation capacity of foragers. By bringing the colonies closer to the field, the distances foragers need to cover might not require as much flight effort. Similarly, bees foraging close to their hive would not need to consume part of the nectar they collect to obtain energy for returning to the hive. Therefore, effects on foragers might be underestimated.

Pesticide exposure has been shown to hinder homing flight and affect foraging behaviour (Vandame *et al.*, 1995; Bortolotti *et al.*, 2003; Colin *et al.*, 2004; Karise *et al.*, 2007; Yang *et al.*, 2008; Decourtye *et al.*, 2011; Henry *et al.*, 2012; Scheinder *et al.*, 2012). Therefore, field trials should be complemented with methodologies specifically evaluating these behaviours. For further information on the protocols to run these tests, see Scheiner *et al.*, 2013. Specific methods can evaluate the impact of pesticide exposure on fecundity, growth and development of individual honey bees (Dai *et al.*, 2010). The development of the colony can be assessed per Delaplane *et al.*, 2013b.

4.3.5.2.5.4. Distance between trial plots
The distance between treatment plots and control ones should be enough to avoid the exposure of the latter to the AI. Therefore, a distance of at least 6 km is desirable. Otherwise, environmental conditions should remain comparable for all plots.

Should the minimum distance of 3 km not be achieved, residue analyses of the contents of the honey stomach of foragers or pollen clusters returning to the hive would provide information about the existence of cross foraging (i.e. bees foraging on the plots not designated for them). Palynological studies can as well help in this task. For method on recovering the honey stomach, see section 4.3.4.1.2.1. or Carreck *et al.*, 2013. Potentially, the same procedure could be developed for the study of the exposure of bee colonies to pesticides in water sources around the apiary.

4.3.5.2.6 Colonies used
Queen-right colonies are used for the trial. Queens should be daughters of one queen of the same age. Ideally, colonies with no remarkable problems (i.e. free of pests/diseases/hive abnormalities) for at least one brood cycle previous to the beginning of the trial should be used.

4.3.5.2.6.1. Colony health status
Colonies should be regularly monitored for the occurrence of diseases (including varroa infestation level, see Dietemann *et al.*, 2013) and any clinical sign should be noted. Prior to the exposure to pesticides, no clinical signs should be observed. Colonies should not be taken if they have received a treatment against varroa in the last 4 weeks prior to the trial. If the varroa treatment is administered during the trial period, the treatment protocol (date of the treatment, product, duration, quantity applied and efficacy observations) should be noted.

Delaplane *et al.*, 2013b describes recommendations concerning colony size, which should be as homogeneous as possible. As field tests should resemble as much as possible realistic conditions, colonies' population would differ depending on the time of the year in which the trial would occur. Colonies of 15,000 individuals would be characteristic of a beginning of the season or overwintering period, while colonies of approx. 50,000-60,000 individuals would be characteristic of the middle of the season (EFSA, 2012). These

estimations however, might vary geographically. The evolution of the colony health status along the trial is one of the observations described later in this method.

4.3.5.2.6.2. Number of colonies/replicates – statistical power
6 to 10 colonies per treatment group (exposure/control) should overcome the inter-colony variability (EFSA, 2012). The number of replicates per trial depends on the magnitude of effects that the test should detect. The statistical power of the test should always be calculated (see Pirk *et al.*, 2013).

4.3.5.2.6.3. Colony placement and equipment
Colonies will be placed all together at an environment free of pesticides where they will be monitored at least 7 days before flowering. If necessary, colonies can be fed with syrup to avoid starvation. The colony should not be exposed to contaminants in syrup. Residue analyses or tracking the syrup origin may help providing this information.

When the crop starts blooming (5 to 15% flowers of the crop have bloomed), colonies will be placed on the edge of the plots. Observations of the colonies will start 7 days before the expected time of flowering.

Pollen traps can be installed in 3 or 4 colonies per treatment group. Each colony should have dead bee traps. Devices like colony scales, bee counters or bee-tracking systems (e.g. Floriade, etc.) may provide extra information on the evolution of the colony throughout the trial (see Human *et al.*, for information on using pollen traps, dead bee traps, and for weighing colonies).

4.3.5.2.7. Duration of the test
Colonies remain on the edge of the field for the period of blossom. However, observations of the their evolution will be extended up to at least 42 days after the placement on the edge of the fields under study. This is the time of two complete brood cycles.

After blooming, they should be moved to an environment where they would overwinter together on the reserves they have accumulated during the trial period. The environment of the colonies should provide enough sources of pollen and nectar to survive. If necessary, colonies can be fed with syrup. This can be done making sure that the colony has consumed first its reserves collected during the exposure period. The colony should not be exposed to further contaminants contained in syrup. Residue analyses or tracking the syrup origin may help providing this information.

The colonies should be monitored through the following season. In the event that pesticide residues are still present in the colony at this time, the monitoring should be extended in the new season. A residue analysis of beekeeping matrices would enable one to know when the exposure of the colonies to the AI has occurred over the winter. It should be noted that these are test conditions. In reality, colonies might be exposed to larger amounts of AI over longer periods or to a mixture of AI.

4.3.5.2.8. Bees' exposure

The exposure of bees to AI following SSST is more difficult to control than that following spraying of non-systemic products. This is because blooming does not occur in the whole surface at the same time and because during the blooming period one cannot say if bees are only going to forage in the treated crop. Therefore, special manipulations need to be performed to ensure the level of exposure achieved by the colony as a whole. The control of the colony's food intake is one parameter that can be achieved.

For this purpose, pollen pellets should be collected with pollen traps installed at the entrance of the colonies prior the blooming of the first flowers of the crop and every 2-3 days during the blooming period (see Human *et al.*, 2013). Samples of at least 5 g of pollen should be collected and kept in hermetic conditions, adequately labelled and immediately frozen. Samples are stored at least at -18°C before analysis.

Pollen from the comb should be collected once before the beginning of the crop bloom and once a week following it. If the samples were taken by cutting a piece of comb, wax samples would be readily available. Otherwise, wax samples should be taken as well on the same days and immediately frozen. Samples are stored at least at -18°C before analysis.

Foragers returning to their hive should be collected (see section 4.3.4.1.2.) at the entrance of the colony to undergo residue analysis of the content of their honey sac. Approximately 50 foragers should be collected prior to the blooming and every 2-3 days during the blooming period. Samples should be kept in hermetic conditions, adequately labelled and immediately frozen. Samples are stored at least at -18°C before analysis.

Honey samples should be collected once before the blooming of the crop and once a week after.

Dead bees should be counted daily from the period starting before the bloom and 42 days after it. Dead bee traps (Human *et al.*, 2013) will be cleaned every evening and samples of bees should be collected from the bee traps before sunrise. The collection period goes from just before the start of blooming and is conducted every 2 days during the blooming period. Samples should be kept in hermetic bags, appropriately labelled and immediately frozen (stored at least at -18°C before analysis).

The quantity of sample per beekeeping matrix hereby proposed is indicative. It should be checked with the laboratory in charge of residue analyses prior to the beginning of the test.

Prepupae should be counted daily, in the same way as dead bees. Bee traps will be cleaned every evening and samples of bees should be collected from the bee traps before sunrise. They can be collected from the bee traps every 2 days and kept in hermetic bags, appropriately labelled and immediately frozen. Another option is the sampling of larvae directly from the comb once before the blooming of the crop and once a week after. Again samples should be kept in hermetic bags, appropriately labelled and frozen in case analyses should be delayed.

4.3.5.2.8.1. Pollen analyses

The origin of pollen in the pollen pellets can be identified through their colour and their palynologic analysis (see Delaplane *et al.*, 2013a). Pollen provides a good tool to monitor the environment of the colony. Palynologic analysis should as well be carried out in honey samples. Therefore, in the week previous to the expected blooming of the treated and control crop and once weekly during this period, pollen samples should be taken with the help of pollen traps (see Human *et al.*, 2013). Pollen origin analysis can be used to complete the information on the environment collected from the satellite images.

4.3.5.2.8.2. Residue analyses

Residue analyses of the previously mentioned matrices should be performed for both treatment and control colonies. Two different analyses could be envisaged, one specific on the AI under study for which the lowest possible LOD and LOQ should be used, and a multi-residue analysis of the most common AI used in the area. The former should be systematically performed when conducting field studies. We do not provide a method for residue analyses as such analyses are typically outsourced to analytical labs.

4.3.5.2.8.3. Reserves of the colonies at the beginning of the trial

It is necessary to reduce as much as possible the content of previous food reserves in hives so that the exposure to the AI present in the field can be maximised. That is why one could remove the frames containing mainly food reserves from colonies before the crop blooms. This could lead colonies to starve in the days immediately following the removal of the food. Consequently, the health of the colony should be monitored closely.

4.3.5.2.9. Observations

4.3.5.2.9.1. Controls

The experimental design allows two kinds of controls: internal and external ones. Each colony serves as its own control (internal control), by comparing its evolution before the exposure to the AI and after it. Additionally, the evolution of the treatment colonies would be compared to that of the control ones (external control).

4.3.5.2.9.2. Brood and reserves content

The surface of brood and reserves should be monitored before, during and after the trial (see Delaplane *et al.*, 2013b). Estimation of colony strength parameters should be performed close before the crop bloom and one week after. Given that the reserve frames should have been removed before the study, there should be visual controls of the food content of the colony. The observation should be repeated once weekly up to the 42 days of the duration of the trial. In case a more intensive data gathering method is used (e.g. the Liebefelder method presented first in Imdorf, 1987 and described in Delaplane *et al.*, 2013b), one could reduce the data collection to every three weeks.

4.3.5.2.9.3. Interpretation of residual information
The information of the residue content in the nectar and pollen brought back to the control and treatment colonies allows one to determine the quality of the control. Additionally, it would provide an estimation of the level of exposure and the comparison of the level of contaminated and non-contaminated food arriving to each colony.

The results of the residue analyses of larvae and dead bees from the trap would provide an indication of the level of exposure that in-hive individuals face. The result of the residue analyses of in-hive stored pollen and honey and the wax would provide an indication of the level of exposure of in-hive bees and of a potential long-term exposure.

4.3.5.2.9.4. Toxicological endpoints
In this section we focus only on the colony as experimental unit. Therefore, the endpoints chosen in this section are directly linked with colony status. Further methodologies could be developed in the field to complement these observations, as is the case of homing flight tests or fecundity tests.

4.3.5.2.9.4.1. Mortality trend
Dead bees can be counted using bee traps placed in front of the hive (Human *et al.*, 2013). If a bee counter is used instead (an electronic device that counts bees exiting and entering the hive), the number of bees leaving the colony and not returning should be determined. These observations should be compared at a certain time of the day with a specific duration (e.g. every morning from 7 to 8 am).

These observations should be done on a daily basis from one week before the colonies are placed in the field until the end of blossom of the treated/control crop. Afterwards, the observations can be done on a weekly basis up to the 42 days.

4.3.5.2.9.4.2. General evolution of the colony during the test
Special attention should be put on the strength and vitality of the colony (see Delaplane *et al.*, 2013b). Should scales be placed on the colonies of study, weight evolution could be used as well as variable to compare treatment and control colonies (Human *et al.*, 2013). The same could be done in case bee counters are installed.

These observations should be done on a daily basis from one week before the day the colonies are placed into the field and until the end of blossom of the treated/control crop. Afterwards, the observations can be done on a weekly basis up to the 42 days.

4.3.5.2.9.4.3. Behavioural observations
The aim of the present protocol is not to evaluate effects on specific behaviours (e.g. homing flight, thermoregulation, etc.), but to observe any alterations on the general behaviour of the colony during the test and after the test. For this reason, any qualitative modification as trembling, aggressiveness, disorientation, apathy, etc. observed at the flight board, outside or within the hive during the test should be noted. Additionally, during a longer period (until next season), abnormalities in the reproduction cycle of the colony should be noted (e.g. supersedure of the queen, problems on egg-laying capacity, etc.). Finally, observations of the flight activity and the foraging behaviour around the hive should be done and alterations should be noted.

There is a wide room for improvement of the behavioural observations that could be done in field test. Namely, specific behavioural traits would increase the accuracy of the observations. The present protocol should be modified in the future as soon as there are advances in methodologies.

4.3.5.2.9.4.4. Colony health
In principle, only colonies not showing disease signs should be included into the experiment. Then pathological signs, their date of appearance and severity should be noted (see *BEEBOOK* Volume II for information on this). The health status of the colony should be monitored from one week before the day the colonies are placed into the field and extended up to the overwintering. The appearance of pathological signs in the treatment colonies, but their absence in the control ones, could be due to a synergic effect pathogens-pesticide.

4.3.5.2.9.4.5. Brood surface and quality
The different observations developed on the brood surface should allow identifying eventual deficiencies in the egg-laying capacity of the queen or the brood success. Any alteration (e.g. mosaic brood, dead larvae/nymphs, increase of pathologies affecting brood, etc.) should be noted, both in quantity and quality. Protocols for brood evolution and monitoring are described in Delaplane *et al.*, 2013b. The assessment of the duration of a brood cycle would be indeed, very interesting from the point of view of the interactions between the pesticide and pathologies. Dead larvae in the bee trap should as well be noted.

4.3.5.2.10. Validity of the trial
Positive residue analyses in samples of pollen or nectar brought back to the control colonies would render the test as invalid. Negative residue analyses in samples of pollen or nectar brought back to the treatment colonies would render the test invalid. Prior to the treatment (before the blooming period) the mortality and behaviour of the colony (incl. foraging activity) should be not statistically differ between

treatment and control groups. Should this not be the case, the study would be invalid.

The evolution of mortality and the different observations described above do not change in the case of the control fields both before and after exposure to flowers. Different crops are susceptible to being treated with the same AI. This could extend the exposure of the colonies under study in time and quantity. Similarly, the different blooms happening in the surroundings of the colonies under testing may dilute the exposure quantities. The purpose of this protocol is to evaluate the effect of on bee colonies of a specific AI applied to a specific crop at a specific time in the year. The uncertainty of the representativeness of the results of the trial to reality is therefore high.

5. Effects of toxic substances on honey bee brood

5.1. Introduction

Honey bee brood may be exposed to pesticides through nectar and pollen collected by foragers. Effects on brood may vary according to the nature of the compound and its concentration in pollen and nectar (Aupinel *et al.*, 2007a, 2007b). Lethal or sublethal effects can be expected throughout the colony life, according to the number of larvae affected, the mode of action and its consequences on bees. Considering that colony survival depends on the adult population directly linked to brood health, it is evident that the effects of pesticides on brood have to be seriously considered.

5.2. *in vivo* larval tests

5.2.1. Oomen test

This test, even if never ring-tested, is a requirement in Europe and it is based on the method outlined in Oomen *et al.* (1992).

- In this in-hive method, experimental units are free flying colonies.
- The artificial contamination with AI is ensured using a syrup feeder of 1 litre fitted to the hive for 24 hours.
- Brood development is followed by weekly inspection of individual brood cells.
- Due to environmental variations, this method may not be easily reproducible since the test product may be stored in the combs and not immediately dispensed to the brood by nurse bees. It may also be diluted by external nectar. No quantitative data can be provided by this test due to the fact that exposure is not controlled.

5.2.2. Semi field test

This in hive method was devised by Schur *et al.* (2003) and is recommended by OECD.

5.2.2.1. Introduction

The European regulatory framework (Directive 91/414/EEC, Regulation 1107/2009/EC) requires data to evaluate the risk of pesticides on the honey bee brood. Beside the possibility to run studies under laboratory conditions, there are 2 publications available to run higher tier studies (e.g. semi-field and field) in order to evaluate the potential impact of a pesticide on the honey bee brood development.

The "in-hive field test" published by Oomen *et al.* (1992), is carried out with free-flying bee colonies, which are fed with contaminated sugar solution. One litre of sugar solution is mixed with a certain amount of pesticide and offered to the bee colonies over a short time period. The brood development is followed by weekly assessments of individual marked brood cells. Such kinds of tests are qualitative test methods or screening tests in order to evaluate the question, whether PPPs are causing harmful effects on the bee brood or not.

A quantitative test method closer to the real field scenario is the semi-field brood test according to the OECD Guidance Document 75 (OECD, 2007). Within this test design a PPP is sprayed directly on a flowering crop and the bee colonies are forced to forage for nectar and pollen in tunnel tents. Thus the bee brood contacts contaminated food and the development of the bee brood in single cells is followed regularly over one complete brood cycle from an egg to a worker bee.

A third possibility to evaluate the risk of PPPs to the bee brood under field conditions is a honey bee field study based on the EPPO 170 (EPPO, 2010a) guideline in combination with detailed brood assessments according to the OECD Guidance Document 75. In the following paragraphs the main focus will be directed to the test method under semi-field conditions.

5.2.2.2. Material and methods of a semi-field brood test

1. Similar as for standard studies based on the EPPO 170 guideline; small healthy honey bee colonies are initially placed in tunnel tents (herein after named tunnels) shortly before full flowering of the crop, a few days before application of the test chemical.

2. Following exposure of the bees in the tunnel for the period of flowering of the crop (e.g. at least 7 days after application of the product), the hives are placed outside the tunnels for the remaining time of the study and are free to forage in the field.

3. It is important to check that the neighbouring environment within a radius of 3 km is free from bee attractive main crops (e.g. sunflower, maize, oil seed rape, fruit orchards) as well as the test substance or other compounds.

4. Mortality of honey bees, flight activity (Human *et al.*, 2013), and condition of the colonies and development of the bee brood (Delaplane *et al.*, 2013b) are evaluated several times over a period of at least 4 weeks after the initial brood assessment.

5. Results are evaluated by comparing the treated colonies with the water-treated colonies and with the reference chemical-treated colonies.

5.2.2.2.1. *Design of the test*

1. A test includes at least 3 treatments:
   - Test chemical
   - Reference chemical or positive reference: An IGR known to produce adverse effects on honey bee brood (e.g. Fenoxycarb (CAS. 121-75-5), rate: at least 150 g/ha)
   - Control: The plants are treated with tap water (water volume: 200-400 L/ha in case of *Phacelia* as test plant)

2. All spray applications should be done with the same water volume. It is suggested to run the test with at least three replicates for better statistical analysis. Thus, in total at least nine tunnels are established for one test. However, it is also possible to increase the number of replicates to four per treatment group in order to increase the stability of the test.

5.2.2.2.2. *Preparation of the colonies*

1. The OECD 75 recommends using small healthy honey bee colonies (e.g. Mini Plus, nuclei, etc.) for the test, but it is also possible to use small commercial bee colonies. However, the size of the colonies should be adapted to the size of the crop area within the tunnels.

2. All colonies of one set or study have to be produced at the same time from colonies headed by sister queens to guarantee that the colonies in all variants are uniform as far as possible (Delaplane *et al.*, 2013b). The colonies must be headed by sister queens which are the progeny of the same queen and mated at the same place in order to minimise genetic variability.

3. The bee colonies should be free of clinical symptoms of disease (e.g. *nosema*, *Amoeba*, chalkbrood, sacbrood, and American or European foulbrood) or pests (*Varroa destructor*): see *BEEBOOK* Volume II. The colonies should be free of unusual occurrences (e.g. presence of dead bees, dark-"bald"-bees, "crawlers" or flightless bees, unusual brood distribution patterns or brood age structure).

4. After establishment of the colonies within the tunnels, all hives are equipped with a dead bee trap at the entrance to count the number of dead bees (Human *et al.*, 2013).

5. The colonies should be established in the tunnels shortly before full flowering of the crop and at least three days before application in order to allow the bees to adapt to the conditions in the tunnels.

6. The colonies should be exposed to the treated crop in the tunnels for a period of at least 7 days after the application.

5.2.2.2.3. *Test conditions*

1. As mentioned in section 5.2.2.2.2., the size of the tunnels should be adapted to the size of the used colonies, but a minimum size of 40 m² floor space is recommended in the OECD 75 guidance document. The minimum height of the tunnels should be 2.5m, to guarantee an unhindered flight of the bees. The covering gauze should have a maximal mesh size of 3mm. The test crop should be attractive to honey bees. Suitable are for example *Phacelia tanacetifolia*, *Sinapis arvensis* and *Brassica napus*.

2. During the whole testing period, the colonies should be supplied with water. A water feeder should be placed into each tunnel as water supply for the bees. During product application, the water feeder should be removed from the tunnel.

5.2.2.2.4. *Application*

1. The applications should be performed with a boom sprayer with calibrated nozzles according to GAP.

2. The spraying should normally be performed at the time of full flowering of the crop and during high bee flight for worst case conditions or, if required (e.g. for testing of residual or delayed action), in accordance with the intended use pattern of the product.

3. The wind speed should not exceed 2m/sec outside the tunnel.

4. Test products should normally be applied at the highest field rate (ml or g/ha) intended for the registration of the product in order to produce a worst-case exposure for the bees.

5. During the applications in the tunnels the water containers should be taken out of the respective tunnels and the bee colonies should be covered with a plastic sheet until the end of application to avoid direct contamination.

5.2.2.2.5. *Assessments*

The total observation period of the colonies is at least 28 days.

5.2.2.2.5.1. **Meteorological data**

During the whole testing period, the following meteorological data should be recorded daily (ideally inside the tunnel):
- temperature (min, max and mean)
- relative humidity (min, max and mean)
- rainfall (total daily)
- wind speed (only during application inside and outside the tunnel)
- cloudiness (during assessment).

Table 5. Time schedule for hive mortality assessment in semi-field brood tests:

DBA = days before application, DAA = days after application.

Timing	Evaluation of number of dead honey bees
At least 3DBA to 1DBA	Once a day, if possible at about the same time
0DBA	Once shortly before application
0DAA	2 hours after application 6 hours after application
1 to 7DAA	Once a day, if possible at about the same time
Outside the tunnels:	
8 to 27(±2)DAA	Once a day, if possible at about the same time at monitoring site (dead bee trap only)

5.2.2.2.5.2. Mortality of honey bees

1. Mortality of honey bees should be assessed on sheets suitable for the collection of dead bees (e.g. linen sheets) which are spread out in front of the hives and at the front, middle and back of the tunnels. From experiences with semi-field studies in general, it is known that most bees which are dying in the crop area can be found in the front and back corner of the tunnels. The middle linen is necessary as a path for walking during the application.

2. Before the start of the test, such paths should be created in each tunnel by removing of the plants and by smoothing the ground. Subsequently, the paths are covered with the aforementioned sheets in order to facilitate the collection of the dead bees in the crop area.

3. Additionally the dead bees are noted and counted in the dead bee traps which are fixed at the entrance of the hives. The assessments could be done according to the Table 5.

4. The assessments of the number of dead bees should be conducted at approximately the same time in the morning in order to cover the same time span from one day to another. During each assessment, the number of dead bees should be differentiated into adult worker bees, drones, freshly emerged bees, pupae and larvae.

Table 6. Time schedule for flight activity assessment in semi-field brood tests:

DBA = days before application, DAA = days after application.

Timing	Evaluation of number of forager honey bees/1 m² and observation of behaviour
At least 3DBA to 1DBA	Once a day during flight activity of the bees
0DBA	Once shortly before application
0DAA	4 times during the first hour after application 2 hours after application 4 hours after application 6 hours after application
1DAA	Three times during flight activity of the bees (preferably in the morning, midday and afternoon)
2 to 7DAA	Once a day during flight activity of the bees

5.2.2.2.5.3. Flight activity and behaviour

1. Flight activity could be recorded on a 1 m² area, at 3 different places in each tunnel according to the time table presented in Table 6.

2. At each assessment time, the number of bees that are both foraging on flowering plants and flying around the crop are counted for a short time period (for example 10-15 seconds depending on the crop) per marked area.

3. During the assessments of flight intensity, the behaviour of the honey bees in the crop and around the hive should be observed with respect to the following criteria:
   - aggressiveness towards the observer
   - guard bees attacking and/or preventing returning bees from entering the hive
   - intensive flying activity in front of the hives without entering the hive
   - intoxication symptoms (e.g. cramping, locomotion problems)
   - clustering of large numbers of bees at the hive entrance.

5.2.2.2.5.4. Brood assessments

5.2.2.2.5.4.1. Condition of the colonies

1. The condition of the colonies is assessed once before the application and several times after the application according to the following time schedule:
   - BFD (brood area fixing day), first assessment
   - Application at +2 days (±1 day) after BFD
   - + 5 days (±1 day) after BFD
   - + 10 days (±1 day) after BFD
   - + 16 days (±1 day) after BFD
   - + 22 days (±1 day) after BFD
   - + 28 days (±1 day) after BFD.

2. For the condition of the colonies the following parameters are assessed in order to record effects of the test chemical:
   - Colony strength (number of bees per Delaplane *et al.*, 2013b)
   - Presence of a healthy queen (e.g. presence of eggs)
   - Pollen storage area and area with nectar or honey (per Delaplane *et al.*, 2013b)
   - Area containing cells with eggs, larvae and capped cells (per Delaplane *et al.*, 2013b).

The coverage of a comb can be estimated assuming that a comb is covered by 120 bees per 100 cm² if bees are sitting very close to each other (Imdorf and Gerig, 1999; Imdorf *et al.*, 1987). The estimations will be done for all combs (both sides) in each hive. The assessment of the areas containing brood and food can be done by estimating subareas of 100 cm². Afterwards the number of cells per brood stage/food stock is calculated assuming that 100 cm² of the comb comprise 400 cells (Imdorf and Gerig, 1999; Imdorf *et al.*, 1987). These estimations will be done for all combs (both sides) in each hive.

5.2.2.2.5.4.2. Development of the bee brood in single cells

The time schedule of the brood assessment days was chosen in order to check the bee brood at different expected stages during the development as mentioned in the Table 7.

1. The application in the tunnels should be performed shortly after BFD (within 2 days afterwards).

2. In contrast to the method described in the OECD Guidance Document 75, it is now common to use the digital photo method (Jeker *et al.*, 2011 but see section 5.2.3. of the present manuscript) to follow the development from an egg to the adult honey bee. In the following text, this method will be used to describe the system.

3. The development of bee brood is assessed in individual marked brood cells of all colonies within a study. At the assessment before the application (BFD) one or more brood combs should be taken out of each colony, marked with the study code, treatment group, hive number, comb number, comb side and BFD date, and photographed with a digital camera. In the laboratory, all photos are transferred to a personal computer and areas with at least 100 cells containing eggs are marked on the screen. The exact position of the markers and of each cell and its content should be stored in a computer file that serves as a template for later assessments. The same cells are assessed on each of the following assessment dates (Table 7). Thus, the development of each individually marked cell throughout the duration of the study can be determined (pre-imaginal development period of worker honey bees typically averages 21 days).

4. For the evaluation of the different brood stages of single marked cells, the recorded growth stages are transformed into values counting from 0 to 5 as listed below:
   - 0: termination/breakup of the development (e.g. nectar or pollen found in a cell, if in the previous assessments the presence of brood was recorded)
   - 1: egg stage
   - 2: young larvae (L1 or L2)
   - 3: old larvae (L3 to L5)
   - 4: pupal stage (capped cell)
   - 5: empty after hatching or again filled with brood (eggs and small larvae)
   - N: cell containing nectar
   - P: cell containing pollen

   Cells filled with nectar and pollen after the termination of brood development in the respective cell (counted 0) may be identified by an "N" and "P" in the following assessments; the respective cells have to be excluded from further calculations, but should be included in the overall evaluation in the end.

Table 7. Time schedule of the brood assessment in semi-field brood tests:

BFD = brood area fixing day. *Assessments will be performed outside the tunnels at the monitoring location.

Timing	Determined brood stage in marked cells
BFD (1-2 days before application)	Egg
Timing	Expected brood stage in marked cells
5(±1) days after BFD	Young to old larvae
10(±1) days after BFD*	Capped cells
16(±1) days after BFD*	Capped cells shortly before hatch
22(±1) days after BFD*	Empty cells or cells containing eggs, young larvae, nectar or pollen

5. Based on the numbering described above, mean values (indices) can be calculated for each colony and assessment day.

6. Assuming that at the first assessment only eggs will be marked, the index is one. An increase of the brood index during the following assessment can be observed, if a normal development of the brood is presumed. This increase is caused by the development from eggs to larval stages, from larvae to pupae and from pupae to adults. Details of the evaluation of the results are presented by Schur *et al.* (2003).

5.2.2.3. Evaluation of the results of the semi-field test

The influence of the test product can be evaluated by comparing the results in the test chemical treatment to the water-treated control and to the reference chemical treatment, and furthermore by comparing the pre- and post-application data regarding:

1. Mortality (dead adult bees, pupae and larvae) within the crop area (linen sheets) and in the dead bee traps (per day and over time after application during bee exposure).It is of interest if an increase in the number of dead pupae is noticed or if malformations of the dead pupae or young dead bees are observed. In case of fenoxycarb in the reference treatment group, an increase in the number of dead pupae can be observed 10-12 days after application. This factor should be considered when demonstrating its sensitivity to bees.

2. Flight intensity in the crop (mean number of forager bees/m² flowering *P. tanacetifolia* after application)

3. Behaviour of the bees on the crop and around the hive

4. Condition of the colonies (strength (number of bees) of the colonies, presence of a healthy queen, mean values of the different brood stages per colony and assessment date, per Delaplane *et al.*, 2013b)

5. Development of the bee brood (brood indices) in > 100 cells:
   - Brood-index:
     The brood-index is an indicator of bee brood development and facilitates comparison between different treatments. It is calculated for each assessment day and colony. For all cells containing the expected brood stage at the respective day, the assessed value (1-5) could be used.

For all cells that do not contain the expected brood stage,
0 is used for calculation. All values per hive and assessment
day are summed and divided by the number of observed
cells in order to obtain the average brood-index.

- Compensation-index:
  The compensation-index is an indicator for recovery of the
  colony. It is calculated for each assessment day and colony.
  The values of all individual cells in each treatment, assessed
  at the respective day for each hive, could be summed and
  divided by the number of observed cells in order to obtain
  the average compensation-index. By that, the compensation
  of bee brood losses is included in the calculation.
- Brood termination rate:
  Percentage of marked cells where a break (i.e. no
  successful development) of the bee brood development is
  recorded, i.e. the bee brood did not reach the expected
  brood stage at one of the assessment days or food was
  stored in the cell during BFD +5 to +15.

  Specific statistical analysis for bee trials in semi-field and
  field conditions are still under development. In general, it
  is recommended to follow the OECD guidelines (OECD,
  2006) and Becker *et al.*, 2011.

5.2.2.4. Discussion and conclusion

Based on the OECD Guidance Document 75 (OECD, 2007), numerous
studies were performed and it became obvious that the brood
termination rate (= mortality of bee brood in selected cells on combs)
was subject to a certain degree of variation, e.g. resulting in replicates
with increased rates up to 100% in the control and reduced rates in
the reference item group down to 21% (Pistorius *et al.*, 2011).
Additionally, a high variation between replicates within a respective
treatment group occurred sometimes. The variability which was
distinctly more present under semi-field conditions compared to a field
method (Oomen *et al.*, 1992) complicates the interpretation of results
regarding potential brood effects of a test item with the outcome that
some studies were regarded as invalid. The time between BFD and
the following assessment on BFD +5 days turned out to be the most
critical for such variations. Due to these variances, no definite
conclusions regarding potential brood effects were possible in such
cases, and the studies needed to be repeated.

In 2011, possible causes and improvements for the existing method
were shown by Pistorius *et al.* (2011) and at the ICPBR (now ICPPR)
meeting in Wageningen. Attempts to improve the methodology were
initiated by the Working Group "Honey bee brood" of the German AG
Bienenschutz. In 2011, honey bee brood studies adapted to these
identified possible improvements, resulting in better results compared
to historical data (for details see Pistorius *et al.*, 2011).

Based on the analysed results, the working group recommended to
improve the method by using bigger colonies with more brood, using
4 instead of 3 replicates for better interpretation of data, starting the
study early in the season, avoiding major modifications of the colonies
shortly before application and using larger tunnels with effective crop
areas preferably > 80 m². To carry out quicker brood cell assessments
to reduce stress for the colonies, it is recommended to use digital
photo brood assessment as described in section 5.2.3., which allows
marking a higher amount of cells (e.g. 200 to 400 cells).

In the overall outcome of the studies of the German working group,
the combination of the suggested improvements showed a reduction
in the breakup rate of the brood development in single cells and in the
variability of the results in the control group (Pistorius *et al.*, 2011).
However, it also showed that even when fulfilling all the described
improvements, it may happen that the brood mortality increases to
such a high level, that an evaluation of the test product data still is
not possible.

Since the bee colonies are kept under semi-field conditions with
restriction in their normal collection and flying behaviour, they generally
are sensitive to any interference from outside. Therefore, one should
avoid stressing the bees too much during the assessments as well as
before set-up of the colonies in the tunnels.

For this reason, it is important to analyse the importance of
additional factors in the future in order to be able to improve semi-
field studies and studies under field conditions, where the detailed
brood assessments are integrated into the study design.

5.2.3. Evaluation of honey bee brood development by using digital image processing

5.2.3.1. Introduction

Evaluations of potential effects on honey bee brood are an important
part of the registration process of PPPs. The recently used methodology
to investigate bee brood development under realistic exposure
conditions are semi-field studies according to Schur *et al.* (2003) (see
section 5.2.2. in this manuscript) superseded by the OECD Guidance
Document No. 75 or field studies according to Oomen *et al.* (1992)
(see section 5.2.1. in this manuscript). Originally, at least 100 brood
cells have to be marked and evaluated on acetate sheets with overhead
markers for both methods. This is time consuming. The disadvantages
of the "acetate method" are the restricted number of cells that can be
marked and the long "off-hive-time" of the brood combs. Therefore a
digital image processing method was developed (Wang & Claßen, 2011,
Jeker *et al.*, 2012;) to reduce the "off-hive-time" of the single brood
combs and therefore the stress for the whole honey bee colony. In
principle, the use of digital image processing allows one to evaluate
the development of an unlimited number of brood cells resulting in
increased statistical power. Further, the digital method allows one to

re-evaluate the brood development of single cells in the case of uncertainties.

5.2.3.2. Material and methods

5.2.3.2.1. Photographing of the brood combs at the field site

1. Before taking photos, each brood comb must be marked with the hive description, treatment group, study code, comb number & side and BFD date (BFD0 is the day of the first photographing, one to two days before treatment application).
2. Further (depending on the image processing software), markers have to be defined that allow the program to recover the single brood cells or it has to be ensured that fixed points of the comb (e. g. the edges of the comb) are photographed at the BFDs.
3. After marking the combs, the photos should be taken with a high resolution camera. To standardise the photos of the different combs at the different BFDs, a "photo box" should be used which allows photographing the combs under the same parameters (e.g. distance, focal length). Additionally the camera should support a "live view mode" which is useful to ensure that the photos are of a high quality and facilitate the setting of the camera. The results are most favourable when the photographed combs are located in the centre of the brood area.

5.2.3.2.2. Evaluation of the brood combs at the laboratory

1. The first step at the laboratory is to set the markers or fixed points with the respective image processing program.
2. Afterwards brood cells containing eggs are chosen. To achieve better results, the cells of choice should be on combs containing nectar and pollen and located close to the centre of the combs and not near the edges. At the following BFDs, the image processing program is able to recover the cells marked at BFD0 by use of the markers or fixed points.
3. At the following BFDs (BFD5, 10, 16, 22), the contents of the brood cells are evaluated according to the respective test method (for a demonstration see the online demo video at Rifcon, 2012).
4. During and after the study, the image processing programs are able to calculate all relevant parameters such as brood termination rate, compensation index and brood index (see section 5.2.2.3.). The results of the single cells are presented tabular or in an image gallery for an easier comparison of the respective brood cells.

5.2.3.3. Discussion and conclusion

The digital image processing (Wang & Claßen, 2011, Jeker *et al.*, 2012) improves the evaluation of the honey bee brood development. It reduces the stress for the honey bee colony as well as unnatural influences on the brood development caused by long lasting manual assessments. Due to the fast and standardised photo taking procedure, a high photo

quality can be guaranteed and the number of brood cells to be evaluated is almost unlimited. Nevertheless, practical experience has proved that the evaluation of a high number of brood cells is time-consuming and thus it was suggested that the evaluation of 200 to 400 brood cells should be sufficient (Pistorius *et al.*, 2012). Future innovations could produce a more automated evaluation (e.g. automatic determination of the brood stages) and also the exact determination of the brood and food status on colony level.

5.3. *in vitro* larval tests

Aupinel *et al.* (2005) devised a standard *in vitro* test usable for any research topic on larvae (Crailsheim *et al.*, 2013) and more specifically for brood risk assessment (Aupinel *et al*, 2007b). This test has already been ring-tested (Aupinel *et al*, 2009) with the participation of 7 laboratories originating from 6 countries that satisfied the 2 criteria of validity: control mortality lower than 15% at D6 and successful emergence of worker adults in at least the control group. This test, based on an individual rearing method permits one to control exactly the individual exposure with a high reproducibility. It provides quantitative oral toxicity data on honey bee brood. It is designed for *in vitro* treatments of AIs or formulated pesticides. Adopted in France by the CEB, it was validated at OECD and will be recommended in the near future as a guideline for acute exposure at D4 and lethal effect at D7. Chronic exposure and observations on pupae and adult stages will be referenced as guidance.

5.3.1. The rearing method

The rearing method used for this test is detailed in Crailsheim *et al.* (2013), summarised in Fig. 9, and outlined in the steps below.

1. For one replicate, larvae are collected preferably from a unique colony. If two colonies are necessary, larvae originated from both colonies must be distributed in two samples of equal size (24 larvae) in each plate. The colonies have to be healthy and must not show any visible clinical symptoms of pests, pathogens (see *BEEBOOK* Volume II) and/or toxin stress.
2. Tests are performed with summer larvae during a period from the middle spring to the middle autumn (the exact time of year varies by location).
3. In case of sanitary treatment (i.e. products added to the hive for purposes of disease/pest control), the date of application and the kind of product has to be noted. No treatment should be applied within the 4 weeks preceding the beginning of experiments.
4. The queen is confined in its own colony in an excluder cage containing a comb with emerging worker brood and empty cells for less than 30 hours in order to obtain a large number of fresh laid eggs. According to queen vigour, the queen's isolation time can be reduced in order to minimize variability in larval size (age).

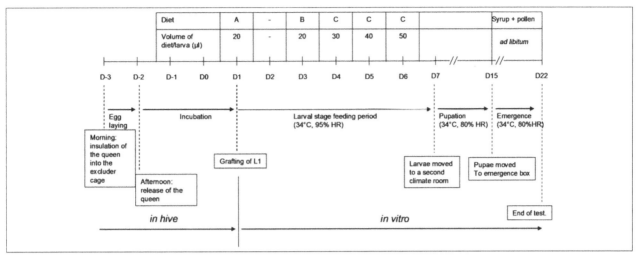

Fig. 9. Steps of a brood *in vitro* test.

5. To ensure one obtains enough larvae, it is recommended to isolate the queens in 2 or 3 colonies in the eventuality that one queen lays few or no eggs.

6. The queen is removed from the cage and the caged comb is left in the hive for 3 days until the larvae hatch.

7. At day 1 (D1, Fig. 9), the comb containing fresh laid eggs is carried from the hive to the laboratory (regulated at a constant temperature of 25°C if possible), in a special wooden container in order to avoid temperature variation and to transfer the larvae into individual rearing cells. We recommend crystal polystyrene grafting cells (ref CNE/3, NICOPLAST Society), having an internal diameter of 9 mm.

8. Before use, the cells are submerged for 30 min in 0.4% MBC (methyl benzethonium chloride) in water, and then dried in a laminar-flow hood. MBC can be replaced by chloride tablets generally used for nursing bottle sterilisation.

9. Each cell is placed into a well of a 48-well tissue culture plate, which was previously half filled with a piece of dental roll wetted with 15.5% glycerol in 0.4% MBC.

10. The young larvae are transferred with a grafting tool (a thin paint brush for example) from the frame into individual plastic cells previously filled with 20 μl of diet A (Table 8).

11. The larvae are fed once a day (except day 2) with a micro-pipette. Diet composition varies according to larval age (Fig. 9, Table 8). The diet is warmed at 34°C prior to each use.

12. The plates are placed into a hermetic Plexiglas desiccator (NALGENE 5314-0120 or 5317-0180 or similar, according to the required volume), provided with a dish filled with K2SO4 saturated solution in order to maintain a water-saturated atmosphere.

13. The desiccator is placed into an incubator at 34 ± 0.5°C. This parameter is crucial considering that susceptibility to a compound may vary significantly according to temperature (Medrzycki *et al.*, 2010).

14. At D7 (pre pupa stage), the plates are transferred into a hermetic container containing a dish filled with a saturated NaCl solution in order to maintain 80% relative humidity. The container is then placed into an incubator at 34°C.

15. At D15, each plate is transferred into a crystal polypropylene box (11 x 15 x 12 cm) with a cover aerated with a wire mesh, and containing a piece of comb with a small plastic royal pheromone diffuser in its centre (Bee Boost®), fixed with a wire.

16. Emerging bees are fed with syrup and pollen powder delivered using bird feeders or similar structures. The boxes are kept in the hermetic container.

Table 8. Composition of the diets provided to larvae (Aupinel *et al.*, 2005, summarised in Crailsheim *et al.* 2013). (Example: to prepare 20 g of diet A (Crailsheim et al., 2013). - Mix 1.2 g glucose, 1.2 g fructose and 0.2 g yeast extract into 7 ml water, and then adjust until 10 ml with water. Mix 10 g of this solution with 10 g of royal jelly.

Diet	A	B	C
Royal jelly (%)	50	50	50
Yeast extract (%)	1.0	1.5	2.0
D glucose (%)	6.0	7.5	9.0
D fructose (%)	6.0	7.5	9.0
Dry matter (%)	29.6	33.1	36.6

5.3.2. Toxicity testing

1. The experimental unit is the 48-larvae plate. For each test, the following treatments should be used:
   - control without solvent (1 plate),
   - control with solvent (1 plate) if necessary,
   - 5 treatments, i.e. the 5 doses or concentrations to be tested (1 plate per treatment),
   - reference treatment with dimethoate (1 plate).
   - 1 additional plate (totally or partially filled with larvae, according to the number of available, remaining larvae) can be used at D4 in the acute toxicity test to replace the larvae which died before D4.

One test has a minimum of three replicates with different larvae origin and new tested solutions for each replicate.

2. The tested pesticide is preferably dissolved in water. If it is not soluble in water at the experimental concentrations, one can use another solvent such as acetone. In that case, it is necessary to prepare a second negative reference fed with diet containing the solvent at the same concentration as in the treated samples.

3. Dilutions of the stock solutions are made with osmosed water, using disposable pipette tips equipped with a filter.

4. The rate of the tested solution in the diet must not exceed 10% of the final volume. In all cases, it is necessary to use a constant volume for the different treatments in order to have a constant rate between the diet and the test pesticide solution.

5. The toxic reference is dimethoate:
   - in acute toxicity test: 3 µg/larva mixed with diet C and provided at D4,
   - in chronic toxicity test: mixed with the three diets at the constant concentration of 20,000 µg/kg diet.

6. In an acute toxicity test, larvae are treated at D4 with diet C containing the preparation to test at the suitable concentration. For a chronic toxicity test, larvae are treated every day (except D2) with the diets containing the preparation to test at a constant concentration.

7. In order to assess the adequate LD_{50} range, it is recommended to run a preliminary experiment where doses of the test preparation may vary according to a geometrical ratio from 5 to 10.

5.3.3. Results

1. Mortality can be defined according to the following criteria:
   - Larva: an immobile larva or a larva which does not react to the contact of the paintbrush is noted as dead.
   - Pupa: a non-emerged individual at D22 is noted as dead during pupal stage.
   - Adult: an immobile adult which does not react to a tactile stimulation is noted as dead.

2. Mortality is checked at the following moments:
   - Larva: At the feeding moment, dead larvae are systematically removed for sanitary reasons. Specific mortality checks are made according to the type of test. In the test where exposure is at D4 (acute toxicity), a first mortality check is made at D4 in order to replace the dead larvae before they have started consuming the diet containing the insecticide. Then one should note the mortality at D5, D6 and D7. In the test with chronic exposure, mortality is noted at D7.
   - Pupa: Non emerged bees are counted at D22.
   - Adult: Alive adult bees and dead adults which have left their cell and show a normal development are both counted at D22.

3. Sublethal effects such as development length, prepupa weight, wing malformation, adult survival, etc. can be noted. It is recommended to weigh prepupa without removing them from the rearing plastic cell. Adults can be kept in the emergence boxes with *ad libitum* food for behaviour observations or longevity assessment.

5.3.4. Statistical analysis

1. The validity of a test depends on some data validity range.

2. In negative reference samples, larval mortality (number of dead larvae/48), pupal mortality (number of dead pupae at D22/number of alive pre pupae at D7) and adult mortality (number of dead emerged bees at D22/total number of emerged bees) must be lower or equal to 15% for the assessment of LD_{50} or LC_{50}, or 20% for the assessment of NOAEL or NOAEC. In case of higher mortality in the control sample, the replicate is invalidated.

3. The mortality rate with positive reference (dimethoate) must be:
   - higher than or equal to 50% at D6 for larvae exposed to 3µg/larva at D4
   - higher than or equal to 50% at D7 in chronic exposure of larvae to the concentration 20,000µg/kg diet.

4. The calculated LD_{50} and LC_{50} must in each case be between the two extreme tested doses. They must not be extrapolated out of the tested limits.

5. Any deviation from the above conditions will invalidate the test.

6. LD_{50} and LC_{50} are calculated from mortalities expressed in percentage of the reference populations after an adjustment according to the Abbott or Shneider-Orelli formula (see section 8.4.1.).

7. The results will be analysed using regression model with high adjustment level, which can be checked with the determination coefficient value (Abbott, 1925).

8. Basing on the same raw mortality data, the NOAEL and NOAEC are assessed (see section 8.4.3.).

5.3.5. General discussion

More research has been published on *in vitro* brood feeding test. Descriptions of laboratory methods have been provided over almost half a century (Weaver, 1955; Rembold and Lackner, 1981; Wittmann and Engels, 1981; Vandenberg and Shimanuki, 1987; Davis *et al.*, 1988; Czoppelt, 1990; Engels, 1990; Peng *et al.*, 1992; Malone *et al.*, 2002; Brodsgaard *et al.*, 2003). These methods generally provide LD_{50} or LC_{50} for the treated larval stage. In 1981, Wittmann and Engels suggested to use the *in vitro* brood feeding test as a routine method for screening insecticides and classifying chemicals according to their toxicity to larvae. Considering both the laboratory toxicity of a product to larvae and exposure data of brood to this product in natural conditions, the *in vitro* larval feeding test seems an appropriate starting point of the brood risk assessment, in other terms a tier 1 study. However, objections have been raised against the *in vitro* method and its regulatory use, in particular doubts on the standardisation of the protocol, criticisms on the frequent high mortality and the presence of intercasts in the control samples. The difference of food quality and mode of dispensing between natural (Haydack, 1968) and artificial conditions described by authors may account for these weaknesses. See a detailed review of *in vitro* larval rearing in Crailsheim *et al.*, 2013.

6. Effects of toxic substances on queen bees and drones

6.1. Introduction

Although the honey bee queen is the only reproductive female in a colony, therefore responsible for the colony sustainability, very few toxicological studies are dedicated to this key member of the social structure. The scientific literature devoted to poisoning of drones is nearly non-existent.

6.2. Mortality and poisoning signs in honey bee queens

Most of the information on pesticide impacts on colonies comes from experimental protocols performed in field conditions, protocols not focused on the effects of pesticides on the queens. In such studies, standardized colonies are fed with sugar syrup or pollen patties contaminated with different pesticides at different concentrations. The administration of contaminated food was regularly repeated over a period of several weeks on colonies in the field.

When pollen patties were contaminated with micro-encapsulated methyl-parathion (Penncap-M), an organophosphate insecticide, and given to colonies in field conditions, Stoner and Wilson (1983) noticed that queens were superseded or died more frequently in the treated groups than in untreated ones (43.3% versus 25%, respectively), without clear relation between concentration and queen problems. When colonies were fed with sugar syrup contaminated with 10 ppm dimethoate, another organophosphate insecticide, Stoner *et al.* (1983) observed that queens died but were not replaced.

Two hypotheses involving the nurse bees were proposed to explain the queen death. The toxin, carried by the sugar syrup, contaminated the crop of the workers and particularly that of the nurse bees. When they offered the glandular secretions to the young larvae or to the queens, they regurgitated contaminated matters at the same time (Davis and Shuel, 1988). Consequently, the queen can be poisoned directly (fed contaminated food) or the queen can reject the contaminated food and suffer from malnutrition. Both hypotheses could result in a situation where the queen drastically decreases egg production. A reduction in egg production generally triggers queen elimination (supersedure) by worker bees. In the case of carbofuran, a carbamate insecticide (Stoner *et al.*, 1982), heavy losses of young bees by poisoning occurred.

6.3. Reduction in egg production

Although often neglected, plant foodstuffs harvested by workers can harm colonies and potentially impact queen physiology. When the nectar and pollen of *Aesculus californica* (California buckeye) is intensively harvested, returned to the hive and consumed, queens lay only male eggs and can be superseded. The poisoning stops generally at the end of buckeye bloom (Vansell, 1926). A deleterious compound of the nectar was suspected but not isolated.

Johansen (1977) mentioned that queens may be affected by insecticides and behave abnormally. For instance, they may produce a an abnormal brood pattern. This was the case with ovicidal effects of certain herbicides. When package bees containing a laying queen were fed with the 2, 4, 5 T and 2, 4 D herbicides at 100 mg/kg, some of the eggs were unable to hatch, thus presenting as a bad brood pattern (Morton and Moffett, 1972).

Bendahou *et al.* (1999) suggested a reduction in the amount of vitellogenin in eggs (see: Tufail and Takeda, 2008) explained a low hatch rate of eggs, and consequently, the resulting high frequency of supersedure observed in colonies fed weekly with sugar syrup including 12.5 µg/l of cypermethrin, a pyrethroid insecticide.

Dai *et al.* (2010) validated that the hatch rate of eggs can be reduced when queens are fed sublethal doses of bifenthrin and deltamethrin, both pyrethroid insecticides. Moreover, the daily number of laid egg was reduced 30 to 50% for bifenthrin and deltamethrin, respectively.

Ovicidal effects, suggested by egg replacement in the cells, can occur after exposure to IGR insecticides such as fenoxycarb or diflubenzuron (Thompson *et al.*, 2005). The maximum replacement rate measured in the first week after treatment was 60% and 90% for fenoxycarb- and diflubenzuron-treated colonies respectively. No queens successfully mated and laid eggs when treated with fenoxycarb.

Other IGR insecticides acting on the Juvenile Hormone III titre in the haemolymph, were shown to inhibit vitellogenin synthesis (Pinto *et al.*, 2000).

The questions of side-effects of acaricide treatments on queen egg laying success were investigated for fluvalinate and coumaphos. After treating queens and attendant bees placed in Benton mailing cage with specially designed strips of fluvalinate for three days, Pettis *et al.* (1991) observed no differences in colony acceptance of queens, brood viability or supersedure rates. After moderate queen larvae exposure to fluvalinate in a starter/finisher colony, Haarmann *et al.* (2002) confirmed the statistical absence of differences compared with the control group of newly mated queens, with queen weight, ovary weight and the number of sperm.

Coumaphos, another acaricide/insecticide, was shown to be more toxic than fluvalinate by Haarmann *et al.* (2002). They contaminated frames of grafted cells placed in starter colonies for 24 h, with two plastic strips each containing 1.360 g of coumaphos. Afterwards, queen cells were raised in finisher colonies. At the end of the experiment, queen cells contained 8 to 28 mg/kg coumaphos depending on the presence or absence of contact of the strips with the grafted cell frames. In coumaphos treated groups, the queen and ovary weights were significantly lower. After artificial contamination of the wax of queen cups with 100 mg/kg of coumaphos, Pettis *et al.* (2004) showed a negative effect on young queen acceptance and on their weights.

6.4. Inability to requeen

In cases where supersedures failed, some authors focused their experiments on the ability of orphan colonies to rear new queens. Before aerial application of fenthion, an organophosphate insecticide, Nunamaker *et al.* (1984) placed orphan colonies in a pasture due to be treated. After treatment, they noticed that some new queens emerged at a later date, compared with control colonies, but neither egg-laying queens nor eggs were found in the exposed colonies.

When Stoner *et al.* (1985) fed nurse colonies for queen rearing purposes with sugar syrup contaminated at 5 mg/kg of acephate, an organophosphate insecticide, for several weeks, most of the queen cells aborted. To observe the effects of 4 insecticides (fenoxycarb, diflubenzuron, tebufenozide, azadirachtin), known as IGR insecticides, on newly emerged queens, Thompson *et al.* (2005), transferred queen cells in nuclei containing about 1000 worker bees and supplied them with contaminated fondant. In the fenoxycarb treated group, the emerged queens showed virgin queen characteristics but none of them successfully mated or laid eggs. These authors were also interested in the effects of the molecules on the drones. They concluded that the number of mature drones was reduced in the diflubenzuron treated colonies and even absent from some fenoxycarb ones.

6.5. Conclusion

Studies are needed to assess pesticide impacts on reproductive activity in the colony, that is to say, the physical and physiological integrity of the queen and drone bees. Methods using a strict control of the toxin exposure of queens and drones must be preferred to field conditions where the exposure of the foragers is always questionable because of the difficulty to locate the foraging sites. Effects on daily egg-laying rates, egg hatch rates, number and viability of the spermatozoa in the queen spermathaeca (see Cobey *et al.*, 2013), and in the seminal vesicles of the mature drone should not be overlooked and may be captured in overall risk assessments of brood and population development in higher tier testing. Nevertheless, specific guidelines may be needed to take into account these criteria in the evaluation of toxicity of any AI or commercial formulations.

7. Evaluation of synergistic effects

7.1. Laboratory testing for interactions between agents

7.1.1. Introduction

The theoretical basis for interpreting interactions between agents is rooted in the history of testing combinations of chemical poisons, such as pesticides, but this theoretical framework is broadly applicable to many biotic and abiotic factors that may interact in bees (section 3 of this manuscript). Bliss (1939) recognized three basic types of interactions between agents that can be observed: Independent Joint Action, Additive Joint Action and Synergistic Action (Robertson *et al.*, 2007).

The simplest interaction between agents, and the implicit null hypothesis in experiments testing for interactions, is termed "Independent Joint Action". In independent joint action, the different agents act on bees through different modes of action and no combinatorial effects are observed. The more highly toxic agent in a combination is understood to cause the observed mortality (or other toxicological endpoint) and the observed mortality is indistinguishable from mortality when the more toxic agent is administered alone.

An agonistic interaction occurs when the toxicity of two agents applied together is higher than that of either agent when applied alone. If an agonistic interaction is observed and agents are known to work through similar modes of action, then the term additive toxicity is used. For example, if bees are exposed to different pyrethroid pesticides which share the same mode of action, then the observed toxicity is a sum of the doses of the different pyrethroid pesticides (e.g. tau-fluvalinate and bifenthrin, Ellis and Baxendale, 1997). Differential potencies between different agents with similar modes of action may need to be taken into account (Robertson *et al.*, 2007).

Agonistic interactions may also be synergistic in nature when the toxicity of a combination of agents cannot be predicted from knowledge of the toxicity of each agent alone. Synergistic interactions do not generally occur at the active site (but see Liu and Plapp, 1992), but instead occur when one agent affects the absorption, distribution, metabolism or excretion of the other agent, rendering it more toxic to bees. For example, piperonyl butoxide acts synergistically with both thiacloprid (Iwasa *et al.*, 2004) and tau-fluvalinate (Johnson *et al.*, 2006) by inhibiting the metabolism of these pesticides and greatly increasing their toxicity to bees.

Antagonistic interactions, where a combination of agents is less toxic than each agent alone, may also be observed.

The potency of an interaction can be substantially affected by the ratio of the different agents, for example the level of exposure to coumaphos affects bees' susceptibility to tau-fluvalinate (Johnson *et al.*, 2009). A range of ratios between agents can be explored using the methods described.

7.1.2. Model synergists

Model synergists are chemical tools that are useful for determining the biological basis of synergistic interactions. Model synergists are not overtly toxic to bees at the doses used, but can greatly alter the toxicity of other agents by changing the absorption, disposition, metabolism or excretion of the second agent.

Commonly used inhibitors of detoxicative metabolism include piperonyl butoxide (PBO), which inhibits cytochrome P450 monooxygenase enzyme activity, S,S,S-tributylphosphorotrithioate (DEF), which inhibits carboxylesterase activity and diethyl maleate (DEM), which inhibits glutathione S-transferase activity. These inhibitors are applied topically to the thoracic notum at sublethal doses of 10 µg (PBO and DEF) or 100 µg (DEM) dissolved in 1 µl of acetone 1h prior to treatment with a second chemical agent (Iwasa *et al.*, 2004; Johnson *et al.*, 2006).

The membrane-bound Multi Drug Resistance transporter can be inhibited by feeding bees verapamil at a concentration of 1mM dissolved in 50% sucrose syrup (Hawthorne and Dively, 2011).

7.1.3. Response variables

Acute mortality is the most commonly used response variable when looking for interactions between agents (section 3). Acute mortality is appropriate when one of the agents to be tested is an insecticide that will reliably kill bees using standard acute testing protocols (Section 3.1-3.3). The protocols listed all assume that mortality is the response to be measured, but this may not be an appropriate response if the agent under study is not acutely toxic to bees or if a binary sublethal effect is of interest.

7.1.4. Experiments testing for interactions
7.1.4.1. Discriminating dose bioassay

The simplest experiment involves treating bees with a single dose, termed the discriminating dose, in the presence and absence of another agent. It is important that an appropriate discriminating dose is chosen that will allow for any changes in toxicity to be detected. Discriminating dose experiments have been extensively conducted in *Varroa destructor* to determine acaricide resistance (Elzen *et al.*, 1998), and have been used in honey bees as well (Hawthorne and Dively, 2011). A significant drawback to the discriminating dose approach is that the full dose-response curve is not explored and it is impossible to differentiate

between interactions affecting the slope and the intercept of the dose-response curve.

1. Preliminary toxicity bioassays are performed singly on both agents to be tested. This bioassay can use adults treated through oral exposure (section 3.2.1.), topical exposure (section 3.2.2.) or exposure on foliage (section 3.2.3.).

2. The dose of the first, less toxic, agent should be chosen using the dose-response curve generated in step 1. Either this "non-killing" dose should be chosen so that it is the maximum dose that can be delivered that does not cause mortality different from control, or it should be an environmentally relevant dose determined through chemical analysis or predicted exposure.

3. The discriminating dose of the second, "killing" agent is chosen using the dose-response curves generated in step 1. The appropriate discriminating dose depends on the expected outcome of the interaction between the two agents – if antagonism is expected, then the LD_{90} or LC_{90} of the more toxic agent should be used. If synergism is expected, then the LD_{10} or LC_{10} is appropriate. If there are no *a priori* expectations the LD_{50} or LC_{50} should be used. An environmentally relevant dose, based results of chemical analysis or predicted exposure, may also be used.

4. To test for interactions bees are treated as recommended for oral, topical or foliage exposure (sections 3.2.1.-3.2.3.), except that only four groups of bees are used. Bees are then exposed to either the "non-killing" dose of the first agent (Step 2) or a control in combination with, or followed by, the discriminating dose of the second "killing" agent (Step 3), or a control. If the two agents cannot be delivered in combination (e.g. an oral "non-killing" agent and a topical "killing" agent) then the "non-killing" agent should be administered 1 h (topical or foliage) or 24 h (oral) prior to administration of the "killing" agent.

5. Testing in Step 4 is repeated to produce 5 replicates. The proportion of bees dying is transformed using the arcsine square root method, then a simple t-test or ANOVA is used to determine the statistical significance of observed differences in mortality (Hawthorne and Dively, 2011).

7.1.4.2. Comparison of dose-response curves

A superior method for detecting interactions can also be detected by comparing the complete dose-response curves of an agent in the presence and absence of a second agent. This approach allows complete characterization of the dose-response curve, including slope, intercept and LD_{50} or LC_{50} (Johnson *et al.*, 2006, 2009).

1. Preliminary toxicity bioassays are performed and the "non-killing" dose of the first agent is determined (steps 1-2 in the section 7.1.4.1.).

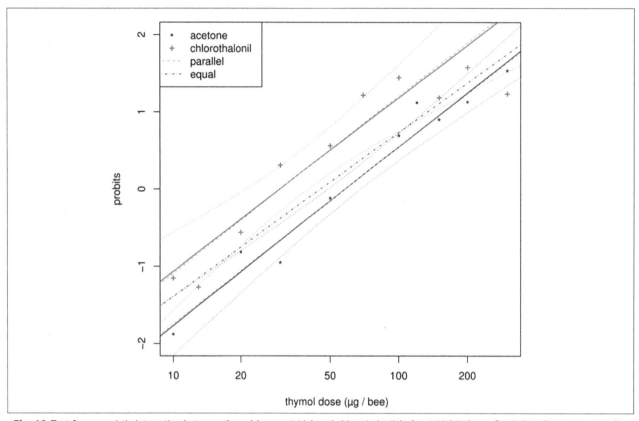

Fig. 10. Test for synergistic interaction between thymol (an acaricide) and chlorothalonil (a fungicide) in bees. Symbols indicate raw mortality data for groups of bees treated with acetone ("*", control, N = 864) or chlorothalonil ("*", N = 467). Solid black and red lines are fit independently to data for acetone and chlorothalonil treatments, respectively. Curved dotted lines correspond to 95% confidence intervals. Dashed green lines were generated using a model where the slope is identical for both lines. The "Test of Parallelism" is a likelihood ratio test between the green lines and the red and black lines (deviance = 0.035, df = 1,17, p-value = 1). The single dashed blue line represents a model fit to pooled data for both treatment groups. The "Test of Equality" is a likelihood ratio test between the blue line and the red and black lines (deviance = 10.449, df = 2,18, p-value < 0.0001).

2. The dose-response of the second "killing" agent is determined by treating bees as recommended for oral, topical or foliage exposure (sections 3.2.1.-3.2.3.), with the exception that all bees are treated with a uniform dose of the "non-killing" agent before, or simultaneous with, administration of a the recommended series of doses of the "killing" agent. A control dose-response series, in which bees are not exposed to the "non-killing" agent at all, is also performed for comparison.

3. Each dose-response series should be repeated at least 3 times.

4. For analysis, the doses are transformed on a log scale and the mortality is transformed on a probit scale, and a dose-response line is fit (Fig. 10). Comparison of the dose-response curves can be performed using commercially available software such as PoloPC (Robertson *et al.*, 2007) or using 'glm' in the R statistical package (R Development Core Team, 2010) (see section 7.3. for a sample script).

5. Three different tests are available to determine the presence of a significant interaction between agents by comparing dose-response curves.

- Comparison of the overlap of 95% confidence intervals around the calculated the LD_{50} or LC_{50}. The LD_{50} or LC_{50} values, and accompanying 95% confidence intervals, are calculated from the log-probit lines using Fieller's method, with correction for heterogeneity where appropriate (Finney, 1971). If the confidence intervals do not overlap, then the treatments are deemed significantly different. However, this test has been criticized for being overly conservative (Payton *et al.*, 2003), it does not generate p-values and there is no method for correcting for multiple comparisons.

- A ratio test comparing the ratio of the LD_{50} or LC_{50} derived from the pair of dose-response curves can be performed.

This test will produce the synergism or antagonism ratio and the associated 95% confidence interval. If the confidence intervals do not overlap "1", then the treatments are deemed significantly different (Robertson *et al.*, 2007). The ratio test does not generate a p-value and there is no method to correct for multiple comparisons.

- Interactions can be determined by comparing the dose-response lines using a test analogous to ANCOVA (Johnson *et al.*, 2013). Models are fit using 'glm' in R with all data from both dose-response curves. For the full model, the second "killing" agent serves as the covariate, and the presence or absence of the "non-killing" agent serves as a categorical factor. The interaction between the "killing" agent dose and "non-killing" agent is then compared using two simplified models with the explanatory power of the terms in the models assessed through a process of model simplification in reference to the likelihood ratio (Savin *et al.*, 1977). The first simplified model leaves out the interaction term and, when compared with the full model, tests for differences in slope between the dose-response lines. The second simplified model leaves out the "non-killing" factor entirely and tests for evidence of an agonistic or antagonistic interaction between the two agents. Model comparison using the likelihood ratio generates a p-value which may be adjusted for multiple comparisons using the Bonferroni correction for multiple comparisons.

7.2. Laboratory approach to study toxico-pathological interactions in honey bees

7.2.1. Introduction

Pesticides and pathogens are two categories of environmental stressors that may contribute to the decline of honey bee populations (vanEngelsdorp and Meixner, 2010). However, if their separate impacts on the honey bee are relatively well studied, knowledge on their interactions are somewhat lacking. Pioneer studies on toxico-pathological interactions have been conducted on the association of *Nosema* and chronic bee paralysis virus (CBPV) with organophosphate, organochlorine and pyrethroid insecticides (Ladas, 1972; Bendahou *et al.*, 1997). These studies focused on the acute exposure to insecticides regardless of their chronic toxicity. However, the introduction of systemic insecticides, such as phenylpyrazoles and neonicotinoids in the mid 1990's renders more relevant the studies on chronic exposures to pesticides by oral route.

A new laboratory approach to study the chronic toxicity of insecticide has offered the possibility to explore the interactions between pathogens and pesticides during chronic exposures (Suchail *et al.*, 2001). Studies on the joint exposure to *Nosema* and systemic insecticides have revealed that toxico-pathological interactions may elicit damaging effects on the bees, even when both stressors have no or limited effects on bee mortality (Alaux *et al.*, 2010; Vidau *et al.*, 2011). Two approaches have been used to study the effects of pesticide-pathogen associations. The first carries out simultaneous exposures to the pathogen and the pesticide and is particularly suitable to reveal antagonistic, additive and synergistic effects (Alaux *et al.*, 2010). The second involves sequential exposures to the pathogen and the pesticide and is particularly relevant to investigate the sensitization to one stressor by another (Vidau *et al.*, 2011; Aufauvre *et al.*, 2012).

The toxico-pathological interactions have been observed in laboratory conditions but the few attempts to demonstrate them in field conditions were not always as successful as expected (Wehling *et al.*, 2009; Pettis *et al.*, 2012). However, workers reared in brood frames containing high levels of pesticide residues exhibited a higher sensitivity to *Nosema* infection (Wu *et al.*, 2012). Hence, since such interactions were observed for humans and other species in their living environment, there is no reason to think that they do not occur in field conditions (Arkoosh *et al.*, 1998; Lewis *et al.*, 2002, Bauer *et al.*, 2012). Thus, in many cases, colony diseases could have been triggered by pollutants in healthy carriers.

7.2.2. Materials

7.2.2.1. Honey bees

Traditionally, the effects of pesticides are investigated in honey bee foragers that are the individuals first exposed to pesticides. Considering the contamination of pollen and honey by systemic insecticides, all individuals may be potentially exposed by ingestion of a contaminated food. Thus, the exploration of the toxico-pathological interactions has also been studied in cohorts of young isolated bees of known age, which represent a relatively homogeneous biological material. A sufficient amount of honey bee colonies not infected by *Nosema*, as confirmed by PCR and using primers previously described (Martin-Hernandez *et al.*, 2007), must be selected in order to obtain the desired number of emerging bees. To make the collection of emerging bees easier, queens can be isolated 20 days before the start of the experiment, using a queen excluder grid during 24 hours.

To fully sustain their physiological maturation after emergence, bees ingest pollen during the first days of their life. Pollen is the natural source of proteins for bees but the risk of contamination by pesticides cannot be ruled out (Chauzat *et al.*, 2006; Mullin *et al.*, 2010). A chemical analysis should normally yield information on the pesticide residues present in the pollen. However, the limit of detection of pesticides achieved with multi-residue methods are above 2 µg/kg for a large number of substances. Thus, a substance may be not detected but might still induce toxicity below its limit of detection. In addition, pathogens, notably *Nosema* and viruses, can be found in the pollen (Higes *et al.*, 2008; Singh *et al.*, 2010). For this reason, pollen is replaced by yeast extracts for protein supply. Commercial protein supplies can be used.

The day before starting the study, frames of sealed brood are sampled from colonies, put in boxes and placed in an incubator in the dark at 34°C with 80% relative humidity.

The day of the study, emerging honey bees (0-1 day) present in the boxes are collected, confined to laboratory cages (e.g. Pain type, 10.5 x 7.5 x 11.5 cm) in groups of 30-50 (see Williams *et al.*, 2013), and maintained in the incubator for different periods of time at 30-32°C and 70-80% relative humidity. To mimic the hive environment, a little piece of wax and a Beeboost® (Pherotech; Delta, BC, Canada) releasing one queen-equivalent of queen mandibular pheromone per day, are placed in each cage.

7.2.2.2. Pesticide

Stock solutions of pesticides in 100% DMSO will be diluted to obtain the required concentration of pesticide and 0.1% DMSO final concentration in 50% (w/v) sucrose syrup.

7.2.2.3. Food supply

Sucrose solution for experimental treatments (pathogens and pesticides) is made with sucrose and distilled water (50%; w/v). Proteins (Provita'bee) and candy (Apifonda®) can be purchased from beekeeping suppliers.

For more details on laboratory rearing methods see Williams *et al.*, 2013.

7.2.3. Joint action of pathogens and pesticides

1. The day of the study, emerging honey bees (0-1 day) present in the boxes are collected and distributed in different experimental groups: (i) uninfected controls, (ii) infected with the pathogen only (e.g. *N. ceranae*), (iii) uninfected and chronically exposed to the pesticide at different doses, and (iv) infected with the pathogen and chronically exposed to the pesticide at different doses. Emerging bees can be handled relatively easily because they are quiet and neither sting or fly.

2. Honey bees are first individually infected by feeding with 3 μl of a freshly prepared 50% (w/v) sucrose solution containing the appropriate inoculum of the pathogen. Feeding is performed by holding each bee with its mouthparts touching the sucrose droplet at the tip of a micropipette (Malone and Gatehouse, 1998). This induces the extension of the proboscis and allows the bees consuming the entire droplet. Non-infected bees are similarly treated with the sucrose solution devoid of pathogen.

3. Bee are then confined to laboratory cages in groups of 30-50, and maintained in the incubator at 30-32°C and 80% relative humidity.

4. Honey bees are chronically exposed to pesticides for different periods of time by ingesting *ad libitum*, 10 h per day, 50% sucrose syrup containing, 1% (w/v) proteins, the pesticide at the appropriate concentration and 0.1% DMSO. The remaining 14 h, bees are fed with Candy and water *ad libitum*.

5. During the experiment, each cage is checked every morning and dead honey bees are removed and counted. The food,

containing or not the pesticide, is freshly prepared and renewed daily. The actual insecticide consumption is quantified by measuring the daily amount of sucrose syrup consumed per bee.

7.2.4. Sensitization to pesticides by a previous exposure to pathogens

1. Bees are distributed in different experimental groups:
   * uninfected controls,
   * infected with the pathogen only (e.g. *N. ceranae*),
   * uninfected and chronically exposed to the pesticide at different doses 10 days post-infection (d.p.i.),
   * infected with the pathogen and chronically exposed to the pesticide at different doses 10 d.p.i.

2. Honey bees are first individually infected with the pathogen (see section 7.2.3.). If studies are conducted on emerging bees, go to step 3. If studies are performed on aged bees, go to step 5.

3. Studies on emerging bees. Honey bees are individually infected by feeding with 3 μl of a freshly prepared 50% (w/v) sucrose solution containing the appropriate inoculum of pathogen. Emerging honey bees are then fed during 10 days with 50% (w/v) sucrose syrup supplemented with 1% (w/v) protein 10 h per day and thereafter with candy and water *ad libitum* 14 h per day. Each day, feeders are replaced and the daily sucrose consumption is quantified.

4. Ten days after infection, honey bees are chronically exposed for 10 days to the pesticide by ingesting *ad libitum*, 10 h per day, 50% (w/v) sucrose syrup containing 1% proteins, the pesticide at the appropriate concentration and 0.1% DMSO. Honey bees not exposed to insecticides are fed *ad libitum* with sucrose syrup containing 1% proteins and 0.1% DMSO. Then, bees are fed with candy and water *ad libitum* 14 h per day.

5. Studies on aged bees. At a given post-emergence time, caged bees are CO2-anaesthetized, put individually in infection boxes consisting of ventilated compartments (3.5x4x2 cm) and starved for 2 h. Each compartment is supplied with a tip containing the appropriate inoculum of pathogen in 3 μL of sucrose syrup (non-infected bees are similarly treated with sucrose syrup devoid of pathogen).

6. Infection boxes are placed in the incubator and 1 h later, bees that have consumed the total pathogen solution are again encaged (50 bees per cage). Bees are then fed during 10 days with 50% (w/v) sucrose syrup supplemented with 1% (w/v) proteins 10 h per day and thereafter with candy and water *ad libitum* 14 h per day. Each day, feeders are replaced and the daily sucrose consumption is quantified.

7. Ten days after infection, honey bees are then exposed for 10 days to the pesticide (see step 4 above).

8. Throughout both types of experiments, each cage is checked every morning and dead honey bees removed and counted. The food, containing or not the pesticide, is freshly prepared

and renewed daily. The actual insecticide consumption is quantified by measuring the daily amount of sucrose syrup consumed per bee.

9. At the end of the experiment (20 d.p.i.), surviving honey bees can be subjected to investigations or may be quickly frozen and set aside for subsequent analysis.

7.2.5. Notes

- To analyse honey bees at a second post-infection time, the number of cages for each modality must be multiplied by two.
- To avoid any bias due to the weather or season on bee physiology, mortality, physiological and chemical investigations should be performed at the same time.
- Honey bees must be handled with a soft insect holding forceps to avoid physiological damages.
- The experimental design may be modified to change the day of infection, the starting day and the duration of exposure to pesticide, and the sequence of exposure to stressors.
- It is proposed to expose the bees to the pesticide 10 h per day in order to avoid overexposure not compatible with environmental exposures (Suchail *et al.*, 2001). However, bees can be exposed continuously to the pesticide.
- The levels of exposure to pesticides are relatively easy to determine on the basis of pesticide residues in pollen, nectar and honey. However, for the pathogens, it is impossible to determine an infectious level that could be representative of an environmental exposure or a pathological situation. Thus, the inoculum has to be determined by the experimenter on the basis of the objectives intended.

7.3. R script for testing synergistic interactions

See online Supplementary Material.
(http://www.ibra.org.uk/downloads/20130809/download)

8. Introduction to the use of statistical methods in honey bee studies

This paper is not written to describe all the possible statistical tests but to provide some information on common statistics used on honey bee toxicological studies. For more information on using statistics in honey bee studies, see Pirk *et al.*, 2013.

8.1. Foreword

Statistics for experimental design are performed to describe the results and to help clarify a conclusion giving a probability to accept or reject a hypothesis which is in many cases a hypothesis of no differentiation.

For most bee study plans or protocols, the variables are mainly counting. Very few are issued from a quantified continue measure such as weight, length, etc. These measured variables can be mortality counts, foraging counts, behavioural counts such as toxicity signs or brood development, etc. These observed counts are raw data issued from experimenter observations in a laboratory box or cage, in a tunnel (semi-field condition), in a field, or directly in a hive. For these counts, two main situations are observed. In the first case, the size is exactly known as when a LD_{50} study is performed in cages with ten or twenty bees, or in a hive for a brood development study, 100 individual brood cells per hive are identified. In the second case, the size is not known. An estimation of population is made in the hive, and the counting is performed on the foraging activities or a counting of the dead bees is performed in the tunnel or in the field.

For most situations, several dose modalities are studied. The experimental design at a minimum includes a negative reference group as a sentinel to measure the experimental background noise (untreated or water treated control). A positive reference group is also often included to measure an experimental bias of no response (i.e. dimethoate). These two kinds of control permit one to validate (or invalidate) the study. Formal criteria are predefined in protocols.

An experimental test item modality is included in the experimental design. At least one modality is studied. The experimental design will include at minimum two or three groups, or product modalities, and up to ten or more product modalities. These modalities are usually independent. The same hive is not observed under several doses or product modalities but the hives are observed several times; then the counting is repeated. If the same modality is studied several times, replicates are observed and can be compared.

8.2. Statistical tests and situations

8.2.1. Honey bee tunnel study

In this study, one hive is observed during several days and several times a day, before and after product applications. The hive population is estimated before its introduction into the tunnel and at the end of the study. Foraging activity and mortality are counted. Indexes are computed as mortality index or forager mortality index for each treatment group: negative reference, positive reference and sponsor's product groups.

If they are no replicates in the study design, the best statistical approach is to compare study index with an historical positive reference index in a database. A control chart with statistical intervals at two levels of significance can be executed and study computed index can be positioned in this control graphic. A decision can be taken about the sponsor's product classification. It is in or outside the statistical bars.

If the study design includes replicates, indexes can be computed in each treatment group at one or several days and index results become study data for parametric or non-parametric analysis of variance.

Table 9. Example of BFD values for a numerical example (see section 8.2.2.).

		Before Exposure	3 days after	7 days after	14 days after	19 days after	Total
Control Group	**H1,1**	1.0	1.9	3.7	3.8	4.7	15.1
	H1,2	1.0	2.2	3.5	3.7	4.4	14.8
	H1,3	1.0	2.1	2.7	2.9	3.2	11.9
	H1,4	1.0	1.9	3.7	3.8	4.7	15.1
	H1,5	1.0	1.8	3.0	3.6	4.5	13.9
	H1,6	1.0	2.0	3.1	3.5	4.0	13.6
	TOTAL	6.0	11.5	18.9	21.0	25.0	82.4
Test Group	**H2,1**	1.0	1.5	2.5	3.1	3.8	11.9
	H2,2	1.0	1.4	2.7	3.0	3.6	11.7
	H2,3	1.0	1.7	2.0	3.3	3.5	11.5
	H2,4	1.0	1.3	2.1	3.4	3.7	11.5
	H2,5	1.0	1.5	2.5	3.0	3.9	11.9
	H2,6	1.0	1.8	2.6	2.9	3.5	11.8
	TOTAL	6.0	9.2	14.4	18.7	22.0	70.3
Total Groups		12.0	20.7	33.3	39.7	47.0	152.7

8.2.1.1. Honey bee brood development

The study is performed usually in field conditions or in semi-field conditions and the study design includes replicates: several hives are observed under the same modality. Indexes are computed from at least a 10 x 10 section of capped brood cells for each hive and for several days during the brood development as a repeated measure.

In this case, a repeated measures ANOVA can be performed to compare results between negative reference and one or several test item modalities. The statistical design is a factor group (modality) and a factor time (repeated measures). Each hive is a basic unit. This statistical analysis permits one to assess factors as group factors but also interactions between factors which could be interesting for the experimenter to assess a slow rate in the brood development.

A second statistical approach is to perform the statistical analysis on the raw data of each cell. In every modality and every hive, each cell among the 100 selected cells is observed during the brood development. A quotation of the development status is assessed by the experimenter. Each cell is a basic unit. The statistical design is a factor group (modality), a factor time (repeated measures), and a factor hive. Multiple interactions between the factors can be computed and statistically assessed. This study design which includes each cell quotation in the statistics permits to increase the statistical power (statistical packages are available to perform this kind of analysis). ANOVA parametric or non-parametric without or with transformation on the data can be performed.

8.2.1.2. LD_{50} determination

The study design is clearly defined in EPPO (2010b), OECD (1998a), or CEB (2011) guidelines. Well known statistical regression analysis from BLISS and LITCHFIELD and WILCOXON (Siegel and Castellan, 1988) and more recent publications lead to perform regressions with dose transformation as logarithm and probit or logit transformation on the response rate.

Dose-response curves at each recommended observation time should be plotted and the slopes of the curves and the median lethal doses (LD_{50}) with 95% confidence limits are calculated (Abbott, 1925). The LD_{50} is determined by the equation of the linear regression. Raw data provide dispersed values which need to be corrected by the control (see section 8.4.1.), then the 50% mortality is calculated with the equation type $y = ax + b$.

In some cases a lack of fit can be observed due to no dose related response. It depends on S shape component or an asymptotic data trend (Winer *et al.*, 1991). Non-linear standard or modified GOMPERTZ regression may give a better fit on experimental data.

Generally for the LDs calculation, different statistical softwares (both commercial and open source) are used. The computer-aided procedure performs the calculations automatically, thus helping to prevent errors.

8.2.2. Brood development index (numerical example)

The numerical example is a factorial experiment in which the factor product has two levels (p): control level and test level. The factor repeated measures has five levels (q): before exposure, three days after exposure, seven days, fourteen days, and nineteen days after exposure. There are six hives (n) in each product modality. In this design, each hive is observed under one modality of the factor product. There are 6 independent hives in every treatment group. The number of hives is twelve (2 x 6). The statistical model has npq = 60 data: n = 6, p = 2; q = 5. Example data are reported in Table 9.

8.2.2.1. Analysis of variance for numerical example

The test calculations are reported in the Table 10. In this example, factor group and factor repeated measures show a *P* value via a Fisher less than the classical level of significance (0.05): Group (p = 0.0019) and repeated measures R (p < 0.00001). These observed probabilities do not permit one to accept a null hypothesis of equality between the

Table 10. Analysis of variance for the example reported in Table 9. Formulae used: (1)= G^2/npq= $152.7^2/60$; (2)= $\Sigma\, x^2$= $1^2+ 1.9^2+......+3.5^2$; (3)= $(\Sigma\, Ai^2)/nq$= $(82.4^2 + 70.3^2)/30$; (4)=$(\Sigma Rj^2)/np$= $(12.0^2+20.7^2+......+47.0^2)/12$; (5)=$[\Sigma\, (ARij^2)]/n$= $(6.0^2+11.5^2+.....+22.0^2)/6$; (6)=$(\Sigma\, Hk^2)/q$= $(15.1^2+14.8^2+......11.8^2)/5$

Source of variation	Computational formula	Sum of square	df	MS	F	(probability)
Between Hives	(6)-(1)	3.84	(pn-1) = 11	0.35		
Group (Product)	(3)-(1)	2.44	(p-1) = 1	2.44	17.48	(p = 0.0019)
Hives within groups	(6)-(3)	1.40	p(n-1) = 10	0.14		
Within Hives	(2)-(6)	70.23	pn(q-1) = 48			
Repeated	(4)-(1)	66.92	(q-1) = 4	16.73	274.70	(p < 0.0001)
Interaction Group x R	(5)-(3)-(4)+(1)	0.88	(p-1)(q-1) = 4	0.22	3.61	(p = 0.0132)
R x Hives within groups	(2)-(5)-(6)+(3)	2.44	p(n-1)(q-1) = 40	0.061		

levels inside each factor. However the experimenter is not authorised to conclude the main factors because the interaction between the factors is significant (p = 0.0132). This statistical observation shows that the mean time profiles are not parallel between both groups (control and test product). The experimenter does analyse this interaction for instance with comparisons between groups at each time of measure.

8.2.2.2. Interaction statistical analysis

An analysis of variance is performed at each time, using a variance error which is computed from both the variance error of the main ANOVA described previously in the table (hives within group, [R x hives] within groups). This computation is performed for comparisons between groups. This combined mean square error with pq(n-1) = 50 degrees of freedom is 0.0766. This degree of freedom must be corrected because this common error comes from two sources of heterogeneity. This correction from SATTERTHWAITE gives the degree of freedom of 43 instead of 50 theoretical degrees.

All kinds of comparisons between both groups will be performed with the same common variance error.

- The comparisons at each level (time of measure) give the statistical results:
- 3 days after exposure: (MS = 0.4408; F = 5.75, observed probability p = 0.0209).
- 7 days after exposure: (MS = 1.6875; F = 22.03, observed probability p < 0.0001).
- 14 days after exposure: (MS = 0.4408; F = 5.75, observed probability p = 0.0209).
- 19 days after exposure: (MS = 0.7500; F = 9.79, observed probability p = 0.0031).
- 7 days after exposure, the comparison between means conduct to reject the null hypothesis with a probability < 0.0001. This observed probability is between 0.01 and 0.05 after 3 days and 14 days. 19 days after exposure, this observed probability is between 0.001 and 0.01.

All the statistical conditions for this statistical model are assumed to be obtained.

8.3. Conclusion

The experimenter needs to use statistical tests to help him make a decision (Fig. 11). A statistical analysis can be conducted only if it is included in the experimental design defined during the drafting of the study protocol. Without *a priori* conception, the statistical performance is frequently poor and the conclusions can be biased.

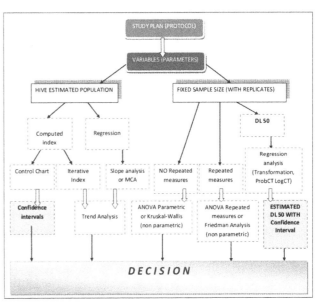

Fig. 11. Statistical decision chart.

8.4. Formulas and procedures frequently used in toxicological studies

8.4.1. Correction of the mortality rates

The mortality ratio is corrected on control mortality with the Henderson-Tilton formula.

$$The\ Henderson-Tilton\ formula: \left\{ 1 - \frac{Ta}{Ca} * \frac{Cb}{Tb} \right\}$$

If the parameter comprises live individuals and uniform numbers of bees per treatment (test and control), the Abbott formula is used.

$$The\ Abbott\ formula: \left\{ \frac{Ca - Ta}{Ca} \right\}$$

If the parameter comprises mortality ratios and a uniform start, the Schneider-Orelli formula should be applied.

$$The\ Schneider-Orelli\ formula: \left\{\frac{b-k}{1-k}\right\}$$

Abbreviations:

Tb = number of live bees before treatment

Ta = number of live bees after treatment

Cb = number of live bees in control before treatment

Ca = number of live bees in control after treatment

b = ratio of dead bees in treatment

k = ratio of dead bees in control

8.4.1.1. Example correction for control mortality

Tb	number of live test bees before treatment		10
Ta	number of live test bees after treatment		5
Cb	number of live control bees before treatment		10
Ca	number of live control bees after treatment		9
b	ratio not corrected test mortality		0.50
k	ratio control mortality		0.10

correction formulas for control mortality	not corrected mortality ratio	formula	corrected mortality ratio	percentage (*100)
Abbott's formula	0.50	((Ca-Ta)/Ca)	0.44	44.44%
Schneider-Orelli	0.50	((b-k)/(1-k))	0.44	44.44%
Henderson-Tilton	0.50	1-((Ta/Ca)* (Cb/Tb))	0.44	44.44%

8.4.2. Calculation of the HQ and RQ

8.4.2.1. Hazard Quotient HQ (EPPO, 2010b)

$$HQ = \frac{application\ rate\ (g\ AI\ /\ ha)}{acute\ LD_{50}\ (\mu g\ AI\ /\ bee)}$$

The critical HQ < 50 indicates low risk.

8.4.2.2. Risk Quotient RQ (EPHC, 2009)

$$RQ = \frac{application\ rate\ (g\ AI\ /\ cm^2)}{acute\ LD_{50}\ (\mu g\ AI\ /\ bee)}$$

Assuming the surface area of a honey bee is 1 cm^2

8.4.3. NOAEL and NOAEC

In individual laboratory assays, The NOAEL and NOAEC are the highest dose (in acute toxicity tests) and concentration (in chronic toxicity tests), respectively, which do not induce mortality significantly higher than that observed in controls. The statistical comparison between uncorrected mortality in the treated sample and in the control is performed using the Chi^2 test. The highest dose/concentration where bee mortality is not significantly different (p = 0.05) from the control is considered as NOAEL/NOAEC (respectively).

8.4.4. Power of a test

The power of a statistical test is the probability that the test will reject the null hypothesis when the null hypothesis is false (Type II error). Conventionally, statisticians require that the power of a test to detect a treatment effect of a specified magnitude is 80% but it may depend on the magnitude of the effects that it is required to detect.

9. Acknowledgements

The COLOSS (Prevention of honey bee COlony LOSSes) network aims to explain and prevent massive honey bee colony losses. It was funded through the COST Action FA0803. COST (European Cooperation in Science and Technology) is a unique means for European researchers to jointly develop their own ideas and new initiatives across all scientific disciplines through trans-European networking of nationally funded research activities. Based on a pan-European intergovernmental framework for cooperation in science and technology, COST has contributed since its creation more than 40 years ago to closing the gap between science, policy makers and society throughout Europe and beyond. COST is supported by the EU Seventh Framework Programme for research, technological development and demonstration activities (Official Journal L 412, 30 December 2006). The European Science Foundation as implementing agent of COST provides the COST Office through an EC Grant Agreement. The Council of the European Union provides the COST Secretariat. The COLOSS network is now supported by the Ricola Foundation - Nature & Culture.

10. References

ABBOTT, W (1925) A method of computing the effectiveness of an insecticide. *Journal of Economic Entomology* 18: 265-267.

ADAMS, S J; HIENRICH, K; HETMANSKI, M; FUSSELL, R J; WILKINS, S; THOMPSON, H M; SHARMAN, M (2007) Study of the depletion of tylosin residues in honey extracted from treated honey bee (*Apis mellifera*) colonies and the effect of the shook swarm procedure. *Apidologie* 38: 315-322.

AFSSA (2009) *Avis de l'Agence française de sécurité sanitaire des aliments relatif à la rédaction d'un projet de document guide de fixation des LMR de pesticides dans le miel dans le cadre du règlement (CE) n° 396/2005.*

ALAUX, C; BRUNET, J L; DUSSAUBAT, C; MONDET, F; TCHAMITCHAN, S; COUSIN, M; BRILLARD, J; BALDY, A; BELZUNCES, L P; LE CONTE, Y (2010) Interactions between *Nosema* microspores and a neonicotinoid weaken honey bees (*Apis mellifera*). *Environmental Microbiology* 12(3): 774-782.

ALIX, A; VERGNET, C; MERCIER, T (2009) Risks to bees from dusts emitted at sowing of coated seeds: concerns, risk assessment and risk management. *Julius- Kühn-Archive* 423: 131–132.

APENET (2009) *Effects of coated maize seed on honey bees. Report based on results obtained from the first year of activity of the APENET project.* 30 pp. http://www.reterurale.it/flex/cm/pages/ ServeBLOB.php/L/IT/IDPagina/4600.

APENET (2010) *Effects of coated maize seed on honey bees. Report based on results obtained from the second year (2010) activity of the APENET project.* 100 pp. http://www.reterurale.it/flex/cm/ pages/ServeBLOB.php/L/IT/IDPagina/5773.

ARKOOSH, M R; CASILLAS, E; CLEMONS, E; KAGLEY, A N; OLSON, R; RENO, P; STEIN, J E (1998). Effect of pollution on fish diseases: potential impacts on salmonid populations. *Journal of Aquatic Animal Health* 10: 182-190.

ARZONE, A; VIDANO, C (1980) Methods for testing pesticide toxicity to honey bees. *Bollettino del Laboratorio di Entomologia Agraria "Filippo Silvestri"* 37: 161-165.

ATKINS, E L; GREYWOOD, E A; MACDONALD, R L (1973) *Toxicity of pesticides and other agricultural chemicals to honey bees: Laboratory studies.* University of California, Agricultural extension service; California, USA. 38 pp.

ATKINS, E L; KELLUM, D (1986) Comparative morphogenic and toxicity studies on the effect of pesticides on honey bee brood. *Journal of Apicultural Research* 25: 242-255.

AUFAUVRE, J; BIRON, D G; VIDAU, C; FONTBONNE, R; ROUDEL, M; DIOGON, M; VIGUÈS, B; BELZUNCES, L P; DELBAC, F; BLOT, N (2012) Parasite-insecticide interactions: a case study of *Nosema ceranae* and fipronil synergy on honey bee. *Scientific Reports* 2: 326. http://dx.doi.org/10.1038/srep00326

AUPINEL, P; FORTINI, D; DUFOUR, H; TASEI, J N; MICHAUD, B; ODOUX, J F; PHAM-DELÈGUE, M H (2005) Improvement of artificial feeding in a standard *in vitro* method for rearing *Apis mellifera* larvae. *Bulletin of Insectology* 58(2): 107-111.

AUPINEL, P; FORTINI, D; MICHAUD, B; MAROLLEAU, F; TASEI, J N; ODOUX, J F (2007a) Toxicity of dimethoate and fenoxycarb to honey bee brood (*Apis mellifera*), using a new *in vitro* standardized feeding method. *Pest Management Science* 63: 1090-1094.

AUPINEL, P; FORTINI, D; MICHAUD, D; MEDRZYCKI, P; PADOVANI, E; PRZYGODA, D; MAUS, C; CHARRIERE, J D; KILCHENMANN, V; RIESSBERGER-GALLE, U; VOLLMANN, J J; JEKER, L; JANKE, M; ODOUX, J F; TASEI, J N (2009) Honey bee brood ring-test: method for testing pesticide toxicity on honey bee brood in laboratory conditions. *Julius-Kühn-Archiv* 423: 96-102.

AUPINEL, P; MEDRZYCKI, P; FORTINI, D; MICHAUD, B; TASEI, J N; ODOUX, J F (2007b) A new larval *in vitro* rearing method to test effects of pesticides on honey bee brood. *REDIA* 90: 91-94.

AVNI, D; DAG, A; SHAFIR, S (2009) The effect of surface area of pollen patties fed to honey bee (*Apis mellifera*) colonies on their consumption, brood production and honey yields. *Journal of Apicultural Research* 48: 23-28.

BARKER, R J; TABER, S (1977) Effects of diflubenzuron fed to caged honey bees. *Environmental Entomology* 6: 167-168.

BAUER, R N; DIAZ-SANCHEZ, D; JASPERS, I (2012) Effects of air pollutants on innate immunity: the role of Toll-like receptors and nucleotide-binding oligomerization domain–like receptors. *The Journal of Allergy and Clinical Immunology* 129: 14-24.

BBA (Biologische Bundesanstalt für Land- und Forstwirtschaft) (1992) Messung der direkten Abtrift beim Ausbringen von flüssigen Pflanzenschutzmitteln im Freiland. *Richtlinien für die Prüfung von Pflanzenschutzgeräten* VII: 2-1.1.

BECHER, M A; SCHARPENBERG, H; MORITZ, R F A (2009) Pupal developmental temperature and behavioural specialization of honey bee workers (*Apis mellifera* L.). *Journal of Comparative Physiology* A195: 673–679. http://dx.doi.org/10.1007/s00359-009-0442-7

BECKER, R; SCHMITZER, S; BARTH, M; BARGEN, H; KAATZ, H-H; RATTE, H T; SCHUR, A (2011) Statistical evaluation of regulatory honey bee trials – a pragmatic approach. In *Proceeding of the 11th International Symposium Hazards of Pesticides to Bees (ICPBR), Wageningen, November 2-4, 2011*.

BENDAHOU, N; BOUNIAS, M; FLÉCHÉ, C (1997) Acute toxicity of cypermethrin and fenithrothion on honey bees according to age, formulations and (chronic paralysis virus) insecticide interaction. *Journal of Environmental Biology* 18: 55-65.

BENDAHOU, N; FLÉCHÉ, C; BOUNIAS, M (1999) Biological and biochemical effects of chronic exposure to very low levels of dietary cypermethrin (Cymbush) on honey bee colonies. *Ecotoxicology and Environmental Safety* 44: 147-153.

BERNARD, C B; PHILOGÈNE, B J (1993) Insecticide synergists: role, importance and perspectives. *Journal of Toxicology and Environmental Health* 38: 199-223.

BLISS, C (1934) The method of probits. *Science* 79: 38-39.

BLISS, C (1939) The toxicity of poisons applied jointly. *Annals of Applied Biology* 26: 585–615.

BORTOLOTTI, L; MONTANARI, R; MARCELINO, J; MEDRZYCKI, P; MAINI, S; PORRINI, C (2003) Effects of sub-lethal imidacloprid doses on the homing rate and foraging activity of honey bees. *Bulletin of Insectology* 56: 63-67.

BORTOLOTTI, L; SABATINI, A G; MUTINELLI, F; ASTUTI, M; LAVAZZA, A; PIRO, R; TESORIERO, D; MEDRZYCKI, P; SGOLASTRA, F; PORRINI, C (2009) Spring honey bee losses in Italy. *Julius-Kühn-Archiv* 423: 148-145.

BRITTAIN, C; POTTS, S (2011) The potential impact of insecticides on the life-history traits of bees and the consequences for pollination. *Basic and Applied Ecology* 12: 321-331.

BRODSGAARD, H F; BRODSGAARD, C J; HANSEN, H; LOVE, I G L (2003) Environmental risk assessment of transgene products using honey bee (*Apis mellifera*) larvae. *Apidologie* 34: 139-145.

BRUNET, J L; FAIVRE D'ARCIER, F; TCHAMITCHIAN, S; CERRUTI, N; BADIOU, A; COUSIN, M; ALAUX, C; DUSSAUBAT, C; MONDET, F; LE CONTE, Y; BELZUNCES, L (2009) Effets différentiels entre hautes doses et basses doses et entre les expositions uniques et chroniques des pesticides chez l'abeille (Round Table 3 - Intoxication in bees due to pesticides: results from scientists). In *Proceedings of the 41th Congress of Apimondia, Montpellier, France, September 15-20.* P 128.

CALABRESE, E J (2005) Paradigm lost, paradigm found: the re-emergence of hormesis as a fundamental dose response model in the toxicological sciences. *Environmental Pollution* 138: 378-411.

CARRECK, N L; ANDREE, M; BRENT, C S; COX-FOSTER, D; DADE, H
A; ELLIS, J D; HATJINA, F; VANENGELSDORP, D (2013) Standard
methods for *Apis mellifera* anatomy and dissection. In *V Dietemann;
J D Ellis; P Neumann (Eds) The COLOSS* BEEBOOK, *Volume I:
standard methods for* Apis mellifera *research. Journal of Apicultural
Research* 52(4): http://dx.doi.org/10.3896/IBRA.1.52.4.03

CEB (2011) *Methode n°230: Method for the evaluation of side-effects
of plant protection products on honey bees (*Apis mellifera *L.).*
Association Française de Protection des Plantes. 43pp.

CELLI, G; MACCAGNANI, B (2003) Honey bees as bioindicators of
environmental pollution. *Bulletin of Insectology* 56 (1): 137-139.

CHAUZAT, M P; FAUCON, J P; MARTEL, A C; LACHAIZE, J; COUGOULE,
N; AUBERT, M (2006) A survey of pesticide residues in pollen
loads collected by honey bees in France. *Journal of Economic
Entomology* 99: 253-262.

CLAUDIANOS, C; RANSON, H; JOHNSON, R M; BISWAS, S; SCHULER,
M A; BERENBAUM, M R; FEYEREISEN, R; OAKESCHOTT, J G (2006)
A deficit of detoxification enzymes: pesticide sensitivity and
environmental response in the honey bee. *Insect Molecular Biology*
15: 615-636.

COLIN, M E; BELZUNCES, L (1992) Evidence of synergy between
prochloraz and deltamethrin in *Apis mellifera*: a convenient
biological approach. *Pesticide Science* 36: 115-119.

COLIN, M; BONMATIN, J; MOINEAU, I; GAIMON, C; BRUN, S;
VERMANDERE, J (2004) A method to quantify an analyse the
foraging activity of honey bees: relevance to the sublethal effects
induced by systemic insecticides. *Environmental Contamination
and Toxicology* 47; 387-395.

CRAILSHEIM, K (1990) The protein balance of the honey bee worker.
Apidologie 21: 417–429.

CRAILSHEIM, K; BRODSCHNEIDER, R; AUPINEL, P; BEHRENS, D;
GENERSCH, E; VOLLMANN, J; RIESSBERGER-GALLÉ, U (2013)
Standard methods for artificial rearing of *Apis mellifera* larvae. In
V Dietemann; J D Ellis; P Neumann (Eds) The COLOSS BEEBOOK,
Volume I: standard methods for Apis mellifera *research. Journal
of Apicultural Research* 52(1):
http://dx.doi.org/10.3896/IBRA.1.52.1.05

CRANE, E (1984) Bees, honey and pollen as indicators of metals in the
environment. *Bee World* 55: 47-49.

CRESSWELL, J E (2011) A meta-analysis of experiments testing the
effects of a neonicotinoid insecticide (imidacloprid) on honey bees.
Ecotoxicology 20(1): 149-57.

CZOPPELT, C (1990) Effect of insect growth inhibitors and pesticides
on honey bee larvae (*Apis mellifera* L.) in contact poison and
feeding poison tests *in vitro*. In *Proceedings of the Fourth
International Symposium on the Harmonization of Methods for
Testing the Toxicity of Pesticides to Bees, May 15-18, 1990, Rez
near Prague, Czechoslovakia.* pp 76-83.

DAI, P L; WANG, Q; SUN, J H; LIU, F; WANG, X; WU, Y Y; ZHOU, T
(2010) Effects of sublethal concentrations of bifenthrin and
deltamethrin on fecundity, growth, and development of the honey
bee *Apis mellifera ligustica. Environmental Toxicology and Chemistry*
29: 644-649.

DAVIS, A R; SOLOMON, K R; SHUEL, R W (1988) Laboratory studies
of honey bee larval growth and development as affected by
systemic insecticides at adult-sublethal levels. *Journal of Apicultural
Research* 27: 146-161.

DAVIS, A; SHUEL, R (1988) Distribution of 14C-labelled coarbofuran
and dimethoate in royal jelly, queen larvae, and nurse honey bees.
Apidologie 19: 37-50.

DECHAUME-MONCHARMONT, F X; DECOURTYE, A; HENNEQUET-
HANTIER, C; PONS, O; PHAM-DELÈGUE, M H (2003) Statistical
analysis of honey bee survival after chronic exposure to insecticides.
Environmental Toxicology and Chemistry 22: 3088-3094.

DECOURTYE, A; DEVILLERS, J; AUPINEL, P; BRUN, F; BAGNIS, C;
FOURRIER, J; GAUTHIER, M (2011) Honey bee tracking with
microchips: a new methodology to measure the effects of pesticides.
Ecotoxicology 20(2): 429-437.

DEGRANDI-HOFFMAN, G; WARDELL, G; AHUMADA-SEGURA, F;
RINDERER, T; DANKA, R; PETTIS, J (2008) Comparisons of pollen
substitute diets for honey bees: consumption rates by colonies
and effects on brood and adult populations. *Journal of Apicultural
Research* 47: 265-270.

DELAPLANE, K S; DAG, A; DANKA, R G; FREITAS, B M; GARIBALDI, L A;
GOODWIN, R M; HORMAZA, J I (2013) Standard methods for
pollination research with *Apis mellifera*. In *V Dietemann; J D Ellis;
P Neumann (Eds) The COLOSS* BEEBOOK, *Volume I: standard
methods for* Apis mellifera *research. Journal of Apicultural Research*
52(4): http://dx.doi.org/10.3896/IBRA.1.52.4.12

DELAPLANE, K S; VAN DER STEEN, J; GUZMAN, E (2013) Standard
methods for estimating strength parameters of *Apis mellifera*
colonies. In *V Dietemann; J D Ellis; P Neumann (Eds) The
COLOSS* BEEBOOK, *Volume I: standard methods for* Apis mellifera
research. Journal of Apicultural Research 52(1):
http://dx.doi.org/10.3896/IBRA.1.52.1.03

DESNEUX, N; DECOURTYE, A; DELPUECH, J M (2007) The sublethal
effects of pesticides on beneficial arthropods. *Annual Review of
Entomology* 52: 81-106.

DIECKMANN, Y; ISHAQUE, M; MUENSTER, I; PICARD, L; BENZ, A;
LANGEWALD, J; KREUZ, K; KOEHLE, H; GOERTH, F C; RAETHER,
R B; MONTAG, J; HUBER-MOULLIET, U; KERL, W (2010) *Patent
application publication for systemicity enhancers.* Publication No.
US 2010/0204045 A1.

DIETEMANN, V; NAZZI, F; MARTIN, S J; ANDERSON, D; LOCKE, B; DELAPLANE, K S; WAUQUIEZ, Q; TANNAHILL, C; FREY, E; ZIEGELMANN, B; ROSENKRANZ, P; ELLIS, J D (2013) Standard methods for varroa research. In *V Dietemann; J D Ellis; P Neumann (Eds) The COLOSS* BEEBOOK, *Volume II: standard methods for* Apis mellifera *pest and pathogen research. Journal of Apicultural Research* 52(1): http://dx.doi.org/10.3896/IBRA.1.52.1.09

EFSA Panel on Plant Protection Products and their Residues (PPR) (2012) Scientific opinion on the science behind the development of a risk assessment of Plant Protection Products on bees (*Apis mellifera, Bombus* spp. and solitary bees). *EFSA Journal* 10(5): 2668. http://www.efsa.europa.eu/en/efsajournal/doc/2668.pdf

ELLIS, M D; BAXENDALE, F P (1997) Toxicity of seven monoterpenoids to tracheal mites (Acari: Tarsonemidae) and their honey bee (Hymenoptera: Apidae) hosts when applied as fumigants. *Journal of Economic Entomology* 90: 1087– 1091.

ELZEN, P J; EISCHEN, F A; BAXTER, J B; PETTIS, J; ELZEN, G W; WILSON, W T (1998) Fluvalinate resistance in *Varroa jacobsoni* from several geographic locations. *American Bee Journal* 138: 674–686.

ENGELS, W (1990) Testing of insect growth regulators and of varroacides by the *Apis*-larvae-test. In *Proceedings of the Fourth International Symposium on the Harmonization of Methods for Testing the Toxicity of Pesticides to Bees, May 15-18, 1990, Rez near Prague, Czechoslovakia.* pp 84-87.

EPHC (2009) *Environmental risk assessment guidance manual for agricultural and veterinary chemicals.* Environmental Protection and Heritage Council; Canberra, Australia.

EPPO (2010a) PP 1/170 (4): Side-effects on honey bees. *EPPO Bulletin* 40(3): 313- 319.

EPPO (2010b) Environmental risk assessment scheme for plant protection products. Chapter 10: Honey bees. *EPPO Bulletin* 40(3): 323-331.

FAUCON, J P; AURIERES, C; DRAJNUDEL, P; MATHIEU, L; RIBIERE, M; MARTEL, A C; ZEGGANE, S; CHAUZAT, M-P; AUBERT, M (2005) Experimental study on the toxicity of imidacloprid given in syrup to honey bee (*Apis mellifera*) colonies. *Pest Management Science* 61: 111-125.

FAUCON, J P; FLAMINI, C; COLIN, M E (1985) Evaluation de l'incidence de la deltamethrine sur les problèmes de cheptel apicole. *Bulletin Vétérinaire* 17: 49-65.

FFRENCH-CONSTANT, R H; ROUCH, R T (1992) Resistance detection and documentation: the relative roles of pesticidal and biochemical aspects. In *Roush, R T; Tabashnik, B E (Eds). Pesticide resistance in Arthropods. Chapman & Hall; New York, USA.* pp. 4-38.

FINNEY, D J (1971) *Probit analysis (3rd edition).* Cambridge University Press; New York, USA. 333 pp.

FLURI, P; LÜSCHER, M; WILLE, H; GERIG, L (1982) Changes in weight of the pharyngeal gland and haemolymph titres of juvenile hormone, protein and vitellogenin in worker honey bees. *Journal of Insect Physiology* 28: 61–68.

FRIES, I; CHAUZAT, M-P; CHEN, Y-P; DOUBLET, V; GENERSCH, E; GISDER, S; HIGES, M; MCMAHON, D P; MARTÍN-HERNÁNDEZ, R; NATSOPOULOU, M; PAXTON, R J; TANNER, G; WEBSTER, T C; WILLIAMS, G R (2013) Standard methods for nosema research. In *V Dietemann; J D Ellis; P Neumann (Eds) The COLOSS* BEEBOOK, *Volume II: Standard methods for* Apis mellifera *pest and pathogen research. Journal of Apicultural Research* 52(1): http://dx.doi.org/10.3896/IBRA.1.52.1.14

GIROLAMI, V; MAZZON, L; SQUARTINI, A; MORI, N; MARZARO, M; DI BERNARDO, A; GREATTI, M; GIORIO, C; TAPPARO, A (2009) Translocation of neonicotinoid insecticides from coated seeds to seedling guttation drops: a novel way of intoxication for bees. *Journal of Economical Entomology* 102: 1808-1815.

GIROLAMI, V; MARZARO, M; VIVAN, L; MAZZON, L; GIORIO, C; MARTON, D; TAPPARO, A (2012b) Aerial powdering of bees inside mobile cages and the extent of neonicotinoid cloud surrounding corn drillers. *Journal of Applied Entomology,* 137(1-2): 35-44. http://dx.doi.org/10.1111/j.1439-0418.2012.01718.x

GIROLAMI, V; MARZARO, M; VIVAN, L; MAZZON, L; GREATTI, M; GIORIO, C; MARTON, D; TAPPARO, A (2012a) Fatal powdering of bees in flight with particulates of neonicotinoids seed coating and humidity implication. *Journal of Applied Entomology* 136: 17-26.

GOUGH, H J; MCINDOE, E C; LEWIS, G B; (1994) The use of dimethoate as a reference compound in laboratory acute toxicity tests on honey bees (*Apis mellifera* L. 1981-1992) *Journal of Apicultural Research* 33(2): 119-125.

GREATTI, M; BARBATTINI, R; STRAVISI, A; SABATINI, A G; ROSSI, S (2006) Presence of the a.i. imidacloprid on vegetation near corn fields sown with Gaucho® dressed seeds. *Bulletin of Insectology* 59(2): 99-103.

GREATTI, M; SABATINI, A G; BARBATTINI, R; ROSSI, S; STRAVISI, A (2003) Risk of environmental contamination by AI imidacloprid used for corn seed dressing. Preliminary results. *Bulletin of Insectology* 59(1): 69-72.

GROH, C; TAUTZ, J; ROSSLER, W (2004) Synaptic organization in the adult honey bee brain is influenced by brood-temperature control during pupal development. *Proceedings of the National Academy of Sciences* 101: 4268-4273. http://dx.doi.org/10.1073/PNAS.0400773101

HAARMANN, T; SPIVAK, M; WEAVER, D; WEAVER, B; GLENN, T (2002) Effects of fluvalinate and coumaphos on queen honey bees in two commercial queen rearing operations. *Journal of Economic Entomology* 95: 28-35.

HAWTHORNE, D J; DIVELY, G P (2011) Killing them with kindness? In -hive medications may inhibit xenobiotic efflux transporters and endanger honey bees. *PLoS One* 6(11): e26796. http://dx.doi.org/10.1371/journal.pone.0026796

HAYDAK, M H (1968) Nutrition des larves d'abeilles, In *Chauvin, R (Eds). Traité de biologie de l'abeille Vol. 1. Masson et Cie; Paris, France.* pp 302-333.

HEIMBACH, U (2008) *Heubach method to determine the particulate matter of maize seeds treated with insecticides*. JKI Institute for Plant Protection in Agriculture and Grassland; Braunschweig, Germany. http://www.jki.bund.de/fileadmin/dam_uploads/_A/pdf/Heubach%20Method%20english.pdf

HENRY, M; BEGUIN, M; REQUIER, F; ROLLIN, O; ODOUX, J F; AUPINEL, P; APTEL, J; TCHAMITCHIAN, S; DECOURTYE, A (2012) A common pesticide decreases foraging success and survival in honey bees. *Science* 336(6079): 348-350. http://dx.doi.org/10.1126/science.1215039

HIGES, M; MARTIN-HERNANDEZ, R; GARRIDO-BAILON, E; GARCIA-PALENCIA, P; MEANA, A (2008) Detection of infective *Nosema ceranae* (Microsporidia) spores in corbicular pollen of forager honey bees. *Journal of Invertebrate Pathology* 97: 76-78.

HODGSON, E (2004) A textbook of modern toxicology (3rd Ed.). John Wiley and sons Inc; UK. 584 pp.

HUMAN, H; BRODSCHNEIDER, R; DIETEMANN, V; DIVELY, G; ELLIS, J; FORSGREN, E; FRIES, I; HATJINA, F; HU, F-L; JAFFÉ, R; JENSEN, A B; KÖHLER, A; MAGYAR, J; ÖZIKRIM, A; PIRK, C W W; ROSE, R; STRAUSS, U; TANNER, G; TARPY, D R; VAN DER STEEN, J J M; VAUDO, A; VEJSNÆS, F; WILDE, J; WILLIAMS, G R; ZHENG, H-Q (2013) Miscellaneous standard methods for *Apis mellifera* research. In *V Dietemann; J D Ellis; P Neumann (Eds) The COLOSS BEEBOOK, Volume I: standard methods for* Apis mellifera *research. Journal of Apicultural Research* 52(4): http://dx.doi.org/10.3896/IBRA.1.52.4.10

IMDORF, A; BUEHLMANN, G; GERIG, L; KILCHMANN, V; WILLE, H (1987) Überprüfung der Schätzmethode zur Ermittlung der Brutfläche und der Anzahl Arbeiterinnen in freifliegenden Bienenvölkern. *Apidologie* 18(2): 137-146.

IMDORF, A; GERIG, L (1999) *Lehrgang zur Erfassung der Volksstärke*. Schweizerisches Zentrum für Bienenforschung. Available from: http://www.agroscope.admin.ch

IWASA, T; MOTOYAMA, N; AMBROSE, J; ROE, R (2004) Mechanism for the differential toxicity of neonicotinoid insecticides in the honey bee, *Apis mellifera*. *Crop Protection* 23: 371–378.

JEKER, L; MESCHBERGER, T; SCHMID, L; CANDOLFI, M; MAGYAR, J P (2011) Digital image analysis tool to improve the assessment and evaluation of brood development in higher tier honey bee studies. In *Proceedings of the 11th International Symposium Hazards of Pesticides to Bees (ICPBR), Wageningen, November 2-4, 2011*.

JEKER, L; SCHMID, L; MESCHBERGER, T; CANDOLFI, M; PUDENZ, S; MAGYAR, J P (2012) Computer-assisted digital image analysis and evaluation of brood development in honey bee combs. *Journal of Apicultural Research* 51(1): 63-73.

JOHANSEN, C (1977) Pesticides and pollinators. *Annual Review of Entomology* 22: 177-192.

JOHANSEN, C (1978) Bee poisoning test protocols for the United States. In *Proceedings of the EPA Conference, 8-9 November 1978*.

JOHANSEN, C A; MAYER, D F; EVES, J D; KIOUS, C W (1983) Pesticides and bees. *Environmental Entomology* 12(5): 1513-1518.

JOHNSON, R M; DAHLGREN L; SIEGFRIED, B D; ELLIS, M D (2013) Acaricide, fungicide and drug interactions in honey bees (*Apis mellifera*). *PLoS One*: e54092. http://dx.doi.org/10.1371/journal.pone.0054092

JOHNSON, R M; POLLOCK, H S; BERENBAUM, M R (2009) Synergistic interactions between in-hive miticides in *Apis mellifera*. *Journal of Economic Entomology* 102: 474-479.

JOHNSON, R M; WEN, Z; SCHULER, M A; BERENBAUM, M R (2006) Mediation of pyrethroid insecticide toxicity to honey bees (Hymenoptera: Apidae) by cytochrome P450 monooxygenases. *Journal of Economic Entomology* 99: 1046–1050.

JONES, J C; HELLIWELL, P; BEEKMAN, M; MALESZKA, R; OLDROYD, B P (2005) The effects of rearing temperature on developmental stability and learning and memory in the honey bee, *Apis mellifera*. *Journal of Comparative Physiology* A191: 1121–1129. http://dx.doi.org/10.1007/s00359-005-0035-z

KARISE, R; VIIK, E; MÄND, M (2007) Impact of alpha-cypermethrin on honey bees foraging on spring oilseed rape (*Brassica napus*) flowers in field conditions. *Pest Management Science* 63: 1085-1089.

KONWICK, B J; FISK, A T; GARRISON, A W; AVANTS, J K; BLACK, M C (2005) Acute enantioselective toxicity of fipronil and its desulfinyl photoproduct to *Ceriodaphnia dubia*. *Environmental Toxicology and Chemistry* 24: 2350-2355.

KRUPKE, C H; HUNT, G J; EITZER, B D; ANDINO, G; GIVEN, K (2012) Multiple routes of pesticide exposure for honey bees living near agricultural fields. *PLoS One* 7(1): e29268.

LADAS, A (1972) Der Einfluss verschiedener Konstitutions-und Umweltfaktoren auf die Anfälligkeit der Honigbiene (*Apis mellifica* L.) gegenüber zwei Insektiziden Pflanzenschutzmitteln. *Apidologie* 3: 55-78.

LEWIS, D; GATTIE, D; NOVAK, M; SANCHEZ, S; PUMPHREY, C (2002) Interactions of pathogens and irritant chemicals in land-applied sewage sludges (biosolids). *BMC Public Health* 2: 11.

LIU, M-Y; PLAPP, F W (1992) Mechanism of formamidine synergism of pyrethroids. *Pesticide Biochemistry and Physiology* 43: 134–140.

LUTTRELL, R G (1985) Efficacy of insecticides applied ultra-low volume in vegetable oils. In *T M Kaneko; L D Spicer (Eds). Pesticide formulations and application systems: Fourth symposium. ASTM International*. pp 67- 77.

MAINI, S; MEDRZYCKI, P; PORRINI, C (2010) The puzzle of honey bee losses: a brief review. *Bulletin of Insectology* 63(1): 153-160.

MALONE, L A; GATEHOUSE, H S (1998) Effects of *Nosema apis* infection on honey bee (*Apis mellifera*) digestive proteolytic enzyme activity. *Journal of Invertebrate Pathology* 71: 169-174.

MALONE, L A; TREGIDGA, E L; TODD, J H; BURGESS, E; PHILIP, B A; MARKWICK, N P; POULTON, J; CHRISTELLER, J T; LESTER, M T; GATEHOUSE, H S (2002) Effects of ingestion of a biotin-binding protein on adult and larval honey bees. *Apidologie* 33: 447-458.

MARTEL, A C; ZEGGANE, S; DRAJNUDEL, P; FAUCON, J P; AUBERT, M (2006) Tetracyclines residues in honey after hive treatment. *Food Additives and Contaminants* 23(3): 265-273.

MARTIN-HERNANDEZ, R; MEANA, A; PRIETO, L; SALVADOR, A M; GARRIDO-BAILON, E; HIGES, M (2007) Outcome of colonization of *Apis mellifera* by *Nosema ceranae*. *Applied and Environmental Microbiology* 73: 6331-6338.

MARZARO, M; VIVAN, L; TARGA, A; MAZZON, L; MORI, N; GREATTI, M; PETRUCCO TOFFOLO, E; DI BERNARDO, A; GIORIO, C; MARTON, D; TAPPARO, A; GIROLAMI, V (2011) Lethal aerial powdering of honey bees with neonicotinoids from fragments of maize seed coat. *Bulletin of Insectology* 64: 119-126.

MATTILA, H R; OTIS, G W (2006a) Effects of pollen availability and *Nosema* infection during the spring on division of labour and survival of worker honey bees (Hymenoptera: Apidae). *Environmental Entomology* 35: 708-717.

MATTILA, H R; OTIS, G W (2006b) The effects of pollen availability during larval development on the behaviour and physiology of spring-reared honey bee workers. *Apidologie* 37: 533-546.

MATTILA, H R; OTIS, G W (2007) Manipulating pollen supply in honey bee colonies during the fall does not affect the performance of winter bees. *Canadian Entomologist* 139: 554-563.

MATTILA, H R; SMITH, B H (2008) Learning and memory in workers reared by nutritionally stressed honey bee (*Apis mellifera* L.) colonies. *Physiology and Behaviour* 95(5): 609-16. http://dx.doi.org/10.1016/j.physbeh.2008.08.003

MAURIZIO, A (1950) The influence of pollen feeding and brood rearing on the length of life and physiological conditions of the honey bee. *Bee World* 31: 9-12.

McMULLAN, J B; BROWN, M J F (2005) Brood pupation temperature affects the susceptibility of honey bees (*Apis mellifera*) to infestation by tracheal mites (*Acarapis woodi*). *Apidologie* 36: 97–105. http://dx.doi.org/10.1051/apido:2004073

MEDRZYCKI, P; SGOLASTRA, F; BORTOLOTTI, L; BOGO, G; TOSI, S; PADOVANI, E; PORRINI, C; SABATINI, A G (2010) Influence of brood rearing temperature on honey bee development and susceptibility to poisoning by pesticides. *Journal of Apicultural Research* 49(1): 52-59.

MORÉTEAU, B (1991) Etude de certains aspects de la physiotoxicologie d'insecticides de synthèse chez le criquet migrateur: *Locusta migratoria*. In *Aupelf-Urek (Eds). La Lutte Anti-acridienne. John Libbey Eurotext; Paris.* pp 167-178.

MORTON, H; MOFFETT, J (1972) Ovicidal and larvicidal effects of certain herbicides on honey bees. *Environmental Entomology* 1: 611-614.

MULLIN, C A; FRAZIER, M; FRAZIER, J L; ASHCRAFT, S; SIMONDS, R; VANENGELSDORP, D; PETTIS, J S (2010) High levels of miticides and agrochemicals in North American apiaries: Implications for honey bee health. *PLoS One* 5: e9754.

NAUEN, R; TIETJEN, K; WAGNER, K; ELBERT, A (1998) Efficacy of plant metabolites of imidacloprid against *Myzus persicae* and *Aphis gossypii*. *Pesticide Science* 52: 53-57.

NAUMANN, K; ISMAN, M (1996) Toxicity of a neem (*Azadirachta indica* A. Juss) insecticide to larval honey bees. *American Bee Journal* 136: 518-520.

NRCC (National Research Council Canada) (1981) *Pesticide-pollinators interactions*. NRCC Report N° 18471, Associate Committee on Scientific Criteria for Environmental Quality; Ottava, National Research Council of Canada. 190 pp.

NUNAMAKER, R; HARVEY, J; WILSON, W (1984) Inability of honey bee colonies to rear queens following exposure to fenthion. *American Bee Journal* 124: 308-309.

OECD (1998a) OECD guideline for testing of chemicals. Test No 213: Honey bees, acute oral toxicity test.

OECD (1998b) OECD guideline for testing of chemicals. Test No 214: Honey bees, acute contact toxicity test.

OECD (2006) Current approaches in the statistical analysis of ecotoxicity data: A guidance to application. *Environment Health and Safety Publications. Series on Testing and Assessment* No. 54.

OECD (2007) Guidance document on the honey bee (*Apis mellifera* L.) brood test under semi-field conditions. *Environment Health and Safety Publications. Series on Testing and Assessment.* No. 75.

OOMEN, P A; DE RUIJTER, A; VAN DER STEEN, J J M (1992) Method for honey bee brood feeding tests with insect growth-regulating insecticides. *EPPO Bulletin* 22(4): 613-616.

PAYTON, M E; GREENSTONE, M H; SCHENKER, N (2003) Overlapping confidence intervals or standard error intervals: What do they mean in terms of statistical significance? *Journal of Insect Science* 3: 34.

PENG, Y S C; MUSSEN, E; FONG, A; MONTAGUE, M A; TYLER, T (1992) Effects of chlortetracycline on honey-bee worker larvae reared *in vitro*. *Journal of Invertebrate Pathology* 60: 127-133.

PETTIS, J S; COLLINS, A; WILBANKS, R; FELDLAUFER, M (2004) Effects of coumaphos on queen rearing in the honey bee, *Apis mellifera*. *Apidologie* 35: 605-610.

PETTIS, J S; VANENGELSDORP, D; JOHNSON, J; DIVELY, G (2012) Pesticide exposure in honey bees results in increased levels of the gut pathogen *Nosema*. *Naturwissenschaften* 99: 153-158.

PETTIS, J S; WILSON, W T; SHIMANUKI, S; TEEL, P D (1991) Fluvalinate treatment of queen and worker honey bees and effects on subsequent mortality, queen acceptance and supersedure. *Apidologie* 22: 1-7.

PINTO, L; BITONDI, M; SIMOES, Z (2000) Inhibition of vitellogenin synthesis in *Apis mellifera* workers by a juvenile hormone analogue, pyriproxyfen. *Journal of Insect Physiology* 46: 153-160.

PIRK, C W W; DE MIRANDA, J R; FRIES, I; KRAMER, M; PAXTON, R; MURRAY, T; NAZZI, F; SHUTLER, D; VAN DER STEEN, J J M; VAN DOOREMALEN, C (2013) Statistical guidelines for *Apis mellifera* research. In *V Dietemann; J D Ellis; P Neumann (Eds) The COLOSS* BEEBOOK, *Volume I: standard methods for* Apis mellifera *research. Journal of Apicultural Research* 52(4): http://dx.doi.org/10.3896/IBRA.1.52.4.13

PISTORIUS, J; BECKER, R; LÜCKMANN, J; SCHUR, A; BARTH, M; JEKER, L; SCHMITZER, S; VON DER OHE, W (2011) Effectiveness of method improvements to reduce variability of brood termination rate in honey bee brood studies under semi-field conditions. In *Proceedings of the 11th International Symposium Hazards of Pesticides to Bees (ICPBR), Wageningen, November 2-4, 2011.*

PISTORIUS, J; BISCHOFF, G; HEIMBACH, U; STÄHLER, M (2009) Bee poisoning incidents in Germany in spring 2008 caused by abrasion of active substance from treated seeds during sowing of maize. *Julius-Kühn-Archiv* 423: 118-126.

PORRINI, C (1995) L'organismo alveare e i fitofarmaci. *Inf.tore Fitopat.* 6: 7-12.

R DEVELOPMENT CORE TEAM (2010) *R: A language and environment for statistical computing.* R Foundation for Statistical Computing; Vienna, Austria.

RAMADE, F (1992) *Précis d'écotoxicologie.* Masson; Paris, France. 300 pp.

REETZ, J E; ZÜHLKE, S; SPITELLER, M; WALLNER, K (2011) Neonicotinoid insecticides translocated in guttated droplets of see-treated maize and wheat: a threat to honey bees? *Apidologie* 42: 797. http://dx.doi.org/10.1007/s13592-011-0049-1

REETZ, J E; ZÜHLKE, S; SPITELLER, M; WALLNER, K (2012) A method for identifying water foraging bees by refractometer analysis: a spotlight on daily and seasonal water collecting activities of *Apis mellifera* L. *Journal of Consumer Protection and Food Safety* 7; S. 283-290.

REMBOLD, H; LACKNER, B (1981) Rearing of honey bee larvae *in vitro*: effect of yeast extract on queen differentiation. *Journal of Apicultural Research* 20: 165-171.

RIFCON (2012): http://www.rifcon.de/consultancy/honey bees/brood-assessment-software

ROBERTSON, J L; SAVIN, N E; PREISLER, H K; RUSSELL, R M (2007) *Bioassays with Arthropods (2nd Ed.).* CRC Press; Boca Raton, FL, USA. 224 pp.

ROBERTSON, J L; SMITH, K C; SAVIN, N E; LAVIGNE, RJ (1984) Effects of dose-selection and sample size on the precision of lethal dose estimates in dose-mortality regression. *Journal of Economic Entomology* 77: 833-837.

ROESSINK, I; VAN DER STEEN, J J M; KASINA, M; GIKUNGU, M; NOCELLI, R (2011) Is the European honey bee (*Apis mellifera mellifera*) a good representative for other pollinator species? In *Proceedings of the SETAC Europe 21st annual meeting, abstract book.* p 35.

RORTAIS, A; ARNOLD, G; HALM, M P; TOUFFET-BRIENS F (2005) Modes of honey bees exposure to systemic insecticides: estimated amounts of contaminated pollen and nectar consumed by different categories of bees. *Apidologie* 36: 71-83.

SAUTET, J; ALDIGHIERI, J; QUILICI, M (1968) L'effet de groupe dans les tests de résistance aux insecticides chez *Aedes aegypti. Bulletin of World Health Organisation* 38: 967-970.

SAVIN, N E; ROBERTSON, J L; RUSSELL, R M (1977) A critical evaluation of bioassay in insecticide research: likelihood ratio tests of dose-mortality regression. *Bulletin of the ESA* 23: 257-266.

SCHEINER, R; ABRAMSON, C I; BRODSCHNEIDER, R; CRAILSHEIM, K; FARINA, W; FUCHS, S; GRÜNEWALD, B; HAHSHOLD, S; KARRER, M; KOENIGER, G; KOENIGER, N; MENZEL, R; MUJAGIC, S; RADSPIELER, G; SCHMICKLI, T; SCHNEIDER, C; SIEGEL, A J; SZOPEK, M; THENIUS, R (2013) Standard methods for behavioural studies of *Apis mellifera*. In *V Dietemann; J D Ellis; P Neumann (Eds) The COLOSS* BEEBOOK, *Volume I: standard methods for* Apis mellifera *research. Journal of Apicultural Research* 52(4): http://dx.doi.org/10.3896/IBRA.1.52.4.04

SCHNEIDER, C W; TAUTZ, J; GRÜNEWALD, B; FUCHS, S (2012) RFID tracking of sublethal effects of two neonicotinoid insecticides on the foraging behaviour of *Apis mellifera. PLoS One* 7(1): e30023. http://dx.doi.org/10.1371/journal.pone.0030023

SCHUR, A; TORNIER, I; BRASSE, D; MÜHLEN, W; VON DER OHE, W; WALLNER, K; WEHLING, M (2003) Honey bee brood ring-test in 2002: method for the assessment of side-effects of plant protection products on the honey bee brood under semi-field conditions. *Bulletin of Insectology* 56(1): 91-96.

SCHUR, A; WALLNER, K (1998) Gathering of non-toxic pesticides by forager bees after treatment of blooming rape seed. *Apidologie* 29: 417-419.

SEBER, G; WILD, C (1989) *Nonlinear regression.* John Wiley and Sons; New York, USA. 768 pp.

SEELEY, T D (1985) *Honey bee ecology: a study of adaptation in social life.* Princeton University Press; Princeton, NJ, USA. 216 pp.

SEELEY, T D (1995) *The wisdom of the hive, the social physiology of honey bee colonies.* Harvard University Press; Cambridge, MA, USA. 295 pp.

SGOLASTRA, F; RENZI, T; DRAGHETTI, S; MEDRZYCKI, P; LODESANI, M; MAINI, S; PORRINI, C (2012). Effects of neonicotinoid dust from maize seed-dressing on honey bees. *Bulletin of Insectology* 65(2): 273-280.

SIEGEL, S; CASTELLAN JR, N J (1988) *Non parametric statistics for the behavioural sciences (2nd Ed.)*. McGraw-Hill Book Company; New York, USA. 399 pp.

SINGH, R; LEVITT, A L; RAJOTTE, E G; HOLMES, E C; OSTIGUY, N; VANENGELSDORP, D; LIPKIN, W I; DEPAMPHILIS, C W; TOTH, A L; COX-FOSTER, D L (2010) RNA Viruses in hymenopteran pollinators: evidence of inter-taxa virus transmission via pollen and potential impact on non-*Apis* hymenopteran species. *PLoS One* 5(12): e14357. http://dx.doi.org/10.1371/journal.pone.0014357

SODERLUND, D M; BLOOMQUIST, J R (1989) Neurotoxic actions of pyrethroid insecticides. *Annual Review of Entomology* 34: 77-96.

STABENTHEINER, A; PRESS, H; PAPST, T; HRASSNIGG, N; CRAILSHEIM, K (2003) Endothermic heat production in honey bee winter clusters. *Journal of Experimental Biology* 206: 353-358.

STEFFAN-DEWENTER, I; KUHN, A (2003) Honey bee foraging in differentially structured landscapes. *Proceedings of the Royal Society London* B270: 569- 575.

STONER, A; RHODES, H A; WILSON, W T (1979) Case histories of the effects of microencapsulated methyl parathion (Penncap M) applied to fields near honey bee colonies. *American Bee Journal* 119: 648-654.

STONER, A; WILSON, W (1983) Microencapsulated methyl parathion (Penncap-M): effect of long term feeding of low doses in pollen on honey bees in standard-size field colonies. *Journal of the Kansas Entomological Society* 56: 234-240.

STONER, A; WILSON, W; HARVEY, J (1983) Dimethoate (Cygon): effect of long-term feeding of low doses on honey bees in standard-size field colonies. *Southwestern Entomologist* 8: 174-177.

STONER, A; WILSON, W; HARVEY, J (1985) Acephate (Orthene): effects on honey bee queen, brood and worker survival. *American Bee Journal* 12: 448-450.

STONER, A; WILSON, W; RHODES, H (1982) Carbofuran: effect of long-term feeding of low doses in sucrose syrup on honey bees in standard size field colonies. *Environmental Entomology* 11: 53-59.

STUTE, K (1991) *Auswirkungen von Pflanzenschutzmitteln auf Honigbiene. Richtlinien für die Prüfung von Pflanzenschutzmitteln im Zulassungsverfahren* Teil VI: 23- 1. Biologische Bundesanstalt für Land und Forstwirtschaft (BBA); Braunschweig, Germany.

SUCHAIL, S; GUEZ, D; BELZUNCES, L (2000) Characteristics of imidacloprid toxicity in two *Apis mellifera* subspecies. *Environmental Toxicology and Chemistry* 19: 1901-1905.

SUCHAIL, S; GUEZ, D; BELZUNCES, L (2001) Discrepancy between acute and chronic toxicity induced by imidacloprid and its metabolites in *Apis mellifera*. *Environmental Toxicology and Chemistry* 20: 2482-2486.

TAPPARO, A; MARTON, D; GIORIO, C; ZANELLA, A; SOLDÀ, L; MARZARO, M; VIVAN, L; GIROLAMI, V (2012) Assessment of the environmental exposure of honey bees to particulate matter containing neonicotinoid insecticides coming from corn coated seeds. *Environmental Science and Technology* 46(5): 2592-2599.

TAUTZ, J (2009) *L'étonnante abeille* (French translation of Phaenomen Honigbiene). De Boeck Université; Bruxelles. Belguim. 278 pp.

THOMPSON, H; WILKINS, S; BATTERSBY, A; WAITE, R; WILKINSON, D (2005) The effects of four growth-regulators-insecticide on honey bee colony development, queen rearing and drone sperm production. *Ecotoxicology* 14: 757-769.

TINGLE, C C; ROTHER, J A; DEWHURST, C F; LAUER, S; KING, W J (2003) Fipronil: environmental fate, ecotoxicology, and human health concerns. *Review of Environmental Contamination and Toxicology* 176: 1-66.

TUFAIL, M; TAKEDA, M (2008) Molecular characteristics of insect vitellogenins. *Journal of Insect Physiology* 54: 1447-1458.

US EPA (1996) *Ecological effects test guidelines OPPTS 850.3020 honey bee acute contact toxicity* (No. EPA 712-C-96-147).

VANDAME, R; BELZUNCES, L (1998) Joint action of deltamethrin and azole fungicides on honey bee thermoregulation. *Neuroscience Letters* 251: 57-60.

VANDAME, R; MELED, M; COLIN, M; BELZUNCES, L (1995) Alteration of the homing-flight in the honey bee *Apis mellifera* L. exposed to sublethal dose of deltamethrin. *Environmental Toxicology and Chemistry* 14: 855-860.

VANDENBERG, J D; SHIMANUKI, H (1987) Technique for rearing worker honey bees in the laboratory. *Journal of Apicultural Research* 26: 90-97.

VANENGELSDORP, D; MEIXNER, M D (2010) A historical review of managed honey bee populations in Europe and the United States and the factors that may affect them. *Journal of Invertebrate Pathology* 103: S80-S95.

VANSELL, G H (1926) Buckeye poisoning of the honey bee. *California Agriculture Experimental Station Circular* 301. 12 pp.

VIDAU, C; DIOGON, M; AUFAUVRE, J; FONTBONNE, R; VIGUES, B; BRUNET, J-L; TEXIER, C; BIRON, D G; BLOT, N; EL ALAOUI, H; BELZUNCES, L P; DELBAC, F (2011) Exposure to sublethal doses of fipronil and thiacloprid highly increases mortality of honey bees previously infected by *Nosema ceranae*. *PLoS One* 6(6): e21550. http://dx.doi.org/10.1371/journal.pone.0021550

WAHL, O; ULM, K (1983) Influence of pollen feeding and physiological condition on pesticide sensitivity of the honey bee *Apis mellifera carnica*. *Oecologia* 59: 106-128.

WALLNER, K (1997) Pesticide gathering after plant protection measures against the fire blight disease. *Apidologie* 28: 172-173.

WALLNER, K (2009) Sprayed and seed dressed pesticides in pollen, nectar and honey of oilseed rape. *Julius Kühn-Archiv* 423: 152-153.

WANG, M; CLAßEN, C (2011) Automated evaluation of honey bee brood trials using digital image processing according to OECD 75, Oomen (1992) and beyond. *Poster presentation at SETAC North America, November 2011, Boston, USA*.

WEAVER, N (1955) Rearing of honey bee larvae on royal jelly in the laboratory. *Science* 121: 509-510.

WEHLING, M; OHE, W V D; BRASSE, D; FORSTER, R (2009) Colony losses - interactions of plant protection products and other factors. Hazards of pesticides to bees. *10th International Symposium of the ICP-Bee Protection Group. Bucharest, Romania, 8-10 October, 2008.* pp 153-154. Julius-Kuhn-Archiv.

WESSTEIN, E http://mathworld.wolfram.com/NormalDistribution.html.

WILLIAMS, G R; ALAUX, C; COSTA, C; CSÁKI, T; DOUBLET, V; EISENHARDT, D; FRIES, I; KUHN, R; MCMAHON, D P; MEDRZYCKI, P; MURRAY, T E; NATSOPOULOU, M E; NEUMANN, P; OLIVER, R; PAXTON, R J; PERNAL, S F; SHUTLER, D; TANNER, G; VAN DER STEEN, J J M; BRODSCHNEIDER, R (2013) Standard methods for maintaining adult *Apis mellifera* in cages under *in vitro* laboratory conditions. In *V Dietemann; J D Ellis; P Neumann (Eds) The COLOSS BEEBOOK, Volume I: standard methods for* Apis mellifera *research. Journal of Apicultural Research* 52(1): http://dx.doi.org/10.3896/IBRA.1.52.1.04

WINER, B J; BROWN, D R; MICHELS, K M (1991) *Statistical principles in experimental design (3rd Ed.).* McGraw-Hill; New York, USA. 928 pp.

WINSTON, M (1987) *The biology of the honey bee.* Harvard University Press; Cambridge, MA, USA. 281 pp.

WITTMANN, D (1982) Determination of the LC_{50} of Dimilin 25 WP for honey bee brood on free flying colonies as an example for the use of a new *Apis*-larvae-test. *Apidologie* 13: 104-107.

WITTMANN, D; ENGELS, W (1981) Development of test procedures for insecticide-induced brood damage in honey bees. *Mittelungen der Deutschen Gesellschaft fur Allgemeine und Angewandte Entomologie* 3: 187-190.

WU, J Y; SMART, M D; ANELLI, C M; SHEPPARD, W S (2012) Honey bees (*Apis mellifera*) reared in brood combs containing high levels of pesticide residues exhibit increased susceptibility to *Nosema* (Microsporidia) infection. *Journal of Invertebrate Pathology* 109: 326-329.

YANG, E C; CHUANG, Y C; CHEN, Y L; CHANG, L H (2008) Abnormal foraging behavior induced by sublethal dosage of imidacloprid in the honey bee (Hymenoptera: Apidae). *Journal of Economic Entomology* 101(6): 1743-1748.

ZBINDEN, G; FLURY-ROVERSI, M (1981) Significance of the LD_{50}-test for the toxicological evaluation of chemical substances. *Archives of Toxicology* 47: 71-99.

CPSIA information can be obtained
at www.ICGtesting.com
Printed in the USA
BVHW02s0553100518
515795BV00024B/115/P